Triangles and Angles

Right Triangle
Triangle has one 90°
(right) angle.

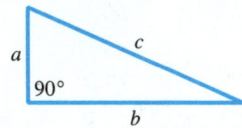

Pythagorean Formula
(*for right triangles*)
$c^2 = a^2 + b^2$

Right Angle
Measure is 90°.

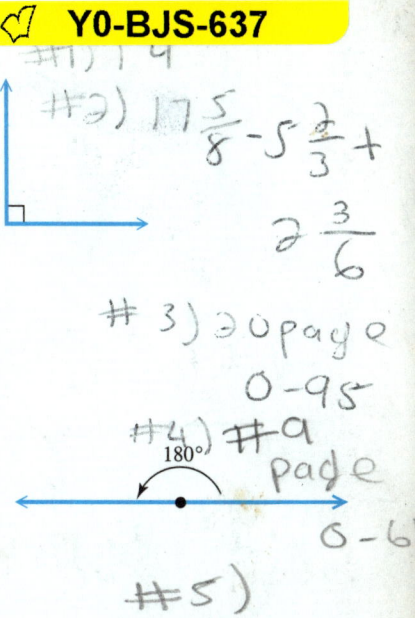

Isosceles Triangle
Two sides are equal.
$AB = BC$

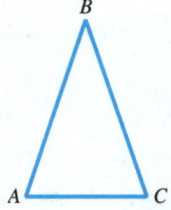

Straight Angle
Measure is 180°.

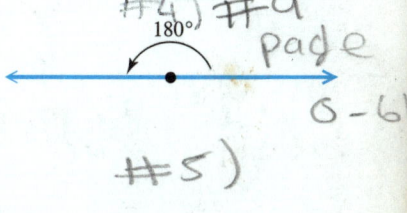

Equilateral Triangle
All sides are equal.
$AB = BC = CA$

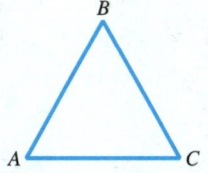

Complementary Angles
The sum of the measures of
two complementary
angles is 90°.

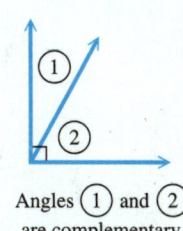

Angles ① and ②
are complementary.

Sum of the Angles of Any Triangle
$A + B + C = 180°$

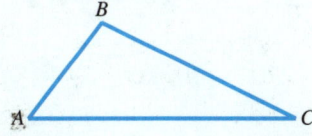

Supplementary Angles
The sum of the
measures of two
supplementary
angles is 180°.

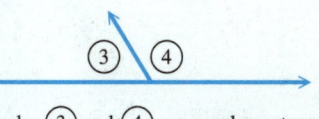

Angles ③ and ④ are supplementary.

Similar Triangles
Corresponding angles are
equal; corresponding sides
are proportional.
$A = D, B = E, C = F$
$\dfrac{AB}{DE} = \dfrac{AC}{DF} = \dfrac{BC}{EF}$

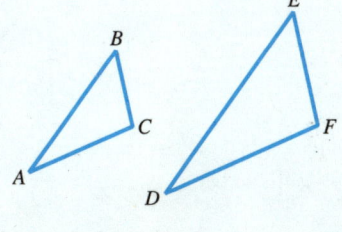

Vertical Angles
Vertical angles have
equal measures.

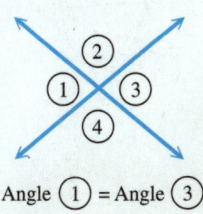

Angle ① = Angle ③
Angle ② = Angle ④

Beginning Algebra

Taken from

Beginning Algebra, Ninth Edition
by Margaret L. Lial, John Hornsby, and Terry McGinnis

Basic College Mathematics, Sixth Edition
by Margaret L. Lial, Stanley A. Salzman, and Diana L. Hestwood

Intermediate Algebra, Ninth Edition
by Margaret L. Lial, John Hornsby, and Terry McGinnis

Cover art: Fractal image © Copyright Sylvie Gallet, 2001.
http://www.fractalus.com/sylvie/homepage.htm.

Book: 2004360352-Lial MDC
Rights: NN

Taken from:

Beginning Algebra, Ninth Edition
by Margaret L. Lial, John Hornsby, and Terry McGinnis
Copyright © 2004 by Pearson Education, Inc.
Published by Addison Wesley
Boston, Massachusetts 02116

Basic College Mathematics, Sixth Edition
by Margaret L. Lial, Stanley A. Salzman, and Diana L. Hestwood
Copyright © 2002 by Pearson Education, Inc.
Published by Addison Wesley

Intermediate Algebra, Ninth Edition
by Margaret L. Lial, John Hornsby, and Terry McGinnis
Copyright © 2004 by Pearson Education, Inc.
Published by Addison Wesley

Copyright © 2004 by Pearson Custom Publishing
All rights reserved.

This copyright covers material written expressly for this volume by the editor(s)/author(s) as well as the compilation itself. It does not cover the individual selections herein that first appeared elsewhere. Permission to reprint these has been obtained by Pearson Custom Publishing for this edition only. Further reproduction by any means, electronic or mechanical, including photocopying and recording, or by any information storage or retrieval system, must be arranged with the individual copyright holders noted.

Printed in the United States of America

10 9 8 7 6 5 4 3 2 1

ISBN 0-536-836140

2004360352

AP/JS

Please visit our web site at *www.pearsoncustom.com*

PEARSON CUSTOM PUBLISHING
75 Arlington Street, Suite 300, Boston, MA 02116
A Pearson Education Company

Contents

List of Applications		ix
Preface		xiii
Feature Walkthrough		xxi

Chapter 0 Topics in Arithmetic — 0-1

Whole Numbers

0.1	Reading and Writing Whole Numbers	0-2
0.2	Adding Whole Numbers	0-5
0.3	Subtracting Whole Numbers	0-10
0.4	Multiplying Whole Numbers	0-17
0.5	Dividing Whole Numbers	0-24
0.6	Long Division	0-31
	Test on Sections 0.1–0.6	0-36

Fractions (Part 1)

0.7	Basics of Fractions	0-37
0.8	Mixed Numbers	0-41
0.9	Factors	0-45
0.10	Writing a Fraction in Lowest Terms	0-50
0.11	Multiplying Fractions	0-57
0.12	Dividing Fractions	0-62
0.13	Multiplying and Dividing Mixed Numbers	0-67
	Test on Sections 0.7–0.13	0-73

Fractions (Part 2)

0.14	Adding and Subtracting Like Fractions	0-74
0.15	Least Common Multiples	0-78
0.16	Adding and Subtracting Unlike Fractions	0-85
0.17	Adding and Subtracting Mixed Numbers	0-90
	Test on Sections 0.14–0.17	0-96

Decimals

0.18	Reading and Writing Decimals	0-97
0.19	Rounding Decimals	0-104
0.20	Adding and Subtracting Decimals	0-108
0.21	Multiplying Decimals	0-112
0.22	Dividing Decimals	0-114
0.23	Writing Fractions as Decimals	0-120
	Test on Sections 0.18–	

Ratio and Proportion

0.24	Ratios	0-126
0.25	Proportions	0-137
0.26	Solving Proportions	0-141
	Test on Sections 0.24–0.26	0-149

Unit Analysis

0.27	Analyzing Units with English Measurement	0-151
0.28	The Metric System—Length	0-158
0.29	The Metric System—Capacity and Weight (Mass)	0-164
0.30	Metric-English Conversions	0-169
	Test on Sections 0.27–0.30	0-172

Percent

0.31	Basics of Percent	0-173
0.32	Percents and Fractions	0-178
0.33	Using the Percent Proportion	0-184
0.34	Using Proportions to Solve Percent Problems	0-191
0.35	Using the Percent Equation	0-197
	Test on Sections 0.31–0.35	0-203

Chapter 1   The Real Number System   1

1.1	Fractions	2
1.2	Exponents, Order of Operations, and Inequality	14
1.3	Variables, Expressions, and Equations	23
1.4	Real Numbers and the Number Line	29
1.5	Adding and Subtracting Real Numbers	38
1.6	Multiplying and Dividing Real Numbers	50
Summary Exercises on Operations with Real Numbers		63
1.7	Properties of Real Numbers	64
1.8	Simplifying Expressions	73
Chapter 1 Group Activity *Comparing Floor Plans of Houses*		**79**
Chapter 1 Summary		**80**
Chapter 1 Review Exercises		**84**
Chapter 1 Test		**89**

Chapter 2   Linear Equations and Inequalities in One Variable   93

2.1	The Addition Property of Equality	94
2.2	The Multiplication Property of Equality	100
2.3	More on Solving Linear Equations	106
Summary Exercises on Solving Linear Equations		114
2.4	An Introduction to Applications of Linear Equations	114
2.5	Formulas and Applications from Geometry	128
2.6	Ratios and Proportions	137
2.7	More about Problem Solving	145
2.8	Solving Linear Inequalities	157
Chapter 2 Group Activity *Are You a Race-Walker?*		**170**

Chapter 2 Summary		**171**
Chapter 2 Review Exercises		**175**
Chapter 2 Test		**179**
Cumulative Review Exercises Chapters 1–2		**181**

Chapter 3 Linear Equations and Inequalities in Two Variables; Functions — 183

3.1	Reading Graphs; Linear Equations in Two Variables	184
3.2	Graphing Linear Equations in Two Variables	196
3.3	The Slope of a Line	208
3.4	Equations of a Line	220
3.5	Graphing Linear Inequalities in Two Variables	230
3.6	Introduction to Functions	238
Chapter 3 Group Activity *Determining Business Profit*		**248**
Chapter 3 Summary		**249**
Chapter 3 Review Exercises		**253**
Chapter 3 Test		**257**
Cumulative Review Exercises Chapters 1–3		**258**

Chapter 4 Systems of Linear Equations and Inequalities — 261

4.1	Solving Systems of Linear Equations by Graphing	262
4.2	Solving Systems of Linear Equations by Substitution	272
4.3	Solving Systems of Linear Equations by Elimination	279
Summary Exercises on Solving Systems of Linear Equations		285
4.4	Applications of Linear Systems	287
4.5	Solving Systems of Linear Inequalities	297
Chapter 4 Group Activity *Top Concert Tours*		**302**
Chapter 4 Summary		**303**
Chapter 4 Review Exercises		**306**
Chapter 4 Test		**309**
Cumulative Review Exercises Chapters 1–4		**311**

Chapter 5 Exponents and Polynomials — 313

5.1	The Product Rule and Power Rules for Exponents	314
5.2	Integer Exponents and the Quotient Rule	321
Summary Exercises on the Rules for Exponents		329
5.3	An Application of Exponents: Scientific Notation	330
5.4	Adding and Subtracting Polynomials; Graphing Simple Polynomials	337
5.5	Multiplying Polynomials	348
5.6	Special Products	356
5.7	Dividing Polynomials	362
Chapter 5 Group Activity *Measuring the Flight of a Rocket*		**372**
Chapter 5 Summary		**374**
Chapter 5 Review Exercises		**377**
Chapter 5 Test		**381**
Cumulative Review Exercises Chapters 1–5		**382**

Chapter 6 Factoring and Applications — 385

6.1	The Greatest Common Factor; Factoring by Grouping	386
6.2	Factoring Trinomials	394
6.3	More on Factoring Trinomials	399
6.4	Special Factoring Rules	407
	Summary Exercises on Factoring	416
6.5	Solving Quadratic Equations by Factoring	418
6.6	Applications of Quadratic Equations	425

Chapter 6 Group Activity   *Factoring Trinomials Made Easy* — 436
Chapter 6 Summary — 438
Chapter 6 Review Exercises — 440
Chapter 6 Test — 444
Cumulative Review Exercises Chapters 1–6 — 445

Chapter 7 Rational Expressions and Applications — 447

7.1	The Fundamental Property of Rational Expressions	448
7.2	Multiplying and Dividing Rational Expressions	458
7.3	Least Common Denominators	464
7.4	Adding and Subtracting Rational Expressions	470
7.5	Complex Fractions	479
7.6	Solving Equations with Rational Expressions	486
	Summary Exercises on Operations and Equations with Rational Expressions	495
7.7	Applications of Rational Expressions	497
7.8	Variation	506

Chapter 7 Group Activity   *Buying a Car* — 513
Chapter 7 Summary — 514
Chapter 7 Review Exercises — 519
Chapter 7 Test — 522
Cumulative Review Exercises Chapters 1–7 — 524

Chapter 8 Roots and Radicals — 527

8.1	Evaluating Roots	528
8.2	Multiplying, Dividing, and Simplifying Radicals	539
8.3	Adding and Subtracting Radicals	547
8.4	Rationalizing the Denominator	550
8.5	More Simplifying and Operations with Radicals	556
	Summary Exercises on Operations with Radicals	564
8.6	Solving Equations with Radicals	565
8.7	Using Rational Numbers as Exponents	576

Chapter 8 Group Activity   *Comparing Television Sizes* — 581
Chapter 8 Summary — 582
Chapter 8 Review Exercises — 584
Chapter 8 Test — 588
Cumulative Review Exercises Chapters 1–8 — 589

Chapter 9 Quadratic Equations — 593

9.1	Solving Quadratic Equations by the Square Root Property	594
9.2	Solving Quadratic Equations by Completing the Square	600
9.3	Solving Quadratic Equations by the Quadratic Formula	607
	Summary Exercises on Quadratic Equations	614
9.4	Complex Numbers	615
9.5	More on Graphing Quadratic Equations; Quadratic Functions	621

Chapter 9 Group Activity *How Is a Radio Telescope Like a Wok?* — 631
Chapter 9 Summary — 632
Chapter 9 Review Exercises — 636
Chapter 9 Test — 638
Cumulative Review Exercises Chapters 1–9 — 639

Appendix A	An Introduction to Calculators	643
Appendix B	Review of Decimals and Percents	649
Appendix C	Sets	655
Appendix D	Mean, Median, and Mode	661
Appendix E	Solving Quadratic Inequalities	669

Answers to Selected Exercises — A-1
Glossary — G-1
Index — I-1

List of Applications

Astronomy/Aerospace
Aerospace industry sales, 336
Altitude of a plane, 50
Circumference of the solar orbit, 335
Cosmic rays, 335, 377
Distance from the sun to a planet, 125, 384
Flight of a rocket, 372–373
Galaxy count, 335
Helium usage at Kennedy Space Center, 332
Large Magellanic Cloud, 381
Light year, 381
NASA's budget, 343–344, 347–348
Object propelled upward on the surface of the moon, 613
Radio signals from Pluto to Earth, 336
Radio telescope, 593, 626, 629, 630, 631, 637
Satellites, 126
Solar luminosity, 335
Speed of light, 332
Velocity of a meteorite, 563

Biology
Bees making honey, 446
DNA, 378
Dolphin's rest time, 383
Fish tagged in a lake, 143
Growth of a plant, 191
Humpback whales, 49
Insecticide, 125
Length of a bass, 597
Migrating hawk, 504
Mouse vs. elephant breaths, 383
Plant species on a Galápagos Island, 580
Population of prairie-chickens, 235
Weight of a female polar bear, 50

Business
Advertising, 311, 336
Annual revenue for eBay, 442
Break-even point, 169, 248, 278
Buying property to expand a business, 14
Car sales, 184
Coffeehouse orders, 117–118
Convention spending, 590
Customers helped at an order desk, 663
Deliveries by a newspaper distributor, 666
Express mail company, 239
Food and drink sales at U.S. restaurants, 257–258
Growth in retail square footage, 219
Investing in a business, 652
Largest restaurant chain, 307
Manufacturing, 237, 509
On-line retail spending, 195
Packaging services, 125
Pharmaceutical income, 12
Printing job, 505
Production, 22, 192, 589
Profit, 166, 169, 248, 294
Purchasing a copier, 229
Revenue, 166, 278, 294
Sales, 166, 661, 665, 666, 667
Shipping merchandise, 237
Sporting goods sales, 206–207, 287–288
Start-up costs, 195, 228, 278
Video rentals, 256
Waitress collecting checks, 665

Chemistry
Inversion of raw sugar, 512
Mixing metals, 146, 153
Mixing solutions, 118, 142, 145, 146–147, 152, 153, 154, 178, 179, 289–290, 292, 295, 308, 310, 312
Pressure of a gas, 511

Construction
Board feet of lumber cut from a log, 613
Bridge span, 179
Dividing a board or pipe, 12, 119
Fencing, 356
Floor plans of houses, 79
Fourth tallest structure in London, 527, 574
Height of a structure, 8–9, 293, 537, 543
Installing tool sheds, 382
Ladder leaning against a wall, 434, 532
Length of a support, 444, 536
Measurements, 664
Mixing concrete, 125
Pitch of a roof, 217
Roofing, 505
Skydome in Toronto, 135
Socket wrenches, 12
Surveying, 536
Tallest buildings in Houston, Texas, 307
Welding, 50

Consumer
Average daily parking costs, 662
Best buy, 137–138, 141, 177, 179, 180
Buying a car, 513
Buying a duplex, 652
Buying gifts, 294
Cost of gasoline, 140, 142, 169, 191, 192, 253, 348, 506
Discount, 651
Monthly bills, 661, 665, 667
Mortgage shopping, 308
Phone costs, 169
Prices, 663
Purchasing cost, 142, 155, 292, 640
Rental costs, 169, 195, 247, 348
Sale price, 651

Economics
Commercial bank assets, 336
Consumer Price Index, 144
Depreciation, 221–222
Dow Jones Industrial Average, 36, 87
Exchange rate, 142, 143
Increase in the value of a home, 653
Monetary value, 113, 152, 292
NASDAQ stock exchange, 1, 12, 36
Number of banks in the United States, 36
Producer Price Index, 35, 38, 45–46
Sales tax, 142, 177, 239
Stock price, 22
Supply and demand, 207, 269
Value of a coin, 124

ix

List of Applications

Education
Annual earnings by level of education, 259
Average annual cost at public or private four-year colleges, 224–225, 229, 246
Average test score, 163–164, 168, 178, 180, 661
Credits per student, 665
Four-year college students graduating within five years, 195–196
Free lunches served in public schools, 237
Funding for Head Start programs, 127
Grade point average, 663, 666
Points on a screening exam, 664
Public school enrollment, 218
SAT scores, 36
Science Literacy Index, 21
Students enrolled in algebra, 666
Students per class, 666
Students per computer, 218
Test or quiz scores, 664, 665
U.S. children educated at home, 243

Environment
Altitude, 50
Broken tree, 537
Earthquake intensity, 333
Elevation, 30, 36, 45, 48, 49, 90
Fertilizer, 177, 296, 371
Heavy-metal nuclear waste, 229
Heights of waterfalls, 176
Land area, 176, 180
Sodding a lawn, 356
Temperature, 30, 45, 49, 87, 89, 169, 228
Volcano, 259

Finance
Annual savings of a family, 653
Bank teller, 114
Borrowing money, 87, 177
Checking account balance, 30, 87
Compound interest, 320
Credit card debt, 50, 183, 202–203
Inheritance, 154, 259
Interest earned, 145, 152, 511
Interest rate, 599, 652
Investment, 147–148, 154, 174, 178, 180, 294, 308, 382, 563
Life insurance policies, 665
Number of bills or coins of each denomination, 148–149, 154, 155, 294, 307

Geometry
Angle measurement, 120, 126, 130–131, 176, 177, 180, 445, 639
Angles of a triangle, 126, 311
Area of a circular region, 135, 176, 512
Area of a rectangular region, 135
Area of a trapezoid, 135
Area of a triangle, 511
Base and height of a triangular object, 130, 432, 444, 606
Base of a parallelogram, 521
Circumference of a circle, 135, 507, 525
Diagonal of a rectangle, 532, 536, 585
Diameter of a circle, 176
Dimensions of a box-shaped object, 136, 431, 432, 441, 442
Dimensions of a cube, 546
Dimensions of a pyramid, 443
Dimensions of a rectangular region, 129, 176, 307, 312, 426, 431, 432, 441, 442, 444, 511, 522, 524, 536, 537, 606, 639, 640
Height of a parallelogram, 521
Length of the side of a square, 179, 433, 613
Lengths of the sides of a triangle, 179, 308, 312, 383, 426, 428, 433, 434, 440, 443, 446
Perimeter of a rectangular region, 135, 169, 478
Perimeter of a triangle, 14, 169, 478
Radius of a circle, 182, 599
Radius of a cylinder, 562
Radius of a sphere, 182, 546, 599
Right triangle, 636, 638, 641
Volume of a box, 136

Government
Acquitting William Jefferson Clinton, 311
Defense budget, 19–20, 91
Electoral votes, 307
Federal budget outlays for treasury department, 48
Social security numbers, 336
U.S. House of Representatives, 49, 123, 176
U.S. Senate, 123
U.S. trade deficit, 435–436

Health/Life Sciences
Age, 113, 664, 665
Body containing copper, 337, 640
Calories, 179
Cost of a doctor's visit, 190–191
Cost of parenthood, 153
Deaths caused by smoking, 457
Emergency calls in California, 573–574
HIV-infected children and teenagers, 235
Height of a woman, 205
Hospital buying glucose solution, 309
Impulses fired after a nerve is stimulated, 429
Increases in drug prices, 430
Medicare beneficiaries, 235
Patients admitted to a hospital, 667
Pharmacist's prescriptions, 124
Pregnant women with hepatitis, 235
Samples taken, 667
Target heart rate zone, 196, 206
Weight of a man, 205–206

Labor
Age of retirees, 667
Ages of part-time employees, 664
Annual salaries, 665
Civilian federal employees, 237
Employee entering tax data, 505
Hours worked, 125, 666
Monthly commissions of sales people, 664
Unemployment, 153
Union members, 153
Working alone to complete a job, 521, 606
Working together to complete a job, 501–502, 505, 518, 521, 523, 525

Miscellaneous
Americans with guns at home, 235
Barrels of pickles, 295
Comparing television sizes, 581
Consecutive integers, 121–122, 127, 176, 427, 433, 442, 443, 444
Contribution to a food pantry, 662
Creating a sundae, 124
Cutting a submarine sandwich, 125
Cutting yarn, 181
Distance one can see from an airplane to the horizon, 572, 580
Distance one can see from the top of a building or structure, 574, 575

Drawing balls out of a box, 378
Everyday items taken for granted, 445
Farming, 22, 336
Filling a sink, 506
Filling a swimming pool, 505, 506
Googol, 378
Grade of a hill, 217
Height of a chair, 144
Height of a kite, 536
Holes in Swiss cheese, 12
Largest box of popcorn, 136
Mixing food or drinks, 125, 155, 295, 307
Most popular automobile colors, 152
Numbers, 115-116, 123, 126, 168, 172–173, 178, 293, 305, 498, 502, 503, 520, 524
Page numbers, 127
Prehistoric ceremonial site, 135
Pumping water out of a basement, 505
Recipes, 13, 182
Serving size, 11
Sheep ranch in Australia, 124
Soft drink consumption, 206
Statues, 235
Survey responses, 84
Tallest candle constructed, 144
Time it takes for fruit to ripen, 523
Trash compactor, 508
Upholsterer covering a chair, 8

Physics

Current in an electrical circuit, 511
Electromagnetic attraction, 378
Force of attraction, 512
Force required to compress a spring, 512
Free-falling objects, 385, 418, 424, 443, 599
Frog's leap, 612
Hooke's law for an elastic spring, 511
Illumination produced by a light source, 512, 599
Impedance of an alternating series circuit, 588
Object dropped, 512
Object propelled upward, 348, 434, 442, 604, 606, 636, 638, 673
Period of a pendulum, 563
Pressure exerted on a scuba diver, 511
Speed of a pulley, 591
Speed of sound, 150

Sports/Entertainment

Athletes training program, 179
Average movie ticket prices, 288, 295
Baseball, 90, 113, 124, 179, 637, 665
Broadway season, 336
Cable TV channels, 445
Car racing, 150, 155, 498, 503, 504
Concours d'elegance competition, 479
Drive-in movies, 123
Freestyle swimming, 150, 503
Husky running the Iditarod, 124
Major broadcast TV networks, 123
Money spent on attending movies, 337
Most popular amusement parks, 310
Most popular magazines, 307
Movie attendance, 284–285
Music production, 270–271
NBA regular season, 124
Net yardage, 87, 113
Network share, 270
Olympic games, 93, 143
Olympic medals, 116–117, 125, 177, 445
Playing cards, 640
Race-walking, 170
Rock band concerts, 124
Running race, 156, 503
Serious sports injuries, 128
Skydiver falling, 507
Softball league batting statistics, 13
Speed skating, 259, 498, 503
Tennis player's serve, 429
Ticket prices to a sporting event, 312
Top-grossing concert tours, 293, 302
Top-grossing movies, 261, 288–289, 293, 337
Top-selling Disney videos, 287
Tourist attractions, 309
U.S. radio stations, 293, 446
Weights of soccer players, 667
World record in 100-m dash, 150

Statistics/Demographics

Asian-American population, 243–244
Growth of U.S. foreign-born population, 246
Population, 36, 152
U.S. citizens 65 years or older, 36
U.S. immigrants, 13

Technology

Cable TV systems in the United States, 36
Cellular phones, 336, 435
Computer parts, 309
Computer service calls, 666
Computer user group meeting, 118
E-mail messages, 666
Personal computers, 185, 270
Satellite-TV home subscribers, 9–10
Speed of a computer, 524
Surfing the Web, 445
U.S. exports and imports of electronics, 570, 573

Transportation

Aircraft carrier, 378
Alternative-fueled vehicles, 443
Cars in a parking lot, 666
Cost of a plane ticket, 114
Distance a plane travels without refueling, 666
Distance between cities, 142, 155, 177, 180, 310, 505, 517, 573
Distances, 150–151, 191, 239, 428–429, 433, 442, 606
Fall speed of a vehicle, 587
Gasoline mileage, 665
Highway grade, 209
Intercity transportation, 89
Miles driven on a vacation, 116
Skidding distance, 313, 347, 572
Speed (Rate), 151–152, 156, 157, 178, 179, 182, 291–292, 296, 297, 308, 309, 310, 498–500, 504, 511, 521, 590, 667
Speed of a current, 291, 296, 504, 523, 641
Speed of the wind, 296, 308, 521
Sport utility vehicle value, 207
Steel-belted radial tires, 653
Time, 155, 156, 174, 178, 180, 371
Top-selling passenger car, 124
Traffic intensity, 457
Traveling by air, 652
U.S. airline industry, 348
Vintage automobiles, 447
Worldwide airline fatalities, 85

Preface

The ninth edition of *Beginning Algebra* continues our ongoing commitment to provide the best possible text and supplements package to help instructors teach and students succeed. To that end, we have tried to address the diverse needs of today's students through a more open design, updated figures and graphs, helpful features, careful explanations of topics, and a comprehensive package of supplements and study aids. We have also taken special care to respond to the suggestions of users and reviewers and have added many new examples and exercises based on their feedback. Students who have never studied algebra—as well as those who require further review of basic algebraic concepts before taking additional courses in mathematics, business, science, nursing, or other fields—will benefit from the text's student-oriented approach.

After many years of benefiting from her behind-the-scenes assistance, we are pleased to welcome Terry McGinnis as coauthor of this series, which includes this text as well as the following books:

- *Intermediate Algebra,* Ninth Edition, by Lial, Hornsby, and McGinnis
- *Beginning and Intermediate Algebra,* Third Edition, by Lial, Hornsby, and McGinnis
- *Algebra for College Students,* Fifth Edition, by Lial, Hornsby, and McGinnis

WHAT'S NEW IN THIS EDITION?

We believe students and instructors will welcome the following new features.

New Real-Life Applications   We are always on the lookout for interesting data to use in real-life applications. As a result, we have included many new or updated examples and exercises throughout the text that focus on real-life applications of mathematics. These applied problems provide a modern flavor that will appeal to and motivate students. (See pp. 184, 430, and 503.) A comprehensive List of Applications appears at the beginning of the text. (See pp. ix–xi.)

New Figures and Photos   Today's students are more visually oriented than ever. Thus, we have made a concerted effort to add mathematical figures, diagrams, tables, and graphs whenever possible. (See pp. 134, 152, and 508.) Many of the graphs use a style similar to that seen by students in today's print and electronic media. Photos have been incorporated to enhance applications in examples and exercises. (See pp. 155, 224, and 574.)

Increased Emphasis on Problem Solving   Introduced in Chapter 2, our six-step problem-solving method has been refined and integrated throughout the text. The six steps, *Read, Assign a Variable, Write an Equation, Solve, State the Answer,* and *Check,* are emphasized in boldface type and repeated in examples and exercises to reinforce the problem-solving process for students. (See pp. 115, 287, and 431.) Special boxes that include additional problem-solving information and tips are interspersed throughout the text. (See pp. 44, 121, and 145.)

Chapter Openers New chapter openers feature real-world applications of mathematics that are relevant to students and tied to specific material within the chapters. Examples of topics include the stock market, the Olympics, and credit card debt. (See pp. 1, 93, and 183—Chapters 1, 2, and 3.)

Now Try Exercises To actively engage students in the learning process, each example now concludes with a reference to one or more parallel exercises from the corresponding exercise set. In this way, students are able to immediately apply and reinforce the concepts and skills presented in the examples. (See pp. 10, 95, and 203.)

Summary Exercises Based on user feedback, we have more than doubled the number of in-chapter summary exercises. These special exercise sets provide students with the all-important *mixed* review problems they need to master topics. Summaries of solution methods or additional examples are often included. (See pp. 63, 285, and 416.)

Glossary A comprehensive glossary of key terms from throughout the text is included at the back of the book. (See pp. G-1 to G-6.)

WHAT FAMILIAR FEATURES HAVE BEEN RETAINED?

We have retained the popular features of previous editions of the text, some of which follow.

Learning Objectives Each section begins with clearly stated, numbered objectives, and the included material is directly keyed to these objectives so that students know exactly what is covered in each section. (See pp. 94, 314, and 528.)

Cautions and Notes One of the most popular features of previous editions, **CAUTION** and **NOTE** boxes warn students about common errors and emphasize important ideas throughout the exposition. (See pp. 14, 108, and 117.) There are more of these in the ninth edition than in the eighth, and the new text design makes them easier to spot.

Connections Connections boxes have been streamlined. They continue to provide connections to the real world or to other mathematical concepts, historical background, and thought-provoking questions for writing or class discussion. (See pages 166, 333, and 534.)

Ample and Varied Exercise Sets The text contains a wealth of exercises to provide students with opportunities to practice, apply, connect, and extend the algebraic skills they are learning. Numerous illustrations, tables, graphs, and photos have been added to the exercise sets to help students visualize the problems they are solving. Problem types include writing ✏, estimation, graphing calculator 📊, and challenging exercises that go beyond the examples as well as applications and multiple-choice, matching, true/false, and fill-in-the-blank problems. (See pp. 192, 292, and 536.)

Relating Concepts Exercises These sets of exercises help students tie together topics and develop problem-solving skills as they compare and contrast ideas, identify and describe patterns, and extend concepts to new situations. (See pp. 168, 256, and 435.) These exercises make great collaborative activities for pairs or small groups of students.

Technology Insights Exercises We assume that all students of this text have access to scientific calculators. *While graphing calculators are not required for this text,* some students may go on to courses that use them. For this reason, we have included Technology Insights exercises in selected exercise sets. These exercises provide an opportunity for students to interpret typical results seen on graphing calculator screens. Actual calculator screens from the Texas Instruments TI-83 Plus graphing calculator are featured. (See pp. 219, 271, and 425.)

Group Activities Appearing at the end of each chapter, these real-data activities allow students to apply the mathematical content of the chapter in a collaborative setting. (See pp. 170, 302, and 513.)

Ample Opportunity for Review Each chapter concludes with an extensive Chapter Summary that features Key Terms, New Symbols, Test Your Word Power, and a Quick Review of each section's content with additional examples. A comprehensive set of Chapter Review Exercises, keyed to individual sections, is included, as are Mixed Review Exercises and a Chapter Test. Beginning with Chapter 2, each chapter concludes with a set of Cumulative Review Exercises that cover material going back to Chapter 1. (See pp. 171, 249, and 303.)

WHAT CONTENT CHANGES HAVE BEEN MADE?

We have worked hard to fine-tune and polish presentations of topics throughout the text based on user and reviewer feedback. Some of the content changes include the following:

- Former Section 2.1 on the addition and multiplication properties of equality has been split into two sections to allow each property to be treated individually before combining their use in Section 2.3.

- All topics on graphing linear equations and inequalities in two variables are now treated in Chapter 3.

- Systems of linear equations and inequalities, presented earlier than in the previous edition, are covered in Chapter 4.

- Chapter 5 on exponents and polynomials has been reorganized so that the sections on exponents are covered early in the chapter.

- Variation is covered in a new Section 7.8.

- Three new appendices have been included. Appendix A provides an introduction to calculators; Appendix D includes all new material on mean, median, and mode; and Appendix E covers quadratic inequalities, formerly included in the chapter on factoring.

WHAT SUPPLEMENTS ARE AVAILABLE?

Our extensive supplements package includes an Annotated Instructor's Edition, testing materials, solutions manuals, tutorial software, videotapes, and a state-of-the-art Web site. For more information about any of the following supplements, please contact your Addison-Wesley sales consultant.

For the Student

Student's Solutions Manual (ISBN 0-321-15714-1) The *Student's Solutions Manual* by Jeff Cole, Anoka-Ramsey Community College, provides detailed solutions to the odd-numbered section and summary exercises and to all Relating Concepts, Chapter Review, Chapter Test, and Cumulative Review exercises.

Addison-Wesley Math Tutor Center The Addison-Wesley Math Tutor Center is staffed by qualified college mathematics instructors who tutor students on examples and exercises from the textbook. Tutoring is provided via toll-free telephone, toll-free fax, e-mail, and the Internet. Interactive Web-based technology allows students and tutors to view and listen to live instruction—in real-time over the Internet. The Math Tutor Center is accessed through a registration number that can be packaged with a new textbook or purchased separately. (*Note:* MyMathLab students obtain access to the Math Tutor Center through their MyMathLab access code.)

 InterAct Math® Tutorial Software (ISBN 0-321-15715-X)   Available on CD-ROM, this interactive tutorial software provides algorithmically generated practice exercises that are correlated at the objective level to the odd-numbered exercises in the text. Every exercise in the program is accompanied by an example and a guided solution designed to involve students in the solution process. Selected problems also include a video clip to help students visualize concepts. The software tracks student activity and scores and can generate printed summaries of students' progress.

MathXL   MathXL is an on-line testing, homework, and tutorial system that uses algorithmically generated exercises correlated to the textbook. Instructors can assign tests and homework provided by Addison-Wesley or create and customize their own tests and homework assignments. Instructors can also track their students' results and tutorial work in an on-line gradebook. Students can take chapter tests and receive personalized study plans that diagnose weaknesses and link students to areas they need to study and retest. Students can also work unlimited practice problems and receive tutorial instruction for areas in which they need improvement. MathXL can be packaged with new copies of *Beginning Algebra,* Ninth Edition. Please contact your Addison-Wesley sales representative for details.

 Videotape Series (ISBN 0-321-15717-6)   This series of videotapes, created specifically for *Beginning Algebra,* Ninth Edition, features an engaging team of lecturers who provide comprehensive lessons on every objective in the text. The videos include a stop-the-tape feature that encourages students to pause the video, work through the example presented on their own, and then resume play to watch the video instructor go over the solution.

 Digital Video Tutor (ISBN 0-321-16823-2)   This supplement provides the entire set of videotapes for the text in digital format on CD-ROM, making it easy and convenient for students to watch video segments from a computer, either at home or on campus. Available for purchase with the text at minimal cost, the Digital Video Tutor is ideal for distance learning and supplemental instruction.

 MyMathLab   MyMathLab is a complete on-line course for Addison-Wesley mathematics textbooks that provides interactive, multimedia instruction correlated to the textbook content. MyMathLab is easily customizable to suit the needs of students and instructors and provides a comprehensive and efficient on-line course-management system that allows for diagnosis, assessment, and tracking of students' progress.

MyMathLab features:

- Chapter and section folders in the on-line course mirror the textbook Table of Contents and contain a wide range of multimedia instruction, including video lectures, tutorial software, and electronic supplements.

- Actual pages of the textbook are loaded into MyMathLab, and as students work through a section of on-line text, they can link to multimedia resources—such as video and audio clips, tutorial exercises, and interactive animations—that are correlated directly to examples and exercises in the text.

- Hyperlinks take the user directly to on-line testing, diagnosis, tutorials, and tracking in MathXL—Addison-Wesley's tutorial and testing system for mathematics and statistics.
- Instructors can create, copy, edit, assign, and track all tests and homework for their course as well as track students' results and practice work.
- With push-button ease, instructors can remove, hide, or annotate Addison-Wesley preloaded content, add their own course documents, or change the order in which material is presented.
- Using the communication tools found in MyMathLab, instructors can hold on-line office hours, host a discussion board, create communication groups within their class, send e-mail, and maintain a course calendar.
- Print supplements are available on-line, side-by-side with the textbook.

For more information, visit our Web site at www.mymathlab.com or contact your Addison-Wesley sales representative for a live demonstration.

For the Instructor

CLASSROOM EXAMPLE

Simplify $\dfrac{2(7 + 8) + 2}{3 \cdot 5 + 1}$.

Answer: 2

TEACHING TIP Warn students that $\dfrac{4}{15} + \dfrac{5}{9}$ is NOT EQUAL TO $\dfrac{9}{24}$.

Annotated Instructor's Edition (ISBN 0-321-12712-9)   For immediate access, the Annotated Instructor's Edition provides answers to all text exercises in the margin or next to the corresponding exercise, as well as Classroom Examples (formerly Chalkboard Examples) and Teaching Tips, printed in blue for easy visibility. Based on user feedback, we have increased the number of Classroom Examples and Teaching Tips.

Exercises designed for writing and graphing calculator use are indicated in both the Student Edition and the Annotated Instructor's Edition.

Instructor's Solutions Manual (ISBN 0-321-15711-7)   The *Instructor's Solutions Manual*, by Jeff Cole, Anoka-Ramsey Community College, provides complete solutions to all text exercises.

Answer Book (ISBN 0-321-15713-3)   The *Answer Book* provides answers to all the exercises in the text.

Printed Test Bank (ISBN 0-321-15712-5)   Written by Jon Becker, Indiana University Northwest, the *Printed Test Bank* contains two diagnostic pretests, four free-response and two multiple-choice test forms per chapter, and two final exams. Additional practice exercises for almost every objective of every section of the text are also included. A conversion guide from the eighth to the ninth edition is also included.

Adjunct Support Manual (ISBN 0-321-19649-X)   This manual includes resources designed to help both new and adjunct faculty with course preparation and classroom management as well as offering helpful teaching tips.

Adjunct Support Center   The Adjunct Support Center offers consultation on suggested syllabi, helpful tips on using the textbook support package, assistance with textbook content, and advice on classroom strategies from qualified mathematics instructors with over fifty years of combined teaching experience. The Adjunct Support Center is available Sunday through Thursday evenings from 5 P.M. to midnight. Phone: 1-800-435-4084; E-mail: adjunctsupport@awl.com; Fax: 1-877-262-9774.

TestGen with QuizMaster **(ISBN 0-321-15709-5)** TestGen enables instructors to build, edit, print, and administer tests using a computerized bank of questions developed to cover all the objectives of the text. Instructors can modify test bank questions or add new questions by using the built-in question editor, which allows users to create graphs, import graphics, insert math notation, and insert variable numbers or text. Tests can be printed or administered on-line via the Web or other network. TestGen comes packaged with QuizMaster, which allows students to take tests on a local area network. The software is available on a dual-platform Windows/Macintosh CD-ROM.

MathXL MathXL is an on-line testing, homework, and tutorial system that uses algorithmically generated exercises correlated to the textbook. Instructors can assign tests and homework provided by Addison-Wesley or create and customize their own tests and homework assignments. Instructors can also track their students' results and tutorial work in an on-line gradebook. Students can take chapter tests and receive personalized study plans that diagnose weaknesses and link students to areas they need to study and retest. Students can also work unlimited practice problems and receive tutorial instruction for areas in which they need improvement. MathXL can be packaged with new copies of *Beginning Algebra,* Ninth Edition. Please contact your Addison-Wesley sales representative for details.

MyMathLab MyMathLab is a complete on-line course for Addison-Wesley mathematics textbooks that provides interactive, multimedia instruction correlated to the textbook content. MyMathLab is easily customizable to suit the needs of students and instructors and provides a comprehensive and efficient on-line course-management system that allows for diagnosis, assessment, and tracking of students' progress.

MyMathLab features:

- Chapter and section folders in the on-line course mirror the textbook Table of Contents and contain a wide range of multimedia instruction, including video lectures, tutorial software, and electronic supplements.
- Actual pages of the textbook are loaded into MyMathLab, and as students work through a section of on-line text, they can link to multimedia resources—such as video and audio clips, tutorial exercises, and interactive animations—that are correlated directly to examples and exercises in the text.
- Hyperlinks take the user directly to on-line testing, diagnosis, tutorials, and tracking in MathXL—Addison-Wesley's tutorial and testing system for mathematics and statistics.
- Instructors can create, copy, edit, assign, and track all tests and homework for their course as well as track students' results and practice work.
- With push-button ease, instructors can remove, hide, or annotate Addison-Wesley preloaded content, add their own course documents, or change the order in which material is presented.
- Using the communication tools found in MyMathLab, instructors can hold on-line office hours, host a discussion board, create communication groups within their class, send e-mail, and maintain a course calendar.
- Print supplements are available on-line, side-by-side with the textbook.

For more information, visit our Web site at www.mymathlab.com or contact your Addison-Wesley sales representative for a live demonstration.

ACKNOWLEDGMENTS

The comments, criticisms, and suggestions of users, nonusers, instructors, and students have positively shaped this textbook over the years, and we are most grateful for the many responses we have received. The feedback gathered for this revision of the text was particularly helpful, and we especially wish to thank the following individuals who provided invaluable suggestions:

James J. Ball, *Indiana State University*
Dixilee Blackinton, *Weber State University*
Marc Campbell, *Daytona Beach Community College*
Carol Flakus, *Lower Columbia College*
Reginald W. Fulwood, *Palm Beach Community College*
Nancy R. Johnson, *Manatee Community College*
Mike Kirby, *Tidewater Community College, Virginia Beach*
Linda Kodama, *Kapi'olani Community College*
Marilyn Platt, *Gaston College*
Larry Pontaski, *Pueblo Community College*
Timothy Thompson, *Oregon Institute of Technology*
Mark Tom, *College of the Sequoias*
Virginia Urban, *Fashion Institute of Technology*
Mansoor Vejdani, *University of Cincinnati*

Special thanks are due all those instructors at Broward Community College for their insightful comments.

Over the years, we have come to rely on an extensive team of experienced professionals. Our sincere thanks go to these dedicated individuals at Addison-Wesley, who worked long and hard to make this revision a success: Maureen O'Connor, Suzanne Alley, Dona Kenly, Kathy Manley, Dennis Schaefer, Susan Raymond, Jolene Lehr, and Lindsay Skay.

Abby Tanenbaum did an outstanding job checking the answers to exercises and also provided invaluable assistance during the production process. Steven Pusztai provided his customary excellent production work. Thanks are due Jeff Cole, who supplied accurate, helpful solutions manuals, and Jon Becker, who provided the comprehensive *Printed Test Bank*. We are most grateful to Paul Van Erden for yet another accurate, useful index; Becky Troutman for preparing the comprehensive List of Applications; and Ellen Sawyer, Joyce Nemeth, Shannon d'Hemecourt, and Abraham Biggs for accuracy checking page proofs.

As an author team, we are committed to the goal stated earlier in this Preface—to provide the best possible text and supplements package to help instructors teach and students succeed. We are most grateful to all those over the years who have aspired to this goal with us. As we continue to work toward it, we would welcome any comments or suggestions you might have. Please feel free to send your comments via e-mail to math@aw.com.

Margaret L. Lial
John Hornsby
Terry McGinnis

Feature Walkthrough

Chapter Opener

Each chapter opens with an application and section outline. The application in the opener is tied to specific material within the chapter.

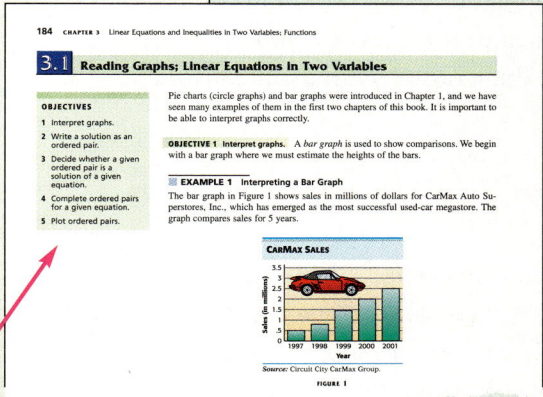

Learning Objectives

Each section opens with a highlighted list of clearly stated, numbered learning objectives. These learning objectives are reinforced throughout the section by restating the learning objective where appropriate so that students always know exactly what is being covered.

xxi

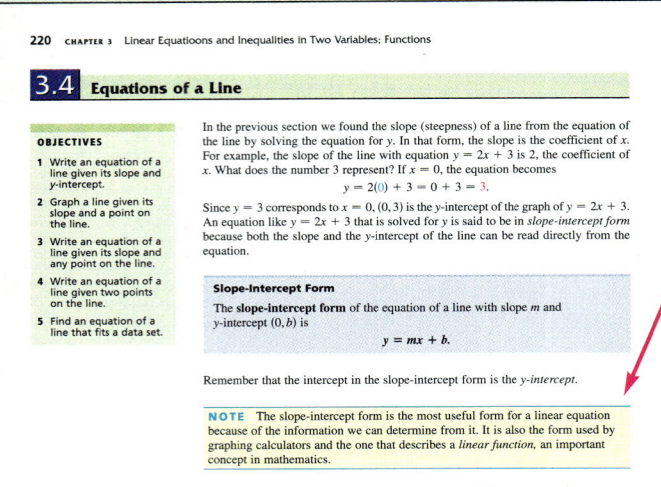

Notes

Important ideas are emphasized in *Note* boxes that appear throughout the text.

Cautions

Students are warned of common errors through the use of *Caution* boxes that are found throughout the text.

New! Now Try Exercises

Now Try exercises are found after each example to encourage active learning. This feature asks students to work exercises in the exercise sets that parallel the example just studied.

Classroom Examples and Teaching Tips

The Annotated Instructor's Edition provides answers to all text exercises and Group Activities in color in the margin or next to the corresponding exercise. In addition, *Classroom Examples* and *Teaching Tips* are included to assist instructors in creating examples to use in class that are different from what students have in their textbooks. *Teaching Tips* offer guidance on presenting the material at hand.

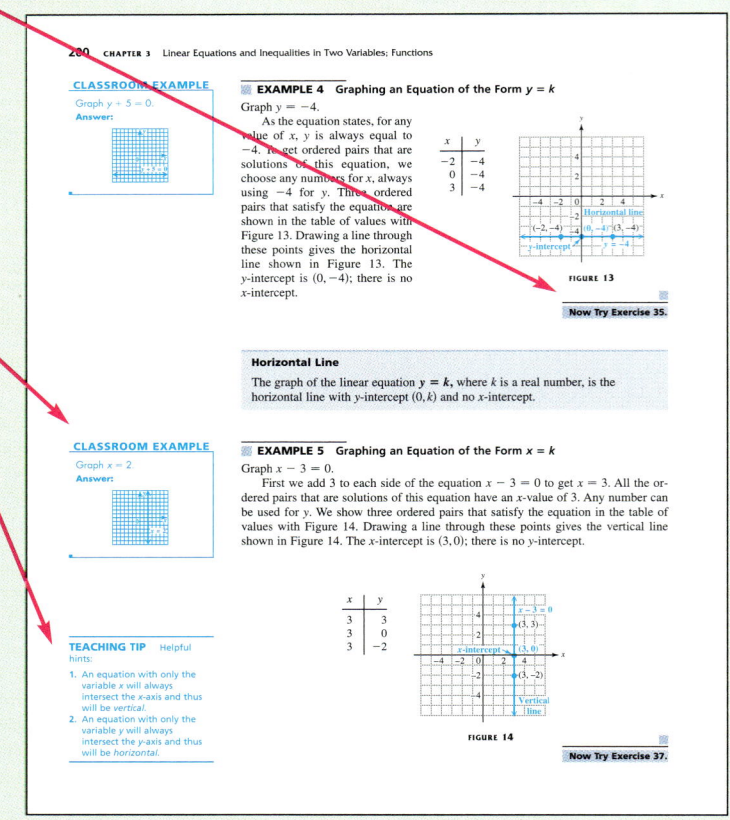

xxii

Connections

Connections boxes provide connections to the real world or to other mathematical concepts, historical background, and thought-provoking questions for writing or class discussion.

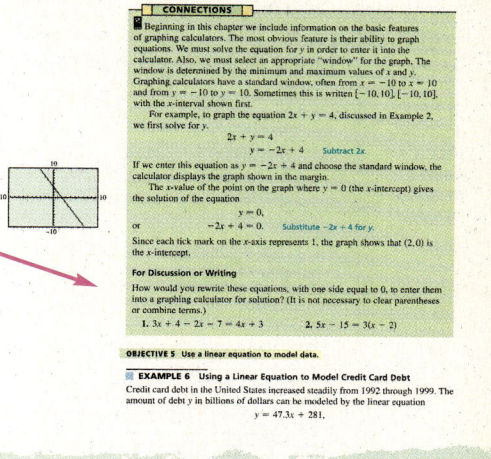

Writing Exercises

Writing exercises abound in the Lial series through the *Connections* boxes and also in the exercise sets. Some writing exercises require only short written answers, and others require lengthier journal-type responses where students are asked to fully explain terminology, procedures and methods, document their understanding using examples, or make connections between topics.

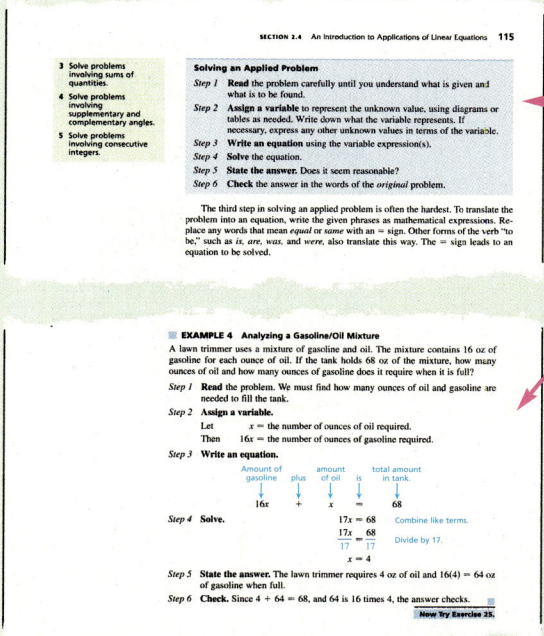

Problem Solving

The Lial *six-step problem-solving* method is clearly explained in Chapter 2 and is then continually reinforced in examples and exercises throughout the text to aid students in solving application problems.

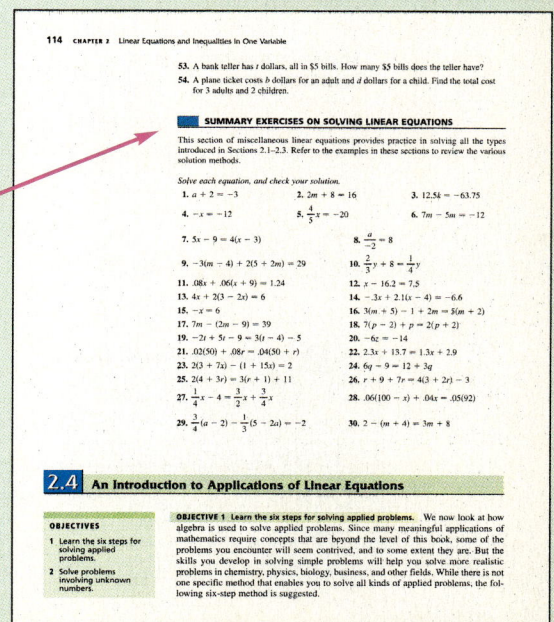

Summary Exercises

Summary Exercises appear in many chapters to provide students with *mixed* practice problems needed to master topics.

xxiii

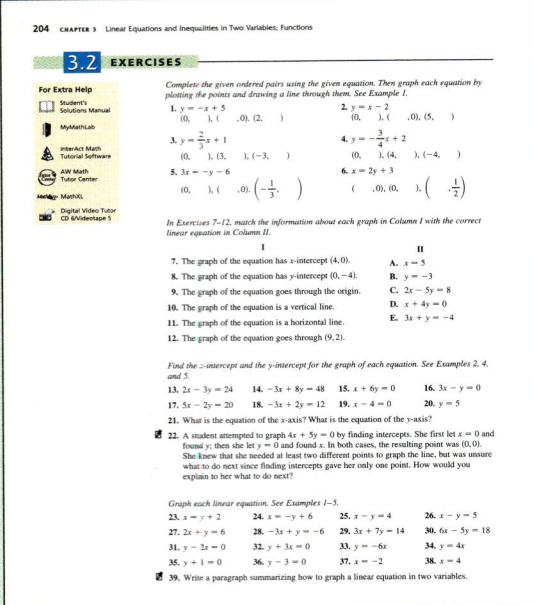

Ample and Varied Exercise Sets

Algebra students require a large number of varied practice exercises to master the material they have just learned. This text contains over 7700 exercises, including over 2100 review exercises, plus numerous conceptual and writing exercises and challenging exercises that go beyond the examples. Multiple-choice, matching, true/false, and completion exercises help to provide variety. Exercises suitable for graphing calculator use are marked with an icon.

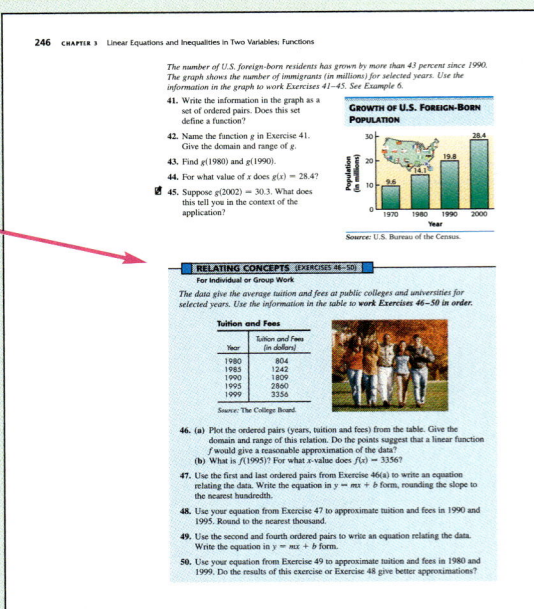

Relating Concepts

Found in selected exercise sets, these exercises tie together topics and highlight the relationships among various concepts and skills. For example, they may show how algebra and geometry are related or how a graph of a linear equation in two variables is related to the solution of the corresponding linear equation in one variable. These sets of exercises make great collaborative activities for small groups of students.

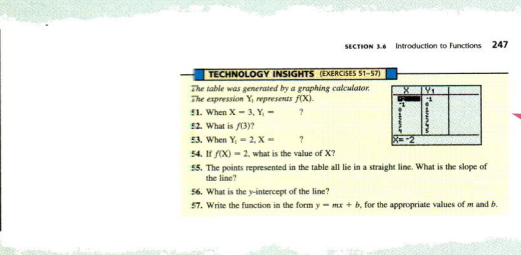

Technology Insights

Technology Insights exercises are found in selected exercise sets throughout the text. These exercises illustrate the power of graphing calculators and provide an opportunity for students to interpret typical results seen on graphing calculator screens. (A graphing calculator is *not* required to complete these exercises.)

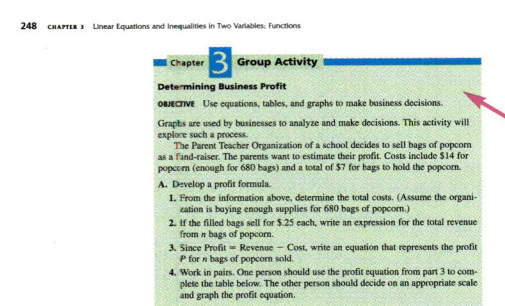

Group Activities

Appearing at the end of each chapter, these activities allow students to work collaboratively to solve a problem related to the chapter material.

Ample Opportunity for Review

One of the most popular features of the Lial textbooks is the extensive and well thought-out end-of-chapter material. At the end of each chapter, students will find:

Key Terms and New Symbols that are keyed back to the appropriate section for easy reference and study

Test Your Word Power to help students understand and master mathematical vocabulary; Key terms from the chapter are presented with four possible definitions in multiple-choice format.

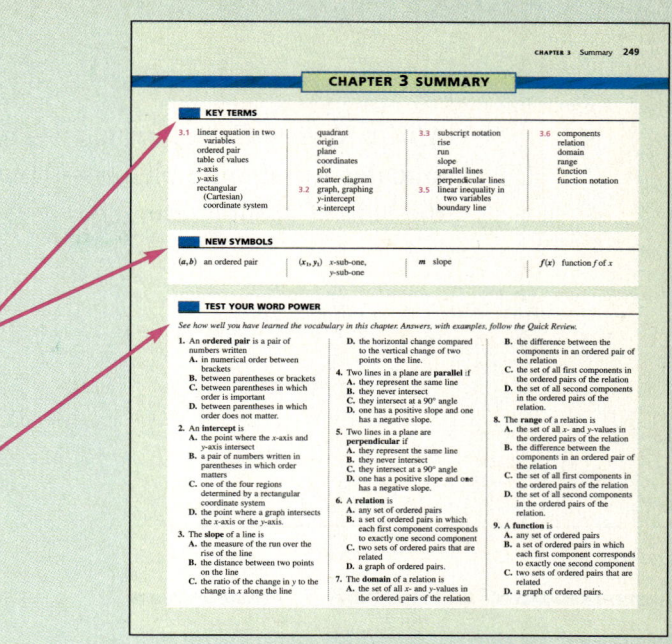

Quick Review sections to give students not only the main concepts from the chapter (referenced back to the appropriate section) but also an adjacent example of each concept

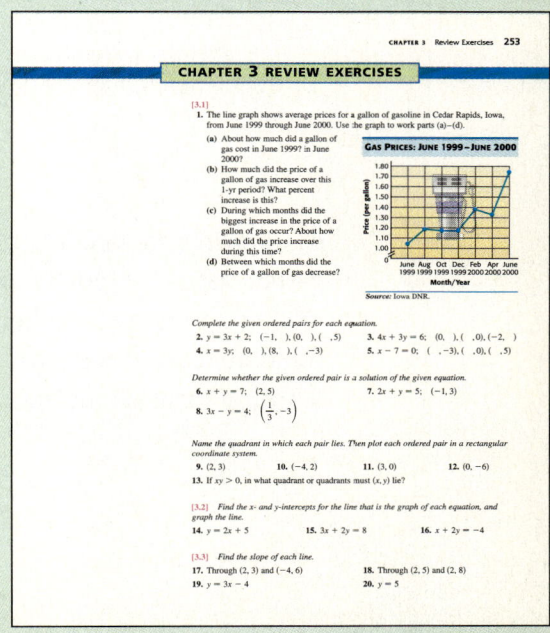

Review Exercises keyed to the appropriate sections so that students can refer to examples of that type of problem if they need help

XXV

Mixed Review Exercises that require students to solve problems without the help of section references

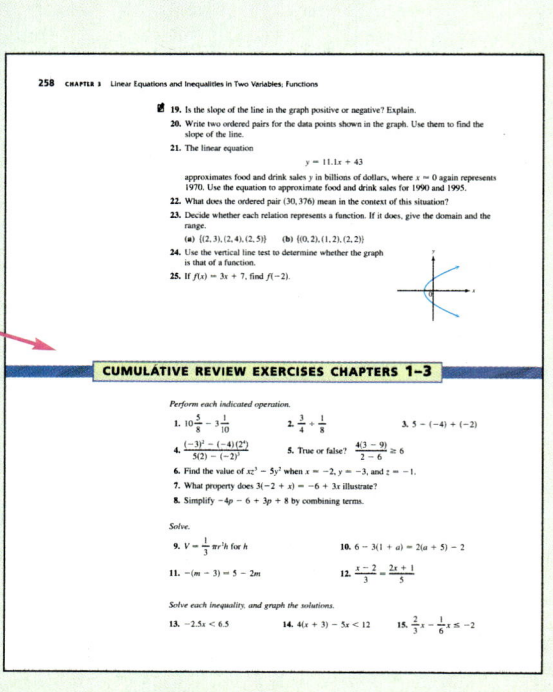

Chapter Test to help students practice for the real thing

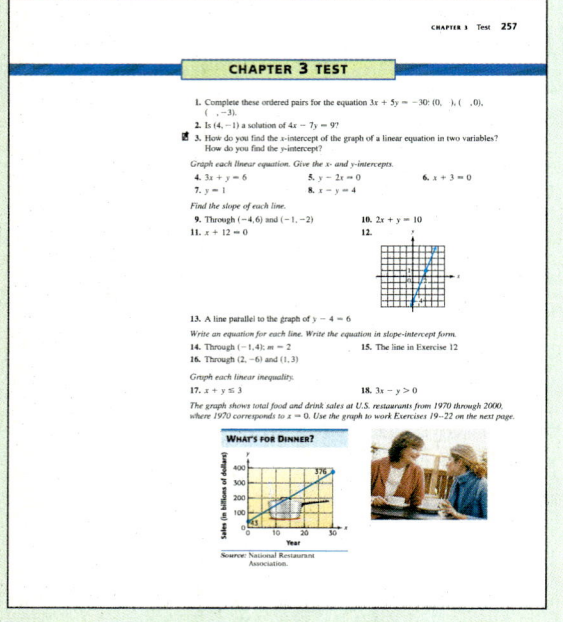

Cumulative Review Exercises that gather various types of exercises from preceding chapters to help students remember and retain what they are learning throughout the course

Topics in Arithmetic

A typical Starbucks location has between 20 and 25 employees. Each of these employees is involved with work schedules, keeping track of their work hours, computing sales and sales taxes, and ordering inventory. An understanding of whole numbers, fractions, decimals, and percent is essential because each of these employees will be using arithmetic as part of their jobs.

Whole Numbers
- **0.1** Reading and Writing Whole Numbers
- **0.2** Adding Whole Numbers
- **0.3** Subtracting Whole Numbers
- **0.4** Multiplying Whole Numbers
- **0.5** Dividing Whole Numbers
- **0.6** Long Division
  Test on Sections 0.1–0.6

Fractions (Part 1)
- **0.7** Basics of Fractions
- **0.8** Mixed Numbers
- **0.9** Factors
- **0.10** Writing a Fraction in Lowest Terms
- **0.11** Multiplying Fractions
- **0.12** Dividing Fractions
- **0.13** Multiplying and Dividing Mixed Numbers
  Test on Sections 0.7–0.13

Fractions (Part 2)
- **0.14** Adding and Subtracting Like Fractions
- **0.15** Least Common Multiples
- **0.16** Adding and Subtracting Unlike Fractions
- **0.17** Adding and Subtracting Mixed Numbers
  Test on Sections 0.14–0.17

Decimals
- **0.18** Reading and Writing Decimals
- **0.19** Rounding Decimals
- **0.20** Adding and Subtracting Decimals
- **0.21** Multiplying Decimals
- **0.22** Dividing Decimals
- **0.23** Writing Fractions as Decimals
  Test on Sections 0.18–0.23

Ratio and Proportion
- **0.24** Ratios
- **0.25** Proportions
- **0.26** Solving Proportions
  Test on Sections 0.24–0.26

Unit Analysis
- **0.27** Analyzing Units with English Measurement
- **0.28** The Metric System—Length
- **0.29** The Metric System—Capacity and Weight (Mass)
- **0.30** Metric-English Conversions
  Test on Sections 0.27–0.30

Percent
- **0.31** Basics of Percent
- **0.32** Percents and Fractions
- **0.33** Using the Percent Proportion
- **0.34** Using Proportions to Solve Percent Problems
- **0.35** Using the Percent Equation
  Test on Sections 0.31–0.35

0.1 Reading and Writing Whole Numbers

OBJECTIVES

1. Identify whole numbers.
2. Give the place value of a digit.

Knowing how to read and write numbers is an important step in learning mathematics.

OBJECTIVE 1 Identify whole numbers. The **decimal system** of writing numbers uses the ten digits

0, 1, 2, 3, 4, 5, 6, 7, 8, 9

to write any number. For example, these digits can be used to write the **whole numbers:**

0, 1, 2, 3, 4, 5, 6, 7, 8, 9, 10, 11, 12, 13 . . .

The three dots indicate that the list goes on forever.

OBJECTIVE 2 Give the place value of a digit. Each digit in a whole number has a **place value,** depending on its position in the whole number. The following place value chart shows the names of the different places used most often and has the whole number 29,022 entered.

1. Identify the place value of the 4 in each whole number.

   (a) 341

   (b) 714

   (c) 479

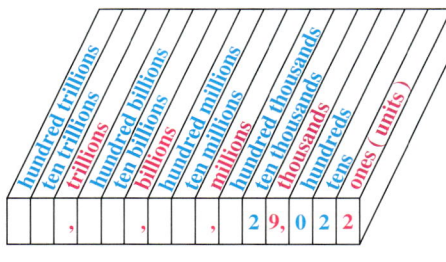

At 29,022 ft, Mt. Everest is the world's tallest mountain.

The tallest mountain in the world is Mt. Everest at 29,022 ft. Each of the 2s in 29,022 represents a different amount because of its position, or *place,* within the number. The *place value* of the 2 on the left is 2 ten thousands (20,000). The place value of the middle 2 is 2 tens (20), and the place value of the 2 on the right is 2 ones (2).

EXAMPLE 1 Identifying Place Values

Identify the place value of 8 in each whole number.

(a) 28   (b) 85   (c) 869
    ↳ 8 ones   ↳ 8 tens   ↳ 8 hundreds

Notice that the value of 8 in each number is different, depending on its location (place) in the number.

Work Problem 1 at the Side.

ANSWERS
1. (a) tens (b) ones (c) hundreds

SECTION 0.1 Reading and Writing Whole Numbers   0-3

2. Identify the place value of each digit.

   (a) 24,386

   (b) 371,942

EXAMPLE 2 Identifying Place Values

Identify the place value of each digit in the number 725,283.

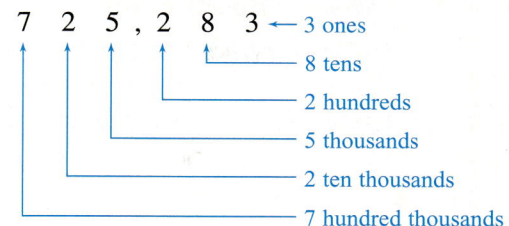

Notice the comma between the hundreds and thousands position in the number 725,283 in Example 2.

Work Problem 2 at the Side.

3. In the number 3,251,609,328 identify the digits in each period (group).

   (a) billions period

   (b) millions period

   (c) thousands period

   (d) ones period

Using Commas

Commas are used to separate each group of three digits, starting from the right. This makes numbers easier to read. (An exception: commas are frequently omitted in four-digit numbers such as 9748 or 1329.) Each three-digit group is called a **period.** Some instructors prefer to just call them **groups.**

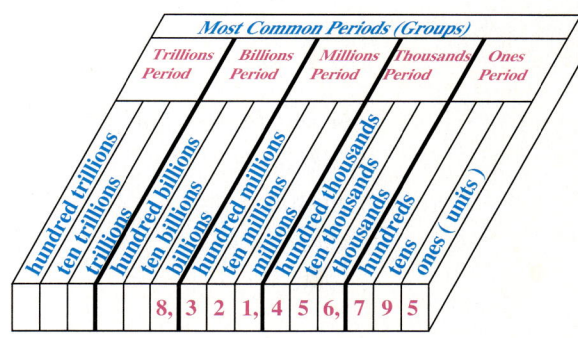

EXAMPLE 3 Knowing the Period or Group Names

Write the digits in each period of 8,321,456,795.

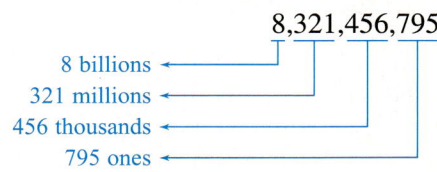

Work Problem 3 at the Side.

ANSWERS
2. (a) 2: ten thousands
       4: thousands
       3: hundreds
       8: tens
       6: ones
   (b) 3: hundred thousands
       7: ten thousands
       1: thousands
       9: hundreds
       4: tens
       2: ones
3. (a) 3 (b) 251 (c) 609
   (d) 328

0-4 CHAPTER 0 Topics in Arithmetic

Use the following rule to read a number with more than three digits.

> **Writing Numbers in Words**
>
> Start at the left when writing a number in words or saying it aloud. Write or say the digit names in each period (group), followed by the name of the period, except for the period name "ones," which is *not* used.

0.1 EXERCISES

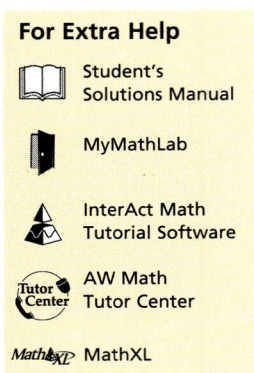

For Extra Help
- Student's Solutions Manual
- MyMathLab
- InterAct Math Tutorial Software
- AW Math Tutor Center
- MathXL

*Write the digit for the given **place value** in each whole number. See Examples 1 and 2.*

1. 2078
   thousands
   tens

2. 5139
   thousands
   ones

3. 18,015
   ten thousands
   hundreds

4. 75,229
   ten thousands
   ones

5. 7,628,592,183
   millions
   thousands

6. 1,700,225,016
   billions
   millions

*Write the digits for the given **period** (group) in each whole number. See Example 3.*

7. 2,768,543
   millions
   thousands
   ones

8. 28,785,203
   millions
   thousands
   ones

9. 60,000,502,109
   billions
   millions
   thousands
   ones

10. 100,258,100,006
    billions
    millions
    thousands
    ones

11. Do you think the fact that humans have four fingers and a thumb on each hand explains why we use a number system based on ten digits? Explain.

12. The decimal system uses ten digits. Fingers and toes are often referred to as digits. In your opinion, is there a relationship here? Explain.

0.2 Adding Whole Numbers

OBJECTIVES

1. Add two single-digit numbers.
2. Add more than two numbers.
3. Add when carrying is not required.
4. Add with carrying.

There are four dollar bills at the left and two at the right. In all, there are six dollar bills.

The process of finding the total is called **addition**. Here 4 and 2 were added to get 6. Addition is written with a + sign, so that

$$4 + 2 = 6.$$

OBJECTIVE 1 Add two single-digit numbers. In addition, the numbers being added are called **addends**, and the resulting answer is called the **sum** or **total**.

$$\begin{array}{r} 4 \leftarrow \text{Addend} \\ +\,2 \leftarrow \text{Addend} \\ \hline 6 \leftarrow \text{Sum (total)} \end{array}$$

Addition problems can also be written horizontally, as follows.

$$\underset{\text{Addend}}{4} \; + \; \underset{\text{Addend}}{2} \; = \; \underset{\text{Sum}}{6}$$

1. Add, and then change the order of numbers to write another addition problem.

    (a) 6 + 3

    (b) 7 + 9

    (c) 7 + 8

    (d) 6 + 9

Commutative Property of Addition

By the **commutative property of addition,** changing the order of the addends in an addition problem does not change the sum.

For example, the sum of 4 + 2 is the same as the sum of 2 + 4. This allows the addition of the same numbers in a different order.

EXAMPLE 1 Adding Two Single-Digit Numbers

Add, and then change the order of numbers to write another addition problem.

(a) 6 + 2 = 8 and 2 + 6 = 8
(b) 5 + 9 = 14 and 9 + 5 = 14
(c) 8 + 3 = 11 and 3 + 8 = 11
(d) 8 + 8 = 16

Work Problem 1 at the Side.

Associative Property of Addition

By the **associative property of addition,** changing the grouping of the addends in an addition problem does not change the sum.

ANSWERS

1. (a) 9; 3 + 6 = 9
   (b) 16; 9 + 7 = 16
   (c) 15; 8 + 7 = 15
   (d) 15; 9 + 6 = 15

CHAPTER 0 Topics in Arithmetic

2. Add each column of numbers.

(a) 6
     5
     3
     2
    +9

(b) 4
     6
     5
     3
    +2

(c) 8
     7
     9
     2
    +1

(d) 3
     8
     6
     4
    +8

For example, the sum of $3 + 5 + 6$ may be found as follows.

$$(3 + 5) + 6 = 8 + 6 = 14 \quad \text{Parentheses tell what to do first.}$$

Another way to add the same numbers is

$$3 + (5 + 6) = 3 + 11 = 14.$$

Either grouping gives the answer 14.

OBJECTIVE 2 Add more than two numbers. To add several numbers, first write them in a column. Add the first number to the second. Add this sum to the third digit; continue until all the digits are used.

EXAMPLE 2 Adding More Than Two Numbers

Add 2, 5, 6, 1, and 4.

$$\begin{array}{r} 2 \\ 5 \\ 6 \\ 1 \\ +4 \\ \hline 18 \end{array}$$

$2 + 5 = 7$
$7 + 6 = 13$
$13 + 1 = 14$
$14 + 4 = 18$

NOTE By the commutative and associative properties of addition, numbers may also be added starting at the bottom of a column. Adding from the top or adding from the bottom will give the same answer.

Work Problem 2 at the Side.

OBJECTIVE 3 Add when carrying is not required. If numbers have two or more digits, you must arrange the numbers in columns so that the ones digits are in the same column, tens are in the same column, hundreds are in the same column, and so on. Next, you add column by column starting at the right.

EXAMPLE 3 Adding without Carrying

Add $511 + 23 + 154 + 10$.

First line up the numbers in columns, with the ones column at the right.

```
        Hundreds in a column
         Tens in a column
          Ones in a column
    5 1 1
      2 3
    1 5 4      Ones digits at
  +   1 0          the right
```

ANSWERS
2. (a) 25   (b) 20   (c) 27
     (d) 29

SECTION 0.2 Adding Whole Numbers 0-7

3. Add.

 (a) 34
     + 62

 (b) 478
     + 221

 (c) 42,305
     + 11,563

Now start at the right and add the ones digits. Add the tens digits next, and finally, the hundreds digits.

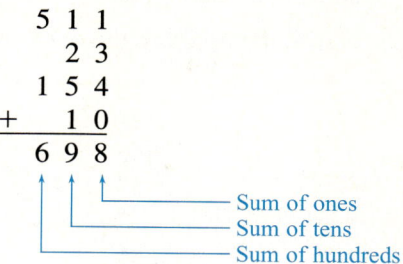
— Sum of ones
— Sum of tens
— Sum of hundreds

The sum of the four numbers is 698.

Work Problem 3 at the Side.

OBJECTIVE 4 Add with carrying. If the sum of the digits in any column is more than 9, use **carrying**.

4. Add by carrying.

 (a) 76
     + 19

 (b) 76
     + 18

 (c) 56
     + 37

 (d) 34
     + 49

EXAMPLE 4 Adding with Carrying

Add 47 and 29.

Add ones.

$$\begin{array}{r} 47 \\ + 29 \\ \hline \end{array}$$

— Sum of ones is 16.

Because 16 is 1 ten plus 6 ones, place 6 in the ones column and carry 1 to the tens column.

$$\begin{array}{r} 1 \\ 47 \\ + 29 \\ \hline 6 \end{array} \quad 7 + 9 = 16$$

Add the tens column, including the carried 1.

$$\begin{array}{r} 1 \\ 47 \\ + 29 \\ \hline 76 \end{array}$$

— Sum of digits in tens column

Work Problem 4 at the Side.

ANSWERS
3. (a) 96 (b) 699 (c) 53,868
4. (a) 95 (b) 94 (c) 93 (d) 83

0-8 CHAPTER 0 Topics in Arithmetic

5. Add by carrying as necessary.

(a) 396
      87
      42
   + 651

(b) 4271
      372
     8976
   + 162

(c)   57
       4
     392
     804
      51
   + 27

(d) 7821
      435
       72
      305
   + 1693

EXAMPLE 5 Adding with Carrying

Add 324 + 7855 + 23 + 7 + 86.

Add the digits in the ones column.

```
     2
   324
  7855
    23
     7
 +  86
     5
```

Carry 2 to the tens column.

Sum of the ones column is 25.

Write 5 in the ones column.

In 25, the 5 represents 5 ones and is written in the ones column, while 2 represents 2 tens and is carried to the tens column.

Now add the digits in the tens column, including the carried 2.

```
    12
   324
  7855
    23
     7
 +  86
    95
```

Carry 1 to the hundreds column.

Sum of the tens column is 19.

Write 9 in the tens column.

Add the hundreds column, including the carried 1.

```
   112
   324
  7855
    23
     7
 +  86
   295
```

Carry 1 to the thousands column.

Sum of the hundreds column is 12.

Write 2 in the hundreds column.

Add the thousands column, including the carried 1.

```
   112
   324
  7855
    23
     7
 +  86
  8295
```

Sum of the thousands column is 8.

Finally, 324 + 7855 + 23 + 7 + 86 = 8295.

ANSWERS

5. (a) 1176 (b) 13,781 (c) 1335
     (d) 10,326

Work Problem 5 at the Side.

SECTION 0.2 Adding Whole Numbers 0-9

6. Add by carrying mentally.

```
   816
   363
    17
     2
     5
+ 7654
```

NOTE For additional speed, try to carry mentally. Do not write the carried number, but just remember it as you move to the top of the next column. Try this method. If it works for you, use it.

Work Problem 6 at the Side.

ANSWERS
6. 8857

0.2 EXERCISES

For Extra Help

 Student's Solutions Manual

 MyMathLab

 InterAct Math Tutorial Software

 AW Math Tutor Center

MathXL MathXL

Add. See Examples 1–3.

1. 34
    + 45

2. 14
    + 13

3. 56
    + 33

4. 83
    + 15

5. 317
    + 572

6. 738
    + 261

7. 318
     151
    + 420

8. 135
     253
    + 410

9. 6310
      252
    + 1223

10. 121
      5705
    + 3163

11. 721 + 23 + 834

12. 517 + 131 + 250

13. 1251 + 4311 + 2114

14. 3241 + 1513 + 2014

15. 12,142 + 43,201 + 23,103

16. 41,124 + 12,302 + 23,500

17. 3213 + 5715

18. 6344 + 1655

19. 38,204 + 21,020

20. 63,251 + 36,305

Add, carrying as necessary. See Examples 4 and 5.

21. 36
    + 78

22. 91
    + 29

23. 86
    + 69

24. 37
    + 85

25. 47
    + 74

26. 79
    + 79

27. 67
    + 78

28. 96
    + 47

29. 73
    + 29

30. 68
    + 37

31. 746
    + 905

32. 621
    + 359

33. 306
    + 848

34. 798
    + 206

35. 278
    + 135

36. 172
    + 156

37. 928
    + 843

38. 686
    + 726

39. 526
    + 884

40. 116
    + 897

0-10   CHAPTER 0   Topics in Arithmetic

41.	7968 + 1285	42.	1768 + 8275	43.	7896 + 3728	44.	9382 + 7586	45.	9625 + 7986
46.	5718 5623 + 7436	47.	9056 78 6089 + 731	48.	4022 709 8621 + 37	49.	18 708 9286 + 636	50.	1708 321 61 + 8926
51.	422 6074 435 + 8663	52.	6505 173 1044 + 168	53.	321 9603 8 21 + 1604	54.	7631 5983 7 36 + 505	55.	2109 63 16 3 + 9887
56.	322 6508 93 745 18 + 2005	57.	553 97 2772 437 63 + 328	58.	3187 810 527 76 2665 + 317	59.	413 85 9919 602 31 + 1218	60.	576 7934 60 781 5968 + 371

0.3 Subtracting Whole Numbers

OBJECTIVES

1. Change addition problems to subtraction and subtraction problems to addition.
2. Identify the minuend, subtrahend, and difference.
3. Subtract when no borrowing is needed.
4. Check subtraction answers by adding.
5. Subtract by borrowing.

Suppose you have $8, and you spend $5 for gasoline. You then have $3 left. There are two different ways of looking at these numbers.

As an addition problem:

$$\underset{\text{Amount spent}}{\$5} + \underset{\text{Amount left}}{\$3} = \underset{\text{Original amount}}{\$8}$$

As a subtraction problem:

$$\underset{\text{Original amount}}{\$8} \underset{\text{Subtraction symbol}}{-} \underset{\text{Amount spent}}{\$5} = \underset{\text{Amount left}}{\$3}$$

OBJECTIVE 1 Change addition problems to subtraction and subtraction problems to addition. As shown in the preceding box, an addition problem can be changed to a subtraction problem and a subtraction problem can be changed to an addition problem.

SECTION 0.3 Subtracting Whole Numbers 0-11

1. Write two subtraction problems for each addition problem.

(a) $6 + 1 = 7$

(b) $7 + 4 = 11$

(c) $15 + 22 = 37$

(d) $23 + 55 = 78$

2. Write an addition problem for each subtraction problem.

(a) $6 - 4 = 2$

(b) $9 - 4 = 5$

(c) $21 - 15 = 6$

(d) $58 - 42 = 16$

■ **EXAMPLE 1** **Changing Addition Problems to Subtraction**

Change each addition problem to a subtraction problem.

(a) $4 + 1 = 5$

Two subtraction problems are possible:

$$5 - 1 = 4 \quad \text{or} \quad 5 - 4 = 1$$

These figures show each subtraction problem.

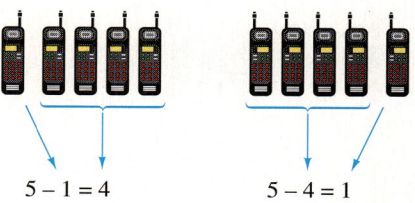

$5 - 1 = 4 \qquad 5 - 4 = 1$

(b) $8 + 7 = 15$

$$15 - 7 = 8 \quad \text{or} \quad 15 - 8 = 7$$

Work Problem 1 at the Side.

■ **EXAMPLE 2** **Changing Subtraction Problems to Addition**

Change each subtraction problem to an addition problem.

(a) $8 - 3 = 5$

$8 = 3 + 5$

It is also correct to write $8 = 5 + 3$.

(b) $18 - 13 = 5$
$18 = 13 + 5$

(c) $29 - 13 = 16$
$29 = 13 + 16$

Work Problem 2 at the Side.

OBJECTIVE 2 Identify the minuend, subtrahend, and difference. In subtraction, as in addition, the numbers in a problem have names. For example, in the problem $8 - 5 = 3$, the number 8 is the **minuend,** 5 is the **subtrahend,** and 3 is the **difference** or answer.

$8 \; - \; 5 \; = 3 \leftarrow$ Difference (answer)
↑   ↑
Minuend   Subtrahend

$8 \leftarrow$ Minuend
$-5 \leftarrow$ Subtrahend
$3 \leftarrow$ Difference

ANSWERS
1. (a) $7 - 1 = 6$ or $7 - 6 = 1$
(b) $11 - 4 = 7$ or $11 - 7 = 4$
(c) $37 - 22 = 15$ or $37 - 15 = 22$
(d) $78 - 55 = 23$ or $78 - 23 = 55$
2. (a) $6 = 4 + 2$ (b) $9 = 4 + 5$
(c) $21 = 15 + 6$
(d) $58 = 42 + 16$

3. Subtract.

(a)   63
    − 22
    ─────

(b)   47
    − 35
    ─────

(c)   429
    − 318
    ─────

(d)   3927
    − 2614
    ─────

(e)   5464
    − 324
    ─────

ANSWERS
3. (a) 41 (b) 12 (c) 111
     (d) 1313 (e) 5140

OBJECTIVE 3 Subtract when no borrowing is needed. Subtract two numbers by lining up the numbers in columns so the digits in the ones place are in the same column. Next subtract by columns, starting at the right with the ones column.

EXAMPLE 3 Subtracting Two Numbers
Subtract.

(a)   53   ← Ones digits are lined up in the same column.
    − 21
    ─────
     32
          3 − 1 = 2
          5 − 2 = 3

(b)   385   ← Ones digits are lined up.
    − 161
    ─────
     224   5 − 1 = 4
             8 − 6 = 2
             3 − 1 = 2

(c)   9431
    − 210
    ─────
     9221   ← 1 − 0 = 1
             3 − 1 = 2
             4 − 2 = 2
             9 − 0 = 9

Work Problem 3 at the Side.

OBJECTIVE 4 Check subtraction answers by adding. Use addition to check your answer to a subtraction problem. For example, check $8 - 3 = 5$ by *adding* 3 and 5:

$$3 + 5 = 8, \text{ so } 8 - 3 = 5 \text{ is correct.}$$

EXAMPLE 4 Checking Subtraction by Using Addition
Check each answer.

(a)   89
    − 47
    ─────
     42

Rewrite as an addition problem, as shown in Example 2.

Subtraction problem $\begin{cases} 89 \\ -47 \\ \overline{42} \end{cases}$ Addition problem $\begin{array}{r} 47 \\ + 42 \\ \hline 89 \end{array}$

Because $47 + 42 = 89$, the subtraction was done correctly.

4. Use addition to determine whether each answer is correct. If incorrect, what should it be?

(a) 76
   -45
   31

(b) 53
   -22
   21

(c) 374
   -251
   113

(d) 7531
   -4301
   3230

5. Subtract.

(a) 58
   -19

(b) 86
   -38

(c) 41
   -27

(d) 863
   -47

(e) 762
   -157

ANSWERS
4. (a) correct
   (b) incorrect, should be 31
   (c) incorrect, should be 123
   (d) correct
5. (a) 39 (b) 48 (c) 14 (d) 816
   (e) 605

(b) $72 - 41 = 21$

Rewrite as an addition problem.

$$72 = 41 + 21$$

But, $41 + 21 = 62$, not 72, so the subtraction was done incorrectly. Rework the original subtraction to get the correct answer, 31.

(c) 374 ← Match
   -141
   233 $141 + 233 = 374$

The answer checks.

Work Problem 4 at the Side.

OBJECTIVE 5 Subtract by borrowing. When a digit in the minuend is less than the one directly below it we subtract by using **borrowing**.

EXAMPLE 5 Subtracting with Borrowing

Subtract 19 from 57.

Write the problem.

$$57$$
$$-19$$

In the ones column, 7 is **less** than 9, so to subtract, we must **borrow a 10** from the 5 (which represents 5 tens, or 50).

$50 - 10 = 40$ 4 17 $10 + 7 = 17$
 $\cancel{5}\cancel{7}$
 -19

Now we can subtract $17 - 9$ in the ones column and then $4 - 1$ in the tens column.

4 17
$\cancel{5}\cancel{7}$
-19
38 Difference

Finally, $57 - 19 = 38$. Check by adding 19 and 38; you should get 57.

Work Problem 5 at the Side.

EXAMPLE 6 Subtracting with Borrowing

Subtract by borrowing as necessary.

(a) 7856
   -137

0-14 CHAPTER 0 Topics in Arithmetic

6. Subtract.

(a) 536
    − 75

(b) 348
    − 79

(c) 477
    − 389

(d) 1437
    − 988

(e) 8739
    − 3892

ANSWERS
6. (a) 461 (b) 269 (c) 88
   (d) 449 (e) 4847

There is no need to borrow, as 4 is greater than 3.

$$10 + 6 = 16$$
$$4\ 16$$
$$7\ 8\ \cancel{5}\ \cancel{6}$$
$$-\ \ 1\ 3\ 7$$
$$\overline{7\ 7\ 1\ 9}\quad \text{Difference}$$

(b) 635
    − 546

$$600 - 100 = 500 \qquad 100 + 20 = 120\ (12\ \text{tens} = 120)$$
$$10 + 5 = 15$$
$$5\ 12\ 15$$
$$\cancel{6}\ \cancel{3}\ \cancel{5}$$
$$-5\ 4\ 6$$
$$\overline{8\ 9}\quad \text{Difference}$$

(c) 3648
    − 1769

$$2\ 15\ 13\ 18$$
$$\cancel{3}\ \cancel{6}\ \cancel{4}\ \cancel{8}$$
$$-1\ 7\ 6\ 9$$
$$\overline{1\ 8\ 7\ 9}$$

▮ **Work Problem 6 at the Side.**

Sometimes a minuend has 0s (zeros) in some of the positions. In such cases, borrowing may be a little more complicated than what we have shown so far.

▬ **EXAMPLE 7 Borrowing with Zeros**

Subtract.

$$4607$$
$$-\ 3168$$

It is not possible to borrow from the tens position. Instead we must first borrow from the hundreds position.

$$600 - 100 = 500 \qquad 100 + 0 = 100$$
$$5\ 10$$
$$4\ \cancel{6}\ \cancel{0}\ 7$$
$$-\ 3\ 1\ 6\ 8$$

Now we may borrow from the tens position.

$$9 \qquad\qquad 100 - 10 = 90$$
$$5\ \cancel{10}\ 17 \qquad 10 + 7 = 17$$
$$4\ \cancel{6}\ \cancel{0}\ \cancel{7}$$
$$-\ 3\ 1\ 6\ 8$$
$$\overline{9}$$

SECTION 0.3 Subtracting Whole Numbers **0-15**

7. Subtract.

(a) 405
-363

(b) 304
-237

(c) 5073
-1632

8. Subtract.

(a) 308
-159

(b) 480
-275

(c) 1570
-983

(d) 7001
-5193

(e) 4000
-1782

Complete the problem.

$$\begin{array}{r} \overset{9}{}\\ 5\,\overset{}{10}\,17 \\ 4\,\cancel{6}\,\cancel{0}\,\cancel{7} \\ -\;3\;1\;6\;8 \\ \hline 1\;4\;3\;9 \end{array}$$ Difference

Check by adding 1439 and 3168; you should get 4607.

Work Problem 7 at the Side.

EXAMPLE 8 Borrowing with Zeros

Subtract.

(a) 708
-149

$100 + 0 = 100$ ⎯⎯⎯ $100 - 10 = 90$ (9 tens = 90)
$700 - 100 = 600$ ⎯⎯⎯ $10 + 8 = 18$

$$\begin{array}{r} \overset{9}{} \\ 6\,\overset{}{10}\,18 \\ \cancel{7}\,\cancel{0}\,\cancel{8} \\ -\;1\;4\;9 \\ \hline 5\;5\;9 \end{array}$$

(b) 380
-276

$80 - 10 = 70$ (7 tens = 70) $10 + 0 = 10$

$$\begin{array}{r} 7\;10 \\ 3\,\cancel{8}\,\cancel{0} \\ -\;2\;7\;6 \\ \hline 1\;0\;4 \end{array}$$

(c) 9000
-6999

$$\begin{array}{r} 9\;9 \\ 8\,\overset{}{10}\,\overset{}{10}\,10 \\ \cancel{9}\,\cancel{0}\,\cancel{0}\,\cancel{0} \\ -\;6\;9\;9\;9 \\ \hline 2\;0\;0\;1 \end{array}$$

Work Problem 8 at the Side.

ANSWERS
7. (a) 42 (b) 67 (c) 3441
8. (a) 149 (b) 205 (c) 587
(d) 1808 (e) 2218

0.3 EXERCISES

For Extra Help
 Student's Solutions Manual
 MyMathLab
 InterAct Math Tutorial Software
 AW Math Tutor Center
MathXL

Work each subtraction problem. Use addition to check each answer. See Examples 3 and 4.

1. 54 − 23	2. 19 − 12	3. 86 − 53	4. 78 − 35	5. 77 − 60
6. 95 − 71	7. 335 − 122	8. 602 − 301	9. 552 − 451	10. 888 − 215
11. 6821 − 610	12. 4420 − 310	13. 5546 − 2134	14. 1875 − 1362	15. 6259 − 4148
16. 9654 − 4323	17. 24,392 − 11,232	18. 57,921 − 34,801	19. 46,253 − 5 143	20. 75,904 − 3 702

Use addition to check each subtraction problem. If an answer is not correct, find the correct answer. See Example 4.

21. 43
− 31
  12
	22. 76
	− 43
	  33
		23. 89
		− 27
		  63
			24. 47
			− 35
			  13
				25. 382
				− 261
				  131

26. 754
− 342
  412
	27. 4683
	− 3542
	  1141
		28. 5217
		− 4105
		  1132
			29. 8643
			− 1421
			  7212
				30. 9428
				− 3124
				  6324

Subtract by borrowing as necessary. See Examples 5–8.

31. 45 − 27	32. 86 − 28	33. 94 − 49	34. 98 − 69	35. 57 − 38
36. 83 − 55	37. 828 − 547	38. 916 − 618	39. 771 − 252	40. 973 − 788
41. 9861 − 684	42. 6171 − 1182	43. 9988 − 2399	44. 3576 − 1658	45. 38,335 − 29,476
46. 61,278 − 3 559	47. 40 − 37	48. 80 − 73	49. 60 − 37	50. 70 − 27

51.	308 − 289	52.	600 − 599	53.	4041 − 1208	54.	4602 − 2063	55.	9305 − 1530
56.	7120 − 6033	57.	1580 − 1077	58.	3068 − 2105	59.	2006 − 1850	60.	8203 − 5365
61.	8240 − 6056	62.	7050 − 6045	63.	8503 − 2816	64.	16,004 − 5 087	65.	80,705 − 61,667
66.	81,000 − 55,456	67.	66,000 − 34,444	68.	77,000 − 65,308	69.	20,080 − 13,496	70.	80,056 − 23,869

0.4 Multiplying Whole Numbers

OBJECTIVES

1. Identify the parts of a multiplication problem.
2. Do chain multiplications.
3. Multiply by single-digit numbers.
4. Use multiplication shortcuts for numbers ending in zeros.
5. Multiply by numbers having more than one digit.

Suppose we want to know the total number of computers in a computer lab. The stations are arranged in four rows with three stations in each row. Adding the number 3 a total of 4 times gives 12.

$$3 + 3 + 3 + 3 = 12$$

This result can also be shown with a figure.

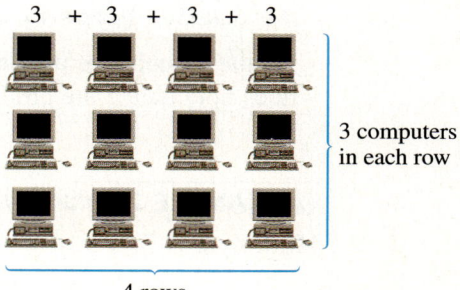

3 computers in each row

4 rows

OBJECTIVE 1 Identify the parts of a multiplication problem. Multiplication is a shortcut for repeated addition. In the computer lab example, instead of *adding* 3 + 3 + 3 + 3 to get 12, we can *multiply* 3 by 4 to get 12. The numbers being multiplied are called **factors**. The answer is called the **product**. For example, the product of 3 and 4 can be written with the symbol ×, a raised dot, or parentheses, as follows.

3   ← Factor (also called *multiplicand*)
× 4 ← Factor (also called *multiplier*)
12 ← Product (answer)

$3 \times 4 = 12$ or $3 \cdot 4 = 12$ or $(3)(4) = 12$

1. Identify the factors and the product in each multiplication problem.

   (a) 2 × 5 = 10
   (b) 6 × 4 = 24
   (c) 7 · 6 = 42
   (d) (3)(9) = 27

ANSWERS

1. (a) factors: 2, 5; product: 10
   (b) factors: 6, 4; product: 24
   (c) factors: 7, 6; product: 42
   (d) factors: 3, 9; product: 27

Work Problem 1 at the Side.

CHAPTER 0 Topics in Arithmetic

2. Multiply.

(a) 3×8

(b) 0×9

(c) $7 \cdot 5$

(d) $6 \cdot 5$

(e) $(3)(8)$

Commutative Property of Multiplication

By the **commutative property of multiplication,** the answer or product remains the same when the order of the factors is changed. For example,

$$3 \times 5 = 15 \quad \text{and} \quad 5 \times 3 = 15$$

CAUTION Recall that addition also has a commutative property. Remember that $4 + 2$ is the same as $2 + 4$. Subtraction, however, is ***not*** commutative.

EXAMPLE 1 Multiplying Two Numbers

Multiply. (Remember that a raised dot or parentheses means to multiply.)

(a) $3 \times 4 = 12$

(b) $6 \cdot 0 = 0$ (The product of any number and 0 is 0; if you give no money to each of 6 relatives, you give no money.)

(c) $(4)(8) = 32$

Work Problem 2 at the Side.

OBJECTIVE 2 Do chain multiplications. Some multiplications contain more than two factors.

Associative Property of Multiplication

By the **associative property of multiplication,** grouping the factors differently does not change the product.

EXAMPLE 2 Multiplying Three Numbers

Multiply $2 \times 3 \times 5$.

$$(2 \times 3) \times 5 \quad \text{Parentheses tell what to do first.}$$
$$6 \times 5 = 30$$

Also,

$$2 \times (3 \times 5)$$
$$2 \times 15 = 30.$$

Either grouping results in the same product.

Calculator Tip The calculator approach to Example 2 uses chain calculations.

$$2 \,\boxtimes\, 3 \,\boxtimes\, 5 \,\boxminus\, 30$$

ANSWERS

2. (a) 24 (b) 0 (c) 35 (d) 30
(e) 24

SECTION 0.4 Multiplying Whole Numbers **0-19**

3. Multiply.

   (a) $2 \times 5 \times 3$

   (b) $7 \cdot 1 \cdot 4$

   (c) $(6)(4)(0)$

A problem with more than two factors, such as the one in Example 2, is called a **chain multiplication**.

> **Work Problem 3 at the Side.**

OBJECTIVE 3 Multiply by single-digit numbers. Carrying may be needed in multiplication problems with larger factors.

EXAMPLE 3   Multiplying with Carrying

Multiply.

(a) 53
     $\times4$

Start by multiplying in the ones column.

$$\begin{array}{r} 1 \\ 53 \\ \times4 \\ \hline 2 \end{array} \quad 4 \times 3 = 12$$

Carry the 1 to the tens column.
Write 2 in the ones column.

Next, multiply 4 ones and 5 tens.

$$\begin{array}{r} 1 \\ 53 \\ \times4 \\ \hline 2 \end{array} \quad 4 \times 5 = \mathbf{20}\text{ tens}$$

Add the 1 that was carried to the tens column.

$$\begin{array}{r} 1 \\ 53 \\ \times4 \\ \hline 212 \end{array} \quad 20 + 1 = \mathbf{21}\text{ tens}$$

(b) 724
     $\times5$

Work as shown.

$$\begin{array}{r} 12 \\ 724 \\ \times5 \\ \hline 3620 \end{array}$$

← $5 \times 4 = \mathbf{20}$ ones; write 0 ones and carry 2 tens.

$5 \times 2 = \mathbf{10}$ tens; add the 2 tens to get 12 tens; write 2 tens and carry 1 hundred.

$5 \times 7 = \mathbf{35}$ hundreds; add the 1 hundred to get 36 hundreds.

4. Multiply.

   (a) 62
       $\times4$

   (b) 98
       $\times0$

   (c) 758
       $\times8$

   (d) 2831
       $\times7$

   (e) 4714
       $\times8$

> **Work Problem 4 at the Side.**

OBJECTIVE 4 Use multiplication shortcuts for numbers ending in zeros. The product of two whole number factors is also called a **multiple** of either factor. For example, since $4 \cdot 2 = 8$, the whole number 8 is a multiple of both 4 and 2. *Multiples of 10* are very useful when multiplying. A **multiple of 10** is a whole number that ends in 0,

ANSWERS
3. (a) 30   (b) 28   (c) 0
4. (a) 248   (b) 0   (c) 6064
   (d) 19,817   (e) 37,712

0-20 CHAPTER 0 Topics in Arithmetic

5. Multiply.

(a) 52×10

(b) 305×100

(c) 418×1000

6. Multiply.

(a) 17×40

(b) 58×300

(c) 180
$\underline{\times30}$

(d) 4200
$\underline{\times80}$

(e) 700
$\underline{\times 400}$

ANSWERS
5. (a) 520 (b) 30,500
   (c) 418,000
6. (a) 680 (b) 17,400 (c) 5400
   (d) 336,000 (e) 280,000

such as 10, 20, or 30; 100, 200, or 300; 1000, 2000, or 3000. There is a short way to multiply by these multiples of 10. Look at the following examples.

$$26 \times 1 = 26$$
$$26 \times 10 = 260$$
$$26 \times 100 = 2600$$
$$26 \times 1000 = 26{,}000$$

Do you see a pattern? These examples suggest the following rule.

Multiplying by Multiples of 10

To multiply a whole number by 10, 100, or 1000, attach one, two, or three 0s to the right of the whole number.

EXAMPLE 4 Using Multiples of 10 to Multiply

Multiply.

(a) $59 \times 10 = 590$ ← Attach 0.

(b) $74 \times 100 = 7400$ ← Attach 00.

(c) $803 \times 1000 = 803{,}000$ ← Attach 000.

Work Problem 5 at the Side.

You can also find the product of other multiples of 10 by attaching 0s.

EXAMPLE 5 Using Multiples of 10 to Multiply

Multiply.

(a) 75×3000

Multiply 75 by 3, and then attach three 0s.

75
$\underline{\times3}$
225

$75 \times 3000 = 225{,}000$ ← Attach 000.

(b) 150×70

Multiply 15 by 7, and then attach two 0s.

15
$\underline{\times7}$
105

$150 \times 70 = 10{,}500$ ← Attach 00.

Work Problem 6 at the Side.

7. Complete each multiplication.

(a) 35
    × 54
    ‾‾‾‾
    140
    175_

(b) 76
    × 49
    ‾‾‾‾
    684
    304_

OBJECTIVE 5 Multiply by numbers having more than one digit. The next example shows multiplication when both factors have more than one digit.

EXAMPLE 6 Multiplying with More Than One Digit

Multiply 46 and 23.

First multiply 46 by 3.

$$\begin{array}{r} 1 \\ 46 \\ \times\ 3 \\ \hline 138 \end{array}$$ ← 46 × 3 = 138

Now multiply 46 by 20.

$$\begin{array}{r} 1 \\ 46 \\ \times\ 20 \\ \hline 920 \end{array}$$ ← 46 × 20 = 920

Add the results.

$$\begin{array}{r} 46 \\ \times\ 23 \\ \hline 138 \\ +\ 920 \\ \hline 1058 \end{array}$$ ← 46 × 3
← 46 × 20
← Add.

Both 138 and 920 are called **partial products.** To save time, the 0 in 920 is usually not written.

$$\begin{array}{r} 46 \\ \times\ 23 \\ \hline 138 \\ 92 \\ \hline 1058 \end{array}$$ ← 0 not written. Be very careful to place the 2 in the tens column.

Work Problem 7 at the Side.

EXAMPLE 7 Using Partial Products

Multiply.

(a)   233
    × 132
    ‾‾‾‾‾‾
      466
     699   (Tens lined up)
    233   (Hundreds lined up)
    ‾‾‾‾‾‾
    30,756 Product

ANSWERS
7. (a) 1890 (b) 3724

CHAPTER 0 Topics in Arithmetic

8. Multiply.

(a) 46
    × 14

(b) 41
    × 38

(c) 75
    × 63

(d) 234
    × 73

(e) 835
    × 189

9. Multiply.

(a) 36
    × 50

(b) 635
    × 40

(c) 562
    × 109

(d) 3526
    × 6002

ANSWERS
8. (a) 644 (b) 1558 (c) 4725
   (d) 17,082 (e) 157,815
9. (a) 1800 (b) 25,400
   (c) 61,258 (d) 21,163,052

(b)   538
    × 46

First multiply by 6.

```
      2 4
    5 3 8
  ×   4 6
    3 2 2 8
```
Carrying is needed here.

Now multiply by 4, being careful to line up the tens.

```
    1 3
    2 4
    5 3 8
  ×   4 6
    3 2 2 8
    2 1 5 2
  2 4,7 4 8
```
Finally, add the results.

Work Problem 8 at the Side.

When 0 appears in the multiplier, be sure to move the partial products to the left to account for the position held by the 0.

EXAMPLE 8 Multiplying with Zeros

Multiply.

(a)
```
      1 3 7
    × 3 0 6
      8 2 2
    0 0 0        (Tens lined up)
  4 1 1          (Hundreds lined up)
  4 1,9 2 2
```

(b)
```
        1 4 0 6
      × 2 0 0 1
        1 4 0 6
      0 0 0 0      ← (0s to line up tens)
    0 0 0 0        ← (0s to line up hundreds)
    2 8 1 2
    2,8 1 3,4 0 6
```

```
        1 4 0 6
      × 2 0 0 1
        1 4 0 6
    2 8 1 2 0 0
    2,8 1 3,4 0 6
```
Zeros are written so this partial product starts in the thousands column.

> **NOTE** In Example 8(b) in the alternative method on the right, 0s were inserted so that thousands were placed in the thousands column. This is a commonly used shortcut.

Work Problem 9 at the Side.

0.4 EXERCISES

For Extra Help

 Student's Solutions Manual

 MyMathLab

 InterAct Math Tutorial Software

 AW Math Tutor Center

MathXL

Work each chain multiplication. See Example 2.

1. $4 \times 1 \times 4$
2. $3 \times 5 \times 3$
3. $8 \times 6 \times 1$
4. $2 \times 4 \times 5$
5. $7 \cdot 8 \cdot 0$
6. $9 \cdot 0 \cdot 5$
7. $4 \cdot 1 \cdot 6$
8. $1 \cdot 5 \cdot 7$
9. $(3)(2)(5)$
10. $(4)(1)(9)$
11. $(3)(0)(7)$
12. $(0)(9)(4)$

13. Explain in your own words the commutative property of multiplication. How do the commutative properties of addition and multiplication compare to each other?

14. Explain in your own words the associative property of multiplication. How do the associative properties of addition and multiplication compare to each other?

Multiply. See Example 3.

15. 25 × 6
16. 62 × 8
17. 34 × 7
18. 76 × 5
19. 642 × 5
20. 472 × 4
21. 624 × 3
22. 852 × 7
23. 2153 × 4
24. 1137 × 3
25. 2521 × 4
26. 2544 × 3
27. 2561 × 8
28. 7326 × 5
29. 36,921 × 7
30. 28,116 × 4

Multiply. See Examples 4 and 5.

31. 30 × 5
32. 20 × 7
33. 80 × 6
34. 70 × 5
35. 740 × 3
36. 200 × 7
37. 600 × 6
38. 860 × 7
39. 125 × 30
40. 246 × 50
41. 1485 × 30
42. 8522 × 50
43. 900 × 300
44. 400 × 700
45. 43,000 × 2 000
46. 11,000 × 9 000
47. $970 \cdot 50$
48. $730 \cdot 40$
49. $500 \cdot 700$
50. $850 \cdot 700$
51. $9700 \cdot 200$
52. $10,050 \cdot 300$

Multiply. See Examples 6–8.

53. 36 × 15
54. 18 × 47
55. 75 × 32
56. 82 × 32
57. 83 × 45
58. $(62)(31)$
59. $(58)(41)$
60. $(82)(67)$
61. $(67)(92)$
62. $(26)(33)$

63. (28)(564)   64. (58)(312)   65. (619)(35)   66. (681)(47)   67. (55)(286)

68. 286   69. 735   70. 621   71. 538   72. 3228
   × 574   × 112   × 415   × 342   × 751

73. 9352   74. 528   75. 215   76. 218   77. 428
   × 264   × 106   × 307   × 106   × 201

78. 3706   79. 6310   80. 3533   81. 2195   82. 1502
   × 208   × 3078   × 5001   × 1038   × 2009

83. A classmate of yours is not clear on how to use a shortcut to multiply a whole number by 10, by 100, or by 1000. Write a short note explaining how this can be done.

84. Show two ways to multiply when a 0 is in the factor that is multiplying. Use the problem 291 × 307 to show this.

0.5 Dividing Whole Numbers

OBJECTIVES

1. Write division problems in three ways.
2. Identify the parts of a division problem.
3. Divide 0 by a number.
4. Recognize that a number cannot be divided by 0.
5. Divide a number by itself.
6. Divide a number by 1.
7. Use short division.

Suppose the cost of a fast-food lunch is $12 and is to be divided equally by three friends. Each person would pay $4, as shown here.

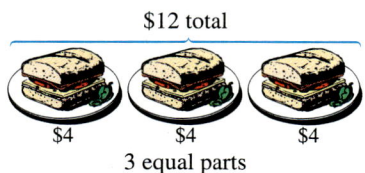

$12 total
$4 $4 $4
3 equal parts

OBJECTIVE 1 Write division problems in three ways. Just as 3 • 4, 3 × 4, and (3)(4) are different ways of indicating the multiplication of 3 and 4, there are several ways to write 12 divided by 3.

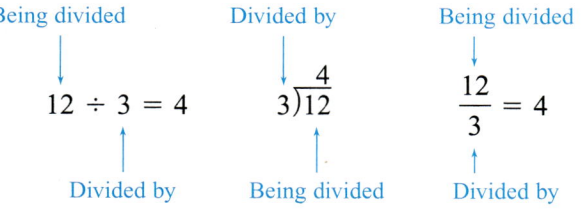

We will use all three division symbols, ÷, $\overline{)}$, and —. In algebra, a slash symbol, /, or a fraction bar, —, is most often used.

SECTION 0.5 Dividing Whole Numbers   0-25

1. Write each division problem using two other symbols.

   (a) $48 \div 6 = 8$

   (b) $24 \div 6 = 4$

   (c) $9\overline{)36}^{\,4}$

   (d) $\dfrac{42}{6} = 7$

■ **EXAMPLE 1**   Using Division Symbols

Write each division problem using two other symbols.

(a) $12 \div 4 = 3$

This division can also be written as

$$4\overline{)12}^{\,3} \quad \text{or} \quad \dfrac{12}{4} = 3.$$

(b) $\dfrac{15}{5} = 3$

$$15 \div 5 = 3 \quad \text{or} \quad 5\overline{)15}^{\,3}$$

(c) $5\overline{)20}^{\,4}$

$$20 \div 5 = 4 \quad \text{or} \quad \dfrac{20}{5} = 4$$

Work Problem 1 at the Side.

OBJECTIVE 2 Identify the parts of a division problem.   In division, the number being divided is the **dividend**, the number divided by is the **divisor**, and the answer is the **quotient**.

$$\text{dividend} \div \text{divisor} = \text{quotient}$$

$$\text{divisor}\overline{)\text{dividend}}^{\,\text{quotient}} \qquad \dfrac{\text{dividend}}{\text{divisor}} = \text{quotient}$$

■ **EXAMPLE 2**   Identifying the Parts of a Division Problem

Identify the dividend, divisor, and quotient.

(a) $35 \div 7 = 5$

$35 \div 7 = 5 \leftarrow$ Quotient
  ↗ ↖
Dividend Divisor

(b) $\dfrac{100}{20} = 5$

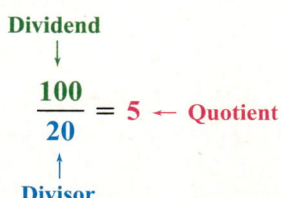

ANSWERS

1. (a) $6\overline{)48}^{\,8}$ and $\dfrac{48}{6} = 8$

   (b) $6\overline{)24}^{\,4}$ and $\dfrac{24}{6} = 4$

   (c) $36 \div 9 = 4$ and $\dfrac{36}{9} = 4$

   (d) $6\overline{)42}^{\,7}$ and $42 \div 6 = 7$

2. Identify the dividend, divisor, and quotient.

   (a) $20 \div 5 = 4$

   (b) $18 \div 6 = 3$

   (c) $\dfrac{28}{7} = 4$

   (d) $2\overline{)36}$ (quotient 18)

3. Divide.

   (a) $0 \div 10$

   (b) $\dfrac{0}{6}$

   (c) $\dfrac{0}{24}$

   (d) $37\overline{)0}$

4. Write each division problem as a multiplication problem.

   (a) $5\overline{)15}$ (quotient 3)

   (b) $\dfrac{32}{4} = 8$

   (c) $48 \div 8 = 6$

(c) $12\overline{)72}$ (quotient 6)

$12\overline{)72}$ ← Quotient 6, ← Dividend 72, ↑ Divisor

Work Problem 2 at the Side.

OBJECTIVE 3 Divide 0 by a number. If no money, or $0, is divided equally among five people, each person gets $0. The general rule for dividing 0 follows.

Dividing 0 by a Number

The number **0** divided by any nonzero number is **0**.

EXAMPLE 3 Dividing 0 by a Number

Divide.

(a) $0 \div 12 = 0$

(b) $0 \div 1728 = 0$

(c) $\dfrac{0}{375} = 0$

(d) $129\overline{)0}$

Work Problem 3 at the Side.

Just as a subtraction such as $8 - 3 = 5$ can be written as the addition $8 = 3 + 5$, any division can be written as a multiplication. For example, $12 \div 3 = 4$ can be written as

$$3 \times 4 = 12 \quad \text{or} \quad 4 \times 3 = 12$$

EXAMPLE 4 Changing Division Problems to Multiplication

Change each division problem to a multiplication problem.

(a) $\dfrac{20}{4} = 5$ becomes $4 \cdot 5 = 20$

(b) $8\overline{)48}$ (quotient 6) becomes $8 \cdot 6 = 48$

(c) $72 \div 9 = 8$ becomes $9 \cdot 8 = 72$

Work Problem 4 at the Side.

ANSWERS
2. (a) dividend: 20; divisor: 5; quotient: 4
   (b) dividend: 18; divisor: 6; quotient: 3
   (c) dividend: 28; divisor: 7; quotient: 4
   (d) dividend: 36; divisor: 2; quotient: 18
3. all 0
4. (a) $5 \cdot 3 = 15$ (b) $4 \cdot 8 = 32$
   (c) $8 \cdot 6 = 48$

SECTION 0.5 Dividing Whole Numbers **0-27**

5. Divide. If the division is not possible, write "undefined."

(a) $\dfrac{6}{0}$

(b) $\dfrac{0}{6}$

(c) $0\overline{)28}$

(d) $28\overline{)0}$

(e) $100 \div 0$

(f) $0 \div 100$

OBJECTIVE 4 Recognize that a number cannot be divided by 0. Division by 0 cannot be done. To see why, try to find

$$9 \div 0 = ?$$

As we have just seen, any division problem can be converted to a multiplication problem so that

$$\text{divisor} \cdot \text{quotient} = \text{dividend}.$$

If you convert the preceding problem to its multiplication counterpart, it reads

$$0 \cdot ? = 9.$$

You already know that 0 times any number must always be 0. Try any number you like to replace the "?" and you'll always get 0 instead of 9. Therefore, the division problem $9 \div 0$ cannot be done. Mathematicians say it is *undefined* and have agreed never to divide by 0. However, $0 \div 9$ *can* be done. Check by rewriting it as a multiplication problem.

$$0 \div 9 = 0 \quad \text{because} \quad 0 \cdot 9 = 0 \text{ is true}.$$

Dividing by 0

Since dividing by 0 cannot be done, we say that division by **0** is *undefined*. It is impossible to compute an answer.

EXAMPLE 5   Dividing Numbers by 0

All the following are undefined.

(a) $\dfrac{6}{0}$ is undefined.

(b) $0\overline{)8}$ is undefined.

(c) $18 \div 0$ is undefined.

(d) $\dfrac{0}{0}$ is undefined.

Division Involving 0

$$0 \div \text{nonzero number} = 0 \quad \text{and} \quad \dfrac{0}{\text{nonzero number}} = 0$$

but

$$\dfrac{\text{nonzero number}}{0} \quad \text{and} \quad \text{nonzero number} \div 0 \text{ are undefined}.$$

CAUTION When 0 is the divisor in a problem, you write "undefined" as the answer. Never divide by 0.

Work Problem 5 at the Side.

ANSWERS
5. (a) undefined   (b) 0   (c) undefined
(d) 0   (e) undefined   (f) 0

0-28 CHAPTER 0 Topics in Arithmetic

6. Divide.

(a) $6 \div 6$

(b) $15 \overline{)15}$

(c) $\dfrac{37}{37}$

7. Divide.

(a) $6 \div 1$

(b) $1 \overline{)18}$

(c) $\dfrac{36}{1}$

Calculator Tip Try these two problems on your calculator. Jot down your answers.

$9 \; \boxed{\div} \; 0 \; \boxed{=}$ _____ $0 \; \boxed{\div} \; 9 =$ _____

When you try to divide by 0, the calculator cannot do it, so it shows the word "Error" or the letter "E" (for error) in the display. But, when you divide 0 by 9 the calculator displays 0, which is the correct answer.

OBJECTIVE 5 Divide a number by itself. What happens when a number is divided by itself? For example, what is $4 \div 4$ or $97 \div 97$?

> **Dividing a Number by Itself**
>
> Any nonzero number divided by itself is **1**.

EXAMPLE 6 Dividing a Nonzero Number by Itself

Divide.

(a) $16 \div 16 = 1$

(b) $32 \overline{)32}^{\;1}$

(c) $\dfrac{57}{57} = 1$

Work Problem 6 at the Side.

OBJECTIVE 6 Divide a number by 1. What happens when a number is divided by 1? For example, what is $5 \div 1$ or $86 \div 1$?

> **Dividing a Number by 1**
>
> Any number divided by 1 is itself.

EXAMPLE 7 Dividing Numbers by 1

Divide.

(a) $8 \div 1 = 8$

(b) $1 \overline{)26}^{\;26}$

(c) $\dfrac{41}{1} = 41$

Work Problem 7 at the Side.

OBJECTIVE 7 Use short division. **Short division** is a method of dividing a number by a one-digit divisor.

ANSWERS
6. all 1
7. (a) 6 (b) 18 (c) 36

SECTION 0.5 Dividing Whole Numbers   0-29

8. Divide.

(a) $2\overline{)18}$

(b) $3\overline{)39}$

(c) $4\overline{)88}$

(d) $2\overline{)462}$

9. Divide.

(a) $2\overline{)125}$

(b) $3\overline{)215}$

(c) $4\overline{)538}$

(d) $\dfrac{819}{5}$

EXAMPLE 8 Using Short Division

Divide: $3\overline{)96}$.

First, divide 9 by 3.

$$3\overline{)96}^{\,3} \quad \leftarrow \dfrac{9}{3} = 3$$

Next, divide 6 by 3.

$$3\overline{)96}^{\,32} \quad \leftarrow \dfrac{6}{3} = 2$$

Work Problem 8 at the Side.

When two numbers do not divide exactly, the leftover portion is called the **remainder**.

EXAMPLE 9 Using Short Division with a Remainder

Divide 147 by 4.

Write the problem.

$$4\overline{)147}$$

Because 1 cannot be divided by 4, divide 14 by 4.

$$4\overline{)14^{2}7}^{\,3} \qquad \dfrac{14}{4} = 3 \text{ with 2 left over}$$

Next, divide 27 by 4. The final number left over is the remainder. Use R to indicate the remainder, and write the remainder to the side.

$$4\overline{)14^{2}7}^{\,3\ 6\ \mathbf{R3}} \qquad \dfrac{27}{4} = 6 \text{ with 3 left over}$$

Work Problem 9 at the Side.

ANSWERS
8. (a) 9   (b) 13   (c) 22   (d) 231
9. (a) 62 R1   (b) 71 R2
   (c) 134 R2   (d) 163 R4

0.5 EXERCISES

For Extra Help
 Student's Solutions Manual
 MyMathLab
 InterAct Math Tutorial Software
 AW Math Tutor Center
MathXL

Write each division problem using two other symbols. See Example 1.

1. $15 \div 3 = 5$
2. $20 \div 5 = 4$
3. $\dfrac{45}{9} = 5$
4. $\dfrac{56}{8} = 7$
5. $2\overline{)16}^{\,8}$
6. $8\overline{)48}^{\,6}$

Divide. If the division is not possible, write "undefined." See Examples 3, 5, 6, and 7.

7. $7 \div 7$
8. $42 \div 6$
9. $\dfrac{14}{2}$
10. $\dfrac{8}{0}$

11. $36 \div 0$
12. $6 \div 6$
13. $\dfrac{24}{1}$
14. $\dfrac{12}{1}$

15. $12\overline{)0}$
16. $\dfrac{0}{7}$
17. $0\overline{)21}$
18. $\dfrac{2}{0}$

19. $\dfrac{15}{1}$
20. $\dfrac{0}{0}$
21. $\dfrac{8}{1}$
22. $\dfrac{0}{5}$

Divide by using short division. Use multiplication to check each answer. See Examples 8 and 9.

23. $5\overline{)75}$
24. $4\overline{)84}$
25. $8\overline{)192}$
26. $6\overline{)168}$

27. $4\overline{)1216}$
28. $5\overline{)2305}$
29. $4\overline{)2509}$
30. $8\overline{)1335}$

31. $6\overline{)9137}$
32. $9\overline{)8371}$
33. $6\overline{)1854}$
34. $8\overline{)856}$

35. $24{,}040 \div 8$
36. $8012 \div 4$
37. $15{,}018 \div 3$
38. $32{,}008 \div 8$

39. $4867 \div 6$
40. $5993 \div 7$
41. $12{,}947 \div 5$
42. $33{,}285 \div 9$

43. $29,298 \div 4$ 44. $17,937 \div 6$ 45. $12,630 \div 4$ 46. $46,560 \div 7$

47. $\dfrac{22,088}{4}$ 48. $\dfrac{8199}{9}$ 49. $\dfrac{74,751}{6}$ 50. $\dfrac{72,543}{5}$

51. $\dfrac{71,776}{7}$ 52. $\dfrac{77,621}{3}$ 53. $\dfrac{128,645}{7}$ 54. $\dfrac{172,255}{4}$

0.6 Long Division

OBJECTIVES

1. Do long division.
2. Divide numbers ending in 0 by numbers ending in 0.

If the total cost of 42 computer modems is $3066, we can find the cost of each modem using **long division**. Long division is used to divide by a number with more than one digit.

OBJECTIVE 1 Do long division. In long division, estimate the various numbers by using a **trial divisor,** which is used to get a **trial quotient.**

EXAMPLE 1 Using a Trial Divisor and a Trial Quotient

Divide: $42\overline{)3066}$.

Because 42 is closer to 40 than to 50, use the first digit of the divisor as a trial divisor.

$$42 \leftarrow \text{Trial divisor}$$

Try to divide the first digit of the dividend by 4. Since 3 cannot be divided by 4, use the first *two* digits, 30.

$$\dfrac{30}{4} = 7 \text{ with remainder 2}$$

$$42\overline{)3066}^{\,7} \leftarrow \text{Trial quotient}$$

7 goes over the 6, because $\dfrac{306}{42}$ is about 7.

Multiply 7 and 42 to get 294; next, subtract 294 from 306.

$$\begin{array}{r} 7 \\ 42\overline{)3066} \\ \underline{294} \leftarrow 7 \times 42 \\ 12 \leftarrow 306 - 294 \end{array}$$

1. Divide.

(a) $14\overline{)1148}$

(b) $32\overline{)2048}$

(c) $61\overline{)4392}$

(d) $\dfrac{5394}{93}$

Bring down the 6 at the right.

$$\begin{array}{r} 7 \\ 42\overline{)3066} \\ 294\downarrow \\ \hline 126 \end{array}$$ ← 6 brought down

Use the trial divisor, 4.

First two digits of 126 → $\dfrac{12}{4} = 3$

$$\begin{array}{r} 73 \\ 42\overline{)3066} \\ 294 \\ \hline 126 \\ 126 \\ \hline 0 \end{array}$$ ← 3 × 42 = 126

Check the answer by multiplying 42 and 73. The product should be 3066.

CAUTION The *first digit* of the quotient in long division must be placed in the proper position over the dividend.

Work Problem 1 at the Side.

EXAMPLE 2 Dividing to Find a Trial Quotient

Divide: $58\overline{)2730}$.

Use 6 as a trial divisor, since 58 is closer to 60 than to 50.

First two digits of dividend → $\dfrac{27}{6}$ = 4 with 3 left over

4 ← Trial quotient

$$\begin{array}{r} 4 \\ 58\overline{)2730} \\ 232 \\ \hline 41 \end{array}$$

← 4 × 58 = 232
← 273 − 232 = 41 (smaller than 58, the divisor)

Bring down the 0.

$$\begin{array}{r} 4 \\ 58\overline{)2730} \\ 232\downarrow \\ \hline 410 \end{array}$$ ← 0 brought down

First two digits of 410 → $\dfrac{41}{6}$ = 6 with 5 left over

46 ← Trial quotient

$$\begin{array}{r} 46 \\ 58\overline{)2730} \\ 232 \\ \hline 410 \\ 348 \\ \hline 62 \end{array}$$

← 6 × 58 = 348
← Greater than 58

ANSWERS
1. (a) 82 (b) 64 (c) 72 (d) 58

2. Divide.

(a) $48\overline{)2688}$

(b) $36\overline{)2236}$

(c) $65\overline{)5416}$

(d) $89\overline{)6649}$

ANSWERS
2. (a) 56 (b) 62 R4
   (c) 83 R21 (d) 74 R63

The remainder, 62, is greater than the divisor, 58, so 7 should be used instead of 6.

$$\begin{array}{r} 47\ \text{R4} \\ 58\overline{)2730} \\ \underline{232} \\ 410 \\ \underline{406} \leftarrow 7 \times 58 = 406 \\ 4 \leftarrow 410 - 406 \end{array}$$

Work Problem 2 at the Side.

Sometimes it is necessary to insert a 0 in the quotient.

EXAMPLE 3   Inserting 0s in the Quotient

Divide: $42\overline{)8734}$.
Start as above.

$$\begin{array}{r} 2 \\ 42\overline{)8734} \\ \underline{84} \leftarrow 2 \times 42 = 84 \\ 3 \leftarrow 87 - 84 = 3 \end{array}$$

Bring down the 3.

$$\begin{array}{r} 2 \\ 42\overline{)8734} \\ \underline{84\downarrow} \\ 33 \leftarrow 3 \text{ brought down} \end{array}$$

Since 33 cannot be divided by 42, place a 0 in the quotient as a placeholder.

$$\begin{array}{r} 20 \leftarrow 0 \text{ in quotient} \\ 42\overline{)8734} \\ \underline{84} \\ 33 \end{array}$$

Bring down the final digit, the 4.

$$\begin{array}{r} 20 \\ 42\overline{)8734} \\ \underline{84\downarrow} \\ 334 \leftarrow 4 \text{ brought down} \end{array}$$

Complete the problem.

$$\begin{array}{r} 207\ \text{R40} \\ 42\overline{)8734} \\ \underline{84} \\ 334 \\ \underline{294} \\ 40 \end{array}$$

The answer is 207 **R40**.

0-34 CHAPTER 0 Topics in Arithmetic

3. Divide.

(a) $34\overline{)3645}$

> **CAUTION** There *must be a digit* in the quotient (answer) above every digit in the dividend once the answer has begun. Notice in Example 3 that a **0** was used to assure an answer digit above every digit in the dividend.

Work Problem 3 at the Side.

(b) $28\overline{)5768}$

OBJECTIVE 2 Divide numbers ending in 0 by numbers ending in 0. When the divisor and dividend both contain 0s at the far right, recall that these numbers are multiples of 10. As with multiplication, there is a short way to divide these multiples of 10. Look at the following examples.

(c) $39\overline{)15,933}$

$$26,000 \div 1 = 26,000$$
$$26,000 \div 10 = 2600$$
$$26,000 \div 100 = 260$$
$$26,000 \div 1000 = 26$$

Do you see a pattern? These examples suggest the following rule.

(d) $78\overline{)23,462}$

> **Dividing a Whole Number by 10, 100, or 1000**
>
> Divide a whole number by 10, 100, or 1000 by dropping the appropriate number of 0s from the whole number.

4. Divide.

(a) $90 \div 10$

EXAMPLE 4 Dividing by Multiples of 10

Divide.

(a) $60 \div 10 = 6$ — One 0 in divisor; 0 dropped

(b) $2400 \div 100$

(b) $3500 \div 100 = 35$ — Two 0s in divisor; 00 dropped

(c) $206,000 \div 1000$

(c) $915,000 \div 1000 = 915$ — Three 0s in divisor; 000 dropped

Work Problem 4 at the Side.

ANSWERS
3. (a) 107 **R**7 (b) 206
   (c) 408 **R**21 (d) 300 **R**62
4. (a) 9 (b) 24 (c) 206

In the next example, we find the quotient for other multiples of 10 by dropping 0s.

SECTION 0.6 Long Division 0-35

5. Divide.

(a) $50\overline{)6250}$

(b) $130\overline{)131{,}040}$

(c) $2600\overline{)195{,}000}$

■ **EXAMPLE 5 Dividing by Multiples of 10**

Divide.

(a) $40\overline{)11{,}000}$ Drop one 0 from the divisor and the dividend.

$$\begin{array}{r} 275 \\ 4\overline{)1100} \\ \underline{8} \\ 30 \\ \underline{28} \\ 20 \\ \underline{20} \\ 0 \end{array}$$

Since $1100 \div 4$ is 275, then $11{,}000 \div 40$ is also 275.

(b) $3500\overline{)31{,}500}$ Drop two 0s from the divisor and the dividend.

$$\begin{array}{r} 9 \\ 35\overline{)315} \\ \underline{315} \\ 0 \end{array}$$

Since $315 \div 35$ is 9, then $31{,}500 \div 3500$ is also 9.

NOTE Dropping 0s when dividing by multiples of 10 *does not* change the quotient (answer).

ANSWERS
5. (a) 125 (b) 1008 (c) 75

Work Problem 5 at the Side.

0.6 EXERCISES

For Extra Help

- Student's Solutions Manual
- MyMathLab
- InterAct Math Tutorial Software
- AW Math Tutor Center
- MathXL

Decide where the first digit in the quotient would be located. Then without finishing the division, you can tell which of the three choices is the correct answer. Circle your choice. See Examples 1 and 2.

1. $25\overline{)550}$
      2   (22)   220

2. $14\overline{)476}$
      3   34   304

3. $18\overline{)4500}$
      2   25   250

4. $42\overline{)7560}$
      18   180   1800

5. $86\overline{)10{,}327}$
      12   120 R7   1200

6. $46\overline{)24{,}026}$
      5   52   522 R14

7. $52\overline{)68{,}025}$
      13   130 R1   (1308 R9)

8. $12\overline{)116{,}953}$
      974 R2   9746 R1   97,460

9. $21\overline{)149{,}826}$
      71   713   7134 R12

10. $64\overline{)208{,}138}$
       325 R2   3252 R10   32,521

11. $523\overline{)470{,}800}$ **12.** $230\overline{)253{,}230}$

9 R100   90 R100   900 R100   11   110   1101

Divide by using long division. Use multiplication to check each answer. See Examples 1–5.

13. $21\overline{)2272}$ **14.** $29\overline{)1827}$ **15.** $56\overline{)10{,}270}$

16. $83\overline{)39{,}692}$ **17.** $26\overline{)62{,}583}$ **18.** $28\overline{)84{,}249}$

19. $74\overline{)84{,}819}$ **20.** $238\overline{)186{,}948}$ **21.** $153\overline{)509{,}725}$

22. $308\overline{)26{,}796}$ **23.** $420\overline{)357{,}000}$ **24.** $900\overline{)153{,}000}$

0.1–0.6 TEST

Give the place value of 0 in each number.

1. 6106 **2.** 85,055

3. Use digits to write four hundred twenty-six thousand, five.

Add.

4. 762
    72
  5186
 + 2694

5. 17,063
        7
       12
    1 505
   93,710
 +   333

Subtract.

6. 6006
     − 4953

7. 5062
     − 1978

Multiply.

8. $6 \times 8 \times 2$ **9.** $57 \cdot 3000$

10. $(95)(18)$ **11.** 7381
                          × 603

Divide. If the division is not possible, write "undefined."

12. $16\overline{)112{,}752}$ **13.** $\dfrac{835}{0}$

14. $19{,}241 \div 42$ **15.** $280\overline{)44{,}800}$

0.7 Basics of Fractions

OBJECTIVES

1. Use a fraction to show which part of a whole is shaded.
2. Identify the numerator and denominator.
3. Identify proper and improper fractions.

In Sections 0.1–0.6 we discussed whole numbers. Many times, however, we find that parts of whole numbers are considered. One way to write parts of a whole is with **fractions.** Another way is with decimals, which is discussed later.

OBJECTIVE 1 Use a fraction to show which part of a whole is shaded. The number $\frac{1}{8}$ is a fraction that represents 1 of 8 equal parts. Read $\frac{1}{8}$ as "one eighth."

EXAMPLE 1 Identifying Fractions

Use fractions to represent the shaded portions and the unshaded portions of each figure.

(a) The figure on the left has 4 equal parts. The 1 shaded part is represented by the fraction $\frac{1}{4}$. The *un*shaded part is $\frac{3}{4}$.

1. Write fractions for the shaded portions and the unshaded portions of each figure.

   (a)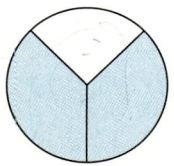

   (b)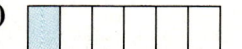

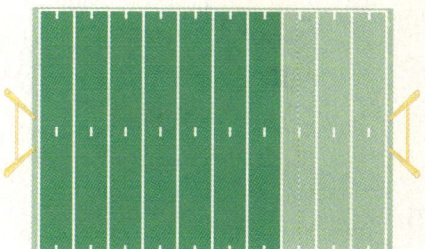

(b) The 3 shaded parts of the 10-part figure on the right are represented by the fraction $\frac{3}{10}$. The *un*shaded part is $\frac{7}{10}$.

Work Problem 1 at the Side.

Fractions can be used to show more than one whole object.

EXAMPLE 2 Representing Fractions Greater Than 1

Use a fraction to represent the shaded part of each figure.

(a)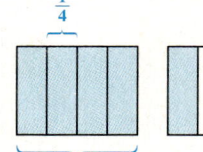

Whole object

An area equal to 5 of the $\frac{1}{4}$ parts is shaded. Write this as $\frac{5}{4}$.

(b)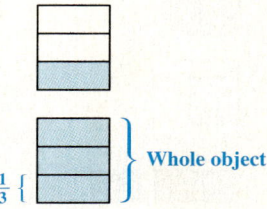

Whole object

An area equal to 4 of the $\frac{1}{3}$ parts is shaded, so $\frac{4}{3}$ is shaded.

2. Write a fraction for the shaded portion of the figure.

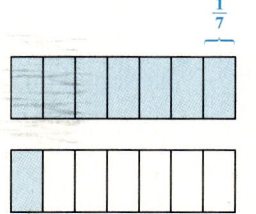

ANSWERS

1. (a) $\frac{2}{3}$; $\frac{1}{3}$ (b) $\frac{1}{6}$, $\frac{5}{6}$ 2. $\frac{8}{7}$

Work Problem 2 at the Side.

0-38 CHAPTER 0 Topics in Arithmetic

3. Identify the numerator and the denominator. Draw a picture with shaded parts to show each fraction. Your drawings may vary, but they should have the correct number of shaded parts.

   (a) $\dfrac{2}{3}$

   (b) $\dfrac{1}{4}$

   (c) $\dfrac{8}{5}$

   (d) $\dfrac{5}{2}$

ANSWERS
3. (a) N: 2; D: 3

   (b) N: 1; D: 4

   (c) N: 8; D: 5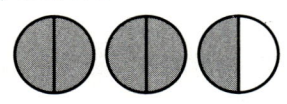

   (d) N: 5; D: 2

OBJECTIVE 2 Identify the numerator and denominator. In the fraction $\tfrac{2}{3}$, the number 2 is the **numerator** and 3 is the **denominator**. The bar between the numerator and the denominator is the *fraction bar*.

Fraction bar → $\dfrac{2}{3}$ ← Numerator
← Denominator

Numerators and Denominators

The **denominator** of a fraction shows the number of equivalent parts in the whole, and the **numerator** shows how many parts are being considered.

NOTE Remember that a bar, —, is one of the division symbols, and that division by 0 is undefined. A fraction with a denominator of 0 is also undefined.

EXAMPLE 3 Identifying Numerators and Denominators
Identify the numerator and denominator in each fraction.

(a) $\dfrac{5}{8}$

(b) $\dfrac{9}{4}$

$\dfrac{5}{8}$ ← Numerator
← Denominator

$\dfrac{9}{4}$ ← Numerator
← Denominator

Work Problem 3 at the Side.

OBJECTIVE 3 Identify proper and improper fractions. Fractions are sometimes called *proper* or *improper* fractions.

Proper and Improper Fractions

If the numerator of a fraction is *smaller* than the denominator, the fraction is a **proper fraction**.

If the numerator is *greater than or equal to* the denominator, the fraction is an **improper fraction**.

Proper Fractions	Improper Fractions
$\dfrac{1}{2}, \dfrac{5}{11}, \dfrac{35}{36}$	$\dfrac{9}{7}, \dfrac{126}{125}, \dfrac{7}{7}$

EXAMPLE 4 Classifying Types of Fractions

(a) Identify all proper fractions in this list.

$\dfrac{3}{4}, \dfrac{5}{9}, \dfrac{17}{5}, \dfrac{9}{7}, \dfrac{3}{3}, \dfrac{12}{25}, \dfrac{1}{9}, \dfrac{5}{3}$

SECTION 0.7 Basics of Fractions 0-39

4. From the following group of fractions:

$$\frac{2}{3}, \frac{4}{3}, \frac{3}{4}, \frac{8}{8}, \frac{3}{1}, \frac{1}{3}$$

(a) list all proper fractions;

(b) list all improper fractions.

Proper fractions have a numerator that is smaller than the denominator. The proper fractions are

$$\frac{3}{4}, \quad \text{← 3 is smaller than 4.} \quad \frac{5}{9}, \frac{12}{25}, \text{ and } \frac{1}{9}.$$

(b) Identify all improper fractions in the list in part (a).
Improper fractions have a numerator that is equal to or greater than the denominator. The improper fractions are

$$\frac{17}{5}, \quad \text{← 17 is greater than 5.} \quad \frac{9}{7}, \frac{3}{3}, \text{ and } \frac{5}{3}.$$

Work Problem 4 at the Side.

ANSWERS

4. (a) $\frac{2}{3}, \frac{3}{4}, \frac{1}{3}$ (b) $\frac{4}{3}, \frac{8}{8}, \frac{3}{1}$

0.7 EXERCISES

For Extra Help
- Student's Solutions Manual
- MyMathLab
- InterAct Math Tutorial Software
- AW Math Tutor Center
- MathXL

Write fractions to represent the shaded and unshaded portions of each figure. See Examples 1 and 2.

1.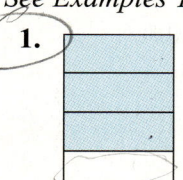

2.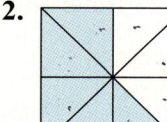

3.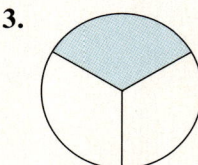

4.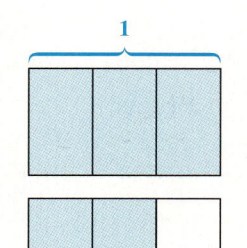

5.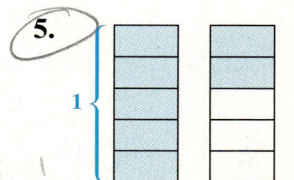

6.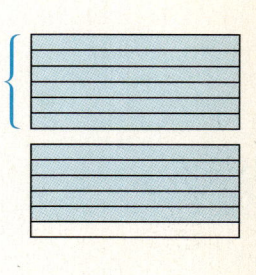

7. What fraction of these 11 coins are dimes?

0-40 CHAPTER 0 Topics in Arithmetic

8. What fraction of these 8 recording artists are men?

9. In an American Sign Language (ASL) class of 25 students, 8 are hearing impaired. What fraction of the students are hearing impaired?

10. Of 98 bicycles in a bike rack, 67 are mountain bikes. What fraction of the bicycles are not mountain bikes?

11. Of 134 employees at D. J. Diecast Company, 85 are part-time employees. What fraction of the employees are not part-time employees?

12. A college cheerleading squad has 12 members. If 5 of the cheerleaders are sophomores and the rest are freshmen, find the fraction of the members that are freshmen.

Identify the numerator and denominator. See Example 3.

	Numerator	Denominator		Numerator	Denominator
13. $\frac{3}{8}$	3	8	14. $\frac{5}{6}$		
15. $\frac{11}{10}$	11	10	16. $\frac{7}{5}$		

List the proper and improper fractions in each group. See Example 4.

                                                                         Proper   Improper

17. $\frac{8}{5}, \frac{1}{3}, \frac{5}{8}, \frac{6}{6}, \frac{12}{2}, \frac{7}{16}$

18. $\frac{1}{3}, \frac{3}{8}, \frac{16}{12}, \frac{10}{8}, \frac{6}{6}, \frac{3}{4}$

19. $\frac{3}{4}, \frac{3}{2}, \frac{5}{5}, \frac{9}{11}, \frac{7}{15}, \frac{19}{18}$

20. $\frac{12}{12}, \frac{15}{11}, \frac{13}{12}, \frac{11}{8}, \frac{17}{17}, \frac{19}{12}$

SECTION 0.8 Mixed Numbers 0-41

21. Write a fraction of your own choice. Label the parts of the fraction and write a sentence describing what each part represents. Draw a picture with shaded parts showing your fraction.

22. Give one example of a proper fraction and one example of an improper fraction. What determines whether a fraction is proper or improper? Draw pictures with shaded parts showing these fractions.

Fill in the blanks to complete each sentence.

23. The fraction $\frac{7}{8}$ represents _____ of the _____ equal parts into which a whole is divided.

24. The fraction $\frac{15}{16}$ represents _____ of the _____ equal parts into which a whole is divided.

25. The fraction $\frac{4}{32}$ represents _____ of the _____ equal parts into which a whole is divided.

26. The fraction $\frac{7}{64}$ represents _____ of the _____ equal parts into which a whole is divided.

0.8 Mixed Numbers

OBJECTIVES

1. Identify mixed numbers.
2. Write mixed numbers as improper fractions.
3. Write improper fractions as mixed numbers.

Suppose you had three whole cases of soft drink cans and half of another case. You would state this as a whole number and a fraction.

OBJECTIVE 1 Identify mixed numbers. When a whole number and a fraction are written together, the result is a **mixed number**. For example, the mixed number

$$3\frac{1}{2} \quad \text{represents} \quad 3 + \frac{1}{2},$$

or 3 wholes and $\frac{1}{2}$ of a whole. Read $3\frac{1}{2}$ as "three and one half." As this figure shows, the mixed number $3\frac{1}{2}$ is equal to the improper fraction $\frac{7}{2}$.

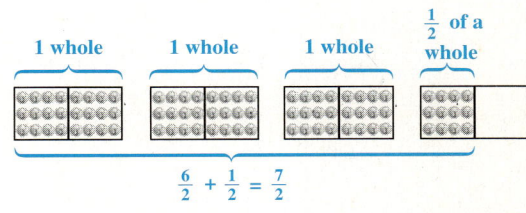

1. Use these diagrams to write $2\frac{1}{4}$ as an improper fraction.

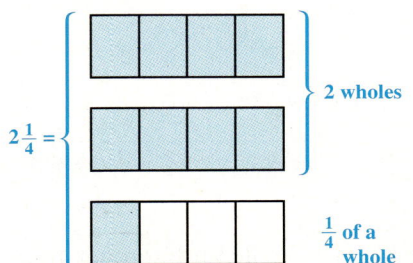

Work Problem 1 at the Side.

ANSWERS

1. $\frac{9}{4}$

CHAPTER 0 Topics in Arithmetic

2. Write as improper fractions.

   (a) $4\dfrac{1}{2}$

   (b) $5\dfrac{3}{4}$

   (c) $4\dfrac{7}{8}$

   (d) $8\dfrac{5}{6}$

OBJECTIVE 2 Write mixed numbers as improper fractions. Use the following steps to write $3\dfrac{1}{2}$ as an improper fraction without drawing a figure.

Step 1 Multiply 3 and 2.

$$3\dfrac{1}{2} \qquad 3 \cdot 2 = 6$$

Step 2 Add 1 to the product.

$$3\dfrac{1}{2} \qquad 6 + 1 = 7$$

Step 3 Use 7, from Step 2, as the numerator and 2 as the denominator.

$$3\dfrac{1}{2} = \dfrac{7}{2}$$

Same denominator

In summary, use the following steps to *write a mixed number as an improper fraction.*

Writing a Mixed Number as an Improper Fraction

Step 1 *Multiply* the denominator of the fraction and the whole number.

Step 2 *Add* to this product the numerator of the fraction.

Step 3 Write the result of Step 2 as the *numerator* and the original denominator as the *denominator.*

EXAMPLE 1 Writing a Mixed Number as an Improper Fraction

Write $7\dfrac{2}{3}$ as an improper fraction (numerator greater than denominator).

Step 1 $7\dfrac{2}{3} \qquad 7 \cdot 3 = 21 \qquad$ Multiply 7 and 3.

Step 2 $7\dfrac{2}{3} \qquad 21 + 2 = 23 \qquad$ Add 2. The numerator is 23.

Step 3 $7\dfrac{2}{3} = \dfrac{23}{3} \qquad$ Use the same denominator.

Work Problem 2 at the Side.

OBJECTIVE 3 Write improper fractions as mixed numbers. Write an improper fraction as a mixed number as follows.

ANSWERS

2. (a) $\dfrac{9}{2}$ (b) $\dfrac{23}{4}$ (c) $\dfrac{39}{8}$ (d) $\dfrac{53}{6}$

SECTION 0.8 Mixed Numbers 0-43

3. Write as whole or mixed numbers.

(a) $\dfrac{4}{3}$

(b) $\dfrac{9}{4}$

(c) $\dfrac{35}{5}$

(d) $\dfrac{85}{8}$

Writing an Improper Fraction as a Mixed Number

Write an **improper fraction** as a mixed number by dividing the numerator by the denominator. The quotient is the whole number (of the mixed number), the remainder is the numerator of the fraction part, and the denominator remains unchanged.

EXAMPLE 2 Writing Improper Fractions as Mixed Numbers

Write each improper fraction as a mixed number.

(a) $\dfrac{17}{5}$

Divide 17 by 5.

$$\begin{array}{r} 3 \\ 5\overline{)17} \\ \underline{15} \\ 2 \end{array}$$ ← Whole number part

← Remainder

The quotient **3** is the whole number part of the mixed number. The remainder **2** is the numerator of the fraction, and the denominator remains as **5**.

$$\dfrac{17}{5} = 3\dfrac{2}{5}$$ ← Remainder

Same denominator

We can verify this by using a diagram in which $\frac{17}{5}$ is shaded.

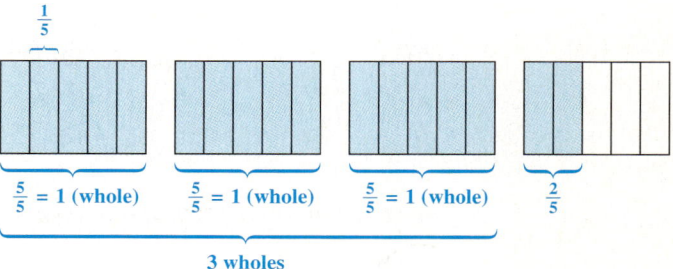

(b) $\dfrac{24}{4}$

Divide 24 by 4.

$$\begin{array}{r} 6 \\ 4\overline{)24} \\ \underline{24} \\ 0 \end{array} \quad \text{so} \quad \dfrac{24}{4} = 6$$

← No remainder

NOTE A proper fraction has a value that is less than 1, while an improper fraction has a value that is 1 or greater.

ANSWERS

3. (a) $1\dfrac{1}{3}$ (b) $2\dfrac{1}{4}$ (c) 7 (d) $10\dfrac{5}{8}$

Work Problem 3 at the Side.

0-44 CHAPTER 0 Topics in Arithmetic

0.8 EXERCISES

For Extra Help

Student's Solutions Manual

MyMathLab

InterAct Math Tutorial Software

AW Math Tutor Center

MathXL

Write each mixed number as an improper fraction. See Example 1.

1. $1\frac{1}{2}$ 2. $2\frac{1}{3}$ 3. $3\frac{2}{5}$

4. $4\frac{3}{4}$ 5. $6\frac{1}{2}$ 6. $5\frac{3}{5}$

7. $7\frac{1}{3}$ 8. $8\frac{1}{2}$ 9. $1\frac{7}{11}$

10. $4\frac{3}{7}$ 11. $6\frac{1}{3}$ 12. $8\frac{2}{3}$

13. $10\frac{1}{8}$ 14. $12\frac{2}{3}$ 15. $10\frac{3}{4}$

16. $5\frac{4}{5}$ 17. $3\frac{3}{8}$ 18. $2\frac{8}{9}$

19. $6\frac{5}{6}$ 20. $3\frac{4}{7}$ 21. $4\frac{10}{11}$

22. $12\frac{3}{8}$ 23. $32\frac{3}{4}$ 24. $15\frac{3}{10}$

25. $17\frac{12}{13}$ 26. $19\frac{8}{11}$ 27. $17\frac{14}{15}$

28. $7\frac{3}{16}$ 29. $7\frac{19}{24}$ 30. $9\frac{7}{12}$

Write each improper fraction as a whole or mixed number. See Example 2.

31. $\frac{5}{2}$ 32. $\frac{8}{5}$ 33. $\frac{9}{4}$ 34. $\frac{13}{3}$

35. $\frac{48}{6}$ 36. $\frac{64}{8}$ 37. $\frac{38}{5}$ 38. $\frac{33}{7}$

39. $\frac{39}{8}$ 40. $\frac{40}{9}$ 41. $\frac{27}{3}$ 42. $\frac{78}{6}$

43. $\frac{63}{4}$ 44. $\frac{19}{5}$ 45. $\frac{47}{9}$ 46. $\frac{65}{9}$

47. $\frac{65}{8}$ 48. $\frac{37}{6}$ 49. $\frac{84}{5}$ 50. $\frac{92}{3}$

51. $\frac{108}{5}$ 52. $\frac{149}{8}$ 53. $\frac{183}{7}$ 54. $\frac{212}{11}$

0.9 Factors

OBJECTIVES

1. Find factors and factor pairs of a number.
2. Identify prime numbers.
3. Find prime factorizations.

OBJECTIVE 1 Find factors and factor pairs of a number. You will recall that numbers multiplied to give a product are called **factors**. Because 2 • 5 = 10, both 2 and 5 are factors of 10. The numbers 1 and 10 are also factors of 10, because

$$1 \cdot 10 = 10.$$

The whole numbers 1, 2, 5, and 10 are the only whole number factors of 10. The products 2 • 5 and 1 • 10 are called **factorizations** of 10, and the pairs 2, 5 and 1, 10 are called **factor pairs**.

1. Find all the whole number factors of each number. List them as factor pairs.

   (a) 10

   (b) 16

   (c) 36

EXAMPLE 1 Finding Factor Pairs

Find all possible factor pairs of each number.

(a) 12

$$1 \cdot 12 = 12 \quad 2 \cdot 6 = 12 \quad 3 \cdot 4 = 12$$

The factor pairs of 12 are 1, 12 and 2, 6 and 3, 4.

(b) 60

$$1 \cdot 60 = 60 \qquad 2 \cdot 30 = 60$$
$$3 \cdot 20 = 60 \qquad 4 \cdot 15 = 60$$
$$5 \cdot 12 = 60 \qquad 6 \cdot 10 = 60$$

The factor pairs of 60 are 1, 60 and 2, 30 and 3, 20 and 4, 15 and 5, 12 and 6, 10.

Work Problem 1 at the Side.

Composite Numbers

A number with a factor other than itself or 1 is called a **composite number**.

2. Which of these numbers are composite?

   2, 4, 5, 6, 8, 10, 11, 13, 19, 21, 27, 28, 33, 36, 42

EXAMPLE 2 Identifying Composite Numbers

Which of the following numbers are composite?

(a) 6

Because 6 has factors of 2 and 3, numbers other than 6 or 1, the number 6 is composite.

(b) 17

The number 17 has only two factors, 17 and 1. It is not composite.

(c) 25

A factor of 25 is 5, so 25 is composite.

Work Problem 2 at the Side.

ANSWERS

1. (a) 1, 10 and 2, 5 (b) 1, 16 and 2, 8 and 4, 4 (c) 1, 36 and 2, 18 and 3, 12 and 4, 9 and 6, 6
2. 4, 6, 8, 10, 21, 27, 28, 33, 36, 42

APTER 0 Topics in Arithmetic

...ich of the following are prime?

4, 7, 9, 13, 17, 19, 29, 33

OBJECTIVE 2 Identify prime numbers. Whole numbers that are not composite are called **prime numbers,** except 0 and 1, which are neither prime nor composite.

> **Prime Numbers**
>
> A **prime number** is a whole number that has exactly *two different* factors, *itself* and *1*.

The number 3 is a prime number, since it can be divided evenly only by itself and 1. The number 8 is not a prime number (it is composite), since 8 can be divided evenly by 2 and 4, as well as by itself and 1.

> **CAUTION** A prime number has only two different factors, itself and 1. The number 1 is not a prime number because it does not have two different factors; the only factor of 1 is 1.

EXAMPLE 3 Finding Prime Numbers

Which of the following numbers are prime?

$$2 \quad 5 \quad 11 \quad 15 \quad 27$$

The number 15 can be divided by 3 and 5, so it is not prime. Also, because 27 can be divided by 3 and 9, 27 is not prime. All the other numbers in the list are divisible by only themselves and 1, so they are prime.

Work Problem 3 at the Side.

OBJECTIVE 3 Find prime factorizations. For reference, here are the prime numbers smaller than 100.

2, 3, 5, 7, 11,
13, 17, 19, 23, 29,
31, 37, 41, 43, 47,
53, 59, 61, 67, 71,
73, 79, 83, 89, 97

> **CAUTION** All prime numbers are odd numbers except the number 2. Be careful though, because *not all odd numbers are prime numbers.* For example, 9, 15, and 21 are odd numbers, but are *not* prime numbers.

The **prime factorization** of a number can be especially useful when we are adding or subtracting fractions and need to find a common denominator or write a fraction in lowest terms.

ANSWERS
3. 7, 13, 17, 19, 29

SECTION 0.9 Factors **0-47**

4. Find the prime factorization of each number.

   (a) 8

   (b) 42

   (c) 18

   (d) 40

5. Find the prime factorization of each number. Write the factorization with exponents.

   (a) 36

   (b) 54

   (c) 60

   (d) 81

Prime Factorization

The **prime factorization** of a number is the factorization in which every factor is a *prime number*.

EXAMPLE 4 Determining the Prime Factorization

Find the prime factorization of 12.
   Try to divide 12 by the first prime, 2.

$$12 \div 2 = 6,$$

     ↑ First prime

so

$$12 = 2 \cdot 6.$$

Try to divide 6 by the prime, 2.

$$6 \div 2 = 3,$$

so

$$12 = 2 \cdot \underbrace{2 \cdot 3}_{\text{Factorization of 6}}.$$

Because all factors are prime, the prime factorization of 12 is

$$2 \cdot 2 \cdot 3.$$

Work Problem 4 at the Side.

EXAMPLE 5 Factoring by Using the Division Method

Find the prime factorization of 48.

All prime factors:
$2 \overline{)48}$ Divide 48 by 2 (first prime).
$2 \overline{)24}$ Divide 24 by 2.
$2 \overline{)12}$ Divide 12 by 2.
$2 \overline{)6}$ Divide 6 by 2.
$3 \overline{)3}$ Divide 3 by 3.
1 Continue to divide until the quotient is 1.

Because all factors (divisors) are prime, the prime factorization of 48 is

$$2 \cdot 2 \cdot 2 \cdot 2 \cdot 3.$$

Using exponents, $2 \cdot 2 \cdot 2 \cdot 2$ is written as 2^4, so the prime factorization of 48 can be written

$$2 \cdot 2 \cdot 2 \cdot 2 \cdot 3 = 2^4 \cdot 3.$$

Work Problem 5 at the Side.

ANSWERS
4. (a) $2 \cdot 2 \cdot 2$ (b) $2 \cdot 3 \cdot 7$
    (c) $2 \cdot 3 \cdot 3$ (d) $2 \cdot 2 \cdot 2 \cdot 5$
5. (a) $2^2 \cdot 3^2$ (b) $2 \cdot 3^3$
    (c) $2^2 \cdot 3 \cdot 5$ (d) 3^4

CHAPTER 0 Topics in Arithmetic

6. Write the prime factorization of each number by using exponents.

   (a) 50

   (b) 88

   (c) 90

   (d) 120

   (e) 180

NOTE When using the division method of factoring, the last quotient found is 1. The "1" is never used as a prime factor because 1 is neither prime nor composite. Besides, 1 times any number is the number itself.

EXAMPLE 6 Using Exponents with Prime Factorization

Find the prime factorization of 225.

$$3\overline{)225}$$ 225 is not divisible by 2; use 3.
$$3\overline{)75}$$ Divide 75 by 3.
$$5\overline{)25}$$ 25 is not divisible by 3; use 5.
$$5\overline{)5}$$ Divide by 5.
$$1$$ Continue to divide until the quotient is 1.

All prime factors

Write the prime factorization,

$$3 \cdot 3 \cdot 5 \cdot 5,$$

with exponents as

$$3^2 \cdot 5^2.$$

Work Problem 6 at the Side.

Another method of factoring is a *factor tree*.

EXAMPLE 7 Factoring by Using a Factor Tree

Find the prime factorization of each number.

(a) 30

Try to divide by the first prime, 2. Write the factors under the 30. Circle the 2, since it is a prime.

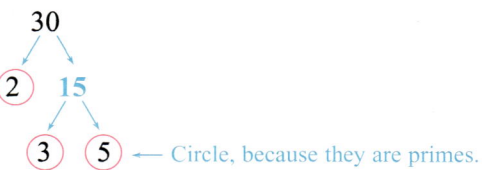

Since 15 cannot be divided evenly by 2, try the next prime, 3.

$$30$$
② 15
③ ⑤ ← Circle, because they are primes.

No uncircled factors remain, so the prime factorization (the circled factors) has been found.

$$30 = 2 \cdot 3 \cdot 5$$

ANSWERS
6. (a) $2 \cdot 5^2$ (b) $2^3 \cdot 11$
   (c) $2 \cdot 3^2 \cdot 5$ (d) $2^3 \cdot 3 \cdot 5$
   (e) $2^2 \cdot 3^2 \cdot 5$

SECTION 0.9 Factors 0-49

7. Complete each factor tree and give the prime factorization.

(a)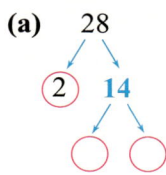

28

(b) 78

ANSWERS

7. (a)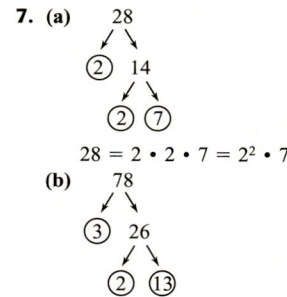

28 = 2 • 2 • 7 = 2² • 7

(b) 78

78 = 2 • 3 • 13

(b) 24
Divide by 2.

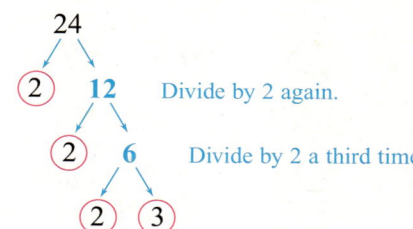

24 = 2 • 2 • 2 • 3 or, using exponents, 24 = 2³ • 3.

(c) 45
Because 45 cannot be divided by 2, try 3.

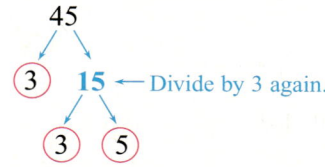

45 = 3 • 3 • 5 or, using exponents, 45 = 3² • 5.

NOTE The diagrams used in Example 7 look like tree branches, and that is why this method is referred to as using a factor tree.

Work Problem 7 at the Side.

0.9 EXERCISES

For Extra Help

 Student's Solutions Manual

 MyMathLab

 InterAct Math Tutorial Software

 AW Math Tutor Center

MathXL

Find all the factors of each number. List them as factor pairs. See Example 1.

1. 6   2. 15
3. 8   4. 28
5. 48   6. 30
7. 36   8. 20
9. 40   10. 60
11. 64   12. 84

Decide whether each number is prime or composite. See Examples 2 and 3.

13. 4   14. 6   15. 11   16. 7

17. 8   18. 9   19. 13   20. 65

21. 13   22. 17   23. 25   24. 26

25. 42   26. 47   27. 45   28. 53

Find the prime factorization of each number. Write answers with exponents when repeated factors appear. See Examples 4–7.

29. 10   30. 15   31. 20

32. 40   33. 25   34. 18

35. 36   36. 56   37. 68

38. 70   39. 72   40. 64

41. 44   42. 104   43. 100

44. 112   45. 125   46. 135

47. 180   48. 300   49. 320

50. 350   51. 360   52. 400

0.10 Writing a Fraction in Lowest Terms

OBJECTIVES

1. Tell whether a fraction is written in lowest terms.
2. Write a fraction in lowest terms using common factors.
3. Write a fraction in lowest terms using prime factors.
4. Tell whether two fractions are equivalent.

When working problems involving fractions, we must often compare two fractions to determine whether they represent the same portion of a whole. Consider the following diagrams.

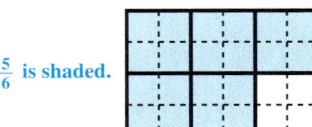
$\frac{5}{6}$ is shaded.

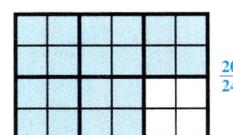
$\frac{20}{24}$ is shaded.

The figures show areas that are $\frac{5}{6}$ shaded and $\frac{20}{24}$ shaded. Because the shaded areas are equivalent (the same portion), the fractions $\frac{5}{6}$ and $\frac{20}{24}$ are **equivalent fractions.**

$$\frac{5}{6} = \frac{20}{24}$$

SECTION 0.10 Writing a Fraction in Lowest Terms 0-51

1. Decide whether the given factor is a common factor of both numbers.

   (a) 8, 18; 2

   (b) 32, 64; 8

   (c) 16, 34; 16

   (d) 56, 73; 1

Because the numbers 20 and 24 both have 4 as a factor, 4 is called a **common factor** of the numbers. Other common factors of 20 and 24 are 1 and 2.

Work Problem 1 at the Side.

OBJECTIVE 1 Tell whether a fraction is written in lowest terms. The fraction $\frac{5}{6}$ is written in *lowest terms* because the numerator and denominator have no common factor other than 1. However, the fraction $\frac{20}{24}$ is *not* in lowest terms because its numerator and denominator have common factors of 4, 2, and 1.

> **Writing a Fraction in Lowest Terms**
>
> A fraction is written in lowest terms when the numerator and denominator have no common factor other than 1.

EXAMPLE 1 Understanding Lowest Terms

Are the following fractions in lowest terms?

(a) $\frac{3}{8}$

The numerator and denominator have no common factor other than 1, so the fraction is in lowest terms.

(b) $\frac{21}{36}$

The numerator and denominator have a common factor of 3, so the fraction is not in lowest terms.

Work Problem 2 at the Side.

2. Are the following fractions in lowest terms?

   (a) $\frac{3}{4}$

   (b) $\frac{5}{15}$

   (c) $\frac{9}{15}$

   (d) $\frac{17}{46}$

OBJECTIVE 2 Write a fraction in lowest terms using common factors. There are two common methods for writing a fraction in lowest terms. These methods are shown in the next examples. The first method works best when the numerator and denominator are small numbers.

EXAMPLE 2 Writing Fractions in Lowest Terms

Write each fraction in lowest terms.

(a) $\frac{20}{24}$

The largest common factor of 20 and 24 is 4. Divide both numerator and denominator by **4**.

$$\frac{20}{24} = \frac{20 \div 4}{24 \div 4} = \frac{5}{6}$$

(b) $\frac{30}{50} = \frac{30 \div 10}{50 \div 10} = \frac{3}{5}$ Divide both numerator and denominator by 10.

ANSWERS
1. (a) yes (b) yes (c) no (d) yes
2. (a) yes (b) no (c) no (d) yes

0-52 CHAPTER 0 Topics in Arithmetic

3. Write in lowest terms.

(a) $\dfrac{6}{12}$

(b) $\dfrac{6}{8}$

(c) $\dfrac{48}{60}$

(d) $\dfrac{30}{80}$

(e) $\dfrac{16}{40}$

(c) $\dfrac{24}{42} = \dfrac{24 \div 6}{42 \div 6} = \dfrac{4}{7}$ Divide both numerator and denominator by 6.

(d) $\dfrac{60}{72}$

Suppose we made an error and thought 4 was the largest common factor of 60 and 72. Dividing by 4 would give

$$\dfrac{60}{72} = \dfrac{60 \div 4}{72 \div 4} = \dfrac{15}{18}.$$

But $\dfrac{15}{18}$ is not in lowest terms, because 15 and 18 have a common factor of 3. So we divide by 3.

$$\dfrac{15}{18} = \dfrac{15 \div 3}{18 \div 3} = \dfrac{5}{6}$$

The fraction $\dfrac{60}{72}$ could have been written in lowest terms in one step by dividing by 12, the largest common factor of 60 and 72.

$$\dfrac{60}{72} = \dfrac{60 \div 12}{72 \div 12} = \dfrac{5}{6}$$

NOTE Dividing the numerator and denominator by the same number results in an equivalent fraction.

In Example 2, we wrote fractions in lowest terms by dividing by a common factor. This method is summarized in the following steps.

The Method of Dividing by a Common Factor

Step 1 Find the largest number that will divide evenly into both the numerator and denominator. This number is a *common factor*.

Step 2 *Divide* both numerator and denominator by the common factor.

Step 3 *Check* to see whether the new fraction has any common factors (besides 1). If it does, repeat Steps 2 and 3. If the only common factor is 1, the fraction is in lowest terms.

Work Problem 3 at the Side.

OBJECTIVE 3 Write a fraction in lowest terms using prime factors. The method of writing a fraction in lowest terms by division works well for fractions with small numerators and denominators. For larger numbers, it is common to use the method of *prime factors,* which is shown in the next example.

ANSWERS

3. (a) $\dfrac{1}{2}$ (b) $\dfrac{3}{4}$ (c) $\dfrac{4}{5}$ (d) $\dfrac{3}{8}$ (e) $\dfrac{2}{5}$

EXAMPLE 3 Using Prime Factors

Write each fraction in lowest terms.

(a) $\dfrac{24}{42}$

Write the prime factorization of both numerator and denominator. See **Section 0.9** for help.

$$\frac{24}{42} = \frac{2 \cdot 2 \cdot 2 \cdot 3}{2 \cdot 3 \cdot 7}$$

Just as with the other method, divide both numerator and denominator by any common factors. Write a **1** by each factor that has been divided.

$$\frac{24}{42} = \frac{\cancel{2} \cdot 2 \cdot 2 \cdot \cancel{3}}{\cancel{2} \cdot \cancel{3} \cdot 7}$$

Multiply the remaining factors in both numerator and denominator.

$$\frac{24}{42} = \frac{1 \cdot 2 \cdot 2 \cdot 1}{1 \cdot 1 \cdot 7} = \frac{4}{7}$$

Finally, $\frac{24}{42}$ written in lowest terms is $\frac{4}{7}$.

(b) $\dfrac{162}{54}$

Write the prime factorization of both numerator and denominator.

$$\frac{162}{54} = \frac{2 \cdot 3 \cdot 3 \cdot 3 \cdot 3}{2 \cdot 3 \cdot 3 \cdot 3}$$

Now divide by the common factors. ***Do not forget to write the 1s.***

$$\frac{162}{54} = \frac{\cancel{2} \cdot \cancel{3} \cdot \cancel{3} \cdot \cancel{3} \cdot 3}{\cancel{2} \cdot \cancel{3} \cdot \cancel{3} \cdot \cancel{3}}$$

$$= \frac{1 \cdot 1 \cdot 1 \cdot 1 \cdot 3}{1 \cdot 1 \cdot 1 \cdot 1} = \frac{3}{1} = 3$$

(c) $\dfrac{18}{90}$

$$\frac{18}{90} = \frac{\cancel{2} \cdot \cancel{3} \cdot \cancel{3}}{\cancel{2} \cdot \cancel{3} \cdot \cancel{3} \cdot 5} = \frac{1 \cdot 1 \cdot 1}{1 \cdot 1 \cdot 1 \cdot 5} = \frac{1}{5}$$

CAUTION In Example 3(c), all factors of the numerator were divided. But $1 \cdot 1 \cdot 1$ is still 1, so the final answer is $\frac{1}{5}$ (***not*** 5).

0-54 CHAPTER 0 Topics in Arithmetic

4. Use the method of prime factors to write each fraction in lowest terms.

(a) $\dfrac{12}{36}$

(b) $\dfrac{32}{56}$

(c) $\dfrac{74}{111}$

(d) $\dfrac{124}{340}$

In Example 3, we wrote fractions in lowest terms using prime factors. This method is summarized as follows.

The Method of Prime Factors

Step 1 Write the *prime factorization* of both numerator and denominator.

Step 2 *Divide* both numerator and denominator by any common factors.

Step 3 *Multiply* the remaining factors in the numerator and denominator.

Work Problem 4 at the Side.

OBJECTIVE 4 Tell whether two fractions are equivalent. The next example shows how to tell whether two fractions are equivalent.

EXAMPLE 4 Determining whether Two Fractions Are Equivalent

Determine whether each pair of fractions is equivalent.

(a) $\dfrac{16}{48}$ and $\dfrac{24}{72}$

Use the method of prime factors to write each fraction in lowest terms.

$$\dfrac{16}{48} = \dfrac{\cancel{2}\cdot\cancel{2}\cdot\cancel{2}\cdot\cancel{2}}{\cancel{2}\cdot\cancel{2}\cdot\cancel{2}\cdot\cancel{2}\cdot 3} = \dfrac{1\cdot 1\cdot 1\cdot 1}{1\cdot 1\cdot 1\cdot 1\cdot 3} = \dfrac{1}{3}$$

$$\dfrac{24}{72} = \dfrac{\cancel{2}\cdot\cancel{2}\cdot\cancel{2}\cdot\cancel{3}}{\cancel{2}\cdot\cancel{2}\cdot\cancel{2}\cdot\cancel{3}\cdot 3} = \dfrac{1\cdot 1\cdot 1\cdot 1}{1\cdot 1\cdot 1\cdot 1\cdot 3} = \dfrac{1}{3}$$

Equivalent $\left(\dfrac{1}{3} = \dfrac{1}{3}\right)$

(b) $\dfrac{32}{52}$ and $\dfrac{64}{112}$

$$\dfrac{32}{52} = \dfrac{\cancel{2}\cdot\cancel{2}\cdot 2\cdot 2\cdot 2}{\cancel{2}\cdot\cancel{2}\cdot 13} = \dfrac{1\cdot 1\cdot 2\cdot 2\cdot 2}{1\cdot 1\cdot 13} = \dfrac{8}{13}$$

$$\dfrac{64}{112} = \dfrac{\cancel{2}\cdot\cancel{2}\cdot\cancel{2}\cdot\cancel{2}\cdot 2\cdot 2}{\cancel{2}\cdot\cancel{2}\cdot\cancel{2}\cdot\cancel{2}\cdot 7} = \dfrac{1\cdot 1\cdot 1\cdot 1\cdot 2\cdot 2}{1\cdot 1\cdot 1\cdot 1\cdot 7} = \dfrac{4}{7}$$

Not equivalent $\left(\dfrac{8}{13} \neq \dfrac{4}{7}\right)$

ANSWERS

4. (a) $\dfrac{1}{3}$ (b) $\dfrac{4}{7}$ (c) $\dfrac{2}{3}$ (d) $\dfrac{31}{85}$

SECTION 0.10 Writing a Fraction in Lowest Terms 0-55

5. Is each pair of fractions equivalent?

(a) $\dfrac{24}{48}$ and $\dfrac{36}{72}$

(b) $\dfrac{45}{60}$ and $\dfrac{50}{75}$

(c) $\dfrac{20}{4}$ and $\dfrac{110}{22}$

(d) $\dfrac{120}{220}$ and $\dfrac{180}{320}$

(c) $\dfrac{75}{15}$ and $\dfrac{60}{12}$

$$\dfrac{75}{15} = \dfrac{\overset{1}{\cancel{3}} \cdot \overset{1}{\cancel{5}} \cdot 5}{\underset{1}{\cancel{3}} \cdot \underset{1}{\cancel{5}}} = \dfrac{1 \cdot 1 \cdot 5}{1 \cdot 1} = 5$$

$$\dfrac{60}{12} = \dfrac{\overset{1}{\cancel{2}} \cdot \overset{1}{\cancel{2}} \cdot \overset{1}{\cancel{3}} \cdot 5}{\underset{1}{\cancel{2}} \cdot \underset{1}{\cancel{2}} \cdot \underset{1}{\cancel{3}}} = \dfrac{1 \cdot 1 \cdot 1 \cdot 5}{1 \cdot 1 \cdot 1} = 5$$

Equivalent (5 = 5)

Work Problem 5 at the Side.

ANSWERS

5. (a) equivalent (b) not equivalent
   (c) equivalent (d) not equivalent

0.10 EXERCISES

For Extra Help
- Student's Solutions Manual
- MyMathLab
- InterAct Math Tutorial Software
- AW Math Tutor Center
- MathXL

Put a ✓ mark in the blank if the number at the left is divisible by the number at the top. Put an X in the blank if the number is not divisible by the number at the top. (For help, see Section 0.5.)

	2	3	5	10
1. 30	15	10	6	3
2. 60	30	20	12	6
3. 48	24	16	X	X
4. 36	16	12	X	X
5. 160	80	X	X	16
6. 175	X	X	35	X
7. 138	69	46	X	
8. 120	60	40	24	12

Write each fraction in lowest terms. See Example 2.

9. $\dfrac{9}{12}$ 10. $\dfrac{5}{10}$ 11. $\dfrac{16}{24}$ 12. $\dfrac{4}{12}$

13. $\dfrac{25}{40}$ 14. $\dfrac{32}{48}$ 15. $\dfrac{36}{42}$ 16. $\dfrac{22}{33}$

17. $\dfrac{63}{70}$ 18. $\dfrac{21}{35}$ 19. $\dfrac{180}{210}$ 20. $\dfrac{72}{80}$

0-56 CHAPTER 0 Topics in Arithmetic

21. $\dfrac{36}{63}$ 22. $\dfrac{73}{146}$ 23. $\dfrac{12}{600}$ 24. $\dfrac{8}{400}$

25. $\dfrac{96}{132}$ 26. $\dfrac{165}{180}$ 27. $\dfrac{60}{108}$ 28. $\dfrac{112}{128}$

Write the numerator and denominator of each fraction as a product of prime factors and divide by the common factors. Then write the fraction in lowest terms. See Example 3.

29. $\dfrac{18}{24}$ 30. $\dfrac{16}{64}$ 31. $\dfrac{35}{40}$

32. $\dfrac{20}{32}$ 33. $\dfrac{90}{180}$ 34. $\dfrac{36}{48}$

35. $\dfrac{36}{12}$ 36. $\dfrac{192}{48}$ 37. $\dfrac{72}{225}$

38. $\dfrac{65}{234}$

Tell whether each pair of fractions is equivalent or not equivalent. See Example 4.

39. $\dfrac{2}{4}$ and $\dfrac{16}{32}$ 40. $\dfrac{4}{10}$ and $\dfrac{20}{50}$ 41. $\dfrac{10}{24}$ and $\dfrac{12}{30}$

42. $\dfrac{22}{32}$ and $\dfrac{32}{48}$ 43. $\dfrac{15}{24}$ and $\dfrac{35}{52}$ 44. $\dfrac{21}{33}$ and $\dfrac{9}{12}$

45. $\dfrac{14}{16}$ and $\dfrac{35}{40}$ 46. $\dfrac{27}{90}$ and $\dfrac{24}{80}$ 47. $\dfrac{48}{6}$ and $\dfrac{72}{8}$

48. $\dfrac{45}{15}$ and $\dfrac{96}{32}$ 49. $\dfrac{25}{30}$ and $\dfrac{65}{78}$ 50. $\dfrac{24}{72}$ and $\dfrac{30}{90}$

51. What does it mean when a fraction is expressed in lowest terms? Give three examples.

52. Explain what equivalent fractions are and give an example of a pair of equivalent fractions. Show that they are equivalent.

Write each fraction in lowest terms.

53. $\dfrac{224}{256}$ 54. $\dfrac{363}{528}$ 55. $\dfrac{356}{178}$ 56. $\dfrac{525}{105}$

0.11 Multiplying Fractions

OBJECTIVES

1. Multiply fractions.
2. Use a multiplication shortcut.
3. Multiply a fraction and a whole number.

OBJECTIVE 1 Multiply fractions. Suppose that you give $\frac{1}{2}$ of your Energy Bar to your kickboxing partner Jennifer. Then Jennifer gives $\frac{1}{2}$ of her share to Tony. How much of the Energy Bar does Tony get to eat?

Start with a sketch showing the Energy Bar cut in half (2 equal pieces).

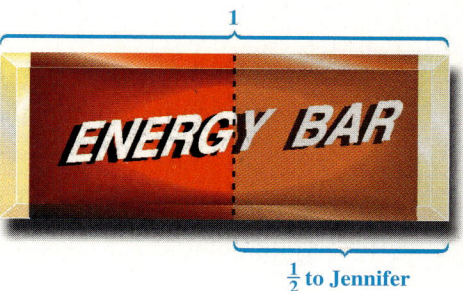

1. Use these figures to find $\frac{1}{4}$ of $\frac{1}{2}$.

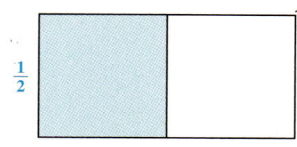

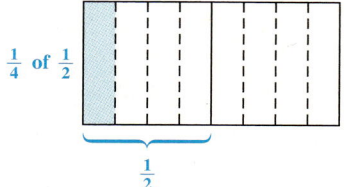

Next, take $\frac{1}{2}$ of the shaded area. (Here we are dividing $\frac{1}{2}$ into 2 equal parts and shading one darker than the other.)

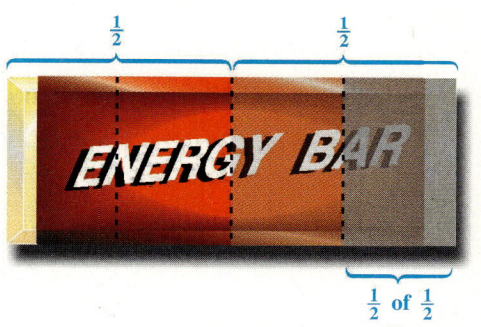

The sketch shows that Tony gets $\frac{1}{4}$ of the Energy Bar.

Tony gets $\frac{1}{2}$ of $\frac{1}{2}$ of the Energy Bar. When used between two fractions, the word **of** tells us to multiply.

$$\frac{1}{2} \text{ of } \frac{1}{2} \quad \text{means} \quad \frac{1}{2} \cdot \frac{1}{2}$$

Tony's share of the Energy Bar is

$$\frac{1}{2} \cdot \frac{1}{2} = \frac{1}{4}.$$

Work Problem 1 at the Side.

The rule for multiplying fractions follows.

Multiplying Fractions

Multiply two fractions by multiplying the numerators and multiplying the denominators.

ANSWERS

1. $\frac{1}{8}$

CHAPTER 0 Topics in Arithmetic

2. Multiply. Write answers in lowest terms.

(a) $\dfrac{3}{4} \cdot \dfrac{1}{2}$

(b) $\dfrac{1}{3} \cdot \dfrac{3}{5}$

(c) $\dfrac{1}{8} \cdot \dfrac{5}{6} \cdot \dfrac{1}{2}$

(d) $\dfrac{1}{2} \cdot \dfrac{3}{4} \cdot \dfrac{3}{8}$

ANSWERS

2. (a) $\dfrac{3}{8}$ (b) $\dfrac{1}{5}$ (c) $\dfrac{5}{96}$ (d) $\dfrac{9}{64}$

Use this rule to find the product of $\dfrac{2}{3}$ and $\dfrac{1}{3}$ (multiply $\dfrac{2}{3}$ by $\dfrac{1}{3}$).

$$\dfrac{2}{3} \cdot \dfrac{1}{3} = \dfrac{2 \cdot 1}{3 \cdot 3} \begin{array}{l} \longleftarrow \text{Multiply numerators.} \\ \longleftarrow \text{Multiply denominators.} \end{array}$$

$$= \dfrac{2}{9}$$

Finish multiplying.

$$\dfrac{2}{3} \cdot \dfrac{1}{3} = \dfrac{2 \cdot 1}{3 \cdot 3} = \dfrac{2}{9} \begin{array}{l} \longleftarrow 2 \cdot 1 = 2 \\ \longleftarrow 3 \cdot 3 = 9 \end{array}$$

Check that the final result is in lowest terms. $\dfrac{2}{9}$ is in lowest terms because 2 and 9 have no common factor other than 1.

EXAMPLE 1 Multiplying Fractions

Multiply. Write answers in lowest terms.

(a) $\dfrac{5}{8} \cdot \dfrac{3}{4}$

Multiply the numerators and multiply the denominators.

$$\dfrac{5}{8} \cdot \dfrac{3}{4} = \dfrac{5 \cdot 3}{8 \cdot 4} = \dfrac{15}{32}$$

Notice that 15 and 32 have no common factors other than 1, so the answer is in lowest terms.

(b) $\dfrac{4}{7} \cdot \dfrac{2}{5}$

$$\dfrac{4}{7} \cdot \dfrac{2}{5} = \dfrac{4 \cdot 2}{7 \cdot 5} = \dfrac{8}{35}$$

(c) $\dfrac{5}{8} \cdot \dfrac{3}{4} \cdot \dfrac{1}{2}$

$$\dfrac{5}{8} \cdot \dfrac{3}{4} \cdot \dfrac{1}{2} = \dfrac{5 \cdot 3 \cdot 1}{8 \cdot 4 \cdot 2} = \dfrac{15}{64}$$

Work Problem 2 at the Side.

OBJECTIVE 2 Use a multiplication shortcut. A multiplication shortcut that can be used with fractions is shown in Example 2.

EXAMPLE 2 Using the Multiplication Shortcut

Multiply $\dfrac{5}{6}$ and $\dfrac{9}{10}$. Write the answer in lowest terms.

$$\dfrac{5}{6} \cdot \dfrac{9}{10} = \dfrac{5 \cdot 9}{6 \cdot 10} = \dfrac{45}{60} \quad \text{Not in lowest terms}$$

The numerator and denominator have a common factor other than 1, so write the prime factorization of each number.

$$\frac{5}{6} \cdot \frac{9}{10} = \frac{5 \cdot 9}{6 \cdot 10} = \frac{5 \cdot 3 \cdot 3}{2 \cdot 3 \cdot 2 \cdot 5}$$

Next, divide by the common factors of 5 and 3.

$$\frac{5}{6} \cdot \frac{9}{10} = \frac{5 \cdot 9}{6 \cdot 10} = \frac{\cancel{5} \cdot \cancel{3} \cdot 3}{2 \cdot \cancel{3} \cdot 2 \cdot \cancel{5}}$$

Finally, multiply the remaining factors in the numerator and in the denominator.

$$\frac{5}{6} \cdot \frac{9}{10} = \frac{1 \cdot 1 \cdot 3}{2 \cdot 1 \cdot 2 \cdot 1} = \frac{3}{4} \quad \text{Lowest terms}$$

As a shortcut, instead of writing the prime factorization of each number, find the product of $\frac{5}{6}$ and $\frac{9}{10}$ as follows.

First, divide both 5 and 10 by 5. $\quad \dfrac{\cancel{5}^{1}}{6} \cdot \dfrac{9}{\cancel{10}_{2}}$

Next, divide both 6 and 9 by 3. $\quad \dfrac{\cancel{5}^{1}}{\cancel{6}_{2}} \cdot \dfrac{\cancel{9}^{3}}{\cancel{10}_{2}}$

Finally, multiply. $\quad \dfrac{1 \cdot 3}{2 \cdot 2} = \dfrac{3}{4}$

CAUTION When using the multiplication shortcut, you are dividing a numerator and a denominator. Be certain that you divide a numerator and a denominator **by the same number**. If you do all possible divisions, your answer will be in lowest terms.

EXAMPLE 3 Using the Multiplication Shortcut

Use the multiplication shortcut to find each product. Write the answers in lowest terms and as mixed numbers where possible.

(a) $\dfrac{6}{11} \cdot \dfrac{7}{8}$

Divide both 6 and 8 by 2. Next, multiply.

$$\dfrac{\cancel{6}^{3}}{11} \cdot \dfrac{7}{\cancel{8}_{4}} = \dfrac{3 \cdot 7}{11 \cdot 4} = \dfrac{21}{44} \quad \text{Lowest terms}$$

0-60 CHAPTER 0 Topics in Arithmetic

3. Use the multiplication shortcut to find each product.

(a) $\dfrac{3}{4} \cdot \dfrac{2}{3}$

(b) $\dfrac{6}{11} \cdot \dfrac{33}{21}$

(c) $\dfrac{20}{4} \cdot \dfrac{3}{40} \cdot \dfrac{1}{3}$

(d) $\dfrac{18}{17} \cdot \dfrac{1}{36} \cdot \dfrac{2}{3}$

(b) $\dfrac{7}{10} \cdot \dfrac{20}{21}$

Divide 7 and 21 by 7, and divide 10 and 20 by 10.

$$\dfrac{\overset{1}{\cancel{7}}}{\underset{1}{\cancel{10}}} \cdot \dfrac{\overset{2}{\cancel{20}}}{\underset{3}{\cancel{21}}} = \dfrac{1 \cdot 2}{1 \cdot 3} = \dfrac{2}{3} \quad \text{Lowest terms}$$

(c) $\dfrac{35}{12} \cdot \dfrac{32}{25}$

$$\dfrac{\overset{7}{\cancel{35}}}{\underset{3}{\cancel{12}}} \cdot \dfrac{\overset{8}{\cancel{32}}}{\underset{5}{\cancel{25}}} = \dfrac{7 \cdot 8}{3 \cdot 5} = \dfrac{56}{15} \quad \text{or} \quad 3\dfrac{11}{15} \quad \text{Mixed number}$$

(d) $\dfrac{2}{3} \cdot \dfrac{8}{15} \cdot \dfrac{3}{4}$

$$\dfrac{\overset{1}{\cancel{2}}}{\underset{1}{\cancel{3}}} \cdot \dfrac{\overset{4}{\cancel{8}}}{15} \cdot \dfrac{\overset{1}{\cancel{3}}}{\underset{2}{\cancel{4}}} = \dfrac{1 \cdot 4 \cdot 1}{1 \cdot 15 \cdot 1} = \dfrac{4}{15} \quad \text{Lowest terms}$$

This shortcut is especially helpful when the fractions involve large numbers.

NOTE There is no specific order that must be used when dividing numerators and denominators as long as both numerator and denominator are divided by the same number.

Work Problem 3 at the Side.

OBJECTIVE 3 Multiply a fraction and a whole number. The rule for multiplying a fraction and a whole number follows.

Multiplying a Whole Number and a Fraction

Multiply a whole number and a fraction by writing the whole number as a fraction with a denominator of 1.

For example, write the whole numbers 8, 10, and 25 as follows.

$$8 = \dfrac{8}{1}, \quad 10 = \dfrac{10}{1}, \quad \text{and} \quad 25 = \dfrac{25}{1}$$

ANSWERS

3. (a) $\dfrac{\overset{1}{\cancel{3}}}{\underset{2}{\cancel{4}}} \cdot \dfrac{\overset{1}{\cancel{2}}}{\underset{1}{\cancel{3}}} = \dfrac{1}{2}$ (b) $\dfrac{\overset{2}{\cancel{6}}}{\underset{1}{\cancel{11}}} \cdot \dfrac{\overset{3}{\cancel{33}}}{\underset{7}{\cancel{21}}} = \dfrac{6}{7}$

(c) $\dfrac{\overset{1}{\cancel{20}}}{4} \cdot \dfrac{\overset{1}{\cancel{3}}}{\underset{2}{\cancel{40}}} \cdot \dfrac{1}{\underset{1}{\cancel{3}}} = \dfrac{1}{8}$

(d) $\dfrac{\overset{1}{\cancel{18}}}{17} \cdot \dfrac{1}{\underset{\underset{1}{2}}{\cancel{36}}} \cdot \dfrac{\overset{1}{\cancel{2}}}{3} = \dfrac{1}{51}$

SECTION 0.11 Multiplying Fractions   0-61

4. Multiply. Write answers in lowest terms.

(a) $5 \cdot \dfrac{1}{5}$

(b) $\dfrac{5}{3} \cdot 12 \cdot \dfrac{3}{4}$

(c) $40 \cdot \dfrac{3}{10}$

(d) $\dfrac{3}{25} \cdot \dfrac{5}{11} \cdot 99$

EXAMPLE 4 Multiplying by Whole Numbers

Multiply. Write answers in lowest terms and as whole numbers where possible.

(a) $8 \cdot \dfrac{3}{4}$

Write 8 as $\dfrac{8}{1}$ and multiply.

$$8 \cdot \dfrac{3}{4} = \dfrac{\overset{2}{\cancel{8}}}{1} \cdot \dfrac{3}{\underset{1}{\cancel{4}}} = \dfrac{2 \cdot 3}{1 \cdot 1} = \dfrac{6}{1} = 6$$

(b) $12 \cdot \dfrac{5}{6}$

$$12 \cdot \dfrac{5}{6} = \dfrac{\overset{2}{\cancel{12}}}{1} \cdot \dfrac{5}{\underset{1}{\cancel{6}}} = \dfrac{2 \cdot 5}{1 \cdot 1} = \dfrac{10}{1} = 10$$

Work Problem 4 at the Side.

ANSWERS
4. (a) 1   (b) 15   (c) 12
   (d) $\dfrac{27}{5}$ or $5\dfrac{2}{5}$

0.11 EXERCISES

For Extra Help

- Student's Solutions Manual
- MyMathLab
- InterAct Math Tutorial Software
- AW Math Tutor Center
- MathXL

Multiply. Write answers in lowest terms. See Examples 1–3.

1. $\dfrac{3}{4} \cdot \dfrac{1}{2}$

2. $\dfrac{1}{2} \cdot \dfrac{2}{3}$

3. $\dfrac{2}{3} \cdot \dfrac{5}{8}$

4. $\dfrac{2}{5} \cdot \dfrac{3}{4}$

5. $\dfrac{8}{5} \cdot \dfrac{15}{32}$

6. $\dfrac{4}{9} \cdot \dfrac{12}{7}$

7. $\dfrac{2}{3} \cdot \dfrac{7}{12} \cdot \dfrac{9}{14}$

8. $\dfrac{7}{8} \cdot \dfrac{16}{21} \cdot \dfrac{1}{2}$

9. $\dfrac{3}{4} \cdot \dfrac{5}{6} \cdot \dfrac{2}{3}$

10. $\dfrac{2}{5} \cdot \dfrac{3}{8} \cdot \dfrac{2}{3}$

11. $\dfrac{9}{22} \cdot \dfrac{11}{16}$

12. $\dfrac{5}{12} \cdot \dfrac{7}{10}$

13. $\dfrac{5}{8} \cdot \dfrac{16}{25}$

14. $\dfrac{6}{11} \cdot \dfrac{22}{15}$

15. $\dfrac{14}{25} \cdot \dfrac{65}{48} \cdot \dfrac{15}{28}$

16. $\dfrac{35}{64} \cdot \dfrac{32}{15} \cdot \dfrac{27}{72}$

17. $\dfrac{16}{25} \cdot \dfrac{35}{32} \cdot \dfrac{15}{64}$

18. $\dfrac{39}{42} \cdot \dfrac{7}{13} \cdot \dfrac{7}{24}$

0-62 CHAPTER 0 Topics in Arithmetic

Multiply. Write answers in lowest terms and as whole or mixed numbers where possible. See Example 4.

19. $6 \cdot \dfrac{5}{6} = \dfrac{5}{1} = 5$

20. $40 \cdot \dfrac{3}{4} = \dfrac{30}{1} = 30$

21. $\dfrac{5}{8} \cdot 64 = \dfrac{40}{1} = 40$

22. $\dfrac{3}{5} \cdot 45 = \dfrac{27}{1} = 27$

23. $28 \cdot \dfrac{3}{4} = \dfrac{21}{1} = 21$

24. $30 \cdot \dfrac{3}{10} = \dfrac{9}{1} = 9$

25. $36 \cdot \dfrac{5}{8} \cdot \dfrac{9}{15} =$

26. $35 \cdot \dfrac{3}{5} \cdot \dfrac{1}{2} = \dfrac{21}{2}$

27. $100 \cdot \dfrac{21}{50} \cdot \dfrac{3}{4}$

28. $200 \cdot \dfrac{7}{8}$

29. $\dfrac{3}{5} \cdot 400 = \dfrac{240}{1} = 240$

30. $\dfrac{3}{7} \cdot 490 = \dfrac{210}{1} = 210$

31. $\dfrac{2}{3} \cdot 284$

32. $\dfrac{12}{25} \cdot 430 =$

33. $\dfrac{28}{21} \cdot 640 \cdot \dfrac{15}{32}$

34. $\dfrac{21}{13} \cdot 520 \cdot \dfrac{7}{20}$

35. $\dfrac{54}{38} \cdot 684 \cdot \dfrac{5}{6}$

36. $\dfrac{76}{43} \cdot 473 \cdot \dfrac{5}{19}$

0.12 Dividing Fractions

OBJECTIVES

1. Find the reciprocal of a fraction.
2. Divide fractions.

OBJECTIVE 1 Find the reciprocal of a fraction. To divide fractions, we need to know how to find the **reciprocal** of a fraction.

Reciprocal of a Fraction

Two numbers are reciprocals of each other if their product is 1. To find the reciprocal of a fraction, interchange the numerator and denominator.

For example, the reciprocal of $\dfrac{3}{4}$ is $\dfrac{4}{3}$.

Fraction $\dfrac{3}{4}$ ✕ $\dfrac{4}{3}$ Reciprocal

NOTE Notice that you invert, or "flip" a fraction to find its reciprocal.

SECTION 0.12 Dividing Fractions

1. Find the reciprocal of each fraction.

   (a) $\dfrac{5}{8}$

   (b) $\dfrac{2}{5}$

   (c) $\dfrac{9}{4}$

   (d) 5

EXAMPLE 1 Finding Reciprocals

Find the reciprocal of each fraction.

(a) The reciprocal of $\dfrac{1}{4}$ is $\dfrac{4}{1}$ because $\dfrac{1}{4} \cdot \dfrac{4}{1} = \dfrac{4}{4} = 1$.

(b) The reciprocal of $\dfrac{2}{3}$ is $\dfrac{3}{2}$ because $\dfrac{2}{3} \cdot \dfrac{3}{2} = \dfrac{6}{6} = 1$.

(c) The reciprocal of $\dfrac{3}{5}$ is $\dfrac{5}{3}$ because $\dfrac{3}{5} \cdot \dfrac{5}{3} = \dfrac{15}{15} = 1$.

(d) The reciprocal of 8 is $\dfrac{1}{8}$ because $\dfrac{8}{1} \cdot \dfrac{1}{8} = \dfrac{8}{8} = 1$. Think of 8 as $\dfrac{8}{1}$.

Work Problem 1 at the Side.

> **NOTE** Every number has a reciprocal except 0. The number 0 has no reciprocal because there is no number that can be multiplied by 0 to get 1.
>
> $$0 \cdot (\text{reciprocal}) = 1$$
>
> There is no number to use here that will give an answer of 1. When you multiply by 0, you always get 0.

Earlier we saw that the division problem $12 \div 3$ asks how many 3s are in 12. In the same way, the division problem $\tfrac{2}{3} \div \tfrac{1}{6}$ asks how many $\tfrac{1}{6}$s are in $\tfrac{2}{3}$. The figure illustrates $\tfrac{2}{3} \div \tfrac{1}{6}$.

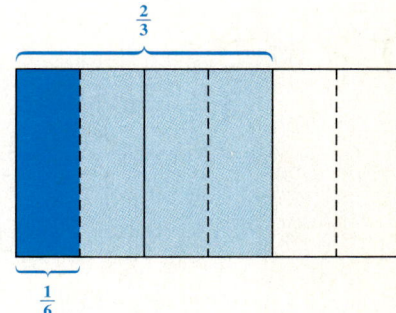

The figure shows that there are 4 of the $\tfrac{1}{6}$s in $\tfrac{2}{3}$, or

$$\dfrac{2}{3} \div \dfrac{1}{6} = 4.$$

ANSWERS

1. (a) $\dfrac{8}{5}$ (b) $\dfrac{5}{2}$ (c) $\dfrac{4}{9}$ (d) $\dfrac{1}{5}$

OBJECTIVE 2 Divide fractions. We will use reciprocals to divide fractions.

Dividing Fractions

To divide two fractions, multiply the first fraction by the reciprocal of the second fraction.

EXAMPLE 2 Dividing One Fraction by Another

Divide. Write answers in lowest terms.

(a) $\dfrac{7}{8} \div \dfrac{15}{16}$

The reciprocal of $\dfrac{15}{16}$ is $\dfrac{16}{15}$.

$$\dfrac{7}{8} \div \dfrac{15}{16} = \dfrac{7}{8} \cdot \dfrac{16}{15}$$

Change division to multiplication.

$$= \dfrac{7}{\overset{1}{\cancel{8}}} \cdot \dfrac{\overset{2}{\cancel{16}}}{15} \quad \text{Divide the numerator and denominator by 8.}$$

$$= \dfrac{7 \cdot 2}{1 \cdot 15} \quad \text{Multiply.}$$

$$= \dfrac{14}{15}$$

(b) $\dfrac{\tfrac{4}{5}}{\tfrac{3}{10}}$

$$\dfrac{\tfrac{4}{5}}{\tfrac{3}{10}} = \dfrac{4}{5} \div \dfrac{3}{10} \quad \text{Rewrite by using the} \div \text{symbol for division.}$$

$$= \dfrac{4}{\underset{1}{\cancel{5}}} \cdot \dfrac{\overset{2}{\cancel{10}}}{3} \quad \text{The reciprocal of } \tfrac{3}{10} \text{ is } \tfrac{10}{3}. \text{ Change ``}\div\text{'' to ``}\cdot\text{'', and divide the numerator and denominator by 5.}$$

$$= \dfrac{4 \cdot 2}{1 \cdot 3} \quad \text{Multiply.}$$

$$= \dfrac{8}{3} = 2\dfrac{2}{3} \quad \text{Mixed number}$$

2. Divide. Write answers in lowest terms.

   (a) $\dfrac{1}{2} \div \dfrac{3}{4}$

   (b) $\dfrac{5}{8} \div \dfrac{7}{8}$

   (c) $\dfrac{\frac{2}{3}}{\frac{4}{5}}$

   (d) $\dfrac{\frac{5}{6}}{\frac{7}{12}}$

3. Divide. Write answers in lowest terms and as whole or mixed numbers where possible.

   (a) $12 \div \dfrac{3}{4}$

   (b) $8 \div \dfrac{8}{9}$

   (c) $\dfrac{4}{5} \div 6$

   (d) $\dfrac{3}{8} \div 4$

CAUTION Be certain that the divisor fraction is changed to its reciprocal *before* you divide numerators and denominators by common factors.

Work Problem 2 at the Side.

EXAMPLE 3 Dividing with a Whole Number

Divide. Write all answers in lowest terms and as whole or mixed numbers where possible.

(a) $5 \div \dfrac{1}{4}$

Write 5 as $\frac{5}{1}$. Next, use the reciprocal of $\frac{1}{4}$, which is $\frac{4}{1}$.

$$5 \div \dfrac{1}{4} = \dfrac{5}{1} \cdot \dfrac{4}{1} \qquad \text{Reciprocal of } \tfrac{1}{4} \text{ is } \tfrac{4}{1}.$$

$$= \dfrac{5 \cdot 4}{1 \cdot 1} \qquad \text{Multiply.}$$

$$= \dfrac{20}{1} = 20 \qquad \text{Whole number}$$

(b) $\dfrac{2}{3} \div 6$

Write 6 as $\frac{6}{1}$. The reciprocal of $\frac{6}{1}$ is $\frac{1}{6}$.

$$\dfrac{2}{3} \div \dfrac{6}{1} = \dfrac{2}{3} \cdot \dfrac{1}{6}$$

$$= \dfrac{\overset{1}{\cancel{2}}}{3} \cdot \dfrac{1}{\underset{3}{\cancel{6}}} \qquad \text{Divide the numerator and denominator by 2, then multiply.}$$

$$= \dfrac{1 \cdot 1}{3 \cdot 3} = \dfrac{1}{9}$$

Work Problem 3 at the Side.

ANSWERS

2. (a) $\dfrac{2}{3}$ (b) $\dfrac{5}{7}$ (c) $\dfrac{5}{6}$ (d) $1\dfrac{3}{7}$

3. (a) 16 (b) 9 (c) $\dfrac{2}{15}$ (d) $\dfrac{3}{32}$

0-66 CHAPTER 0 Topics in Arithmetic

0.12 EXERCISES

For Extra Help

 Student's Solutions Manual

 MyMathLab

 InterAct Math Tutorial Software

Tutor Center — AW Math Tutor Center

MathXL — MathXL

Find the reciprocal of each number. See Example 1.

1. $\dfrac{2}{3}$ 2. $\dfrac{3}{4}$ 3. $\dfrac{8}{5}$ 4. $\dfrac{12}{7}$

5. $\dfrac{5}{6}$ 6. $\dfrac{13}{20}$ 7. 4 8. 10

Divide. Write answers in lowest terms and as whole or mixed numbers where possible. See Examples 2 and 3.

9. $\dfrac{1}{4} \div \dfrac{3}{4}$ 10. $\dfrac{3}{8} \div \dfrac{5}{8}$ 11. $\dfrac{7}{8} \div \dfrac{1}{3}$ 12. $\dfrac{7}{8} \div \dfrac{3}{4}$

13. $\dfrac{3}{4} \div \dfrac{5}{3}$ 14. $\dfrac{4}{5} \div \dfrac{9}{4}$ 15. $\dfrac{7}{9} \div \dfrac{7}{36}$ 16. $\dfrac{5}{8} \div \dfrac{5}{16}$

17. $\dfrac{15}{32} \div \dfrac{5}{64}$ 18. $\dfrac{7}{12} \div \dfrac{14}{15}$ 19. $\dfrac{\frac{13}{20}}{\frac{4}{5}}$ 20. $\dfrac{\frac{9}{10}}{\frac{3}{5}}$

21. $\dfrac{\frac{5}{6}}{\frac{25}{24}}$ 22. $\dfrac{\frac{28}{15}}{\frac{21}{5}}$ 23. $12 \div \dfrac{2}{3}$ 24. $7 \div \dfrac{1}{4}$

25. $\dfrac{\frac{15}{2}}{\frac{2}{3}}$ 26. $\dfrac{\frac{9}{3}}{\frac{3}{4}}$ 27. $\dfrac{\frac{4}{7}}{\frac{7}{8}}$ 28. $\dfrac{\frac{7}{10}}{3}$

0.13 Multiplying and Dividing Mixed Numbers

In Section 0.8 we worked with mixed numbers—a whole number and a fraction written together. Many of the fraction problems you encounter in everyday life involve mixed numbers.

OBJECTIVES

1. Estimate the answer and multiply mixed numbers.
2. Estimate the answer and divide mixed numbers.

OBJECTIVE 1 Estimate the answer and multiply mixed numbers. When multiplying mixed numbers, it is a good idea to estimate the answer first. Then multiply the mixed numbers by using the following steps.

Multiplying Mixed Numbers

Step 1 *Change* each mixed number to an improper fraction.

Step 2 *Multiply* as fractions.

Step 3 *Simplify* the answer, which means to write it in *lowest terms*, and change it to a mixed number or whole number where possible.

To estimate the answer, round each mixed number to the nearest whole number. If the numerator is *half* of the denominator or *more,* round up the whole number part. If the numerator is *less* than half the denominator, leave the whole number as it is.

$1\frac{5}{8}$ ← 5 is more than 4.
$\phantom{1\frac{5}{8}}$ ← Half of 8 is 4. $\quad\Big\}\quad 1\frac{5}{8}$ rounds up to 2

$3\frac{2}{5}$ ← 2 is less than $2\frac{1}{2}$.
$\phantom{3\frac{2}{5}}$ ← Half of 5 is $2\frac{1}{2}$. $\quad\Big\}\quad 3\frac{2}{5}$ rounds to 3

Work Problem 1 at the Side.

1. Round each mixed number to the nearest whole number.

(a) $3\frac{2}{3}$

(b) $5\frac{2}{5}$

(c) $2\frac{3}{4}$

(d) $4\frac{7}{12}$

(e) $6\frac{1}{2}$

(f) $1\frac{4}{9}$

■ **EXAMPLE 1 Multiplying Mixed Numbers**

First estimate the answer. Then multiply to get an exact answer. Simplify your answers.

(a) $2\frac{1}{2} \cdot 3\frac{1}{5}$

Estimate the answer by rounding the mixed numbers.

$2\frac{1}{2}$ rounds to 3 and $3\frac{1}{5}$ rounds to 3

$3 \cdot 3 = 9$ Estimated answer

To find the exact answer, change each mixed number to an improper fraction.

Step 1 $2\frac{1}{2} = \frac{5}{2}$ and $3\frac{1}{5} = \frac{16}{5}$

ANSWERS

1. (a) 4   (b) 5   (c) 3   (d) 5
    (e) 7   (f) 1

0-68 CHAPTER 0 Topics in Arithmetic

2. First estimate the answer. Then multiply to find the exact answer. Write answers in lowest terms. Simplify your answers.

(a) $3\frac{1}{2} \cdot 6\frac{1}{3}$

   $\underline{\ 4\ } \cdot \underline{\ 6\ }$

   $= \underline{\ 24\ }$ estimate

(b) $4\frac{2}{3} \cdot 2\frac{3}{4}$

   $\underline{\ 5\ } \cdot \underline{\ 3\ }$

   $= \underline{\ 15\ }$ estimate

(c) $3\frac{3}{5} \cdot 4\frac{4}{9}$

   $\underline{\ 4\ } \cdot \underline{\ 4\ }$

   $= \underline{\ 16\ }$ estimate

(d) $5\frac{1}{4} \cdot 3\frac{2}{5}$

   $\underline{\ 5\ } \cdot \underline{\ 3\ }$

   $= \underline{\ 15\ }$ estimate

ANSWERS

2. (a) Estimate: $4 \cdot 6 = 24$; Exact: $22\frac{1}{6}$

   (b) Estimate: $5 \cdot 3 = 15$; Exact: $12\frac{5}{6}$

   (c) Estimate: $4 \cdot 4 = 16$; Exact: 16

   (d) Estimate: $5 \cdot 3 = 15$; Exact: $17\frac{17}{20}$

Next, multiply.

$$2\frac{1}{2} \cdot 3\frac{1}{5} = \frac{5}{2} \cdot \frac{16}{5} = \frac{\overset{1}{\cancel{5}}}{\underset{1}{\cancel{2}}} \cdot \frac{\overset{8}{\cancel{16}}}{\underset{1}{\cancel{5}}} = \frac{1 \cdot 8}{1 \cdot 1} = \frac{8}{1} = 8$$

The estimated answer is 9 and the exact answer is 8. The exact answer is reasonable.

(b) $3\frac{5}{8} \cdot 4\frac{4}{5}$

$3\frac{5}{8}$ rounds to 4   and $4\frac{4}{5}$ rounds to 5

$4 \cdot 5 = 20$ Estimated answer

Now find the exact answer.

$$3\frac{5}{8} \cdot 4\frac{4}{5} = \frac{29}{8} \cdot \frac{24}{5} = \frac{29}{\underset{1}{\cancel{8}}} \cdot \frac{\overset{3}{\cancel{24}}}{5} = \frac{29 \cdot 3}{1 \cdot 5} = \frac{87}{5}$$

As a mixed number,

$$\frac{87}{5} = 17\frac{2}{5}.$$ Simplified answer

The estimate was 20, so the exact answer is reasonable.

(c) $1\frac{3}{5} \cdot 3\frac{1}{3}$

$1\frac{3}{5}$ rounds to 2   and $3\frac{1}{3}$ rounds to 3

$2 \cdot 3 = 6$ Estimated answer

The exact answer is

$$1\frac{3}{5} \cdot 3\frac{1}{3} = \frac{8}{5} \cdot \frac{\overset{2}{\cancel{10}}}{3} = \frac{8 \cdot 2}{1 \cdot 3} = \frac{16}{3} = 5\frac{1}{3}.$$

The estimate was 6, so the exact answer is reasonable.

Work Problem 2 at the Side.

OBJECTIVE 2 Estimate the answer and divide mixed numbers. Just as you did when multiplying mixed numbers, it is also a good idea to estimate the answer when dividing mixed numbers. To divide mixed numbers, use the following steps.

Dividing Mixed Numbers

Step 1 *Change* each mixed number to an improper fraction.

Step 2 Use the *reciprocal* of the second fraction (divisor).

Step 3 *Multiply.*

Step 4 Simplify the answer, which means to write it in *lowest terms*, and change it to a mixed number or whole number where possible.

NOTE Recall that the reciprocal of a fraction is found by interchanging the numerator and the denominator.

EXAMPLE 2 Dividing Mixed Numbers

First estimate the answer. Then divide to find the exact answer. Simplify your answers.

(a) $2\frac{2}{5} \div 1\frac{1}{2}$

First estimate the answer by rounding each mixed number to the nearest whole number.

$$2\frac{2}{5} \div 1\frac{1}{2}$$

Rounded

$$2 \div 2 = 1 \quad \text{Estimated answer}$$

To find the exact answer, first change each mixed number to an improper fraction.

$$2\frac{2}{5} \div 1\frac{1}{2} = \frac{12}{5} \div \frac{3}{2} \quad \text{Step 1}$$

Next, use the reciprocal of the second fraction and multiply.

$$\frac{12}{5} \div \frac{3}{2} = \frac{\overset{4}{\cancel{12}}}{5} \cdot \frac{2}{\cancel{3}} = \frac{4 \cdot 2}{5 \cdot 1} = \frac{8}{5} = 1\frac{3}{5} \quad \text{Simplified answer}$$

Reciprocals

Steps 2, 3, 4

The estimate was 1, so the exact answer is reasonable.

3. First estimate the answer. Then divide to find the exact answer. Simplify all answers.

(a) $3\dfrac{1}{8} \div 6\dfrac{1}{4}$

$\underline{} \div \underline{}$

$= \underline{}$ estimate

(b) $5\dfrac{1}{3} \div 1\dfrac{1}{4}$

$\underline{} \div \underline{}$

$= \underline{}$ estimate

(c) $8 \div 5\dfrac{1}{3}$

$\underline{} \div \underline{}$

$= \underline{}$ estimate

(d) $13\dfrac{1}{2} \div 18$

$\underline{} \div \underline{}$

$= \underline{}$ estimate

(b) $8 \div 3\dfrac{3}{5}$

$8 \div 3\dfrac{3}{5}$

↓ Rounded ↓

$8 \div 4 = 2$ Estimate

Now find the exact answer.

$$8 \div 3\dfrac{3}{5} = \dfrac{8}{1} \div \dfrac{18}{5} = \dfrac{\cancel{8}^{4}}{1} \cdot \dfrac{5}{\cancel{18}_{9}} = \dfrac{20}{9} = 2\dfrac{2}{9}$$

Write 8 as $\dfrac{8}{1}$.

The estimate was 2, so the exact answer is reasonable.

(c) $4\dfrac{3}{8} \div 5$

$4\dfrac{3}{8} \div 5$

↓ Rounded ↓ Reciprocals

$4 \div 5 = \dfrac{4}{1} \div \dfrac{5}{1} = \dfrac{4}{1} \cdot \dfrac{1}{5} = \dfrac{4}{5}$ Estimate

The exact answer is

$$4\dfrac{3}{8} \div 5 = \dfrac{35}{8} \div \dfrac{5}{1} = \dfrac{\cancel{35}^{7}}{8} \cdot \dfrac{1}{\cancel{5}_{1}} = \dfrac{7}{8}.$$

Write 5 as $\dfrac{5}{1}$.

The estimate was $\dfrac{4}{5}$, so the exact answer is reasonable.

Work Problem 3 at the Side.

ANSWERS

3. (a) *Estimate:* $3 \div 6 = \dfrac{1}{2}$; *Exact:* $\dfrac{1}{2}$

(b) *Estimate:* $5 \div 1 = 5$; *Exact:* $4\dfrac{4}{15}$

(c) *Estimate:* $8 \div 5 = 1\dfrac{3}{5}$; *Exact:* $1\dfrac{1}{2}$

(d) *Estimate:* $14 \div 18 = \dfrac{7}{9}$; *Exact:* $\dfrac{3}{4}$

SECTION 0.13 Multiplying and Dividing Mixed Numbers 0-71

0.13 EXERCISES

For Extra Help

 Student's Solutions Manual

 MyMathLab

 InterAct Math Tutorial Software

 AW Math Tutor Center

 MathXL

First estimate the answer. Then multiply to find the exact answer. Simplify all answers. See Example 1.

1. Exact:

   $3\dfrac{1}{4} \cdot 2\dfrac{1}{2}$

   Estimate:

   ____ • ____ = ____

2. Exact:

   $3\dfrac{1}{2} \cdot 1\dfrac{1}{4}$

   Estimate:

   ____ • ____ = ____

3. Exact:

   $1\dfrac{2}{3} \cdot 2\dfrac{7}{10}$

   Estimate:

   ____ • ____ = ____

4. Exact:

   $4\dfrac{1}{2} \cdot 2\dfrac{1}{4}$

   Estimate:

   ____ • ____ = ____

5. Exact:

   $3\dfrac{1}{9} \cdot 1\dfrac{2}{7}$

   Estimate:

   ____ • ____ = ____

6. Exact:

   $6\dfrac{1}{4} \cdot 3\dfrac{1}{5}$

   Estimate:

   ____ • ____ = ____

7. Exact:

   $8 \cdot 6\dfrac{1}{4}$

   Estimate:

   ____ • ____ = ____

8. Exact:

   $6 \cdot 2\dfrac{1}{3}$

   Estimate:

   ____ • ____ = ____

9. Exact:

   $4\dfrac{1}{2} \cdot 2\dfrac{1}{5} \cdot 5$

   Estimate:

   ____ • ____ • ____ = ____

10. Exact:

    $5\dfrac{1}{2} \cdot 1\dfrac{1}{3} \cdot 2\dfrac{1}{4}$

    Estimate:

    ____ • ____ • ____ = ____

11. Exact:

    $3 \cdot 1\dfrac{1}{2} \cdot 2\dfrac{2}{3}$

    Estimate:

    ____ • ____ • ____ = ____

12. Exact:

    $\dfrac{2}{3} \cdot 3\dfrac{2}{3} \cdot \dfrac{6}{11}$

    Estimate:

    ____ • ____ • ____ = ____

0-72　CHAPTER 0　Topics in Arithmetic

First estimate the answer. Then divide to find the exact answer. Simplify all answers. See Example 2.

13. Exact:

$2\frac{1}{2} \div 7\frac{1}{2}$

Estimate:

$\underline{3} \div \underline{8} = \underline{3}$

14. Exact:

$1\frac{1}{8} \div 2\frac{1}{4}$

Estimate:

$\underline{\qquad} \div \underline{\qquad} = \underline{\qquad}$

15. Exact:

$2\frac{1}{2} \div 3$

Estimate:

$\underline{\qquad} \div \underline{\qquad} = \underline{\qquad}$

16. Exact:

$2\frac{3}{4} \div 2$

Estimate:

$\underline{\qquad} \div \underline{\qquad} = \underline{\qquad}$

17. Exact:

$9 \div 2\frac{1}{2}$

Estimate:

$\underline{\qquad} \div \underline{\qquad} = \underline{\qquad}$

18. Exact:

$5 \div 1\frac{7}{8}$

Estimate:

$\underline{\qquad} \div \underline{\qquad} = \underline{\qquad}$

19. Exact:

$\frac{3}{4} \div 1\frac{3}{4}$

Estimate:

$\underline{\qquad} \div \underline{\qquad} = \underline{\qquad}$

20. Exact:

$\frac{3}{4} \div 2\frac{1}{2}$

Estimate:

$\underline{\qquad} \div \underline{\qquad} = \underline{\qquad}$

21. Exact:

$1\frac{7}{8} \div 6\frac{1}{4}$

Estimate:

$\underline{\qquad} \div \underline{\qquad} = \underline{\qquad}$

22. Exact:

$8\frac{2}{5} \div 3\frac{1}{2}$

Estimate:

$\underline{\qquad} \div \underline{\qquad} = \underline{\qquad}$

23. Exact:

$5\frac{2}{3} \div 6$

Estimate:

$\underline{\qquad} \div \underline{\qquad} = \underline{\qquad}$

24. Exact:

$5\frac{3}{4} \div 2$

Estimate:

$\underline{\qquad} \div \underline{\qquad} = \underline{\qquad}$

0.7–0.13 TEST

SECTIONS 0.7–0.13 Test 0-73

Write a fraction to represent each shaded portion.

1. 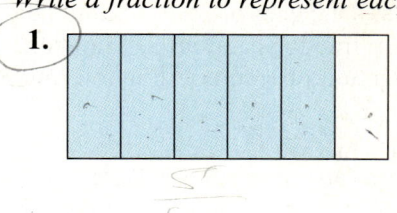   2.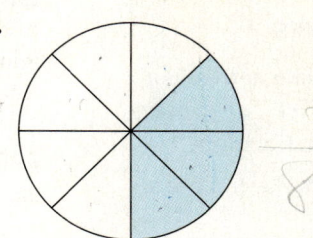

3. Identify all the proper fractions in this list: $\frac{2}{3}, \frac{4}{4}, \frac{6}{7}, \frac{5}{2}, \frac{1}{4}, \frac{5}{8}, \frac{30}{18}$.

4. Write $6\frac{2}{3}$ as an improper fraction.   5. Write $\frac{118}{5}$ as a mixed number.

6. Find all factors of 18.

Find the prime factorization of each number. Write the answers using exponents.

7. 63   8. 96   9. 500

Write each fraction in lowest terms.

10. $\frac{27}{36}$ 11. $\frac{45}{54}$

12. The method of prime factors is used to write a fraction in lowest terms. Briefly explain how this is done. Use the fraction $\frac{56}{84}$ to show how this works.

13. Explain how to multiply fractions. What additional step must be taken when dividing fractions?

Multiply or divide. Write answers in lowest terms, and as mixed numbers or whole numbers where possible.

14. $\frac{2}{5} \cdot \frac{3}{8}$ 15. $36 \cdot \frac{4}{9}$ 16. $\frac{3}{4} \div \frac{5}{6}$ 17. $\frac{\frac{7}{4}}{\frac{4}{9}}$ 18. $\frac{\frac{2}{3}}{6}$

First estimate the answer. Then either multiply or divide to find the exact answer. Simplify all answers.

19. $4\frac{1}{8} \cdot 3\frac{1}{2}$ 20. $1\frac{5}{6} \cdot 4\frac{1}{3}$ 21. $9\frac{3}{5} \div 2\frac{1}{4}$ 22. $\frac{8\frac{1}{2}}{1\frac{2}{3}}$

0.14 Adding and Subtracting Like Fractions

OBJECTIVES

1. Define like and unlike fractions.
2. Add like fractions.
3. Subtract like fractions.

Earlier we looked at the basics of fractions and then practiced with multiplication and division of common fractions and mixed numbers. In this chapter we will work with addition and subtraction of common fractions and mixed numbers.

OBJECTIVE 1 **Define like and unlike fractions.** Fractions with the same denominators are **like fractions**. Fractions with different denominators are **unlike fractions**.

1. Next to each pair of fractions write *like* or *unlike*.

   (a) $\frac{2}{3}$ $\frac{1}{3}$ _____

   (b) $\frac{3}{4}$ $\frac{3}{8}$ _____

   (c) $\frac{6}{15}$ $\frac{11}{15}$ _____

   (d) $\frac{5}{12}$ $\frac{5}{7}$ _____

EXAMPLE 1 Identifying Like and Unlike Fractions

(a) $\frac{3}{4}, \frac{1}{4}, \frac{5}{4}, \frac{6}{4}$, and $\frac{4}{4}$ are **like** fractions.
— All denominators are the same.

(b) $\frac{7}{12}$ and $\frac{12}{7}$ are **unlike** fractions.
— All denominators are different.

Work Problem 1 at the Side.

NOTE Like fractions have the *same* denominator.

OBJECTIVE 2 **Add like fractions.** The following figures show you how to add the fractions $\frac{2}{7}$ and $\frac{4}{7}$.

$$\frac{2}{7} + \frac{4}{7} = \frac{6}{7}$$

As the figures show,

$$\frac{2}{7} + \frac{4}{7} = \frac{6}{7}.$$

Add like fractions as follows.

ANSWERS
1. (a) like (b) unlike (c) like
   (d) unlike

SECTION 0.14 Adding and Subtracting Like Fractions   **0-75**

2. Add and write the answers in lowest terms.

(a) $\dfrac{1}{5} + \dfrac{3}{5}$

Adding Like Fractions

Step 1 Add the numerators to find the numerator of the sum.

Step 2 Write the denominator of the like fractions as the denominator of the sum.

Step 3 Write the answer in lowest terms.

(b) $\dfrac{1}{9} + \dfrac{5}{9}$

■ **EXAMPLE 2** Adding Like Fractions

Add and write the answer in lowest terms.

(a) $\dfrac{1}{5} + \dfrac{2}{5}$

$$\dfrac{1}{5} + \dfrac{2}{5} = \dfrac{\overbrace{1+2}^{\text{Add numerators.}}}{5} = \dfrac{3}{5} \leftarrow \text{Same denominator}$$

(b) $\dfrac{1}{12} + \dfrac{7}{12} + \dfrac{1}{12}$

(c) $\dfrac{5}{8} + \dfrac{1}{8}$

Step 1 $\dfrac{\overbrace{1+7+1}^{\text{Add numerators.}}}{12}$

Step 2 $= \dfrac{9}{12}$ ← Sum
← Same denominator

Step 3 $= \dfrac{9 \div 3}{12 \div 3} = \dfrac{3}{4}$ In lowest terms

(d) $\dfrac{3}{10} + \dfrac{1}{10} + \dfrac{4}{10}$

CAUTION Fractions may be added *only* if they have like denominators.

Work Problem 2 at the Side.

OBJECTIVE 3 Subtract like fractions. The figures show $\dfrac{7}{8}$ broken into $\dfrac{4}{8}$ and $\dfrac{3}{8}$.

Subtracting $\dfrac{3}{8}$ from $\dfrac{7}{8}$ gives the answer $\dfrac{4}{8}$, or

$$\dfrac{7}{8} - \dfrac{3}{8} = \dfrac{4}{8}.$$

ANSWERS

2. (a) $\dfrac{4}{5}$ (b) $\dfrac{2}{3}$ (c) $\dfrac{3}{4}$ (d) $\dfrac{4}{5}$

0-76　CHAPTER 0　Topics in Arithmetic

3. Subtract and simplify.

(a) $\dfrac{7}{9} - \dfrac{5}{9}$

(b) $\dfrac{13}{16} - \dfrac{5}{16}$

(c) $\dfrac{15}{3} - \dfrac{5}{3}$

(d) $\dfrac{25}{32} - \dfrac{6}{32}$

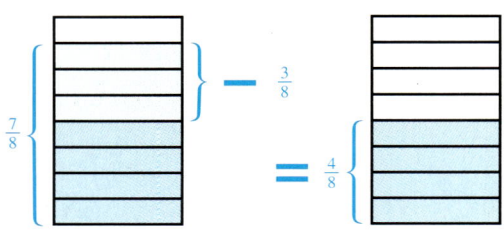

Write $\dfrac{4}{8}$ in lowest terms.

$$\dfrac{7}{8} - \dfrac{3}{8} = \dfrac{4 \div 4}{8 \div 4} = \dfrac{1}{2}$$

The steps for subtracting like fractions are very similar to those for adding like fractions.

Subtracting Like Fractions

Step 1　Subtract the numerators to find the numerator of the difference.

Step 2　Write the denominator of the like fractions as the denominator of the difference.

Step 3　Write the answer in lowest terms.

EXAMPLE 3　Subtracting Like Fractions

Subtract and simplify the answer.

(a) $\dfrac{15}{16} - \dfrac{3}{16}$　　Subtract numerators.

Step 1　$\dfrac{15}{16} - \dfrac{3}{16} = \dfrac{15 - 3}{16}$

Step 2　$= \dfrac{12}{16}$ ← Difference
← Same denominator

Step 3　$= \dfrac{12 \div 4}{16 \div 4} = \dfrac{3}{4}$　In lowest terms

(b) $\dfrac{13}{4} - \dfrac{6}{4}$　　Subtract numerators.

$\dfrac{13}{4} - \dfrac{6}{4} = \dfrac{13 - 6}{4}$ ← Same denominator

$= \dfrac{7}{4}$

CAUTION　Fractions may be subtracted *only* if they have like denominators.

ANSWERS

3. (a) $\dfrac{2}{9}$　(b) $\dfrac{1}{2}$　(c) $3\dfrac{1}{3}$　(d) $\dfrac{19}{32}$

Work Problem 3 at the Side.

0.14 EXERCISES

SECTION 0.14 Adding and Subtracting Like Fractions 0-77

For Extra Help

 Student's Solutions Manual

 MyMathLab

 InterAct Math Tutorial Software

 AW Math Tutor Center

MathXL

Add and simplify the answer. See Example 2.

1. $\dfrac{3}{5} + \dfrac{1}{5}$ 2. $\dfrac{5}{8} + \dfrac{2}{8}$ 3. $\dfrac{2}{6} + \dfrac{3}{6}$

4. $\dfrac{9}{11} + \dfrac{1}{11}$ 5. $\dfrac{1}{6} + \dfrac{1}{6}$ 6. $\dfrac{1}{12} + \dfrac{1}{12}$

7. $\dfrac{9}{10} + \dfrac{3}{10}$ 8. $\dfrac{13}{12} + \dfrac{5}{12}$ 9. $\dfrac{2}{9} + \dfrac{1}{9}$

10. $\dfrac{3}{14} + \dfrac{5}{14}$ 11. $\dfrac{6}{20} + \dfrac{4}{20} + \dfrac{3}{20}$ 12. $\dfrac{1}{7} + \dfrac{2}{7} + \dfrac{3}{7}$

13. $\dfrac{4}{15} + \dfrac{2}{15} + \dfrac{5}{15}$ 14. $\dfrac{5}{11} + \dfrac{1}{11} + \dfrac{4}{11}$ 15. $\dfrac{3}{8} + \dfrac{7}{8} + \dfrac{2}{8}$

16. $\dfrac{4}{9} + \dfrac{1}{9} + \dfrac{7}{9}$ 17. $\dfrac{2}{54} + \dfrac{8}{54} + \dfrac{12}{54}$ 18. $\dfrac{7}{64} + \dfrac{15}{64} + \dfrac{20}{64}$

Subtract and simplify the answer. See Example 3.

19. $\dfrac{7}{8} - \dfrac{4}{8}$ 20. $\dfrac{2}{3} - \dfrac{1}{3}$ 21. $\dfrac{10}{11} - \dfrac{4}{11}$ 22. $\dfrac{4}{5} - \dfrac{3}{5}$ 23. $\dfrac{9}{10} - \dfrac{3}{10}$

24. $\dfrac{7}{8} - \dfrac{3}{8}$ 25. $\dfrac{31}{21} - \dfrac{7}{21}$ 26. $\dfrac{43}{24} - \dfrac{13}{24}$ 27. $\dfrac{27}{40} - \dfrac{19}{40}$ 28. $\dfrac{38}{55} - \dfrac{16}{55}$

29. $\dfrac{47}{36} - \dfrac{5}{36}$ 30. $\dfrac{76}{45} - \dfrac{21}{45}$ 31. $\dfrac{73}{60} - \dfrac{7}{60}$ 32. $\dfrac{181}{100} - \dfrac{31}{100}$

0.15 Least Common Multiples

OBJECTIVES

1. Find the least common multiple.
2. Find the least common multiple by using multiples of the largest number.
3. Find the least common multiple by using prime factorization.
4. Find the least common multiple by using an alternative method.
5. Write a fraction with an indicated denominator.

1. **(a)** List the multiples of 5.

   5, ___, ___, ___,

   ___, ___, ___,

   ___, . . .

   (b) List the multiples of 8.

   8, ___, ___, ___,

   ___, ___, ___, . . .

   (c) Find the least common multiple of 5 and 8.

ANSWERS

1. (a) 10, 15, 20, 25, 30, 35, 40, . . .
   (b) 16, 24, 32, 40, 48, 56, . . .
   (c) 40

Only *like* fractions can be added or subtracted. Because of this, we must rewrite *unlike* fractions as *like* fractions before we can add or subtract.

OBJECTIVE 1 Find the least common multiple. We can rewrite unlike fractions as like fractions by finding the *least common multiple* of the denominators.

> **Least Common Multiple**
>
> The **least common multiple (LCM)** of two whole numbers is the smallest whole number divisible by both those numbers.

EXAMPLE 1 Finding the Least Common Multiple

Find the least common multiple of 6 and 9.
   This list shows the multiples of 6.

$$\underbrace{6 \cdot 1}_{6,} \quad \underbrace{6 \cdot 2}_{12,} \quad \underbrace{6 \cdot 3}_{18,} \quad \underbrace{6 \cdot 4}_{24,} \quad \underbrace{6 \cdot 5}_{30,} \quad \underbrace{6 \cdot 6}_{36,} \quad \underbrace{6 \cdot 7}_{42,} \quad \underbrace{6 \cdot 8}_{48,} \ldots$$

(The three dots at the end of the list show that the list continues in the same pattern without stopping.) The next list shows multiples of 9.

$$\underbrace{9 \cdot 1}_{9,} \quad \underbrace{9 \cdot 2}_{18,} \quad \underbrace{9 \cdot 3}_{27,} \quad \underbrace{9 \cdot 4}_{36,} \quad \underbrace{9 \cdot 5}_{45,} \quad \underbrace{9 \cdot 6}_{54,} \quad \underbrace{9 \cdot 7}_{63,} \quad \underbrace{9 \cdot 8}_{72,} \ldots$$

The smallest number found in *both* lists is 18, so 18 is the **least common multiple** of 6 and 9; the number 18 is the smallest whole number divisible by both 6 and 9.

Multiples of 6:   6, 12, **18**, 24, 30, 36, 42, 48, . . .
Multiples of 9:   9, **18**, 27, 36, 45, 54, 63, 72, . . .

18 is the smallest number found in both lists. **18** is the least common multiple of 6 and 9.

Work Problem 1 at the Side.

OBJECTIVE 2 Find the least common multiple by using multiples of the largest number.
There are several ways to find the least common multiple of a number. If the numbers are small, the least common multiple can often be found by inspection. Can you think of a number that can be divided evenly by both 3 and 4? What about 6 or 8, or perhaps 10 or 12? The number 12 will work; it is the least common multiple of the numbers 3 and 4. A method that works well to find the least common multiple is to write multiples of the larger number.
   In this case, write the multiples of 4:

$$4, 8, 12, 16, 20, \ldots$$

SECTION 0.15 Least Common Multiples **0-79**

2. Use multiples of the larger number to find the least common multiple in each set of numbers.

   (a) 2 and 3

   (b) 4 and 5

   (c) 6 and 8

   (d) 5 and 9

Now, check each multiple of 4 to see if it is divisible by 3.

4 is *not* divisible by 3.
8 is *not* divisible by 3.
12 *is* divisible by 3.

The first multiple of 4 that is divisible by 3 is 12, so 12 is the least common multiple of 3 and 4.

EXAMPLE 2 Finding the Least Common Multiple

Use multiples of the larger number to find the least common multiple of 6 and 9. We start by writing the first few multiples of 9.

Multiples of 9
9, 18, 27, 36, 45, 54, . . .

Now, we check each multiple of 9 to see if it is divisible by 6. The first multiple of 9 that is divisible by 6 is 18.

9, **18**, 27, 36, 45, 54, . . .

First multiple divisible by 6 (18 ÷ 6 = 3)

The least common multiple of the numbers 6 and 9 is 18.

Work Problem 2 at the Side.

3. Use prime factorization to find the LCM for each pair of numbers.

   (a) 15 and 18

   (b) 12 and 20

OBJECTIVE 3 Find the least common multiple by using prime factorization. Example 2 shows how to find the least common multiple of two numbers by making a list of the multiples of the *larger* number. Although this method works for smaller numbers, it is usually easier to find the least common multiple for larger numbers by using *prime factorization*, as shown in the next example.

EXAMPLE 3 Applying Prime Factorization Knowledge

Use prime factorization to find the least common multiple of 9 and 12.
We start by finding the prime factorization of each number.

Factors of 9

$9 = 3 \cdot 3$
$12 = 2 \cdot 2 \cdot 3$

$LCM = 3 \cdot 3 \cdot 2 \cdot 2 = 36$

Factors of 12

Check to see that 36 is divisible by 9 (yes) and by 12 (yes). The smallest whole number divisible by both 9 and 12 is 36.

CAUTION Notice that we did *not* have to repeat the factors that 9 and 12 have in common. In this case, the **3** in $2 \cdot 2 \cdot 3 = 12$ was *not* used because 3 is already included in $3 \cdot 3 = 9$.

ANSWERS
2. (a) 6 (b) 20 (c) 24 (d) 45
3. (a) $15 = 3 \cdot 5$
      $18 = 2 \cdot 3 \cdot 3$

      15

      $LCM = 2 \cdot 3 \cdot 3 \cdot 5 = 90$

      18

   (b) $12 = 2 \cdot 2 \cdot 3$
       $20 = 2 \cdot 2 \cdot 5$

       12

       $LCM = 2 \cdot 2 \cdot 3 \cdot 5 = 60$

       20

Work Problem 3 at the Side.

0-80　CHAPTER 0　Topics in Arithmetic

4. Find the least common mul-tiple of the denominators in each set of fractions.

   (a) $\dfrac{3}{8}$ and $\dfrac{6}{5}$

   (b) $\dfrac{4}{9}, \dfrac{5}{18},$ and $\dfrac{7}{24}$

5. Find the least common multiple for each set of numbers.

   (a) 12, 15

   (b) 8, 9, 12

   (c) 18, 20, 30

   (d) 15, 20, 30, 40

ANSWERS

4. (a) $8 = 2 \cdot 2 \cdot 2$
   $5 = 1 \cdot 5$
   LCM $= 2 \cdot 2 \cdot 2 \cdot 5 = 40$

   (b) $9 = 3 \cdot 3$
   $18 = 2 \cdot 3 \cdot 3$
   $24 = 2 \cdot 2 \cdot 2 \cdot 3$
   LCM $= 2 \cdot 2 \cdot 2 \cdot 3 \cdot 3 = 72$

5. (a) 60 (b) 72 (c) 180 (d) 120

EXAMPLE 4 Using Prime Factorization

Find the least common multiple of 12, 18, and 20.

Find the prime factorization of each number. Then use the prime factors to build the LCM.

$12 = 2 \cdot 2 \cdot 3$
$18 = 2 \cdot 3 \cdot 3$
$20 = 2 \cdot 2 \cdot 5$

LCM $= 2 \cdot 2 \cdot 3 \cdot 3 \cdot 5 = 180$

Check to see that 180 is divisible by 12 (yes), and by 18 (yes) and by 20 (yes). The smallest whole number divisible by 12, 18, and 20 is 180.

Note: We did *not* repeat the factors that 12, 18, and 20 have in common.

Work Problem 4 at the Side.

EXAMPLE 5 Finding the Least Common Multiple

Find the least common multiple for each set of numbers.

(a) 5, 6, 35

Find the prime factorization for each number.

$5 = 5$
$6 = 2 \cdot 3$
$35 = 5 \cdot 7$

LCM $= 2 \cdot 3 \cdot 5 \cdot 7 = 210$

The least common multiple of 5, 6, and 35 is 210.

(b) 10, 20, 24

Find the prime factorization for each number.

$10 = 2 \cdot 5$
$20 = 2 \cdot 2 \cdot 5$
$24 = 2 \cdot 2 \cdot 2 \cdot 3$

LCM $= 2 \cdot 2 \cdot 2 \cdot 3 \cdot 5 = 120$

The least common multiple of 10, 20, and 24 is 120.

Work Problem 5 at the Side.

SECTION 0.15 Least Common Multiples **0-81**

OBJECTIVE 4 **Find the least common multiple by using an alternative method.** Some people like the following *alternative method* for finding the least common multiple for larger numbers. Try both methods, and *use the one you prefer*. As a review, a list of the first few prime numbers follows.

$$2, 3, 5, 7, 11, 13, 17$$

EXAMPLE 6 Alternative Method for Finding the Least Common Multiple

Find the least common multiple of each set of numbers.

(a) 14 and 21

Start by trying to divide 14 and 21 by the preceding list of prime numbers, 2, 3, 5, 7, 11, 13, and 17.

Use the following shortcut.

Divide by 2, the first prime

$$\begin{array}{r|rr} 2 & 14 & \cancel{21} \\ \hline & 7 & 21 \end{array}$$

Because 21 cannot be divided evenly by 2, cross 21 out and bring it down. Divide by 3, the second prime.

$$\begin{array}{r|rr} 2 & 14 & \cancel{21} \\ \hline 3 & \cancel{7} & 21 \\ \hline & 7 & 7 \end{array}$$

Since 7 cannot be divided evenly by the third prime, 5, skip 5 and divide by the next prime, 7.

Divide by 7, the fourth prime.

$$\begin{array}{r|rr} 2 & 14 & \cancel{21} \\ \hline 3 & \cancel{7} & 21 \\ \hline 7 & 7 & 7 \\ \hline & 1 & 1 \end{array}$$ All quotients are 1.

When all quotients are 1, multiply the prime numbers on the left side.

$$\text{least common multiple} = 2 \cdot 3 \cdot 7 = 42$$

The least common multiple of 14 and 21 is 42.

(b) 6, 15, 18

Divide by 2.

$$\begin{array}{r|rrr} 2 & 6 & \cancel{15} & 18 \\ \hline & 3 & 15 & 9 \end{array}$$ Cross out 15 and bring it down.

Divide by 3.

$$\begin{array}{r|rrr} 2 & 6 & \cancel{15} & 18 \\ \hline 3 & 3 & 15 & 9 \\ \hline & 1 & 5 & 3 \end{array}$$

0-82 CHAPTER 0 Topics in Arithmetic

6. In the following problems, the divisions have already been worked out. Multiply the prime numbers on the left to find the least common multiple.

(a) 2 | 6   15
    3 | 3   15
    5 | 1   5
        1   1

(b) 2 | 20   36
    2 | 10   18
    3 | 5   9
    3 | 5   3
    5 | 5   1
         1   1

7. Find the least common multiple of each set of numbers.

(a) 4, 8, 12

(b) 25 and 30

(c) 9, 24

(d) 8, 21, 24

ANSWERS
6. (a) 30   (b) 180
7. (a) 24   (b) 150   (c) 72
    (d) 168

Divide by 3 again, since the remaining 3 can be divided.

2 | 6   15   18
3 | 3   15   9
3 | 1   5   3
    1   5   1

Finally, divide by 5.

2 | 6   15   18
3 | 3   15   9
3 | 1   5   3
5 | 1   5   1
    1   1   1   All quotients are 1.

Multiply the prime numbers on the left side.

$2 \cdot 3 \cdot 3 \cdot 5 = 90$ Least common multiple

■ **Work Problem 6 at the Side.**

OBJECTIVE 5 Write a fraction with an indicated denominator. Before we can add or subtract unlike fractions, we must find the least common multiple, which is then used as the denominator of the fractions.

■ **EXAMPLE 7** Writing a Fraction with an Indicated Denominator

Write the fraction $\frac{2}{3}$ with a denominator of 15.
Find a numerator, so that

$$\frac{2}{3} = \frac{?}{15}.$$

To find the new numerator, first divide **15** by **3**.

$$\frac{2}{3} = \frac{?}{15} \qquad 15 \div 3 = 5$$

Multiply both numerator and denominator of the fraction $\frac{2}{3}$ by 5.

$$\frac{2}{3} = \frac{2 \cdot 5}{3 \cdot 5} = \frac{10}{15}$$

■ **Work Problem 7 at the Side.**

This process is just the opposite of writing a fraction in lowest terms. Check the answer by writing $\frac{10}{15}$ in lowest terms; you should get $\frac{2}{3}$.

■ **EXAMPLE 8** Writing Fractions with a New Denominator

Rewrite each fraction with the indicated denominator.

(a) $\frac{3}{8} = \frac{?}{48}$

SECTION 0.15 Least Common Multiples 0-83

8. Rewrite each fraction with the indicated denominator.

(a) $\dfrac{1}{3} = \dfrac{?}{12}$

(b) $\dfrac{3}{4} = \dfrac{?}{8}$

(c) $\dfrac{4}{5} = \dfrac{?}{25}$

(d) $\dfrac{6}{11} = \dfrac{?}{33}$

Divide 48 by 8, getting 6. Now multiply both numerator and denominator of $\tfrac{3}{8}$ by 6.

$$\dfrac{3}{8} = \dfrac{3 \cdot 6}{8 \cdot 6} = \dfrac{18}{48}$$ Multiply numerator and denominator by 6.

That is, $\tfrac{3}{8} = \tfrac{18}{48}$. As a check, write $\tfrac{18}{48}$ in lowest terms.

(b) $\dfrac{5}{6} = \dfrac{?}{42}$

Divide 42 by 6, getting 7. Next, multiply both numerator and denominator of $\tfrac{5}{6}$ by 7.

$$\dfrac{5}{6} = \dfrac{5 \cdot 7}{6 \cdot 7} = \dfrac{35}{42}$$ Multiply numerator and denominator by 7.

This shows that $\tfrac{5}{6} = \tfrac{35}{42}$. As a check, write $\tfrac{35}{42}$ in lowest terms.

> **NOTE** In Example 7, the fraction $\tfrac{2}{3}$ was multiplied by $\tfrac{5}{5}$. In Example 8, the fraction $\tfrac{3}{8}$ was multiplied by $\tfrac{6}{6}$ and the fraction $\tfrac{5}{6}$ was multiplied by $\tfrac{7}{7}$. The fractions $\tfrac{5}{5}, \tfrac{6}{6},$ and $\tfrac{7}{7}$ are all equal to 1.
>
> $$\dfrac{5}{5} = 1 \qquad \dfrac{6}{6} = 1 \qquad \dfrac{7}{7} = 1$$
>
> Recall that any number multiplied by 1 is the number itself.

Work Problem 8 at the Side.

ANSWERS

8. (a) $\dfrac{4}{12}$ (b) $\dfrac{6}{8}$ (c) $\dfrac{20}{25}$ (d) $\dfrac{18}{33}$

0-84　CHAPTER 0　Topics in Arithmetic

0.15 EXERCISES

For Extra Help

　Student's Solutions Manual

　MyMathLab

　InterAct Math Tutorial Software

　AW Math Tutor Center

MathXL

Use multiples of the larger number to find the least common multiple in each set of numbers. See Examples 1 and 2.

1. 2 and 4
2. 6 and 12
3. 3 and 5
4. 3 and 7
5. 4 and 9
6. 4 and 10
7. 2 and 7
8. 6 and 8
9. 6 and 10
10. 12 and 16
11. 20 and 50
12. 25 and 75

Find the least common multiple of each set of numbers. Use any method. See Examples 2–6.

13. 4, 10
14. 8, 10
15. 18, 24
16. 9, 15
17. 6, 9, 12
18. 20, 24, 30
19. 4, 6, 8, 10
20. 8, 9, 12, 18
21. 12, 15, 18, 20
22. 6, 9, 27, 36
23. 8, 12, 16, 36
24. 5, 6, 25, 30

Rewrite each fraction with a denominator of 24. See Examples 7 and 8.

25. $\dfrac{1}{3} = \dfrac{8}{24}$
26. $\dfrac{3}{8} = \dfrac{9}{24}$
27. $\dfrac{3}{4} = \dfrac{18}{24}$
28. $\dfrac{5}{12} = \dfrac{10}{24}$
29. $\dfrac{5}{6} = \dfrac{20}{24}$
30. $\dfrac{7}{8} = \dfrac{21}{24}$

Rewrite each fraction with the indicated denominators. See Example 8.

31. $\dfrac{1}{4} = \dfrac{2}{8}$
32. $\dfrac{2}{3} = \dfrac{}{12}$
33. $\dfrac{5}{8} = \dfrac{15}{24}$
34. $\dfrac{7}{10} = \dfrac{}{30}$
35. $\dfrac{7}{8} = \dfrac{}{32}$
36. $\dfrac{5}{12} = \dfrac{10}{48}$
37. $\dfrac{5}{6} = \dfrac{55}{66}$
38. $\dfrac{7}{8} = \dfrac{84}{96}$
39. $\dfrac{8}{5} = \dfrac{32}{20}$
40. $\dfrac{7}{8} = \dfrac{}{40}$
41. $\dfrac{9}{7} = \dfrac{}{56}$
42. $\dfrac{3}{2} = \dfrac{96}{64}$

0.16 Adding and Subtracting Unlike Fractions

OBJECTIVES

1. Add unlike fractions.
2. Add fractions vertically.
3. Subtract unlike fractions.

OBJECTIVE 1 Add unlike fractions. In this section, we add and subtract unlike fractions. To add unlike fractions, we must first change them to like fractions (fractions with the same denominator). For example, the diagrams show $\frac{3}{8}$ and $\frac{1}{4}$.

These fractions can be added by changing them to like fractions. Make like fractions by changing $\frac{1}{4}$ to the equivalent fraction $\frac{2}{8}$.

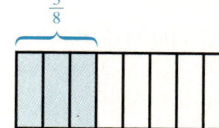
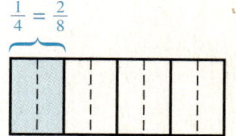

Next, add.

$$\frac{3}{8} + \frac{1}{4} = \frac{3}{8} + \frac{2}{8} = \frac{5}{8}$$

Becomes

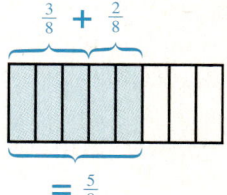

Use the following steps to add or subtract unlike fractions.

Adding or Subtracting Unlike Fractions

Step 1 Rewrite the *unlike fractions* as *like fractions* with the least common multiple as their new denominator. This new denominator is called the **least common denominator (LCD)**.

Step 2 Add or subtract as with like fractions.

Step 3 Simplify the answer by writing in lowest terms and as a whole or mixed number where possible.

1. Add.

    (a) $\dfrac{1}{2} + \dfrac{3}{8}$

    (b) $\dfrac{3}{4} + \dfrac{1}{8}$

    (c) $\dfrac{3}{5} + \dfrac{3}{10}$

    (d) $\dfrac{1}{12} + \dfrac{5}{6}$

ANSWERS

1. (a) $\dfrac{7}{8}$ (b) $\dfrac{7}{8}$ (c) $\dfrac{9}{10}$ (d) $\dfrac{11}{12}$

EXAMPLE 1 Adding Unlike Fractions

Add $\dfrac{2}{3}$ and $\dfrac{1}{9}$.

The least common multiple of 3 and 9 is 9, so first write the fractions as like fractions with a denominator of 9. This is the **least common denominator** of 3 and 9.

Step 1 $\qquad \dfrac{2}{3} = \dfrac{?}{9}$

Divide 9 by 3, getting 3. Next, multiply numerator and denominator by 3.

$$\dfrac{2}{3} = \dfrac{2 \cdot 3}{3 \cdot 3} = \dfrac{6}{9}$$

Now, add the like fractions $\dfrac{6}{9}$ and $\dfrac{1}{9}$.

Step 2 $\qquad \dfrac{2}{3} + \dfrac{1}{9} = \dfrac{6}{9} + \dfrac{1}{9} = \dfrac{6+1}{9} = \dfrac{7}{9}$

Step 3 Step 3 is not needed because $\dfrac{7}{9}$ is already in lowest terms.

Work Problem 1 at the Side.

EXAMPLE 2 Adding Fractions

Add the following fractions using the three steps. Simplify all answers.

(a) $\dfrac{1}{3} + \dfrac{1}{6}$

The least common multiple of 3 and 6 is 6. Rewrite both fractions as fractions with a least common denominator of 6.

Step 1 $\qquad \dfrac{1}{3} + \dfrac{1}{6} = \dfrac{2}{6} + \dfrac{1}{6}$ — Rewritten as like fractions

Step 2 $\qquad \dfrac{2}{6} + \dfrac{1}{6} = \dfrac{2+1}{6} = \dfrac{3}{6}$

Step 3 $\qquad \dfrac{3}{6} = \dfrac{1}{2}$ ← In lowest terms

(b) $\dfrac{6}{15} + \dfrac{3}{10}$

The least common multiple of 15 and 10 is 30, so rewrite both fractions with a least common denominator of 30.

Step 1 $\qquad \dfrac{6}{15} + \dfrac{3}{10} = \dfrac{12}{30} + \dfrac{9}{30}$ — Rewritten as like fractions

SECTION 0.16 Adding and Subtracting Unlike Fractions 0-87

2. Add. Simplify all answers.

(a) $\dfrac{3}{10} + \dfrac{1}{5}$

Step 2 $\dfrac{12}{30} + \dfrac{9}{30} = \dfrac{\overbrace{12+9}^{\text{Add numerators.}}}{30} = \dfrac{21}{30}$

Step 3 $\dfrac{21}{30} = \dfrac{7}{10}$ ← In lowest terms

Work Problem 2 at the Side.

(b) $\dfrac{5}{8} + \dfrac{1}{3}$

OBJECTIVE 2 Add fractions vertically. Fractions can also be added vertically (up and down).

EXAMPLE 3 Vertical Addition of Fractions

Add the following fractions vertically.

(c) $\dfrac{1}{10} + \dfrac{1}{3} + \dfrac{1}{6}$

(a)
$$\begin{aligned}
\dfrac{3}{8} &= \dfrac{3 \cdot 3}{8 \cdot 3} = \dfrac{9}{24}\\
+\dfrac{7}{12} &= \dfrac{7 \cdot 2}{12 \cdot 2} = \dfrac{14}{24}\\
&\qquad\qquad\qquad \dfrac{23}{24}
\end{aligned}$$

← Rewritten as like fractions

← Add the numerators.

3. Add the following fractions vertically.

(a) $\begin{aligned}&\dfrac{3}{4}\\&+\dfrac{3}{16}\end{aligned}$

(b)
$$\begin{aligned}
\dfrac{2}{9} &= \dfrac{2 \cdot 4}{9 \cdot 4} = \dfrac{8}{36}\\
+\dfrac{1}{4} &= \dfrac{1 \cdot 9}{4 \cdot 9} = \dfrac{9}{36}\\
&\qquad\qquad\qquad \dfrac{17}{36}
\end{aligned}$$

← Rewritten as like fractions

← Add the numerators.

Work Problem 3 at the Side.

(b) $\begin{aligned}&\dfrac{5}{9}\\&+\dfrac{1}{3}\end{aligned}$

OBJECTIVE 3 Subtract unlike fractions. The next example shows subtraction of unlike fractions.

EXAMPLE 4 Subtracting Unlike Fractions

Subtract. Simplify all answers.
  As with addition, rewrite unlike fractions with a least common denominator.

(a) $\dfrac{3}{4} - \dfrac{3}{8}$

← Rewritten as like fractions

ANSWERS

2. (a) $\dfrac{1}{2}$ (b) $\dfrac{23}{24}$ (c) $\dfrac{3}{5}$

3. (a) $\dfrac{15}{16}$ (b) $\dfrac{8}{9}$

Step 1 $\dfrac{3}{4} - \dfrac{3}{8} = \dfrac{6}{8} - \dfrac{3}{8}$

0-88   CHAPTER 0   Topics in Arithmetic

4. Subtract. Simplify all answers.

(a) $\dfrac{5}{8} - \dfrac{1}{4}$

(b) $\dfrac{4}{5} - \dfrac{3}{4}$

(c) $\begin{array}{r} \dfrac{7}{8} \\ -\dfrac{2}{3} \\ \hline \end{array}$

Step 2   Subtract numerators.
$\dfrac{6}{8} - \dfrac{3}{8} = \dfrac{6-3}{8} = \dfrac{3}{8}$

Step 3   Not needed. ($\dfrac{3}{8}$ is in lowest terms.)

(b) $\dfrac{3}{4} - \dfrac{5}{9}$

Step 1   Rewritten as like fractions
$\dfrac{3}{4} - \dfrac{5}{9} = \dfrac{27}{36} - \dfrac{20}{36}$

Step 2   Subtract numerators.
$\dfrac{27}{36} - \dfrac{20}{36} = \dfrac{27-20}{36} = \dfrac{7}{36}$

Step 3   Not needed. ($\dfrac{7}{36}$ is in lowest terms.)

Work Problem 4 at the Side.

ANSWERS

4. (a) $\dfrac{3}{8}$ (b) $\dfrac{1}{20}$ (c) $\dfrac{5}{24}$

0.16   EXERCISES

For Extra Help

- Student's Solutions Manual
- MyMathLab
- InterAct Math Tutorial Software
- AW Math Tutor Center
- MathXL

Add the following fractions. Simplify all answers. See Examples 1–3.

1. $\dfrac{1}{6} + \dfrac{2}{3}$ 2. $\dfrac{1}{4} + \dfrac{1}{8}$ 3. $\dfrac{2}{3} + \dfrac{2}{9}$

4. $\dfrac{3}{7} + \dfrac{1}{14}$ 5. $\dfrac{9}{20} + \dfrac{3}{10}$ 6. $\dfrac{5}{8} + \dfrac{1}{4}$

7. $\dfrac{3}{5} + \dfrac{3}{8}$ 8. $\dfrac{5}{7} + \dfrac{3}{14}$ 9. $\dfrac{2}{9} + \dfrac{5}{12}$

10. $\dfrac{5}{8} + \dfrac{1}{12}$ 11. $\dfrac{1}{3} + \dfrac{3}{5}$ 12. $\dfrac{2}{5} + \dfrac{3}{7}$

13. $\dfrac{1}{4} + \dfrac{2}{9} + \dfrac{1}{3}$ 14. $\dfrac{3}{7} + \dfrac{2}{5} + \dfrac{1}{10}$ 15. $\dfrac{3}{10} + \dfrac{2}{5} + \dfrac{3}{20}$

16. $\dfrac{1}{3} + \dfrac{3}{8} + \dfrac{1}{4}$ 17. $\dfrac{4}{15} + \dfrac{1}{6} + \dfrac{1}{3}$ 18. $\dfrac{5}{12} + \dfrac{2}{9} + \dfrac{1}{6}$

SECTION 0.16 Adding and Subtracting Unlike Fractions   **0-89**

19. $\dfrac{1}{3}$
    $+\dfrac{1}{4}$

20. $\dfrac{7}{12}$
    $+\dfrac{1}{8}$

21. $\dfrac{5}{12}$
    $+\dfrac{1}{16}$

22. $\dfrac{3}{7}$
    $+\dfrac{1}{3}$

Subtract the following fractions. Simplify all answers. See Example 4.

23. $\dfrac{7}{8} - \dfrac{1}{2}$

24. $\dfrac{7}{8} - \dfrac{1}{4}$

25. $\dfrac{2}{3} - \dfrac{1}{6}$

26. $\dfrac{3}{4} - \dfrac{5}{8}$

27. $\dfrac{2}{3} - \dfrac{1}{5}$

28. $\dfrac{5}{6} - \dfrac{7}{9}$

29. $\dfrac{5}{12} - \dfrac{1}{4}$

30. $\dfrac{5}{7} - \dfrac{1}{3}$

31. $\dfrac{8}{9} - \dfrac{7}{15}$

32. $\dfrac{4}{5}$
    $-\dfrac{1}{3}$

33. $\dfrac{7}{8}$
    $-\dfrac{4}{5}$

34. $\dfrac{5}{8}$
    $-\dfrac{1}{3}$

35. $\dfrac{5}{12}$
    $-\dfrac{1}{16}$

36. $\dfrac{7}{12}$
    $-\dfrac{1}{3}$

Solve each application problem.

37. To prepare her flower bed, Katie Lim ordered $\dfrac{1}{4}$ cubic yard of sand, $\dfrac{3}{8}$ cubic yard of mulch, and $\dfrac{1}{3}$ cubic yard of peat moss. How many cubic yards of material did she order?

38. Sergy Shibko has used his savings to repair his car. He spent $\dfrac{1}{4}$ of his savings for new tires, $\dfrac{1}{6}$ of it for brakes, $\dfrac{1}{10}$ of it for a tune-up, and $\dfrac{1}{12}$ of it for new belts and hoses. What fraction of his total savings has he spent?

0.17 Adding and Subtracting Mixed Numbers

OBJECTIVES

1. Estimate an answer, then add or subtract mixed numbers.
2. Estimate an answer, then subtract mixed numbers by borrowing.
3. Add or subtract mixed numbers using an alternate method.

Recall that a mixed number is the sum of a whole number and a fraction. For example,

$$3\frac{2}{5} \text{ means } 3 + \frac{2}{5}.$$

OBJECTIVE 1 Estimate an answer, then add or subtract mixed numbers. Add or subtract mixed numbers by adding or subtracting the fraction parts and then the whole number parts. It is a good idea to estimate the answer first, as we did when multiplying and dividing mixed numbers.

Work Problem 1 at the Side.

1. As a review of mixed numbers, convert each mixed number to an improper fraction and each improper fraction to a mixed number.

   (a) $\frac{7}{4}$

   (b) $\frac{8}{3}$

   (c) $5\frac{3}{5}$

   (d) $3\frac{7}{8}$

EXAMPLE 1 Adding and Subtracting Mixed Numbers

First estimate the answer. Then add or subtract to find the exact answer.

(a) $16\frac{1}{8} + 5\frac{5}{8}$

Estimate: Rounds to 16
           +Rounds to + 6
                       ——
                        22 ← Sum of whole numbers

Exact: $16\frac{1}{8}$
        $+ 5\frac{5}{8}$
        ————
        $21\frac{6}{8}$ ← Sum of fractions

In lowest terms $\frac{6}{8}$ is $\frac{3}{4}$, so the final answer is $21\frac{3}{4}$. This is reasonable because it is close to the estimate of 22.

(b) $8\frac{5}{8} - 3\frac{1}{12}$

Estimate: 9 Rounds to
            − 3 Rounds to
            ——
              6

Exact: $8\frac{5}{8} = 8\frac{15}{24}$ ← Least common denominator
        $-3\frac{1}{12} = -3\frac{2}{24}$
        ————————
        $5\frac{13}{24}$ ← Subtract fractions

Subtract whole numbers.

$5\frac{13}{24}$ is reasonable because it is close to the estimated answer of 6. Just as before, check by adding $5\frac{13}{24}$ and $3\frac{1}{12}$; the sum should be $8\frac{5}{8}$.

ANSWERS

1. (a) $1\frac{3}{4}$ (b) $2\frac{2}{3}$ (c) $\frac{28}{5}$
   (d) $\frac{31}{8}$

SECTION 0.17 Adding and Subtracting Mixed Numbers 0-91

2. First estimate, and then add or subtract to find the exact answer.

(a) *Estimate:*   *Exact:*

$$7 \xleftarrow{\text{Rounds to}} \left\{ 6\frac{7}{8} \right.$$
$$+\, 2 \xleftarrow{\text{Rounds to}} \left\{ +\, 2\frac{1}{4} \right.$$

(b) *Estimate:*   *Exact:*

$$\xleftarrow{\text{Rounds to}} \left\{ 4\frac{7}{9} \right.$$
$$-\ \xleftarrow{\text{Rounds to}} \left\{ -\, 2\frac{2}{3} \right.$$

3. First estimate, and then add to find the exact answer.

(a) *Estimate:*   *Exact:*

$$\xleftarrow{\text{Rounds to}} \left\{ 9\frac{3}{4} \right.$$
$$+\ \xleftarrow{\text{Rounds to}} \left\{ +\, 7\frac{1}{2} \right.$$

(b) *Estimate:*   *Exact:*

$$\xleftarrow{\text{Rounds to}} \left\{ 15\frac{4}{5} \right.$$
$$+\ \xleftarrow{\text{Rounds to}} \left\{ +\, 12\frac{2}{3} \right.$$

ANSWERS

2. (a) $7 + 2 = 9; \ 9\frac{1}{8}$

   (b) $5 - 3 = 2; \ 2\frac{1}{9}$

3. (a) $10 + 8 = 18; \ 17\frac{1}{4}$

   (b) $16 + 13 = 29; \ 28\frac{7}{15}$

NOTE When estimating, if the numerator is *half* of the denominator or *more*, round up the whole number part. If the numerator is *less* than *half* the denominator, leave the whole number part as it is.

Work Problem 2 at the Side.

When you add the fraction parts of mixed numbers, the sum may be greater than 1. If this happens, **carrying** from the fraction column to the whole number column is the best procedure. (You wouldn't want a whole number along with an improper fraction as the answer.)

EXAMPLE 2   Carrying When Adding Mixed Numbers

First estimate, and then add $9\frac{5}{8} + 13\frac{7}{8}$.

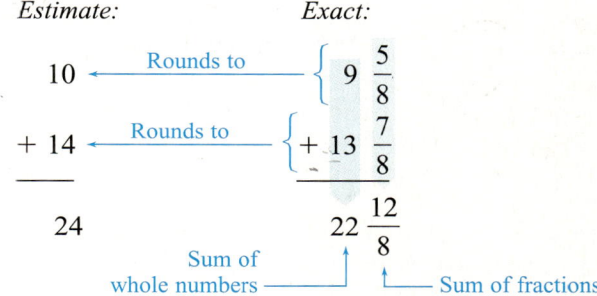

Estimate: *Exact:*

$$10 \xleftarrow{\text{Rounds to}} \left\{ 9\frac{5}{8} \right.$$
$$+\, 14 \xleftarrow{\text{Rounds to}} \left\{ +\, 13\frac{7}{8} \right.$$
$$\overline{24} \qquad\qquad\qquad\qquad\overline{22\frac{12}{8}}$$

Sum of whole numbers   Sum of fractions

The improper fraction $\frac{12}{8}$ can be written in lowest terms as $\frac{3}{2}$. Because $\frac{3}{2} = 1\frac{1}{2}$, the sum is

Becomes

$$22\frac{12}{8} = 22 + \frac{12}{8} = 22 + 1\frac{1}{2} = 23\frac{1}{2}.$$

The estimate was 24, so the exact answer is reasonable.

NOTE When adding mixed numbers, first add the fraction parts, then add the whole number parts. Then combine the two answers simplifying the fraction part when necessary.

Work Problem 3 at the Side.

OBJECTIVE 2 Estimate an answer, then subtract mixed numbers by borrowing. Borrowing is sometimes necessary when subtracting mixed numbers.

EXAMPLE 3   Borrowing When Subtracting Mixed Numbers

First estimate, and then subtract.

(a) $7 - 2\frac{5}{6}$

Estimate: *Exact:*

$$7 \xleftarrow{\text{Rounds to}} \begin{cases} 7 \\ -2\dfrac{5}{6} \end{cases} \leftarrow \text{There is no fraction here from which to subtract } \dfrac{5}{6}.$$

$$\underline{-3} \xleftarrow{\text{Rounds to}}$$

$$\overline{4}$$

It is not possible to subtract $\dfrac{5}{6}$ without borrowing from the whole number 7 first.

$$7 = 6 + \overset{\text{Borrow 1.}}{1}$$

$$= 6 + \underset{\uparrow\; 1 = \frac{6}{6}}{\dfrac{6}{6}}$$

$$= 6\dfrac{6}{6}$$

Now, subtract.

$$\begin{aligned} 7 &= 6\dfrac{6}{6} \\ -2\dfrac{5}{6} &= -2\dfrac{5}{6} \\ \hline & \;4\dfrac{1}{6} \end{aligned}$$

The estimate was 4, so the exact answer is reasonable.

(b) $8\dfrac{1}{3} - 4\dfrac{3}{5}$

Estimate: *Exact:*

$$8 \xleftarrow{\text{Rounds to}} \begin{cases} 8\dfrac{1}{3} = 8\dfrac{5}{15} \\ -4\dfrac{3}{5} = -4\dfrac{9}{15} \end{cases} \leftarrow \text{Least common denominator}$$

$$\underline{-5} \xleftarrow{\text{Rounds to}}$$

$$\overline{3}$$

It is not possible to subtract $\dfrac{9}{15}$ from $\dfrac{5}{15}$, so borrow from the whole number **8**.

$$8\dfrac{5}{15} = 8 + \dfrac{5}{15} = 7 + \overset{\text{Borrow 1.}}{1} + \dfrac{5}{15}$$

$$\phantom{8\dfrac{5}{15}} = 7 + \underset{1 = \frac{15}{15}.}{\dfrac{15}{15}} + \dfrac{5}{15}$$

$$\phantom{8\dfrac{5}{15}} = 7 + \dfrac{20}{15} \leftarrow \dfrac{15}{15} + \dfrac{5}{15}$$

$$\phantom{8\dfrac{5}{15}} = 7\dfrac{20}{15}$$

SECTION 0.17 Adding and Subtracting Mixed Numbers **0-93**

4. First estimate and then subtract to find the exact answer.

(a) *Estimate:*   *Exact:*

$\underset{\text{Rounds to}}{\longleftarrow} \left\{ 7\dfrac{1}{3} \right.$

$\underset{\text{Rounds to}}{\longleftarrow} \left\{ -4\dfrac{5}{6} \right.$

(b) *Estimate:*   *Exact:*

$\underset{\text{Rounds to}}{\longleftarrow} \left\{ 15 \right.$

$\underset{\text{Rounds to}}{\longleftarrow} \left\{ -6\dfrac{4}{9} \right.$

Now, subtract.

$$8\dfrac{1}{3} = 8\dfrac{5}{15} = 7\dfrac{20}{15}$$
$$-4\dfrac{3}{5} = 4\dfrac{9}{15} = 4\dfrac{9}{15}$$
$$\phantom{-4\dfrac{3}{5} = 4\dfrac{9}{15} =} 3\dfrac{11}{15}$$

The exact answer is $3\dfrac{11}{15}$ (lowest terms), which is reasonable because it is close to the estimate of 3.

■ **Work Problem 4 at the Side.**

OBJECTIVE 3 Add or subtract mixed numbers using an alternate method. An alternate method for adding or subtracting mixed numbers is to first change the mixed numbers to improper fractions. Then rewrite the unlike fractions as like fractions. Finally, add or subtract the numerators and write the answer in lowest terms.

■ **EXAMPLE 4** Adding or Subtracting Mixed Numbers

Add or subtract.

(a) $2\dfrac{3}{8} = \dfrac{19}{8} = \dfrac{19}{8}$ ← Least common denominator
$+3\dfrac{3}{4} = \dfrac{15}{4} = \dfrac{30}{8}$
$\phantom{+3\dfrac{3}{4} = \dfrac{15}{4} =} \dfrac{49}{8} = 6\dfrac{1}{8}$ Answer as mixed number

↑ Change to improper fractions.

5. Add or subtract by changing mixed numbers to improper fractions. Write answers as mixed numbers in lowest terms.

(a) $3\dfrac{3}{8}$ (b) $5\dfrac{3}{5}$
$+2\dfrac{1}{2}$ $+3\dfrac{1}{2}$

(c) $6\dfrac{3}{4}$ (d) $9\dfrac{7}{8}$
$-4\dfrac{2}{3}$ $-4\dfrac{3}{4}$

(b) $4\dfrac{2}{3} = \dfrac{14}{3} = \dfrac{70}{15}$ ← Least common denominator
$-2\dfrac{1}{5} = \dfrac{11}{5} = \dfrac{33}{15}$
$\phantom{-2\dfrac{1}{5} = \dfrac{11}{5} =} \dfrac{37}{15} = 2\dfrac{7}{15}$ Answer as mixed number

↑ Improper fractions

■ **Work Problem 5 at the Side.**

ANSWERS

4. (a) $7 - 5 = 2$; $2\dfrac{1}{2}$

(b) $15 - 6 = 9$; $8\dfrac{5}{9}$

5. (a) $\dfrac{47}{8} = 5\dfrac{7}{8}$ (b) $\dfrac{91}{10} = 9\dfrac{1}{10}$

(c) $\dfrac{25}{12} = 2\dfrac{1}{12}$ (d) $\dfrac{41}{8} = 5\dfrac{1}{8}$

NOTE The advantage of this alternate method of adding or subtracting mixed numbers is that it eliminates the need to carry when adding or to borrow when subtracting. However, if the mixed numbers are large, then the numerators of the improper fractions become so large that they are difficult to work with. In such cases, you may want to keep the numbers as mixed numbers.

0-94 CHAPTER 0 Topics in Arithmetic

0.17 EXERCISES

For Extra Help

 Student's Solutions Manual

 MyMathLab

 InterAct Math Tutorial Software

 AW Math Tutor Center

MathXL

First estimate the answer. Then add to find the exact answer. Write answers as mixed numbers. See Examples 1 and 2.

1. Estimate: Exact:
   ←Rounds to $6\frac{1}{3}$
   + ←Rounds to $+ 2\frac{1}{2}$
   ___   ___

2. Estimate: Exact:
                    $8\frac{3}{10}$
   + $+ 1\frac{3}{5}$
   ___   ___

3. Estimate: Exact:
                    $7\frac{1}{3}$
   + $+ 4\frac{1}{6}$
   ___   ___

4. Estimate: Exact:
                    $10\frac{1}{4}$
   + $+ 5\frac{5}{8}$
   ___   ___

5. Estimate: Exact:
                    $\frac{5}{8}$
   + $+ 3\frac{7}{12}$
   ___   ___

6. Estimate: Exact:
                    $12\frac{4}{5}$
   + $+ \frac{7}{10}$
   ___   ___

7. Estimate: Exact:
                    $24\frac{5}{6}$
   + $+ 18\frac{5}{6}$
   ___   ___

8. Estimate: Exact:
                    $14\frac{6}{7}$
   + $+ 15\frac{1}{2}$
   ___   ___

9. Estimate: Exact:
                    $33\frac{3}{5}$
   + $+ 18\frac{1}{2}$
   ___   ___

10. Estimate: Exact:
                    $18\frac{5}{8}$
    + $+ 6\frac{2}{3}$
    ___   ___

11. Estimate: Exact:
                    $22\frac{3}{4}$
    + $+ 15\frac{3}{7}$
    ___   ___

12. Estimate: Exact:
                    $7\frac{1}{4}$
    + $+ 25\frac{7}{8}$
    ___   ___

13. Estimate: Exact:
                    $10\frac{3}{5}$
                    $17\frac{7}{10}$
    + $+ 15\frac{8}{15}$
    ___   ___

14. Estimate: Exact:
                    $14\frac{9}{10}$
                    $8\frac{1}{4}$
    + $+ 13\frac{3}{5}$
    ___   ___

SECTION 0.17 Adding and Subtracting Mixed Numbers **0-95**

First estimate the answer. Then subtract to find the exact answer. Simplify all answers. See Examples 1 and 3.

15. *Estimate:* *Exact:*
$12\frac{3}{8}$
$-10\frac{1}{4}$

16. *Estimate:* *Exact:*
$14\frac{3}{4}$
$-11\frac{3}{8}$

17. *Estimate:* *Exact:*
$12\frac{2}{3}$
$-1\frac{1}{5}$

18. *Estimate:* *Exact:*
$11\frac{9}{20}$
$-4\frac{3}{5}$

19. *Estimate:* *Exact:*
$28\frac{3}{10}$
$-6\frac{1}{15}$

20. *Estimate:* *Exact:*
$15\frac{7}{20}$
$-6\frac{1}{8}$

21. *Estimate:* *Exact:*
17
$-6\frac{5}{8}$

22. *Estimate:* *Exact:*
22
$-4\frac{5}{8}$

23. *Estimate:* *Exact:*
$18\frac{3}{4}$
$-5\frac{4}{5}$

24. *Estimate:* *Exact:*
$17\frac{5}{8}$
$-5\frac{2}{3}$

25. *Estimate:* *Exact:*
$16\frac{3}{8}$
$-10\frac{1}{2}$

26. *Estimate:* *Exact:*
$20\frac{3}{5}$
$-12\frac{7}{15}$

Add or subtract by changing mixed numbers to improper fractions. Write answers as mixed numbers. See Example 4.

27. $3\frac{1}{2}$
$+5\frac{7}{8}$

28. $8\frac{3}{4}$
$+1\frac{5}{8}$

29. $4\frac{2}{3}$
$+6\frac{5}{6}$

30. $7\frac{5}{12}$
$+6\frac{2}{3}$

CHAPTER 0 Topics in Arithmetic

31. $2\frac{2}{3}$
    $+ 1\frac{1}{6}$

32. $4\frac{1}{2}$
    $+ 2\frac{3}{4}$

33. $3\frac{1}{4}$
    $+ 3\frac{2}{3}$

34. $2\frac{4}{5}$
    $+ 5\frac{1}{3}$

35. $1\frac{3}{8}$
    $+ 6\frac{3}{4}$

36. $1\frac{5}{12}$
    $+ 1\frac{7}{8}$

37. $3\frac{1}{2}$
    $- 2\frac{2}{3}$

38. $4\frac{1}{4}$
    $- 3\frac{7}{12}$

0.14–0.17 TEST

Add or subtract. Write answers in lowest terms.

1. $\frac{3}{8} + \frac{3}{8}$

2. $\frac{3}{16} + \frac{5}{16}$

3. $\frac{7}{10} - \frac{3}{10}$

4. $\frac{7}{12} - \frac{5}{12}$

Find the least common multiple of each set of numbers.

5. 2, 3, 4

6. 6, 3, 5, 15

7. 6, 9, 27, 36

Add or subtract. Write answers in lowest terms.

8. $\frac{3}{8} + \frac{1}{4}$

9. $\frac{2}{9} + \frac{5}{12}$

10. $\frac{7}{8} - \frac{2}{3}$

11. $\frac{2}{5} - \frac{3}{8}$

First estimate the answer. Then add or subtract to find the exact answer; simplify exact answers.

12. $7\frac{2}{3} + 4\frac{5}{6}$

13. $16\frac{2}{5} - 11\frac{2}{3}$

14. $18\frac{3}{4} + 9\frac{2}{5} + 12\frac{1}{3}$

15. $24 - 18\frac{3}{8}$

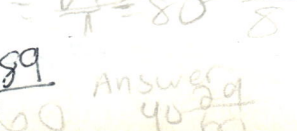

0.18 Reading and Writing Decimals

OBJECTIVES

1. Write parts of a whole using decimals.
2. Identify the place value of a digit.
3. Read and write decimals in words.
4. Write decimals as fractions or mixed numbers.

1. There are 10 dimes in one dollar. Each dime is $\frac{1}{10}$ of a dollar. Write the yellow shaded portion of each dollar as a fraction, as a decimal, and in words.

   (a)

   (b)

   (c)

ANSWERS

1. (a) $\frac{1}{10}$; 0.1; one tenth
   (b) $\frac{3}{10}$; 0.3; three tenths
   (c) $\frac{9}{10}$; 0.9; nine tenths

Fractions are used to represent parts of a whole. In this chapter, **decimals** are used as another way to show parts of a whole. For example, our money system is based on decimals. One dollar is divided into 100 equivalent parts. One cent ($0.01) is one of the parts, and a dime ($0.10) is 10 of the parts.

OBJECTIVE 1 Write parts of a whole using decimals. Decimals are used when a whole is divided into 10 equivalent parts, or into 100 or 1000 or 10,000 equivalent parts. In other words, decimals are fractions with denominators that are a power of 10. For example, the square below is cut into 10 equivalent parts. Written as a fraction, each part is $\frac{1}{10}$ of the whole. Written as a decimal, each part is 0.1. Both $\frac{1}{10}$ and 0.1 are read as *"one tenth."*

One-tenth of the square is shaded.

The dot in 0.1 is called the **decimal point.**

$$0.1 \uparrow \text{Decimal point}$$

The square at the right has **7** of its 10 parts shaded.

Written as a *fraction*, $\frac{7}{10}$ of the square is shaded.

Written as a *decimal*, 0.7 of the square is shaded.

Both $\frac{7}{10}$ and 0.7 are read as *"seven tenths."*

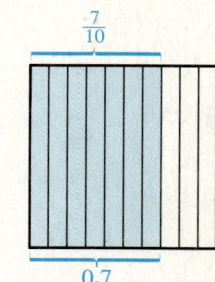

Seven-tenths of the square is shaded.

Work Problem 1 at the Side.

The square below is cut into 100 equivalent parts. Written as a *fraction*, each part is $\frac{1}{100}$ of the whole.

Written as a decimal, each part is **0.01** of the whole.
Both $\frac{1}{100}$ and 0.01 are read as "one hundredth."

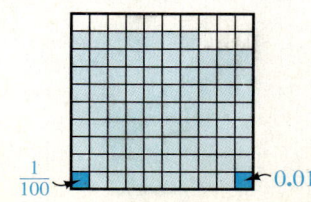

0-98 CHAPTER 0 Topics in Arithmetic

2. Write the portion of each square that is shaded as a fraction, as a decimal, and in words.

(a)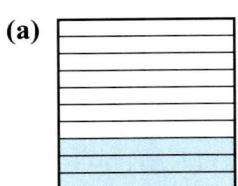

(b)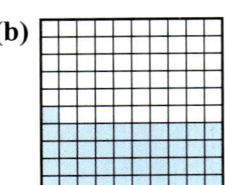

3. Write each decimal as a fraction.

(a) 0.7

(b) 0.2

(c) 0.03

(d) 0.69

(e) 0.047

(f) 0.351

The square at the bottom of the preceding page has 87 parts shaded.

Written as a fraction, $\frac{87}{100}$ of the total area is shaded.

Written as a decimal, **0.87** of the total area is shaded.

Both $\frac{87}{100}$ and 0.87 are read as "*eighty-seven hundredths.*"

Work Problem 2 at the Side.

Example 1 shows several numbers written as both fractions and decimals.

EXAMPLE 1 Using the Decimal Forms of Fractions

	Fraction	Decimal	Read As
(a)	$\frac{4}{10}$	0.4	four tenths
(b)	$\frac{9}{100}$	0.09	nine hundredths
(c)	$\frac{71}{100}$	0.71	seventy-one hundredths
(d)	$\frac{8}{1000}$	0.008	eight thousandths
(e)	$\frac{45}{1000}$	0.045	forty-five thousandths
(f)	$\frac{832}{1000}$	0.832	eight hundred thirty-two thousandths

Work Problem 3 at the Side.

OBJECTIVE 2 Identify the place value of a digit. The decimal point separates the *whole number part* from the *fractional part* in a decimal number. In the chart below, you see that the **place value names** for fractional parts are similar to those on the whole number side, but end in "*ths.*"

Decimal Place Value Chart

hundred thousands | ten thousands | thousands | hundreds | tens | ones • tenths | hundredths | thousandths | ten-thousandths | hundred-thousandths

100,000   10,000   1000   100   10   1   • $\frac{1}{10}$ $\frac{1}{100}$ $\frac{1}{1000}$ $\frac{1}{10,000}$ $\frac{1}{100,000}$

Whole number part   Decimal point (Read "and")   Fractional part

ANSWERS

2. (a) $\frac{3}{10}$; 0.3; three tenths

(b) $\frac{41}{100}$; 0.41; forty-one hundredths

3. (a) $\frac{7}{10}$ (b) $\frac{2}{10}$ (c) $\frac{3}{100}$

(d) $\frac{69}{100}$ (e) $\frac{47}{1000}$ (f) $\frac{351}{1000}$

SECTION 0.18 Reading and Writing Decimals **0-99**

4. Identify the place value of each digit.

(a) 971.54

Notice that the **ones** place is at the center. (There is no "oneths" place.) Also notice that each place is 10 times the value of the place to its right.

> **CAUTION** In this chapter, if a number does *not* have a decimal point, it is a *whole number*. A whole number has no fractional part. If you want to show the decimal point in a whole number, it is just to the *right* of the digit in the ones place. For example:
>
> $$8 = 8. \qquad 306 = 306.$$
> ↑ Decimal point   ↑ Decimal point

(b) 0.4

EXAMPLE 2 Identifying the Place Value of a Digit

Identify the place value of each digit.

(a) 178.36

(b) 0.00935

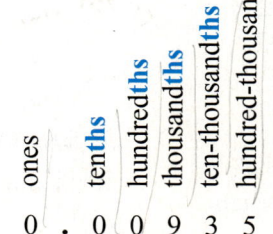

(c) 5.60

1 7 8 . 3 6
hundreds, tens, ones, tenths, hundredths

0 . 0 0 9 3 5
ones, tenths, hundredths, thousandths, ten-thousandths, hundred-thousandths

Notice in Example 2(b) that we do *not* use commas on the right side of the decimal point.

Work Problem 4 at the Side.

(d) 0.0835

OBJECTIVE 3 Read and write decimals in words. A decimal is read according to its form as a fraction. We read 0.9 as "nine tenths" because 0.9 is the same as $\frac{9}{10}$. Notice that 0.9 ends in the tenths place.

0. 9
ones, **tenths**

We read 0.02 as "two hundredths" because 0.02 is the same as $\frac{2}{100}$. Notice that 0.02 ends in the hundredths place.

0. 0 2
ones, tenths, **hundredths**

ANSWERS

4. (a) 9 7 1 . 5 4
hundreds, tens, ones, tenths, hundredths
(b) 0 . 4
ones, tenths
(c) 5 . 6 0
ones, tenths, hundredths
(d) 0 . 0 8 3 5
ones, tenths, hundredths, thousandths, ten-thousandths

EXAMPLE 3 Reading Decimal Numbers

Tell how to read each decimal in words.

(a) 0.3

Because $0.3 = \frac{3}{10}$, read the decimal as three ten**ths**.

(b) 0.49   Read it as: forty-nine hundred**ths**.

(c) 0.08   Read it as: eight hundred**ths**.

CHAPTER 0 Topics in Arithmetic

5. Tell how to read each decimal in words.

(a) 0.6

(b) 0.46

(c) 0.05

(d) 0.409

(e) 0.0003

(f) 0.0703

(g) 0.088

6. Tell how to read each decimal in words.

(a) 3.8

(b) 15.001

(c) 0.0073

(d) 64.309

ANSWERS

5. (a) six tenths
   (b) forty-six hundredths
   (c) five hundredths
   (d) four hundred nine thousandths
   (e) three ten-thousandths
   (f) seven hundred three ten-thousandths
   (g) eighty-eight thousandths
6. (a) three and eight tenths
   (b) fifteen and one thousandth
   (c) seventy-three ten-thousandths
   (d) sixty-four and three hundred nine thousandths

(d) 0.918   Read it as: nine hundred eighteen thousand**ths**.

(e) 0.0106   Read it as: one hundred six ten-thousand**ths**.

Work Problem 5 at the Side.

Reading Decimal Numbers

Step 1 Read any whole number part to the *left* of the decimal point as you normally would.

Step 2 Read the decimal point as "*and*."

Step 3 Read the part of the number to the *right* of the decimal point as if it were an ordinary whole number.

Step 4 Finish with the place value name of the rightmost digit; these names all end in "*ths*."

NOTE If there is *no whole number part,* you will use only Steps 3 and 4.

EXAMPLE 4   Reading Decimals

Read each decimal.

(a)   9 is in tenths place.

16.9

sixteen **and** nine **tenths**

16.9 is read "sixteen and nine tenths."

(b)   5 is in hundredths place.

482.35

four hundred eighty-two **and** thirty-five **hundredths**

482.35 is read "four hundred eighty-two and thirty-five hundredths."

        3 is in thousandths place.

(c) 0.063 is "sixty-three **thousandths**." (No whole number part.)

(d) 11.1085 is "eleven **and** one thousand eighty-five **ten-thousandths**."

CAUTION Use "and" *only* when reading a decimal point. A common mistake is to read the whole number 405 as "four hundred *and* five." But there is *no decimal point* shown in 405, so it is read "four hundred five."

Work Problem 6 at the Side.

OBJECTIVE 4 Write decimals as fractions or mixed numbers. Knowing how to read decimals will help you when writing decimals as fractions.

SECTION 0.18 Reading and Writing Decimals

7. Write each decimal as a fraction or mixed number.

   (a) 0.7

   (b) 12.21

   (c) 0.101

   (d) 0.007

   (e) 1.3717

Writing Decimals as Fractions or Mixed Numbers

Step 1 The digits to the right of the decimal point are the numerator of the fraction.

Step 2 The denominator is 10 for tenths, 100 for hundredths, 1000 for thousandths, 10,000 for ten-thousandths, and so on.

Step 3 If the decimal has a whole number part, the fraction will be a mixed number with the same whole number part.

EXAMPLE 5 Writing Decimals as Fractions or Mixed Numbers

Write each decimal as a fraction or mixed number.

(a) 0.19

The digits to the right of the decimal point, 19, are the numerator of the fraction. The denominator is 100 for hundredths because the right-most digit is in the hundredths place.

$$0.19 = \frac{19}{100} \quad \leftarrow 100 \text{ for hundredths}$$

Hundredths place

(b) 0.863

$$0.863 = \frac{863}{1000} \quad \leftarrow 1000 \text{ for thousandths}$$

Thousandths place

(c) 4.0099

The whole number part stays the same.

$$4.0099 = 4\frac{99}{10{,}000} \quad \leftarrow 10{,}000 \text{ for ten-thousandths}$$

Ten-thousandths place

Work Problem 7 at the Side.

CAUTION After you write a decimal as a fraction or a mixed number, make sure the fraction is in lowest terms.

EXAMPLE 6 Writing Decimals as Fractions or Mixed Numbers in Lowest Terms

Write each decimal as a fraction or mixed number in lowest terms.

(a) $0.4 = \frac{4}{10} \quad \leftarrow 10 \text{ for tenths}$

Write $\frac{4}{10}$ in lowest terms. $\quad \frac{4}{10} = \frac{4 \div 2}{10 \div 2} = \frac{2}{5} \quad \leftarrow$ Lowest terms

(b) $0.75 = \frac{75}{100} = \frac{75 \div 25}{100 \div 25} = \frac{3}{4} \quad \leftarrow$ Lowest terms

ANSWERS

7. (a) $\frac{7}{10}$ (b) $12\frac{21}{100}$ (c) $\frac{101}{1000}$
   (d) $\frac{7}{1000}$ (e) $1\frac{3717}{10{,}000}$

0-102 CHAPTER 0 Topics in Arithmetic

8. Write each decimal as a fraction or mixed number in lowest terms.

(a) 0.5

(b) 12.6

(c) 0.85

(d) 3.05

(e) 0.225

(f) 420.0802

(c) $18.105 = 18\dfrac{105}{1000} = 18\dfrac{105 \div 5}{1000 \div 5} = 18\dfrac{21}{200}$ ← Lowest terms

(d) $42.8085 = 42\dfrac{8085}{10{,}000} = 42\dfrac{8085 \div 5}{10{,}000 \div 5} = 42\dfrac{1617}{2000}$ ← Lowest terms

Work Problem 8 at the Side.

Calculator Tip In this chapter you'll notice that we use a 0 in the ones place for decimal fractions. We write **0**.45 instead of just .45, to emphasize that there is no whole number. Your scientific (not graphing) calculator shows these 0s also. Enter ⊙ ④ ⑤. Notice that the display automatically shows 0.45 even though you did not press 0. For comparison, enter the whole number 45 by pressing ④ ⑤ ⊕ and notice where the decimal point is shown in the display. (It automatically appears to the *right* of the 5.)

ANSWERS

8. (a) $\dfrac{1}{2}$ (b) $12\dfrac{3}{5}$ (c) $\dfrac{17}{20}$ (d) $3\dfrac{1}{20}$

(e) $\dfrac{9}{40}$ (f) $420\dfrac{401}{5000}$

0.18 EXERCISES

For Extra Help

- Student's Solutions Manual
- MyMathLab
- InterAct Math Tutorial Software
- AW Math Tutor Center
- MathXL

Identify the digit that has the given place value. See Example 2.

1. 70.489
   tens
   ones
   tenths

2. 135.296
   ones
   tenths
   tens

3. 0.2518
   hundredths
   thousandths
   ten-thousandths

4. 0.9347
   hundredths
   thousandths
   ten-thousandths

5. 93.01472
   thousandths
   ten-thousandths
   tenths

6. 0.51968
   tenths
   ten-thousandths
   hundredths

7. 314.658
   tens
   tenths
   hundreds

8. 51.325
   tens
   tenths
   hundredths

9. 149.0832
   hundreds
   hundredths
   ones

10. 3458.712
    hundreds
    hundredths
    tenths

11. 6285.7125
    thousands
    thousandths
    hundredths

12. 5417.6832
    thousands
    thousandths
    ones

SECTION 0.18   Reading and Writing Decimals   0-103

Write the decimal number that has the specified place values. See Example 2.

13. 0 ones, 5 hundredths, 1 ten, 4 hundreds, 2 tenths

14. 7 tens, 9 tenths, 3 ones, 6 hundredths, 8 hundreds

15. 3 thousandths, 4 hundredths, 6 ones, 2 ten-thousandths, 5 tenths

16. 8 ten-thousandths, 4 hundredths, 0 ones, 2 tenths, 6 thousandths

17. 4 hundredths, 4 hundreds, 0 tens, 0 tenths, 5 thousandths, 5 thousands, 6 ones

18. 7 tens, 7 tenths, 6 thousands, 6 thousandths, 3 hundreds, 3 hundredths, 2 ones

Write each decimal as a fraction or mixed number in lowest terms. See Examples 1, 5, and 6.

19. 0.7   20. 0.1   21. 13.4   22. 9.8   23. 0.25

24. 0.55   25. 0.66   26. 0.33   27. 10.17 28. 31.99

29. 0.06   30. 0.08   31. 0.205   32. 0.805

33. 5.002 34. 4.008 35. 0.686   36. 0.492

Tell how to read each decimal in words. See Examples 3 and 4.

37. 0.5   38. 0.2

39. 0.78   40. 0.55

41. 0.105   42. 0.609

43. 12.04   44. 86.09

45. 1.075   46. 4.025

Write each decimal in numbers. See Examples 3 and 4.

47. six and seven tenths

48. eight and twelve hundredths

49. thirty-two hundredths

50. one hundred eleven thousandths

51. four hundred twenty and eight thousandths

52. two hundred and twenty-four thousandths

53. seven hundred three ten-thousandths

54. eight hundred and six hundredths

55. seventy-five and thirty thousandths

56. sixty and fifty hundredths

0.19 Rounding Decimals

OBJECTIVES

1. Learn the rules for rounding decimals.
2. Round decimals to any given place.

OBJECTIVE 1 Learn the rules for rounding decimals.   It is also important to be able to **round** decimals. For example, a store is selling 2 candy mints for $0.75 but you want only one mint. The price of each mint is $0.75 ÷ 2, which is $0.375, but you cannot pay part of a cent. Is $0.375 closer to $0.37 or to $0.38? Actually, it's exactly halfway between. When this happens in everyday situations, the rule is to round *up*. The store will charge you $0.38 for the mint.

Rounding Decimals

Step 1 Find the place to which the rounding is being done. Draw a "cut-off" line *after* that place to show that you are cutting off and dropping the rest of the digits.

Step 2 Look *only* at the *first* digit you are cutting off.

Step 3(a) If this digit is **4 or less,** the part of the number you are keeping *stays the same.*

Step 3(b) If this digit is **5 or more,** you must **round up** the part of the number you are keeping.

Step 4 You can use the "≈" sign to indicate that the rounded number is now an approximation (close, but not exact). "≈" means "is approximately equal to."

CAUTION   Do *not* move the decimal point when rounding.

OBJECTIVE 2 Round decimals to any given place.   The following examples show you how to round decimals.

SECTION 0.19 Rounding Decimals

1. Round to the nearest thousandth.

(a) 0.33492

(b) 8.00851

(c) 265.42038

(d) 10.70180

EXAMPLE 1 Rounding a Decimal Number

Round 14.39652 to the nearest thousandth. Is it closer to 14.396 or to 14.397?

Step 1 Draw a "cut-off" line after the thousandths place.

$$14.396 \,|\, 52$$

Thousandths

You are cutting off the 5 and 2. They will be dropped.

Step 2 Look *only* at the *first* digit you are cutting off. Ignore the other digits you are cutting off.

$$14.396 \,|\, 5\,2$$

Look only at the 5. Ignore the 2.

Step 3 If the first digit you are cutting off is 5 or more, round up the part of the number you are keeping.

$$\begin{array}{r} 14.396\,|\,5\,2 \\ +\ 0.001 \\ \hline 14.397 \end{array}$$

First digit cut is 5 or more, so round up by adding 1 thousandth to the part you are keeping.

So, 14.39652 rounded to the nearest thousandth is 14.397. We can write 14.39652 ≈ 14.397.

CAUTION When rounding whole numbers, we keep all the digits but change some to 0s. With decimals, you cut off and *drop the extra digits*. In Example 1 above, 14.39652 rounds to 14.397 *not* 14.39700.

Work Problem 1 at the Side.

In Example 1, the rounded number 14.397 had *three decimal places*. **Decimal places** are the number of digits to the *right* of the decimal point. The first decimal place is tenths, the second is hundredths, the third is thousandths, and so on.

EXAMPLE 2 Rounding Decimals to Different Places

Round to the place indicated.

(a) 5.3496 to the nearest tenth (Is it closer to 5.3 or to 5.4?)

Step 1 Draw a cut-off line after the tenths place.

$$5.3 \,|\, 4\,9\,6$$

Tenths

You are cutting off the 4, 9, and 6. They will be dropped.

Step 2

$$5.3 \,|\, \underline{4}\,9\,6$$

Look only at the 4.

Ignore these digits.

ANSWERS

1. (a) 0.335 **(b)** 8.009
 (c) 265.420 **(d)** 10.702

Step 3   5 . 3 ⧸4̸ 9 6   ── First digit cut is 4 or less, so the part you are keeping stays the same.

5 . 3 ← Stays the same

5.3496 rounded to the nearest tenth is 5.3 (*one* decimal place for *tenths*). We can write 5.3496 ≈ 5.3. Notice that 5.3496 does ***not*** round to 5.3000, which would be ten-thousandths.

(b) 0.69738 to the nearest hundredth (Is it closer to 0.69 or to 0.70?)

Step 1   0 . 6 9 | 7 3 8   Draw a cut-off line after the hundredths place.
             ↑
         Hundredths

Step 2   0 . 6 9 | 7 3 8   ── Look only at the 7.

Step 3   0 . 6 9 | 7 3 8   ── First digit cut is 5 or more, so round up by adding 1 hundredth to the part you are keeping.

    0 . 6 9   ← Keep this part.
  + 0 . 0 1   ← To round up, add 1 hundredth.
    0 . 7 0   ← 9 + 1 is 10; write 0 and carry 1 to the 6 in the tenths place.

0.69738 rounded to the nearest hundredth is 0.70. Hundredths is *two* decimal places so you *must* write the 0 in the hundredths place. We can write 0.69738 ≈ 0.70.

(c) 0.01806 to the nearest thousandth (Is it closer to 0.018 or to 0.019?)

    0 . 0 1 8 | 0 6   ── First digit cut is 4 or less, so the part you are keeping stays the same.
    0 . 0 1 8

0.01806 rounded to the nearest thousandth is 0.018 (*three* decimal places for *thousandths*). We can write 0.01806 ≈ 0.018.

(d) 57.976 to the nearest tenth (Is it closer to 57.9 or to 58.0?)

    57.9 | 76   ── First digit cut is 5 or more, so round up by adding 1 tenth to the part you are keeping.
    57.9
  + 0.1
    58.0   ← 9 + 1 is 10; write the 0 and carry 1 to the 7 in the ones place.

57.976 rounded to the nearest tenth is 58.0. We can write 57.976 ≈ 58.0. You *must* write the 0 in the tenths place to show that the number was rounded to the nearest tenth.

2. Round to the place indicated.

(a) 0.8988 to the nearest hundredth

(b) 5.8903 to the nearest hundredth

(c) 11.0299 to the nearest thousandth

(d) 0.545 to the nearest tenth

CAUTION Check that your rounded answer shows *exactly* the number of decimal places called for, even if a 0 is in that place. See Examples 2(b) and 2(d). Be sure your answer shows *one* decimal place if you rounded to *tenths*, *two* decimal places for *hundredths*, or *three* decimal places for *thousandths*.

Work Problem 2 at the Side.

ANSWERS
2. (a) 0.90 (b) 5.89 (c) 11.030
   (d) 0.5

0.19 EXERCISES

For Extra Help
- Student's Solutions Manual
- MyMathLab
- InterAct Math Tutorial Software
- AW Math Tutor Center
- MathXL

Round each number to the place indicated. See Examples 1 and 2.

1. 16.8974 to the nearest tenth

2. 193.845 to the nearest hundredth

3. 0.95647 to the nearest thousandth

4. 96.81584 to the nearest ten-thousandth

5. 0.799 to the nearest hundredth

6. 0.952 to the nearest tenth

7. 3.66062 to the nearest thousandth

8. 1.5074 to the nearest hundredth

9. 793.988 to the nearest tenth

10. 476.1196 to the nearest thousandth

11. 0.09804 to the nearest ten-thousandth

12. 176.004 to the nearest tenth

13. 48.512 to the nearest one

14. 3.385 to the nearest one

15. 9.0906 to the nearest hundredth

16. 30.1290 to the nearest thousandth

17. 82.000151 to the nearest ten-thousandth

18. 0.400594 to the nearest ten-thousandth

0.20 Adding and Subtracting Decimals

OBJECTIVES

1. Add decimals.
2. Subtract decimals.

OBJECTIVE 1 Add decimals. When adding or subtracting *whole* numbers (Sections 0.2 and 0.3), you lined up the numbers in columns so that you were adding ones to ones, tens to tens, and so on. A similar idea applies to adding or subtracting *decimal* numbers. With decimals, you line up the decimal points to make sure you are adding tenths to tenths, hundredths to hundredths, and so on.

1. Find each sum.

    (a) 2.86 + 7.09

    (b) 13.761 + 8.325

    (c) 0.319 + 56.007 + 8.252

    (d) 39.4 + 0.4 + 177.2

Adding and Subtracting Decimals

Step 1 Write the numbers in columns with the decimal points lined up.

Step 2 If necessary, write in 0s so both numbers have the same number of decimal places. Then add or subtract as if they were whole numbers.

Step 3 Line up the decimal point in the answer directly below the decimal points in the problem.

EXAMPLE 1 Adding Decimal Numbers

Add.

(a) 16.92 and 48.34

Step 1 Write the numbers in columns with the decimal points lined up.

```
  tens
   ones
     tenths
      hundredths
   1 6 . 9 2
 + 4 8 . 3 4
```
— Decimal points are lined up.

Step 2 Add as if these were whole numbers.

```
    1 1
   1 6 . 9 2
 + 4 8 . 3 4
```

Step 3
```
   6 5 . 2 6
```
Decimal point in answer is lined up under decimal points in problem.

(b) 5.897 + 4.632 + 12.174

Write the numbers vertically with decimal points lined up. Then add.

```
   1 1   2 1
     5 . 8 9 7
     4 . 6 3 2
 + 1 2 . 1 7 4
   2 2 . 7 0 3
```
— Decimal points are lined up.

Work Problem 1 at the Side.

ANSWERS

1. (a) 9.95   (b) 22.086
   (c) 64.578   (d) 217.0

SECTION 0.20 Adding and Subtracting Decimals

2. Find each sum.

(a) 6.54 + 9.8

In Example 1(a), both numbers had *two* decimal places (two digits to the right of the decimal point). In Example 1(b), all the numbers had *three decimal places* (three digits to the right of the decimal point). That made it easy to add tenths to tenths, hundredths to hundredths, and so on.

If the number of decimal places does *not* match, you can write in 0s as placeholders to make them match. This is shown in Example 2.

EXAMPLE 2 Writing 0s as Placeholders before Adding

Add.

(a) 7.3 + 0.85

(b) 0.831 + 222.2 + 10

There are two decimal places in 0.85 (tenths and hundredths), so write a 0 in the hundredths place in 7.3 so that it has two decimal places also.

$$\begin{array}{r} 7.30 \\ + 0.85 \\ \hline 8.15 \end{array}$$ ← One 0 is written in.

7.30 is equivalent to 7.3 because

$7\dfrac{30}{100}$ in lowest terms is $7\dfrac{3}{10}$

(b) 6.42 + 9 + 2.576

Write in 0s so that all the addends have three decimal places. Notice how the whole number 9 is written with the decimal point at the *far right* side. (If you put the decimal point on the *left* side of the 9, you would turn it into the decimal fraction 0.9.)

(c) 8.64 + 39.115 + 3.0076

$$\begin{array}{r} 6.4\,2\,0 \\ 9.0\,0\,0 \\ + 2.5\,7\,6 \\ \hline 17.9\,9\,6 \end{array}$$

← One 0 is written in.
← 9 is a whole number; decimal point and three 0s are written in.
← No 0s are needed.

Decimal points are lined up.

(d) 5 + 429.823 + 0.76

NOTE Writing 0s to the right of a *decimal* number does *not* change the value of the number.

Work Problem 2 at the Side.

OBJECTIVE 2 Subtract decimals. Subtraction of decimals is done in much the same way as addition of decimals. You can check the answers to subtraction problems using addition, as you did with whole numbers (see **Section 0.3**).

EXAMPLE 3 Subtracting Decimal Numbers

Subtract. Check your answers using addition.

(a) 15.82 from 28.93

Step 1

$$\begin{array}{r} 28.93 \\ -15.82 \end{array}$$

Line up decimal points. Then you will be subtracting hundredths from hundredths and tenths from tenths.

ANSWERS
2. (a) 16.34 (b) 233.031
(c) 50.7626 (d) 435.583

0-110 CHAPTER 0 Topics in Arithmetic

3. Subtract. Check your answers using addition.

(a) 22.7 from 72.9

(b) 6.425 from 11.813

(c) $20.15 - 19.67$

Step 2

$$\begin{array}{r} 28.93 \\ -15.82 \\ \hline 13\ 11 \end{array}$$

Both numbers have two decimal places; no need to write in 0s.

Subtract as if they were whole numbers.

Step 3

$$\begin{array}{r} 28.93 \\ -15.82 \\ \hline 13.11 \end{array}$$

⎯ Decimal point in answer is lined up.

Check the answer by adding 13.11 and 15.82. If the subtraction is done correctly, the sum will be 28.93.

(b) 146.35 minus 58.98
Borrowing is needed here.

$$\begin{array}{r} 0\ 13\ 15\ \ 12\ 15 \\ \cancel{1}\cancel{4}\cancel{6}.\cancel{3}\cancel{5} \\ -\ \ 5\ 8\ .\ 9\ 8 \\ \hline \ \ \ 8\ 7\ .\ 3\ 7 \end{array}$$

⎯ Line up decimal points.

4. Subtract. Check your answers using addition.

(a) 18.651 from 25.3

(b) $5.816 - 4.98$

(c) 40 less 3.66

(d) $1 - 0.325$

Check the answer by adding 87.37 and 58.98. If you did the subtraction correctly, the sum will be 146.35. (If it *isn't*, you need to rework the problem.)

Work Problem 3 at the Side.

■ **EXAMPLE 4** Writing 0s as Placeholders before Subtracting

Subtract.

(a) 16.5 from 28.362
Use the same steps as in Example 3 above, remembering to write in 0s so both numbers have three decimal places.

⎯ Line up decimal points.

$$\begin{array}{r} 28.362 \\ -16.500 \\ \hline 11.862 \end{array}$$ ← Write two 0s.
← Subtract as usual.

Check the answer by adding.

$$\begin{array}{r} 16.500 \\ +11.862 \\ \hline 28.362 \end{array}$$ ← Matches minuend in original problem.

(b) $59.7 - 38.914$

$$\begin{array}{r} 59.700 \\ -38.914 \\ \hline 20.786 \end{array}$$ ← Write two 0s
← Subtract as usual.

(c) 12 less 5.83

$$\begin{array}{r} 12.00 \\ -\ 5.83 \\ \hline \ \ 6.17 \end{array}$$ ← Write a decimal point and two 0s.
← Subtract as usual.

Work Problem 4 at the Side.

ANSWERS
3. (a) 50.2; $50.2 + 22.7 = 72.9$
(b) 5.388; $5.388 + 6.425 = 11.813$
(c) 0.48; $0.48 + 19.67 = 20.15$
4. (a) 6.649; $6.649 + 18.651 = 25.3$
(b) 0.836; $0.836 + 4.98 = 5.816$
(c) 36.34; $36.34 + 3.66 = 40$
(d) 0.675; $0.675 + 0.325 = 1$

0.20 EXERCISES

For Extra Help

 Student's Solutions Manual

 MyMathLab

 InterAct Math Tutorial Software

 AW Math Tutor Center

MathXL

Find each sum or difference. See Examples 1–4.

1. 5.69 + 11.79
2. 372.1 − 33.7
3. 24.008 − 0.995

4. 0.7759 + 9.8883
5. 8.263 − 0.5
6. 47.658 − 20.9

7. 76.5 + 0.506
8. 1.87 + 9.749
9. 21 − 0.896

10. 9 − 1.183
11. 0.4 − 0.291
12. 0.35 − 0.088

13. 39.76005 + 182 + 4.799 + 98.31 + 5.9999

14. 489.76 + 0.9993 + 38 + 8.55087 + 80.697

This drawing of a human skeleton shows the average length of the longest bones, in inches. Use the drawing to answer Exercises 15–18.

15. (a) What is the combined length of the humerus and radius bones?
    (b) What is the difference in the lengths of these two bones?

16. (a) What is the total length of the femur and tibia bones?
    (b) How much longer is the femur than the tibia?

17. (a) Find the sum of the lengths of the humerus, ulna, femur, and tibia.
    (b) How much shorter is the 8th rib than the 7th rib?

18. (a) What is the difference in the lengths of the two bones in the lower arm?
    (b) What is the difference in the lengths of the two bones in the lower leg?

7th rib 9.45 in.
Humerus 14.35 in.
Radius 10.4 in.
8th rib 9.06 in.
Ulna 11.1 in.
Femur 19.88 in.
Tibia 16.94 in.
Fibula 15.94 in.

Source: *The Top 10 of Everything 2000.*

0.21 Multiplying Decimals

OBJECTIVE

1. Multiply decimals.

1. Multiply.

   (a) 2.6×0.4

   (b) 45.2×0.25

   (c) 0.104×7

   (d) 3.18×2.23

   (e) 611×3.7

ANSWERS

1. (a) 1.04 (b) 11.300 (c) 0.728
   (d) 7.0914 (e) 2260.7

OBJECTIVE 1 Multiply decimals. The decimals 0.3 and 0.07 can be multiplied by writing them as fractions.

$$0.3 \times 0.07 = \frac{3}{10} \times \frac{7}{100} = \frac{3 \times 7}{10 \times 100} = \frac{21}{1000} = 0.021$$

1 decimal place + 2 decimal places → 3 decimal places

Can you see a way to multiply decimals without writing them as fractions? Try these steps. Remember that each number in a multiplication problem is called a *factor*, and the answer is called the *product*.

Multiplying Decimals

Step 1 Multiply the numbers (the factors) as if they were whole numbers.

Step 2 Find the *total* number of decimal places in *both* factors.

Step 3 Write the decimal point in the product (the answer) so it has the same number of decimal places as the total from Step 2. You may need to write in extra 0s on the left side of the product to get the correct number of decimal places.

NOTE When multiplying decimals, you do *not* need to line up decimal points. (You *do* need to line up decimal points when adding or subtracting decimals.)

EXAMPLE 1 Multiplying Decimal Numbers

Multiply: 8.34 times 4.2.

Step 1 Multiply the numbers as if they were whole numbers.

```
   8.34
 × 4.2
  1668
 3336
 35028
```

Step 2 Count the total number of decimal places in both factors.

```
   8.34  ← 2 decimal places
 × 4.2   ← 1 decimal place
  1668     3 total decimal places
 3336
 35028
```

SECTION 0.21 Multiplying Decimals 0-113

2. Multiply.

(a) 0.04 × 0.09

(b) 0.2 · 0.008

(c) (0.063)(0.04)

(d) 0.0081 · 0.003

(e) (0.11)(0.0005)

ANSWERS
2. (a) 0.0036 (b) 0.0016
   (c) 0.00252 (d) 0.0000243
   (e) 0.000055

Step 3 Count over 3 places in the product and write the decimal point. Count from right to left.

$$\begin{array}{r} 8.3\,4 \\ \times\ \ 4.2 \\ \hline 1\,6\,6\,8 \\ 3\,3\,3\,6\ \ \ \\ \hline 3\,5.0\,2\,8 \end{array}$$ ← 2 decimal places
← 1 decimal place
3 total decimal places

← 3 decimal places in product

Count over 3 places from right to left to position the decimal point.

Work Problem 1 on p. 0-112.

EXAMPLE 2 Writing 0s as Placeholders in the Product

Multiply 0.042 by 0.03.

Start by multiplying, then count decimal places.

$$\begin{array}{r} 0.0\,4\,2 \\ \times\ \ 0.0\,3 \\ \hline 1\,2\,6 \end{array}$$ ← 3 decimal places
← 2 decimal places
← 5 decimal places needed in product

After multiplying, the answer has only three decimal places, but five are needed. So write two 0s on the *left* side of the answer.

$$\begin{array}{r} 0.0\,4\,2 \\ \times\ \ 0.0\,3 \\ \hline 0\,0\,1\,2\,6 \end{array}$$

Write two 0s on *left* side of answer.

$$\begin{array}{r} 0.0\,4\,2 \\ \times\ \ 0.0\,3 \\ \hline .0\,0\,1\,2\,6 \end{array}$$ ← 3 decimal places
← 2 decimal places
← 5 decimal places

Now count over 5 places and write in the decimal point.

The final product is 0.00126, which has five decimal places.

Work Problem 2 at the Side.

0.21 EXERCISES

For Extra Help

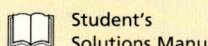 Student's Solutions Manual

 MyMathLab

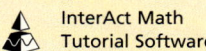 InterAct Math Tutorial Software

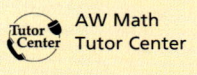 AW Math Tutor Center

MathXL

Multiply. See Example 1.

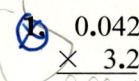
1.   0.042
   × 3.2

2.   0.571
   × 2.9

3.   21.5
   × 7.4

4.   85.4
   × 3.5

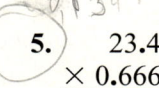
5.   23.4
   × 0.666

6.   0.896
   × 0.799

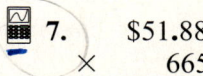
7. $51.88
      ×   665

8. $736.75
      ×   118

0-114　CHAPTER 0　Topics in Arithmetic

Use the fact that 72 × 6 = 432 to help you answer Exercises 9–16 by simply counting decimal places. See Examples 1 and 2.

9. 72 × 0.6 = 4 3 2　　**10.** 7.2 × 6 = 4 3 2　　**11.** (7.2)(0.06) = 4 3 2

12. (0.72)(0.6) = 4 3 2　　**13.** 0.72 × 0.06 = 4 3 2　　**14.** 72 × 0.0006 = 4 3 2

15. 0.0072 × 0.6 = 4 3 2　　**16.** 0.072 × 0.006 = 4 3 2

Multiply. See Example 2.

17. (0.006)(0.0052)　　　　　　　**18.** (0.0052)(0.009)

19. 0.003 · 0.002　　　　　　　　**20.** 0.0079 · 0.006

RELATING CONCEPTS (EXERCISES 21–22)

For Individual or Group Work

*Look for patterns in the multiplications as you **work Exercises 21 and 22 in order.***

21. Do these multiplications:

(5.96)(10)　　(3.2)(10)
(0.476)(10)　　(80.35)(10)
(722.6)(10)　　(0.9)(10)

What pattern do you see? Write a "rule" for multiplying by 10. What do you think the rule is for multiplying by 100? by 1000? Write the rules and try them out on the numbers above.

22. Do these multiplications:

(59.6)(0.1)　　(3.2)(0.1)
(0.476)(0.1)　　(80.35)(0.1)
(65)(0.1)　　(523)(0.1)

What pattern do you see? Write a "rule" for multiplying by 0.1. What do you think the rule is for multiplying by 0.01? by 0.001? Write the rules and try them out on the numbers above.

0.22　Dividing Decimals

OBJECTIVES

1. Divide a decimal by a whole number.
2. Divide a decimal by a decimal.

There are two kinds of decimal division problems; those in which a decimal is divided by a whole number, and those in which a decimal is divided by a decimal. First recall the parts of a division problem from **Section 0.5**.

Divisor → 4)33 ← Dividend
　　　　　　8 ← Quotient
　　　　　　32
　　　　　　 1 ← Remainder

SECTION 0.22 Dividing Decimals 0-115

1. Divide. Check your answers by multiplying.

   (a) $4\overline{)93.6}$

   (b) $6\overline{)6.804}$

   (c) $11\overline{)278.3}$

   (d) $0.51835 \div 5$

   (e) $213.45 \div 15$

OBJECTIVE 1 Divide a decimal by a whole number. When the divisor is a whole number, use these steps.

Dividing Decimals by Whole Numbers

Step 1 Write the decimal point in the quotient (answer) directly above the decimal point in the dividend.

Step 2 Divide as if both numbers were whole numbers.

EXAMPLE 1 Dividing Decimals by Whole Numbers

Divide.

(a) 21.93 by 3

   Dividend — 21.93   Divisor — 3

   Rewrite the division problem. $3\overline{)21.93}$

   Step 1 Write the decimal point in the quotient directly above the decimal point in the dividend.

   Decimal points lined up
   $3\overline{)21.93}$ with decimal point above

   Step 2 Divide as if the numbers were whole numbers.

   $\begin{array}{r} 7.31 \\ 3\overline{)21.93} \end{array}$

   Check by multiplying the quotient times the divisor.

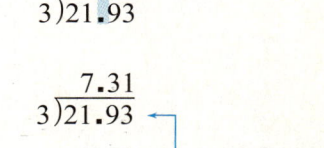

   Matches, so 7.31 is correct.

   The quotient (answer) is 7.31.

(b) $9\overline{)470.7}$
     Divisor — Dividend

   Write the decimal point in the quotient above the decimal point in the dividend. Then divide as if the numbers were whole numbers.

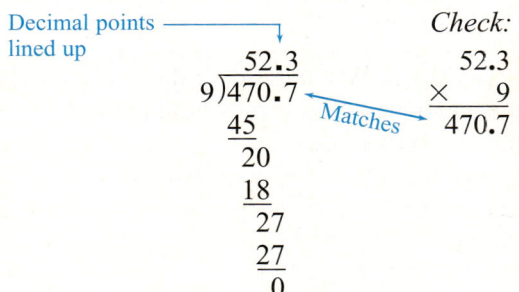

   The quotient is 52.3.

ANSWERS
1. (a) 23.4; (23.4)(4) = 93.6
   (b) 1.134; (1.134)(6) = 6.804
   (c) 25.3; (25.3)(11) = 278.3
   (d) 0.10367; (0.10367)(5) = 0.51835
   (e) 14.23; (14.23)(15) = 213.45

Work Problem 1 at the Side.

CHAPTER 0 Topics in Arithmetic

2. Divide. Check your answers by multiplying.

(a) $5 \overline{)6.4}$

(b) $30.87 \div 14$

(c) $\dfrac{259.5}{30}$

(d) $0.3 \div 8$

EXAMPLE 2 Writing Extra 0s to Complete a Division

Divide 1.5 by 8.

Keep dividing until the remainder is 0, or until the digits in the quotient begin to repeat in a pattern. In Example 1(b), you ended up with a remainder of 0. But sometimes you run out of digits in the dividend before that happens. If so, write extra 0s on the right side of the dividend so you can continue dividing.

$$\begin{array}{r} 0.1 \\ 8\overline{)1.5} \\ \underline{8} \\ 7 \end{array}$$
← All digits have been used.

← Remainder is not yet 0.

Write a 0 after the 5 in the dividend so you can continue dividing. Keep writing more 0s in the dividend if needed. Recall that writing 0s to the *right* of a decimal number does *not* change its value.

$$\begin{array}{r} 0.1875 \\ 8\overline{)1.5000} \\ \underline{8} \\ 70 \\ \underline{64} \\ 60 \\ \underline{56} \\ 40 \\ \underline{40} \\ 0 \end{array}$$

← Three 0s needed to complete the division.

← Stop dividing when the remainder is 0.

Check:

$$\begin{array}{r} 0.1875 \\ \times 8 \\ \hline 1.5000 \end{array}$$

Matches dividend, so 0.1875 is correct.

Calculator Tip When *multiplying* numbers, you can enter them in any order because multiplication is commutative (see **Section 0.4**). But division is *not* commutative. It *does* matter which number you enter first. Try Example 2 both ways; jot down your answers.

$$1.5 \div 8 = \underline{} \qquad 8 \div 1.5 = \underline{}$$

Notice that the first answer, 0.1875, matches the result from Example 2. But the second answer is much different: 5.333333333. Be careful to enter the dividend first.

CAUTION When dividing decimals, notice that the dividend may *not* be the larger number, as it was in whole numbers. In Example 2 the dividend is 1.5, which is *smaller* than 8.

Work Problem 2 at the Side.

The next example shows a quotient (answer) that must be rounded because you will never get a remainder of 0.

ANSWERS

2. (a) 1.28; (1.28)(5) = 6.40 or 6.4
(b) 2.205; (2.205)(14) = 30.870 or 30.87
(c) 8.65; (8.65)(30) = 259.50 or 259.5
(d) 0.0375; (0.0375)(8) = 0.3000 or 0.3

SECTION 0.22 Dividing Decimals **0-117**

3. Divide. Round answers to the nearest thousandth. If it is a repeating decimal, also write the answer using a bar. Check your answers by multiplying.

(a) $13\overline{)267.01}$

(b) $6\overline{)20.5}$

(c) $\dfrac{10.22}{9}$

(d) $16.15 \div 3$

(e) $116.3 \div 7$

EXAMPLE 3 Rounding a Decimal Quotient

Divide 4.7 by 3. Round the quotient to the nearest thousandth.

Write extra 0s in the dividend so you can continue dividing.

```
      1.5 6 6 6
   3)4.7 0 0 0    ← Three 0s added so far
     3
     ―
     1 7
     1 5
       ―
       2 0
       1 8
         ―
         2 0
         1 8
           ―
           2 0
           1 8
             ―
             2    ← Remainder is still not 0.
```

Notice that the digit 6 in the answer is repeating. It will continue to do so. The remainder will *never be 0*. There are two ways to show that the answer is a **repeating decimal** that goes on forever. You can write three dots after the answer, or you can write a bar above the digits that repeat (in this case, the 6).

$1.5666\ldots$ (Three dots) or $1.5\overline{6}$ ← Bar above repeating digit

When repeating decimals occur, round the answer according to the directions in the problem. In this example, to round to thousandths, divide out one *more* place, to ten-thousandths.

$4.7 \div 3 = 1.5666\ldots$ rounds to 1.567

Check the answer by multiplying 1.567 by 3. Because 1.567 is a rounded answer, the check will not give exactly 4.7, but it should be very close.

$(1.567)(3) = 4.701$ ← Does not equal exactly 4.7 because 1.567 was rounded.

CAUTION When checking answers that you've rounded, the check will *not* match the dividend exactly, but it should be very close.

Work Problem 3 at the Side.

OBJECTIVE 2 Divide a decimal by a decimal. To divide by a *decimal* divisor, first change the divisor to a whole number. Then divide as before. To see how this is done, write the problem in fraction form. Here is an example.

$1.2\overline{)6.36}$ can be written $\dfrac{6.36}{1.2}$

ANSWERS

3. (a) 20.539 (rounded); no repeating digits visible on calculator; (20.539)(13) = 267.007
(b) 3.417 (rounded); 3.41$\overline{6}$; (3.417)(6) = 20.502
(c) 1.136 (rounded); 1.13$\overline{5}$; (1.136)(9) = 10.224
(d) 5.383 (rounded); 5.38$\overline{3}$; (5.383)(3) = 16.149
(e) 16.614 (rounded); starts repeating in eighth decimal place as 16.6$\overline{142857}$; (16.614)(7) = 116.298

In **Section 0.15** you learned that multiplying the numerator and denominator by the same number gives an equivalent fraction. We want the divisor (1.2) to be a whole number. Multiplying by 10 will accomplish that.

$$\text{Decimal divisor} \rightarrow \frac{6.36}{1.2} = \frac{6.36 \cdot 10}{1.2 \cdot 10} = \frac{63.6}{12} \leftarrow \text{Whole number divisor}$$

The short way to multiply by 10 is to move the decimal point *one place* to the *right* in both the divisor and the dividend.

$$1.2 \overline{)6.3\,6} \quad \text{is equivalent to} \quad 12 \overline{)63.6}$$

NOTE Moving the decimal points the *same* number of places in *both* the divisor and dividend will *not* change the answer.

Dividing by Decimals

Step 1 Count the number of decimal places in the divisor and move the decimal point that many places to the *right*. (This changes the divisor to a whole number.)

Step 2 Move the decimal point in the dividend the *same* number of places to the *right*. (Write in extra 0s if needed.)

Step 3 Write the decimal point in the quotient directly above the decimal point in the dividend. Then divide as usual.

EXAMPLE 4 Dividing by Decimals

(a) $0.003\overline{)27.69}$

Move the decimal point in the divisor *three* places to the *right* so 0.003 becomes the whole number 3. To move the decimal point in the dividend the same number of places, write in an extra 0.

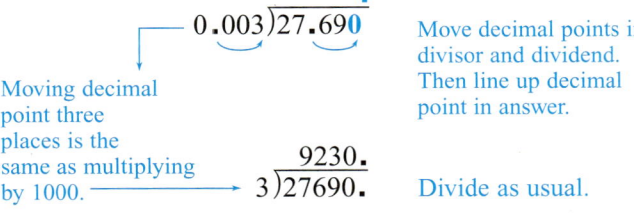

Moving decimal point three places is the same as multiplying by 1000.

Move decimal points in divisor and dividend. Then line up decimal point in answer.

Divide as usual.

SECTION 0.22 Dividing Decimals 0-119

4. Divide. If the quotient does not come out even, round to the nearest hundredth.

   (a) $0.2\overline{)1.04}$

   (b) $0.06\overline{)1.8072}$

   (c) $0.005\overline{)32}$

   (d) $8.1 \div 0.025$

   (e) $\dfrac{7}{1.3}$

   (f) $5.3091 \div 6.2$

ANSWERS
4. (a) 5.2   (b) 30.12   (c) 6400
   (d) 324   (e) 5.38 (rounded)
   (f) 0.86 (rounded)

(b) Divide 5 by 4.2. Round to the nearest hundredth.

Move the decimal point in the divisor one place to the right so 4.2 becomes the whole number 42. The decimal point in the dividend starts on the right side of 5 and is also moved one place to the right.

$$\begin{array}{r} 1.190 \\ 4.2\overline{)5.0000} \\ \underline{4\,2} \\ 80 \\ \underline{42} \\ 380 \\ \underline{378} \\ 20 \end{array}$$

← To round to hundredths, divide out one *more* place, to thousandths.

Round the quotient. It is 1.19 (rounded to the nearest hundredth).

Work Problem 4 at the Side.

0.22 EXERCISES

For Extra Help
- Student's Solutions Manual
- MyMathLab
- InterAct Math Tutorial Software
- AW Math Tutor Center
- MathXL

Divide. See Examples 1 and 4.

1. $7\overline{)27.3}$  2. $8\overline{)50.4}$ 3. $\dfrac{4.23}{9}$    4. $\dfrac{1.62}{6}$

5. $0.05\overline{)20.01}$ 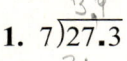 6. $0.08\overline{)16.04}$ 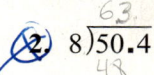 7. $1.5\overline{)54}$ 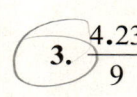 8. $2.4\overline{)132}$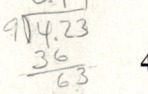

Use the fact that $108 \div 18 = 6$ to work Exercises 9–12 simply by moving decimal points. See Examples 2 and 4.

9. $0.108 \div 1.8$ 10. $10.8 \div 18$ 11. $0.018\overline{)108}$ 12. $0.18\overline{)1.08}$

Divide. Round quotients to the nearest hundredth if necessary. See Examples 3 and 4.

13. $4.6\overline{)116.38}$ 14. $2.6\overline{)4.992}$ 15. $\dfrac{3.1}{0.006}$ 16. $\dfrac{1.7}{0.09}$

 Divide. Round quotients to the nearest thousandth. See Example 4.

17. $240 \div 9.88$ 18. $7643 \div 5.36$ 19. $0.034\overline{)342.81}$ 20. $0.043\overline{)1748.4}$

> **RELATING CONCEPTS** (EXERCISES 21–22)
>
> **For Individual or Group Work**
>
> *Look back at your work in Exercises 21 and 22 in Section 0.21. Then* **work Exercises 21 and 22 in order.**
>
> **21.** Do these division problems:
>
> 3.77 ÷ 10   9.1 ÷ 10
> 0.886 ÷ 10   30.19 ÷ 10
> 406.5 ÷ 10   6625.7 ÷ 10
>
> What pattern do you see? Write a "rule" for dividing by 10. What do you think the rule is for dividing by 100? by 1000? Write the rules and try them out on the numbers above.
>
> **22.** Do these division problems:
>
> 40.2 ÷ 0.1   7.1 ÷ 0.1
> 0.339 ÷ 0.1   15.77 ÷ 0.1
> 46 ÷ 0.1   873 ÷ 0.1
>
> What pattern do you see? Write a "rule" for dividing by 0.1. What do you think the rule is for dividing by 0.01? by 0.001? Write the rules and try them out on the numbers above.

0.23 Writing Fractions as Decimals

OBJECTIVES

1 Write a fraction as a decimal.

Writing fractions as equivalent decimals can help you do calculations more easily or compare the size of two numbers.

OBJECTIVE 1 Write a fraction as a decimal. Recall that a fraction is one way to show division (see **Section 0.5**). For example, $\frac{3}{4}$ means 3 ÷ 4. If you are doing the division by hand, write it as $4\overline{)3}$. When you do the division, you will get the decimal equivalent of $\frac{3}{4}$, which is 0.75.

Writing Fractions as Decimals
Step 1 Divide the numerator of the fraction by the denominator.
Step 2 If necessary, round the answer to the place indicated.

Work Problem 1 at the Side.

1. Rewrite each fraction so you could do the division by hand. Do *not* complete the division.

   (a) $\dfrac{1}{9}$ is written $9\overline{)}$

   (b) $\dfrac{2}{3}$ is written $\overline{)}$

   (c) $\dfrac{5}{4}$ is written $\overline{)}$

   (d) $\dfrac{3}{10}$ is written $\overline{)}$

   (e) $\dfrac{21}{16}$ is written $\overline{)}$

   (f) $\dfrac{1}{50}$ is written $\overline{)}$

EXAMPLE 1 Writing Fractions or Mixed Numbers as Decimals

(a) Write $\dfrac{1}{8}$ as a decimal.

$\dfrac{1}{8}$ means $1 \div 8$. Write it as $8\overline{)1}$. The decimal point in the dividend is on the right side of the 1. Write extra 0s in the dividend so you can continue dividing until the remainder is 0.

$$\dfrac{1}{8} = 1 \div 8 = 8\overline{)1} \quad\Rightarrow\quad \begin{array}{r} 0.125 \\ 8\overline{)1.000} \\ \underline{8} \\ 20 \\ \underline{16} \\ 40 \\ \underline{40} \\ 0 \end{array}$$

— Decimal points lined up
← Three extra 0s needed
← Remainder is 0.

Therefore, $\dfrac{1}{8} = 0.125$. To check this, write 0.125 as a fraction, then change it to lowest terms.

$$0.125 = \dfrac{125}{1000} \qquad \text{In lowest terms} \qquad \dfrac{125 \div 125}{1000 \div 125} = \dfrac{1}{8} \quad\leftarrow\text{Original fraction}$$

Calculator Tip When changing fractions to decimals on your calculator, enter the numbers from the top down. Remember that the order in which you enter the numbers *does* matter in division. Example 1(a) works like this:

$\dfrac{1}{8}$ ↓ Top down   Enter 1 ÷ 8 =   Answer is 0.125.

What happens if you enter 8 ÷ 1 = ? Do you see why that cannot possibly be correct? (Answer: $8 \div 1 = 8$, and a proper fraction like $\dfrac{1}{8}$ *cannot* be equivalent to a whole number.)

(b) Write $2\dfrac{3}{4}$ as a decimal.

One method is to divide 3 by 4 to get 0.75 for the fraction part. Then add the whole number part to 0.75.

$$\dfrac{3}{4} \quad\Rightarrow\quad \begin{array}{r} 0.75 \\ 4\overline{)3.00} \\ \underline{2\,8} \\ 20 \\ \underline{20} \\ 0 \end{array} \qquad \begin{array}{r} 2.00 \leftarrow \text{Whole number part} \\ +\,0.75 \\ \hline 2.75 \end{array}$$

— Fraction part

So, $2\dfrac{3}{4} = 2.75$.   Check: $2.75 = 2\dfrac{75}{100} = 2\dfrac{3}{4}$ ← Lowest terms

Whole number parts match.

ANSWERS

1. (a) $9\overline{)1}$ (b) $3\overline{)2}$ (c) $4\overline{)5}$ (d) $10\overline{)3}$ (e) $16\overline{)21}$ (f) $50\overline{)1}$

2. Write each fraction or mixed number as a decimal.

(a) $\dfrac{1}{4}$

(b) $2\dfrac{1}{2}$

(c) $\dfrac{5}{8}$

(d) $4\dfrac{3}{5}$

(e) $\dfrac{7}{8}$

A second method is to first write $2\dfrac{3}{4}$ as an improper fraction and then divide numerator by denominator.

$$2\dfrac{3}{4} = \dfrac{11}{4}$$

$$\dfrac{11}{4} = 11 \div 4 = 4\overline{)11} \quad \Longrightarrow \quad \begin{array}{r} 2.75 \\ 4\overline{)11.00} \\ \underline{8} \\ 30 \\ \underline{28} \\ 20 \\ \underline{20} \\ 0 \end{array}$$ ← Two extra 0s needed

Whole number parts match.

So, $2\dfrac{3}{4} = 2.75$

$\dfrac{3}{4}$ is equivalent to $\dfrac{75}{100}$ or 0.75.

Work Problem 2 at the Side.

■ EXAMPLE 2 Writing a Fraction as a Decimal with Rounding

Write $\dfrac{2}{3}$ as a decimal and round to the nearest thousandth.

$\dfrac{2}{3}$ means $2 \div 3$. To round to thousandths, divide out one *more* place, to ten-thousandths.

$$\dfrac{2}{3} = 2 \div 3 = 3\overline{)2}. \quad \Longrightarrow \quad \begin{array}{r} 0.6666 \\ 3\overline{)2.0000} \\ \underline{18} \\ 20 \\ \underline{18} \\ 20 \\ \underline{18} \\ 20 \\ \underline{18} \\ 2 \end{array}$$ ← Four 0s needed for ten-thousandths

Written as a repeating decimal, $\dfrac{2}{3} = 0.\overline{6}$. ← Bar above repeating digit.

Rounded to the nearest thousandth, $\dfrac{2}{3} \approx 0.667$.

▦ Calculator Tip Try Example 2 on your calculator. Enter 2 ÷ 3. Which answer do you get?

| 0.666666667 | or | 0.6666666 |

Many scientific calculators will show a 7 as the last digit. Because the 6s keep on repeating forever, the calculator automatically rounds in the last decimal place it

---

ANSWERS
2. (a) 0.25 (b) 2.5 (c) 0.625
   (d) 4.6 (e) 0.875

3. Write as decimals. Round to the nearest thousandth.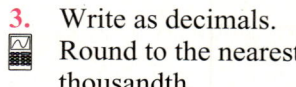

(a) $\dfrac{1}{3}$ (b) $2\dfrac{7}{9}$

(c) $\dfrac{10}{11}$ (d) $\dfrac{3}{7}$

(e) $3\dfrac{5}{6}$

ANSWERS

3. All answers are rounded.
(a) 0.333 (b) 2.778 (c) 0.909
(d) 0.429 (e) 3.833

has room to show. If you have a 10-digit display space, the calculator is rounding like this:

0.6666666666 (11 digits)   rounds to   0.666666667.

Other calculators, especially standard, four-function ones, may *not* round. They just cut off, or *truncate,* the extra digits. Such a calculator would show 0.6666666 in the display.

Would this difference in calculators show up when changing $\tfrac{1}{3}$ to a decimal? Why not? (Answer: The repeating digit is 3, which is 4 or less, so it stays as 3 whether it's rounded or not.)

Work Problem 3 at the Side.

0.23 EXERCISES

For Extra Help

- Student's Solutions Manual
- MyMathLab
- InterAct Math Tutorial Software
- AW Math Tutor Center
- MathXL

Write each fraction or mixed number as a decimal. Round to the nearest thousandth if necessary. See Examples 1 and 2.

1. $\dfrac{1}{2}$ 2. $\dfrac{1}{4} = .25$ 3. $\dfrac{3}{4} = .75$ 4. $\dfrac{1}{10}$ 5. $\dfrac{3}{10}$

6. $\dfrac{7}{10} = .7$ 7. $\dfrac{9}{10}$ 8. $\dfrac{4}{5}$ 9. $\dfrac{3}{5}$ 10. $\dfrac{2}{5}$

11. $\dfrac{7}{8}$ 12. $\dfrac{3}{8}$ 13. $2\dfrac{1}{4}$ 14. $1\dfrac{1}{2}$ 15. $14\dfrac{7}{10}$

16. $23\dfrac{3}{5}$ 17. $3\dfrac{5}{8}$ 18. $2\dfrac{7}{8}$ 19. $\dfrac{1}{3}$ 20. $\dfrac{2}{3}$

21. $\dfrac{5}{6}$ 22. $\dfrac{1}{6}$ 23. $1\dfrac{8}{9}$ 24. $5\dfrac{4}{7}$

0-124 CHAPTER 0 Topics in Arithmetic

RELATING CONCEPTS (EXERCISES 25–28)

For Individual or Group Work

Use your knowledge of fractions and decimals to work Exercises 25–28 in order.

25. (a) Explain how you can tell that Keith made an error *just by looking at his final answer*. Here is his work.

$$\frac{5}{9} = 5\overline{)9.0}^{\;1.8} \quad \text{so} \quad \frac{5}{9} = 1.8$$

(b) Show the correct way to change $\frac{5}{9}$ to a decimal. Explain why your answer makes sense.

26. (a) How can you prove to Sandra that $2\frac{7}{20}$ is *not* equivalent to 2.035? Here is her work.

$$2\frac{7}{20} = 20\overline{)7.00}^{\;0.35} \quad \text{so} \quad 2\frac{7}{20} = 2.035$$

(b) What is the correct answer? Show how to prove that it is correct.

27. Ving knows that $\frac{3}{8} = 0.375$. How can he write $1\frac{3}{8}$ as a decimal *without* having to do a division? How can he write $3\frac{3}{8}$ as a decimal? $295\frac{3}{8}$? Explain your answer.

28. Iris has found a shortcut for writing mixed numbers as decimals:

$$2\frac{7}{10} = 2.7 \qquad 1\frac{13}{100} = 1.13$$

Does her shortcut work for all mixed numbers? Explain when it works and why it works.

Find each decimal or fraction equivalent. Write fractions in lowest terms.

	Fraction	Decimal		Fraction	Decimal
29.	_____	0.4	30.	_____	0.75
31.	_____	0.625	32.	_____	0.111
33.	_____	0.35	34.	_____	0.9
35.	$\frac{7}{20}$	_____	36.	$\frac{1}{40}$	_____
37.	_____	0.04	38.	_____	0.52

	Fraction	Decimal		Fraction	Decimal
39.	_____	0.15	40.	_____	0.85
41.	$\frac{1}{5}$	_____	42.	$\frac{1}{8}$	_____
43.	_____	0.09	44.	_____	0.02

0.18–0.23 TEST

Write each decimal as a fraction or mixed number in lowest terms.

1. 18.4
2. 0.075

Write each decimal in words.

3. 60.007
4. 0.0208

Round each decimal to the place indicated.

5. 725.6089 to the nearest tenth
6. 0.62951 to the nearest thousandth
7. $1.4945 to the nearest cent
8. $7859.51 to the nearest dollar

First use front end rounding to round each number and estimate the answer. Then find the exact answer.

9. 7.6 + 82.0128 + 39.59
10. 79.1 − 3.602
11. 5.79 • 1.2
12. 20.04 ÷ 4.8

Add, subtract, multiply, or divide as indicated.

13. 53.1 + 4.631 + 782 + 0.031
14. 670 − 0.996
15. (0.0069)(0.007)
16. $0.15 \overline{)72}$

17. Write $2\frac{5}{8}$ as a decimal. Round to the nearest thousandth, if necessary.

0.24 Ratios

OBJECTIVES

1. Write ratios as fractions.
2. Solve ratio problems involving decimals or mixed numbers.
3. Solve ratio problems after converting units.
4. Write rates as fractions.
5. Find unit rates.
6. Find the best buy based on cost per unit.

A **ratio** compares two quantities. You can compare two numbers, such as 8 and 4, or two measurements that have the same type of units, such as 3 days and 12 days.

Ratios can help you see important relationships. For example, if the ratio of your monthly expenses to your monthly income is 10 to 9, then you are spending $10 for every $9 you earn, going deeper into debt.

OBJECTIVE 1 Write ratios as fractions. A ratio can be written in three ways.

Writing a Ratio

The ratio of $7 **to** $3 can be written:

$$7 \text{ to } 3 \quad \text{or} \quad 7{:}3 \quad \text{or} \quad \frac{7}{3} \leftarrow \text{Fraction bar indicates "to."}$$

":" indicates "to."

Writing a ratio as a fraction is the most common method, and the one we will use here. All three ways are read, "the ratio of 7 **to** 3." The word **to** separates the quantities being compared.

Writing a Ratio as a Fraction

Order is important when writing a ratio. The quantity mentioned **first** is the **numerator**. The quantity mentioned **second** is the **denominator**. For example:

$$\text{The ratio of } 5 \text{ to } 12 \text{ is written } \frac{5}{12}.$$

EXAMPLE 1 Writing Ratios

The Anasazi, ancestors of the Pueblo Indians, built multistory apartment towns in New Mexico about 1100 years ago. A room might measure 14 ft long, 11 ft wide, and 15 ft high.

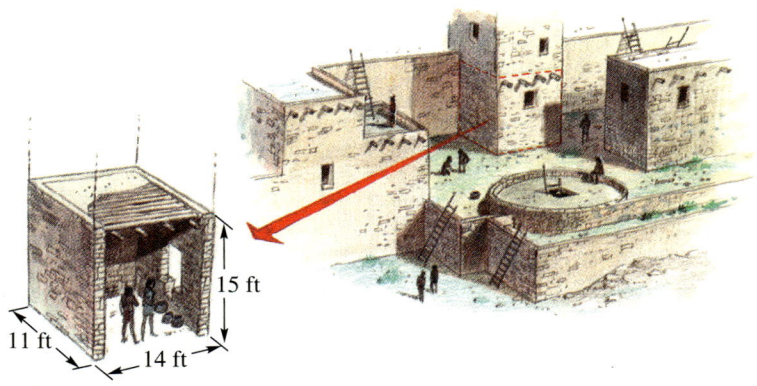

SECTION 0.24　Ratios　0-127

1. Shane spent $14 on meat, $5 on milk, and $7 on fresh fruit. Write each ratio as a fraction.

(a) The ratio of amount spent on fruit to amount spent on milk.

Write each ratio as a fraction, using these room measurements.

(a) Ratio of length to width

The ratio of **length** to **width** is $\dfrac{14 \text{ ft}}{11 \text{ ft}} = \dfrac{14}{11}$.

Numerator (mentioned first) — Denominator (mentioned second)

You can divide out common *units* just like you divided out common *factors* when writing fractions in lowest terms. (See **Section 0.10**.)

(b) Ratio of width to height

The ratio of width **to** height is $\dfrac{11 \text{ ft}}{15 \text{ ft}} = \dfrac{11}{15}$.

(b) The ratio of amount spent on milk to amount spent on meat.

CAUTION　Remember, the *order* of the numbers is important in a ratio. Look for the words "ratio of **a** to **b**." Write the ratio as $\dfrac{a}{b}$, **not** $\dfrac{b}{a}$. The quantity mentioned first is the numerator.

Work Problem 1 at the Side.

Any ratio can be written as a fraction. Therefore, you can write a ratio in *lowest terms*, just as you do with any fraction.

EXAMPLE 2 Writing Ratios in Lowest Terms

Write each ratio in lowest terms.

(a) 60 days to 20 days

The ratio is $\dfrac{60}{20}$. Write this ratio in lowest terms by dividing the numerator and denominator by 20.

(c) The ratio of amount spent on meat to amount spent on milk.

$$\dfrac{60}{20} = \dfrac{60 \div 20}{20 \div 20} = \dfrac{3}{1} \quad \leftarrow \left\{ \begin{array}{l} \text{Ratio in} \\ \text{lowest terms} \end{array} \right.$$

CAUTION　In the fractions sections you would have rewritten $\dfrac{3}{1}$ as 3. But a *ratio* compares *two* quantities, so you need to keep both parts of the ratio and write it as $\dfrac{3}{1}$.

(b) 50 ounces of medicine to 120 ounces of medicine

The ratio is $\dfrac{50}{120}$. Divide the numerator and denominator by 10.

ANSWERS

1. (a) $\dfrac{7}{5}$　(b) $\dfrac{5}{14}$　(c) $\dfrac{14}{5}$

$$\dfrac{50}{120} = \dfrac{50 \div 10}{120 \div 10} = \dfrac{5}{12} \quad \leftarrow \left\{ \begin{array}{l} \text{Ratio in} \\ \text{lowest terms} \end{array} \right.$$

2. Write each ratio as a fraction in lowest terms.

(a) 9 hours to 12 hours

(b) 100 meters to 50 meters

(c) Write the ratio of width to length for this rectangle.

Length 48 ft
Width 24 ft

3. Write each ratio as a ratio of whole numbers in lowest terms.

(a) The price of Tamar's favorite brand of lipstick increased from $5.50 to $7.00. Find the ratio of the increase in price to the original price.

(b) Last week, Lance worked 4.5 hours each day. This week he cut back to 3 hours each day. Find the ratio of the decrease in hours to the original number of hours.

ANSWERS

2. (a) $\frac{3}{4}$ (b) $\frac{2}{1}$ (c) $\frac{1}{2}$

3. (a) $\frac{1.50 \times 100}{5.50 \times 100} = \frac{150 \div 50}{550 \div 50} = \frac{3}{11}$

(b) $\frac{1.5 \times 10}{4.5 \times 10} = \frac{15 \div 15}{45 \div 15} = \frac{1}{3}$

(c) 15 people in a large van to 6 people in a small van

The ratio is $\frac{15}{6} = \frac{15 \div 3}{6 \div 3} = \frac{5}{2}$ ← Ratio in lowest terms

NOTE Although $\frac{5}{2} = 2\frac{1}{2}$, ratios are *not* written as mixed numbers. Nevertheless, in Example 2(c), the ratio $\frac{5}{2}$ does mean the large van holds $2\frac{1}{2}$ times as many people as the small van.

Work Problem 2 at the Side.

OBJECTIVE 2 Solve ratio problems involving decimals or mixed numbers. Sometimes a ratio compares two decimal numbers or two fractions. It is easier to understand if we rewrite the ratio as a ratio of two whole numbers.

EXAMPLE 3 Using Decimal Numbers in a Ratio

The price of a Sunday newspaper increased from $1.50 to $1.75. Find the ratio of the increase in price to the original price.

The words increase in price are mentioned first, so the increase will be the numerator. How much did the price go up? Use subtraction.

new price − original price = increase
$1.75 − $1.50 = $0.25

The words the original price are mentioned second, so the original price of $1.50 is the denominator.

The ratio of increase in price to original price is

$\frac{0.25}{1.50}$ ← increase
← original price

Now rewrite the ratio as a ratio of whole numbers. Recall that if you multiply both the numerator and denominator of a fraction by the same number, you get an equivalent fraction. The decimals in this example are hundredths, so multiply by 100 to get whole numbers. (If the decimals are tenths, multiply by 10. If thousandths, multiply by 1000.) Then write the ratio in lowest terms.

$\frac{0.25}{1.50} = \frac{0.25 \cdot 100}{1.50 \cdot 100} = \frac{25}{150} = \frac{25 \div 25}{150 \div 25} = \frac{1}{6}$ ← Ratio in lowest terms

Ratio as two whole numbers

Work Problem 3 at the Side.

EXAMPLE 4 Using Mixed Numbers in Ratios

Write each ratio as a comparison of whole numbers in lowest terms.

(a) 2 days to $2\frac{1}{4}$ days

SECTION 0.24 Ratios   0-129

4. Write each ratio as a ratio of whole numbers in lowest terms.

   (a) $3\frac{1}{2}$ to 4

   (b) $5\frac{5}{8}$ pounds to $3\frac{3}{4}$ pounds

   (c) $3\frac{1}{2}$ in. to $\frac{7}{8}$ in.

Write the ratio as follows. Divide out the common units.

$$\frac{2 \text{ days}}{2\frac{1}{4} \text{ days}} = \frac{2}{2\frac{1}{4}}$$

Next, write 2 as $\frac{2}{1}$ and $2\frac{1}{4}$ as the improper fraction $\frac{9}{4}$.

$$\frac{2}{2\frac{1}{4}} = \frac{\frac{2}{1}}{\frac{9}{4}}$$

Now rewrite the problem in horizontal format, using the "÷" symbol for division. Finally, multiply by the reciprocal of the divisor, as you did in **Section 0.12**.

$$\frac{\frac{2}{1}}{\frac{9}{4}} = \frac{2}{1} \div \frac{9}{4} = \frac{2}{1} \cdot \frac{4}{9} = \frac{8}{9}$$

The ratio, in lowest terms, is $\frac{8}{9}$.

(b) $3\frac{1}{4}$ to $1\frac{1}{2}$

Write the ratio as $\frac{3\frac{1}{4}}{1\frac{1}{2}}$. Then write $3\frac{1}{4}$ and $1\frac{1}{2}$ as improper fractions.

$$3\frac{1}{4} = \frac{13}{4} \quad \text{and} \quad 1\frac{1}{2} = \frac{3}{2}$$

The ratio is shown below.

$$\frac{3\frac{1}{4}}{1\frac{1}{2}} = \frac{\frac{13}{4}}{\frac{3}{2}}$$

Rewrite as a division problem in horizontal format, using the "÷" symbol. Then multiply by the reciprocal of the divisor.

$$\frac{13}{4} \div \frac{3}{2} = \frac{13}{4} \cdot \frac{2}{3} = \frac{13}{6} \leftarrow \left\{ \begin{array}{l} \text{Ratio in} \\ \text{lowest terms} \end{array} \right.$$

Work Problem 4 at the Side.

OBJECTIVE 3 Solve ratio problems after converting units. When a ratio compares measurements, both measurements must be in the *same* units. For example, *feet* must be compared to *feet*, *hours* to *hours*, *pints* to *pints*, and *inches* to *inches*.

ANSWERS

4. (a) $\frac{7}{8}$ (b) $\frac{3}{2}$ (c) $\frac{4}{1}$

EXAMPLE 5 Ratio Applications Using Measurement

(a) Write the ratio of the length of the board on the left to the length of the board on the right. Compare in inches.

First, express 2 ft in inches. Because 1 ft has 12 in., 2 ft is

$$2 \cdot 12 \text{ in.} = 24 \text{ in.}$$

The length of the board on the left is 24 in., so the ratio of the lengths is

$$\frac{24 \text{ in.}}{30 \text{ in.}} = \frac{24}{30}$$

Write the ratio in lowest terms

$$\frac{24}{30} = \frac{24 \div 6}{30 \div 6} = \frac{4}{5} \quad \leftarrow \left\{ \begin{array}{l} \text{Ratio in} \\ \text{lowest terms} \end{array} \right.$$

The shorter board on the left is $\frac{4}{5}$ the length of the longer board on the right.

NOTE Notice that we wrote the ratio using the smaller unit (inches are smaller than feet). Using the smaller unit will help you avoid working with fractions. If we wrote the ratio using feet, then

$$30 \text{ in.} = 2\frac{1}{2} \text{ ft.}$$

So the ratio in feet is shown below.

$$\frac{2 \text{ ft}}{2\frac{1}{2} \text{ ft}} = \frac{2}{1} \div \frac{5}{2} = \frac{2}{1} \cdot \frac{2}{5} = \frac{4}{5} \quad \leftarrow \text{Same result.}$$

The ratio is the same, but it takes more steps to get the answer. Using the smaller unit is usually easier.

(b) Write the ratio of 28 days to 3 weeks.

Since it is easier to write the ratio using the smaller measurement unit, compare in *days* because days are shorter than weeks.

First express 3 weeks in days. Because 1 week has 7 days, 3 weeks is

$$3 \cdot 7 \text{ days} = 21 \text{ days.}$$

So the ratio in days is shown below.

$$\frac{28 \text{ days}}{21 \text{ days}} = \frac{28}{21} = \frac{28 \div 7}{21 \div 7} = \frac{4}{3} \quad \leftarrow \text{Lowest terms}$$

5. Write each ratio as a fraction in lowest terms. (*Hint:* Recall that it is usually easier to write the ratio using the smaller measurement unit.)

(a) 9 in. to 6 ft

(b) 2 days to 8 hours

(c) 7 yd to 14 ft

(d) 3 quarts to 3 gallons

(e) 25 minutes to 2 hours

(f) 4 pounds to 12 ounces

The following table will help you set up ratios that compare measurements.

Measurement Comparisons

Length	Capacity (Volume)
1 foot = 12 inches	1 pint = 2 cups
1 yard = 3 feet	1 quart = 2 pints
1 mile = 5280 feet	1 gallon = 4 quarts

Weight	Time
1 pound = 16 ounces	1 minute = 60 seconds
1 ton = 2000 pounds	1 hour = 60 minutes
	1 day = 24 hours
	1 week = 7 days

Work Problem 5 at the Side.

OBJECTIVE 4 Write rates as fractions. A *ratio* compares two measurements with the same type of units, such as 9 feet to 12 feet (both length measurements). But many of the comparisons we make use measurements with different types of units, such as the following.

$$160 \text{ dollars for } 8 \text{ hours (money to time)}$$
$$450 \text{ miles on } 15 \text{ gallons (distance to capacity)}$$

This type of comparison is called a **rate**.

Suppose you hiked 18 miles in 4 hours. The *rate* at which you hiked can be written as a fraction in lowest terms.

$$\frac{18 \text{ miles}}{4 \text{ hours}} = \frac{18 \text{ miles} \div 2}{4 \text{ hours} \div 2} = \frac{9 \text{ miles}}{2 \text{ hours}} \leftarrow \text{Lowest terms}$$

In a rate, you often find these words separating the quantities you are comparing.

$$\text{in} \quad \text{for} \quad \text{on} \quad \text{per} \quad \text{from}$$

CAUTION When writing a rate, always include the units, such as miles, hours, dollars, and so on. Because the units in a rate are different, the units do *not* divide out.

EXAMPLE 6 Writing Rates in Lowest Terms

Write each rate as a fraction in lowest terms.

(a) 5 gallons of chemical for $60.

$$\frac{5 \text{ gallons} \div 5}{60 \text{ dollars} \div 5} = \frac{1 \text{ gallon}}{12 \text{ dollars}} \quad \text{Write the units: gallons and dollars}$$

ANSWERS

5. (a) $\frac{1}{8}$ (b) $\frac{6}{1}$ (c) $\frac{3}{2}$ (d) $\frac{1}{4}$
    (e) $\frac{5}{24}$ (f) $\frac{16}{3}$

6. Write each rate as a fraction in lowest terms.

(a) $6 for 30 packages

(b) 500 miles in 10 hours

(c) 4 teachers for 90 students

(d) 1270 bushels on 30 acres

7. Find each unit rate.

(a) $4.35 for 3 pounds of cheese

(b) 304 miles on 9.5 gallons of gas

(c) $850 in 5 days

(d) 24-pound turkey for 15 people

(b) $1500 wages **in** 10 weeks

$$\frac{1500 \text{ dollars} \div 10}{10 \text{ weeks} \div 10} = \frac{150 \text{ dollars}}{1 \text{ week}}$$

(c) 2225 miles **on** 75 gallons of gas

$$\frac{2225 \text{ miles} \div 25}{75 \text{ gallons} \div 25} = \frac{89 \text{ miles}}{3 \text{ gallons}}$$

Work Problem 6 at the Side.

OBJECTIVE 5 Find unit rates. When the *denominator* of a rate is 1, it is called a **unit rate.** We use unit rates frequently. For example, you earn $12.75 for *1 hour* of work. This unit rate is written:

$$\$12.75 \text{ per hour} \quad \text{or} \quad \$12.75/\text{hour.}$$

Or, you drive 28 miles on *1 gallon* of gas. This unit rate is written

$$28 \text{ miles per gallon} \quad \text{or} \quad 28 \text{ miles/gallon.}$$

Use **per** or a slash mark (/) when writing unit rates.

EXAMPLE 7 Finding Unit Rates

Find each unit rate.

(a) 337.5 miles on 13.5 gallons of gas
Write the rate as a fraction.

$$\frac{337.5 \text{ miles}}{13.5 \text{ gallons}} \quad \leftarrow \text{The fraction bar indicates division.}$$

Divide 337.5 by 13.5 to find the unit rate.

$$13.5\overline{)337.5} \quad \text{quotient } 25.$$

$$\frac{337.5 \text{ miles} \div 13.5}{13.5 \text{ gallons} \div 13.5} = \frac{25 \text{ miles}}{1 \text{ gallon}}$$

The unit rate is 25 miles **per** gallon, or 25 miles/gallon.

(b) 549 miles in 18 hours

$$\frac{549 \text{ miles}}{18 \text{ hours}} \qquad \text{Divide: } 18\overline{)549.0} \quad 30.5$$

The unit rate is 30.5 miles/hour.

(c) $810 in 6 days

$$\frac{810 \text{ dollars}}{6 \text{ days}} \qquad \text{Divide: } 6\overline{)810} \quad 135$$

The unit rate is $135/day.

Work Problem 7 at the Side.

ANSWERS

6. (a) $\dfrac{\$1}{5 \text{ packages}}$ (b) $\dfrac{50 \text{ miles}}{1 \text{ hour}}$
   (c) $\dfrac{2 \text{ teachers}}{45 \text{ students}}$ (d) $\dfrac{127 \text{ bushels}}{3 \text{ acres}}$

7. (a) $1.45/pound
   (b) 32 miles/gallon   (c) $170/day
   (d) 1.6 pounds/person

SECTION 0.24 Ratios 0-133

8. Find the best buy (lowest cost per unit) for each purchase.

(a) 2 quarts for $3.25
3 quarts for $4.95
4 quarts for $6.48

OBJECTIVE 6 Find the best buy based on cost per unit. When shopping for groceries, household supplies, and health and beauty items, you will find many different brands and package sizes. You can save money by finding the lowest *cost per unit*.

Cost per Unit

Cost per unit is a rate that tells how much you pay for *one* item or *one* unit. Examples are $1.65 per gallon, $47 per shirt, and $2.98 per pound.

■ **EXAMPLE 8** Determining the Best Buy

The local store charges the following prices for pancake syrup. Find the best buy.

(b) 6 cans of cola for $1.99
12 cans of cola for $3.49
24 cans of cola for $7

The best buy is the container with the *lowest* cost per unit. All the containers are measured in *ounces* (oz), so you first need to find the *cost per ounce* for each one. Divide the price of the container by the number of ounces in it. Round to the nearest thousandth, if necessary.

Size	Cost per Unit (Rounded)
12 ounces	$\frac{\$1.28}{12 \text{ ounces}} \approx \$0.107$ per ounce (highest)
24 ounces	$\frac{\$1.81}{24 \text{ ounces}} \approx \$0.075$ per ounce (lowest)
36 ounces	$\frac{\$2.73}{36 \text{ ounces}} \approx \$0.076$ per ounce

The lowest cost per ounce is $0.075, so the 24-ounce container is the best buy. ■

NOTE Earlier we rounded money amounts to the nearest hundredth (nearest cent). But when comparing unit costs, rounding to the nearest thousandth will help you see the difference between very similar unit costs. Notice that the 24-ounce and 36-ounce syrup containers would both have rounded to $0.08 per ounce if we had rounded to hundredths.

ANSWERS
3. (a) 4 quarts, at $1.62 per quart
   (b) 12 cans, at $0.291 per can (rounded)

Work Problem 8 at the Side.

0.24 EXERCISES

For Extra Help

 Student's Solutions Manual

 MyMathLab

 InterAct Math Tutorial Software

 AW Math Tutor Center

MathXL

Write each ratio as a fraction in lowest terms. See Examples 1 and 2.

1. 8 to 9
2. 11 to 15
3. $100 to $50
4. 35¢ to 7¢

5. 30 minutes to 90 minutes
6. 9 pounds to 36 pounds

7. 80 miles to 50 miles
8. 300 people to 450 people

9. 6 hours to 16 hours
10. 45 books to 35 books

Write each ratio as a ratio of whole numbers in lowest terms. See Examples 3 and 4.

11. $4.50 to $3.50
12. $0.08 to $0.06
13. 15 to $2\frac{1}{2}$

14. 5 to $1\frac{1}{4}$
15. $1\frac{1}{4}$ to $1\frac{1}{2}$
16. $2\frac{1}{3}$ to $2\frac{2}{3}$

Write each ratio as a fraction in lowest terms. For help, use the table of measurement relationships. See Example 5.

17. 4 ft to 30 in.
18. 8 ft to 4 yd

19. 5 minutes to 1 hour
20. 8 quarts to 5 pints

21. 15 hours to 2 days
22. 3 pounds to 6 ounces

23. 5 gallons to 5 quarts
24. 3 cups to 3 pints

SECTION 0.24 Ratios 0-135

The table shows the number of greeting cards that Americans buy for various occasions. Use the information to answer Exercises 25–30. Write each ratio as a fraction in lowest terms.

Holiday/Event	Cards
Valentine's Day	900 million
Mother's Day	150 million
Father's Day	95 million
Graduation	60 million
Thanksgiving	30 million
Halloween	25 million

Source: Hallmark Cards.

25. Find the ratio of Thanksgiving cards to graduation cards.

26. Find the ratio of Halloween cards to Mother's Day cards.

27. Find the ratio of Valentine's Day cards to Halloween cards.

28. Find the ratio of Mother's Day cards to Father's Day cards.

29. Explain how you might use the information in the table if you owned a shop selling gifts and greeting cards.

30. Why is the ratio of Valentine's Day cards to graduation cards $\frac{15}{1}$? Give two possible reasons.

The bar graph shows the number of Americans who play various instruments. Use the graph to complete Exercises 31–34. Write each ratio as a fraction in lowest terms.

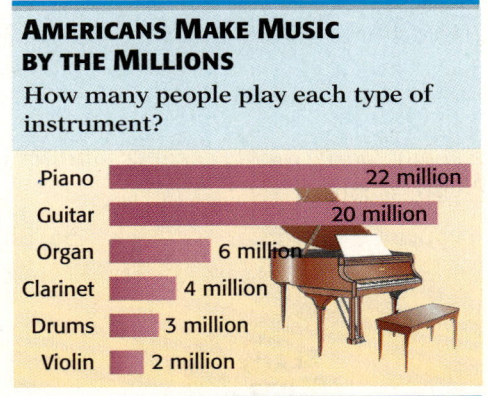

AMERICANS MAKE MUSIC BY THE MILLIONS
How many people play each type of instrument?

- Piano — 22 million
- Guitar — 20 million
- Organ — 6 million
- Clarinet — 4 million
- Drums — 3 million
- Violin — 2 million

Source: America by the Numbers.

31. Write six ratios that compare the least popular instrument to each of the other instruments.

32. Which two instruments give each of these ratios:
(a) $\frac{5}{1}$; (b) $\frac{2}{1}$. There may be more than one correct answer.

33. Why might the ratio of guitar players to drum players be $\frac{20}{3}$? Give two possible explanations.

34. If you ran a music school, explain how you might use the information in the graph to decide how many teachers to hire.

Write each rate as a fraction in lowest terms. See Example 6.

35. 10 cups for 6 people

36. $12 for 30 pens

37. 15 feet in 35 seconds

38. 100 miles in 30 hours

39. 14 people for 28 dresses

40. 12 wagons for 48 horses

41. 25 letters in 5 minutes

42. 68 pills for 17 people

43. $63 for 6 visits

44. 25 doctors for 310 patients

45. 72 miles on 4 gallons

46. 132 miles on 8 gallons

Find each unit rate. See Example 7.

47. $60 in 5 hours

48. $2500 in 20 days

49. 50 eggs from 10 chickens

50. 36 children from 12 families

51. 7.5 pounds for 6 people

52. 44 bushels from 8 trees

53. $413.20 for 4 days

54. $74.25 for 9 hours

Find the best buy (based on the cost per unit) for each item. See Example 8. (Source: Piggly Wiggly.)

55. Black pepper

56. Shampoo

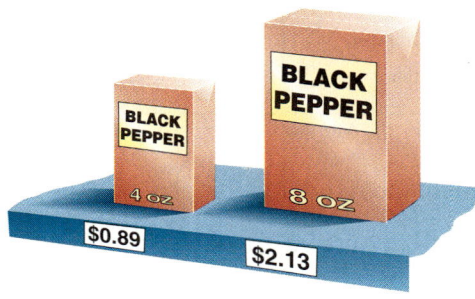

57. Cereal
13 ounces for $2.80
15 ounces for $3.15
18 ounces for $3.98

58. Soup
2 cans for $0.95
3 cans for $1.45
5 cans for $2.29

59. Chunky peanut butter
12 ounces for $1.29
18 ounces for $1.79
28 ounces for $3.39
40 ounces for $4.39

60. Pork and beans
8 ounces for $0.37
16 ounces for $0.77
21 ounces for $0.99
31 ounces for $1.50

0.25 Proportions

OBJECTIVES

1. Write proportions.
2. Determine whether proportions are true or false.
3. Find diagonal products.

1. Write each proportion.

   (a) $7 is to 3 cans as $28 is to 12 cans

   (b) 9 meters is to 16 meters as 18 meters is to 32 meters

   (c) 5 is to 7 as 35 is to 49

   (d) 10 is to 30 as 60 is to 180

ANSWERS

1. (a) $\dfrac{\$7}{3 \text{ cans}} = \dfrac{\$28}{12 \text{ cans}}$

   (b) $\dfrac{9}{16} = \dfrac{18}{32}$

   (c) $\dfrac{5}{7} = \dfrac{35}{49}$ (d) $\dfrac{10}{30} = \dfrac{60}{180}$

OBJECTIVE 1 Write proportions. A **proportion** states that two ratios (or rates) are equivalent. For example,

$$\dfrac{\$20}{4 \text{ hours}} = \dfrac{\$40}{8 \text{ hours}}$$

is a proportion that says the rate $\dfrac{\$20}{4 \text{ hours}}$ is equivalent to the rate $\dfrac{\$40}{8 \text{ hours}}$. As the amount of money doubles, the number of hours also doubles. This proportion is read:

20 dollars **is to** 4 hours **as** 40 dollars **is to** 8 hours.

EXAMPLE 1 Writing Proportions

Write each proportion.

(a) 6 ft is to 11 ft **as** 18 ft is to 33 ft.

$$\dfrac{6 \text{ ft}}{11 \text{ ft}} = \dfrac{18 \text{ ft}}{33 \text{ ft}} \quad \text{so} \quad \dfrac{6}{11} = \dfrac{18}{33} \quad \text{The common units (ft) divide out and are not written.}$$

(b) $9 is to 6 liters **as** $3 is to 2 liters.

$$\dfrac{\$9}{6 \text{ liters}} = \dfrac{\$3}{2 \text{ liters}} \quad \text{Units must be written.}$$

Work Problem 1 at the Side.

OBJECTIVE 2 Determine whether proportions are true or false. There are two ways to see whether a proportion is true. One way is to *write both of the ratios in lowest terms.*

0-138 CHAPTER 0 Topics in Arithmetic

2. Determine whether each proportion is true or false by writing both ratios in lowest terms.

(a) $\dfrac{6}{12} = \dfrac{15}{30}$

(b) $\dfrac{20}{24} = \dfrac{3}{4}$

(c) $\dfrac{25}{40} = \dfrac{30}{48}$

(d) $\dfrac{35}{45} = \dfrac{12}{18}$

(e) $\dfrac{21}{45} = \dfrac{56}{120}$

ANSWERS
2. (a) true (b) false (c) true
   (d) false (e) true

EXAMPLE 2 Writing Both Ratios in Lowest Terms

Are the following proportions true?

(a) $\dfrac{5}{9} = \dfrac{18}{27}$

Write each ratio in lowest terms.

$\dfrac{5}{9}$ ← Already in lowest terms $\dfrac{18 \div 9}{27 \div 9} = \dfrac{2}{3}$ ← Lowest terms

Because $\dfrac{5}{9}$ is *not* equivalent to $\dfrac{2}{3}$, the proportion is *false*.

(b) $\dfrac{16}{12} = \dfrac{28}{21}$

Write each ratio in lowest terms.

$\dfrac{16 \div 4}{12 \div 4} = \dfrac{4}{3}$ and $\dfrac{28 \div 7}{21 \div 7} = \dfrac{4}{3}$

Both ratios are equivalent to $\dfrac{4}{3}$, so the proportion is *true*.

Work Problem 2 at the Side.

OBJECTIVE 3 Find diagonal products. Another way to test whether the ratios in a proportion are equivalent is to compare *diagonal products*.

Using Diagonal Products to Determine Whether a Proportion Is True

To see whether a proportion is true, first multiply along one diagonal, then multiply along the other diagonal, as shown here.

$5 \cdot 4 = 20$

$\dfrac{2}{5} = \dfrac{4}{10}$ Diagonal products are equal.

$2 \cdot 10 = 20$

In this case the **diagonal products** are both 20. When diagonal products are *equal*, the proportion is *true*. If the diagonal products are *unequal*, the proportion is *false*.

The diagonal products test is based on rewriting both fractions with the common denominator of $5 \cdot 10$, or 50.

$\dfrac{2 \cdot 10}{5 \cdot 10} = \dfrac{20}{50}$ and $\dfrac{4 \cdot 5}{10 \cdot 5} = \dfrac{20}{50}$

We see that $\dfrac{2}{5}$ and $\dfrac{4}{10}$ are equivalent because both can be rewritten as $\dfrac{20}{50}$. The diagonal product test takes a shortcut by comparing only the two numerators (20 = 20).

3. Diagonal multiply to see whether each proportion is true or false.

(a) $\dfrac{5}{9} = \dfrac{10}{18}$

(b) $\dfrac{32}{15} = \dfrac{16}{8}$

(c) $\dfrac{10}{17} = \dfrac{20}{34}$

(d) $\dfrac{2.4}{6} = \dfrac{5}{12}$ $6 \cdot 5 =$ $2.4 \cdot 12 =$

(e) $\dfrac{3}{4.25} = \dfrac{24}{34}$

(f) $\dfrac{1\frac{1}{6}}{2\frac{1}{3}} = \dfrac{4}{8}$

EXAMPLE 3   Using Diagonal Products

Use diagonal products to see whether each proportion is true or false.

(a) $\dfrac{3}{5} = \dfrac{12}{20}$

Multiply along one diagonal and then along the other diagonal.

$5 \cdot 12 = 60$
$3 \cdot 20 = 60$ Equal

The diagonal products are *equal*, so the proportion is *true*.

(b) $\dfrac{2\frac{1}{3}}{3\frac{1}{3}} = \dfrac{9}{16}$

Diagonal multiply.

Changed to improper fractions

$3\dfrac{1}{3} \cdot 9 = \dfrac{10}{\cancel{3}} \cdot \dfrac{\cancel{9}^{3}}{1} = \dfrac{30}{1} = 30$

$2\dfrac{1}{3} \cdot 16 = \dfrac{7}{3} \cdot \dfrac{16}{1} = \dfrac{112}{3} = 37\dfrac{1}{3}$ Unequal

The diagonal products are *unequal*, so the proportion is *false*.

NOTE   The numbers in a proportion do *not* have to be whole numbers. They can be fractions, mixed numbers, decimal numbers, and so on.

Work Problem 3 at the Side.

ANSWERS
3. (a) true   (b) false   (c) true
 (d) $6 \cdot 5 = 30$; $2.4 \cdot 12 = 28.8$; false
 (e) true   (f) true

0.25 EXERCISES

For Extra Help

 Student's Solutions Manual

 MyMathLab

 InterAct Math Tutorial Software

 AW Math Tutor Center

MathXL

Write each proportion. See Example 1.

1. $9 is to 12 cans as $18 is to 24 cans.

2. 28 people is to 7 cars as 16 people is to 4 cars.

3. 200 adults is to 450 children as 4 adults is to 9 children.

4. 150 trees is to 1 acre as 1500 trees is to 10 acres.

5. 120 ft is to 150 ft as 8 ft is to 10 ft.

6. $6 is to $9 as $10 is to $15.

Determine whether each proportion is true or false by writing the ratios in lowest terms. See Example 2.

7. $\dfrac{6}{10} = \dfrac{3}{5}$

8. $\dfrac{1}{4} = \dfrac{9}{36}$

9. $\dfrac{5}{8} = \dfrac{25}{40}$

10. $\dfrac{2}{3} = \dfrac{20}{27}$

11. $\dfrac{150}{200} = \dfrac{200}{300}$

12. $\dfrac{100}{120} = \dfrac{75}{100}$

13. $\dfrac{42}{15} = \dfrac{28}{10}$

14. $\dfrac{18}{16} = \dfrac{36}{32}$

15. $\dfrac{32}{18} = \dfrac{48}{27}$

16. $\dfrac{15}{48} = \dfrac{10}{24}$

17. $\dfrac{7}{6} = \dfrac{54}{48}$

18. $\dfrac{28}{21} = \dfrac{44}{33}$

Use diagonal multiplication to determine whether each proportion is true or false. See Example 3.

19. $\dfrac{2}{9} = \dfrac{6}{27}$

20. $\dfrac{20}{25} = \dfrac{4}{5}$

21. $\dfrac{20}{28} = \dfrac{12}{16}$

22. $\dfrac{16}{40} = \dfrac{22}{55}$

23. $\dfrac{110}{18} = \dfrac{160}{27}$

24. $\dfrac{600}{420} = \dfrac{20}{14}$

25. $\dfrac{3.5}{4} = \dfrac{7}{8}$

26. $\dfrac{36}{23} = \dfrac{9}{5.75}$

27. $\dfrac{18}{16} = \dfrac{2.8}{2.5}$

28. $\dfrac{0.26}{0.39} = \dfrac{1.3}{1.9}$

29. $\dfrac{6}{3\frac{2}{3}} = \dfrac{18}{11}$

30. $\dfrac{16}{13} = \dfrac{2}{1\frac{5}{8}}$

31. $\dfrac{2\frac{5}{8}}{3\frac{1}{4}} = \dfrac{21}{26}$ **32.** $\dfrac{28}{17} = \dfrac{9\frac{1}{3}}{5\frac{2}{3}}$ **33.** $\dfrac{\frac{2}{3}}{2} = \dfrac{2.7}{8}$

34. $\dfrac{3.75}{1\frac{1}{4}} = \dfrac{7.5}{2\frac{1}{2}}$ **35.** $\dfrac{2\frac{3}{10}}{8.05} = \dfrac{\frac{1}{4}}{0.9}$ **36.** $\dfrac{\frac{3}{5}}{\frac{5}{6}} = \dfrac{1.5}{\frac{7}{12}}$

37. Suppose Jerome Walton of the Atlanta Braves had 16 hits in 50 times at bat and Mariano Duncan of the New York Yankees was at bat 400 times and got 128 hits. Paul is trying to convince Jamie that the two men hit equally well. Show how you could use a proportion and diagonal products to see whether Paul is correct.

38. Jay worked 3.5 hours and packed 91 cartons. Craig packed 126 cartons in 5.25 hours. To see whether the men worked equally fast, Barry set up this proportion:
$$\dfrac{3.5}{91} = \dfrac{126}{5.25}$$
Explain what is wrong with Barry's proportion and write a correct one. Is the correct proportion true or false?

0.26 Solving Proportions

OBJECTIVES

1. Find the unknown number in a proportion.
2. Find the unknown number in a proportion with mixed numbers or decimals.
3. Use proportions to solve application problems.

OBJECTIVE 1 Find the unknown number in a proportion. Four numbers are used in a proportion. If any three of these numbers are known, the fourth can be found. For example, find the unknown number that will make this proportion true.

$$\dfrac{3}{5} = \dfrac{x}{40}$$

The x represents the unknown number. Start by finding the diagonal products.

$$\dfrac{3}{5} = \dfrac{x}{40} \quad \text{Diagonal products: } 5 \cdot x,\ 3 \cdot 40$$

To make the proportion true, the diagonal products must be equal.

$$5 \cdot x = 3 \cdot 40$$
$$5 \cdot x = 120$$

The equal sign says that 5 • x and 120 are equivalent. If 5 • x and 120 are both divided by 5, the results will still be equivalent.

$$\frac{5 \cdot x}{5} = \frac{120}{5} \leftarrow \text{Divide each side by 5.}$$

Divide out 5 in numerator and denominator.

$$\frac{\overset{1}{\cancel{5}} \cdot x}{\underset{1}{\cancel{5}}} = 24 \qquad \text{On the right side, divide 120 by 5 to get 24.}$$

Multiplying by 1 does *not* change a number, so in the numerator on the left side, 1 • x is the same as x.

$$\frac{x}{1} = 24$$

Dividing by 1 does *not* change a number, so on the left side, $\frac{x}{1}$ is the same as x.

$$x = 24$$

The unknown number in the proportion is 24. The complete proportion is shown below.

$$\frac{3}{5} = \frac{24}{40} \leftarrow x \text{ is 24.}$$

Check by finding the diagonal products. If they are equal, you solved the problem correctly. If they are unequal, rework the problem.

$$5 \cdot 24 = 120$$
$$\frac{3}{5} = \frac{24}{40} \qquad \text{Equal; proportion is true.}$$
$$3 \cdot 40 = 120$$

The diagonal products are equal, so the solution, $x = 24$, is correct.

CAUTION The solution is 24, which is the unknown number in the proportion. 120 is *not* the solution; it is the diagonal product you get when *checking* the solution.

Solve a proportion for an unknown number by using the following steps.

Finding an Unknown Number in a Proportion

Step 1 Find the diagonal products.

Step 2 Show that the diagonal products are equivalent.

Step 3 Divide both products by the number multiplied by *x* (the number next to *x*).

Step 4 Check by writing the solution in the proportion and finding the diagonal products.

EXAMPLE 1 Solving for Unknown Numbers

Find the unknown number in each proportion. Round to hundredths, if necessary.

(a) $\dfrac{16}{x} = \dfrac{32}{20}$

Recall that ratios can be rewritten in lowest terms. If desired, you can do that before finding the diagonal products. In this example, write $\tfrac{32}{20}$ in lowest terms $\left(\tfrac{8}{5}\right)$ to get $\dfrac{16}{x} = \dfrac{8}{5}$.

Step 1 $\dfrac{16}{x} = \dfrac{8}{5}$ ⟵ Find the diagonal products: $x \cdot 8$ and $16 \cdot 5$.

Step 2 $x \cdot 8 = 16 \cdot 5$ ⟵ Show that diagonal products are equivalent.
$x \cdot 8 = 80$

Step 3 $\dfrac{x \cdot \cancel{8}}{\cancel{8}} = \dfrac{80}{8}$ ⟵ Divide each side by 8.

$x = 10$ ⟵ Find x. (No rounding necessary.)

Step 4   Write the solution in the proportion and check by finding diagonal products.

x is 10. → $\dfrac{16}{10} = \dfrac{8}{5}$

$10 \cdot 8 = 80$
$16 \cdot 5 = 80$

Equal; proportion is true.

The diagonal products are equal, so 10 is the correct solution.

NOTE It is not necessary to write the ratios in lowest terms before solving. However, if you do, you will have smaller numbers to work with.

(b) $\dfrac{7}{12} = \dfrac{15}{x}$

$\dfrac{7}{12} = \dfrac{15}{x}$

$12 \cdot 15 = 180$
$7 \cdot x$

Find the diagonal products.

Show that diagonal products are equivalent.

$7 \cdot x = 180$

Divide each side by 7.

$\dfrac{\cancel{7} \cdot x}{\cancel{7}} = \dfrac{180}{7}$

$x \approx 25.71$ ⟵ Rounded to nearest hundredth

0-144 CHAPTER 0 Topics in Arithmetic

1. Find the unknown numbers. Round to hundredths, if necessary. Check your answers by finding the diagonal products.

   (a) $\dfrac{1}{2} = \dfrac{x}{12}$

   (b) $\dfrac{6}{10} = \dfrac{15}{x}$

   (c) $\dfrac{28}{x} = \dfrac{21}{9}$

   (d) $\dfrac{x}{8} = \dfrac{3}{5}$

   (e) $\dfrac{14}{11} = \dfrac{x}{3}$

ANSWERS

1. (a) $x = 6$ (b) $x = 25$
   (c) $x = 12$ (d) $x = 4.8$
   (e) $x \approx 3.82$ (rounded to nearest hundredth)

When the division does not come out even, check for directions on how to round your answer. Divide out one more place, then round.

$$\begin{array}{r} 25.714 \\ 7\overline{)180.000} \end{array}$$ ← Divide out to thousandths.
Round to hundredths.

Write the solution in the proportion and check by finding diagonal products.

$$\dfrac{7}{12} = \dfrac{15}{25.71}$$

$12 \cdot 15 = 180$
$7 \cdot 25.71 = 179.97$ Very close, but not equal

The diagonal products are slightly different because you rounded the value of x. However, they are close enough to see that the problem was done correctly and 25.71 is the solution.

Work Problem 1 at the Side.

OBJECTIVE 2 Find the unknown number in a proportion with mixed numbers or decimals. The following examples show how to work with mixed numbers or decimals in a proportion.

EXAMPLE 2 Solving Proportions with Mixed Numbers and Decimals

Find the unknown number in each proportion.

(a) $\dfrac{2\frac{1}{5}}{6} = \dfrac{x}{10}$

$\dfrac{2\frac{1}{5}}{6} = \dfrac{x}{10}$ $2\frac{1}{5} \cdot 10$ Find the diagonal products.
                                             $6 \cdot x$

Find $2\frac{1}{5} \cdot 10$.

$$2\frac{1}{5} \cdot 10 = \dfrac{11}{5} \cdot \dfrac{10}{1} = \dfrac{11}{\cancel{5}} \cdot \dfrac{\cancel{10}^2}{1} = \dfrac{22}{1} = 22$$

Changed to improper fraction

Show that the diagonal products are equivalent.

$$6 \cdot x = 22$$

Divide each side by 6.

$$\dfrac{\cancel{6} \cdot x}{\cancel{6}} = \dfrac{22}{6}$$

Write the answer as a mixed number in lowest terms.

$$x = \dfrac{22 \div 2}{6 \div 2} = \dfrac{11}{3} = 3\dfrac{2}{3}$$

SECTION 0.26 Solving Proportions   0-145

2. Find the unknown numbers. Round to hundredths on the decimal problems, if necessary. Check your answers by finding the diagonal products.

(a) $\dfrac{3\frac{1}{4}}{2} = \dfrac{x}{8}$

(b) $\dfrac{x}{3} = \dfrac{1\frac{2}{3}}{5}$

(c) $\dfrac{0.06}{x} = \dfrac{0.3}{0.4}$

(d) $\dfrac{2.2}{5} = \dfrac{13}{x}$

(e) $\dfrac{x}{6} = \dfrac{0.5}{1.2}$

(f) $\dfrac{0}{2} = \dfrac{x}{7.092}$

ANSWERS
2. (a) $x = 13$ (b) $x = 1$
 (c) $x = 0.08$
 (d) $x \approx 29.55$
    (rounded to nearest hundredth)
 (e) $x = 2.5$ (f) $x = 0$

Write the solution in the proportion and check by finding diagonal products.

$$6 \cdot 3\tfrac{2}{3} = \dfrac{\cancel{6}^{2}}{1} \cdot \dfrac{11}{\cancel{3}_{1}} = \dfrac{22}{1} = 22$$

$$\dfrac{2\tfrac{1}{5}}{6} \bowtie \dfrac{3\tfrac{2}{3}}{10}$$

Equal

$$2\tfrac{1}{5} \cdot 10 = \dfrac{11}{\cancel{5}_{1}} \cdot \dfrac{\cancel{10}^{2}}{1} = \dfrac{22}{1} = 22$$

The diagonal products are equal, so $3\tfrac{2}{3}$ is the correct solution.

(b) $\dfrac{1.5}{0.6} = \dfrac{2}{x}$ 0.8

Show that diagonal products are equivalent.

$$1.5 \cdot x = 0.6 \cdot 2$$
$$1.5 \cdot x = 1.2$$

Divide each side by 1.5.

$$\dfrac{\cancel{1.5}^{1} \cdot x}{\cancel{1.5}_{1}} = \dfrac{1.2}{1.5}$$

$$x = \dfrac{1.2}{1.5}$$

Complete the division.

$$1.5\overline{)1.20} \; .8$$

So the unknown number is 0.8. Check by finding the diagonal products.

$$0.6 \cdot 2 = 1.2$$
$$\dfrac{1.5}{0.6} \bowtie \dfrac{2}{0.8}$$
Equal
$$1.5 \cdot 0.8 = 1.2$$

The diagonal products are equal, so 0.8 is the correct solution.

Work Problem 2 at the Side.

OBJECTIVE 3 Use proportions to solve application problems. Proportions can be used to solve a wide variety of problems. Watch for problems in which you are given a ratio or rate and then asked to find part of a corresponding ratio or rate. Remember that a ratio or rate compares two quantities and often includes one of the following indicator words.

in   for   on   per   from   to

0-146 CHAPTER 0 Topics in Arithmetic

3. Set up and solve a proportion for each problem.

   (a) If 2 pounds of fertilizer will cover 50 square feet of garden, how many pounds are needed for 225 square feet?

   (b) A U.S. map has a scale of 1 inch to 75 miles. Lake Superior is 4.75 inches long on the map. What is the lake's actual length in miles?

   (c) Cough syrup is to be given at the rate of 30 milliliters for each 100 pounds of body weight. How much should be given to a 34-pound child? Round to the nearest whole milliliter.

EXAMPLE 3 Solving a Proportion Application

Mike's car can travel 163 **miles** on 6.4 **gallons** of gas. How far can it travel on a full tank of 14 **gallons** of gas? Round to the nearest whole mile.

Step 1 **Read** the problem. The problem asks for the number of miles the car can travel on 14 gallons of gas.

Step 2 **Work out a plan.** Decide what is being compared. This example compares **miles** to **gallons**. Write a proportion using the two rates. Be sure that *both* rates compare miles to gallons in the same order. In other words, miles is in both numerators and gallons is in both denominators. Use a letter to represent the unknown number.

Step 3 **Estimate** a reasonable answer. To estimate the answer, notice that 14 gallons is a little more than *twice as much* as 6.4 gallons, so the car should travel a little more than *twice as far*; 2 • 163 miles = 326 miles.

Step 4 **Solve** the problem. With the proportion set up correctly, solve for the unknown number.

This rate compares miles to gallons. $\dfrac{163 \text{ miles}}{6.4 \text{ gallons}} = \dfrac{x \text{ miles}}{14 \text{ gallons}}$ This rate compares miles to gallons.

(Matching units)

Ignore the units while finding the diagonal products and dividing each side by **6.4**.

$6.4 \cdot x = 163 \cdot 14$ Show that diagonal products are equivalent.

$6.4 \cdot x = 2282$

$\dfrac{6.4 \cdot x}{6.4} = \dfrac{2282}{6.4}$ Divide each side by **6.4**.

$x = 356.5625$

Step 5 **State the answer.** Rounded to the nearest mile, the car can travel 357 miles on a full tank of gas.

Step 6 **Check** your work. The answer, 357 miles, is a little more than the estimate of 326 miles, so it is reasonable.

Work Problem 3 at the Side.

ANSWERS

3. (a) $\dfrac{2 \text{ pounds}}{50 \text{ square feet}} = \dfrac{x \text{ pounds}}{225 \text{ square feet}}$
   $x = 9$ pounds

   (b) $\dfrac{1 \text{ inch}}{75 \text{ miles}} = \dfrac{4.75 \text{ inches}}{x \text{ miles}}$
   $x = 356.25$ miles or $x \approx 356$ miles

   (c) $\dfrac{30 \text{ milliliters}}{100 \text{ pounds}} = \dfrac{x \text{ milliliters}}{34 \text{ pounds}}$
   $x \approx 10$ milliliters

SECTION 0.26 Solving Proportions 0-147

0.26 EXERCISES

For Extra Help

Student's Solutions Manual

MyMathLab

InterAct Math Tutorial Software

AW Math Tutor Center

MathXL

Find the unknown number in each proportion. Round your answers to hundredths, if necessary. Check your answers by finding the diagonal products. See Examples 1 and 2.

1. $\dfrac{1}{3} = \dfrac{x}{12}$ 2. $\dfrac{x}{6} = \dfrac{15}{18}$ [handwritten: x=5, 18x=90, =5] 3. $\dfrac{15}{10} = \dfrac{3}{x}$

4. $\dfrac{5}{x} = \dfrac{20}{8}$ 5. $\dfrac{x}{11} = \dfrac{32}{4}$ 6. $\dfrac{12}{9} = \dfrac{8}{x}$

7. $\dfrac{42}{x} = \dfrac{18}{39}$ 8. $\dfrac{49}{x} = \dfrac{14}{18}$ 9. $\dfrac{x}{25} = \dfrac{4}{20}$

10. $\dfrac{6}{x} = \dfrac{4}{8}$ 11. $\dfrac{8}{x} = \dfrac{24}{30}$ 12. $\dfrac{32}{5} = \dfrac{x}{10}$

 13. $\dfrac{99}{55} = \dfrac{44}{x}$ 14. $\dfrac{x}{12} = \dfrac{101}{147}$ 15. $\dfrac{0.7}{9.8} = \dfrac{3.6}{x}$

16. $\dfrac{x}{3.6} = \dfrac{4.5}{6}$ 17. $\dfrac{250}{24.8} = \dfrac{x}{1.75}$ 18. $\dfrac{4.75}{17} = \dfrac{43}{x}$

Find the unknown number in each proportion. Write your answers as whole or mixed numbers when possible. See Example 2.

19. $\dfrac{15}{1\frac{2}{3}} = \dfrac{9}{x}$ 20. $\dfrac{x}{\frac{3}{10}} = \dfrac{2\frac{2}{9}}{1}$

21. $\dfrac{2\frac{1}{3}}{1\frac{1}{2}} = \dfrac{x}{2\frac{1}{4}}$ 22. $\dfrac{1\frac{5}{6}}{x} = \dfrac{\frac{3}{14}}{\frac{6}{7}}$

Solve each proportion two different ways. First change all the numbers to decimal form and solve. Then change all the numbers to fraction form and solve; write your answers in lowest terms.

23. $\dfrac{\frac{1}{2}}{x} = \dfrac{2}{0.8}$

24. $\dfrac{\frac{3}{20}}{0.1} = \dfrac{0.03}{x}$

25. $\dfrac{x}{\frac{3}{50}} = \dfrac{0.15}{1\frac{4}{5}}$

26. $\dfrac{8\frac{4}{5}}{1\frac{1}{10}} = \dfrac{x}{0.4}$

RELATING CONCEPTS (EXERCISES 27–28)

For Individual or Group Work

Work Exercises 27–28 in order. *First prove that the proportions are* **not** *true. Then create four true proportions for each exercise by changing one number at a time.*

27. $\dfrac{10}{4} = \dfrac{5}{3}$

28. $\dfrac{6}{8} = \dfrac{24}{30}$

Set up and solve a proportion for each application problem. See Example 3.

29. Caroline can sketch four cartoon strips in five hours. How long will it take her to sketch 18 strips?

30. The Cosmic Toads recorded eight songs on their first CD in 26 hours. How long will it take them to record 14 songs for their second CD?

31. Sixty newspapers cost $27. Find the cost of 16 newspapers.

32. Twenty-two guitar lessons cost $396. Find the cost of 12 lessons.

33. If three pounds of fescue grass seed cover about 350 square feet of ground, how many pounds are needed for 4900 square feet?

34. Anna earns $1242.08 in 14 days. How much does she earn in 260 days?

SECTIONS 0.24–0.26   Test   0-149

35. Tom makes $455.75 in 5 days. How much does he make in 3 days?

36. If 5 ounces of a medicine must be mixed with 8 ounces of water, how many ounces of medicine would be mixed with 20 ounces of water?

37. The bag of rice noodles shown below makes 7 servings. (*Source:* Everfresh Foods.) At that rate, how many ounces of noodles do you need for 12 servings, to the nearest ounce?

38. This can of sweet potatoes is enough for 4 servings. (*Source:* Moody Dunbar, Inc.) How many ounces are needed for 9 servings, to the nearest ounce?

39. Three quarts of a latex enamel paint will cover about 270 square feet of wall surface. (*Source:* Thompson and Formby, Inc.) How many quarts will you need to cover 350 square feet of wall surface in your kitchen and 100 square feet of wall surface in your bathroom?

40. One gallon of clear gloss wood finish covers about 550 square feet of surface. (*Source:* The Flecto Company.) If you need to apply three coats of finish to 400 square feet of surface, how many gallons do you need, to the nearest tenth?

0.24–0.26   TEST

Write each rate or ratio as a fraction in lowest terms. Change to the same units when necessary.

1. 16 fish to 20 fish

2. 300 miles on 15 gallons

3. $15 for 75 minutes

4. The little theater has 320 seats. The auditorium has 1200 seats. Find the ratio of auditorium seats to theater seats.

5. 3 quarts to 60 gallons

6. 3 hours to 40 minutes

7. Find the best buy on spaghetti sauce.
   28 ounces of Brand X for $3.89
   18 ounces of Brand Y for $1.89
   13 ounces of Brand Z for $1.29

Determine whether each proportion is true or false.

8. $\dfrac{6}{14} = \dfrac{18}{45}$

9. $\dfrac{8.4}{2.8} = \dfrac{2.1}{0.7}$

Find the unknown number in each proportion. Round the answers to the nearest hundredth, if necessary.

10. $\dfrac{5}{9} = \dfrac{x}{45}$

11. $\dfrac{3}{1} = \dfrac{8}{x}$

12. $\dfrac{x}{20} = \dfrac{6.5}{0.4}$

13. $\dfrac{2\frac{1}{3}}{x} = \dfrac{\frac{8}{9}}{4}$

Set up and solve a proportion for each application problem.

14. Pedro types 240 words in five minutes. At that rate, how many words can he type in 12 minutes?

15. Just 0.8 ounce of wildflower seeds is enough for 50 square feet of ground. What weight of seeds is needed for a garden with 225 square feet? (*Source:* White Swan Ltd.)

16. A medication is given at the rate of 8.2 grams for every 50 pounds of body weight. How much should be given to a 145-pound person? Round to the nearest tenth.

0.27 Analyzing Units with English Measurement

OBJECTIVES

1. Learn the basic measurement units in the English system.
2. Convert among measurement units using multiplication or division.
3. Convert among measurement units using unit fractions.

1. After memorizing the measurement conversions, answer these questions.

   (a) 1 c = __8__ fl oz

   (b) __4__ qt = 1 gal

   (c) 1 wk = __7__ days

   (d) __3__ ft = 1 yd

   (e) 1 ft = __12__ in.

   (f) __16__ oz = 1 lb

   (g) 1 T = _____ lb

   (h) __60__ min = 1 hr

   (i) 1 pt = __2__ c

   (j) __24__ hr = 1 day

   (k) 1 min = __60__ sec

   (l) 1 qt = __2__ pt

   (m) __5,280__ ft = 1 mi

ANSWERS

1. (a) 8 (b) 4 (c) 7 (d) 3
   (e) 12 (f) 16 (g) 2000
   (h) 60 (i) 2 (j) 24 (k) 60
   (l) 2 (m) 5280

We measure things all the time: the distance traveled on vacation, the floor area we want to cover with carpet, the amount of milk in a recipe, the weight of the bananas we buy at the store, the number of hours we work, and many more.

In the United States we still use the **English system** of measurement for many everyday activities. Examples of English units are inches, feet, quarts, ounces, and pounds. However, the fields of science, medicine, sports, and manufacturing use the **metric system** (meters, liters, and grams). And, because the rest of the world uses only the metric system, U.S. businesses are beginning to change to the metric system in order to compete internationally.

OBJECTIVE 1 Learn the basic measurement units in the English system. Until the switch to the metric system is complete, we still need to know how to use the English system of measurement. The table below lists the relationships you should memorize. The time relationships are used in both the English and metric systems.

English Measurement Relationships

Length	Weight
1 foot (ft) = 12 inches (in.)	1 pound (lb) = 16 ounces (oz)
1 yard (yd) = 3 feet (ft)	1 ton (T) = 2000 pounds (lb)
1 mile (mi) = 5280 feet (ft)	

Capacity	Time
1 cup (c) = 8 fluid ounces (fl oz)	1 week (wk) = 7 days
1 pint (pt) = 2 cups (c)	1 day = 24 hours (hr)
1 quart (qt) = 2 pints (pt)	1 hour (hr) = 60 minutes (min)
1 gallon (gal) = 4 quarts (qt)	1 minute (min) = 60 seconds (sec)

As you can see, there is no simple or "natural" way to convert among these various measures. The units evolved over hundreds of years and were based on a variety of "standards." For example, one yard was the distance from the tip of a king's nose to his thumb when his arm was outstretched. An inch was three dried barleycorns laid end to end.

EXAMPLE 1 Knowing English Measurement Units

Memorize the English measurement conversions. Then answer these questions.

(a) 24 hr = _____ day   Answer: 1 day

(b) 1 yd = _____ ft   Answer: 3 ft

Work Problem 1 at the Side.

OBJECTIVE 2 Convert among measurement units using multiplication or division. You often need to convert from one unit of measure to another. Two methods of converting measurements are shown here. Study each way and use the method you prefer. The first method involves deciding whether to multiply or divide.

CHAPTER 0 Topics in Arithmetic

2. Convert each measurement using multiplication or division.

   (a) $5\frac{1}{2}$ ft to inches

   (b) 64 oz to pounds

   (c) 6 yd to feet

   (d) 2 T to pounds

   (e) 35 pt to quarts

   (f) 20 min to hours

   (g) 4 wk to days

Converting among Measurement Units

1. *Multiply* when converting from a larger unit to a smaller unit.
2. *Divide* when converting from a smaller unit to a larger unit.

EXAMPLE 2 Converting from One Unit of Measure to Another

Convert each measurement.

(a) 7 ft to inches

You are converting from a *larger* unit to a *smaller* unit (feet to inches), so multiply.
Because *1 ft = 12 in.*, multiply by 12.

$$7 \text{ ft} = 7 \cdot 12 = 84 \text{ in.}$$

(b) $3\frac{1}{2}$ lb to ounces

You are converting from a *larger* unit to a *smaller* unit (pounds to ounces), so multiply.
Because *1 lb = 16 oz*, multiply by 16.

$$3\frac{1}{2} \text{ lb} = 3\frac{1}{2} \cdot 16 = \frac{7}{2} \cdot \frac{\overset{8}{\cancel{16}}}{1} = \frac{56}{1} = 56 \text{ oz}$$

(c) 20 qt to gallons

You are converting from a *smaller* unit to a *larger* unit (quarts to gallons), so divide.
Because *4 qt = 1 gal*, divide by 4.

$$20 \text{ qt} = \frac{20}{4} = 5 \text{ gal}$$

Divide by 4.

(d) 45 min to hours

You are converting from a *smaller* unit to a *larger* unit (minutes to hours), so divide.
Because *60 min = 1 hr*, divide by 60 and write the fraction in lowest terms.

$$45 \text{ min} = \frac{45}{60} = \frac{45 \div 15}{60 \div 15} = \frac{3}{4} \text{ hr} \leftarrow \text{Lowest terms}$$

Divide by 60.

Work Problem 2 at the Side.

OBJECTIVE 3 Convert among measurement units using unit fractions. If you have trouble deciding whether to multiply or divide when converting measurements, use *unit fractions* to solve the problem. You'll also find this method useful in science classes. A **unit fraction** is equivalent to 1. Here is an example.

$$\frac{12 \text{ in.}}{12 \text{ in.}} = \frac{\overset{1}{\cancel{12}} \text{ in.}}{\underset{1}{\cancel{12}} \text{ in.}} = 1$$

ANSWERS

2. (a) 66 in. (b) 4 lb (c) 18 ft
   (d) 4000 lb (e) $17\frac{1}{2}$ qt (f) $\frac{1}{3}$ hr
   (g) 28 days

Use the table of measurement relationships on the first page of this section to find that 12 in. is the same as 1 ft. So you can substitute 1 ft for 12 in. in the numerator, or you can substitute 1 ft for 12 in. in the denominator. This makes two useful unit fractions.

$$\frac{1 \text{ ft}}{12 \text{ in.}} = 1 \quad \text{or} \quad \frac{12 \text{ in.}}{1 \text{ ft}} = 1$$

To convert from one measurement unit to another, just multiply by the appropriate unit fraction. Remember, a unit fraction is equivalent to 1. Multiplying something by 1 does *not* change its value.

Use these guidelines to choose the correct unit fraction.

Choosing a Unit Fraction

The *numerator* should use the measurement unit you want in the *answer*.

The *denominator* should use the measurement unit you want to *change*.

EXAMPLE 3 Using Unit Fractions with Length Measurements

(a) Convert 60 in. to feet.

Use a unit fraction with feet (the unit for your answer) in the numerator, and inches (the unit being changed) in the denominator. Because *1 ft = 12 in.*, the necessary unit fraction is

$$\frac{1 \text{ ft}}{12 \text{ in.}} \quad \begin{array}{l} \leftarrow \text{Unit for your answer is feet.} \\ \leftarrow \text{Unit being changed is inches.} \end{array}$$

Next, multiply 60 in. times this unit fraction. Write 60 in. as the fraction $\frac{60 \text{ in.}}{1}$. Then divide out common units and factors wherever possible.

$$60 \text{ in.} \cdot \frac{1 \text{ ft}}{12 \text{ in.}} = \frac{\overset{5}{\cancel{60} \text{ in.}}}{1} \cdot \frac{1 \text{ ft}}{\underset{1}{\cancel{12} \text{ in.}}} = \frac{5 \cdot 1 \text{ ft}}{1} = 5 \text{ ft}$$

These units should match. — Divide out inches. Divide 60 and 12 by 12.

(b) Convert 9 ft to inches.

Select the correct unit fraction to change 9 ft to inches.

$$\frac{12 \text{ in.}}{1 \text{ ft}} \quad \begin{array}{l} \leftarrow \text{Unit for your answer is inches.} \\ \leftarrow \text{Unit being changed is feet.} \end{array}$$

Multiply 9 ft times the unit fraction.

$$9 \text{ ft} \cdot \frac{12 \text{ in.}}{1 \text{ ft}} = \frac{9 \cancel{\text{ft}}}{1} \cdot \frac{12 \text{ in.}}{1 \cancel{\text{ft}}} = \frac{9 \cdot 12 \text{ in.}}{1} = 108 \text{ in.}$$

These units should match. — Divide out feet.

0-154 CHAPTER 0 Topics in Arithmetic

3. First write the unit fraction needed to make each conversion. Then complete the conversion.

   (a) 36 in. to feet

   unit fraction } $\dfrac{1\ \text{ft}}{12\ \text{in.}}$

   (b) 14 ft to inches

   unit fraction } $\dfrac{\text{in.}}{\text{ft}}$

   (c) 39 ft to yards

   unit fraction } _____

   (d) 2 mi to feet

   unit fraction } _____

4. Convert using unit fractions.

   (a) 16 qt to gallons

   (b) 3 c to pints

   (c) $3\dfrac{1}{2}$ T to pounds

   (d) $1\dfrac{3}{4}$ lb to ounces

ANSWERS

3. (a) 3 ft   (b) $\dfrac{12\ \text{in.}}{1\ \text{ft}}$; 168 in.
   (c) $\dfrac{1\ \text{yd}}{3\ \text{ft}}$; 13 yd   (d) $\dfrac{5280\ \text{ft}}{1\ \text{mi}}$; 10,560 ft

4. (a) 4 gal   (b) $1\dfrac{1}{2}$ pt or 1.5 pt
   (c) 7000 lb   (d) 28 oz

CAUTION If no units will divide out, you made a mistake in choosing the unit fraction.

Work Problem 3 at the Side.

■ **EXAMPLE 4** Using Unit Fractions with Capacity and Weight Measurements

(a) Convert 9 pt to quarts.

First select the correct unit fraction.

$\dfrac{1\ \text{qt}}{2\ \text{pt}}$ ← Unit for your answer is quarts.
← Unit being changed is pints.

Next multiply.

$$9\ \text{pt} \cdot \dfrac{1\ \text{qt}}{2\ \text{pt}} = \dfrac{9\ \text{pt}}{1} \cdot \dfrac{1\ \text{qt}}{2\ \text{pt}} = \dfrac{9}{2}\ \text{qt} = 4\dfrac{1}{2}\ \text{qt}$$

These units should match.   Divide out pints.

(b) Convert $7\dfrac{1}{2}$ gal to quarts.

$$\dfrac{7\dfrac{1}{2}\ \text{gal}}{1} \cdot \dfrac{4\ \text{qt}}{1\ \text{gal}} = \dfrac{15}{2} \cdot \dfrac{4}{1}\ \text{qt}$$

$$= \dfrac{15}{2} \cdot \dfrac{4}{1}\ \text{qt}$$

$$= 30\ \text{qt}$$

(c) Convert 36 oz to pounds.

$$\dfrac{36\ \text{oz}}{1} \cdot \dfrac{1\ \text{lb}}{16\ \text{oz}} = \dfrac{9}{4}\ \text{lb} = 2\dfrac{1}{4}\ \text{lb}$$

NOTE In Example 4(c) you get $\dfrac{9}{4}$ lb. Recall that $\dfrac{9}{4}$ means 9 ÷ 4. If you do 9 ÷ 4 on your calculator, you get 2.25 lb. English measurements usually use fractions or mixed numbers, like $2\dfrac{1}{4}$ lb. However, 2.25 lb is also correct and is the way grocery stores often show weights of produce, meat, and cheese.

Work Problem 4 at the Side.

SECTION 0.27 Analyzing Units with English Measurement 0-155

5. Convert using two or three unit fractions.

(a) 4 T to ounces

(b) 3 mi to inches

(c) 36 pt to gallons

(d) 2 wk to minutes

■ **EXAMPLE 5** Using Several Unit Fractions

Sometimes you may need to use two or three unit fractions in problems like these.

(a) Convert 63 in. to yards.

Use the unit fraction $\frac{1 \text{ ft}}{12 \text{ in.}}$ to change inches to feet and the unit fraction $\frac{1 \text{ yd}}{3 \text{ ft}}$ to change feet to yards. Notice how all the units divide out except yards, which is the unit you want in the answer.

$$\frac{63 \text{ in.}}{1} \cdot \frac{1 \text{ ft}}{12 \text{ in.}} \cdot \frac{1 \text{ yd}}{3 \text{ ft}} = \frac{63}{36} \text{ yd} = 1\frac{3}{4} \text{ yd}$$

You can also divide out common factors in the numbers.

$$\frac{\cancel{63}^{21\ 7}}{1} \cdot \frac{1}{\cancel{12}_{4}} \cdot \frac{1}{\cancel{3}_{1}} = \frac{7}{4} = 1\frac{3}{4} \text{ yd}$$

Instead of changing $\frac{7}{4}$ to $1\frac{3}{4}$, you can enter $7 \div 4$ on your calculator to get 1.75 yd. Both answers are correct because 1.75 is equivalent to $1\frac{3}{4}$.

(b) Convert 2 days to seconds.

Use three unit fractions. The first one changes days to hours, the second one changes hours to minutes, and the third one changes minutes to seconds. All the units divide out except seconds, which is what you want in your answer.

$$\frac{2 \text{ days}}{1} \cdot \frac{24 \text{ hr}}{1 \text{ day}} \cdot \frac{60 \text{ min}}{1 \text{ hr}} \cdot \frac{60 \text{ seconds}}{1 \text{ min}} = 172,800 \text{ seconds}$$

Divide out **days**.
Divide out **hr**.
Divide out **min**.

6. Convert the following. Round your answers to the nearest tenth.

(a) 50 miles/hour to feet/second

(b) 10 feet/second to miles/hour

> Work Problem 5 at the Side.

■ **EXAMPLE 6** Using Several Unit Fractions

The maximum speed limit on urban interstate highways in Florida is 65 miles/hour. (*Source: The World Almanac and Book of Facts, 2004.*) Convert 65 miles/hour to feet/second.

To solve this problem, use several unit fractions to change $\frac{65 \text{ mi}}{1 \text{ hr}}$ to $\frac{? \text{ ft}}{1 \text{ sec}}$. Remember these relationships.

$$1 \text{ mile} = 5280 \text{ feet}$$
$$1 \text{ hour} = 60 \text{ minutes}$$
$$1 \text{ minute} = 60 \text{ seconds}$$

$$\frac{65 \text{ mi}}{1 \text{ hr}} \cdot \frac{5280 \text{ ft}}{1 \text{ mi}} \cdot \frac{1 \text{ hr}}{60 \text{ min}} \cdot \frac{1 \text{ min}}{60 \text{ sec}} = \frac{65 \cdot 5280 \text{ ft}}{60 \cdot 60 \text{ sec}}$$
$$\approx 95.3 \text{ ft/sec}$$

Thus, 65 miles/hour is approximately equal to 95.3 feet/second.

> Work Problem 6 at the Side.

ANSWERS
5. (a) 128,000 oz (b) 190,080 in.
  (c) $4\frac{1}{2}$ gal or 4.5 gal
  (d) 20,160 min
6. (a) 73.3 ft/sec (b) 6.8 mi/hr

0.27 EXERCISES

For Extra Help
 Student's Solutions Manual
 MyMathLab
 InterAct Math Tutorial Software
 AW Math Tutor Center
MathXL

Fill in the blanks with the measurement relationships you have memorized. See Example 1.

1. 1 yd = _____ ft
2. 1 ft = _____ in.
3. _____ fl oz = 1 c
4. _____ qt = 1 gal
5. 1 mi = 5280 ft
6. 1 wk = _____ days
7. 2000 lb = 1 T
8. 16 oz = 1 lb
9. 1 min = 60 sec
10. 1 day = 24 hr

Convert each measurement using unit fractions. See Examples 3 and 4.

11. 120 sec = _____ min
12. 180 min = 3 hr
13. 8 qt = _____ gal
14. 6 gal = _____ qt

15. An adult sperm whale may weigh 38 to 40 tons. How many pounds could it weigh? (*Source: Grolier Multimedia Encyclopedia.*)

16. Recent fossil finds in Argentina indicate that the largest meat-eating dinosaur may have been 45 ft long. How many yards long was this dinosaur? (*Source: Washington Post.*)

45 ft

17. 9 yd = _____ ft
18. 20,000 lb = 10 T
19. 7 lb = _____ oz
20. 96 oz = _____ lb
21. 5 qt = _____ pt
22. 26 pt = _____ qt
23. 90 min = _____ hr
24. 45 sec = _____ min
25. 3 in. = _____ ft
26. 30 in. = _____ ft
27. 24 oz = _____ lb
28. 36 oz = _____ lb
29. 5 c = _____ pt
30. 15 qt = _____ gal

31. Mr. Kashpaws and his son worked for 12 hours doing traditional harvesting of wild rice. What part of a day did they work?

32. Michelle prepares 4-oz hamburgers at a fast-food restaurant. Each hamburger is what part of a pound?

33. $2\frac{1}{2}$ T = _____ lb

34. $4\frac{1}{2}$ pt = _____ c

35. $4\frac{1}{4}$ gal = _____ qt

36. $2\frac{1}{4}$ hr = _____ min

37. Our premature baby weighed $2\frac{3}{4}$ lb at birth. How many ounces did our baby weigh?

38. Gheorge Muresan, the tallest person playing basketball in the NBA, is $7\frac{2}{3}$ ft tall. What is his height in inches? (*Source:* NBA.)

Use two or three unit fractions to convert the following. See Example 5.

39. 6 yd = _____ in.

40. 2 T = _____ oz

41. 112 c = _____ qt

42. 336 hr = _____ wk

43. 6 days = _____ sec

44. 5 gal = _____ c

45. $1\frac{1}{2}$ T = _____ oz

46. $3\frac{1}{3}$ yd = _____ in.

47. The statement 8 = 2 is *not* true. But with appropriate measurement units, it *is* true.

$$8 \text{ quarts} = 2 \text{ gallons}$$

Attach measurement units to these numbers to make the statements true.

(a) 1 _____ = 16 _____
(b) 10 _____ = 20 _____
(c) 120 _____ = 2 _____
(d) 2 _____ = 24 _____
(e) 6000 _____ = 3 _____
(f) 35 _____ = 5 _____

48. Explain in your own words why you can add 2 feet + 12 inches to get 3 feet, but you cannot add 2 feet + 12 pounds.

CHAPTER 0 Topics in Arithmetic

Convert the following. See Example 5.

49. $2\frac{3}{4}$ mi = _____ in.

50. $5\frac{3}{4}$ tons = _____ oz

51. $6\frac{1}{4}$ gal = _____ fl oz

52. $3\frac{1}{2}$ days = _____ sec

53. 24,000 oz = _____ T

54. 57,024 in. = _____ mi

Convert the following. Round your answers to the nearest tenth. See Example 6.

55. 35 miles/hour to feet/second

56. 75 miles/hour to feet/second

57. 25 feet/second to miles/hour

58. 40 feet/second to miles/hour

59. 10 yards/second to miles/hour

60. 18 yards/second to miles/hour

0.28 The Metric System—Length

OBJECTIVES

1. Learn the basic metric units of length.
2. Use unit fractions to convert among units.
3. Move the decimal point to convert among units.

Around 1790, a group of French scientists developed the metric system of measurement. It is an organized system based on multiples of 10, like our number system and our money. After you are familiar with metric units, you will see that they are easier to use than the hodgepodge of English measurement relationships you used in **Section 0.27.**

OBJECTIVE 1 Learn the basic metric units of length. The basic unit of length in the metric system is the **meter** (also spelled *metre*). Use the symbol **m** for meter; do not put a period after it. If you put five of the pages from this textbook side by side, they would measure about 1 meter. Or, look at a yardstick—a meter is just a little longer. A yard is 36 inches long; a meter is about 39 inches long.

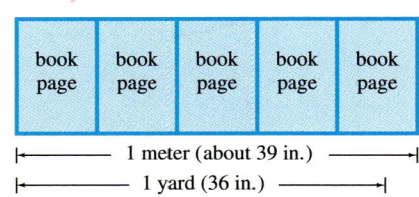

SECTION 0.28 The Metric System—Length 0-159

To make longer or shorter length units in the metric system, **prefixes** are written in front of the word *meter*. For example, the prefix *kilo* means 1000, so a *kilo*meter is 1000 meters. The table below shows how to use the prefixes for length measurements. It is helpful to memorize the prefixes because they are also used with weight and capacity measurements. The blue boxes are the units you will use most often in daily life.

Prefix	kilo-meter	hecto-meter	deka-meter	meter	deci-meter	centi-meter	milli-meter
Meaning	1000 meters	100 meters	10 meters	1 meter	$\frac{1}{10}$ of a meter	$\frac{1}{100}$ of a meter	$\frac{1}{1000}$ of a meter
Symbol	km	hm	dam	m	dm	cm	mm

Here are some comparisons to help you get acquainted with the commonly used length units: km, m, cm, mm.

*Kilo*meters are used instead of miles. A kilometer is 1000 meters. It is about 0.6 mile (a little more than half a mile) or about 5 to 6 city blocks. If you participate in a 10 km run, you'll run about 6 miles.

A meter is divided into 100 smaller pieces called *centi*meters. Each centimeter is $\frac{1}{100}$ of a meter. Centimeters are used instead of inches. A centimeter is a little shorter than $\frac{1}{2}$ inch. The cover of this textbook is about 21 cm wide. A nickel is about 2 cm across.

A meter is divided into 1000 smaller pieces called *milli*meters. Each millimeter is $\frac{1}{1000}$ of a meter. It takes 10 mm to equal 1 cm, so it is a very small length. The thickness of a dime is about 1 mm.

OBJECTIVE 2 Use unit fractions to convert among units. You can convert among metric length units using unit fractions. Keep these relationships in mind when setting up the unit fractions.

Metric Length Relationships

1 km = 1000 m so the unit fractions are:	**1 m = 1000 mm** so the unit fractions are:
$\frac{1 \text{ km}}{1000 \text{ m}}$ or $\frac{1000 \text{ m}}{1 \text{ km}}$	$\frac{1 \text{ m}}{1000 \text{ mm}}$ or $\frac{1000 \text{ mm}}{1 \text{ m}}$
1 m = 100 cm so the unit fractions are:	**1 cm = 10 mm** so the unit fractions are:
$\frac{1 \text{ m}}{100 \text{ cm}}$ or $\frac{100 \text{ cm}}{1 \text{ m}}$	$\frac{1 \text{ cm}}{10 \text{ mm}}$ or $\frac{10 \text{ mm}}{1 \text{ cm}}$

0-160 CHAPTER 0 Topics in Arithmetic

1. First write the unit fraction needed to make each conversion. Then complete the conversion.

   (a) 3.67 m to cm

   $\left.\begin{matrix}\text{unit}\\\text{fraction}\end{matrix}\right\}\dfrac{100\text{ cm}}{1\text{ m}}$

   (b) 92 cm to m

   $\left.\begin{matrix}\text{unit}\\\text{fraction}\end{matrix}\right\}\dfrac{\text{m}}{\text{cm}}$

   (c) 432.7 cm to m

   $\left.\begin{matrix}\text{unit}\\\text{fraction}\end{matrix}\right\}\rule{2cm}{0.4pt}$

   (d) 65 mm to cm

   $\left.\begin{matrix}\text{unit}\\\text{fraction}\end{matrix}\right\}\rule{2cm}{0.4pt}$

   (e) 0.9 m to mm

   $\left.\begin{matrix}\text{unit}\\\text{fraction}\end{matrix}\right\}\rule{2cm}{0.4pt}$

2. Convert each of the following. Round your answers to the nearest tenth as necessary.

   (a) 120 kilometers/hour to meters/second

   (b) 18 meters/second to kilometers/hour

ANSWERS

1. (a) 367 cm (b) $\dfrac{1\text{ m}}{100\text{ cm}}$; 0.92 m

   (c) $\dfrac{1\text{ m}}{100\text{ cm}}$; 4.327 m

   (d) $\dfrac{1\text{ cm}}{10\text{ mm}}$; 6.5 cm

   (e) $\dfrac{1000\text{ mm}}{1\text{ m}}$; 900 mm

2. (a) 33.3 m/sec (b) 64.8 km/hr

EXAMPLE 1 Using Unit Fractions to Convert Length Measurements

Convert the following.

(a) 5 km to m

Put the unit for the answer (meters) in the numerator of the unit fraction; put the unit you want to change (km) in the denominator.

$$\text{Unit fraction equivalent to 1}\left\{\dfrac{1000\text{ m}}{1\text{ km}}\begin{matrix}\leftarrow\text{ Unit for answer}\\\leftarrow\text{ Unit being changed}\end{matrix}\right.$$

Multiply. Divide out common units where possible.

$$5\text{ km}\cdot\dfrac{1000\text{ m}}{1\text{ km}}=\dfrac{5\text{ km}}{1}\cdot\dfrac{1000\text{ m}}{1\text{ km}}=\dfrac{5\cdot1000\text{ m}}{1}=5000\text{ m}$$

These units should match.

The answer makes sense because a kilometer is much longer than a meter, so 5 km will contain many meters.

(b) 18.6 cm to m

Multiply by a unit fraction that allows you to divide out centimeters.

$$\dfrac{18.6\text{ cm}}{1}\cdot\overbrace{\dfrac{1\text{ m}}{100\text{ cm}}}^{\text{Unit fraction}}=\dfrac{18.6}{100}\text{ m}=0.186\text{ m}$$

There are 100 cm in a meter, so 18.6 cm will be a small part of a meter. The answer makes sense.

Work Problem 1 at the Side.

EXAMPLE 2 Using Several Unit Fractions

Convert 100 kilometers/hour to meters/second.

As in the previous section, use several unit fractions to make this conversion. Remember the following relationships.

$$1\text{ kilometer} = 1000\text{ meters}$$
$$1\text{ hour} = 60\text{ minutes}$$
$$1\text{ minute} = 60\text{ seconds}$$

$$\dfrac{100\text{ km}}{1\text{ hr}}\cdot\dfrac{1000\text{ m}}{1\text{ km}}\cdot\dfrac{1\text{ hr}}{60\text{ min}}\cdot\dfrac{1\text{ min}}{60\text{ sec}}=\dfrac{100\cdot1000\text{ m}}{60\cdot60\text{ sec}}$$

$$\approx 27.8\text{ m/sec}$$

Work Problem 2 at the Side.

OBJECTIVE 3 Move the decimal point to convert among units. By now you have probably noticed that conversions among metric units are made by multiplying or dividing by 10, by 100, or by 1000. A quick way to *multiply* by 10 is to move the decimal point one place to the *right*. Move it two places to the right to multiply by

SECTION 0.28 The Metric System—Length 0-161

3. Do each multiplication or division by hand or on a calculator. Compare your answer to the one you get by moving the decimal point.

(a) 43.5 • 10 = _____

43.5 gives 435.

(b) 43.5 ÷ 10 = _____

43.5 gives _____.

(c) 28 • 100 = _____

28.00 gives _____.

(d) 28 ÷ 100 = _____

28. gives _____.

(e) 0.7 • 1000 = _____

0.700 gives _____.

(f) 0.7 ÷ 1000 = _____

000.7 gives _____.

ANSWERS
3. (a) 435 (b) 4.35 (c) 2800
   (d) 0.28 (e) 700 (f) 0.0007

100, three places to multiply by 1000. *Dividing* is done by moving the decimal point to the *left* in the same manner.

Work Problem 3 at the Side.

An alternate conversion method to unit fractions is moving the decimal point using this **metric conversion line.**

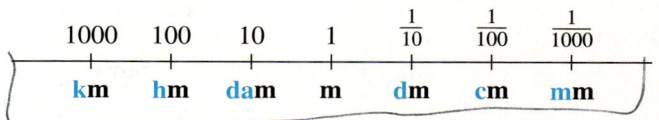

Here are the steps for using the conversion line.

Using the Metric Conversion Line

Step 1 Find the unit you are given on the metric conversion line.

Step 2 Count the number of places to get from the unit you are given to the unit you want in the answer.

Step 3 Move the decimal point the *same number of places* and in the *same direction* as you did on the conversion line.

EXAMPLE 3 Using the Metric Conversion Line

Use the metric conversion line to make the following conversions.

(a) 5.702 km to m

Find km on the metric conversion line. To get to m , you move *three places* to the *right*. So move the decimal point in 5.702 *three places* to the *right*.

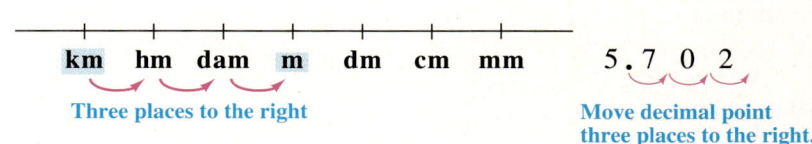

5.702 km = 5702 m

(b) 69.5 cm to m

Find cm on the conversion line. To get to m , move *two places* to the *left*.

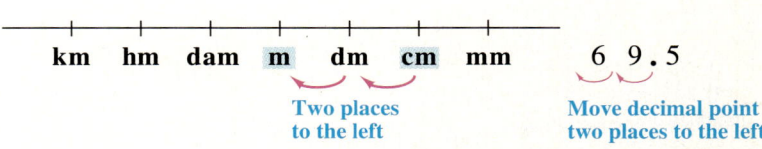

69.5 cm = 0.695 m

0-162 CHAPTER 0 Topics in Arithmetic

4. Convert using the metric conversion line.

   (a) 12.008 km to m

   (b) 561.4 m to km

   (c) 20.7 cm to m

   (d) 20.7 cm to mm

   (e) 4.66 m to cm

   (f) 85.6 mm to cm

5. Convert using the metric conversion line.

   (a) 9 m to mm

   (b) 3 cm to m

   (c) 14.6 km to m

   (d) 5 mm to cm

   (e) 70 m to km

   (f) 0.8 m to cm

ANSWERS
4. (a) 12,008 m (b) 0.5614 km
   (c) 0.207 m (d) 207 mm
   (e) 466 cm (f) 8.56 cm
5. (a) 9000 mm (b) 0.03 m
   (c) 14,600 m (d) 0.5 cm
   (e) 0.07 km (f) 80 cm

(c) 8.1 cm to mm

From **cm** to **mm** is *one place* to the *right*.

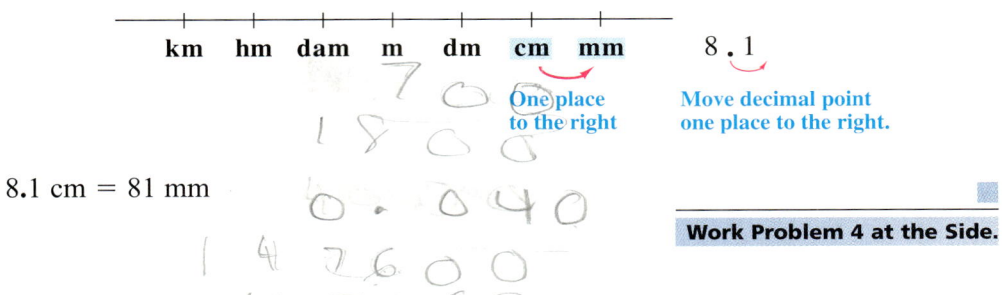

8.1 cm = 81 mm

Work Problem 4 at the Side.

EXAMPLE 4 Practicing Length Conversions

Convert using the metric conversion line.

(a) 1.28 m to mm

Moving from **m** to **mm** is going *three places to the right*. In order to move the decimal point in 1.28 three places to the right, you must write a 0 as a placeholder.

1.28**0**   Zero is written in as a placeholder.

Move decimal point three places to the right.

1.28 m = 1280 mm

(b) 60 cm to m

From **cm** to **m** is two places to the left. The decimal point in 60 starts at the *far right side* because 60 is a whole number. Then move it two places to the left.

60 cm = 0.60 m, which is equivalent to 0.6 m.

(c) 8 m to km

From **m** to **km** is three places to the left. The decimal point in 8 starts at the far right side. In order to move it three places to the left, you must write two 0s as placeholders.

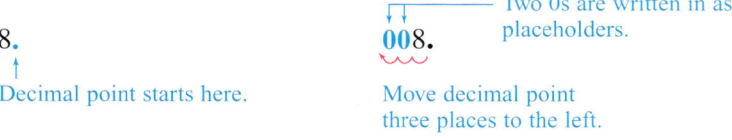

8 m = 0.008 km.

Work Problem 5 at the Side.

0.28 EXERCISES

For Extra Help
 Student's Solutions Manual
 MyMathLab
 InterAct Math Tutorial Software
 AW Math Tutor Center
MathXL

Use your knowledge of the meaning of metric prefixes to fill in the blanks.

1. *kilo* means __1000__ so

   1 km = __1000__ m

2. *deka* means __10__ so

   1 dam = __10__ m

3. *milli* means __$\frac{1}{1000}$__ so

   1 mm = __$\frac{1}{1000}$__ m

4. *deci* means __$\frac{1}{10}$__ so

   1 dm = __$\frac{1}{10}$__ m

5. *centi* means __$\frac{1}{100}$__ so

   1 cm = __$\frac{1}{100}$__ m

6. *hecto* means __100__ so

   1 hm = __100__ m

Convert each measurement. Use unit fractions or the metric conversion line. See Examples 1, 3, and 4.

7. 7 m to cm

8. 18 m to cm

9. 40 mm to m   0.040

10. 6 mm to m

11. 9.4 km to m

12. 0.7 km to m

13. 509 cm to m

14. 30 cm to m

15. 400 mm to cm

16. 25 mm to cm

17. 0.91 m to mm

18. 4 m to mm

19. Use two unit fractions to convert 5.6 mm to km.

20. Use two unit fractions to convert 16.5 km to mm.

Convert the following. Round your answers to the nearest tenth as necessary. See Example 2.

21. 60 kilometers/hour to meters/second

22. 75 kilometers/hour to meters/second

23. 88 kilometers/hour to meters/second

24. 110 kilometers/hour to meters/second

25. 25 meters/second to kilometers/hour

26. 35 meters/second to kilometers/hour

27. 45 meters/second to kilometers/hour

28. 20 meters/second to kilometers/hour

0-164 CHAPTER 0 Topics in Arithmetic

0.29 The Metric System—Capacity and Weight (Mass)

OBJECTIVES

1. Learn the basic metric units of capacity.
2. Convert among metric capacity units.
3. Learn the basic metric units of weight (mass).
4. Convert among metric weight (mass) units.

We use capacity units to measure liquids, such as the amount of milk in a recipe, the gasoline in our car tank, and the water in an aquarium. (The English capacity units we've been using are cups, pints, quarts, and gallons.) The basic metric unit for capacity is the **liter** (also spelled *litre*). The capital letter **L** is the symbol for liter, to avoid confusion with the numeral 1.

OBJECTIVE 1 Learn the basic metric units of capacity.   The liter is related to metric length in this way: a box that measures 10 cm on every side holds exactly one liter. (The volume of the box is 1000 cubic centimeters.) A liter is just a little more than 1 quart.

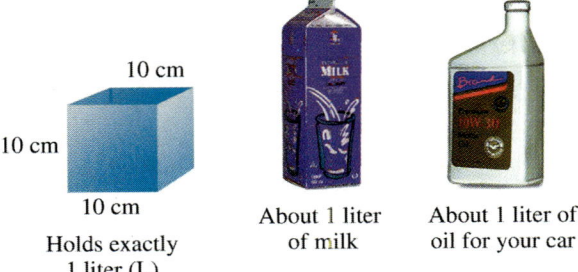

Holds exactly 1 liter (L) About 1 liter of milk About 1 liter of oil for your car

A liter is a little more than one quart (just $\frac{1}{4}$ cup more).

To make larger or smaller capacity units, we use the same **prefixes** as we did with length units. For example, *kilo* means 1000 so a *kilo*meter is 1000 meters. In the same way, a *kilo*liter is 1000 liters.

Prefix	kilo- liter	hecto- liter	deka- liter	liter	deci- liter	centi- liter	milli- liter
Meaning	1000 liters	100 liters	10 liters	1 liter	$\frac{1}{10}$ of a liter	$\frac{1}{100}$ of a liter	$\frac{1}{1000}$ of a liter
Symbol	kL	hL	daL	L	dL	cL	mL

The capacity units you will use most often in daily life are liters (L) and *milli-liters* (mL). A tiny box that measures 1 cm on every side holds exactly one milliliter. (In medicine, this small amount is also called 1 cubic centimeter, or 1 cc for short.) It takes 1000 mL to make 1 L. Here are some other useful comparisons.

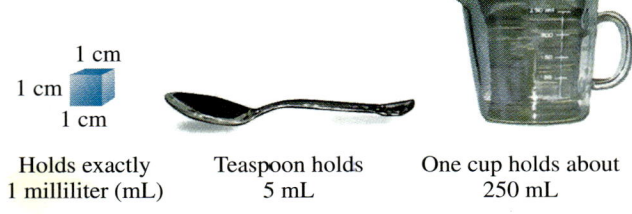

Holds exactly 1 milliliter (mL)   Teaspoon holds 5 mL   One cup holds about 250 mL

1. Convert.

   (a) 9 L to mL

   (b) 0.75 L to mL

   (c) 500 mL to L

   (d) 5 mL to L

   (e) 2.07 L to mL

   (f) 3275 mL to L

OBJECTIVE 2 Convert among metric capacity units. Just as with length units, you can convert between milliliters and liters using unit fractions.

Metric Capacity Relationships

$1\ L = 1000\ mL$, so the unit fractions are:

$$\frac{1\ L}{1000\ mL} \quad \text{or} \quad \frac{1000\ mL}{1\ L}$$

Or you can use a metric conversion line to decide how to move the decimal point.

```
  1000   100    10     1    1/10   1/100  1/1000
   |      |      |      |     |      |      |
   kL    hL    daL     L    dL     cL     mL
```

EXAMPLE 1 Converting among Metric Capacity Units

Convert using unit fractions or the metric conversion line.

(a) 2.5 L to mL

Using unit fractions:

Multiply by a unit fraction that allows you to divide out liters.

$$\frac{2.5\ \cancel{L}}{1} \cdot \frac{1000\ mL}{1\ \cancel{L}} = 2500\ mL$$

Using the metric conversion line:

From **L** to **mL** is *three places* to the *right*.

2.500   Write two 0s as placeholders.

2.5 L = 2500 mL

(b) 80 mL to L

Using unit fractions:

Multiply by a unit fraction that allows you to divide out mL.

$$\frac{80\ \cancel{mL}}{1} \cdot \frac{1\ L}{1000\ \cancel{mL}}$$

$$= \frac{80}{1000}\ L = 0.08\ L$$

Using the metric conversion line:

From **mL** to **L** is *three places* to the *left*.

80.   080.
↑
Decimal point   Move three
starts here.   places left.

80 mL = 0.080 L or 0.08 L

Work Problem 1 at the Side.

OBJECTIVE 3 Learn the basic metric units of weight (mass). The **gram** is the basic metric unit for *mass*. Although we often call it "weight," there is a difference. Weight is a measure of the pull of gravity; the farther you are from the center of the earth, the less you weigh. In outer space you become weightless, but your mass, the amount

ANSWERS
1. (a) 9000 mL (b) 750 mL
   (c) 0.5 L (d) 0.005 L
   (e) 2070 mL (f) 3.275 L

of matter in your body, stays the same regardless of where you are. In science courses, it will be important to distinguish between the weight of an object and its mass. But for everyday purposes, we will use the word *weight*.

The gram is related to metric length in this way: The weight of the water in a box measuring 1 cm on every side is 1 gram. This is a very tiny amount of water (1 mL) and a very small weight. One gram is also the weight of a dollar bill or a single raisin. A nickel weighs 5 grams.

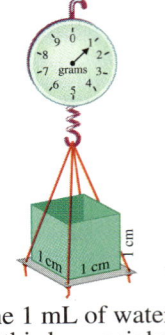

The 1 mL of water in this box weighs 1 gram.

A dollar bill weighs 1 gram.

To make larger or smaller weight units, we use the same **prefixes** as we did with length and capacity units. For example, *kilo* means 1000 so a *kilo*meter is 1000 meters, a *kilo*liter is 1000 liters, and a *kilo*gram is 1000 grams.

Prefix	kilo-gram	hecto-gram	deka-gram	gram	deci-gram	centi-gram	milli-gram
Meaning	1000 grams	100 grams	10 grams	1 gram	$\frac{1}{10}$ of a gram	$\frac{1}{100}$ of a gram	$\frac{1}{1000}$ of a gram
Symbol	kg	hg	dag	g	dg	cg	mg

The units you will use most often in daily life are kilograms (kg), grams (g), and milligrams (mg). *Kilo*grams are used instead of pounds. A kilogram is 1000 grams. It is about 2.2 pounds. Two packages of butter plus one stick of butter weigh about 1 kg. An average newborn baby weighs 3 to 4 kg; a college football player might weigh 100 to 130 kg.

Extremely small weights are measured in *milli*grams. It takes 1000 mg to make 1 g. Recall that a dollar bill weighs about 1 g. Imagine cutting it into 1000 pieces; the weight of one tiny piece would be 1 mg. Dosages of medicine and vitamins are given in milligrams. You will also use milligrams in science classes.

Cut a dollar bill into 1000 pieces. One tiny piece weighs 1 milligram.

SECTION 0.29 The Metric System—Capacity and Weight (Mass) 0-167

2. Convert.

(a) 10 kg to g

OBJECTIVE 4 Convert among metric weight (mass) units. As with length and capacity, you can convert among metric weight units by using unit fractions. The unit fractions you need are shown here.

Metric Weight (Mass) Relationships

1 kg = 1000 g so the unit fractions are:

$$\frac{1 \text{ kg}}{1000 \text{ g}} \quad \text{or} \quad \frac{1000 \text{ g}}{1 \text{ kg}}$$

1 g = 1000 mg so the unit fractions are:

$$\frac{1 \text{ g}}{1000 \text{ mg}} \quad \text{or} \quad \frac{1000 \text{ mg}}{1 \text{ g}}$$

(b) 45 mg to g

Or you can use a metric conversion line to decide how to move the decimal point.

(c) 6.3 kg to g

```
   1000   100    10     1    1/10   1/100  1/1000
    |      |      |     |     |      |      |
    kg     hg    dag    g     dg     cg     mg
```

EXAMPLE 2 Converting among Metric Weight Units

Convert using unit fractions or the metric conversion line.

(d) 0.077 g to mg

(a) 7 mg to g

Using unit fractions:

Multiply by a unit fraction that allows you to divide out mg.

$$\frac{7 \text{ mg}}{1} \cdot \frac{1 \text{ g}}{1000 \text{ mg}} = \frac{7}{1000} \text{ g}$$

$$= 0.007 \text{ g}$$

Using the metric conversion line:

From **mg** to **g** is *three places* to the *left*.

7.   007.
↑
Decimal point   Move three
starts here.   places left.

7 mg = 0.007 g

(e) 5630 g to kg

(b) 13.72 kg to g

Using unit fractions:

Multiply by a unit fraction that allows you to divide out kg.

$$\frac{13.72 \text{ kg}}{1} \cdot \frac{1000 \text{ g}}{1 \text{ kg}} = 13,720 \text{ g}$$
↑
A comma
(not a decimal point)

(f) 90 g to kg

Using the metric conversion line:

From **kg** to **g** is *three* places to the *right*.

13.720   Decimal point moves
           three places to
           the right.

13.72 kg = 13,720 g
           ↑
        A comma
        (not a decimal point)

ANSWERS
2. (a) 10,000 g (b) 0.045 g
   (c) 6300 g   (d) 77 mg
   (e) 5.63 kg   (f) 0.09 kg

Work Problem 2 at the Side.

0.29 EXERCISES

For Extra Help

 Student's Solutions Manual

 MyMathLab

 InterAct Math Tutorial Software

 AW Math Tutor Center

MathXL

1. Describe how you decide which unit fraction to use when converting 6.5 kg to grams.

2. Explain in your own words how the meter, liter, and gram are related.

Convert each measurement. Use unit fractions or the metric conversion line. See Examples 2 and 4.

3. 15 L to mL

4. 6 L to mL

5. 3000 mL to L

6. 18,000 mL to L

7. 925 mL to L

8. 200 mL to L

9. 8 mL to L

10. 25 mL to L

11. 4.15 L to mL

12. 11.7 L to mL

13. 8000 g to kg

14. 25,000 g to kg

15. 5.2 kg to g

16. 12.42 kg to g

17. 0.85 g to mg

18. 0.2 g to mg

19. 30,000 mg to g

20. 7500 mg to g

21. 598 mg to g

22. 900 mg to g

23. 60 mL to L

24. 6.007 kg to g

25. 3 g to kg

26. 12 mg to g

27. 0.99 L to mL

28. 13,700 mL to L

0.30 Metric–English Conversions

OBJECTIVE

1. Use unit fractions to convert between metric and English units.

OBJECTIVE 1 Use unit fractions to convert between metric and English units. Until the United States has switched completely from the English system to the metric system, it will be necessary to make conversions from one system to the other. *Approximate* conversions can be made with the help of the following table, in which the values have been rounded to the nearest hundredth or thousandth. (The only value that is exact, not rounded, is 1 inch = 2.54 cm.)

Metric to English	English to Metric
1 kilometer ≈ 0.62 mile	1 mile ≈ 1.61 kilometers
1 meter ≈ 1.09 yards	1 yard ≈ 0.91 meter
1 meter ≈ 3.28 feet	1 foot ≈ 0.30 meter
1 centimeter ≈ 0.39 inch	1 inch = 2.54 centimeters
1 liter ≈ 0.26 gallon	1 gallon ≈ 3.78 liters
1 liter ≈ 1.06 quarts	1 quart ≈ 0.95 liter
1 kilogram ≈ 2.20 pounds	1 pound ≈ 0.45 kilogram
1 gram ≈ 0.035 ounce	1 ounce ≈ 28.35 grams

EXAMPLE 1 Converting between Metric and English Length Units

Convert 10 m to yards using unit fractions. Round your answer to the nearest tenth if necessary.

We're changing from a metric unit to an English unit. In the "Metric to English" side of the table, you see that 1 meter ≈ 1.09 yards. Two unit fractions can be written using that information.

$$\frac{1 \text{ m}}{1.09 \text{ yd}} \quad \text{or} \quad \frac{1.09 \text{ yd}}{1 \text{ m}}$$

Multiply by the unit fraction that allows you to divide out meters (that is, meters is in the denominator).

$$10 \text{ m} \cdot \frac{1.09 \text{ yd}}{1 \text{ m}} = \frac{10 \text{ m}}{1} \cdot \frac{1.09 \text{ yd}}{1 \text{ m}} = \frac{(10)(1.09 \text{ yd})}{1} = 10.9 \text{ yd}$$

These units should match.

$$10 \text{ m} \approx 10.9 \text{ yd}$$

1. Convert using unit fractions. Round your answers to the nearest tenth.

   (a) 23 m to yards

   (b) 40 cm to inches

   (c) 5 mi to kilometers (Look at the "English to Metric" side of the table.)

   (d) 12 in. to centimeters

2. Convert. Use the values from the table on the previous page to make unit fractions. Round answers to the nearest tenth.

   (a) 17 kg to pounds

   (b) 5 L to quarts

   (c) 90 g to ounces

   (d) 3.5 gal to liters

   (e) 145 lb to kilograms

   (f) 8 oz to grams

NOTE In Example 1, you could also use the other numbers from the table involving meters and yards: 1 yard ≈ 0.91 meter.

$$\frac{10 \text{ m}}{1} \cdot \frac{1 \text{ yd}}{0.91 \text{ m}} = \frac{10}{0.91} \text{ yd} \approx 10.99 \text{ yd}$$

The answer is slightly different because the values in the table are approximate. Also, you have to divide instead of multiply, which is usually more difficult to do without a calculator. We will use the first method in this section.

Work Problem 1 at the Side.

EXAMPLE 2 Converting between Metric and English Weight and Capacity Units

Convert using unit fractions. Round your answers to the nearest tenth.

(a) 3.5 kg to pounds

Look in the "Metric to English" side of the table on the previous page to see that 1 kilogram ≈ 2.20 pounds. Use this information to write a unit fraction that allows you to divide out kilograms.

$$\frac{3.5 \text{ kg}}{1} \cdot \frac{2.20 \text{ lb}}{1 \text{ kg}} = \frac{(3.5)(2.20 \text{ lb})}{1} = 7.7 \text{ lb}$$

3.5 kg ≈ 7.7 lb

(b) 18 gal to liters

Look in the "English to Metric" side of the table to see that 1 gallon ≈ 3.78 liters. Write a unit fraction that will allow you to divide out gallons.

$$\frac{18 \text{ gal}}{1} \cdot \frac{3.78 \text{ L}}{1 \text{ gal}} = \frac{(18)(3.78 \text{ L})}{1} = 68.04 \text{ L}$$

68.04 rounded to the nearest tenth is 68.0.
18 gal ≈ 68.0 L

(c) 300 g to ounces

In the "Metric to English" side of the table, 1 gram ≈ 0.035 ounce.

$$\frac{300 \text{ g}}{1} \cdot \frac{0.035 \text{ oz}}{1 \text{ g}} = 10.5 \text{ oz}$$

300 g ≈ 10.5 oz

CAUTION Because the metric and English systems were developed independently, almost all comparisons are approximate. Your answers should be written with the "≈" symbol to show they are approximate.

Work Problem 2 at the Side.

ANSWERS

1. (a) 23 m ≈ 25.1 yd
   (b) 40 cm ≈ 15.6 in.
   (c) 5 mi ≈ 8.1 km
   (d) 12 in. ≈ 30.5 cm
2. (a) 17 kg ≈ 37.4 lb
   (b) 5 L ≈ 5.3 qt
   (c) 90 g ≈ 3.2 oz
   (d) 3.5 gal ≈ 13.2 L
   (e) 145 lb ≈ 65.3 kg
   (f) 8 oz ≈ 226.8 g

0.30 EXERCISES

For Extra Help
- Student's Solutions Manual
- MyMathLab
- InterAct Math Tutorial Software
- AW Math Tutor Center
- MathXL

Use the table on the first page of this section and unit fractions to make approximate conversions from metric to English or English to metric. Round your answers to the nearest tenth. See Examples 1 and 2.

1. 20 m to yards
2. 8 km to miles
3. 80 m to feet

4. 85 cm to inches
5. 16 ft to meters
6. 3.2 yd to meters

 150 g to ounces
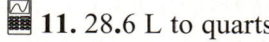 8. 2.5 oz to grams
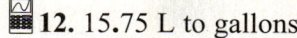 9. 248 lb to kilograms

10. 7.68 kg to pounds
11. 28.6 L to quarts
12. 15.75 L to gallons

13. The 3M Company used 5 g of pure gold to coat Michael Johnson's Olympic track shoes.
    (a) How many ounces of gold were used, to the nearest tenth?
    (b) Was this enough extra weight to slow him down?

14. The fastest nerve signals in the human body travel 120 m per second. (*Source: Big Book of Knowledge.*) How many feet per second do the signals travel?

15. The heavy duty wash cycle in a dishwasher uses 8.4 gal of water. How many liters does it use, to the nearest tenth? (*Source:* Frigidaire.)

16. The rinse-and-hold cycle in a dishwasher uses only 4.5 L of water. How many gallons does it use, to the nearest tenth? (*Source:* Frigidaire.)

17. The smallest pet fish are dwarf gobies, which are half an inch long. (*Source: Big Book of Knowledge.*) How many centimeters long is a dwarf gobie, to the nearest tenth? (*Hint:* Write half an inch in decimal form.)

18. A Toshiba Protégé laptop computer weighs 4.4 lb. How many kilograms does it weigh, to the nearest tenth? (*Source:* Toshiba.)

0.27–0.30 TEST

Convert the following measurements.

1. 9 gal = _____ qt
2. 45 ft = _____ yd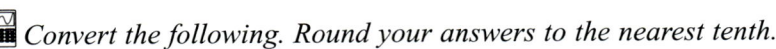
3. 135 min = _____ hr
4. 9 in. = _____ ft
5. $3\frac{1}{2}$ lb = _____ oz
6. 5 days = _____ min

Convert the following. Round your answers to the nearest tenth.

7. 15 miles/hour to feet/second

8. 48 kilometers/hour to meters/second

Convert the following measurements.

9. 250 cm to meters
10. 4.6 km to meters
11. 5 mm to centimeters
12. 325 mg to grams
13. 16 L to milliliters
14. 0.4 kg to grams
15. 10.55 m to centimeters
16. 95 mL to liters

*Use the table from **Section 0.30** and unit fractions to convert the following measurements. Round your answers to the nearest tenth if necessary.*

17. 6 ft to meters
18. 125 lb to kilograms

19. 50 L to gallons
20. 8.1 km to miles

0.31 Basics of Percent

OBJECTIVES

1. Learn the meaning of percent.
2. Write percents as decimals.
3. Write decimals as percents.

Notice that this figure has one hundred squares of equal size. Eleven of the squares are shaded. The shaded portion is $\frac{11}{100}$, or 0.11, of the total figure.

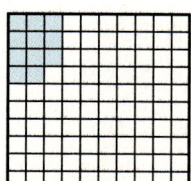

The shaded portion is also 11% of the total, or "eleven parts out of 100 parts." Read **11%** as "eleven percent."

OBJECTIVE 1 Learn the meaning of percent. As we just saw, a percent is a ratio with a denominator of 100.

> **The Meaning of Percent**
>
> **Percent** means *per one hundred*. The "%" sign is used to show the number of parts out of one hundred parts.

1. Write as percents.

    (a) In a group of 100 people, 63 are unmarried. What percent are unmarried?

    (b) The sales tax is $5 per $100. What percent is this?

EXAMPLE 1 Understanding Percent

(a) If 43 out of 100 students are men, then 43 per (out of) 100 or $\frac{43}{100}$ or **43%** of the students are men.

(b) If a person pays a tax of $7 on every $100 of purchases, then the tax rate is $7 per $100. The ratio is $\frac{7}{100}$ and the percent of tax is **7%**.

Work Problem 1 at the Side.

(c) Out of 100 students, 68 are working part-time. What percent are working part-time?

OBJECTIVE 2 Write percents as decimals. If 8% means 8 parts out of 100 parts or $\frac{8}{100}$, then $p\%$ means p parts out of 100 parts or $\frac{p}{100}$. Because $\frac{p}{100}$ is another way to write the division $p \div 100$, we have

$$p\% = \frac{p}{100} = p \div 100.$$

> **Writing a Percent as a Decimal**
>
> $p\% = \dfrac{p}{100}$ or $p\% = p \div 100$
>
> As a fraction   As a decimal

ANSWERS

1. (a) 63% (b) 5% (c) 68%

0-174　CHAPTER 0　Topics in Arithmetic

2. Write each percent as a decimal.

   (a) 68%

   (b) 34%

   (c) 58.5%

   (d) 175%

■ **EXAMPLE 2**　Writing Percents as Decimals

Write each percent as a decimal.

(a) 47%

$$p\% = p \div 100$$
$$47\% = 47 \div 100 = 0.47 \quad \text{Decimal form}$$

(b) 76%　　$76\% = 76 \div 100 = 0.76$　Decimal form

(c) 28.2%　$28.2\% = 28.2 \div 100 = 0.282$　Decimal form

(d) 100%　$100\% = 100 \div 100 = 1.00$　Decimal form

CAUTION　In Example 2(d), notice that 100% is 1.00, or 1, which is a whole number. Whenever you have a percent that is 100% or higher, the equivalent decimal number will be a number of 1 or higher.

Work Problem 2 at the Side.

The resulting answers in Example 2 suggest the following procedure for writing a percent as a decimal.

Writing a Percent as a Decimal

Step 1　Drop the percent sign.

Step 2　Divide by 100.

NOTE　A quick way to divide a number by 100 is to move the decimal point two places to the left.

■ **EXAMPLE 3**　Writing Percents as Decimals by Moving the Decimal Point

Write each percent as a decimal by moving the decimal point two places to the left.

(a) 17%

$17\% = 17.\%$　Decimal point starts at far right side.

　　　　0.17　← Percent sign is dropped. (Step 1)

　　　　　　　Decimal point is moved two places to the left. (Step 2)

$17\% = 0.17$

(b) 160%

$160\% = 160.\% = 1.60$ or 1.6　Decimal starts at far right side. 1.60 is equivalent to 1.6.

(c) 4.9%

　　$.049\%$　0 is attached so the decimal point can be moved two places to the left.

$4.9\% = 0.049$

ANSWERS

2. (a) 0.68　(b) 0.34　(c) 0.585
   (d) 1.75

SECTION 0.31 Basics of Percent 0-175

3. Write each percent as a decimal.

(a) 96%

(b) 6%

(c) 24.8%

(d) 0.9%

(d) 0.6%

$0.6\% = 0.006$ 0s are attached so the decimal point can be moved.

CAUTION Look at Example 3(d), where 0.6% is less than 1%. Because 1% is equivalent to 0.01 or $\frac{1}{100}$, any fraction of a percent smaller than 1% is less than 0.01.

Work Problem 3 at the Side.

OBJECTIVE 3 Write decimals as percents. You can write any decimal as a percent. For example, the decimal 0.78 is the same as the fraction

$$\frac{78}{100}.$$

This fraction means 78 of 100 parts, or 78%. The following steps give the same result.

Writing a Decimal as a Percent

Step 1 Multiply by 100.

Step 2 Attach a percent sign.

NOTE A quick way to divide or multiply a number by 100 is to move the decimal point two places to the left or two places to the right, respectively.

Multiply decimal by 100; move decimal point two places to the *right*.

Decimal ⟶ Percent

Divide percent by 100; move decimal point two places to the *left*.

EXAMPLE 4 Writing Decimals as Percents by Moving the Decimal Point

Write each decimal as a percent by moving the decimal point two places to the right.

(a) 0.21

0.21% ← Percent sign is attached. (Step 1)

⎿— Decimal point is moved two places to the right. (Step 2)

0.21 = 21%

⎿— Decimal point is not written with whole number percents.

(b) 0.529 = 52.9%

(c) 1.92 = 192%

ANSWERS

3. (a) 0.96 (b) 0.06 (c) 0.248
   (d) 0.009

0-176 CHAPTER 0 Topics in Arithmetic

4. Write each decimal as a percent.

(a) 0.95   (b) 0.18

(c) 0.09   (d) 0.617

(e) 0.834   (f) 5.34

(g) 2.8   (h) 4

ANSWERS
4. (a) 95% (b) 18% (c) 9%
   (d) 61.7% (e) 83.4% (f) 534%
   (g) 280% (h) 400%

(d) 2.5

2.50%   0 is attached so the decimal point can be moved two places to the right.

2.5 = 250%

(e) 3

3. = 3.00%   Two 0s are attached.

so 3 = 300%

CAUTION Look at Examples 4(c), 4(d), and 4(e) where 1.92, 2.5, and 3 are greater than 1. Because the number 1 is equivalent to 100%, all numbers greater than 1 will be greater than 100%.

Work Problem 4 at the Side.

0.31 EXERCISES

For Extra Help

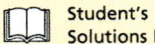 Student's Solutions Manual

 MyMathLab

 InterAct Math Tutorial Software

 AW Math Tutor Center

MathXL

Write each percent as a decimal. See Examples 2 and 3.

1. 15%   2. 41%   3. 60%   4. 40%

5. 25%   6. 35%   7. 140%   8. 250%

9. 5.5%   10. 6.7%   11. 100%   12. 600%

13. 0.5%   14. 0.25%   15. 0.35%   16. 0.75%

Write each decimal as a percent. See Example 4.

17. 0.8   18. 0.4   19. 0.58   20. 0.25

21. 0.01   22. 0.07   23. 0.125   24. 0.875

25. 0.375   26. 0.625   27. 2   28. 5

29. 3.7   30. 2.2   31. 0.0312   32. 0.0625

33. 4.162   34. 8.715   35. 0.0028   36. 0.0064

SECTION 0.31   Basics of Percent   0-177

37. Fractions, decimals, and percents are all used to describe a part of something. The use of percents is much more common than fractions and decimals. Why do you suppose this is true?

38. List five uses of percent that are or will be part of your life. Consider the activities of working, shopping, saving, and planning for the future.

Write each percent as a decimal and each decimal as a percent. See Examples 2–4.

39. In Lincoln, 45% of the refuse is recycled.   0.45

40. At College of DuPage, 82% of the students work part-time.   0.82

41. At Bett's Boutique, 18% of the items sold are returned.   0.18

42. There was a 43.2% voter turnout at the election.   0.432

43. The property tax rate in Alpine County is 0.035.   3.5

44. A church building fund has 0.49 of the money needed.   49%

45. The number of people successfully completing CPR training this session is 2 times that of the last session.

46. The number of newspaper subscribers was 4 times as great as last quarter.

47. Only 0.005 of the total population has this genetic defect.

48. The return rate of defective keyboards is 0.0075 of total output.

49. The patient's blood pressure was 153.6% of normal.

50. Success with the diet was 248.7% greater than anticipated.

0.32 Percents and Fractions

OBJECTIVES

1. Write percents as fractions.
2. Write fractions as percents.

OBJECTIVE 1 Write percents as fractions. Percents can be written as fractions by using what we learned in the previous section.

Writing a Percent as a Fraction

$$p\% = \frac{p}{100}, \text{ as a fraction.}$$

---

1. Write each percent as a fraction or mixed number in lowest terms.

   (a) 50%

   (b) 35%

   (c) 52%

   (d) 23%

   (e) 125%

   (f) 250%

EXAMPLE 1 Writing Percents as Fractions

Write each percent as a fraction or mixed number in lowest terms.

(a) 25%

As we saw in the last section, 25% can be written as a decimal.

$$25\% = 25 \div 100 = 0.25 \quad \text{Percent sign dropped}$$

Because 0.25 means 25 hundredths,

$$0.25 = \frac{25 \div 25}{100 \div 25} = \frac{1}{4}. \quad \text{Lowest terms}$$

It is not necessary, however, to write 25% as a decimal first. Just write

$$25\% = \frac{25}{100} \quad \text{25 per 100}$$

$$= \frac{1}{4}. \quad \text{Lowest terms}$$

(b) 76%

Write 76% as $\frac{76}{100}$. ⟵ The percent becomes the numerator.

⟵ The *denominator* is always 100 because percent means *parts per 100*.

Write $\frac{76}{100}$ in lowest terms.

$$\frac{76 \div 4}{100 \div 4} = \frac{19}{25} \quad \text{Lowest terms}$$

(c) 150%

$$150\% = \frac{150}{100} = \frac{3}{2} = 1\frac{1}{2} \quad \text{Mixed number}$$

NOTE Remember that percent means *per 100*.

ANSWERS

1. (a) $\frac{1}{2}$ (b) $\frac{7}{20}$ (c) $\frac{13}{25}$ (d) $\frac{23}{100}$
   (e) $1\frac{1}{4}$ (f) $2\frac{1}{2}$

Work Problem 1 at the Side.

SECTION 0.32 Percents and Fractions 0-179

2. Write each percent as a fraction in lowest terms.

(a) 37.5%

(b) 62.5%

(c) 4.5%

(d) $66\frac{2}{3}\%$

(e) $10\frac{1}{3}\%$

(f) $87\frac{1}{2}\%$

The next example shows how to write decimal and fraction percents as fractions.

EXAMPLE 2 Writing Decimal or Fraction Percents as Fractions

Write each percent as a fraction in lowest terms.

(a) 15.5%

Write 15.5 over 100.

$$15.5\% = \frac{15.5}{100}$$

To get a whole number in the numerator, multiply the numerator and denominator by 10. (Recall that multiplying by $\frac{10}{10}$ is the same as multiplying by 1.)

$$\frac{15.5}{100} = \frac{15.5 \cdot 10}{100 \cdot 10} = \frac{155}{1000}$$

Write in lowest terms.

$$\frac{155 \div 5}{1000 \div 5} = \frac{31}{200}$$

(b) $33\frac{1}{3}\%$

Write $33\frac{1}{3}$ over 100.

$$33\frac{1}{3}\% = \frac{33\frac{1}{3}}{100}$$

When we have a mixed number in the numerator, we must write the mixed number as an improper fraction.

Here, we write $33\frac{1}{3}$ as the improper fraction $\frac{100}{3}$. Now,

$$\frac{33\frac{1}{3}}{100} = \frac{\frac{100}{3}}{100}.$$

We rewrite the division problem in a horizontal form. Then we multiply by the reciprocal of the divisor.

Multiply by the reciprocal.

$$\frac{\frac{100}{3}}{100} = \frac{100}{3} \div 100 = \frac{100}{3} \div \frac{100}{1} = \frac{\cancel{100}}{3} \cdot \frac{1}{\cancel{100}} = \frac{1}{3}$$

NOTE In Example 2(a) we could have changed 15.5% to $15\frac{1}{2}\%$ and then written it as the improper fraction $\frac{31}{2}$ over 100, as was done in part (b). It is usually easier not to change decimals to fractions but to leave decimal percents as they are.

ANSWERS

2. (a) $\frac{3}{8}$ (b) $\frac{5}{8}$ (c) $\frac{9}{200}$ (d) $\frac{2}{3}$
   (e) $\frac{31}{300}$ (f) $\frac{7}{8}$

Work Problem 2 at the Side.

OBJECTIVE 2 Write fractions as percents. We will use the formula given at the beginning of this section to write fractions as percents.

$$p\% = \frac{p}{100}$$

■ EXAMPLE 3 Writing Fractions as Percents

Write each fraction as a percent. Round to the nearest tenth if necessary.

(a) $\dfrac{3}{5}$

Write this fraction as a percent by solving for p in the proportion

$$\frac{3}{5} = \frac{p}{100}$$

Find diagonal products and show that they are equivalent.

$$5 \cdot p = 3 \cdot 100$$
$$5 \cdot p = 300$$

Divide each side by 5.

$$\frac{\overset{1}{\cancel{5}} \cdot p}{\underset{1}{\cancel{5}}} = \frac{300}{5}$$

$$p = 60$$

This result means that $\frac{3}{5} = \frac{60}{100}$ or 60%.

NOTE Solving proportions can be reviewed in **Section 0.26.**

(b) $\dfrac{7}{8}$

Write a proportion.

$$\frac{7}{8} = \frac{p}{100}$$

$8 \cdot p = 7 \cdot 100$ Show that diagonal products
$8 \cdot p = 700$ are equivalent.

$$\frac{\overset{1}{\cancel{8}} \cdot p}{\underset{1}{\cancel{8}}} = \frac{700}{8} \qquad \text{Divide each side by 8.}$$

$$p = 87.5$$

Finally, $\frac{7}{8} = 87.5\%$.

NOTE If you think of $\frac{700}{8}$ as an improper fraction, changing it to a mixed number gives an answer of $87\frac{1}{2}$. So $\frac{7}{8} = 87.5\%$ or $87\frac{1}{2}\%$.

SECTION 0.32 Percents and Fractions 0-181

3. Write as percents. Round to the nearest tenth if necessary.

(a) $\dfrac{1}{2}$ (b) $\dfrac{7}{10}$

(c) $\dfrac{6}{25}$ (d) $\dfrac{5}{8}$

(e) $\dfrac{1}{6}$ (f) $\dfrac{2}{9}$

(c) $\dfrac{5}{6}$

Start with a proportion.

$$\dfrac{5}{6} = \dfrac{p}{100}$$

$6 \cdot p = 5 \cdot 100$ — Show that diagonal products are equivalent.
$6 \cdot p = 500$

$\dfrac{\cancel{6} \cdot p}{\cancel{6}} = \dfrac{500}{6}$ — Divide each side by 6.

$p = 83.\overline{3}$ — The answer is a repeating decimal.
$p \approx 83.3$ — Round to the nearest tenth.

Solving this proportion shows

$$\dfrac{5}{6} = 83.\overline{3}\% \approx 83.3\% \quad \text{(rounded)}.$$

NOTE You can treat $\dfrac{500}{6}$ as an improper fraction to get an exact answer of $83\dfrac{1}{3}\%$.

ANSWERS
3. (a) 50% (b) 70%
   (c) 24% (d) 62.5%
   (e) 16.7% (rounded); $16\dfrac{2}{3}\%$ (exact)
   (f) 22.2% (rounded); $22\dfrac{2}{9}\%$ (exact)

Work Problem 3 at the Side.

0.32 EXERCISES

For Extra Help
- Student's Solutions Manual
- MyMathLab
- InterAct Math Tutorial Software
- AW Math Tutor Center
- MathXL

Write each percent as a fraction or mixed number in lowest terms. See Examples 1 and 2.

1. 25%   2. 60%   3. 75%   4. 80%

5. 85%   6. 45%   7. 62.5%   8. 87.5%

9. 6.25%   10. 43.75%   11. $16\dfrac{2}{3}\%$ 12. $83\dfrac{1}{3}\%$

13. $6\dfrac{2}{3}\%$ 14. $46\dfrac{2}{3}\%$ 15. 0.5%   16. 0.8%

17. 120%   18. 140%   19. 375%   20. 225%

Write each fraction as a percent. Round percents to the nearest tenth if necessary. See Example 3.

21. $\dfrac{1}{2}$ 22. $\dfrac{4}{10}$ 23. $\dfrac{4}{5}$ 24. $\dfrac{3}{10}$

25. $\dfrac{1}{4}$ 26. $\dfrac{3}{4}$ 27. $\dfrac{37}{100}$ 28. $\dfrac{63}{100}$

29. $\dfrac{5}{8}$ 30. $\dfrac{1}{8}$ 31. $\dfrac{7}{8}$ 32. $\dfrac{3}{8}$

33. $\dfrac{9}{25}$ 34. $\dfrac{15}{25}$ 35. $\dfrac{23}{50}$ 36. $\dfrac{18}{50}$

37. $\dfrac{3}{20}$ 38. $\dfrac{9}{20}$ 39. $\dfrac{5}{6}$ 40. $\dfrac{1}{6}$

41. $\dfrac{5}{9}$ 42. $\dfrac{7}{9}$ 43. $\dfrac{1}{7}$ 44. $\dfrac{5}{7}$

Complete the chart. Round decimals to the nearest thousandth and percents to the nearest tenth if necessary.

	Fraction	Decimal	Percent
45.	_____	0.5	_____
46.	$\dfrac{3}{50}$	_____	_____
47.	_____	_____	87.5%
48.	$\dfrac{3}{4}$	_____	_____
49.	_____	0.8	_____
50.	_____	_____	60%

SECTION 0.32   Percents and Fractions   **0-183**

	Fraction	Decimal	Percent
51.	$\frac{1}{6}$	_____	_____
52.	$\frac{1}{3}$	_____	_____
53.	_____	0.25	_____
54.	_____	_____	37.5%
55.	_____	_____	12.5%
56.	$\frac{5}{8}$	_____	_____
57.	$\frac{2}{3}$	_____	_____
58.	_____	0.833 (rounded)	_____
59.	$\frac{2}{5}$	_____	_____
60.	$\frac{3}{10}$	_____	_____
61.	$\frac{8}{100}$	_____	_____
62.	_____	_____	100%
63.	$\frac{1}{200}$	_____	_____
64.	$\frac{1}{400}$	_____	_____

0.33 Using the Percent Proportion

OBJECTIVES

1. Learn the percent proportion.
2. Solve for an unknown value in a proportion.
3. Identify the percent.
4. Identify the whole.
5. Identify the part.

There are two ways to solve percent problems. One method uses proportions and is discussed in this and the next section. The other method uses the percent equation and is explained in **Section 0.35**.

OBJECTIVE 1 Learn the percent proportion. We have seen that a statement of two equivalent ratios is called a proportion.

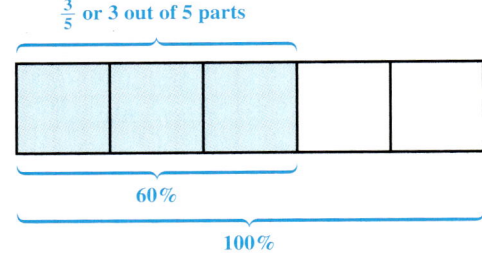

For example, the fraction $\frac{3}{5}$ is the same as the ratio 3 to 5, and 60% is the same as the ratio 60 to 100. As the figure shows, these two ratios are equivalent and make a proportion.

Work Problem 1 at the Side.

1. As a review of proportions, use the method of diagonal products to decide whether each proportion is *true* or *false*.

    (a) $\dfrac{1}{2} = \dfrac{25}{50}$

    (b) $\dfrac{3}{5} = \dfrac{75}{125}$

    (c) $\dfrac{7}{8} = \dfrac{180}{200}$

    (d) $\dfrac{32}{53} = \dfrac{160}{265}$

    (e) $\dfrac{112}{41} = \dfrac{332}{123}$

The **percent proportion** can be used to solve percent problems.

Percent Proportion

Part is to *whole* as percent is to 100.

$$\frac{\text{part}}{\text{whole}} = \frac{\text{percent}}{100} \quad \leftarrow \text{Always 100 because percent means per 100}$$

In the figure at the top of the page, the **whole** is 5 (the entire quantity), the **part** is 3 (the part of the whole), and the **percent** is 60. Write the percent proportion as follows.

$$\frac{\text{part} \rightarrow 3}{\text{whole} \rightarrow 5} = \frac{60}{100} \begin{array}{l}\leftarrow \text{percent}\\ \leftarrow 100\end{array}$$

OBJECTIVE 2 Solve for an unknown value in a proportion. As shown in **Section 0.26,** if any two of the three values (part, whole, or percent) in the percent proportion are known, the third can be found by solving the proportion.

ANSWERS
1. (a) true   (b) true   (c) false
   (d) true   (e) false

EXAMPLE 1 Using the Percent Proportion

Use the percent proportion and solve for the unknown value. Let x represent the unknown value.

(a) part = 12, percent = 25; find the whole.

$$\frac{\text{part}}{\text{whole}} = \frac{\text{percent}}{100} \quad \text{Percent proportion}$$

$$\text{Part} \rightarrow \frac{12}{x} = \frac{25}{100} \quad \text{or} \quad \frac{12}{x} = \frac{1}{4} \quad \text{Lowest terms}$$
Whole (unknown)

Find the diagonal products to solve this proportion.

$$\frac{12}{x} = \frac{1}{4}$$

$x \cdot 1$ and $12 \cdot 4$

Show that the diagonal products are equivalent.

$$x \cdot 1 = 12 \cdot 4$$
$$x = 48$$

The whole is 48.

CAUTION You cannot divide out common factors from a numerator and denominator in a proportion. This can be done *only* when the fractions are being multiplied.

(b) part = 30, whole = 50; find the percent.
Use the percent proportion.

$$\text{Part} \rightarrow \frac{30}{50} = \frac{x}{100} \quad \text{Percent proportion}$$
Whole

$$\frac{3}{5} = \frac{x}{100} \quad \text{Lowest terms}$$

$$5 \cdot x = 3 \cdot 100 \quad \text{Diagonal products}$$
$$5 \cdot x = 300$$

$$\frac{\cancel{5} \cdot x}{\cancel{5}} = \frac{300}{5} \quad \text{Divide each side by 5.}$$

$$x = 60$$

The percent is 60, written as 60%.

CHAPTER 0 Topics in Arithmetic

2. Use the percent proportion $\left(\dfrac{\text{part}}{\text{whole}} = \dfrac{\text{percent}}{100}\right)$ and solve for the unknown value.

(a) part = 12, percent = 16

(b) part = 30, whole = 120

(c) whole = 210, percent = 20

(d) whole = 4000, percent = 32

(e) part = 74, whole = 185

ANSWERS
2. (a) whole = 75 **(b)** percent = 25 (so, the percent is 25%) **(c)** part = 42 **(d)** part = 1280 **(e)** percent = 40 (so, the percent is 40%)

(c) whole = 150, percent = 18; find the part.

Part (unknown) → $\dfrac{x}{150} = \dfrac{18}{100}$ ← Percent   or $\dfrac{x}{150} = \dfrac{9}{50}$ Lowest terms
Whole →

$x \cdot 50 = 150 \cdot 9$ Diagonal products

$x \cdot 50 = 1350$

$\dfrac{x \cdot \cancel{50}}{\cancel{50}} = \dfrac{1350}{50}$ Divide each side by 50.

$x = 27$

The part is 27.

Work Problem 2 at the Side.

As a help in solving percent problems, keep in mind this basic idea.

Percent Problems

All percent problems involve a comparison between a part of something and the whole.

Solving these problems requires identifying the three components of a percent proportion: part, whole, and percent.

OBJECTIVE 3 Identify the percent. Look for the percent first. It is the easiest to identify.

Percent

The **percent** is the ratio of a part to a whole, with 100 as the denominator. In a problem, the percent appears with the word *percent* or with the symbol "%" after it.

EXAMPLE 2 Finding the Percent in Percent Problems

Find the percent in the following.

(a) 32% of the 900 men were retired.
       ↓
     Percent

The percent is 32. The number 32 appears with the symbol %.

SECTION 0.33 Using the Percent Proportion   0-187

3. Identify the percent.
   (a) Of the $1800, 18% will be spent on window coverings.

   (b) Find the amount of sales tax by multiplying sales of $590 and $6\frac{1}{2}$ percent.

   (c) 105 is 3% of what number?

   (d) What percent of the 380 guests will return this year?

4. Identify the whole.
   (a) Of the $1800, 18% will be spent on window coverings.

   (b) Find the amount of sales tax by multiplying sales of $590 and $6\frac{1}{2}$ percent.

   (c) $105 is 3% of what number?

   (d) What percent of the 380 guests will return this year?

ANSWERS

3. (a) 18 (b) $6\frac{1}{2}$ (c) 3
   (d) The percent is unknown.
4. (a) $1800 (b) $590 (c) what number (an unknown) (d) 380

(b) $150 is 25 percent of what number?
        ↓
      Percent

The percent is 25 because 25 appears with the word *percent*.

(c) What percent of the 350 women will go?
         ↓
    Percent (unknown)

The word *percent* has no number with it, so the percent is the unknown part of the problem.

Work Problem 3 at the Side.

OBJECTIVE 4 Identify the whole. Next, look for the whole.

Whole

The **whole** is the entire quantity. In a percent problem, the whole often appears after the word **of**.

EXAMPLE 3 Finding the Whole in Percent Problems

Identify the whole in the following.

(a) 32% **of** the 900 men were too large for the imported car.
                ↓
              Whole

The whole is 900. The number 900 appears after the word *of*.

(b) $150 is 25 percent **of** what number?
                        ↓
                      Whole The whole is the unknown part of the problem.

(c) 85% **of** 7000 is what number?
              ↓
            Whole

Work Problem 4 at the Side.

OBJECTIVE 5 Identify the part. Finally, look for the part.

Part

The **part** is the portion being compared with the whole.

NOTE If you have trouble identifying the part, find the whole and percent first. The remaining number is the part.

5. Identify the part.

(a) Of the $1800, 18%, or $324 will be spent on window coverings.

(b) Find the sales tax by multiplying $590 and $6\frac{1}{2}$ percent.

(c) $105 is 3% of what number?

(d) 80% of the 380 guests will return this year.

ANSWERS
5. (a) $324 (b) unknown
   (c) $105 (d) unknown

EXAMPLE 4 Finding the Part in Percent Problems

Identify the part in the following.

(a) 54% of 700 students is 378 students.
First find the percent and the whole.

54% of 700 students is 378 students.
↓   ↓
Percent; with % sign   Whole; follows "of"

The remaining number is the part.

54% of 700 students is 378 students.
↓   ↓   ↓
Percent Whole   Part

The part is 378.

(b) $150 is 25% of what number?
        ↓   ↓
      Percent   Whole (unknown)

$150 is the remaining number, so the part is $150.

(c) 85% of $7000 is what number?
     ↓   ↓   ↓
   Percent Whole   Part (unknown)

Work Problem 5 at the Side.

0.33 EXERCISES

For Extra Help

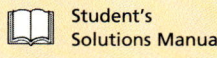 Student's Solutions Manual

 MyMathLab

 InterAct Math Tutorial Software

 AW Math Tutor Center

MathXP MathXL

Find the unknown value in the percent proportion $\dfrac{part}{whole} = \dfrac{percent}{100}$. Round to the nearest tenth if necessary. If the answer is a percent, be sure to include a percent sign (%). See Example 1.

1. part = 5, percent = 10
2. part = 20, percent = 25
3. part = 30, percent = 20
4. part = 25, percent = 25
5. part = 28, percent = 40
6. part = 11, percent = 5
7. part = 15, whole = 60
8. part = 105, whole = 35
9. part = 36, whole = 24
10. part = 1.5, whole = 4.5
11. part = 9.25, whole = 27.75
12. part = 12.8, whole = 9.6
13. whole = 52, percent = 50
14. whole = 160, percent = 35

SECTION 0.33   Using the Percent Proportion   0-189

15. whole = 72, percent = 30

16. whole = 115, percent = 38

17. whole = 94.4, part = 25

18. whole = 89.6, part = 50

Solve each problem. If the answer is a percent, be sure to include a percent sign (%). See Examples 2–4.

19. Find the whole if the part is 46 and the percent is 40.

20. The percent is 45 and the whole is 160. Find the part.

21. The whole is 5000 and the part is 20. Find the percent.

22. Suppose the part is 15 and the whole is 2500. Find the percent.

23. Find the percent if the whole is 4300 and the part is $107\frac{1}{2}$.

24. What is the part, if the percent is $12\frac{3}{4}$ and the whole is 5600?

25. The whole is 6480 and the part is 19.44. Find the percent.

26. Suppose the part is 281.25 and the percent is $1\frac{1}{4}$. Find the whole.

Identify the percent, whole, and part in the following. Do not try to solve for any unknowns. See Examples 2–4.

	Percent	Whole	Part
27. 10% of how many bicycles is 60 bicycles?	_____	_____	_____
28. 58% of how many preschoolers is 203 preschoolers?	_____	_____	_____
29. 75% of $800 is $600.	75%	800	600
30. 93% of $1500 is $1395.	_____	_____	_____
31. What is 25% of $970?	_____	_____	_____
32. What is 61% of 830 homes?	_____	_____	_____
33. 12 injections is 20% of what number of injections?	_____	_____	_____

	Percent	Whole	Part
34. 92 servings is 26% of what number of servings?	_____	_____	_____
35. 34 trophies is 50% of 68 trophies.	_____	_____	_____
36. 410 pallets is $33\frac{1}{3}$% of 1230 pallets.	_____	_____	_____
37. What percent of $296 is $177.60?	_____	_____	_____
38. What percent of $120.80 is $30.20?	_____	_____	_____
39. 54.34 is 3.25% of what number?	_____	_____	_____
40. 16.74 is 11.9% of what number?	_____	_____	_____
41. 0.68% of $487 is what amount?	_____	_____	_____
42. What amount is 6.21% of $704.35?	_____	_____	_____

0.34 Using Proportions to Solve Percent Problems

OBJECTIVES

1. Use the percent proportion to find the part.
2. Find the whole using the percent proportion.
3. Find the percent using the percent proportion.

This is the percent proportion.

$$\frac{\text{part}}{\text{whole}} = \frac{\text{percent}}{100}$$

As discussed in **Section 0.33**, if any two of the three values are known, the third can be found by solving the percent proportion.

OBJECTIVE 1 Use the percent proportion to find the part. The first example shows how to use the percent proportion to find the part.

EXAMPLE 1 Finding the Part with the Percent Proportion

Find 15% of $160.

Here the percent is 15 and the whole is 160. (Recall that the whole often comes after the word *of*.) Now find the part. Let x represent the unknown part.

$$\frac{\text{part}}{\text{whole}} = \frac{\text{percent}}{100} \quad \text{so} \quad \frac{x}{160} = \frac{15}{100} \quad \text{or} \quad \frac{x}{160} = \frac{3}{20} \quad \text{Lowest terms}$$

Find the diagonal products in the proportion.

$$x \cdot 20 = 160 \cdot 3 \quad \text{Diagonal products}$$
$$x \cdot 20 = 480$$

$$\frac{x \cdot \cancel{20}^1}{\cancel{20}_1} = \frac{480}{20} \quad \text{Divide each side by 20.}$$

$$x = 24 \quad \text{Part}$$

15% of $160 is **$24**.

Work Problem 1 at the Side.

Just as with the application problems given earlier, the word *of* is an indicator word meaning *multiply*. For example:

15% of 160
↓
15% · 160.

Because of this, there is another way to find the part.

Finding the Part Using Multiplication

To find the part:

Step 1 Identify the percent. Write the percent as a decimal.

Step 2 Multiply this decimal by the whole.

---

1. Use the percent proportion to find the part.

   (a) 10% of 1250 sailboats

   (b) 15% of $3220

   (c) 7% of 2700 miles

   (d) 39% of 1220 meters

ANSWERS

1. (a) 125 sailboats (b) $483
   (c) 189 miles (d) 475.8 meters

0-192 CHAPTER 0 Topics in Arithmetic

2. Use multiplication to find the part.

(a) 55% of 10,000 injections

(b) 16% of 120 miles

(c) 135% of 60 dosages

(d) 0.5% of $238

EXAMPLE 2 Finding the Part by Using Multiplication

Use multiplication to find the part.

(a) Find 42% of 830 yards.

Step 1 Here, the percent is 42. Write 42% as the decimal 0.42.

Step 2 Multiply 0.42 and the whole, which is 830.

$$\text{part} = 0.42 \cdot 830$$
$$= 348.6 \text{ yd}$$

It is a good idea to estimate the answer, to make sure no mistakes were made with decimal points. Round 42% to 40% or 0.4, and round 830 as 800. Next, 40% of 800 is

$$0.4 \cdot 800 = 320, \leftarrow \text{Estimate.}$$

so 348.6 is a reasonable answer.

(b) Find 25% of 1680 cars.

Identify the percent as 25. Write 25% in decimal form as 0.25. Now, multiply 0.25 and 1680.

$$\text{part} = 0.25 \cdot 1680 = 420 \text{ cars} \quad \text{Multiply.}$$

You can use a shortcut to estimate the answer. Since 25% means 25 parts out of 100 parts, this is the same as $\frac{1}{4}$ of the whole ($\frac{25}{100} = \frac{1}{4}$). Do you see a shortcut here? You can find $\frac{1}{4}$ of a number by dividing the number by 4. So, this shortcut gives us the exact answer, $1680 \div 4 = 420$.

(c) Find 140% of 60 miles.

In this problem, the percent is 140. Write 140% as the decimal 1.40. Next, multiply 1.40 and 60.

$$\text{part} = 1.40 \cdot 60 = 84 \text{ miles} \quad \text{Multiply.}$$

You can estimate the answer by realizing that 140% is close to 150% (which is $1\frac{1}{2}$) and $1\frac{1}{2}$ times 60 is 90. So, 84 miles is a reasonable answer.

(d) Find 0.4% of 50 kilometers.

$$\text{part} = 0.004 \cdot 50 = 0.2 \text{ kilometers} \quad \text{Multiply.}$$
$$\uparrow$$
Write 0.4% as a decimal.

Estimate the answer. 0.4% is less than 1%.

$$1\% \text{ of } 50 \text{ miles} = 50. = 0.5 \text{ miles}$$

So our answer should be *less than* 0.5 miles, and 0.2 miles fits this requirement.

Work Problem 2 at the Side.

ANSWERS

2. (a) 5500 injections (b) 19.2 miles
(c) 81 dosages (d) $1.19

SECTION 0.34 Using Proportions to Solve Percent Problems   **0-193**

3. Use the six problem-solving steps to solve each problem.

   (a) One day on Jacob's mail route there were 2920 pieces of mail. If 45% of those were advertising pieces, find the number of advertising pieces.

■ **EXAMPLE 3** Solving for the Part in an Application Problem

Raley's Markets has 850 employees. Of these employees, 28% are students. How many of the employees are students?

Step 1 **Read** the problem. The problem asks us to find the number of employees who are students.

Step 2 **Work out a plan.** Look for the word *of* as an indicator word for multiplication.

28% **of** the employees are students.
　　　└── Indicator word

The total number of employees is 850, so the whole is 850. The percent is 28. To find the number of students, find the part.

Step 3 **Estimate** a reasonable answer. You can estimate the answer by rounding 28% to 25% and 850 to 900. Remember 25% is 25 parts out of 100, which is equivalent to $\frac{1}{4}$. So divide 900 by 4.

$$900 \div 4 = 225 \text{ students} \leftarrow \text{Estimate}$$

Step 4 **Solve** the problem.

$$\text{part} = 0.28 \cdot 850 = 238 \quad \text{Multiply.}$$
　　　　　　　└── Write 28% as a decimal.

Step 5 **State the answer.** Raley's Markets has 238 student employees.

Step 6 **Check.** The answer, 238 students, is close to our estimate of 225 students.

■

Work Problem 3 at the Side.

   (b) There are 8550 students attending a college. If 28% of the students drink coffee, find the number of coffee drinkers.

Calculator Tip If you are using a calculator, you could solve Example 3 like this.

$$0.28 \; \times \; 850 \; = \; 238$$

Or, you can use this alternate approach on calculators with a % key.

$$850 \; \times \; 28 \; \% \; = \; 238$$

OBJECTIVE 2 Find the whole using the percent proportion. The next example shows how to use the percent proportion to find the whole.

NOTE Remember, the whole is the entire quantity.

■ **EXAMPLE 4** Finding the Whole with the Percent Proportion

(a) 8 tables is 4% of what number of tables?

Here the percent is 4, the whole is unknown, and the part is 8. Use the percent proportion to find the whole. Let x represent the unknown whole.

ANSWERS
3. (a) 1314 advertising pieces
   (b) 2394 coffee drinkers

$$\frac{\text{part}}{\text{whole}} = \frac{\text{percent}}{100} \quad \text{so} \quad \frac{8}{x} = \frac{4}{100} \quad \text{or} \quad \frac{8}{x} = \frac{1}{25} \quad \text{Lowest terms}$$

0-194 CHAPTER 0 Topics in Arithmetic

4. Use the percent proportion to find the unknown whole.

(a) 150 tickets is 25% of what number of tickets?

(b) 28 antiques is 35% of what number of antiques?

(c) 387 customers is 36% of what number of customers?

(d) 292.5 miles is 37.5% of what number of miles?

ANSWERS
4. (a) 600 tickets (b) 80 antiques
   (c) 1075 customers (d) 780 miles

Find the diagonal products.

$$x \cdot 1 = 8 \cdot 25$$
$$x = 200$$

8 tables is 4% of **200 tables**.

(b) 135 tourists is 15% of what number of tourists?
The percent is 15 and part is 135, so

Part → $\dfrac{135}{x} = \dfrac{15}{100}$ ← Percent
Whole (unknown) →

$\dfrac{135}{x} = \dfrac{3}{20}$ Lowest terms

$x \cdot 3 = 135 \cdot 20$ Diagonal products
$x \cdot 3 = 2700$

$\dfrac{x \cdot \cancel{3}}{\cancel{3}} = \dfrac{2700}{3}$ Divide each side by 3.

$x = 900.$

135 tourists is 15% of **900 tourists**.

Work Problem 4 at the Side.

EXAMPLE 5 Applying the Percent Proportion

At Newark Salt Works, 78 employees are absent because of illness. If this is 5% of the total number of employees, how many employees does the company have?

Step 1 **Read** the problem. The problem asks for the total number of employees.

Step 2 **Work out a plan.** From the information in the problem, the percent is 5 and the part of the total number of employees is 78. The total number of employees or entire quantity, which is the whole, is the unknown.

Step 3 **Estimate** a reasonable answer. Round the number of employees from 78 to 80. Then, 5% is equivalent to the fraction $\frac{1}{20}$. Since 80 is $\frac{1}{20}$ of the total number of employees,

$$80 \cdot 20 = 1600 \text{ employees} \leftarrow \text{Estimate}$$

Step 4 **Solve** the problem. Use the percent proportion to find the whole (the total number of employees).

Part → $\dfrac{78}{x} = \dfrac{5}{100}$ ← Percent
Whole (unknown) →

$\dfrac{78}{x} = \dfrac{1}{20}$ Lowest terms

Find the diagonal products.

$$x \cdot 1 = 78 \cdot 20$$
$$x = 1560$$

SECTION 0.34 Using Proportions to Solve Percent Problems **0-195**

5. Use the six problem-solving steps and the percent proportion to solve each problem.

    (a) A freeze resulted in a loss of 52% of an avocado crop. If the loss was 182 tons, find the total number of tons in the crop.

    (b) A metal alloy contains 450 pounds of zinc, which is 8% of the alloy. Find the total weight of the alloy.

Step 5 **State the answer.** The company has **1560 employees**.

Step 6 **Check.** The answer, 1560 employees, is close to our estimate of 1600 employees.

NOTE To estimate the answer to Example 5, the 5% was changed to its fraction equivalent $\frac{1}{20}$. Because 80 (rounded) is $\frac{1}{20}$ of the total employees, 80 was multiplied by 20 to get 1600, the estimated answer.

Work Problem 5 at the Side.

OBJECTIVE 3 Find the percent using the percent proportion. Finally, if the part and the whole are known, the percent proportion can be used to find the percent.

EXAMPLE 6 Using the Percent Proportion to Find the Percent

(a) 13 roofs is what percent of 52 roofs?

The whole is 52 (follows *of*) and the part is 13. Next, find the percent.

$$\frac{\text{part}}{\text{whole}} = \frac{\text{percent}}{100}$$

Part → $\frac{13}{52} = \frac{x}{100}$ ← Percent (unknown)
Whole →

$\frac{1}{4} = \frac{x}{100}$ Lowest terms

Find the diagonal products.

$4 \cdot x = 1 \cdot 100$

$\frac{\cancel{4} \cdot x}{\cancel{4}} = \frac{100}{4}$ Divide each side by 4.

$x = 25$

13 roofs is **25%** of 52 roofs.

(b) What percent of $500 is $100?

The whole is 500 (follows *of*) and the part is 100, so

$\frac{100}{500} = \frac{x}{100}$ ← Percent (unknown)

$\frac{1}{5} = \frac{x}{100}$ Lowest terms

$5 \cdot x = 1 \cdot 100$ Diagonal products

$5 \cdot x = 100$

$\frac{\cancel{5} \cdot x}{\cancel{5}} = \frac{100}{5}$ Divide each side by 5.

$x = 20$

20% of $500 is $100.

ANSWERS
5. (a) 350 tons (b) 5625 pounds

0.34 EXERCISES

For Extra Help

 Student's Solutions Manual

 MyMathLab

 InterAct Math Tutorial Software

 AW Math Tutor Center

MathXL

Find the part using the multiplication shortcut. See Example 2.

1. 20% of 80 guests
2. 25% of 3500 salespeople
3. 45% of 4080 military personnel
4. 12% of 3650 Web sites
5. 4% of 120 ft
6. 9% of $150
7. 150% of 210 files
8. 130% of 60 trees
9. 52.5% of 1560 trucks
10. 38.2% of 4250 loads
11. 2% of $164
12. 6% of $434
13. 225% of 680 tables
14. 135% of 800 commuters
15. 17.5% of 1040 homes
16. 46.1% of 843 kilograms
17. 0.9% of $2400
18. 0.3% of $1400

Find the whole using the percent proportion. See Example 4.

19. 80 e-mails is 25% of what number of e-mails?
20. 32 medical exams is 5% of what number of medical exams?
21. 30% of what number of hay bales is 48 hay bales?
22. 55% of what number of experiments is 209 experiments?
23. 495 successful students is 90% of what number of students?
24. 84 letters is 28% of what number of letters?
25. 1496 graduates is 110% of what number of graduates?
26. 154 bicycles is 140% of what number of bicycles?
27. $12\frac{1}{2}$% of what number is 350?
    ($Hint$: $12\frac{1}{2}$% = 12.5%)
28. $5\frac{1}{2}$% of what number is 176?

Find the percent using the percent proportion. Round your answers to the nearest tenth if necessary. See Example 6.

29. 18 bean burritos is what percent of 36 bean burritos?

30. 62 rooms is what percent of 248 rooms?

31. 26 desks is what percent of 50 desks?

32. 650 liters is what percent of 1000 liters?

33. 32 patients is what percent of 400 patients?

34. 7 bridges is what percent of 350 bridges?

35. 9 rolls is what percent of 600 rolls?

36. 60 cartons is what percent of 2400 cartons?

37. What percent of $344 is $64?

38. What percent of $398 is $14?

39. What percent of 250 tires is 23 tires?

40. What percent of 105 employees is 54 employees?

0.35 Using the Percent Equation

OBJECTIVES

1. Use the percent equation to find the part.
2. Find the whole using the percent equation.
3. Find the percent using the percent equation.

In the last section you were shown how to use a proportion to solve percent problems. In this section we show another way to solve these problems by using the **percent equation**. The percent equation is just a rearrangement of the percent proportion.

Percent Equation

$$\text{part} = \text{percent} \cdot \text{whole}$$

Be sure to write the percent as a decimal before using the equation.

When using the percent proportion, we did *not* have to write the percent as a decimal because of the 100 in the denominator of the proportion. However, because there is no 100 in the percent *equation,* we *must* first write the percent as a decimal by dividing by 100.

Some of the examples solved earlier will be reworked by using the percent equation. If you want to, you can look back at **Section 0.34** to see how some of these same problems were solved using proportions. This will give you a comparison of the two methods.

CHAPTER 0 Topics in Arithmetic

1. Use the percent equation to find the part.

   (a) 14% of 550 magazine titles

   (b) 23% of 840 gallons

   (c) 120% of $220

   (d) 135% of $1080

   (e) 0.5% of 1200 test tubes

   (f) 0.25% of 1600 lab tests

OBJECTIVE 1 Use the percent equation to find the part. The first example shows how to find the part.

EXAMPLE 1 Finding the Part

(a) Find 15% of $160.

Write 15% as the decimal 0.15. The whole, which comes after the word *of*, is 160. Next, use the percent equation. Let x represent the unknown part.

$$\text{part} = \text{percent} \cdot \text{whole}$$
$$x = 0.15 \cdot 160$$

Multiply 0.15 and 160 to get

$$x = 24.$$

15% of $160 is **$24**.

(b) Find 110% of 80 cases.

Write 110% as the decimal 1.10. The whole is 80. Let x represent the unknown part.

$$\text{part} = \text{percent} \cdot \text{whole}$$
$$x = 1.10 \cdot 80$$
$$x = 88$$

110% of 80 cases is **88** cases.

(c) Find 0.4% of 250 patients.

Write 0.4% as the decimal 0.004. The whole is 250. Let x represent the unknown part.

$$\text{part} = \text{percent} \cdot \text{whole}$$
$$x = 0.004 \cdot 250$$
$$x = 1$$

0.4% of 250 patients is **1 patient**.

Estimate the answer. 0.4% is approximately 0.5 or $\frac{1}{2}$ of 1%. Because 1% is $\frac{1}{100}$, 1% of 250 is

$$250 \div 100 = 2.5.$$

Since 2.5 is equal to 1%, half of 2.5 is equal to 1.25 (2.5 ÷ 2 = 1.25). So, 1 patient is a reasonable answer.

CAUTION When using the percent equation, the percent must always be *changed to a decimal* before multiplying.

Work Problem 1 at the Side.

OBJECTIVE 2 Find the whole using the percent equation. The next example shows how to use the percent equation to find the whole.

ANSWERS
1. (a) 77 magazine titles
   (b) 193.2 gallons (c) $264
   (d) $1458 (e) 6 test tubes
   (f) 4 lab tests

SECTION 0.35 Using the Percent Equation **0-199**

> **NOTE** Remember that the word *of* is an indicator word for *multiply*.

EXAMPLE 2 Solving for the Whole

(a) 8 tables is 4% of what number of tables?

The part is 8 and the percent is 4% or the decimal 0.04. The whole is unknown.

8   is   4% **of** what number?
            └── Indicator word

Next, use the percent equation

$$\text{part} = \text{percent} \cdot \text{whole}$$
$$8 = 0.04 \cdot x \quad \text{Let } x \text{ represent the unknown whole.}$$

$$\frac{8}{0.04} = \frac{\cancel{0.04} \cdot x}{\cancel{0.04}} \quad \text{Divide each side by 0.04.}$$

$$200 = x \longleftarrow \text{Whole}$$

8 tables is 4% of **200 tables**.

(b) 135 tourists is 15% of what number of tourists?

Write 15% as 0.15. The part is 135. Next, use the percent equation to find the whole.

$$\text{part} = \text{percent} \cdot \text{whole}$$
$$135 = 0.15 \cdot x \quad \text{Let } x \text{ represent the unknown whole.}$$

$$\frac{135}{0.15} = \frac{\cancel{0.15} \cdot x}{\cancel{0.15}} \quad \text{Divide each side by 0.15.}$$

$$900 = x \longleftarrow \text{Whole}$$

135 tourists is 15% of **900 tourists**.

(c) $8\frac{1}{2}\%$ of what number is 102?

Write $8\frac{1}{2}\%$ as 8.5%, or the decimal 0.085. The part is 102. Use the percent equation.

$$\text{part} = \text{percent} \cdot \text{whole}$$
$$102 = 0.085 \cdot x \quad \text{Let } x \text{ represent the unknown whole.}$$

$$\frac{102}{0.085} = \frac{\cancel{0.085} \cdot x}{\cancel{0.085}} \quad \text{Divide each side by 0.085.}$$

$$1200 = x \longleftarrow \text{Whole}$$

102 is $8\frac{1}{2}\%$ of **1200**.

Estimate the answer. Notice that $8\frac{1}{2}\%$ is close to 10%. If 102 is 10% of a number, then the number is 10 times 102, or 1020. So the answer, 1200, is reasonable.

CHAPTER 0 Topics in Arithmetic

2. Find the whole using the percent equation.

(a) 36 dance contestants is 18% of what number of dance contestants?

(b) 67.5 containers is 27% of what number of containers?

(c) 666 inoculations is 45% of what number of inoculations?

(d) $5\frac{1}{2}$% of what number of policies is 66 policies?

ANSWERS
2. (a) 200 dance contestants
   (b) 250 containers
   (c) 1480 inoculations
   (d) 1200 policies

CAUTION In Example 2(c) the $8\frac{1}{2}$% was changed to 8.5%, which is the decimal form of $8\frac{1}{2}$% ($8\frac{1}{2}$ = 8.5). The percent sign still remained in 8.5%. Then 8.5% was changed to the decimal 0.085 before dividing.

> Work Problem 2 at the Side.

OBJECTIVE 3 Find the percent using the percent equation. The final example shows how to use the percent equation to find the percent.

EXAMPLE 3 Finding the Percent

(a) 13 roofs is what percent of 52 roofs?

Because 52 follows *of*, the whole is 52. The part is 13, and the percent is unknown. Use the percent equation.

$$\text{part} = \text{percent} \cdot \text{whole}$$
$$13 = x \cdot 52 \quad \text{Let } x \text{ represent the unknown percent.}$$
$$\frac{13}{52} = \frac{x \cdot \cancel{52}}{\cancel{52}} \quad \text{Divide each side by 52.}$$
$$0.25 = x$$
$$0.25 \text{ is } 25\% \quad \text{Write the decimal as a percent.}$$

13 roofs is **25%** of 52 roofs.

The equation can also be set up by using *of* as an indicator word for multiplication, and *is* as an indicator word for "is equal to."

13 is what percent of 52?
$$13 = \quad x \quad \cdot 52$$
$$13 = x \cdot 52 \quad \text{Same equation as above}$$

(b) What percent of $500 is $100?

The whole is 500 and the part is 100. Let *x* represent the unknown percent.

$$\text{part} = \text{percent} \cdot \text{whole}$$
$$100 = x \cdot 500 \quad \text{Let } x \text{ represent the unknown percent.}$$
$$\frac{100}{500} = \frac{x \cdot \cancel{500}}{\cancel{500}} \quad \text{Divide each side by 500.}$$
$$0.20 = x$$
$$0.20 \text{ is } 20\% \quad \longleftarrow \text{Write the decimal as a percent.}$$

20% of $500 is $100.

SECTION 0.35 Using the Percent Equation   0-201

3. Find the percent using the percent equation.

(a) What percent of 70 keyboards is 14 keyboards?

(b) 34 post office boxes is what percent of 85 post office boxes?

(c) What percent of 920 invitations is 1288 invitations?

(d) 9 world class runners is what percent of 1125 runners?

(c) What percent of $300 is $390?
The whole is 300 and the part is 390. Let x represent the unknown percent.

$$\text{part} = \text{percent} \cdot \text{whole}$$
$$390 = x \cdot 300 \quad \text{Let } x \text{ represent the unknown percent.}$$
$$\frac{390}{300} = \frac{x \cdot \cancel{300}}{\cancel{300}} \quad \text{Divide each side by 300.}$$
$$1.3 = x$$
1.3 is 130% ⟵ Write the decimal as a percent.

130% of $300 is $390.

(d) 6 ladders is what percent of 1200 ladders?
Since 1200 follows *of*, the whole is 1200. The part is 6. Let x represent the unknown percent.

$$\text{part} = \text{percent} \cdot \text{whole}$$
$$6 = x \cdot 1200 \quad \text{Let } x \text{ represent the unknown percent.}$$
$$\frac{6}{1200} = \frac{x \cdot \cancel{1200}}{\cancel{1200}} \quad \text{Divide each side by 1200.}$$
$$0.005 = x$$
0.005 is 0.5% ⟵ Write the decimal as a percent.

6 ladders is **0.5%** of 1200 ladders.

You can estimate the answer because 1% of 1200 ladders is found by moving the decimal point two places to the left in 1200, resulting in 12. Since 6 ladders is half of 12 ladders, our answer should be $\frac{1}{2}$ of 1% or 0.5%. Our answer is reasonable.

> **CAUTION** When you use the percent equation to solve for an unknown percent, the answer will always be in decimal form. Notice that in Example 3(a), (b), (c), and (d) the decimal answer had to be changed to a percent by multiplying by 100 and attaching the percent sign. The answers became: (a) 0.25 = 25%; (b) 0.20 = 20%; (c) 1.3 = 130%; and (d) 0.005 = 0.5%.

Work Problem 3 at the Side.

ANSWERS
3. (a) 20% (b) 40% (c) 140%
   (d) 0.8%

0.35 EXERCISES

Find the part using the percent equation. See Example 1.

1. 25% of 540 hamburgers
2. 19% of 700 pages
3. 45% of 1500 garments
4. 75% of 360 dosages
5. 32% of 260 quarts
6. 44% of 430 liters
7. 140% of 500 tablets
8. 145% of 580 donors
9. 12.4% of 8300 meters
10. 26.4% of 4700 miles
11. 0.8% of $520
12. 0.3% of $480

Find the whole using the percent equation. See Example 2.

13. 24 patients is 15% of what number of patients?
14. 32 classrooms is 20% of what number of classrooms?
15. 40% of what number of salads is 130 salads?
16. 75% of what number of wrenches is 675 wrenches?
17. 476 circuits is 70% of what number of circuits?
18. 270 lab tests is 45% of what number of lab tests?
19. $12\frac{1}{2}$% of what number of people is 135 people?
20. $18\frac{1}{2}$% of what number of circuit breakers is 370 circuit breakers?
21. $1\frac{1}{4}$% of what number of gallons is 3.75 gallons?
22. $2\frac{1}{4}$% of what number of files is 9 files?

Find the percent using the percent equation. See Example 3.

23. 35 rail cars is what percent of 70 rail cars?
24. 90 mail carriers is what percent of 225 mail carriers?
25. 114 tuxedos is what percent of 150 tuxedos?
26. 75 offices is what percent of 125 offices?
27. What percent of $264 is $330?
28. What percent of $480 is $696?
29. What percent of 160 liters is 2.4 liters?
30. What percent of 600 meters is 7.5 meters?
31. 170 cartons is what percent of 68 cartons?
32. 612 orders is what percent of 425 orders?

The Real Number System

1

- 1.1 Fractions
- 1.2 Exponents, Order of Operations, and Inequality
- 1.3 Variables, Expressions, and Equations
- 1.4 Real Numbers and the Number Line
- 1.5 Adding and Subtracting Real Numbers
- 1.6 Multiplying and Dividing Real Numbers

Summary Exercises on Operations with Real Numbers

- 1.7 Properties of Real Numbers
- 1.8 Simplifying Expressions

The table indicates the figures for several stocks listed on the NASDAQ stock exchange for Wednesday, May 29, 2002. The final column represents the change in the price of the stock from the previous day. These changes are indicated by signed numbers. A positive number, such as $+.75$, means that the price of one share of the stock rose 75 cents from the previous day's closing; a negative number, such as -2.24, means that the price fell $2.24. In the exercises for Sections 1.1 and 1.4, we apply the concepts of this chapter to investigate situations involving stock prices.

Stock	Vol.	Close	Chg.
AAONs	14	31.85	+.75
ABC Bcp	7	14.55	
AC Moore	58	40.35	−2.24
ACT Tele	400	3.30	+.14
ADC Tel	4752	3.49	−.12
AFC Ent	225	30.40	−.96
AHL Sev	79	2.03	−.21

Source: Times Picayune.

1

2 CHAPTER 1 The Real Number System

1.1 Fractions

OBJECTIVES

1. Learn the definition of *factor.*
2. Write fractions in lowest terms.
3. Multiply and divide fractions.
4. Add and subtract fractions.
5. Solve applied problems that involve fractions.
6. Interpret data in a circle graph.

As preparation for the study of algebra, we begin this section with a brief review of arithmetic. In everyday life, the numbers seen most often are the **natural numbers,**

$$1, 2, 3, 4, \ldots,$$

the **whole numbers,**

$$0, 1, 2, 3, 4, \ldots,$$

and **fractions,** such as

$$\frac{1}{2}, \quad \frac{2}{3}, \quad \text{and} \quad \frac{15}{7}.$$

The parts of a fraction are named as follows.

$$\text{Fraction bar} \rightarrow \frac{4 \leftarrow \text{Numerator}}{7 \leftarrow \text{Denominator}}$$

As we will see later, the fraction bar represents division $\left(\frac{a}{b} = a \div b\right)$ and also serves as a grouping symbol.

If the numerator of a fraction is less than the denominator, we call it a **proper fraction.** A proper fraction has a value less than 1. If the numerator is greater than or equal to the denominator, the fraction is an **improper fraction.** An improper fraction that has a value greater than 1 is often written as a **mixed number.** For example, $\frac{12}{5}$ may be written as $2\frac{2}{5}$. In algebra, we prefer to use the improper form because it is easier to work with. In applications, we usually convert answers to mixed form, which is more meaningful.

OBJECTIVE 1 Learn the definition of *factor.* In the statement $2 \times 9 = 18$, the numbers 2 and 9 are called **factors** of 18. Other factors of 18 include 1, 3, 6, and 18. The result of the multiplication, 18, is called the **product.**

There are several ways of representing the product of two numbers. For example,

$$6 \times 3, \quad 6 \cdot 3, \quad (6)(3), \quad 6(3), \quad \text{and} \quad (6)3$$

all represent the product of 6 and 3.

The number 18 is **factored** by writing it as the product of two or more numbers. For example, 18 can be factored as $6 \cdot 3$, $18 \cdot 1$, $9 \cdot 2$, or $3 \cdot 3 \cdot 2$. (In algebra, a raised dot $\cdot$ is often used instead of the $\times$ symbol to indicate multiplication.)

A natural number (except 1) is **prime** if it has only itself and 1 as factors. "Factors" are understood here to mean natural number factors. (By agreement, the number 1 is not a prime number.) The first dozen primes are

$$2, 3, 5, 7, 11, 13, 17, 19, 23, 29, 31, 37.$$

A natural number (except 1) that is not prime, such as 4, 6, 8, 9, and 12, is a **composite** number.

It is often useful to find all **prime factors** of a number—those factors that are prime numbers. For example, the only prime factors of 18 are 2 and 3.

EXAMPLE 1 Factoring Numbers

Write each number as the product of prime factors.

(a) 35

Write 35 as the product of the prime fractors 5 and 7, or as

$$35 = 5 \cdot 7.$$

(b) 24

One way to begin is to divide by the smallest prime, 2, to get

$$24 = 2 \cdot 12.$$

Now divide 12 by 2 to find factors of 12.

$$24 = 2 \cdot 2 \cdot 6$$

Since 6 can be written as $2 \cdot 3$,

$$24 = 2 \cdot 2 \cdot 2 \cdot 3,$$

where all factors are prime.

Now Try Exercises 9 and 19.

NOTE It is not necessary to start with the smallest prime factor, as shown in Example 1(b). In fact, no matter which prime factor we start with, we will *always* obtain the same prime factorization.

OBJECTIVE 2 Write fractions in lowest terms. We use prime numbers to write fractions in *lowest terms*. A fraction is in **lowest terms** when the numerator and denominator have no factors in common (other than 1). We use the **basic principle of fractions** to write a fraction in lowest terms.

Basic Principle of Fractions

If the numerator and denominator of a fraction are multiplied or divided by the same nonzero number, the value of the fraction is not changed.

For example, $\frac{12}{16}$ can be written in lowest terms as follows:

$$\frac{12}{16} = \frac{3 \cdot 4}{4 \cdot 4} = \frac{3}{4} \cdot \frac{4}{4} = \frac{3}{4} \cdot 1 = \frac{3}{4}.$$

This procedure uses the rule for multiplying fractions (covered in the next objective) and the multiplication property of 1 (covered in Section 1.7).

To write a fraction in lowest terms, use these steps.

Writing a Fraction in Lowest Terms

Step 1 Write the numerator and the denominator as the product of prime factors.

Step 2 Divide the numerator and the denominator by the **greatest common factor,** the product of all factors common to both.

EXAMPLE 2 Writing Fractions in Lowest Terms

Write each fraction in lowest terms.

(a) $\dfrac{10}{15} = \dfrac{2 \cdot 5}{3 \cdot 5} = \dfrac{2 \cdot 1}{3 \cdot 1} = \dfrac{2}{3}$

Since 5 is the greatest common factor of 10 and 15, dividing both numerator and denominator by 5 gives the fraction in lowest terms.

(b) $\dfrac{15}{45} = \dfrac{3 \cdot 5}{3 \cdot 3 \cdot 5} = \dfrac{1 \cdot 1}{3 \cdot 1 \cdot 1} = \dfrac{1}{3}$

The factored form shows that 3 and 5 are the common factors of both 15 and 45. Dividing both 15 and 45 by $3 \cdot 5 = 15$ gives $\dfrac{15}{45}$ in lowest terms as $\dfrac{1}{3}$.

Now Try Exercises 25 and 29.

We can simplify this process by finding the greatest common factor in the numerator and denominator by inspection. For instance, in Example 2(b), we can use 15 rather than $3 \cdot 5$.

$$\dfrac{15}{45} = \dfrac{15}{3 \cdot 15} = \dfrac{1}{3 \cdot 1} = \dfrac{1}{3}$$

CAUTION Errors may occur when writing fractions in lowest terms if the factor 1 is not included. To see this, refer to Example 2(b). In the equation

$$\dfrac{3 \cdot 5}{3 \cdot 3 \cdot 5} = \dfrac{?}{3},$$

if 1 is not written in the numerator when dividing out common factors, you may make an error. The ? should be replaced by 1.

OBJECTIVE 3 Multiply and divide fractions. We multiply two fractions by first multiplying their numerators and then multiplying their denominators. This rule is written in symbols as follows.

Multiplying Fractions

If $\dfrac{a}{b}$ and $\dfrac{c}{d}$ are fractions, then $\dfrac{a}{b} \cdot \dfrac{c}{d} = \dfrac{a \cdot c}{b \cdot d}$.

EXAMPLE 3 Multiplying Fractions

Find each product, and write it in lowest terms.

(a) $\dfrac{3}{8} \cdot \dfrac{4}{9} = \dfrac{3 \cdot 4}{8 \cdot 9}$ Multiply numerators.
    Multiply denominators.

$= \dfrac{3 \cdot 4}{2 \cdot 4 \cdot 3 \cdot 3}$ Factor.

$= \dfrac{1}{2 \cdot 3} = \dfrac{1}{6}$ Write in lowest terms.

(b) $2\dfrac{1}{3} \cdot 5\dfrac{1}{2} = \dfrac{7}{3} \cdot \dfrac{11}{2}$ Write as improper fractions.

$= \dfrac{77}{6}$ or $12\dfrac{5}{6}$ Multiply numerators and denominators.

Now Try Exercises 35 and 39.

Two fractions are **reciprocals** of each other if their product is 1. For example, $\dfrac{3}{4}$ and $\dfrac{4}{3}$ are reciprocals since

$$\dfrac{3}{4} \cdot \dfrac{4}{3} = \dfrac{12}{12} = 1.$$

Also, $\dfrac{7}{11}$ and $\dfrac{11}{7}$ are reciprocals of each other, as are $\dfrac{1}{6}$ and 6. Because division is the opposite (or inverse) of multiplication, we use reciprocals to divide fractions. To divide two fractions, multiply the first fraction by the reciprocal of the second fraction.

Dividing Fractions

For the fractions $\dfrac{a}{b}$ and $\dfrac{c}{d}$, $\dfrac{a}{b} \div \dfrac{c}{d} = \dfrac{a}{b} \cdot \dfrac{d}{c}$.

That is, to divide by a fraction, multiply by its reciprocal.

We will explain why this method works in Chapter 7. However, as an example, we know that $20 \div 10 = 2$ and $20 \cdot \dfrac{1}{10} = 2$. The answer to a division problem is called a **quotient**. For example, the quotient of 20 and 10 is 2.

EXAMPLE 4 Dividing Fractions

Find each quotient, and write it in lowest terms.

(a) $\dfrac{3}{4} \div \dfrac{8}{5} = \dfrac{3}{4} \cdot \dfrac{5}{8} = \dfrac{3 \cdot 5}{4 \cdot 8} = \dfrac{15}{32}$

Multiply by the reciprocal of the second fraction.

(b) $\dfrac{3}{4} \div \dfrac{5}{8} = \dfrac{3}{4} \cdot \dfrac{8}{5} = \dfrac{3 \cdot 8}{4 \cdot 5} = \dfrac{3 \cdot 4 \cdot 2}{4 \cdot 5} = \dfrac{6}{5}$ or $1\dfrac{1}{5}$

(c) $\dfrac{5}{8} \div 10 = \dfrac{5}{8} \div \dfrac{10}{1} = \dfrac{5}{8} \cdot \dfrac{1}{10} = \dfrac{1}{16}$

Write 10 as $\dfrac{10}{1}$.

(d) $1\dfrac{2}{3} \div 4\dfrac{1}{2} = \dfrac{5}{3} \div \dfrac{9}{2}$ Write as improper fractions.

$= \dfrac{5}{3} \cdot \dfrac{2}{9}$ Multiply by the reciprocal of the second fraction.

$= \dfrac{10}{27}$

Now Try Exercises 43, 45, and 47.

OBJECTIVE 4 Add and subtract fractions. The result of adding two numbers is called the **sum** of the numbers. For example, $2 + 3 = 5$, so 5 is the sum of 2 and 3. To find the sum of two fractions having the same denominator, add the numerators and keep the same denominator.

> **Adding Fractions**
>
> If $\dfrac{a}{b}$ and $\dfrac{c}{b}$ are fractions, then $\dfrac{a}{b} + \dfrac{c}{b} = \dfrac{a+c}{b}.$

■ **EXAMPLE 5** Adding Fractions with the Same Denominator

Add.

(a) $\dfrac{3}{7} + \dfrac{2}{7} = \dfrac{3+2}{7} = \dfrac{5}{7}$ Add numerators; keep the same denominator.

(b) $\dfrac{2}{10} + \dfrac{3}{10} = \dfrac{2+3}{10} = \dfrac{5}{10} = \dfrac{1}{2}$ Write in lowest terms.

Now Try Exercise 51.

If the fractions to be added do not have the same denominators, we can still use the procedure above, but only *after* we rewrite the fractions with a common denominator. For example, to rewrite $\tfrac{3}{4}$ as a fraction with denominator 32,

$$\dfrac{3}{4} = \dfrac{?}{32},$$

we find the number that can be multiplied by 4 to give 32. Since $4 \cdot 8 = 32$, we use the number 8. By the basic principle of fractions, we multiply the numerator and the denominator by 8.

$$\dfrac{3}{4} = \dfrac{3 \cdot 8}{4 \cdot 8} = \dfrac{24}{32}$$

> **Finding the Least Common Denominator**
>
> To add or subtract fractions with different denominators, find the **least common denominator (LCD)** as follows.
>
> *Step 1* Factor both denominators.
>
> *Step 2* For the LCD, use every factor that appears in any factored form. If a factor is repeated, use the largest number of repeats in the LCD.

The next example shows this procedure.

■ **EXAMPLE 6** Adding Fractions with Different Denominators

Add. Write the sums in lowest terms.

(a) $\dfrac{4}{15} + \dfrac{5}{9}$

To find the least common denominator, first factor both denominators.

$$15 = 5 \cdot 3 \quad \text{and} \quad 9 = 3 \cdot 3$$

Since 5 and 3 appear as factors, and 3 is a factor of 9 twice, the LCD is

$$\overset{15}{5 \cdot 3} \cdot \overset{9}{3} \quad \text{or} \quad 45.$$

Write each fraction with 45 as denominator.

$$\frac{4}{15} = \frac{4 \cdot 3}{15 \cdot 3} = \frac{12}{45} \quad \text{and} \quad \frac{5}{9} = \frac{5 \cdot 5}{9 \cdot 5} = \frac{25}{45}$$

Now add the two equivalent fractions.

$$\frac{4}{15} + \frac{5}{9} = \frac{12}{45} + \frac{25}{45} = \frac{37}{45}$$

(b) $3\frac{1}{2} + 2\frac{3}{4}$

We add mixed numbers using either of two methods.

Method 1

Rewrite both numbers as improper fractions.

$$3\frac{1}{2} = 3 + \frac{1}{2} = \frac{3}{1} + \frac{1}{2} = \frac{6}{2} + \frac{1}{2} = \frac{6+1}{2} = \frac{7}{2}$$

$$2\frac{3}{4} = 2 + \frac{3}{4} = \frac{8}{4} + \frac{3}{4} = \frac{8+3}{4} = \frac{11}{4}$$

Now add. The common denominator is 4.

$$3\frac{1}{2} + 2\frac{3}{4} = \frac{7}{2} + \frac{11}{4} = \frac{14}{4} + \frac{11}{4} = \frac{25}{4} \quad \text{or} \quad 6\frac{1}{4}$$

Method 2

Write $3\frac{1}{2}$ as $3\frac{2}{4}$. Then add vertically.

$$\begin{array}{r} 3\frac{1}{2} \\ + 2\frac{3}{4} \\ \hline \end{array} \quad \rightarrow \quad \begin{array}{r} 3\frac{2}{4} \\ + 2\frac{3}{4} \\ \hline 5\frac{5}{4} \end{array}$$

Since $\frac{5}{4} = 1\frac{1}{4}$,

$$5\frac{5}{4} = 5 + 1\frac{1}{4} = 6\frac{1}{4} \quad \text{or} \quad \frac{25}{4}.$$

Now Try Exercises 53 and 55.

The **difference** between two numbers is found by subtracting the numbers. For example, $9 - 5 = 4$, so the difference between 9 and 5 is 4. Subtraction of fractions is similar to addition. Just subtract the numerators instead of adding them; again, keep the same denominator.

8 CHAPTER 1 The Real Number System

> **Subtracting Fractions**
>
> If $\dfrac{a}{b}$ and $\dfrac{c}{b}$ are fractions, then $\dfrac{a}{b} - \dfrac{c}{b} = \dfrac{a-c}{b}$.

EXAMPLE 7 Subtracting Fractions

Subtract. Write the differences in lowest terms.

(a) $\dfrac{15}{8} - \dfrac{3}{8} = \dfrac{15-3}{8}$ Subtract numerators; keep the same denominator.

$= \dfrac{12}{8} = \dfrac{3}{2}$ or $1\dfrac{1}{2}$ Lowest terms

(b) $\dfrac{7}{18} - \dfrac{4}{15}$

Here, $18 = 2 \cdot 3 \cdot 3$ and $15 = 3 \cdot 5$, so the LCD is $2 \cdot 3 \cdot 3 \cdot 5 = 90$.

$\dfrac{7}{18} - \dfrac{4}{15} = \dfrac{7 \cdot 5}{2 \cdot 3 \cdot 3 \cdot 5} - \dfrac{4 \cdot 2 \cdot 3}{2 \cdot 3 \cdot 3 \cdot 5} = \dfrac{35}{90} - \dfrac{24}{90} = \dfrac{11}{90}$

(c) $\dfrac{15}{32} - \dfrac{11}{45}$

Since $32 = 2 \cdot 2 \cdot 2 \cdot 2 \cdot 2$ and $45 = 3 \cdot 3 \cdot 5$, there are no common factors, and the LCD is $32 \cdot 45 = 1440$.

$\dfrac{15}{32} - \dfrac{11}{45} = \dfrac{15 \cdot 45}{32 \cdot 45} - \dfrac{11 \cdot 32}{45 \cdot 32}$ Get a common denominator.

$= \dfrac{675}{1440} - \dfrac{352}{1440}$

$= \dfrac{323}{1440}$ Subtract.

Now Try Exercises 57 and 59.

OBJECTIVE 5 Solve applied problems that involve fractions.   Applied problems often require work with fractions. For example, when a carpenter reads diagrams and plans, he or she often must work with fractions whose denominators are 2, 4, 8, 16, or 32, as shown in the next example.

EXAMPLE 8 Adding Fractions to Solve an Applied Problem

The diagram in Figure 1 appears in the book *Woodworker's 39 Sure-Fire Projects*. It is the front view of a corner bookcase/desk. Add the fractions shown in the diagram to find the height of the bookcase/desk to the top of the writing surface.

We must add the following measures (″ means inches):

$\dfrac{3}{4},\ 4\dfrac{1}{2},\ 9\dfrac{1}{2},\ \dfrac{3}{4},\ 9\dfrac{1}{2},\ \dfrac{3}{4},\ 4\dfrac{1}{2}.$

We begin by changing $4\dfrac{1}{2}$ to $4\dfrac{2}{4}$ and $9\dfrac{1}{2}$ to $9\dfrac{2}{4}$, since the common denominator is 4. Then we use Method 2 from Example 6(b).

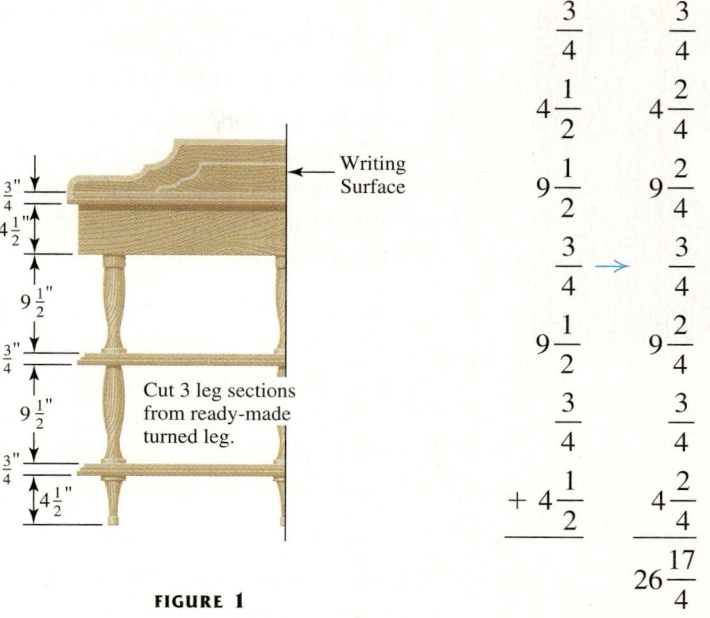

FIGURE 1

Since $\frac{17}{4} = 4\frac{1}{4}$, $26\frac{17}{4} = 26 + 4\frac{1}{4} = 30\frac{1}{4}$. The height is $30\frac{1}{4}$ in. It is best to give answers as mixed numbers in applications like this.

Now Try Exercise 69.

OBJECTIVE 6 Interpret data in a circle graph. A **circle graph** or **pie chart** is often used to give a pictorial representation of data. A circle is used to indicate the total of all the categories represented. The circle is divided into sectors, or wedges (like pieces of pie), whose sizes show the relative magnitudes of the categories. The sum of all the fractional parts must be 1 (for 1 whole circle).

EXAMPLE 9 Using a Circle Graph (Pie Chart) to Interpret Information

The 1999 market share for satellite-TV home subscribers is shown in the circle graph in Figure 2. The number of subscribers reached 12 million in August 1999.

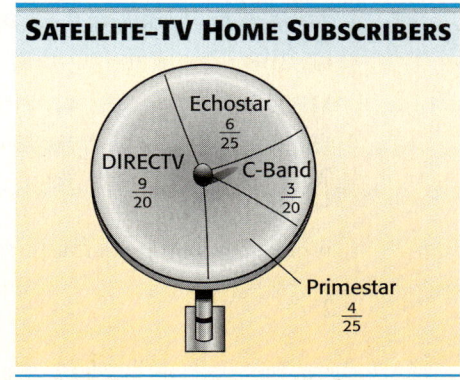

Source: Skyreport.com; *USA Today.*

FIGURE 2

10 CHAPTER 1 The Real Number System

(a) Which provider had the largest share of the home subscriber market in August 1999? What was that share?

In the circle graph, the sector for DIRECTV is the largest, so DIRECTV had the largest market share, $\frac{9}{20}$.

(b) Estimate the number of home subscribers to DIRECTV in August 1999.

A market share of $\frac{9}{20}$ can be rounded to $\frac{10}{20}$, or $\frac{1}{2}$. We multiply $\frac{1}{2}$ by the total number of subscribers, 12 million. A good estimate for the number of DIRECTV subscribers would be

$$\frac{1}{2}(12) = 6 \text{ million.}$$

(c) How many actual home subscribers to DIRECTV were there?

To find the answer, we multiply the actual fraction from the graph for DIRECTV, $\frac{9}{20}$, by the number of subscribers, 12 million:

$$\frac{9}{20}(12) = \frac{9}{20} \cdot \frac{12}{1} = \frac{27}{5} = 5\frac{2}{5}.$$

Thus, $5\frac{2}{5}$ million, or 5,400,000, homes subscribed to DIRECTV. This is reasonable given our estimate in part (b).

Now Try Exercises 75 and 77.

1.1 EXERCISES

For Extra Help

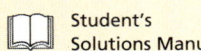 Student's Solutions Manual

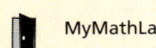 MyMathLab

InterAct Math Tutorial Software

AW Math Tutor Center

MathXL

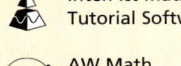 Digital Video Tutor CD 1/Videotape 1

Decide whether each statement is true or false. If it is false, say why.

1. In the fraction $\frac{3}{7}$, 3 is the numerator and 7 is the denominator.
2. The mixed number equivalent of $\frac{41}{5}$ is $8\frac{1}{5}$.
3. The fraction $\frac{17}{51}$ is in lowest terms.
4. The reciprocal of $\frac{8}{2}$ is $\frac{4}{1}$.
5. The product of 8 and 2 is 10.
6. The difference between 12 and 2 is 6.

Identify each number as prime, composite, or neither. If the number is composite, write it as the product of prime factors. See Example 1.

7. 19
8. 31
9. 64
10. 99
11. 3458
12. 1025
13. 1
14. 0
15. 30
16. 40
17. 500
18. 700
19. 124
20. 120
21. 29
22. 83

Write each fraction in lowest terms. See Example 2.

23. $\frac{8}{16}$
24. $\frac{4}{12}$
25. $\frac{15}{18}$
26. $\frac{16}{20}$
27. $\frac{18}{60}$
28. $\frac{16}{64}$
29. $\frac{144}{120}$
30. $\frac{132}{77}$

31. One of the following is the correct way to write $\frac{16}{24}$ in lowest terms. Which one is it?

A. $\frac{16}{24} = \frac{8+8}{8+16} = \frac{8}{16} = \frac{1}{2}$ **B.** $\frac{16}{24} = \frac{4 \cdot 4}{4 \cdot 6} = \frac{4}{6}$

C. $\frac{16}{24} = \frac{8 \cdot 2}{8 \cdot 3} = \frac{2}{3}$ **D.** $\frac{16}{24} = \frac{14+2}{21+3} = \frac{2}{3}$

32. For the fractions $\frac{p}{q}$ and $\frac{r}{s}$, which one of the following can serve as a common denominator?

A. $q \cdot s$ **B.** $q + s$ **C.** $p \cdot r$ **D.** $p + r$

Find each product or quotient, and write it in lowest terms. See Examples 3 and 4.

33. $\frac{4}{5} \cdot \frac{6}{7}$ **34.** $\frac{5}{9} \cdot \frac{10}{7}$ **35.** $\frac{1}{10} \cdot \frac{12}{5}$ **36.** $\frac{6}{11} \cdot \frac{2}{3}$

37. $\frac{15}{4} \cdot \frac{8}{25}$ **38.** $\frac{4}{7} \cdot \frac{21}{8}$ **39.** $2\frac{2}{3} \cdot 5\frac{4}{5}$ **40.** $3\frac{3}{5} \cdot 7\frac{1}{6}$

41. $\frac{5}{4} \div \frac{3}{8}$ **42.** $\frac{7}{6} \div \frac{9}{10}$ **43.** $\frac{32}{5} \div \frac{8}{15}$ **44.** $\frac{24}{7} \div \frac{6}{21}$

45. $\frac{3}{4} \div 12$ **46.** $\frac{2}{5} \div 30$ **47.** $2\frac{5}{8} \div 1\frac{15}{32}$ **48.** $2\frac{3}{10} \div 7\frac{4}{5}$

49. Write a summary explaining how to multiply and divide two fractions. Give examples.

50. Write a summary explaining how to add and subtract two fractions. Give examples.

Find each sum or difference, and write it in lowest terms. See Examples 5–7.

51. $\frac{7}{12} + \frac{1}{12}$ **52.** $\frac{3}{16} + \frac{5}{16}$ **53.** $\frac{5}{9} + \frac{1}{3}$ **54.** $\frac{4}{15} + \frac{1}{5}$

55. $3\frac{1}{8} + 2\frac{1}{4}$ **56.** $5\frac{3}{4} + 3\frac{1}{3}$ **57.** $\frac{13}{15} - \frac{3}{15}$ **58.** $\frac{11}{12} - \frac{3}{12}$

59. $\frac{7}{12} - \frac{1}{9}$ **60.** $\frac{11}{16} - \frac{1}{12}$ **61.** $6\frac{1}{4} - 5\frac{1}{3}$ **62.** $8\frac{4}{5} - 7\frac{4}{9}$

Solve each applied problem. See Example 8.

Use the table, which appears on a package of Quaker Quick Grits, to answer Exercises 63 and 64.

	Microwave		Stove Top	
Servings	1	1	4	6
Water	$\frac{3}{4}$ cup	1 cup	3 cups	4 cups
Grits	3 Tbsp	3 Tbsp	$\frac{3}{4}$ cup	1 cup
Salt (optional)	Dash	Dash	$\frac{1}{4}$ tsp	$\frac{1}{2}$ tsp

63. How many cups of water would be needed for eight microwave servings?

64. How many tsp of salt would be needed for five stove top servings? (*Hint:* 5 is halfway between 4 and 6.)

12 CHAPTER 1 The Real Number System

Stock prices are now reported using decimal numerals. However, prior to April 9, 2001, fractions were used for stock prices. With this in mind, answer Exercises 65 and 66.

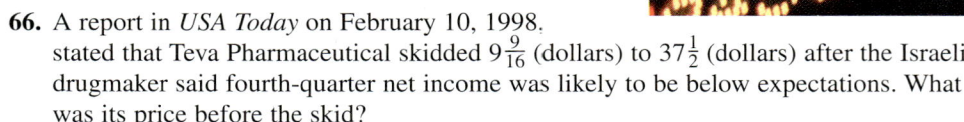

65. On Tuesday, February 10, 1998, Earthlink stock on the NASDAQ exchange closed the day at $4\frac{5}{8}$ (dollars) ahead of where it had opened. It closed at $38\frac{5}{8}$ (dollars). What was its opening price?

66. A report in *USA Today* on February 10, 1998, stated that Teva Pharmaceutical skidded $9\frac{9}{16}$ (dollars) to $37\frac{1}{2}$ (dollars) after the Israeli drugmaker said fourth-quarter net income was likely to be below expectations. What was its price before the skid?

67. A hardware store sells a 40-piece socket wrench set. The measure of the largest socket is $\frac{3}{4}$ in., while the measure of the smallest socket is $\frac{3}{16}$ in. What is the difference between these measures?

68. Two sockets in a socket wrench set have measures of $\frac{9}{16}$ in. and $\frac{3}{8}$ in. What is the difference between these two measures?

69. A motel owner has decided to expand his business by buying a piece of property next to the motel. The property has an irregular shape, with five sides as shown in the figure. Find the total distance around the piece of property. (This is called the *perimeter* of the figure.)

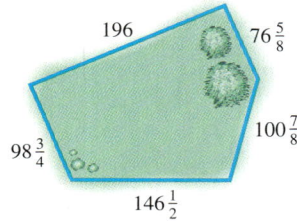

Measurements in feet

70. Find the perimeter of the triangle in the figure.

71. A piece of board is $15\frac{5}{8}$ in. long. If it must be divided into 3 pieces of equal length, how long must each piece be?

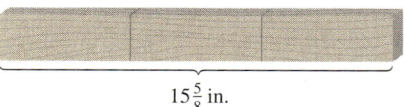

72. Under existing standards, most of the holes in Swiss cheese must have diameters between $\frac{11}{16}$ and $\frac{13}{16}$ in. To accommodate new high-speed slicing machines, the USDA wants to reduce the minimum size to $\frac{3}{8}$ in. How much smaller is $\frac{3}{8}$ in. than $\frac{11}{16}$ in.? (*Source:* U.S. Department of Agriculture.)

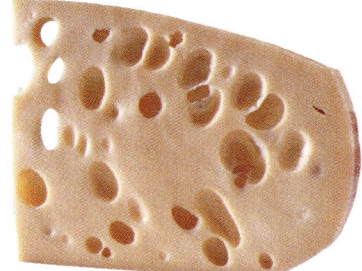

73. Tex's favorite recipe for barbecue sauce calls for $2\frac{1}{3}$ cups of tomato sauce. The recipe makes enough barbecue sauce to serve 7 people. How much tomato sauce is needed for 1 serving?

74. A cake recipe calls for $1\frac{3}{4}$ cups of sugar. A caterer has $15\frac{1}{2}$ cups of sugar on hand. How many cakes can he make?

More than 8 million immigrants were admitted to the United States between 1990 and 1997. The circle graph gives the fractional number from each region of birth for these immigrants. Use the graph to answer the following questions. See Example 9.

75. What fractional part of the immigrants were from Other regions?

76. What fractional part of the immigrants were from Latin America or Asia?

77. How many (in millions) were from Europe?

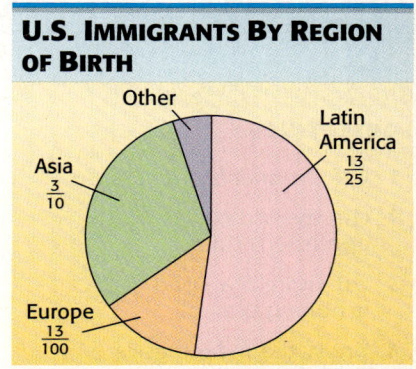

U.S. IMMIGRANTS BY REGION OF BIRTH

Other
Latin America $\frac{13}{25}$
Asia $\frac{3}{10}$
Europe $\frac{13}{100}$

Source: U.S. Bureau of the Census.

78. At the conclusion of the Addison Wesley softball league season, batting statistics for five players were as follows.

Player	At-Bats	Hits	Home Runs
Alison Romike	40	9	2
Jennifer Crum	36	12	3
Jason Jordan	11	5	1
Greg Tobin	16	8	0
Greg Erb	20	10	2

Use the table to answer each question, using estimation skills as necessary.

(a) Which player got a hit in exactly $\frac{1}{3}$ of his or her at-bats?
(b) Which player got a hit in just less than $\frac{1}{2}$ of his or her at-bats?
(c) Which player got a home run in just less than $\frac{1}{10}$ of his or her at-bats?
(d) Which player got a hit in just less than $\frac{1}{4}$ of his or her at-bats?
(e) Which two players got hits in exactly the same fractional parts of their at-bats? What was the fractional part, expressed in lowest terms?

79. For each description, write a fraction in lowest terms that represents the region described.

(a) The dots in the rectangle as a part of the dots in the entire figure
(b) The dots in the triangle as a part of the dots in the entire figure
(c) The dots in the overlapping region of the triangle and the rectangle as a part of the dots in the triangle alone
(d) The dots in the overlapping region of the triangle and the rectangle as a part of the dots in the rectangle alone

80. Estimate the best approximation for the following sum:

$$\frac{14}{26} + \frac{98}{99} + \frac{100}{51} + \frac{90}{31} + \frac{13}{27}.$$

A. 6 **B.** 7 **C.** 5 **D.** 8

1.2 Exponents, Order of Operations, and Inequality

OBJECTIVES

1. Use exponents.
2. Use the order of operations rules.
3. Use more than one grouping symbol.
4. Know the meanings of $\neq, <, >, \leq,$ and $\geq$.
5. Translate word statements to symbols.
6. Write statements that change the direction of inequality symbols.
7. Interpret data in a bar graph.

OBJECTIVE 1 Use exponents. In the prime factored form of 81, written

$$81 = 3 \cdot 3 \cdot 3 \cdot 3,$$

the factor 3 appears four times. In algebra, repeated factors are written with an *exponent*. For example, in $3 \cdot 3 \cdot 3 \cdot 3$, the number 3 appears as a factor four times, so the product is written as 3^4, and is read "3 to the fourth power."

$$\underbrace{3 \cdot 3 \cdot 3 \cdot 3}_{\text{4 factors of 3}} = 3^{\overset{\text{Exponent}}{4}}$$
$$\text{Base}$$

The number 4 is the **exponent** or **power** and 3 is the **base** in the **exponential expression** 3^4. A natural number exponent, then, tells how many times the base is used as a factor. A number raised to the first power is simply that number. For example, $5^1 = 5$ and $\left(\frac{1}{2}\right)^1 = \frac{1}{2}$.

EXAMPLE 1 Evaluating Exponential Expressions

Find the value of each exponential expression.

(a) $5^2 = \underline{5 \cdot 5} = 25$

              5 is used as a factor 2 times.

Read 5^2 as "5 squared" or "the square of 5."

(b) $6^3 = \underline{6 \cdot 6 \cdot 6} = 216$

              6 is used as a factor 3 times.

Read 6^3 as "6 cubed" or "the cube of 6."

(c) $2^5 = 2 \cdot 2 \cdot 2 \cdot 2 \cdot 2 = 32$ 2 is used as a factor 5 times.

Read 2^5 as "2 to the fifth power."

(d) $\left(\frac{2}{3}\right)^3 = \frac{2}{3} \cdot \frac{2}{3} \cdot \frac{2}{3} = \frac{8}{27}$ $\frac{2}{3}$ is used as a factor 3 times.

Now Try Exercises 5 and 17.

CAUTION Squaring, or raising a number to the second power, is *not* the same as doubling the number. For example,

$$3^2 \quad \text{means} \quad 3 \cdot 3, \quad not \quad 2 \cdot 3.$$

Thus $3^2 = 9$, not 6. Similarly, cubing, or raising a number to the third power, does *not* mean tripling the number.

SECTION 1.2 Exponents, Order of Operations, and Inequality   **15**

OBJECTIVE 2 Use the order of operations rules. Many problems involve more than one operation. To indicate the order in which the operations should be performed, we often use **grouping symbols.** If no grouping symbols are used, we apply the order of operations rules discussed below.

Consider the expression $5 + 2 \cdot 3$. To show that the multiplication should be performed before the addition, we can use parentheses to write

$$5 + (2 \cdot 3) = 5 + 6 = 11.$$

If addition is to be performed first, the parentheses should group $5 + 2$ as follows.

$$(5 + 2) \cdot 3 = 7 \cdot 3 = 21$$

Other grouping symbols used in more complicated expressions are brackets [ ], braces { }, and fraction bars. (For example, in $\frac{8-2}{3}$, the expression $8 - 2$ is considered to be grouped in the numerator.)

To work problems with more than one operation, we use the following **order of operations.** This order is used by most calculators and computers.

Order of Operations

If grouping symbols are present, simplify within them, innermost first (and above and below fraction bars separately), in the following order.

Step 1 Apply all **exponents.**

Step 2 Do any **multiplications** or **divisions** in the order in which they occur, working from left to right.

Step 3 Do any **additions** or **subtractions** in the order in which they occur, working from left to right.

If no grouping symbols are present, start with Step 1.

When no operation is indicated, as in $3(7)$ or $(-5)(-4)$, multiplication is understood.

EXAMPLE 2   Using the Order of Operations

Find the value of each expression.

(a) $9(6 + 11)$

Using the order of operations given above, work first inside the parentheses.

$$\begin{aligned} 9(6 + 11) &= 9(17) &&\text{Work inside parentheses.} \\ &= 153 &&\text{Multiply.} \end{aligned}$$

(b) $6 \cdot 8 + 5 \cdot 2$

Do the multiplications, working from left to right, and then add.

$$\begin{aligned} 6 \cdot 8 + 5 \cdot 2 &= 48 + 10 &&\text{Multiply.} \\ &= 58 &&\text{Add.} \end{aligned}$$

(c) $\begin{aligned} 2(5 + 6) + 7 \cdot 3 &= 2(11) + 7 \cdot 3 &&\text{Work inside parentheses.} \\ &= 22 + 21 &&\text{Multiply.} \\ &= 43 &&\text{Add.} \end{aligned}$

16   CHAPTER 1   The Real Number System

(d) $9 - 2^3 + 5$

Find 2^3 first.

$$
\begin{aligned}
9 - 2^3 + 5 &= 9 - 2 \cdot 2 \cdot 2 + 5 && \text{Use the exponent.} \\
&= 9 - 8 + 5 && \text{Multiply.} \\
&= 1 + 5 && \text{Subtract.} \\
&= 6 && \text{Add.}
\end{aligned}
$$

(e)
$$
\begin{aligned}
72 \div 2 \cdot 3 + 4 \cdot 2^3 - 3^3 &= 72 \div 2 \cdot 3 + 4 \cdot 8 - 27 && \text{Use the exponents.} \\
&= 36 \cdot 3 + 4 \cdot 8 - 27 && \text{Divide.} \\
&= 108 + 32 - 27 && \text{Multiply.} \\
&= 140 - 27 && \text{Add.} \\
&= 113 && \text{Subtract.}
\end{aligned}
$$

Notice that the multiplications and divisions are performed from left to right *as they appear;* then the additions and subtractions should be done from left to right, *as they appear.*

Now Try Exercises 23, 27, and 29.

OBJECTIVE 3 Use more than one grouping symbol. An expression with double (or *nested*) parentheses, such as $2(8 + 3(6 + 5))$, can be confusing. For clarity, square brackets, [ ], often are used in place of one pair of parentheses. Fraction bars also act as grouping symbols.

EXAMPLE 3   Using Brackets and Fraction Bars as Grouping Symbols

Simplify each expression.

(a) $2[8 + 3(6 + 5)]$

Work first within the parentheses, and then simplify inside the brackets until a single number remains.

$$
\begin{aligned}
2[8 + 3(6 + 5)] &= 2[8 + 3(11)] \\
&= 2[8 + 33] \\
&= 2[41] \\
&= 82
\end{aligned}
$$

(b) $\dfrac{4(5 + 3) + 3}{2(3) - 1}$

The expression can be written as the quotient

$$[4(5 + 3) + 3] \div [2(3) - 1],$$

which shows that the fraction bar groups the numerator and denominator separately. Simplify both numerator and denominator, then divide, if possible.

$$
\begin{aligned}
\frac{4(5 + 3) + 3}{2(3) - 1} &= \frac{4(8) + 3}{2(3) - 1} && \text{Work inside parentheses.} \\
&= \frac{32 + 3}{6 - 1} && \text{Multiply.}
\end{aligned}
$$

$$= \frac{35}{5} \qquad \text{Add and subtract.}$$

$$= 7 \qquad \text{Divide.}$$

Now Try Exercises 31 and 35.

NOTE Parentheses and fraction bars are used as grouping symbols to indicate an expression that represents a single number. That is why we must first simplify within parentheses and above and below fraction bars.

OBJECTIVE 4 Know the meanings of $\neq, <, >, \leq,$ and $\geq$. So far, we have used the symbols for the operations of arithmetic and the symbol for equality ($=$). The symbols $\neq, <, >, \leq,$ and $\geq$ are used to express an **inequality,** a statement that two expressions are not equal. The equality symbol with a slash through it, $\neq$, means "is not equal to." For example,

$$7 \neq 8$$

indicates that 7 is not equal to 8.

If two numbers are not equal, then one of the numbers must be less than the other. The symbol $<$ represents "is less than," so that "7 is less than 8" is written

$$7 < 8.$$

The word *is* in the phrase "is less than" is a verb, which actually makes $7 < 8$ a sentence. This is not the case for expressions, like $7 + 8, 7 - 8,$ and so on.

The symbol $>$ means "is greater than." Write "8 is greater than 2" as

$$8 > 2.$$

To keep the meanings of the symbols $<$ and $>$ clear, remember that the symbol always *points to the lesser number*. For example, write "8 is less than 15" by pointing the symbol toward the 8:

Lesser number $\rightarrow$ $8 < 15$.

Two other symbols, $\leq$ and $\geq$, also represent the idea of inequality. The symbol $\leq$ means "is less than or equal to," so

$$5 \leq 9$$

means "5 is less than or equal to 9." This statement is true, since $5 < 9$ is true. *If either the $<$ part or the $=$ part is true, then the inequality $\leq$ is true.*

The symbol $\geq$ means "is greater than or equal to." Again,

$$9 \geq 5$$

is true because $9 > 5$ is true. Also, $8 \leq 8$ is true since $8 = 8$ is true. But it is not true that $13 \leq 9$ because neither $13 < 9$ nor $13 = 9$ is true.

EXAMPLE 4 Using Inequality Symbols

Determine whether each statement is true or false.

(a) $6 \neq 5 + 1$

The statement is false because 6 *is equal to* $5 + 1$.

(b) $5 + 3 < 19$

Since $5 + 3$ represents a number (8) that is less than 19, this statement is true.

(c) $15 \leq 20 \cdot 2$

The statement $15 \leq 20 \cdot 2$ is true, since $15 < 40$.

(d) $25 \geq 30$

Both $25 > 30$ and $25 = 30$ are false. Because of this, $25 \geq 30$ is false.

(e) $12 \geq 12$

Since $12 = 12$, this statement is true.

(f) $9 < 9$

Since $9 = 9$, this statement is false.

> **Now Try Exercises 41 and 43.**

OBJECTIVE 5 Translate word statements to symbols. An important part of algebra deals with translating words into algebraic notation.

> **PROBLEM SOLVING**
>
> As we will see throughout this book, the ability to solve problems using mathematics is based on translating the words of the problem into symbols. The next example is the first of many that illustrate such translations.

EXAMPLE 5 Translating from Words to Symbols

Write each word statement in symbols.

(a) Twelve equals ten plus two.

$12 = 10 + 2$

(b) Nine is less than ten.

$9 < 10$

(c) Fifteen is not equal to eighteen.

$15 \neq 18$

(d) Seven is greater than four.

$7 > 4$

(e) Thirteen is less than or equal to forty.

$13 \leq 40$

(f) Eleven is greater than or equal to eleven.

$11 \geq 11$

> **Now Try Exercises 55 and 57.**

OBJECTIVE 6 Write statements that change the direction of inequality symbols. Any statement with $<$ can be converted to one with $>$, and any statement with $>$ can be converted to one with $<$. We do this by reversing the order of the numbers and the direction of the symbol. For example, the statement $6 < 10$ can be written with $>$ as $10 > 6$. Similarly, the statement $4 \leq 10$ can be changed to $10 \geq 4$.

EXAMPLE 6 Converting between Inequality Symbols

The following examples show the same statement written in two equally correct ways.

(a) $9 < 16$ $16 > 9$

(b) $5 > 2$ $2 < 5$

(c) $3 \leq 8$ $8 \geq 3$

(d) $12 \geq 5$ $5 \leq 12$

Note that in each pair of inequalities, the inequality symbol points toward the smaller number.

Now Try Exercise 71.

Here is a summary of the symbols discussed in this section.

Symbols of Equality and Inequality

=	is equal to	≠	is not equal to
<	is less than	>	is greater than
≤	is less than or equal to	≥	is greater than or equal to

CAUTION The equality and inequality symbols are used to write mathematical *sentences* that describe the relationship between two numbers. On the other hand, the symbols for operators ($+$, $-$, $\times$, $\div$) are used to write mathematical *expressions* that represent a single number. For example, compare the sentence $4 < 10$ with the expression $4 + 10$, which represents the number 14.

OBJECTIVE 7 Interpret data in a bar graph. **Bar graphs** are often used to summarize data in a concise manner.

EXAMPLE 7 Interpreting Inequality Concepts Using a Bar Graph

The bar graph in Figure 3 shows the federal budget outlays for national defense during the 1990s.

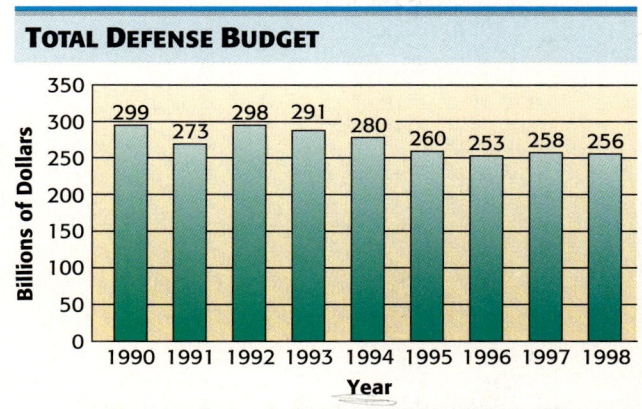

TOTAL DEFENSE BUDGET

Source: U.S. Office of Management and Budget.

FIGURE 3

(a) Which years had outlays greater than 280 billion dollars?

Look for numbers greater than 280 at the tops of the columns. In 1990, 1992, and 1993, the outlays were 299 billion, 298 billion, and 291 billion dollars, respectively.

20 CHAPTER 1 The Real Number System

(b) Which years had outlays greater than 300 billion dollars?

Since the tops of all the bars are below the line representing 300 billion, there were no such years. This is reinforced by the fact that all of the numbers at the tops of the bars are less than 300 (billion).

Now Try Exercise 79.

1.2 EXERCISES

For Extra Help

Student's Solutions Manual

MyMathLab

InterAct Math Tutorial Software

AW Math Tutor Center

MathXL

Digital Video Tutor CD 1/Videotape 1

Decide whether each statement is true or false. If it is false, explain why.

1. When evaluated, $4 + 3(8 - 2)$ is equal to 42.
2. $3^3 = 9$
3. The statement "4 is 12 less than 16" is interpreted $4 = 12 - 16$.
4. The statement "6 is 4 less than 10" is interpreted $6 < 10 - 4$.

Find the value of each exponential expression. See Example 1.

5. 7^2
6. 4^2
7. 12^2
8. 14^2
9. 4^3
10. 5^3
11. 10^3
12. 11^3
13. 3^4
14. 6^4
15. 4^5
16. 3^5
17. $\left(\dfrac{2}{3}\right)^4$
18. $\left(\dfrac{3}{4}\right)^3$
19. $(.04)^3$
20. $(.05)^4$

21. Explain in your own words how to evaluate a power of a number, such as 6^3.
22. Explain why any power of 1 must be equal to 1.

Find the value of each expression. See Examples 2 and 3.

23. $64 \div 4 \times 2$
24. $250 \div 5 \times 2$
25. $\dfrac{1}{4} \cdot \dfrac{2}{3} + \dfrac{2}{5} \cdot \dfrac{11}{3}$
26. $\dfrac{9}{4} \cdot \dfrac{2}{3} + \dfrac{4}{5} \cdot \dfrac{5}{3}$
27. $9 \cdot 4 - 8 \cdot 3$
28. $11 \cdot 4 + 10 \cdot 3$
29. $3(4 + 2) + 8 \cdot 3$
30. $9(1 + 7) + 2 \cdot 5$
31. $5[3 + 4(2^2)]$
32. $6[2 + 8(3^3)]$
33. $3^2[(11 + 3) - 4]$
34. $4^2[(13 + 4) - 8]$
35. $\dfrac{6(3^2 - 1) + 8}{8 - 2^2}$
36. $\dfrac{2(8^2 - 4) + 8}{29 - 3^3}$
37. $\dfrac{4(6 + 2) + 8(8 - 3)}{6(4 - 2) - 2^2}$
38. $\dfrac{6(5 + 1) - 9(1 + 1)}{5(8 - 6) - 2^3}$
39. $-3^2 + 5[12 \div (2 - 8)]$
40. $-5^2 - 3[36 \div (4 - 10)]$
41. $48 \div 3[12 - (10 - 6)] - 2^4$
42. $96 \div 6[20 - (18 - 2)] + 3^3$
43. $-32 \div 4 - 2^5 + (-2)^4$
44. $-27 \div 9 + 2^3 - (-3)^2$
45. $30 + 5 \cdot \dfrac{4^2 - 6}{3^2 - 7} + (-4)^3$
46. $45 - 8 \cdot \dfrac{3^3 - 2}{8^2 - 59} - (-5)^3$
47. $3.2^2 - 4.56 + 2.87$
48. $4.7^2 + 8.32 - 1.44$

SECTION 1.2 Exponents, Order of Operations, and Inequality **21**

First simplify both sides of each inequality. Then tell whether the given statement is true *or* false. *See Examples 2–4.*

49. $9 \cdot 3 - 11 \leq 16$
50. $6 \cdot 5 - 12 \leq 18$
51. $5 \cdot 11 + 2 \cdot 3 \leq 60$
52. $9 \cdot 3 + 4 \cdot 5 \geq 48$
53. $0 \geq 12 \cdot 3 - 6 \cdot 6$
54. $10 \leq 13 \cdot 2 - 15 \cdot 1$
55. $45 \geq 2[2 + 3(2 + 5)]$
56. $55 \geq 3[4 + 3(4 + 1)]$
57. $[3 \cdot 4 + 5(2)] \cdot 3 > 72$
58. $2 \cdot [7 \cdot 5 - 3(2)] \leq 58$
59. $\dfrac{3 + 5(4 - 1)}{2 \cdot 4 + 1} \geq 3$
60. $\dfrac{7(3 + 1) - 2}{3 + 5 \cdot 2} \leq 2$
61. $3 \geq \dfrac{2(5 + 1) - 3(1 + 1)}{5(8 - 6) - 4 \cdot 2}$
62. $7 \leq \dfrac{3(8 - 3) + 2(4 - 1)}{9(6 - 2) - 11(5 - 2)}$

Write each word statement in symbols. See Example 5.

63. Fifteen is equal to five plus ten.
64. Twelve is equal to twenty minus eight.
65. Nine is greater than five minus four.
66. Ten is greater than six plus one.
67. Sixteen is not equal to nineteen.
68. Three is not equal to four.
69. Two is less than or equal to three.
70. Five is less than or equal to nine.

Write each statement in words and decide whether it is true *or false.*

71. $7 < 19$
72. $9 < 10$
73. $3 \neq 6$
74. $9 \neq 13$
75. $8 \geq 11$
76. $4 \leq 2$

Write each statement with the inequality symbol reversed while keeping the same meaning. See Example 6.

77. $5 < 30$
78. $8 > 4$
79. $12 \geq 3$
80. $25 \leq 41$

81. What English-language phrase is used to express the fact that one person's age *is less than* another person's age?

82. What English-language phrase is used to express the fact that one person's height *is greater than* another person's height?

83. $12 \geq 12$ is a true statement. Suppose that someone tells you the following: "$12 \geq 12$ is false, because even though 12 is equal to 12, 12 is not greater than 12." How would you respond to this?

84. The table shows results of a science literacy survey by Jon Miller of the International Center for the Advancement of Science Literacy in Chicago.

    (a) Which countries scored more than 50?
    (b) Which countries scored at most 40?
    (c) For which countries were scores not less than 50?

Country	Science Literacy Index
United States	56
Netherlands	52
France	50
Canada	45
Greece	38
Japan	36

The bar graph shows world coal production by year for the years 1993 through 1999. Use the graph to answer Exercises 85 and 86. See Example 7.

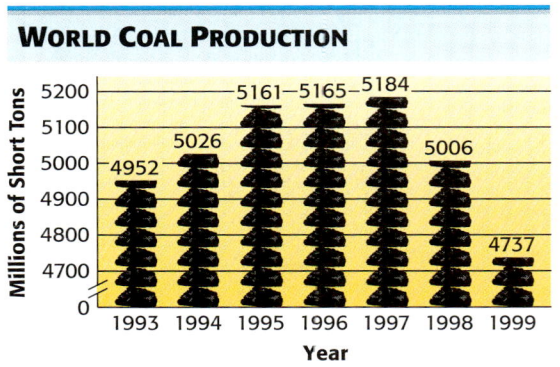

Source: U.S. Energy Information Administration, *International Energy Review,* 1999.

85. In two of the years represented, coal production was less than production in the previous year. What were these two years?

86. What was the first year in which production was greater than 5100 millions of short tons?

87. According to an article in the April 1, 2002 issue of *USA Today,* the stock for Clear Channel, the largest radio station owner in the country, closed at $36.80 per share on September 20, 2001. A year prior to that date, the same stock had closed at $53.76.

  (a) By how much had the stock price decreased?
  (b) To find the percent of decrease, divide the amount from part (a) by 53.76 and convert to a percent. What was the percent of decrease? (Round to the nearest tenth of a percent.)

88. In March 2001, farmers were paid $15.00 per hundred pounds of lettuce. Due to a cold snap in Arizona and California, one year later the price had soared to $86.50. (*Source:* "Prices Soar as Cold Snap Shreds Iceberg Lettuce Supply," *USA Today,* April 1, 2001.)

  (a) By how much had the price increased?
  (b) What was the percent of increase? (Round to the nearest percent.) (*Hint:* See Exercise 87(b).)

Insert one pair of parentheses to make the left side of each equation equal to the right side.

89. $3 \cdot 6 + 4 \cdot 2 = 60$

90. $2 \cdot 8 - 1 \cdot 3 = 42$

91. $10 - 7 - 3 = 6$

92. $15 - 10 - 2 = 7$

93. $8 + 2^2 = 100$

94. $4 + 2^2 = 36$

1.3 Variables, Expressions, and Equations

OBJECTIVES

1. Define *variable*, and find the value of an algebraic expression, given the values of the variables.
2. Convert phrases from words to algebraic expressions.
3. Identify solutions of equations.
4. Identify solutions of equations from a set of numbers.
5. Distinguish between an *expression* and an *equation*.

OBJECTIVE 1 Define *variable*, and find the value of an algebraic expression, given the values of the variables. A **variable** is a symbol, usually a letter such as x, y, or z, used to represent any unknown number. An **algebraic expression** is a collection of numbers, variables, operation symbols, and grouping symbols (such as parentheses). For example,

$$6(x + 5), \quad 2m - 9, \quad \text{and} \quad 8p^2 + 6p + 2$$

are all algebraic expressions. In the algebraic expression $2m - 9$, the expression $2m$ indicates the product of 2 and m, just as $8p^2$ shows the product of 8 and p^2. Also, $6(x + 5)$ means the product of 6 and $x + 5$. An algebraic expression has different numerical values for different values of the variable.

EXAMPLE 1 Evaluating Expressions

Find the numerical value of each algebraic expression when $m = 5$.

(a) $8m$

Replace m with 5, to get
$$8m = 8 \cdot 5 = 40.$$

(b) $3m^2$

For $m = 5$,
$$3m^2 = 3 \cdot 5^2 = 3 \cdot 25 = 75.$$

Now Try Exercises 15 and 17.

CAUTION In Example 1(b), notice that $3m^2$ means $3 \cdot m^2$; it *does not* mean $3m \cdot 3m$. The product $3m \cdot 3m$ is indicated by $(3m)^2$.

EXAMPLE 2 Evaluating Expressions

Find the value of each expression when $x = 5$ and $y = 3$.

(a) $2x + 7y$

Replace x with 5 and y with 3. Follow the order of operations: Multiply first, then add.

$$2x + 7y = 2 \cdot 5 + 7 \cdot 3 \qquad \text{Let } x = 5 \text{ and } y = 3.$$
$$= 10 + 21 \qquad \text{Multiply.}$$
$$= 31 \qquad \text{Add.}$$

(b) $\dfrac{9x - 8y}{2x - y} = \dfrac{9 \cdot 5 - 8 \cdot 3}{2 \cdot 5 - 3} \qquad \text{Let } x = 5 \text{ and } y = 3.$

$$= \dfrac{45 - 24}{10 - 3} \qquad \text{Multiply.}$$
$$= \dfrac{21}{7} \qquad \text{Subtract.}$$
$$= 3 \qquad \text{Divide.}$$

24 CHAPTER 1 The Real Number System

(c) $x^2 - 2y^2 = 5^2 - 2 \cdot 3^2$ Let $x = 5$ and $y = 3$.
$ = 25 - 2 \cdot 9$ Use the exponents.
$ = 25 - 18$ Multiply.
$ = 7$ Subtract.

Now Try Exercises 27, 35, and 37.

OBJECTIVE 2 Convert phrases from words to algebraic expressions.

PROBLEM SOLVING

Sometimes variables must be used to change word phrases into algebraic expressions. Such translations are used in problem solving.

EXAMPLE 3 Using Variables to Change Word Phrases to Algebraic Expressions

Change each word phrase to an algebraic expression. Use x as the variable.

(a) The sum of a number and 9

"Sum" is the answer to an addition problem. This phrase translates as

$x + 9$ or $9 + x$.

(b) 7 minus a number

"Minus" indicates subtraction, so the answer is $7 - x$.

(c) 7 less than a number

Write 7 less than a number as $x - 7$. In this case $7 - x$ would not be correct, because "less than" means "subtracted from."

(d) The product of 11 and a number

$11 \cdot x$ or $11x$

As mentioned earlier, $11x$ means 11 times x. No symbol is needed to indicate the product of a number and a variable.

(e) 5 divided by a number

This translates as $\frac{5}{x}$. The expression $\frac{x}{5}$ would *not* be correct here.

(f) The product of 2, and the sum of a number and 8

We are multiplying 2 times another number. This number is the sum of x and 8, written $x + 8$. Using parentheses for this sum, the final expression is

$2(x + 8)$.

Now Try Exercises 43, 49, and 53.

CAUTION In Example 3(b), the response $x - 7$ would *not* be correct; this statement translates as "a number minus 7," not "7 minus a number." The expressions $7 - x$ and $x - 7$ are rarely equal. For example, if $x = 10$, $10 - 7 \neq 7 - 10$. ($7 - 10$ is a *negative number,* discussed in Section 1.4.)

SECTION 1.3 Variables, Expressions, and Equations   25

OBJECTIVE 3 Identify solutions of equations. An **equation** is a statement that two algebraic expressions are equal. Therefore, an equation *always* includes the equality symbol, =. Examples of equations are

$$x + 4 = 11, \quad 2y = 16, \quad \text{and} \quad 4p + 1 = 25 - p.$$

To **solve** an equation means to find the values of the variable that make the equation true. Such values of the variable are called the **solutions** of the equation.

EXAMPLE 4 Deciding Whether a Number Is a Solution of an Equation

Decide whether the given number is a solution of the equation.

(a) $5p + 1 = 36; \quad 7$
Replace p with 7.

$$5p + 1 = 36$$
$$5 \cdot 7 + 1 = 36 \quad ? \quad \text{Let } p = 7.$$
$$35 + 1 = 36 \quad ?$$
$$36 = 36 \quad \text{True}$$

The number 7 is a solution of the equation.

(b) $9m - 6 = 32; \quad 4$

$$9m - 6 = 32$$
$$9 \cdot 4 - 6 = 32 \quad ? \quad \text{Let } m = 4.$$
$$36 - 6 = 32 \quad ?$$
$$30 = 32 \quad \text{False}$$

The number 4 is not a solution of the equation.

Now Try Exercise 61.

OBJECTIVE 4 Identify solutions of equations from a set of numbers. A **set** is a collection of objects. In mathematics, these objects are most often numbers. The objects that belong to the set, called **elements** of the set, are written between **set braces.** For example, the set containing the numbers 1, 2, 3, 4, and 5 is written as

$$\{1, 2, 3, 4, 5\}.$$

For more information about sets, see Appendix C at the back of this book.

In some cases, the set of numbers from which the solutions of an equation must be chosen is specifically stated. One way of determining solutions is direct substitution of all possible replacements. The ones that lead to a true statement are solutions.

EXAMPLE 5 Finding a Solution from a Given Set

Change each word statement to an equation. Use x as the variable. Then find all solutions of the equation from the set

$$\{0, 2, 4, 6, 8, 10\}.$$

(a) The sum of a number and four is six.

The word *is* suggests "equals." If x represents the unknown number, then translate as follows.

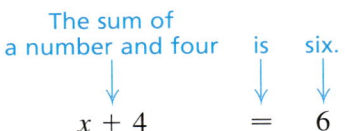

$$x + 4 = 6$$

Try each number from the given set $\{0, 2, 4, 6, 8, 10\}$, in turn, to see that 2 is the only solution of $x + 4 = 6$.

(b) 9 more than five times a number is 49.

Use x to represent the unknown number. Start with $5x$ and then add 9 to it. The word *is* translates as $=$.

$$5x + 9 = 49$$

Try each number from $\{0, 2, 4, 6, 8, 10\}$. The solution is 8, since $5 \cdot 8 + 9 = 49$.

(c) The sum of a number and 12 is equal to four times the number.

If x represents the number, "the sum of a number and 12," is represented by $x + 12$. The translation is

$$x + 12 = 4x.$$

Trying each replacement leads to a true statement when $x = 4$, since $4 + 12 = 4(4) = 16$.

Now Try Exercise 69.

OBJECTIVE 5 Distinguish between an *expression* and an *equation*. Students often have trouble distinguishing between equations and expressions. Remember that an equation is a sentence; an expression is a phrase.

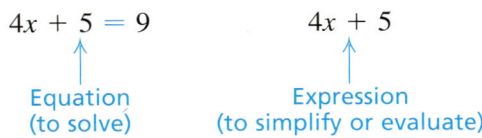

■ **EXAMPLE 6** Distinguishing between Equations and Expressions

Decide whether each of the following is an equation or an expression.

(a) $2x - 5y$

There is no equals sign, so this is an expression.

(b) $2x = 5y$

Because an equals sign is present, this is an equation.

Now Try Exercises 77 and 81.

1.3 EXERCISES

For Extra Help

 Student's Solutions Manual

 MyMathLab

 InterAct Math Tutorial Software

 AW Math Tutor Center

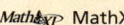 MathXL

 Digital Video Tutor CD 1/Videotape 1

Fill in each blank with the correct response.

1. If $x = 3$, then the value of $x + 7$ is ____.
2. If $x = 1$ and $y = 2$, then the value of $4xy$ is ____.
3. The sum of 12 and x is represented by the expression ____. If $x = 9$, the value of that expression is ____.
4. If x can be chosen from the set $\{0, 1, 2, 3, 4, 5\}$, the only solution of $x + 5 = 9$ is ____.
5. Will the equation $x = x + 4$ ever have a solution? ____.
6. $2x + 3$ is an _____, while $2x + 3 = 8$ is an _____.
   (equation/expression)   (equation/expression)

Exercises 7–12 cover some of the concepts introduced in this section. Give a short explanation for each.

7. Explain why $2x^3$ is not the same as $2x \cdot 2x \cdot 2x$.
8. Why are "5 less than a number" and "5 is less than a number" translated differently?
9. Explain why, when evaluating the expression $4x^2$ for $x = 3$, 3 must be squared *before* multiplying by 4.
10. What value of x would cause the expression $2x + 3$ to equal 9? Explain your reasoning.
11. There are many pairs of values of x and y for which $2x + y$ will equal 6. Name two such pairs and describe how you determined them.
12. Suppose that for the equation $3x - y = 9$, the value of x is given as 4. What would be the corresponding value of y? How do you know this?

Find the numerical value if (a) $x = 4$ and (b) $x = 6$. See Example 1.

13. $x + 9$
14. $x - 1$
15. $5x$
16. $7x$
17. $4x^2$
18. $5x^2$
19. $\dfrac{x + 1}{3}$
20. $\dfrac{x - 2}{5}$
21. $\dfrac{3x - 5}{2x}$
22. $\dfrac{4x - 1}{3x}$
23. $3x^2 + x$
24. $2x + x^2$
25. $6.459x$
26. $.74x^2$

Find the numerical value if (a) $x = 2$ and $y = 1$ and (b) $x = 1$ and $y = 5$. See Example 2.

27. $8x + 3y + 5$
28. $4x + 2y + 7$
29. $3(x + 2y)$
30. $2(2x + y)$
31. $x + \dfrac{4}{y}$
32. $y + \dfrac{8}{x}$
33. $\dfrac{x}{2} + \dfrac{y}{3}$
34. $\dfrac{x}{5} + \dfrac{y}{4}$
35. $\dfrac{2x + 4y - 6}{5y + 2}$
36. $\dfrac{4x + 3y - 1}{x}$
37. $2y^2 + 5x$
38. $6x^2 + 4y$
39. $\dfrac{3x + y^2}{2x + 3y}$
40. $\dfrac{x^2 + 1}{4x + 5y}$
41. $.841x^2 + .32y^2$
42. $.941x^2 + .2y^2$

Change each word phrase to an algebraic expression. Use x as the variable to represent the number. See Example 3.

43. Twelve times a number
44. Nine times a number
45. Seven added to a number
46. Thirteen added to a number

28 CHAPTER 1 The Real Number System

47. Two subtracted from a number
48. Eight subtracted from a number
49. A number subtracted from seven
50. A number subtracted from fourteen
51. The difference between a number and 6
52. The difference between 6 and a number
53. 12 divided by a number
54. A number divided by 12
55. The product of 6, and four less than a number.
56. The product of 9, and five more than a number.

57. In the phrase "Four more than the product of a number and 6," does the word *and* signify the operation of addition? Explain.

58. Suppose that the directions on a test read "Solve the following expressions." How would you politely correct the person who wrote these directions? What alternative directions might you suggest?

Decide whether the given number is a solution of the equation. See Example 4.

59. $5m + 2 = 7$; 1
60. $3r + 5 = 8$; 1
61. $2y + 3(y - 2) = 14$; 3
62. $6a + 2(a + 3) = 14$; 2
63. $6p + 4p + 9 = 11$; $\frac{1}{5}$
64. $2x + 3x + 8 = 20$; $\frac{12}{5}$
65. $3r^2 - 2 = 46$; 4
66. $2x^2 + 1 = 19$; 3
67. $\frac{z + 4}{2 - z} = \frac{13}{5}$; $\frac{1}{3}$
68. $\frac{x + 6}{x - 2} = \frac{37}{5}$; $\frac{13}{4}$

Change each word statement to an equation. Use x as the variable. Find all solutions from the set {2, 4, 6, 8, 10}. See Example 5.

69. The sum of a number and 8 is 18.
70. A number minus three equals 1.
71. Sixteen minus three-fourths of a number is 13.
72. The sum of six-fifths of a number and 2 is 14.
73. One more than twice a number is 5.
74. The product of a number and 3 is 6.
75. Three times a number is equal to 8 more than twice the number.
76. Twelve divided by a number equals $\frac{1}{3}$ times that number.

Identify each as an expression or an equation. See Example 6.

77. $3x + 2(x - 4)$
78. $5y - (3y + 6)$
79. $7t + 2(t + 1) = 4$
80. $9r + 3(r - 4) = 2$
81. $x + y = 3$
82. $x + y - 3$

A **mathematical model** is an equation that describes the relationship between two quantities. For example, based on data from the United States Olympic Committee, the winning distances for the men's discus throw can be approximated by the equation
$$y = 1.2304x - 2224.5,$$
where x is the year (between 1896 and 1996) and y is in feet. Use this model to approximate the winning distances for the following years.

83. 1912
84. 1936
85. 1960
86. 1992

1.4 Real Numbers and the Number Line

OBJECTIVES

1. Classify numbers and graph them on number lines.
2. Tell which of two different real numbers is smaller.
3. Find additive inverses and absolute values of real numbers.
4. Interpret the meanings of real numbers from a table of data.

OBJECTIVE 1 Classify numbers and graph them on number lines. In Section 1.1 we introduced two important sets of numbers, the *natural numbers* and the *whole numbers*.

> **Natural Numbers***
>
> $\{1, 2, 3, 4, \ldots\}$ is the set of **natural numbers**.

> **Whole Numbers**
>
> $\{0, 1, 2, 3, 4, \ldots\}$ is the set of **whole numbers**.

> **NOTE** The three dots show that the list of numbers continues in the same way indefinitely.

These numbers, along with many others, can be represented on **number lines** like the one in Figure 4. The **graph** of a number is the point on the number line associated with that number. The number is called the **coordinate** of the point. We draw a number line by choosing any point on the line and labeling it 0. Then we choose any point to the right of 0 and label it 1. The distance between 0 and 1 gives a unit of measure used to locate other points, as shown in Figure 4. The points labeled in Figure 4 correspond to the first few whole numbers.

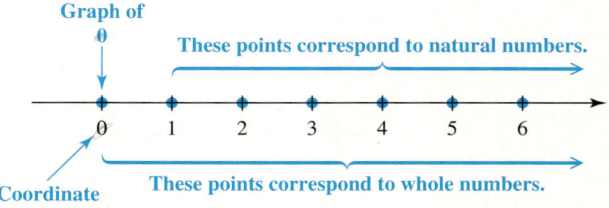

FIGURE 4

The natural numbers are located to the right of 0 on the number line. But numbers may also be placed to the left of 0. For each natural number we can place a corresponding number to the left of 0. These numbers, written $-1, -2, -3, -4$, and so on, are shown in Figure 5 on the next page. Each is the **opposite** or **negative** of a natural number. The natural numbers, their opposites, and 0 form a new set of numbers called the *integers*.

> **Integers**
>
> $\{\ldots, -3, -2, -1, 0, 1, 2, 3, \ldots\}$ is the set of **integers**.

---

*The symbols { and } are braces used in conjunction with sets. See Appendix C.

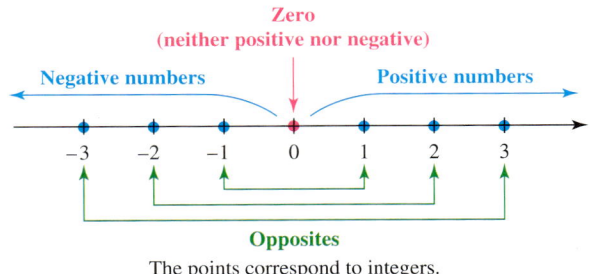

FIGURE 5

Positive numbers and negative numbers are called **signed numbers.** There are many practical applications of negative numbers. For example, a Fahrenheit temperature on a cold January day might be $-10°$, and a business that spends more than it takes in has a negative "profit."

EXAMPLE 1 Using Negative Numbers in Applications

Use an integer to express the number in italics in each application.

(a) The lowest Fahrenheit temperature ever recorded in meteorological records was *129°* below zero at Vostok, Antarctica, on July 21, 1983. (*Source: World Almanac and Book of Facts,* 2000).
Use $-129°$ because "below zero" indicates a negative number.

(b) The shore surrounding the Dead Sea is *1340* ft below sea level. (*Source: Microsoft Encarta Encyclopedia 2000*).
Again, "below sea level" indicates a negative number, -1340.

Now Try Exercises 3 and 5.

Fractions, introduced in Section 1.1, are examples of *rational numbers.*

Rational Numbers

$\{x \mid x$ is a quotient of two integers, with denominator not $0\}$ is the set of **rational numbers.**

(Read the part in the braces as "the set of all numbers x such that x is a quotient of two integers, with denominator not 0.")

NOTE
The set symbolism used in the definition of rational numbers,

$$\{x \mid x \text{ has a certain property}\},$$

is called **set-builder notation.** This notation is convenient to use when it is not possible to list all the elements of a set.

Since any integer can be written as the quotient of itself and 1, all integers also are rational numbers. Figure 6 shows a number line with the graphs of several rational numbers.

SECTION 1.4 Real Numbers and the Number Line 31

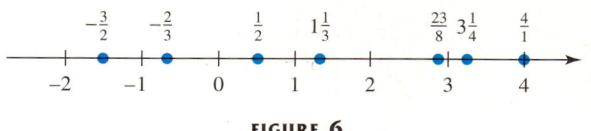

FIGURE 6

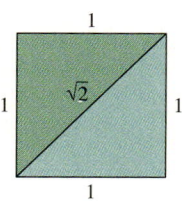

FIGURE 7

Although many numbers are rational, not all are. For example, a square that measures one unit on a side has a diagonal whose length is the square root of 2, written $\sqrt{2}$. See Figure 7. It can be shown that $\sqrt{2}$ cannot be written as a quotient of integers. Because of this, $\sqrt{2}$ is not rational; it is *irrational*. Other examples of **irrational numbers** are $\sqrt{3}$, $\sqrt{7}$, $-\sqrt{10}$, and π (the ratio of the *circumference* of a circle to its diameter).

Both rational and irrational numbers can be represented by points on the number line and together form the set of *real numbers*.

> **Real Numbers**
>
> $\{x \mid x$ is a rational or an irrational number$\}$ is the set of **real numbers.**

Real numbers can be written as decimals. Any rational number will have a decimal that either comes to an end (terminates) or repeats in a fixed "block" of digits. For example, $\frac{2}{5} = .4$ and $\frac{27}{100} = .27$ are rational numbers with terminating decimals; $\frac{1}{3} = .333\ldots$ and $\frac{3}{11} = .27272727\ldots$ are repeating decimals, often written as $.\overline{3}$ and $.\overline{27}$ with a bar over the repeating digit(s). The decimal representation of an irrational number will neither terminate nor repeat.

An example of a number that is not a real number is the square root of a negative number like $\sqrt{-5}$. These numbers are discussed in Chapter 9.

Two ways to represent the relationships among the various types of numbers are shown in Figure 8. Part (a) also gives some examples. Notice that every real number is either a rational number or an irrational number.

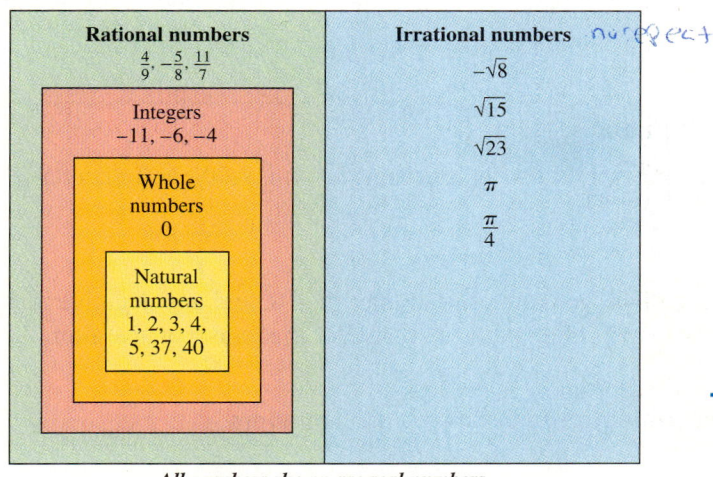

All numbers shown are real numbers.

(a)

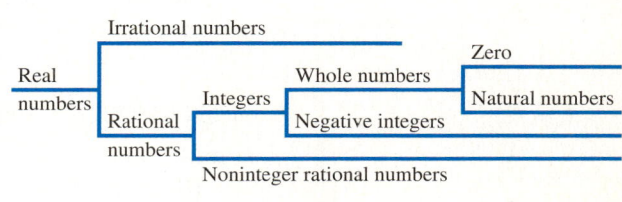

(b)

FIGURE 8

EXAMPLE 2 Determining Whether a Number Belongs to a Set

List the numbers in the set

$$\left\{-5, -\frac{2}{3}, 0, \sqrt{2}, 3\frac{1}{4}, 5, 5.8\right\}$$

that belong to each set of numbers.

(a) Natural numbers
The only natural number in the set is 5.

(b) Whole numbers
The whole numbers consist of the natural numbers and 0. So the elements of the set that are whole numbers are 0 and 5.

(c) Integers
The integers in the set are -5, 0, and 5.

(d) Rational numbers
The rational numbers are $-5, -\frac{2}{3}, 0, 3\frac{1}{4}, 5, 5.8$, since each of these numbers *can* be written as the quotient of two integers. For example, $3\frac{1}{4} = \frac{13}{4}$ and $5.8 = \frac{58}{10}$.

(e) Irrational numbers
The only irrational number in the set is $\sqrt{2}$.

(f) Real numbers
All the numbers in the set are real numbers.

Now Try Exercise 19.

OBJECTIVE 2 Tell which of two different real numbers is smaller. Given any two whole numbers, you probably can tell which number is smaller. But what happens with negative numbers, as in the set of integers? Positive numbers decrease as the corresponding points on the number line go to the left. For example, $8 < 12$ because 8 is to the left of 12 on the number line. This ordering is extended to all real numbers by definition.

Ordering of Real Numbers

For any two real numbers a and b, **a is less than b** if a is to the left of b on the number line.

This means that any negative number is smaller than 0, and any negative number is smaller than any positive number. Also, 0 is smaller than any positive number.

EXAMPLE 3 Determining the Order of Real Numbers

Is it true that $-3 < -1$?

To decide whether the statement is true, locate both numbers, -3 and -1, on a number line, as shown in Figure 9. Since -3 is to the left of -1 on the number line, -3 is smaller than -1. The statement $-3 < -1$ is true.

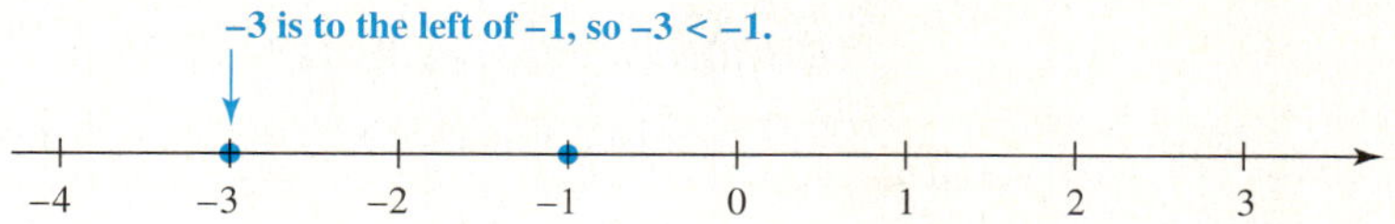

SECTION 1.4 Real Numbers and the Number Line   33

−3 is to the left of −1, so −3 < −1.

FIGURE 9

Now Try Exercise 53.

NOTE In Section 1.2 we saw how it is possible to rewrite a statement involving < as an equivalent statement involving >. The question in Example 2 can also be worded as follows: Is it true that $-1 > -3$? This is, or course, also a true statement.

We can also say that for any two real numbers a and b, **a is greater than b** if a is to the right of b on the number line.

OBJECTIVE 3 Find additive inverses and absolute values of real numbers. By a property of the real numbers, for any real number x (except 0), there is exactly one number on the number line the same distance from 0 as x but on the opposite side of 0. For example, Figure 10 shows that the numbers 1 and -1 are each the same distance from 0 but are on opposite sides of 0. The numbers 1 and -1 are called *additive inverses,* or *opposites,* of each other.

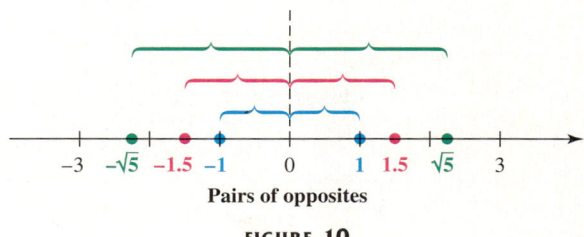

Pairs of opposites

FIGURE 10

Additive Inverse

The **additive inverse** of a number x is the number that is the same distance from 0 on the number line as x, but on the opposite side of 0.

The additive inverse of a number can be indicated by writing the symbol − in front of the number. With this symbol, the additive inverse of 7 is written -7. The additive inverse of -3 is written $-(-3)$, and can be read "the opposite of -3" or "the negative of -3." Figure 10 suggests that 3 is the additive inverse of -3. A number can have only one additive inverse, and so the symbols 3 and $-(-3)$ must represent the same number, which means that

$$-(-3) = 3.$$

This idea can be generalized as follows.

34 CHAPTER 1 The Real Number System

> **Double Negative Rule***
>
> For any real number x, $\qquad -(-x) = x.$

Number	Additive Inverse
-4	$-(-4)$ or 4
0	0
19	-19
$-\frac{2}{3}$	$\frac{2}{3}$
.52	$-.52$

The table in the margin shows several numbers and their additive inverses. It suggests that the additive inverse of a number is found by changing the sign of the number. An important property of additive inverses will be studied later in this chapter: $a + (-a) = (-a) + a = 0$ for all real numbers a.

As previously mentioned, additive inverses are numbers that are the same distance from 0 on the number line. See Figure 10. We can also express this idea by saying that a number and its additive inverse have the same absolute value. The **absolute value** of a real number can be defined as the distance between 0 and the number on the number line. The symbol for the absolute value of the number x is $|x|$, read "the absolute value of x." For example, the distance between 2 and 0 on the number line is 2 units, so

$$|2| = 2.$$

Because the distance between -2 and 0 on the number is also 2 units,

$$|-2| = 2.$$

Since distance is a physical measurement, which is never negative, *the absolute value of a number is never negative.* For example, $|12| = 12$ and $|-12| = 12$, since both 12 and -12 lie at a distance of 12 units from 0 on the number line. Also, since 0 is a distance of 0 units from 0, $|0| = 0$.

In symbols, the absolute value of x is defined as follows.

> **Absolute Value**
>
> $$|x| = \begin{cases} x & \text{if } x \geq 0 \\ -x & \text{if } x < 0 \end{cases}$$

By this definition, if x is a positive number or 0, then its absolute value is x itself. For example, since 8 is a positive number, $|8| = 8$. However, if x is a negative number, then its absolute value is the additive inverse of x. This means that if $x = -9$, then $|-9| = -(-9) = 9$, since the additive inverse of -9 is 9.

CAUTION The definition of absolute value can be confusing if it is not read carefully. The "$-x$" in the second part of the definition *does not* represent a negative number. Since x is negative in the second part, $-x$ represents the opposite of a negative number, that is, a positive number. *The absolute value of a number is never negative.*

---

*This rule is justified by interpreting $-(-x)$ as $-1 \cdot (-x) = [(-1)(-1)]x = 1 \cdot x = x$. This requires concepts covered in later sections of this chapter.

EXAMPLE 4 Finding Absolute Value

Simplify by finding the absolute value.

(a) $|5| = 5$
(b) $|-5| = -(-5) = 5$
(c) $-|5| = -(5) = -5$
(d) $-|-14| = -(14) = -14$
(e) $|8 - 2| = |6| = 6$
(f) $-|8 - 2| = -|6| = -6$

Now Try Exercises 35, 37, and 39.

Parts (e) and (f) of Example 4 show that absolute value bars are also grouping symbols. You must perform any operations that appear inside absolute value symbols before finding the absolute value.

OBJECTIVE 4 Interpret the meanings of real numbers from a table of data. The Producer Price Index is the oldest continuous statistical series published by the Bureau of Labor Statistics. It measures the average changes in prices received by producers of all commodities produced in the United States. The next example shows how signed numbers can be used to interpret such data.

EXAMPLE 5 Interpreting Data

The table shows the percent change in the Producer Price Index for selected commodities from 1998 to 1999 and from 1999 to 2000. Use the table to answer each question.

Commodity	Change from 1998 to 1999	Change from 1999 to 2000
Fresh fruits and melons	13.1	−12.8
Gasoline	11.3	30.1
X-ray and electromedical equipment	−2.4	−2.5
Prepared animal feeds	−9.6	4.6
Office and store machines and equipment	0	.6

Source: U.S. Bureau of Labor Statistics.

(a) What commodity in which year represents the greatest percent decrease?

We must find the negative number with the greatest absolute value. The number that satisfies this condition is −12.8; the greatest percent decrease was shown by fresh fruits and melons from 1999 to 2000.

(b) Which commodity in which year represents the least change?

In this case, we must find the number (either positive, negative, or zero) with the least absolute value. From 1998 to 1999, office and store machines and equipment showed no change, as represented by 0.

Now Try Exercises 65 and 67.

1.4 EXERCISES

For Extra Help

 Student's Solutions Manual

 MyMathLab

 InterAct Math Tutorial Software

 AW Math Tutor Center

MathXL MathXL

 Digital Video Tutor CD 2/Videotape 1

In the applications in Exercises 1–6, use an integer to express each number in italics representing a change. In Exercises 7 and 8, use a rational number. See Example 1.

1. Between 1995 and 2000, the number of U.S. citizens 65 years of age or older increased by *1,198,000*. (*Source:* U.S. Bureau of the Census.)

2. Between 1990 and 2000, the mean SAT verbal score for Florida residents increased by *3*, while the mathematics score increased by *7*. (*Source:* The College Board.)

3. From 1990 to 2000, the number of cable TV systems in the United States went from 9575 to 10,500, representing an increase of *925*. (*Source: Television and Cable Factbook.*)

4. From 1990 to 1998, the population of Glen Ellyn, Illinois, went from 25,956 to 24,919, representing a decrease of *1037* people. (*Source:* U.S. Bureau of the Census.)

5. In 1935, there were 15,295 banks in the United States. By 1999, this number was 10,221, representing a decrease of *5074* banks. (*Source:* Federal Deposit Insurance Corporation.)

6. Death Valley lies *282* feet below sea level. (Sea level is considered 0 feet.) (*Source:* U.S. Geological Survey, Department of the Interior.)

7. On Thursday, May 30, 2002, the Dow Jones Industrial Average closed at 9911.69. On the previous day it had closed at 9923.04. Thus on Thursday it closed down *11.35*. (*Source: Times Picayune.*)

8. On Thursday, May 30, 2002, the NASDAQ closed at 1631.92. On the previous day it had closed at 1624.39. Thus on Thursday it closed up *7.53*. (*Source: Times Picayune.*)

In Exercises 9–14, give a number that satisfies the given condition.

9. An integer between 3.5 and 4.5

10. A rational number between 3.8 and 3.9

11. A whole number that is not positive and is less than 1

12. A whole number greater than 4.5

13. An irrational number that is between $\sqrt{11}$ and $\sqrt{13}$

14. A real number that is neither negative nor positive

In Exercises 15–18, decide whether each statement is true *or* false.

15. Every natural number is positive.
16. Every whole number is positive.
17. Every integer is a rational number.
18. Every rational number is a real number.

For Exercises 19 and 20, see Example 2.

19. List all numbers from the set

$$\left\{-9, -\sqrt{7}, -1\frac{1}{4}, -\frac{3}{5}, 0, \sqrt{5}, 3, 5.9, 7\right\}$$

that are

(a) natural numbers; (b) whole numbers; (c) integers;
(d) rational numbers; (e) irrational numbers; (f) real numbers.

20. List all numbers from the set

$$\left\{-5.3, -5, -\sqrt{3}, -1, -\frac{1}{9}, 0, 1.2, 1.8, 3, \sqrt{11}\right\}$$

that are

(a) natural numbers; (b) whole numbers; (c) integers;
(d) rational numbers; (e) irrational numbers; (f) real numbers.

21. Explain in your own words the different sets of numbers introduced in this section, and give an example of each kind.

22. What two possible situations exist for the decimal representation of a rational number?

Graph each group of numbers on a number line. See Figures 5 and 6.

23. 0, 3, −5, −6

24. 2, 6, −2, −1

25. −2, −6, −4, 3, 4

26. −5, −3, −2, 0, 4

27. $\frac{1}{4}, 2\frac{1}{2}, -3\frac{4}{5}, -4, -1\frac{5}{8}$

28. $5\frac{1}{4}, 4\frac{5}{9}, -2\frac{1}{3}, 0, -3\frac{2}{5}$

29. Match each expression in Column I with its value in Column II. Choices in Column II may be used once, more than once, or not at all.

I	II
(a) $\|-7\|$	A. 7
(b) $-(-7)$	B. −7
(c) $-\|-7\|$	C. Neither A nor B
(d) $-\|-(-7)\|$	D. Both A and B

30. Fill in the blanks with the correct values: The opposite of −2 is ____, while the absolute value of −2 is ____. The additive inverse of −2 is ____, while the additive inverse of the absolute value of −2 is ____.

Find (a) the opposite (or additive inverse) of each number and (b) the absolute value of each number.

31. −2

32. −8

33. 6

34. 11

Simplify by finding the absolute value. See Example 4.

35. $|-6|$

36. $|-12|$

37. $-|-12|$

38. $-|-6|$

39. $|6-3|$

40. $-|6-3|$

Select the smaller of the two given numbers. See Examples 3 and 4.

41. −12, −4

42. −9, −14

43. −8, −1

44. −15, −16

45. 3, $|-4|$

46. 5, $|-2|$

47. $|-3|, |-4|$

48. $|-8|, |-9|$

49. $-|-6|, -|-4|$

50. $-|-2|, -|-3|$

51. $|5-3|, |6-2|$

52. $|7-2|, |8-1|$

38 CHAPTER 1 The Real Number System

Decide whether each statement is true or false. See Examples 3 and 4.

53. $-5 < -2$
54. $-8 > -2$
55. $-4 \leq -(-5)$
56. $-6 \leq -(-3)$
57. $|-6| < |-9|$
58. $|-12| < |-20|$
59. $-|8| > |-9|$
60. $-|12| > |-15|$
61. $-|-5| \geq -|-9|$
62. $-|-12| \leq -|-15|$
63. $|6-5| \geq |6-2|$
64. $|13-8| \leq |7-4|$

The table shows the percent change in the Producer Price Index for selected construction commodities from 1998 to 1999 and from 1999 to 2000. Use the table to answer Exercises 65–68. See Example 5.*

Commodity	Change from 1998 to 1999	Change from 1999 to 2000
Softwood plywood	32.1	−33.2
Outdoor lighting equipment	−.5	2.4
Steel pipe and tubes	−6.9	4.3
Gypsum products	30.4	−6.6
Paving mixtures and blocks	.4	17.4

*2000 data are preliminary.
Source: U.S. Bureau of Labor Statistics.

65. Which commodity in which year represents the greatest percentage increase?
66. Which commodity in which year represents the greatest percentage decrease?
67. Which commodity in which year represents the least change?
68. Which commodity represents an increase for both years?

Give three numbers between −6 and 6 that satisfy each given condition.

69. Positive real numbers but not integers
70. Real numbers but not positive numbers
71. Real numbers but not whole numbers
72. Rational numbers but not integers
73. Real numbers but not rational numbers
74. Rational numbers but not negative numbers
75. Students often say "Absolute value is always positive." Is this true? Explain.
76. True or false: If a is negative, $|a| = -a$.

1.5 Adding and Subtracting Real Numbers

OBJECTIVES

1. Add two numbers with the same sign.
2. Add positive and negative numbers.

In this and the next section, we extend the rules for operations with positive numbers to the negative numbers.

OBJECTIVE 1 Add two numbers with the same sign. A number line can be used to illustrate adding real numbers.

SECTION 1.5 Adding and Subtracting Real Numbers   39

3 Use the definition of subtraction.

4 Use the order of operations with real numbers.

5 Interpret words and phrases involving addition and subtraction.

6 Use signed numbers to interpret data.

■ **EXAMPLE 1**   Adding Numbers on a Number Line

(a) Use a number line to find the sum $2 + 3$.

Add the positive numbers 2 and 3 on the number line by starting at 0 and drawing an arrow 2 units to the *right*, as shown in Figure 11. This arrow represents the number 2 in the sum $2 + 3$. Then, from the right end of this arrow draw another arrow 3 units to the right. The number below the end of this second arrow is 5, so $2 + 3 = 5$.

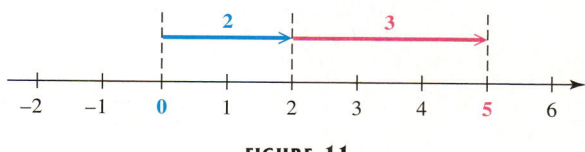

FIGURE 11

(b) Use a number line to find the sum $-2 + (-4)$. (Parentheses are placed around the -4 to avoid the confusing use of $+$ and $-$ next to each other.)

Add the negative numbers -2 and -4 on the number line by starting at 0 and drawing an arrow 2 units to the *left*, as shown in Figure 12. The arrow is drawn to the left to represent the addition of a *negative* number. From the left end of the first arrow, draw a second arrow 4 units to the left. The number below the end of this second arrow is -6, so $-2 + (-4) = -6$.

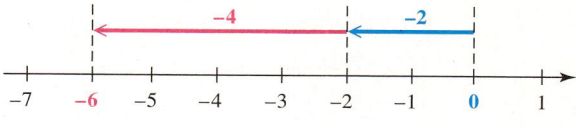

FIGURE 12

Now Try Exercise 1.

In Example 1, the sum of the two negative numbers -2 and -4 is a negative number whose distance from 0 is the sum of the distance of -2 from 0 and the distance of -4 from 0. That is, *the sum of two negative numbers is the negative of the sum of their absolute values.*

$$-2 + (-4) = -(|-2| + |-4|) = -(2 + 4) = -6$$

Adding Numbers with the Same Sign

To add two numbers with the *same* sign, add the absolute values of the numbers. The sum has the same sign as the numbers being added.

■ **EXAMPLE 2**   Using the Rule to Add Two Negative Numbers

Find each sum.

(a) $-2 + (-9) = -(|-2| + |-9|) = -(2 + 9) = -11$

(b) $-8 + (-12) = -20$ **(c)** $-15 + (-3) = -18$

Now Try Exercise 7.

OBJECTIVE 2 Add positive and negative numbers. We can use a number line to illustrate the sum of a positive number and a negative number.

EXAMPLE 3 Adding Numbers with Different Signs

Use a number line to find the sum $-2 + 5$.

Find the sum $-2 + 5$ on the number line by starting at 0 and drawing an arrow 2 units to the left. From the left end of this arrow, draw a second arrow 5 units to the right, as shown in Figure 13. The number below the end of the second arrow is 3, so $-2 + 5 = 3$.

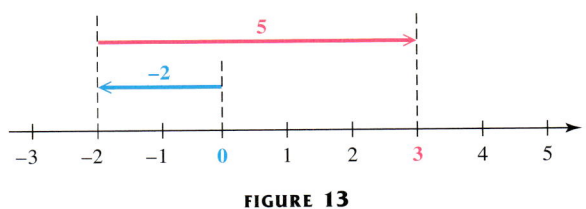

FIGURE 13

Now Try Exercise 3.

Adding Numbers with Different Signs

To add two numbers with *different* signs, subtract the smaller absolute value from the larger absolute value. The answer has the sign of the number with the larger absolute value.

For example, to add -12 and 5, find their absolute values: $|-12| = 12$ and $|5| = 5$. Then find the difference between these absolute values: $12 - 5 = 7$. Since $|-12| > |5|$, the sum will be negative, so the final answer is $-12 + 5 = -7$.

While a number line is useful in showing the rules for addition, it is important to be able to find sums mentally.

EXAMPLE 4 Adding Mentally

Check each answer, trying to work the addition mentally. If you have trouble, use a number line.

(a) $7 + (-4) = 3$ **(b)** $-8 + 12 = 4$

(c) $-\dfrac{1}{2} + \dfrac{1}{8} = -\dfrac{4}{8} + \dfrac{1}{8} = -\dfrac{3}{8}$ Remember to find a common denominator first.

(d) $\dfrac{5}{6} + \left(-\dfrac{4}{3}\right) = -\dfrac{1}{2}$ **(e)** $-4.6 + 8.1 = 3.5$

(f) $-16 + 16 = 0$ **(g)** $42 + (-42) = 0$

Now Try Exercises 15 and 27.

Parts (f) and (g) in Example 4 suggest that the sum of a number and its additive inverse is 0. This is always true, and this property is discussed further in Section 1.7.

The rules for adding signed numbers are summarized as follows.

Adding Signed Numbers

Same sign Add the absolute values of the numbers. The sum has the same sign as the given numbers.

Different signs Find the difference between the larger absolute value and the smaller. The sum has the sign of the number with the larger absolute value.

OBJECTIVE 3 Use the definition of subtraction. We can illustrate subtraction of 4 from 7, written $7 - 4$, using a number line. As seen in Figure 14, we begin at 0 and draw an arrow 7 units to the right. From the right end of this arrow, we draw an arrow 4 units to the left. The number at the end of the second arrow shows that $7 - 4 = 3$.

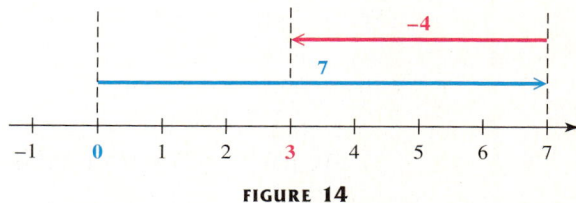

FIGURE 14

The procedure used to find $7 - 4$ is exactly the same procedure that would be used to find $7 + (-4)$, so

$$7 - 4 = 7 + (-4).$$

This suggests that *subtracting* a positive number from a larger positive number is the same as *adding* the additive inverse of the smaller number to the larger. This result leads to the definition of subtraction for all real numbers.

Subtraction

For any real numbers x and y,

$$x - y = x + (-y).$$

That is, to *subtract* y from x, add the additive inverse (or opposite) of y to x.

The definition gives the following procedure for subtracting signed numbers.

Subtracting Signed Numbers

Step 1 Change the subtraction symbol to the addition symbol.
Step 2 Change the sign of the number being subtracted.
Step 3 Add.

EXAMPLE 5   Using the Definition of Subtraction

Subtract.

(a) $12 - 3 = 12 + (-3) = 9$

(b) $5 - 7 = 5 + (-7) = -2$

(c) $-3 - (-5) = -3 + (5) = 2$

(d) $-6 - 9 = -6 + (-9) = -15$

(e) $\dfrac{4}{3} - \left(-\dfrac{1}{2}\right) = \dfrac{4}{3} + \dfrac{1}{2} = \dfrac{8}{6} + \dfrac{3}{6} = \dfrac{11}{6}$ or $1\dfrac{5}{6}$

>>> **Now Try Exercises 43, 51, and 55.**

We have now used the symbol $-$ for three purposes:

1. to represent subtraction, as in $9 - 5 = 4$;
2. to represent negative numbers, such as $-10, -2,$ and -3;
3. to represent the opposite (or negative) of a number, as in "the opposite (or negative) of 8 is -8."

We may see more than one use of $-$ in the same problem, such as $-6 - (-9)$, where -9 is subtracted from -6. The meaning of the $-$ symbol depends on its position in the algebraic expression.

OBJECTIVE 4   Use the order of operations with real numbers.   As before, with problems that have grouping symbols, first do any operations inside the parentheses and brackets. Work within the innermost set of grouping symbols first, and then work outward.

EXAMPLE 6   Adding and Subtracting with Grouping Symbols

Perform each indicated operation.

(a) $-6 - [2 - (8 + 3)] = -6 - [2 - 11]$ Add.

$\qquad\qquad\qquad\qquad\quad = -6 - [2 + (-11)]$ Definition of subtraction

$\qquad\qquad\qquad\qquad\quad = -6 - (-9)$ Add.

$\qquad\qquad\qquad\qquad\quad = -6 + (9)$ Definition of subtraction

$\qquad\qquad\qquad\qquad\quad = 3$ Add.

(b) $5 + [(-3 - 2) - (4 - 1)] = 5 + [(-3 + (-2)) - 3]$
$= 5 + [(-5) - 3]$
$= 5 + [(-5) + (-3)]$
$= 5 + (-8)$
$= -3$

(c) $\dfrac{2}{3} - \left[\dfrac{1}{12} - \left(-\dfrac{1}{4}\right)\right] = \dfrac{8}{12} - \left[\dfrac{1}{12} - \left(-\dfrac{3}{12}\right)\right]$ Find a common denominator.

$= \dfrac{8}{12} - \left[\dfrac{1}{12} + \dfrac{3}{12}\right]$ Definition of subtraction

$= \dfrac{8}{12} - \dfrac{4}{12}$ Add.

$= \dfrac{4}{12}$ Subtract.

$= \dfrac{1}{3}$ Lowest terms

(d) $|4 - 7| + 2|6 - 3| = |-3| + 2|3|$ Work within absolute value bars.
$= 3 + 2 \cdot 3$ Evaluate absolute values.
$= 3 + 6$ Multiply.
$= 9$ Add.

Now Try Exercises 65, 75, and 79.

OBJECTIVE 5 Interpret words and phrases involving addition and subtraction.

PROBLEM SOLVING

As we mentioned earlier, problem solving often requires translating words and phrases into symbols. The word *sum* indicates addition. The table lists some of the words and phrases that also signify addition.

Word or Phrase	Example	Numerical Expression and Simplification
Sum of	The *sum of* −3 and 4	−3 + 4 = 1
Added to	5 *added to* −8	−8 + 5 = −3
More than	12 *more than* −5	−5 + 12 = 7
Increased by	−6 *increased by* 13	−6 + 13 = 7
Plus	3 *plus* 14	3 + 14 = 17

44 CHAPTER 1 The Real Number System

EXAMPLE 7 Interpreting Words and Phrases Involving Addition

Write a numerical expression for each phrase and simplify the expression.

(a) The sum of -8 and 4 and 6

$$-8 + 4 + 6 = -4 + 6 = 2 \quad \text{Add in order from left to right.}$$

(b) 3 more than -5, increased by 12

$$-5 + 3 + 12 = -2 + 12 = 10$$

Now Try Exercise 107.

PROBLEM SOLVING

To solve problems that involve subtraction, we must be able to interpret key words and phrases that indicate subtraction. *Difference* is one of them. Some of these are given in the table.

Word, Phrase, or Sentence	Example	Numerical Expression and Simplification
Difference between	The difference between -3 and -8	$-3 - (-8) = -3 + 8 = 5$
Subtracted from	12 subtracted from 18	$18 - 12 = 6$
From..., subtract....	From 12, subtract 8.	$12 - 8 = 12 + (-8) = 4$
Less	6 less 5	$6 - 5 = 1$
Less than	6 less than 5	$5 - 6 = 5 + (-6) = -1$
Decreased by	9 decreased by -4	$9 - (-4) = 9 + 4 = 13$
Minus	8 minus 5	$8 - 5 = 3$

CAUTION When you are subtracting two numbers, it is important that you write them in the correct order, because, in general, $a - b \neq b - a$. For example, $5 - 3 \neq 3 - 5$. For this reason, *think carefully before interpreting an expression involving subtraction.* (This difficulty did not arise for addition.)

EXAMPLE 8 Interpreting Words and Phrases Involving Subtraction

Write a numerical expression for each phrase and simplify the expression.

(a) The difference between -8 and 5

In this book, we will write the numbers in the order they are given when "difference between" is used.*

$$-8 - 5 = -8 + (-5) = -13$$

(b) 4 subtracted from the sum of 8 and -3

Here addition is also used, as indicated by the word *sum*. First, add 8 and -3. Next, subtract 4 *from* this sum.

$$[8 + (-3)] - 4 = 5 - 4 = 1$$

---
*In some cases, people interpret "the difference between" (at least for two positive numbers) to represent the larger minus the smaller. However, we will not do so in this book.

SECTION 1.5 Adding and Subtracting Real Numbers   **45**

(c) 4 less than −6

Be careful with order. Here, 4 must be taken *from* −6.

$$-6 - 4 = -6 + (-4) = -10$$

Notice that "4 less than −6" differs from "4 *is less than* −6." The second of these is symbolized 4 < −6 (which is a false statement).

(d) 8, decreased by 5 less than 12

First, write "5 less than 12" as 12 − 5. Next, subtract 12 − 5 from 8.

$$8 - (12 - 5) = 8 - 7 = 1$$

Now Try Exercises 111 and 117.

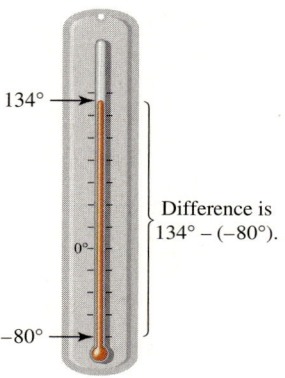

FIGURE 15

■ **EXAMPLE 9**   Solving a Problem Involving Subtraction

The record high temperature in the United States was 134° Fahrenheit, recorded at Death Valley, California in 1913. The record low was −80°F, at Prospect Creek, Alaska, in 1971. See Figure 15. What is the difference between these highest and lowest temperatures? (*Source:* World Almanac and Book of Facts, 2000.)

We must subtract the lowest temperature from the highest temperature.

$$134 - (-80) = 134 + 80 \quad \text{Definition of subtraction}$$
$$= 214 \quad \text{Add.}$$

The difference between the two temperatures is 214°F.

Now Try Exercise 131.

OBJECTIVE 6 Use signed numbers to interpret data.

■ **EXAMPLE 10**   Using a Signed Number in Data Interpretation

The bar graph in Figure 16 gives the Producer Price Index (PPI) for crude materials between 1994 and 1999.

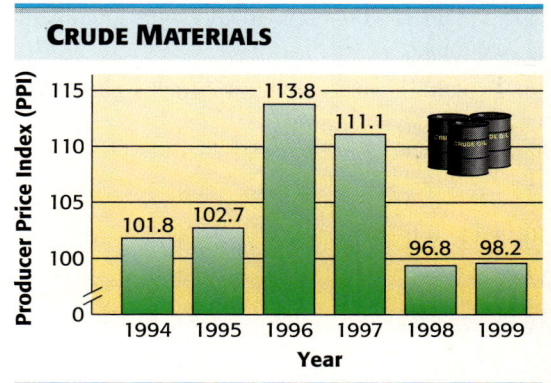

Source: U.S. Bureau of Labor Statistics, Producer Price Indexes, monthly and annual.

FIGURE 16

(a) Use a signed number to represent the change in the PPI from 1995 to 1996.

46 CHAPTER 1 The Real Number System

To find this change, we start with the index number from 1996 and subtract from it the index number from 1995.

$$\underbrace{113.8}_{\text{1996 index}} - \underbrace{102.7}_{\text{1995 index}} = \underbrace{+11.1}_{\substack{\text{A positive number} \\ \text{indicates an increase.}}}$$

(b) Use a signed number to represent the change in the PPI from 1996 to 1997. We use the same procedure as in part (a).

$$\underbrace{111.1}_{\text{1997 index}} - \underbrace{113.8}_{\text{1996 index}} = \underbrace{111.1 + (-113.8) = -2.7}_{\substack{\text{A negative number} \\ \text{indicates a decrease.}}}$$

Now Try Exercises 119 and 121.

1.5 EXERCISES

For Extra Help

- Student's Solutions Manual
- MyMathLab
- InterAct Math Tutorial Software
- AW Math Tutor Center
- MathXL
- Digital Video Tutor CD 2/Videotape 2

Fill in each blank with the correct response.

1. The sum of two negative numbers will always be a _____ number.
   (positive/negative)
   Give a number-line illustration using the sum $-2 + (-3)$.

2. The sum of a number and its opposite will always be _____.

3. If I am adding a positive number and a negative number, and the negative number has the larger absolute value, the sum will be a _____ number. Give
   (positive/negative)
   a number-line illustration using the sum $-4 + 2$.

4. To simplify the expression $8 + [-2 + (-3 + 5)]$, I should begin by adding _____ and _____, according to the rule for order of operations.

5. Explain in words how to add signed numbers. Consider the various cases and give examples.

6. Explain in words how to subtract signed numbers.

Find each sum. See Examples 1–6.

7. $-6 + (-2)$
8. $-8 + (-3)$
9. $-3 + (-9)$
10. $-11 + (-5)$
11. $5 + (-3)$
12. $11 + (-8)$
13. $6 + (-8)$
14. $3 + (-7)$
15. $12 + (-8)$
16. $10 + (-2)$
17. $4 + [13 + (-5)]$
18. $6 + [2 + (-13)]$
19. $8 + [-2 + (-1)]$
20. $12 + [-3 + (-4)]$
21. $-2 + [5 + (-1)]$
22. $-8 + [9 + (-2)]$
23. $-6 + [6 + (-9)]$
24. $-3 + [11 + (-8)]$
25. $[(-9) + (-3)] + 12$
26. $[(-8) + (-6)] + 10$
27. $-\dfrac{1}{6} + \dfrac{2}{3}$
28. $\dfrac{9}{10} + \left(-\dfrac{3}{5}\right)$
29. $\dfrac{5}{8} + \left(-\dfrac{17}{12}\right)$
30. $-\dfrac{6}{25} + \dfrac{19}{20}$

SECTION 1.5 Adding and Subtracting Real Numbers **47**

31. $2\frac{1}{2} + \left(-3\frac{1}{4}\right)$
32. $-4\frac{3}{8} + 6\frac{1}{2}$
33. $-6.1 + [3.2 + (-4.8)]$
34. $-9.4 + [-5.8 + (-1.4)]$
35. $[-3 + (-4)] + [5 + (-6)]$
36. $[-8 + (-3)] + [-7 + (-6)]$
37. $[-4 + (-3)] + [8 + (-1)]$
38. $[-5 + (-9)] + [16 + (-21)]$
39. $[-4 + (-6)] + [(-3) + (-8)] + [12 + (-11)]$
40. $[-2 + (-11)] + [12 + (-2)] + [18 + (-6)]$

Find each difference. See Examples 1–6.

41. $3 - 6$
42. $7 - 12$
43. $5 - 9$
44. $8 - 13$
45. $-6 - 2$
46. $-11 - 4$
47. $-9 - 5$
48. $-12 - 15$
49. $6 - (-3)$
50. $12 - (-2)$
51. $-6 - (-2)$
52. $-7 - (-5)$
53. $2 - (3 - 5)$
54. $-3 - (4 - 11)$
55. $\frac{1}{2} - \left(-\frac{1}{4}\right)$
56. $\frac{1}{3} - \left(-\frac{4}{3}\right)$
57. $-\frac{3}{4} - \frac{5}{8}$
58. $-\frac{5}{6} - \frac{1}{2}$
59. $\frac{5}{8} - \left(-\frac{1}{2} - \frac{3}{4}\right)$
60. $\frac{9}{10} - \left(\frac{1}{8} - \frac{3}{10}\right)$
61. $3.4 - (-8.2)$
62. $5.7 - (-11.6)$
63. $-6.4 - 3.5$
64. $-4.4 - 8.6$

Perform each indicated operation. See Examples 1–6.

65. $(4 - 6) + 12$
66. $(3 - 7) + 4$
67. $(8 - 1) - 12$
68. $(9 - 3) - 15$
69. $6 - (-8 + 3)$
70. $8 - (-9 + 5)$
71. $2 + (-4 - 8)$
72. $6 + (-9 - 2)$
73. $|-5 - 6| + |9 + 2|$
74. $|-4 + 8| + |6 - 1|$
75. $|-8 - 2| - |-9 - 3|$
76. $|-4 - 2| - |-8 - 1|$
77. $-9 + [(3 - 2) - (-4 + 2)]$
78. $-8 - [(-4 - 1) + (9 - 2)]$
79. $-3 + [(-5 - 8) - (-6 + 2)]$
80. $-4 + [(-12 + 1) - (-1 - 9)]$
81. $\left(\frac{3}{4} + \frac{2}{5}\right) + \left(\frac{1}{10} + \frac{3}{2}\right)$
82. $\left(\frac{4}{9} + \frac{2}{5}\right) + \left(\frac{4}{3} + \frac{2}{15}\right)$
83. $\left(\frac{1}{8} - \frac{3}{4}\right) + \left(\frac{5}{4} + \frac{3}{8}\right)$
84. $\left(\frac{1}{6} - \frac{5}{3}\right) + \left(\frac{5}{6} + \frac{1}{3}\right) + (-2)$
85. $\left(5\frac{1}{2} + 4\frac{1}{4}\right) - \left(2\frac{3}{4} + 1\frac{1}{2}\right)$
86. $\left(3\frac{1}{6} + 1\frac{3}{8}\right) - \left(5\frac{1}{2} - 2\frac{1}{4}\right)$
87. $\left|\frac{1}{3} - \frac{2}{7} - \left(\frac{1}{2} + \frac{4}{7}\right)\right|$
88. $\left|\frac{4}{7} - \frac{3}{2} - \left(\frac{1}{3} + \frac{4}{21}\right)\right|$
89. $\left|\frac{8}{3} + \frac{5}{6} - \frac{1}{2}\right| - \left|\frac{5}{12} - \frac{3}{2} + \frac{1}{6}\right|$
90. $\left|-\frac{7}{3} - \frac{5}{12} + \frac{5}{2}\right| - \left|\frac{7}{12} + \frac{7}{2} - \frac{1}{6}\right|$
91. $(3.45 - 4.25) + (8.23 - 7.46) - (9.11 + 10.87)$
92. $(6.82 + 3.15) - (12.46 + 1.33) - (4.88 + 9.20)$
93. $7.5 - [(2.3 - 6.8) + (4.4 + 9.3)]$

94. $13.5 - [(6.2 - 7.5) + (8.8 + 2.4)]$
95. $|7.40 - 8.65 - 2.25| - |8.60 + 2.45 - 6.75|$
96. $|8.55 - 7.75 - 1.30| - |10.95 + 8.75 - 7.50|$
97. $(5.67 - 8.34) - (1.15 - 7.91) - (-2.15 - 3.67)$
98. $(2.34 - 10.43) - (3.66 - 8.99) - (-7.13 - 9.88)$
99. $-3.4 - [(-2.1 + 6.8) - (-9.2 + 3.2)]$
100. $-7.3 - [(-1.4 + 12.5) - (-3.0 + 10.1)]$
101. $-9.1237 + [(-4.8099 - 3.2516) + 11.27903]$
102. $-7.6247 - [(-3.9928 + 1.42773) - (-2.80981)]$

Write a numerical expression for each phrase and simplify. See Example 7.

103. The sum of -5 and 12 and 6
104. The sum of -3 and 5 and -12
105. 14 added to the sum of -19 and -4
106. -2 added to the sum of -18 and 11
107. The sum of -4 and -10, increased by 12
108. The sum of -7 and -13, increased by 14
109. 4 more than the sum of 8 and -18
110. 10 more than the sum of -4 and -6

Write a numerical expression for each phrase and simplify. See Example 8.

111. The difference between 4 and -8
112. The difference between 7 and -14
113. 8 less than -2
114. 9 less than -13
115. The sum of 9 and -4, decreased by 7
116. The sum of 12 and -7, decreased by 14
117. 12 less than the difference between 8 and -5
118. 19 less than the difference between 9 and -2

The bar graph shows federal budget outlays for the U.S. Treasury Department for the years 1998 through 2001. Use a signed number to represent the change in outlay for each time period. See Example 10.

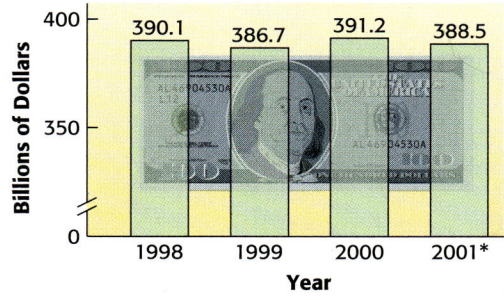

119. 1998 to 1999
120. 1999 to 2000
121. 2000 to 2001
122. 1998 to 2001

Use the information given to answer Exercises 123–128.

Mountain	Height (in feet)	Trench	Depth (in feet, as a negative number)
Foraker	17,400	Philippine	−32,995
Wilson	14,246	Cayman	−24,721
Pikes Peak	14,110	Java	−23,376

Source: World Almanac and Book of Facts, 2000.

123. What is the difference between the height of Mt. Foraker and the depth of the Philippine Trench?

124. What is the difference between the height of Pikes Peak and the depth of the Java Trench?

125. How much deeper is the Cayman Trench than the Java Trench?

126. How much deeper is the Philippine Trench than the Cayman Trench?

127. How much higher is Mt. Wilson than Pikes Peak?

128. If Mt. Wilson and Pikes Peak were stacked one on top of the other, how much higher would they be than Mt. Foraker?

Solve each problem. See Example 9.

129. On January 23, 1943, the temperature rose 49°F in two minutes in Spearfish, South Dakota. The starting temperature was −4°F. What was the temperature two minutes later? (*Source: Guinness Book of World Records, 2002.*)

130. The lowest temperature ever recorded in Arkansas was −29°F. The highest temperature ever recorded there was 149°F more than the lowest. What was this highest temperature? (*Source: World Almanac and Book of Facts, 2000.*)

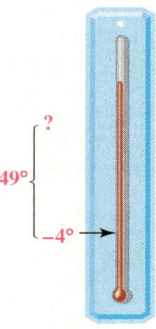
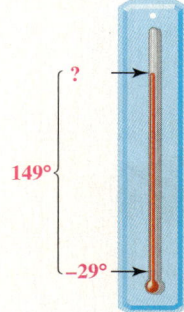

131. The coldest temperature recorded in Chicago, Illinois, was −35°F in 1996. The record low in South Dakota was set in 1936 and was 23°F lower than Chicago's record low. What was the record low in South Dakota? (*Source: World Almanac and Book of Facts, 2000.*)

132. The top of Mt. Whitney, visible from Death Valley, has an altitude of 14,494 ft above sea level. The bottom of Death Valley is 282 ft below sea level. Using 0 as sea level, find the difference between these two elevations. (*Source: World Almanac and Book of Facts, 2000.*)

133. No one knows just why humpback whales heave their 45-ton bodies out of the water, but leap they do. (This activity is called *breaching*.) Mark and Debbie, two researchers based on the island of Maui, noticed that one of their favorite whales, "Pineapple," leaped 15 ft above the surface of the ocean while her mate cruised 12 ft below. Find the difference between these two distances.

134. The surface, or rim, of a canyon is at altitude 0. On a hike down into the canyon, a party of hikers stops for a rest at 130 m below the surface. They then descend another 54 m. Write the new altitude as a signed number.

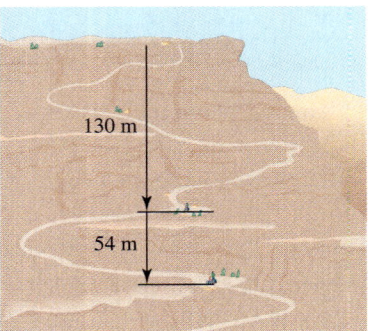

135. A pilot announces to the passengers that the current altitude of their plane is 34,000 ft. Because of turbulence, the pilot is forced to descend 2100 ft. Write the new altitude as a signed number.

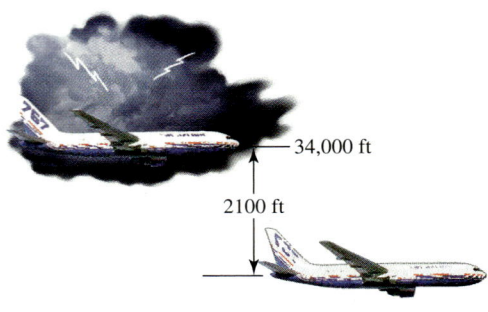

136. A welder working with stainless steel must use precise measurements. Suppose a welder attaches two pieces of steel that are each 3.60 in. long, and then attaches an additional three pieces that are each 9.10 in. long. She finally cuts off a piece that is 7.60 in. long. Find the length of the welded piece of steel.

137. Jennifer owes $153 to a credit card company. She makes a $14 purchase with the card and then pays $60 on the account. What is her current balance as a signed number?

138. A female polar bear weighed 660 lb when she entered her winter den. She lost 45 lb during each of the first two months of hibernation, and another 205 lb before leaving the den with her two cubs in March. How much did she weigh when she left the den?

1.6 Multiplying and Dividing Real Numbers

OBJECTIVES

1. Find the product of a positive number and a negative number.
2. Find the product of two negative numbers.
3. Identify factors of integers.

In this section we learn how to multiply with positive and negative numbers. We already know how to multiply positive numbers and that the product of two positive numbers is positive. We also know that the product of 0 and any positive number is 0, so we extend that property to all real numbers.

Multiplication by Zero

For any real number x, $\quad x \cdot 0 = 0.$

SECTION 1.6 Multiplying and Dividing Real Numbers

4 Use the reciprocal of a number to apply the definition of division.

5 Use the order of operations when multiplying and dividing signed numbers.

6 Evaluate expressions involving variables.

7 Interpret words and phrases involving multiplication and division.

8 Translate simple sentences into equations.

OBJECTIVE 1 Find the product of a positive number and a negative number. To define the product of a positive and a negative number so that the result is consistent with the multiplication of two positive numbers, look at the following pattern.

$$3 \cdot 5 = 15$$
$$3 \cdot 4 = 12$$
$$3 \cdot 3 = 9$$
$$3 \cdot 2 = 6$$
$$3 \cdot 1 = 3$$
$$3 \cdot 0 = 0$$
$$3 \cdot (-1) = \;?$$

The products decrease by 3.

What should $3(-1)$ equal? The product $3(-1)$ represents the sum

$$-1 + (-1) + (-1) = -3,$$

so the product should be -3. Also,

$$3(-2) = -2 + (-2) + (-2) = -6$$

and

$$3(-3) = -3 + (-3) + (-3) = -9.$$

These results maintain the pattern in the list, which suggests the following rule.

Multiplying Numbers with Different Signs

For any positive real numbers x and y,

$$x(-y) = -(xy) \quad \text{and} \quad (-x)y = -(xy).$$

That is, the product of two numbers with opposite signs is negative.

EXAMPLE 1 Multiplying a Positive Number and a Negative Number

Find each product using the multiplication rule given in the box.

(a) $8(-5) = -(8 \cdot 5) = -40$ (b) $5(-4) = -(5 \cdot 4) = -20$
(c) $-7(2) = -(7 \cdot 2) = -14$ (d) $-9(3) = -(9 \cdot 3) = -27$

Now Try Exercise 13.

OBJECTIVE 2 Find the product of two negative numbers. The product of two positive numbers is positive, and the product of a positive and a negative number is negative. What about the product of two negative numbers? Look at another pattern.

$$-5(4) = -20$$
$$-5(3) = -15$$
$$-5(2) = -10$$
$$-5(1) = -5$$
$$-5(0) = 0$$
$$-5(-1) = \;?$$

The products increase by 5.

The numbers on the left of the equals sign (in color) decrease by 1 for each step down the list. The products on the right increase by 5 for each step down the list. To maintain this pattern, $-5(-1)$ should be 5 more than $-5(0)$, or 5 more than 0, so

$$-5(-1) = 5.$$

The pattern continues with

$$-5(-2) = 10$$
$$-5(-3) = 15$$
$$-5(-4) = 20$$
$$-5(-5) = 25,$$

and so on, which suggests the next rule.

Multiplying Two Negative Numbers

For any positive real numbers x and y,

$$-x(-y) = xy.$$

That is, the product of two negative numbers is positive.

EXAMPLE 2 Multiplying Two Negative Numbers

Find each product using the multiplication rule given in the box.

(a) $-9(-2) = 9 \cdot 2 = 18$ **(b)** $-6(-12) = 6 \cdot 12 = 72$
(c) $-8(-1) = 8 \cdot 1 = 8$ **(d)** $-15(-2) = 15 \cdot 2 = 30$

Now Try Exercise 11.

A summary of multiplying signed numbers is given here.

Multiplying Signed Numbers

The product of two numbers having the *same* sign is *positive,* and the product of two numbers having *different* signs is *negative.*

OBJECTIVE 3 Identify factors of integers. In Section 1.1 the definition of a *factor* was given for whole numbers. (For example, since $9 \cdot 5 = 45$, both 9 and 5 are factors of 45.) The definition can now be extended to integers.

If the product of two integers is a third integer, then each of the two integers is a *factor* of the third. For example, $(-3)(-4) = 12$, so -3 and -4 are both factors of 12. The integer factors of 12 are $-12, -6, -4, -3, -2, -1, 1, 2, 3, 4, 6,$ and 12.

The table on the next page shows several integers and the factors of those integers.

Integer	18	20	15	7	1
Pairs of factors	1, 18	1, 20	1, 15	1, 7	1, 1
	2, 9	2, 10	3, 5	−1, −7	−1, −1
	3, 6	4, 5	−1, −15		
	−1, −18	−1, −20	−3, −5		
	−2, −9	−2, −10			
	−3, −6	−4, −5			

Now Try Exercise 29.

OBJECTIVE 4 Use the reciprocal of a number to apply the definition of division. In Section 1.5 we saw that the difference between two numbers is found by adding the additive inverse of the second number to the first. Similarly, the *quotient* of two numbers is found by *multiplying* by the *reciprocal,* or *multiplicative inverse.* By definition, since

$$8 \cdot \frac{1}{8} = \frac{8}{8} = 1 \quad \text{and} \quad \frac{5}{4} \cdot \frac{4}{5} = \frac{20}{20} = 1,$$

the reciprocal or multiplicative inverse of 8 is $\frac{1}{8}$, and of $\frac{5}{4}$ is $\frac{4}{5}$.

> **Reciprocal or Multiplicative Inverse**
>
> Pairs of numbers whose product is 1 are **reciprocals,** or **multiplicative inverses,** of each other.

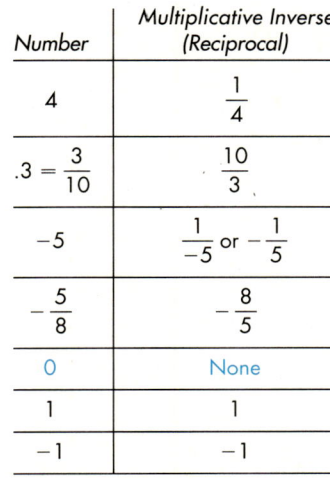

The table in the margin shows several numbers and their multiplicative inverses. Why is there no multiplicative inverse for the number 0? Suppose that k is to be the multiplicative inverse of 0. Then $k \cdot 0$ should equal 1. But $k \cdot 0 = 0$ for any number k. Since there is no value of k that is a solution of the equation $k \cdot 0 = 1$, the following statement can be made.

0 has no multiplicative inverse.

By definition, the quotient of x and y is the product of x and the multiplicative inverse of y.

> **Division**
>
> For any real numbers x and y, with $y \neq 0$, $\quad \dfrac{x}{y} = x \cdot \dfrac{1}{y}.$

The definition of division indicates that y, the number to divide by, cannot be 0. The reason is that 0 has no multiplicative inverse, so that $\frac{1}{0}$ is not a number. **Because 0 has no multiplicative inverse,** *division by 0 is undefined.* **If a division problem turns out to involve division by 0, write "undefined."**

54 CHAPTER 1 The Real Number System

> **NOTE** While division *by* 0 $\left(\frac{a}{0}\right)$ is undefined, we may divide 0 by any nonzero number. In fact, if $a \neq 0$,
> $$\frac{0}{a} = 0.$$

Since division is defined in terms of multiplication, all the rules for multiplying signed numbers also apply to dividing them.

EXAMPLE 3 Using the Definition of Division

Find each quotient using the definition of division.

(a) $\dfrac{12}{3} = 12 \cdot \dfrac{1}{3} = 4$ (b) $\dfrac{-10}{2} = -10 \cdot \dfrac{1}{2} = -5$

(c) $\dfrac{-14}{-7} = -14\left(\dfrac{1}{-7}\right) = 2$ (d) $-\dfrac{2}{3} \div \left(-\dfrac{4}{5}\right) = -\dfrac{2}{3} \cdot \left(-\dfrac{5}{4}\right) = \dfrac{5}{6}$

(e) $\dfrac{0}{13} = 0\left(\dfrac{1}{13}\right) = 0$ $\dfrac{0}{a} = 0$ $(a \neq 0)$

(f) $\dfrac{-10}{0}$ Undefined

Now Try Exercises 33, 39, and 45.

When dividing fractions, multiplying by the reciprocal works well. However, using the definition of division directly with integers is awkward. It is easier to divide in the usual way, then determine the sign of the answer.

Dividing Signed Numbers

The quotient of two numbers having the *same* sign is *positive;* the quotient of two numbers having *different* signs is *negative.*

Note that these are the same as the rules for multiplication.

EXAMPLE 4 Dividing Signed Numbers

Find each quotient.

(a) $\dfrac{8}{-2} = -4$ (b) $\dfrac{-4.5}{-.09} = 50$

(c) $-\dfrac{1}{8} \div \left(-\dfrac{3}{4}\right) = -\dfrac{1}{8} \cdot \left(-\dfrac{4}{3}\right) = \dfrac{1}{6}$

Now Try Exercises 35, 41, and 43.

From the definitions of multiplication and division of real numbers,

$$\frac{-40}{8} = -40 \cdot \frac{1}{8} = -5, \quad \text{and} \quad \frac{40}{-8} = 40\left(\frac{1}{-8}\right) = -5, \quad \text{so}$$

$$\frac{-40}{8} = \frac{40}{-8}.$$

Based on this example, the quotient of a positive and a negative number can be expressed in any of the following three forms.

> For any positive real numbers x and y, $\quad \dfrac{-x}{y} = \dfrac{x}{-y} = -\dfrac{x}{y}.$

Similarly, the quotient of two negative numbers can be expressed as a quotient of two positive numbers.

> For any positive real numbers x and y, $\quad \dfrac{-x}{-y} = \dfrac{x}{y}.$

OBJECTIVE 5 Use the order of operations when multiplying and dividing signed numbers.

EXAMPLE 5 Using the Order of Operations

Perform each indicated operation.

(a) $-9(2) - (-3)(2)$

First find all products, working from left to right.

$$-9(2) - (-3)(2) = -18 - (-6)$$
$$= -18 + 6$$
$$= -12$$

(b) $-6(-2) - 3(-4) = 12 - (-12)$
$$= 12 + 12$$
$$= 24$$

(c) $-5(-2 - 3) = -5(-5) = 25$

(d) $\dfrac{5(-2) - 3(4)}{2(1 - 6)}$

Simplify the numerator and denominator separately. Then write in lowest terms.

$$\frac{5(-2) - 3(4)}{2(1 - 6)} = \frac{-10 - 12}{2(-5)} = \frac{-22}{-10} = \frac{11}{5}$$

> **Now Try Exercises 53 and 77.**

We summarize the rules for operations with signed numbers on the next page.

Operations with Signed Numbers

ADDITION

Same sign   Add the absolute values of the numbers. The sum has the same sign as the numbers.
$$-4 + (-6) = -10$$

Different signs   Subtract the number with the smaller absolute value from the one with the larger. Give the sum the sign of the number having the larger absolute value.
$$4 + (-6) = -(6 - 4) = -2$$

SUBTRACTION

Add the opposite of the second number to the first number.
$$8 - (-3) = 8 + 3 = 11$$

MULTIPLICATION AND DIVISION

Same sign   The product or quotient of two numbers with the same sign is positive.
$$-5(-6) = 30 \quad \text{and} \quad \frac{-36}{-12} = 3$$

Different signs   The product or quotient of two numbers with different signs is negative.
$$-5(6) = -30 \quad \text{and} \quad \frac{18}{-6} = -3$$

Division by 0 is undefined.

OBJECTIVE 6 Evaluate expressions involving variables. The next examples show numbers substituted for variables where the rules for multiplying and dividing signed numbers must be used.

EXAMPLE 6   Evaluating Expressions for Numerical Values

Evaluate each expression, given that $x = -1$, $y = -2$, and $m = -3$.

(a) $(3x + 4y)(-2m)$

First substitute the given values for the variables. Then use the order of operations to find the value of the expression.

$(3x + 4y)(-2m) = [3(-1) + 4(-2)][-2(-3)]$ Put parentheses around the value for each variable.

$ = [-3 + (-8)][6]$ Multiply.

$ = (-11)(6)$ Add inside the parentheses.

$ = -66$ Multiply.

(b) $2x^2 - 3y^2$

Use parentheses as shown.

$$2(-1)^2 - 3(-2)^2 = 2(1) - 3(4) \quad \text{Apply the exponents.}$$
$$= 2 - 12 \quad \text{Multiply.}$$
$$= -10 \quad \text{Subtract.}$$

(c) $\dfrac{4y^2 + x}{m}$

$$\dfrac{4(-2)^2 + (-1)}{-3} = \dfrac{4(4) + (-1)}{-3} \quad \text{Apply the exponent.}$$
$$= \dfrac{16 + (-1)}{-3} \quad \text{Multiply.}$$
$$= \dfrac{15}{-3} \quad \text{Add.}$$
$$= -5 \quad \text{Divide.}$$

Notice how the fraction bar was used as a grouping symbol.

Now Try Exercises 93 and 101.

OBJECTIVE 7 Interpret words and phrases involving multiplication and division. Just as there are words and phrases that indicate addition and subtraction, certain words also indicate multiplication and division.

PROBLEM SOLVING

The word *product* refers to multiplication. The table gives other key words and phrases that indicate multiplication.

Word or Phrase	Example	Numerical Expression and Simplification
Product of	The *product of* -5 and -2	$-5(-2) = 10$
Times	13 *times* -4	$13(-4) = -52$
Twice (meaning "2 times")	*Twice* 6	$2(6) = 12$
Of (used with fractions)	$\frac{1}{2}$ *of* 10	$\frac{1}{2}(10) = 5$
Percent of	12% *of* -16	$.12(-16) = -1.92$
As much as	$\frac{2}{3}$ *as much as* 30	$\frac{2}{3}(30) = 20$

EXAMPLE 7 Interpreting Words and Phrases Involving Multiplication

Write a numerical expression for each phrase and simplify. Use the order of operations.

(a) The *product of* 12 and the sum of 3 and -6

Here, 12 is multiplied by "the sum of 3 and -6."

$$12[3 + (-6)] = 12(-3) = -36$$

(b) Twice the difference between 8 and -4

$$2[8 - (-4)] = 2[8 + 4] = 2(12) = 24$$

(c) Two-thirds *of* the sum of -5 and -3

$$\frac{2}{3}[-5 + (-3)] = \frac{2}{3}[-8] = -\frac{16}{3}$$

(d) 15% *of* the difference between 14 and -2
Remember that $15\% = .15$.

$$.15[14 - (-2)] = .15[14 + 2] = .15(16) = 2.4$$

(e) Double the product of 3 and 4

$$2 \cdot (3 \cdot 4) = 2 \cdot 12 = 24$$

> **Now Try Exercises 105, 109, and 117.**

PROBLEM SOLVING

The word *quotient* refers to the answer in a division problem. In algebra, quotients are usually represented with a fraction bar; the symbol $\div$ is seldom used. When translating applied problems involving division, use a fraction bar. The table gives some key phrases associated with division.

Phrase	Example	Numerical Expression and Simplification
Quotient of	The *quotient of* -24 and 3	$\frac{-24}{3} = -8$
Divided by	-16 *divided by* -4	$\frac{-16}{-4} = 4$
Ratio of	The *ratio of* 2 to 3	$\frac{2}{3}$

It is customary to write the first number named as the numerator and the second as the denominator when interpreting a phrase involving division, as shown in the next example.

EXAMPLE 8 Interpreting Words and Phrases Involving Division

Write a numerical expression for each phrase and simplify the expression.

(a) The quotient of 14 and the sum of -9 and 2

"Quotient" indicates division. The number 14 is the numerator and "the sum of -9 and 2" is the denominator.

$$\frac{14}{-9 + 2} = \frac{14}{-7} = -2$$

(b) The product of 5 and -6, divided by the difference between -7 and 8

The numerator of the fraction representing the division is obtained by multiplying 5 and -6. The denominator is found by subtracting -7 and 8.

$$\frac{5(-6)}{-7 - 8} = \frac{-30}{-15} = 2$$

> **Now Try Exercise 111.**

OBJECTIVE 8 Translate simple sentences into equations. In this section and the previous one, important words and phrases involving the four operations of arithmetic have been introduced. We can use these words and phrases to interpret sentences that translate into equations.

EXAMPLE 9 Translating Words into Equations

Write each sentence in symbols, using x as the variable, and guess or use trial and error to find the solution. All solutions come from the list of integers between -12 and 12, inclusive.

(a) Three *times* a number *is* -18.

The word *times* indicates multiplication, and the word *is* translates as the equals sign ($=$).

$$3x = -18$$

Since the integer between -12 and 12 inclusive that makes this statement true is -6, the solution of the equation is -6.

(b) The *sum* of a number and 9 *is* 12.

$$x + 9 = 12$$

Since $3 + 9 = 12$, the solution of this equation is 3.

(c) The *difference between* a number and 5 *is* 0.

$$x - 5 = 0$$

Since $5 - 5 = 0$, the solution of this equation is 5.

(d) The *quotient* of 24 and a number *is* -2.

$$\frac{24}{x} = -2$$

Here, x must be a negative number, since the numerator is positive and the quotient is negative. Since $\frac{24}{-12} = -2$, the solution is -12.

Now Try Exercises 119 and 123.

CAUTION It is important to recognize the distinction between the types of problems found in Examples 7 and 8 and Example 9. In Examples 7 and 8, the *phrases* translate as *expressions,* while in Example 9, the *sentences* translate as *equations.* Remember that an equation is a sentence with an $=$ sign, while an expression is a phrase.

$$\underbrace{\frac{5(-6)}{-7-8}}_{\text{Expression}} \qquad \underbrace{3x = -18}_{\text{Equation}}$$

1.6 EXERCISES

For Extra Help

Student's Solutions Manual

MyMathLab

InterAct Math Tutorial Software

AW Math Tutor Center

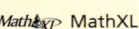
MathXL

Digital Video Tutor CD 2/Videotape 2

Fill in each blank with one of the following: greater than 0, less than 0, equal to 0.

1. A positive number is _____.
2. A negative number is _____.
3. The product or the quotient of two numbers with the same sign is _____.
4. The product or the quotient of two numbers with different signs is _____.
5. If three negative numbers are multiplied together, the product is _____.
6. If two negative numbers are multiplied and then their product is divided by a negative number, the result is _____.
7. If a negative number is squared and the result is added to a positive number, the final answer is _____.
8. The reciprocal of a negative number is _____.
9. If three positive numbers, five negative numbers, and zero are multiplied, the product is _____.
10. The fifth power of a negative number is _____.

Find each product. See Examples 1 and 2.

11. $-3(-4)$
12. $-3(4)$
13. $3(-4)$
14. $-2(-8)$
15. $-10(-12)$
16. $9(-5)$
17. $3(-11)$
18. $3(-15)$
19. $15(-11)$
20. $-9(-4)$
21. $-\dfrac{3}{8} \cdot \left(-\dfrac{10}{9}\right)$
22. $-\dfrac{5}{4} \cdot \left(-\dfrac{5}{8}\right)$
23. $\left(-1\dfrac{1}{4}\right)\left(\dfrac{2}{15}\right)$
24. $\left(\dfrac{3}{7}\right)\left(-1\dfrac{5}{9}\right)$
25. $(-8)\left(-\dfrac{3}{4}\right)$
26. $(-6)\left(-\dfrac{5}{3}\right)$

Find all integer factors of each number.

27. 32
28. 36
29. 40
30. 50
31. 31
32. 17

Find each quotient. See Examples 3 and 4.

33. $\dfrac{15}{5}$
34. $\dfrac{25}{5}$
35. $\dfrac{-30}{6}$
36. $\dfrac{-28}{14}$
37. $\dfrac{-28}{-4}$
38. $\dfrac{-35}{-7}$
39. $\dfrac{96}{-16}$
40. $\dfrac{38}{-19}$
41. $-\dfrac{4}{3} \div \left(-\dfrac{1}{8}\right)$
42. $-\dfrac{5}{6} \div \left(-\dfrac{15}{7}\right)$
43. $\dfrac{-8.8}{2.2}$
44. $\dfrac{-4.6}{-.23}$
45. $\dfrac{0}{-2}$
46. $\dfrac{0}{-8}$
47. $\dfrac{12}{0}$
48. $\dfrac{6}{0}$

Perform each indicated operation. See Examples 5(a), (b), and (c).

49. $7 - 3 \cdot 6$
50. $8 - 2 \cdot 5$
51. $-10 - (-4)(2)$
52. $-11 - (-3)(6)$
53. $-7(3 - 8)$
54. $-5(4 - 7)$
55. $(12 - 14)(1 - 4)$
56. $(8 - 9)(4 - 12)$
57. $(7 - 10)(10 - 4)$
58. $(5 - 12)(19 - 4)$
59. $(-2 - 8)(-6) + 7$
60. $(-9 - 4)(-2) + 10$
61. $3(-5) + |3 - 10|$
62. $4(-8) + |4 - 15|$

SECTION 1.6 Multiplying and Dividing Real Numbers

63. $\left(\dfrac{2}{3} - \dfrac{4}{5}\right) \cdot \left(\dfrac{1}{3} + \dfrac{2}{5}\right)$

64. $\left(\dfrac{4}{7} - \dfrac{4}{3}\right) \cdot \left(\dfrac{3}{7} + \dfrac{8}{3}\right)$

65. $\dfrac{3}{5}\left(-\dfrac{1}{8}\right) + \left|\dfrac{5}{6}\left(-\dfrac{1}{2}\right)\right|$

66. $-\dfrac{7}{5}\left(-\dfrac{4}{7}\right) + \left|\dfrac{10}{7}\left(-\dfrac{3}{5}\right)\right|$

67. $2.4(-3.5) - (-1.8)(-4.7)$

68. $7.3(-6.6) - (-4.8)(3.3)$

69. $|-1.3(-4.5)| - |5.7(-1.4)|$

70. $|-8.84(-1.43)| - |-1.32(-2.56)|$

71. $\left(-4\dfrac{1}{2}\right)\left(-3\dfrac{1}{4}\right) - (-6.5)(1.25)$

72. $\left(-5\dfrac{1}{3}\right)\left(-2\dfrac{3}{4}\right) - (-3.5)(5.75)$

Perform each indicated operation. See Example 5(d).

73. $\dfrac{-5(-6)}{9 - (-1)}$

74. $\dfrac{-12(-5)}{7 - (-5)}$

75. $\dfrac{-21(3)}{-3 - 6}$

76. $\dfrac{-40(3)}{-2 - 3}$

77. $\dfrac{-10(2) + 6(2)}{-3 - (-1)}$

78. $\dfrac{8(-1) - |(-4)(-3)|}{-6 - (-1)}$

79. $\dfrac{-27(-2) - |6 \cdot 4|}{-2(3) - 2(2)}$

80. $\dfrac{-13(-4) - (-8)(-2)}{(-10)(2) - 4(-2)}$

81. $\dfrac{-5(2) + [3(-2) - 4]}{-3 - (-1)}$

82. $\dfrac{(-2.5)(4.5)}{-5.625}$

83. $\dfrac{(-3.75)(4.0)}{-7.5}$

84. $\dfrac{|(-1.5)(4.0)|}{-.75}$

85. $\dfrac{|(4.5)(-12.0)|}{-1.50}$

86. $\dfrac{|(-1.5)(4.0)|}{-.75}$

87. $\dfrac{3.5^2(-6.25) - 3.5(-12.5)}{-2.5^2}$

88. $\dfrac{4.5^2(-8.5) - 4.5(-10.25)}{-3.0^2}$

89. $\dfrac{\left|\left(\tfrac{2}{3}\right)\left(-\tfrac{9}{8}\right)\right|}{\left|\left(-\tfrac{1}{2}\right) \div \tfrac{1}{4}\right|}$

90. $\dfrac{\left|\left(-\tfrac{2}{9}\right)\left(-\tfrac{3}{4}\right)\right|}{\left|\left(-\tfrac{3}{2}\right) \div \left(-\tfrac{1}{6}\right)\right|}$

Evaluate each expression if $x = 6$, $y = -4$, and $a = 3$. See Example 6.

91. $5x - 2y + 3a$

92. $6x - 5y + 4a$

93. $(2x + y)(3a)$

94. $(5x - 2y)(-2a)$

95. $\left(\dfrac{1}{3}x - \dfrac{4}{5}y\right)\left(-\dfrac{1}{5}a\right)$

96. $\left(\dfrac{5}{6}x + \dfrac{3}{2}y\right)\left(-\dfrac{1}{3}a\right)$

97. $(-5 + x)(-3 + y)(3 - a)$

98. $(6 - x)(5 + y)(3 + a)$

99. $-2y^2 + 3a$

100. $5x - 4a^2$

101. $\dfrac{2y^2 - x}{a + 10}$

102. $\dfrac{xy + 8a}{x - y}$

Write a numerical expression for each phrase and simplify. See Examples 7 and 8.

103. The product of −9 and 2, added to 9

104. The product of 4 and −7, added to −12

105. Twice the product of −1 and 6, subtracted from −4

106. Twice the product of −8 and 2, subtracted from −1

107. Nine subtracted from the product of 1.5 and −3.2

108. Three subtracted from the product of 4.2 and −8.5

109. The product of 12 and the difference between 9 and −8

110. The product of −3 and the difference between 3 and −7

111. The quotient of −12 and the sum of −5 and −1

112. The quotient of −20 and the sum of −8 and −2

62 CHAPTER 1 The Real Number System

113. The sum of 15 and −3, divided by the product of 4 and −3

114. The sum of −18 and −6, divided by the product of 2 and −4

115. Twice the sum of 8 and 9

116. Three-fourths of the sum of −8 and 12

117. 20% of the product of −5 and 6

118. $\frac{2}{3}$ as much as the difference between 8 and −1

Write each statement in symbols, using x as the variable, and find the solution by guessing or by using trial and error. All solutions come from the set of integers between −12 and 12, inclusive. See Example 9.

119. The quotient of a number and 3 is −3.

120. The quotient of a number and 4 is −1.

121. 6 less than a number is 4.

122. 7 less than a number is 2.

123. When 5 is added to a number, the result is −5.

124. When 6 is added to a number, the result is −3.

To find the **average** *of a group of numbers, we add the numbers and then divide the sum by the number of terms added. For example, to find the average of 14, 8, 3, 9, and 1, we add them and then divide by 5:*

$$\frac{14 + 8 + 3 + 9 + 1}{5} = \frac{35}{5} = 7.$$

The average of these numbers is 7.

Exercises 125–128 involve finding the average of a group of numbers.

125. Find the average of 23, 18, 13, −4, and −8.

126. Find the average of 18, 12, 0, −4, and −10.

127. What is the average of all integers between −10 and 14, inclusive of both?

128. What is the average of all even integers between −18 and 4, inclusive of both?

The operation of division is used in divisibility *tests. A divisibility test allows us to determine whether a given number is divisible (without remainder) by another number. For example, a number is divisible by 2 if its last digit is divisible by 2, and not otherwise.*

129. Tell why **(a)** 3,473,986 is divisible by 2 and **(b)** 4,336,879 is not divisible by 2.

130. An integer is divisible by 3 if the sum of its digits is divisible by 3, and not otherwise. Show that **(a)** 4,799,232 is divisible by 3 and **(b)** 2,443,871 is not divisible by 3.

131. An integer is divisible by 4 if its last two digits form a number divisible by 4, and not otherwise. Show that **(a)** 6,221,464 is divisible by 4 and **(b)** 2,876,335 is not divisible by 4.

132. An integer is divisible by 5 if its last digit is divisible by 5, and not otherwise. Show that **(a)** 3,774,595 is divisible by 5 and **(b)** 9,332,123 is not divisible by 5.

133. An integer is divisible by 6 if it is divisible by both 2 and 3, and not otherwise. Show that **(a)** 1,524,822 is divisible by 6 and **(b)** 2,873,590 is not divisible by 6.

134. An integer is divisible by 8 if its last three digits form a number divisible by 8, and not otherwise. Show that **(a)** 2,923,296 is divisible by 8 and **(b)** 7,291,623 is not divisible by 8.

135. An integer is divisible by 9 if the sum of its digits is divisible by 9, and not otherwise. Show that **(a)** 4,114,107 is divisible by 9 and **(b)** 2,287,321 is not divisible by 9.

136. An integer is divisible by 12 if it is divisible by both 3 and 4, and not otherwise. Show that **(a)** 4,253,520 is divisible by 12 and **(b)** 4,249,474 is not divisible by 12.

SUMMARY EXERCISES ON OPERATIONS WITH REAL NUMBERS

Addition

Same sign Add the absolute values of the numbers. The sum has the same sign as the numbers.

Different signs Subtract the number with the smaller absolute value from the one with the larger. Give the sum the sign of the number having the larger absolute value.

Subtraction

Add the opposite of the second number to the first number.

Multiplication and Division

Same sign The product or quotient of two numbers with the same sign is positive.

Different signs The product or quotient of two numbers with different signs is negative.

Division by 0 is undefined.

Perform each indicated operation.

1. $14 - 3 \cdot 10$
2. $-3(8) - 4(-7)$
3. $(3 - 8)(-2) - 10$
4. $-6(7 - 3)$
5. $7 - (-3)(2 - 10)$
6. $-4[(-2)(6) - 7]$
7. $(-4)(7) - (-5)(2)$
8. $-5[-4 - (-2)(-7)]$
9. $40 - (-2)[8 - 9]$
10. $\dfrac{5(-4)}{-7 - (-2)}$
11. $\dfrac{-3 - (-9 + 1)}{-7 - (-6)}$
12. $\dfrac{5(-8 + 3)}{13(-2) + (-7)(-3)}$
13. $\dfrac{6^2 - 8}{-2(2) + 4(-1)}$
14. $\dfrac{16(-8 + 5)}{15(-3) + (-7 - 4)(-3)}$
15. $\dfrac{9(-6) - 3(8)}{4(-7) + (-2)(-11)}$
16. $\dfrac{2^2 + 4^2}{5^2 - 3^2}$
17. $\dfrac{(2 + 4)^2}{(5 - 3)^2}$
18. $\dfrac{4^3 - 3^3}{-5(-4 + 2)}$
19. $\dfrac{-9(-6) + (-2)(27)}{3(8 - 9)}$
20. $|-4(9)| - |-11|$
21. $\dfrac{6(-10 + 3)}{15(-2) - 3(-9)}$
22. $\dfrac{(-9)^2 - 9^2}{3^2 - 5^2}$
23. $\dfrac{(-10)^2 + 10^2}{-10(5)}$
24. $-\dfrac{3}{4} \div \left(-\dfrac{5}{8}\right)$
25. $\dfrac{1}{2} \div \left(-\dfrac{1}{2}\right)$
26. $\dfrac{8^2 - 12}{(-5)^2 + 2(6)}$
27. $\left[\dfrac{5}{8} - \left(-\dfrac{1}{16}\right)\right] + \dfrac{3}{8}$
28. $\left(\dfrac{1}{2} - \dfrac{1}{3}\right) - \dfrac{5}{6}$
29. $-.9(-3.7)$
30. $-5.1(-.2)$
31. $-3^2 - 2^2$
32. $|-2(3) + 4| - |-2|$
33. $40 - (-2)[-5 - 3]$

Evaluate each expression if $x = -2$, $y = 3$, and $a = 4$.

34. $-x + y - 3a$
35. $(x + 6)^3 - y^3$
36. $(x - y) - (a - 2y)$
37. $\left(\dfrac{1}{2}x + \dfrac{2}{3}y\right)\left(-\dfrac{1}{4}a\right)$
38. $\dfrac{2x + 3y}{a - xy}$
39. $\dfrac{x^2 - y^2}{x^2 + y^2}$
40. $-x^2 + 3y$

64 CHAPTER 1 The Real Number System

1.7 Properties of Real Numbers

OBJECTIVES

1. Use the commutative properties.
2. Use the associative properties.
3. Use the identity properties.
4. Use the inverse properties.
5. Use the distributive property.

If you were asked to find the sum $3 + 89 + 97$, you might mentally add $3 + 97$ to get 100, and then add $100 + 89$ to get 189. While the rule for order of operations says to add from left to right, we may change the order of the terms and group them in any way we choose without affecting the sum. These are examples of shortcuts that we use in everyday mathematics. These shortcuts are justified by the basic properties of addition and multiplication, discussed in this section. In these properties, a, b, and c represent real numbers.

OBJECTIVE 1 Use the commutative properties. The word *commute* means to go back and forth. Many people commute to work or to school. If you travel from home to work and follow the same route from work to home, you travel the same distance each time. The **commutative properties** say that if two numbers are added or multiplied in any order, the result is the same.

Commutative Properties

$$a + b = b + a$$
$$ab = ba$$

EXAMPLE 1 Using the Commutative Properties

Use a commutative property to complete each statement.

(a) $-8 + 5 = 5 + \underline{}$

By the commutative property for addition, the missing number is -8, since $-8 + 5 = 5 + (-8)$.

(b) $(-2)7 = \underline{}(-2)$

By the commutative property for multiplication, the missing number is 7, since $(-2)7 = 7(-2)$.

Now Try Exercises 1 and 3.

OBJECTIVE 2 Use the associative properties. When we *associate* one object with another, we tend to think of those objects as being grouped together. The **associative properties** say that when we add or multiply three numbers, we can group the first two together or the last two together and get the same answer.

Associative Properties

$$(a + b) + c = a + (b + c)$$
$$(ab)c = a(bc)$$

EXAMPLE 2 Using the Associative Properties

Use an associative property to complete each statement.

(a) $8 + (-1 + 4) = (8 + \underline{\quad}) + 4$

The missing number is -1.

(b) $[2 \cdot (-7)] \cdot 6 = 2 \cdot \underline{\quad}$

The completed expression on the right should be $2 \cdot [(-7) \cdot 6]$.

Now Try Exercises 5 and 7.

By the associative property of addition, the sum of three numbers will be the same no matter how the numbers are "associated" in groups. For this reason, parentheses can be left out in many addition problems. For example, both

$$(-1 + 2) + 3 \quad \text{and} \quad -1 + (2 + 3)$$

can be written as

$$-1 + 2 + 3.$$

In the same way, parentheses also can be left out of many multiplication problems.

EXAMPLE 3 Distinguishing between the Associative and Commutative Properties

(a) Is $(2 + 4) + 5 = 2 + (4 + 5)$ an example of the associative property or the commutative property?

The order of the three numbers is the same on both sides of the equals sign. The only change is in the *grouping,* or association, of the numbers. Therefore, this is an example of the associative property.

(b) Is $6(3 \cdot 10) = 6(10 \cdot 3)$ an example of the associative property or the commutative property?

The same numbers, 3 and 10, are grouped on each side. On the left, the 3 appears first, but on the right, the 10 appears first. Since the only change involves the *order* of the numbers, this statement is an example of the commutative property.

(c) Is $(8 + 1) + 7 = 8 + (7 + 1)$ an example of the associative property or the commutative property?

In the statement, both the order and the grouping are changed. On the left the order of the three numbers is 8, 1, and 7. On the right it is 8, 7, and 1. On the left the 8 and 1 are grouped, and on the right the 7 and 1 are grouped. Therefore, both the associative and the commutative properties are used.

Now Try Exercises 15 and 23.

EXAMPLE 4 Using the Commutative and Associative Properties

Find the sum $23 + 41 + 2 + 9 + 25$.

The commutative and associative properties make it possible to choose pairs of numbers whose sums are easy to add.

$$23 + 41 + 2 + 9 + 25 = (41 + 9) + (23 + 2) + 25$$
$$= 50 + 25 + 25$$
$$= 100$$

Now Try Exercise 35.

OBJECTIVE 3 Use the identity properties. If a child wears a costume on Halloween, the child's appearance is changed, but his or her *identity* is unchanged. The identity of a real number is left unchanged when identity properties are applied. The **identity properties** say that the sum of 0 and any number equals that number, and the product of 1 and any number equals that number.

Identity Properties

$$a + 0 = a \quad \text{and} \quad 0 + a = a$$
$$a \cdot 1 = a \quad \text{and} \quad 1 \cdot a = a$$

The number 0 leaves the identity, or value, of any real number unchanged by addition. For this reason, 0 is called the **identity element for addition,** or the **additive identity.** Since multiplication by 1 leaves any real number unchanged, 1 is the **identity element for multiplication,** or the **multiplicative identity.**

EXAMPLE 5 Using the Identity Properties

These statements are examples of the identity properties.

(a) $-3 + 0 = -3$ (b) $1 \cdot \dfrac{1}{2} = \dfrac{1}{2}$

Now Try Exercise 31.

We use the identity property for multiplication to write fractions in lowest terms and to find common denominators.

EXAMPLE 6 Using the Identity Property to Simplify Expressions

Simplify each expression.

(a) $\dfrac{49}{35} = \dfrac{7 \cdot 7}{5 \cdot 7}$ Factor.

$= \dfrac{7}{5} \cdot \dfrac{7}{7}$ Write as a product.

$= \dfrac{7}{5} \cdot 1$ Divide.

$= \dfrac{7}{5}$ Identity property

(b) $\dfrac{3}{4} + \dfrac{5}{24} = \dfrac{3}{4} \cdot 1 + \dfrac{5}{24}$ Identity property

$= \dfrac{3}{4} \cdot \dfrac{6}{6} + \dfrac{5}{24}$ Find a common denominator.

$= \dfrac{18}{24} + \dfrac{5}{24}$ Multiply.

$= \dfrac{23}{24}$ Add.

Now Try Exercise 27.

OBJECTIVE 4 Use the inverse properties. Each day before you go to work or school, you probably put on your shoes before you leave. Before you go to sleep at night, you probably take them off, and this leads to the same situation that existed before you put them on. These operations from everyday life are examples of *inverse* operations. The **inverse properties** of addition and multiplication lead to the additive and multiplicative identities, respectively. Recall that $-a$ is the **additive inverse, or opposite,** of a and $\frac{1}{a}$ is the **multiplicative inverse, or reciprocal,** of the nonzero number a. The sum of the numbers a and $-a$ is 0, and the product of the nonzero numbers a and $\frac{1}{a}$ is 1.

Inverse Properties

$$a + (-a) = 0 \quad \text{and} \quad -a + a = 0$$

$$a \cdot \frac{1}{a} = 1 \quad \text{and} \quad \frac{1}{a} \cdot a = 1 \quad (a \neq 0)$$

EXAMPLE 7 Using the Inverse Properties

The following statements are examples of the inverse properties.

(a) $\dfrac{2}{3} \cdot \dfrac{3}{2} = 1$ (b) $-5\left(-\dfrac{1}{5}\right) = 1$

(c) $-\dfrac{1}{2} + \dfrac{1}{2} = 0$ (d) $4 + (-4) = 0$

Now Try Exercise 19.

In the next example, we show how the various properties are used to simplify an expression.

EXAMPLE 8 Using Properties to Simplify an Expression

Simplify $-2x + 10 + 2x$, using the properties discussed in this section.

$$\begin{aligned}
-2x + 10 + 2x &= (-2x + 10) + 2x &&\text{Order of operations}\\
&= [10 + (-2x)] + 2x &&\text{Commutative property}\\
&= 10 + [(-2x) + 2x] &&\text{Associative property}\\
&= 10 + 0 &&\text{Inverse property}\\
&= 10 &&\text{Identity property}
\end{aligned}$$

Note that for *any* value of x, $-2x$ and $2x$ are additive inverses; that is why we can use the inverse property in this simplification.

Now Try Exercise 47.

NOTE The detailed procedure shown in Example 8 is seldom, if ever, used in practice. We include the example to show how the properties of this section apply, even though steps may be skipped when actually doing the simplification.

68 CHAPTER 1 The Real Number System

OBJECTIVE 5 Use the distributive property. The everyday meaning of the word *distribute* is "to give out from one to several." An important property of real number operations involves this idea.

Look at the value of the following expressions.
$$2(5 + 8) = 2(13) = 26$$
$$2(5) + 2(8) = 10 + 16 = 26$$

Since both expressions equal 26,
$$2(5 + 8) = 2(5) + 2(8).$$

This result is an example of the *distributive property*, the only property involving *both* addition and multiplication. With this property, a product can be changed to a sum or difference. This idea is illustrated by the divided rectangle in Figure 17.

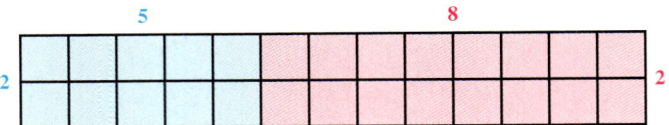

The area of the left part is 2(5) = 10.
The area of the right part is 2(8) = 16.
The total area is 2(5 + 8) = 2(13) = 26 or the total area is
2(5) + 2(8) = 10 + 16 = 26.
Thus, 2(5 + 8) = 2(5) + 2(8).

FIGURE 17

The **distributive property** says that multiplying a number a by a sum of numbers $b + c$ gives the same result as multiplying a by b and a by c and then adding the two products.

Distributive Property

$$a(b + c) = ab + ac \quad \text{and} \quad (b + c)a = ba + ca$$

As the arrows show, the a outside the parentheses is "distributed" over the b and c inside. Another form of the distributive property is valid for subtraction.
$$a(b - c) = ab - ac \quad \text{and} \quad (b - c)a = ba - ca$$

The distributive property also can be extended to more than two numbers.
$$a(b + c + d) = ab + ac + ad$$

The distributive property can be used "in reverse." For example, we can write
$$ac + bc = (a + b)c.$$

EXAMPLE 9 Using the Distributive Property

Use the distributive property to rewrite each expression.

(a) $5(9 + 6) = 5 \cdot 9 + 5 \cdot 6$ Distributive property
$ = 45 + 30$ Multiply.
$ = 75$ Add.

(b) $4(x + 5 + y) = 4x + 4 \cdot 5 + 4y$ Distributive property
$\qquad\qquad\qquad = 4x + 20 + 4y$ Multiply.

(c) $-2(x + 3) = -2x + (-2)(3)$ Distributive property
$\qquad\qquad = -2x - 6$ Multiply.

(d) $3(k - 9) = 3k - 3 \cdot 9$ Distributive property
$\qquad\qquad = 3k - 27$ Multiply.

(e) $8(3r + 11t + 5z) = 8(3r) + 8(11t) + 8(5z)$ Distributive property
$\qquad\qquad\qquad\quad = (8 \cdot 3)r + (8 \cdot 11)t + (8 \cdot 5)z$ Associative property
$\qquad\qquad\qquad\quad = 24r + 88t + 40z$ Multiply.

(f) $6 \cdot 8 + 6 \cdot 2 = 6(8 + 2)$ Distributive property
$\qquad\qquad\quad = 6(10) = 60$ Add, then multiply.

(g) $4x - 4m = 4(x - m)$ Distributive property

(h) $6x - 12 = 6 \cdot x - 6 \cdot 2 = 6(x - 2)$ Distributive property

> **Now Try Exercises 65, 71, 73, and 77.**

The symbol $-a$ may be interpreted as $-1 \cdot a$. Similarly, when a negative sign precedes an expression within parentheses, it may also be interpreted as a factor of -1. Thus, we can use the distributive property to remove parentheses from expressions such as $-(2y + 3)$. We do this by first writing $-(2y + 3)$ as $-1 \cdot (2y + 3)$.

$-(2y + 3) = -1 \cdot (2y + 3)$
$\qquad\qquad = -1 \cdot 2y + (-1) \cdot 3$ Distributive property
$\qquad\qquad = -2y - 3$ Multiply.

EXAMPLE 10 Using the Distributive Property to Remove Parentheses

Write without parentheses.

(a) $-(7r - 8) = -1(7r - 8)$
$\qquad\qquad\quad = -1(7r) + (-1)(-8)$ Distributive property
$\qquad\qquad\quad = -7r + 8$ Multiply.

(b) $-(-9w + 2) = -1(-9w + 2)$
$\qquad\qquad\qquad = 9w - 2$

> **Now Try Exercise 93.**

The properties discussed here are the basic properties that justify how we do algebra. Here is a summary of these properties.

Properties of Addition and Multiplication

For any real numbers a, b, and c, the following properties hold.

Commutative Properties $a + b = b + a \qquad ab = ba$

Associative Properties $(a + b) + c = a + (b + c)$
$\qquad\qquad\qquad\qquad\quad (ab)c = a(bc)$

(continued)

70 CHAPTER 1 The Real Number System

Identity Properties	There is a real number 0 such that $a + 0 = a$ and $0 + a = a.$ There is a real number 1 such that $a \cdot 1 = a$ and $1 \cdot a = a.$
Inverse Properties	For each real number a, there is a single real number $-a$ such that $a + (-a) = 0$ and $(-a) + a = 0.$ For each nonzero real number a, there is a single real number $\frac{1}{a}$ such that $a \cdot \frac{1}{a} = 1$ and $\frac{1}{a} \cdot a = 1.$
Distributive Properties	$a(b + c) = ab + ac$ $(b + c)a = ba + ca$

1.7 EXERCISES

For Extra Help

- Student's Solutions Manual
- MyMathLab
- InterAct Math Tutorial Software
- AW Math Tutor Center
- MathXL
- Digital Video Tutor CD 3/Videotape 2

Use the commutative or the associative property to complete each statement. State which property is used. See Examples 1 and 2.

1. $-12 + 6 = 6 + \underline{}$
2. $8 + (-4) = -4 + \underline{}$
3. $-6 \cdot 3 = \underline{} \cdot (-6)$
4. $-12 \cdot 6 = 6 \cdot \underline{}$
5. $(4 + 7) + 8 = 4 + (\underline{} + 8)$
6. $(-2 + 3) + 6 = -2 + (\underline{} + 6)$
7. $8 \cdot (3 \cdot 6) = (\underline{} \cdot 3) \cdot 6$
8. $6 \cdot (4 \cdot 2) = (6 \cdot \underline{}) \cdot 2$

9. Match each item in Column I with the correct choice(s) from Column II. Choices may be used once, more than once, or not at all.

I	II
(a) Identity element for addition	A. $(5 \cdot 4) \cdot 3 = 5 \cdot (4 \cdot 3)$
(b) Identity element for multiplication	B. 0
(c) Additive inverse of a	C. $-a$
(d) Multiplicative inverse, or reciprocal, of the nonzero number a	D. -1
(e) The number that is its own additive inverse	E. $5 \cdot 4 \cdot 3 = 60$
(f) The two numbers that are their own multiplicative inverses	F. 1
(g) The only number that has no multiplicative inverse	G. $(5 \cdot 4) \cdot 3 = 3 \cdot (5 \cdot 4)$
(h) An example of the associative property	H. $5(4 + 3) = 5 \cdot 4 + 5 \cdot 3$
(i) An example of the commutative property	I. $\frac{1}{a}$
(j) An example of the distributive property	

10. Explain the difference between the commutative and associative properties.

Decide whether each statement is an example of the commutative, associative, identity, inverse, or distributive property. See Examples 1, 2, 3, 5, 6, 7, and 9.

11. $7 + 18 = 18 + 7$
12. $13 + 12 = 12 + 13$
13. $5 \cdot (13 \cdot 7) = (5 \cdot 13) \cdot 7$
14. $-4 \cdot (2 \cdot 6) = (-4 \cdot 2) \cdot 6$
15. $-6 + (12 + 7) = (-6 + 12) + 7$
16. $(-8 + 13) + 2 = -8 + (13 + 2)$
17. $-6 + 6 = 0$
18. $12 + (-12) = 0$

19. $\frac{2}{3}\left(\frac{3}{2}\right) = 1$ Inverse
20. $\frac{5}{8}\left(\frac{8}{5}\right) = 1$
21. $2.34 + 0 = 2.34$ identity
22. $-8.456 + 0 = -8.456$
23. $(4 + 17) + 3 = 3 + (4 + 17)$ C
24. $(-8 + 4) + 12 = 12 + (-8 + 4)$
25. $6(x + y) = 6x + 6y$ Distributive
26. $14(t + s) = 14t + 14s$
27. $-\frac{5}{9} = -\frac{5}{9} \cdot \frac{3}{3} = -\frac{15}{27}$ identity
28. $\frac{13}{12} = \frac{13}{12} \cdot \frac{7}{7} = \frac{91}{84}$
29. $5(2x) + 5(3y) = 5(2x + 3y)$ distributive
30. $3(5t) - 3(7r) = 3(5t - 7r)$

31. The following conversation actually took place between one of the authors of this book and his son, Jack, when Jack was 4 years old:

    DADDY: "Jack, what is 3 + 0?"
    JACK: "3."
    DADDY: "Jack, what is 4 + 0?"
    JACK: "4. And Daddy, *string* plus zero equals *string*!"

    What property of addition did Jack recognize? identity

32. The distributive property holds for multiplication with respect to addition. Is there a distributive property for addition with respect to multiplication? If not, give an example to show why.

Find each sum. Use the commutative and associative properties to make your work easier. See Example 4.

33. $97 + 13 + 3 + 37 = 150$
34. $49 + 199 + 1 + 1$
35. $1999 + 2 + 1 + 8 = 2010$
36. $2998 + 3 + 2 + 17$
37. $159 + 12 + 141 + 88 = 300 + 100 = 400$
38. $106 + 8 + (-6) + (-8)$
39. $843 + 627 + (-43) + (-27) = 1400$
40. $1846 + 1293 + (-46) + (-93)$
41. $(-7.8) + 3.5 + (-2.2) + 6.5 = 0$
42. $(-4.4) + 9.1 + (-5.6) + .9$
43. $\left(-\frac{1}{5} + \frac{9}{2} - \frac{4}{7}\right) + \left(\frac{1}{2} + \frac{6}{5} + \frac{11}{7}\right)$
44. $\left(-\frac{11}{3} + \frac{2}{7} - \frac{4}{5}\right) + \left(-\frac{1}{5} - \frac{1}{3} + \frac{5}{7}\right)$
45. $\left(2\frac{5}{8} - 3.7\right) + \left(3.7 + \frac{3}{8}\right) - \left(2.5 - \frac{1}{4}\right) + \left(\frac{5}{2} - .25\right) = 3$
46. $\left(7\frac{1}{6} - 5.3\right) + \left(15.3 + \frac{5}{6}\right) + \left(7.5 + \frac{1}{10}\right) - \left(\frac{15}{2} + .1\right)$

Use the properties of this section to simplify each expression. See Examples 7 and 8.

47. $6t + 8 - 6t + 3 = 11$
48. $9r + 12 - 9r + 1$
49. $\frac{2}{3}x - 11 + 11 - \frac{2}{3}x = 0$
50. $\frac{1}{5}y + 4 - 4 - \frac{1}{5}y$
51. $\left(\frac{9}{7}\right)(-.38)\left(\frac{7}{9}\right) = -.38$
52. $\left(\frac{4}{5}\right)(-.73)\left(\frac{5}{4}\right)$
53. $t + (-t) + \frac{1}{2}(2) = 1$
54. $w + (-w) + \frac{1}{4}(4)$
55. $2.87w + \frac{1}{2}(2) + (-2.87w) - 3$
56. $5.99x + \frac{1}{4}(4) + (-5.99x) + 7$
57. $\left(-\frac{2}{3}\right)(6.548)\left(-\frac{3}{2}\right) - 3(6.548)\left(\frac{1}{3}\right) = 0$
58. $\left(-\frac{17}{4}\right)(3.1416)\left(-\frac{4}{17}\right) - (2.5)(3.1416)\left(\frac{2}{5}\right)$

59. Evaluate $25 - (6 - 2)$ and evaluate $(25 - 6) - 2$. Do you think subtraction is associative? No, subtraction is not associative.

60. Evaluate $180 \div (15 \div 3)$ and evaluate $(180 \div 15) \div 3$. Do you think division is associative?

61. Suppose that a student shows you the following work.
$$-3(4 - 6) = -3(4) - 3(6) = -12 - 18 = -30$$
The student has made a very common error. Explain the student's mistake, and work the problem correctly.

62. Explain how the procedure of changing $\frac{3}{4}$ to $\frac{9}{12}$ requires the use of the multiplicative identity element, 1.

Use the distributive property to rewrite each expression. Simplify if possible. See Example 9.

63. $5(9 + 8)$ **64.** $6(11 + 8)$ **65.** $4(t + 3)$
66. $5(w + 4)$ **67.** $-8(r + 3)$ **68.** $-11(x + 4)$
69. $-5(y - 4)$ **70.** $-9(g - 4)$ **71.** $-\frac{4}{3}(12y + 15z)$
72. $-\frac{2}{5}(10b + 20a)$ **73.** $8z + 8w$ **74.** $4s + 4r$
75. $7(2v) + 7(5r)$ **76.** $13(5w) + 13(4p)$ **77.** $8(3r + 4s - 5y)$
78. $2(5u - 3v + 7w)$ **79.** $-3(8x + 3y + 4z)$ **80.** $-5(2x - 5y + 6z)$
81. $5x + 15$ **82.** $9p + 18$ **83.** $\frac{4}{5}x - \frac{4}{5}y + \frac{4}{5}t$
84. $\frac{9}{4}w + \frac{9}{4}x - \frac{9}{4}z$ **85.** $-\frac{7}{2}\left(-\frac{2}{7}m + \frac{2}{7}n\right)$ **86.** $-\frac{8}{3}\left(-\frac{3}{8}a + \frac{3}{8}b\right)$
87. $1.5xy - 1.5ab + 1.5mn$ **88.** $8.25pq + 8.25rs - 8.25yz$
89. $-\frac{1}{8}(16p + 80q - 40r - 64s)$ **90.** $-\frac{1}{10}(30a + 50b - 20c - 90d)$
91. $-6.7(10r + 100s - 1000t)$ **92.** $-9.9(100a + 1000b - 10,000c)$

Use the distributive property to write each expression without parentheses. See Example 10.

93. $-(4t + 3m)$ **94.** $-(9x + 12y)$ **95.** $-(-5c - 4d)$
96. $-(-13x - 15y)$ **97.** $-(-3q + 5r - 8s)$ **98.** $-(-4z + 5w - 9y)$

99. The operations of "getting out of bed" and "taking a shower" are not commutative. Give an example of another pair of everyday operations that are not commutative.

100. The phrase "dog biting man" has two different meanings, depending on how the words are associated:

(dog biting) man   dog (biting man)

Give another example of a three-word phrase that has different meanings depending on how the words are associated.

RELATING CONCEPTS (EXERCISES 101–104)

For Individual or Group Work

In Section 1.6 we used a pattern to see that the product of two negative numbers is a positive number. In the group of exercises that follows, we show another justification for determining the sign of the product of two negative numbers. **Work Exercises 101–104 in order.**

101. Evaluate the expression $-3[5 + (-5)]$ by using the order of operations.

102. Write the expression in Exercise 101 using the distributive property. Do not simplify the products.

103. The product $-3(5)$ should be part of the answer you wrote for Exercise 102. Based on the results in Section 1.6, what is this product?

104. In Exercise 101, you should have obtained 0 as an answer. Now, consider the following, using the results of Exercises 101 and 103.

$$-3[5 + (-5)] = -3(5) + (-3)(-5)$$
$$0 = -15 + \,?$$

The question mark represents the product $(-3)(-5)$. When added to -15, it must give a sum of 0. Therefore, how must we interpret $(-3)(-5)$?

1.8 Simplifying Expressions

OBJECTIVES

1. Simplify expressions.
2. Identify terms and numerical coefficients.
3. Identify like terms.
4. Combine like terms.
5. Simplify expressions from word phrases.

OBJECTIVE 1 Simplify expressions. In this section we show how to simplify expressions using the properties of addition and multiplication introduced in the previous section.

EXAMPLE 1   Simplifying Expressions

Simplify each expression.

(a) $4x + 8 + 9$
Since $8 + 9 = 17$,
$$4x + 8 + 9 = 4x + 17.$$

(b) $4(3m - 2n)$
Use the distributive property first.

$$4(3m - 2n) = 4(3m) - 4(2n) \quad \text{Arrows denote distributive property.}$$
$$= (4 \cdot 3)m - (4 \cdot 2)n \quad \text{Associative property}$$
$$= 12m - 8n$$

(c) $6 + 3(4k + 5) = 6 + 3(4k) + 3(5) \quad \text{Distributive property}$
$$= 6 + (3 \cdot 4)k + 3(5) \quad \text{Associative property}$$
$$= 6 + 12k + 15$$
$$= 6 + 15 + 12k \quad \text{Commutative property}$$
$$= 21 + 12k$$

(d) $5 - (2y - 8) = 5 - 1 \cdot (2y - 8) \quad -a = -1 \cdot a$
$$= 5 - 1(2y) - 1(-8) \quad \text{Distributive property}$$
$$= 5 - 2y + 8$$
$$= 5 + 8 - 2y \quad \text{Commutative property}$$
$$= 13 - 2y$$

Now Try Exercises 3 and 5.

NOTE In Examples 1(c) and 1(d), a different use of the commutative property would have resulted in answers of $12k + 21$ and $-2y + 13$. These answers also would be acceptable.

OBJECTIVE 2 Identify terms and numerical coefficients. A **term** is a number, a variable, or a product or quotient of numbers and variables raised to powers.* Examples of terms include

$$-9x^2, \quad 15y, \quad -3, \quad 8m^2n, \quad \frac{2}{p}, \quad \text{and} \quad k.$$

The **numerical coefficient** of the term $9m$ is 9, the numerical coefficient of $-15x^3y^2$ is -15, the numerical coefficient of x is 1, and the numerical coefficient of 8 is 8. In the expression $\frac{x}{3}$, the numerical coefficient of x is $\frac{1}{3}$ since $\frac{x}{3} = \frac{1x}{3} = \frac{1}{3}x$.

Now Try Exercises 9 and 15.

Term	Numerical Coefficient
$-7y$	-7
$34r^3$	34
$-26x^5yz^4$	-26
$-k$	-1
r	1
$\frac{3x}{8} = \frac{3}{8}x$	$\frac{3}{8}$

NOTE It is important to be able to distinguish between *terms* and *factors*. For example, in the expression $8x^3 + 12x^2$, there are two *terms*, $8x^3$ and $12x^2$. On the other hand, in the one-term expression $(8x^3)(12x^2)$, $8x^3$ and $12x^2$ are *factors*.

Examples of terms and their numerical coefficients are shown in the table in the margin.

OBJECTIVE 3 Identify like terms. Terms with exactly the same variables that have the same exponents are **like terms.** For example, $9m$ and $4m$ have the same variable and are like terms. Also, $6x^3$ and $-5x^3$ are like terms. The terms $-4y^3$ and $4y^2$ have different exponents and are **unlike terms.**

Here are some additional examples.

$5x$ and $-12x$ $3x^2y$ and $5x^2y$ Like terms
$4xy^2$ and $5xy$ $-7w^3z^3$ and $2xz^3$ Unlike terms

Now Try Exercises 19 and 25.

OBJECTIVE 4 Combine like terms. Recall the distributive property.

$$x(y + z) = xy + xz.$$

As seen in the previous section, this statement can also be written "backward" as

$$xy + xz = x(y + z).$$

This form of the distributive property may be used to find the sum or difference of like terms. For example,

$$3x + 5x = (3 + 5)x = 8x.$$

This process is called **combining like terms.**

NOTE Remember that *only like terms may be combined.*

---

*Another name for certain terms, **monomial**, is introduced in Chapter 5.

EXAMPLE 2 Combining Like Terms

Combine like terms in each expression.

(a) $9m + 5m$

Use the distributive property as given above.

$$9m + 5m = (9 + 5)m = 14m$$

(b) $6r + 3r + 2r = (6 + 3 + 2)r = 11r$ Distributive property

(c) $4x + x = 4x + 1x = (4 + 1)x = 5x$ (Note: $x = 1x$.)

(d) $16y^2 - 9y^2 = (16 - 9)y^2 = 7y^2$

(e) $32y + 10y^2$ cannot be combined because $32y$ and $10y^2$ are unlike terms. We cannot use the distributive property here to combine coefficients.

Now Try Exercises 31, 35, and 45.

When an expression involves parentheses, the distributive property is used both "forward" and "backward" to combine like terms, as shown in the following example.

EXAMPLE 3 Simplifying Expressions Involving Like Terms

Combine like terms in each expression.

(a) $14y + 2(6 + 3y) = 14y + 2(6) + 2(3y)$ Distributive property
$\qquad\qquad\qquad\quad = 14y + 12 + 6y$ Multiply.
$\qquad\qquad\qquad\quad = 20y + 12$ Combine like terms.

(b) $9k - 6 - 3(2 - 5k) = 9k - 6 - 3(2) - 3(-5k)$ Distributive property
$\qquad\qquad\qquad\qquad = 9k - 6 - 6 + 15k$ Multiply.
$\qquad\qquad\qquad\qquad = 24k - 12$ Combine like terms.

(c) $-(2 - r) + 10r = -1(2 - r) + 10r$ $-(2 - r) = -1(2 - r)$
$\qquad\qquad\qquad = -1(2) - 1(-r) + 10r$ Distributive property
$\qquad\qquad\qquad = -2 + 1r + 10r$ Multiply.
$\qquad\qquad\qquad = -2 + 11r$ Combine like terms.

(d) $100[.03(x + 4)] = [(100)(.03)](x + 4)$ Associative property
$\qquad\qquad\qquad = 3(x + 4)$ Multiply.
$\qquad\qquad\qquad = 3x + 12$ Distributive property

(e) $5(2a - 6) - 3(4a - 9) = 10a - 30 - 12a + 27$ Distributive property
$\qquad\qquad\qquad\qquad = -2a - 3$ Combine like terms.

Now Try Exercises 49, 51, 53, and 55.

Example 3(e) suggests that the commutative property can be used with subtraction by treating the subtracted terms as the addition of their additive inverses.

NOTE Examples 2 and 3 suggest that like terms may be combined by adding or subtracting the coefficients of the terms and keeping the same variable factors.

76 CHAPTER 1 The Real Number System

OBJECTIVE 5 Simplify expressions from word phrases. Earlier we saw how to translate words, phrases, and statements into expressions and equations. Now we can simplify translated expressions by combining like terms.

EXAMPLE 4 Translating Words to a Mathematical Expression

Translate to a mathematical expression, and simplify: The sum of 9, five times a number, four times the number, and six times the number.

The word "sum" indicates that the terms should be added. Use x to represent the number. Then the phrase translates as

$$9 + 5x + 4x + 6x,$$ Write with symbols.

which simplifies to

$$9 + 15x.$$ Combine like terms.

Now Try Exercise 77.

CAUTION In Example 4, we are dealing with an expression to be simplified, *not* an equation to be solved.

1.8 EXERCISES

For Extra Help

- Student's Solutions Manual
- MyMathLab
- InterAct Math Tutorial Software
- AW Math Tutor Center
- MathXL
- Digital Video Tutor CD 3/Videotape 2

Simplify each expression. See Example 1.

1. $4r + 19 - 8$
2. $7t + 18 - 4$
3. $5 + 2(x - 3y)$
4. $8 + 3(s - 6t)$
5. $-2 - (5 - 3p)$
6. $-10 - (7 - 14r)$
7. $6 + (4 - 3x) - 8$
8. $-12 + (7 - 8x) + 6$

Give the numerical coefficient of each term.

9. $-12k$
10. $-23y$
11. $5m^2$
12. $-3n^6$
13. xw
14. pq
15. $-x$
16. $-t$
17. 74
18. 98

Identify each group of terms as like or unlike.

19. $8r, -13r$
20. $-7a, 12a$
21. $5z^4, 9z^3$
22. $8x^5, -10x^3$
23. $4, 9, -24$
24. $7, 17, -83$
25. x, y
26. t, s

Simplify each expression by combining like terms. See Examples 1–3.

27. $9y + 8y$
28. $15m + 12m$
29. $-4a - 2a$
30. $-3z - 9z$
31. $12b + b$
32. $30x + x$
33. $2k + 9 + 5k + 6$
34. $2 + 17z + 1 + 2z$
35. $-5y + 3 - 1 + 5 + y - 7$
36. $2k - 7 - 5k + 7k - 3 - k$
37. $-2x + 3 + 4x - 17 + 20$
38. $r - 6 - 12r - 4 + 6r$
39. $16 - 5m - 4m - 2 + 2m$
40. $6 - 3z - 2z - 5 + z - 3z$

SECTION 1.8 Simplifying Expressions 77

41. $-10 + x + 4x - 7 - 4x$
42. $-p + 10p - 3p - 4 - 5p$
43. $1 + 7x + 11x - 1 + 5x$
44. $-r + 2 - 5r + 3 + 4r$
45. $6y^2 + 11y^2 - 8y^2$
46. $-9m^3 + 3m^3 - 7m^3$
47. $2p^2 + 3p^2 - 8p^3 - 6p^3$
48. $5y^3 + 6y^3 - 3y^2 - 4y^2$
49. $2(4x + 6) + 3$
50. $4(6y - 9) + 7$
51. $100[.05(x + 3)]$
52. $100[.06(x + 5)]$
53. $-4(y - 7) - 6$
54. $-5(t - 13) - 4$
55. $-5(5y - 9) + 3(3y + 6)$
56. $-3(2t + 4) + 8(2t - 4)$
57. $-3(2r - 3) + 2(5r + 3)$
58. $-4(5y - 7) + 3(2y - 5)$
59. $8(2k - 1) - (4k - 3)$
60. $6(3p - 2) - (5p + 1)$
61. $-2(-3k + 2) - (5k - 6) - 3k - 5$
62. $-2(3r - 4) - (6 - r) + 2r - 5$
63. $-4(-3k + 3) - (6k - 4) - 2k + 1$
64. $-5(8j + 2) - (5j - 3) - 3j + 17$
65. $-7.5(2y + 4) - 2.9(3y - 6)$
66. $8.4(6t - 6) + 2.4(9 - 3t)$
67. $5.2[3y - 4(7 + 2x)] + 6.1[-2x + 3(4 - 2y)]$
68. $6.6[2y + 3(10 - 9x)] + 4.4[x + 2(9 - 4x)]$
69. $\dfrac{1}{2}[6(9y - 4x) + 2(3y - 6x)] + \dfrac{2}{3}[9(6x + 3y) - 12(5x - 8y)]$
70. $\dfrac{1}{4}[8(2y + 3x) - 4(6y - 3x)] + \dfrac{1}{5}[15(x + 2y) + 5(4x - 12y)]$
71. $-1.55(4s + 5t) - 2.68(3s - 2t) + 7.25(s - 6t)$
72. $-8.34(8r + 2s) + 9.15(4r - 3s) - 2.45(2r - 7s)$
73. $\dfrac{1}{3}(6a - 9b + 12c) + 3.5(30a - 18b - 20c)$
74. $\dfrac{1}{2}(8x + 10y + 12z) - 6.5(16x - 10y + 8z)$
75. $-12\{4[2.5(x - 6) + (y + 4)]\} - 10(4.5x + 6.5y)$
76. $-14\{6[3.5(x + 2) + (2y - 8)]\} + 16(2.5x - 3.5y)$

Translate each phrase into a mathematical expression. Use x as the variable. Combine like terms when possible. See Example 4.

77. Five times a number, added to the sum of the number and three
78. Six times a number, added to the sum of the number and six
79. A number multiplied by -7, subtracted from the sum of 13 and six times the number
80. A number multiplied by 5, subtracted from the sum of 14 and eight times the number
81. Six times a number added to -4, subtracted from twice the sum of three times the number and 4 (*Hint: Twice* means two times.)
82. Nine times a number added to 6, subtracted from triple the sum of 12 and 8 times the number (*Hint: Triple* means three times.)

RELATING CONCEPTS (EXERCISES 83–90)

For Individual or Group Work

Work Exercises 83–90 in order. *They will help prepare you for graphing later in the text.*

83. Evaluate the expression $x + 2$ for the values of x shown in the table.

x	$x + 2$
0	
1	
2	
3	

84. Based on your results from Exercise 83, complete the following statement: For every increase of 1 unit for x, the value of $x + 2$ increases by _____ unit(s).

85. Repeat Exercise 83 for these expressions:
   (a) $x + 1$ (b) $x + 3$ (c) $x + 4$

86. Based on your results from Exercises 83 and 85, make a conjecture (an educated guess) about what happens to the value of an expression of the form $x + b$ for any value of b, as x increases by 1 unit.

87. Repeat Exercise 83 for these expressions:
   (a) $2x + 2$ (b) $3x + 2$ (c) $4x + 2$

88. Based on your results from Exercise 87, complete the following statement: For every increase of 1 unit for x, the value of $mx + 2$ increases by _____ units.

89. Repeat Exercise 83 and compare your results to those in Exercise 87 for these expressions:
   (a) $2x + 7$ (b) $3x + 5$ (c) $4x + 1$

90. Based on your results from Exercises 83–89, complete the following statement: For every increase of 1 unit for x, the value of $mx + b$ increases by _____ units.

91. There is an old saying, "You can't add apples and oranges." Explain how this saying can be applied to the goal of Objective 4 in this section.

92. Explain how the distributive property is used in combining $6t + 5t$ to get $11t$.

93. Write the expression $9x - (x + 2)$ using words, as in Exercises 77–82.

94. Write the expression $2(3x + 5) - 2(x + 4)$ using words, as in Exercises 77–82.

Chapter Group Activity

Comparing Floor Plans of Houses

OBJECTIVE Use arithmetic skills to make comparisons.

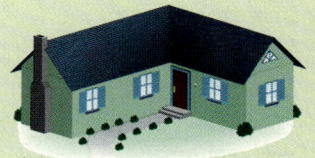

This activity explores perimeters and areas of different shaped homes. Floor plans for three different homes are given below.

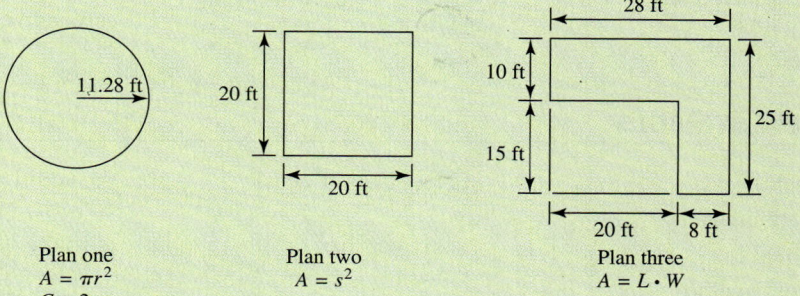

Plan one
$A = \pi r^2$
$C = 2\pi r$

Plan two
$A = s^2$

Plan three
$A = L \cdot W$

A. As a group, look at the dimensions of the given floor plans. Considering only the dimensions, which plan do you think has the greatest area?

B. Now have each student in your group pick one floor plan. For each plan, find the following. (Round all answers to the nearest whole number. In Plan one, let $\pi = 3.14$.)

   1. The area of the plan

   2. The perimeter or circumference (the distance around the outside) of the plan

C. Share your findings with the group and answer the following questions.

   1. What did you determine about the areas of the three floor plans?

   2. Which plan has the smallest perimeter? Which has the largest perimeter?

   3. Why do you think houses with round floor plans might be more energy efficient?

   4. What advantages do you think houses with square or rectangular floor plans have? Why do you think floor plans with these shapes are most common for homes today?

CHAPTER 1 SUMMARY

KEY TERMS

1.1
natural numbers
whole numbers
fractions
numerator
denominator
proper fraction
improper fraction
mixed number
factor
product
factored
prime number
composite number
greatest common factor
lowest terms
basic principle of fractions

reciprocal
quotient
sum
least common denominator (LCD)
difference
1.2 exponent (power)
base
exponential expression
grouping symbols
order of operations
inequality
1.3 variable
algebraic expression
equation
solution
set
element

1.4 number line
graph
coordinate
negative numbers
positive numbers
signed numbers
integers
rational numbers
set-builder notation
irrational numbers
real numbers
additive inverse (opposite)
absolute value
1.6 multiplicative inverse (reciprocal)
1.7 commutative property
associative property

identity property
identity element for addition (additive identity)
identity element for multiplication (multiplicative identity)
inverse property
distributive property
1.8 term
numerical coefficient
like terms
unlike terms
combining like terms

NEW SYMBOLS

a^n n factors of a
[ ]   square brackets
=   is equal to
≠   is not equal to
<   is less than
>   is greater than

≤   is less than or equal to
≥   is greater than or equal to
{ }   set braces
$\{x \mid x$ has a certain property$\}$ set-builder notation

$-x$ the additive inverse, or opposite, of x
$|x|$ absolute value of x
$\dfrac{1}{x}$ the multiplicative inverse, or reciprocal, of the nonzero number x

$a(b), (a)b, (a)(b), a \cdot b,$ or ab a times b
$\dfrac{a}{b}$ or a/b a divided by b

TEST YOUR WORD POWER

See how well you have learned the vocabulary in this chapter. Answers, with examples, follow the Quick Review.

1. A **factor** is
   A. the answer in an addition problem
   B. the answer in a multiplication problem
   C. one of two or more numbers that are added to get another number
   D. one of two or more numbers that are multiplied to get another number.

2. A number is **prime** if
   A. it cannot be factored
   B. it has just one factor
   C. it has only itself and 1 as factors
   D. it has at least two different factors.

3. An **exponent** is
   A. a symbol that tells how many numbers are being multiplied
   B. a number raised to a power
   C. a number that tells how many times a factor is repeated
   D. one of two or more numbers that are multiplied.

4. A **variable** is
   A. a symbol used to represent an unknown number
   B. a value that makes an equation true
   C. a solution of an equation
   D. the answer in a division problem.

5. An **integer** is
   A. a positive or negative number
   B. a natural number, its opposite, or zero

CHAPTER 1 Summary **81**

C. any number that can be graphed on a number line
D. the quotient of two numbers.

6. A coordinate is
A. the number that corresponds to a point on a number line
B. the graph of a number
C. any point on a number line
D. the distance from 0 on a number line.

7. The **absolute value** of a number is
A. the graph of the number
B. the reciprocal of the number
C. the opposite of the number
D. the distance between 0 and the number on a number line.

8. A term is
A. a numerical factor
B. a number or a product or quotient of numbers and variables raised to powers

C. one of several variables with the same exponents
D. a sum of numbers and variables raised to powers.

9. A numerical coefficient is
A. the numerical factor in a term
B. the number of terms in an expression
C. a variable raised to a power
D. the variable factor in a term.

QUICK REVIEW

CONCEPTS	EXAMPLES

1.1 FRACTIONS

Operations with Fractions

Addition/Subtraction:
1. To add/subtract fractions with the same denominator, add/subtract the numerators and keep the same denominator.
2. To add/subtract fractions with different denominators, find the LCD and write each fraction with this LCD. Then follow the procedure above.

Multiplication: Multiply numerators and multiply denominators.

Division: Multiply the first fraction by the reciprocal of the second fraction.

Perform each operation.

$$\frac{2}{5} + \frac{7}{5} = \frac{2+7}{5} = \frac{9}{5}$$

$$\frac{2}{3} - \frac{1}{2} = \frac{4}{6} - \frac{3}{6} \qquad \text{6 is the LCD.}$$

$$= \frac{4-3}{6} = \frac{1}{6}$$

$$\frac{4}{3} \cdot \frac{5}{6} = \frac{20}{18} = \frac{10}{9}$$

$$\frac{6}{5} \div \frac{1}{4} = \frac{6}{5} \cdot \frac{4}{1} = \frac{24}{5}$$

1.2 EXPONENTS, ORDER OF OPERATIONS, AND INEQUALITY

Order of Operations
Simplify within parentheses or above and below fraction bars first, in the following order.

Step 1 Apply all exponents.

Step 2 Do any multiplications or divisions from left to right.

Step 3 Do any additions or subtractions from left to right.

If no grouping symbols are present, start with Step 1.

Simplify $36 - 4(2^2 + 3)$.

$$36 - 4(2^2 + 3) = 36 - 4(4 + 3)$$
$$= 36 - 4(7)$$
$$= 36 - 28$$
$$= 8$$

(continued)

CONCEPTS	EXAMPLES
1.3 VARIABLES, EXPRESSIONS, AND EQUATIONS	
Evaluate an expression with a variable by substituting a given number for the variable.	Evaluate $2x + y^2$ if $x = 3$ and $y = -4$. $$2x + y^2 = 2(3) + (-4)^2$$ $$= 6 + 16$$ $$= 22$$
Values of a variable that make an equation true are solutions of the equation.	Is 2 a solution of $5x + 3 = 18$? $5(2) + 3 = 18$? $13 = 18$ False 2 is not a solution.
1.4 REAL NUMBERS AND THE NUMBER LINE	
The Ordering of Real Numbers a is less than b if a is to the left of b on the number line.	Graph -2, 0, and 3. $-2 < 3$ $3 > 0$ $0 < 3$
The additive inverse of x is $-x$.	$-(5) = -5$ $-(-7) = 7$ $-0 = 0$
The absolute value of x, $\|x\|$, is the distance between x and 0 on the number line.	$\|13\| = 13$ $\|0\| = 0$ $\|-5\| = 5$
1.5 ADDING AND SUBTRACTING REAL NUMBERS	
Rules for Adding Signed Numbers To add two numbers with the same sign, add their absolute values. The sum has that same sign. To add two numbers with different signs, subtract their absolute values. The sum has the sign of the number with larger absolute value.	Add. $9 + 4 = 13$ $-8 + (-5) = -13$ $7 + (-12) = -5$ $-5 + 13 = 8$
Definition of Subtraction $x - y = x + (-y)$	Subtract. $5 - (-2) = 5 + 2 = 7$
Rules for Subtracting Signed Numbers 1. Change the subtraction symbol to the addition symbol. 2. Change the sign of the number being subtracted. 3. Add, using the rules for addition.	$-3 - 4 = -3 + (-4) = -7$ $-2 - (-6) = -2 + 6 = 4$ $13 - (-8) = 13 + 8 = 21$

CONCEPTS	EXAMPLES

1.6 MULTIPLYING AND DIVIDING REAL NUMBERS

Rules for Multiplying and Dividing Signed Numbers The product (or quotient) of two numbers having the *same sign* is *positive;* the product (or quotient) of two numbers having *different signs* is *negative*.	Multiply or divide. $6 \cdot 5 = 30 \qquad -7(-8) = 56 \qquad \dfrac{20}{4} = 5$ $\dfrac{-24}{-6} = 4 \qquad -6(5) = -30 \qquad 6(-5) = -30$ $\dfrac{-18}{9} = -2 \qquad \dfrac{49}{-7} = -7$
Definition of Division $\dfrac{x}{y} = x \cdot \dfrac{1}{y}, \quad y \neq 0$	$\dfrac{10}{2} = 10 \cdot \dfrac{1}{2} = 5$
Division by 0 is undefined.	$\dfrac{5}{0}$ is undefined.
0 divided by a nonzero number equals 0.	$\dfrac{0}{5} = 0$

1.7 PROPERTIES OF REAL NUMBERS

Commutative Properties $a + b = b + a$ $ab = ba$	$7 + (-1) = -1 + 7$ $5(-3) = (-3)5$
Associative Properties $(a + b) + c = a + (b + c)$ $(ab)c = a(bc)$	$(3 + 4) + 8 = 3 + (4 + 8)$ $[-2(6)]4 = -2[(6)4]$
Identity Properties $a + 0 = a \qquad 0 + a = a$ $a \cdot 1 = a \qquad 1 \cdot a = a$	$-7 + 0 = -7 \qquad 0 + (-7) = -7$ $9 \cdot 1 = 9 \qquad 1 \cdot 9 = 9$
Inverse Properties $a + (-a) = 0 \qquad -a + a = 0$ $a \cdot \dfrac{1}{a} = 1 \qquad \dfrac{1}{a} \cdot a = 1 \quad (a \neq 0)$	$7 + (-7) = 0 \qquad -7 + 7 = 0$ $-2\left(-\dfrac{1}{2}\right) = 1 \qquad -\dfrac{1}{2}(-2) = 1$
Distributive Properties $a(b + c) = ab + ac$ $(b + c)a = ba + ca$ $a(b - c) = ab - ac$	$5(4 + 2) = 5(4) + 5(2)$ $(4 + 2)5 = 4(5) + 2(5)$ $9(5 - 4) = 9(5) - 9(4)$

(continued)

84 CHAPTER 1 The Real Number System

CONCEPTS	EXAMPLES
1.8 SIMPLIFYING EXPRESSIONS	
Only like terms may be combined.	$-3y^2 + 6y^2 + 14y^2 = 17y^2$
	$4(3 + 2x) - 6(5 - x)$
	$\quad = 12 + 8x - 30 + 6x$ *Distributive property*
	$\quad = 14x - 18$

Answers to Test Your Word Power
1. D; *Example:* Since $2 \times 5 = 10$, the numbers 2 and 5 are factors of 10; other factors of 10 are $-10, -5, -2, -1, 1,$ and 10.
2. C; *Examples:* 2, 3, 11, 41, 53   3. C; *Example:* In 2^3, the number 3 is the exponent (or power), so 2 is a factor three times; $2^3 = 2 \cdot 2 \cdot 2 = 8$.   4. A; *Examples:* a, b, c 5. B; *Examples:* $-9, 0, 6$ 6. A; *Example:* The point graphed 3 units to the right of 0 on a number line has coordinate 3.   7. D; *Examples:* $|2| = 2$ and $|-2| = 2$ 8. B; *Examples:* $6, \frac{x}{2}, -4ab^2$ 9. A; *Examples:* The term 3 has numerical coefficient 3, $8z$ has numerical coefficient 8, and $-10x^4y$ has numerical coefficient -10.

CHAPTER 1 REVIEW EXERCISES

[1.1] *Perform each indicated operation.**

1. $\dfrac{8}{5} \div \dfrac{32}{15}$

2. $\dfrac{3}{8} + 3\dfrac{1}{2} - \dfrac{3}{16}$

3. The circle graph illustrates how 400 people responded to a survey that asked "Do you think that the CEOs of corporations guilty of accounting fraud should go to jail?" What fractional part of the group did not have an opinion?

4. Based on the graph in Exercise 3, how many people responded "yes"?

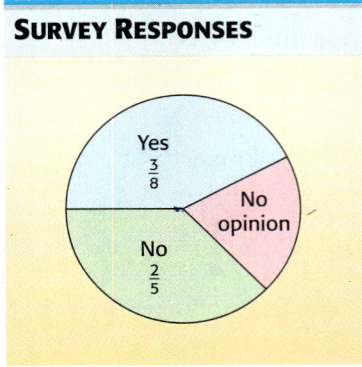

SURVEY RESPONSES

Yes $\frac{3}{8}$
No opinion
No $\frac{2}{5}$

[1.2] *Find the value of each exponential expression.*

5. 5^4

6. $\left(\dfrac{3}{5}\right)^3$

7. $(.02)^5$

8. $(.001)^3$

---

*For help with the Review Exercises in this text, refer to the appropriate section given in brackets.

Find the value of each expression.

9. $8 \cdot 5 - 13$
10. $7[3 + 6(3^2)]$
11. $\dfrac{9(4^2 - 3)}{4 \cdot 5 - 17}$
12. $\dfrac{6(5 - 4) + 2(4 - 2)}{3^2 - (4 + 3)}$

Tell whether each statement is true or false.

13. $12 \cdot 3 - 6 \cdot 6 \leq 0$
14. $3[5(2) - 3] > 20$
15. $9 \leq 4^2 - 8$

Write each word statement in symbols.

16. Thirteen is less than seventeen.
17. Five plus two is not equal to ten.
18. The bar graph shows the number of worldwide airline fatalities for the years 1990 to 1999.

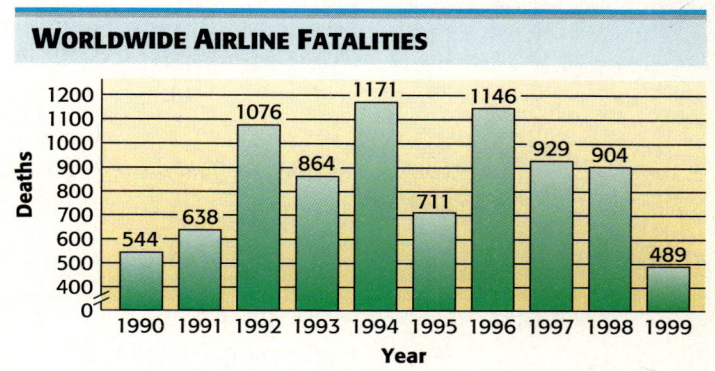

WORLDWIDE AIRLINE FATALITIES

Source: International Civil Aviation Organization.

(a) In which years were *fewer than* 700 people killed?
(b) In which years were *at least* 905 people killed?
(c) How many people were killed in the 5 yr having the largest numbers of deaths?

[1.3] *Find the numerical value of each expression if $x = 6$ and $y = 3$.*

19. $2x + 6y$
20. $4(3x - y)$
21. $\dfrac{x}{3} + 4y$
22. $\dfrac{x^2 + 3}{3y - x}$

Change each word phrase to an algebraic expression. Use x as the variable to represent the number.

23. Six added to a number
24. A number subtracted from eight
25. Nine subtracted from six times a number
26. Three-fifths of a number added to 12

Decide whether the given number is a solution of the given equation.

27. $5x + 3(x + 2) = 22;\ 2$
28. $\dfrac{t + 5}{3t} = 1;\ 6$

Change each word statement to an equation. Use x as the variable. Then find the solution from the set $\{0, 2, 4, 6, 8, 10\}$.

29. Six less than twice a number is 10.
30. The product of a number and 4 is 8.

86 CHAPTER 1 The Real Number System

[1.4] *Graph each group of numbers on a number line.*

31. $-4, -\dfrac{1}{2}, 0, 2.5, 5$

32. $-2, |-3|, -3, |-1|$

Classify each number, using the sets natural numbers, whole numbers, integers, rational numbers, irrational numbers, real numbers.

33. $\dfrac{4}{3}$

34. $\sqrt{6}$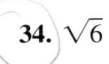

Select the smaller number in each pair.

35. $-10, 5$ **36.** $-8, -9$ **37.** $-\dfrac{2}{3}, -\dfrac{3}{4}$ **38.** $0, -|23|$

Decide whether each statement is **true** *or* **false.**

39. $12 > -13$ **40.** $0 > -5$ **41.** $-9 < -7$ **42.** $-13 \geq -13$

For each number, **(a)** *find the opposite of the number and* **(b)** *find the absolute value of the number.*

43. -9 **44.** 0 **45.** 6 **46.** $-\dfrac{5}{7}$

Simplify each number by removing absolute value symbols.

47. $|-12|$ **48.** $-|3|$ **49.** $-|-19|$ **50.** $-|9-2|$

[1.5] *Perform each indicated operation.*

51. $-10 + 4$ **52.** $14 + (-18)$ **53.** $-8 + (-9)$

54. $\dfrac{4}{9} + \left(-\dfrac{5}{4}\right)$ **55.** $-13.5 + (-8.3)$ **56.** $(-10 + 7) + (-11)$

57. $[-6 + (-8) + 8] + [9 + (-13)]$ **58.** $(-4 + 7) + (-11 + 3) + (-15 + 1)$

59. $-7 - 4$ **60.** $-12 - (-11)$

61. $5 - (-2)$ **62.** $-\dfrac{3}{7} - \dfrac{4}{5}$

63. $2.56 - (-7.75)$ **64.** $(-10 - 4) - (-2)$

65. $(-3 + 4) - (-1)$ **66.** $-(-5 + 6) - 2$

Write a numerical expression for each phrase and simplify the expression.

67. 19 added to the sum of -31 and 12   **68.** 13 more than the sum of -4 and -8

69. The difference between -4 and -6 **70.** Five less than the sum of 4 and -8

Find the solution of each equation from the set $\{-3, -2, -1, 0, 1, 2, 3\}$ *by guessing or by trial and error.*

71. $x + (-2) = -4$ **72.** $12 + x = 11$

CHAPTER 1 Review Exercises **87**

Solve each problem.

73. Like many people, Otis Taylor neglects to keep up his checkbook balance. When he finally balanced his account, he found the balance was −$23.75, so he deposited $50.00. What is his new balance?

74. The low temperature in Yellowknife, in the Canadian Northwest Territories, one January day was −26°F. It rose 16° that day. What was the high temperature?

75. Eric owed his brother $28. He repaid $13 but then borrowed another $14. What positive or negative amount represents his present financial status?

76. If the temperature drops 7° below its previous level of −3°, what is the new temperature?

77. A football team gained 3 yd on the first play from scrimmage, lost 12 yd on the second play, and then gained 13 yd on the third play. How many yards did the team gain or lose altogether?

78. On Wednesday, May 29, 2002, the Dow Jones Industrial Average closed at 9923.04, down 58.54 from the previous day. What was the closing value the previous day? (*Source: Times Picayune.*)

[1.6] *Perform each indicated operation.*

79. $(-12)(-3)$ 80. $15(-7)$ 81. $-\frac{4}{3}\left(-\frac{3}{8}\right)$ 82. $(-4.8)(-2.1)$

83. $5(8 - 12)$ 84. $(5 - 7)(8 - 3)$ 85. $2(-6) - (-4)(-3)$

86. $3(-10) - 5$ 87. $\frac{-36}{-9}$ 88. $\frac{220}{-11}$

89. $-\frac{1}{2} \div \frac{2}{3}$ 90. $-33.9 \div (-3)$ 91. $\frac{-5(3) - 1}{8 - 4(-2)}$

92. $\frac{5(-2) - 3(4)}{-2[3 - (-2)] - 1}$ 93. $\frac{10^2 - 5^2}{8^2 + 3^2 - (-2)}$ 94. $\frac{(.6)^2 + (.8)^2}{(-1.2)^2 - (-.56)}$

Evaluate each expression if $x = -5$, $y = 4$, and $z = -3$.

95. $6x - 4z$ 96. $5x + y - z$ 97. $5x^2$ 98. $z^2(3x - 8y)$

Write a numerical expression for each phrase and simplify the expression.

99. Nine less than the product of −4 and 5

100. Five-sixths of the sum of 12 and −6

101. The quotient of 12 and the sum of 8 and −4

102. The product of −20 and 12, divided by the difference between 15 and −15

Write each sentence in symbols, using x as the variable, and find the solution by guessing or by trial and error. All solutions come from the list of integers between −12 and 12.

103. 8 times a number is −24.

104. The quotient of a number and 3 is −2.

88 CHAPTER 1 The Real Number System

Find the average of each group of numbers.

105. 26, 38, 40, 20, 4, 14, 96, 18

106. $-12, 28, -36, 0, 12, -10$

[1.7] *Decide whether each statement is an example of the* commutative, associative, identity, inverse, *or* distributive *property.*

107. $6 + 0 = 6$

108. $5 \cdot 1 = 5$

109. $-\dfrac{2}{3}\left(-\dfrac{3}{2}\right) = 1$

110. $17 + (-17) = 0$

111. $5 + (-9 + 2) = [5 + (-9)] + 2$

112. $w(xy) = (wx)y$

113. $3x + 3y = 3(x + y)$

114. $(1 + 2) + 3 = 3 + (1 + 2)$

Use the distributive property to rewrite each expression. Simplify if possible.

115. $7y + 14$ **116.** $-12(4 - t)$ **117.** $3(2s) + 3(5y)$ **118.** $-(-4r + 5s)$

119. Evaluate $25 - (5 - 2)$ and $(25 - 5) - 2$. Use this example to explain why subtraction is not associative.

120. Evaluate $180 \div (15 \div 5)$ and $(180 \div 15) \div 5$. Use this example to explain why division is not associative.

[1.8] *Combine like terms whenever possible.*

121. $2m + 9m$

122. $15p^2 - 7p^2 + 8p^2$

123. $5p^2 - 4p + 6p + 11p^2$

124. $-2(3k - 5) + 2(k + 1)$

125. $7(2m + 3) - 2(8m - 4)$

126. $-(2k + 8) - (3k - 7)$

In Exercises 127–130, choose the letter of the correct response.

127. Which one of the following is true for all real numbers x?

A. $6 + 2x = 8x$ **B.** $6 - 2x = 4x$
C. $6x - 2x = 4x$ **D.** $3 + 8(4x - 6) = 11(4x - 6)$

128. Which one of the following is an example of a pair of like terms?

A. $6t, 6w$ **B.** $-8x^2y, 9xy^2$ **C.** $5ry, 6yr$ **D.** $-5x^2, 2x^3$

129. Which one of the following is an example of a term with numerical coefficient 5?

A. $5x^3y^7$ **B.** x^5 **C.** $\dfrac{x}{5}$ **D.** 5^2xy^3

130. Which one of the following is a correct translation for "six times a number, subtracted from the product of eleven and the number" (if x represents the number)?

A. $6x - 11x$ **B.** $11x - 6x$ **C.** $(11 + x) - 6x$ **D.** $6x - (11 + x)$

MIXED REVIEW EXERCISES*

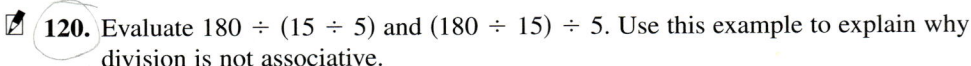

Perform each indicated operation.

131. $[(-2) + 7 - (-5)] + [-4 - (-10)]$

132. $\left(-\dfrac{5}{6}\right)^2$

---

*The order of exercises in this final group does not correspond to the order in which topics occur in the chapter. This random ordering should help you prepare for the chapter test in yet another way.

133. $\dfrac{6(-4) + 2(-12)}{5(-3) + (-3)}$ 134. $\dfrac{3}{8} - \dfrac{5}{12}$

135. $\dfrac{8^2 + 6^2}{7^2 + 1^2}$ 136. $-16(-3.5) - 7.2(-3)$

137. $2\dfrac{5}{6} - 4\dfrac{1}{3}$ 138. $-8 + [(-4 + 17) - (-3 - 3)]$

139. $-\dfrac{12}{5} \div \dfrac{9}{7}$ 140. $(-8 - 3) - 5(2 - 9)$

141. $5x^2 - 12y^2 + 3x^2 - 9y^2$ 142. $-4(2t + 1) - 8(-3t + 4)$

143. Write a sentence or two explaining the special considerations involving 0 when dividing.

144. "Two negatives give a positive" is often heard from students. Is this correct? Use more precise language in explaining what this means.

145. Use x as the variable and write an expression for "the product of 5, and the sum of a number and 7." Then use the distributive property to rewrite the expression.

146. The highest temperature ever recorded in Albany, New York, was 99°F, while the lowest was 112° less than the highest. What was the lowest temperature ever recorded in Albany? (*Source:* World Almanac and Book of Facts, 2000.)

CHAPTER 1 TEST

1. Write $\dfrac{63}{99}$ in lowest terms. 2. Add: $\dfrac{5}{8} + \dfrac{11}{12} + \dfrac{7}{15}$. 3. Divide: $\dfrac{19}{15} \div \dfrac{6}{5}$.

4. The circle graph indicates the market share of different means of intercity transportation in a recent year, based on 1230 million passengers carried.

    (a) How many of these passengers used air travel?
    (b) How many of these passengers did not use the bus?

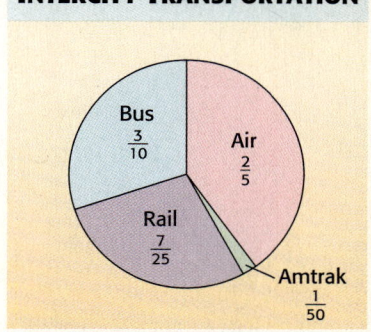

INTERCITY TRANSPORTATION

Bus $\dfrac{3}{10}$
Air $\dfrac{2}{5}$
Rail $\dfrac{7}{25}$
Amtrak $\dfrac{1}{50}$

Source: Eno Transportation Foundation, Inc.

5. Decide whether $4[-20 + 7(-2)] \le 135$ is true or false.

6. Graph the group of numbers $-1, -3, |-4|, |-1|$ on a number line.

7. To which of the following sets does $-\dfrac{2}{3}$ belong: natural numbers, whole numbers, integers, rational numbers, irrational numbers, real numbers?

8. Explain how a number line can be used to show that -8 is less than -1.

9. Write in symbols: The quotient of -6 and the sum of 2 and -8. Simplify the expression.

Perform each indicated operation.

10. $-2 - (5 - 17) + (-6)$

11. $-5\dfrac{1}{2} + 2\dfrac{2}{3}$

12. $-6 - [-7 + (2 - 3)]$

13. $4^2 + (-8) - (2^3 - 6)$

14. $(-5)(-12) + 4(-4) + (-8)^2$

15. $\dfrac{-7 - (-6 + 2)}{-5 - (-4)}$

16. $\dfrac{30(-1 - 2)}{-9[3 - (-2)] - 12(-2)}$

Find the solution for each equation from the set $\{-6, -4, -2, 0, 2, 4, 6\}$ by guessing or by trial and error.

17. $-x + 3 = -3$

18. $-3x = -12$

Evaluate each expression, given $x = -2$ and $y = 4$.

19. $3x - 4y^2$

20. $\dfrac{5x + 7y}{3(x + y)}$

Solve each problem.

21. The highest elevation in Argentina is Mt. Aconcagua, which is 6960 m above sea level. The lowest point in Argentina is the Valdes Peninsula, 40 m below sea level. Find the difference between the highest and lowest elevations.

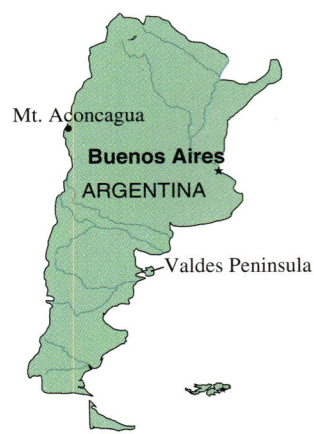

22. For a certain system of rating major league baseball relief pitchers, 3 points are awarded for a save, 3 points are awarded for a win, 2 points are subtracted for a loss, and 2 points are subtracted for a blown save. If Mariano Rivera of the New York Yankees has 4 saves, 3 wins, 2 losses, and 1 blown save, how many points does he have?

23. The bar graph shows the federal budget outlays for national defense for the years 1990 through 1998. Use a signed number to represent the change in outlays for each time period. For example, the change from 1995 to 1996 was $253.3 − $259.6 = −$6.3 billion.

(a) 1991–1992
(b) 1993–1994
(c) 1996–1997
(d) 1997–1998

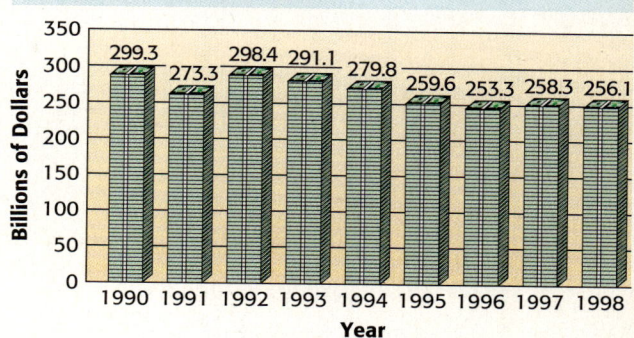

Source: U.S. Office of Management and Budget.

Match each property in Column I with the example of it in Column II.

I	II
24. Commutative property	**A.** $3x + 0 = 3x$
25. Associative property	**B.** $(5 + 2) + 8 = 8 + (5 + 2)$
26. Inverse property	**C.** $-3(x + y) = -3x + (-3y)$
27. Identity property	**D.** $-5 + (3 + 2) = (-5 + 3) + 2$
28. Distributive property	**E.** $-\dfrac{5}{3}\left(-\dfrac{3}{5}\right) = 1$

29. What property is used to show that $3(x + 1) = 3x + 3$?

30. Consider the expression $-6[5 + (-2)]$.
 (a) Evaluate it by first working within the brackets.
 (b) Evaluate it by using the distributive property.
 (c) Why must the answers in items (a) and (b) be the same?

Simplify by combining like terms.

31. $8x + 4x - 6x + x + 14x$

32. $5(2x - 1) - (x - 12) + 2(3x - 5)$

Linear Equations and Inequalities in One Variable

2

2.1 The Addition Property of Equality

2.2 The Multiplication Property of Equality

2.3 More on Solving Linear Equations

Summary Exercises on Solving Linear Equations

2.4 An Introduction to Applications of Linear Equations

2.5 Formulas and Applications from Geometry

2.6 Ratios and Proportions

2.7 More about Problem Solving

2.8 Solving Linear Inequalities

From a gathering of 13 nations in Athens in 1896, the Olympic Games have truly become a worldwide event. The XIX Olympic Winter Games in Salt Lake City in 2002 attracted athletes from 82 countries and entertained 3.5 billion viewers around the world.

One ceremonial aspect of the games is the flying of the Olympic flag with its five interlocking rings of different colors on a white background. First introduced at the 1920 Games in Antwerp, Belgium, the five rings on the flag symbolize unity among the nations of Africa, the Americas, Asia, Australia, and Europe. (*Source: Microsoft Encarta Encyclopedia 2002;* www.olympic-usa.org) Throughout this chapter we use linear equations to solve applications about the Olympics.

2.1 The Addition Property of Equality

OBJECTIVES

1. Identify linear equations.
2. Use the addition property of equality.
3. Simplify equations, and then use the addition property of equality.

Recall from Section 1.3 that an *equation* is a statement that two algebraic expressions are equal. The simplest type of equation is a *linear equation*.

OBJECTIVE 1 Identify linear equations.

Linear Equation in One Variable

A **linear equation in one variable** can be written in the form

$$Ax + B = C$$

for real numbers A, B, and C, with $A \neq 0$.

For example,

$$4x + 9 = 0, \quad 2x - 3 = 5, \quad \text{and} \quad x = 7$$

are linear equations in one variable (x). The final two can be written in the specified form using properties developed in this chapter. However,

$$x^2 + 2x = 5, \quad \frac{1}{x} = 6, \quad \text{and} \quad |2x + 6| = 0$$

are *not* linear equations.

As we saw in Section 1.3, a *solution* of an equation is a number that makes the equation true when it replaces the variable. An equation is solved by finding its **solution set**, the set of all solutions. Equations that have exactly the same solution sets are **equivalent equations.** Linear equations are solved by using a series of steps to produce a simpler equivalent equation of the form

$$x = \text{a number} \quad \text{or} \quad \text{a number} = x.$$

OBJECTIVE 2 Use the addition property of equality. In the equation $x - 5 = 2$, both $x - 5$ and 2 represent the same number because this is the meaning of the equals sign. To solve the equation, we change the left side from $x - 5$ to just x. We do this by adding 5 to $x - 5$. We use 5 because 5 is the opposite (additive inverse) of -5, and $-5 + 5 = 0$. To keep the two sides equal, we must also add 5 to the right side.

$x - 5 = 2$	Given equation
$x - 5 + 5 = 2 + 5$	Add 5 to each side.
$x + 0 = 7$	Additive inverse property
$x = 7$	Additive identity property

The solution of the given equation is 7. We check by replacing x with 7 in the original equation.

Check:
$x - 5 = 2$	Original equation
$7 - 5 = 2$?	Let $x = 7$.
$2 = 2$	True

Since the final equation is true, 7 checks as the solution and {7} is the solution set.

To solve the equation, we added the same number to each side. The **addition property of equality** justifies this step.

> **Addition Property of Equality**
>
> If A, B, and C are real numbers, then the equations
>
> $$A = B \quad \text{and} \quad A + C = B + C$$
>
> are equivalent equations.
>
> That is, we can add the same number to each side of an equation without changing the solution.

In the addition property, C represents a real number. This means that any quantity that represents a real number can be added to each side of an equation to change it to an equivalent equation.

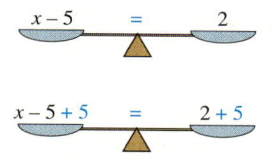

FIGURE 1

> **NOTE** Equations can be thought of in terms of a balance. Thus, adding the same quantity to each side does not affect the balance. See Figure 1.

EXAMPLE 1 Using the Addition Property of Equality

Solve $x - 16 = 7$.

Our goal is to get an equivalent equation of the form $x =$ a number. To do this, we use the addition property of equality and add 16 to each side.

$$x - 16 = 7$$
$$x - 16 + 16 = 7 + 16 \qquad \text{Add 16 to each side.}$$
$$x = 23 \qquad \text{Combine like terms.}$$

Note that we combined the steps that change $x - 16 + 16$ to $x + 0$ and $x + 0$ to x. We will combine these steps from now on. We check by substituting 23 for x in the *original* equation.

Check:
$$x - 16 = 7 \qquad \text{Original equation}$$
$$23 - 16 = 7 \quad ? \qquad \text{Let } x = 23.$$
$$7 = 7 \qquad \text{True}$$

Since a true statement results, $\{23\}$ is the solution set.

Now Try Exercise 7.

EXAMPLE 2 Using the Addition Property of Equality

Solve $x - 2.9 = -6.4$.

To get x alone on the left side, we must eliminate the -2.9. We use the addition property of equality and add 2.9 to each side.

$$x - 2.9 = -6.4$$
$$x - 2.9 + 2.9 = -6.4 + 2.9 \qquad \text{Add 2.9 to each side.}$$
$$x = -3.5$$

Check:
$$x - 2.9 = -6.4 \quad \text{Original equation}$$
$$-3.5 - 2.9 = -6.4 \quad ? \quad \text{Let } x = -3.5.$$
$$-6.4 = -6.4 \quad \text{True}$$

Since a true statement results, the solution set is $\{-3.5\}$.

Now Try Exercise 21.

The addition property of equality says that the same number may be *added* to each side of an equation. In Chapter 1, subtraction was defined as addition of the opposite. Thus, we can also use the following rule when solving an equation.

> The same number may be subtracted from each side of an equation without changing the solution.

EXAMPLE 3 Using the Addition Property of Equality

Solve $-7 = x + 22$.

Here the variable x is on the right side of the equation. To get x alone on the right, we must eliminate the 22 by subtracting 22 from each side.

$$-7 = x + 22$$
$$-7 - 22 = x + 22 - 22 \quad \text{Subtract 22 from each side.}$$
$$-29 = x \quad \text{or} \quad x = -29$$

Check:
$$-7 = x + 22 \quad \text{Original equation}$$
$$-7 = -29 + 22 \quad ? \quad \text{Let } x = -29.$$
$$-7 = -7 \quad \text{True}$$

Thus, the solution set is $\{-29\}$.

Now Try Exercise 17.

EXAMPLE 4 Subtracting a Variable Expression

Solve $\frac{3}{5}k + 17 = \frac{8}{5}k$.

To get all terms with variables on the same side of the equation, subtract $\frac{3}{5}k$ from each side.

$$\frac{3}{5}k + 17 = \frac{8}{5}k$$
$$\frac{3}{5}k + 17 - \frac{3}{5}k = \frac{8}{5}k - \frac{3}{5}k \quad \text{Subtract } \frac{3}{5}k \text{ from each side.}$$
$$17 = 1k \quad \text{Combine like terms; } \frac{5}{5}k = 1k.$$
$$17 = k \quad \text{Multiplicative identity property}$$

(From now on we will skip the step that changes $1k$ to k.) Check the solution by replacing k with 17 in the original equation. The solution set is $\{17\}$.

Now Try Exercise 25.

What happens if we solve the equation in Example 4 by first subtracting $\frac{8}{5}k$ from each side?

$$\frac{3}{5}k + 17 = \frac{8}{5}k$$

$$\frac{3}{5}k + 17 - \frac{8}{5}k = \frac{8}{5}k - \frac{8}{5}k \qquad \text{Subtract } \tfrac{8}{5}k \text{ from each side.}$$

$$17 - k = 0 \qquad \text{Combine like terms; } -\tfrac{5}{5}k = -1k = -k.$$

$$17 - k - 17 = 0 - 17 \qquad \text{Subtract 17 from each side.}$$

$$-k = -17 \qquad \text{Combine like terms; additive inverse}$$

This result gives the value of $-k$, but not of k itself. However, it does say that the additive inverse of k is -17, which means that k must be 17, the same result we obtained in Example 4.

$$-k = -17$$
$$k = 17$$

(This result can also be justified using the multiplication property of equality, covered in Section 2.2.)

OBJECTIVE 3 Simplify equations, and then use the addition property of equality. Sometimes an equation must be simplified as a first step in its solution.

EXAMPLE 5 Simplifying an Equation before Solving

Solve $3k - 12 + k + 2 = 5 + 3k + 2$.

Begin by combining like terms on each side of the equation to get

$$4k - 10 = 7 + 3k.$$

Next, get all terms that contain variables on the same side of the equation and all terms without variables on the other side. One way to start is to subtract $3k$ from each side.

$$4k - 10 - 3k = 7 + 3k - 3k \qquad \text{Subtract } 3k \text{ from each side.}$$
$$k - 10 = 7 \qquad \text{Combine like terms.}$$
$$k - 10 + 10 = 7 + 10 \qquad \text{Add 10 to each side.}$$
$$k = 17 \qquad \text{Combine like terms.}$$

Check: Substitute 17 for k in the original equation.

$$3k - 12 + k + 2 = 5 + 3k + 2 \qquad \text{Original equation}$$
$$3(17) - 12 + 17 + 2 = 5 + 3(17) + 2 \quad ? \quad \text{Let } k = 17.$$
$$51 - 12 + 17 + 2 = 5 + 51 + 2 \quad ? \quad \text{Multiply.}$$
$$58 = 58 \qquad \text{True}$$

The check results in a true statement, so the solution set is $\{17\}$.

Now Try Exercise 41.

98 CHAPTER 2 Linear Equations and Inequalities in One Variable

> **EXAMPLE 6** Using the Distributive Property to Simplify an Equation
> Solve $3(2 + 5x) - (1 + 14x) = 6$.
>
> $$3(2 + 5x) - (1 + 14x) = 6$$
> $$3(2 + 5x) - 1(1 + 14x) = 6 \quad -(1 + 14x) = -1(1 + 14x)$$
> $$3(2) + 3(5x) - 1(1) - 1(14x) = 6 \quad \text{Distributive property}$$
> $$6 + 15x - 1 - 14x = 6 \quad \text{Multiply.}$$
> $$x + 5 = 6 \quad \text{Combine like terms.}$$
> $$x + 5 - 5 = 6 - 5 \quad \text{Subtract 5 from each side.}$$
> $$x = 1 \quad \text{Combine like terms.}$$
>
> Check by substituting 1 for x in the original equation. The solution set is $\{1\}$.
>
> **Now Try Exercise 49.**

> **CAUTION** Be careful to apply the distributive property correctly in a problem like that in Example 6, or a sign error may result.

2.1 EXERCISES

For Extra Help

 Student's Solutions Manual

 MyMathLab

 InterAct Math Tutorial Software

 AW Math Tutor Center

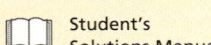 MathXL

 Digital Video Tutor CD 3/Videotape 3

1. Which pairs of equations are equivalent equations?
   A. $x + 2 = 6$ and $x = 4$ **B.** $10 - x = 5$ and $x = -5$
   C. $x + 3 = 9$ and $x = 6$ **D.** $4 + x = 8$ and $x = -4$

2. Decide whether each of the following is an expression or an equation. If it is an expression, simplify it. If it is an equation, solve it.
   (a) $5x + 8 - 4x + 7$ **(b)** $-6y + 12 + 7y - 5$
   (c) $5x + 8 - 4x = 7$ **(d)** $-6y + 12 + 7y = -5$

3. In your own words, state the addition property of equality. Give an example.

4. Explain how to check a solution of an equation.

Solve each equation, and check your solution. See Examples 1–4.

5. $x - 4 = 8$ 6. $x - 8 = 9$ 7. $y - 12 = 19$
8. $t - 15 = 25$ 9. $x - 5 = -8$ 10. $x - 7 = -9$
11. $r + 9 = 13$ 12. $y + 6 = 10$ 13. $x + 26 = 17$
14. $x + 45 = 24$ 15. $7 + r = -3$ 16. $8 + k = -4$
17. $2 = p + 15$ 18. $3 = z + 17$ 19. $-2 = x - 12$
20. $-6 = x - 21$ 21. $x - 8.4 = -2.1$ 22. $y - 15.5 = -5.1$
23. $t + 12.3 = -4.6$ 24. $x + 21.5 = -13.4$ 25. $\frac{2}{5}w - 6 = \frac{7}{5}w$
26. $-\frac{2}{7}z + 2 = \frac{5}{7}z$ 27. $5.6x + 2 = 4.6x$ 28. $9.1x - 5 = 8.1x$

SECTION 2.1 The Addition Property of Equality   99

29. $3p + 6 = 10 + 2p$ 30. $8b - 4 = -6 + 7b$ 31. $1.2y - 4 = .2y - 4$

32. $7.7r + 6 = 6.7r + 6$ 33. $\frac{1}{2}x + 2 = -\frac{1}{2}x$ 34. $\frac{1}{5}x - 7 = -\frac{4}{5}x$

35. $3x + 7 - 2x = 0$ 36. $5x + 4 - 4x = 0$

37. Which of the following are *not* linear equations in one variable?
    A. $x^2 - 5x + 6 = 0$ B. $x^3 = x$
    C. $3x - 4 = 0$ D. $7x - 6x = 3 + 9x$

Define a linear equation in one variable in words.

38. Refer to the definition of linear equation in one variable given in this section. Why is the restriction $A \neq 0$ necessary?

Solve each equation, and check your solution. See Examples 5 and 6.

39. $5t + 3 + 2t - 6t = 4 + 12$ 40. $4x + 3x - 6 - 6x = 10 + 3$

41. $6x + 5 + 7x + 3 = 12x + 4$ 42. $4x - 3 - 8x + 1 = -5x + 9$

43. $5.2q - 4.6 - 7.1q = -.9q - 4.6$ 44. $-4.0x + 2.7 - 1.6x = -4.6x + 2.7$

45. $\frac{5}{7}x + \frac{1}{3} = \frac{2}{5} - \frac{2}{7}x + \frac{2}{5}$ 46. $\frac{6}{7}s - \frac{3}{4} = \frac{4}{5} - \frac{1}{7}s + \frac{1}{6}$

47. $(5y + 6) - (3 + 4y) = 10$ 48. $(8r - 3) - (7r + 1) = -6$

49. $2(p + 5) - (9 + p) = -3$ 50. $4(k - 6) - (3k + 2) = -5$

51. $-6(2b + 1) + (13b - 7) = 0$ 52. $-5(3w - 3) + (1 + 16w) = 0$

53. $10(-2x + 1) = -19(x + 1)$ 54. $2(2 - 3r) = -5(r - 3)$

55. $-2(8p + 2) - 3(2 - 7p) - 2(4 + 2p) = 0$

56. $-5(1 - 2z) + 4(3 - z) - 7(3 + z) = 0$

57. $4(7x - 1) + 3(2 - 5x) - 4(3x + 5) = -6$

58. $9(2m - 3) - 4(5 + 3m) - 5(4 + m) = -3$

59. Write an equation that requires the use of the addition property of equality, where 6 must be added to each side to solve the equation, and the solution is a negative number.

60. Write an equation that requires the use of the addition property of equality, where $\frac{1}{2}$ must be subtracted from each side, and the solution is a positive number.

Write an equation using the information given in the problem. Use x as the variable. Then solve the equation.

61. Three times a number is 17 more than twice the number. Find the number.

62. One, added to three times a number, is three less than four times the number. Find the number.

63. If six times a number is subtracted from seven times the number, the result is -9. Find the number.

64. If five times a number is added to three times the number, the result is the sum of seven times the number and 9. Find the number.

2.2 The Multiplication Property of Equality

Solve (handwritten annotation)

OBJECTIVES

1. Use the multiplication property of equality.
2. Simplify equations, and then use the multiplication property of equality.

OBJECTIVE 1 Use the multiplication property of equality. The addition property of equality alone is not enough to solve some equations, such as $3x + 2 = 17$.

$$3x + 2 = 17$$
$$3x + 2 - 2 = 17 - 2 \qquad \text{Subtract 2 from each side.}$$
$$3x = 15 \qquad \text{Combine like terms.}$$

Notice that the coefficient of x on the left side is 3, not 1 as desired. Another property is needed to change $3x = 15$ to an equation of the form

$$x = \text{a number.}$$

If $3x = 15$, then $3x$ and 15 both represent the same number. Multiplying both $3x$ and 15 by the same number will also result in an equality. The **multiplication property of equality** states that we can multiply each side of an equation by the same nonzero number without changing the solution.

Multiplication Property of Equality

If A, B, and C ($C \neq 0$) represent real numbers, then the equations

$$A = B \qquad \text{and} \qquad AC = BC$$

are equivalent equations.

That is, we can multiply each side of an equation by the same nonzero number without changing the solution.

This property can be used to solve $3x = 15$. The $3x$ on the left must be changed to $1x$, or x, instead of $3x$. To isolate x, we multiply each side of the equation by $\frac{1}{3}$. We use $\frac{1}{3}$ because $\frac{1}{3}$ is the reciprocal of 3, and $\frac{1}{3} \cdot 3 = \frac{3}{3} = 1$.

$$3x = 15$$
$$\frac{1}{3}(3x) = \frac{1}{3} \cdot 15 \qquad \text{Multiply each side by } \tfrac{1}{3}.$$
$$\left(\frac{1}{3} \cdot 3\right)x = \frac{1}{3} \cdot 15 \qquad \text{Associative property}$$
$$1x = 5 \qquad \text{Multiplicative inverse property}$$
$$x = 5 \qquad \text{Multiplicative identity property}$$

We check by substituting 5 for x in the original equation. The solution set of the equation is $\{5\}$. From now on, we will combine the last two steps shown in the preceding example.

Just as the addition property of equality permits *subtracting* the same number from each side of an equation, the multiplication property of equality permits *dividing* each side of an equation by the same nonzero number. For example, the equation $3x = 15$, which we just solved by multiplying each side by $\frac{1}{3}$, could also be solved by dividing each side by 3.

$$3x = 15$$

$$\frac{3x}{3} = \frac{15}{3} \quad \text{Divide each side by 3.}$$

$$x = 5$$

We can divide each side of an equation by the same nonzero number without changing the solution. Do not, however, divide each side by a variable, as that may result in losing a valid solution.

> **NOTE** In practice, it is usually easier to multiply on each side if the coefficient of the variable is a fraction, and divide on each side if the coefficient is an integer. For example, to solve
>
> $$-\frac{3}{4}x = 12,$$
>
> it is easier to multiply by $-\frac{4}{3}$, the reciprocal of $-\frac{3}{4}$, than to divide by $-\frac{3}{4}$. On the other hand, to solve
>
> $$-5x = -20,$$
>
> it is easier to divide by -5 than to multiply by $-\frac{1}{5}$.

EXAMPLE 1 Dividing Each Side of an Equation by a Nonzero Number

Solve $25p = 30$.

Transform the equation so that p (instead of $25p$) is on the left by using the multiplication property of equality. Divide each side of the equation by 25, the coefficient of p.

$$25p = 30$$

$$\frac{25p}{25} = \frac{30}{25} \quad \text{Divide each side by 25.}$$

$$p = \frac{30}{25} = \frac{6}{5} \quad \text{Write in lowest terms.}$$

To check, substitute $\frac{6}{5}$ for p in the original equation.

Check:
$$25p = 30$$

$$\frac{25}{1}\left(\frac{6}{5}\right) = 30 \quad ? \quad \text{Let } p = \frac{6}{5}.$$

$$30 = 30 \quad \text{True}$$

The solution set is $\left\{\frac{6}{5}\right\}$.

Now Try Exercise 25.

CHAPTER 2 Linear Equations and Inequalities in One Variable

EXAMPLE 2 Solving an Equation with Decimals

Solve $-2.1x = 6.09$.

$$-2.1x = 6.09$$

$$\frac{-2.1x}{-2.1} = \frac{6.09}{-2.1} \qquad \text{Divide each side by } -2.1.$$

A calculator will simplify the work at this point.

$$x = -2.9 \qquad \text{Divide.}$$

Check:
$$-2.1x = 6.09 \qquad \text{Original equation}$$
$$-2.1(-2.9) = 6.09 \quad ? \quad \text{Let } x = -2.9.$$
$$6.09 = 6.09 \qquad \text{True}$$

The solution set is $\{-2.9\}$.

Now Try Exercise 39.

In the next two examples, multiplication produces the solution more quickly than division would.

EXAMPLE 3 Using the Multiplication Property of Equality

Solve $\dfrac{a}{4} = 3$.

Replace $\frac{a}{4}$ by $\frac{1}{4}a$, since dividing by 4 is the same as multiplying by $\frac{1}{4}$. To get a alone on the left, multiply each side by 4, the reciprocal of the coefficient of a.

$$\frac{a}{4} = 3$$

$$\frac{1}{4}a = 3 \qquad \text{Change } \tfrac{a}{4} \text{ to } \tfrac{1}{4}a.$$

$$4 \cdot \frac{1}{4}a = 4 \cdot 3 \qquad \text{Multiply by 4.}$$

$$a = 12 \qquad \text{Multiplicative inverse property; multiplicative identity property}$$

Check that 12 is the solution.

Check:
$$\frac{a}{4} = 3 \qquad \text{Original equation}$$
$$\frac{12}{4} = 3 \quad ? \quad \text{Let } a = 12.$$
$$3 = 3 \qquad \text{True}$$

The solution set is $\{12\}$.

Now Try Exercise 43.

EXAMPLE 4 Using the Multiplication Property of Equality

Solve $\frac{3}{4}h = 6$.

To transform the equation so that h is alone on the left, multiply each side of the equation by $\frac{4}{3}$. Use $\frac{4}{3}$ because $\frac{4}{3} \cdot \frac{3}{4} h = 1 \cdot h = h$.

$$\frac{3}{4}h = 6$$

$$\frac{4}{3}\left(\frac{3}{4}h\right) = \frac{4}{3} \cdot 6 \qquad \text{Multiply by } \tfrac{4}{3}.$$

$$1 \cdot h = \frac{4}{3} \cdot \frac{6}{1} \qquad \text{Multiplicative inverse property}$$

$$h = 8 \qquad \text{Multiplicative identity property; multiply fractions.}$$

Check by substituting 8 for h in the original equation. The solution set is $\{8\}$.

Now Try Exercise 47.

In Section 2.1, we obtained the equation $-k = -17$ in our alternate solution to Example 4. We reasoned that since this equation says that the additive inverse (or opposite) of k is -17, then k must equal 17. We can also use the multiplication property of equality to obtain the same result, as shown in the next example.

EXAMPLE 5 Using the Multiplication Property of Equality When the Coefficient of the Variable Is −1

Solve $-k = -17$.

On the left side, change $-k$ to k by first writing $-k$ as $-1 \cdot k$.

$$-k = -17$$
$$-1 \cdot k = -17 \qquad -k = -1 \cdot k$$
$$-1(-1 \cdot k) = -1(-17) \qquad \text{Multiply by } -1, \text{ since } -1(-1) = 1.$$
$$[-1(-1)] \cdot k = 17 \qquad \text{Associative property; multiply.}$$
$$1 \cdot k = 17 \qquad \text{Multiplicative inverse property}$$
$$k = 17 \qquad \text{Multiplicative identity property}$$

Check: $\quad -k = -17 \qquad$ Original equation
$\quad -(17) = -17 \quad ? \quad$ Let $k = 17$.
$\quad -17 = -17 \qquad$ True

The solution, 17, checks, so $\{17\}$ is the solution set.

Now Try Exercise 53.

From Example 5, we see that the following is true.

$$\text{If} \quad -x = a, \quad \text{then} \quad x = -a.$$

104 CHAPTER 2 Linear Equations and Inequalities in One Variable

OBJECTIVE 2 Simplify equations, and then use the multiplication property of equality. In the next example, it is necessary to simplify the equation before using the multiplication property of equality.

EXAMPLE 6 Simplifying an Equation

Solve $5m + 6m = 33$.

$$5m + 6m = 33$$
$$11m = 33 \quad \text{Combine like terms.}$$
$$\frac{11m}{11} = \frac{33}{11} \quad \text{Divide by 11.}$$
$$1m = 3 \quad \text{Divide.}$$
$$m = 3 \quad \text{Multiplicative identity property}$$

Check this proposed solution. The solution set is $\{3\}$.

Now Try Exercise 57.

CONNECTIONS

The use of algebra to solve equations and applied problems is very old. The 3600-year-old Rhind Papyrus includes the following "word problem." "Aha, its whole, its seventh, it makes 19." This brief sentence describes the equation

$$x + \frac{x}{7} = 19.$$

The solution of this equation is $16\frac{5}{8}$. The word *algebra* is from the work *Hisab al-jabr m'al muquabalah*, written in the ninth century by Muhammed ibn Musa Al-Khowarizmi. The title means "the science of transposition and cancellation." From Latin versions of Khowarizmi's text, "al-jabr" became the broad term covering the art of equation solving.

2.2 EXERCISES

For Extra Help

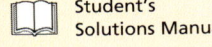
Student's Solutions Manual

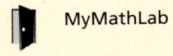
MyMathLab

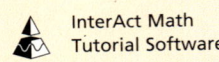
InterAct Math Tutorial Software

AW Math Tutor Center

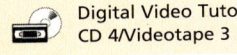
MathXL

Digital Video Tutor CD 4/Videotape 3

1. In your own words, state the multiplication property of equality. Give an example.

2. In the statement of the multiplication property of equality in this section, there is a restriction that $C \neq 0$. What would happen if you multiplied each side of an equation by 0?

3. Which equation does *not* require the use of the multiplication property of equality?

    A. $3x - 5x = 6$ **B.** $-\frac{1}{4}x = 12$ **C.** $5x - 4x = 7$ **D.** $\frac{x}{3} = -2$

4. Tell whether you would use the addition or multiplication property of equality to solve each equation. Explain your answer.

    (a) $3x = 12$ **(b)** $3 + x = 12$

SECTION 2.2 The Multiplication Property of Equality   105

5. A student tried to solve the equation $4x = 8$ by dividing each side by 8. Why is this wrong?

6. State how you would find the solution of a linear equation if your next-to-last step reads "$-x = 5$."

By what number is it necessary to multiply both sides of each equation to get just x on the left side? Do not actually solve these equations.

7. $\frac{2}{3}x = 8$
8. $\frac{4}{5}x = 6$
9. $\frac{x}{10} = 3$
10. $\frac{x}{100} = 8$

11. $-\frac{9}{2}x = -4$
12. $-\frac{8}{3}x = -11$
13. $-x = .36$
14. $-x = .29$

By what number is it necessary to divide both sides of each equation to get just x on the left side? Do not actually solve these equations.

15. $6x = 5$
16. $7x = 10$
17. $-4x = 13$
18. $-13x = 6$
19. $12x = 48$
20. $.21x = 63$
21. $-x = 23$
22. $-x = 49$

Solve each equation, and check your solution. See Examples 1–5.

23. $5x = 30$
24. $7x = 56$
25. $2m = 15$
26. $3m = 10$
27. $3a = -15$
28. $5k = -70$
29. $-3x = 12$
30. $-4x = 36$
31. $10t = -36$
32. $4s = -34$
33. $-6x = -72$
34. $-8x = -64$
35. $2r = 0$
36. $5x = 0$
37. $.2t = 8$
38. $.9x = 18$
39. $-2.1m = 25.62$
40. $-3.9a = 31.2$
41. $\frac{1}{4}y = -12$
42. $\frac{1}{5}p = -3$
43. $\frac{z}{6} = 12$
44. $\frac{x}{5} = 15$
45. $\frac{x}{7} = -5$
46. $\frac{k}{8} = -3$
47. $\frac{2}{7}p = 4$
48. $\frac{3}{8}y = 9$
49. $-\frac{5}{6}t = -15$
50. $-\frac{3}{4}k = -21$
51. $-\frac{7}{9}c = \frac{3}{5}$
52. $-\frac{5}{6}d = \frac{4}{9}$
53. $-y = 12$
54. $-t = 14$
55. $-x = -\frac{3}{4}$
56. $-y = -\frac{1}{2}$

Solve each equation, and check your solution. See Example 6.

57. $4x + 3x = 21$
58. $9x + 2x = 121$
59. $3r - 5r = 10$
60. $9p - 13p = 24$
61. $5m + 6m - 2m = 63$
62. $11r - 5r + 6r = 168$
63. $-6x + 4x - 7x = 0$
64. $-5x + 4x - 8x = 0$
65. $9w - 5w + w = -3$
66. $10y - 6y + 3y = -4$

67. Write an equation that requires the use of the multiplication property of equality, where each side must be multiplied by $\frac{2}{3}$, and the solution is a negative number.

68. Write an equation that requires the use of the multiplication property of equality, where each side must be divided by 100, and the solution is not an integer.

Write an equation using the information given in the problem. Use x as the variable. Then solve the equation.

69. When a number is multiplied by 4, the result is 6. Find the number.
70. When a number is multiplied by −4, the result is 10. Find the number.
71. When a number is divided by −5, the result is 2. Find the number.
72. If twice a number is divided by 5, the result is 4. Find the number.

2.3 More on Solving Linear Equations

OBJECTIVES

1. Learn and use the four steps for solving a linear equation.
2. Solve equations with fractions or decimals as coefficients.
3. Solve equations with no solution or infinitely many solutions.
4. Write expressions for two related unknown quantities.

OBJECTIVE 1 Learn and use the four steps for solving a linear equation. In this section, we use the addition and multiplication properties together to solve more complicated equations. We use the following four-step method.

Solving a Linear Equation

Step 1 **Simplify each side separately.** Clear parentheses using the distributive property, if needed, and combine like terms.

Step 2 **Isolate the variable term on one side.** Use the addition property if necessary so that the variable term is on one side of the equation and a number is on the other.

Step 3 **Isolate the variable.** Use the multiplication property if necessary to get the equation in the form $x = $ a number.

Step 4 **Check.** Substitute the proposed solution into the *original* equation to see if a true statement results.

EXAMPLE 1 Using the Four Steps to Solve an Equation

Solve $3r + 4 - 2r - 7 = 4r + 3$.

Step 1 $3r + 4 - 2r - 7 = 4r + 3$
               $r - 3 = 4r + 3$ Combine like terms.

Step 2 $r - 3 + 3 = 4r + 3 + 3$ Add 3.
               $r = 4r + 6$ Combine like terms.
               $r - 4r = 4r + 6 - 4r$ Subtract 4r.
               $-3r = 6$ Combine like terms.

Step 3 $\dfrac{-3r}{-3} = \dfrac{6}{-3}$ Divide by −3.
               $r = -2$ $\dfrac{-3}{-3} = 1; 1r = r$

Step 4 Substitute -2 for r in the original equation to check.

$$3r + 4 - 2r - 7 = 4r + 3 \quad \text{Original equation}$$
$$3(-2) + 4 - 2(-2) - 7 = 4(-2) + 3 \quad ? \quad \text{Let } r = -2.$$
$$-6 + 4 + 4 - 7 = -8 + 3 \quad ? \quad \text{Multiply.}$$
$$-5 = -5 \quad \text{True}$$

The solution set of the equation is $\{-2\}$.

Now Try Exercise 7.

> **NOTE** In Step 2 of Example 1, we added and subtracted the terms in such a way that the variable term ended up on the left side of the equation. Choosing differently would have put the variable term on the right side of the equation. Either way, the same solution results.

EXAMPLE 2 Using the Four Steps to Solve an Equation

Solve $4(k - 3) - k = k - 6$.

Step 1 Before combining like terms, use the distributive property to simplify $4(k - 3)$.

$$4(k - 3) - k = k - 6$$
$$4(k) + 4(-3) - k = k - 6 \quad \text{Distributive property}$$
$$4k - 12 - k = k - 6$$
$$3k - 12 = k - 6 \quad \text{Combine like terms.}$$

Step 2
$$3k - 12 - k = k - 6 - k \quad \text{Subtract } k.$$
$$2k - 12 = -6 \quad \text{Combine like terms.}$$
$$2k - 12 + 12 = -6 + 12 \quad \text{Add 12.}$$
$$2k = 6 \quad \text{Combine like terms.}$$

Step 3
$$\frac{2k}{2} = \frac{6}{2} \quad \text{Divide by 2.}$$
$$k = 3$$

Step 4 Check by substituting 3 for k in the original equation. Work inside the parentheses first.

$$4(k - 3) - k = k - 6 \quad \text{Original equation}$$
$$4(3 - 3) - 3 = 3 - 6 \quad ? \quad \text{Let } k = 3.$$
$$4(0) - 3 = 3 - 6 \quad ?$$
$$0 - 3 = 3 - 6 \quad ?$$
$$-3 = -3 \quad \text{True}$$

The solution set of the equation is $\{3\}$.

Now Try Exercise 11.

EXAMPLE 3 Using the Four Steps to Solve an Equation

Solve $8a - (3 + 2a) = 3a + 1$.

Step 1 Clear parentheses using the distributive property.

$8a - (3 + 2a) = 3a + 1$

$8a - 1(3 + 2a) = 3a + 1$ Multiplicative identity property

$8a - 3 - 2a = 3a + 1$ Distributive property

$6a - 3 = 3a + 1$

Step 2 $6a - 3 - 3a = 3a + 1 - 3a$ Subtract 3a.

$3a - 3 = 1$

$3a - 3 + 3 = 1 + 3$ Add 3.

$3a = 4$

Step 3 $\dfrac{3a}{3} = \dfrac{4}{3}$ Divide by 3.

$a = \dfrac{4}{3}$

Step 4 Check this solution in the original equation.

$8a - (3 + 2a) = 3a + 1$ Original equation

$8\left(\dfrac{4}{3}\right) - \left[3 + 2\left(\dfrac{4}{3}\right)\right] = 3\left(\dfrac{4}{3}\right) + 1$?   Let $a = \dfrac{4}{3}$.

$\dfrac{32}{3} - \left[3 + \dfrac{8}{3}\right] = 4 + 1$?

$\dfrac{32}{3} - \left[\dfrac{9}{3} + \dfrac{8}{3}\right] = 5$?

$\dfrac{32}{3} - \dfrac{17}{3} = 5$?

$5 = 5$ True

The check shows that $\left\{\dfrac{4}{3}\right\}$ is the solution set.

Now Try Exercise 9.

CAUTION Be very careful with signs when solving an equation like the one in Example 3. When clearing parentheses in the expression

$8 - (3 + 2a),$

remember that the − sign acts like a factor of −1 and affects the sign of *every* term within the parentheses. Thus,

$8 - (3 + 2a) = 8 + (-1)(3 + 2a)$

$= 8 - 3 - 2a.$

   ↑ ↑

Change to − in *both* terms.

EXAMPLE 4 Using the Four Steps to Solve an Equation

Solve $4(8 - 3t) = 32 - 8(t + 2)$.

Step 1
$$4(8 - 3t) = 32 - 8(t + 2)$$
$$32 - 12t = 32 - 8t - 16 \quad \text{Distributive property}$$
$$32 - 12t = 16 - 8t$$

Step 2
$$32 - 12t + 12t = 16 - 8t + 12t \quad \text{Add } 12t.$$
$$32 = 16 + 4t$$
$$32 - 16 = 16 + 4t - 16 \quad \text{Subtract 16.}$$
$$16 = 4t$$

Step 3
$$\frac{16}{4} = \frac{4t}{4} \quad \text{Divide by 4.}$$
$$4 = t \quad \text{or} \quad t = 4$$

Step 4 Check:
$$4(8 - 3t) = 32 - 8(t + 2) \quad \text{Original equation}$$
$$4(8 - 3 \cdot 4) = 32 - 8(4 + 2) \quad ? \quad \text{Let } t = 4.$$
$$4(8 - 12) = 32 - 8(6) \quad ?$$
$$4(-4) = 32 - 48 \quad ?$$
$$-16 = -16 \quad \text{True}$$

Since a true statement results, the solution set is {4}.

Now Try Exercise 13.

OBJECTIVE 2 Solve equations with fractions or decimals as coefficients. We can clear an equation of fractions by multiplying each side by the least common denominator (LCD) of all the fractions in the equation. It is a good idea to do this *before* starting the four-step method to avoid working with fractions.

EXAMPLE 5 Solving an Equation with Fractions as Coefficients

Solve $\frac{2}{3}x - \frac{1}{2}x = -\frac{1}{6}x - 2$.

The LCD of all the fractions in the equation is 6, so multiply each side by 6.

$$\frac{2}{3}x - \frac{1}{2}x = -\frac{1}{6}x - 2$$

$$6\left(\frac{2}{3}x - \frac{1}{2}x\right) = 6\left(-\frac{1}{6}x - 2\right) \quad \text{Multiply by 6.}$$

$$6\left(\frac{2}{3}x\right) + 6\left(-\frac{1}{2}x\right) = 6\left(-\frac{1}{6}x\right) + 6(-2) \quad \text{Distributive property}$$

$$4x - 3x = -x - 12$$

Now use the four steps to solve this equivalent equation.

Step 1 $\qquad\qquad\qquad x = -x - 12 \qquad$ Combine like terms.

110 CHAPTER 2 Linear Equations and Inequalities in One Variable

Step 2 $x + x = -x - 12 + x$ Add x.
$$2x = -12$$ Combine like terms.

Step 3 $\dfrac{2x}{2} = \dfrac{-12}{2}$ Divide by 2.
$$x = -6$$

Step 4   Check:

$$\dfrac{2}{3}x - \dfrac{1}{2}x = -\dfrac{1}{6}x - 2$$ Original equation

$$\dfrac{2}{3}(-6) - \dfrac{1}{2}(-6) = -\dfrac{1}{6}(-6) - 2 \quad ?$$ Let $x = -6$.

$$-4 + 3 = 1 - 2 \quad ?$$

$$-1 = -1$$ True

The solution set of the equation is $\{-6\}$.

Now Try Exercise 23.

> **CAUTION**   When clearing an equation of fractions, be sure to multiply *every* term on each side of the equation by the LCD.

The multiplication property is also used to clear an equation of decimals.

EXAMPLE 6   Solving an Equation with Decimals as Coefficients

Solve $.1t + .05(20 - t) = .09(20)$.

The decimals are expressed as tenths and hundredths. Choose the smallest exponent on 10 needed to eliminate the decimals; in this case, use $10^2 = 100$. A number can be multiplied by 100 by moving the decimal point two places to the right.

$$.10t + .05(20 - t) = .09(20) \quad\quad .1 = .10$$

$$10t + 5(20 - t) = 9(20) \quad\quad \text{Multiply by 100.}$$

Now use the four steps.

Step 1 $10t + 5(20) + 5(-t) = 180$ Distributive property
$$10t + 100 - 5t = 180$$
$$5t + 100 = 180$$ Combine like terms.

Step 2 $5t + 100 - 100 = 180 - 100$ Subtract 100.
$$5t = 80$$ Combine like terms.

Step 3 $\dfrac{5t}{5} = \dfrac{80}{5}$ Divide by 5.
$$t = 16$$

Step 4   Check that $\{16\}$ is the solution set by substituting 16 for t in the original equation.

Now Try Exercise 29.

SECTION 2.3 More on Solving Linear Equations 111

OBJECTIVE 3 Solve equations with no solution or infinitely many solutions. Each equation that we have solved so far has had exactly one solution. An equation with exactly one solution is a **conditional equation** because it is only true under certain conditions. As the next examples show, linear equations may also have no solution or infinitely many solutions. (The four steps are not identified in these examples. See if you can identify them.)

EXAMPLE 7 Solving an Equation That Has Infinitely Many Solutions

Solve $5x - 15 = 5(x - 3)$.

$$5x - 15 = 5(x - 3)$$
$$5x - 15 = 5x - 15 \qquad \text{Distributive property}$$
$$5x - 15 - 5x = 5x - 15 - 5x \qquad \text{Subtract } 5x.$$
$$-15 = -15$$
$$-15 + 15 = -15 + 15 \qquad \text{Add 15.}$$
$$0 = 0 \qquad \text{True}$$

The variable has "disappeared." Since the last statement ($0 = 0$) is true, *any* real number is a solution. (We could have predicted this from the line in the solution that says $5x - 15 = 5x - 15$, which is certainly true for *any* value of x.) An equation with both sides exactly the same, like $0 = 0$, is called an **identity**. An identity is true for all replacements of the variables. We indicate this by writing the solution set as {all real numbers}.

Now Try Exercise 17.

CAUTION When solving an equation like the one in Example 7, do not write {0} as the solution set. Although 0 is a solution, there are infinitely many other solutions.

EXAMPLE 8 Solving an Equation That Has No Solution

Solve $2x + 3(x + 1) = 5x + 4$.

$$2x + 3(x + 1) = 5x + 4$$
$$2x + 3x + 3 = 5x + 4 \qquad \text{Distributive property}$$
$$5x + 3 = 5x + 4$$
$$5x + 3 - 5x = 5x + 4 - 5x \qquad \text{Subtract } 5x.$$
$$3 = 4 \qquad \text{False}$$

Again, the variable has disappeared, but this time a false statement ($3 = 4$) results. Whenever this happens in solving an equation, it is a signal that the equation has no solution. An equation with no solution is called a **contradiction**. Its solution set is the **empty set**, or **null set**, symbolized $\emptyset$.

Now Try Exercise 21.

The table on the next page summarizes the solution sets of the three types of linear equations.

112 CHAPTER 2 Linear Equations and Inequalities in One Variable

Type of Linear Equation	Final Equation in Solution	Number of Solutions	Solution Set
Conditional	$x =$ a number	One	{a number}
Identity	A true statement with no variable, such as $0 = 0$	Infinite	{all real numbers}
Contradiction	A false statement with no variable, such as $3 = 4$	None	$\emptyset$

OBJECTIVE 4 Write expressions for two related unknown quantities. Next, we continue our work with translating from words to symbols.

> **PROBLEM SOLVING**
>
> Often we are given a problem in which the sum of two quantities is a particular number, and we are asked to find the values of the two quantities. Example 9 shows how to express the unknown quantities in terms of a single variable.

EXAMPLE 9 Translating a Phrase into an Algebraic Expression

Two numbers have a sum of 23. If one of the numbers is represented by k, find an expression for the other number.

First, suppose that the sum of two numbers is 23, and one of the numbers is 10. How would you find the other number? You would subtract 10 from 23 to get 13; $23 - 10 = 13$. So instead of using 10 as one of the numbers, use k as stated in the problem. The other number would be obtained in the same way. You must subtract k from 23. Therefore, an expression for the other number is $23 - k$.

Now Try Exercise 47.

NOTE The approach used in Example 9, first writing an expression with a trial number, is also useful when translating applied problems to equations.

2.3 EXERCISES

For Extra Help

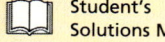 Student's Solutions Manual

 MyMathLab

 InterAct Math Tutorial Software

 AW Math Tutor Center

MathXL MathXL

 Digital Video Tutor CD 4/Videotape 3

1. In your own words, give the four steps used to solve a linear equation. Give an example to demonstrate the steps.

2. After working correctly through several steps of the solution of a linear equation, a student obtains the equation $7x = 3x$. Then the student divides each side by x to get $7 = 3$, and gives $\emptyset$ as the answer. Is this correct? If not, explain why.

3. Which one of the following linear equations does *not* have {all real numbers} as its solution set?

   A. $5x = 4x + x$ **B.** $2(x + 6) = 2x + 12$ **C.** $\frac{1}{2}x = .5x$ **D.** $3x = 2x$

4. Based on the discussion in this section, if an equation has decimals or fractions as coefficients, what additional step will make the work easier?

SECTION 2.3 More on Solving Linear Equations **113**

Solve each equation, and check your solution. See Examples 1–4, 7, and 8.

5. $3x + 8 = 5x + 10$
6. $10p + 6 = 12p - 4$
7. $12h - 5 = 11h + 5 - h$
8. $-4x - 1 = -5x + 1 + 3x$
9. $-2p + 7 = 3 - (5p + 1)$
10. $4x + 9 = 3 - (x - 2)$
11. $3(4x + 2) + 5x = 30 - x$
12. $5(2m + 3) - 4m = 8m + 27$
13. $6(3w + 5) = 2(10w + 10)$
14. $4(2x - 1) = -6(x + 3)$
15. $6(4x - 1) = 12(2x + 3)$
16. $6(2x + 8) = 4(3x - 6)$
17. $3(2x - 4) = 6(x - 2)$
18. $3(6 - 4x) = 2(-6x + 9)$
19. $7r - 5r + 2 = 5r - r$
20. $9p - 4p + 6 = 7p - 3p$
21. $11x - 5(x + 2) = 6x + 5$
22. $6x - 4(x + 1) = 2x + 4$

Solve each equation by clearing fractions or decimals. Check your solution. See Examples 5 and 6.

23. $\dfrac{3}{5}t - \dfrac{1}{10}t = t - \dfrac{5}{2}$
24. $-\dfrac{2}{7}r + 2r = \dfrac{1}{2}r + \dfrac{17}{2}$
25. $-\dfrac{1}{4}(x - 12) + \dfrac{1}{2}(x + 2) = x + 4$
26. $\dfrac{1}{9}(y + 18) + \dfrac{1}{3}(2y + 3) = y + 3$
27. $\dfrac{2}{3}k - \left(k + \dfrac{1}{4}\right) = \dfrac{1}{12}(k + 4)$
28. $-\dfrac{5}{6}q - \left(q - \dfrac{1}{2}\right) = \dfrac{1}{4}(q + 1)$
29. $.20(60) + .05x = .10(60 + x)$
30. $.30(30) + .15x = .20(30 + x)$
31. $1.00x + .05(12 - x) = .10(63)$
32. $.92x + .98(12 - x) = .96(12)$
33. $.06(10,000) + .08x = .072(10,000 + x)$
34. $.02(5000) + .03x = .025(5000 + x)$

Solve each equation, and check your solution. See Examples 1–8.

35. $10(2x - 1) = 8(2x + 1) + 14$
36. $9(3k - 5) = 12(3k - 1) - 51$
37. $-(4y + 2) - (-3y - 5) = 3$
38. $-(6k - 5) - (-5k + 8) = -3$
39. $\dfrac{1}{2}(x + 2) + \dfrac{3}{4}(x + 4) = x + 5$
40. $\dfrac{1}{3}(x + 3) + \dfrac{1}{6}(x - 6) = x + 3$
41. $.10(x + 80) + .20x = 14$
42. $.30(x + 15) + .40(x + 25) = 25$
43. $4(x + 8) = 2(2x + 6) + 20$
44. $4(x + 3) = 2(2x + 8) - 4$
45. $9(v + 1) - 3v = 2(3v + 1) - 8$
46. $8(t - 3) + 4t = 6(2t + 1) - 10$

Write the answer to each problem as an algebraic expression. See Example 9.

47. Two numbers have a sum of 11. One of the numbers is q. Find the other number.
48. The product of two numbers is 9. One of the numbers is k. What is the other number?
49. A football player gained x yd rushing. On the next down he gained 7 yd. How many yards did he gain altogether?
50. A baseball player got 65 hits one season. He got h of the hits in one game. How many hits did he get in the rest of the games?
51. Mary is a yr old. How old will she be in 12 yr? How old was she 5 yr ago?
52. Tom has r quarters. Find the value of the quarters in cents.

114 CHAPTER 2 Linear Equations and Inequalities in One Variable

53. A bank teller has t dollars, all in \$5 bills. How many \$5 bills does the teller have?

54. A plane ticket costs b dollars for an adult and d dollars for a child. Find the total cost for 3 adults and 2 children.

SUMMARY EXERCISES ON SOLVING LINEAR EQUATIONS

This section of miscellaneous linear equations provides practice in solving all the types introduced in Sections 2.1–2.3. Refer to the examples in these sections to review the various solution methods.

Solve each equation, and check your solution.

1. $a + 2 = -3$ **2.** $2m + 8 = 16$ **3.** $12.5k = -63.75$

4. $-x = -12$ **5.** $\dfrac{4}{5}x = -20$ **6.** $7m - 5m = -12$

7. $5x - 9 = 4(x - 3)$ **8.** $\dfrac{a}{-2} = 8$

9. $-3(m - 4) + 2(5 + 2m) = 29$ **10.** $\dfrac{2}{3}y + 8 = \dfrac{1}{4}y$

11. $.08x + .06(x + 9) = 1.24$ **12.** $x - 16.2 = 7.5$

13. $4x + 2(3 - 2x) = 6$ **14.** $-.3x + 2.1(x - 4) = -6.6$

15. $-x = 6$ **16.** $3(m + 5) - 1 + 2m = 5(m + 2)$

17. $7m - (2m - 9) = 39$ **18.** $7(p - 2) + p = 2(p + 2)$

19. $-2t + 5t - 9 = 3(t - 4) - 5$ **20.** $-6z = -14$

21. $.02(50) + .08r = .04(50 + r)$ **22.** $2.3x + 13.7 = 1.3x + 2.9$

23. $2(3 + 7x) - (1 + 15x) = 2$ **24.** $6q - 9 = 12 + 3q$

25. $2(4 + 3r) = 3(r + 1) + 11$ **26.** $r + 9 + 7r = 4(3 + 2r) - 3$

27. $\dfrac{1}{4}x - 4 = \dfrac{3}{2}x + \dfrac{3}{4}x$ **28.** $.06(100 - x) + .04x = .05(92)$

29. $\dfrac{3}{4}(a - 2) - \dfrac{1}{3}(5 - 2a) = -2$ **30.** $2 - (m + 4) = 3m + 8$

2.4 An Introduction to Applications of Linear Equations

OBJECTIVES

1 Learn the six steps for solving applied problems.

2 Solve problems involving unknown numbers.

OBJECTIVE 1 Learn the six steps for solving applied problems. We now look at how algebra is used to solve applied problems. Since many meaningful applications of mathematics require concepts that are beyond the level of this book, some of the problems you encounter will seem contrived, and to some extent they are. But the skills you develop in solving simple problems will help you solve more realistic problems in chemistry, physics, biology, business, and other fields. While there is not one specific method that enables you to solve all kinds of applied problems, the following six-step method is suggested.

SECTION 2.4 An Introduction to Applications of Linear Equations

3 Solve problems involving sums of quantities.

4 Solve problems involving supplementary and complementary angles.

5 Solve problems involving consecutive integers.

Solving an Applied Problem

Step 1 **Read** the problem carefully until you understand what is given and what is to be found.

Step 2 **Assign a variable** to represent the unknown value, using diagrams or tables as needed. Write down what the variable represents. If necessary, express any other unknown values in terms of the variable.

Step 3 **Write an equation** using the variable expression(s).

Step 4 **Solve** the equation.

Step 5 **State the answer.** Does it seem reasonable?

Step 6 **Check** the answer in the words of the *original* problem.

The third step in solving an applied problem is often the hardest. To translate the problem into an equation, write the given phrases as mathematical expressions. Replace any words that mean *equal* or *same* with an $=$ sign. Other forms of the verb "to be," such as *is, are, was,* and *were,* also translate this way. The $=$ sign leads to an equation to be solved.

OBJECTIVE 2 Solve problems involving unknown numbers. Some of the simplest applied problems involve unknown numbers.

■ **EXAMPLE 1** Finding the Value of an Unknown Number

The product of 4, and a number decreased by 7, is 100. What is the number?

Step 1 **Read** the problem carefully. Decide what you are being asked to find.

Step 2 **Assign a variable** to represent the unknown quantity. In this problem, we are asked to find a number, so we write

$$\text{Let } x = \text{the number.}$$

There are no other unknown quantities to find.

Step 3 **Write an equation.**

The product of 4, and a number decreased by 7, is 100.

$$4 \cdot (x - 7) = 100$$

Because of the commas in the given problem, writing the equation as $4x - 7 = 100$ is incorrect. The equation $4x - 7 = 100$ corresponds to the statement "The product of 4 and a number, decreased by 7, is 100."

Step 4 **Solve** the equation.

$$4(x - 7) = 100$$
$$4x - 28 = 100 \quad \text{Distributive property}$$
$$4x - 28 + 28 = 100 + 28 \quad \text{Add 28.}$$
$$4x = 128 \quad \text{Combine like terms.}$$
$$\frac{4x}{4} = \frac{128}{4} \quad \text{Divide by 4.}$$
$$x = 32$$

Step 5 **State the answer.** The number is 32.

Step 6 **Check.** When 32 is decreased by 7, we get $32 - 7 = 25$. If 4 is multiplied by 25, we get 100, as required. The answer, 32, is correct.

Now Try Exercise 5.

OBJECTIVE 3 Solve problems involving sums of quantities. A common type of problem in elementary algebra involves finding two quantities when the sum of the quantities is known. In Example 9 of the previous section, we prepared for this type of problem by writing mathematical expressions for two related unknown quantities.

PROBLEM SOLVING

In general, to solve problems involving sums of quantities, choose a variable to represent one of the unknowns. Then represent the other quantity in terms of the *same* variable, using information from the problem. Write an equation based on the words of the problem.

EXAMPLE 2 Finding Numbers of Olympic Medals

In the 2002 Winter Olympics in Salt Lake City, the United States won 10 more medals than Norway. The two countries won a total of 58 medals. How many medals did each country win? (*Source:* U.S. Olympic Committee.)

Step 1 **Read** the problem. We are given information about the total number of medals and asked to find the number each country won.

Step 2 **Assign a variable.**

Let $\quad x =$ the number of medals Norway won.

Then $\quad x + 10 =$ the number of medals the U.S. won.

Step 3 **Write an equation.**

The total is the number of medals Norway won plus the number of medals the U.S. won.

$$58 = x + (x + 10)$$

Step 4 **Solve** the equation.

$58 = 2x + 10$ Combine like terms.

$58 - 10 = 2x + 10 - 10$ Subtract 10.

$48 = 2x$ Combine like terms.

$\dfrac{48}{2} = \dfrac{2x}{2}$ Divide by 2.

$24 = x$ or $x = 24$

Step 5 **State the answer.** The variable x represents the number of medals Norway won, so Norway won 24 medals. Then the number of medals the United States won is $x + 10 = 24 + 10 = 34$.

Step 6 **Check.** Since the United States won 34 medals and Norway won 24, the total number of medals was $34 + 24 = 58$. Because $34 - 24 = 10$, the United States won 10 more medals than Norway. This information agrees with what is given in the problem, so the answer checks.

Now Try Exercise 11.

NOTE The problem in Example 2 could also be solved by letting x represent the number of medals the United States won. Then $x - 10$ would represent the number of medals Norway won. The equation would be
$$58 = x + (x - 10).$$
The solution of this equation is 34, which is the number of U.S. medals. The number of Norwegian medals would be $34 - 10 = 24$. The answers are the same, whichever approach is used.

EXAMPLE 3 Finding the Number of Orders for Tea

The owner of P. J.'s Coffeehouse found that on one day the number of orders for tea was $\frac{1}{3}$ the number of orders for coffee. If the total number of orders for the two drinks was 76, how many orders were placed for tea?

Step 1 **Read** the problem. It asks for the number of orders for tea.

Step 2 **Assign a variable.** Because of the way the problem is stated, let the variable represent the number of orders for coffee.

   Let $\quad x = $ the number of orders for coffee.

   Then $\quad \frac{1}{3}x = $ the number of orders for tea.

Step 3 **Write an equation.** Use the fact that the total number of orders was 76.

$$\underset{\text{total}}{\text{The}} \; \underset{}{\text{is}} \; \underset{\text{coffee}}{\text{orders for}} \; \underset{}{\text{plus}} \; \underset{\text{for tea.}}{\text{orders}}$$

$$76 \; = \; x \; + \; \frac{1}{3}x$$

Step 4 **Solve** the equation.

$$76 = \frac{4}{3}x \qquad x = 1x = \frac{3}{3}x;\ \text{Combine like terms.}$$

$$\frac{3}{4}(76) = \frac{3}{4}\left(\frac{4}{3}x\right) \qquad \text{Multiply by } \tfrac{3}{4}.$$

$$57 = x$$

Step 5 **State the answer.** In this problem, x does not represent the quantity that we are asked to find. The number of orders for tea was $\frac{1}{3}x$. So $\frac{1}{3}(57) = 19$ is the number of orders for tea.

Step 6 **Check.** The number of coffee orders (x) was 57 and the number of tea orders was 19; 19 is one-third of 57, and $19 + 57 = 76$. Since this agrees with the information given in the problem, the answer is correct.

Now Try Exercise 23.

PROBLEM SOLVING

In Example 3, it was easier to let the variable represent the quantity that was *not* asked for. This required extra work in Step 5 to find the number of orders for tea. In some cases, this approach is easier than letting the variable represent the quantity that we are asked to find. Experience in solving problems will indicate when this approach is useful, and experience comes only from solving many problems.

EXAMPLE 4 Analyzing a Gasoline/Oil Mixture

A lawn trimmer uses a mixture of gasoline and oil. The mixture contains 16 oz of gasoline for each ounce of oil. If the tank holds 68 oz of the mixture, how many ounces of oil and how many ounces of gasoline does it require when it is full?

Step 1 **Read** the problem. We must find how many ounces of oil and gasoline are needed to fill the tank.

Step 2 **Assign a variable.**

Let $x =$ the number of ounces of oil required.
Then $16x =$ the number of ounces of gasoline required.

Step 3 **Write an equation.**

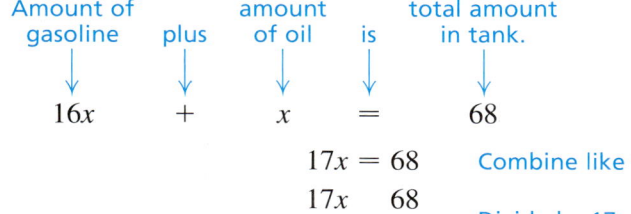

Step 4 **Solve.**

$$17x = 68 \quad \text{Combine like terms.}$$
$$\frac{17x}{17} = \frac{68}{17} \quad \text{Divide by 17.}$$
$$x = 4$$

Step 5 **State the answer.** The lawn trimmer requires 4 oz of oil and $16(4) = 64$ oz of gasoline when full.

Step 6 **Check.** Since $4 + 64 = 68$, and 64 is 16 times 4, the answer checks.

Now Try Exercise 25.

PROBLEM SOLVING

Sometimes it is necessary to find three unknown quantities in an applied problem. Frequently the three unknowns are compared in *pairs*. When this happens, it is usually easiest to let the variable represent the unknown found in both pairs.

SECTION 2.4 An Introduction to Applications of Linear Equations 119

EXAMPLE 5 Dividing a Board into Pieces

The instructions for a woodworking project call for three pieces of wood. The longest piece must be twice the length of the middle-sized piece, and the shortest piece must be 10 in. shorter than the middle-sized piece. Maria Gonzales has a board 70 in. long that she wishes to use. How long can each piece be?

Step 1 **Read** the problem. There will be three answers.

Step 2 **Assign a variable.** Since the middle-sized piece appears in both pairs of comparisons, let x represent the length, in inches, of the middle-sized piece. We have

$x =$ the length of the middle-sized piece,
$2x =$ the length of the longest piece, and
$x - 10 =$ the length of the shortest piece.

A sketch is helpful here. See Figure 2.

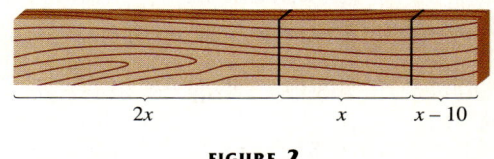

FIGURE 2

Step 3 **Write an equation.**

Longest plus middle-sized plus shortest is total length.
$$2x + x + (x - 10) = 70$$

Step 4 **Solve.**

$$4x - 10 = 70 \quad \text{Combine like terms.}$$
$$4x - 10 + 10 = 70 + 10 \quad \text{Add 10.}$$
$$4x = 80 \quad \text{Combine like terms.}$$
$$\frac{4x}{4} = \frac{80}{4} \quad \text{Divide by 4.}$$
$$x = 20$$

Step 5 **State the answer.** The middle-sized piece is 20 in. long, the longest piece is $2(20) = 40$ in. long, and the shortest piece is $20 - 10 = 10$ in. long.

Step 6 **Check.** The sum of the lengths is 70 in. All conditions of the problem are satisfied.

Now Try Exercise 29.

OBJECTIVE 4 Solve problems involving supplementary and complementary angles. An angle can be measured by a unit called the **degree** (°), which is $\frac{1}{360}$ of a complete rotation. Two angles whose sum is 90° are said to be **complementary,** or complements of each other. An angle that measures 90° is a **right angle.** Two angles whose sum is 180° are said to be **supplementary,** or supplements of each other. One angle *supplements* the other to form a **straight angle** of 180°. See Figure 3. If x represents the degree measure of an angle, then

$$90 - x \text{ represents the degree measure of its complement,}$$

and $$180 - x \text{ represents the degree measure of its supplement.}$$

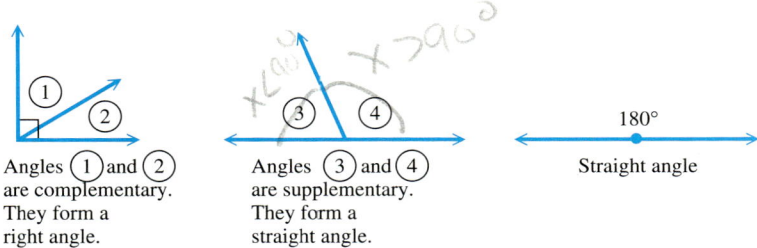

Angles ① and ② are complementary. They form a right angle.

Angles ③ and ④ are supplementary. They form a straight angle.

180° Straight angle

FIGURE 3

EXAMPLE 6 Finding the Measure of an Angle

Find the measure of an angle whose supplement is 10° more than twice its complement.

Step 1 **Read** the problem. We are to find the measure of an angle, given information about its complement and its supplement.

Step 2 **Assign a variable.**

Let x = the degree measure of the angle.
Then $90 - x$ = the degree measure of its complement,
and $180 - x$ = the degree measure of its supplement.

Step 3 **Write an equation.**

Supplement is 10 more than twice its complement.

$$180 - x = 10 + 2 \cdot (90 - x)$$

Step 4 **Solve.**

$$180 - x = 10 + 180 - 2x \quad \text{Distributive property}$$
$$180 - x = 190 - 2x \quad \text{Combine like terms.}$$
$$180 - x + 2x = 190 - 2x + 2x \quad \text{Add } 2x.$$
$$180 + x = 190 \quad \text{Combine like terms.}$$
$$180 + x - 180 = 190 - 180 \quad \text{Subtract 180.}$$
$$x = 10$$

Step 5 **State the answer.** The measure of the angle is 10°.

Step 6 **Check.** The complement of 10° is 80° and the supplement of 10° is 170°. 170° is equal to 10° more than twice 80° ($170 = 10 + 2(80)$ is true); therefore, the answer is correct.

Now Try Exercise 43.

OBJECTIVE 5 Solve problems involving consecutive integers. Two integers that differ by 1 are called **consecutive integers.** For example, 3 and 4, 6 and 7, and -2 and -1

are pairs of consecutive integers. In general, if x represents an integer, $x + 1$ represents the next larger consecutive integer.

Consecutive *even* integers, such as 8 and 10, differ by 2. Similarly, consecutive *odd* integers, such as 9 and 11, also differ by 2. In general, if x represents an even integer, $x + 2$ represents the next larger consecutive even integer. The same holds true for odd integers; that is, if x is an odd integer, $x + 2$ is the next larger odd integer.

PROBLEM SOLVING

When solving consecutive integer problems, if $x =$ the first integer, then for any

two consecutive integers, use $\quad x,\ x + 1;$

two consecutive *even* integers, use $\quad x,\ x + 2;$

two consecutive *odd* integers, use $\quad x,\ x + 2.$

EXAMPLE 7 Finding Consecutive Integers

Two pages that face each other in this book have 569 as the sum of their page numbers. What are the page numbers?

Step 1 **Read** the problem. Because the two pages face each other, they must have page numbers that are consecutive integers.

Step 2 **Assign a variable.**

Let $\qquad x =$ the smaller page number.

Then $\quad x + 1 =$ the larger page number.

Step 3 **Write an equation.** Because the sum of the page numbers is 569, the equation is

$$x + (x + 1) = 569.$$

Step 4 **Solve.** $\qquad 2x + 1 = 569 \qquad$ Combine like terms.

$\qquad\qquad\qquad 2x = 568 \qquad$ Subtract 1.

$\qquad\qquad\qquad x = 284 \qquad$ Divide by 2.

Step 5 **State the answer.** The smaller page number is 284, and the larger page number is $284 + 1 = 285$.

Step 6 **Check.** The sum of 284 and 285 is 569. The answer is correct.

Now Try Exercise 47.

In the final example, we do not number the steps. See if you can identify them.

EXAMPLE 8 Finding Consecutive Odd Integers

If the smaller of two consecutive odd integers is doubled, the result is 7 more than the larger of the two integers. Find the two integers.

Let x be the smaller integer. Since the two numbers are consecutive *odd* integers, then $x + 2$ is the larger. Now we write an equation.

If the smaller is doubled,	the result is	7	more than	the larger.
$2x$	$=$	7	$+$	$(x + 2)$

$2x = 9 + x$ Combine like terms.
$x = 9$ Subtract x.

The first integer is 9 and the second is $9 + 2 = 11$. To check, we see that when 9 is doubled, we get 18, which is 7 more than the larger odd integer, 11. The answers are correct.

Now Try Exercise 49.

CONNECTIONS

George Polya (1888–1985), a native of Budapest, Hungary, wrote the modern classic *How to Solve It*. In this book he proposed a four-step process for problem solving:

1. Understand the problem.
2. Devise a plan.
3. Carry out the plan.
4. Look back and check.

For Discussion or Writing

Compare Polya's four-step process with the six steps given in this section. Identify which of the steps in our list match Polya's four steps. Trial and error is also a useful problem-solving tool. Where does this method fit into Polya's steps?

2.4 EXERCISES

For Extra Help

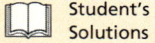 Student's Solutions Manual

 MyMathLab

 InterAct Math Tutorial Software

 AW Math Tutor Center

 MathXL

Digital Video Tutor CD 4/Videotape 3

1. In your own words, write the general procedure for solving applications as outlined in this section.

2. List some of the words that translate as "=" when writing an equation to solve an applied problem.

3. Suppose that a problem requires you to find the number of cars on a dealer's lot. Which one of the following would *not* be a reasonable answer? Justify your answer.

   A. 0   **B.** 45   **C.** 1   **D.** $6\frac{1}{2}$

4. Suppose that a problem requires you to find the number of hours a light bulb is on during a day. Which one of the following would *not* be a reasonable answer? Justify your answer.

   A. 0   **B.** 4.5   **C.** 13   **D.** 25

SECTION 2.4 An Introduction to Applications of Linear Equations   **123**

Solve each problem. See Example 1.

5. If 2 is added to five times a number, the result is equal to 5 more than four times the number. Find the number.

6. If four times a number is added to 8, the result is three times the number added to 5. Find the number.

7. If 2 is subtracted from a number and this difference is tripled, the result is 6 more than the number. Find the number.

8. If 3 is added to a number and this sum is doubled, the result is 2 more than the number. Find the number.

9. The sum of three times a number and 7 more than the number is the same as the difference between −11 and twice the number. What is the number?

10. If 4 is added to twice a number and this sum is multiplied by 2, the result is the same as if the number is multiplied by 3 and 4 is added to the product. What is the number?

Solve each problem. See Example 2.

11. The number of drive-in movie screens has declined steadily in the United States since the 1960s. California and New York were two of the states with the most remaining drive-in movie screens in 2001. California had 11 more screens than New York, and there were 107 screens total in the two states. How many drive-in movie screens remained in each state? (*Source:* National Association of Theatre Owners.)

12. Thursday is the most-watched night for the major broadcast TV networks (ABC, CBS, NBC, and Fox), with 20 million more viewers than Saturday, the least-watched night. The total for the two nights is 102 million viewers. How many viewers of the major networks are there on each of these nights? (*Source:* Nielsen Media Research.)

13. The U.S. Senate has 100 members. During the 106th session (1999–2001), there were 10 more Republicans than Democrats. How many Democrats and Republicans were there in the Senate? (*Source: World Almanac and Book of Facts,* 2000.)

14. The total number of Democrats and Republicans in the U.S. House of Representatives during the 106th session was 434. There were 12 more Republicans than Democrats. How many members of each party were there? (*Source: World Almanac and Book of Facts,* 2000.)

124 CHAPTER 2 Linear Equations and Inequalities in One Variable

15. The rock band U2 generated top revenue on the concert circuit in 2001. U2 and second-place 'N Sync together took in $196.5 million from ticket sales. If 'N Sync took in $22.9 million less than U2, how much revenue did each generate? (*Source:* Pollstar.)

16. The Toyota Camry was the top-selling passenger car in the United States in 2000, followed by the Honda Accord. Honda Accord sales were 18 thousand less than Toyota Camry sales, and 828 thousand of these two cars were sold. How many of each make of car were sold? (*Source:* Ward's Communications.)

17. In the 2001–2002 NBA regular season, the Sacramento Kings won two less than three times as many games as they lost. The Kings played 82 games. How many wins and losses did the team have? (*Source:* nba.com)

18. In the 2000–2001 regular baseball season, the Oakland Athletics won 18 less than twice as many games as they lost. They played 162 regular season games. How many wins and losses did the team have? (*Source:* World Almanac and Book of Facts, 2002.)

Solve each problem. See Examples 3 and 4.

19. A September 2000 issue of *Coin World* listed the value of a "Mint State-65" (uncirculated) 1950 Jefferson nickel minted at Denver at $\frac{7}{6}$ the value of a similar condition 1945 nickel minted at Philadelphia. Together the total value of the two coins is $26.00. What is the value of each coin?

20. The largest sheep ranch in the world is located in Australia. The number of sheep on the ranch is $\frac{8}{3}$ the number of uninvited kangaroos grazing on the pastureland. Together, herds of these two animals number 88,000. How many sheep and how many kangaroos roam the ranch? (*Source: Guinness Book of Records.*)

21. In 1988, a dairy in Alberta, Canada, created a sundae with approximately 1 lb of topping for every 83.2 lb of ice cream. The total of the two ingredients weighed approximately 45,225 lb. To the nearest tenth of a pound, how many pounds of ice cream and how many pounds of topping were there? (*Source: Guinness Book of Records.*)

22. A husky running the Iditarod (a thousand-mile race between Anchorage and Nome, Alaska) burns $5\frac{3}{8}$ calories in exertion for every 1 calorie burned in thermoregulation in extreme cold. According to one scientific study, a husky in top condition burns an amazing total of 11,200 calories per day. How many calories are burned for exertion, and how many are burned for regulation of body temperature? Round answers to the nearest whole number.

23. A pharmacist found that at the end of the day she had $\frac{4}{3}$ as many prescriptions for antibiotics as she did for tranquilizers. She had 42 prescriptions altogether for these two types of drugs. How many did she have for tranquilizers?

SECTION 2.4 An Introduction to Applications of Linear Equations 125

24. In a mixture of concrete, there are 3 lb of cement mix for every 1 lb of gravel. If the mixture contains a total of 140 lb of these two ingredients, how many pounds of gravel are there?

25. A mixture of nuts contains only peanuts and cashews. For every 1 oz of cashews there are 5 oz of peanuts. If the mixture contains a total of 27 oz, how many ounces of each type of nut does the mixture contain?

26. An insecticide contains 95 cg of inert ingredient for every 1 cg of active ingredient. If a quantity of the insecticide weighs 336 cg, how much of each type of ingredient does it contain?

Solve each problem. See Example 5.

27. In one day, Akilah Cadet received 13 packages. Federal Express delivered three times as many as Airborne Express, while United Parcel Service delivered 2 fewer than Airborne Express. How many packages did each service deliver to Akilah?

28. In his job at the post office, Eddie Thibodeaux works a 6.5-hr day. He sorts mail, sells stamps, and does supervisory work. One day he sold stamps twice as long as he sorted mail, and he supervised .5 hr longer than he sorted mail. How many hours did he spend at each task?

29. The United States earned an all-time high of 34 medals at the 2002 Winter Olympics. The number of gold medals earned was 1 less than the number of bronze medals. The number of silver medals earned was 2 more than the number of bronze medals. How many of each kind of medal did the United States earn? (*Source:* U.S. Olympic Committee.)

30. Nagaraj Nanjappa has a party-length submarine sandwich 59 in. long. He wants to cut it into three pieces so that the middle piece is 5 in. longer than the shortest piece and the shortest piece is 9 in. shorter than the longest piece. How long should the three pieces be?

31. Venus is 31.2 million mi farther from the sun than Mercury, while Earth is 57 million mi farther from the sun than Mercury. If the total of the distances from these three planets to the sun is 196.2 million mi, how far away from the sun is Mercury? (All distances given here are *mean* (*average*) distances.) (*Source: Universal Almanac,* 1997.)

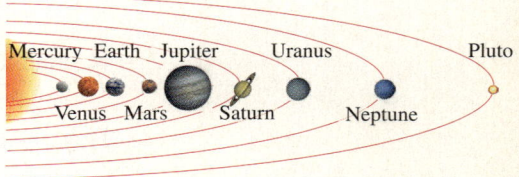

32. Together, Saturn, Jupiter, and Mars have a total of 36 known satellites (moons). Jupiter has 2 fewer satellites than Saturn, and Mars has 16 fewer satellites than Saturn. How many known satellites does Mars have? (*Source: World Almanac and Book of Facts,* 2000.)

33. The sum of the measures of the angles of any triangle is 180°. In triangle *ABC*, angles *A* and *B* have the same measure, while the measure of angle *C* is 60° larger than each of *A* and *B*. What are the measures of the three angles?

34. In triangle *ABC*, the measure of angle *A* is 141° more than the measure of angle *B*. The measure of angle *B* is the same as the measure of angle *C*. Find the measure of each angle. (*Hint:* See Exercise 33.)

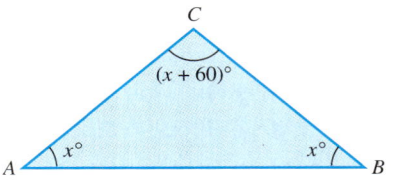

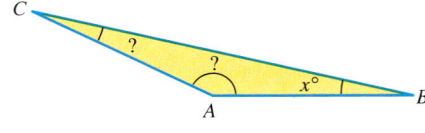

Use the concepts of this section to answer each question.

35. If the sum of two numbers is k, and one of the numbers is m, how can you express the other number?

36. If the product of two numbers is r, and one of the numbers is s ($s \neq 0$), how can you express the other number?

37. Is there an angle whose supplement is equal to its complement? If so, what is the measure of the angle?

38. Is there an angle that is equal to its supplement? Is there an angle that is equal to its complement? If the answer is yes to either question, give the measure of the angle.

39. If x represents an integer, how can you express the next smaller consecutive integer in terms of x?

40. If x represents an integer, how can you express the next smaller even integer in terms of x?

Solve each problem. See Example 6.

41. Find the measure an an angle whose complement is four times its measure.

42. Find the measure of an angle whose supplement is three times its measure.

43. Find the measure of an angle whose supplement measures 39° more than twice its complement.

44. Find the measure of an angle whose supplement measures 38° less than three times its complement.

45. Find the measure of an angle such that the difference between the measures of its supplement and three times its complement is 10°.

46. Find the measure of an angle such that the sum of the measures of its complement and its supplement is 160°.

SECTION 2.4 An Introduction to Applications of Linear Equations 127

Solve each problem. See Examples 7 and 8.

47. The numbers on two consecutively numbered gym lockers have a sum of 137. What are the locker numbers?

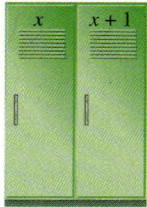

48. The sum of two consecutive checkbook check numbers is 357. Find the numbers.

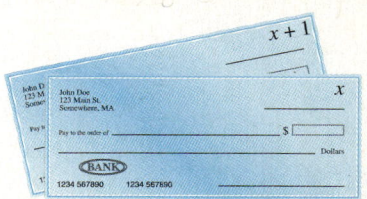

49. Find two consecutive even integers such that the smaller added to three times the larger gives a sum of 46.

50. Find two consecutive odd integers such that twice the larger is 17 more than the smaller.

51. Two pages that are back-to-back in this book have 203 as the sum of their page numbers. What are the page numbers?

52. Two houses on the same side of the street have house numbers that are consecutive even integers. The sum of the integers is 58. What are the two house numbers?

53. When the smaller of two consecutive integers is added to three times the larger, the result is 43. Find the integers.

54. If five times the smaller of two consecutive integers is added to three times the larger, the result is 59. Find the integers.

55. If the sum of three consecutive even integers is 60, what is the smallest even integer? (*Hint:* If x and $x + 2$ represent the first two consecutive even integers, how would you represent the third consecutive even integer?)

56. If the sum of three consecutive odd integers is 69, what is the largest odd integer?

57. If 6 is subtracted from the largest of three consecutive odd integers, with this result multiplied by 2, the answer is 23 less than the sum of the first and twice the second of the integers. Find the integers.

58. If the first and third of three consecutive even integers are added, the result is 22 less than three times the second integer. Find the integers.

Apply the ideas of this section to solve Exercises 59 and 60, which are based on the graphs.

59. In a recent year, the funding for Head Start programs increased by .55 billion dollars from the funding in the previous year. The following year, the increase was .20 billion dollars more. For this 3-year period, the total funding was 9.64 billion dollars. How much was funded in each of these years? (*Source:* U.S. Department of Health and Human Services.)

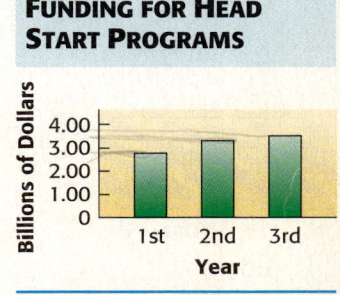

FUNDING FOR HEAD START PROGRAMS

128 CHAPTER 2 Linear Equations and Inequalities in One Variable

60. According to data provided by the National Safety Council for a recent year, the number of serious injuries per 100,000 participants in football, bicycling, and golf is illustrated in the graph. There were 800 more in bicycling than in golf, and there were 1267 more in football than in bicycling. Altogether there were 3179 serious injuries per 100,000 participants. How many such serious injuries were there in each sport?

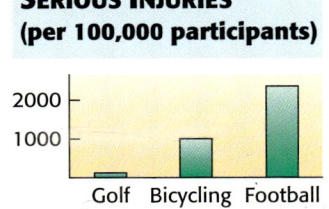

SERIOUS INJURIES
(per 100,000 participants)

2.5 Formulas and Applications from Geometry

OBJECTIVES

1. Solve a formula for one variable, given the values of the other variables.
2. Use a formula to solve an applied problem.
3. Solve problems involving vertical angles and straight angles.
4. Solve a formula for a specified variable.

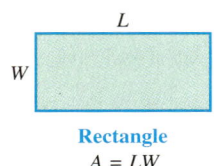

Rectangle
$A = LW$
FIGURE 4

Many applied problems can be solved with formulas. For example, formulas exist for geometric figures such as squares and circles, for distance, for money earned on bank savings, and for converting English measurements to metric measurements. The formulas used in this book are given on the inside covers.

OBJECTIVE 1 Solve a formula for one variable, given the values of the other variables. Given the values of all but one of the variables in a formula, we can find the value of the remaining variable by using the methods introduced in this chapter.

In Example 1, we use the idea of *area*. The **area** of a plane (two-dimensional) geometric figure is a measure of the surface covered by the figure.

EXAMPLE 1 Using Formulas to Evaluate Variables

Find the value of the remaining variable in each formula.

(a) $A = LW$; $A = 64$, $L = 10$

As shown in Figure 4, this formula gives the area of a rectangle with length L and width W. Substitute the given values into the formula and then solve for W.

$$A = LW$$
$$64 = 10W \quad \text{Let } A = 64 \text{ and } L = 10.$$
$$\frac{64}{10} = \frac{10W}{10} \quad \text{Divide by 10.}$$
$$6.4 = W$$

The width of the rectangle is 6.4. Since $10(6.4) = 64$, the given area, the answer checks.

(b) $A = \dfrac{1}{2}h(b + B)$; $A = 210$, $B = 27$, $h = 10$

This formula gives the area of a trapezoid with parallel sides of lengths b and B and distance h between the parallel sides. See Figure 5. Again, begin by substituting the given values into the formula.

$$A = \frac{1}{2}h(b + B)$$
$$210 = \frac{1}{2}(10)(b + 27) \quad A = 210, h = 10, B = 27$$

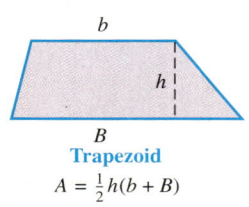

Trapezoid
$A = \frac{1}{2}h(b + B)$
FIGURE 5

Now solve for b.

$$210 = 5(b + 27) \quad \text{Multiply.}$$
$$210 = 5b + 135 \quad \text{Distributive property}$$
$$210 - 135 = 5b + 135 - 135 \quad \text{Subtract 135.}$$
$$75 = 5b \quad \text{Combine like terms.}$$
$$\frac{75}{5} = \frac{5b}{5} \quad \text{Divide by 5.}$$
$$15 = b$$

Check that the length of the shorter parallel side, b, is 15.

Now Try Exercises 19 and 23.

OBJECTIVE 2 Use a formula to solve an applied problem. As the next examples show, formulas are often used to solve applied problems. *It is a good idea to draw a sketch when a geometric figure is involved.* Example 2 uses the idea of perimeter. The **perimeter** of a plane (two-dimensional) geometric figure is the distance around the figure, that is, the sum of the lengths of its sides. We use the six steps introduced in the previous section.

EXAMPLE 2 Finding the Width of a Rectangular Lot

A rectangular lot has perimeter 80 m and length 25 m. Find the width of the lot.

Step 1 **Read.** We are told to find the width of the lot.

Step 2 **Assign a variable.** Let W = the width of the lot in meters. See Figure 6.

Step 3 **Write an equation.** The formula for the perimeter of a rectangle is

$$P = 2L + 2W.$$

We find the width by substituting 80 for P and 25 for L in the formula.

$$80 = 2(25) + 2W \quad P = 80, L = 25$$

Step 4 **Solve** the equation.

$$80 = 50 + 2W \quad \text{Multiply.}$$
$$80 - 50 = 50 + 2W - 50 \quad \text{Subtract 50.}$$
$$30 = 2W \quad \text{Combine like terms.}$$
$$\frac{30}{2} = \frac{2W}{2} \quad \text{Divide by 2.}$$
$$15 = W$$

Step 5 **State the answer.** The width is 15 m.

Step 6 **Check.** If the width is 15 m and the length is 25 m, the distance around the rectangular lot (perimeter) is $2(25) + 2(15) = 50 + 30 = 80$ m, as required.

Now Try Exercise 37.

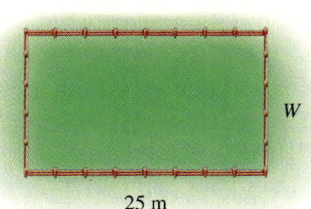

25 m

FIGURE 6

FIGURE 7

EXAMPLE 3 Finding the Height of a Triangular Sail

The area of a triangular sail of a sailboat is 126 ft². (Recall that ft² means "square feet.") The base of the sail is 12 ft. Find the height of the sail.

Step 1 **Read.** We must find the height of the triangular sail.

Step 2 **Assign a variable.** Let h = the height of the sail in feet. See Figure 7.

Step 3 **Write an equation.** The formula for the area of a triangle is $A = \frac{1}{2}bh$, where A is the area, b is the base, and h is the height. Using the information given in the problem, we substitute 126 for A and 12 for b in the formula.

$$A = \frac{1}{2}bh$$

$$126 = \frac{1}{2}(12)h \quad A = 126, b = 12$$

Step 4 **Solve.** $\quad 126 = 6h \quad$ Multiply.

$\quad\quad\quad\quad\quad\quad\quad 21 = h \quad$ Divide by 6.

Step 5 **State the answer.** The height of the sail is 21 ft.

Step 6 **Check** to see that the values $A = 126$, $b = 12$, and $h = 21$ satisfy the formula for the area of a triangle.

Now Try Exercise 39.

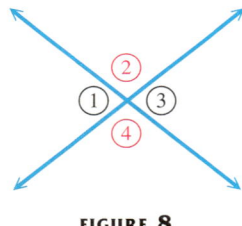

FIGURE 8

OBJECTIVE 3 Solve problems involving vertical angles and straight angles. Figure 8 shows two intersecting lines forming angles that are numbered ①, ②, ③, and ④. Angles ① and ③ lie "opposite" each other. They are called **vertical angles.** Another pair of vertical angles is ② and ④. In geometry, it is shown that vertical angles have equal measures.

Now look at angles ① and ②. When their measures are added, we get 180°, the measure of a straight angle. There are three other such pairs of angles: ② and ③, ③ and ④, and ① and ④.

The next example uses these ideas.

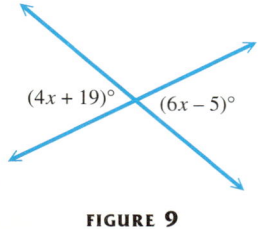

FIGURE 9

EXAMPLE 4 Finding Angle Measures

Refer to the appropriate figure in each part.

(a) Find the measure of each marked angle in Figure 9.

Since the marked angles are vertical angles, they have equal measures. Set $4x + 19$ equal to $6x - 5$ and solve.

$$4x + 19 = 6x - 5$$
$$19 = 2x - 5 \quad \text{Subtract } 4x.$$
$$24 = 2x \quad \text{Add 5.}$$
$$12 = x \quad \text{Divide by 2.}$$

Since $x = 12$, one angle has measure $4(12) + 19 = 67$ degrees. The other has the same measure, since $6(12) - 5 = 67$ as required. Each angle measures 67°.

(b) Find the measure of each marked angle in Figure 10.

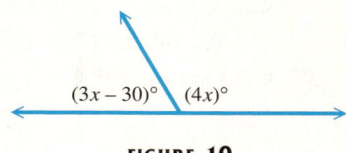

FIGURE 10

The measures of the marked angles must add to 180° because together they form a straight angle. The equation to solve is

$(3x - 30) + 4x = 180.$

$7x - 30 = 180$ Combine like terms.

$7x = 210$ Add 30.

$x = 30$ Divide by 7.

To find the measures of the angles, replace x with 30 in the two expressions.

$3x - 30 = 3(30) - 30 = 90 - 30 = 60$

$4x = 4(30) = 120$

The two angle measures are 60° and 120°.

Now Try Exercises 45 and 47.

CAUTION In Example 4, the answer is *not* the value of x. Remember to substitute the value of the variable into the expression given for each angle.

OBJECTIVE 4 Solve a formula for a specified variable. Sometimes it is necessary to solve a number of problems that use the same formula. For example, a surveying class might need to solve several problems that involve the formula for the area of a rectangle, $A = LW$. Suppose that in each problem the area (A) and the length (L) of a rectangle are given, and the width (W) must be found. Rather than solving for W each time the formula is used, it would be simpler to *rewrite the formula* so that it is solved for W. This process is called **solving for a specified variable.**

In solving a formula for a specified variable, we treat the specified variable as if it were the *only* variable in the equation, and treat the other variables as if they were numbers. We use the same steps to solve the equation for the specified variable that we use to solve equations with just one variable.

EXAMPLE 5 Solving for a Specified Variable

Solve $A = LW$ for W.

Think of undoing what has been done to W. Since W is multiplied by L, undo the multiplication by dividing each side of $A = LW$ by L.

$A = LW$

$\dfrac{A}{L} = \dfrac{LW}{L}$ Divide by L.

$\dfrac{A}{L} = W$ or $W = \dfrac{A}{L}$ $\dfrac{L}{L} = 1; 1W = W$

Now Try Exercise 51.

EXAMPLE 6 Solving for a Specified Variable

Solve $P = 2L + 2W$ for L.

We want to get L alone on one side of the equation. We begin by subtracting $2W$ from each side.

$$P = 2L + 2W$$
$$P - 2W = 2L + 2W - 2W \quad \text{Subtract } 2W.$$
$$P - 2W = 2L \quad \text{Combine like terms.}$$
$$\frac{P - 2W}{2} = \frac{2L}{2} \quad \text{Divide by 2.}$$
$$\frac{P - 2W}{2} = L \quad \tfrac{2}{2} = 1;\ 1L = L$$

or $$L = \frac{P - 2W}{2}$$

The last step gives the formula solved for L, as required.

Now Try Exercise 67.

EXAMPLE 7 Solving for a Specified Variable

Solve $F = \dfrac{9}{5}C + 32$ for C. (This is the formula for converting from Celsius to Fahrenheit.)

We need to isolate C on one side of the equation. First we undo the addition of 32 to $\tfrac{9}{5}C$ by subtracting 32 from each side.

$$F = \frac{9}{5}C + 32$$
$$F - 32 = \frac{9}{5}C + 32 - 32 \quad \text{Subtract 32.}$$
$$F - 32 = \frac{9}{5}C$$

Now we multiply each side by $\tfrac{5}{9}$, using parentheses on the left.

$$\frac{5}{9}(F - 32) = \frac{5}{9} \cdot \frac{9}{5}C \quad \text{Multiply by } \tfrac{5}{9}.$$
$$\frac{5}{9}(F - 32) = C \quad \text{or} \quad C = \frac{5}{9}(F - 32)$$

This last result is the formula for converting temperatures from Fahrenheit to Celsius.

Now Try Exercise 69.

EXAMPLE 8 Solving for a Specified Variable

Solve $A = \dfrac{1}{2}(b + B)h$ for B.

To get B alone, begin by multiplying each side by 2 to clear the fraction.

SECTION 2.5 Formulas and Applications from Geometry

$$A = \frac{1}{2}(b + B)h$$

$$2A = 2 \cdot \frac{1}{2}(b + B)h \quad \text{Multiply by 2. (\textit{Do not distribute the 2 on the right side.})}$$

$$2A = (b + B)h \quad 2 \cdot \frac{1}{2} = \frac{2}{2} = 1$$

$$2A = bh + Bh \quad \text{Distributive property}$$

$$2A - bh = bh + Bh - bh \quad \text{Subtract } bh.$$

$$2A - bh = Bh \quad \text{Combine like terms.}$$

$$\frac{2A - bh}{h} = \frac{Bh}{h} \quad \text{Divide by } h.$$

$$\frac{2A - bh}{h} = B \quad \text{or} \quad B = \frac{2A - bh}{h}$$

Now Try Exercise 73.

NOTE The result in Example 8 can be written in a different form as follows.

$$B = \frac{2A - bh}{h} = \frac{2A}{h} - \frac{bh}{h} = \frac{2A}{h} - b$$

Either form is correct.

2.5 EXERCISES

For Extra Help

 Student's Solutions Manual

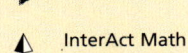 MyMathLab

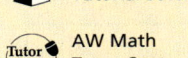 InterAct Math Tutorial Software

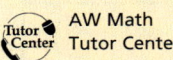 AW Math Tutor Center

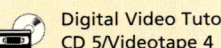 MathXL

Digital Video Tutor CD 5/Videotape 4

1. In your own words, explain what is meant by each term.
   (a) Perimeter of a plane geometric figure
   (b) Area of a plane geometric figure

2. Perimeter is to a polygon as _____ is to a circle.

3. If a formula has exactly five variables, how many values would you need to be given in order to find the value of any one variable?

4. The formula for changing Celsius to Fahrenheit is given in Example 7 as $F = \frac{9}{5}C + 32$. Sometimes it is seen as $F = \frac{9C}{5} + 32$. These are both correct. Why is it true that $\frac{9}{5}C$ is equal to $\frac{9C}{5}$?

Decide whether perimeter or area would be used to solve a problem concerning the measure of the quantity.

5. Sod for a lawn
6. Carpeting for a bedroom
7. Baseboards for a living room
8. Fencing for a yard
9. Fertilizer for a garden
10. Tile for a bathroom
11. Determining the cost of planting rye grass in a lawn for the winter
12. Determining the cost of replacing a linoleum floor with a wood floor

In the following exercises a formula is given, along with the values of all but one of the variables in the formula. Find the value of the variable that is not given. (When necessary, use 3.14 as an approximation for π.) See Example 1.

13. $P = 2L + 2W$ (perimeter of a rectangle); $L = 8$, $W = 5$

14. $P = 2L + 2W$; $L = 6$, $W = 4$

15. $A = \frac{1}{2}bh$ (area of a triangle); $b = 8$, $h = 16$

16. $A = \frac{1}{2}bh$; $b = 10$, $h = 14$

17. $P = a + b + c$ (perimeter of a triangle); $P = 12$, $a = 3$, $c = 5$

18. $P = a + b + c$; $P = 15$, $a = 3$, $b = 7$

19. $d = rt$ (distance formula); $d = 252$, $r = 45$

20. $d = rt$; $d = 100$, $t = 2.5$

21. $I = prt$ (simple interest); $p = 7500$, $r = .035$, $t = 6$

22. $I = prt$; $p = 5000$, $r = .025$, $t = 7$

23. $A = \frac{1}{2}h(b + B)$ (area of a trapezoid); $A = 91$, $h = 7$, $b = 12$

24. $A = \frac{1}{2}h(b + B)$; $A = 75$, $b = 19$, $B = 31$

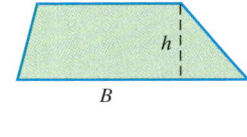

25. $C = 2\pi r$ (circumference of a circle); $C = 16.328$

26. $C = 2\pi r$; $C = 8.164$

27. $A = \pi r^2$ (area of a circle); $r = 4$

28. $A = \pi r^2$; $r = 12$

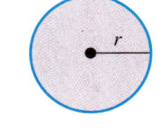

*The **volume** of a three-dimensional object is a measure of the space occupied by the object. For example, we would need to know the volume of a gasoline tank in order to know how many gallons of gasoline it would take to completely fill the tank. In the following exercises, a formula for the volume (V) of a three-dimensional object is given, along with values for the other variables. Evaluate V. (Use 3.14 as an approximation for π.) See Example 1.*

29. $V = LWH$ (volume of a rectangular box); $L = 10$, $W = 5$, $H = 3$

30. $V = LWH$; $L = 12$, $W = 8$, $H = 4$

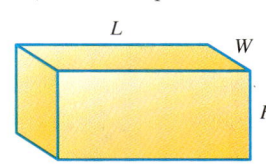

31. $V = \frac{1}{3}Bh$ (volume of a pyramid); $B = 12$, $h = 13$

32. $V = \frac{1}{3}Bh$; $B = 36$, $h = 4$

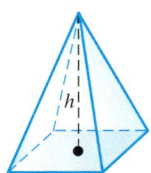

33. $V = \frac{4}{3}\pi r^3$ (volume of a sphere); $r = 12$

34. $V = \frac{4}{3}\pi r^3$; $r = 6$

SECTION 2.5 Formulas and Applications from Geometry   135

Use a formula to write an equation for each application, and then use the problem-solving method of Section 2.4 to solve. (Use 3.14 as an approximation for π.) Formulas are found on the inside covers of this book. See Examples 2 and 3.

35. Recently, a prehistoric ceremonial site dating to about 3000 B.C. was discovered at Stanton Drew in southwestern England. The site, which is larger than Stonehenge, is a nearly perfect circle, consisting of nine concentric rings that probably held upright wooden posts. Around this timber temple is a wide, encircling ditch enclosing an area with a diameter of 443 ft. Find this enclosed area to the nearest thousand square feet. (*Hint:* Find the radius. Then use $A = \pi r^2$.) (*Source: Archaeology,* vol. 51, no. 1, Jan./Feb. 1998.)

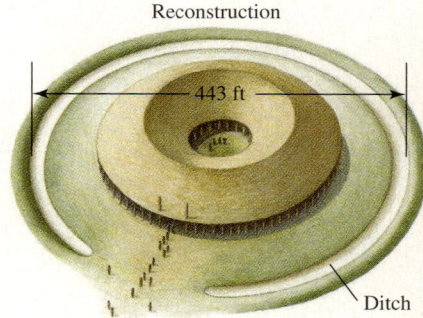

36. The Skydome in Toronto, Canada, is the first stadium with a hard-shell, retractable roof. The steel dome is 630 ft in diameter. To the nearest foot, what is the circumference of this dome? (*Source:* www.4ballparks.com)

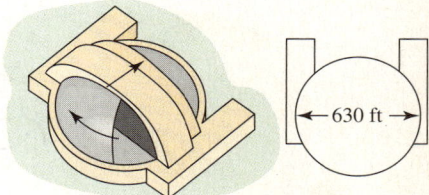

37. The *Daily Banner,* published in Roseberg, Oregon, in the 19th century, had page size 3 in. by 3.5 in. What was the perimeter? What was the area? (*Source: Guinness Book of Records.*)

38. The newspaper *The Constellation,* printed in 1859 in New York City as part of the Fourth of July celebration, had length 51 in. and width 35 in. What was the perimeter? What was the area? (*Source: Guinness Book of Records.*)

39. The largest drum ever constructed was played at the Royal Festival Hall in London in 1987. It had a diameter of 13 ft. What was the area of the circular face of the drum? (*Hint:* $A = \pi r^2$.) (*Source: Guinness Book of Records.*)

40. What was the circumference of the drum described in Exercise 39? (*Hint:* Use $C = 2\pi r$.)

41. The survey plat depicted here shows two lots that form a trapezoid. The measures of the parallel sides are 115.80 ft and 171.00 ft. The height of the trapezoid is 165.97 ft. Find the combined area of the two lots. Round your answer to the nearest hundredth of a square foot.

42. Lot A in the figure is in the shape of a trapezoid. The parallel sides measure 26.84 ft and 82.05 ft. The height of the trapezoid is 165.97 ft. Find the area of Lot A. Round your answer to the nearest hundredth of a square foot.

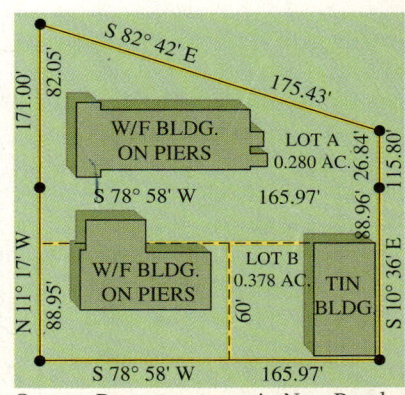

Source: Property survey in New Roads, Louisiana.

136 CHAPTER 2 Linear Equations and Inequalities in One Variable

43. The U.S. Postal Service requires that any box sent through the mail have length plus girth (distance around) totaling no more than 108 in. The maximum volume that meets this condition is contained by a box with a square end 18 in. on each side. What is the length of the box? What is the maximum volume?

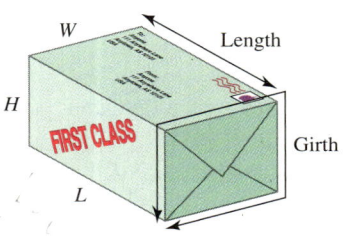

44. The largest box of popcorn was filled by students in Jacksonville, Florida. The box was approximately 40 ft long, $20\frac{2}{3}$ ft wide, and 8 ft high. To the nearest cubic foot, what was the volume of the box? (*Source: Guinness Book of Records.*)

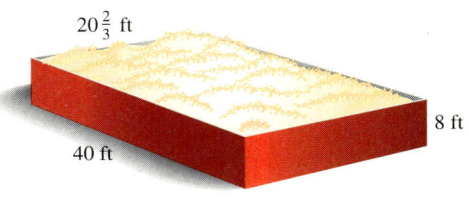

Find the measure of each marked angle. See Example 4.

45. 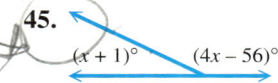 $(x + 1)°$, $(4x - 56)°$

46. $(10x + 7)°$ $(7x + 3)°$

47. 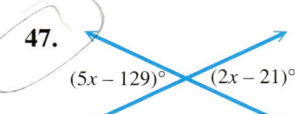 $(5x - 129)°$ $(2x - 21)°$

48. $(3x + 45)°$ $(7x + 5)°$

49. 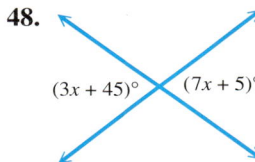 $(10x + 15)°$ $(12x - 3)°$

50. 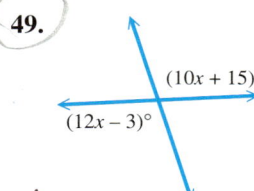 $(11x - 37)°$ $(7x + 27)°$

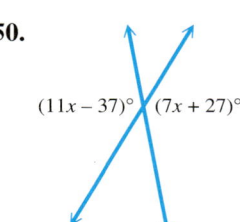

Solve each formula for the specified variable. See Examples 5–8.

51. $d = rt$ for t **52.** $d = rt$ for r **53.** $A = bh$ for b

54. $A = LW$ for L **55.** $C = \pi d$ for d **56.** $P = 4s$ for s

57. $V = LWH$ for H **58.** $V = LWH$ for W **59.** $I = prt$ for r

60. $I = prt$ for p **61.** $A = \frac{1}{2}bh$ for h **62.** $A = \frac{1}{2}bh$ for b

63. $V = \frac{1}{3}\pi r^2 h$ for h **64.** $V = \pi r^2 h$ for h **65.** $P = a + b + c$ for b

66. $P = a + b + c$ for a **67.** $P = 2L + 2W$ for W **68.** $A = p + prt$ for r

69. $y = mx + b$ for m **70.** $y = mx + b$ for x **71.** $Ax + By = C$ for y

72. $Ax + By = C$ for x **73.** $M = C(1 + r)$ for r **74.** $C = \frac{5}{9}(F - 32)$ for F

2.6 Ratios and Proportions

OBJECTIVES
1. Write ratios.
2. Solve proportions.
3. Solve applied problems using proportions.

OBJECTIVE 1 Write ratios. A **ratio** is a comparison of two quantities using a quotient.

> **Ratio**
>
> The ratio of the number a to the number b ($b \neq 0$) is written
>
> $$a \text{ to } b, \quad a{:}b, \quad \text{or} \quad \frac{a}{b}.$$

The last way of writing a ratio is most common in algebra.

Percents are ratios where the second number is always 100. For example, 50% represents the ratio of 50 to 100, 27% represents the ratio of 27 to 100, and so on.

EXAMPLE 1 Writing Word Phrases as Ratios

Write a ratio for each word phrase.

(a) The ratio of 5 hr to 3 hr is

$$\frac{5 \text{ hr}}{3 \text{ hr}} = \frac{5}{3}.$$

(b) To find the ratio of 6 hr to 3 days, first convert 3 days to hours.

$$3 \text{ days} = 3 \cdot 24$$
$$= 72 \text{ hr}$$

The ratio of 6 hr to 3 days is thus

$$\frac{6 \text{ hr}}{3 \text{ days}} = \frac{6 \text{ hr}}{72 \text{ hr}} = \frac{6}{72} = \frac{1}{12}.$$

Now Try Exercises 3 and 7.

An application of ratios is in unit pricing, to see which size of an item offered in different sizes produces the best price per unit. To do this, set up the ratio of the price of the item to the number of units on the label. Then divide to obtain the price per unit.

EXAMPLE 2 Finding Price per Unit

The Winn-Dixie supermarket in Mandeville, Louisiana, charges the prices shown in the table for a box of trash bags. Which size is the best buy? That is, which size has the lowest unit price?

Size	Price
10-count	$1.28
20-count	$2.68
30-count	$3.88

To find the best buy, write ratios comparing the price for each box size to the number of units (bags) per box. The results in the table on the next page are rounded to the nearest thousandth.

138 CHAPTER 2 Linear Equations and Inequalities in One Variable

Size	Unit Cost (dollars per bag)
10-count	$\frac{\$1.28}{10} = \$.128$ ← The best buy
20-count	$\frac{\$2.68}{20} = \$.134$
30-count	$\frac{\$3.88}{30} = \$.129$

Because the 10-count size produces the lowest unit cost, it is the best buy. This example shows that buying the largest size does not always provide the best buy, although this is often true.

Now Try Exercise 15.

OBJECTIVE 2 Solve proportions. A ratio is used to compare two numbers or amounts. A **proportion** says that two ratios are equal, so it is a special type of equation. For example,

$$\frac{3}{4} = \frac{15}{20}$$

is a proportion that says that the ratios $\frac{3}{4}$ and $\frac{15}{20}$ are equal. In the proportion

$$\frac{a}{b} = \frac{c}{d} \quad (b, d \neq 0),$$

a, b, c, and d are the **terms** of the proportion. The a and d terms are called the **extremes,** and the b and c terms are called the **means.** We read the proportion $\frac{a}{b} = \frac{c}{d}$ as "a is to b as c is to d." Beginning with this proportion and multiplying each side by the common denominator, bd, gives

$$\frac{a}{b} = \frac{c}{d}$$

$$bd \cdot \frac{a}{b} = bd \cdot \frac{c}{d}$$

$$\frac{b}{b}(d \cdot a) = \frac{d}{d}(b \cdot c) \quad \text{Associative and commutative properties}$$

$$ad = bc. \quad \text{Commutative and identity properties}$$

We can also find the products ad and bc by multiplying diagonally.

$$\frac{a}{b} \diagup\!\!\!\!\diagdown \frac{c}{d}$$

with bc (top) and ad (bottom)

For this reason, ad and bc are called **cross products.**

Cross Products

If $\dfrac{a}{b} = \dfrac{c}{d}$, then the cross products ad and bc are equal.

Also, if $ad = bc$, then $\dfrac{a}{b} = \dfrac{c}{d}$ $(b, d \neq 0)$.

From this rule, if $\frac{a}{b} = \frac{c}{d}$ then $ad = bc$; that is, the product of the extremes equals the product of the means.

> **NOTE** If $\frac{a}{c} = \frac{b}{d}$, then $ad = cb$, or $ad = bc$. This means that the two proportions are equivalent, and
>
> $$\text{the proportion } \frac{a}{b} = \frac{c}{d} \text{ can also be written as } \frac{a}{c} = \frac{b}{d} \quad (c \neq 0).$$
>
> Sometimes one form is more convenient to work with than the other.

EXAMPLE 3 Deciding Whether Proportions Are True

Decide whether each proportion is true or false.

(a) $\dfrac{3}{4} = \dfrac{15}{20}$

Check to see whether the cross products are equal.

$$4 \cdot 15 = 60$$
$$\frac{3}{4} = \frac{15}{20}$$
$$3 \cdot 20 = 60$$

The cross products are equal, so the proportion is true.

(b) $\dfrac{6}{7} = \dfrac{30}{32}$

The cross products are $6 \cdot 32 = 192$ and $7 \cdot 30 = 210$. The cross products are not equal, so the proportion is false.

Now Try Exercises 23 and 25.

Four numbers are used in a proportion. If any three of these numbers are known, the fourth can be found.

EXAMPLE 4 Finding an Unknown in a Proportion

Solve the proportion

$$\frac{5}{9} = \frac{x}{63}.$$

The cross products must be equal.

$$5 \cdot 63 = 9 \cdot x \quad \text{Cross products}$$
$$315 = 9x \quad \text{Multiply.}$$
$$35 = x \quad \text{Divide by 9.}$$

Check by substituting 35 for x in the proportion. The solution set is $\{35\}$.

Now Try Exercise 29.

> **CAUTION** The cross product method cannot be used directly if there is more than one term on either side of the equals sign.

EXAMPLE 5 Solving an Equation Using Cross Products

Solve the equation
$$\frac{m-2}{5} = \frac{m+1}{3}.$$

Find the cross products.

$3(m-2) = 5(m+1)$ Be sure to use parentheses.
$3m - 6 = 5m + 5$ Distributive property
$3m = 5m + 11$ Add 6.
$-2m = 11$ Subtract 5m.
$m = -\frac{11}{2}$ Divide by -2.

The solution set is $\{-\frac{11}{2}\}$.

Now Try Exercise 37.

NOTE When you set cross products equal to each other, you are really multiplying each ratio in the proportion by a common denominator.

OBJECTIVE 3 Solve applied problems using proportions. Proportions are useful in many practical applications. We continue to use the six-step method, although the steps are not numbered here.

EXAMPLE 6 Applying Proportions

After Lee Ann Spahr pumped 5.0 gal of gasoline, the display showing the price read $7.90. When she finished pumping the gasoline, the price display read $21.33. How many gallons did she pump?

To solve this problem, set up a proportion, with prices in the numerators and gallons in the denominators. Make sure that the corresponding numbers appear together.

Let $x =$ the number of gallons she pumped. Then

Price ⟶ $\dfrac{\$7.90}{5.0} = \dfrac{\$21.33}{x}$ ⟵ Price
Gallons ⟶   ⟵ Gallons

$7.90x = 5.0(21.33)$ Cross products
$7.90x = 106.65$ Multiply.
$x = 13.5.$ Divide by 7.90.

She pumped 13.5 gal. Check this answer. Using a calculator to perform the arithmetic reduces the possibility of errors. Notice that the way the proportion was set up uses the fact that the unit price is the same, no matter how many gallons are purchased.

Now Try Exercise 43.

2.6 EXERCISES

1. Match each ratio in Column I with the ratio equivalent to it in Column II.

I	II
(a) 75 to 100	A. 80 to 100
(b) 5 to 4	B. 50 to 100
(c) $\frac{1}{2}$	C. 3 to 4
(d) 4 to 5	D. 15 to 12

2. Give three different, equivalent forms of the ratio $\frac{4}{3}$.

Write a ratio for each word phrase. In Exercises 7–12, first write the amounts with the same units. Write fractions in lowest terms. See Example 1.

3. 60 ft to 70 ft
4. 40 mi to 30 mi
5. 72 dollars to 220 dollars
6. 120 people to 90 people
7. 30 in. to 8 ft
8. 20 yd to 8 ft
9. 16 min to 1 hr
10. 24 min to 2 hr
11. 5 days to 40 hr
12. 60 in. to 2 yd

A supermarket was surveyed to find the prices charged for items in various sizes. Find the best buy (based on price per unit) for each item. See Example 2.

13. Seasoning mix
8-oz size: $1.75
17-oz size: $2.88

14. Red beans
1-lb package: $.89
2-lb package: $1.79

15. Prune juice
32-oz can: $1.95
48-oz can: $2.89
64-oz can: $3.29

16. Corn oil
24-oz bottle: $2.08
64-oz bottle: $3.94
128-oz bottle: $7.65

17. Artificial sweetener packets
50-count: $1.19
100-count: $1.85
250-count: $3.79
500-count: $6.38

18. Chili (no beans)
7.5-oz can: $1.19
10.5-oz can: $1.29
15-oz can: $1.78
25-oz can: $2.59

19. Extra crunchy peanut butter
12-oz size: $1.49
28-oz size: $1.99
40-oz size: $3.99

20. Tomato ketchup
14-oz size: $.93
32-oz size: $1.19
44-oz size: $2.19

21. Explain the distinction between *ratio* and *proportion*. Give examples.

22. Suppose that someone told you to use cross products to multiply fractions. How would you explain to the person what is wrong with his or her thinking?

Decide whether each proportion is true *or* false. *See Example 3.*

23. $\frac{5}{35} = \frac{8}{56}$

24. $\frac{4}{12} = \frac{7}{21}$

25. $\frac{120}{82} = \frac{7}{10}$

26. $\dfrac{27}{160} = \dfrac{18}{110}$ **27.** $\dfrac{\frac{1}{2}}{5} = \dfrac{1}{10}$ **28.** $\dfrac{\frac{1}{3}}{6} = \dfrac{1}{18}$

Solve each equation. See Examples 4 and 5.

29. $\dfrac{k}{4} = \dfrac{175}{20}$ **30.** $\dfrac{x}{6} = \dfrac{18}{4}$ **31.** $\dfrac{49}{56} = \dfrac{z}{8}$

32. $\dfrac{20}{100} = \dfrac{z}{80}$ **33.** $\dfrac{a}{24} = \dfrac{15}{16}$ **34.** $\dfrac{x}{4} = \dfrac{12}{30}$

35. $\dfrac{z}{2} = \dfrac{z+1}{3}$ **36.** $\dfrac{m}{5} = \dfrac{m-2}{2}$ **37.** $\dfrac{3y-2}{5} = \dfrac{6y-5}{11}$

38. $\dfrac{2r+8}{4} = \dfrac{3r-9}{3}$ **39.** $\dfrac{5k+1}{6} = \dfrac{3k-2}{3}$ **40.** $\dfrac{x+4}{6} = \dfrac{x+10}{8}$

41. $\dfrac{2p+7}{3} = \dfrac{p-1}{4}$ **42.** $\dfrac{3m-2}{5} = \dfrac{4-m}{3}$

Solve each problem. See Example 6.

43. If nine pairs of jeans cost $121.50, find the cost of five pairs. (Assume all are equally priced.)

44. If 7 shirts cost $87.50, find the cost of 11 shirts. (Assume all are equally priced.)

45. If 6 gal of premium unleaded gasoline cost $11.34, how much would it cost to completely fill a 15-gal tank?

46. If sales tax on a $16.00 compact disc is $1.32, how much would the sales tax be on a $120.00 compact disc player?

47. The distance between Kansas City, Missouri, and Denver is 600 mi. On a certain wall map, this is represented by a length of 2.4 ft. On the map, how many feet would there be between Memphis and Philadelphia, two cities that are actually 1000 mi apart?

48. The distance between Singapore and Tokyo is 3300 mi. On a certain wall map, this distance is represented by 11 in. The actual distance between Mexico City and Cairo is 7700 mi. How far apart are they on the same map?

49. A chain saw requires a mixture of 2-cycle engine oil and gasoline. According to the directions on a bottle of Oregon 2-cycle Engine Oil, for a 50 to 1 ratio requirement, approximately 2.5 fluid oz of oil are required for 1 gal of gasoline. For 2.75 gal, how many fluid ounces of oil are required?

50. The directions on the bottle mentioned in Exercise 49 indicate that if the ratio requirement is 24 to 1, approximately 5.5 oz of oil are required for 1 gal of gasoline. If gasoline is to be mixed with 22 oz of oil, how much gasoline is to be used?

51. In a recent year, the average exchange rate between British pounds and U.S. dollars was 1 pound to $1.6762. Margaret went to London and exchanged her U.S. currency for British pounds, and received 400 pounds. How much in U.S. dollars did Margaret exchange? (*Note:* Great Britain and Switzerland (see Exercise 52) are among the European nations not using euros.)

SECTION 2.6 Ratios and Proportions 143

52. If 3 U.S. dollars can be exchanged for 4.5204 Swiss francs, how many Swiss francs can be obtained for $49.20? (Round to the nearest hundredth.)

53. Biologists tagged 500 fish in Willow Lake on October 5. At a later date they found 7 tagged fish in a sample of 700. Estimate the total number of fish in Willow Lake to the nearest hundred.

54. On May 13 researchers at Argyle Lake tagged 840 fish. When they returned a few weeks later, their sample of 1000 fish contained 18 that were tagged. Give an approximation of the fish population in Argyle Lake to the nearest hundred.

The International Olympic Committee has come to rely more and more on television rights and major corporate sponsors to finance the games. The circle graphs show funding plans for the first Olympics in Athens and the 1996 Olympics in Atlanta 100 yr later. Use proportions and the graphs to work Exercises 55 and 56.

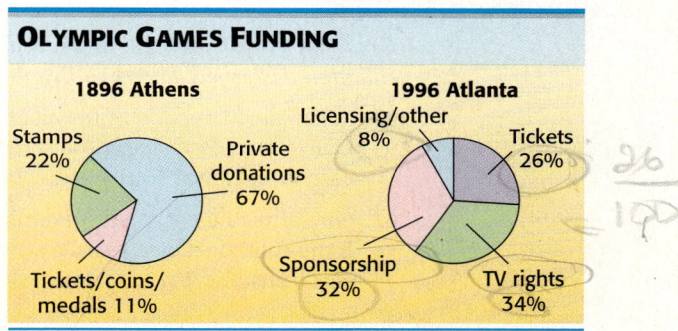

Source: International Olympic Committee.

55. In the 1996 Olympics, total revenue of $350 million was raised. There were 10 major sponsors.

   (a) Write a proportion to find the amount of revenue provided by tickets. Solve it.
   (b) What amount was provided by sponsors? Assuming the sponsors contributed equally, how much was provided per sponsor?
   (c) What amount was raised by TV rights?

56. Suppose the amount of revenue raised in the 1896 Olympics was equivalent to the $350 million in 1996.

   (a) Write a proportion for the amount of revenue provided by stamps and solve it.
   (b) What amount (in dollars) would have been provided by private donations?

Two triangles are **similar** if they have the same shape (but not necessarily the same size). Similar triangles have sides that are proportional. The figure shows two similar triangles. Notice that the ratios of the corresponding sides all equal $\frac{3}{2}$:

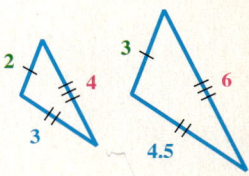

$$\frac{3}{2} = \frac{3}{2} \qquad \frac{4.5}{3} = \frac{3}{2} \qquad \frac{6}{4} = \frac{3}{2}.$$

If we know that two triangles are similar, we can set up a proportion to solve for the length of an unknown side.

(continued)

Use a proportion to find the length x, given that each pair of triangles is similar.

57. 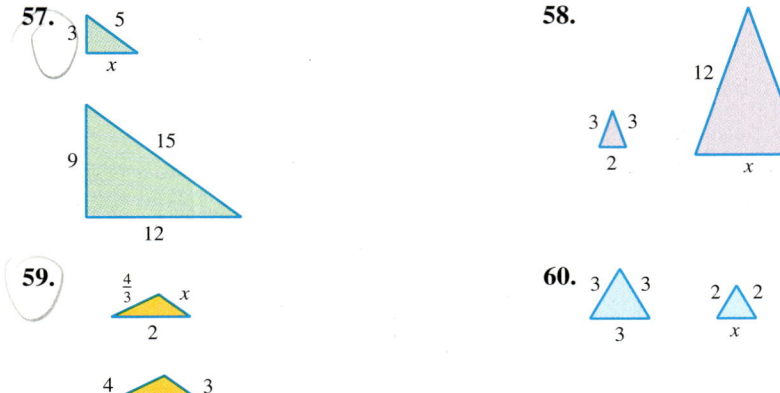 **58.**

59. **60.**

For Exercises 61 and 62, (a) draw a sketch consisting of two right triangles, depicting the situation described, and (b) solve the problem. (Source: Guinness Book of Records.)

61. An enlarged version of the chair used by George Washington at the Constitutional Convention casts a shadow 18 ft long at the same time a vertical pole 12 ft high casts a shadow 4 ft long. How tall is the chair?

62. One of the tallest candles ever constructed was exhibited at the 1897 Stockholm Exhibition. If it cast a shadow 5 ft long at the same time a vertical pole 32 ft high cast a shadow 2 ft long, how tall was the candle?

The Consumer Price Index provides a means of determining the purchasing power of the U.S. dollar from one year to the next. Using the period from 1982 to 1984 as a measure of 100.0, the Consumer Price Index for selected years from 1990 through 2000 is shown in the table. To use the Consumer Price Index to predict a price in a particular year, we set up a proportion and compare it with a known price in another year, as follows:

Year	Consumer Price Index
1990	130.7
1992	140.3
1994	148.2
1996	156.9
1998	163.0
1999	166.6
2000	172.2

Source: U.S. Bureau of Labor Statistics.

$$\frac{\text{price in year } A}{\text{index in year } A} = \frac{\text{price in year } B}{\text{index in year } B}.$$

Use the Consumer Price Index figures in the table to find the amount that would be charged for the use of the same amount of electricity that cost $225 in 1990. Give your answer to the nearest dollar.

63. in 1996   **64.** in 1998   **65.** in 1999   **66.** in 2000

RELATING CONCEPTS (EXERCISES 67–70)

For Individual or Group Work

In Section 2.3 we solved equations with fractions by first multiplying each side of the equation by the common denominator. A proportion with a variable is this kind of equation. **Work Exercises 67–70 in order.** *The steps justify the method of solving a proportion by cross products.*

67. What is the LCD of the fractions in the equation $\dfrac{x}{6} = \dfrac{2}{5}$?

68. Solve the equation in Exercise 67 as follows.
   (a) Multiply each side by the LCD. What equation do you get?
   (b) Solve the equation from part (a) by dividing each side by the coefficient of x.

69. Solve the equation in Exercise 67 using cross products.

70. Compare your solutions from Exercises 68 and 69. What do you notice?

2.7 More about Problem Solving

OBJECTIVES

1. Use percent in problems involving rates.
2. Solve problems involving mixtures.
3. Solve problems involving simple interest.
4. Solve problems involving denominations of money.
5. Solve problems involving distance, rate, and time.

OBJECTIVE 1 Use percent in problems involving rates. Recall that percent means "per hundred." Thus, percents are ratios where the second number is always 100. For example, 50% represents the ratio of 50 to 100 and 27% represents the ratio of 27 to 100.

PROBLEM SOLVING

Percents are often used in problems involving mixing different concentrations of a substance or different interest rates. In each case, to get the amount of pure substance or the interest, we multiply.

Mixture Problems	Interest Problems (annual)
base × rate (%) = percentage	principal × rate (%) = interest
$b \times r = p$	$p \times r = I$

In an equation, *percent is always written as a decimal.* For example, 35% is written .35, *not* 35, and 7% is written .07, *not* 7.

EXAMPLE 1 Using Percents to Find Percentages

(a) If a chemist has 40 L of a 35% acid solution, then the amount of pure acid in the solution is

$$40 \times .35 = 14 \text{ L.}$$

Amount of solution — Rate of concentration — Amount of pure acid

(b) If $1300 is invested for one year at 7% simple interest, the amount of interest earned in the year is

$$\$1300 \times .07 = \$91.$$

Principal — Interest rate — Interest earned

Now Try Exercises 1 and 3.

146 CHAPTER 2 Linear Equations and Inequalities in One Variable

> ### PROBLEM SOLVING
> In the examples that follow, we use tables to organize the information in the problems. A table enables us to more easily set up an equation, which is usually the most difficult step.

OBJECTIVE 2 Solve problems involving mixtures. In the next example, we use percent to solve a mixture problem.

EXAMPLE 2 Solving a Mixture Problem

A chemist needs to mix 20 L of 40% acid solution with some 70% acid solution to get a mixture that is 50% acid. How many liters of the 70% acid solution should be used?

Step 1 **Read** the problem. Note the percent of each solution and of the mixture.

Step 2 **Assign a variable.**

Let x = the number of liters of 70% acid solution needed.

Recall from Example 1(a) that the amount of pure acid in this solution will be given by the product of the percent of strength and the number of liters of solution, or

liters of pure acid in x L of 70% solution = $.70x$.

The amount of pure acid in the 20 L of 40% solution is

liters of pure acid in the 40% solution = $.40(20) = 8$.

The new solution will contain $(x + 20)$ L of 50% solution. The amount of pure acid in this solution is

liters of pure acid in the 50% solution = $.50(x + 20)$.

Figure 11 illustrates this information, which is summarized in the table.

After mixing

+ = ← from 40%
from 70% from 40% 50% ← from 70%

Unknown number   20 L $(x + 20)$ L
of liters, x

FIGURE 11

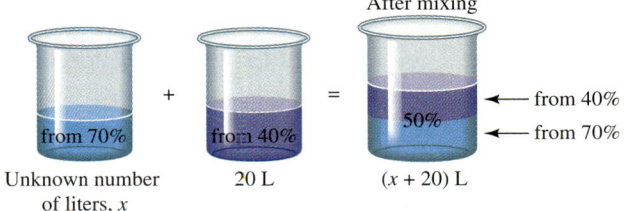

Step 3 **Write an equation.** The number of liters of pure acid in the 70% solution added to the number of liters of pure acid in the 40% solution will equal the number of liters of pure acid in the final mixture, so the equation is

$$\underset{\text{Pure acid in 70%}}{.70x} + \underset{\text{pure acid in 40%}}{.40(20)} = \underset{\text{pure acid in 50%.}}{.50(x + 20).}$$

Step 4 **Solve** the equation. Clear parentheses on the right side, then multiply by 100 to clear decimals.

$.70x + .40(20) = .50x + .50(20)$ Distributive property
$70x + 40(20) = 50x + 50(20)$ Multiply by 100.
$70x + 800 = 50x + 1000$
$20x + 800 = 1000$ Subtract 50x.
$20x = 200$ Subtract 800.
$x = 10$ Divide by 20.

Step 5 **State the answer.** The chemist needs to use 10 L of 70% solution.

Step 6 **Check.** Since

$$.70(10) + .40(20) = 7 + 8 = 15$$

and

$$.50(10 + 20) = .50(30) = 15,$$

the answer checks.

Now Try Exercise 15.

NOTE In a problem such as Example 2, the concentration of the final mixture must be *between* the concentrations of the two solutions making up the mixture.

OBJECTIVE 3 Solve problems involving simple interest. The next example uses the formula for simple interest, $I = prt$. Remember that when $t = 1$, the formula becomes $I = pr$, as shown in the Problem-Solving box at the beginning of this section. Once again, the idea of multiplying the total amount (principal) by the rate (rate of interest) gives the percentage (amount of interest).

■ **EXAMPLE 3** Solving a Simple Interest Problem

Elizabeth Suco receives an inheritance. She plans to invest part of it at 9% and $2000 more than this amount at 10%. To earn $1150 per year in interest, how much should she invest at each rate?

Step 1 **Read** the problem again.

Step 2 **Assign a variable.**

Let $x =$ the amount invested at 9% (in dollars).
Then $x + 2000 =$ the amount invested at 10% (in dollars).

148 CHAPTER 2 Linear Equations and Inequalities in One Variable

Use a table to arrange the information given in this problem.

Amount Invested in Dollars	Rate of Interest	Interest for One Year
x	.09	.09x
x + 2000	.10	.10(x + 2000)

Step 3 **Write an equation.** Multiply amount by rate to get the interest earned. Since the total interest is to be $1150, the equation is

Interest at 9%	plus	interest at 10%	is	total interest.
↓	↓	↓	↓	↓
.09x	+	.10(x + 2000)	=	1150.

Step 4 **Solve** the equation. Clear parentheses; then clear decimals.

$.09x + .10x + .10(2000) = 1150$ Distributive property
$9x + 10x + 10(2000) = 115{,}000$ Multiply by 100.
$9x + 10x + 20{,}000 = 115{,}000$
$19x + 20{,}000 = 115{,}000$ Combine like terms.
$19x = 95{,}000$ Subtract 20,000.
$x = 5000$ Divide by 19.

Step 5 **State the answer.** She should invest $5000 at 9% and $5000 + $2000 = $7000 at 10%.

Step 6 **Check.** Investing $5000 at 9% and $7000 at 10% gives total interest of .09($5000) + .10($7000) = $450 + $700 = $1150, as required in the original problem.

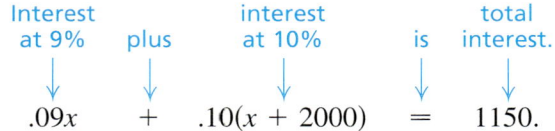

Now Try Exercise 25.

OBJECTIVE 4 Solve problems involving denominations of money.

PROBLEM SOLVING

Problems that involve different denominations of money or items with different monetary values are very similar to mixture and interest problems. To get the total value, we multiply.

Money Problems

number × value of one item = total value

For example, 30 dimes have a monetary value of 30($.10) = $3. Fifteen $5 bills have a value of 15($5) = $75. A table is helpful for these problems, too.

EXAMPLE 4 Solving a Money Problem

A bank teller has 25 more $5 bills than $10 bills. The total value of the money is $200. How many of each denomination of bill does she have?

SECTION 2.7 More about Problem Solving 149

Step 1 **Read** the problem. We must find the number of each denomination of bill that the teller has.

Step 2 **Assign a variable.**
Let x = the number of $10 bills.
Then $x + 25$ = the number of $5 bills.

Organize the given information in a table.

Number of Bills	Denomination	Total Value
x	10	$10x$
$x + 25$	5	$5(x + 25)$

Step 3 **Write an equation.** Multiplying the number of bills by the denomination gives the monetary value. The value of the tens added to the value of the fives must be $200.

Value of tens plus value of fives is $200.

$$10x + 5(x + 25) = 200$$

Step 4 **Solve.**

$10x + 5x + 125 = 200$ Distributive property
$15x + 125 = 200$ Combine like terms.
$15x = 75$ Subtract 125.
$x = 5$ Divide by 15.

Step 5 **State the answer.** The teller has 5 tens and $5 + 25 = 30$ fives.

Step 6 **Check.** The teller has $30 - 5 = 25$ more fives, and the value of the money is $5(\$10) + 30(\$5) = \$200$, as required.

Now Try Exercise 29.

OBJECTIVE 5 Solve problems involving distance, rate, and time. If your car travels at an average rate of 50 mph for 2 hr, then it travels $50 \times 2 = 100$ mi. This is an example of the basic relationship between distance, rate, and time,

$$\text{distance} = \text{rate} \times \text{time},$$

given by the formula $d = rt$. By solving, in turn, for r and t in the formula, we obtain two other equivalent forms of the formula. The three forms are given here.

Distance, Rate, and Time Relationship

$$d = rt \qquad r = \frac{d}{t} \qquad t = \frac{d}{r}$$

The next examples illustrate the uses of these formulas.

EXAMPLE 5 Finding Distance, Rate, or Time

(a) The speed of sound is 1088 ft per sec at sea level at 32°F. In 5 sec under these conditions, sound travels

$$1088 \times 5 = 5440 \text{ ft.}$$
$$\text{Rate} \times \text{Time} = \text{Distance}$$

Here, we found distance given rate and time, using $d = rt$.

(b) The winner of the first Indianapolis 500 race (in 1911) was Ray Harroun, driving a Marmon Wasp at an average speed of 74.59 mph. (*Source: Universal Almanac,* 1997.) To complete the 500 mi, it took him

$$\frac{\text{Distance} \rightarrow 500}{\text{Rate} \rightarrow 74.59} = 6.70 \text{ hr} \quad \text{(rounded)}. \leftarrow \text{Time}$$

Here, we found time given rate and distance using $t = \frac{d}{r}$. To convert .70 hr to minutes, we multiply by 60 to get .70(60) = 42. It took Harroun about 6 hr, 42 min to complete the race.

(c) At the 2000 Olympic Games in Sydney, Australia, Dutch swimmer Inge de Bruijn set a world record in the women's 50-m freestyle swimming event of 24.13 sec. (*Source: World Almanac and Book of Facts,* 2002.) Her rate was

$$\text{Rate} = \frac{\text{Distance} \rightarrow 50}{\text{Time} \rightarrow 24.13} = 2.07 \text{ m per sec (rounded)}.$$

Now Try Exercises 37, 39, and 41.

EXAMPLE 6 Solving a Motion Problem

Two cars leave Baton Rouge, Louisiana, at the same time and travel east on Interstate 10. One travels at a constant speed of 55 mph, and the other travels at a constant speed of 63 mph. In how many hours will the distance between them be 24 mi?

Step 1 **Read** the problem. We must find the time it will take for the distance between the cars to be 24 mi.

Step 2 **Assign a variable.** Since we are looking for time, we let $t =$ the number of hours until the distance between them is 24 mi. The sketch in Figure 12 shows what is happening in the problem.

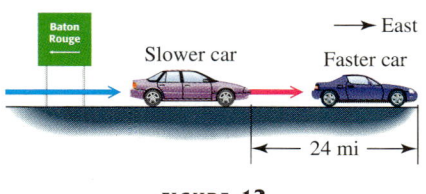

FIGURE 12

Now we construct a table using the information given in the problem and t for the time traveled by each car. We multiply rate by time to get the expressions for distances traveled.

SECTION 2.7 More about Problem Solving   151

	Rate	×	Time	=	Distance
Faster Car	63		t		63t
Slower Car	55		t		55t

Difference is 24 mi.

The quantities 63t and 55t represent the two distances. Refer to Figure 12, and notice that the *difference* between the larger distance and the smaller distance is 24 mi.

Step 3 **Write an equation.**
$$63t - 55t = 24$$

Step 4 **Solve.** $8t = 24$ Combine like terms.
                           $t = 3$ Divide by 8.

Step 5 **State the answer.** It will take the cars 3 hr to be 24 mi apart.

Step 6 **Check.** After 3 hr the faster car will have traveled $63 \times 3 = 189$ mi, and the slower car will have traveled $55 \times 3 = 165$ mi. Since $189 - 165 = 24$, the conditions of the problem are satisfied.

Now Try Exercise 45.

NOTE In motion problems like the one in Example 6, once you have filled in two pieces of information in each row of the table, you should automatically fill in the third piece of information, using the appropriate form of the formula relating distance, rate, and time. Set up the equation based on your sketch and the information in the table.

EXAMPLE 7 Solving a Motion Problem

Two planes leave Los Angeles at the same time. One heads south to San Diego; the other heads north to San Francisco. The San Francisco plane flies 50 mph faster. In $\frac{1}{2}$ hr, the planes are 275 mi apart. What are their speeds?

Step 1 **Read** the problem carefully.

Step 2 **Assign a variable.**

Let $r =$ the speed of the slower plane.
Then $r + 50 =$ the speed of the faster plane.

Fill in a table.

	Rate	Time	Distance
Slower Plane	r	$\frac{1}{2}$	$\frac{1}{2}r$
Faster Plane	r + 50	$\frac{1}{2}$	$\frac{1}{2}(r+50)$

Sum is 275 mi.

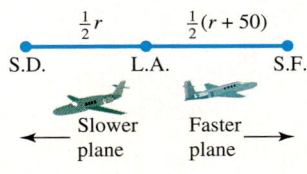

FIGURE 13

Step 3 **Write an equation.** As Figure 13 shows, the planes are headed in *opposite* directions. The *sum* of their distances equals 275 mi, so

$$\frac{1}{2}r + \frac{1}{2}(r + 50) = 275.$$

Step 4 Solve. $\frac{1}{2}r + \frac{1}{2}(r + 50) = 275$

$r + (r + 50) = 550$ Multiply by 2.

$2r + 50 = 550$ Combine like terms.

$2r = 500$ Subtract 50.

$r = 250$ Divide by 2.

Step 5 **State the answer.** The slower plane (headed south) has a speed of 250 mph. The speed of the faster plane is 250 + 50 = 300 mph.

Step 6 **Check.** Verify that $\frac{1}{2}(250) + \frac{1}{2}(300) = 275$ mi.

Now Try Exercise 51.

Another way to solve the problems in this section is given in Chapter 4.

2.7 EXERCISES

For Extra Help

 Student's Solutions Manual

 MyMathLab

 InterAct Math Tutorial Software

 AW Math Tutor Center

 MathXL

Digital Video Tutor CD 5/Videotape 4

Use the concepts of this section to answer each question. See Example 1 and the Problem-Solving box before Example 4.

1. How much pure acid is in 250 mL of a 14% acid solution?
2. How much pure alcohol is in 150 L of a 30% alcohol solution?
3. If $10,000 is invested for 1 yr at 3.5% simple interest, how much interest is earned?
4. If $25,000 is invested at 3% simple interest for 2 yr, how much interest is earned?
5. What is the monetary amount of 283 nickels?
6. What is the monetary amount of 35 half-dollars?

Solve each percent problem. Remember that base × rate = percentage.

7. The 2000 U.S. Census showed that the population of Alabama was 4,447,000, with 26.0% represented by African-Americans. What is the best estimate of the African-American population in Alabama? (*Source:* U.S. Bureau of the Census.)

   A. 500,000   B. 750,000   C. 1,100,000   D. 1,500,000

8. The 2000 U.S. Census showed that the population of New Mexico was 1,819,000, with 42.1% being Hispanic. What is the best estimate of the Hispanic population in New Mexico? (*Source:* U.S. Bureau of the Census.)

   A. 720,000   B. 72,000   C. 650,000
   D. 36,000

9. The graph shows the breakdown, by approximate percents, of colors chosen for new compact/sports cars in the 2000 model year. If about 2.5 million compact/sports cars were sold in 2000, about how many were each color? (*Source:* Ward's Communications.)

   (a) Red   (b) White   (c) Black

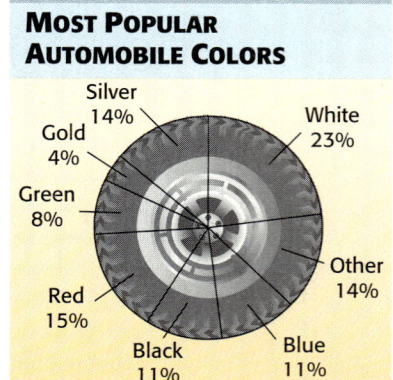

MOST POPULAR AUTOMOBILE COLORS

Silver 14%
White 23%
Gold 4%
Green 8%
Other 14%
Red 15%
Black 11%
Blue 11%

Source: DuPont Automotive Products.

SECTION 2.7 More about Problem Solving 153

10. An average middle-income family will spend $160,140 to raise a child born in 1999 from birth to age 17. The graph shows the breakdown, by approximate percents, for various expense categories. To the nearest dollar, about how much will be spent to provide the following?

    (a) Housing   (b) Food
    (c) Health care

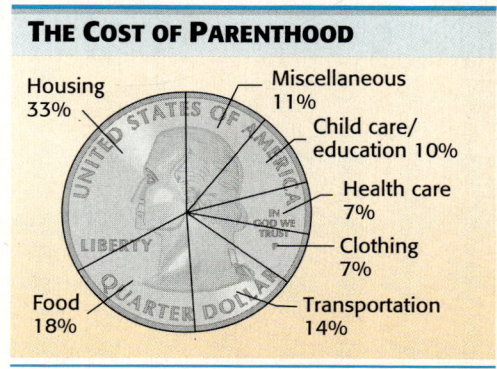

THE COST OF PARENTHOOD

Housing 33%
Miscellaneous 11%
Child care/education 10%
Health care 7%
Clothing 7%
Transportation 14%
Food 18%

Source: U.S. Department of Agriculture.

11. In 1998, the U.S. civilian labor force consisted of 137,673,000 persons. Of this total, 6,210,000 were unemployed. To the nearest tenth, what was the percent of unemployment? (*Source:* U.S. Bureau of Labor Statistics.)

12. In 1998, the U.S. labor force (excluding agricultural employees, self-employed persons, and the unemployed) consisted of 116,730,000 persons. Of this total, 16,211,000 were union members. To the nearest tenth, what percent of this labor force belonged to unions? (*Source:* U.S. Bureau of Labor Statistics.)

Use the concepts of this section to answer each question.

13. Suppose that a chemist is mixing two acid solutions, one of 20% concentration and the other of 30% concentration. Which one of the following concentrations could *not* be obtained?

    A. 22%   B. 24%   C. 28%   D. 32%

14. Suppose that pure alcohol is added to a 24% alcohol mixture. Which one of the following concentrations could *not* be obtained?

    A. 22%   B. 26%   C. 28%   D. 30%

Work each mixture problem. See Example 2.

15. How many gallons of 50% antifreeze must be mixed with 80 gal of 20% antifreeze to get a mixture that is 40% antifreeze?

Gallons of Mixture	Rate	Gallons of Antifreeze
x	.50	.50x
80	.20	.20(80)
x + 80	.40	.40(x + 80)

16. How many liters of 25% acid solution must be added to 80 L of 40% solution to get a solution that is 30% acid?

Liters of Solution	Rate	Liters of Acid
x	.25	.25x
80	.40	.40(80)
x + 80	.30	.30(x + 80)

17. A certain metal is 20% tin. How many kilograms of this metal must be mixed with 80 kg of a metal that is 70% tin to get a metal that is 50% tin?

Kilograms of Metal	Rate	Kilograms of Pure Tin
x	.20	
	.70	
	.50	

18. A pharmacist has 20 L of a 10% drug solution. How many liters of 5% solution must be added to get a mixture that is 8%?

Liters of Solution	Rate	Liters of Pure Drug
20	.10	20(.10)
	.05	
	.08	

19. In a chemistry class, 12 L of a 12% alcohol solution must be mixed with a 20% solution to get a 14% solution. How many liters of the 20% solution are needed?

20. How many liters of a 60% acid solution must be mixed with a 75% acid solution to get 20 L of a 72% solution?

21. How many liters of a 10% alcohol solution must be mixed with 40 L of a 50% solution to get a 40% solution?

22. How many gallons of a 12% indicator solution must be mixed with a 20% indicator solution to get 10 gal of a 14% solution?

23. Minoxidil is a drug that has recently proven to be effective in treating male pattern baldness. A pharmacist wishes to mix a solution that is 2% minoxidil. She has on hand 50 mL of a 1% solution, and she wishes to add some 4% solution to it to obtain the desired 2% solution. How much 4% solution should she add?

24. Water must be added to 20 mL of a 4% minoxidil solution to dilute it to a 2% solution. How many milliliters of water should be used? (*Hint:* Water is 0% minoxidil.)

Work each investment problem using simple interest. See Example 3.

25. Li Nguyen invested some money at 3% and $4000 less than that amount at 5%. The two investments produced a total of $200 interest in 1 yr. How much was invested at each rate?

26. LaShondra Williams inherited some money from her uncle. She deposited part of the money in a savings account paying 2%, and $3000 more than that amount in a different account paying 3%. Her annual interest income was $690. How much did she deposit at each rate?

27. With income earned by selling the rights to his life story, an actor invests some of the money at 3% and $30,000 more than twice as much at 4%. The total annual interest earned from the investments is $5600. How much is invested at each rate?

28. An artist invests her earnings in two ways. Some goes into a tax-free bond paying 6%, and $6000 more than three times as much goes into mutual funds paying 5%. Her total annual interest income from the investments is $825. How much does she invest at each rate?

Work each problem involving monetary values. See Example 4.

29. A bank teller has some $5 bills and some $20 bills. The teller has 5 more twenties than fives. The total value of the money is $725. Find the number of $5 bills that the teller has.

Number of Bills	Denomination	Total Value
x	5	
x + 5	20	

SECTION 2.7 More about Problem Solving 155

30. A coin collector has $1.70 in dimes and nickels. She has 2 more dimes than nickels. How many nickels does she have?

Number of Coins	Denomination	Total Value
x	.05	.05x
	.10	

31. A cashier has a total of 126 bills, made up of fives and tens. The total value of the money is $840. How many of each kind does he have?

32. A convention manager finds that she has $1290, made up of twenties and fifties. She has a total of 42 bills. How many of each kind does she have?

33. A merchant wishes to mix candy worth $5 per lb with 40 lb of candy worth $2 per lb to get a mixture that can be sold for $3 per lb. How many pounds of $5 candy should be used?

34. At Vern's Grill, hamburgers cost $2.70 each, and a bag of french fries costs $1.20. How many hamburgers and how many bags of french fries can a customer buy with $26.40 if he wants twice as many hamburgers as bags of french fries?

Vern's Grill
Hamburgers..........$2.70 each
French fries..........$1.20 a bag
Soda pop..............$ .75 each
Shakes................$2.70 each
Cherry pie...........$2.70 a slice
Ice cream............$1.20 a scoop

Solve each problem involving distance, rate, and time. See Example 5.

35. Which choice is the best estimate for the average speed of a trip of 405 mi that lasted 8.2 hr?

   A. 50 mph   B. 30 mph   C. 60 mph   D. 40 mph

36. Suppose that an automobile averages 45 mph, and travels for 30 min. Is the distance traveled 45 × 30 = 1350 mi? If not, explain why not, and give the correct distance.

37. A driver averaged 53 mph and took 10 hr to travel from Memphis to Chicago. What is the distance between Memphis and Chicago?

38. A small plane traveled from Warsaw to Rome, averaging 164 mph. The trip took 2 hr. What is the distance from Warsaw to Rome?

39. The winner of the 1998 Charlotte 500 (mile) race was Mark Martin, who drove his Ford to victory with a rate of 123.188 mph. What was his time? (*Source: Sports Illustrated 2000 Sports Almanac.*)

40. In 1998, Jeff Gordon drove his Chevrolet to victory in the North Carolina 400 (mile) race. His rate was 128.423 mph. What was his time? (*Source: Sports Illustrated 2000 Sports Almanac.*)

156 CHAPTER 2 Linear Equations and Inequalities in One Variable

In Exercises 41–44, find the rate based on the information provided. Use a calculator and round your answers to the nearest hundredth. All events were at the 2000 Summer Olympics. (Source: http://espn.go.com/oly/summer00)

Event	Participant	Distance	Time
41. 100-m hurdles, Women	Olga Shishigina, Kazakhstan	100 m	12.65 sec
42. 400-m hurdles, Women	Irina Privalova, Russia	400 m	53.02 sec
43. 400-m hurdles, Men	Angelo Taylor, USA	400 m	47.50 sec
44. 400-m dash, Men	Michael Johnson, USA	400 m	43.84 sec

Solve each motion problem. See Examples 6 and 7.

45. St. Louis and Portland are 2060 mi apart. A small plane leaves Portland, traveling toward St. Louis at an average speed of 90 mph. Another plane leaves St. Louis at the same time, traveling toward Portland, averaging 116 mph. How long will it take them to meet?

	r	t	d
Plane Leaving Portland	90	t	90t
Plane Leaving St. Louis	116	t	116t

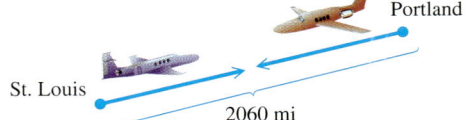

46. Atlanta and Cincinnati are 440 mi apart. John leaves Cincinnati, driving toward Atlanta at an average speed of 60 mph. Pat leaves Atlanta at the same time, driving toward Cincinnati in her antique auto, averaging 28 mph. How long will it take them to meet?

	r	t	d
John	60	t	60t
Pat	28	t	28t

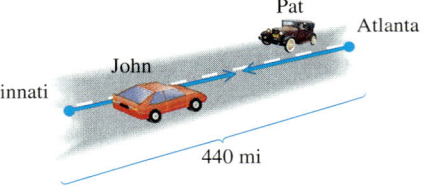

47. Two steamers leave a port on a river at the same time, traveling in opposite directions. Each is traveling 22 mph. How long will it take for them to be 110 mi apart?

48. A train leaves Kansas City, Kansas, and travels north at 85 km per hr. Another train leaves at the same time and travels south at 95 km per hour. How long will it take before they are 315 km apart?

49. At a given hour two steamboats leave a city in the same direction on a straight canal. One travels at 18 mph and the other travels at 25 mph. In how many hours will the boats be 35 mi apart?

50. From a point on a straight road, Lupe and Maria ride bicycles in the same direction. Lupe rides 10 mph and Maria rides 12 mph. In how many hours will they be 5 mi apart?

51. Two trains leave a city at the same time. One travels north and the other travels south at 20 mph faster. In 2 hr the trains are 280 mi apart. Find their speeds.

	r	t	d
Northbound	x	2	
Southbound	x + 20	2	

52. Two planes leave an airport at the same time, one flying east, the other flying west. The eastbound plane travels 150 mph slower. They are 2250 mi apart after 3 hr. Find the speed of each plane.

	r	t	d
Eastbound	x − 150	3	
Westbound	x	3	

53. Two cars leave towns 230 km apart at the same time, traveling directly toward one another. One car travels 15 km per hr slower than the other. They pass one another 2 hr later. What are their speeds?

54. Two cars start from towns 400 mi apart and travel toward each other. They meet after 4 hr. Find the speed of each car if one travels 20 mph faster than the other.

2.8 Solving Linear Inequalities

OBJECTIVES

1. Graph intervals on a number line.
2. Use the addition property of inequality.
3. Use the multiplication property of inequality.
4. Solve linear inequalities using both properties of inequality.
5. Solve applied problems using inequalities.
6. Solve linear inequalities with three parts.

Inequalities are algebraic expressions related by

    <   "is less than,"   ≤   "is less than or equal to,"

    >   "is greater than,"   ≥   "is greater than or equal to."

We solve an inequality by finding all real number solutions for it. For example, the solution set of $x \leq 2$ includes *all real numbers* that are less than or equal to 2, not just the *integers* less than or equal to 2.

OBJECTIVE 1 Graph intervals on a number line. A good way to show the solution set of an inequality is by graphing. We graph all the real numbers satisfying $x \leq 2$ by placing a square bracket at 2 on a number line and drawing an arrow extending from the bracket to the left (to represent the fact that all numbers less than 2 are also part of the graph). The graph is shown in Figure 14.

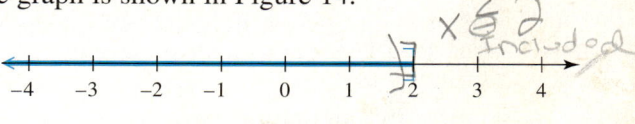

FIGURE 14

The set of numbers less than or equal to 2 is an example of an **interval** on the number line. To write intervals, we use **interval notation.** For example, using this notation, the interval of all numbers less than or equal to 2 is written as $(-\infty, 2]$. The **negative infinity** symbol $-\infty$ does not indicate a number. It is used to show that the interval includes all real numbers less than 2. As on the number line, the square bracket indicates that 2 is part of the solution. A parenthesis is always used next to the infinity symbol. The set of real numbers is written in interval notation as $(-\infty, \infty)$.

EXAMPLE 1 Graphing Intervals Written in Interval Notation on a Number Line

Write each inequality in interval notation and graph the interval.

(a) $x > -5$

The statement $x > -5$ says that x can represent any number greater than -5 but cannot equal -5. The interval is written $(-5, \infty)$. We show this on a graph by placing a parenthesis at −5 and drawing an arrow to the right, as shown in Figure 15. The parenthesis at −5 indicates that −5 is not part of the graph.

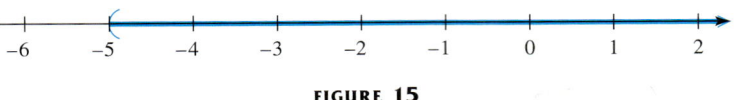

FIGURE 15

(b) $-1 \leq x < 3$

The statement is read "−1 is less than or equal to x and x is less than 3." Thus, we want the set of numbers that are *between* −1 and 3, with −1 included and 3 excluded. In interval notation, we write $[-1, 3)$, using a square bracket at −1 because −1 is part of the graph, and a parenthesis at 3 because 3 is not part of the graph. The graph is shown in Figure 16.

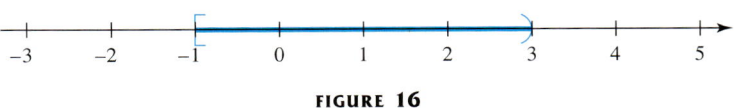

FIGURE 16

Now Try Exercises 7 and 15.

OBJECTIVE 2 Use the addition property of inequality. Solving inequalities is similar to solving equations.

Linear Inequality in One Variable

A **linear inequality in one variable** can be written in the form

$$Ax + B < C,$$

where A, B, and C are real numbers, with $A \neq 0$.

Examples of linear inequalities in one variable include

$$x + 5 < 2, \quad y - 3 \geq 5, \quad \text{and} \quad 2k + 5 \leq 10.$$

(Throughout this section we give definitions and rules only for $<$, but they are also valid for $>$, $\leq$, and $\geq$.)

Consider the inequality $2 < 5$. If 4 is added to each side of this inequality, the result is

$$2 + 4 < 5 + 4$$
$$6 < 9,$$

a true sentence. Now subtract 8 from each side:

$$2 - 8 < 5 - 8$$
$$-6 < -3.$$

The result is again a true sentence. These examples suggest the **addition property of inequality,** which states that the same real number can be added to each side of an inequality without changing the solutions.

> **Addition Property of Inequality**
>
> For any real numbers A, B, and C, the inequalities
>
> $$A < B \quad \text{and} \quad A + C < B + C$$
>
> have exactly the same solutions.
>   That is, the same number may be added to each side of an inequality without changing the solutions.

As with the addition property of equality, the same number may also be *subtracted* from each side of an inequality.

EXAMPLE 2 Using the Addition Property of Inequality

Solve $7 + 3k > 2k - 5$.
  Use the addition property of inequality twice, once to get the terms containing k alone on one side of the inequality and a second time to get the integers together on the other side. (These steps can be done in either order.)

$$7 + 3k > 2k - 5$$
$$7 + 3k - 2k > 2k - 5 - 2k \quad \text{Subtract } 2k.$$
$$7 + k > -5 \quad \text{Combine like terms.}$$
$$7 + k - 7 > -5 - 7 \quad \text{Subtract 7.}$$
$$k > -12 \quad \text{Combine like terms.}$$

The solution set is $(-12, \infty)$. Its graph is shown in Figure 17.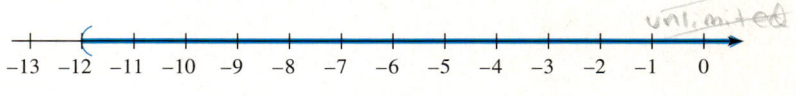

FIGURE 17

Now Try Exercise 21.

OBJECTIVE 3 Use the multiplication property of inequality. The addition property of inequality cannot be used to solve inequalities such as $4y \geq 28$. These inequalities require the *multiplication property of inequality*. To see how this property works, we look at some examples.
  First, start with the inequality $3 < 7$ and multiply each side by the positive number 2.

$$3 < 7$$
$$2(3) < 2(7) \quad \text{Multiply each side by 2.}$$
$$6 < 14 \quad \text{True}$$

Now multiply each side of $3 < 7$ by the negative number -5.

$$3 < 7$$
$$-5(3) < -5(7) \quad \text{Multiply each side by } -5.$$
$$-15 < -35 \quad \text{False}$$

To get a true statement when multiplying each side by -5, we must reverse the direction of the inequality symbol.

$$3 < 7$$
$$-5(3) > -5(7) \quad \text{Multiply by } -5; \text{ reverse the symbol.}$$
$$-15 > -35 \quad \text{True}$$

Take the inequality $-6 < 2$ as another example. Multiply each side by the positive number 4.

$$-6 < 2$$
$$4(-6) < 4(2) \quad \text{Multiply by 4.}$$
$$-24 < 8 \quad \text{True}$$

Multiplying each side of $-6 < 2$ by -5 *and at the same time reversing the direction of the inequality symbol* gives

$$-6 < 2$$
$$-5(-6) > -5(2) \quad \text{Multiply by } -5; \text{ reverse the symbol.}$$
$$30 > -10. \quad \text{True}$$

In summary, the **multiplication property of inequality** has two parts.

Multiplication Property of Inequality

For any real numbers A, B, and C, with $C \neq 0$,

1. if C is *positive*, then the inequalities

$$A < B \quad \text{and} \quad AC < BC$$

have exactly the same solutions;

2. if C is *negative*, then the inequalities

$$A < B \quad \text{and} \quad AC > BC$$

have exactly the same solutions.

That is, each side of an inequality may be multiplied by the same positive number without changing the solutions. If the multiplier is negative, we must reverse the direction of the inequality symbol.

The multiplication property of inequality also permits *division* of each side of an inequality by the same nonzero number.

It is important to remember the differences in the multiplication property for positive and negative numbers.

1. When each side of an inequality is multiplied or divided by a positive number, the direction of the inequality symbol *does not change*. (Adding or subtracting terms on each side also does not change the symbol.)

2. When each side of an inequality is multiplied or divided by a negative number, the direction of the symbol *does change*. **Reverse the direction of the inequality symbol only when multiplying or dividing each side by a negative number.**

EXAMPLE 3 Using the Multiplication Property of Inequality

Solve each inequality and graph the solution set.

(a) $3r < -18$

Using the multiplication property of inequality, we divide each side by 3. Since 3 is a positive number, the direction of the inequality symbol *does not* change.

$$3r < -18$$
$$\frac{3r}{3} < \frac{-18}{3} \quad \text{Divide by 3.}$$
$$r < -6$$

The solution set is $(-\infty, -6)$. The graph is shown in Figure 18.

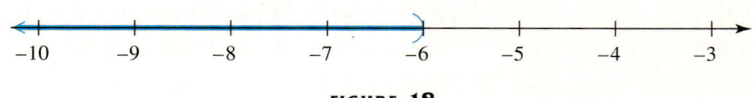

FIGURE 18

(b) $-4t \geq 8$

Here each side of the inequality must be divided by -4, a negative number, which *does* change the direction of the inequality symbol.

$$-4t \geq 8$$
$$\frac{-4t}{-4} \leq \frac{8}{-4} \quad \text{Divide by } -4\text{; reverse the symbol.}$$
$$t \leq -2$$

The solution set $(-\infty, -2]$ is graphed in Figure 19.

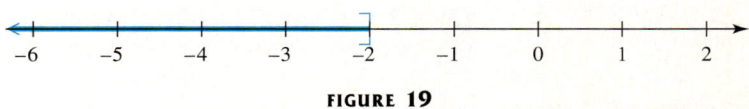

FIGURE 19

Now Try Exercises 31 and 33.

CAUTION Even though the number on the right side of the inequality in Example 3(a) is negative (-18), *do not reverse the direction of the inequality symbol.* Reverse the symbol only when multiplying or dividing by a negative number, as shown in Example 3(b).

OBJECTIVE 4 Solve linear inequalities using both properties of inequality. To solve a linear inequality, follow these steps.

Solving a Linear Inequality

Step 1 **Simplify each side separately.** Use the distributive property to clear parentheses and combine like terms on each side as needed.

(continued)

Step 2 **Isolate the variable terms on one side.** Use the addition property of inequality to get all terms with variables on one side of the inequality and all numbers on the other side.

Step 3 **Isolate the variable.** Use the multiplication property of inequality to change the inequality to the form $x < k$ or $x > k$.

Remember: Reverse the direction of the inequality symbol *only when multiplying or dividing each side of an inequality by a negative number.*

EXAMPLE 4 Solving a Linear Inequality

Solve $3z + 2 - 5 > -z + 7 + 2z$ and graph the solution set.

Step 1 Combine like terms and simplify.

$$3z + 2 - 5 > -z + 7 + 2z$$
$$3z - 3 > z + 7$$

Step 2 Use the addition property of inequality.

$$3z - 3 + 3 > z + 7 + 3 \quad \text{Add 3.}$$
$$3z > z + 10$$
$$3z - z > z + 10 - z \quad \text{Subtract } z.$$
$$2z > 10$$

Step 3 Use the multiplication property of inequality.

$$\frac{2z}{2} > \frac{10}{2} \quad \text{Divide by 2.}$$
$$z > 5$$

Since 2 is positive, the direction of the inequality symbol is not changed in Step 3. The solution set is $(5, \infty)$. Its graph is shown in Figure 20.

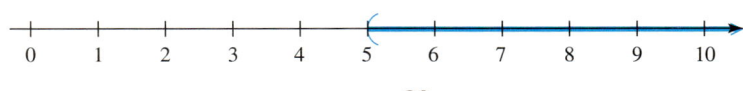

FIGURE 20

Now Try Exercise 43.

EXAMPLE 5 Solving a Linear Inequality

Solve $5(k - 3) - 7k \geq 4(k - 3) + 9$ and graph the solution set.

Step 1 Simplify and combine like terms.

$$5(k - 3) - 7k \geq 4(k - 3) + 9$$
$$5k - 15 - 7k \geq 4k - 12 + 9 \quad \text{Distributive property}$$
$$-2k - 15 \geq 4k - 3 \quad \text{Combine like terms.}$$

Step 2 Use the addition property.

$$-2k - 15 - 4k \geq 4k - 3 - 4k \quad \text{Subtract } 4k.$$
$$-6k - 15 \geq -3$$
$$-6k - 15 + 15 \geq -3 + 15 \quad \text{Add 15.}$$
$$-6k \geq 12$$

Step 3 Divide each side by −6, a negative number. Change the direction of the inequality symbol.

$$\frac{-6k}{-6} \leq \frac{12}{-6}$$ Divide by −6; reverse the symbol.

$$k \leq -2$$

The solution set is $(-\infty, -2]$. Its graph is shown in Figure 21.

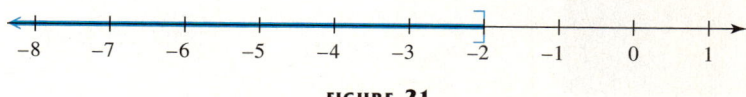

FIGURE 21

Now Try Exercise 47.

OBJECTIVE 5 Solve applied problems using inequalities. Until now, the applied problems that we have studied have all led to equations.

PROBLEM SOLVING

Inequalities can be used to solve applied problems involving phrases that suggest inequality. The table gives some of the more common such phrases along with examples and translations.

Phrase	Example	Inequality
Is more than	A number *is more than* 4.	$x > 4$
Is less than	A number *is less than* −12.	$x < -12$
Is at least	A number *is at least* 6.	$x \geq 6$
Is at most	A number *is at most* 8.	$x \leq 8$

We use the same six problem-solving steps from Section 2.4, changing Step 3 to "Write an inequality" instead of "Write an equation."

CAUTION Do not confuse statements like "5 is more than a number" with the phrase "5 more than a number." The first of these is expressed as $5 > x$ while the second is expressed as $x + 5$ or $5 + x$.

The next example shows an application of algebra that is important to anyone who has ever asked, "What score can I make on my next test and have a (particular grade) in this course?" It uses the idea of finding the average of a number of grades. In general, to find the average of n numbers, add the numbers, then divide by n.

EXAMPLE 6 Finding an Average Test Score

Brent has test grades of 86, 88, and 78 on his first three tests in geometry. If he wants an average of at least 80 after his fourth test, what are the possible scores he can make on his fourth test?

Step 1 **Read** the problem again.

Step 2 **Assign a variable.** Let x = Brent's score on his fourth test.

164 CHAPTER 2 Linear Equations and Inequalities in One Variable

Step 3 **Write an inequality.** To find his average after 4 tests, add the test scores and divide by 4.

$$\underbrace{\frac{86 + 88 + 78 + x}{4}}_{\text{Average}} \underbrace{\geq}_{\text{is at least}} \underbrace{80}_{80.}$$

Step 4 **Solve.**

$$\frac{252 + x}{4} \geq 80 \qquad \text{Add the known scores.}$$

$$4\left(\frac{252 + x}{4}\right) \geq 4(80) \qquad \text{Multiply by 4.}$$

$$252 + x \geq 320$$

$$252 + x - 252 \geq 320 - 252 \qquad \text{Subtract 252.}$$

$$x \geq 68 \qquad \text{Combine like terms.}$$

Step 5 **State the answer.** He must score 68 or more on the fourth test to have an average of *at least* 80.

Step 6 **Check.**

$$\frac{86 + 88 + 78 + 68}{4} = \frac{320}{4} = 80$$

Now Try Exercise 77.

CAUTION Errors often occur when the phrases "at least" and "at most" appear in applied problems. Remember that

        at least   translates as   greater than or equal to

and

        at most   translates as   less than or equal to.

OBJECTIVE 6 Solve linear inequalities with three parts. Inequalities that say that one number is *between* two other numbers are **three-part inequalities.** For example,

$$-3 < 5 < 7$$

says that 5 is between -3 and 7. For some applications, it is necessary to work with an inequality such as

$$3 < x + 2 < 8,$$

where $x + 2$ is between 3 and 8. To solve this inequality, we subtract 2 from each of the three parts of the inequality, giving

$$3 - 2 < x + 2 - 2 < 8 - 2$$

$$1 < x < 6.$$

The idea is to get the inequality in the form

    a number $< x <$ another number,

using "is less than." The solution set (in this case the interval (1, 6)) can then easily be graphed.

SECTION 2.8 Solving Linear Inequalities 165

> **CAUTION** When inequalities have three parts, the order of the parts is important. It would be *wrong* to write an inequality as $8 < x + 2 < 3$, since this would imply that $8 < 3$, a false statement. In general, three-part inequalities are written so that the symbols point in the same direction, and both point toward the smaller number.

EXAMPLE 7 Solving Three-Part Inequalities

Solve each inequality and graph the solution set.

(a)
$$4 \leq 3x - 5 < 6$$
$$4 + 5 \leq 3x - 5 + 5 < 6 + 5 \quad \text{Add 5 to each part.}$$
$$9 \leq 3x < 11$$
$$\frac{9}{3} \leq \frac{3x}{3} < \frac{11}{3} \quad \text{Divide each part by 3.}$$
$$3 \leq x < \frac{11}{3}$$

The solution set is $\left[3, \frac{11}{3}\right)$. Its graph is shown in Figure 22.

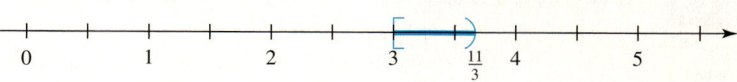

FIGURE 22

(b) $-4 \leq \frac{2}{3}m - 1 < 8$

Fractions as coefficients in equations can be eliminated by multiplying each side by the least common denominator of the fractions. The same is true for inequalities.

$$3(-4) \leq 3\left(\frac{2}{3}m - 1\right) < 3(8) \quad \text{Multiply each part by 3.}$$
$$-12 \leq 2m - 3 < 24 \quad \text{Distributive property}$$
$$-12 + 3 \leq 2m - 3 + 3 < 24 + 3 \quad \text{Add 3 to each part.}$$
$$-9 \leq 2m < 27$$
$$\frac{-9}{2} \leq \frac{2m}{2} < \frac{27}{2} \quad \text{Divide each part by 2.}$$
$$-\frac{9}{2} \leq m < \frac{27}{2}$$

The solution set is $\left[-\frac{9}{2}, \frac{27}{2}\right)$. Its graph is shown in Figure 23.

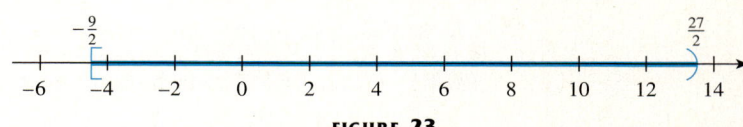

FIGURE 23

Now Try Exercises 59 and 65.

> **NOTE** The inequality in Example 7(b) can also be solved by first adding 1 to each part, and then multiplying each part by $\frac{3}{2}$. Do this and confirm that the same solution set results.

166 CHAPTER 2 Linear Equations and Inequalities in One Variable

---

CONNECTIONS

Many applications from economics involve inequalities rather than equations. For example, a company that produces videocassettes has found that revenue from the sales of the cassettes is $5 per cassette less sales costs of $100. Production costs are $125 plus $4 per cassette. Profit ($P$) is given by revenue ($R$) less cost ($C$), so the company must find the production level x that makes

$$P = R - C > 0.$$

For Discussion or Writing

Write an expression for revenue, letting x represent the production level (number of cassettes to be produced). Write an expression for production costs using x. Write an expression for profit and solve the inequality shown above. Describe the solution in terms of the problem.

2.8 EXERCISES

For Extra Help

- Student's Solutions Manual
- MyMathLab
- InterAct Math Tutorial Software
- AW Math Tutor Center
- MathXL
- Digital Video Tutor CD 5/Videotape 4

1. Explain how to determine whether to use a parenthesis or a square bracket at the endpoint when graphing an inequality on a number line.

2. How does the graph of $t \geq -7$ differ from the graph of $t > -7$?

Write an inequality involving the variable x that describes each set of numbers graphed. See Example 1.

3.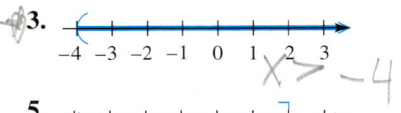

4.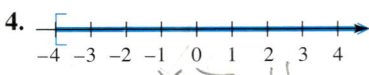

5.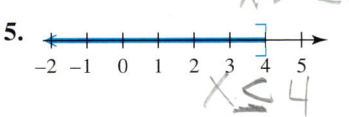

6.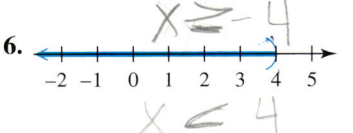

Write each inequality in interval notation and graph the interval. See Example 1.

7. $k \leq 4$ 8. $r \leq -11$ 9. $x < -3$ 10. $y < 3$

11. $t > 4$ 12. $m > 5$ 13. $8 \leq x \leq 10$ 14. $3 \leq x \leq 5$

15. $0 < y \leq 10$ 16. $-3 \leq x < 5$

17. Why is it wrong to write $3 < x < -2$ to indicate that x is between -2 and 3?

18. If $p < q$ and $r < 0$, which one of the following statements is false?

   A. $pr < qr$ B. $pr > qr$ C. $p + r < q + r$ D. $p - r < q - r$

Solve each inequality. Write the solution set in interval notation and graph it. See Example 2.

19. $z - 8 \geq -7$ 20. $p - 3 \geq -11$ 21. $2k + 3 \geq k + 8$

22. $3x + 7 \geq 2x + 11$ 23. $3n + 5 < 2n - 6$ 24. $5x - 2 < 4x - 5$

SECTION 2.8 Solving Linear Inequalities 167

25. Under what conditions must the inequality symbol be reversed when solving an inequality?

26. Explain the steps you would use to solve the inequality $-5x > 20$.

27. Your friend tells you that when solving the inequality $6x < -42$ he reversed the direction of the inequality symbol because of the -42. How would you respond?

28. By what number must you *multiply* each side of $.2x > 6$ to get just x on the left side?

Solve each inequality. Write the solution set in interval notation and graph it. See Example 3.

29. $3x < 18$
30. $5x < 35$
31. $2y \geq -20$
32. $6m \geq -24$
33. $-8t > 24$
34. $-7x > 49$
35. $-x \geq 0$
36. $-k < 0$
37. $-\dfrac{3}{4}r < -15$
38. $-\dfrac{7}{8}t < -14$
39. $-.02x \leq .06$
40. $-.03v \geq -.12$

Solve each inequality. Write the solution set in interval notation and graph it. See Examples 4 and 5.

41. $5r + 1 \geq 3r - 9$
42. $6t + 3 < 3t + 12$
43. $6x + 3 + x < 2 + 4x + 4$
44. $-4w + 12 + 9w \geq w + 9 + w$
45. $-x + 4 + 7x \leq -2 + 3x + 6$
46. $14y - 6 + 7y > 4 + 10y - 10$
47. $5(x + 3) - 6x \leq 3(2x + 1) - 4x$
48. $2(x - 5) + 3x < 4(x - 6) + 1$
49. $\dfrac{2}{3}(p + 3) > \dfrac{5}{6}(p - 4)$
50. $\dfrac{7}{9}(y - 4) \leq \dfrac{4}{3}(y + 5)$
51. $4x - (6x + 1) \leq 8x + 2(x - 3)$
52. $2y - (4y + 3) > 6y + 3(y + 4)$
53. $5(2k + 3) - 2(k - 8) > 3(2k + 4) + k - 2$
54. $2(3z - 5) + 4(z + 6) \geq 2(3z + 2) + 3z - 15$

Write a three-part inequality involving the variable x that describes each set of numbers graphed. See Example 1(b).

55.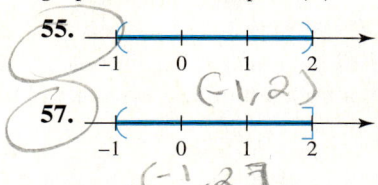
56.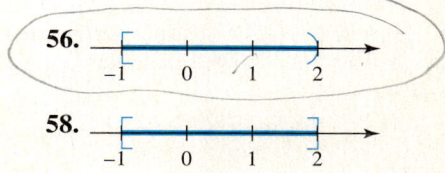
57.
58.

Solve each inequality. Write the solution set in interval notation and graph it. See Example 7.

59. $-5 \leq 2x - 3 \leq 9$
60. $-7 \leq 3x - 4 \leq 8$
61. $5 < 1 - 6m < 12$
62. $-1 \leq 1 - 5q \leq 16$
63. $10 < 7p + 3 < 24$
64. $-8 \leq 3r - 1 \leq -1$

168 CHAPTER 2 Linear Equations and Inequalities in One Variable

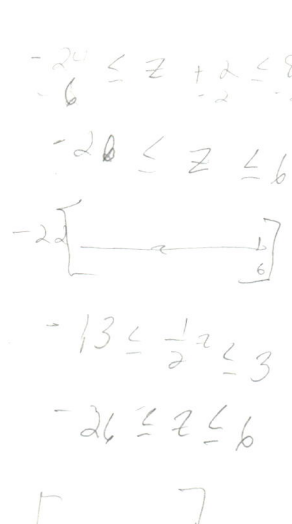

65. $-12 \leq \dfrac{1}{2}z + 1 \leq 4$

66. $-6 \leq 3 + \dfrac{1}{3}a \leq 5$

67. $1 \leq 3 + \dfrac{2}{3}p \leq 7$

68. $2 < 6 + \dfrac{3}{4}y < 12$

69. $-7 \leq \dfrac{5}{4}r - 1 \leq -1$

70. $-12 \leq \dfrac{3}{7}a + 2 \leq -4$

RELATING CONCEPTS (EXERCISES 71–76)

For Individual or Group Work

The methods for solving linear equations and linear inequalities are quite similar. Work Exercises 71–76 in order, to see how the solutions of an inequality are closely connected to the solution of the corresponding equation.

71. Solve the equation $3x + 2 = 14$ and graph the solution set as a single point on the number line.

72. Solve the inequality $3x + 2 > 14$ and graph the solution set as an interval on the number line.

73. Solve the inequality $3x + 2 < 14$ and graph the solution set as an interval on the number line.

74. If you were to graph all the solution sets from Exercises 71–73 on the same number line, what would the graph be? (This is called the *union* of all the solution sets.)

75. Based on your results from Exercises 71–74, if you were to graph the union of the solution sets of

$-4x + 3 = -1,$ $-4x + 3 > -1,$ and $-4x + 3 < -1,$

what do you think the graph would be?

76. Comment on the following statement: *Equality is the boundary between less than and greater than.*

Solve each problem. See Example 6.

77. Inkie Landry has scores of 76 and 81 on her first two algebra tests. If she wants an average of at least 80 after her third test, what possible scores can she make on her third test?

78. Mabimi Pampo has scores of 96 and 86 on his first two geometry tests. What possible scores can he make on his third test so that his average is at least 90?

79. When 2 is added to the difference between six times a number and 5, the result is greater than 13 added to five times the number. Find all such numbers.

80. When 8 is subtracted from the sum of three times a number and 6, the result is less than 4 more than the number. Find all such numbers.

81. The formula for converting Fahrenheit temperature to Celsius is

$$C = \frac{5}{9}(F - 32).$$

If the Celsius temperature on a certain summer day in Toledo is never more than 30°, how would you describe the corresponding Fahrenheit temperatures?

82. The formula for converting Celsius temperature to Fahrenheit is

$$F = \frac{9}{5}C + 32.$$

The Fahrenheit temperature of Key West, Florida, has never exceeded 95°. How would you describe this using Celsius temperature?

83. For what values of x would the rectangle have perimeter of at least 400?

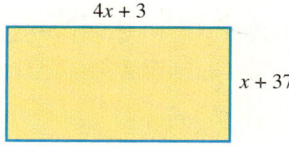

84. For what values of x would the triangle have perimeter of at least 72?

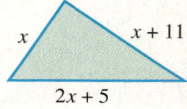

85. A long-distance phone call costs $2.00 for the first three minutes plus $.30 per minute for each minute or fractional part of a minute after the first three minutes. If x represents the number of minutes of the length of the call after the first three minutes, then $2 + .30x$ represents the cost of the call. If Jorge has $5.60 to spend on a call, what is the maximum total time he can use the phone?

86. If the call described in Exercise 85 costs between $5.60 and $6.50, what are the possible total time lengths for the call?

87. At the Speedy Gas'n Go, a car wash costs $3.00, and gasoline is selling for $1.50 per gal. Lee Ann Spahr has $17.25 to spend, and her car is so dirty that she must have it washed. What is the maximum number of gallons of gasoline that she can purchase?

88. A rental company charges $15 to rent a chain saw, plus $2 per hr. Al Ghandi can spend no more than $35 to clear some logs from his yard. What is the maximum amount of time he can use the rented saw?

89. A product will break even or produce a profit if the revenue R from selling the product is at least equal to the cost C of producing it. Suppose that the cost C (in dollars) to produce x units of bicycle helmets is $C = 50x + 5000$, while the revenue R (in dollars) collected from the sale of x units is $R = 60x$. For what values of x does the product break even or produce a profit?

90. (See Exercise 89.) If the cost to produce x units of baseball cards is $C = 100x + 6000$ (in dollars), and the revenue collected from selling x units is $R = 500x$ (in dollars), for what values of x does the product break even or produce a profit?

CHAPTER 2 Group Activity

Are You a Race-Walker?

OBJECTIVE Use proportions to calculate walking speeds.

MATERIALS Students will need a large area for walking. Each group will need a stopwatch.

Race-walking at speeds exceeding 8 mph is a high fitness, long-distance competitive sport. The table contains gold medal winners in Olympic race-walking competition. Complete the table by applying the proportion given below to find the race-walker's steps per minute.

Use 10 km ≈ 6.21 mi. (Round all answers except those for steps per minute to the nearest thousandth. Round steps per minute to the nearest whole number.)

$$\frac{70 \text{ steps per min}}{2 \text{ mph}} = \frac{x \text{ steps per min}}{y \text{ mph}}$$

Event	Gold Medal Winner	Country	Time in Hours: Minutes: Seconds	Time in Minutes	Time in Hours	y Miles per Hour	x Steps per Minute
10-km Walk, Women	Yelena Nikolayeva	Russia	0:41:49				
20-km Walk, Men	Jefferson Perez	Ecuador	1:20:07				
50-km Walk, Men	Robert Korzeniowski	Poland	3:43:30				

Source: World Almanac and Book of Facts, 1999.

A. Using a stopwatch, take turns counting how many steps each member of the group takes in one minute while walking at a normal pace. Record the results in the following table. Then do it again at a fast pace. Record these results.

	Normal Pace		Fast Pace	
Name	x Steps per Minute	y Miles per Hour	x Steps per Minute	y Miles per Hour

B. Use the given proportion to convert the numbers from part A to miles per hour and complete the table.

  1. Find the average speed for the group at a normal pace and at a fast pace.

  2. What is the minimum number of steps per minute you would have to take to be a race-walker?

  3. At a fast pace did anyone in the group walk fast enough to be a race-walker? Explain how you decided.

CHAPTER 2 SUMMARY

KEY TERMS

2.1 linear equation in one variable
solution set
equivalent equations
2.3 conditional equation
identity
contradiction
empty (null) set

2.4 degree
complementary angles
right angle
supplementary angles
straight angle
consecutive integers
2.5 area
perimeter

vertical angles
volume
2.6 ratio
proportion
terms of a proportion
extremes
means
cross products

2.8 interval on the number line
interval notation
linear inequality in one variable
three-part inequality

NEW SYMBOLS

$\emptyset$ empty set
$1°$ one degree

a to b, $a:b$, or $\dfrac{a}{b}$ the ratio of a to b

(a, b) interval notation for $a < x < b$
$[a, b]$ interval notation for $a \leq x \leq b$

∞ infinity
$-\infty$ negative infinity
$(-\infty, \infty)$ set of all real numbers

TEST YOUR WORD POWER

See how well you have learned the vocabulary in this chapter. Answers, with examples, follow the Quick Review.

1. A **solution set** is the set of numbers that
   A. make an expression undefined
   B. make an equation false
   C. make an equation true
   D. make an expression equal to 0.

2. The **empty set** is a set
   A. with 0 as its only element
   B. with an infinite number of elements
   C. with no elements
   D. of ideas.

3. **Complementary angles** are angles
   A. formed by two parallel lines
   B. whose sum is 90°
   C. whose sum is 180°
   D. formed by perpendicular lines.

4. **Supplementary angles** are angles
   A. formed by two parallel lines
   B. whose sum is 90°
   C. whose sum is 180°
   D. formed by perpendicular lines.

5. A **ratio**
   A. compares two quantities using a quotient
   B. says that two quotients are equal
   C. is a product of two quantities
   D. is a difference between two quantities.

6. A **proportion**
   A. compares two quantities using a quotient
   B. says that two quotients are equal
   C. is a product of two quantities
   D. is a difference between two quantities.

7. An **inequality** is
   A. a statement that two algebraic expressions are equal
   B. a point on a number line
   C. an equation with no solutions
   D. a statement with algebraic expressions related by $<$, $\leq$, $>$, or $\geq$.

8. **Interval notation** is
   A. a portion of a number line
   B. a special notation for describing a point on a number line
   C. a way to use symbols to describe an interval on a number line
   D. a notation to describe unequal quantities.

QUICK REVIEW

CONCEPTS	EXAMPLES

2.1 THE ADDITION PROPERTY OF EQUALITY

The same number may be added to (or subtracted from) each side of an equation without changing the solution.

Solve.
$$x - 6 = 12$$
$$x - 6 + 6 = 12 + 6 \quad \text{Add 6.}$$
$$x = 18 \quad \text{Combine like terms.}$$

Solution set: $\{18\}$

2.2 THE MULTIPLICATION PROPERTY OF EQUALITY

Each side of an equation may be multiplied (or divided) by the same nonzero number without changing the solution.

Solve.
$$\frac{3}{4}x = -9$$
$$\frac{4}{3} \cdot \frac{3}{4}x = \frac{4}{3}(-9) \quad \text{Multiply by } \tfrac{4}{3}.$$
$$x = -12$$

Solution set: $\{-12\}$

2.3 MORE ON SOLVING LINEAR EQUATIONS

Step 1 Simplify each side separately.

Step 2 Isolate the variable term on one side.

Step 3 Isolate the variable.

Step 4 Check.

Solve.
$$2x + 2(x + 1) = 14 + x$$
$$2x + 2x + 2 = 14 + x \quad \text{Distributive property}$$
$$4x + 2 = 14 + x \quad \text{Combine like terms.}$$
$$4x + 2 - x - 2 = 14 + x - x - 2 \quad \text{Subtract } x; \text{ subtract 2.}$$
$$3x = 12 \quad \text{Combine like terms.}$$
$$\frac{3x}{3} = \frac{12}{3} \quad \text{Divide by 3.}$$
$$x = 4$$

Check:
$$2(4) + 2(4 + 1) = 14 + 4 \quad ? \quad \text{Let } x = 4.$$
$$18 = 18 \quad \text{True}$$

Solution set: $\{4\}$

2.4 AN INTRODUCTION TO APPLICATIONS OF LINEAR EQUATIONS

One number is 5 more than another. Their sum is 21. What are the numbers?

Step 1 Read.

Step 2 Assign a variable.

We are looking for two numbers.

Let x represent the smaller number. Then $x + 5$ represents the larger number.

CONCEPTS	EXAMPLES
Step 3 Write an equation. *Step 4* Solve the equation.	$x + (x + 5) = 21$ $2x + 5 = 21$ Combine like terms. $2x + 5 - 5 = 21 - 5$ Subtract 5. $2x = 16$ Combine like terms. $\dfrac{2x}{2} = \dfrac{16}{2}$ Divide by 2. $x = 8$
Step 5 State the answer. *Step 6* Check.	The numbers are 8 and 13. 13 is 5 more than 8, and $8 + 13 = 21$. It checks.

Handwritten annotation: $8 + (8+5)$; $8 + 13 = 21$

2.5 FORMULAS AND APPLICATIONS FROM GEOMETRY

To find the value of one of the variables in a formula, given values for the others, substitute the known values into the formula.	Find L if $A = LW$, given that $A = 24$ and $W = 3$. $24 = L \cdot 3$ $A = 24, W = 3$ $\dfrac{24}{3} = \dfrac{L \cdot 3}{3}$ Divide by 3. $8 = L$
To solve a formula for one of the variables, isolate that variable by treating the other variables as numbers and using the steps for solving equations.	Solve $P = 2L + 2W$ for W. $P - 2L = 2L + 2W - 2L$ Subtract $2L$. $P - 2L = 2W$ Combine like terms. $\dfrac{P - 2L}{2} = \dfrac{2W}{2}$ Divide by 2. $\dfrac{P - 2L}{2} = W$ or $W = \dfrac{P - 2L}{2}$

2.6 RATIOS AND PROPORTIONS

To write a ratio, express quantities in the same units.	4 ft to 8 in. = 48 in. to 8 in. = $\dfrac{48}{8} = \dfrac{6}{1}$
To solve a proportion, use the method of cross products.	Solve $\dfrac{x}{12} = \dfrac{35}{60}$. $60x = 12 \cdot 35$ Cross products $60x = 420$ Multiply. $\dfrac{60x}{60} = \dfrac{420}{60}$ Divide by 60. $x = 7$ Solution set: $\{7\}$

(continued)

174 CHAPTER 2 Linear Equations and Inequalities in One Variable

CONCEPTS	EXAMPLES

2.7 MORE ABOUT PROBLEM SOLVING

Step 1 Read.

A sum of money is invested at simple interest in two ways. Part is invested at 12%, and $20,000 less than that amount is invested at 10%. If the total interest for 1 yr is $9000, find the amount invested at each rate.

Step 2 Assign a variable. Make a table to help solve the problems in this section.

Let x = amount invested at 12%.
Then $x - 20,000$ = amount invested at 10%.

Dollars Invested	Rate of Interest	Interest for One Year
x	.12	.12x
$x - 20,000$	.10	.10($x - 20,000$)

Step 3 Write an equation.
Step 4 Solve the equation.

$$.12x + .10(x - 20,000) = 9000$$
$$.12x + .10x + .10(-20,000) = 9000 \quad \text{Distributive property}$$
$$12x + 10x + 10(-20,000) = 900,000 \quad \text{Multiply by 100.}$$
$$12x + 10x - 200,000 = 900,000$$
$$22x - 200,000 = 900,000$$
$$22x = 1,100,000 \quad \text{Add 200,000.}$$
$$x = 50,000 \quad \text{Divide by 22.}$$

Steps 5 and 6 State the answer and check the solution.

$50,000 is invested at 12% and $30,000 is invested at 10%.

The three forms of the formula relating distance, rate, and time are

$$d = rt, \quad r = \frac{d}{t}, \quad \text{and} \quad t = \frac{d}{r}.$$

Two cars leave from the same point, traveling in opposite directions. One travels at 45 mph and the other at 60 mph. How long will it take them to be 210 mi apart?

Let t = time it takes for them to be 210 mi apart.

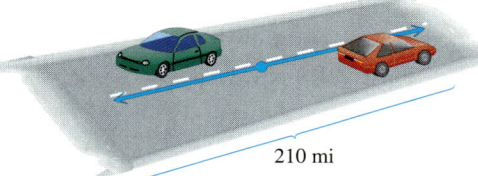

210 mi

The table gives the information from the problem.

	Rate	Time	Distance
One Car	45	t	45t
Other Car	60	t	60t

The sum of the distances, 45t and 60t, must be 210 mi.

$$45t + 60t = 210$$
$$105t = 210 \quad \text{Combine like terms.}$$
$$t = 2 \quad \text{Divide by 105.}$$

It will take them 2 hr to be 210 mi apart.

CHAPTER 2 Review Exercises 175

CONCEPTS	EXAMPLES
2.8 SOLVING LINEAR INEQUALITIES	
Step 1 Simplify each side separately.	Solve and graph the solution set. $$3(1-x)+5-2x>9-6$$ $$3-3x+5-2x>9-6 \quad \text{Clear parentheses.}$$ $$8-5x>3 \quad \text{Combine like terms.}$$
Step 2 Isolate the variable term on one side.	$$8-5x-8>3-8 \quad \text{Subtract 8.}$$ $$-5x>-5 \quad \text{Combine like terms.}$$
Step 3 Isolate the variable. **Be sure to reverse the direction of the inequality symbol when multiplying or dividing by a negative number.**	$$\frac{-5x}{-5} < \frac{-5}{-5} \quad \text{Divide by } -5;$$ $$\text{change} > \text{to} <.$$ $$x<1$$ Solution set: $(-\infty, 1)$ 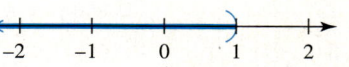
To solve an inequality such as $$4 < 2x+6 < 8$$ work with all three expressions at the same time.	Solve. $$4 < 2x+6 < 8$$ $$4-6 < 2x+6-6 < 8-6 \quad \text{Subtract 6.}$$ $$-2 < 2x < 2 \quad \text{Combine like terms.}$$ $$\frac{-2}{2} < \frac{2x}{2} < \frac{2}{2} \quad \text{Divide by 2.}$$ $$-1 < x < 1$$ Solution set: $(-1, 1)$

Answers to Test Your Word Power
1. C; *Example:* {8} is the solution set of $2x+5=21$.   2. C; *Example:* The empty set $\emptyset$ is the solution set of $5x+3=5x+4$.
3. B; *Example:* Angles with measures 35° and 55° are complementary angles.   4. C; *Example:* Angles with measures 112° and 68° are supplementary angles.   5. A; *Example:* $\frac{7 \text{ in.}}{12 \text{ in.}}$ or $\frac{7}{12}$ 6. B; *Example:* $\frac{2}{3} = \frac{8}{12}$ 7. D; *Examples:* $x < 5, 7 + 2y \geq 11$, $-5 < 2z - 1 \leq 3$ 8. C; *Examples:* $(-\infty, 5], (1, \infty), [-3, 3)$

CHAPTER 2 REVIEW EXERCISES

[2.1–2.3] Solve each equation.

1. $m - 5 = 1$ 2. $y + 8 = -4$ 3. $3k + 1 = 2k + 8$

4. $5k = 4k + \frac{2}{3}$ 5. $(4r - 2) - (3r + 1) = 8$ 6. $3(2y - 5) = 2 + 5y$

7. $7k = 35$ 8. $12r = -48$ 9. $2p - 7p + 8p = 15$

10. $\frac{m}{12} = -1$ 11. $\frac{5}{8}k = 8$ 12. $12m + 11 = 59$

13. $3(2x + 6) - 5(x + 8) = x - 22$ 14. $5x + 9 - (2x - 3) = 2x - 7$

176 CHAPTER 2 Linear Equations and Inequalities in One Variable

15. $\dfrac{1}{2}r - \dfrac{r}{3} = \dfrac{r}{6}$

16. $.10(x + 80) + .20x = 14$

17. $3x - (-2x + 6) = 4(x - 4) + x$

18. $2(y - 3) - 4(y + 12) = -2(y + 27)$

[2.4] *Use the six-step method to solve each problem.*

19. In a recent year, the state of Florida had a total of 120 members in its House of Representatives, consisting of only Democrats and Republicans. There were 30 more Democrats than Republicans. How many representatives from each party were there?

20. The land area of Hawaii is 5213 mi² greater than the area of Rhode Island. Together, the areas total 7637 mi². What is the area of each of the two states?

21. The height of Seven Falls in Colorado is $\dfrac{5}{2}$ the height of Twin Falls in Idaho. The sum of the heights is 420 ft. Find the height of each. (*Source: World Almanac and Book of Facts.*)

22. The supplement of an angle measures 10 times the measure of its complement. What is the measure of the angle?

23. Find two consecutive odd integers such that when the smaller is added to twice the larger, the result is 24 more than the larger integer.

[2.5] *A formula is given along with the values for all but one of the variables. Find the value of the variable that is not given.*

24. $A = \dfrac{1}{2}bh$; $A = 44$, $b = 8$

25. $A = \dfrac{1}{2}h(b + B)$; $b = 3$, $B = 4$, $h = 8$

26. $C = 2\pi r$; $C = 29.83$, $\pi = 3.14$

27. $V = \dfrac{4}{3}\pi r^3$; $r = 6$, $\pi = 3.14$

Solve each formula for the specified variable.

28. $A = bh$ for h

29. $A = \dfrac{1}{2}h(b + B)$ for h

Find the measure of each marked angle.

30. $(8x - 1)°$ $(3x - 6)°$

31. $(3x + 10)°$ $(4x - 20)°$

Solve each problem.

32. The perimeter of a certain rectangle is 16 times the width. The length is 12 cm more than the width. Find the width of the rectangle.

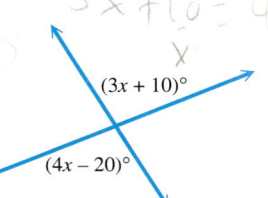

33. The Ziegfield Room in Reno, Nevada, has a circular turntable on which its showgirls dance. The circumference of the table is 62.5 ft. What is the diameter? What is the radius? What is the area? (Use $\pi = 3.14$.) (*Source: Guinness Book of Records.*)

CHAPTER 2 Review Exercises 177

34. A baseball diamond is a square with a side of 90 ft. The pitcher's rubber is located 60.5 ft from home plate, as shown in the figure. Find the measures of the angles marked in the figure. (*Hint:* Recall that the sum of the measures of the angles of any triangle is 180°.)

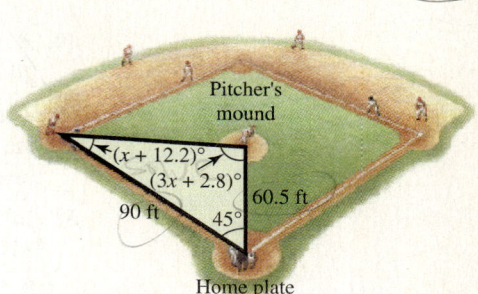

35. What is wrong with the following problem? "The formula for the area of a trapezoid is $A = \frac{1}{2}h(b + B)$. If $h = 12$ and $b = 14$, find the value of A."

[2.6] *Give a ratio for each word phrase, writing fractions in lowest terms.*

36. 60 cm to 40 cm
37. 5 days to 2 weeks
38. 90 in. to 10 ft
39. 3 mo to 3 yr

Solve each equation.

40. $\dfrac{p}{21} = \dfrac{5}{30}$

41. $\dfrac{5 + x}{3} = \dfrac{2 - x}{6}$

42. $\dfrac{y}{5} = \dfrac{6y - 5}{11}$

Use proportions to solve each problem.

43. If 2 lb of fertilizer will cover 150 ft² of lawn, how many pounds would be needed to cover 500 ft²?

44. The tax on a $24.00 item is $2.04. How much tax would be paid on a $36.00 item?

45. The distance between two cities on a road map is 32 cm. The two cities are actually 150 km apart. The distance on the map between two other cities is 80 cm. How far apart are these cities?

46. In the 2000 Olympics held in Sydney, Australia, Russian athletes earned 88 medals. Four of every 11 medals were gold. How many gold medals did Russia earn? (*Source: Times Picayune,* October 2, 2000.)

47. Which is the best buy for a popular breakfast cereal?

    15-oz size: $2.69
    20-oz size: $3.29
    25.5-oz size: $3.49

[2.7] *Solve each problem.*

48. The San Francisco Giants baseball team borrowed $160 million of the $290 million needed to build Pacific Bell Park. Corporate sponsors provided the rest of the cost. What percent of the cost of the park was borrowed? (*Source:* Michael K. Ozanian, "Fields of Debt," *Forbes,* December 15, 1997.)

178 CHAPTER 2 Linear Equations and Inequalities in One Variable

49. A nurse must mix 15 L of a 10% solution of a drug with some 60% solution to get a 20% mixture. How many liters of the 60% solution will be needed?

50. Todd Cardella invested $10,000 from which he earns an annual income of $550 per yr. He invested part of it at 5% annual interest and the remainder in bonds paying 6% interest. How much did he invest at each rate?

51. In 1846, the vessel *Yorkshire* traveled from Liverpool to New York, a distance of 3150 mi, in 384 hr. What was the *Yorkshire*'s average speed? Round your answer to the nearest tenth.

52. Honey Kirk drove from Louisville to Dallas, a distance of 819 mi, averaging 63 mph. What was her driving time?

53. Two planes leave St. Louis at the same time. One flies north at 350 mph and the other flies south at 420 mph. In how many hours will they be 1925 mi apart?

[2.8] *Write each inequality in interval notation and graph it.*

54. $p \geq -4$ **55.** $x < 7$ **56.** $-5 \leq y < 6$

Solve each inequality and graph the solution set.

57. $y + 6 \geq 3$ **58.** $5t < 4t + 2$

59. $-6x \leq -18$ **60.** $8(k - 5) - (2 + 7k) \geq 4$

61. $4x - 3x > 10 - 4x + 7x$

62. $3(2w + 5) + 4(8 + 3w) < 5(3w + 2) + 2w$

63. $-3 \leq 2m + 1 \leq 4$ **64.** $9 < 3m + 5 \leq 20$

Solve each problem by writing an inequality.

65. Carlotta Valdes has grades of 94 and 88 on her first two calculus tests. What possible scores on a third test will give her an average of at least 90?

66. If nine times a number is added to 6, the result is at most 3. Find all such numbers.

MIXED REVIEW EXERCISES*

Solve.

67. $\dfrac{x}{7} = \dfrac{x - 5}{2}$ **68.** $I = prt$ for r **69.** $-2x > -4$

70. $2k - 5 = 4k + 13$ **71.** $.05x + .02x = 4.9$ **72.** $2 - 3(x - 5) = 4 + x$

73. $9x - (7x + 2) = 3x + (2 - x)$ **74.** $\dfrac{1}{3}s + \dfrac{1}{2}s + 7 = \dfrac{5}{6}s + 5 + 2$

75. A student solved $3 - (8 + 4x) = 2x + 7$ and gave the answer as $\{6\}$. Verify that this answer is incorrect by checking it in the equation. Then explain the error and give the correct solution set. (*Hint:* The error involves the subtraction sign.)

---

*The order of the exercises in this final group does not correspond to the order in which topics occur in the chapter. This random ordering should help you prepare for the chapter test.

76. Athletes in vigorous training programs can eat 50 calories per day for every 2.2 lb of body weight. To the nearest hundred, how many calories can a 175 lb athlete consume per day? (*Source: The Gazette,* March 23, 2002.)

77. The Golden Gate Bridge in San Francisco is 2605 ft longer than the Brooklyn Bridge. Together, their spans total 5795 ft. How long is each bridge? (*Source: World Almanac and Book of Facts.*)

78. Which is the best buy for apple juice?

    32-oz size: $1.19
    48-oz size: $1.79
    64-oz size: $1.99

79. If 1 qt of oil must be mixed with 24 qt of gasoline, how much oil would be needed for 192 qt of gasoline?

80. Two trains are 390 mi apart. They start at the same time and travel toward one another, meeting 3 hr later. If the speed of one train is 30 mph more than the speed of the other train, find the speed of each train.

81. The perimeter of a triangle is 96 m. One side is twice as long as another, and the third side is 30 m long. What is the length of the longest side?

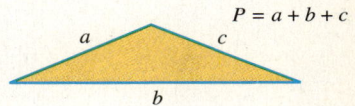

82. The perimeter of a certain square cannot be greater than 200 m. Find the possible values for the length of a side.

CHAPTER 2 TEST

Solve each equation.

1. $5x + 9 = 7x + 21$
2. $-\frac{4}{7}x = -12$
3. $7 - (m - 4) = -3m + 2(m + 1)$
4. $.06(x + 20) + .08(x - 10) = 4.6$
5. $-8(2x + 4) = -4(4x + 8)$

Solve each problem.

6. In the 2000–2001 baseball season, the Seattle Mariners tied a league record set by the 1906 Chicago Cubs for most wins in a season. The Mariners won 24 more than twice as many games as they lost. They played 162 regular season games. How many wins and losses did the Mariners have?

7. The three largest islands in the Hawaiian island chain are Hawaii (the Big Island), Maui, and Kauai. Together, their areas total 5300 mi². The island of Hawaii is 3293 mi² larger than the island of Maui, and Maui is 177 mi² larger than Kauai. What is the area of each island?

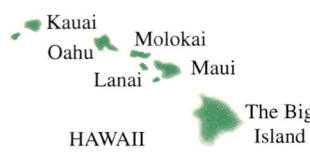

8. Find the measure of an angle if its supplement measures 10° more than three times its complement.

9. The formula for the perimeter of a rectangle is $P = 2L + 2W$.
   (a) Solve for W.
   (b) If $P = 116$ and $L = 40$, find the value of W.

10. Find the measure of each marked angle.

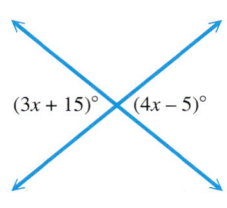

$(3x + 15)°$ $(4x - 5)°$

Solve each proportion.

11. $\dfrac{z}{8} = \dfrac{12}{16}$

12. $\dfrac{x + 5}{3} = \dfrac{x - 3}{4}$

Solve each problem.

13. Which is the better buy for processed cheese slices: 8 slices for $2.19 or 12 slices for $3.30?

14. The distance between Milwaukee and Boston is 1050 mi. On a certain map, this distance is represented by 42 in. On the same map, Seattle and Cincinnati are 92 in. apart. What is the actual distance between Seattle and Cincinnati?

15. Steven Pusztai invested some money at 3% simple interest and $6000 more than that amount at 4.5% simple interest. After 1 yr his total interest from the two accounts was $870. How much did he invest at each rate?

16. Two cars leave from the same point, traveling in opposite directions. One travels at a constant rate of 50 mph while the other travels at a constant rate of 65 mph. How long will it take for them to be 460 mi apart?

Solve each inequality and graph the solution set.

17. $-4x + 2(x - 3) \geq 4x - (3 + 5x) - 7$

18. $-10 < 3k - 4 \leq 14$

19. Twylene Johnson has grades of 76 and 81 on her first two algebra tests. If she wants an average of at least 80 after her third test, what score must she make on her third test?

20. Write a short explanation of the additional (extra) rule that must be remembered when solving an inequality (as opposed to solving an equation).

CUMULATIVE REVIEW EXERCISES CHAPTERS 1–2

1. Write $\frac{108}{144}$ in lowest terms.

Perform each indicated operation.

2. $\frac{5}{6} + \frac{1}{4} - \frac{7}{15}$

3. $\frac{9}{8} \cdot \frac{16}{3} \div \frac{5}{8}$

Translate from words to symbols. Use x as the variable, if necessary.

4. The difference between half a number and 18
5. The quotient of 6 and 12 more than a number is 2.
6. True or false? $\dfrac{8(7) - 5(6 + 2)}{3 \cdot 5 + 1} \geq 1$

Perform each indicated operation.

7. $9 - (-4) + (-2)$
8. $\dfrac{-4(9)(-2)}{-3^2}$
9. $(-7 - 1)(-4) + (-4)$

10. Find the value of $\dfrac{3x^2 - y^3}{-4z}$ when $x = -2$, $y = -4$, and $z = 3$.

Name each property illustrated.

11. $7(k + m) = 7k + 7m$
12. $3 + (5 + 2) = 3 + (2 + 5)$
13. Simplify $-4(k + 2) + 3(2k - 1)$ by combining like terms.

Solve each equation, then check the solution.

14. $2r - 6 = 8r$
15. $4 - 5(a + 2) = 3(a + 1) - 1$
16. $\dfrac{2}{3}x + \dfrac{3}{4}x = -17$
17. $\dfrac{2x + 3}{5} = \dfrac{x - 4}{2}$

Solve each formula for the indicated variable.

18. $3x + 4y = 24$ for y
19. $A = P(1 + ni)$ for n

Solve each inequality. Graph the solution set.

20. $6(r - 1) + 2(3r - 5) \leq -4$
21. $-18 \leq -9z < 9$

Solve each problem.

22. For a woven hanging, Miguel Hidalgo needs three pieces of yarn, which he will cut from a 40-cm piece. The longest piece is to be 3 times as long as the middle-sized piece, and the shortest piece is to be 5 cm shorter than the middle-sized piece. What lengths should he cut?

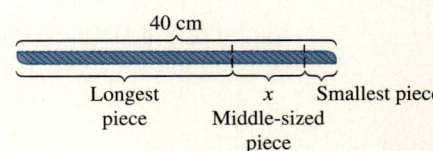

23. A fully inflated professional basketball has a circumference of 78 cm. What is the radius of a circular cross section through the center of the ball? (Use 3.14 as the approximation for π.) Round your answer to the nearest hundredth.

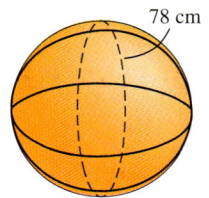

24. A cook wants to increase a recipe that serves 6 to make enough for 20 people. The recipe calls for $1\frac{1}{4}$ cups of grated cheese. How much cheese will be needed to serve 20?

25. Two cars are 400 mi apart. Both start at the same time and travel toward one another. They meet 4 hr later. If the speed of one car is 20 mph faster than the other, what is the speed of each car?

Linear Equations and Inequalities in Two Variables; Functions

3

3.1 Reading Graphs; Linear Equations in Two Variables

3.2 Graphing Linear Equations in Two Variables

3.3 The Slope of a Line

3.4 Equations of a Line

3.5 Graphing Linear Inequalities in Two Variables

3.6 Introduction to Functions

U.S. debt from credit cards continues to increase. In recent years, college campuses have become fertile territory as credit card companies pitch their plastic to students at bookstores, student unions, and sporting events. As a result, three out of four undergrads now have at least one credit card and carry an average balance of $2748. (*Source:* Nellie Mae.) In Example 6 of Section 3.2, we use the concepts of this chapter to investigate credit card debt.

184 CHAPTER 3 Linear Equations and Inequalities in Two Variables; Functions

3.1 Reading Graphs; Linear Equations in Two Variables

OBJECTIVES

1. Interpret graphs.
2. Write a solution as an ordered pair.
3. Decide whether a given ordered pair is a solution of a given equation.
4. Complete ordered pairs for a given equation.
5. Plot ordered pairs.

Pie charts (circle graphs) and bar graphs were introduced in Chapter 1, and we have seen many examples of them in the first two chapters of this book. It is important to be able to interpret graphs correctly.

OBJECTIVE 1 Interpret graphs. A *bar graph* is used to show comparisons. We begin with a bar graph where we must estimate the heights of the bars.

EXAMPLE 1 Interpreting a Bar Graph

The bar graph in Figure 1 shows sales in millions of dollars for CarMax Auto Superstores, Inc., which has emerged as the most successful used-car megastore. The graph compares sales for 5 years.

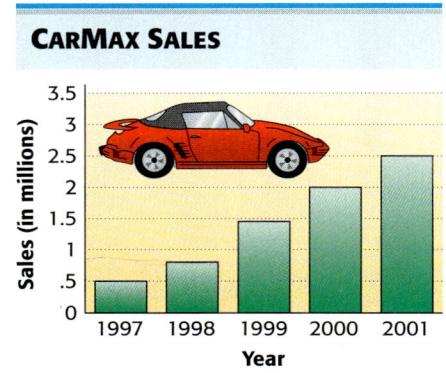

Source: Circuit City CarMax Group.

FIGURE 1

(a) Estimate sales in 1998.

Move horizontally from the top of the bar for 1998 to the scale on the left to see that sales in 1998 were about $.8 million.

(b) In what years were sales greater than $1 million?

Locate 1 on the vertical scale and follow the line across to the right. Three years—1999, 2000, and 2001—have bars that extend above the line for 1, so sales were greater than $1 million in those years.

(c) As the years progress, describe the change in sales.

Sales increase steadily as the years progress, from about $.5 million to $2.5 million.

Now Try Exercises 1 and 3.

A *line graph* is used to show changes or trends in data over time. To form a line graph, we connect a series of points representing data with line segments.

EXAMPLE 2 Interpreting a Line Graph

The line graph in Figure 2 shows average prices for personal computers (PCs) for the years 1993 through 1999.

FIGURE 2

(a) Between which years did the average price of a PC increase?

The line between 1993 and 1994 rises, so PC prices increased from 1993 to 1994.

(b) What has been the general trend in average PC prices since 1994?

The line graph falls from 1994 to 1999, so PC prices have been decreasing over these years.

(c) Estimate average PC prices in 1996 and 1999. About how much did PC prices decline between 1996 and 1999?

Move up from 1996 on the horizontal scale to the point plotted for 1996. Then move across to the vertical scale. The average price of a PC in 1996 was about $2000.

The point for 1999 is a little more than halfway between the lines for $1000 and $1500, so estimate the average price in 1999 at about $1300.

Between 1996 and 1999, PC prices declined about

$$\$2000 - \$1300 = \$700.$$

Now Try Exercises 5 and 7.

Many everyday situations, such as those illustrated in Examples 1 and 2, involve two quantities that are related. The equations and applications we discussed in Chapter 2 had only one variable. In this chapter, we extend those ideas to *linear equations in two variables*.

186 CHAPTER 3 Linear Equations and Inequalities in Two Variables; Functions

> **Linear Equation in Two Variables**
>
> A **linear equation in two variables** is an equation that can be written in the form
>
> $$Ax + By = C,$$
>
> where A, B, and C are real numbers and A and B are not both 0.

> **NOTE** If A or B is 0, we have equations such as $y = -1$ or $x = 3$. We think of these as linear equations in two variables by writing $y = -1$ as $0x + y = -1$. Similarly, $x = 3$ is the same as $x + 0y = 3$.

OBJECTIVE 2 Write a solution as an ordered pair. A solution of a linear equation in *two* variables requires *two* numbers, one for each variable. For example, the equation $y = 4x + 5$ is a true statement if x is replaced with 2 and y is replaced with 13, since

$$13 = 4(2) + 5. \qquad \text{Let } x = 2; y = 13.$$

The pair of numbers $x = 2$ and $y = 13$ gives a solution of the equation $y = 4x + 5$. The phrase "$x = 2$ and $y = 13$" is abbreviated

$$\underbrace{(\overset{\downarrow}{2}, \overset{\downarrow}{13})}_{\text{Ordered pair}} \quad \text{\textit{x}-value} \quad \text{\textit{y}-value}$$

with the *x*-value, 2, and the *y*-value, 13, given as a pair of numbers written inside parentheses. ***The x-value is always given first.*** A pair of numbers such as $(2, 13)$ is called an **ordered pair**. As the name indicates, the order in which the numbers are written is important. The ordered pairs $(2, 13)$ and $(13, 2)$ are not the same. The second pair indicates that $x = 13$ and $y = 2$.

OBJECTIVE 3 Decide whether a given ordered pair is a solution of a given equation. An ordered pair that is a solution of an equation is said to *satisfy* the equation.

EXAMPLE 3 Deciding Whether an Ordered Pair Satisfies an Equation

Decide whether the given ordered pair is a solution of the given equation.

(a) $(3, 2);\quad 2x + 3y = 12$

To see whether $(3, 2)$ is a solution of the equation $2x + 3y = 12$, we substitute 3 for x and 2 for y in the given equation.

$$2x + 3y = 12$$
$$2(3) + 3(2) = 12 \quad ? \qquad \text{Let } x = 3; \text{ let } y = 2.$$
$$6 + 6 = 12 \quad ?$$
$$12 = 12 \qquad \text{True}$$

This result is true, so $(3, 2)$ satisfies $2x + 3y = 12$.

(b) $(-2, -7)$; $m + 5n = 33$

$$m + 5n = 33$$
$$-2 + 5(-7) = 33 \quad ? \quad \text{Let } m = -2; \text{ let } n = -7.$$
$$-2 + (-35) = 33 \quad ?$$
$$-37 = 33 \quad \quad \text{False}$$

This result is false, so $(-2, -7)$ is *not* a solution of $m + 5n = 33$.

Now Try Exercises 17 and 21.

OBJECTIVE 4 Complete ordered pairs for a given equation. Choosing a number for one variable in a linear equation makes it possible to find the value of the other variable.

EXAMPLE 4   Completing an Ordered Pair

Complete the ordered pair $(7, \ )$ for the equation $y = 4x + 5$.

In this ordered pair, $x = 7$. (Remember that x always comes first.) To find the corresponding value of y, replace x with 7 in the equation $y = 4x + 5$.

$$y = 4(7) + 5 = 28 + 5 = 33$$

The ordered pair is $(7, 33)$.

Now Try Exercise 31.

Ordered pairs often are displayed in a **table of values.** Although we usually write tables of values vertically, they may be written horizontally as well.

EXAMPLE 5   Completing Tables of Values

Complete the given table of values for each equation. Then write the results as ordered pairs.

(a) $x - 2y = 8$

x	y
2	
10	
	0
	-2

To complete the first two ordered pairs, let $x = 2$ and $x = 10$, respectively.

If $x = 2,$ If $x = 10,$
then $x - 2y = 8$ then $x - 2y = 8$
becomes $2 - 2y = 8$ becomes $10 - 2y = 8$
$-2y = 6$ $-2y = -2$
$y = -3.$ $y = 1.$

Now complete the last two ordered pairs by letting $y = 0$ and $y = -2$, respectively.

If $y = 0$,	If $y = -2$,
then $x - 2y = 8$	then $x - 2y = 8$
becomes $x - 2(0) = 8$	becomes $x - 2(-2) = 8$
$x - 0 = 8$	$x + 4 = 8$
$x = 8$.	$x = 4$.

The completed table of values is as follows.

x	y
2	−3
10	1
8	0
4	−2

The corresponding ordered pairs are $(2, -3)$, $(10, 1)$, $(8, 0)$, and $(4, -2)$.

(b) $x = 5$

x	y
	−2
	6
	3

The given equation is $x = 5$. No matter which value of y might be chosen, the value of x is always the same, 5.

x	y
5	−2
5	6
5	3

The ordered pairs are $(5, -2)$, $(5, 6)$, and $(5, 3)$.

Now Try Exercises 41 and 45.

OBJECTIVE 5 Plot ordered pairs. Every linear equation in two variables has an infinite number of ordered pairs as solutions. Each choice of a number for one variable leads to a particular real number for the other variable. To graph these solutions, represented as ordered pairs (x, y), we need *two* number lines, one for each variable. These two number lines are drawn as shown in Figure 3. The horizontal number line is called the **x-axis.** The vertical line is called the **y-axis.** Together, the x-axis and y-axis form a **rectangular coordinate system,** also called the **Cartesian coordinate system,** in honor of René Descartes.

The coordinate system is divided into four regions, called **quadrants.** These quadrants are numbered counterclockwise, as shown in Figure 3. Points on the axes themselves are not in any quadrant. The point at which the *x*-axis and *y*-axis meet is called the **origin.** The origin, labeled 0 in Figure 3, is the point corresponding to (0, 0).

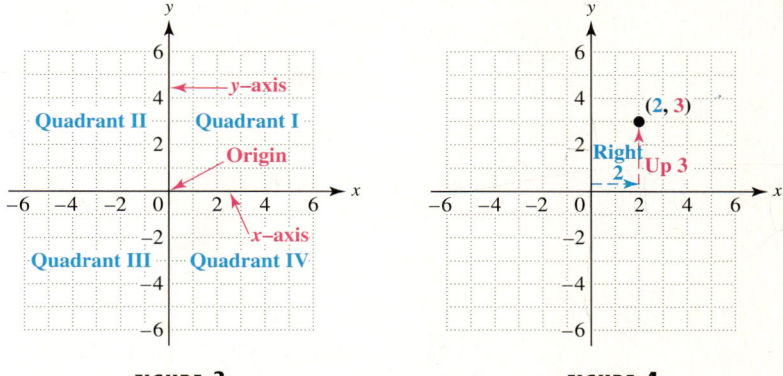

FIGURE 3   **FIGURE 4**

The *x*-axis and *y*-axis in a coordinate system determine a **plane,** a flat surface similar to a sheet of paper. By referring to the two axes, every point on the plane can be associated with an ordered pair. The numbers in the ordered pair are called the **coordinates** of the point. For example, locate the point associated with the ordered pair (2, 3) by starting at the origin. Since the *x*-coordinate is 2, go 2 units to the right along the *x*-axis. Then, since the *y*-coordinate is 3, turn and go up 3 units on a line parallel to the *y*-axis. This is called **plotting** the point (2, 3). (See Figure 4.) From now on we refer to the point with *x*-coordinate 2 and *y*-coordinate 3 as the point (2, 3).

NOTE On a plane, both numbers in the ordered pair are needed to locate a point. The ordered pair is a name for the point.

EXAMPLE 6   Plotting Ordered Pairs

In parts (a)–(f), plot the given points on a coordinate system. Then in part (g), identify the ordered pairs for the points labled *A*–*E*.

(a)   (1, 5)   (b)   (−2, 3)   (c)   (−1, −4)

(d)   (7, −2)   (e) $\left(\dfrac{3}{2}, 2\right)$ (f)   (5, 0)

(g)

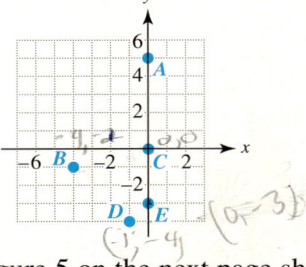

Figure 5 on the next page shows the graphs of the points in parts (a)–(f). Locate the point (−1, −4) in part (c) by first going 1 unit to the left along the *x*-axis. Then turn and go 4 units down, parallel to the *y*-axis. Plot the point $\left(\frac{3}{2}, 2\right)$ in part (e) by going $\frac{3}{2}$ (or $1\frac{1}{2}$) units to the right along the *x*-axis. Then turn and go 2 units up, parallel to the *y*-axis.

190 CHAPTER 3 Linear Equations and Inequalities in Two Variables; Functions

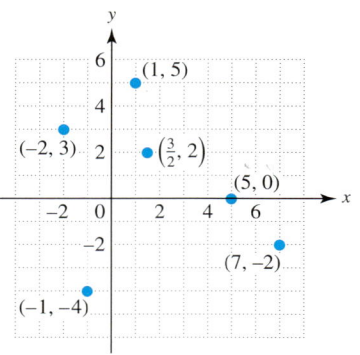

FIGURE 5

In part (g) the ordered pairs are as follows: A: $(0, 5)$; B: $(-4, -1)$; C: $(0, 0)$; D: $(-1, -4)$; E: $(0, -3)$.

Now Try Exercises 47, 49, and 53.

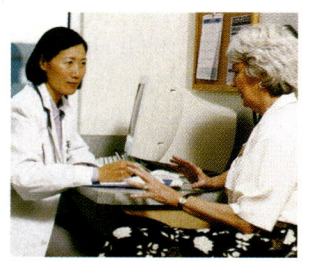

EXAMPLE 7 Completing Ordered Pairs to Estimate Annual Costs of Doctors' Visits

The amount Americans pay annually for doctors' visits has increased steadily from 1990 through 2000. This amount can be closely approximated by the linear equation

Cost ↴   ↴ Year
$$y = 34.3x - 67{,}693,$$

which relates x, the year, and y, the cost in dollars. (*Source:* U.S. Health Care Financing Administration.)

(a) Complete the table of values for this linear equation.

x (Year)	y (Cost)
1990	
1996	
2000	

To find y when $x = 1990$, we substitute into the equation.

$y = 34.3(1990) - 67{,}693$ Let $x = 1990$.
$y = 564$ Use a calculator.

This means that in 1990, Americans each spent about $564 on doctors' visits.

We substitute the years 1996 and 2000 in the same way to complete the table as follows.

x (Year)	y (Cost)
1990	564
1996	770
2000	907

We can write the results from the table of values as ordered pairs (x, y). Each year x is paired with its cost y:

(1990, 564),   (1996, 770),   and   (2000, 907).

(b) Graph the ordered pairs found in part (a).

The ordered pairs are graphed in Figure 6. This graph of ordered pairs of data is called a **scatter diagram.** Notice how the axes are labeled: x represents the year, and y represents the cost in dollars. Different scales are used on the two axes. Here, each square represents two units in the horizontal direction and 100 units in the vertical direction. Because the numbers in the first ordered pair are so large, we show a break in the axes near the origin.

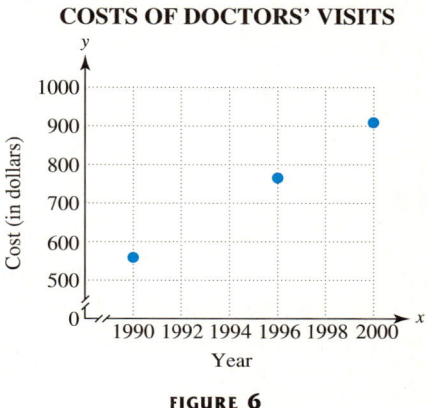

FIGURE 6

A scatter diagram enables us to tell whether two quantities are related to each other. In Figure 6, the plotted points could be connected to form a straight *line,* so the variables x (year) and y (cost) have a *line*ar relationship. The increase in costs is also reflected.

Now Try Exercise 71.

> **CAUTION** The equation in Example 7 is valid only for the years 1990 through 2000 because it was based on data for those years. Do not assume that this equation would provide reliable data for other years since the data for those years may not follow the same pattern.

We can think of ordered pairs as representing an input value x and an output value y. If we input x into the equation, the output is y. We encounter many examples of this type of relationship every day.

- The cost to fill a tank with gasoline depends on how many gallons are needed; the number of gallons is the input, and the cost is the output.
- The distance traveled depends on the traveling time; input a time, and the output is a distance.
- The growth of a plant depends on the amount of sun it gets; the input is the amount of sun, and the output is the growth.

This idea is illustrated in Figure 7 on the next page with an input-output "machine."

192 CHAPTER 3 Linear Equations and Inequalities in Two Variables; Functions

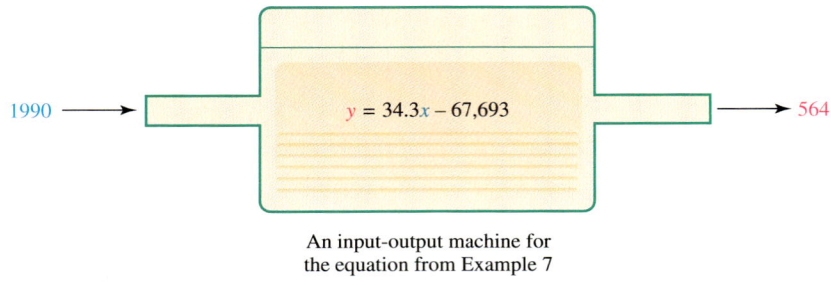

An input-output machine for
the equation from Example 7

FIGURE 7

In Section 3.6, we extend this idea to the concept of a *function*.

3.1 EXERCISES

For Extra Help

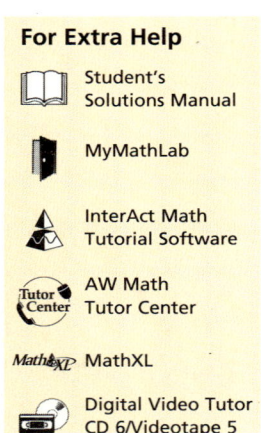

Student's Solutions Manual

MyMathLab

InterAct Math Tutorial Software

AW Math Tutor Center

MathXL

Digital Video Tutor CD 6/Videotape 5

The bar graph compares egg production in millions of eggs for six states in June 1999. Use the bar graph to work Exercises 1–4. See Example 1.

1. Name the top two egg-producing states in June 1999. Estimate their production.

2. Which states had egg production less than 400 million eggs?

3. Which states appear to have had equal production? Estimate this production.

4. How does egg production in Ohio compare to egg production in North Carolina?

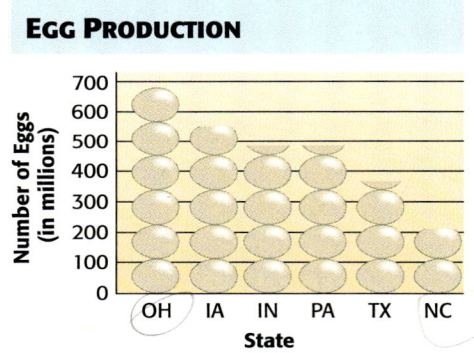

Source: Iowa Agricultural Statistics.

The line graph shows the average price, adjusted for inflation, that Americans have paid for a gallon of gasoline for selected years since 1970. Use the line graph to work Exercises 5–8. See Example 2.

5. Over which period of years did the greatest increase in the price of a gallon of gas occur? About how much was this increase?

6. Estimate the price of a gallon of gas during 1985, 1990, 1995, and 2000.

7. Describe the trend in gas prices from 1980 to 1995.

8. During which year(s) did a gallon of gas cost $1.50?

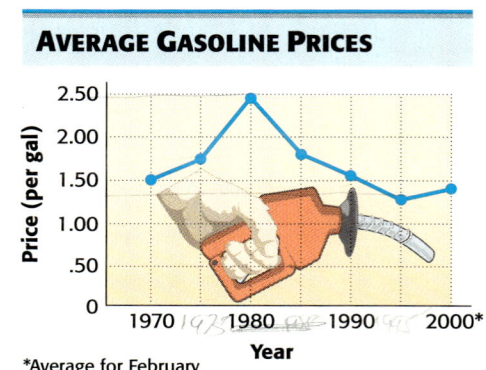

*Average for February

Source: American Petroleum Institute; AP research.

Fill in each blank with the correct response.

9. The symbol (x, y) _____ represent an ordered pair, while the symbols $[x, y]$ and
$\{x, y\}$ _____ represent ordered pairs.
(does/does not)
(do/do not)

10. The ordered pair $(3, 2)$ is a solution of the equation $2x - 5y =$ _____.

11. The point whose graph has coordinates $(-4, 2)$ is in quadrant _____.

12. The point whose graph has coordinates $(0, 5)$ lies along the _____ -axis.

13. The ordered pair $(4, $ _____$)$ is a solution of the equation $y = 3$.

14. The ordered pair $($_____$, -2)$ is a solution of the equation $x = 6$.

15. Define a linear equation in one variable and a linear equation in two variables, and give examples of each.

Decide whether the given ordered pair is a solution of the given equation. See Example 3.

16. $x + y = 9$; $(0, 9)$ **17.** $x + y = 8$; $(0, 8)$ **18.** $2p - q = 6$; $(4, 2)$
19. $2v + w = 5$; $(3, -1)$ **20.** $4x - 3y = 6$; $(2, 1)$ **21.** $5x - 3y = 15$; $(5, 2)$
22. $y = 3x$; $(2, 6)$ **23.** $x = -4y$; $(-8, 2)$ **24.** $x = -6$; $(-6, 5)$
25. $y = 2$; $(4, 2)$ **26.** $x + 4 = 0$; $(-6, 2)$ **27.** $x - 6 = 0$; $(4, 2)$

28. Do $(4, -1)$ and $(-1, 4)$ represent the same ordered pair? Explain.

29. Do the ordered pairs $(3, 4)$ and $(4, 3)$ correspond to the same point on the plane? Explain.

Complete each ordered pair for the equation $y = 2x + 7$. See Example 4.

30. $(2, \quad)$ **31.** $(5, \quad)$ **32.** $(\quad, 0)$ **33.** $(\quad, -3)$

Complete each ordered pair for the equation $y = -4x - 4$. See Example 4.

34. $(0, \quad)$ **35.** $(\quad, 0)$ **36.** $(\quad, 16)$ **37.** $(\quad, 24)$

38. Explain why it would be easier to find the corresponding y-value for $x = \frac{1}{3}$ than for $x = \frac{1}{7}$ in the equation $y = 6x + 2$.

39. For the equation $y = mx + b$, what is the y-value corresponding to $x = 0$ for *any* value of m?

Complete each table of values. See Example 5.

40. $2x + 3y = 12$

x	y
0	
	0
	8

41. $4x + 3y = 24$

x	y
0	
	0
	4

42. $3x - 5y = -15$

x	y
0	
	0
	-6

43. $4x - 9y = -36$

x	y
0	
	0
	8

194 CHAPTER 3 Linear Equations and Inequalities in Two Variables; Functions

44. $x = -9$

x	y
-9	6
-9	2
-9	-3

45. $x = 12$

x	y
12	3
12	8
12	0

46. $y = -6$

x	y
8	-6
4	-6
-2	-6

47. Give the ordered pairs that correspond to the points labeled in the figure. (All coordinates are integers.)

A = (2,5)
B = (-2,2)
C =
D = (-5,5)
E = (-6,-2)
(0,-1)

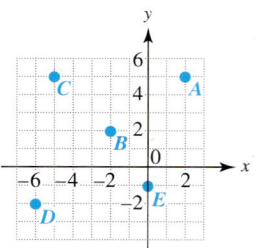

48. What is true about any point that lies on the x-axis? the y-axis?

Plot each ordered pair in a rectangular coordinate system. See Example 6.

49. $(6, 2)$ **50.** $(5, 3)$ **51.** $(-4, 2)$ **52.** $(-3, 5)$

53. $\left(-\dfrac{4}{5}, -1\right)$ **54.** $\left(-\dfrac{3}{2}, -4\right)$ **55.** $(0, 4)$ **56.** $(-3, 0)$

Fill in each blank with the word positive *or the word* negative.

The point with coordinates (x, y) is in

57. quadrant III if x is _____ N and y is _____ N.
58. quadrant II if x is _____ N and y is _____ P.
59. quadrant IV if x is _____ P and y is _____ N.
60. quadrant I if x is _____ P and y is _____ P.

Complete each table of values and then plot the ordered pairs. See Examples 5 and 6.

61. $x - 2y = 6$

x	y
0	
	0
2	
	-1

62. $2x - y = 4$

x	y
0	
	0
1	
	-6

63. $3x - 4y = 12$

x	y
0	3
4	0
-4	
	-4

64. $2x - 5y = 10$

x	y
0	
	0
-5	
	-3

65. $y + 4 = 0$

x	y
0	-4
5	-4
-2	-4
-3	-4

66. $x - 5 = 0$

x	y
5	1
5	0
5	6
5	-4

67. Look at your graphs of the ordered pairs in Exercises 61–66. Describe the pattern indicated by the plotted points.

SECTION 3.1 Reading Graphs; Linear Equations in Two Variables 195

Solve each problem. See Example 7.

68. Suppose that it costs $5000 to start up a business selling snow cones. Furthermore, it costs $.50 per cone in labor, ice, syrup, and overhead. Then the cost to make x snow cones is given by y dollars, where $y = .50x + 5000$. Express as an ordered pair each of the following.

(a) When 100 snow cones are made, the cost is $5050. (*Hint:* What does x represent? What does y represent?)

(b) When the cost is $6000, the number of snow cones made is 2000.

69. It costs a flat fee of $20 plus $5 per day to rent a pressure washer. Therefore, the cost to rent the pressure washer for x days is given by $y = 5x + 20$, where y is in dollars. Express as an ordered pair each of the following.

(a) When the washer is rented for 5 days, the cost is $45.

(b) I paid $50 when I returned the washer, so I must have rented it for 6 days.

Work each problem. See Example 7.

70. The table shows on-line retail spending in billions of dollars.

Year	Spending (in billions)
1998	7.8
1999	14.9
2000	23.1
2001	34.6
2002	53.0

Source: Jupiter Communications.

ON-LINE RETAIL SPENDING

(a) Write the data from the table as ordered pairs (x, y), where x represents the year and y represents on-line spending in billions of dollars.

(b) What does the ordered pair $(2003, 78.0)$ mean in the context of this problem?

(c) Make a scatter diagram of the data using the ordered pairs from part (a) and the given grid.

(d) Describe the pattern indicated by the points on the scatter diagram. What is the trend in on-line spending?

71. The table shows the rate (in percent) at which 4-yr college students graduate within 5 yr.

Year	Rate (%)
1996	53.3
1997	52.8
1998	52.1
1999	51.6

Source: ACT.

4-YEAR COLLEGE STUDENTS GRADUATING WITHIN 5 YEARS

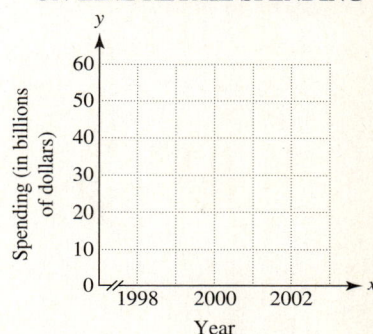

(a) Write the data from the table as ordered pairs (x, y), where x represents the year and y represents graduation rate.

(b) What does the ordered pair $(1995, 54.0)$ mean in the context of this problem?

(c) Make a scatter diagram of the data using the ordered pairs from part (a) and the given grid.

196 CHAPTER 3 Linear Equations and Inequalities in Two Variables; Functions

✍ **(d)** Describe the pattern indicated by the points on the scatter diagram. What is happening to graduation rates for 4-yr college students within 5 yr?

72. The maximum benefit for the heart from exercising occurs if the heart rate is in the target heart rate zone. The lower limit of this target zone can be approximated by the linear equation

$$y = -.7x + 154,$$

where *x* represents age and *y* represents heartbeats per minute. (*Source:* Hockey, R. V., *Physical Fitness: The Pathway to Healthy Living,* Times Mirror/Mosby College Publishing, 1989.)

Age	Heartbeats (per minute)
20	
40	
60	
80	

(a) Complete the table of values for this linear equation.
(b) Write the data from the table of values as ordered pairs.
(c) Make a scatter diagram of the data. Do the points lie in an approximately linear pattern?

73. (See Exercise 72.) The upper limit of the target heart rate zone can be approximated by the linear equation

$$y = -.8x + 186,$$

where *x* represents age and *y* represents heartbeats per minute. (*Source:* Hockey, R. V., *Physical Fitness: The Pathway to Healthy Living,* Times Mirror/Mosby College Publishing, 1989.)

Age	Heartbeats (per minute)
20	
40	
60	
80	

(a) Complete the table of values for this linear equation.
(b) Write the data from the table of values as ordered pairs.
(c) Make a scatter diagram of the data. Describe the pattern indicated by the data.

74. Refer to Exercises 72 and 73. What is the target heart rate zone for age 20? age 40?

3.2 Graphing Linear Equations in Two Variables

OBJECTIVES

1. Graph linear equations.
2. Find intercepts.
3. Graph linear equations of the form $Ax + By = 0$.
4. Graph linear equations of the form $y = k$ or $x = k$.
5. Use a linear equation to model data.

In this section we use a few ordered pairs that satisfy a linear equation to graph the equation.

OBJECTIVE 1 Graph linear equations. We know that infinitely many ordered pairs satisfy a linear equation. Some ordered pairs that are solutions of $x + 2y = 7$ are graphed in Figure 8. Notice that the points plotted in this figure all appear to lie on a straight line as shown in Figure 9. In fact,

every point on the line represents a solution of the equation $x + 2y = 7$, and each solution of the equation corresponds to a point on the line.

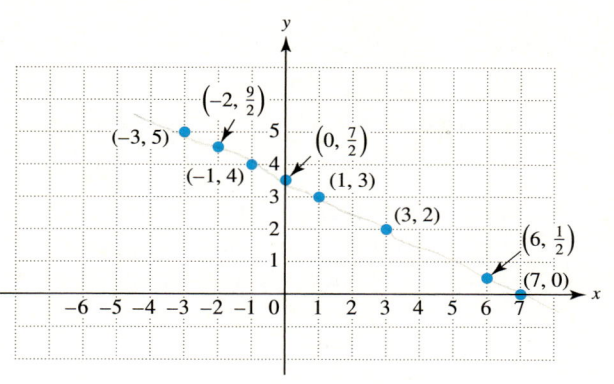

FIGURE 8

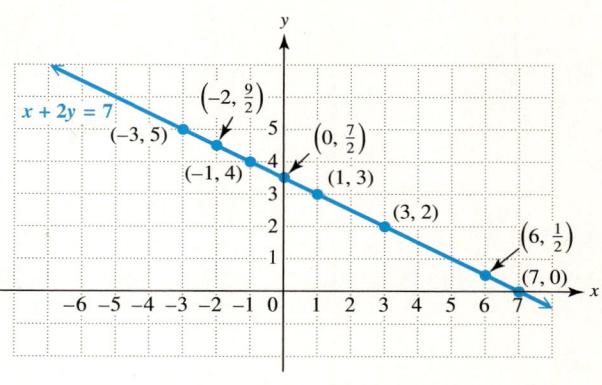

FIGURE 9

The line is a "picture" of all the solutions of the equation $x + 2y = 7$. Only a portion of the line is shown here, but it extends indefinitely in both directions, as suggested by the arrowhead on each end of the line. The line is called the **graph** of the equation, and the process of plotting the ordered pairs and drawing the line through the corresponding points is called **graphing**.

The preceding discussion can be generalized.

> ### Graph of a Linear Equation
> The graph of any linear equation in two variables is a straight line.

Notice that the word *line* appears in the name "*line*ar equation."

Since two distinct points determine a line, we can graph a straight line by finding any two different points on the line. However, it is a good idea to plot a third point as a check.

EXAMPLE 1 Graphing a Linear Equation

Graph the linear equation $2y = -3x + 6$.

Although this equation is not in the form $Ax + By = C$, it *could* be put in that form, and so is a linear equation. For most linear equations, we can find two different points on the graph by first letting $x = 0$, and then letting $y = 0$. Doing this gives the ordered pairs $(0, 3)$ and $(2, 0)$. We get a third ordered pair (as a check) by letting x or y equal some other number. For example, if $x = -2$, we find that $y = 6$, giving the ordered pair $(-2, 6)$. These three ordered pairs are shown in the table of values with Figure 10. The table of values is an alternative way of listing the ordered pairs $(0, 3)$, $(2, 0)$, and $(-2, 6)$.

x	y
0	3
2	0
-2	6

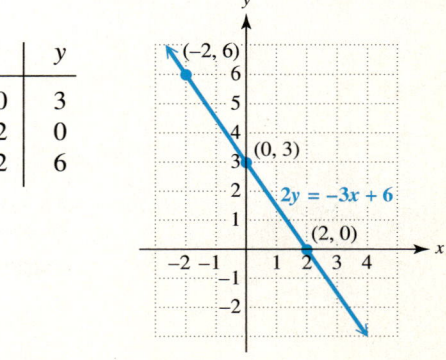

FIGURE 10

198 CHAPTER 3 Linear Equations and Inequalities in Two Variables; Functions

We plot the corresponding points, then draw a line through them. This line, shown in Figure 10, is the graph of $2y = -3x + 6$.

Now Try Exercise 5.

OBJECTIVE 2 Find intercepts. In Figure 10 the graph intersects (crosses) the y-axis at $(0, 3)$ and the x-axis at $(2, 0)$. For this reason $(0, 3)$ is called the **y-intercept** and $(2, 0)$ is called the **x-intercept** of the graph. The intercepts are particularly useful for graphing linear equations.

> **Finding Intercepts**
>
> To find the x-intercept, let $y = 0$ in the given equation and solve for x. Then $(x, 0)$ is the x-intercept.
>
> To find the y-intercept, let $x = 0$ in the given equation and solve for y. Then $(0, y)$ is the y-intercept.

EXAMPLE 2 Finding Intercepts

Find the intercepts for the graph of $2x + y = 4$. Draw the graph.

Find the y-intercept by letting $x = 0$; find the x-intercept by letting $y = 0$.

$$2x + y = 4 \qquad\qquad 2x + y = 4$$
$$2(0) + y = 4 \quad \text{Let } x = 0. \qquad 2x + 0 = 4 \quad \text{Let } y = 0.$$
$$0 + y = 4 \qquad\qquad 2x = 4$$
$$y = 4 \qquad\qquad x = 2$$

The y-intercept is $(0, 4)$. The x-intercept is $(2, 0)$. The graph, with the two intercepts shown in red, is given in Figure 11. Get a third point as a check. For example, choosing $x = 4$ gives $y = -4$. These three ordered pairs are shown in the table with Figure 11. Plot $(0, 4)$, $(2, 0)$, and $(4, -4)$ and draw a line through them. This line, shown in Figure 11, is the graph of $2x + y = 4$.

x	y
0	4
2	0
4	-4

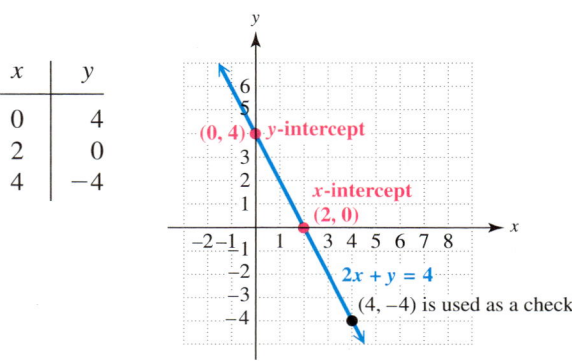

FIGURE 11

Now Try Exercise 13.

SECTION 3.2 Graphing Linear Equations in Two Variables **199**

> **CAUTION** When choosing x- or y-values to find ordered pairs to plot, be careful to choose so that the resulting points are not too close together. For example, using $(-1, -1)$, $(0, 0)$, and $(1, 1)$ may result in an inaccurate line. It is better to choose points where the x-values differ by at least 2.

OBJECTIVE 3 Graph linear equations of the form $Ax + By = 0$. In earlier examples, the x- and y-intercepts were used to help draw the graphs. This is not always possible. Example 3 shows what to do when the x- and y-intercepts are the same point.

EXAMPLE 3 Graphing an Equation of the Form $Ax + By = 0$

Graph $x - 3y = 0$.

If we let $x = 0$, then $y = 0$, giving the ordered pair $(0, 0)$. Letting $y = 0$ also gives $(0, 0)$. This is the same ordered pair, so we choose two *other* values for x or y. Choosing 2 for y gives $x - 3 \cdot 2 = 0$, or $x = 6$, giving the ordered pair $(6, 2)$. For a check point, we choose -6 for x getting -2 for y. We use the ordered pairs $(-6, -2)$, $(0, 0)$, and $(6, 2)$ to get the graph shown in Figure 12.

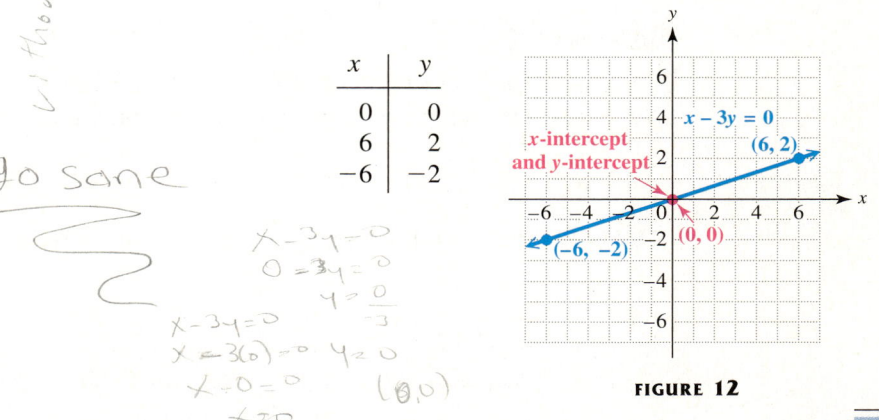

FIGURE 12

Now Try Exercise 31.

Example 3 can be generalized as follows.

> **Line through the Origin**
>
> If A and B are nonzero real numbers, the graph of a linear equation of the form
>
> $$Ax + By = 0$$
>
> goes through the origin $(0, 0)$.

OBJECTIVE 4 Graph linear equations of the form $y = k$ or $x = k$. The equation $y = -4$ is a linear equation in which the coefficient of x is 0. (Write $y = -4$ as $0x + y = -4$ to see this.) Also, $x = 3$ is a linear equation in which the coefficient of y is 0. These equations lead to horizontal or vertical straight lines.

EXAMPLE 4 Graphing an Equation of the Form y = k

Graph $y = -4$.

As the equation states, for any value of x, y is always equal to -4. To get ordered pairs that are solutions of this equation, we choose any numbers for x, always using -4 for y. Three ordered pairs that satisfy the equation are shown in the table of values with Figure 13. Drawing a line through these points gives the horizontal line shown in Figure 13. The y-intercept is $(0, -4)$; there is no x-intercept.

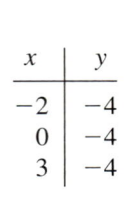
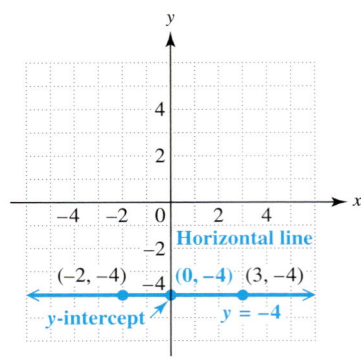

FIGURE 13

Now Try Exercise 35.

Horizontal Line

The graph of the linear equation $y = k$, where k is a real number, is the horizontal line with y-intercept $(0, k)$ and no x-intercept.

EXAMPLE 5 Graphing an Equation of the Form x = k

Graph $x - 3 = 0$.

First we add 3 to each side of the equation $x - 3 = 0$ to get $x = 3$. All the ordered pairs that are solutions of this equation have an x-value of 3. Any number can be used for y. We show three ordered pairs that satisfy the equation in the table of values with Figure 14. Drawing a line through these points gives the vertical line shown in Figure 14. The x-intercept is $(3, 0)$; there is no y-intercept.

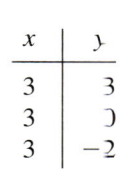
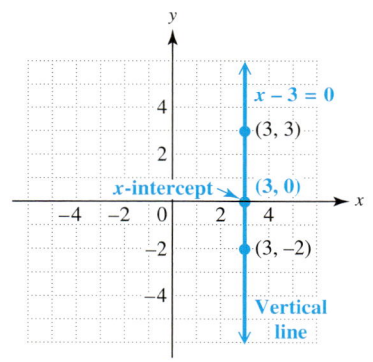

FIGURE 14

Now Try Exercise 37.

Vertical Line

The graph of the linear equation $x = k$, where k is a real number, is the vertical line with x-intercept $(k, 0)$ and no y-intercept.

In particular, notice that the horizontal line $y = 0$ is the x-axis and the vertical line $x = 0$ is the y-axis.

CAUTION The equations of horizontal and vertical lines are often confused with each other. Remember that the graph of $y = k$ is parallel to the x-axis and that of $x = k$ is parallel to the y-axis.

The different forms of straight-line equations and the methods of graphing them are given in the following summary.

Graphing Straight Lines

Equation	To Graph	Example
$y = k$	Draw a horizontal line, through $(0, k)$.	$y = -2$
$x = k$	Draw a vertical line, through $(k, 0)$.	$x = 4$
$Ax + By = 0$	Graph goes through $(0, 0)$. Get additional points that lie on the graph by choosing any value of x or y, except 0.	$x = 2y$

(continued)

$Ax + By = C$ but not of the types above	Find any two points the line goes through. A good choice is to find the intercepts: let $x = 0$, and find the corresponding value of y; then let $y = 0$, and find x. As a check, get a third point by choosing a value of x or y that has not yet been used.	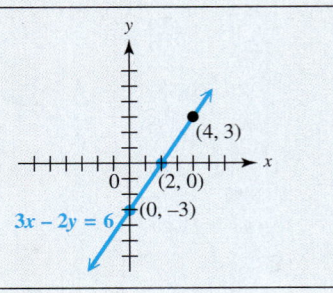

(Another important form, the slope-intercept form, is introduced on page 220.)

CONNECTIONS

Beginning in this chapter we include information on the basic features of graphing calculators. The most obvious feature is their ability to graph equations. We must solve the equation for y in order to enter it into the calculator. Also, we must select an appropriate "window" for the graph. The window is determined by the minimum and maximum values of x and y. Graphing calculators have a standard window, often from $x = -10$ to $x = 10$ and from $y = -10$ to $y = 10$. Sometimes this is written $[-10, 10]$, $[-10, 10]$, with the x-interval shown first.

For example, to graph the equation $2x + y = 4$, discussed in Example 2, we first solve for y.

$$2x + y = 4$$
$$y = -2x + 4 \quad \text{Subtract } 2x.$$

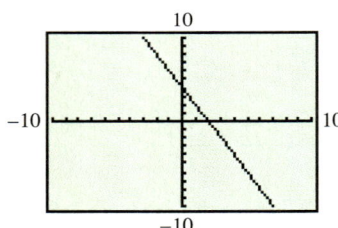

If we enter this equation as $y = -2x + 4$ and choose the standard window, the calculator displays the graph shown in the margin.

The x-value of the point on the graph where $y = 0$ (the x-intercept) gives the solution of the equation

$$y = 0,$$
or $\quad -2x + 4 = 0. \quad$ Substitute $-2x + 4$ for y.

Since each tick mark on the x-axis represents 1, the graph shows that $(2, 0)$ is the x-intercept.

For Discussion or Writing

How would you rewrite these equations, with one side equal to 0, to enter them into a graphing calculator for solution? (It is not necessary to clear parentheses or combine terms.)

1. $3x + 4 - 2x - 7 = 4x + 3$ 2. $5x - 15 = 3(x - 2)$

OBJECTIVE 5 Use a linear equation to model data.

EXAMPLE 6 Using a Linear Equation to Model Credit Card Debt

Credit card debt in the United States increased steadily from 1992 through 1999. The amount of debt y in billions of dollars can be modeled by the linear equation

$$y = 47.3x + 281,$$

where $x = 0$ represents 1992, $x = 1$ represents 1993, and so on. (*Source:* Board of Governors of the Federal Reserve System.)

(a) Use the equation to approximate credit card debt in the years 1992, 1993, and 1999.

For 1992: $y = 47.3(0) + 281$ Replace *x* with 0.
                $y = 281$ billion dollars

For 1993: $y = 47.3(1) + 281$ Replace *x* with 1.
                $y = 328.3$ billion dollars

For 1999: $y = 47.3(7) + 281$ $1999 - 1992 = 7$;
                $y = 612.1$ billion dollars   replace *x* with 7.

(b) Write the information from part (a) as three ordered pairs, and use them to graph the given linear equation.

Since *x* represents the year and *y* represents the debt in billions of dollars, the ordered pairs are (0, 281), (1, 328.3), and (7, 612.1). Figure 15 shows a graph of these ordered pairs and the line through them. (Note that arrowheads are not included with the graphed line since the data are for the years 1992 to 1999 only, that is, from $x = 0$ to $x = 7$.)

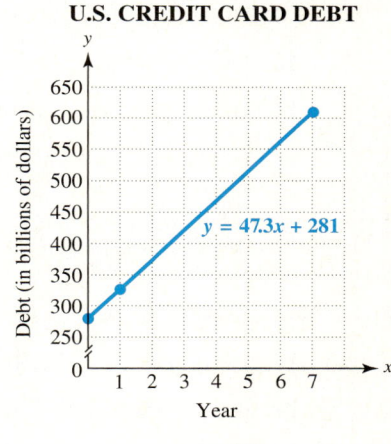

FIGURE 15

(c) Use the graph and then the equation to approximate credit card debt in 1996.

For 1996, $x = 4$. On the graph, find 4 on the horizontal axis and move up to the graphed line, then across to the vertical axis. It appears that credit card debt in 1996 was about 470 billion dollars.

To use the equation, substitute 4 for *x*.

$y = 47.3(4) + 281$ Let $x = 4$.
$y = 470.2$ billion dollars

This result is quite similar to our estimate using the graph.

Now Try Exercise 71.

204 CHAPTER 3 Linear Equations and Inequalities in Two Variables; Functions

 EXERCISES

For Extra Help

 Student's Solutions Manual

 MyMathLab

 InterAct Math Tutorial Software

 AW Math Tutor Center

 MathXL

Digital Video Tutor CD 6/Videotape 5

Complete the given ordered pairs using the given equation. Then graph each equation by plotting the points and drawing a line through them. See Example 1.

1. $y = -x + 5$
(0,), (,0), (2,)

2. $y = x - 2$
(0,), (,0), (5,)

3. $y = \frac{2}{3}x + 1$
(0,), (3,), (-3,)

4. $y = -\frac{3}{4}x + 2$
(0,), (4,), (-4,)

5. $3x = -y - 6$
(0,), (,0), $\left(-\frac{1}{3}, \right)$

6. $x = 2y + 3$
(,0), (0,), $\left(, \frac{1}{2}\right)$

In Exercises 7–12, match the information about each graph in Column I with the correct linear equation in Column II.

I

7. The graph of the equation has *x*-intercept (4, 0).

8. The graph of the equation has *y*-intercept (0, -4).

9. The graph of the equation goes through the origin.

10. The graph of the equation is a vertical line.

11. The graph of the equation is a horizontal line.

12. The graph of the equation goes through (9, 2).

II

A. $x = 5$
B. $y = -3$
C. $2x - 5y = 8$
D. $x + 4y = 0$
E. $3x + y = -4$

Find the x-intercept and the y-intercept for the graph of each equation. See Examples 2, 4, and 5.

13. $2x - 3y = 24$
14. $-3x + 8y = 48$
15. $x + 6y = 0$
16. $3x - y = 0$
17. $5x - 2y = 20$
18. $-3x + 2y = 12$
19. $x - 4 = 0$
20. $y = 5$

21. What is the equation of the *x*-axis? What is the equation of the *y*-axis?

22. A student attempted to graph $4x + 5y = 0$ by finding intercepts. She first let $x = 0$ and found y; then she let $y = 0$ and found x. In both cases, the resulting point was (0, 0). She knew that she needed at least two different points to graph the line, but was unsure what to do next since finding intercepts gave her only one point. How would you explain to her what to do next?

Graph each linear equation. See Examples 1–5.

23. $x = y + 2$
24. $x = -y + 6$
25. $x - y = 4$
26. $x - y = 5$
27. $y = -2x + 6$
28. $y = 3x - 6$
29. $3x + 7y = 14$
30. $6x - 5y = 18$
31. $y - 2x = 0$
32. $y + 3x = 0$
33. $y = -6x$
34. $y = 4x$
35. $y + 1 = 0$
36. $y - 3 = 0$
37. $x = -2$
38. $x = 4$

39. $3y + 2x = 12$ **40.** $2y + 5x = 10$ **41.** $y = \frac{1}{3}x - 2$

42. $y = \frac{1}{4}x - 1$ **43.** $2x = 3y$ **44.** $3x = 2y$

45. $y = 2.5x - 5$ **46.** $y = 1.5x + 2$ **47.** $y = -\frac{1}{2}x + 2$

48. $y = -\frac{1}{3}x + 3$ **49.** $8x - 6y = 24$ **50.** $4x - 3y = -24$

51. $2x = -5y + 20$ **52.** $4x = -6x + 8$ **53.** $-3x = -2y - 20$

54. $-2x = -3y - 9$ **55.** $3y = 15$ **56.** $2y = 12$

57. $-4x = -8$ **58.** $-5x = -10$ **59.** $y = -\frac{3}{4}x + 2$

60. $y = -\frac{2}{3}x + 1$ **61.** $y = \frac{3}{2}x - 3$ **62.** $y = \frac{4}{3}x - 4$

63. $y = .75x + 1$ **64.** $y = 1.5x - 2$ **65.** $2y = -10$

66. $3y = -6$ **67.** $-5x = 10$ **68.** $-3x = 9$

69. $-2x = 4y$ **70.** $-3x = 9y$

Solve each problem. See Example 6.

71. The height *y* (in centimeters) of a woman is related to the length of her radius bone *x* (from the wrist to the elbow) and is approximated by the linear equation
$$y = 3.9x + 73.5.$$
(a) Use the equation to find the approximate heights of women with radius bones of lengths 20 cm, 26 cm, and 22 cm.
(b) Graph the equation using the data from part (a).
(c) Use the graph to estimate the length of the radius bone in a woman who is 167 cm tall. Then use the equation to find the length of this radius bone to the nearest centimeter. (*Hint:* Substitute for *y* in the equation.)

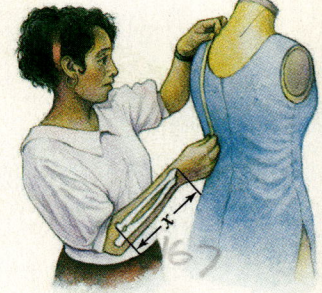

72. The weight *y* (in pounds) of a man taller than 60 in. can be roughly approximated by the linear equation
$$y = 5.5x - 220,$$
where *x* is the height of the man in inches.
(a) Use the equation to approximate the weights of men whose heights are 62 in., 66 in., and 72 in.
(b) Graph the equation using the data from part (a).
(c) Use the graph to estimate the height of a man who weighs 155 lb. Then use the equation to find the height of this man to the nearest inch. (*Hint:* Substitute for *y* in the equation.)

73. Refer to Section 3.1 Exercise 72. Draw a line through the points you plotted in the scatter diagram there.

  (a) Use the graph to estimate the lower limit of the target heart rate zone for age 30.
  (b) Use the linear equation given there to approximate the lower limit for age 30.
  (c) How does the approximation using the equation compare to the estimate from the graph?

74. Refer to Section 3.1 Exercise 73. Draw a line through the points you plotted in the scatter diagram there.

  (a) Use the graph to estimate the upper limit of the target heart rate zone for age 30.
  (b) Use the linear equation given there to approximate the upper limit for age 30.
  (c) How does the approximation using the equation compare to the estimate from the graph?

75. Use the results of Exercises 73(b) and 74(b) to determine the target heart rate zone for age 30.

76. Should the graphs of the target heart rate zone in Section 3.1 Exercises 72 and 73 be used to estimate the target heart rate zone for ages below 20 or above 80? Why or why not?

77. Per capita consumption of carbonated soft drinks increased for the years 1992 through 1997 as shown in the graph. If $x = 0$ represents 1992, $x = 1$ represents 1993, and so on, per capita consumption can be modeled by the linear equation

$$y = .8x + 49,$$

where y is in gallons.

  (a) Use the equation to approximate consumption in 1993, 1995, and 1997.
  (b) Use the graph to estimate consumption for the same years.
  (c) How do the approximations using the equation compare to the estimates from the graph?

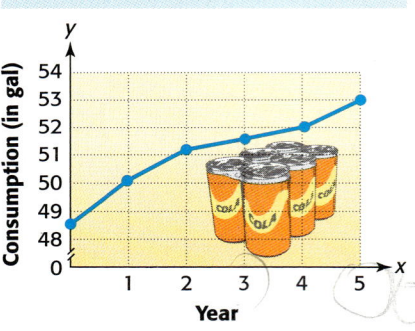

SOFT DRINK CONSUMPTION

Source: U.S. Department of Agriculture.

78. Sporting goods sales y (in billions of dollars) from 1995 through 2000 are modeled by the linear equation

$$y = 3.606x + 41.86,$$

where $x = 0$ corresponds to 1990, $x = 5$ corresponds to 1995, and so on.

SECTION 3.2 Graphing Linear Equations in Two Variables 207

SPORTING GOODS SALES

Source: U.S. Bureau of the Census.

(a) Use the equation to approximate sporting goods sales in 1995, 1997, and 2000. Round your answers to the nearest billion dollars.
(b) Use the graph to estimate sales for the same years.
(c) How do the approximations using the equation compare to the estimates using the graph?

79. The graph shows the value of a certain sport utility vehicle over the first 5 yr of ownership.

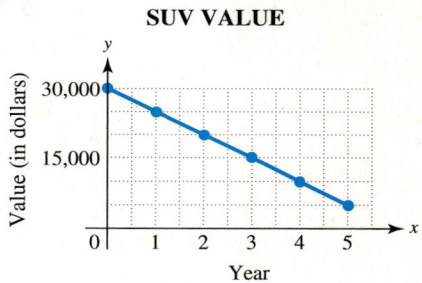

SUV VALUE

Use the graph to do the following.
(a) Determine the initial value of the SUV.
(b) Find the *depreciation* (loss in value) from the original value after the first 3 yr.
(c) What is the annual or yearly depreciation in each of the first 5 yr?
(d) What does the ordered pair (5, 5000) mean in the context of this problem?

80. Demand for an item is often closely related to its price. As price increases, demand decreases, and as price decreases, demand increases. Suppose demand for a video game is 2000 units when the price is $40, and demand is 2500 units when the price is $30.
(a) Let x be the price and y be the demand for the game. Graph the two given pairs of prices and demands.
(b) Assume the relationship is linear. Draw a line through the two points from part (a). From your graph, estimate the demand if the price drops to $20.
(c) Use the graph to estimate the price if the demand is 3500 units.

208 CHAPTER 3 Linear Equations and Inequalities in Two Variables; Functions

3.3 The Slope of a Line

OBJECTIVES

1. Find the slope of a line given two points.
2. Find the slope from the equation of a line.
3. Use slope to determine whether two lines are parallel, perpendicular, or neither.

An important characteristic of the lines we graphed in the previous section is their slant or "steepness." See Figure 16.

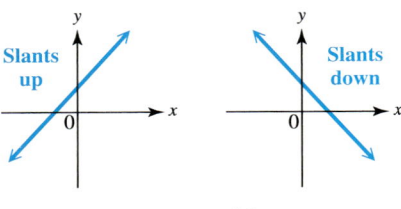

FIGURE 16

One way to measure the steepness of a line is to compare the vertical change in the line to the horizontal change while moving along the line from one fixed point to another. This measure of steepness is called the *slope* of the line.

OBJECTIVE 1 Find the slope of a line given two points. Figure 17 shows a line through two nonspecific points (x_1, y_1) and (x_2, y_2). (This notation is called **subscript notation.** Read x_1 as "x-sub-one" and x_2 as "x-sub-two.")

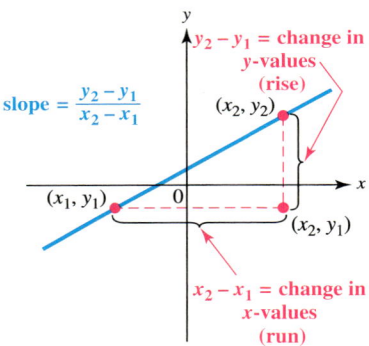

FIGURE 17

Moving along the line from the point (x_1, y_1) to the point (x_2, y_2) causes y to change by $y_2 - y_1$ units. This is the vertical change or **rise.** Similarly, x changes by $x_2 - x_1$ units, which is the horizontal change or **run.** (In both cases, the change is expressed as a *difference.*) Remember from Section 2.6 that one way to compare two numbers is by using a ratio. **Slope** is the ratio of the vertical change in y to the horizontal change in x.

EXAMPLE 1 Comparing Rise to Run

Figure 18 shows two lines. Find the slope ratio of the rise to the run for each line.

We use the two points shown on each line. For the line in Figure 18(a), we find the rise from point Q to point P by determining the vertical change from −4 to 4, the

SECTION 3.3 The Slope of a Line 209

FIGURE 18

difference $4 - (-4) = 8$. Similarly, we find the run by determining the horizontal change from Q to P, $5 - (-5) = 10$. Slope is the ratio

$$\frac{\text{rise}}{\text{run}} = \frac{8}{10} \text{ or } \frac{4}{5}.$$

From Q to P, the line in Figure 18(b) has rise $3 - (-1) = 4$ and run $2 - 6 = -4$, so the slope is the ratio

$$\frac{\text{rise}}{\text{run}} = \frac{4}{-4} \text{ or } -1.$$

To confirm our slope ratios, count grid squares from one point on the line to another. For example, starting at the origin in Figure 18(a), count up 4 squares (the rise in the slope ratio $\frac{4}{5}$) and then 5 squares to the right (the run) to arrive at the point $(5, 4)$ on the line.

Now Try Exercise 3.

The slopes we found for the two lines in Figure 18 suggest that a line with positive slope slants upward from left to right, and a line with negative slope slants downward. These facts can be generalized.

Positive and Negative Slopes

A line with positive slope rises from left to right.

A line with negative slope falls from left to right.

Hill

The idea of slope is used in many everyday situations. For example, because $10\% = \frac{1}{10}$, a highway with a 10% grade (or slope) rises 1 meter for every 10 horizontal meters. A highway sign is used to warn of a downgrade that may be long or steep. Architects specify the pitch of a roof using slope; a $\frac{5}{12}$ roof means that the roof rises 5 ft for every 12 ft in the horizontal direction. See the figure on the next page. The slope of a stairwell also indicates the ratio of the vertical rise to the horizontal run. The slope of the stairs in the figure is $\frac{8}{14}$.

210 CHAPTER 3 Linear Equations and Inequalities in Two Variables; Functions

$\frac{5}{12}$ roof pitch

Rise: 8 ft
Run: 14 ft
Slope of a stairwell

Traditionally, the letter m represents slope. The slope m of a line is defined as follows.

Slope Formula

The **slope** of the line through the points (x_1, y_1) and (x_2, y_2) is

$$m = \frac{\text{change in } y}{\text{change in } x} = \frac{y_2 - y_1}{x_2 - x_1}, \quad \text{if } x_1 \neq x_2.$$

The slope of a line tells how fast y changes for each unit of change in x; that is, the slope gives the rate of change in y for each unit of change in x.

EXAMPLE 2 Finding Slopes of Lines

Find the slope of each line.

(a) The line through $(1, -2)$ and $(-4, 7)$

Use the slope formula. Let $(-4, 7) = (x_2, y_2)$ and $(1, -2) = (x_1, y_1)$. Then

$$\text{slope } m = \frac{\text{change in } y}{\text{change in } x} = \frac{y_2 - y_1}{x_2 - x_1} = \frac{7 - (-2)}{-4 - 1} = \frac{9}{-5} = -\frac{9}{5}.$$

See Figure 19.

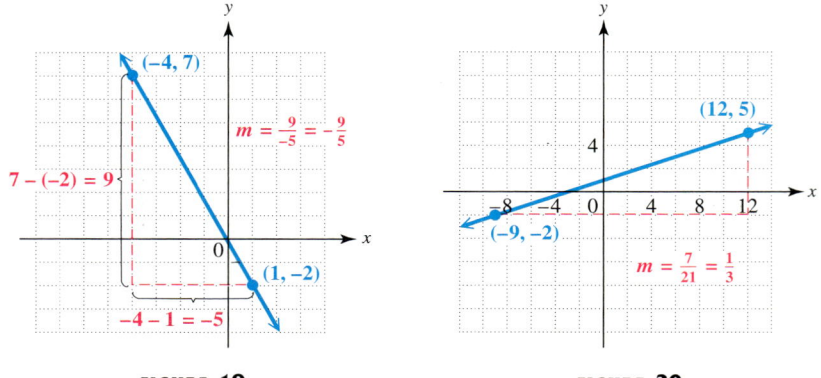

FIGURE 19

FIGURE 20

(b) The line through $(-9, -2)$ and $(12, 5)$

$$m = \frac{y_2 - y_1}{x_2 - x_1} = \frac{5 - (-2)}{12 - (-9)} = \frac{7}{21} = \frac{1}{3}$$

SECTION 3.3 The Slope of

See Figure 20. Note that the same slope is obtained by subtracting in

$$m = \frac{-2 - 5}{-9 - 12} = \frac{-7}{-21} = \frac{1}{3}$$

Now Try Exercises 17 and 21.

CAUTION It makes no difference which point is (x_1, y_1) or (x_2, y_2); however, be consistent. Start with the x- and y-values of one point (either one) and subtract the corresponding values of the other point. Also, the slope of a line is the same for *any* two points on the line.

EXAMPLE 3 Finding the Slope of a Horizontal Line

Find the slope of the line through $(-8, 4)$ and $(2, 4)$.
Use the slope formula.

$$m = \frac{4 - 4}{-8 - 2} = \frac{0}{-10} = 0 \quad \text{Zero slope}$$

As shown in Figure 21, the line through these two points is horizontal, with equation $y = 4$. *All horizontal lines have slope 0,* since the difference in y-values is always 0.

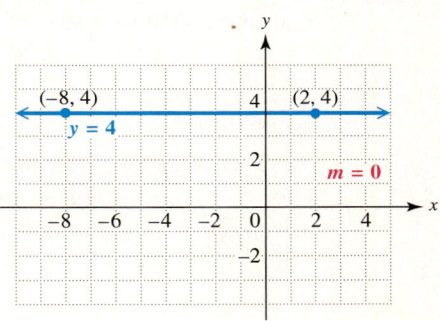

FIGURE 21

Now Try Exercise 23.

EXAMPLE 4 Finding the Slope of a Vertical Line

Find the slope of the line through $(6, 2)$ and $(6, -9)$.

$$m = \frac{2 - (-9)}{6 - 6} = \frac{11}{0} \quad \text{Undefined slope}$$

Since division by 0 is undefined, the slope is undefined. The graph in Figure 22 shows that the line through these two points is vertical with equation $x = 6$. All points on a vertical line have the same x-value, so *the slope of any vertical line is undefined.*

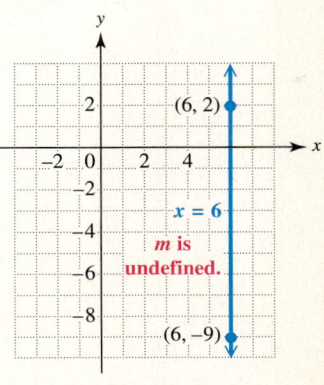

FIGURE 22

Now Try Exercise 25.

CHAPTER 3 Linear Equations and Inequalities in Two Variables; Functions

> **Slopes of Horizontal and Vertical Lines**
>
> **Horizontal lines,** with equations of the form $y = k$, have **slope 0.**
>
> **Vertical lines,** with equations of the form $x = k$, have **undefined slope.**

OBJECTIVE 2 Find the slope from the equation of a line. The slope of a line can be found directly from its equation. For example, the slope of the line

$$y = -3x + 5$$

is found using any two points on the line. We get these two points by first choosing two different values of x and then finding the corresponding values of y. We choose $x = -2$ and $x = 4$.

$y = -3x + 5$		$y = -3x + 5$
$y = -3(-2) + 5$ Let $x = -2$.		$y = -3(4) + 5$ Let $x = 4$.
$y = 6 + 5$		$y = -12 + 5$
$y = 11$		$y = -7$

The ordered pairs are $(-2, 11)$ and $(4, -7)$. Now we use the slope formula.

$$m = \frac{11 - (-7)}{-2 - 4} = \frac{18}{-6} = -3$$

The slope, -3, is the same number as the coefficient of x in the equation $y = -3x + 5$. It can be shown that this always happens, *as long as the equation is solved for y.* This fact is used to find the slope of a line from its equation.

> **Finding the Slope of a Line from Its Equation**
>
> *Step 1*   Solve the equation for y.
>
> *Step 2*   The slope is given by the coefficient of x.

EXAMPLE 5 Finding Slopes from Equations

Find the slope of each line.

(a) $2x - 5y = 4$

Solve the equation for y.

$$2x - 5y = 4$$
$$-5y = -2x + 4 \quad \text{Subtract } 2x \text{ from each side.}$$
$$y = \frac{2}{5}x - \frac{4}{5} \quad \text{Divide by } -5.$$

The slope is given by the coefficient of x, so the slope is $\frac{2}{5}$.

(b) $8x + 4y = 1$

Solve the equation for y.

$$8x + 4y = 1$$
$$4y = -8x + 1 \quad \text{Subtract } 8x.$$
$$y = -2x + \frac{1}{4} \quad \text{Divide by 4.}$$

The slope of this line is given by the coefficient of x, -2.

Now Try Exercise 33.

OBJECTIVE 3 Use slope to determine whether two lines are parallel, perpendicular, or neither. Two lines in a plane that never intersect are **parallel.** We use slopes to tell whether two lines are parallel. For example, Figure 23 shows the graphs of $x + 2y = 4$ and $x + 2y = -6$. These lines appear to be parallel. Solve for y to find that both $x + 2y = 4$ and $x + 2y = -6$ have slope $-\frac{1}{2}$. Nonvertical parallel lines always have equal slopes.

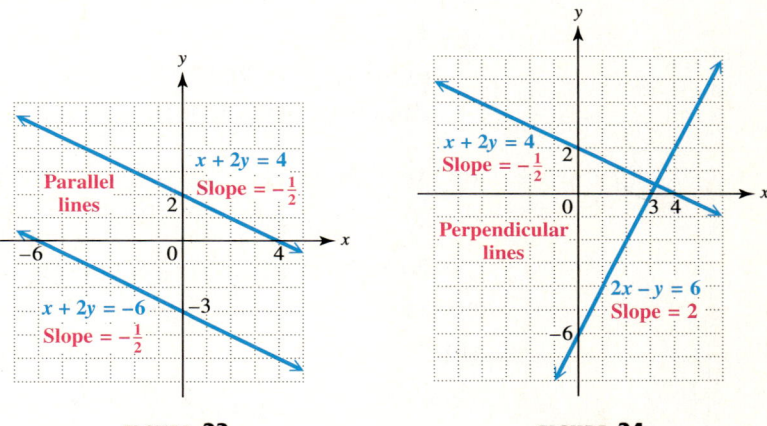

FIGURE 23   FIGURE 24

Figure 24 shows the graphs of $x + 2y = 4$ and $2x - y = 6$. These lines appear to be **perpendicular** (that is, they intersect at a 90° angle). Solving for y shows that the slope of $x + 2y = 4$ is $-\frac{1}{2}$, while the slope of $2x - y = 6$ is 2. The product of $-\frac{1}{2}$ and 2 is

$$-\frac{1}{2}(2) = -1.$$

This is true in general; the product of the slopes of two perpendicular lines, neither of which is vertical, is always -1. This means that the slopes of perpendicular lines are negative reciprocals; if one slope is the nonzero number a, the other is $-\frac{1}{a}$.

Slopes of Parallel and Perpendicular Lines

Two lines with the same slope are parallel.

Two lines whose slopes have a product of -1 are perpendicular.

EXAMPLE 6 Deciding Whether Two Lines Are Parallel or Perpendicular

Decide whether each pair of lines is *parallel, perpendicular,* or *neither.*

(a) $x + 2y = 7$
         $-2x + y = 3$

Find the slope of each line by first solving each equation for *y*.

$x + 2y = 7$ $\qquad$ $-2x + y = 3$
$2y = -x + 7$ $\qquad$ $y = 2x + 3$
$y = -\dfrac{1}{2}x + \dfrac{7}{2}$

Slope: $-\dfrac{1}{2}$ $\qquad$ Slope: 2

Since the slopes are not equal, the lines are not parallel. Check the product of the slopes: $-\dfrac{1}{2}(2) = -1$. The two lines are perpendicular because the product of their slopes is -1, indicating that the slopes are negative reciprocals.

(b) $3x - y = 4$ $\quad$ *Solve for y.* $\quad$ $y = 3x - 4$
       $6x - 2y = -12$ $\longrightarrow$ $y = 3x + 6$

Both lines have slope 3, so the lines are parallel.

(c) $4x + 3y = 6$ $\quad$ *Solve for y.* $\quad$ $y = -\dfrac{4}{3}x + 2$
       $2x - y = 5$ $\longrightarrow$ $y = 2x - 5$

Here the slopes are $-\dfrac{4}{3}$ and 2. These lines are neither parallel nor perpendicular because $-\dfrac{4}{3} \neq 2$ and $-\dfrac{4}{3} \cdot 2 \neq -1$.

(d) $5x - y = 1$ $\quad$ $y = 5x - 1$
       $x - 5y = -10$ $\quad$ *Solve for y.* $\quad$ $y = \dfrac{1}{5}x + 2$

The slopes are 5 and $\dfrac{1}{5}$. The lines are not parallel, nor are they perpendicular. $\left(\text{Be careful! } 5\left(\dfrac{1}{5}\right) = 1, \text{ not } -1.\right)$

Now Try Exercises 51, 55, and 57.

CONNECTIONS

Because the viewing window of a graphing calculator is a rectangle, the graphs of perpendicular lines will not appear perpendicular unless appropriate intervals are used for *x* and *y*. Graphing calculators usually have a key to select a "square" window automatically. In a square window, the *x*-interval is about 1.5 times the *y*-interval because the screen is about 1.5 times as wide as it is high. The equations from Figure 24 are graphed with the standard (nonsquare) window and then with a square window on the next page.

A standard (nonsquare) window
Lines do not appear perpendicular.

A square window
Lines appear perpendicular.

3.3 EXERCISES

For Extra Help

 Student's Solutions Manual

 MyMathLab

 InterAct Math Tutorial Software

 AW Math Tutor Center

 MathXL

Digital Video Tutor CD 6/Videotape 5

1. In the context of the graph of a straight line, what is meant by "rise"? What is meant by "run"?

Use the coordinates of the indicated points to find the ratio of rise to run for each line. See Example 1.

2.

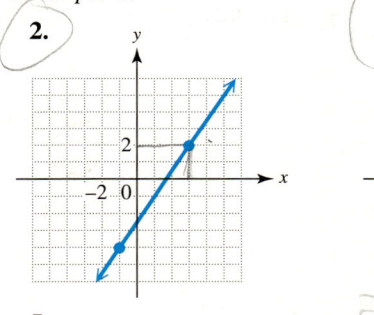

3.

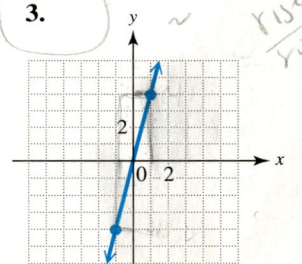

4.

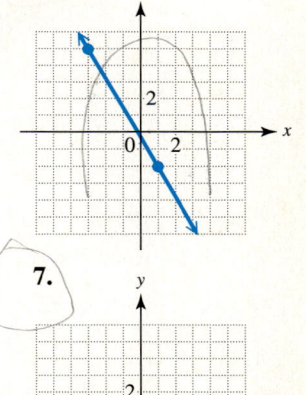

5.

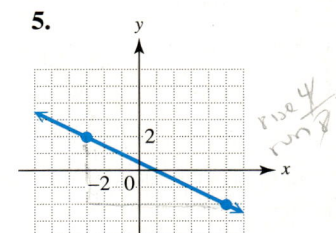

6.

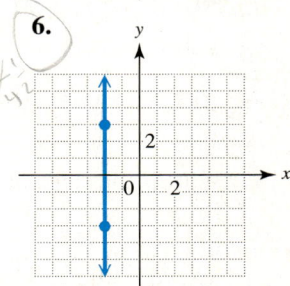

7.

8. Look at the graph in Exercise 2 and answer the following.

(a) Start at the point $(-1, -4)$ and count vertically up to the horizontal line that goes through the other plotted point. What is this vertical change? (Remember: "up" means positive, "down" means negative.)

(b) From this new position, count horizontally to the other plotted point. What is this horizontal change? (Remember: "right" means positive, "left" means negative.)

(c) What is the quotient of the numbers found in parts (a) and (b)? What do we call this number?

216 CHAPTER 3 Linear Equations and Inequalities in Two Variables; Functions

9. Refer to Exercise 8. If we were to *start* at the point $(3, 2)$ and *end* at the point $(-1, -4)$, do you think that the answer to part (c) would be the same? Explain why or why not.

On a pair of axes similar to the one shown, sketch the graph of a straight line having the indicated slope.

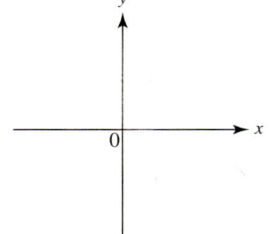

10. Negative

11. Positive

12. Undefined

13. Zero

14. Explain in your own words what is meant by *slope* of a line.

15. A student was asked to find the slope of the line through the points $(2, 5)$ and $(-1, 3)$. His answer, $-\frac{2}{3}$, was incorrect. He showed his work as

$$\frac{3 - 5}{2 - (-1)} = \frac{-2}{3} = -\frac{2}{3}.$$

What was his error? Give the correct slope.

Find the slope of the line through each pair of points. See Examples 2–4.

16. $(4, -1)$ and $(-2, -8)$ 17. $(1, -2)$ and $(-3, -7)$ 18. $(-8, 0)$ and $(0, -5)$

19. $(0, 3)$ and $(-2, 0)$ 20. $(-4, -5)$ and $(-5, -8)$ 21. $(-2, 4)$ and $(-3, 7)$

22. $(6, -5)$ and $(-12, -5)$ 23. $(4, 3)$ and $(-6, 3)$

24. $(-8, 6)$ and $(-8, -1)$ 25. $(-12, 3)$ and $(-12, -7)$

26. $(3.1, 2.6)$ and $(1.6, 2.1)$ 27. $\left(-\frac{7}{5}, \frac{3}{10}\right)$ and $\left(\frac{1}{5}, -\frac{1}{2}\right)$

Find the slope of each line. See Example 5.

28. $y = 2x - 3$ 29. $y = 5x + 12$ 30. $2y = -x + 4$ 31. $4y = x + 1$

32. $-6x + 4y = 4$ 33. $3x - 2y = 3$ 34. $2x + 4y = 5$ 35. $-3x + 2y = 5$

36. $x = -2$ 37. $y = -5$ 38. $y = 4$ 39. $x = 6$

40. What is the slope of a line whose graph is parallel to the graph of $-5x + y = -3$? Perpendicular to the graph of $-5x + y = -3$?

41. What is the slope of a line whose graph is parallel to the graph of $3x + y = 7$? Perpendicular to the graph of $3x + y = 7$?

The figure shows a line that has a positive slope (because it rises from left to right) and a positive y-value for the y-intercept (because it intersects the y-axis above the origin).

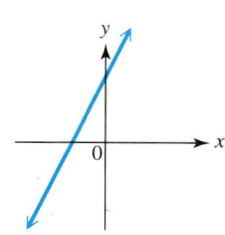

For each line in Exercises 42–47, decide whether (a) the slope is positive, negative, or zero and (b) the y-value of the y-intercept is positive, negative, or zero.

42. 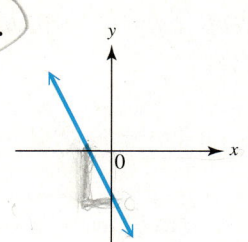 **43.** 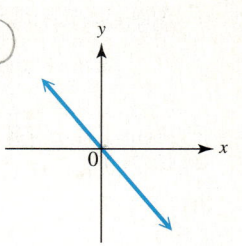 **44.**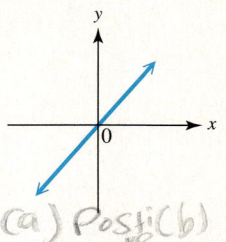

45. 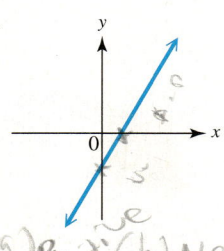 **46.** 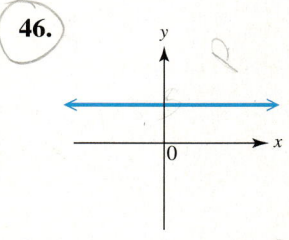 **47.**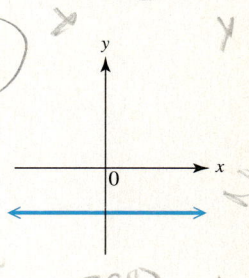

48. If two nonvertical lines are parallel, what do we know about their slopes? If two lines are perpendicular and neither is parallel to an axis, what do we know about their slopes? Why must the lines be nonvertical?

49. If two lines are both vertical or both horizontal, which of the following are they?
    A. Parallel   **B.** Perpendicular   **C.** Neither parallel nor perpendicular

50. If a line is vertical, what is true of any line that is perpendicular to it?

For each pair of equations, give the slopes of the lines and then determine whether the two lines are parallel, perpendicular, or neither parallel nor perpendicular. See Example 6.

51. $2x + 5y = 4$
     $4x + 10y = 1$

52. $-4x + 3y = 4$
     $-8x + 6y = 0$

53. $8x - 9y = 6$
     $8x + 6y = -5$

54. $5x - 3y = -2$
     $3x - 5y = -8$

55. $3x - 2y = 6$
     $2x + 3y = 3$

56. $3x - 5y = -1$
     $5x + 3y = 2$

57. $5x - y = 1$
     $x - 5y = -10$

58. $3x - 4y = 12$
     $4x + 3y = 12$

59. What is the slope (or pitch) of this roof?

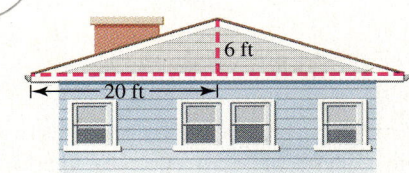

60. What is the slope (or grade) of this hill?

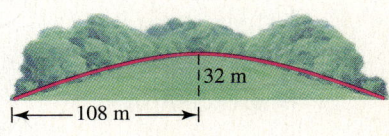

218 CHAPTER 3 Linear Equations and Inequalities in Two Variables; Functions

RELATING CONCEPTS (EXERCISES 61–66)

For Individual or Group Work

Figure A gives public school enrollment (in thousands) in grades 9–12 in the United States. Figure B gives the (average) number of public school students per computer.

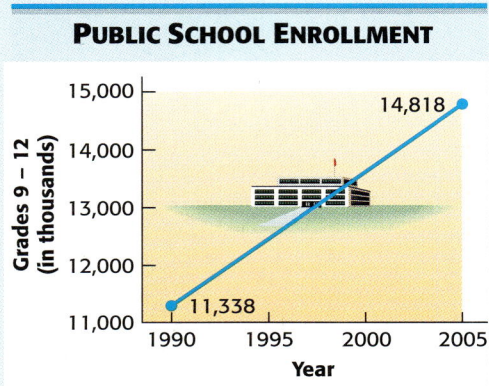

FIGURE A

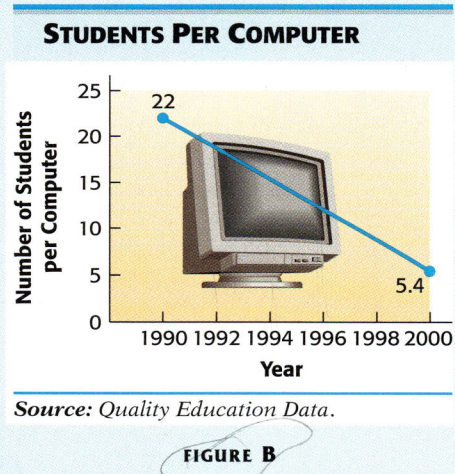

FIGURE B

Work Exercises 61–66 in order.

61. Use the ordered pairs (1990, 11,338) and (2005, 14,818) to find the slope of the line in Figure A.

62. The slope of the line in Figure A is _____ . This means that
(positive/negative)
during the period represented, enrollment _____ .
(increased/decreased)

63. The slope of a line represents its *rate of change*. Based on Figure A, what was the increase in students *per year* during the period shown?

64. Use the given information to find the slope of the line in Figure B.

65. The slope of the line in Figure B is _____. This means that
    (positive/negative)
    during the period represented, the number of students per computer
    _____.
    (increased/decreased)

66. Based on Figure B, what was the decrease in students per computer *per year* during the period shown?

67. The growth in retail square footage, in billions, is shown in the line graph. This graph looks like a straight line. If the change in square footage each year is the same, then it is a straight line. Find the change in square footage for the years shown in the graph. (*Hint:* To find the change in square footage from 1991 to 1992, subtract the *y*-value for 1991 from the *y*-value for 1992.) Is the graph a straight line?

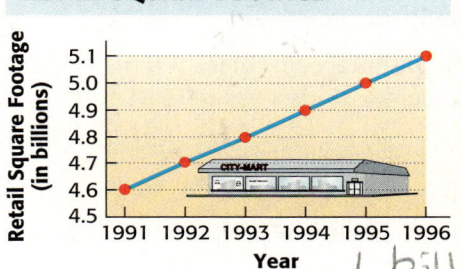

RETAIL SQUARE FOOTAGE

Source: International Council of Shopping Centers.

68. Find the slope of the line in Exercise 67 by using any two of the points shown on the line. How does the slope compare with the yearly change in square footage?

TECHNOLOGY INSIGHTS (EXERCISES 69–72)

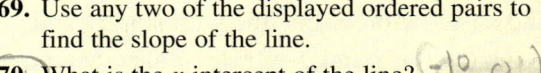

Some graphing calculators have the capability of displaying a table of points for a graph. The table shown here gives several points that lie on a line designated Y_1.

69. Use any two of the displayed ordered pairs to find the slope of the line.

70. What is the *x*-intercept of the line?

71. What is the *y*-intercept of the line?

72. Which one of the two lines shown is the graph of Y_1?

  A.
  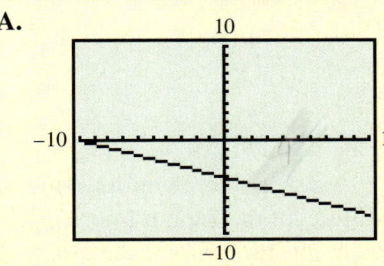

  B.
  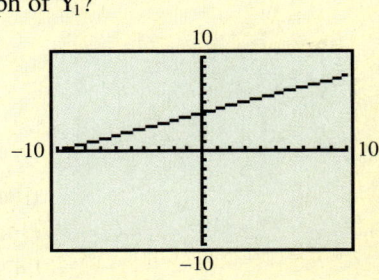

3.4 Equations of a Line

OBJECTIVES

1. Write an equation of a line given its slope and y-intercept.
2. Graph a line given its slope and a point on the line.
3. Write an equation of a line given its slope and any point on the line.
4. Write an equation of a line given two points on the line.
5. Find an equation of a line that fits a data set.

In the previous section we found the slope (steepness) of a line from the equation of the line by solving the equation for y. In that form, the slope is the coefficient of x. For example, the slope of the line with equation $y = 2x + 3$ is 2, the coefficient of x. What does the number 3 represent? If $x = 0$, the equation becomes

$$y = 2(0) + 3 = 0 + 3 = 3.$$

Since $y = 3$ corresponds to $x = 0$, $(0, 3)$ is the y-intercept of the graph of $y = 2x + 3$. An equation like $y = 2x + 3$ that is solved for y is said to be in *slope-intercept form* because both the slope and the y-intercept of the line can be read directly from the equation.

Slope-Intercept Form

The **slope-intercept form** of the equation of a line with slope m and y-intercept $(0, b)$ is

$$y = mx + b.$$

Remember that the intercept in the slope-intercept form is the *y-intercept*.

> **NOTE** The slope-intercept form is the most useful form for a linear equation because of the information we can determine from it. It is also the form used by graphing calculators and the one that describes a *linear function*, an important concept in mathematics.

OBJECTIVE 1 Write an equation of a line given its slope and y-intercept. Given the slope and y-intercept of a line, we can use the slope-intercept form to find an equation of the line.

EXAMPLE 1 Finding an Equation of a Line

Find an equation of the line with slope $\frac{2}{3}$ and y-intercept $(0, -1)$.

Here $m = \frac{2}{3}$ and $b = -1$, so the equation is

$$y = \underset{\text{Slope}}{m}x + \underset{\text{y-intercept}}{b}$$

$$y = \frac{2}{3}x - 1.$$

Now Try Exercise 11.

OBJECTIVE 2 Graph a line given its slope and a point on the line. We can use the slope and y-intercept to graph a line. For example, to graph $y = \frac{2}{3}x - 1$, we first locate the y-intercept, $(0, -1)$, on the y-axis. From the definition of slope and the fact that the slope of the line is $\frac{2}{3}$,

$$m = \frac{\text{rise}}{\text{run}} = \frac{2}{3}.$$

Another point P on the graph of the line can be found by counting from the y-intercept 2 units up and then counting 3 units to the right. We then draw the line through point P and the y-intercept, as shown in Figure 25.

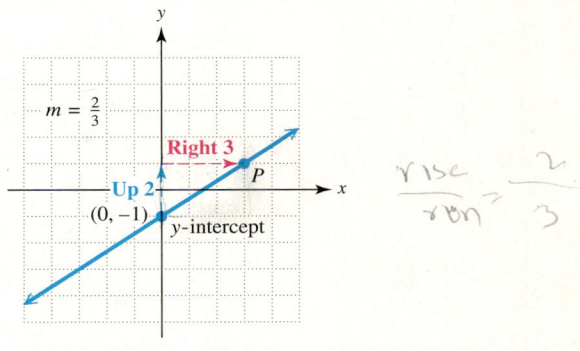

FIGURE 25

This method can be extended to graph a line given its slope and any point on the line, not just the y-intercept.

EXAMPLE 2 Graphing a Line Given a Point and the Slope

Graph the line through $(-2, 3)$ with slope -4.

First, locate the point $(-2, 3)$. Write the slope as

$$m = \frac{\text{rise}}{\text{run}} = -4 = \frac{-4}{1}.$$

Locate another point on the line by counting 4 units down (because of the negative sign) and then 1 unit to the right. Finally, draw the line through this new point P and the given point $(-2, 3)$. See Figure 26.

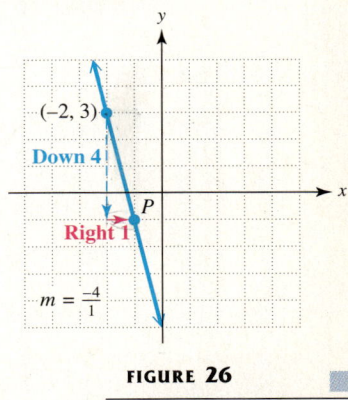

FIGURE 26

Now Try Exercise 19.

NOTE In Example 2, we could have written the slope as $\frac{4}{-1}$ instead. In this case, we would move 4 units up from $(-2, 3)$ and then 1 unit to the left (because of the negative sign). Verify that this produces the same line.

CONNECTIONS

Businesses must consider the amount of value lost, called **depreciation**, during each year of a machine's useful life. The simplest way to calculate depreciation is to assume that an item with a useful life of n years loses $\frac{1}{n}$ of its original

(continued)

value each year. Historically, if the equipment had salvage value, the depreciation was calculated on the **net cost,** the difference between the purchase price and the salvage value. If P represents the purchase price and S the salvage value of an item with a useful life of n years, then the annual depreciation would be

$$D = \frac{1}{n}(P - S).$$

Because this is a linear equation, this is called **straight-line depreciation.**

However, from a practical viewpoint, it is often difficult to determine the salvage value when the equipment is new. Also, for some equipment, such as computers, there is no residual dollar value at the end of the useful life due to obsolescence. In actual practice now, it is customary to find straight-line depreciation by the simpler linear equation

$$D = \frac{1}{n}P,$$

which assumes no salvage value.

For Discussion or Writing

1. Find the depreciation using both methods for a $50,000 (new) asset with a useful life of 10 yr and a salvage value of $15,000. How much is "written off" in each case over the 10-yr period?
2. The depreciation equation is given in slope-intercept form. What does the slope represent here? What does the y-value of the y-intercept represent?

Source: Joel E. Halle, CPA.

OBJECTIVE 3 Write an equation of a line given its slope and any point on the line. Let m represent the slope of a line and let (x_1, y_1) represent a given point on the line. Let (x, y) represent any other point on the line. Then by the definition of slope,

$$\frac{y - y_1}{x - x_1} = m \quad \text{or} \quad y - y_1 = m(x - x_1).$$

This result is the *point-slope form* of the equation of a line.

Point-Slope Form

The **point-slope form** of the equation of a line with slope m going through (x_1, y_1) is

$$y - y_1 = m(x - x_1).$$

EXAMPLE 3 Using the Point-Slope Form to Write Equations

Find an equation of each line. Write the equation in slope-intercept form.

(a) Through $(-2, 4)$, with slope -3

The given point is $(-2, 4)$ so $x_1 = -2$ and $y_1 = 4$. Also, $m = -3$. First substitute these values into the point-slope form to get an equation of the line. Then solve for y to write the equation in slope-intercept form.

$$y - y_1 = m(x - x_1) \qquad \text{Point-slope form}$$
$$y - 4 = -3[x - (-2)] \qquad \text{Let } y_1 = 4, m = -3, x_1 = -2.$$
$$y - 4 = -3(x + 2)$$
$$y - 4 = -3x - 6 \qquad \text{Distributive property}$$
$$y = -3x - 2 \qquad \text{Add 4.}$$

The last equation is in slope-intercept form.

(b) Through $(4, 2)$, with slope $\frac{3}{5}$

$$y - y_1 = m(x - x_1)$$
$$y - 2 = \frac{3}{5}(x - 4) \qquad \text{Let } y_1 = 2, m = \frac{3}{5}, x_1 = 4.$$
$$y - 2 = \frac{3}{5}x - \frac{12}{5} \qquad \text{Distributive property}$$
$$y = \frac{3}{5}x - \frac{12}{5} + \frac{10}{5} \qquad \text{Add } 2 = \frac{10}{5} \text{ to both sides.}$$
$$y = \frac{3}{5}x - \frac{2}{5} \qquad \text{Combine terms.}$$

We did not clear fractions after the substitution step because we want the equation in slope-intercept form; that is, solved for y.

Now Try Exercises 29 and 31.

OBJECTIVE 4 Write an equation of a line given two points on the line. We can also use the point-slope form to find an equation of a line when two points on the line are known.

EXAMPLE 4 Finding the Equation of a Line Given Two Points

Find an equation of the line through the points $(-2, 5)$ and $(3, 4)$. Write the equation in slope-intercept form.

First, find the slope of the line, using the slope formula.

$$\text{slope } m = \frac{y_2 - y_1}{x_2 - x_1} = \frac{5 - 4}{-2 - 3} = \frac{1}{-5} = -\frac{1}{5}$$

Now use either $(-2, 5)$ or $(3, 4)$ and the point-slope form. We choose $(3, 4)$.

$$y - y_1 = m(x - x_1)$$
$$y - 4 = -\frac{1}{5}(x - 3) \qquad \text{Let } y_1 = 4, m = -\frac{1}{5}, x_1 = 3.$$
$$y - 4 = -\frac{1}{5}x + \frac{3}{5} \qquad \text{Distributive property}$$
$$y = -\frac{1}{5}x + \frac{3}{5} + \frac{20}{5} \qquad \text{Add } 4 = \frac{20}{5} \text{ to both sides.}$$
$$y = -\frac{1}{5}x + \frac{23}{5} \qquad \text{Combine terms.}$$

The same result would be found by using $(-2, 5)$ for (x_1, y_1).

Now Try Exercise 39.

Many of the linear equations in Sections 3.1 through 3.3 were given in the form $Ax + By = C$, which is called *standard form*.

> **Standard Form**
>
> A linear equation is in **standard form** if it is written as
>
> $$Ax + By = C,$$
>
> where A, B, and C are integers and $A > 0$, $B \neq 0$.

NOTE The given definition of standard form is not the same in all texts. A linear equation can be written in this form in many different, equally correct, ways. For example, $3x + 4y = 12$, $6x + 8y = 24$, and $9x + 12y = 36$ all represent the same set of ordered pairs. Let us agree that $3x + 4y = 12$ is preferable to the other forms because the greatest common factor of 3, 4, and 12 is 1.

Linear Equations

$x = k$	**Vertical line** Slope is undefined; x-intercept is $(k, 0)$.
$y = k$	**Horizontal line** Slope is 0; y-intercept is $(0, k)$.
$y = mx + b$	**Slope-intercept form** Slope is m; y-intercept is $(0, b)$.
$y - y_1 = m(x - x_1)$	**Point-slope form** Slope is m; line goes through (x_1, y_1).
$Ax + By = C$	**Standard form** Slope is $-\frac{A}{B}$; x-intercept is $\left(\frac{C}{A}, 0\right)$; y-intercept is $\left(0, \frac{C}{B}\right)$.

OBJECTIVE 5 Find an equation of a line that fits a data set. Earlier in this chapter, we gave linear equations that modeled real data, such as annual costs of doctors' visits and amounts of credit card debt, and then used these equations to estimate or predict values. Using the information in this section, we can now develop a procedure to find such an equation if the given set of data fits a linear pattern—that is, its graph consists of points lying close to a straight line.

EXAMPLE 5　Finding the Equation of a Line That Describes Data

The table lists the average annual cost (in dollars) in tuition and fees at public 4-yr colleges for selected years. Year 1 represents 1991, year 3 represents 1993, and so on. Plot the data and find an equation that approximates it.

Letting y represent the cost in year x, we plot the data as shown in Figure 27.

Year	Cost (in dollars)
1	2137
3	2527
5	2860
7	3111
9	3356

Source: The College Board.

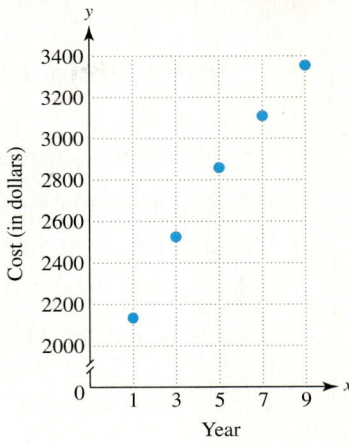

AVERAGE ANNUAL COSTS AT PUBLIC 4-YEAR COLLEGES

FIGURE 27

The points appear to lie approximately in a straight line. We can use two of the data pairs and the point-slope form of the equation of a line to get an equation that describes the relationship between the year and the cost. We choose the ordered pairs $(3, 2527)$ and $(7, 3111)$ from the table and find the slope of the line through these points.

$$m = \frac{y_2 - y_1}{x_2 - x_1} \quad \text{Slope formula}$$

$$m = \frac{3111 - 2527}{7 - 3} \quad \text{Let } (7, 3111) = (x_2, y_2) \text{ and } (3, 2527) = (x_1, y_1).$$

$$m = 146$$

As we might expect, the slope, 146, is positive, indicating that tuition and fees increased $146 each year. Now use this slope and the point $(3, 2527)$ in the point-slope form to find an equation of the line.

$$y - y_1 = m(x - x_1) \quad \text{Point-slope form}$$
$$y - 2527 = 146(x - 3) \quad \text{Substitute for } x_1, y_1, \text{ and } m.$$
$$y - 2527 = 146x - 438 \quad \text{Distributive property}$$
$$y = 146x + 2089 \quad \text{Add 2527.}$$

To see how well this equation approximates the ordered pairs in the data table, let $x = 9$ (for 1999) and find y.

$$y = 146x + 2089 \quad \text{Equation of the line}$$
$$y = 146(9) + 2089 \quad \text{Substitute 9 for } x.$$
$$y = 3403$$

The corresponding value in the table for $x = 9$ is 3356, so the equation approximates the data reasonably well. With caution, the equation could be used to predict values for years that are not included in the table.

Now Try Exercise 61.

NOTE In Example 5, if we had chosen two different data points, we would have gotten a slightly different equation.

Here is a summary of what is needed to find the equation of a line.

Finding the Equation of a Line

To find the equation of a line, you need
1. a point on the line, and
2. the slope of the line.

If two points are known, first find the slope and then use the point-slope form.

3.4 EXERCISES

For Extra Help

 Student's Solutions Manual

 MyMathLab

 InterAct Math Tutorial Software

 AW Math Tutor Center

 MathXL

Digital Video Tutor CD 7/Videotape 6

Match the correct equation in Column II with the description in Column I.

I	II
1. Slope $= -2$, through the point $(4, 1)$	**A.** $y = 4x$
2. Slope $= -2$, y-intercept $(0, 1)$	**B.** $y = \dfrac{1}{4}x$
3. Passing through the points $(0, 0)$ and $(4, 1)$	**C.** $y = -2x + 1$
4. Passing through the points $(0, 0)$ and $(1, 4)$	**D.** $y - 1 = -2(x - 4)$

5. Explain why the equation of a vertical line cannot be written in the form $y = mx + b$.

6. Match each equation with the graph that would most closely resemble its graph.

(a) $y = x + 3$
(b) $y = -x + 3$
(c) $y = x - 3$
(d) $y = -x - 3$

A.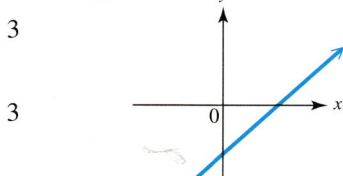

B.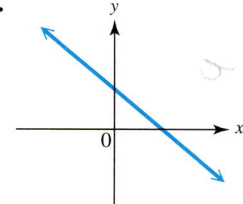

C.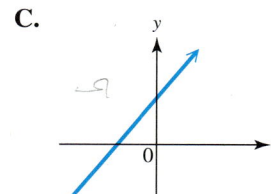

D.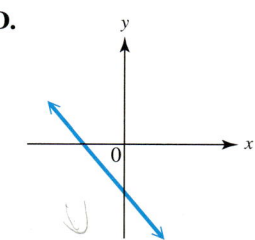

SECTION 3.4 Equations of a Line **227**

Use the geometric interpretation of slope (rise divided by run, from Section 3.3) to find the slope of each line. Then, by identifying the y-intercept from the graph, write the slope-intercept form of the equation of the line.

7. 8.

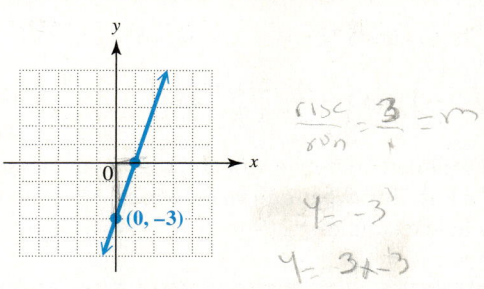
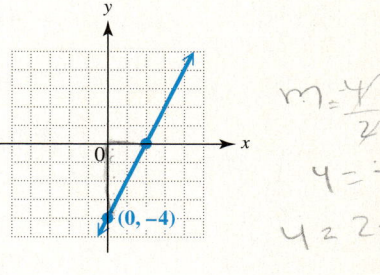

9. 10.

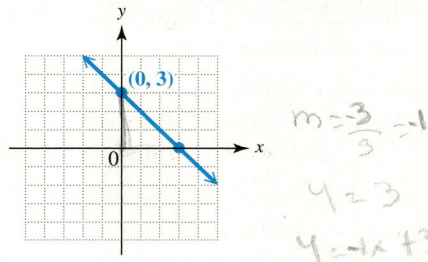
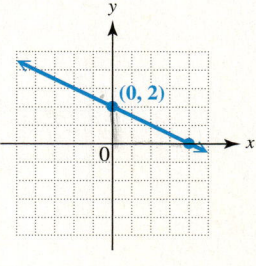

Write the equation of each line with the given slope and y-intercept. See Example 1.

11. $m = 4$, $(0, -3)$
12. $m = -5$, $(0, 6)$
13. $m = 0$, $(0, 3)$
14. $m = 0$, $(0, -4)$
15. Undefined slope, $(0, -2)$
16. Undefined slope, $(0, 5)$

Graph each line passing through the given point and having the given slope. (In Exercises 22–25, recall the types of lines having slope 0 and undefined slope.) See Example 2.

17. $(0, 2)$, $m = 3$
18. $(0, -5)$, $m = -2$
19. $(1, -5)$, $m = -\dfrac{2}{5}$
20. $(2, -1)$, $m = -\dfrac{1}{3}$
21. $(-1, 4)$, $m = \dfrac{2}{5}$
22. $(3, 2)$, $m = 0$
23. $(-2, 3)$, $m = 0$
24. $(3, -2)$, undefined slope
25. $(2, 4)$, undefined slope

26. What is the common name given to a vertical line whose *x*-intercept is the origin?
27. What is the common name given to a line with slope 0 whose *y*-intercept is the origin?

Write an equation for each line passing through the given point and having the given slope. Write the equation in slope-intercept form. See Example 3.

28. $(4, 1)$, $m = 2$
29. $(-1, 3)$, $m = -4$
30. $(2, 7)$, $m = 3$
31. $(-2, 5)$, $m = \dfrac{2}{3}$
32. $(4, 2)$, $m = -\dfrac{1}{3}$
33. $(-4, 1)$, $m = \dfrac{3}{4}$
34. $(6, -3)$, $m = -\dfrac{4}{5}$
35. $(2, 1)$, $m = \dfrac{5}{2}$
36. $(7, -2)$, $m = -\dfrac{7}{2}$

37. If a line passes through the origin and a second point whose *x*- and *y*-coordinates are equal, what is an equation of the line?

Write an equation for the line passing through the given pair of points. Write each equation in slope-intercept form. See Example 4.

38. $(8, 5)$ and $(9, 6)$

39. $(4, 10)$ and $(6, 12)$

40. $(-1, -7)$ and $(-8, -2)$

41. $(-2, -1)$ and $(3, -4)$

42. $(0, -2)$ and $(-3, 0)$

43. $(-4, 0)$ and $(0, 2)$

44. $\left(\dfrac{1}{2}, \dfrac{3}{2}\right)$ and $\left(-\dfrac{1}{4}, \dfrac{5}{4}\right)$

45. $\left(-\dfrac{2}{3}, \dfrac{8}{3}\right)$ and $\left(\dfrac{1}{3}, \dfrac{7}{3}\right)$

46. Describe in your own words the slope-intercept and point-slope forms of the equation of a line. Tell what information must be given to use each form to write an equation. Include examples.

RELATING CONCEPTS (EXERCISES 47–54)

For Individual or Group Work

If we think of ordered pairs of the form (C, F), then the two most common methods of measuring temperature, Celsius and Fahrenheit, can be related as follows: When $C = 0$, $F = 32$, and when $C = 100$, $F = 212$. **Work Exercises 47–54 in order.**

47. Write two ordered pairs relating these two temperature scales.

48. Find the slope of the line through the two points.

49. Use the point-slope form to find an equation of the line. (Your variables should be C and F rather than x and y.)

50. Write an equation for F in terms of C.

51. Use the equation from Exercise 50 to write an equation for C in terms of F.

52. Use the equation from Exercise 50 to find the Fahrenheit temperature when $C = 30$.

53. Use the equation from Exercise 51 to find the Celsius temperature when $F = 50$.

54. For what temperature is $F = C$?

Write an equation of the line satisfying the given conditions. Write the equation in slope-intercept form.

55. Through $(2, -3)$, parallel to $3x = 4y + 5$

56. Through $(-1, 4)$, perpendicular to $2x + 3y = 8$

57. Perpendicular to $x - 2y = 7$, y-intercept $(0, -3)$

58. Parallel to $5x = 2y + 10$, y-intercept $(0, 4)$

The cost to produce x items is, in some cases, expressed as $y = mx + b$. The number b gives the fixed cost *(the cost that is the same no matter how many items are produced), and the number m is the* variable cost *(the cost to produce an additional item). Use this information to work Exercises 59 and 60.*

59. It costs $400 to start up a business selling snow cones. Each snow cone costs $.25 to produce.

    (a) What is the fixed cost?
    (b) What is the variable cost?
    (c) Write the cost equation.
    (d) What will be the cost to produce 100 snow cones, based on the cost equation?
    (e) How many snow cones will be produced if total cost is $775?

60. It costs $2000 to purchase a copier, and each copy costs $.02 to make.

  (a) What is the fixed cost?
  (b) What is the variable cost?
  (c) Write the cost equation.
  (d) What will be the cost to produce 10,000 copies, based on the cost equation?
  (e) How many copies will be produced if total cost is $2600?

Solve each problem. See Example 5.

61. The table lists the average annual cost (in dollars) of tuition and fees at private 4-yr colleges for selected years, where year 1 represents 1991, year 3 represents 1993, and so on.

Year	Cost (in dollars)
1	10,017
3	11,025
5	12,432
7	13,785
9	15,380

Source: The College Board.

  (a) Write five ordered pairs for the data.
  (b) Plot the ordered pairs. Do the points lie approximately in a straight line?
  (c) Use the ordered pairs (3, 11,025) and (9, 15,380) to find the equation of a line that approximates the data. Write the equation in slope-intercept form. (Round the slope to the nearest tenth.)
  (d) Use the equation from part (c) to estimate the average annual cost at private 4-yr colleges in 2003. (*Hint:* What is the value of x for 2003?)

62. The table gives heavy-metal nuclear waste (in thousands of metric tons) from spent reactor fuel now stored temporarily at reactor sites, awaiting permanent storage. (*Source:* "Burial of Radioactive Nuclear Waste Under the Seabed," *Scientific American,* January 1998.)

Year x	Waste y
1995	32
2000	42
2010*	61
2020*	76

*Estimated by the U.S. Department of Energy.

Let $x = 0$ represent 1995, $x = 5$ represent 2000 (since $2000 - 1995 = 5$), and so on.

  (a) For 1995, the ordered pair is (0, 32). Write ordered pairs for the data for the other years given in the table.
  (b) Plot the ordered pairs (x, y). Do the points lie approximately in a straight line?
  (c) Use the ordered pairs (0, 32) and (25, 76) to find the equation of a line that approximates the other ordered pairs. Write the equation in slope-intercept form.
  (d) Use the equation from part (c) to estimate the amount of nuclear waste in 2005. (*Hint:* What is the value of x for 2005?)

230 CHAPTER 3 Linear Equations and Inequalities in Two Variables; Functions

TECHNOLOGY INSIGHTS (EXERCISES 63–66)

In Exercises 63 and 64, two graphing calculator views of the same line are shown. Use the displays at the bottom of the screen to find an equation of the form $y = mx + b$ for each line.

63.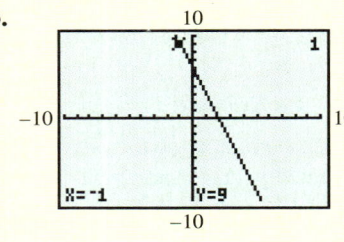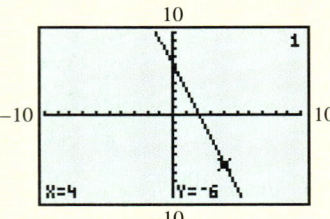

64.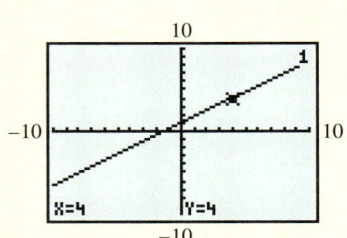

In Exercises 65 and 66, a table of points, generated by a graphing calculator, is shown for a line Y_1.* Use any two points to find the equation of each line. Write the equation in slope-intercept form.

65.

X	Y₁
-2	-.5
-1	.25
0	1
1	1.75
2	2.5
3	3.25
4	4

X = -2

66.

X	Y₁
-4	14
-3	10
-2	6
-1	2
0	-2
1	-6
2	-10

X = -4

*With graphing calculators, we use capital Y_1 and X like the calculator does.

3.5 Graphing Linear Inequalities in Two Variables

OBJECTIVES

1. Graph $\leq$ or $\geq$ linear inequalities.
2. Graph $<$ or $>$ linear inequalities.
3. Graph inequalities with a boundary through the origin.

In Section 3.2 we graphed linear equations, such as $2x + 3y = 6$. Now we extend this work to *linear inequalities in two variables,* such as $2x + 3y \leq 6$. (Recall that $\leq$ is read "is less than or equal to.")

Linear Inequality in Two Variables

An inequality that can be written as
$$Ax + By < C \quad \text{or} \quad Ax + By > C,$$
where A, B, and C are real numbers and A and B are not both 0, is a **linear inequality in two variables.**

The symbols $\leq$ and $\geq$ may replace $<$ and $>$ in the definition.

OBJECTIVE 1 Graph $\leq$ or $\geq$ linear inequalities. The inequality $2x + 3y \leq 6$ means that

$$2x + 3y < 6 \quad \text{or} \quad 2x + 3y = 6.$$

As we found earlier, the graph of $2x + 3y = 6$ is a line. This **boundary line** divides the plane into two regions. The graph of the solutions of the inequality $2x + 3y < 6$ will include only *one* of these regions. We find the required region by solving the given inequality for y.

$$2x + 3y \leq 6$$
$$3y \leq -2x + 6 \quad \text{Subtract } 2x.$$
$$y \leq -\frac{2}{3}x + 2 \quad \text{Divide by 3.}$$

From this last statement, ordered pairs in which *y is less than or equal to* $-\frac{2}{3}x + 2$ will be solutions to the inequality. Ordered pairs in which y is equal to $-\frac{2}{3}x + 2$ are on the boundary line, so pairs in which *y is less than* $-\frac{2}{3}x + 2$ will be *below* that line. To indicate the solution, shade the region below the line, as in Figure 28. The shaded region, along with the line, is the desired graph.

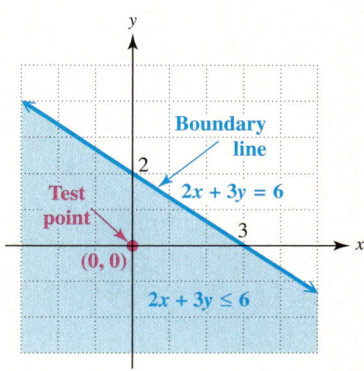

FIGURE 28

Alternatively, a test point gives a quick way to find the correct region to shade. We choose any point *not* on the line. Because $(0, 0)$ is easy to substitute into an inequality, it is often a good choice, and we use it here. We substitute 0 for x and 0 for y in the given inequality to see whether the resulting statement is true or false. In the example above,

$$2x + 3y \leq 6 \quad \text{Original inequality}$$
$$2(0) + 3(0) \leq 6 \quad ? \quad \text{Let } x = 0 \text{ and } y = 0.$$
$$0 + 0 \leq 6 \quad ?$$
$$0 \leq 6. \quad \text{True}$$

Since the last statement is true, we shade the region that includes the test point $(0, 0)$. This agrees with the result shown in Figure 28.

CHAPTER 3 Linear Equations and Inequalities in Two Variables; Functions

> **CAUTION** If the boundary lines passes through the origin, the point $(0,0)$ lies on the line, so choose as test point any ordered pair that does not lie on the line.

OBJECTIVE 2 Graph < or > linear inequalities. Inequalities that do not include the equals sign are graphed in a similar way.

EXAMPLE 1 Graphing a Linear Inequality

Graph $x - y > 5$.

This inequality does not include the equals sign. Therefore, the points on the line $x - y = 5$ do *not* belong to the graph. However, the line still serves as a boundary for two regions, one of which satisfies the inequality. To graph the inequality, first graph the equation $x - y = 5$. Use a *dashed line* to show that the points on the line are *not* solutions of the inequality $x - y > 5$. Now choose a test point, such as $(1, -2)$, to see which side of the line satisfies the inequality.

$$x - y > 5 \quad \text{Original inequality}$$
$$1 - (-2) > 5 \quad ? \quad \text{Let } x = 1 \text{ and } y = -2.$$
$$3 > 5 \quad \text{False}$$

Since $3 > 5$ is false, the graph of the inequality is the region that *does not* contain $(1, -2)$. Shade the other region, as shown in Figure 29. This shaded region is the required graph. To check that the correct region is shaded, select a test point in the shaded region and substitute for x and y in the inequality $x - y > 5$. For example, use $(4, -3)$ from the shaded region as follows.

$$x - y > 5$$
$$4 - (-3) > 5 \quad ?$$
$$7 > 5 \quad \text{True}$$

This verifies that the correct region was shaded in Figure 29.

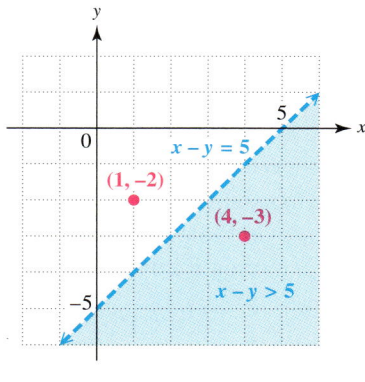

FIGURE 29

Now Try Exercise 21.

A summary of the steps used to graph a linear inequality in two variables follows.

Graphing a Linear Inequality

Step 1 **Graph the boundary.** Graph the line that is the boundary of the region. Use the methods of Section 3.2. Draw a solid line if the inequality has ≤ or ≥; draw a dashed line if the inequality has < or >.

Step 2 **Shade the appropriate side.** Use any point not on the line as a test point. Substitute for *x* and *y* in the *inequality.* If a true statement results, shade the side containing the test point. If a false statement results, shade the other side.

EXAMPLE 2 Graphing a Linear Inequality with a Vertical Boundary Line

Graph $x \leq 3$.

First, we graph $x = 3$, a vertical line going through the point $(3, 0)$. We use a solid line (why?) and choose $(0, 0)$ as a test point.

$x \leq 3$		Original inequality
$0 \leq 3$	?	Let $x = 0$.
$0 \leq 3$		True

Since $0 \leq 3$ is true, we shade the region containing $(0, 0)$, as in Figure 30.

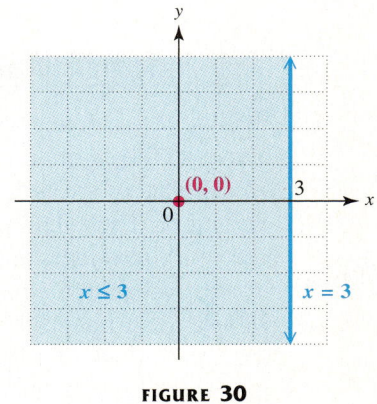

FIGURE 30

Now Try Exercise 25.

OBJECTIVE 3 Graph inequalities with a boundary through the origin. Remember that when we graph an inequality having a boundary line that goes through the origin, $(0, 0)$ cannot be used as a test point.

EXAMPLE 3 Graphing a Linear Inequality with a Boundary Line through the Origin

Graph $x \leq 2y$.

Begin by graphing $x = 2y$. Some ordered pairs that can be used to graph this line are $(0, 0)$, $(6, 3)$, and $(4, 2)$. Draw a solid line. Since $(0, 0)$ is on the line $x = 2y$, it

234 CHAPTER 3 Linear Equations and Inequalities in Two Variables; Functions

cannot be used as a test point. Instead, we choose a test point off the line, $(1, 3)$.

$$x \leq 2y \qquad \text{Original inequality}$$
$$1 \leq 2(3) \quad ? \quad \text{Let } x = 1 \text{ and } y = 3.$$
$$1 \leq 6 \qquad \text{True}$$

Since $1 \leq 6$ is true, shade the side of the graph containing the test point $(1, 3)$. See Figure 31.

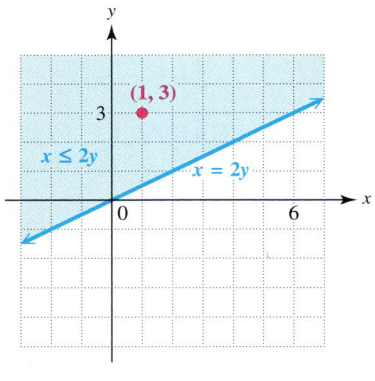

FIGURE 31

Now Try Exercise 29.

CONNECTIONS

Graphing calculators have a feature that allows us to shade regions in the plane, so they can be used to graph a linear inequality in two variables. Since all calculators are different, consult the manual for directions for your calculator. The calculator will not draw the graph as a dashed line, so it is still necessary to understand what is and is not included in the solution set.

To solve the inequalities in one variable, $-2x + 4 > 0$ and $-2x + 4 < 0$, we use the graph of $y = -2x + 4$, shown below. For $y = -2x + 4 > 0$, we want the values of x such that $y > 0$, so that the line is *above* the x-axis. From the graph, we see that this is the case for $x < 2$. Thus, the solution set is $(-\infty, 2)$. Similarly, the solution set of $y = -2x + 4 < 0$ is $(2, \infty)$, because the line is *below* the x-axis when $x > 2$.

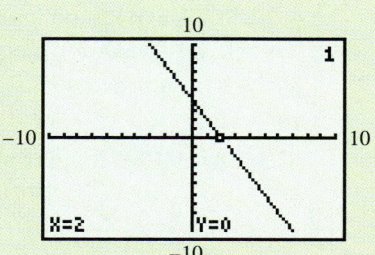

SECTION 3.5 Graphing Linear Inequalities in Two Variables

For Discussion or Writing

1. Discuss the pros and cons of using a calculator to solve a linear inequality in one variable.
2. Use a graphing calculator to solve the following inequalities in one variable from Section 2.8.
   (a) $3x + 2 - 5 > -x + 7 + 2x$ (Example 4)
   (b) $3x + 2 - 5 < -x + 7 + 2x$ (Use the result from part (a).)
   (c) $4 \leq 3x - 5 < 6$ (Example 7)

3.5 EXERCISES

All H.W

For Extra Help

 Student's Solutions Manual

 MyMathLab

 InterAct Math Tutorial Software

 AW Math Tutor Center

 MathXL

Digital Video Tutor CD 7/Videotape 6

Answer true *or* false *to each of the following.*

1. The point $(4, 0)$ lies on the graph of $3x - 4y < 12$. F
2. The point $(4, 0)$ lies on the graph of $3x - 4y \leq 12$. T
3. Both points $(4, 1)$ and $(0, 0)$ lie on the graph of $3x - 2y \geq 0$. T
4. The graph of $y > x$ does not contain points in quadrant IV. T

The following statements were taken from recent articles in newspapers or magazines. Each includes a phrase that can be symbolized with one of the inequality symbols $<$, $\leq$, $>$, or $\geq$. In Exercises 5–10, give the inequality symbol for the bold-faced words.

5. According to the Kaiser Family Foundation, in January 2001, (about) one-quarter of Medicare beneficiaries spend **more than** $2000 a year on drugs. >
6. Treating HIV-infected children and teenagers with new drugs cut the risk of death by two-thirds, to **less than** 1 percent annually. (*Source: Sacramento Bee,* November 22, 2001.) <
7. In one study, 9 percent of pregnant women had active hepatitis, meaning that **at most** 9 percent of children could get it at birth. (*Source: 2000 Dow Jones & Company, Inc.*) ≤
8. Forty percent of Americans keep **at least** one gun at home. (*Source:* Gallup Organization, quoted in *Reader's Digest,* March 2002). ≥
9. By 1937, a population of as many as a million Attwater's prairie-chickens had been cut to **less than** 9000. (*Source: National Geographic,* March 2002). <
10. [Easter Island's] nearly 1000 statues, some **almost** 30 feet tall and weighing **as much as** 80 tons, are still an enigma. (*Source: Smithsonian,* March 2002). ≤

236 CHAPTER 3 Linear Equations and Inequalities in Two Variables; Functions

In Exercises 11–16, the straight-line boundary has been drawn. Complete the graph by shading the correct region. See Examples 1–3.

11. $x + 2y \geq 7$ **12.** $2x + y \geq 5$ **13.** $-3x + 4y > 12$

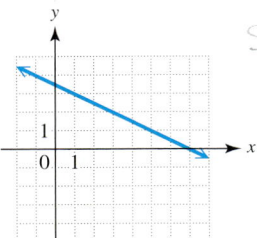 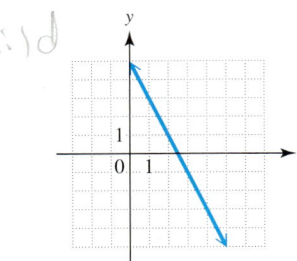 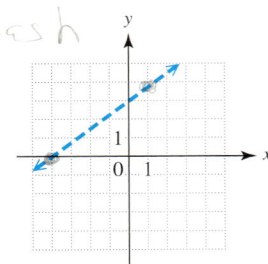

14. $x > 4$ **15.** $y < -1$ **16.** $x \leq 3y$

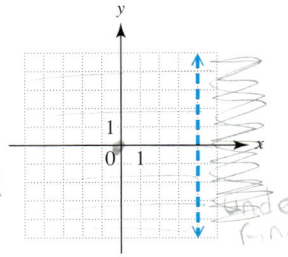 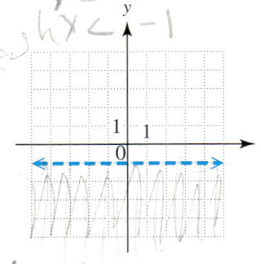 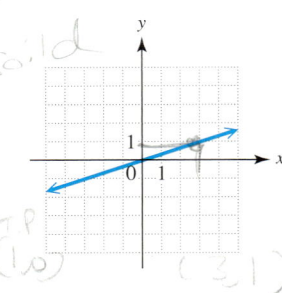

17. Explain how to determine whether to use a dashed line or a solid line when graphing a linear inequality in two variables.

18. Explain why the point $(0, 0)$ is not an appropriate choice for a test point when graphing an inequality whose boundary goes through the origin.

Graph each linear inequality. See Examples 1–3.

19. $x + y \leq 5$ **20.** $x + y \geq 3$ **21.** $2x + 3y > -6$ **22.** $3x + 4y < 12$
23. $y \geq 2x + 1$ **24.** $y < -3x + 1$ **25.** $x \leq -2$ **26.** $x \geq 1$
27. $y < 5$ **28.** $y < -3$ **29.** $y \geq 4x$ **30.** $y \leq 2x$

31. Explain why the graph of $y > x$ cannot lie in quadrant IV.

32. Explain why the graph of $y < x$ cannot lie in quadrant II.

TECHNOLOGY INSIGHTS (EXERCISES 33–36)

A calculator was used to generate the shaded graphs in choices A–D. Each graph has boundary line $y = 3x - 7$ or $y = -3x + 7$.

A. **B.**

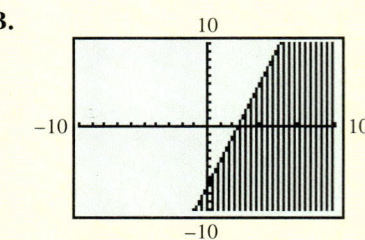

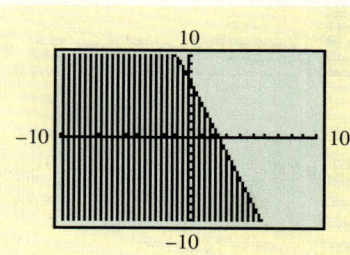

Match each inequality with the appropriate choice.

33. $y \geq 3x - 7$ **34.** $y \leq 3x - 7$ **35.** $y \geq -3x + 7$ **36.** $y \leq -3x + 7$

For the given information (a) graph the inequality. Here $x \geq 0$ and $y \geq 0$, so graph only the part of the inequality in quadrant I. (b) Give some ordered pairs that satisfy the inequality.

37. A company will ship x units of merchandise to outlet I and y units of merchandise to outlet II. The company must ship a total of at least 500 units to these two outlets. This can be expressed by writing
$$x + y \geq 500.$$

38. A toy manufacturer makes stuffed bears and geese. It takes 20 minutes to sew a bear and 30 minutes to sew a goose. There is a total of 480 minutes of sewing time available to make x bears and y geese. These restrictions lead to the inequality
$$20x + 30y \leq 480.$$

39. The number of civilian federal employees (in thousands) is approximated by $y = -51.1x + 3153$, where x represents the year. Here, $x = 1$ corresponds to 1991, $x = 2$ corresponds to 1992, and so on. Thus, the total number of federal employees (including noncivilians) in each year is described by the inequality
$$y \geq -51.1x + 3153.$$

(*Hint:* Use tick marks of 1 to 10 on the x-axis and 2500 to 3500, at intervals of 200, on the y-axis.) (*Source:* U.S. Bureau of the Census.)

40. The equation $y = 57.1x + 1798$ approximates the number of free lunches served in public schools each year in millions. The year 1992 is represented by $x = 2$, 1993 is represented by $x = 3$, and so on. Free school breakfasts are also served in some (but not all) of the same schools. The number of children receiving only breakfast is described by the inequality
$$y \leq 57.1x + 1798.$$

(*Hint:* Use tick marks of 1 to 10 on the x-axis and 1800, 1900, 2000, 2100, 2200, 2300, and 2400 on the y-axis.) (*Source:* U.S. Department of Agriculture, Food and Consumer Service.)

3.6 Introduction to Functions

OBJECTIVES

1. Understand the definition of a relation.
2. Understand the definition of a function.
3. Decide whether an equation defines a function.
4. Find domains and ranges.
5. Use function notation.
6. Apply the function concept in an application.

If gasoline costs $1.45 per gal and you buy 1 gal, then you must pay $1.45(1) = $1.45. If you buy 2 gal, your cost is $1.45(2) = $2.90; for 3 gal, your cost is $1.45(3) = $4.35, and so on. Generalizing, if x represents the number of gallons, then the cost is $1.45x$. If we let y represent the cost, then the equation $y = 1.45x$ *relates* the number of gallons, x, to the cost in dollars, y. The ordered pairs (x, y) that satisfy this equation form a *relation*.

OBJECTIVE 1 Understand the definition of a relation. In an ordered pair (x, y), x and y are called the **components** of the ordered pair. Any set of ordered pairs is called a **relation**. The set of all first components in the ordered pairs of a relation is the **domain** of the relation, and the set of all second components in the ordered pairs is the **range** of the relation.

EXAMPLE 1 Using Ordered Pairs to Define a Relation

(a) The relation $\{(0, 1), (2, 5), (3, 8), (4, 2)\}$ has domain $\{0, 2, 3, 4\}$ and range $\{1, 2, 5, 8\}$. The correspondence between the elements of the domain and the elements of the range is shown in Figure 32.

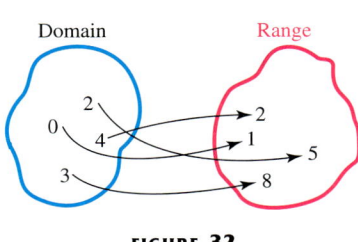

FIGURE 32   FIGURE 33

(b) Figure 33 shows a relation where the domain elements represent four mothers and the range elements represent eight children. The figure shows the correspondence between each mother and her children. The relation also could be written as the set of ordered pairs

$\{(A, 8), (B, 1), (B, 2), (C, 3), (C, 4), (C, 7), (D, 5), (D, 6)\}.$

> **Now Try Exercises 7(a) and 9(a).**

OBJECTIVE 2 Understand the definition of a function. We now investigate an important type of relation, called a *function*.

Function

A **function** is a set of ordered pairs in which each first component corresponds to exactly one second component.

By this definition, the relation in Example 1(a) is a function. But the relation in Example 1(b) is *not* a function, because at least one first component (mother) corresponds to more than one second component (child). Notice that if the ordered pairs in Example 1(b) were interchanged, giving the relation

$$\{(8, A), (1, B), (2, B), (3, C), (4, C), (7, C), (5, D), (6, D)\},$$

the result *would be* a function. In that case, each domain element (first component) corresponds to *exactly one* range element (second component).

EXAMPLE 2 Determining Whether Relations Are Functions

Determine whether each relation is a function.

(a) $\{(-2, 4), (-1, 1), (0, 0), (1, 1), (2, 4)\}$

Notice that each first component appears once and only once. Because of this, the relation is a function.

(b) $\{(9, 3), (9, -3), (4, 2)\}$

The first component 9 appears in two ordered pairs, and corresponds to two different second components. Therefore, this relation is not a function.

Now Try Exercises 7(b) and 9(b).

The simple relations given in Examples 1 and 2 were defined by listing the ordered pairs or by showing the correspondence with a figure. Most useful functions have an infinite number of ordered pairs and are usually defined with equations that tell how to get the second components, given the first. We have been using equations with x and y as the variables, where x represents the first component (input) and y the second component (output) in the ordered pairs.

Here are some everyday examples of functions.

1. The cost y in dollars charged by an express mail company is a function of the weight in pounds x determined by the equation $y = 1.5(x - 1) + 9$.

2. In Cedar Rapids, Iowa, the sales tax is 6% of the price of an item. The tax y on a particular item is a function of the price x, because $y = .06x$.

3. The distance d traveled by a car moving at a constant speed of 45 mph is a function of the time t. Here, $d = 45t$.

As mentioned in Section 3.1, the function concept can be illustrated by an input-output "machine," as seen in Figure 34. It shows how the express mail company equation $y = 1.5(x - 1) + 9$ provides an output (the cost, represented by y) for a given input (the weight in pounds, given by x).

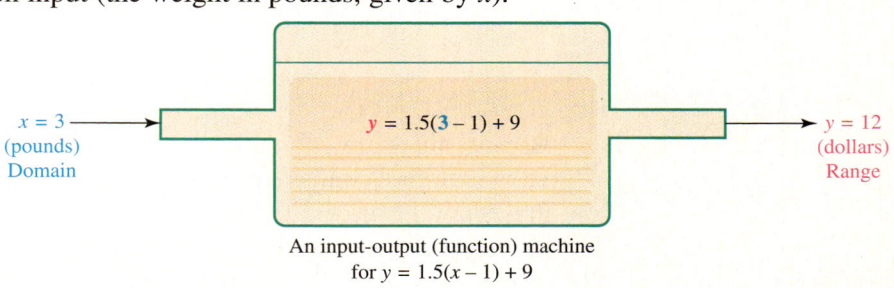

An input-output (function) machine
for $y = 1.5(x - 1) + 9$

FIGURE 34

OBJECTIVE 3 Decide whether an equation defines a function. Given the graph of an equation, the definition of a function can be used to decide whether or not the graph represents a function. By the definition of a function, each *x*-value must lead to exactly one *y*-value. In Figure 35(a), the indicated *x*-value leads to two *y*-values, so this graph is not the graph of a function. A vertical line can be drawn that intersects this graph in more than one point.

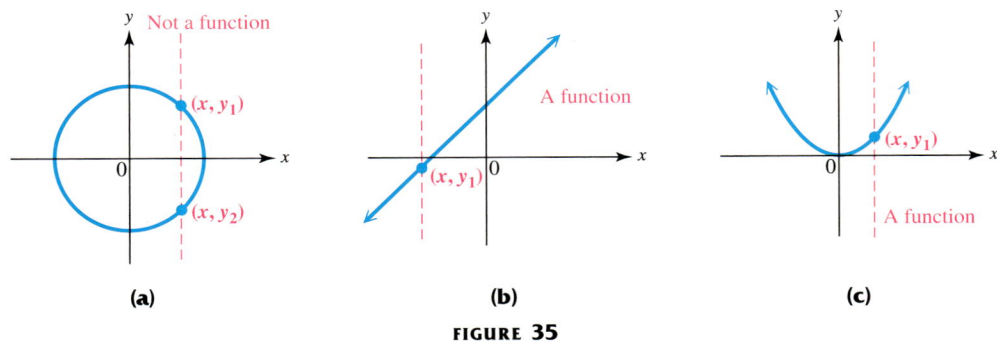

FIGURE 35

On the other hand, in Figures 35(b) and 35(c) any vertical line will intersect each graph in no more than one point. Because of this, the graphs in Figures 35(b) and 35(c) are graphs of functions. This idea leads to the **vertical line test** for a function.

Vertical Line Test

If a vertical line intersects a graph in more than one point, the graph is not the graph of a function.

As Figure 35(b) suggests, any nonvertical line is the graph of a function. For this reason, any linear equation of the form $y = mx + b$ defines a function. (Recall that a vertical line has undefined slope.)

EXAMPLE 3 Deciding Whether Relations Define Functions

Decide whether each relation graphed or defined here is a function.

(a)

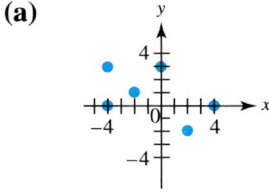

Because there are two ordered pairs with first component -4, this is not the graph of a function.

(b)

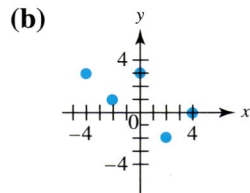

Every first component is matched with one and only one second component, and as a result, no vertical line intersects the graph in more than one point. Therefore, this is the graph of a function.

(c) $y = 2x - 9$

This linear equation is in the form $y = mx + b$. Since the graph of this equation is a line that is not vertical, the equation defines a function.

(d) 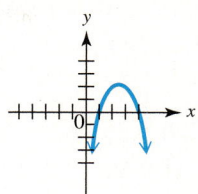   **(e)**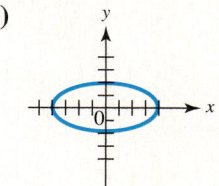

Use the vertical line test. Any vertical line will intersect the graph of a vertical parabola just once, so this is the graph of a function.

The vertical line test shows that this graph is not the graph of a function; a vertical line could intersect the graph twice.

(f) $x = 4$

The graph of $x = 4$ is a vertical line, so the equation does not define a function.

> **Now Try Exercises 11(b), 15, and 17.**

An equation in which y is squared cannot define a function because most x-values will lead to two y-values. This is true for any *even* power of y, such as y^2, y^4, y^6, and so on. Similarly, an equation involving $|y|$ does not define a function, because some x-values lead to more than one y-value.

OBJECTIVE 4 Find domains and ranges. By the definitions of domain and range given for relations, the set of all numbers that can be used as replacements for x in a function is the domain of the function, and the set of all possible values of y is the range of the function.

EXAMPLE 4 Finding the Domain and Range of Functions

Find the domain and range for each function.

(a) $y = 2x - 4$

Any number may be used for x, so the domain is the set of all real numbers. Also, any number may be used for y, so the range is also the set of all real numbers. As indicated in Figure 36, the graph of the equation is a straight line that extends infinitely in both directions, confirming that both the domain and range are $(-\infty, \infty)$.

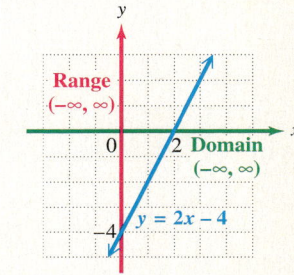

FIGURE 36

(b) $y = x^2$

Any number can be squared, so the domain is the set of all real numbers. However, since the square of a real number cannot be negative and since $y = x^2$, the values of y cannot be negative, making the range the set of all nonnegative numbers, or

$[0, \infty)$ in interval notation. The ordered pairs shown in the table are used to get the graph of the function in Figure 37. While x can take any real number value, notice that y is always greater than or equal to 0.

x	y
0	0
1	1
-1	1
2	4
-2	4
3	9
-3	9

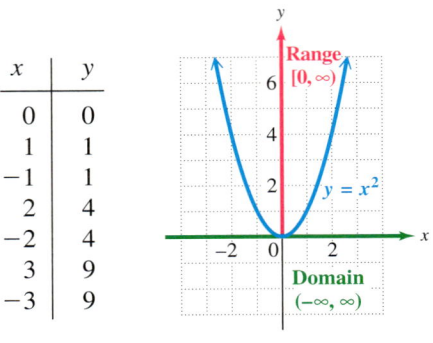

FIGURE 37

Now Try Exercises 23 and 25.

OBJECTIVE 5 Use function notation. The letters f, g, and h are commonly used to name functions. For example, the function $y = 3x + 5$ may be written

$$f(x) = 3x + 5,$$

where $f(x)$ is read "f of x." The notation $f(x)$ is another way of writing y in a function. For the function $f(x) = 3x + 5$, if $x = 7$ then

$$f(7) = 3 \cdot 7 + 5 \qquad \text{Let } x = 7.$$
$$= 21 + 5 = 26.$$

Read this result, $f(7) = 26$, as "f of 7 equals 26." The notation $f(7)$ means the value of y when x is 7. The statement $f(7) = 26$ says that the value of y is 26 when x is 7. It also indicates that the point $(7, 26)$ lies on the graph of f.

Similarly, to find $f(-3)$, substitute -3 for x.

$$f(-3) = 3(-3) + 5 \qquad \text{Let } x = -3.$$
$$= -9 + 5 = -4$$

CAUTION The notation $f(x)$ does *not* mean f times x; $f(x)$ means the function value of x. It represents the y-value that corresponds to x.

Function Notation

In the notation $f(x)$,

f is the name of the function,

x is the domain value,

and $f(x)$ is the range value y for the domain value x.

EXAMPLE 5 Using Function Notation

For the function defined by $f(x) = x^2 - 3$, find the following.

(a) $f(2)$

Substitute 2 for x.

$$f(x) = x^2 - 3$$
$$f(2) = 2^2 - 3 \qquad \text{Let } x = 2.$$
$$= 4 - 3 = 1$$

(b) $f(0) = 0^2 - 3 = 0 - 3 = -3$

(c) $f(-3) = (-3)^2 - 3 = 9 - 3 = 6$

Now Try Exercise 35.

OBJECTIVE 6 Apply the function concept in an application. Because a function assigns each element in its domain to exactly one element in its range, the function concept is used in real-data applications where two quantities are related. Our final example discusses such an application, using a table of values.

EXAMPLE 6 Applying the Function Concept to Population

Asian-American populations (in millions) are shown in the table.

Asian-American Population

Year	Population (in millions)
1996	9.7
1998	10.5
2000	11.2
2002	12

Source: U.S. Bureau of the Census.

(a) Use the table to write a set of ordered pairs that defines a function f.

If we choose the years as the domain elements and the populations as the range elements, the information in the table can be written as a set of four ordered pairs. In set notation, the function f is

$$f = \{(1996, 9.7), (1998, 10.5), (2000, 11.2), (2002, 12)\}.$$

(b) What is the domain of f? What is the range?

The domain is the set of years, or x-values:

$$\{1996, 1998, 2000, 2002\}.$$

The range is the set of populations, in millions, or y-values:

$$\{9.7, 10.5, 11.2, 12\}.$$

(c) Find $f(1996)$ and $f(2000)$.

From the table or the set of ordered pairs in part (a),

$$f(1996) = 9.7 \text{ million} \qquad \text{and} \qquad f(2000) = 11.2 \text{ million}.$$

244 CHAPTER 3 Linear Equations and Inequalities in Two Variables; Functions

(d) For what x-value does $f(x)$ equal 12 million? 10.5 million?

We use the table or the ordered pairs found in part (a) to determine $f(2002) = 12$ million and $f(1998) = 10.5$ million.

Now Try Exercises 41 and 43.

3.6 EXERCISES

For Extra Help

 Student's Solutions Manual

 MyMathLab

 InterAct Math Tutorial Software

 AW Math Tutor Center

 MathXL

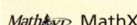 Digital Video Tutor CD 7/Videotape 6

Complete the following table for the linear function defined by $f(x) = x + 2$.

x	$x + 2$	$f(x)$	(x, y)
0	2	2	(0, 2)
1. 1			
2. 2			
3. 3			
4. t			

5. Describe the graph of function f in Exercises 1–4 if the domain is $\{0, 1, 2, 3\}$.

6. Describe the graph of function f in Exercises 1–4 if the domain is the set of real numbers $(-\infty, \infty)$.

*In Exercises 7–12, **(a)** give the domain and range of each relation; **(b)** determine whether the relation is or is not a function. See Examples 1–3.*

7. $\{(-4, 3), (-2, 1), (0, 5), (-2, -8)\}$

8. $\{(3, 7), (1, 4), (0, -2), (-1, -1), (-2, 5)\}$

9.
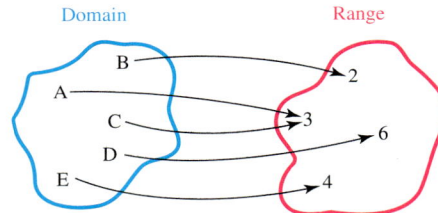

10.
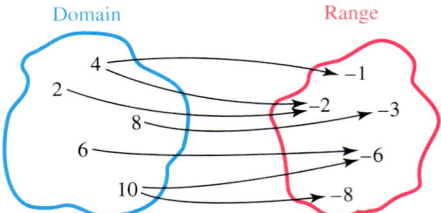

11.
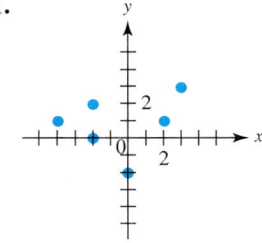

12.
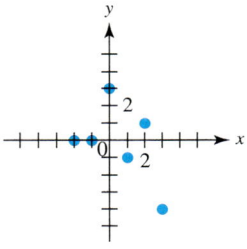

Decide whether each graph is that of a function. See Example 3.

13.
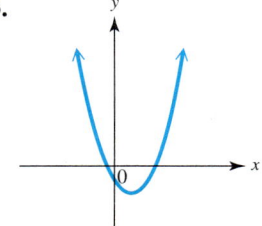

14.
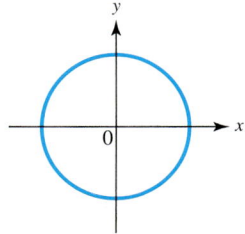

15. 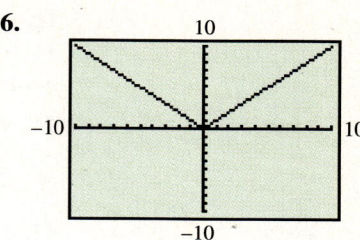 16.

Decide whether each equation defines y as a function of x. (Remember that to be a function, every value of x must give one and only one value of y.)

17. $y = 5x + 3$ **18.** $y = -7x + 12$ **19.** $y = x^2$
20. $y = -3x^2$ **21.** $x = y^2$ **22.** $x = |y|$

Find the domain and the range for each function. See Example 4.

23. $y = 3x - 2$ **24.** $y = x^2 - 3$ **25.** $y = x^2 + 2$
26. $y = -x + 3$ **27.** $f(x) = \sqrt{x}$ **28.** $f(x) = |x|$

RELATING CONCEPTS (EXERCISES 29–32)

For Individual or Group Work

A function defined by the equation of a line, such as $f(x) = 3x - 4$, called a linear function, can be graphed by replacing $f(x)$ with y and then using the methods described earlier. Let us assume that some function is written in the form $f(x) = mx + b$, for particular values of m and b. **Work Exercises 29–32 in order.**

29. If $f(2) = 4$, name the coordinates of one point on the line.

30. If $f(-1) = -4$, name the coordinates of another point on the line.

31. Use the results of Exercises 29 and 30 to find the slope of the line.

32. Use the slope-intercept form of the equation of a line to write the function in the form $f(x) = mx + b$.

For each function f, find (a) $f(2)$, (b) $f(0)$, (c) $f(-3)$. See Example 5.

33. $f(x) = 4x + 3$ **34.** $f(x) = -3x + 5$ **35.** $f(x) = x^2 - x + 2$
36. $f(x) = x^3 + x$ **37.** $f(x) = |x|$ **38.** $f(x) = |x + 7|$

A calculator can be thought of as a function machine. We input a number value (from the domain), and then by pressing the appropriate key, we obtain an output value (from the range). Use your calculator, follow the directions, and then answer each question.

39. Suppose we enter the value 4 and then take the square root (that is, activate the square root *function,* using the key labeled $\sqrt{x}$).

   (a) What is the domain value here?   **(b)** What is the range value obtained?

40. Suppose we enter the value -8 and then activate the squaring function, using the key labeled x^2.

   (a) What is the domain value here?   **(b)** What range value is obtained?

The number of U.S. foreign-born residents has grown by more than 43 percent since 1990. The graph shows the number of immigrants (in millions) for selected years. Use the information in the graph to work Exercises 41–45. See Example 6.

41. Write the information in the graph as a set of ordered pairs. Does this set define a function?

42. Name the function g in Exercise 41. Give the domain and range of g.

43. Find $g(1980)$ and $g(1990)$.

44. For what value of x does $g(x) = 28.4$?

45. Suppose $g(2002) = 30.3$. What does this tell you in the context of the application?

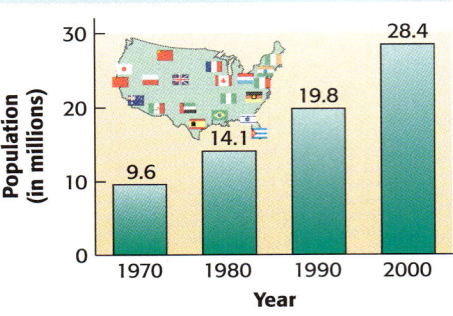

GROWTH OF U.S. FOREIGN-BORN POPULATION

Source: U.S. Bureau of the Census.

RELATING CONCEPTS (EXERCISES 46–50)
For Individual or Group Work

The data give the average tuition and fees at public colleges and universities for selected years. Use the information in the table to **work Exercises 46–50 in order.**

Tuition and Fees

Year	Tuition and Fees (in dollars)
1980	804
1985	1242
1990	1809
1995	2860
1999	3356

Source: The College Board.

46. **(a)** Plot the ordered pairs (years, tuition and fees) from the table. Give the domain and range of this relation. Do the points suggest that a linear function f would give a reasonable approximation of the data?
    (b) What is $f(1995)$? For what x-value does $f(x) = 3356$?

47. Use the first and last ordered pairs from Exercise 46(a) to write an equation relating the data. Write the equation in $y = mx + b$ form, rounding the slope to the nearest hundredth.

48. Use your equation from Exercise 47 to approximate tuition and fees in 1990 and 1995. Round to the nearest dollar.

49. Use the second and fourth ordered pairs to write an equation relating the data. Write the equation in $y = mx + b$ form.

50. Use your equation from Exercise 49 to approximate tuition and fees in 1980 and 1999. Do the results of this exercise or Exercise 48 give better approximations?

SECTION 3.6 Introduction to Functions **247**

TECHNOLOGY INSIGHTS (EXERCISES 51–57)

The table was generated by a graphing calculator. The expression Y_1 represents $f(X)$.

X	Y₁
-2	-1
-1	0
0	1
1	2
2	3
3	4
4	5

X = -2

51. When $X = 3$, $Y_1 = $ _____?

52. What is $f(3)$?

53. When $Y_1 = 2$, $X = $ _____?

54. If $f(X) = 2$, what is the value of X?

55. The points represented in the table all lie in a straight line. What is the slope of the line?

56. What is the y-intercept of the line?

57. Write the function in the form $y = mx + b$, for the appropriate values of m and b.

58. Define *relation* and *function*. Compare the two definitions. How are they alike? How are they different?

59. In your own words, explain the meaning of domain and range of a function.

Not every function can be readily expressed as an equation. Here is an example of such a function from everyday life.

60. A chain-saw rental firm charges $7 per day or fraction of a day to rent a saw, plus a fixed fee of $4 for resharpening the blade. Let y represent the cost of renting a saw for x days. A portion of the graph is shown here.

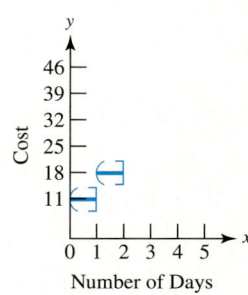

Number of Days

(a) Give the domain and range of this relation.
(b) Explain how the graph can be continued.
(c) Explain how the vertical line test applies here.
(d) What do the numbers in the domain represent here? What do the numbers in the range represent?

Chapter 3 Group Activity

Determining Business Profit

OBJECTIVE Use equations, tables, and graphs to make business decisions.

Graphs are used by businesses to analyze and make decisions. This activity will explore such a process.

The Parent Teacher Organization of a school decides to sell bags of popcorn as a fund-raiser. The parents want to estimate their profit. Costs include $14 for popcorn (enough for 680 bags) and a total of $7 for bags to hold the popcorn.

A. Develop a profit formula.
1. From the information above, determine the total costs. (Assume the organization is buying enough supplies for 680 bags of popcorn.)
2. If the filled bags sell for $.25 each, write an expression for the total revenue from n bags of popcorn.
3. Since Profit = Revenue − Cost, write an equation that represents the profit P for n bags of popcorn sold.
4. Work in pairs. One person should use the profit equation from part 3 to complete the table below. The other person should decide on an appropriate scale and graph the profit equation.

n	P
0	
	0
100	

B. Choose a different price for a bag of popcorn (between $.20 and $.75).
1. Write the profit equation for this cost.
2. Switch roles, that is, if you drew the graph in part A, now make a table of values for this equation. Have your partner graph the profit equation on the same coordinate system you used in part A.

C. Compare your findings and answer the following.
1. What is the break-even point (that is, when profits are 0 or revenue equals costs) for each equation?
2. What does it mean if you end up with a negative value for P?
3. If you sell all 680 bags, what will your profits be for the two different prices? Explain how you would estimate this from your graph.
4. What are the advantages and/or disadvantages of charging a higher price?
5. Together, decide on the price you would charge for a bag of popcorn. Explain why you chose this price.

CHAPTER 3 SUMMARY

KEY TERMS

3.1 linear equation in two variables
ordered pair
table of values
x-axis
y-axis
rectangular (Cartesian) coordinate system

quadrant
origin
plane
coordinates
plot
scatter diagram
3.2 graph, graphing
y-intercept
x-intercept

3.3 subscript notation
rise
run
slope
parallel lines
perpendicular lines
3.5 linear inequality in two variables
boundary line

3.6 components
relation
domain
range
function
function notation

NEW SYMBOLS

(a, b) an ordered pair

(x_1, y_1) x-sub-one, y-sub-one

m slope

$f(x)$ function f of x

TEST YOUR WORD POWER

See how well you have learned the vocabulary in this chapter. Answers, with examples, follow the Quick Review.

1. An **ordered pair** is a pair of numbers written
   A. in numerical order between brackets
   B. between parentheses or brackets
   C. between parentheses in which order is important
   D. between parentheses in which order does not matter.

2. An **intercept** is
   A. the point where the x-axis and y-axis intersect
   B. a pair of numbers written in parentheses in which order matters
   C. one of the four regions determined by a rectangular coordinate system
   D. the point where a graph intersects the x-axis or the y-axis.

3. The **slope** of a line is
   A. the measure of the run over the rise of the line
   B. the distance between two points on the line
   C. the ratio of the change in y to the change in x along the line
   D. the horizontal change compared to the vertical change of two points on the line.

4. Two lines in a plane are **parallel** if
   A. they represent the same line
   B. they never intersect
   C. they intersect at a 90° angle
   D. one has a positive slope and one has a negative slope.

5. Two lines in a plane are **perpendicular** if
   A. they represent the same line
   B. they never intersect
   C. they intersect at a 90° angle
   D. one has a positive slope and one has a negative slope.

6. A **relation** is
   A. any set of ordered pairs
   B. a set of ordered pairs in which each first component corresponds to exactly one second component
   C. two sets of ordered pairs that are related
   D. a graph of ordered pairs.

7. The **domain** of a relation is
   A. the set of all x- and y-values in the ordered pairs of the relation
   B. the difference between the components in an ordered pair of the relation
   C. the set of all first components in the ordered pairs of the relation
   D. the set of all second components in the ordered pairs of the relation.

8. The **range** of a relation is
   A. the set of all x- and y-values in the ordered pairs of the relation
   B. the difference between the components in an ordered pair of the relation
   C. the set of all first components in the ordered pairs of the relation
   D. the set of all second components in the ordered pairs of the relation.

9. A **function** is
   A. any set of ordered pairs
   B. a set of ordered pairs in which each first component corresponds to exactly one second component
   C. two sets of ordered pairs that are related
   D. a graph of ordered pairs.

250 CHAPTER 3 Linear Equations and Inequalities in Two Variables; Functions

QUICK REVIEW

CONCEPTS	EXAMPLES

3.1 READING GRAPHS; LINEAR EQUATIONS IN TWO VARIABLES

An ordered pair is a solution of an equation if it satisfies the equation.

Is $(2, -5)$ or $(0, -6)$ a solution of $4x - 3y = 18$?

$4(2) - 3(-5) = 23 \neq 18$ | $4(0) - 3(-6) = 18$
$(2, -5)$ is not a solution. | $(0, -6)$ is a solution.

If a value of either variable in an equation is given, the other variable can be found by substitution.

Complete the ordered pair $(0, \quad)$ for $3x = y + 4$.

$$3(0) = y + 4$$
$$0 = y + 4$$
$$-4 = y$$

The ordered pair is $(0, -4)$.

Plot the ordered pair $(-3, 4)$ by starting at the origin, going 3 units to the left, then going 4 units up.

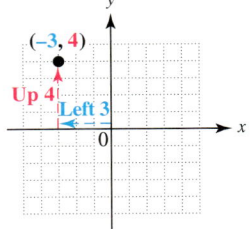

3.2 GRAPHING LINEAR EQUATIONS IN TWO VARIABLES

The graph of $y = k$ is a horizontal line through $(0, k)$.

The graph of $x = k$ is a vertical line through $(k, 0)$.

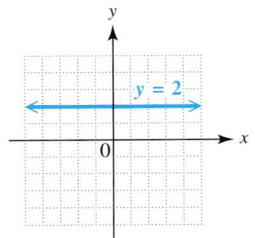 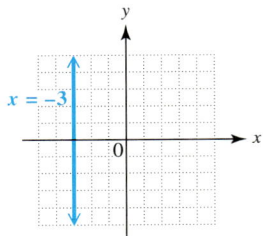

The graph of $Ax + By = 0$ goes through the origin. Find and plot another point that satisfies the equation. Then draw the line through the two points.

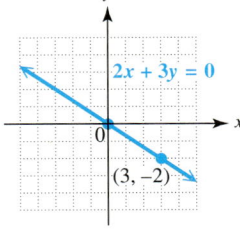

To graph a linear equation:

Step 1 Find at least two ordered pairs that satisfy the equation.

Step 2 Plot the corresponding points.

Step 3 Draw a straight line through the points.

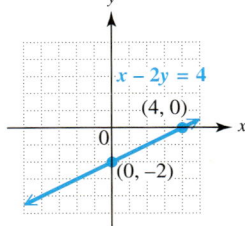

CONCEPTS	EXAMPLES
3.3 THE SLOPE OF A LINE	
The slope of the line through (x_1, y_1) and (x_2, y_2) is $$m = \frac{y_2 - y_1}{x_2 - x_1} \quad (x_1 \neq x_2).$$	The line through $(-2, 3)$ and $(4, -5)$ has slope $$m = \frac{-5 - 3}{4 - (-2)} = \frac{-8}{6} = -\frac{4}{3}.$$
Horizontal lines have slope 0.	The line $y = -2$ has slope 0.
Vertical lines have undefined slope.	The line $x = 4$ has undefined slope.
To find the slope of a line from its equation, solve for y. The slope is the coefficient of x.	Find the slope of $3x - 4y = 12$. $$-4y = -3x + 12$$ $$y = \frac{3}{4}x - 3$$ The slope is $\frac{3}{4}$.
Parallel lines have the same slope.	The lines $y = 3x - 1$ and $y = 3x + 4$ are parallel because both have slope 3.
The slopes of perpendicular lines are negative reciprocals (that is, their product is -1).	The lines $y = -3x - 1$ and $y = \frac{1}{3}x + 4$ are perpendicular because their slopes are -3 and $\frac{1}{3}$, and $-3(\frac{1}{3}) = -1$.
3.4 EQUATIONS OF A LINE	
Slope-Intercept Form $y = mx + b$ m is the slope. $(0, b)$ is the y-intercept.	Find an equation of the line with slope 2 and y-intercept $(0, -5)$. The equation is $y = 2x - 5$.
Point-Slope Form $y - y_1 = m(x - x_1)$ m is the slope. (x_1, y_1) is a point on the line.	Find an equation of the line with slope $-\frac{1}{2}$ through $(-4, 5)$. $$y - 5 = -\frac{1}{2}[x - (-4)]$$ $$y - 5 = -\frac{1}{2}(x + 4)$$ $$y - 5 = -\frac{1}{2}x - 2$$ $$y = -\frac{1}{2}x + 3$$
Standard Form $Ax + By = C$ A, B, and C are integers and $A > 0$, $B \neq 0$.	This equation is written in standard form as $$x + 2y = 6,$$ with $A = 1$, $B = 2$, and $C = 6$.

CONCEPTS	EXAMPLES
3.5 GRAPHING LINEAR INEQUALITIES IN TWO VARIABLES	
Step 1 Graph the line that is the boundary of the region. Make it solid if the inequality is $\leq$ or $\geq$; make it dashed if the inequality is $<$ or $>$.	Graph $2x + y \leq 5$. Graph the line $2x + y = 5$. Make it solid because of $\leq$. 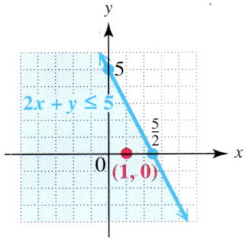
Step 2 Use any point not on the line as a test point. Substitute for x and y in the inequality. If the result is true, shade the side of the line containing the test point; if the result is false, shade the other side.	Use $(1, 0)$ as a test point. $2(1) + 0 \leq 5$? $2 \leq 5$ True Shade the side of the line containing $(1, 0)$.
3.6 INTRODUCTION TO FUNCTIONS	
Vertical Line Test If a vertical line intersects a graph in more than one point, the graph is not the graph of a function.	The graph shown is not the graph of a function. 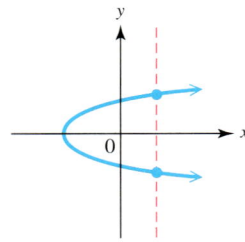
The **domain** of a function is the set of numbers that can replace x in the expression for the function. The **range** is the set of y-values that result as x is replaced by each number in the domain. To find $f(x)$ for a specific value of x, replace x by that value in the expression for the function.	Let $f(x) = x - 1$. Find **(a)** the domain of f; **(b)** the range of f; **(c)** $f(-1)$. **(a)** The domain is $(-\infty, \infty)$. **(b)** The range is $(-\infty, \infty)$. **(c)** $f(-1) = -1 - 1 = -2$

Answers to Test Your Word Power

1. C; *Examples:* $(0, 3), (-3, 8), (4, 0)$ **2.** D; *Example:* The graph of the equation $4x - 3y = 12$ has x-intercept $(3, 0)$ and y-intercept $(0, -4)$.   **3.** C; *Example:* The line through $(3, 6)$ and $(5, 4)$ has slope $\frac{4-6}{5-3} = \frac{-2}{2} = -1$.   **4.** B; *Example:* See Figure 23 in Section 3.3.   **5.** C; *Example:* See Figure 24 in Section 3.3.   **6.** A; *Example:* $\{(0, 2), (2, 4), (3, 6), (-1, 3)\}$ **7.** C; *Example:* The domain in the relation in Answer 6 is the set of x-values, that is, $\{0, 2, 3, -1\}$.   **8.** D; *Example:* The range of the relation in Answer 6 is the set of y-values, that is, $\{2, 4, 6, 3\}$.   **9.** B; *Example:* The relation in Answer 6 is a function since each x-value corresponds to exactly one y-value.

CHAPTER 3 REVIEW EXERCISES

[3.1]
1. The line graph shows average prices for a gallon of gasoline in Cedar Rapids, Iowa, from June 1999 through June 2000. Use the graph to work parts (a)–(d).

   (a) About how much did a gallon of gas cost in June 1999? in June 2000?
   (b) How much did the price of a gallon of gas increase over this 1-yr period? What percent increase is this?
   (c) During which months did the biggest increase in the price of a gallon of gas occur? About how much did the price increase during this time?
   (d) Between which months did the price of a gallon of gas decrease?

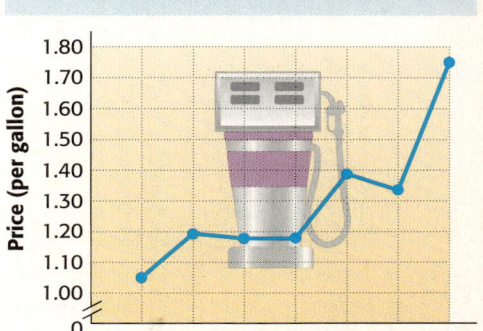

Source: Iowa DNR.

Complete the given ordered pairs for each equation.

2. $y = 3x + 2$; $(-1, \ )$, $(0, \ )$, $(\ , 5)$
3. $4x + 3y = 6$; $(0, \ )$, $(\ , 0)$, $(-2, \ )$
4. $x = 3y$; $(0, \ )$, $(8, \ )$, $(\ , -3)$
5. $x - 7 = 0$; $(\ , -3)$, $(\ , 0)$, $(\ , 5)$

Determine whether the given ordered pair is a solution of the given equation.

6. $x + y = 7$; $(2, 5)$
7. $2x + y = 5$; $(-1, 3)$
8. $3x - y = 4$; $\left(\dfrac{1}{3}, -3\right)$

Name the quadrant in which each pair lies. Then plot each ordered pair in a rectangular coordinate system.

9. $(2, 3)$
10. $(-4, 2)$
11. $(3, 0)$
12. $(0, -6)$
13. If $xy > 0$, in what quadrant or quadrants must (x, y) lie?

[3.2] *Find the x- and y-intercepts for the line that is the graph of each equation, and graph the line.*

14. $y = 2x + 5$
15. $3x + 2y = 8$
16. $x + 2y = -4$

[3.3] *Find the slope of each line.*

17. Through $(2, 3)$ and $(-4, 6)$
18. Through $(2, 5)$ and $(2, 8)$
19. $y = 3x - 4$
20. $y = 5$

21.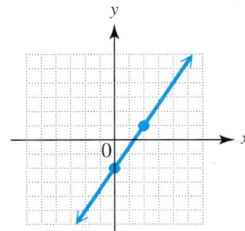

22.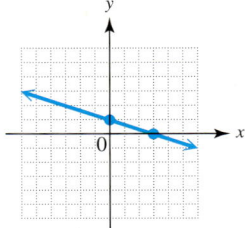

23. The line having these points

x	y
0	1
2	4
6	10

24. (a) A line parallel to the graph of $y = 2x + 3$
(b) A line perpendicular to the graph of $y = -3x + 3$

Decide whether each pair of lines is parallel, perpendicular, *or* neither.

25. $3x + 2y = 6$
$6x + 4y = 8$

26. $x - 3y = 1$
$3x + y = 4$

27. $x - 2y = 8$
$x + 2y = 8$

[3.4] *Write an equation for each line in the form* $y = mx + b$, *if possible.*

28. $m = -1, b = \dfrac{2}{3}$

29. Through $(2, 3)$ and $(-4, 6)$

30. Through $(4, -3)$, $m = 1$

31. Through $(-1, 4)$, $m = \dfrac{2}{3}$

32. Through $(1, -1)$, $m = -\dfrac{3}{4}$

33. $m = -\dfrac{1}{4}, b = \dfrac{3}{2}$

34. Slope 0, through $(-4, 1)$

35. Through $\left(\dfrac{1}{3}, -\dfrac{5}{4}\right)$ with undefined slope

[3.5] *Graph each linear inequality.*

36. $3x + 5y > 9$

37. $2x - 3y > -6$

38. $x - 2y \geq 0$

[3.6] *Decide whether each relation* is *or* is not *a function. In Exercises 39 and 40, give the domain and the range.*

39. $\{(-2, 4), (0, 8), (2, 5), (2, 3)\}$

40. $\{(8, 3), (7, 4), (6, 5), (5, 6), (4, 7)\}$

41.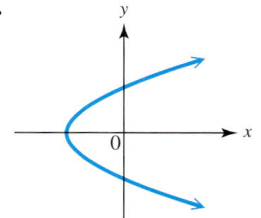

42.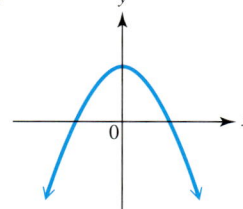

43. $2x + 3y = 12$

44. $y = x^2$

Find (a) $f(2)$ and (b) $f(-1)$.

45. $f(x) = 3x + 2$

46. $f(x) = 2x^2 - 1$

47. $f(x) = |x + 3|$

MIXED REVIEW EXERCISES

In Exercises 48–53, match each statement to the appropriate graph or graphs in A–D. Graphs may be used more than once.

A.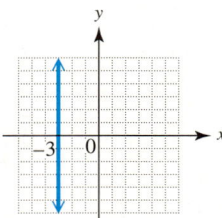

B.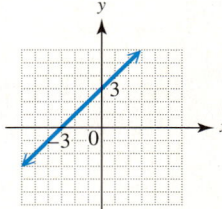

C.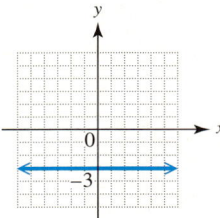

D.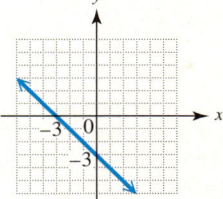

48. The line shown in the graph has undefined slope.

49. The graph of the equation has y-intercept $(0, -3)$.

50. The graph of the equation has x-intercept $(-3, 0)$.

51. The line shown in the graph has negative slope.

52. The graph is that of the equation $y = -3$.

53. The line shown in the graph has slope 1.

Find the intercepts and the slope of each line. Then graph the line.

54. $y = -2x - 5$

55. $x + 3y = 0$

56. $y - 5 = 0$

Write an equation for each line. Write the equation in slope-intercept form.

57. $m = -\dfrac{1}{4}, b = -\dfrac{5}{4}$

58. Through $(8, 6)$, $m = -3$

59. Through $(3, -5)$ and $(-4, -1)$

Graph each inequality.

60. $y < -4x$

61. $x - 2y \leq 6$

RELATING CONCEPTS (EXERCISES 62–70)

For Individual or Group Work

The total amount spent (in billions of dollars) on video rentals in the United States from 1996 through 2000 is shown in the graph. Use the graph to **work Exercises 62–70 in order.**

62. About how much did the amount spent on video rentals decrease during the years shown in the graph?

63. Since the points of the graph lie approximately in a linear pattern, a straight line can be used to model the data. Will this line have positive or negative slope? Explain.

64. The table gives the actual amounts spent on video rentals in 1996 and 2000. Write two ordered pairs for the data.

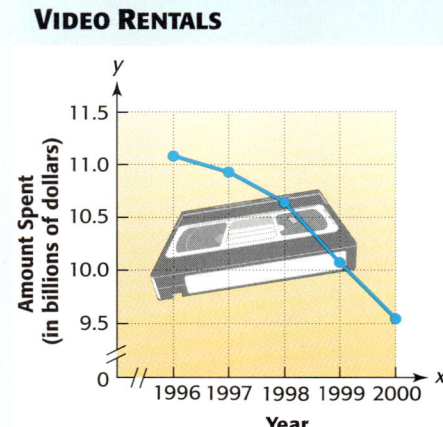

VIDEO RENTALS

Source: Blockbuster.

Year x	Amount y (in billions of dollars)
1996	11.1
2000	9.6

65. Use the ordered pairs from Exercise 64 to find the equation of a line that models the data. Write the equation in slope-intercept form.

66. Based on the equation you found in Exercise 65, what is the slope of the line? Does it agree with your answer in Exercise 63?

67. Use the equation from Exercise 65 to approximate the amount spent on video rentals from 1997 through 1999, and complete the table. Round your answers to the nearest tenth.

x	y
1996	11.1
1997	
1998	
1999	
2000	9.6

68. The actual amounts spent on video rentals are given in the following ordered pairs.

(1996, 11.1), (1997, 10.9), (1998, 10.6), (1999, 10.1), (2000, 9.6)

How do the actual amounts compare to those you found in Exercise 67 using the linear equation?

69. Since the equation in Exercise 65 models the data fairly well, use it to estimate the amount spent on video rentals in 2002.

70. Discuss reasons why the amount spent on video rentals has been decreasing in recent years.

CHAPTER 3 TEST

1. Complete these ordered pairs for the equation $3x + 5y = -30$: (0,), (,0), (,−3).
2. Is $(4, -1)$ a solution of $4x - 7y = 9$?
3. How do you find the x-intercept of the graph of a linear equation in two variables? How do you find the y-intercept?

Graph each linear equation. Give the x- and y-intercepts.

4. $3x + y = 6$
5. $y - 2x = 0$
6. $x + 3 = 0$
7. $y = 1$
8. $x - y = 4$

Find the slope of each line.

9. Through $(-4, 6)$ and $(-1, -2)$
10. $2x + y = 10$
11. $x + 12 = 0$
12.

13. A line parallel to the graph of $y - 4 = 6$

Write an equation for each line. Write the equation in slope-intercept form.

14. Through $(-1, 4)$; $m = 2$
15. The line in Exercise 12
16. Through $(2, -6)$ and $(1, 3)$

Graph each linear inequality.

17. $x + y \leq 3$
18. $3x - y > 0$

The graph shows total food and drink sales at U.S. restaurants from 1970 through 2000, where 1970 corresponds to $x = 0$. Use the graph to work Exercises 19–22 on the next page.

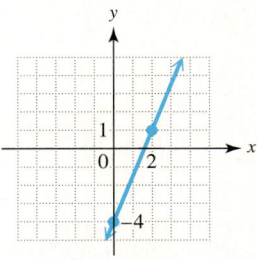

WHAT'S FOR DINNER?

Source: National Restaurant Association.

19. Is the slope of the line in the graph positive or negative? Explain.

20. Write two ordered pairs for the data points shown in the graph. Use them to find the slope of the line.

21. The linear equation
$$y = 11.1x + 43$$
approximates food and drink sales y in billions of dollars, where $x = 0$ again represents 1970. Use the equation to approximate food and drink sales for 1990 and 1995.

22. What does the ordered pair $(30, 376)$ mean in the context of this situation?

23. Decide whether each relation represents a function. If it does, give the domain and the range.

    (a) $\{(2, 3), (2, 4), (2, 5)\}$ (b) $\{(0, 2), (1, 2), (2, 2)\}$

24. Use the vertical line test to determine whether the graph is that of a function.

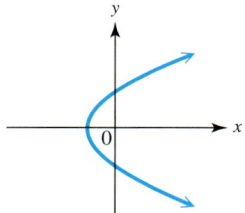

25. If $f(x) = 3x + 7$, find $f(-2)$.

CUMULATIVE REVIEW EXERCISES CHAPTERS 1–3

Perform each indicated operation.

1. $10\dfrac{5}{8} - 3\dfrac{1}{10}$

2. $\dfrac{3}{4} \div \dfrac{1}{8}$

3. $5 - (-4) + (-2)$

4. $\dfrac{(-3)^2 - (-4)(2^4)}{5(2) - (-2)^3}$

5. True or false? $\dfrac{4(3 - 9)}{2 - 6} \geq 6$

6. Find the value of $xz^3 - 5y^2$ when $x = -2$, $y = -3$, and $z = -1$.

7. What property does $3(-2 + x) = -6 + 3x$ illustrate?

8. Simplify $-4p - 6 + 3p + 8$ by combining terms.

Solve.

9. $V = \dfrac{1}{3}\pi r^2 h$ for h

10. $6 - 3(1 + a) = 2(a + 5) - 2$

11. $-(m - 3) = 5 - 2m$

12. $\dfrac{x - 2}{3} = \dfrac{2x + 1}{5}$

Solve each inequality, and graph the solutions.

13. $-2.5x < 6.5$

14. $4(x + 3) - 5x < 12$

15. $\dfrac{2}{3}x - \dfrac{1}{6}x \leq -2$

Cumulative Review Exercises Chapters 1–3 **259**

Solve each problem.

16. The gap in average annual earnings by level of education continues to increase. Based on the most recent statistics available, a person with a bachelor's degree can expect to earn $17,583 more each year than someone with a high school diploma. Together the individuals would earn $63,373. How much can a person at each level of education expect to earn? (*Source:* U.S. Bureau of the Census.)

17. Mount Mayon in the Philippines is the most perfectly shaped conical volcano in the world. Its base is a perfect circle with circumference 80 mi, and it has a height of about 8200 ft. (One mile is 5280 ft.) Find the radius of the circular base to the nearest mile. (*Hint:* This problem has some unneeded information.) (*Source: Microsoft Encarta Encyclopedia 2000.*)

Circumference = 80 mi

18. The winning times in seconds for the women's 1000-m speed skating event in the Winter Olympics for the years 1960 through 1998 can be closely approximated by the linear equation

$$y = -.4685x + 95.07,$$

where x is the number of years since 1960. That is, $x = 4$ represents 1964, $x = 8$ represents 1968, and so on. (*Source: The Universal Almanac, 1998.*)

(a) Use this equation to complete the table of values. Round times to the nearest hundredth of a second.
(b) What does the ordered pair $(20, 85.7)$ mean in the context of the problem?

x	y
12	
28	
36	

19. Baby boomers are expected to inherit $10.4 trillion from their parents over the next 45 yr, an average of $50,000 each. The circle graph shows how they plan to spend their inheritances.

(a) How much of the $50,000 is expected to go toward home purchase?
(b) How much is expected to go toward retirement?
(c) Use the answer from part (b) to estimate the amount expected to go toward paying off debts or funding children's education.

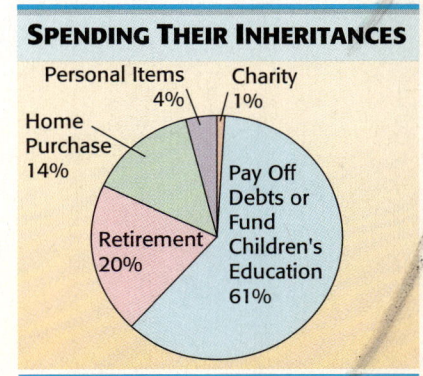

Source: First Interstate Bank Trust and Private Banking Group.

Consider the linear equation $-3x + 4y = 12$. Find the following.

20. The x- and y-intercepts   **21.** The slope   **22.** The graph

23. Are the lines with equations $x + 5y = -6$ and $y = 5x - 8$ *parallel, perpendicular, or neither*?

Write an equation for each line. Write the equation in slope-intercept form, if possible.

24. Through $(2, -5)$ with slope 3   **25.** Through $(0, 4)$ and $(2, 4)$

Systems of Linear Equations and Inequalities

4

4.1 Solving Systems of Linear Equations by Graphing

4.2 Solving Systems of Linear Equations by Substitution

4.3 Solving Systems of Linear Equations by Elimination

Summary Exercises on Solving Systems of Linear Equations

4.4 Applications of Linear Systems

4.5 Solving Systems of Linear Inequalities

Americans continue to flock to movie theaters in record numbers. Box office grosses reached $8.14 billion in 2001, the first time they exceeded $8 billion, as more than 1.4 billion movie tickets were sold. Surprisingly, the top three movies of the year were family films—*Harry Potter and the Sorcerer's Stone, Shrek,* and *Monsters, Inc.* not only attracted scores of adults with kids, but also unaccompanied adults, perhaps wishing to get away from it all for a few hours. (*Source:* ACNielsen EDI.) In Exercise 11 of Section 4.4, we use systems of linear equations to find out just how much money these top films earned.

261

4.1 Solving Systems of Linear Equations by Graphing

OBJECTIVES

1. Decide whether a given ordered pair is a solution of a system.
2. Solve linear systems by graphing.
3. Solve special systems by graphing.
4. Identify special systems without graphing.
5. Use a graphing calculator to solve a linear system.

A **system of linear equations**, often called a **linear system**, consists of two or more linear equations with the same variables. Examples of systems of two linear equations include

$$2x + 3y = 4 \qquad x + 3y = 1 \qquad x - y = 1$$
$$3x - y = -5 \qquad -y = 4 - 2x \qquad y = 3.$$

In the system on the right, think of $y = 3$ as an equation in two variables by writing it as $0x + y = 3$.

OBJECTIVE 1 Decide whether a given ordered pair is a solution of a system. A **solution of a system** of linear equations is an ordered pair that makes both equations true at the same time. A solution of an equation is said to *satisfy* the equation.

EXAMPLE 1 Determining Whether an Ordered Pair Is a Solution

Is $(4, -3)$ a solution of each system?

(a) $x + 4y = -8$
$3x + 2y = 6$

To decide whether $(4, -3)$ is a solution of the system, substitute 4 for x and -3 for y in each equation.

$x + 4y = -8$		$3x + 2y = 6$	
$4 + 4(-3) = -8$?		$3(4) + 2(-3) = 6$?	
$4 + (-12) = -8$?	Multiply.	$12 + (-6) = 6$?	Multiply.
$-8 = -8$	True	$6 = 6$	True

Because $(4, -3)$ satisfies both equations, it is a solution of the system.

(b) $2x + 5y = -7$
$3x + 4y = 2$

Again, substitute 4 for x and -3 for y in both equations.

$2x + 5y = -7$		$3x + 4y = 2$	
$2(4) + 5(-3) = -7$?		$3(4) + 4(-3) = 2$?	
$8 + (-15) = -7$?	Multiply.	$12 + (-12) = 2$?	Multiply.
$-7 = -7$	True	$0 = 2$	False

The ordered pair $(4, -3)$ is not a solution of this system because it does not satisfy the second equation.

Now Try Exercises 3 and 5.

We discuss three methods of solving a system of two linear equations in two variables in this chapter.

OBJECTIVE 2 Solve linear systems by graphing. The set of all ordered pairs that are solutions of a system is its **solution set.** One way to find the solution set of a system of two linear equations is to graph both equations on the same axes. The graph of each line shows points whose coordinates satisfy the equation of that line. Any intersection point would be on both lines and would therefore be a solution of both equations. Thus, the coordinates of any point where the lines intersect give a solution of the system. Because two *different* straight lines can intersect at no more than one point, there can never be more than one solution for such a system. The graph in Figure 1 shows that the solution of the system in Example 1(a) is the intersection point $(4, -3)$.

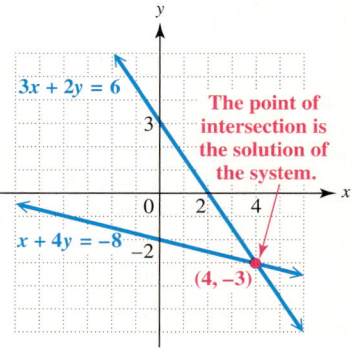

FIGURE 1

EXAMPLE 2 Solving a System by Graphing

Solve the system of equations by graphing both equations on the same axes.

$$2x + 3y = 4$$
$$3x - y = -5$$

To graph these two equations, we choose *any* number for either x or y to get an ordered pair. The intercepts are often convenient choices. It is a good idea to use a third ordered pair as a check.

$2x + 3y = 4$

x	y
0	$\frac{4}{3}$
2	0
-2	$\frac{8}{3}$

$3x - y = -5$

x	y
0	5
$-\frac{5}{3}$	0
-2	-1

The lines in Figure 2 suggest that the graphs intersect at the point $(-1, 2)$. We check this by substituting -1 for x and 2 for y in both equations.

$$2x + 3y = 4$$
$$2(-1) + 3(2) = 4 \quad ?$$
$$4 = 4 \quad \text{True}$$

$$3x - y = -5$$
$$3(-1) - 2 = -5 \quad ?$$
$$-5 = -5 \quad \text{True}$$

Because $(-1, 2)$ satisfies both equations, the solution set of this system is $\{(-1, 2)\}$.

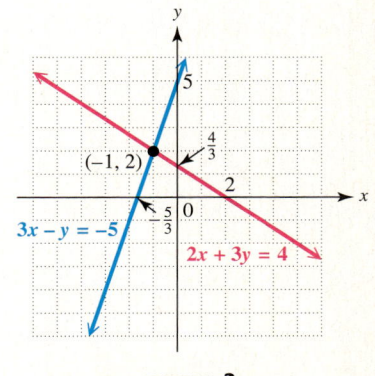

FIGURE 2

Now Try Exercise 13.

To solve a system by graphing, follow these steps.

> **Solving a Linear System by Graphing**
>
> *Step 1* **Graph each equation** of the system on the same coordinate axes.
>
> *Step 2* **Find the coordinates of the point of intersection** of the graphs, if possible. This is the solution of the system.
>
> *Step 3* **Check** the solution in both of the original equations. Then write the solution set.

CAUTION A difficulty with the graphing method of solution is that it may not be possible to determine from the graph the exact coordinates of the point that represents the solution, particularly if these coordinates are not integers. For this reason, algebraic methods of solution are explained later in this chapter. The graphing method does, however, show geometrically how solutions are found and is useful when approximate answers will do.

OBJECTIVE 3 Solve special systems by graphing. Sometimes the graphs of the two equations in a system either do not intersect at all or are the same line, as in the systems of Example 3.

EXAMPLE 3 Solving Special Systems by Graphing

Solve each system by graphing.

(a) $2x + y = 2$
$2x + y = 8$

The graphs of these lines are shown in Figure 3. The two lines are parallel and have no points in common. For such a system, there is no solution; we write the solution set as $\emptyset$.

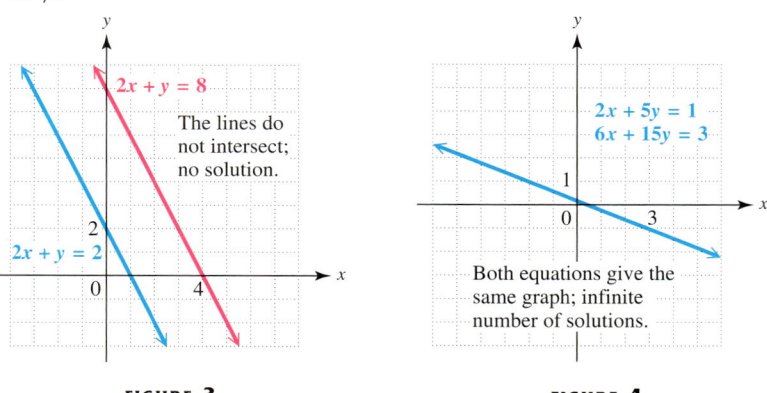

FIGURE 3   FIGURE 4

(b) $2x + 5y = 1$
$6x + 15y = 3$

The graphs of these two equations are the same line. See Figure 4. We can obtain the second equation by multiplying each side of the first equation by 3. In this

SECTION 4.1 Solving Systems of Linear Equations by Graphing   265

case, every point on the line is a solution of the system, and the solution set contains an infinite number of ordered pairs. We write "infinite number of solutions" to indicate this case.

The system in Example 2 has exactly one solution. A system with at least one solution is called a **consistent system.** A system with no solution, such as the one in Example 3(a), is called an **inconsistent system.** The equations in Example 2 are **independent equations** with different graphs. The equations of the system in Example 3(b) have the same graph and are equivalent. Because they are different forms of the same equation, these equations are called **dependent equations.** Examples 2 and 3 show the three cases that may occur when solving a system of equations with two variables.

Possible Types of Solutions

1. The graphs intersect at exactly one point, which gives the (single) ordered pair solution of the system. The **system is consistent** and the **equations are independent.** See Figure 5(a).
2. The graphs are parallel lines, so there is no solution and the solution set is ∅. The **system is inconsistent.** See Figure 5(b).
3. The graphs are the same line. There are an infinite number of solutions. The **equations are dependent.** See Figure 5(c).

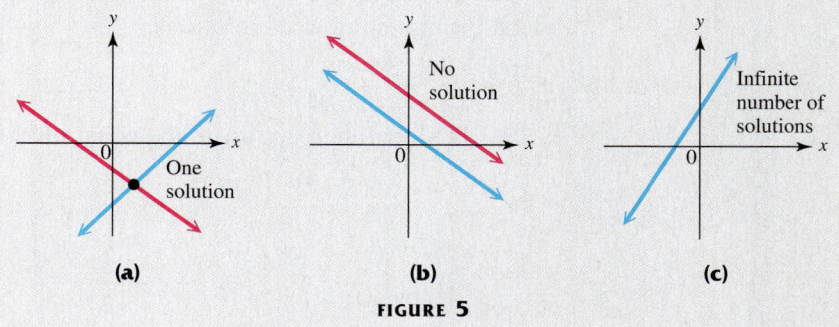

FIGURE 5

Now Try Exercises 23 and 25.

OBJECTIVE 4 Identify special systems without graphing. Example 3 showed that the graphs of an inconsistent system are parallel lines and the graphs of a system of dependent equations are the same line. We can recognize these special kinds of systems without graphing by using slopes.

EXAMPLE 4 Identifying the Three Cases Using Slopes

Describe each system without graphing.

(a) $3x + 2y = 6$
     $-2y = 3x - 5$

Write each equation in slope-intercept form by solving for y.

$$3x + 2y = 6 \qquad\qquad -2y = 3x - 5$$
$$2y = -3x + 6 \qquad\qquad y = -\frac{3}{2}x + \frac{5}{2}$$
$$y = -\frac{3}{2}x + 3$$

Both equations have slope $-\frac{3}{2}$ but they have different y-intercepts, 3 and $\frac{5}{2}$. In Chapter 3 we found that lines with the same slope are parallel, so these equations have graphs that are parallel lines. The system has no solution.

(b) $2x - y = 4$
$x = \frac{y}{2} + 2$

Again, write the equations in slope-intercept form.

$$2x - y = 4 \qquad\qquad x = \frac{y}{2} + 2$$
$$-y = -2x + 4 \qquad\qquad \frac{y}{2} + 2 = x$$
$$y = 2x - 4 \qquad\qquad \frac{y}{2} = x - 2$$
$$\qquad\qquad\qquad\qquad y = 2x - 4$$

The equations are exactly the same; their graphs are the same line. The system has an infinite number of solutions.

(c) $x - 3y = 5$
$2x + y = 8$

In slope-intercept form, the equations are as follows.

$$x - 3y = 5 \qquad\qquad 2x + y = 8$$
$$-3y = -x + 5 \qquad\qquad y = -2x + 8$$
$$y = \frac{1}{3}x - \frac{5}{3}$$

The graphs of these equations are neither parallel nor the same line since the slopes are different. This system has exactly one solution.

Now Try Exercises 31, 33, and 35.

OBJECTIVE 5 Use a graphing calculator to solve a linear system. In the next example we use a graphing calculator to find the solution of the system in Example 1(a).

EXAMPLE 5 Finding the Solution Set of a System with a Graphing Calculator

Solve the system by graphing with a calculator.

$$x + 4y = -8$$
$$3x + 2y = 6$$

SECTION 4.1 Solving Systems of Linear Equations by Graphing 267

To enter the equations in a graphing calculator, first solve each equation for *y*.

$$x + 4y = -8$$
$$4y = -x - 8$$
$$y = -\frac{1}{4}x - 2$$

$$3x + 2y = 6$$
$$2y = -3x + 6$$
$$y = -\frac{3}{2}x + 3$$

The calculator allows us to enter several equations to be graphed at the same time. We designate the first one Y_1 and the second one Y_2. See Figure 6(a). Notice the careful use of parentheses with the fractions. We graph the two equations using a standard window and then use the capability of the calculator to find the coordinates of the point of intersection of the graphs. The display at the bottom of Figure 6(b) indicates that the solution set is $\{(4, -3)\}$.

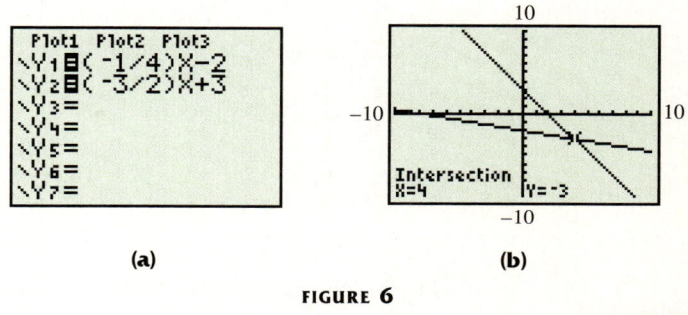

(a) (b)

FIGURE 6

Now Try Exercise 51.

4.1 EXERCISES

For Extra Help

  Student's Solutions Manual

  MyMathLab

  InterAct Math Tutorial Software

  AW Math Tutor Center

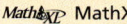 MathXL

  Digital Video Tutor CD 8/Videotape 7

1. Each ordered pair in (a)–(d) is a solution of one of the systems graphed in A–D. Because of the location of the point of intersection, you should be able to determine the correct system for each system. Match each system from A–D with its solution from (a)–(d).

(a) (3, 4)

(b) (−2, 3)

(c) (−4, −1)

(d) (5, −2)

A.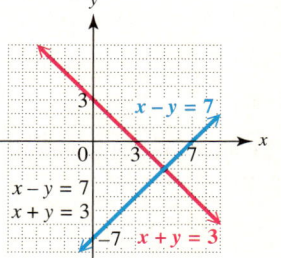

B.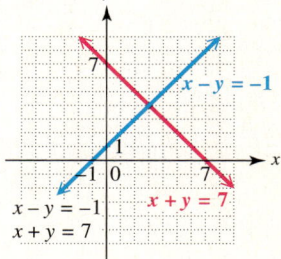

C.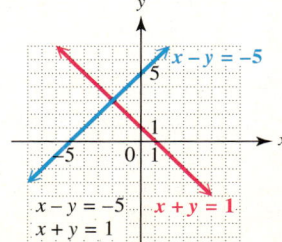

D.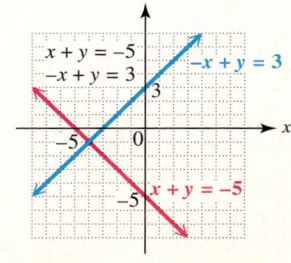

✏️ **2.** When a student was asked to determine whether the ordered pair $(1, -2)$ is a solution of the system

$$x + y = -1$$
$$2x + y = 4,$$

he answered "yes." His reasoning was that the ordered pair satisfies the equation $x + y = -1$ since $1 + (-2) = -1$ is true. Why is the student's answer wrong?

Decide whether the given ordered pair is a solution of the given system. See Example 1.

3. $(2, -3)$
$x + y = -1$
$2x + 5y = 19$

4. $(4, 3)$
$x + 2y = 10$
$3x + 5y = 3$

5. $(-1, -3)$
$3x + 5y = -18$
$4x + 2y = -10$

6. $(-9, -2)$
$2x - 5y = -8$
$3x + 6y = -39$

7. $(7, -2)$
$4x = 26 - y$
$3x = 29 + 4y$

8. $(9, 1)$
$2x = 23 - 5y$
$3x = 24 + 3y$

9. $(6, -8)$
$-2y = x + 10$
$3y = 2x + 30$

10. $(-5, 2)$
$5y = 3x + 20$
$3y = -2x - 4$

11. Which ordered pair could not possibly be a solution of the system graphed? Why is it the only valid choice?

A. $(-4, -4)$ **B.** $(-2, 2)$
C. $(-4, 4)$ **D.** $(-3, 3)$

12. Which ordered pair could possibly be a solution of the system graphed? Why is it the only valid choice?

A. $(2, 0)$ **B.** $(0, 2)$
C. $(-2, 0)$ **D.** $(0, -2)$

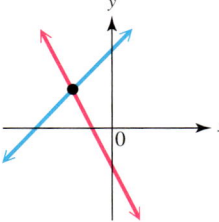
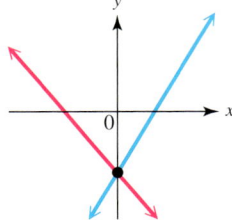

Solve each system of equations by graphing. If the system is inconsistent or the equations are dependent, say so. See Examples 2 and 3.

13. $x - y = 2$
$x + y = 6$

14. $x - y = 3$
$x + y = -1$

15. $x + y = 4$
$y - x = 4$

16. $x + y = -5$
$x - y = 5$

17. $x - 2y = 6$
$x + 2y = 2$

18. $2x - y = 4$
$4x + y = 2$

19. $3x - 2y = -3$
$-3x - y = -6$

20. $2x - y = 4$
$2x + 3y = 12$

21. $2x - 3y = -6$
$y = -3x + 2$

22. $x + 2y = 4$
$2x + 4y = 12$

23. $2x - y = 6$
$4x - 2y = 8$

24. $2x - y = 4$
$4x = 2y + 8$

25. $3x = 5 - y$
$6x + 2y = 10$

26. $-3x + y = -3$
$y = x - 3$

27. $3x - 4y = 24$
$y = -\dfrac{3}{2}x + 3$

28. Solve the system

$$2x + 3y = 6$$
$$x - 3y = 5$$

by graphing. Can you check your solution? Why or why not?

29. Explain one of the drawbacks of solving a system of equations graphically.

30. Explain the three situations that may occur (regarding the number of solutions) when solving a system of two linear equations in two variables by graphing.

Without graphing, answer the following questions for each linear system. See Example 4.

(a) *Is the system inconsistent, are the equations dependent, or neither?*
(b) *Is the graph a pair of intersecting lines, a pair of parallel lines, or one line?*
(c) *Does the system have one solution, no solution, or an infinite number of solutions?*

31. $y - x = -5$
$x + y = 1$

32. $2x + y = 6$
$x - 3y = -4$

33. $x + 2y = 0$
$4y = -2x$

34. $y = 3x$
$y + 3 = 3x$

35. $5x + 4y = 7$
$10x + 8y = 4$

36. $2x + 3y = 12$
$2x - y = 4$

37. $x - 3y = 5$
$2x + y = 8$

38. $2x - y = 4$
$x = \dfrac{1}{2}y + 2$

An application of mathematics in economics deals with **supply and demand.** *Typically, as the price of an item increases, the demand for the item decreases, while the supply increases. (There are exceptions to this, however.) If supply and demand can be described by straight-line equations, the point at which the lines intersect determines the* **equilibrium supply** *and* **equilibrium demand.**

Suppose that an economist has studied the supply and demand for aluminum siding and has concluded that the price per unit, p, and the demand, x, are related by the demand equation $p = 60 - \tfrac{3}{4}x$, while the supply is given by the equation $p = \tfrac{3}{4}x$. The graphs of these two equations are shown in the figure.

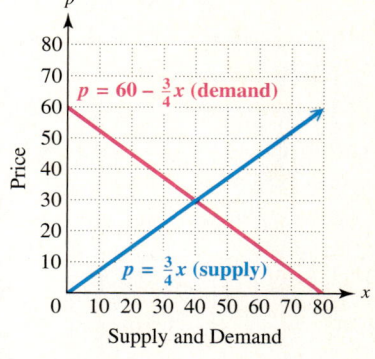
Supply and Demand

Use the graph to answer Exercises 39–42.

39. At what value of x does supply equal demand?

40. At what value of p does supply equal demand?

41. What are the coordinates of the point of intersection of the two lines?

42. When $x > 40$, does demand exceed supply or does supply exceed demand?

The personal computer market share for two manufacturers is shown in the graph.

43. For which years was Hewlett-Packard's share less than Packard Bell, NEC's share?

44. Estimate the year in which market share for Hewlett-Packard and Packard Bell, NEC was the same. About what was this share?

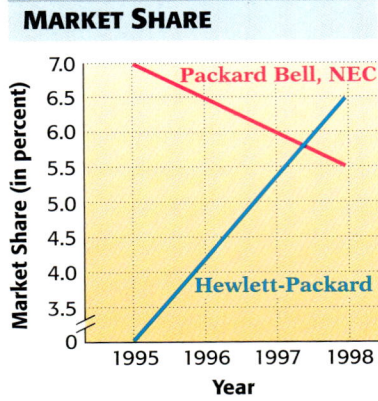

Source: Intelliquest; IDC.

45. The graph shows network share (the percentage of TV sets in use) for the early evening news programs for the three major broadcast networks from 1986 through 2000.

(a) Between what years did the ABC early evening news dominate?

(b) During what year did ABC's dominance end? Which network equaled ABC's share that year? What was that share?

(c) During what years did ABC and CBS have equal network share? What was the share for each of these years?

(d) Which networks most recently had equal share? Write their share as an ordered pair of the form (year, share).

(e) Describe the general trend in viewership for the three major networks during these years.

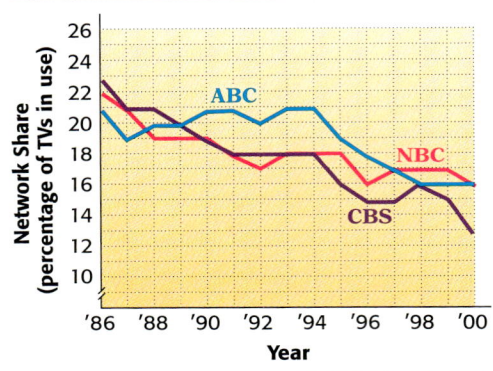

Source: Nielsen Media Research.

46. The graph shows how the production of vinyl LPs, audiocassettes, and compact discs (CDs) changed over the years from 1986 through 1998.

(a) In what year did cassette production and CD production reach equal levels? What was that level?

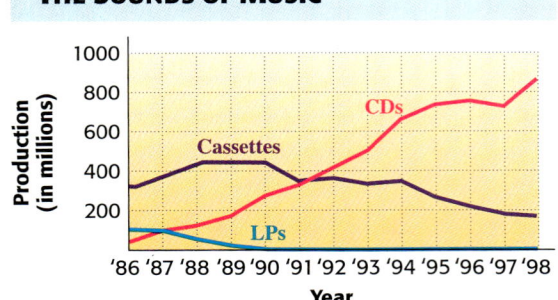

Source: Recording Industry Association of America.

(b) Express the point of intersection of the graphs of LP production and CD production as an ordered pair of the form (year, production level).
(c) Between what years did cassette production first stabilize and remain fairly constant?
(d) Describe the trend in CD production from 1986 through 1998. If a straight line were used to approximate its graph, would the line have positive, negative, or 0 slope?
(e) If a straight line were used to approximate the graph of cassette production from 1990 through 1998, would the line have positive, negative, or 0 slope? Explain.

TECHNOLOGY INSIGHTS (EXERCISES 47–50)

Match the graphing calculator screens in choices A–D with the appropriate system in Exercises 47–50. See Example 5.

A.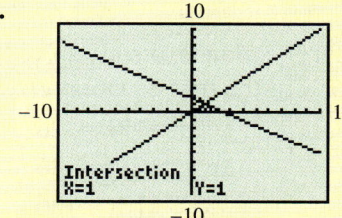

B.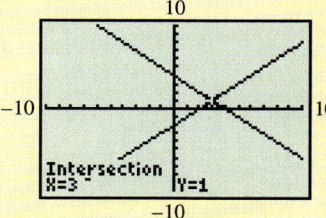

C.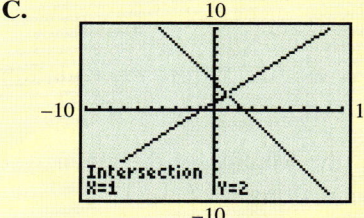

D.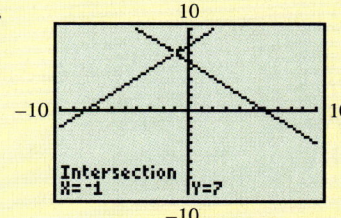

47. $x + y = 4$
$x - y = 2$

48. $x + y = 6$
$x - y = -8$

49. $2x + 3y = 5$
$x - y = 0$

50. $3x + 2y = 7$
$-x + y = 1$

Use a graphing calculator to solve each system. See Example 5.

51. $3x + y = 2$
$2x - y = -7$

52. $x + 2y = -2$
$2x - y = 11$

53. $8x + 4y = 0$
$4x - 2y = 2$

54. $3x + 3y = 0$
$4x + 2y = 3$

272 CHAPTER 4 Systems of Linear Equations and Inequalities

4.2 Solving Systems of Linear Equations by Substitution

OBJECTIVES

1. Solve linear systems by substitution.
2. Solve special systems by substitution.
3. Solve linear systems with fractions.

OBJECTIVE 1 Solve linear systems by substitution. Graphing to solve a system of equations has a serious drawback: It is difficult to accurately find a solution such as $\left(\frac{1}{3}, -\frac{5}{6}\right)$ from a graph. One algebraic method for solving a system of equations is the **substitution method.** This method is particularly useful for solving systems where one equation is already solved, or can be solved quickly, for one of the variables.

EXAMPLE 1 Using the Substitution Method

Solve the system by the substitution method.

$$3x + 5y = 26$$
$$y = 2x$$

The second equation is already solved for y. This equation says that $y = 2x$. Substituting $2x$ for y in the first equation gives

$$3x + 5y = 26$$
$$3x + 5(2x) = 26 \quad \text{Let } y = 2x.$$
$$3x + 10x = 26 \quad \text{Multiply.}$$
$$13x = 26 \quad \text{Combine like terms.}$$
$$x = 2. \quad \text{Divide by 13.}$$

Because $x = 2$, we find y from the equation $y = 2x$ by substituting 2 for x.

$$y = 2(2) = 4 \quad \text{Let } x = 2.$$

We check that the solution of the given system is $(2, 4)$ by substituting 2 for x and 4 for y in *both* equations.

Check:
$$3x + 5y = 26 \qquad\qquad y = 2x$$
$$3(2) + 5(4) = 26 \enspace ? \qquad 4 = 2(2) \enspace ?$$
$$6 + 20 = 26 \enspace ? \qquad\qquad 4 = 4 \quad \text{True}$$
$$26 = 26 \quad \text{True}$$

Since $(2, 4)$ satisfies both equations, the solution set is $\{(2, 4)\}$.

Now Try Exercise 3.

EXAMPLE 2 Using the Substitution Method

Solve the system

$$2x + 5y = 7$$
$$x = -1 - y.$$

The second equation gives x in terms of y. Substitute $-1 - y$ for x in the first equation.

$$2x + 5y = 7$$
$$2(-1 - y) + 5y = 7 \quad \text{Let } x = -1 - y.$$
$$-2 - 2y + 5y = 7 \quad \text{Distributive property}$$

SECTION 4.2 Solving Systems of Linear Equations by Substitution 273

$$-2 + 3y = 7 \quad \text{Combine like terms.}$$
$$3y = 9 \quad \text{Add 2.}$$
$$y = 3 \quad \text{Divide by 3.}$$

To find x, substitute 3 for y in the equation $x = -1 - y$ to get

$$x = -1 - 3 = -4.$$

Check that the solution set of the given system is $\{(-4, 3)\}$.

Now Try Exercise 5.

> **CAUTION** Be careful when you write the ordered pair solution of a system. Even though we found y first in Example 2, the x-coordinate is *always* written first in the ordered pair.

To solve a system by substitution, follow these steps.

Solving a Linear System by Substitution

Step 1 **Solve one equation for either variable.** If one of the variables has coefficient 1 or -1, choose it, since the substitution method is usually easier this way.

Step 2 **Substitute** for that variable in the other equation. The result should be an equation with just one variable.

Step 3 **Solve** the equation from Step 2.

Step 4 **Substitute** the result from Step 3 into the equation from Step 1 to find the value of the other variable.

Step 5 **Check** the solution in both of the original equations. Then write the solution set.

EXAMPLE 3 Using the Substitution Method

Use substitution to solve the system

$$2x = 4 - y \quad (1)$$
$$5x + 3y = 10. \quad (2)$$

Step 1 We must solve one of the equations for either x or y. Because the coefficient of y in equation (1) is -1, we avoid fractions by solving this equation for y.

$$2x = 4 - y \quad (1)$$
$$2x - 4 = -y \quad \text{Subtract 4.}$$
$$-2x + 4 = y \quad \text{Multiply by } -1.$$

Step 2 Now substitute $-2x + 4$ for y in equation (2).

$$5x + 3y = 10 \quad (2)$$
$$5x + 3(-2x + 4) = 10 \quad \text{Let } y = -2x + 4.$$

Step 3 Solve the equation from Step 2.

$$5x - 6x + 12 = 10 \quad \text{Distributive property}$$
$$-x + 12 = 10 \quad \text{Combine like terms.}$$
$$-x = -2 \quad \text{Subtract 12.}$$
$$x = 2 \quad \text{Multiply by } -1.$$

Step 4 Since $y = -2x + 4$ and $x = 2$,
$$y = -2(2) + 4 = 0,$$
and the solution is $(2, 0)$.

Step 5 Check:

$$2x = 4 - y \quad (1) \qquad\qquad 5x + 3y = 10 \quad (2)$$
$$2(2) = 4 - 0 \quad ? \qquad\qquad 5(2) + 3(0) = 10 \quad ?$$
$$4 = 4 \quad \text{True} \qquad\qquad 10 = 10 \quad \text{True}$$

Since both results are true, the solution set of the system is $\{(2, 0)\}$.

Now Try Exercise 7.

OBJECTIVE 2 Solve special systems by substitution. In the previous section we solved inconsistent systems with graphs that are parallel lines and systems of dependent equations with graphs that are the same line. We can also solve these special systems with the substitution method.

EXAMPLE 4 Solving an Inconsistent System by Substitution

Use substitution to solve the system

$$x = 5 - 2y \quad (1)$$
$$2x + 4y = 6. \quad (2)$$

Substitute $5 - 2y$ for x in equation (2).

$$2x + 4y = 6 \quad (2)$$
$$2(5 - 2y) + 4y = 6 \quad \text{Let } x = 5 - 2y.$$
$$10 - 4y + 4y = 6 \quad \text{Distributive property}$$
$$10 = 6 \quad \text{False}$$

This false result means that the equations in the system have graphs that are parallel lines. The system is inconsistent and has no solution, so the solution set is $\emptyset$. See Figure 7.

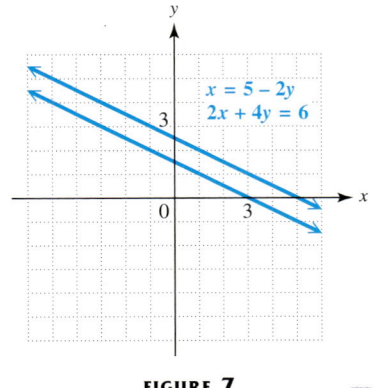

FIGURE 7

Now Try Exercise 17.

CAUTION It is a common error to give "false" as the solution of an inconsistent system. The correct response is ∅.

EXAMPLE 5 Solving a System with Dependent Equations by Substitution

Solve the system by the substitution method.

$$3x - y = 4 \quad (1)$$
$$-9x + 3y = -12 \quad (2)$$

Begin by solving equation (1) for y to get $y = 3x - 4$. Substitute $3x - 4$ for y in equation (2) and solve the resulting equation.

$$-9x + 3y = -12 \quad (2)$$
$$-9x + 3(3x - 4) = -12 \quad \text{Let } y = 3x - 4.$$
$$-9x + 9x - 12 = -12 \quad \text{Distributive property}$$
$$0 = 0 \quad \text{Add 12; combine like terms.}$$

This true result means that every solution of one equation is also a solution of the other, so the system has an infinite number of solutions: all the ordered pairs corresponding to points that lie on the common graph. A graph of the equations of this system is shown in Figure 8.

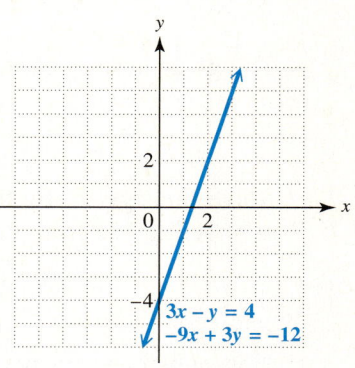

FIGURE 8

Now Try Exercise 19.

CAUTION It is a common error to give "true" as the solution of a system of dependent equations. The correct response is "infinite number of solutions."

OBJECTIVE 3 Solve linear systems with fractions. When a system includes an equation with fractions as coefficients, eliminate the fractions by multiplying each side of the equation by a common denominator. Then solve the resulting system.

EXAMPLE 6 Using the Substitution Method with Fractions as Coefficients

Solve the system by the substitution method.

$$3x + \frac{1}{4}y = 2 \quad (1)$$

$$\frac{1}{2}x + \frac{3}{4}y = -\frac{5}{2} \quad (2)$$

Clear equation (1) of fractions by multiplying each side by 4.

$$4\left(3x + \frac{1}{4}y\right) = 4(2) \quad \text{Multiply by 4.}$$

$$4(3x) + 4\left(\frac{1}{4}y\right) = 4(2) \quad \text{Distributive property}$$

$$12x + y = 8 \quad (3)$$

Now clear equation (2) of fractions by multiplying each side by the common denominator 4.

$$4\left(\frac{1}{2}x + \frac{3}{4}y\right) = 4\left(-\frac{5}{2}\right) \quad \text{Multiply by 4.}$$

$$4\left(\frac{1}{2}x\right) + 4\left(\frac{3}{4}y\right) = 4\left(-\frac{5}{2}\right) \quad \text{Distributive property}$$

$$2x + 3y = -10 \quad (4)$$

The given system of equations has been simplified to the equivalent system

$$12x + y = 8 \quad (3)$$
$$2x + 3y = -10. \quad (4)$$

Solve this system by the substitution method. Equation (3) can be solved for y by subtracting $12x$ from each side.

$$12x + y = 8 \quad (3)$$
$$y = -12x + 8 \quad \text{Subtract } 12x.$$

Now substitute this result for y in equation (4).

$$2x + 3y = -10 \quad (4)$$
$$2x + 3(-12x + 8) = -10 \quad \text{Let } y = -12x + 8.$$
$$2x - 36x + 24 = -10 \quad \text{Distributive property}$$
$$-34x = -34 \quad \text{Combine like terms; subtract 24.}$$
$$x = 1 \quad \text{Divide by } -34.$$

Substitute 1 for x in $y = -12x + 8$ to get

$$y = -12(1) + 8 = -4.$$

Check by substituting 1 for x and -4 for y in both of the original equations. The solution set is $\{(1, -4)\}$.

Now Try Exercise 25.

4.2 EXERCISES

For Extra Help

 Student's Solutions Manual

 MyMathLab

 InterAct Math Tutorial Software

 AW Math Tutor Center

 MathXL

Digital Video Tutor CD 8/Videotape 7

1. A student solves the system
$$5x - y = 15$$
$$7x + y = 21$$
and finds that $x = 3$, which is the correct value for x. The student gives the solution set as $\{3\}$. Is this correct? Explain.

2. A student solves the system
$$x + y = 4$$
$$2x + 2y = 8$$
and obtains the equation $0 = 0$. The student gives the solution set as $\{(0, 0)\}$. Is this correct? Explain.

Solve each system by the substitution method. Check each solution. See Examples 1–5.

3. $x + y = 12$
   $y = 3x$

4. $x + 3y = -28$
   $y = -5x$

5. $3x + 2y = 27$
   $x = y + 4$

6. $4x + 3y = -5$
   $x = y - 3$

7. $3x + 5y = 25$
   $x - 2y = -10$

8. $5x + 2y = -15$
   $2x - y = -6$

9. $3x + 4 = -y$
   $2x + y = 0$

10. $2x - 5 = -y$
    $x + 3y = 0$

11. $7x + 4y = 13$
    $x + y = 1$

12. $3x - 2y = 19$
    $x + y = 8$

13. $3x - y = 5$
    $y = 3x - 5$

14. $4x - y = -3$
    $y = 4x + 3$

15. $2x + y = 0$
    $4x - 2y = 2$

16. $x + y = 0$
    $4x + 2y = 3$

17. $2x + 8y = 3$
    $x = 8 - 4y$

18. $2x + 10y = 3$
    $x = 1 - 5y$

19. $2x = -12 + y$
    $2y = 4x + 24$

20. $3x = 7 - y$
    $2y = 14 - 6x$

21. When you use the substitution method, how can you tell that a system has
    (a) no solution?
    (b) an infinite number of solutions?

22. Solve each system.
    (a) $5x - 4y = 7$
        $x = 3$
    (b) $5x - 4y = 7$
        $y = -3$

    Why are these systems easier to solve than the examples in this section?

Solve each system by the substitution method. First clear all fractions. Check each solution. See Example 6.

23. $\frac{1}{2}x + \frac{1}{3}y = 3$
    $y = 3x$

24. $\frac{1}{4}x - \frac{1}{5}y = 9$
    $y = 5x$

25. $\frac{1}{2}x + \frac{1}{3}y = -\frac{1}{3}$
    $\frac{1}{2}x + 2y = -7$

26. $\frac{1}{6}x + \frac{1}{6}y = 1$
    $-\frac{1}{2}x - \frac{1}{3}y = -5$

27. $\frac{x}{5} + 2y = \frac{8}{5}$
    $\frac{3x}{5} + \frac{y}{2} = -\frac{7}{10}$

28. $\frac{x}{2} + \frac{y}{3} = \frac{7}{6}$
    $\frac{x}{4} - \frac{3y}{2} = \frac{9}{4}$

29. $\dfrac{x}{5} + y = \dfrac{6}{5}$
$\dfrac{x}{10} + \dfrac{y}{3} = \dfrac{5}{6}$

30. $\dfrac{1}{2}x - \dfrac{1}{8}y = -\dfrac{1}{4}$
$-4x + y = 2$

31. $\dfrac{1}{6}x + \dfrac{1}{3}y = 8$
$\dfrac{1}{4}x + \dfrac{1}{2}y = 12$

32. One student solved the system

$$\dfrac{1}{3}x - \dfrac{1}{2}y = 7$$
$$\dfrac{1}{6}x + \dfrac{1}{3}y = 0$$

and wrote as his answer "$x = 12$," while another solved it and wrote as her answer "$y = -6$." Who, if either, was correct? Why?

RELATING CONCEPTS (EXERCISES 33–36)

For Individual or Group Work

A system of linear equations can be used to model the cost and the revenue of a business. **Work Exercises 33–36 in order.**

33. Suppose that you start a business manufacturing and selling bicycles, and it costs you $5000 to get started. You determine that each bicycle will cost $400 to manufacture. Explain why the linear equation $y_1 = 400x + 5000$ gives your *total* cost to manufacture x bicycles (y_1 in dollars).

34. You decide to sell each bike for $600. What expression in x represents the revenue you will take in if you sell x bikes? Write an equation using y_2 to express your revenue when you sell x bikes (y_2 in dollars).

35. Form a system from the two equations in Exercises 33 and 34 and then solve the system.

36. The value of x from Exercise 35 is the number of bikes it takes to *break even*. Fill in the blanks: When _____ bikes are sold, the break-even point is reached. At that point, you have spent _____ dollars and taken in _____ dollars.

Solve each system by substitution. Then graph both lines in the standard viewing window of a graphing calculator and use the intersection feature to support your answer. See Example 5 in Section 4.1. (In Exercises 41 and 42, you will need to solve each equation for y first before graphing.)

37. $y = 6 - x$
$y = 2x$

38. $y = 4x - 4$
$y = -3x - 11$

39. $y = -\dfrac{4}{3}x + \dfrac{19}{3}$
$y = \dfrac{15}{2}x - \dfrac{5}{2}$

40. $y = -\dfrac{15}{2}x + 10$
$y = \dfrac{25}{3}x - \dfrac{65}{3}$

41. $4x + 5y = 5$
$2x + 3y = 1$

42. $6x + 5y = 13$
$3x + 3y = 4$

43. If the point of intersection does not appear on your screen when solving a linear system using a graphing calculator, how can you find the point of intersection?

44. Suppose that you were asked to solve the system

$$y = 1.73x + 5.28$$
$$y = -2.94x - 3.85.$$

Why would it probably be easier to solve this system using a graphing calculator than using the substitution method?

4.3 Solving Systems of Linear Equations by Elimination

OBJECTIVES

1. Solve linear systems by elimination.
2. Multiply when using the elimination method.
3. Use an alternative method to find the second value in a solution.
4. Use the elimination method to solve special systems.

OBJECTIVE 1 Solve linear systems by elimination. An algebraic method that depends on the addition property of equality can also be used to solve systems. As mentioned earlier, adding the same quantity to each side of an equation results in equal sums.

$$\text{If} \quad A = B, \quad \text{then} \quad A + C = B + C.$$

We can take this addition a step further. Adding *equal* quantities, rather than the *same* quantity, to each side of an equation also results in equal sums.

$$\text{If} \quad A = B \quad \text{and} \quad C = D, \quad \text{then} \quad A + C = B + D.$$

Using the addition property to solve systems is called the **elimination method.** When using this method, the idea is to *eliminate* one of the variables. To do this, one pair of variable terms in the two equations must have coefficients that are opposites.

EXAMPLE 1 Using the Elimination Method

Use the elimination method to solve the system

$$x + y = 5$$
$$x - y = 3.$$

Each equation in this system is a statement of equality, so the sum of the right sides equals the sum of the left sides. Adding in this way gives

$$(x + y) + (x - y) = 5 + 3.$$

Combine terms and simplify to get

$$2x = 8$$
$$x = 4. \quad \text{Divide by 2.}$$

Notice that y has been eliminated. The result, $x = 4$, gives the x-value of the solution of the given system. To find the y-value of the solution, substitute 4 for x in either of the two equations of the system. Choosing the first equation, $x + y = 5$, gives

$$x + y = 5$$
$$4 + y = 5 \quad \text{Let } x = 4.$$
$$y = 1. \quad \text{Subtract 4.}$$

Check the solution, (4, 1), by substituting 4 for *x* and 1 for *y* in both equations of the given system.

$$x + y = 5 \qquad\qquad x - y = 3$$
$$4 + 1 = 5 \text{ ?} \qquad\qquad 4 - 1 = 3 \text{ ?}$$
$$5 = 5 \quad \text{True} \qquad\qquad 3 = 3 \quad \text{True}$$

Since both results are true, the solution set of the system is {(4, 1)}.

Now Try Exercise 5.

CAUTION A system is not completely solved until values for *both x* and *y* are found. Do not stop after finding the value of only one variable. Remember to write the solution set as a set containing an ordered pair.

In general, use the following steps to solve a linear system of equations by the elimination method.

Solving a Linear System by Elimination

Step 1 **Write both equations in standard form** $Ax + By = C$.

Step 2 **Transform so that the coefficients of one pair of variable terms are opposites.** Multiply one or both equations by appropriate numbers so that the sum of the coefficients of either the *x*- or *y*-terms is 0.

Step 3 **Add** the new equations to eliminate a variable. The sum should be an equation with just one variable.

Step 4 **Solve** the equation from Step 3 for the remaining variable.

Step 5 **Substitute** the result from Step 4 into either of the original equations and solve for the other variable.

Step 6 **Check** the solution in both of the original equations. Then write the solution set.

It does not matter which variable is eliminated first. Usually we choose the one that is more convenient to work with.

EXAMPLE 2 Using the Elimination Method

Solve the system

$$y + 11 = 2x$$
$$5x = y + 26.$$

Step 1 Rewrite both equations in standard form $Ax - By = C$ to get the system

$$-2x + y = -11 \qquad \text{Subtract } 2x \text{ and } 11.$$
$$5x - y = 26. \qquad \text{Subtract } y.$$

Step 2 Because the coefficients of *y* are 1 and −1, adding will eliminate *y*. It is not necessary to multiply either equation by a number.

Step 3 Add the two equations. This time we use vertical addition.
$$-2x + y = -11$$
$$\underline{5x - y = 26}$$
$$3x = 15 \qquad \text{Add in columns.}$$

Step 4 Solve the equation.
$$3x = 15$$
$$x = 5 \qquad \text{Divide by 3.}$$

Step 5 Find the value of y by substituting 5 for x in either of the original equations. Choosing the first gives
$$y + 11 = 2x$$
$$y + 11 = 2(5)$$
$$y + 11 = 10 \qquad \text{Let } x = 5.$$
$$y = -1. \qquad \text{Subtract 11.}$$

Step 6 Check the solution by substituting $x = 5$ and $y = -1$ in both of the original equations.

Check:
$$y + 11 = 2x \qquad\qquad 5x = y + 26$$
$$(-1) + 11 = 2(5) \ ? \qquad 5(5) = -1 + 26 \ ?$$
$$10 = 10 \quad \text{True} \qquad\quad 25 = 25 \quad \text{True}$$

Since $(5, -1)$ is a solution of *both* equations, the solution set is $\{(5, -1)\}$.

Now Try Exercise 9.

OBJECTIVE 2 Multiply when using the elimination method. In both of the preceding examples, a variable was eliminated by adding the equations. Sometimes we need to multiply each side of one or both equations in a system by some number before adding will eliminate a variable.

EXAMPLE 3 Multiplying Both Equations When Using the Elimination Method

Solve the system
$$2x + 3y = -15 \qquad (1)$$
$$5x + 2y = 1. \qquad (2)$$

Adding the two equations gives $7x + 5y = -14$, which does not eliminate either variable. However, we can multiply each equation by a suitable number so that the coefficients of one of the two variables are opposites. For example, to eliminate x, we multiply each side of equation (1) by 5, and each side of equation (2) by -2.

$$10x + 15y = -75 \qquad \text{Multiply equation (1) by 5.}$$
$$\underline{-10x - 4y = -2} \qquad \text{Multiply equation (2) by } -2.$$
$$11y = -77 \qquad \text{Add.}$$
$$y = -7 \qquad \text{Divide by 11.}$$

Substituting -7 for y in either equation (1) or (2) gives $x = 3$. Check that the solution set of the system is $\{(3, -7)\}$.

Now Try Exercise 19.

CAUTION When using the elimination method, remember to *multiply both sides* of an equation by the same nonzero number.

OBJECTIVE 3 Use an alternative method to find the second value in a solution. Sometimes it is easier to find the value of the second variable in a solution by using the elimination method twice. The next example shows this approach.

EXAMPLE 4 Finding the Second Value Using an Alternative Method

Solve the system

$$4x = 9 - 3y \quad (1)$$
$$5x - 2y = 8. \quad (2)$$

Rearrange the terms in equation (1) so that like terms are aligned in columns. Add $3y$ to each side to get the system

$$4x + 3y = 9 \quad (3)$$
$$5x - 2y = 8. \quad (2)$$

One way to proceed is to eliminate y by multiplying each side of equation (3) by 2 and each side of equation (2) by 3, and then adding.

$$\begin{aligned} 8x + 6y &= 18 & \text{Multiply equation (3) by 2.}\\ 15x - 6y &= 24 & \text{Multiply equation (2) by 3.}\\ \hline 23x &= 42 & \text{Add.}\\ x &= \frac{42}{23} & \text{Divide by 23.} \end{aligned}$$

Substituting $\frac{42}{23}$ for x in one of the given equations would give y, but the arithmetic involved would be messy. Instead, solve for y by starting again with the original equations and eliminating x. Multiply each side of equation (3) by 5 and each side of equation (2) by -4, and then add.

$$\begin{aligned} 20x + 15y &= 45 & \text{Multiply equation (3) by 5.}\\ -20x + 8y &= -32 & \text{Multiply equation (2) by } -4.\\ \hline 23y &= 13 & \text{Add.}\\ y &= \frac{13}{23} & \text{Divide by 23.} \end{aligned}$$

Check that the solution set is $\left\{\left(\frac{42}{23}, \frac{13}{23}\right)\right\}$.

Now Try Exercise 27.

NOTE When the value of the first variable is a fraction, the method used in Example 4 helps avoid arithmetic errors. Of course, this method could be used to solve any system of equations.

OBJECTIVE 4 Use the elimination method to solve special systems. The next example shows the elimination method when a system is inconsistent or the equations of the system are dependent. To contrast the elimination method with the substitution method, in part (b) we use the same system solved in Example 5 of the previous section.

EXAMPLE 5 Using the Elimination Method for an Inconsistent System or Dependent Equations

Solve each system by the elimination method.

(a) $2x + 4y = 5$
$4x + 8y = -9$

Multiply each side of $2x + 4y = 5$ by -2; then add to $4x + 8y = -9$.

$$-4x - 8y = -10$$
$$4x + 8y = -9$$
$$0 = -19 \quad \text{False}$$

The false statement $0 = -19$ indicates that the given system has solution set $\emptyset$.

(b) $3x - y = 4$
$-9x + 3y = -12$

Multiply each side of the first equation by 3; then add the two equations.

$$9x - 3y = 12$$
$$-9x + 3y = -12$$
$$0 = 0 \quad \text{True}$$

A true statement occurs when the equations are equivalent. As before, this result indicates that every solution of one equation is also a solution of the other; there are an infinite number of solutions.

Now Try Exercises 35 and 37.

4.3 EXERCISES

For Extra Help

 Student's Solutions Manual

 MyMathLab

 InterAct Math Tutorial Software

 AW Math Tutor Center

MathXL MathXL

 Digital Video Tutor CD 8/Videotape 7

Answer true *or* false *for each statement. If false, tell why.*

1. To eliminate the *y*-terms in the system

$$2x + 12y = 7$$
$$3x + 4y = 1,$$

we should multiply the bottom equation by 3 and then add.

2. The ordered pair $(0, 0)$ *must* be a solution of a system of the form

$$Ax + By = 0$$
$$Cx + Dy = 0.$$

3. The system

$$x + y = 1$$
$$x + y = 2$$

has $\emptyset$ as its solution set.

4. The ordered pair (4, −5) cannot be a solution of a system that contains the equation $5x - 4y = 0$.

Solve each system by the elimination method. Check each solution. See Examples 1 and 2.

5. $x - y = -2$
   $x + y = 10$

6. $x + y = 10$
   $x - y = -6$

7. $2x + y = -5$
   $x - y = 2$

8. $2x + y = -15$
   $-x - y = 10$

9. $2y = -3x$
   $-3x - y = 3$

10. $5x = y + 5$
    $-5x + 2y = 0$

11. $6x - y = -1$
    $5y = 17 + 6x$

12. $y = 9 - 6x$
    $-6x + 3y = 15$

Solve each system by the elimination method. (Hint: In Exercises 33 and 34, first clear all fractions.) Check each solution. See Examples 3–5.*

13. $2x - y = 12$
    $3x + 2y = -3$

14. $x + y = 3$
    $-3x + 2y = -19$

15. $x + 4y = 16$
    $3x + 5y = 20$

16. $2x + y = 8$
    $5x - 2y = -16$

17. $2x - 8y = 0$
    $4x + 5y = 0$

18. $3x - 15y = 0$
    $6x + 10y = 0$

19. $3x + 3y = 33$
    $5x - 2y = 27$

20. $4x - 3y = -19$
    $3x + 2y = 24$

21. $5x + 4y = 12$
    $3x + 5y = 15$

22. $2x + 3y = 21$
    $5x - 2y = -14$

23. $5x - 4y = 15$
    $-3x + 6y = -9$

24. $4x + 5y = -16$
    $5x - 6y = -20$

25. $6x - 2y = -21$
    $-3x + 4y = 36$

26. $6x - 2y = -21$
    $3x + 4y = 36$

27. $3x - 7y = 1$
    $-5x + 4y = 4$

28. $-4x + 3y = 2$
    $5x - 2y = -3$

29. $2x + 3y = 0$
    $4x + 12 = 9y$

30. $-4x + 3y = 2$
    $5x + 3 = -2y$

31. $24x + 12y = -7$
    $16x - 17 = 18y$

32. $9x + 4y = -3$
    $6x + 7 = -6y$

33. $3x = 3 + 2y$
    $-\dfrac{4}{3}x + y = \dfrac{1}{3}$

34. $3x = 27 + 2y$
    $x - \dfrac{7}{2}y = -25$

35. $-x + 3y = 4$
    $-2x + 6y = 8$

36. $6x - 2y = 24$
    $-3x + y = -12$

37. $5x - 2y = 3$
    $10x - 4y = 5$

38. $3x - 5y = 1$
    $6x - 10y = 4$

RELATING CONCEPTS (EXERCISES 39–44)

For Individual or Group Work

The graph on the next page shows movie attendance from 1991 through 1999. In 1991, attendance was 1141 million, as represented by the point P(1991, 1141). In 1999, attendance was 1465 million, as represented by the point Q(1999, 1465). We can find an equation of line segment PQ using a system of equations and then use the equation to approximate the attendance in any of the years between 1991 and 1999. **Work Exercises 39–44 in order.**

---

*The authors thank Mitchel Levy of Broward Community College for his suggestions for this group of exercises.

MOVIE BOX OFFICE ATTENDANCE/ADMISSIONS

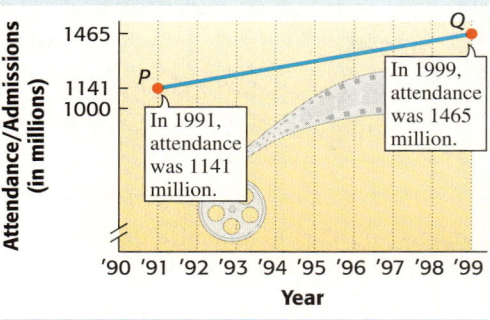

Source: Motion Picture Association of America.

39. The line segment has an equation that can be written in the form $y = ax + b$. Using the coordinates of point P with $x = 1991$ and $y = 1141$, write an equation in the variables a and b.

40. Using the coordinates of point Q with $x = 1999$ and $y = 1465$, write a second equation in the variables a and b.

41. Write the system of equations formed from the two equations in Exercises 39 and 40, and solve the system using the elimination method.

42. What is the equation of the segment PQ?

43. Let $x = 1998$ in the equation of Exercise 42, and solve for y. How does the result compare with the actual figure of 1481 million?

44. The actual data points for the years 1991 through 1999 do not lie in a perfectly straight line. Explain the pitfalls of relying too heavily on using the equation in Exercise 42 to approximate attendance.

SUMMARY EXERCISES ON SOLVING SYSTEMS OF LINEAR EQUATIONS

The exercises in this summary include a variety of problems on solving systems of linear equations. Since we do not usually specify the method of solution, use the following guidelines to help you decide whether to use substitution or elimination.

Choosing a Method When Solving a System of Linear Equations

1. If one of the equations of the system is already solved for one of the variables, as in the systems

$$3x + 4y = 9 \qquad \text{or} \qquad -5x + 3y = 9$$
$$y = 2x - 6 \qquad\qquad\qquad x = 3y - 7,$$

the substitution method is the better choice.

2. If both equations are in standard $Ax + By = C$ form, as in

$$4x - 11y = 3$$
$$-2x + 3y = 4,$$

and none of the variables has coefficient -1 or 1, the elimination method is the better choice.

(continued)

3. If one or both of the equations are in standard form and the coefficient of one of the variables is -1 or 1, as in the systems

$$3x + y = -2 \qquad \text{or} \qquad -x + 3y = -4$$
$$-5x + 2y = 4 \qquad \qquad 3x - 2y = 8,$$

use the elimination method, or solve for the variable with coefficient -1 or 1 and then use the substitution method.

Use the information in the preceding box to solve each problem.

1. Assuming you want to minimize the amount of work required, tell whether you would use the substitution or elimination method to solve each system. Explain your answers. *Do not actually solve.*

   (a) $3x + 5y = 69$ (b) $3x + y = -7$ (c) $3x - 2y = 0$
          $y = 4x$ $x - y = -5$ $9x + 8y = 7$

2. Which one of the following systems would be easier to solve using the substitution method? Why?

   $$5x - 3y = 7 \qquad\qquad 7x + 2y = 4$$
   $$2x + 8y = 3 \qquad\qquad y = -3x + 1$$

In Exercises 3 and 4, (a) solve the system by the elimination method, (b) solve the system by the substitution method, and (c) tell which method you prefer for that particular system and why.

3. $4x - 3y = -8$ 4. $2x + 5y = 0$
    $x + 3y = 13$ $x = -3y + 1$

Solve each system by the method of your choice. (For Exercises 5–7, see your answers for Exercise 1.)

5. $3x + 5y = 69$ 6. $3x + y = -7$ 7. $3x - 2y = 0$
    $y = 4x$ $x - y = -5$ $9x + 8y = 7$

8. $x + y = 7$ 9. $6x + 7y = 4$ 10. $6x - y = 5$
    $x = -3 - y$ $5x + 8y = -1$ $y = 11x$

11. $4x - 6y = 10$ 12. $3x - 5y = 7$ 13. $5x = 7 + 2y$
     $-10x + 15y = -25$ $2x + 3y = 30$ $5y = 5 - 3x$

14. $4x + 3y = 1$ 15. $2x - 3y = 7$ 16. $2x + 3y = 10$
     $3x + 2y = 2$ $-4x + 6y = 14$ $-3x + y = 18$

17. $2x + 5y = 4$ 18. $x - 3y = 7$
     $x + y = -1$ $4x + y = 5$

Solve each system by any method. First clear all fractions.

19. $\dfrac{1}{3}x - \dfrac{1}{2}y = \dfrac{1}{6}$ 20. $\dfrac{1}{5}x + \dfrac{2}{3}y = -\dfrac{8}{5}$ 21. $\dfrac{1}{6}x + \dfrac{1}{6}y = 2$
      $3x - 2y = 9$ $3x - y = 9$ $-\dfrac{1}{2}x - \dfrac{1}{3}y = -8$

22. $\dfrac{x}{2} - \dfrac{y}{3} = 9$ 23. $\dfrac{x}{3} - \dfrac{3y}{4} = -\dfrac{1}{2}$ 24. $\dfrac{x}{5} + 2y = \dfrac{16}{5}$
      $\dfrac{x}{5} - \dfrac{y}{4} = 5$ $\dfrac{x}{6} + \dfrac{y}{8} = \dfrac{3}{4}$ $\dfrac{3x}{5} + \dfrac{y}{2} = -\dfrac{7}{5}$

4.4 Applications of Linear Systems

OBJECTIVES

1. Solve problems about unknown numbers.
2. Solve problems about quantities and their costs.
3. Solve problems about mixtures.
4. Solve problems about distance, rate (or speed), and time.

Recall from Chapter 2 the six-step method for solving applied problems. We modify those steps slightly to allow for two variables and two equations.

Solving an Applied Problem with Two Variables

Step 1 **Read** the problem carefully until you understand what is given and what is to be found.

Step 2 **Assign variables** to represent the unknown values, using diagrams or tables as needed. Write down what each variable represents.

Step 3 **Write two equations** using both variables.

Step 4 **Solve** the system of two equations.

Step 5 **State the answer** to the problem. Is the answer reasonable?

Step 6 **Check** the answer in the words of the original problem.

OBJECTIVE 1 Solve problems about unknown numbers. Use the modified six-step method to solve problems about unknown numbers.

EXAMPLE 1 Solving a Problem about Two Unknown Numbers

In 1999, sales of sports clothing were $179 million more than sales of sports footwear. Together, the total sales for these items amounted to $26,601 million. (*Source:* National Sporting Goods Association.) What were the sales for each?

Step 1 **Read** the problem carefully. We must find the 1999 sales (in millions of dollars) for sports clothing and sports footwear. We know how much more sports clothing sales were than sports footwear sales. Also, we know the total sales.

Step 2 **Assign variables.**

Let x = sales of sports clothing in millions of dollars,

and y = sales of sports footwear in millions of dollars.

Step 3 **Write two equations.**

$x = 179 + y$ *Sales of sports clothing were $179 million more than sales of sports footwear.*

$x + y = 26{,}601$ *Total sales were $26,601 million.*

Step 4 **Solve** the system from Step 3. We use the substitution method since the first equation is already solved for x.

$x + y = 26{,}601$ *Second equation*

$(179 + y) + y = 26{,}601$ *Let $x = 179 + y$.*

$179 + 2y = 26{,}601$ *Combine like terms.*

$2y = 26{,}422$ *Subtract 179.*

$y = 13{,}211$ *Divide by 2.*

We substitute 13,211 for y in either equation of the system to find that $x = 13,390$.

Step 5 **State the answer.** Clothing sales were $13,390 million and footwear sales were $13,211 million.

Step 6 **Check** the answer in the original problem. Since

$$13,390 - 13,211 = 179 \quad \text{and} \quad 13,390 + 13,211 = 26,601,$$

the answer satisfies the information in the problem.

Now Try Exercise 9.

CAUTION If an applied problem asks for *two* values as in Example 1, be sure to give both of them in your answer.

OBJECTIVE 2 Solve problems about quantities and their costs. We can also use a linear system to solve an applied problem involving two quantities and their costs.

EXAMPLE 2 Solving a Problem about Quantities and Costs

The 1997 box office smash *Titanic,* the all-time top-grossing American movie, earned more in Europe than in the United States. This may be because average movie prices in Europe exceed those in the United States. (*Source: Parade,* September 13, 1998; *World Almanac and Book of Facts,* 2002.)

For example, while the average movie ticket (to the nearest dollar) in 1997–1998 cost $5 in the United States, it cost an equivalent of $11 in London. Suppose that a group of 41 Americans and Londoners who paid these average prices spent a total of $307 for tickets. How many from each country were in the group?

Step 1 **Read** the problem again.

Step 2 **Assign variables.**

Let $x =$ the number of Americans in the group,

and $y =$ the number of Londoners in the group.

Summarize the information given in the problem in a table. The entries in the first two rows of the Total Value column were found by multiplying the number of tickets sold by the price per ticket.

	Number of Tickets	Price per Ticket (in dollars)	Total Value (in dollars)
Americans	x	5	$5x$
Londoners	y	11	$11y$
Total	41		307

Step 3 **Write two equations.** The total number of tickets was 41, so one equation is

$$x + y = 41. \quad \text{Total number of tickets}$$

Since the total value was $307, the final column leads to

$$5x + 11y = 307.$$ Total value of tickets

These two equations form the system

$$x + y = 41 \quad (1)$$
$$5x + 11y = 307. \quad (2)$$

Step 4 **Solve** the system of equations using the elimination method. To eliminate the x-terms, multiply each side of equation (1) by -5. Then add this result to equation (2).

$$\begin{aligned} -5x - 5y &= -205 \quad &\text{Multiply equation (1) by } -5. \\ 5x + 11y &= 307 \quad &(2) \\ \hline 6y &= 102 \quad &\text{Add.} \\ y &= 17 \quad &\text{Divide by 6.} \end{aligned}$$

Substitute 17 for y in equation (1) to get

$$x + y = 41 \quad (1)$$
$$x + 17 = 41 \quad \text{Let } y = 17.$$
$$x = 24. \quad \text{Subtract 17.}$$

Step 5 **State the answer.** There were 24 Americans and 17 Londoners in the group.

Step 6 **Check.** The sum of 24 and 17 is 41, so the number of movie-goers is correct. Since 24 Americans paid $5 each and 17 Londoners paid $11 each, the total of the admission prices is $5(24) + $11(17) = $307, which agrees with the total amount stated in the problem.

Now Try Exercise 21.

OBJECTIVE 3 Solve problems about mixtures. In Chapter 2 we solved mixture problems using one variable. Many mixture problems can also be solved using a system of two equations in two variables.

EXAMPLE 3 Solving a Mixture Problem Involving Percent

A pharmacist needs 100 L of 50% alcohol solution. She has on hand 30% alcohol solution and 80% alcohol solution, which she can mix. How many liters of each will be required to make the 100 L of 50% alcohol solution?

Step 1 **Read** the problem. Note the percent of each solution and of the mixture.

Step 2 **Assign variables.**

Let x = the number of liters of 30% alcohol needed,

and y = the number of liters of 80% alcohol needed.

Summarize the information in a table. Percents are represented in decimal form.

290 CHAPTER 4 Systems of Linear Equations and Inequalities

Liters of Solution	Percent	Liters of Pure Alcohol
x	.30	.30x
y	.80	.80y
100	.50	.50(100)

Figure 9 gives an idea of what is actually happening in this problem.

FIGURE 9

Step 3 **Write two equations.** Since the total number of liters in the final mixture will be 100, the first equation is
$$x + y = 100.$$
To find the amount of pure alcohol in each mixture, multiply the number of liters by the concentration. The amount of pure alcohol in the 30% solution added to the amount of pure alcohol in the 80% solution will equal the amount of pure alcohol in the final 50% solution. This gives the second equation,
$$.30x + .80y = .50(100).$$
These two equations form the system
$$x + y = 100$$
$$.30x + .80y = 50. \qquad .50(100) = 50$$

Step 4 **Solve** this system by the substitution method. Solving the first equation of the system for x gives $x = 100 - y$. Substitute $100 - y$ for x in the second equation to get

$.30(100 - y) + .80y = 50$	Let $x = 100 - y$.
$30 - .30y + .80y = 50$	Distributive property
$30 + .50y = 50$	Combine like terms.
$.50y = 20$	Subtract 30.
$y = 40.$	Divide by .50.

Then $x = 100 - y = 100 - 40 = 60$.

Step 5 **State the answer.** The pharmacist should use 60 L of the 30% solution and 40 L of the 80% solution.

Step 6 **Check.** Since $60 + 40 = 100$ and $.30(60) + .80(40) = 50$, this mixture will give the 100 L of 50% solution, as required in the original problem.

Now Try Exercise 23.

SECTION 4.4 Applications of Linear Systems

NOTE The system in Example 3 could have been solved by the elimination method. Also, we could have cleared decimals by multiplying each side of the second equation by 10.

OBJECTIVE 4 Solve problems about distance, rate (or speed), and time. Problems that use the distance formula $d = rt$ were first introduced in Chapter 2. In many cases, these problems can be solved with systems of two linear equations. Keep in mind that setting up a table and drawing a sketch will help you solve such problems.

EXAMPLE 4 Solving a Problem about Distance, Rate, and Time

Two executives in cities 400 mi apart drive to a business meeting at a location on the line between their cities. They meet after 4 hr. Find the speed of each car if one travels 20 mph faster than the other.

Step 1 **Read** the problem carefully.

Step 2 **Assign variables.**

$$\text{Let } x = \text{the speed of the faster car,}$$
$$\text{and } y = \text{the speed of the slower car.}$$

Use the formula $d = rt$. Since each car travels for 4 hr, the time, t, for each car is 4. This information is shown in the table. The distance is found by using the formula $d = rt$ and the expressions already entered in the table.

	r	t	d
Faster Car	x	4	4x
Slower Car	y	4	4y

Find d using $d = rt$.

Draw a sketch showing what is happening in the problem. See Figure 10.

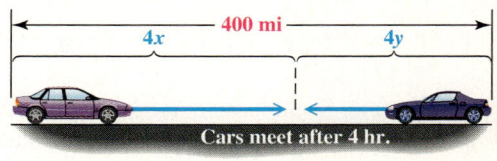

FIGURE 10

Step 3 **Write two equations.** As shown in the figure, since the total distance traveled by both cars is 400 mi, one equation is

$$4x + 4y = 400.$$

Because the faster car goes 20 mph faster than the slower car, the second equation is

$$x = 20 + y.$$

Step 4 **Solve** the system of equations

$$4x + 4y = 400 \quad (1)$$
$$x = 20 + y. \quad (2)$$

292 CHAPTER 4 Systems of Linear Equations and Inequalities

Use substitution. Replace x with $20 + y$ in equation (1) and solve for y.

$$4(20 + y) + 4y = 400 \quad \text{Let } x = 20 + y.$$
$$80 + 4y + 4y = 400 \quad \text{Distributive property}$$
$$80 + 8y = 400 \quad \text{Combine like terms.}$$
$$8y = 320 \quad \text{Subtract 80.}$$
$$y = 40 \quad \text{Divide by 8.}$$

Then $x = 20 + y = 20 + 40 = 60$.

Step 5 **State the answer.** The speeds of the two cars are 40 mph and 60 mph.

Step 6 **Check.** Since each car travels for 4 hr, the total distance traveled is

$$4(60) + 4(40) = 240 + 160 = 400 \text{ mi},$$

as required.

Now Try Exercise 29.

The problems in Examples 1–4 also could be solved using only one variable. Many students find that the solution is simpler with two variables.

CAUTION Be careful! When you use two variables to solve a problem, you must write two equations.

4.4 EXERCISES

For Extra Help

Student's Solutions Manual

MyMathLab

InterAct Math Tutorial Software

AW Math Tutor Center

MathXL MathXL

Digital Video Tutor CD 9/Videotape 7

Choose the correct response in Exercises 1–5.

1. Which expression represents the monetary value of x 20-dollar bills?

  A. $\dfrac{x}{20}$ dollars   **B.** $\dfrac{20}{x}$ dollars   **C.** $20 + x$ dollars   **D.** $20x$ dollars

2. Which expression represents the cost of t pounds of candy that sells for \$1.95 per lb?

  A. \$1.95t **B.** $\dfrac{\$1.95}{t}$ **C.** $\dfrac{t}{\$1.95}$ **D.** \$1.95 + t

3. Suppose that x liters of a 40% acid solution are mixed with y liters of a 35% solution to obtain 100 L of a 38% solution. One equation in a system for solving this problem is $x + y = 100$. Which one of the following is the other equation?

  A. $.35x + .40y = .38(100)$ **B.** $.40x + .35y = .38(100)$
  C. $35x + 40y = 38$ **D.** $40x + 35y = .38(100)$

4. What is the speed of a plane that travels at a rate of 560 mph *into* a wind of r mph?

  A. $560 + r$ mph   **B.** $\dfrac{560}{r}$ mph   **C.** $560 - r$ mph   **D.** $r - 560$ mph

5. What is the speed of a plane that travels at a rate of 560 mph *with* a wind of r mph?

  A. $\dfrac{r}{560}$ mph   **B.** $560 - r$ mph   **C.** $560 + r$ mph   **D.** $r - 560$ mph

6. Using the list of steps for solving an applied problem with two variables, write a short paragraph describing the general procedure you will use to solve the problems that follow in this exercise set.

SECTION 4.4 Applications of Linear Systems **293**

Exercises 7 and 8 are good warm-up problems. In each case, refer to the six-step problem-solving method, fill in the blanks for Steps 2 and 3, and then complete the solution by applying Steps 4–6.

7. The sum of two numbers is 98 and the difference between them is 48. Find the two numbers.

    Step 1 **Read** the problem carefully.

    Step 2 **Assign variables.**
    Let x = the first number and let y = _____.

    Step 3 **Write two equations.**
    First equation: $x + y = 98$
    Second equation: _____

8. The sum of two numbers is 201 and the difference between them is 11. Find the two numbers.

    Step 1 **Read** the problem carefully.

    Step 2 **Assign variables.**
    Let x = the first number and let y = _____.

    Step 3 **Write two equations.**
    First equation: $x + y = 201$
    Second equation: _____

Write a system of equations for each problem, and then solve the problem. See Example 1.

9. During 2000, two of the top-grossing concert tours were KISS and the Dave Matthews Band. Together the two tours visited 163 cities. KISS visited 77 more cities than the Dave Matthews Band. How many cities did each group visit? (*Source:* Pollstar.)

10. In 2001, the top two formats of U.S. commercial radio stations were country and news/talk/sports. There were 1051 fewer news/talk/sports stations than country stations, and together they totaled 3329 stations. How many stations of each format were there? (*Source:* M Street Corporation.)

11. The two top-grossing movies of 2001 were *Harry Potter and the Sorcerer's Stone* and *Shrek*. *Shrek* grossed $26 million less than *Harry Potter and the Sorcerer's Stone*, and together the two films took in $562 million. (*Source:* ACNielsen EDI.)

    (a) How much did each of these movies earn?
    (b) If *Shrek* earned $27 million more than *Monsters, Inc.*, how much did *Monsters, Inc.* earn? (*Hint:* Use your answer from part (a).)
    (c) What is the total amount these top three films earned?

12. The Terminal Tower in Cleveland, Ohio, is 242 ft shorter than the Key Tower, also in Cleveland. The total of the heights of the two buildings is 1658 ft. Find the heights of the buildings. (*Source:* World Almanac and Book of Facts, 2000.)

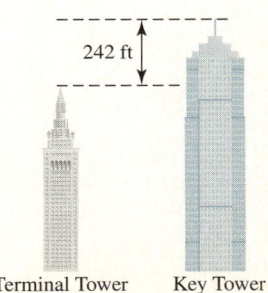

Terminal Tower   Key Tower

294 CHAPTER 4 Systems of Linear Equations and Inequalities

If x units of a product cost C dollars to manufacture and earn revenue of R dollars, the value of x where the expressions for C and R are equal is called the break-even quantity, *the number of units that produce 0 profit. In Exercises 13 and 14,* **(a)** *find the break-even quantity, and* **(b)** *decide whether the product should be produced based on whether it will earn a profit. (Profit = Revenue − Cost.)*

13. $C = 85x + 900$; $R = 105x$; no more than 38 units can be sold.

14. $C = 105x + 6000$; $R = 255x$; no more than 400 units can be sold.

Write a system of equations for each problem, and then solve the system. See Example 2.

15. A motel clerk counts his $1 and $10 bills at the end of a day. He finds that he has a total of 74 bills having a combined monetary value of $326. Find the number of bills of each denomination that he has.

Number of Bills	Denomination	Total Value
x	$1	1x
y	$10	10y
74		$326

16. Letarsha is a bank teller. At the end of a day, she has a total of 69 $5 and $10 bills. The total value of the money is $590. How many of each denomination does she have?

Number of Bills	Denomination	Total Value
x	$5	$5x
y	$10	

17. A newspaper advertised videocassettes and CDs. Christine May went shopping and bought each of her seven nephews a gift, either a copy of the movie *Harry Potter and the Sorcerer's Stone* or the latest Backstreet Boys CD. The movie cost $14.95 and the CD cost $16.88, and she spent a total of $114.30. How many movies and how many CDs did she buy?

18. Terry Wong saw the ad (see Exercise 17) and he, too, went shopping. He bought each of his five nieces a gift, either a copy of *Monsters, Inc.* or the CD soundtrack to *Scooby-Doo*. The movie cost $14.99 and the soundtrack cost $13.88, and he spent a total of $70.51. How many movies and how many soundtracks did he buy?

19. Maria Lopez has twice as much money invested at 5% simple annual interest as she does at 4%. If her yearly income from these two investments is $350, how much does she have invested at each rate?

20. Charles Miller invested his textbook royalty income in two accounts, one paying 3% annual simple interest and the other paying 2% interest. He earned a total of $11 interest. If he invested three times as much in the 3% account as he did in the 2% account, how much did he invest at each rate?

SECTION 4.4 Applications of Linear Systems

21. Average movie ticket prices in the United States are, in general, lower than in other countries. It would cost $77.87 to buy three tickets in Japan plus two tickets in Switzerland. Three tickets in Switzerland plus two tickets in Japan would cost $73.83. How much does an average movie ticket cost in each of these countries? (*Source: Business Traveler International.*)

22. (See Exercise 21.) Four movie tickets in Germany plus three movie tickets in France would cost $62.27. Three tickets in Germany plus four tickets in France would cost $62.19. How much does an average movie ticket cost in each of these countries? (*Source: Business Traveler International.*)

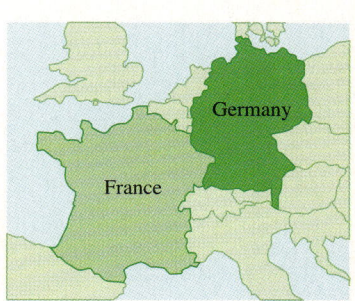

Write a system of equations for each problem, and then solve the system. See Example 3.

23. A 40% dye solution is to be mixed with a 70% dye solution to get 120 L of a 50% solution. How many liters of the 40% and 70% solutions will be needed?

Liters of Solution	Percent (as a decimal)	Liters of Pure Dye
x	.40	
y	.70	
120	.50	

24. A 90% antifreeze solution is to be mixed with a 75% solution to make 120 L of a 78% solution. How many liters of the 90% and 75% solutions will be used?

Liters of Solution	Percent (as a decimal)	Liters of Pure Antifreeze
x	.90	
y	.75	
120	.78	

25. A merchant wishes to mix coffee worth $6 per lb with coffee worth $3 per lb to get 90 lb of a mixture worth $4 per lb. How many pounds of the $6 and the $3 coffees will be needed?

Pounds	Dollars per Pound	Cost
x	6	
y		
90		

26. A grocer wishes to blend candy selling for $1.20 per lb with candy selling for $1.80 per lb to get a mixture that will be sold for $1.40 per lb. How many pounds of the $1.20 and the $1.80 candies should be used to get 45 lb of the mixture?

Pounds	Dollars per Pound	Cost
x		
y	1.80	
45		

27. How many barrels of pickles worth $40 per barrel and pickles worth $60 per barrel must be mixed to obtain 50 barrels of a mixture worth $48 per barrel?

28. The owner of a nursery wants to mix some fertilizer worth $70 per bag with some worth $90 per bag to obtain 40 bags of mixture worth $77.50 per bag. How many bags of each type should she use?

Write a system of equations for each problem, and then solve the system. See Example 4.

29. Two cars start from points 420 mi apart and travel toward each other. They meet after 3.5 hr. Find the average speed of each car if one travels 30 mph slower than the other.

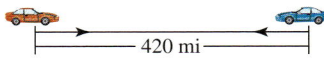

30. Two cars that were 450 mi apart traveled toward each other. They met after 5 hr. If one car traveled twice as fast as the other, what were their speeds?

31. Toledo and Cincinnati are 200 mi apart. A car leaves Toledo traveling toward Cincinnati, and another car leaves Cincinnati at the same time, traveling toward Toledo. The car leaving Toledo averages 15 mph faster than the other, and they meet after 1 hr and 36 min. What are the rates of the cars?

32. Kansas City and Denver are 600 mi apart. Two cars start from these cities, traveling toward each other. They meet after 6 hr. Find the rate of each car if one travels 30 mph slower than the other.

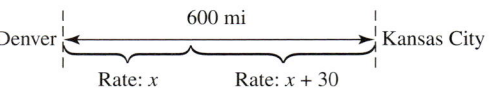

33. A boat takes 3 hr to go 24 mi upstream. It can go 36 mi downstream in the same time. Find the speed of the current and the speed of the boat in still water if x = the speed of the boat in still water and y = the speed of the current. (*Hint:* Because the current pushes the boat when it is going downstream, the speed of the boat downstream is the *sum* of the speed of the boat and the speed of the current. The current slows down the boat when it is going upstream, so the speed of the boat upstream is the *difference* between the speed of the boat and the speed of the current.)

	r	t	d
Downstream	$x + y$	3	36
Upstream	$x - y$	3	24

34. It takes a boat $1\frac{1}{2}$ hr to go 12 mi downstream, and 6 hr to return. Find the speed of the boat in still water and the speed of the current. Let x = the speed of the boat in still water and y = the speed of the current.

	r	t	d
Downstream	$x + y$	$\frac{3}{2}$	12
Upstream	$x - y$	6	12

35. If a plane can travel 440 mph into the wind and 500 mph with the wind, find the speed of the wind and the speed of the plane in still air.

36. A small plane travels 200 mph with the wind and 120 mph against it. Find the speed of the wind and the speed of the plane in still air.

37. At the beginning of a bicycle ride for charity, Roberto and Juana are 30 mi apart. If they leave at the same time and ride in the same direction, Roberto overtakes Juana in 6 hr. If they ride toward each other, they meet in 1 hr. What are their speeds?

38. Mr. Abbot left Farmersville in a plane at noon to travel to Exeter. Mr. Baker left Exeter in his automobile at 2 P.M. to travel to Farmersville. It is 400 mi from Exeter to Farmersville. If the sum of their speeds was 120 mph, and if they crossed paths at 4 P.M., find the speed of each.

4.5 Solving Systems of Linear Inequalities

OBJECTIVES

1 Solve systems of linear inequalities by graphing.

2 Use a graphing calculator to solve a system of linear inequalities.

We graphed the solutions of a linear inequality in Section 3.5. Recall that to graph the solutions of $x + 3y > 12$, for example, we first graph $x + 3y = 12$ by finding and plotting a few ordered pairs that satisfy the equation. Because the points on the line do *not* satisfy the inequality, we use a dashed line. To decide which region includes the points that are solutions, we choose a test point not on the line, such as (0, 0). Substituting these values for x and y in the inequality gives

$$x + 3y > 12$$
$$0 + 3(0) > 12$$
$$0 > 12,$$

a false result. This indicates that the solutions are those points on the side of the line that does not include (0, 0), as shown in Figure 11.

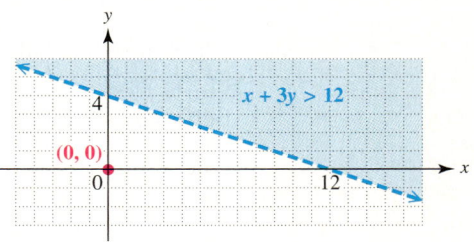

FIGURE 11

Now we use the same techniques to solve systems of linear inequalities.

OBJECTIVE 1 Solve systems of linear inequalities by graphing. A **system of linear inequalities** consists of two or more linear inequalities. The **solution set of a system of linear inequalities** includes all ordered pairs that make all inequalities of the system true at the same time. To solve a system of linear inequalities, use the following steps.

Solving a System of Linear Inequalities

Step 1 **Graph the inequalities.** Graph each inequality using the method of Section 3.5.

Step 2 **Choose the intersection.** Indicate the solution set of the system by shading the intersection of the graphs (the region where the graphs overlap).

298 CHAPTER 4 Systems of Linear Equations and Inequalities

EXAMPLE 1 Solving a System of Linear Inequalities

Graph the solution set of the system
$$3x + 2y \leq 6$$
$$2x - 5y \geq 10.$$

Step 1 To graph $3x + 2y \leq 6$, graph the solid boundary line $3x + 2y = 6$ and shade the region containing $(0, 0)$, as shown in Figure 12(a). Then graph $2x - 5y \geq 10$ with the solid boundary line $2x - 5y = 10$. The test point $(0, 0)$ makes this inequality false, so shade the region on the other side of the boundary line. See Figure 12(b).

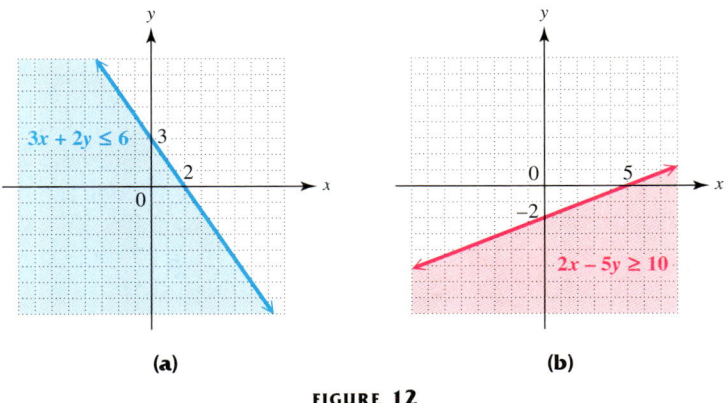

FIGURE 12

Step 2 The solution set of this system includes all points in the intersection (overlap) of the graphs of the two inequalities. It includes the gray shaded region and portions of the two boundary lines that surround it, as shown in Figure 13.

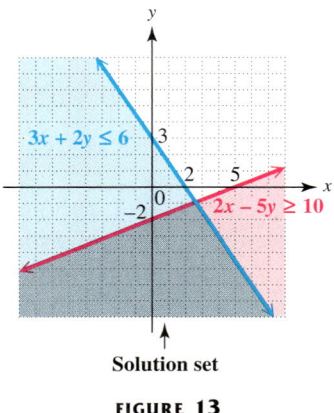

Solution set

FIGURE 13

Now Try Exercise 5.

NOTE We usually do all the work on one set of axes. In the following examples, only one graph is shown. Be sure that the region of the final solution is clearly indicated.

EXAMPLE 2 Solving a System of Linear Inequalities

Graph the solution set of the system

$$x - y > 5$$
$$2x + y < 2.$$

Figure 14 shows the graphs of both $x - y > 5$ and $2x + y < 2$. Dashed lines show that the graphs of the inequalities do not include their boundary lines. The solution set of the system is the region with the gray shading. The solution set does not include either boundary line.

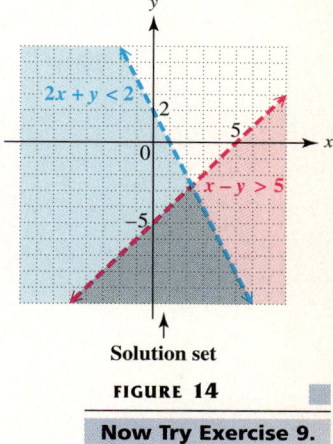

FIGURE 14

Now Try Exercise 9.

EXAMPLE 3 Solving a System of Three Linear Inequalities

Graph the solution set of the system

$$4x - 3y \leq 8$$
$$x \geq 2$$
$$y \leq 4.$$

Recall that $x = 2$ is a vertical line through the point $(2, 0)$, and $y = 4$ is a horizontal line through the point $(0, 4)$. The graph of the solution set is the shaded region in Figure 15, including all boundary lines.

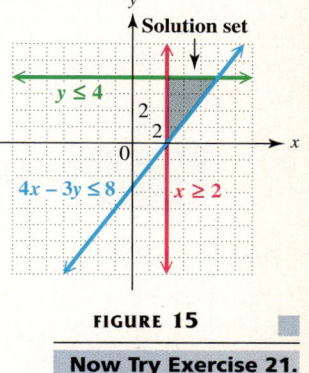

FIGURE 15

Now Try Exercise 21.

OBJECTIVE 2 Use a graphing calculator to solve a system of linear inequalities.

EXAMPLE 4 Finding the Solution Set of a System of Linear Inequalities with a Graphing Calculator

Graph the solution set of the system with a calculator.

$$y < 3x + 2$$
$$y > -2x - 5$$

To graph the first inequality, we direct the calculator to shade *below* the line

$$Y_1 = 3X + 2$$

(because of the $<$ symbol). To graph the second inequality, we direct the calculator to shade *above* the line

$$Y_2 = -2X - 5$$

(because of the $>$ symbol). Figure 16(a) on the next page shows the appropriate screen for these directions on a TI-83 Plus calculator. Graphing the inequalities in the standard viewing window gives the screen in Figure 16(b). The cross-hatched region is the intersection of the solution sets of the two individual inequalities and represents the solution set of the system.

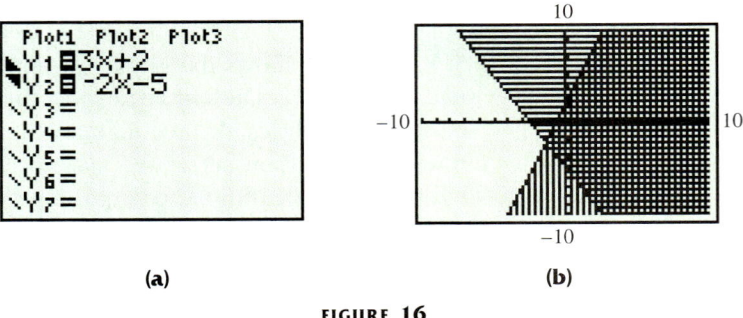

FIGURE 16

If the inequalities in a system are not solved for *y* as in this example, we must do so in order to enter them into the calculator. Notice that we cannot determine from the screen in Figure 16(b) whether the boundary lines are included or excluded in the solution set. For this system, neither boundary line is included.

Now Try Exercise 25.

4.5 EXERCISES

For Extra Help

- Student's Solutions Manual
- MyMathLab
- InterAct Math Tutorial Software
- AW Math Tutor Center
- MathXL
- Digital Video Tutor CD 9/Videotape 7

Match each system of inequalities with the correct graph from choices A–D.

1. $x \geq 5$
   $y \leq -3$

2. $x \leq 5$
   $y \geq -3$

3. $x > 5$
   $y < -3$

4. $x < 5$
   $y > -3$

A.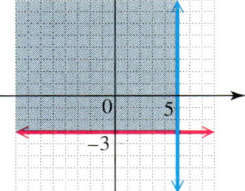

B.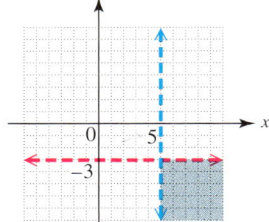

C.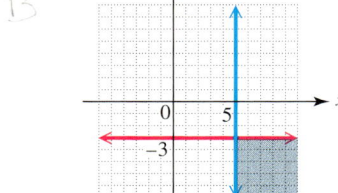

D.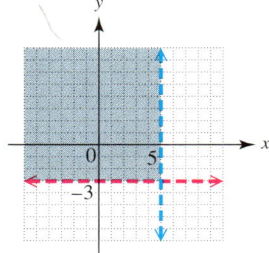

Graph the solution set of each system of linear inequalities. See Examples 1 and 2.

5. $x + y \leq 6$
   $x - y \geq 1$

6. $x + y \leq 2$
   $x - y \geq 3$

7. $4x + 5y \geq 20$
   $x - 2y \leq 5$

SECTION 4.5 Solving Systems of Linear Inequalities **301**

8. $x + 4y \leq 8$
   $2x - y \geq 4$

9. $2x + 3y < 6$
   $x - y < 5$

10. $x + 2y < 4$
    $x - y < -1$

11. $y \leq 2x - 5$
    $x < 3y + 2$

12. $x \geq 2y + 6$
    $y > -2x + 4$

13. $4x + 3y < 6$
    $x - 2y > 4$

14. $3x + y > 4$
    $x + 2y < 2$

15. $x \leq 2y + 3$
    $x + y < 0$

16. $x \leq 4y + 3$
    $x + y > 0$

17. $-3x + y \geq 1$
    $6x - 2y \geq -10$

18. $2x + 3y < 6$
    $4x + 6y > 18$

19. $x - 3y \leq 6$
    $x \geq -4$

20. Suppose that your friend was absent from class (because of a bad headache) on the day that systems of inequalities were covered. He asked you to write a short explanation of the procedure. Do this for him.

Graph the solution set of each system. See Example 3.

21. $4x + 5y < 8$
    $y > -2$
    $x > -4$

22. $x + y \geq -3$
    $x - y \leq 3$
    $y \leq 3$

23. $3x - 2y \geq 6$
    $x + y \leq 4$
    $x \geq 0$
    $y \geq -4$

24. How can we determine that the point (0, 0) is a solution of the system in Exercise 21 without actually graphing the system?

TECHNOLOGY INSIGHTS (EXERCISES 25–28)

Match each system of inequalities with its solution set. See Example 4.

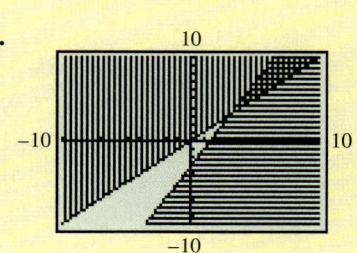

A.

B.

C.

D.

25. $y \geq x$
    $y \leq 2x - 3$

26. $y \leq x$
    $y \geq 2x - 3$

27. $y \geq -x$
    $y \leq 2x - 3$

28. $y \leq -x$
    $y \geq 2x - 3$

Chapter 4 Group Activity

Top Concert Tours

OBJECTIVE Use systems of equations to analyze data.

This activity will use systems of equations to analyze the top-grossing concert tours of 1997 and 2000.

- In 1997 the top-grossing North American concert tours were The Rolling Stones and U2. Together they grossed $169.2 million. The Rolling Stones grossed $9.4 million more than U2. The two groups performed a total of 79 shows, and U2 performed 13 more shows than The Rolling Stones. (*Source:* Pollstar.)

- In 2000 the top-grossing North American concert tours were Tina Turner and 'N Sync. Together they grossed $156.6 million. Tina Turner grossed $3.8 million more than 'N Sync. The two groups performed 181 shows, and 'N Sync performed 9 less shows than Tina Turner. (*Source:* Pollstar.)

Using the given information, complete the table. One student should complete the information for 1997 and the other student for 2000.

A. Write a system of equations to find the total gross for each group.

B. Write a system of equations to find the number of shows per year for each group.

C. Use the information from Exercises A and B to find the gross per show.

Year	Group	Total Gross	Shows per Year	Gross per Show
1997	Rolling Stones			
1997	U2			
2000	Tina Turner			
2000	'N Sync			

D. Once you have completed the table, compare the results. Answer these questions.

  1. What differences do you notice between 1997 and 2000?
  2. Which group had the highest total gross?
  3. Which group had the highest gross per show?

CHAPTER 4 SUMMARY

KEY TERMS

4.1 system of linear equations (linear system)
solution of a system
solution set of a system

consistent system
inconsistent system
independent equations
dependent equations

4.5 system of linear inequalities
solution set of a system of linear inequalities

TEST YOUR WORD POWER

See how well you have learned the vocabulary in this chapter. Answers, with examples, follow the Quick Review.

1. A **system of linear equations** consists of
   A. at least two linear equations with different variables
   B. two or more linear equations that have an infinite number of solutions
   C. two or more linear equations with the same variables
   D. two or more linear inequalities.

2. A **solution of a system** of linear equations is
   A. an ordered pair that makes one equation of the system true

   B. an ordered pair that makes all the equations of the system true at the same time
   C. any ordered pair that makes one or the other or both equations of the system true
   D. the set of values that make all the equations of the system false.

3. A **consistent system** is a system of equations
   A. with one solution
   B. with no solution
   C. with an infinite number of solutions

   D. that have the same graph.

4. An **inconsistent system** is a system of equations
   A. with one solution
   B. with no solution
   C. with an infinite number of solutions
   D. that have the same graph.

5. **Dependent equations**
   A. have different graphs
   B. have no solution
   C. have one solution
   D. are different forms of the same equation.

QUICK REVIEW

CONCEPTS	EXAMPLES

4.1 SOLVING SYSTEMS OF LINEAR EQUATIONS BY GRAPHING

An ordered pair is a solution of a system if it makes all equations of the system true at the same time.

Is $(4, -1)$ a solution of the system $\begin{array}{l} x + y = 3 \\ 2x - y = 9? \end{array}$

Yes, because $4 + (-1) = 3$, and $2(4) - (-1) = 9$ are both true.

To solve a linear system by graphing,

Step 1 Graph each equation of the system on the same axes.

Step 2 Find the coordinates of the point of intersection.

Step 3 Check. Write the solution set.

Solve the system by graphing: $\begin{array}{l} x + y = 5 \\ 2x - y = 4. \end{array}$

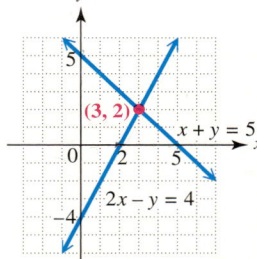

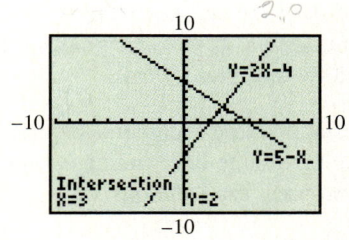

A graphing calculator can find the solution of a system by locating the point of intersection of the graphs.

The solution $(3, 2)$ checks, so $\{(3, 2)\}$ is the solution set.

(continued)

CONCEPTS	EXAMPLES

4.2 SOLVING SYSTEMS OF LINEAR EQUATIONS BY SUBSTITUTION

Step 1 Solve one equation for either variable.

Solve by substitution.
$$x + 2y = -5 \quad (1)$$
$$y = -2x - 1 \quad (2)$$

Equation (2) is already solved for y.

Step 2 Substitute for that variable in the other equation to get an equation in one variable.

Substitute $-2x - 1$ for y in equation (1).
$$x + 2(-2x - 1) = -5$$

Step 3 Solve the equation from Step 2.

Solve to get $x = 1$.

Step 4 Substitute the result into the equation from Step 1 to get the value of the other variable.

To find y, let $x = 1$ in equation (2).
$$y = -2(1) - 1 = -3$$

Step 5 Check. Write the solution set.

The solution, $(1, -3)$, checks so $\{(1, -3)\}$ is the solution set.

4.3 SOLVING SYSTEMS OF LINEAR EQUATIONS BY ELIMINATION

Step 1 Write both equations in standard form $Ax + By = C$.

Solve by elimination.
$$x + 3y = 7 \quad (1)$$
$$3x - y = 1 \quad (2)$$

Step 2 Multiply to make the coefficients of one pair of variable terms opposites.

Multiply equation (1) by -3 to eliminate the x-terms.

Step 3 Add the equations to get an equation with only one variable.

$$-3x - 9y = -21$$
$$\underline{3x - y = 1}$$

Step 4 Solve the equation from Step 3.

$$-10y = -20 \quad \text{Add.}$$
$$y = 2 \quad \text{Divide by } -10.$$

Step 5 Substitute the solution from Step 4 into either of the original equations to find the value of the remaining variable.

Substitute to get the value of x.
$$x + 3y = 7 \quad (1)$$
$$x + 3(2) = 7 \quad \text{Let } y = 2.$$
$$x + 6 = 7 \quad \text{Multiply.}$$
$$x = 1 \quad \text{Subtract 6.}$$

Step 6 Check. Write the solution set.

Since $1 + 3(2) = 7$ and $3(1) - 2 = 1$, the solution set is $\{(1, 2)\}$.

If the result of the addition step (Step 3) is a false statement, such as $0 = 4$, the graphs are parallel lines and *the solution set is* $\emptyset$.

$$x - 2y = 6$$
$$\underline{-x + 2y = -2}$$
$$0 = 4 \quad \text{Solution set: } \emptyset$$

If the result is a true statement, such as $0 = 0$, the graphs are the same line, and an *infinite number of ordered pairs are solutions*.

$$x - 2y = 6$$
$$\underline{-x + 2y = -6}$$
$$0 = 0 \quad \text{Infinite number of solutions}$$

CONCEPTS	EXAMPLES
4.4 APPLICATIONS OF LINEAR SYSTEMS	
Use the modified six-step method.	The sum of two numbers is 30. Their difference is 6. Find the numbers.
Step 1 Read.	
Step 2 Assign variables.	Let $x =$ one number; let $y =$ the other number.
Step 3 Write two equations using both variables.	$x + y = 30$
	$\underline{x - y = 6}$
Step 4 Solve the system.	$2x = 36 \quad$ Add.
	$x = 18 \quad$ Divide by 2.
	Let $x = 18$ in the top equation: $18 + y = 30$. Solve to get $y = 12$.
Step 5 State the answer.	The two numbers are 18 and 12.
Step 6 Check.	$18 + 12 = 30$ and $18 - 12 = 6$, so the solution checks.
4.5 SOLVING SYSTEMS OF LINEAR INEQUALITIES	
To solve a system of linear inequalities, graph each inequality on the same axes. (This was explained in Section 3.5.) The solution set of the system is formed by the overlap of the regions of the two graphs.	The shaded region shows the solution set of the system $$2x + 4y \geq 5$$ $$x \geq 1.$$ 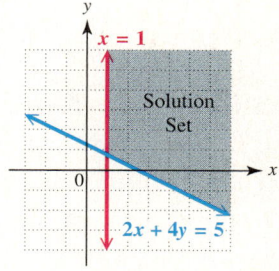
Graphing calculators also can represent solution sets of systems of inequalities.	The cross-hatched region shows the solution set of the system $$y > 2x + 1$$ $$y < 3.$$ The boundary lines are not included. 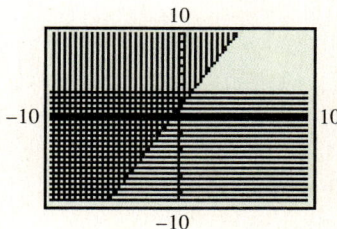

Answers to Test Your Word Power

1. C; *Example:* $2x + y = 7$, $3x - y = 3$ **2.** B; *Example:* The ordered pair (2, 3) satisfies both equations of the system in Answer 1, so it is a solution of the system.   **3.** A; *Example:* The system in Answer 1 is consistent. The graphs of the equations intersect at exactly one point, in this case the solution (2, 3).   **4.** B; *Example:* The equations of two parallel lines make up an inconsistent system; their graphs never intersect, so there is no solution to the system.   **5.** D; *Example:* The equations $4x - y = 8$ and $8x - 2y = 16$ are dependent because their graphs are the same line.

CHAPTER 4 REVIEW EXERCISES

[4.1] *Decide whether the given ordered pair is a solution of the given system.*

1. $(3, 4)$
   $4x - 2y = 4$
   $5x + y = 19$

2. $(-5, 2)$
   $x - 4y = -13$
   $2x + 3y = 4$

Solve each system by graphing.

3. $x + y = 4$
   $2x - y = 5$

4. $x - 2y = 4$
   $2x + y = -2$

5. $2x + 4 = 2y$
   $y - x = -3$

[4.2]

6. A student solves the system $\begin{array}{l} 2x + y = 6 \\ -2x - y = 4 \end{array}$ and gets the equation $0 = 10$. The student gives the solution set as $\{(0, 10)\}$. Is this correct? Explain.

7. Can a system of two linear equations in two variables have exactly three solutions? Explain.

Solve each system by the substitution method.

8. $3x + y = 7$
   $x = 2y$

9. $2x - 5y = -19$
   $y = x + 2$

10. $5x + 15y = 30$
    $x + 3y = 6$

11. After solving a system of linear equations by the substitution method, a student obtained the equation "$0 = 0$." He gave the solution set of the system as $\{(0, 0)\}$. Was his answer correct? Why or why not?

12. Suppose that you were asked to solve the system $\begin{array}{l} 5x - 3y = 7 \\ -x + 2y = 4 \end{array}$ by substitution. Which variable in which equation would be easiest to solve for in your first step? Why?

[4.3]

13. Only one of the following systems does not require that we multiply one or both equations by a constant to solve the system by the elimination method. Which one is it?

    A. $-4x + 3y = 7$
       $3x - 4y = 4$

    B. $5x + 8y = 13$
       $12x + 24y = 36$

    C. $2x + 3y = 5$
       $x - 3y = 12$

    D. $x + 2y = 9$
       $3x - y = 6$

14. For the system

    $2x + 12y = 7$
    $3x + 4y = 1,$

    if we were to multiply the first equation by -3, by what number would we have to multiply the second equation to
    (a) eliminate the *x*-terms when solving by the elimination method?
    (b) eliminate the *y*-terms when solving by the elimination method?

Solve each system by the elimination method.

15. $2x - y = 13$
    $x + y = 8$

16. $-4x + 3y = 25$
    $6x - 5y = -39$

17. $3x - 4y = 9$
    $6x - 8y = 18$

18. $2x + y = 3$
    $-4x - 2y = 6$

[4.1–4.3] *Solve each system by any method.*

19. $2x + 3y = -5$
    $3x + 4y = -8$

20. $6x - 9y = 0$
    $2x - 3y = 0$

21. $x - 2y = 5$
    $y = x - 7$

22. $\dfrac{x}{2} + \dfrac{y}{3} = 7$
    $\dfrac{x}{4} + \dfrac{2y}{3} = 8$

23. What are the three methods of solving systems discussed in this chapter? Choose one and discuss its advantages and disadvantages.

24. Why would it be easier to solve system B by the substitution method than system A?

    A: $-5x + 6y = 7$ **B:** $2x + 9y = 13$
        $2x + 5y = -5$ $y = 3x - 2$

[4.4] *Solve each problem using a system of equations.*

25. At the end of 2001, Subway topped McDonald's as the largest restaurant chain in the United States. Subway operated 148 more restaurants than McDonald's, and together the two chains had 26,346 restaurants. How many restaurants did each company operate? (*Source: USA Today,* February 4, 2002.)

26. Two of the most popular magazines in the United States are *Modern Maturity* and *Reader's Digest.* Together, the average total circulation for these two magazines during a recent 6-month period was 35.6 million. *Reader's Digest* circulation was 5.4 million less than that of *Modern Maturity.* What were the circulation figures for each magazine? (*Source:* Audit Bureau of Circulations and Magazine Publishers of America.)

27. The two tallest buildings in Houston, Texas, are the Texas Commerce Tower and the First Interstate Plaza. The first of these is 4 stories taller than the second. Together the buildings have 146 stories. How many stories tall are the individual buildings? (*Source: World Almanac and Book of Facts,* 2002.)

28. In the 2000 presidential election, George W. Bush received 5 more electoral votes than Al Gore. Together, the two candidates received a total of 537 electoral votes. How many electoral votes did each receive? (*Source: World Almanac and Book of Facts,* 2002.)

29. The perimeter of a rectangle is 90 m. Its length is $1\tfrac{1}{2}$ times its width. Find the length and width of the rectangle.

30. A cashier has 20 bills, all of which are $10 or $20 bills. The total value of the money is $330. How many of each type does the cashier have?

31. Candy that sells for $1.30 per lb is to be mixed with candy selling for $.90 per lb to get 100 lb of a mix that will sell for $1 per lb. How much of each type should be used?

32. A 40% antifreeze solution is to be mixed with a 70% solution to get 90 L of a 50% solution. How many liters of the 40% and 70% solutions will be needed?

Number of Liters	Percent (as a decimal)	Amount of Pure Antifreeze
x	.40	
y	.70	
90	.50	

33. Ms. Cesar invested $18,000. Part of it was invested at 3% annual simple interest and the rest was invested at 4%. Her interest income for the first year was $650. How much did she invest at each rate?

Amount of Principal	Rate	Interest
x	.03	
y	.04	
$18,000		

34. A certain plane flying with the wind travels 540 mi in 2 hr. Later, flying against the same wind, the plane travels 690 mi in 3 hr. Find the speed of the plane in still air and the speed of the wind.

[4.5] *Graph the solution set of each system of linear inequalities.*

35. $x + y \geq 2$
$x - y \leq 4$

36. $y \geq 2x$
$2x + 3y \leq 6$

37. $x + y < 3$
$2x > y$

38. $3x - y \leq 3$
$x \geq -2$
$y \leq 2$

■ MIXED REVIEW EXERCISES

Solve each problem using the methods of this chapter.

39. Eboni Perkins compared the monthly payments she would incur for two types of mortgages: fixed-rate and variable-rate. Her observations led to the following graph.

(a) For which years would the monthly payment be more for the fixed-rate mortgage than for the variable-rate mortgage?

(b) In what year would the payments be the same, and what would those payments be?

40. $\dfrac{2x}{3} + \dfrac{y}{4} = \dfrac{14}{3}$
$\dfrac{x}{2} + \dfrac{y}{12} = \dfrac{8}{3}$

41. $x = y + 6$
$2y - 2x = -12$

42. $3x + 4y = 6$
$4x - 5y = 8$

43. $\dfrac{3x}{2} + \dfrac{y}{5} = -3$
$4x + \dfrac{y}{3} = -11$

44. $x + y < 5$
$x - y \geq 2$

45. $y \leq 2x$
$x + 2y > 4$

46. The perimeter of an isosceles triangle measures 29 in. One side of the triangle is 5 in. longer than each of the two equal sides. Find the lengths of the sides of the triangle.

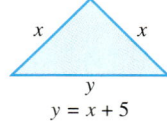
$y = x + 5$

47. In 2000, a total of 9.3 million people visited the Statue of Liberty and the Vietnam Veterans Memorial, two popular tourist attractions. The Statue of Liberty had 1.7 million more visitors than the Vietnam Veterans Memorial. How many visitors did each of these sites have? (*Source:* National Park Service, Department of the Interior.)

48. Two cars leave from the same place and travel in opposite directions. One car travels 30 mph faster than the other. After $2\frac{1}{2}$ hr they are 265 mi apart. What are the rates of the cars?

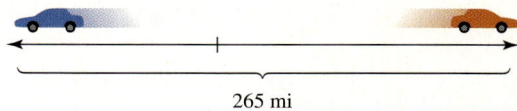

265 mi

49. A hospital bought a total of 146 bottles of glucose solution. Small bottles cost $2 each, and large ones cost $3 each. The total cost was $336. How many of each size bottle were bought?

50. Can the problem in Exercise 46 be worked using a single variable? If so, explain how it can be done, and then solve it.

CHAPTER 4 TEST

1. The graph shows a company's costs to produce computer parts and the revenue from the sale of those parts.

   (a) At what production level does the cost equal the revenue?
   (b) What is the revenue at that point?

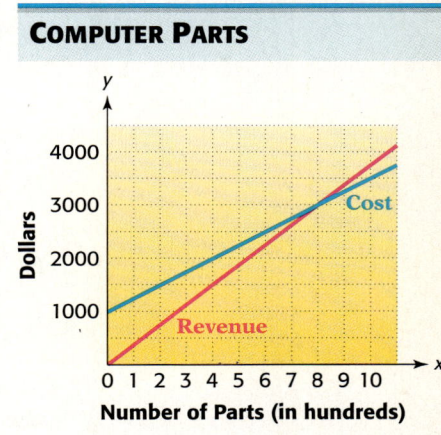

2. Decide whether each ordered pair is a solution of the system $\begin{array}{r} 2x + y = -3 \\ x - y = -9. \end{array}$

   (a) $(1, -5)$ **(b)** $(1, 10)$ **(c)** $(-4, 5)$

3. Solve the system by graphing: $\begin{array}{r} x + 2y = 6 \\ -2x + y = -7. \end{array}$

Solve each system by substitution.

4. $2x + y = -4$
    $x = y + 7$

5. $4x + 3y = -35$
    $x + y = 0$

Solve each system by elimination.

6. $2x - y = 4$
    $3x + y = 21$

7. $4x + 2y = 2$
    $5x + 4y = 7$

8. $3x + 4y = 9$
    $2x + 5y = 13$

9. $4x + 5y = 2$
    $-8x - 10y = 6$

10. $6x - 5y = 0$
    $-2x + 3y = 0$

11. $\frac{6}{5}x - \frac{1}{3}y = -20$
    $-\frac{2}{3}x + \frac{1}{6}y = 11$

12. Solve the system by any method: $4y = -3x + 5$
    $6x = -8y + 10.$

13. Suppose that the graph of a system of two linear equations consists of lines that have the same slope but different *y*-intercepts. How many solutions does the system have?

Solve each problem using a system of equations.

14. The distance between Memphis and Atlanta is 300 mi less than the distance between Minneapolis and Houston. Together the two distances total 1042 mi. How far is it between Memphis and Atlanta? How far is it between Minneapolis and Houston?

15. In 2000, the two most popular amusement parks in the United States were Disneyland and the Magic Kingdom at Walt Disney World. Disneyland had 1.5 million fewer visitors than the Magic Kingdom, and together they had 29.3 million visitors. How many visitors did each park have? (*Source: Amusement Business,* August 2001.)

16. A 25% solution of alcohol is to be mixed with a 40% solution to get 50 L of a final mixture that is 30% alcohol. How much of each of the original solutions should be used?

17. Two cars leave from Perham, Minnesota, at the same time and travel in the same direction. One car travels one and one-third times as fast as the other. After 3 hr they are 45 mi apart. What are the speeds of the cars?

Graph the solution set of each system of inequalities.

18. $2x + 7y \leq 14$
    $x - y \geq 1$

19. $2x - y > 6$
    $4y + 12 \geq -3x$

20. Without actually graphing, determine which one of the following systems of inequalities has no solution.

    A. $x \geq 4$
    $y \leq 3$

    B. $x + y > 4$
    $x + y < 3$

    C. $x > 2$
    $y < 1$

    D. $x + y > 4$
    $x - y < 3$

CUMULATIVE REVIEW EXERCISES CHAPTERS 1–4

1. List all integer factors of 40.

2. Find the value of the expression $\dfrac{3x^2 + 2y^2}{10y + 3}$ if $x = 1$ and $y = 5$.

Name the property that justifies each statement.

3. $5 + (-4) = (-4) + 5$

4. $r(s - k) = rs - rk$

5. $-\dfrac{2}{3} + \dfrac{2}{3} = 0$

6. Evaluate $-2 + 6[3 - (4 - 9)]$.

Solve each linear equation.

7. $2 - 3(6x + 2) = 4(x + 1) + 18$

8. $\dfrac{3}{2}\left(\dfrac{1}{3}x + 4\right) = 6\left(\dfrac{1}{4} + x\right)$

9. Solve the formula $P = \dfrac{kT}{V}$ for T.

Solve each linear inequality.

10. $-\dfrac{5}{6}x < 15$

11. $-8 < 2x + 3$

12. A recent survey measured public recognition of the most popular contemporary advertising slogans. Complete the results shown in the table if 2500 people were surveyed.

Slogan (product or company)	Percent Recognition (nearest tenth of a percent)	Actual Number Who Recognized Slogan (nearest whole number)
Please Don't Squeeze the . . . (Charmin)	80.4%	
The Breakfast of Champions (Wheaties)	72.5%	
The King of Beers (Budweiser)		1570
Like a Good Neighbor (State Farm)		1430

(Other slogans included "You're in Good Hands" (Allstate), "Snap, Crackle, Pop" (Rice Krispies), and "The Un-Cola" (7-Up).)
Source: Department of Integrated Marketing Communications, Northwestern University.

Solve each problem.

13. On February 12, 1999, the U.S. Senate voted to acquit William Jefferson Clinton on both counts of impeachment (perjury and obstruction of justice). Of the 200 "guilty" or "not guilty" votes cast that day, there were 10 more "not guilty" votes than "guilty" votes. How many of each vote were there? (*Source:* MSNBC Web site, February 13, 1999.)

14. Two angles of a triangle have the same measure. The measure of the third angle is 4° less than twice the measure of each of the equal angles. Find the measures of the three angles.

Measures are in degrees. (Triangle with angles x, x, and $2x - 4$.)

15. No baseball fan should be without a copy of *The Sports Encyclopedia: Baseball 2000* by David S. Neft and Richard M. Cohen. The 20th edition provides exhaustive statistics for professional baseball dating back to 1876. This book has a perimeter of 38 in., and its width measures 2.5 in. less than its length. What are the dimensions of the book?

Graph each linear equation.

16. $x - y = 4$

17. $3x + y = 6$

Find the slope of each line.

18. Through $(-5, 6)$ and $(1, -2)$

19. Perpendicular to the line $y = 4x - 3$

Find an equation for each line. Write it in slope-intercept form.

20. Through $(-4, 1)$ with slope $\frac{1}{2}$

21. Through the points $(1, 3)$ and $(-2, -3)$

22. (a) Write an equation of the vertical line through $(9, -2)$.
    (b) Write an equation of the horizontal line through $(4, -1)$.

Solve each system by any method.

23. $2x - y = -8$
    $x + 2y = 11$

24. $4x + 5y = -8$
    $3x + 4y = -7$

25. $3x + 5y = 1$
    $x = y + 3$

26. $3x + 4y = 2$
    $6x + 8y = 1$

Use a system of equations to solve each problem.

27. Admission prices at a football game were $6 for adults and $2 for children. The total value of the tickets sold was $2528, and 454 tickets were sold. How many adults and how many children attended the game?

Kind of Ticket	Number Sold	Cost of Each (in dollars)	Total Value (in dollars)
Adult	x	6	6x
Child	y		
Total	454		

28. The perimeter of a triangle measures 53 in. If two sides are of equal length, and the third side measures 4 in. less than each of the equal sides, what are the lengths of the three sides?

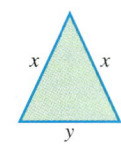

29. A chemist needs 12 L of a 40% alcohol solution. She must mix a 20% solution and a 50% solution. How many liters of each will be required to obtain what she needs?

30. Graph the solution set of the system $\begin{array}{l} x + 2y \leq 12 \\ 2x - y \leq 8. \end{array}$

Exponents and Polynomials

5

5.1 The Product Rule and Power Rules for Exponents

5.2 Integer Exponents and the Quotient Rule

Summary Exercises on the Rules for Exponents

5.3 An Application of Exponents: Scientific Notation

5.4 Adding and Subtracting Polynomials; Graphing Simple Polynomials

5.5 Multiplying Polynomials

5.6 Special Products

5.7 Dividing Polynomials

The expression $100x - 13x^2$ gives the distance in feet a car going approximately 68 mph will skid in x sec. This expression in x is an example of a *polynomial,* one topic of this chapter. Accident investigators use polynomials like this to determine the length of a skid or the elapsed time during a skid. In Exercises 95 and 96 of Section 5.4, we use this polynomial to approximate skidding distance.

5.1 The Product Rule and Power Rules for Exponents

OBJECTIVES

1. Identify bases and exponents.
2. Use the product rule for exponents.
3. Use the rule $(a^m)^n = a^{mn}$.
4. Use the rule $(ab)^m = a^m b^m$.
5. Use the rule $\left(\dfrac{a}{b}\right)^m = \dfrac{a^m}{b^m}$.
6. Use combinations of rules.

OBJECTIVE 1 Identify bases and exponents. Recall from Section 1.2 that in the expression 5^2, the number 5 is the *base* and 2 is the *exponent* or *power*. The expression 5^2 is called an *exponential expression*. Usually we do not write the exponent when it is 1; however, sometimes it is convenient to do so. In general, for any quantity a,

$$a^1 = a.$$

EXAMPLE 1 Determining the Base and Exponent in Exponential Expressions

Evaluate each exponential expression. Name the base and the exponent.

	Base	Exponent
(a) $5^4 = 5 \cdot 5 \cdot 5 \cdot 5 = 625$	5	4
(b) $-5^4 = -1 \cdot 5^4 = -1 \cdot (5 \cdot 5 \cdot 5 \cdot 5) = -625$	5	4
(c) $(-5)^4 = (-5)(-5)(-5)(-5) = 625$	-5	4

Now Try Exercises 15 and 17.

> **CAUTION** Note the differences between parts (b) and (c) of Example 1. In -5^4 the lack of parentheses shows that the exponent 4 refers only to the base 5, not -5; in $(-5)^4$ the parentheses show that the exponent 4 refers to the base -5. In summary, $-a^n$ and $(-a)^n$ are not necessarily the same.
>
> | Expression | Base | Exponent | Example |
> |---|---|---|---|
> | $-a^n$ | a | n | $-3^2 = -(3 \cdot 3) = -9$ |
> | $(-a)^n$ | $-a$ | n | $(-3)^2 = (-3)(-3) = 9$ |

OBJECTIVE 2 Use the product rule for exponents. By the definition of exponents,

$$2^4 \cdot 2^3 = \underbrace{(2 \cdot 2 \cdot 2 \cdot 2)}_{4 \text{ factors}} \underbrace{(2 \cdot 2 \cdot 2)}_{3 \text{ factors}}$$

$$= \underbrace{2 \cdot 2 \cdot 2 \cdot 2 \cdot 2 \cdot 2 \cdot 2}_{4 + 3 = 7 \text{ factors}}$$

$$= 2^7.$$

Also,

$$6^2 \cdot 6^3 = (6 \cdot 6)(6 \cdot 6 \cdot 6)$$

$$= 6 \cdot 6 \cdot 6 \cdot 6 \cdot 6$$

$$= 6^5.$$

Generalizing from these examples, $2^4 \cdot 2^3 = 2^{4+3} = 2^7$ and $6^2 \cdot 6^3 = 6^{2+3} = 6^5$, suggests the **product rule for exponents**.

SECTION 5.1 The Product Rule and Power Rules for Exponents **315**

> **Product Rule for Exponents**
>
> For any positive integers m and n, $\quad a^m \cdot a^n = a^{m+n}$.
> (Keep the same base; add the exponents.)
> *Example:* $6^2 \cdot 6^5 = 6^{2+5} = 6^7$

> **CAUTION** Avoid the common error of multiplying the bases when using the product rule:
> $$6^2 \cdot 6^5 \neq 36^7.$$
> **Keep the same base and add the exponents.**

EXAMPLE 2 Using the Product Rule

Use the product rule for exponents to find each result when possible.

(a) $6^3 \cdot 6^5 = 6^{3+5} = 6^8$

(b) $(-4)^7(-4)^2 = (-4)^{7+2} = (-4)^9$

(c) $x^2 \cdot x = x^2 \cdot x^1 = x^{2+1} = x^3$

(d) $m^4 m^3 m^5 = m^{4+3+5} = m^{12}$

(e) $2^3 \cdot 3^2$

The product rule does not apply to the product $2^3 \cdot 3^2$, since the bases are different.
$$2^3 \cdot 3^2 = 8 \cdot 9 = 72$$

(f) $2^3 + 2^4$

The product rule does not apply to $2^3 + 2^4$, since this is a *sum*, not a *product*.
$$2^3 + 2^4 = 8 + 16 = 24$$

(g) $(2x^3)(3x^7)$

Since $2x^3$ means $2 \cdot x^3$ and $3x^7$ means $3 \cdot x^7$, we use the associative and commutative properties to get
$$(2x^3)(3x^7) = (2 \cdot 3) \cdot (x^3 \cdot x^7) = 6x^{10}.$$

> Now Try Exercises 25, 29, 31, 35, and 39.

> **CAUTION** Be sure you understand the difference between *adding* and *multiplying* exponential expressions. For example,
> $$8x^3 + 5x^3 = (8 + 5)x^3 = 13x^3,$$
> but
> $$(8x^3)(5x^3) = (8 \cdot 5)x^{3+3} = 40x^6.$$

OBJECTIVE 3 Use the rule $(a^m)^n = a^{mn}$. We simplify an expression such as $(8^3)^2$ with the product rule for exponents.
$$(8^3)^2 = (8^3)(8^3) = 8^{3+3} = 8^6$$

The exponents in $(8^3)^2$ are multiplied to give the exponent in 8^6. As another example,

$$(5^2)^4 = 5^2 \cdot 5^2 \cdot 5^2 \cdot 5^2 \quad \text{Definition of exponent}$$
$$= 5^{2+2+2+2} \quad \text{Product rule}$$
$$= 5^8$$

and $2 \cdot 4 = 8$. These examples suggest **power rule (a) for exponents.**

CHAPTER 5 Exponents and Polynomials

Power Rule (a) for Exponents

For any positive integers m and n, $\quad (a^m)^n = a^{mn}$.
(Raise a power to a power by multiplying exponents.)
Example: $(3^2)^4 = 3^{2 \cdot 4} = 3^8$

EXAMPLE 3 Using Power Rule (a)

Use power rule (a) for exponents to simplify each expression.

(a) $(2^5)^3 = 2^{5 \cdot 3} = 2^{15}$ **(b)** $(5^7)^2 = 5^{7(2)} = 5^{14}$ **(c)** $(x^2)^5 = x^{2(5)} = x^{10}$

> **Now Try Exercises 43 and 45.**

OBJECTIVE 4 Use the rule $(ab)^m = a^m b^m$. We can use the properties studied in Chapter 1 to develop two more rules for exponents. Using the definition of an exponential expression and the commutative and associative properties, we can rewrite the expression $(4x)^3$ as follows.

$$(4x)^3 = (4x)(4x)(4x) \qquad \text{Definition of exponent}$$
$$= (4 \cdot 4 \cdot 4)(x \cdot x \cdot x) \qquad \text{Commutative and associative properties}$$
$$= 4^3 \cdot x^3 \qquad \text{Definition of exponent}$$

This example suggests **power rule (b) for exponents.**

Power Rule (b) for Exponents

For any positive integer m, $\quad (ab)^m = a^m b^m$.
(Raise a product to a power by raising each factor to the power.)
Example: $(2p)^5 = 2^5 p^5$

EXAMPLE 4 Using Power Rule (b)

Use power rule (b) for exponents to simplify each expression.

(a) $(3xy)^2 = 3^2 x^2 y^2 = 9x^2 y^2 \qquad$ Power rule (a)

(b) $5(pq)^2 = 5(p^2 q^2) \qquad$ Power rule (b)
$ = 5p^2 q^2 \qquad$ Multiply.

(c) $3(2m^2 p^3)^4 = 3[2^4(m^2)^4(p^3)^4] \qquad$ Power rule (b)
$ = 3 \cdot 2^4 m^8 p^{12} \qquad$ Power rule (a)
$ = 48 m^8 p^{12} \qquad$ Multiply.

(d) $(-5^6)^3 = (-1 \cdot 5^6)^3$
$ = (-1)^3 \cdot (5^6)^3$
$ = -1 \cdot 5^{18}$
$ = -5^{18}$

> **Now Try Exercises 49 and 53.**

> **CAUTION** Power rule (b) *does not* apply to a *sum*.
> $$(x+4)^2 \neq x^2 + 4^2$$

OBJECTIVE 5 Use the rule $\left(\dfrac{a}{b}\right)^m = \dfrac{a^m}{b^m}$. Since the quotient $\dfrac{a}{b}$ can be written as $a\left(\dfrac{1}{b}\right)$, we use power rule (b), together with some of the properties of real numbers, to get **power rule (c) for exponents.**

Power Rule (c) for Exponents

For any positive integer m,

$$\left(\dfrac{a}{b}\right)^m = \dfrac{a^m}{b^m} \quad (b \neq 0).$$

(Raise a quotient to a power by raising both numerator and denominator to the power.)

Example: $\left(\dfrac{5}{3}\right)^2 = \dfrac{5^2}{3^2}$

EXAMPLE 5 Using Power Rule (c)

Use power rule (c) for exponents to simplify each expression.

(a) $\left(\dfrac{2}{3}\right)^5 = \dfrac{2^5}{3^5}$ **(b)** $\left(\dfrac{m}{n}\right)^3 = \dfrac{m^3}{n^3}$ $(n \neq 0)$

> **Now Try Exercises 59 and 61.**

In the following box, we list the rules for exponents discussed in this section. These rules are basic to the study of algebra.

Rules for Exponents

For positive integers m and n:

		Examples
Product rule	$a^m \cdot a^n = a^{m+n}$	$6^2 \cdot 6^5 = 6^{2+5} = 6^7$
Power rules (a)	$(a^m)^n = a^{mn}$	$(3^2)^4 = 3^{2 \cdot 4} = 3^8$
(b)	$(ab)^m = a^m b^m$	$(2p)^5 = 2^5 p^5$
(c)	$\left(\dfrac{a}{b}\right)^m = \dfrac{a^m}{b^m}$ $(b \neq 0)$	$\left(\dfrac{5}{3}\right)^2 = \dfrac{5^2}{3^2}$

OBJECTIVE 6 Use combinations of rules. As shown in the next example, more than one rule may be needed to simplify an expression with exponents.

CHAPTER 5 Exponents and Polynomials

EXAMPLE 6 Using Combinations of Rules

Use the rules for exponents to simplify each expression.

(a) $\left(\dfrac{2}{3}\right)^2 \cdot 2^3 = \dfrac{2^2}{3^2} \cdot \dfrac{2^3}{1}$ Power rule (c)

$= \dfrac{2^2 \cdot 2^3}{3^2 \cdot 1}$ Multiply fractions.

$= \dfrac{2^5}{3^2}$ Product rule

(b) $(5x)^3(5x)^4 = (5x)^7$ Product rule
$= 5^7 x^7$ Power rule (b)

(c) $(2x^2y^3)^4(3xy^2)^3 = 2^4(x^2)^4(y^3)^4 \cdot 3^3x^3(y^2)^3$ Power rule (b)
$= 2^4 x^8 y^{12} \cdot 3^3 x^3 y^6$ Power rule (a)
$= 2^4 \cdot 3^3 x^8 x^3 y^{12} y^6$ Commutative and associative properties
$= 16 \cdot 27 x^{11} y^{18}$ Product rule
$= 432 x^{11} y^{18}$

(d) $(-x^3y)^2(-x^5y^4)^3 = (-1 \cdot x^3 y)^2(-1 \cdot x^5 y^4)^3$
$= (-1)^2(x^3)^2 y^2 \cdot (-1)^3(x^5)^3(y^4)^3$ Power rule (b)
$= (-1)^2(x^6)(y^2)(-1)^3(x^{15})(y^{12})$ Power rule (a)
$= (-1)^5(x^{21})(y^{14})$ Product rule
$= -x^{21}y^{14}$

Now Try Exercises 63, 67, 75, and 77.

CAUTION Refer to Example 6(c). Notice that
$$(2x^2y^3)^4 = 2^4 x^{2\cdot4} y^{3\cdot4}, \quad \text{not} \quad (2 \cdot 4)x^{2\cdot4} y^{3\cdot4}.$$
Do not multiply the coefficient 2 and the exponent 4.

5.1 EXERCISES

For Extra Help

- Student's Solutions Manual
- MyMathLab
- InterAct Math Tutorial Software
- AW Math Tutor Center
- MathXL
- Digital Video Tutor CD 9/Videotape 8

Decide whether each statement is true *or* false.

1. $3^3 = 9$ 2. $(-2)^4 = 2^4$ 3. $(a^2)^3 = a^5$ 4. $\left(\dfrac{1}{4}\right)^2 = \dfrac{1}{4^2}$

Write each expression using exponents.

5. $w \cdot w \cdot w \cdot w \cdot w \cdot w$ 6. $t \cdot t \cdot t \cdot t \cdot t \cdot t \cdot t$

7. $\dfrac{1}{4 \cdot 4 \cdot 4 \cdot 4}$ 8. $\dfrac{1}{3 \cdot 3 \cdot 3}$

9. $(-7x)(-7x)(-7x)(-7x)$ 10. $(-8p)(-8p)$

11. $\left(\dfrac{1}{2}\right)\left(\dfrac{1}{2}\right)\left(\dfrac{1}{2}\right)\left(\dfrac{1}{2}\right)\left(\dfrac{1}{2}\right)\left(\dfrac{1}{2}\right)$ 12. $\left(-\dfrac{1}{4}\right)\left(-\dfrac{1}{4}\right)\left(-\dfrac{1}{4}\right)\left(-\dfrac{1}{4}\right)\left(-\dfrac{1}{4}\right)$

13. Explain how the expressions $(-3)^4$ and -3^4 are different.

14. Explain how the expressions $(5x)^3$ and $5x^3$ are different.

Identify the base and the exponent for each exponential expression. In Exercises 15–18, also evaluate each expression. See Example 1.

15. 3^5 **16.** 2^7 **17.** $(-3)^5$ **18.** $(-2)^7$

19. $(-6x)^4$ **20.** $(-8x)^4$ **21.** $-6x^4$ **22.** $-8x^4$

23. Explain why the product rule does not apply to the expression $5^2 + 5^3$. Then evaluate the expression by finding the individual powers and adding the results.

24. Repeat Exercise 23 for the expression $(-4)^3 + (-4)^4$.

Use the product rule, if possible, to simplify each expression. Write each answer in exponential form. See Example 2.

25. $5^2 \cdot 5^6$ **26.** $3^6 \cdot 3^7$ **27.** $4^2 \cdot 4^7 \cdot 4^3$

28. $5^3 \cdot 5^8 \cdot 5^2$ **29.** $(-7)^3(-7)^6$ **30.** $(-9)^8(-9)^5$

31. $t^3 \cdot t^8 \cdot t^{13}$ **32.** $n^5 \cdot n^6 \cdot n^9$ **33.** $(-8r^4)(7r^3)$

34. $(10a^7)(-4a^3)$ **35.** $(-6p^5)(-7p^5)$ **36.** $(-5w^8)(-9w^8)$

37. $(5x^2)(-2x^3)(3x^4)$ **38.** $(12y^3)(4y)(-3y^5)$ **39.** $3^8 + 3^9$

40. $4^{12} + 4^5$ **41.** $5^8 \cdot 3^9$ **42.** $6^3 \cdot 8^9$

Use the power rules for exponents to simplify each expression. Write each answer in exponential form. See Examples 3–5.

43. $(4^3)^2$ **44.** $(8^3)^6$ **45.** $(t^4)^5$

46. $(y^6)^5$ **47.** $(7r)^3$ **48.** $(11x)^4$

49. $(5xy)^5$ **50.** $(9pq)^6$ **51.** $(-5^2)^6$

52. $(-9^4)^8$ **53.** $(-8^3)^5$ **54.** $(-7^5)^7$

55. $8(qr)^3$ **56.** $4(vw)^5$ **57.** $\left(\dfrac{1}{2}\right)^3$

58. $\left(\dfrac{1}{3}\right)^5$ **59.** $\left(\dfrac{a}{b}\right)^3$ $(b \neq 0)$ **60.** $\left(\dfrac{r}{t}\right)^4$ $(t \neq 0)$

61. $\left(\dfrac{9}{5}\right)^8$ **62.** $\left(\dfrac{12}{7}\right)^3$

Use a combination of the rules of exponents to simplify each expression. See Example 6.

63. $\left(\dfrac{5}{2}\right)^3 \cdot \left(\dfrac{5}{2}\right)^2$ **64.** $\left(\dfrac{3}{4}\right)^5 \cdot \left(\dfrac{3}{4}\right)^6$ **65.** $\left(\dfrac{9}{8}\right)^3 \cdot 9^2$

66. $\left(\dfrac{8}{5}\right)^4 \cdot 8^3$ **67.** $(2x)^9(2x)^3$ **68.** $(6y)^5(6y)^8$

69. $(-6p)^4(-6p)$ **70.** $(-13q)^3(-13q)$ **71.** $(6x^2y^3)^5$

72. $(5r^5t^6)^7$ **73.** $(x^2)^3(x^3)^5$ **74.** $(y^4)^5(y^3)^5$

75. $(2w^2x^3y)^2(x^4y)^5$ **76.** $(3x^4y^2z)^3(yz^4)^5$ **77.** $(-r^4s)^2(-r^2s^3)^5$

78. $(-ts^6)^4(-t^3s^5)^3$

79. $\left(\dfrac{5a^2b^5}{c^6}\right)^3$ $(c \neq 0)$

80. $\left(\dfrac{6x^3y^9}{z^5}\right)^4$ $(z \neq 0)$

81. A student tried to simplify $(10^2)^3$ as 1000^6. Is this correct? If not, how is it simplified using the product rule for exponents?

82. Explain why $(3x^2y^3)^4$ is *not* equivalent to $(3 \cdot 4)x^8y^{12}$.

Find the area of each figure. Use the formulas found on the inside covers. (The small squares in the figures indicate 90° right angles.)

83.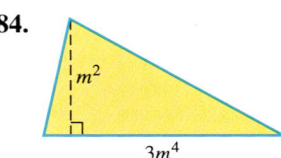
rectangle with sides $3x^2$ and $4x^3$

84.
triangle with height m^2 and base $3m^4$

85.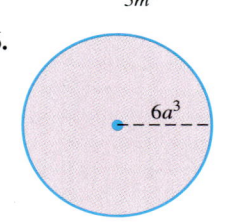
parallelogram with height $3p^2$ and base $2p^5$

86.
circle with radius $6a^3$

Find the volume of each figure. Use the formulas found on the inside covers.

87.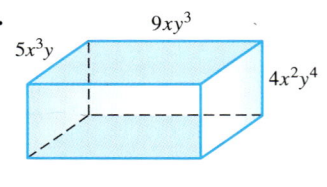
cube with side $5x^2$

88.
rectangular box with dimensions $9xy^3$, $5x^3y$, $4x^2y^4$

89. Assume a is a positive number greater than 1. Arrange the following terms in order from smallest to largest: $-(-a)^3$, $-a^3$, $(-a)^4$, $-a^4$. Explain how you decided on the order.

90. Devise a rule to tell whether an exponential expression with a negative base is positive or negative.

In Chapter 2 we used the formula for simple interest, $I = prt$, which deals with interest paid only on the principal. With **compound interest,** interest is paid on the principal and the interest earned earlier. The formula for compound interest, which involves an exponential expression, is

$$A = P(1 + r)^n.$$

Here A is the amount accumulated from a principal of P dollars left untouched for n years with an annual interest rate r (expressed as a decimal).

In Exercises 91–94, use this formula and a calculator to find A to the nearest cent.

91. $P = \$250$, $r = .04$, $n = 5$

92. $P = \$400$, $r = .04$, $n = 3$

93. $P = \$1500$, $r = .035$, $n = 6$

94. $P = \$2000$, $r = .025$, $n = 4$

5.2 Integer Exponents and the Quotient Rule

OBJECTIVES

1. Use 0 as an exponent.
2. Use negative numbers as exponents.
3. Use the quotient rule for exponents.
4. Use combinations of rules.

In Section 5.1 we studied the product rule for exponents. In all our earlier work, exponents were positive integers. Now we want to develop meaning for exponents that are not positive integers.

Consider the following list of exponential expressions.

$$2^4 = 16$$
$$2^3 = 8$$
$$2^2 = 4$$

Do you see the pattern in the values? Each time we reduce the exponent by 1, the value is divided by 2 (the base). Using this pattern, we can continue the list to smaller and smaller integer exponents.

$$2^1 = 2$$
$$2^0 = 1$$
$$2^{-1} = \frac{1}{2}$$
$$2^{-2} = \frac{1}{4}$$
$$2^{-3} = \frac{1}{8}$$

From the preceding list, it appears that we should define 2^0 as 1 and negative exponents as reciprocals.

OBJECTIVE 1 Use 0 as an exponent. We want the definitions of 0 and negative exponents to satisfy the rules for exponents from Section 5.1. For example, if $6^0 = 1$,

$$6^0 \cdot 6^2 = 1 \cdot 6^2 = 6^2 \quad \text{and} \quad 6^0 \cdot 6^2 = 6^{0+2} = 6^2,$$

so the product rule is satisfied. Check that the power rules are also valid for a 0 exponent. Thus, we define a 0 exponent as follows.

Zero Exponent

For any nonzero real number a, $a^0 = 1.$

Example: $17^0 = 1$

EXAMPLE 1 Using Zero Exponents

Evaluate each exponential expression.

(a) $60^0 = 1$ (b) $(-60)^0 = 1$
(c) $-60^0 = -(1) = -1$ (d) $y^0 = 1 \quad (y \neq 0)$
(e) $6y^0 = 6(1) = 6 \quad (y \neq 0)$ (f) $(6y)^0 = 1 \quad (y \neq 0)$

Now Try Exercises 1 and 5.

322 CHAPTER 5 Exponents and Polynomials

> **CAUTION** Notice the difference between parts (b) and (c) of Example 1. In $(-60)^0$, the base is -60 and the exponent is 0. Any nonzero base raised to the 0 exponent is 1. But in -60^0, the base is 60. Then $60^0 = 1$, and $-60^0 = -1$.

OBJECTIVE 2 Use negative numbers as exponents. From the lists at the beginning of this section, since $2^{-2} = \frac{1}{4}$ and $2^{-3} = \frac{1}{8}$, we can deduce that 2^{-n} should equal $\frac{1}{2^n}$. Is the product rule valid in such cases? For example, if we multiply 6^{-2} by 6^2, we get

$$6^{-2} \cdot 6^2 = 6^{-2+2} = 6^0 = 1.$$

The expression 6^{-2} behaves as if it were the reciprocal of 6^2 because their product is 1. The reciprocal of 6^2 is also $\frac{1}{6^2}$, leading us to define 6^{-2} as $\frac{1}{6^2}$. This is a particular case of the definition of negative exponents.

Negative Exponents

For any nonzero real number a and any integer n, $\quad a^{-n} = \dfrac{1}{a^n}.$

Example: $3^{-2} = \dfrac{1}{3^2}$

By definition, a^{-n} and a^n are reciprocals, since

$$a^n \cdot a^{-n} = a^n \cdot \frac{1}{a^n} = 1.$$

Since $1^n = 1$, the definition of a^{-n} can also be written

$$a^{-n} = \frac{1}{a^n} = \frac{1^n}{a^n} = \left(\frac{1}{a}\right)^n.$$

For example,

$$6^{-3} = \left(\frac{1}{6}\right)^3 \quad \text{and} \quad \left(\frac{1}{3}\right)^{-2} = 3^2.$$

EXAMPLE 2 Using Negative Exponents

Simplify by writing each expression with positive exponents.

(a) $3^{-2} = \dfrac{1}{3^2} = \dfrac{1}{9}$

(b) $5^{-3} = \dfrac{1}{5^3} = \dfrac{1}{125}$

(c) $\left(\dfrac{1}{2}\right)^{-3} = 2^3 = 8 \quad$ $\frac{1}{2}$ and 2 are reciprocals.

Notice that we can change the base to its reciprocal if we also change the sign of the exponent.

(d) $\left(\dfrac{2}{5}\right)^{-4} = \left(\dfrac{5}{2}\right)^4 \quad$ $\frac{2}{5}$ and $\frac{5}{2}$ are reciprocals.

(e) $\left(\dfrac{4}{3}\right)^{-5} = \left(\dfrac{3}{4}\right)^5$

(f) $4^{-1} - 2^{-1} = \dfrac{1}{4} - \dfrac{1}{2} = \dfrac{1}{4} - \dfrac{2}{4} = -\dfrac{1}{4}$

We apply the exponents first, then subtract.

(g) $p^{-2} = \dfrac{1}{p^2} \quad (p \neq 0)$

(h) $\dfrac{1}{x^{-4}} \quad (x \neq 0)$

$$\dfrac{1}{x^{-4}} = \dfrac{1^{-4}}{x^{-4}} \qquad 1^{-4} = 1$$

$$= \left(\dfrac{1}{x}\right)^{-4} \qquad \text{Power rule (c)}$$

$$= x^4 \qquad \dfrac{1}{x} \text{ and } x \text{ are reciprocals.}$$

Now Try Exercises 19, 21, 23, and 27.

> **CAUTION** A negative exponent does not indicate a negative number; negative exponents lead to reciprocals.
>
> | Expression | Example | |
> |---|---|---|
> | a^{-n} | $3^{-2} = \dfrac{1}{3^2} = \dfrac{1}{9}$ | Not negative |
> | $-a^{-n}$ | $-3^{-2} = -\dfrac{1}{3^2} = -\dfrac{1}{9}$ | Negative |

The definition of negative exponents allows us to move factors across a fraction bar if we also change the signs of the exponents. For example,

$$\dfrac{2^{-3}}{3^{-4}} = \dfrac{\frac{1}{2^3}}{\frac{1}{3^4}} = \dfrac{1}{2^3} \cdot \dfrac{3^4}{1} = \dfrac{3^4}{2^3},$$

so

$$\dfrac{2^{-3}}{3^{-4}} = \dfrac{3^4}{2^3}.$$

> **Changing from Negative to Positive Exponents**
>
> For any nonzero numbers a and b, and any integers m and n,
>
> $$\dfrac{a^{-m}}{b^{-n}} = \dfrac{b^n}{a^m} \quad \text{and} \quad \left(\dfrac{a}{b}\right)^{-m} = \left(\dfrac{b}{a}\right)^m.$$
>
> *Examples:* $\dfrac{3^{-5}}{2^{-4}} = \dfrac{2^4}{3^5} \quad \text{and} \quad \left(\dfrac{4}{5}\right)^{-3} = \left(\dfrac{5}{4}\right)^3$

■ EXAMPLE 3 Changing from Negative to Positive Exponents

Write with only positive exponents. Assume all variables represent nonzero real numbers.

324 CHAPTER 5 Exponents and Polynomials

(a) $\dfrac{4^{-2}}{5^{-3}} = \dfrac{5^3}{4^2}$ or $\dfrac{125}{16}$

(b) $\dfrac{m^{-5}}{p^{-1}} = \dfrac{p^1}{m^5} = \dfrac{p}{m^5}$

(c) $\dfrac{a^{-2}b}{3d^{-3}} = \dfrac{bd^3}{3a^2}$

Notice that b in the numerator and 3 in the denominator were not affected.

(d) $x^3 y^{-4} = \dfrac{x^3 y^{-4}}{1} = \dfrac{x^3}{y^4}$

(e) $\left(\dfrac{x}{2y}\right)^{-4} = \left(\dfrac{2y}{x}\right)^4 = \dfrac{2^4 y^4}{x^4}$

Now Try Exercises 31, 35, and 47.

CAUTION Be careful. We cannot change negative exponents to positive exponents using this rule if the exponents occur in a sum of terms. For example,

$$\dfrac{5^{-2} + 3^{-1}}{7 - 2^{-3}}$$

cannot be written with positive exponents using the rule given here. We would have to use the definition of a negative exponent to rewrite this expression with positive exponents, as

$$\dfrac{\dfrac{1}{5^2} + \dfrac{1}{3}}{7 - \dfrac{1}{2^3}}.$$

OBJECTIVE 3 Use the quotient rule for exponents. How should we handle the quotient of two exponential expressions with the same base? We know that

$$\dfrac{6^5}{6^3} = \dfrac{6 \cdot 6 \cdot 6 \cdot 6 \cdot 6}{6 \cdot 6 \cdot 6} = 6^2.$$

Notice that the difference between the exponents, $5 - 3 = 2$, is the exponent in the quotient. Also,

$$\dfrac{6^2}{6^4} = \dfrac{6 \cdot 6}{6 \cdot 6 \cdot 6 \cdot 6} = \dfrac{1}{6^2} = 6^{-2}.$$

Here, $2 - 4 = -2$. These examples suggest the **quotient rule for exponents.**

Quotient Rule for Exponents

For any nonzero real number a and any integers m and n,

$$\dfrac{a^m}{a^n} = a^{m-n}.$$

(Keep the base and subtract the exponents.)

Example: $\dfrac{5^8}{5^4} = 5^{8-4} = 5^4$

CAUTION A common **error** is to write $\dfrac{5^8}{5^4} = 1^{8-4} = 1^4$. Notice that by the quotient rule, the quotient should have the *same base*, 5. That is,

$$\dfrac{5^8}{5^4} = 5^{8-4} = 5^4.$$

If you are not sure, use the definition of an exponent to write out the factors:

$$5^8 = 5 \cdot 5 \cdot 5 \cdot 5 \cdot 5 \cdot 5 \cdot 5 \cdot 5 \quad \text{and} \quad 5^4 = 5 \cdot 5 \cdot 5 \cdot 5.$$

Then it is clear that the quotient is 5^4.

EXAMPLE 4 Using the Quotient Rule

Simplify, using the quotient rule for exponents. Write answers with positive exponents.

(a) $\dfrac{5^8}{5^6} = 5^{8-6} = 5^2$ or 25

(b) $\dfrac{4^2}{4^9} = 4^{2-9} = 4^{-7} = \dfrac{1}{4^7}$

(c) $\dfrac{5^{-3}}{5^{-7}} = 5^{-3-(-7)} = 5^4$ or 625

(d) $\dfrac{q^5}{q^{-3}} = q^{5-(-3)} = q^8 \quad (q \neq 0)$

(e) $\dfrac{3^2 x^5}{3^4 x^3} = \dfrac{3^2}{3^4} \cdot \dfrac{x^5}{x^3} = 3^{2-4} \cdot x^{5-3} = 3^{-2} x^2 = \dfrac{x^2}{3^2} \quad (x \neq 0)$

(f) $\dfrac{(m+n)^{-2}}{(m+n)^{-4}} = (m+n)^{-2-(-4)} = (m+n)^{-2+4} = (m+n)^2 \quad (m \neq -n)$

(g) $\dfrac{7x^{-3}y^2}{2^{-1}x^2y^{-5}} = \dfrac{7 \cdot 2^1 y^2 y^5}{x^2 x^3} = \dfrac{14y^7}{x^5} \quad (x, y \neq 0)$

Now Try Exercises 29, 41, 45, and 49.

The definitions and rules for exponents given in this section and Section 5.1 are summarized here.

Definitions and Rules for Exponents

For any integers m and n:

		Examples
Product rule	$a^m \cdot a^n = a^{m+n}$	$7^4 \cdot 7^5 = 7^9$
Zero exponent	$a^0 = 1 \quad (a \neq 0)$	$(-3)^0 = 1$
Negative exponent	$a^{-n} = \dfrac{1}{a^n} \quad (a \neq 0)$	$5^{-3} = \dfrac{1}{5^3}$
Quotient rule	$\dfrac{a^m}{a^n} = a^{m-n} \quad (a \neq 0)$	$\dfrac{2^2}{2^5} = 2^{2-5} = 2^{-3} = \dfrac{1}{2^3}$
Power rules (a)	$(a^m)^n = a^{mn}$	$(4^2)^3 = 4^6$
(b)	$(ab)^m = a^m b^m$	$(3k)^4 = 3^4 k^4$
(c)	$\left(\dfrac{a}{b}\right)^m = \dfrac{a^m}{b^m} \quad (b \neq 0)$	$\left(\dfrac{2}{3}\right)^2 = \dfrac{2^2}{3^2}$

(continued)

326 CHAPTER 5 Exponents and Polynomials

Negative to positive rules	$\dfrac{a^{-m}}{b^{-n}} = \dfrac{b^n}{a^m}$ $(a \neq 0, b \neq 0)$	$\dfrac{2^{-4}}{5^{-3}} = \dfrac{5^3}{2^4}$
	$\left(\dfrac{a}{b}\right)^{-m} = \left(\dfrac{b}{a}\right)^m$	$\left(\dfrac{4}{7}\right)^{-2} = \left(\dfrac{7}{4}\right)^2$

OBJECTIVE 4 Use combinations of rules. We may sometimes need to use more than one rule to simplify an expression.

EXAMPLE 5 Using Combinations of Rules

Use the rules for exponents to simplify each expression. Assume all variables represent nonzero real numbers.

(a) $\dfrac{(4^2)^3}{4^5} = \dfrac{4^6}{4^5}$ Power rule (a)

$\phantom{\dfrac{(4^2)^3}{4^5}} = 4^{6-5}$ Quotient rule

$\phantom{\dfrac{(4^2)^3}{4^5}} = 4^1 = 4$

(b) $(2x)^3(2x)^2 = (2x)^5$ Product rule

$ = 2^5 x^5$ or $32x^5$ Power rule (b)

(c) $\left(\dfrac{2x^3}{5}\right)^{-4} = \left(\dfrac{5}{2x^3}\right)^4$ Negative to positive rule

$\phantom{\left(\dfrac{2x^3}{5}\right)^{-4}} = \dfrac{5^4}{2^4 x^{12}}$ or $\dfrac{625}{16x^{12}}$ Power rules (a)–(c)

(d) $\left(\dfrac{3x^{-2}}{4^{-1}y^3}\right)^{-3} = \dfrac{3^{-3}x^6}{4^3 y^{-9}}$ Power rules (a)–(c)

$\phantom{\left(\dfrac{3x^{-2}}{4^{-1}y^3}\right)^{-3}} = \dfrac{x^6 y^9}{4^3 \cdot 3^3}$ or $\dfrac{x^6 y^9}{1728}$ Negative to positive rule

(e) $\dfrac{(4m)^{-3}}{(3m)^{-4}} = \dfrac{4^{-3} m^{-3}}{3^{-4} m^{-4}}$ Power rule (b)

$\phantom{\dfrac{(4m)^{-3}}{(3m)^{-4}}} = \dfrac{3^4 m^4}{4^3 m^3}$ Negative to positive rule

$\phantom{\dfrac{(4m)^{-3}}{(3m)^{-4}}} = \dfrac{3^4 m^{4-3}}{4^3}$ Quotient rule

$\phantom{\dfrac{(4m)^{-3}}{(3m)^{-4}}} = \dfrac{3^4 m}{4^3}$ or $\dfrac{81m}{64}$

Now Try Exercises 57, 63, 65, and 67.

NOTE Since the steps can be done in several different orders, there are many equally correct ways to simplify expressions like those in Examples 5(d) and 5(e).

5.2 EXERCISES

For Extra Help

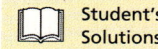 Student's Solutions Manual

 MyMathLab

 InterAct Math Tutorial Software

 AW Math Tutor Center

 MathXL

Digital Video Tutor CD 9/Videotape 8

Each expression is either equal to 0, 1, or −1. Decide which is correct. See Example 1.

1. 9^0
2. 5^0
3. $(-4)^0$
4. $(-10)^0$
5. -9^0
6. -5^0
7. $(-2)^0 - 2^0$
8. $(-8)^0 - 8^0$
9. $\dfrac{0^{10}}{10^0}$
10. $\dfrac{0^5}{5^0}$

Match each expression in Column I with the equivalent expression in Column II. Choices in Column II may be used once, more than once, or not at all.

I		II
11. -2^{-4}		A. 8
12. $(-2)^{-4}$		B. 16
13. 2^{-4}		C. $-\dfrac{1}{16}$
14. $\dfrac{1}{2^{-4}}$		D. -8
15. $\dfrac{1}{-2^{-4}}$		E. -16
16. $\dfrac{1}{(-2)^{-4}}$		F. $\dfrac{1}{16}$

Evaluate each expression. See Examples 1 and 2.

17. $7^0 + 9^0$
18. $8^0 + 6^0$
19. 4^{-3}
20. 5^{-4}
21. $\left(\dfrac{1}{2}\right)^{-4}$
22. $\left(\dfrac{1}{3}\right)^{-3}$
23. $\left(\dfrac{6}{7}\right)^{-2}$
24. $\left(\dfrac{2}{3}\right)^{-3}$
25. $(-3)^{-4}$
26. $(-4)^{-3}$
27. $5^{-1} + 3^{-1}$
28. $6^{-1} + 2^{-1}$

Use the quotient rule to simplify each expression. Write the expression with positive exponents. Assume that all variables represent nonzero real numbers. See Examples 2–4.

29. $\dfrac{5^8}{5^5}$
30. $\dfrac{11^6}{11^3}$
31. $\dfrac{3^{-2}}{5^{-3}}$
32. $\dfrac{4^{-3}}{3^{-2}}$
33. $\dfrac{5}{5^{-1}}$
34. $\dfrac{6}{6^{-2}}$
35. $\dfrac{x^{12}}{x^{-3}}$
36. $\dfrac{y^4}{y^{-6}}$
37. $\dfrac{1}{6^{-3}}$
38. $\dfrac{1}{5^{-2}}$
39. $\dfrac{2}{r^{-4}}$
40. $\dfrac{3}{s^{-8}}$
41. $\dfrac{4^{-3}}{5^{-2}}$
42. $\dfrac{6^{-2}}{5^{-4}}$
43. $p^5 q^{-8}$
44. $x^{-8} y^4$
45. $\dfrac{r^5}{r^{-4}}$
46. $\dfrac{a^6}{a^{-4}}$
47. $\dfrac{x^{-3} y}{4 z^{-2}}$
48. $\dfrac{p^{-5} q^{-4}}{9 r^{-3}}$
49. $\dfrac{(a+b)^{-3}}{(a+b)^{-4}}$
50. $\dfrac{(x+y)^{-8}}{(x+y)^{-9}}$
51. $\dfrac{(x+2y)^{-3}}{(x+2y)^{-5}}$
52. $\dfrac{(p-3q)^{-2}}{(p-3q)^{-4}}$

CHAPTER 5 Exponents and Polynomials

RELATING CONCEPTS (EXERCISES 53–56)

For Individual or Group Work

In Objective 1, we showed how 6^0 acts as 1 when it is applied to the product rule, thus motivating the definition for 0 as an exponent. We can also use the quotient rule to motivate this definition. **Work Exercises 53–56 in order.**

53. Consider the expression $\frac{25}{25}$. What is its simplest form?

54. Because $25 = 5^2$, the expression $\frac{25}{25}$ can be written as the quotient of powers of 5. Write the expression in this way.

55. Apply the quotient rule for exponents to the expression you wrote in Exercise 54. Give the answer as a power of 5.

56. Your answers in Exercises 53 and 55 must be equal because they both represent $\frac{25}{25}$. Write this equality. What definition does this result support?

Use a combination of the rules for exponents to simplify each expression. Write answers with only positive exponents. Assume that all variables represent nonzero real numbers. See Example 5.

57. $\dfrac{(7^4)^3}{7^9}$

58. $\dfrac{(5^3)^2}{5^2}$

59. $x^{-3} \cdot x^5 \cdot x^{-4}$

60. $y^{-8} \cdot y^5 \cdot y^{-2}$

61. $\dfrac{(3x)^{-2}}{(4x)^{-3}}$

62. $\dfrac{(2y)^{-3}}{(5y)^{-4}}$

63. $\left(\dfrac{x^{-1}y}{z^2}\right)^{-2}$

64. $\left(\dfrac{p^{-4}q}{r^{-3}}\right)^{-3}$

65. $(6x)^4(6x)^{-3}$

66. $(10y)^9(10y)^{-8}$

67. $\dfrac{(m^7n)^{-2}}{m^{-4}n^3}$

68. $\dfrac{(m^8n^{-4})^2}{m^{-2}n^5}$

69. $\dfrac{(x^{-1}y^2z)^{-2}}{(x^{-3}y^3z)^{-1}}$

70. $\dfrac{(a^{-2}b^{-3}c^{-4})^{-5}}{(a^2b^3c^4)^5}$

71. $\left(\dfrac{xy^{-2}}{x^2y}\right)^{-3}$

72. $\left(\dfrac{wz^{-5}}{w^{-3}z}\right)^{-2}$

73. $\dfrac{(4a^2b^3)^{-2}(2ab^{-1})^3}{(a^3b)^{-4}}$

74. $\dfrac{(m^6n)^{-2}(m^2n^{-2})^3}{m^{-1}n^{-2}}$

75. $\dfrac{(2y^{-1}z^2)^2(3y^{-2}z^{-3})^3}{(y^3z^2)^{-1}}$

76. $\dfrac{(3p^{-2}q^3)^2(5p^{-1}q^{-4})^{-1}}{(p^2q^{-2})^{-3}}$

77. $\dfrac{(9^{-1}z^{-2}x)^{-1}(4z^2x^4)^{-2}}{(5z^{-2}x^{-3})^2}$

78. $\dfrac{(4^{-1}a^{-1}b^{-2})^{-2}(5a^{-3}b^4)^{-2}}{(3a^{-3}b^{-5})^2}$

79. Consider the following typical student **error**:

$$\dfrac{16^3}{2^2} = \left(\dfrac{16}{2}\right)^{3-2} = 8^1 = 8.$$

Explain what the student did incorrectly, and then give the correct answer.

80. Consider the following typical student **error**:

$$-5^4 = (-5)^4 = 625.$$

Explain what the student did incorrectly, and then give the correct answer.

SUMMARY EXERCISES ON THE RULES FOR EXPONENTS

Use the rules for exponents to simplify each expression. Use only positive exponents in your answers. Assume all variable expressions represent positive numbers.

1. $\left(\dfrac{6x^2}{5}\right)^{12}$

2. $\left(\dfrac{rs^2t^3}{3t^4}\right)^6$

3. $(10x^2y^4)^2(10xy^2)^3$

4. $(-2ab^3c)^4(-2a^2b)^3$

5. $\left(\dfrac{9wx^3}{y^4}\right)^3$

6. $(4x^{-2}y^{-3})^{-2}$

7. $\dfrac{c^{11}(c^2)^4}{(c^3)^3(c^2)^{-6}}$

8. $\left(\dfrac{k^4t^2}{k^2t^{-4}}\right)^{-2}$

9. $5^{-1} + 6^{-1}$

10. $\dfrac{(3y^{-1}z^3)^{-1}(3y^2)}{(y^3z^2)^{-3}}$

11. $\dfrac{(2xy^{-1})^3}{2^3x^{-3}y^2}$

12. $-8^0 + (-8)^0$

13. $(z^4)^{-3}(z^{-2})^{-5}$

14. $\left(\dfrac{r^2st^5}{3r}\right)^{-2}$

15. $\dfrac{(3^{-1}x^{-3}y)^{-1}(2x^2y^{-3})^2}{(5x^{-2}y^2)^{-2}}$

16. $\left(\dfrac{5x^2}{3x^{-4}}\right)^{-1}$

17. $\left(\dfrac{-2x^{-2}}{2x^2}\right)^{-2}$

18. $\dfrac{(x^{-4}y^2)^3(x^2y)^{-1}}{(xy^2)^{-3}}$

19. $\dfrac{(a^{-2}b^3)^{-4}}{(a^{-3}b^2)^{-2}(ab)^{-4}}$

20. $(2a^{-30}b^{-29})(3a^{31}b^{30})$

21. $5^{-2} + 6^{-2}$

22. $\left[\dfrac{(x^{47}y^{23})^2}{x^{-26}y^{-42}}\right]^0$

23. $\left(\dfrac{7a^2b^3}{2}\right)^3$

24. $-(-12^0)$

25. $-(-12)^0$

26. $\dfrac{0^{12}}{12^0}$

27. $\dfrac{(2xy^{-3})^{-2}}{(3x^{-2}y^4)^{-3}}$

28. $\left(\dfrac{a^2b^3c^4}{a^{-2}b^{-3}c^{-4}}\right)^{-2}$

29. $(6x^{-5}z^3)^{-3}$

30. $(2p^{-2}qr^{-3})(2p)^{-4}$

31. $\dfrac{(xy)^{-3}(xy)^5}{(xy)^{-4}}$

32. $42^0 - (-12)^0$

33. $\dfrac{(7^{-1}x^{-3})^{-2}(x^4)^{-6}}{7^{-1}x^{-3}}$

34. $\left(\dfrac{3^{-4}x^{-3}}{3^{-3}x^{-6}}\right)^{-2}$

35. $(5p^{-2}q)^{-3}(5pq^3)^4$

36. $8^{-1} + 6^{-1}$

37. $\left[\dfrac{4r^{-6}s^{-2}t}{2r^8s^{-4}t^2}\right]^{-1}$

38. $(13x^{-6}y)(13x^{-6}y)^{-1}$

39. $\dfrac{(8pq^{-2})^4}{(8p^{-2}q^{-3})^3}$

40. $\left(\dfrac{mn^{-2}p}{m^2np^4}\right)^{-2}\left(\dfrac{mn^{-2}p}{m^2np^4}\right)^3$

41. $-(-3^0)^0$

42. $5^{-1} - 8^{-1}$

330 CHAPTER 5 Exponents and Polynomials

5.3 An Application of Exponents: Scientific Notation

OBJECTIVES

1. Express numbers in scientific notation.
2. Convert numbers in scientific notation to numbers without exponents.
3. Use scientific notation in calculations.

OBJECTIVE 1 Express numbers in scientific notation. One example of the use of exponents comes from science. The numbers occurring in science are often extremely large (such as the distance from Earth to the sun, 93,000,000 mi) or extremely small (the wavelength of yellow-green light, approximately .0000006 m). Because of the difficulty of working with many zeros, scientists often express such numbers with exponents. Each number is written as $a \times 10^n$, where $1 \leq |a| < 10$ and n is an integer. This form is called **scientific notation**. There is always one nonzero digit before the decimal point. This is shown in the following examples. (In work with scientific notation, the times symbol, $\times$, is commonly used.)

$3.19 \times 10^1 = 3.19 \times 10 = 31.9$ Decimal point moves 1 place to the right.
$3.19 \times 10^2 = 3.19 \times 100 = 319.$ Decimal point moves 2 places to the right.
$3.19 \times 10^3 = 3.19 \times 1000 = 3190.$ Decimal point moves 3 places to the right.
$3.19 \times 10^{-1} = 3.19 \times .1 = .319$ Decimal point moves 1 place to the left.
$3.19 \times 10^{-2} = 3.19 \times .01 = .0319$ Decimal point moves 2 places to the left.
$3.19 \times 10^{-3} = 3.19 \times .001 = .00319$ Decimal point moves 3 places to the left.

A number in scientific notation is always written with the decimal point after the first nonzero digit and then multiplied by the appropriate power of 10. For example, 35 is written 3.5×10^1, or 3.5×10; 56,200 is written 5.62×10^4, since

$$56,200 = 5.62 \times 10,000 = 5.62 \times 10^4.$$

To write a number in scientific notation, follow these steps.

Writing a Number in Scientific Notation

Step 1 Move the decimal point to the right of the first nonzero digit.

Step 2 Count the number of places you moved the decimal point.

Step 3 The number of places in Step 2 is the absolute value of the exponent on 10.

Step 4 The exponent on 10 is positive if the original number is larger than the number in Step 1; the exponent is negative if the original number is smaller than the number in Step 1. If the decimal point is not moved, the exponent is 0.

EXAMPLE 1 Using Scientific Notation

Write each number in scientific notation.

(a) 93,000,000

The number will be written in scientific notation as 9.3×10^n. To find the value of n, first compare the original number, 93,000,000, with 9.3. Here 93,000,000 is *larger* than 9.3. Therefore, multiply by a *positive* power of 10 so the product 9.3×10^n will equal the larger number.

SECTION 5.3 An Application of Exponents: Scientific Notation 331

Move the decimal point to follow the first nonzero digit (the 9). Count the number of places the decimal point was moved.

$$93{,}000{,}000 \quad \text{7 places}$$

Since the decimal point was moved 7 places, and since n is positive,

$$93{,}000{,}000 = 9.3 \times 10^7.$$

(b) $63{,}200{,}000{,}000 = 6.3200000000 = 6.32 \times 10^{10}$

10 places

(c) .00462

Move the decimal point to the right of the first nonzero digit, and count the number of places the decimal point was moved.

$$.00462 \quad \text{3 places}$$

Since .00462 is *smaller* than 4.62, the exponent must be *negative*.

$$.00462 = 4.62 \times 10^{-3}$$

(d) $.0000762 = 7.62 \times 10^{-5}$

> Now Try Exercises 15 and 19.

NOTE To choose the exponent when writing a number in scientific notation, think: If the original number is "large," like 93,000,000, use a *positive* exponent on 10, since positive is larger than negative. However, if the original number is "small," like .00462, use a *negative* exponent on 10, since negative is smaller than positive.

OBJECTIVE 2 Convert numbers in scientific notation to numbers without exponents. To convert a number written in scientific notation to a number without exponents, work in reverse. Multiplying by a positive power of 10 will make the number larger; multiplying by a negative power of 10 will make the number smaller.

EXAMPLE 2 Writing Numbers without Exponents

Write each number without exponents.

(a) 6.2×10^3

Since the exponent is positive, make 6.2 larger by moving the decimal point 3 places to the right, inserting zeros as needed.

$$6.2 \times 10^3 = 6.200 = 6200$$

(b) $4.283 \times 10^5 = 4.28300 = 428{,}300$ Move 5 places to the right.

(c) $7.04 \times 10^{-3} = .00704$ Move 3 places to the left.

The exponent tells the number of places and the direction that the decimal point is moved.

> Now Try Exercises 23 and 29.

OBJECTIVE 3 Use scientific notation in calculations. The next example uses scientific notation with products and quotients.

EXAMPLE 3   Multiplying and Dividing with Scientific Notation

Write each product or quotient without exponents.

(a) $(6 \times 10^3)(5 \times 10^{-4}) = (6 \times 5)(10^3 \times 10^{-4})$ Commutative and associative properties

$= 30 \times 10^{-1}$ Product rule

$= 3$ Write without exponents.

(b) $\dfrac{4 \times 10^{-5}}{2 \times 10^3} = \dfrac{4}{2} \times \dfrac{10^{-5}}{10^3} = 2 \times 10^{-8} = .00000002$

Now Try Exercises 33 and 43.

NOTE Multiplying or dividing numbers written in scientific notation may produce an answer in the form $a \times 10^0$. Since $10^0 = 1$, $a \times 10^0 = a$. For example,

$$(8 \times 10^{-4})(5 \times 10^4) = 40 \times 10^0 = 40.$$

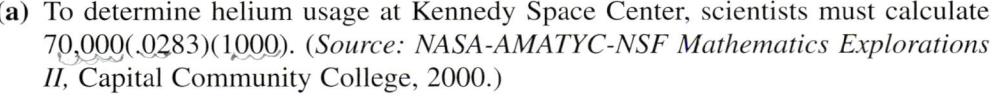

EXAMPLE 4   Calculating Using Scientific Notation

Convert to scientific notation, perform each computation, then give the result without scientific notation.

(a) To determine helium usage at Kennedy Space Center, scientists must calculate $70,000(.0283)(1000)$. (*Source:* NASA-AMATYC-NSF *Mathematics Explorations II*, Capital Community College, 2000.)

$70,000(.0283)(1000) = (7 \times 10^4)(2.83 \times 10^{-2})(1 \times 10^3)$

$= (7 \times 2.83 \times 1)(10^{4-2+3})$

$= 19.81 \times 10^5$

$= 1,981,000$

(b) The ratio of the tidal force exerted by the moon compared to that exerted by the sun is given by

$$\dfrac{73.5 \times 10^{21} \times (1.5 \times 10^8)^3}{1.99 \times 10^{30} \times (3.84 \times 10^5)^3}.$$

(*Source:* Kastner, Bernice, *Space Mathematics,* NASA.)

$\dfrac{7.35 \times 10^1 \times 10^{21} \times 1.5^3 \times 10^{24}}{1.99 \times 10^{30} \times 3.84^3 \times 10^{15}} \approx .22 \times 10^{1+21+24-30-15}$

$= .22 \times 10^1$

$= 2.2$

Now Try Exercises 61 and 65.

CONNECTIONS

Charles F. Richter devised a scale in 1935 to compare the intensities, or relative power, of earthquakes. The *intensity* of an earthquake is measured relative to the intensity of a standard *zero-level* earthquake of intensity I_0. The relationship is equivalent to $I = I_0 \times 10^R$, where R is the *Richter scale* measure. For example, if an earthquake has magnitude 5.0 on the Richter scale, then its intensity is calculated as $I = I_0 \times 10^{5.0} = I_0 \times 100,000$, which is 100,000 times as intense as a zero-level earthquake. The following diagram illustrates the intensities of earthquakes and their Richter scale magnitudes.

Intensity	$I_0 \times 10^0$	$I_0 \times 10^1$	$I_0 \times 10^2$	$I_0 \times 10^3$	$I_0 \times 10^4$	$I_0 \times 10^5$	$I_0 \times 10^6$	$I_0 \times 10^7$	$I_0 \times 10^8$
Richter Scale	0	1	2	3	4	5	6	7	8

To compare two earthquakes to each other, a ratio of the intensities is calculated. For example, to compare an earthquake that measures 8.0 on the Richter scale to one that measures 5.0, simply find the ratio of the intensities:

$$\frac{\text{intensity 8.0}}{\text{intensity 5.0}} = \frac{I_0 \times 10^{8.0}}{I_0 \times 10^{5.0}} = \frac{10^8}{10^5} = 10^{8-5} = 10^3 = 1000.$$

Therefore an earthquake that measures 8.0 on the Richter Scale is 1000 times as intense as one that measures 5.0.

For Discussion or Writing

The table gives Richter scale measurements for several earthquakes.

Earthquake	Richter Scale Measurement
1960 Concepción, Chile	9.5
1906 San Francisco, California	8.3
1939 Erzincan, Turkey	8.0
1998 Sumatra, Indonesia	7.0
1998 Adana, Turkey	6.3

Source: World Almanac and Books of Facts, 2000.

1. Compare the intensity of the 1939 Erzincan earthquake to the 1998 Sumatra earthquake.
2. Compare the intensity of the 1998 Adana earthquake to the 1906 San Francisco earthquake.
3. Compare the intensity of the 1939 Erzincan earthquake to the 1998 Adana earthquake.
4. Suppose an earthquake measures 7.2 on the Richter scale. How would the intensity of a second earthquake compare if its Richter scale measure differed by $+3.0$? By -1.0?

334 CHAPTER 5 Exponents and Polynomials

5.3 EXERCISES

For Extra Help

 Student's Solutions Manual

 MyMathLab

 InterAct Math Tutorial Software

 AW Math Tutor Center

MathXL

 Digital Video Tutor CD 9/Videotape 8

Match each number written in scientific notation in Column I with the correct choice from Column II.

I	II
1. 4.6×10^{-4}	A. .00046
2. 4.6×10^{4}	B. 46,000
3. 4.6×10^{5}	C. 460,000
4. 4.6×10^{-5}	D. .000046

Determine whether or not each number is written in scientific notation as defined in Objective 1. If it is not, write it as such.

5. 4.56×10^3 6. 7.34×10^5 7. 5,600,000   8. 34,000

9. $.8 \times 10^2$ 10. $.9 \times 10^3$ 11. .004   12. .0007

13. Explain in your own words what it means for a number to be written in scientific notation. Give examples.

14. Explain how to multiply a number by a positive power of ten. Then explain how to multiply a number by a negative power of ten.

Write each number in scientific notation. See Example 1.

15. 5,876,000,000   16. 9,994,000,000   17. 82,350   18. 78,330

19. .000007   20. .0000004   21. .00203   22. .0000578

Write each number without exponents. See Example 2.

23. 7.5×10^5 24. 8.8×10^6 25. 5.677×10^{12} 26. 8.766×10^9

27. 6.21×10^0 28. 8.56×10^0 29. 7.8×10^{-4} 30. 8.9×10^{-5}

31. 5.134×10^{-9} 32. 7.123×10^{-10}

Use properties and rules for exponents to perform the indicated operations, and write each answer without exponents. See Example 3.

33. $(2 \times 10^8) \times (3 \times 10^3)$ 34. $(4 \times 10^7) \times (3 \times 10^3)$

35. $(5 \times 10^4) \times (3 \times 10^2)$ 36. $(8 \times 10^5) \times (2 \times 10^3)$

37. $(3 \times 10^{-4}) \times (2 \times 10^8)$ 38. $(4 \times 10^{-3}) \times (2 \times 10^7)$

39. $\dfrac{9 \times 10^{-5}}{3 \times 10^{-1}}$ 40. $\dfrac{12 \times 10^{-4}}{4 \times 10^{-3}}$ 41. $\dfrac{8 \times 10^3}{2 \times 10^2}$

42. $\dfrac{5 \times 10^4}{1 \times 10^3}$ 43. $\dfrac{2.6 \times 10^{-3}}{2 \times 10^2}$ 44. $\dfrac{9.5 \times 10^{-1}}{5 \times 10^3}$

TECHNOLOGY INSIGHTS (EXERCISES 45–50)

Graphing calculators such as the TI-83 Plus can display numbers in scientific notation (when in scientific mode), using the format shown in the screen at the top left on the next page. For example, the calculator displays 5.4E3 to represent 5.4×10^3, *the scientific notation form for 5400. The display 5.4E−4 means 5.4 × 10⁻⁴. It will also perform operations with numbers entered in scientific notation, as shown in the screen on the right. Notice how the rules for exponents are applied.*

```
5400
            5.4E3
5.4*10^(-4)
           5.4E-4
```

```
(2E5)*(4E-2)
              8E3
(2E5)/(4E-2)
              5E6
(2E5)³
             8E15
```

Predict the display the calculator would give for the expression shown in each screen.

45. `.00000047`

46. `.000021`

47. `(8E5)/(4E-2)`

48. `(9E-4)/(3E3)`

49. `(2E6)*(2E-3)/(4E2)`

50. `(5E-3)*(1E9)/(5E3)`

Each statement comes from Astronomy! A Brief Edition *by James B. Kaler (Addison-Wesley, 1997). If the number in italics is in scientific notation, write it without exponents. If the number is written without exponents, write it in scientific notation.*

51. Multiplying this view over the whole sky yields a galaxy count of more than *10 billion.* (page 496)

52. The circumference of the solar orbit is . . . about *4.7 million* km . . . . (in reference to the orbit of Jupiter, page 395)

53. The solar luminosity requires that *2 × 10⁹* kg of mass be converted into energy every second. (page 327)

54. At maximum, a cosmic ray particle—a mere atomic nucleus of only *10⁻¹³* cm across—can carry the energy of a professionally pitched baseball. (page 445)

Each of the following statements contains a number in italics. Write the number in scientific notation.

55. During the 1999–2000 Broadway season, gross receipts were *$603,000,000*. (*Source:* The League of American Theatres and Producers, Inc.)

56. In 1999, the leading U.S. advertiser was the General Motors Corporation, which spent approximately *$2,900,000,000*. (*Source:* Competitive Media Reporting and Publishers Information Bureau.)

57. In 2000, service revenue of the cellular telephone industry was *$52,466,000,000*. (*Source:* Cellular Telecommunications & Internet Association.)

58. Assets of the insured commercial banks in the state of New York totaled *$1,304,300,000*. (*Source:* U.S. Federal Deposit Insurance Corporation.)

Use scientific notation to calculate the answer to each problem. See Example 4.

59. The distance to Earth from the planet Pluto is 4.58×10^9 km. In April 1983, Pioneer 10 transmitted radio signals from Pluto to Earth at the speed of light, 3.00×10^5 km per sec. How long (in seconds) did it take for the signals to reach Earth?

60. In Exercise 59, how many hours did it take for the signals to reach Earth?

61. In a recent year, the state of Texas had about 1.3×10^6 farms with an average of 7.1×10^2 acres per farm. What was the total number of acres devoted to farmland in Texas that year? (*Source:* National Agricultural Statistics Service, U.S. Department of Agriculture.)

62. The graph depicts aerospace industry sales. The figures at the tops of the bars represent billions of dollars.

    (a) For each year, write the figure in scientific notation.
    (b) If a line segment is drawn between the tops of the bars for 1997 and 1999, what is its slope?

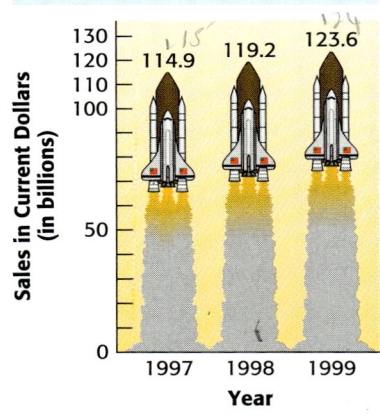

63. There are 10^9 Social Security numbers. The population of the United States is about 3×10^8. How many Social Security numbers are available for each person? (*Source:* U.S. Bureau of the Census.)

SECTION 5.4 Adding and Subtracting Polynomials; Graphing Simple Polynomials 337

64. The top-grossing movie of 1997 was *Titanic*, with box office receipts of about 6×10^8 dollars. That amount represented a fraction of about 9.5×10^{-3} of the total receipts for motion pictures in that year. What were the total receipts? (*Source:* U.S. Bureau of the Census.)

65. The body of a 150-lb person contains about 2.3×10^{-4} lb of copper. How much copper is contained in the bodies of 1200 such people?

66. There were 6.3×10^{10} dollars spent to attend motion pictures in a recent year. Approximately 1.3×10^8 adults attended a motion picture theater at least once. What was the average amount spent per person that year? (*Source:* U.S. National Endowment for the Arts.)

5.4 Adding and Subtracting Polynomials; Graphing Simple Polynomials

OBJECTIVES

1. Identify terms and coefficients.
2. Add like terms.
3. Know the vocabulary for polynomials.
4. Evaluate polynomials.
5. Add and subtract polynomials.
6. Graph equations defined by polynomials of degree 2.

OBJECTIVE 1 Identify terms and coefficients. In Chapter 1 we saw that in an expression such as

$$4x^3 + 6x^2 + 5x + 8,$$

the quantities $4x^3$, $6x^2$, $5x$, and 8 are called *terms*. As mentioned earlier, in the term $4x^3$, the number 4 is called the *numerical coefficient*, or simply the *coefficient*, of x^3. In the same way, 6 is the coefficient of x^2 in the term $6x^2$, 5 is the coefficient of x in the term $5x$, and 8 is the coefficient in the term 8. A constant term, like 8 in the expression above, can be thought of as $8 \cdot 1 = 8x^0$, since $x^0 = 1$.

EXAMPLE 1 Identifying Coefficients

Name the (numerical) coefficient of each term in these expressions.

(a) $4x^3$

The coefficient is 4.

(b) $x - 6x^4$

The coefficient of x is 1 because $x = 1 \cdot x$ or $1x$. The coefficient of x^4 is -6 since we can write $x - 6x^4$ as the sum $x + (-6x^4)$.

(c) $5 - v^3$

The coefficient of the term 5 is 5 because $5 = 5v^0$. By writing $5 - v^3$ as a sum, $5 + (-v^3)$, or $5 + (-1v^3)$, we can identify the coefficient of v^3 as -1.

Now Try Exercises 9 and 13.

OBJECTIVE 2 Add like terms. Recall from Section 1.8 that *like terms* have exactly the same combination of variables with the same exponents on the variables. Only

338 CHAPTER 5 Exponents and Polynomials

the coefficients may differ. Examples of like terms are

$$19m^5 \quad \text{and} \quad 14m^5,$$
$$6y^9, \quad -37y^9, \quad \text{and} \quad y^9,$$
$$3pq \quad \text{and} \quad -2pq,$$
$$2xy^2 \quad \text{and} \quad -xy^2.$$

Using the distributive property, we combine, or add, like terms by adding their coefficients.

EXAMPLE 2 Adding Like Terms

Simplify each expression by adding like terms.

(a) $-4x^3 + 6x^3 = (-4 + 6)x^3 = 2x^3$ Distributive property

(b) $9x^6 - 14x^6 + x^6 = (9 - 14 + 1)x^6 = -4x^6$

(c) $12m^2 + 5m + 4m^2 = (12 + 4)m^2 + 5m = 16m^2 + 5m$

(d) $3x^2y + 4x^2y - x^2y = (3 + 4 - 1)x^2y = 6x^2y$

Now Try Exercises 17, 23, and 27.

In Example 2(c), we cannot combine $16m^2$ and $5m$. These two terms are unlike because the exponents on the variables are different. *Unlike terms* have different variables or different exponents on the same variables.

OBJECTIVE 3 Know the vocabulary for polynomials. A **polynomial in x** is a term or the sum of a finite number of terms of the form ax^n, for any real number a and any whole number n. For example,

$$16x^8 - 7x^6 + 5x^4 - 3x^2 + 4$$

is a polynomial in x. (The 4 can be written as $4x^0$.) This polynomial is written in **descending powers** of the variable, since the exponents on x decrease from left to right. On the other hand,

$$2x^3 - x^2 + \frac{4}{x}$$

is not a polynomial in x, since a variable appears in a denominator. Of course, we could define *polynomial* using any variable and not just x, as in Example 2(c). In fact, polynomials may have terms with more than one variable, as in Example 2(d).

The **degree of a term** is the sum of the exponents on the variables. For example, $3x^4$ has degree 4, while $6x^{17}$ has degree 17. The term $5x$ has degree 1, -7 has degree 0 (since -7 can be written as $-7x^0$), and $2x^2y$ has degree $2 + 1 = 3$. (y has an exponent of 1.) The **degree of a polynomial** is the greatest degree of any nonzero term of the polynomial. For example, $3x^4 - 5x^2 + 6$ is of degree 4, the polynomial $5x + 7$ is of degree 1, 3 (or $3x^0$) is of degree 0, and $x^2y + xy - 5xy^2$ is of degree 3.

Three types of polynomials are very common and are given special names. A polynomial with exactly three terms is called a **trinomial**. (*Tri-* means "three," as in *tri*angle.) Examples are

$$9m^3 - 4m^2 + 6, \quad 19y^2 + 8y + 5, \quad \text{and} \quad -3m^5n^2 + 2n^3 - m^4.$$

SECTION 5.4 Adding and Subtracting Polynomials; Graphing Simple Polynomials 339

A polynomial with exactly two terms is called a **binomial**. (*Bi-* means "two," as in *bi*cycle.) Examples are

$$-9x^4 + 9x^3, \quad 8m^2 + 6m, \quad \text{and} \quad 3m^5n^2 - 9m^2n^4.$$

A polynomial with only one term is called a **monomial**. (*Mon(o)-* means "one," as in *mono*rail.) Examples are

$$9m, \quad -6y^5, \quad a^2b^2, \quad \text{and} \quad 6.$$

EXAMPLE 3 Classifying Polynomials

For each polynomial, first simplify if possible by combining like terms. Then give the degree and tell whether it is a monomial, a binomial, a trinomial, or none of these.

(a) $2x^3 + 5$

The polynomial cannot be simplified. The degree is 3. The polynomial is a binomial.

(b) $4xy - 5xy + 2xy$

Add like terms to simplify: $4xy - 5xy + 2xy = xy$, which is a monomial of degree 2.

Now Try Exercises 29 and 31.

OBJECTIVE 4 Evaluate polynomials. A polynomial usually represents different numbers for different values of the variable, as shown in the next example.

EXAMPLE 4 Evaluating a Polynomial

Find the value of $3x^4 + 5x^3 - 4x - 4$ when $x = -2$ and when $x = 3$.

First, substitute -2 for x.

$$3x^4 + 5x^3 - 4x - 4 = 3(-2)^4 + 5(-2)^3 - 4(-2) - 4$$
$$= 3 \cdot 16 + 5 \cdot (-8) - 4(-2) - 4 \quad \text{Apply exponents.}$$
$$= 48 - 40 + 8 - 4 \quad \text{Multiply.}$$
$$= 12 \quad \text{Add and subtract.}$$

Next, replace x with 3.

$$3x^4 + 5x^3 - 4x - 4 = 3(3)^4 + 5(3)^3 - 4(3) - 4$$
$$= 3 \cdot 81 + 5 \cdot 27 - 12 - 4$$
$$= 362$$

Now Try Exercise 37.

CAUTION Notice the use of parentheses around the numbers that are substituted for the variable in Example 4. This is particularly important when substituting a negative number for a variable that is raised to a power, so that the sign of the product is correct.

OBJECTIVE 5 Add and subtract polynomials. Polynomials may be added, subtracted, multiplied, and divided.

Adding Polynomials

To add two polynomials, add like terms.

EXAMPLE 5 Adding Polynomials Vertically

Add $6x^3 - 4x^2 + 3$ and $-2x^3 + 7x^2 - 5$.

Write like terms in columns.

$$\begin{array}{r} 6x^3 - 4x^2 + 3 \\ -2x^3 + 7x^2 - 5 \end{array}$$

Now add, column by column.

$$\begin{array}{ccc} 6x^3 & -4x^2 & 3 \\ -2x^3 & 7x^2 & -5 \\ \hline 4x^3 & 3x^2 & -2 \end{array}$$

Add the three sums together.

$$4x^3 + 3x^2 + (-2) = 4x^3 + 3x^2 - 2$$

Now Try Exercise 45.

The polynomials in Example 5 also can be added horizontally.

EXAMPLE 6 Adding Polynomials Horizontally

Add $6x^3 - 4x^2 + 3$ and $-2x^3 + 7x^2 - 5$.

Write the sum as

$$(6x^3 - 4x^2 + 3) + (-2x^3 + 7x^2 - 5).$$

Use the associative and commutative properties to rewrite this sum with the parentheses removed and with the subtractions changed to additions of inverses.

$$6x^3 + (-4x^2) + 3 + (-2x^3) + 7x^2 + (-5)$$

Place like terms together.

$$6x^3 + (-2x^3) + (-4x^2) + 7x^2 + 3 + (-5)$$

Combine like terms to get

$$4x^3 + 3x^2 + (-2), \quad \text{or} \quad 4x^3 + 3x^2 - 2,$$

the same answer found in Example 5.

Now Try Exercise 61.

Earlier, we defined the difference $x - y$ as $x + (-y)$. (We find the difference $x - y$ by adding x and the opposite of y.) For example,

$$7 - 2 = 7 + (-2) = 5 \quad \text{and} \quad -8 - (-2) = -8 + 2 = -6.$$

A similar method is used to subtract polynomials.

SECTION 5.4 Adding and Subtracting Polynomials; Graphing Simple Polynomials   341

> **Subtracting Polynomials**
>
> To subtract two polynomials, change all the signs in the second polynomial and add the result to the first polynomial.

EXAMPLE 7   Subtracting Polynomials

(a) Perform the subtraction $(5x - 2) - (3x - 8)$.
By the definition of subtraction,
$$(5x - 2) - (3x - 8) = (5x - 2) + [-(3x - 8)].$$
As shown in Chapter 1, the distributive property gives
$$-(3x - 8) = -1(3x - 8) = -3x + 8,$$
so
$$(5x - 2) - (3x - 8) = (5x - 2) + (-3x + 8)$$
$$= 2x + 6.$$

(b) Subtract $6x^3 - 4x^2 + 2$ from $11x^3 + 2x^2 - 8$.
Write the problem.
$$(11x^3 + 2x^2 - 8) - (6x^3 - 4x^2 + 2)$$
Change all signs in the second polynomial and add.
$$(11x^3 + 2x^2 - 8) + (-6x^3 + 4x^2 - 2) = 5x^3 + 6x^2 - 10$$

To check a subtraction problem, use the fact that if $a - b = c$, then $a = b + c$. For example, $6 - 2 = 4$, so check by writing $6 = 2 + 4$, which is correct. Check the polynomial subtraction above by adding $6x^3 - 4x^2 + 2$ and $5x^3 + 6x^2 - 10$. Since the sum is $11x^3 + 2x^2 - 8$, the subtraction was performed correctly.

Now Try Exercise 59.

Subtraction also can be done in columns (vertically). We use vertical subtraction in Section 5.7 when we divide polynomials.

EXAMPLE 8   Subtracting Polynomials Vertically

Use the method of subtracting by columns to find
$$(14y^3 - 6y^2 + 2y - 5) - (2y^3 - 7y^2 - 4y + 6).$$

Arrange like terms in columns.
$$\begin{array}{r} 14y^3 - 6y^2 + 2y - 5 \\ 2y^3 - 7y^2 - 4y + 6 \end{array}$$

Change all signs in the second row, and then add.
$$\begin{array}{r} 14y^3 - 6y^2 + 2y - 5 \\ -2y^3 + 7y^2 + 4y - 6 \\ \hline 12y^3 + y^2 + 6y - 11 \end{array}$$
Change all signs.
Add.

Now Try Exercise 55.

342 CHAPTER 5 Exponents and Polynomials

Either the horizontal or the vertical method may be used to add and subtract polynomials.

We add and subtract polynomials in more than one variable by combining like terms, just as with single variable polynomials.

EXAMPLE 9 Adding and Subtracting Polynomials with More Than One Variable

Add or subtract as indicated.

(a) $(4a + 2ab - b) + (3a - ab + b) = 4a + 2ab - b + 3a - ab + b$
$= 7a + ab$ Combine like terms.

(b) $(2x^2y + 3xy + y^2) - (3x^2y - xy - 2y^2)$
$= 2x^2y + 3xy + y^2 - 3x^2y + xy + 2y^2$
$= -x^2y + 4xy + 3y^2$

Now Try Exercises 83 and 85.

OBJECTIVE 6 Graph equations defined by polynomials of degree 2. In Chapter 3 we introduced graphs of straight lines. These graphs were defined by linear equations (which are actually polynomial equations of degree 1). By selective point-plotting, we can graph polynomial equations of degree 2.

EXAMPLE 10 Graphing Equations Defined by Polynomials with Degree 2

Graph each equation.

(a) $y = x^2$ *Parabola*

Select several values for x; then find the corresponding y-values. For example, selecting $x = 2$ gives

$$y = 2^2 = 4,$$

and so the point $(2, 4)$ is on the graph of $y = x^2$. (Recall that in an ordered pair such as $(2, 4)$, the x-value comes first and the y-value second.) We show some ordered pairs that satisfy $y = x^2$ in the table next to Figure 1. If we plot the ordered pairs from the table on a coordinate system and draw a smooth curve through them, we obtain the graph shown in Figure 1.

x	y
3	9
2	4
1	1
0	0
-1	1
-2	4
-3	9

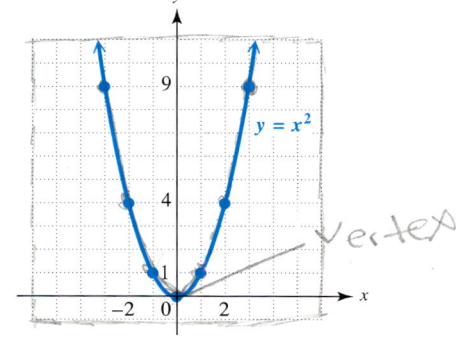

FIGURE 1

SECTION 5.4 Adding and Subtracting Polynomials; Graphing Simple Polynomials 343

The graph of $y = x^2$ is the graph of a function, since each input x is related to just one output y. The curve in Figure 1 is called a **parabola**. The point $(0, 0)$, the lowest point on this graph, is called the **vertex** of the parabola. The vertical line through the vertex (the y-axis here) is called the **axis** of the parabola. The axis of a parabola is a **line of symmetry** for the graph. If the graph is folded on this line, the two halves will match.

(b) $y = -x^2 + 3$

Once again plot points to obtain the graph. For example, if $x = -2$,
$$y = -(-2)^2 + 3 = -4 + 3 = -1.$$

This point and several others are shown in the table that accompanies the graph in Figure 2. The vertex of this parabola is $(0, 3)$. This time the vertex is the *highest* point on the graph. The graph opens downward because x^2 has a negative coefficient.

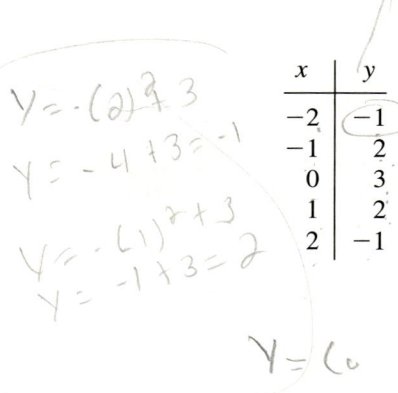
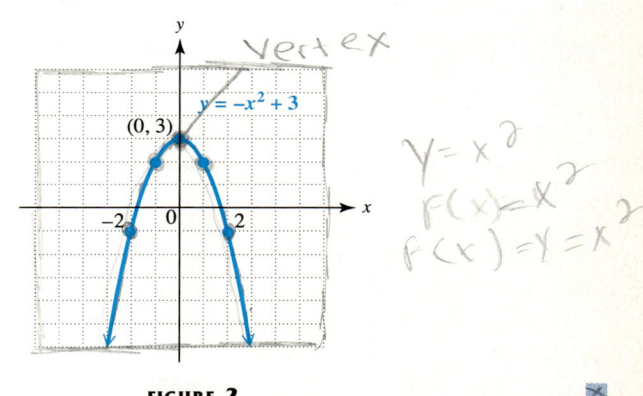

FIGURE 2

Now Try Exercises 107 and 111.

NOTE All polynomials of degree 2 have parabolas as their graphs. When graphing by plotting points, it is necessary to continue finding points until the vertex and points on either side of it are located. (In this section, all parabolas have their vertices on the x-axis or the y-axis.)

CONNECTIONS

In Section 3.6 we introduced the idea of a function: for every input x, there is one output y. For example, according to the U.S. National Aeronautics and Space Administration (NASA), the budget in millions of dollars for space station research for 1996–2001 can be approximated by the polynomial equation
$$y = -10.25x^2 - 126.04x + 5730.21,$$
where $x = 0$ represents 1996, $x = 1$ represents 1997, and so on, up to $x = 5$ representing 2001. The actual budget for 1998 was 5327 million dollars; an input of $x = 2$ (for 1998) in the equation gives approximately $y = 5437$. Considering the magnitude of the numbers, this is a good approximation.

5.4 EXERCISES

For Extra Help

 Student's Solutions Manual

 MyMathLab

 InterAct Math Tutorial Software

AW Math Tutor Center

MathXL

Digital Video Tutor CD 10/Videotape 8

Fill in each blank with the correct response.

1. In the term $7x^5$, the coefficient is ____7____ and the exponent is ____5____.
2. The expression $5x^3 - 4x^2$ has ____2____ term(s). (how many?)
3. The degree of the term $-4x^8$ is ____8____.
4. The polynomial $4x^2 - y^2$ ____is not____ an example of a trinomial. (is/is not)
5. When $x^2 + 10$ is evaluated for $x = 4$, the result is _____.
6. $5x^{\_\_} + 3x^3 - 7x$ is a trinomial of degree 4.
7. $3xy + 2xy - 5xy = $ _____.
8. _____ is an example of a monomial with coefficient 5, in the variable x, having degree 9.

For each polynomial, determine the number of terms and name the coefficients of the terms. See Example 1.

9. $6x^4$
10. $-9y^5$
11. t^4
12. s^7
13. $-19r^2 - r$
14. $2y^3 - y$
15. $x + 8x^2 + 5x^3$
16. $v - 2v^3 - v^7$

In each polynomial, add like terms whenever possible. Write the result in descending powers of the variable. See Example 2.

17. $-3m^5 + 5m^5$
18. $-4y^3 + 3y^3$
19. $2r^5 + (-3r^5)$
20. $-19y^2 + 9y^2$
21. $.2m^5 - .5m^2$
22. $-.9y + .9y^2$
23. $-3x^5 + 2x^5 - 4x^5$
24. $6x^3 - 8x^3 + 9x^3$
25. $-4p^7 + 8p^7 + 5p^9$
26. $-3a^8 + 4a^8 - 3a^2$
27. $-4xy^2 + 3xy^2 - 2xy^2 + xy^2$
28. $3pr^5 - 8pr^5 + pr^5 + 2pr^5$

For each polynomial, first simplify, if possible, and write it in descending powers of the variable. Then give the degree of the resulting polynomial and tell whether it is a monomial, binomial, trinomial, or none of these. See Example 3.

29. $6x^4 - 9x$
30. $7t^3 - 3t$
31. $5m^4 - 3m^2 + 6m^4 - 7m^3$
32. $6p^5 + 4p^3 - 8p^5 + 10p^2$
33. $\dfrac{5}{3}x^4 - \dfrac{2}{3}x^4$
34. $\dfrac{4}{5}r^6 + \dfrac{1}{5}r^6$
35. $.8x^4 - .3x^4 - .5x^4 + 7$
36. $1.2t^3 - .9t^3 - .3t^3 + 9$

Find the value of each polynomial when (a) $x = 2$ and when (b) $x = -1$. See Example 4.

37. $2x^5 - 4x^4 + 5x^3 - x^2$
38. $2x^2 + 5x + 1$
39. $-3x^2 + 14x - 2$
40. $-2x^2 + 3$
41. $2x^2 - 3x - 5$
42. $x^2 + 5x - 10$

Add. See Example 5.

43. $2x^2 - 4x$
    $3x^2 + 2x$

44. $-5y^3 + 3y$
    $8y^3 - 4y$

45. $3m^2 + 5m + 6$
    $2m^2 - 2m - 4$

SECTION 5.4 Adding and Subtracting Polynomials; Graphing Simple Polynomials **345**

46. $4a^3 - 4a^2 - 4$
    $\underline{6a^3 + 5a^2 - 8}$

47. $\frac{2}{3}x^2 + \frac{1}{5}x + \frac{1}{6}$
    $\underline{\frac{1}{2}x^2 - \frac{1}{3}x + \frac{2}{3}}$

48. $\frac{4}{7}y^2 - \frac{1}{5}y + \frac{7}{9}$
    $\underline{\frac{1}{3}y^2 - \frac{1}{3}y + \frac{2}{5}}$

49. $9m^3 - 5m^2 + 4m - 8$
    $\underline{-3m^3 + 6m^2 + 8m - 6}$

50. $12r^5 + 11r^4 - 7r^3 - 2r^2$
    $\underline{-8r^5 + 10r^4 + 3r^3 + 2r^2}$

Subtract. See Example 8.

51. $5y^3 - 3y^2$
    $\underline{2y^3 + 8y^2}$

52. $-6t^3 + 4t^2$
    $\underline{8t^3 - 6t^2}$

53. $12x^4 - x^2 + x$
    $\underline{8x^4 + 3x^2 - 3x}$

54. $13y^5 - y^3 - 8y^2$
    $\underline{7y^5 + 5y^3 + y^2}$

55. $12m^3 - 8m^2 + 6m + 7$
    $\underline{-3m^3 + 5m^2 - 2m - 4}$

56. $5a^4 - 3a^3 + 2a^2 - a + 6$
    $\underline{-6a^4 + a^3 - a^2 + a - 1}$

57. After reading Examples 5–8, do you have a preference regarding horizontal or vertical addition and subtraction of polynomials? Explain your answer.

58. Write a paragraph explaining how to add and subtract polynomials. Give an example using addition.

Perform each indicated operation. See Examples 6 and 7.

59. $(8m^2 - 7m) - (3m^2 + 7m - 6)$
60. $(x^2 + x) - (3x^2 + 2x - 1)$
61. $(16x^3 - x^2 + 3x) + (-12x^3 + 3x^2 + 2x)$
62. $(-2b^6 + 3b^4 - b^2) + (b^6 + 2b^4 + 2b^2)$
63. $(7y^4 + 3y^2 + 2y) - (18y^4 - 5y^2 + y)$
64. $(8t^5 + 3t^3 + 5t) - (19t^5 - 6t^3 + t)$
65. $(9a^4 - 3a^2 + 2) + (4a^4 - 4a^2 + 2) + (-12a^4 + 6a^2 - 3)$
66. $(4m^2 - 3m + 2) + (5m^2 + 13m - 4) - (16m^2 + 4m - 3)$
67. $[(8m^2 + 4m - 7) - (2m^2 - 5m + 2)] - (m^2 + m + 1)$
68. $[(9b^3 - 4b^2 + 3b + 2) - (-2b^3 - 3b^2 + b)] - (8b^3 + 6b + 4)$
69. $[(3x^2 - 2x + 7) - (4x^2 + 2x - 3)] - [(9x^2 + 4x - 6) + (-4x^2 + 4x + 4)]$
70. $[(6t^2 - 3t + 1) - (12t^2 + 2t - 6)] - [(4t^2 - 3t - 8) + (-6t^2 + 10t - 12)]$
71. $(3.47x^2 - 2.51x + 7.77) + (9.54x^2 + 7.44x - 8.33)$
72. $(10.45x^2 + 4.53x - 9.87) + (8.82x^2 - 6.54x + 2.10)$
73. $(4.5a^3 - 2.4a^2 + 9.0a - 4.7) - (8.3a^3 + 2.7a^2 - 7.3a + 12.1) + (5.3a^3 - 6.4a^2 + 2.1a - 5.7)$
74. $(8.2a^3 + 7.5a^2 - 10.1a + 6.9) - (5.3a^3 - 2.6a^2 - 9.3a + 4.4) + (4.8a^3 - 8.2a^2 + 3.9a - 7.7)$
75. $(-3.2x^2 + 4.5x - 7.1) - [(3.3x^2 + 2.5x - 8.2) - (4.0x^2 + 2.3x - 7.9)]$
76. $(-9.9x^2 - 3.1x + 2.6) - [(8.0x^2 - 4.6x + 6.9) - (1.3x^2 - 6.5x + 5.3)]$
77. $\left(\frac{2}{3}y^3 - \frac{1}{6}y^2 + \frac{3}{5}y - \frac{8}{7}\right) + \left(\frac{4}{3}y^3 - \frac{5}{6}y^2 + \frac{2}{5}y + \frac{1}{7}\right)$
78. $\left(\frac{6}{7}y^3 + \frac{2}{9}y^2 + \frac{8}{5}y - \frac{4}{3}\right) + \left(\frac{1}{7}y^3 - \frac{11}{9}y^2 + \frac{7}{5}y + \frac{1}{3}\right)$
79. $\left[\left(\frac{4}{3}x^2 - \frac{5}{6}x + \frac{2}{3}\right) - \left(\frac{5}{12}x^2 + \frac{2}{9}x - \frac{7}{12}\right)\right] - \left(\frac{5}{12}x^2 + \frac{7}{18}x + \frac{11}{12}\right)$
80. $\left[\left(\frac{7}{12}x^2 + \frac{5}{9}x + \frac{1}{5}\right) - \left(\frac{7}{3}x^2 + \frac{4}{3}x - \frac{3}{10}\right)\right] - \left(\frac{7}{12}x^2 + \frac{1}{9}x + \frac{11}{10}\right)$

81. Without actually performing the operations, determine mentally the coefficient of x^2 in the simplified form of $(-4x^2 + 2x - 3) - (-2x^2 + x - 1) + (-8x^2 + 3x - 4)$.

346 CHAPTER 5 Exponents and Polynomials

82. Without actually performing the operations, determine mentally the coefficient of x in the simplified form of $(-8x^2 - 3x + 2) - (4x^2 - 3x + 8) - (-2x^2 - x + 7)$.

Add or subtract as indicated. See Example 9.

83. $(6b + 3c) + (-2b - 8c)$
84. $(-5t + 13s) + (8t - 3s)$
85. $(4x + 2xy - 3) - (-2x + 3xy + 4)$
86. $(8ab + 2a - 3b) - (6ab - 2a + 3b)$
87. $(5x^2y - 2xy + 9xy^2) - (8x^2y + 13xy + 12xy^2)$
88. $(16t^3s^2 + 8t^2s^3 + 9ts^4) - (-24t^3s^2 + 3t^2s^3 - 18ts^4)$
89. $(4.6x + 3.5y - 6.1z) + (9.5x - 8.8y - 4.0z)$
90. $(7.5a - 6.9b + 4.1c) + (3.2a - 12.1b + 1.7c)$
91. $(9.66s + 7.11st - 8.90) - (-7.44s - 10.13st - 8.55)$
92. $(12.24m + 9.33mn - 2.38) - (-6.34m - 9.32mn - 1.15)$
93. $\left(\frac{2}{3}x^2y^3 + \frac{5}{8}x^3y^2 - \frac{1}{9}x^4y\right) + \left(-\frac{5}{6}x^2y^3 + \frac{3}{4}x^3y^2 + \frac{2}{3}x^4y\right)$
94. $\left(\frac{4}{9}x^2y^2 - \frac{6}{5}x^3y - \frac{2}{5}x^4\right) + \left(-\frac{8}{9}x^2y^2 + \frac{3}{10}x^3y + \frac{7}{10}x^4y\right)$
95. $\left(\frac{1}{4}a^2 - \frac{5}{3}ab + \frac{5}{12}b^2\right) - \left[\left(\frac{3}{4}a^2 + \frac{2}{3}ab - \frac{1}{12}b^2\right) - \left(\frac{1}{4}a^2 + \frac{7}{3}ab + \frac{7}{12}b^2\right)\right]$
96. $\left(\frac{1}{3}x^2 - \frac{1}{12}xy + \frac{1}{4}y^2\right) - \left[\left(\frac{2}{3}x^2 + \frac{7}{12}xy - \frac{3}{4}y^2\right) - \left(\frac{10}{3}x^2 + \frac{11}{12}xy - \frac{10}{3}y^2\right)\right]$
97. $(-3.5a^3 + 2.5a^2 + 1.4a + 6.0) + \left(\frac{5}{2}a^3 + \frac{3}{10}a^2 - \frac{3}{5}a + \frac{5}{2}\right)$
98. $(-4.5x^3 - 12.5x^2 - 2.8x + 3.5) + \left(\frac{7}{2}x^3 + \frac{7}{10}x^2 - \frac{7}{10}x + \frac{7}{5}\right)$

For Exercises 99–102, use the formulas found on the inside covers. Find the perimeter of each rectangle.

99.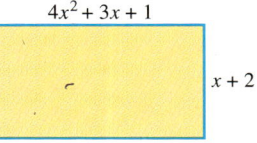
$4x^2 + 3x + 1$
$x + 2$

100.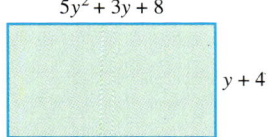
$5y^2 + 3y + 8$
$y + 4$

*Find **(a)** a polynomial representing the perimeter of each triangle and **(b)** the measures of the angles of the triangle.*

101.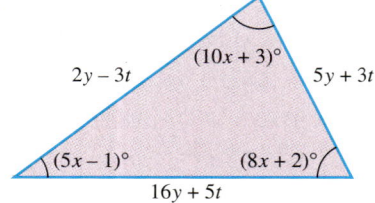
$2y - 3t$, $(10x + 3)°$, $5y + 3t$, $(5x - 1)°$, $(8x + 2)°$, $16y + 5t$

102.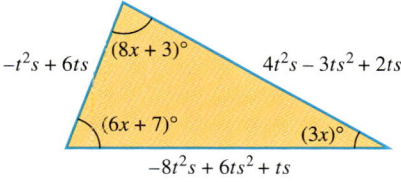
$-t^2s + 6ts$, $(8x + 3)°$, $4t^2s - 3ts^2 + 2ts$, $(6x + 7)°$, $(3x)°$, $-8t^2s + 6ts^2 + ts$

The concepts required to work Exercises 103–106 have been covered, but the usual wording of the problem has been changed. Perform the indicated operations.

103. Subtract $9x^2 - 6x + 5$ from $3x^2 - 2$.

104. Find the difference when $9x^4 + 3x^2 + 5$ is subtracted from $8x^4 - 2x^3 + x - 1$.

105. Find the difference between the sum of $5x^2 + 2x - 3$ and $x^2 - 8x + 2$ and the sum of $7x^2 - 3x + 6$ and $-x^2 + 4x - 6$.

106. Subtract the sum of $9t^3 - 3t + 8$ and $t^2 - 8t + 4$ from the sum of $12t + 8$ and $t^2 - 10t + 3$.

SECTION 5.4 Adding and Subtracting Polynomials; Graphing Simple Polynomials   **347**

Graph each equation by completing the table of values. See Example 10.

107. $y = x^2 - 4$

x	y
-2	
-1	
0	
1	
2	

108. $y = x^2 - 6$

x	y
-2	
-1	
0	
1	
2	

109. $y = 2x^2 - 1$

x	y
-2	
-1	
0	
1	
2	

110. $y = 2x^2 + 2$

x	y
-2	
-1	
0	
1	
2	

111. $y = -x^2 + 4$

x	y
-2	
-1	
0	
1	
2	

112. $y = -x^2 + 2$

x	y
-2	
-1	
0	
1	
2	

113. $y = (x + 3)^2$

x	-5	-4	-3	-2	-1
y					

114. $y = (x - 4)^2$

x	2	3	4	5	6
y					

In the introduction to this chapter, we gave a polynomial that models the distance in feet that a car going approximately 68 mph will skid in x seconds. If we let D represent this distance, then

$$D = 100x - 13x^2.$$

Each time we evaluate this polynomial for a value of x, we get one and only one output value D. This idea is basic to the concept of a function, introduced in Section 3.6. Exercises 115 and 116 illustrate the idea with this polynomial.

115. Use the given polynomial equation to approximate the skidding distance in feet if $x = 5$ sec.

116. Use the given polynomial equation to find the distance the car will skid in 1 sec. Write an ordered pair of the form (x, D).

RELATING CONCEPTS (EXERCISES 117–120)

For Individual or Group Work

As explained earlier in this section, the polynomial equation

$$y = -10.25x^2 - 126.04x + 5730.21$$

gives a good approximation of NASA's budget for space station research, in millions of dollars, for 1996 through 2001, where $x = 0$ represents 1996, $x = 1$ represents

(continued)

1997, and so on. If we evaluate the polynomial for a specific input value x, we will get one and only one output value y as a result. This idea is basic to the study of functions, one of the most important concepts in mathematics. **Work Exercises 117–120 in order.**

117. If gasoline costs $1.25 per gallon, then the monomial $1.25x$ gives the cost of x gallons. Evaluate this monomial for 4, and then use the result to fill in the blanks: If _____ gallons are purchased, the cost is _____.

118. If it costs $15 to rent a chain saw plus $2 per day, the binomial $2x + 15$ gives the cost to rent the chain saw for x days. Evaluate this polynomial for 6 and then use the result to fill in the blanks: If the saw is rented for _____ days, the cost is _____.

119. If an object is thrown upward under certain conditions, its height in feet is given by the trinomial $-16x^2 + 60x + 80$, where x is in seconds. Evaluate this polynomial for 2.5 and then use the result to fill in the blanks: If _____ seconds have elapsed, the height of the object is _____ feet.

120. The polynomial $2.69x^2 + 4.75x + 452.43$ gives a good approximation for the number of revenue passenger miles, in billions, for the U.S. airline industry during the period from 1990 to 1995, where $x = 0$ represents 1990. Use this polynomial to approximate the number of revenue passenger miles in 1991. (*Hint:* Any power of 1 is equal to 1, so simply add the coefficients and the constant.) (*Source:* Air Transportation Association of America.)

5.5 Multiplying Polynomials

OBJECTIVES

1. Multiply a monomial and a polynomial.
2. Multiply two polynomials.
3. Multiply binomials by the FOIL method.

OBJECTIVE 1 Multiply a monomial and a polynomial. As shown earlier, we find the product of two monomials by using the rules for exponents and the commutative and associative properties. For example,

$$(-8m^6)(-9n^6) = (-8)(-9)(m^6)(n^6) = 72m^6n^6.$$

CAUTION Do not confuse *addition* of terms with *multiplication* of terms. For example,

$$7q^5 + 2q^5 = 9q^5, \quad \text{but} \quad (7q^5)(2q^5) = 7 \cdot 2q^{5+5} = 14q^{10}.$$

To find the product of a monomial and a polynomial with more than one term, we use the distributive property and multiplication of monomials.

EXAMPLE 1 Multiplying Monomials and Polynomials

Use the distributive property to find each product.

(a) $4x^2(3x + 5)$

$$4x^2(3x + 5) = 4x^2(3x) + 4x^2(5) \quad \text{Distributive property}$$
$$= 12x^3 + 20x^2 \quad \text{Multiply monomials.}$$

(b) $-8m^3(4m^3 + 3m^2 + 2m - 1)$
$$= -8m^3(4m^3) + (-8m^3)(3m^2)$$
$$+ (-8m^3)(2m) + (-8m^3)(-1) \quad \text{Distributive property}$$
$$= -32m^6 - 24m^5 - 16m^4 + 8m^3 \quad \text{Multiply monomials.}$$

Now Try Exercises 11 and 19.

OBJECTIVE 2 Multiply two polynomials. We use the distributive property repeatedly to find the product of any two polynomials. For example, to find the product of the polynomials $x^2 + 3x + 5$ and $x - 4$, think of $x - 4$ as a single quantity and use the distributive property as follows.

$$(x^2 + 3x + 5)(x - 4) = x^2(x - 4) + 3x(x - 4) + 5(x - 4)$$

Now use the distributive property three times to find $x^2(x - 4)$, $3x(x - 4)$, and $5(x - 4)$.

$$x^2(x - 4) + 3x(x - 4) + 5(x - 4)$$
$$= x^2(x) + x^2(-4) + 3x(x) + 3x(-4) + 5(x) + 5(-4)$$
$$= x^3 - 4x^2 + 3x^2 - 12x + 5x - 20 \quad \text{Multiply monomials.}$$
$$= x^3 - x^2 - 7x - 20 \quad \text{Combine like terms.}$$

This example suggests the following rule.

Multiplying Polynomials

To multiply two polynomials, multiply each term of the second polynomial by each term of the first polynomial and add the products.

EXAMPLE 2 Multiplying Two Polynomials

Multiply $(m^2 + 5)(4m^3 - 2m^2 + 4m)$.

Multiply each term of the second polynomial by each term of the first.

$$(m^2 + 5)(4m^3 - 2m^2 + 4m)$$
$$= m^2(4m^3) + m^2(-2m^2) + m^2(4m) + 5(4m^3) + 5(-2m^2) + 5(4m)$$
$$= 4m^5 - 2m^4 + 4m^3 + 20m^3 - 10m^2 + 20m$$
$$= 4m^5 - 2m^4 + 24m^3 - 10m^2 + 20m \quad \text{Combine like terms.}$$

Now Try Exercise 25.

When at least one of the factors in a product of polynomials has three or more terms, the multiplication can be simplified by writing one polynomial above the other vertically.

EXAMPLE 3 Multiplying Polynomials Vertically

Multiply $(x^3 + 2x^2 + 4x + 1)(3x + 5)$ using the vertical method.

Write the polynomials as follows.

$$\begin{array}{r} x^3 + 2x^2 + 4x + 1 \\ 3x + 5 \\ \hline \end{array}$$

It is not necessary to line up terms in columns, because any terms may be multiplied (not just like terms). Begin by multiplying each of the terms in the top row by 5.

$$\begin{array}{r} x^3 + 2x^2 + 4x + 1 \\ 3x + 5 \\ \hline 5x^3 + 10x^2 + 20x + 5 \end{array} \qquad 5(x^3 + 2x^2 + 4x + 1)$$

Notice how this process is similar to multiplication of whole numbers. Now multiply each term in the top row by $3x$. Be careful to place like terms in columns, since the final step will involve addition (as in multiplying two whole numbers).

$$\begin{array}{r} x^3 + 2x^2 + 4x + 1 \\ 3x + 5 \\ \hline 5x^3 + 10x^2 + 20x + 5 \\ 3x^4 + 6x^3 + 12x^2 + 3x \end{array} \qquad 3x(x^3 + 2x^2 + 4x + 1)$$

Add like terms.

$$\begin{array}{r} x^3 + 2x^2 + 4x + 1 \\ 3x + 5 \\ \hline 5x^3 + 10x^2 + 20x + 5 \\ 3x^4 + 6x^3 + 12x^2 + 3x \\ \hline 3x^4 + 11x^3 + 22x^2 + 23x + 5 \end{array}$$

The product is $3x^4 + 11x^3 + 22x^2 + 23x + 5$.

Now Try Exercise 29.

EXAMPLE 4 Multiplying Polynomials with Fractional Coefficients Vertically

Find the product of $4m^3 - 2m^2 + 4m$ and $\frac{1}{2}m^2 + \frac{5}{2}$.

$$\begin{array}{r} 4m^3 - 2m^2 + 4m \\ \frac{1}{2}m^2 + \frac{5}{2} \\ \hline 10m^3 - 5m^2 + 10m \\ 2m^5 - m^4 + 2m^3 \\ \hline 2m^5 - m^4 + 12m^3 - 5m^2 + 10m \end{array}$$

Terms of top row multiplied by $\frac{5}{2}$
Terms of top row multiplied by $\frac{1}{2}m^2$
Add.

Now Try Exercise 35.

SECTION 5.5 Multiplying Polynomials **351**

We can use a rectangle to model polynomial multiplication. For example, to find the product

$$(2x + 1)(3x + 2),$$

label a rectangle with each term as shown here.

	$3x$	2
$2x$		
1		

Now put the product of each pair of monomials in the appropriate box.

	$3x$	2
$2x$	$6x^2$	$4x$
1	$3x$	2

The product of the original binomials is the sum of these four monomial products.

$$(2x + 1)(3x + 2) = 6x^2 + 4x + 3x + 2$$
$$= 6x^2 + 7x + 2$$

This approach can be extended to polynomials with any number of terms.

OBJECTIVE 3 Multiply binomials by the FOIL method. In algebra, many of the polynomials to be multiplied are both binomials (with just two terms). For these products, the **FOIL method** reduces the rectangle method to a systematic approach without the rectangle. To develop the FOIL method, we use the distributive property to find $(x + 3)(x + 5)$.

$$(x + 3)(x + 5) = (x + 3)x + (x + 3)5$$
$$= x(x) + 3(x) + x(5) + 3(5)$$
$$= x^2 + 3x + 5x + 15$$
$$= x^2 + 8x + 15$$

Here is where the letters of the word FOIL originate.

$(x + 3)(x + 5)$ Multiply the **First** terms: $x(x)$.   **F**

$(x + 3)(x + 5)$ Multiply the **Outer** terms: $x(5)$.   **O**
This is the **outer product.**

$(x + 3)(x + 5)$ Multiply the **Inner** terms: $3(x)$.   **I**
This is the **inner product.**

$(x + 3)(x + 5)$ Multiply the **Last** terms: $3(5)$.   **L**

The inner product and the outer product should be added mentally so that the three terms of the answer can be written without extra steps as

$$(x + 3)(x + 5) = x^2 + 8x + 15.$$

A summary of the steps in the FOIL method follows.

352 CHAPTER 5 Exponents and Polynomials

> **Multiplying Binomials by the FOIL Method**
>
> *Step 1* Multiply the two **F**irst terms of the binomials to get the first term of the answer.
>
> *Step 2* Find the **O**uter product and the **I**nner product and add them (when possible) to get the middle term of the answer.
>
> *Step 3* Multiply the two **L**ast terms of the binomials to get the last term of the answer.
>
> $$\mathbf{F} = x^2 \qquad \mathbf{L} = 15$$
> $$(x + 3)(x + 5)$$
> $$\mathbf{I} \quad 3x$$
> $$\mathbf{O} \quad 5x$$
> $$\overline{8x} \quad \text{Add.}$$

EXAMPLE 5 Using the FOIL Method

Use the FOIL method to find the product $(x + 8)(x - 6)$.

Step 1 **F** Multiply the *first* terms.
$$x(x) = x^2$$

Step 2 **O** Find the *outer* product.
$$x(-6) = -6x$$

I Find the *inner* product.
$$8(x) = 8x$$

Add the outer and inner products mentally.
$$-6x + 8x = 2x$$

Step 3 **L** Multiply the *last* terms.
$$8(-6) = -48$$

The product of $x + 8$ and $x - 6$ is the sum of the four terms found in three steps above, so
$$(x + 8)(x - 6) = x^2 + 2x - 48.$$

As a shortcut, this product can be found in the following manner.

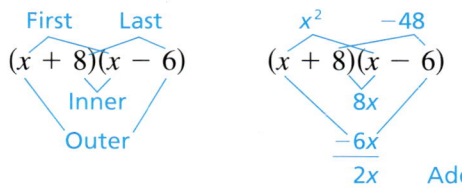

Now Try Exercise 37.

It is not possible to add the inner and outer products of the FOIL method if unlike terms result, as shown in the next example.

SECTION 5.5 Multiplying Polynomials **353**

EXAMPLE 6 Using the FOIL Method

Multiply $(9x - 2)(3y + 1)$.

First	$(9x - 2)(3y + 1)$	$27xy$
Outer	$(9x - 2)(3y + 1)$	$9x$
Inner	$(9x - 2)(3y + 1)$	$-6y$
Last	$(9x - 2)(3y + 1)$	-2

Unlike terms

$$\begin{matrix} & \text{F} & \text{O} & \text{I} & \text{L} \\ (9x - 2)(3y + 1) = & 27xy + 9x - 6y - 2 \end{matrix}$$

Now Try Exercise 51.

EXAMPLE 7 Using the FOIL Method

Find each product.

$$\begin{matrix} & & \text{F} & \text{O} & \text{I} & \text{L} \end{matrix}$$

(a) $(2k + 5y)(k + 3y) = 2k(k) + 2k(3y) + 5y(k) + 5y(3y)$
$= 2k^2 + 6ky + 5ky + 15y^2$
$= 2k^2 + 11ky + 15y^2$

(b) $(7p + 2q)(3p - q) = 21p^2 - pq - 2q^2$ FOIL

(c) $2x^2(x - 3)(3x + 4) = 2x^2(3x^2 - 5x - 12)$ FOIL
$= 6x^4 - 10x^3 - 24x^2$ Distributive property

Now Try Exercises 53 and 55.

NOTE Example 7(c) showed one way to multiply three polynomials. We could have multiplied $2x^2$ and $x - 3$ first, then multiplied that product and $3x + 4$ as follows.
$$2x^2(x - 3)(3x + 4) = (2x^3 - 6x^2)(3x + 4)$$
$$= 6x^4 - 10x^3 - 24x^2 \qquad \text{FOIL}$$

5.5 EXERCISES

For Extra Help

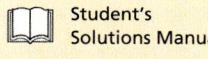
Student's
Solutions Manual

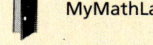
MyMathLab

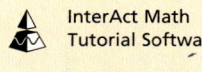
InterAct Math
Tutorial Software

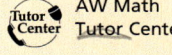
AW Math
Tutor Center

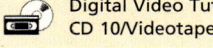
MathXL

Digital Video Tutor
CD 10/Videotape 9

1. Match each product in Column I with the correct monomial in Column II.

I	II
(a) $(5x^3)(6x^5)$	A. $125x^{15}$
(b) $(-5x^5)(6x^3)$	B. $30x^8$
(c) $(5x^5)^3$	C. $-216x^9$
(d) $(-6x^3)^3$	D. $-30x^8$

2. Match each product in Column I with the correct polynomial in Column II.

I	II
(a) $(x - 5)(x + 3)$	A. $x^2 + 8x + 15$
(b) $(x + 5)(x + 3)$	B. $x^2 - 8x + 15$
(c) $(x - 5)(x - 3)$	C. $x^2 - 2x - 15$
(d) $(x + 5)(x - 3)$	D. $x^2 + 2x - 15$

354 CHAPTER 5 Exponents and Polynomials

Find each product. Use the rules for exponents discussed earlier in the chapter.

3. $(5y^4)(3y^7)$
4. $(10p^2)(5p^3)$
5. $(-15a^4)(-2a^5)$
6. $(-3m^6)(-5m^4)$
7. $(5p)(3q^2)$
8. $(4a^3)(3b^2)$
9. $(-6m^3)(3n^2)$
10. $(9r^3)(-2s^2)$

Find each product. See Example 1.

11. $2m(3m + 2)$
12. $4x(5x + 3)$
13. $3p(-2p^3 + 4p^2)$
14. $4x(3 + 2x + 5x^3)$
15. $-8z(2z + 3z^2 + 3z^3)$
16. $-7y(3 + 5y^2 - 2y^3)$
17. $2y^3(3 + 2y + 5y^4)$
18. $2m^4(3m^2 + 5m + 6)$
19. $-4r^3(-7r^2 + 8r - 9)$
20. $-9a^5(-3a^6 - 2a^4 + 8a^2)$
21. $3a^2(2a^2 - 4ab + 5b^2)$
22. $4z^3(8z^2 + 5zy - 3y^2)$
23. $7m^3n^2(3m^2 + 2mn - n^3)$
24. $2p^2q(3p^2q^2 - 5p + 2q^2)$

Find each product. See Examples 2–4.

25. $(6x + 1)(2x^2 + 4x + 1)$
26. $(9y - 2)(8y^2 - 6y + 1)$
27. $(9a + 2)(9a^2 + a + 1)$
28. $(2r - 1)(3r^2 + 4r - 4)$
29. $(4m + 3)(5m^3 - 4m^2 + m - 5)$
30. $(y + 4)(3y^4 - 2y^2 + 1)$
31. $(2x - 1)(3x^5 - 2x^3 + x^2 - 2x + 3)$
32. $(2a + 3)(a^4 - a^3 + a^2 - a + 1)$
33. $(5x^2 + 2x + 1)(x^2 - 3x + 5)$
34. $(2m^2 + m - 3)(m^2 - 4m + 5)$
35. $(6x^4 - 4x^2 + 8x)\left(\frac{1}{2}x + 3\right)$
36. $(8y^6 + 4y^4 - 12y^2)\left(\frac{3}{4}y^2 + 2\right)$

Find each product. Use the FOIL method. See Examples 5–7.

37. $(m + 7)(m + 5)$
38. $(n - 1)(n + 4)$
39. $(x + 5)(x - 5)$
40. $(y + 8)(y - 8)$
41. $(2x + 3)(6x - 4)$
42. $(4m + 3)(4m + 3)$
43. $(3x - 2)(3x - 2)$
44. $(b + 8)(6b - 2)$
45. $(5a + 1)(2a + 7)$
46. $(8 - 3a)(2 + a)$
47. $(6 - 5m)(2 + 3m)$
48. $(-4 + k)(2 - k)$
49. $(5 - 3x)(4 + x)$
50. $(2m - 3n)(m + 5n)$
51. $(4x + 3)(2y - 1)$
52. $(5x + 7)(3y - 8)$
53. $(3x + 2y)(5x - 3y)$
54. $x(2x - 5)(x + 3)$
55. $3y^3(2y + 3)(y - 5)$
56. $5t^4(t + 3)(3t - 1)$
57. $-8r^3(5r^2 + 2)(5r^2 - 2)$

58. Find a polynomial that represents the area of this square.

59. Find a polynomial that represents the area of this rectangle.

$6x + 2$

$3y + 7$

$y + 1$

60. Perform the indicated multiplications, and then describe the pattern that you observe in the products.

(a) $(x + 4)(x - 4)$; $(y + 2)(y - 2)$; $(r + 7)(r - 7)$
(b) $(x + 4)(x + 4)$; $(y - 2)(y - 2)$; $(r + 7)(r + 7)$

Find each product. In Exercises 71–74 and 78–80, apply the meaning of exponents.

61. $\left(3p + \dfrac{5}{4}q\right)\left(2p - \dfrac{5}{3}q\right)$

62. $\left(-x + \dfrac{2}{3}y\right)\left(3x - \dfrac{3}{4}y\right)$

63. $(x + 7)^2$

64. $(m + 6)^2$

65. $(a - 4)(a + 4)$

66. $(b - 10)(b + 10)$

67. $(2p - 5)^2$

68. $(3m + 1)^2$

69. $(5k + 3q)^2$

70. $(8m - 3n)^2$

71. $(m - 5)^3$

72. $(p + 3)^3$

73. $(2a + 1)^3$

74. $(3m - 1)^3$

75. $7(4m - 3)(2m + 1)$

76. $-4r(3r + 2)(2r - 5)$

77. $-3a(3a + 1)(a - 4)$

78. $(k + 1)^4$

79. $(3r - 2s)^4$

80. $(2z + 5y)^4$

81. $3p^3(2p^2 + 5p)(p^3 + 2p + 1)$

82. $5k^2(k^2 - k + 4)(k^3 - 3)$

83. $-2x^5(3x^2 + 2x - 5)(4x + 2)$

84. $-4x^3(3x^4 + 2x^2 - x)(-2x + 1)$

Find a polynomial that represents the area of each shaded region. In Exercises 87 and 88 leave π in your answer. Use the formulas found on the inside covers.

85.
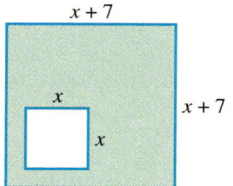

86.
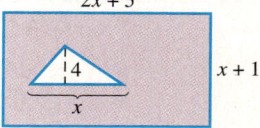

87.
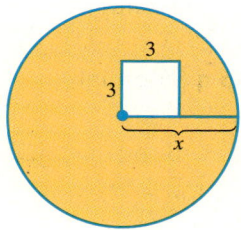

88.
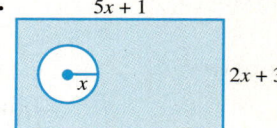

RELATING CONCEPTS (EXERCISES 89–96)

For Individual or Group Work

Work Exercises 89–96 in order. *Refer to the figure as necessary.*

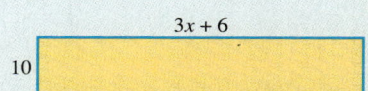

89. Find a polynomial that represents the area of the rectangle.

(continued)

90. Suppose you know that the area of the rectangle is 600 yd². Use this information and the polynomial from Exercise 89 to write an equation that allows you to solve for x.

91. Solve for x.

92. What are the dimensions of the rectangle (assume units are all in yards)?

93. Suppose the rectangle represents a lawn and it costs $3.50 per yd² to lay sod on the lawn. How much will it cost to sod the entire lawn?

94. Use the result of Exercise 92 to find the perimeter of the lawn.

95. Again, suppose the rectangle represents a lawn and it costs $9.00 per yd to fence the lawn. How much will it cost to fence the lawn?

96. (a) Suppose that it costs k dollars per yd² to sod the lawn. Determine a polynomial in the variables x and k that represents the cost to sod the entire lawn.

(b) Suppose that it costs r dollars per yd to fence the lawn. Determine a polynomial in the variables x and r that represents the cost to fence the lawn.

97. Explain the FOIL method for multiplying two binomials. Give an example.

98. Why does the FOIL method not apply to the product of a binomial and a trinomial? Give an example.

5.6 Special Products

OBJECTIVES

1 Square binomials.

2 Find the product of the sum and difference of two terms.

3 Find higher powers of binomials.

In this section, we develop shortcuts to find certain binomial products that occur frequently.

OBJECTIVE 1 Square binomials. The square of a binomial can be found quickly by using the method suggested by Example 1.

EXAMPLE 1 Squaring a Binomial

Find $(m + 3)^2$.

Squaring $m + 3$ by the FOIL method gives
$$(m + 3)(m + 3) = m^2 + 3m + 3m + 9$$
$$= m^2 + 6m + 9.$$

Now Try Exercise 1.

The result above has the squares of the first and the last terms of the binomial:
$$m^2 = m^2 \quad \text{and} \quad 3^2 = 9.$$

The middle term is twice the product of the two terms of the binomial, since the outer and inner products are $m(3)$ and $3(m)$, and
$$m(3) + 3(m) = 2(m)(3) = 6m.$$

This example suggests the following rules.

Square of a Binomial

The square of a binomial is a trinomial consisting of the square of the first term of the binomial, plus twice the product of the two terms, plus the square of the last term of the binomial. For x and y,

$$(x + y)^2 = x^2 + 2xy + y^2.$$

Also,
$$(x - y)^2 = x^2 - 2xy + y^2.$$

EXAMPLE 2 Squaring Binomials

Use the rules to square each binomial.

$$(x - y)^2 = x^2 - 2 \cdot x \cdot y + y^2$$

(a) $(5z - 1)^2 = (5z)^2 - 2(5z)(1) + (1)^2$
$= 25z^2 - 10z + 1$ $(5z)^2 = 5^2 z^2 = 25z^2$

(b) $(3b + 5r)^2 = (3b)^2 + 2(3b)(5r) + (5r)^2$
$= 9b^2 + 30br + 25r^2$

(c) $(2a - 9x)^2 = 4a^2 - 36ax + 81x^2$

(d) $\left(4m + \dfrac{1}{2}\right)^2 = (4m)^2 + 2(4m)\left(\dfrac{1}{2}\right) + \left(\dfrac{1}{2}\right)^2$

$= 16m^2 + 4m + \dfrac{1}{4}$

> Now Try Exercises 5, 7, 9, and 15.

Notice that in the square of a sum all of the terms are positive, as in Examples 2(b) and (d). In the square of a difference, the middle term is negative, as in Examples 2(a) and (c).

CAUTION A common error when squaring a binomial is to forget the middle term of the product. In general,
$$(x + y)^2 \neq x^2 + y^2.$$

OBJECTIVE 2 Find the product of the sum and difference of two terms. Binomial products of the form $(x + y)(x - y)$ also occur frequently. In these products, one binomial is the sum of two terms, and the other is the difference of the same two terms. For example, the product of $x + 2$ and $x - 2$ is

$$(x + 2)(x - 2) = x^2 - 2x + 2x - 4$$
$$= x^2 - 4.$$

As this example suggests, the product of $x + y$ and $x - y$ is the difference between two squares.

Product of the Sum and Difference of Two Terms

$$(x + y)(x - y) = x^2 - y^2$$

EXAMPLE 3 Finding the Product of the Sum and Difference of Two Terms

Find each product.

(a) $(x + 4)(x - 4)$

Use the rule for the product of the sum and difference of two terms.
$$(x + 4)(x - 4) = x^2 - 4^2$$
$$= x^2 - 16$$

(b) $\left(\dfrac{2}{3} - w\right)\left(\dfrac{2}{3} + w\right)$

By the commutative property, this product is the same as $\left(\dfrac{2}{3} + w\right)\left(\dfrac{2}{3} - w\right)$.

$$\left(\dfrac{2}{3} - w\right)\left(\dfrac{2}{3} + w\right) = \left(\dfrac{2}{3} + w\right)\left(\dfrac{2}{3} - w\right)$$
$$= \left(\dfrac{2}{3}\right)^2 - w^2$$
$$= \dfrac{4}{9} - w^2$$

Now Try Exercises 23 and 37.

EXAMPLE 4 Finding the Product of the Sum and Difference of Two Terms

Find each product.

$(x + y)\ (x - y)$
↓ ↓ ↓ ↓

(a) $(5m + 3)(5m - 3)$

Use the rule for the product of the sum and difference of two terms.
$$(5m + 3)(5m - 3) = (5m)^2 - 3^2$$
$$= 25m^2 - 9$$

(b) $(4x + y)(4x - y) = (4x)^2 - y^2$
$$= 16x^2 - y^2$$

(c) $\left(z - \dfrac{1}{4}\right)\left(z + \dfrac{1}{4}\right) = z^2 - \dfrac{1}{16}$

Now Try Exercises 31 and 33.

The product rules of this section will be important later, particularly in Chapters 6 and 7. Therefore, it is essential to learn these rules and practice using them.

OBJECTIVE 3 Find higher powers of binomials. The methods used in the previous section and this section can be combined to find higher powers of binomials.

SECTION 5.6 Special Products 359

EXAMPLE 5 Finding Higher Powers of Binomials

Find each product.

(a) $(x + 5)^3 = (x + 5)^2(x + 5)$ $a^3 = a^2 \cdot a$
$\quad = (x^2 + 10x + 25)(x + 5)$ Square the binomial.
$\quad = x^3 + 10x^2 + 25x + 5x^2 + 50x + 125$ Multiply polynomials.
$\quad = x^3 + 15x^2 + 75x + 125$ Combine like terms.

(b) $(2y - 3)^4 = (2y - 3)^2(2y - 3)^2$ $a^4 = a^2 \cdot a^2$
$\quad = (4y^2 - 12y + 9)(4y^2 - 12y + 9)$ Square each binomial.
$\quad = 16y^4 - 48y^3 + 36y^2 - 48y^3 + 144y^2$ Multiply polynomials.
$\quad - 108y + 36y^2 - 108y + 81$
$\quad = 16y^4 - 96y^3 + 216y^2 - 216y + 81$ Combine like terms.

Now Try Exercises 41 and 49.

5.6 EXERCISES

For Extra Help

Student's Solutions Manual

MyMathLab

InterAct Math Tutorial Software

AW Math Tutor Center

MathXL

Digital Video Tutor CD 11/Videotape 9

1. Consider the square $(2x + 3)^2$.
   (a) What is the square of the first term, $(2x)^2$?
   (b) What is twice the product of the two terms, $2(2x)(3)$?
   (c) What is the square of the last term, 3^2?
   (d) Write the final product, which is a trinomial, using your results in parts (a)–(c).

 2. Explain in your own words how to square a binomial. Give an example.

Find each product. See Examples 1 and 2.

3. $(m + 2)^2$ 4. $(x + 8)^2$ 5. $(r - 3)^2$
6. $(z - 5)^2$ 7. $(x + 2y)^2$ 8. $(3m - p)^2$
9. $(5p + 2q)^2$ 10. $(8a - 3b)^2$ 11. $(4a + 5b)^2$
12. $(9y + z)^2$ 13. $(7t + s)^2$ 14. $\left(5x + \dfrac{2}{5}y\right)^2$
15. $\left(6m - \dfrac{4}{5}n\right)^2$ 16. $x(2x + 5)^2$ 17. $t(3t - 1)^2$
18. $-(4r - 2)^2$ 19. $-(3y - 8)^2$

20. Consider the product $(7x + 3y)(7x - 3y)$.
    (a) What is the product of the first terms, $(7x)(7x)$?
    (b) Multiply the outer terms, $(7x)(-3y)$. Then multiply the inner terms, $(3y)(7x)$. Add the results. What is this sum?
    (c) What is the product of the last terms, $(3y)(-3y)$?
    (d) Write the complete product using your answers in parts (a) and (c). Why is the sum found in part (b) omitted here?

 21. Explain in your own words how to find the product of the sum and the difference of two terms. Give an example.

22. The square of a binomial leads to a polynomial with how many terms? The product of the sum and difference of two terms leads to a polynomial with how many terms?

Find each product. See Examples 3 and 4.

23. $(a + 8)(a - 8)$ **24.** $(k + 5)(k - 5)$ **25.** $(2 + p)(2 - p)$

26. $(4 - 3t)(4 + 3t)$ **27.** $(2m + 5)(2m - 5)$ **28.** $(5x + 2)(5x - 2)$

29. $(3x + 4y)(3x - 4y)$ **30.** $(6a - p)(6a + p)$ **31.** $(5y + 3x)(5y - 3x)$

32. $(10x + 3y)(10x - 3y)$ **33.** $(13r + 2z)(13r - 2z)$ **34.** $(2x^2 - 5)(2x^2 + 5)$

35. $(9y^2 - 2)(9y^2 + 2)$ **36.** $\left(7x + \dfrac{3}{7}\right)\left(7x - \dfrac{3}{7}\right)$ **37.** $\left(9y + \dfrac{2}{3}\right)\left(9y - \dfrac{2}{3}\right)$

38. $p(3p + 7)(3p - 7)$ **39.** $q(5q - 1)(5q + 1)$

40. Does $(a + b)^3$ equal $a^3 + b^3$ in general? Explain.

Find each product. See Example 5.

41. $(x + 1)^3$ **42.** $(y + 2)^3$ **43.** $(t - 3)^3$

44. $(m - 5)^3$ **45.** $(r + 5)^3$ **46.** $(p + 3)^3$

47. $(2a + 1)^3$ **48.** $(3m - 1)^3$ **49.** $(3r - 2t)^4$

50. $(2z + 5y)^4$

RELATING CONCEPTS (EXERCISES 51–60)

For Individual or Group Work

*Special products can be illustrated by using areas of rectangles. Use the figure, and **work Exercises 51–56 in order** to justify the special product*

$$(a + b)^2 = a^2 + 2ab + b^2.$$

51. Express the area of the large square as the square of a binomial.

52. Give the monomial that represents the area of the red square.

53. Give the monomial that represents the sum of the areas of the blue rectangles.

54. Give the monomial that represents the area of the yellow square.

55. What is the sum of the monomials you obtained in Exercises 52–54?

56. Explain why the binomial square you found in Exercise 51 must equal the polynomial you found in Exercise 55.

*To understand how the special product $(a + b)^2 = a^2 + 2ab + b^2$ can be applied to a purely numerical problem, **work Exercises 57–60 in order.***

57. Evaluate 35^2 using either traditional paper-and-pencil methods or a calculator.

58. The number 35 can be written as $30 + 5$. Therefore, $35^2 = (30 + 5)^2$. Use the special product for squaring a binomial with $a = 30$ and $b = 5$ to write an expression for $(30 + 5)^2$. Do not simplify at this time.

59. Use the order of operations to simplify the expression you found in Exercise 58.

60. How do the answers in Exercises 57 and 59 compare?

The special product

$$(a + b)(a - b) = a^2 - b^2$$

can be used to perform some multiplication problems. For example,

$$51 \times 49 = (50 + 1)(50 - 1)$$
$$= 50^2 - 1^2$$
$$= 2500 - 1$$
$$= 2499.$$

$$102 \times 98 = (100 + 2)(100 - 2)$$
$$= 100^2 - 2^2$$
$$= 10{,}000 - 4$$
$$= 9996.$$

Once these patterns are recognized, multiplications of this type can be done mentally. Use this method to calculate each product mentally.

61. 101×99

62. 103×97

63. 201×199

64. 301×299

65. $20\frac{1}{2} \times 19\frac{1}{2}$

66. $30\frac{1}{3} \times 29\frac{2}{3}$

Determine a polynomial that represents the area of each figure. Use the formulas found on the inside covers.

67.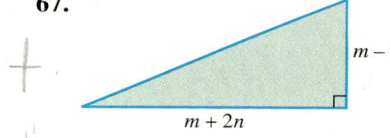

68.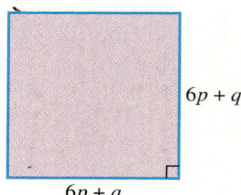

69.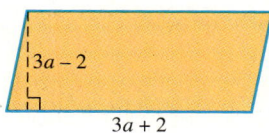

70.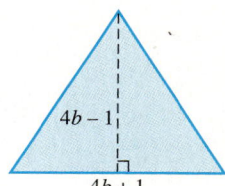

71.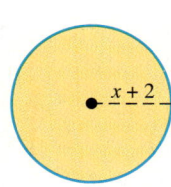

72.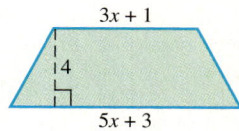

In Exercises 73 and 74, refer to the figure shown here.

73. Find a polynomial that represents the volume of the cube.

74. If the value of x is 6, what is the volume of the cube?

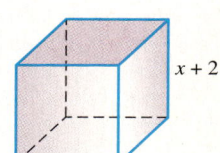

362 CHAPTER 5 Exponents and Polynomials

5.7 Dividing Polynomials

OBJECTIVES

1. Divide a polynomial by a monomial.
2. Divide a polynomial by a polynomial.

OBJECTIVE 1 Divide a polynomial by a monomial. We add two fractions with a common denominator as follows.

$$\frac{a}{c} + \frac{b}{c} = \frac{a+b}{c}$$

Looking at this statement in reverse gives us a rule for dividing a polynomial by a monomial.

Dividing a Polynomial by a Monomial

To divide a polynomial by a monomial, divide each term of the polynomial by the monomial:

$$\frac{a+b}{c} = \frac{a}{c} + \frac{b}{c} \quad (c \neq 0).$$

Examples: $\dfrac{2+5}{3} = \dfrac{2}{3} + \dfrac{5}{3}$ and $\dfrac{x+3z}{2y} = \dfrac{x}{2y} + \dfrac{3z}{2y}$

The parts of a division problem are named here.

Dividend $\rightarrow$ $\dfrac{12x^2 + 6x}{6x} = 2x + 1$ $\leftarrow$ Quotient

Divisor $\rightarrow$

EXAMPLE 1 Dividing a Polynomial by a Monomial

Divide $5m^5 - 10m^3$ by $5m^2$.

Use the preceding rule, with $+$ replaced by $-$. Then use the quotient rule.

$$\frac{5m^5 - 10m^3}{5m^2} = \frac{5m^5}{5m^2} - \frac{10m^3}{5m^2} = m^3 - 2m$$

Check by multiplying: $5m^2(m^3 - 2m) = 5m^5 - 10m^3$.

Because division by 0 is undefined, the quotient

$$\frac{5m^5 - 10m^3}{5m^2}$$

is undefined if $m = 0$. From now on, we assume that no denominators are 0.

Now Try Exercise 15.

EXAMPLE 2 Dividing a Polynomial by a Monomial

Divide $\dfrac{16a^5 - 12a^4 + 8a^2}{4a^3}$.

Divide each term of $16a^5 - 12a^4 + 8a^2$ by $4a^3$.

$$\frac{16a^5 - 12a^4 + 8a^2}{4a^3} = \frac{16a^5}{4a^3} - \frac{12a^4}{4a^3} + \frac{8a^2}{4a^3}$$

$$= 4a^2 - 3a + \frac{2}{a} \qquad \text{Quotient rule}$$

The quotient is not a polynomial because of the expression $\frac{2}{a}$, which has a variable in the denominator. While the sum, difference, and product of two polynomials are always polynomials, the quotient of two polynomials may not be. Again, check by multiplying.

$$4a^3\left(4a^2 - 3a + \frac{2}{a}\right) = 4a^3(4a^2) + 4a^3(-3a) + 4a^3\left(\frac{2}{a}\right)$$

$$= 16a^5 - 12a^4 + 8a^2$$

Now Try Exercise 13.

EXAMPLE 3 Dividing a Polynomial by a Monomial with a Negative Coefficient

Divide $-7x^3 + 12x^4 - 4x$ by $-4x$.

The polynomial should be written in descending powers before dividing. Write it as $12x^4 - 7x^3 - 4x$; then divide by $-4x$.

$$\frac{12x^4 - 7x^3 - 4x}{-4x} = \frac{12x^4}{-4x} + \frac{-7x^3}{-4x} + \frac{-4x}{-4x}$$

$$= -3x^3 + \frac{7x^2}{4} + 1 \quad \text{or} \quad -3x^3 + \frac{7}{4}x^2 + 1$$

Check by multiplying.

Now Try Exercise 23.

CAUTION In Example 3, notice the quotient $\frac{-4x}{-4x} = 1$. It is a common error to leave this term out of the answer. A check by multiplying will show that the answer $-3x^3 + \frac{7}{4}x^2$ is not correct.

EXAMPLE 4 Dividing a Polynomial by a Monomial

Divide the polynomial

$$180x^4y^{10} - 150x^3y^8 + 120x^2y^6 - 90xy^4 + 100y$$

by the monomial $30xy^2$.

$$\frac{180x^4y^{10} - 150x^3y^8 + 120x^2y^6 - 90xy^4 + 100y}{30xy^2}$$

$$= \frac{180x^4y^{10}}{30xy^2} - \frac{150x^3y^8}{30xy^2} + \frac{120x^2y^6}{30xy^2} - \frac{90xy^4}{30xy^2} + \frac{100y}{30xy^2}$$

$$= 6x^3y^8 - 5x^2y^6 + 4xy^4 - 3y^2 + \frac{10}{3xy}$$

Now Try Exercise 31.

364 CHAPTER 5 Exponents and Polynomials

OBJECTIVE 2 Divide a polynomial by a polynomial. We use a method of "long division" to divide a polynomial by a polynomial (other than a monomial). This method is similar to the method of long division used for two whole numbers. For comparison, the division of whole numbers is shown alongside the division of polynomials. Both polynomials must first be written in descending powers.

Dividing Whole Numbers	**Dividing Polynomials**
Step 1 Divide 6696 by 27. $27\overline{)6696}$	Divide $8x^3 - 4x^2 - 14x + 15$ by $2x + 3$. $2x + 3\overline{)8x^3 - 4x^2 - 14x + 15}$
Step 2 66 divided by 27 = **2**; $2 \cdot 27 = $ **54**. $\;\;\mathbf{2}$ $27\overline{)6696}$ $\;\;\mathbf{54}$	$8x^3$ divided by $2x = \mathbf{4x^2}$; $4x^2(2x + 3) = \mathbf{8x^3 + 12x^2}$. $\;\;\mathbf{4x^2}$ $2x + 3\overline{)8x^3 - 4x^2 - 14x + 15}$ $\;\;\mathbf{8x^3 + 12x^2}$
Step 3 Subtract; then bring down the next digit. $\;\;2$ $27\overline{)6696}$ $\;\;54\downarrow$ $\;\;12\mathbf{9}$	Subtract; then bring down the next term. $\;\;4x^2$ $2x + 3\overline{)8x^3 - 4x^2 - 14x + 15}$ $\;\;\underline{8x^3 + 12x^2}\downarrow$ $\;\;\mathbf{-16x^2 - 14x}$

(To subtract two polynomials, change the signs of the second and then add.)

Step 4 129 divided by 27 = **4**; $4 \cdot 27 = $ **108**. $\;\;24$ $27\overline{)6696}$ $\;\;54$ $\;\;129$ $\;\;108$	$-16x^2$ divided by $2x = \mathbf{-8x}$; $-8x(2x+3) = \mathbf{-16x^2 - 24x}$. $\;\;4x^2 - \mathbf{8x}$ $2x + 3\overline{)8x^3 - 4x^2 - 14x + 15}$ $\;\;\underline{8x^3 + 12x^2}$ $\;\;-16x^2 - 14x$ $\;\;\mathbf{-16x^2 - 24x}$
Step 5 Subtract; then bring down the next digit. $\;\;24$ $27\overline{)6696}$ $\;\;54$ $\;\;129$ $\;\;108\downarrow$ $\;\;21\mathbf{6}$	Subtract; then bring down the next term. $\;\;4x^2 - 8x$ $2x + 3\overline{)8x^3 - 4x^2 - 14x + 15}$ $\;\;\underline{8x^3 + 12x^2}$ $\;\;-16x^2 - 14x$ $\;\;\underline{-16x^2 - 24x}\downarrow$ $\;\;\mathbf{10x + 15}$

Step 6

216 divided by 27 = **8**;
8 · 27 = 216.

$$\begin{array}{r} 24\mathbf{8} \\ 27\overline{)6696} \\ \underline{54} \\ 129 \\ \underline{108} \\ 216 \\ \underline{216} \\ 0 \end{array}$$

6696 divided by 27 is 248. There is no remainder.

10x divided by 2x = **5**;
5(2x + 3) = 10x + 15.

$$\begin{array}{r} 4x^2 - 8x + \mathbf{5} \\ 2x + 3\overline{)8x^3 - 4x^2 - 14x + 15} \\ \underline{8x^3 + 12x^2} \\ -16x^2 - 14x \\ \underline{-16x^2 - 24x} \\ 10x + 15 \\ \underline{10x + 15} \\ 0 \end{array}$$

$8x^3 - 4x^2 - 14x + 15$ divided by $2x + 3$ is $4x^2 - 8x + 5$. There is no remainder.

Step 7

Check by multiplying.

27 · 248 = 6696

Check by multiplying.

$(2x + 3)(4x^2 - 8x + 5)$
$= 8x^3 - 4x^2 - 14x + 15$

Now Try Exercise 53.

EXAMPLE 5 Dividing a Polynomial by a Polynomial

Divide $\dfrac{5x + 4x^3 - 8 - 4x^2}{2x - 1}$.

The first polynomial must be written with the exponents in descending order as $4x^3 - 4x^2 + 5x - 8$. Then begin the division process.

Divide $4x^3 - 4x^2 + 5x - 8$ by $2x - 1$.

$$\begin{array}{r} 2x^2 - x + 2 \\ 2x - 1\overline{)4x^3 - 4x^2 + 5x - 8} \\ \underline{4x^3 - 2x^2} \\ -2x^2 + 5x \\ \underline{-2x^2 + x} \\ 4x - 8 \\ \underline{4x - 2} \\ -6 \leftarrow \text{Remainder} \end{array}$$

Step 1 $4x^3$ divided by $2x = 2x^2$; $2x^2(2x - 1) = 4x^3 - 2x^2$.

Step 2 Subtract; bring down the next term.

Step 3 $-2x^2$ divided by $2x = -x$; $-x(2x - 1) = -2x^2 + x$.

Step 4 Subtract; bring down the next term.

Step 5 $4x$ divided by $2x = 2$; $2(2x - 1) = 4x - 2$.

Step 6 Subtract. The remainder is -6. Thus $4x^3 - 4x^2 + 5x - 8$ divided by $2x - 1$ has a quotient of $2x^2 - x + 2$ and a remainder of -6. Write the remainder as the numerator of a fraction that has $2x - 1$ as its denominator. The answer is not a polynomial because of the remainder.

$$\frac{4x^3 - 4x^2 + 5x - 8}{2x - 1} = 2x^2 - x + 2 + \frac{-6}{2x - 1}$$

Step 7 Check by multiplying.

$$(2x - 1)\left(2x^2 - x + 2 + \frac{-6}{2x - 1}\right)$$
$$= (2x - 1)(2x^2) + (2x - 1)(-x) + (2x - 1)(2) + (2x - 1)\left(\frac{-6}{2x - 1}\right)$$
$$= 4x^3 - 2x^2 - 2x^2 + x + 4x - 2 - 6$$
$$= 4x^3 - 4x^2 + 5x - 8$$

Now Try Exercise 59.

EXAMPLE 6 Dividing into a Polynomial with Missing Terms

Divide $x^3 - 1$ by $x - 1$.

Here the polynomial $x^3 - 1$ is missing the x^2-term and the x-term. When terms are missing, use 0 as the coefficient for each missing term. (Zero acts as a placeholder here, just as it does in our number system.)

$$x^3 - 1 = x^3 + 0x^2 + 0x - 1$$

Now divide.

$$\begin{array}{r} x^2 + x + 1 \\ x - 1 \overline{\smash{\big)}\, x^3 + 0x^2 + 0x - 1} \\ \underline{x^3 - x^2} \\ x^2 + 0x \\ \underline{x^2 - x} \\ x - 1 \\ \underline{x - 1} \\ 0 \end{array}$$

The remainder is 0. The quotient is $x^2 + x + 1$. Check by multiplying.

$$(x^2 + x + 1)(x - 1) = x^3 - 1$$

Now Try Exercise 65.

EXAMPLE 7 Dividing by a Polynomial with Missing Terms

Divide $x^4 + 2x^3 + 2x^2 - x - 1$ by $x^2 + 1$.

Since $x^2 + 1$ has a missing x-term, write it as $x^2 + 0x + 1$. Then go through the division process as follows.

$$\begin{array}{r}
x^2 + 2x + 1\\
x^2 + 0x + 1 \overline{)x^4 + 2x^3 + 2x^2 - x - 1}\\
\underline{x^4 + 0x^3 + x^2}\\
2x^3 + x^2 - x\\
\underline{2x^3 + 0x^2 + 2x}\\
x^2 - 3x - 1\\
\underline{x^2 + 0x + 1}\\
-3x - 2 \quad \leftarrow \text{Remainder}
\end{array}$$

When the result of subtracting ($-3x - 2$, in this case) is a polynomial of smaller degree than the divisor ($x^2 + 0x + 1$), that polynomial is the remainder. Write the answer as

$$x^2 + 2x + 1 + \frac{-3x - 2}{x^2 + 1}.$$

Multiply to check that this is correct.

Now Try Exercise 69.

EXAMPLE 8 Dividing a Polynomial When the Quotient Has Fractional Coefficients

Divide $4x^3 + 2x^2 + 3x + 1$ by $4x - 4$.

$$\begin{array}{r}
x^2 + \frac{3}{2}x + \frac{9}{4}\\
4x - 4 \overline{)4x^3 + 2x^2 + 3x + 1}\\
\underline{4x^3 - 4x^2}\\
6x^2 + 3x\\
\underline{6x^2 - 6x}\\
9x + 1\\
\underline{9x - 9}\\
10
\end{array}$$

The answer is $x^2 + \frac{3}{2}x + \frac{9}{4} + \frac{10}{4x - 4}$.

Now Try Exercise 73.

CONNECTIONS

In Section 5.4, we found the value of a polynomial in x for a given value of x by substituting that number for x. Surprisingly, we can accomplish the same thing by division. Suppose we want to find the value of $2x^3 - 4x^2 + 3x - 5$ for $x = -3$. Instead of substituting -3 for x in the polynomial, we divide the polynomial by $x - (-3) = x + 3$. The remainder will give the value of the

(continued)

368 CHAPTER 5 Exponents and Polynomials

polynomial for $x = -3$. In general, when a polynomial P is divided by $x - r$, the remainder is equal to P evaluated at $x = r$.

For Discussion or Writing

1. Evaluate $2x^3 - 4x^2 + 3x - 5$ for $x = -3$.
2. Divide $2x^3 - 4x^2 + 3x - 5$ by $x + 3$. Give the remainder.
3. Compare the answers to Exercises 1 and 2. What do you notice?
4. Choose another polynomial and evaluate it both ways at some value of the variable. Do the answers agree?

5.7 EXERCISES

For Extra Help

Student's Solutions Manual

MyMathLab

InterAct Math Tutorial Software

AW Math Tutor Center

MathXL

Digital Video Tutor
CD 11/Videotape 9

Fill in each blank with the correct response.

1. In the statement $\dfrac{6x^2 + 8}{2} = 3x^2 + 4$, _____ is the dividend, _____ is the divisor, and _____ is the quotient.

2. The expression $\dfrac{3x + 12}{x}$ is undefined if $x =$ _____.

3. To check the division shown in Exercise 1, multiply _____ by _____ and show that the product is _____.

4. The expression $5x^2 - 3x + 6 + \dfrac{2}{x}$ _____ a polynomial.
    (is/is not)

5. Explain why the division problem $\dfrac{16m^3 - 12m^2}{4m}$ can be performed using the methods of this section, while the division problem $\dfrac{4m}{16m^3 - 12m^2}$ cannot.

6. Suppose that a polynomial in the variable x has degree 5 and it is divided by a monomial in the variable x having degree 3. What is the degree of the quotient?

Perform each division. See Examples 1–3.

7. $\dfrac{60x^4 - 20x^2 + 10x}{2x}$

8. $\dfrac{120x^6 - 60x^3 + 80x^2}{2x}$

9. $\dfrac{20m^5 - 10m^4 + 5m^2}{5m^2}$

10. $\dfrac{12t^5 - 6t^3 + 6t^2}{6t^2}$

11. $\dfrac{8t^5 - 4t^3 + 4t^2}{2t}$

12. $\dfrac{8r^4 - 4r^3 + 6r^2}{2r}$

13. $\dfrac{4a^5 - 4a^2 + 8}{4a}$

14. $\dfrac{5t^8 + 5t^7 + 15}{5t}$

Divide each polynomial by $3x^2$. See Examples 1–3.

15. $12x^5 - 9x^4 + 6x^3$

16. $24x^6 - 12x^5 + 30x^4$

17. $3x^2 + 15x^3 - 27x^4$

18. $3x^2 - 18x^4 + 30x^5$

19. $36x + 24x^2 + 6x^3$

20. $9x - 12x^2 + 9x^3$

21. $4x^4 + 3x^3 + 2x$

22. $5x^4 - 6x^3 + 8x$

Perform each division. See Examples 1–4.

23. $\dfrac{-27r^4 + 36r^3 - 6r^2 - 26r + 2}{-3r}$

24. $\dfrac{-8k^4 + 12k^3 + 2k^2 - 7k + 3}{-2k}$

25. $\dfrac{2m^5 - 6m^4 + 8m^2}{-2m^3}$

26. $\dfrac{6r^5 - 8r^4 + 10r^2}{-2r^4}$

27. $(20a^4 - 15a^5 + 25a^3) \div (5a^4)$

28. $(16y^5 - 8y^2 + 12y) \div (4y^2)$

29. $(120x^{11} - 60x^{10} + 140x^9 - 100x^8) \div (10x^{12})$

30. $(120x^{12} - 84x^9 + 60x^8 - 36x^7) \div (12x^9)$

31. $(120x^5y^4 - 80x^2y^3 + 40x^2y^4 - 20x^5y^3) \div (20xy^2)$

32. $(200a^5b^6 - 160a^4b^7 - 120a^3b^9 + 40a^2b^2) \div (40a^2b)$

33. The area of the rectangle is given by the polynomial $15x^3 + 12x^2 - 9x + 3$. What polynomial expresses the length?

34. The area of the triangle is given by the polynomial $24m^3 + 48m^2 + 12m$. What polynomial expresses the length of the base?

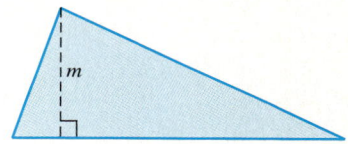

35. The quotient of a certain polynomial and $-7m^2$ is $9m^2 + 3m + 5 - \frac{2}{m}$. Find the polynomial.

36. Suppose that a polynomial of degree n is divided by a monomial of degree m to get a *polynomial* quotient.

   (a) How do m and n compare in value?
   (b) What is the expression that gives the degree of the quotient?

RELATING CONCEPTS (EXERCISES 37–40)

For Individual or Group Work

Our system of numeration is called a decimal system. It is based on powers of ten. In a whole number such as 2846, *each digit is understood to represent the number of powers of ten for its place value. The 2 represents two thousands* (2×10^3), *the 8 represents eight hundreds* (8×10^2), *the 4 represents four tens* (4×10^1), *and the 6 represents six ones (or units)* (6×10^0). *In expanded form we write*

$$2846 = (2 \times 10^3) + (8 \times 10^2) + (4 \times 10^1) + (6 \times 10^0).$$

Keeping this information in mind, **work Exercises 37–40 in order.**

37. Divide 2846 by 2, using paper-and-pencil methods: $2\overline{)2846}$.

38. Write your answer from Exercise 37 in expanded form.

39. Use the methods of this section to divide the polynomial $2x^3 + 8x^2 + 4x + 6$ by 2.

40. Compare your answers in Exercises 38 and 39. How are they similar? How are they different? For what value of x does the answer in Exercise 39 equal the answer in Exercise 38?

370 CHAPTER 5 Exponents and Polynomials

Perform each division. See Example 5.

41. $\dfrac{x^2 - x - 6}{x - 3}$

42. $\dfrac{m^2 - 2m - 24}{m - 6}$

43. $\dfrac{2y^2 + 9y - 35}{y + 7}$

44. $\dfrac{2y^2 + 9y + 7}{y + 1}$

45. $\dfrac{p^2 + 2p + 20}{p + 6}$

46. $\dfrac{x^2 + 11x + 16}{x + 8}$

47. $(r^2 - 8r + 15) \div (r - 3)$

48. $(t^2 + 2t - 35) \div (t - 5)$

49. $\dfrac{12m^2 - 20m + 3}{2m - 3}$

50. $\dfrac{12y^2 + 20y + 7}{2y + 1}$

51. $\dfrac{4a^2 - 22a + 32}{2a + 3}$

52. $\dfrac{9w^2 + 6w + 10}{3w - 2}$

53. $\dfrac{8x^3 - 10x^2 - x + 3}{2x + 1}$

54. $\dfrac{12t^3 - 11t^2 + 9t + 18}{4t + 3}$

55. $\dfrac{8k^4 - 12k^3 - 2k^2 + 7k - 6}{2k - 3}$

56. $\dfrac{27r^4 - 36r^3 - 6r^2 + 26r - 24}{3r - 4}$

57. $\dfrac{5y^4 + 5y^3 + 2y^2 - y - 3}{y + 1}$

58. $\dfrac{2r^3 - 5r^2 - 6r + 15}{r - 3}$

59. $\dfrac{3k^3 - 4k^2 - 6k + 10}{k - 2}$

60. $\dfrac{5z^3 - z^2 + 10z + 2}{z + 2}$

61. $\dfrac{6p^4 - 16p^3 + 15p^2 - 5p + 10}{3p + 1}$

62. $\dfrac{6r^4 - 11r^3 - r^2 + 16r - 8}{2r - 3}$

Perform each division. See Examples 5–8.

63. $\dfrac{5 - 2r^2 + r^4}{r^2 - 1}$

64. $\dfrac{4t^2 + t^4 + 7}{t^2 + 1}$

65. $\dfrac{y^3 + 1}{y + 1}$

66. $\dfrac{y^3 - 1}{y - 1}$

67. $\dfrac{a^4 - 1}{a^2 - 1}$

68. $\dfrac{a^4 - 1}{a^2 + 1}$

69. $\dfrac{x^4 - 4x^3 + 5x^2 - 3x + 2}{x^2 + 3}$

70. $\dfrac{3t^4 + 5t^3 - 8t^2 - 13t + 2}{t^2 - 5}$

71. $\dfrac{2x^5 + 9x^4 + 8x^3 + 10x^2 + 14x + 5}{2x^2 + 3x + 1}$

72. $\dfrac{4t^5 - 11t^4 - 6t^3 + 5t^2 - t + 3}{4t^2 + t - 3}$

73. $(3a^2 - 11a + 17) \div (2a + 6)$

74. $(4x^2 + 11x - 8) \div (3x + 6)$

75. Suppose that one of your classmates asks you the following question: "How do I know when to stop the division process in a problem like the one in Exercise 69?" How would you respond?

76. Suppose that someone asks you if the following division problem is correct:

$$(6x^3 + 4x^2 - 3x + 9) \div (2x - 3) = 4x^2 + 9x - 3.$$

Tell how, by looking only at the *first term* of the quotient, you immediately know that the problem has been worked incorrectly.

Find a polynomial that describes each quantity required. Use the formulas found on the inside covers.

77. Give the length of the rectangle.

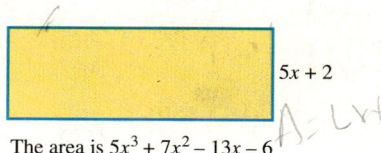

The area is $5x^3 + 7x^2 - 13x - 6$ sq. units.

78. Find the measure of the base of the parallelogram.

The area is $2x^3 + 2x^2 - 3x - 1$ sq. units.

79. If the distance traveled is $5x^3 - 6x^2 + 3x + 14$ mi and the rate is $x + 1$ mph, what is the time traveled?

80. If it costs $4x^5 + 3x^4 + 2x^3 + 9x^2 - 29x + 2$ dollars to fertilize a garden, and fertilizer costs $x + 2$ dollars per yd^2, what is the area of the garden?

RELATING CONCEPTS (EXERCISES 81–84)

For Individual or Group Work

Students often would like to know quickly whether the quotient obtained in a polynomial division problem is actually correct or whether an error was made. While the method described here is not 100% foolproof, it will at least give a fairly good idea as to the accuracy of the result. To illustrate, suppose that $4x^4 + 2x^3 - 14x^2 + 19x + 10$ is divided by $2x + 5$. Lakeisha and Stan obtain the following answers:

Lakeisha	Stan
$2x^3 - 4x^2 + 3x + 2$	$2x^3 - 4x^2 - 3x + 2.$

As a "quick check" we can evaluate Lakeisha's answer for $x = 1$ and Stan's answer for $x = 1$:

Lakeisha: When $x = 1$, her answer gives $2(1)^3 - 4(1)^2 + 3(1) + 2 = 3$.
Stan: When $x = 1$, his answer gives $2(1)^3 - 4(1)^2 - 3(1) + 2 = -3$.

Now, if the original quotient of the two polynomials is evaluated for $x = 1$, we get

$$\frac{4x^4 + 2x^3 - 14x^2 + 19x + 10}{2x + 5} = \frac{4(1)^4 + 2(1)^3 - 14(1)^2 + 19(1) + 10}{2(1) + 5}$$

$$= \frac{21}{7}$$

$$= 3.$$

Because Stan's answer, -3, is different from the quotient 3 just obtained, Stan can conclude his answer is incorrect. Lakeisha's answer, 3, agrees with the quotient just obtained, and while this does not *guarantee* that she is correct, at least she can feel better and go on to the next problem.

(continued)

In Exercises 81–84, a division problem is given, along with two possible answers. One is correct and one is incorrect. Use the method just described to determine which one is correct and which one is not.

81. Problem

$$\frac{2x^2 + 3x - 14}{x - 2}$$

Possible answers

A. $2x + 7$  **B.** $2x - 7$

82. Problem

$$\frac{x^4 + 4x^3 - 5x^2 - 12x + 6}{x^2 - 3}$$

Possible answers

A. $x^2 - 4x - 2$ **B.** $x^2 + 4x - 2$

83. Problem

$$\frac{2y^3 + 17y^2 + 37y + 7}{2y + 7}$$

Possible answers

A. $y^2 + 5y + 1$ **B.** $y^2 - 5y + 1$

84. In the explanation preceding Exercise 81 we used 1 for the value of x to check our work. This is because a polynomial is easy to evaluate for 1. Why is this so? Why would we not be able to use 1 if the divisor is $x - 1$?

Chapter 5 Group Activity

Measuring the Flight of a Rocket

OBJECTIVE Graph quadratic equations to solve an application problem.

A physics class observes the launch of two rockets that are slightly different. One has an initial velocity of 48 ft per sec. The second one has an initial velocity of 64 ft per sec. Follow the steps below to determine the maximum height each rocket will achieve and how long it will take it to reach that height.

The following formula is used to find the height of a projectile when shot vertically into the air:

$$H = -16t^2 + Vt,$$

where H = height above the launch pad in feet, t = time in seconds, and V = initial velocity.

A. One student should complete the table on the next page for the first rocket. The other student should do the same for the second rocket. Use the formula for the height of a projectile and the values provided.

Initial Velocity = 48 Feet per Second	
Time (seconds)	Height (feet)
0	
.5	
1	
1.5	
2	
2.5	
3	

Initial Velocity = 64 Feet per Second	
Time (seconds)	Height (feet)
0	
.5	
1	
1.5	
2	
2.5	
3	
3.5	
4	

B. Graph the results. You must determine the scales for the two axes on the coordinate system. Clearly label each graph and its scale.

C. Answer the following questions for each rocket.

1. Find the time in seconds when the rocket is at height 0 ft. Why does this happen twice?
2. What is the maximum height the rocket reached?
3. At what time (in seconds) did the rocket reach its maximum height?

D. Compare the data from both rockets and answer the following questions.

1. Which rocket had the greater maximum height? Explain how you decided.
2. Which rocket reached its maximum height first? Explain how you decided.
3. Consider the answers for Exercises D1 and D2 and discuss why this occurred.
4. Do both rockets reach a height of 48 ft? If yes, at what times? What observations about projectiles might this lead to?
5. What are some other possible uses for the formula of the height of a projectile?

374 CHAPTER 5 Exponents and Polynomials

CHAPTER 5 SUMMARY

KEY TERMS

5.3 scientific notation
5.4 polynomial
    descending powers
    degree of a term

degree of a
    polynomial
trinomial
binomial

monomial
parabola
vertex
axis

line of symmetry
5.5 outer product
    inner product
    FOIL

NEW SYMBOLS

x^{-n} x to the negative n power

TEST YOUR WORD POWER

See how well you have learned the vocabulary in this chapter. Answers, with examples, follow the Quick Review.

1. A **polynomial** is an algebraic expression made up of
   A. a term or a finite product of terms with positive coefficients and exponents
   B. a term or a finite sum of terms with real coefficients and whole number exponents
   C. the product of two or more terms with positive exponents
   D. the sum of two or more terms with whole number coefficients and exponents.

2. The **degree of a term** is
   A. the number of variables in the term

   B. the product of the exponents on the variables
   C. the smallest exponent on the variables
   D. the sum of the exponents on the variables.

3. A **trinomial** is a polynomial with
   A. only one term
   B. exactly two terms
   C. exactly three terms
   D. more than three terms.

4. A **binomial** is a polynomial with
   A. only one term
   B. exactly two terms
   C. exactly three terms

   D. more than three terms.

5. A **monomial** is a polynomial with
   A. only one term
   B. exactly two terms
   C. exactly three terms
   D. more than three terms.

6. **FOIL** is a method for
   A. adding two binomials
   B. adding two trinomials
   C. multiplying two binomials
   D. multiplying two trinomials.

QUICK REVIEW

CONCEPTS	EXAMPLES
5.1 THE PRODUCT RULE AND POWER RULES FOR EXPONENTS	

For any integers m and n with no denominators zero:
 	
 Product Rule $a^m \cdot a^n = a^{m+n}$
 Power Rules (a) $(a^m)^n = a^{mn}$
                (b) $(ab)^m = a^m b^m$
                (c) $\left(\dfrac{a}{b}\right)^m = \dfrac{a^m}{b^m}$ $(b \neq 0)$
	Perform the operations by using rules for exponents.
	
	 $2^4 \cdot 2^5 = 2^9$
	 $(3^4)^2 = 3^8$
	 $(6a)^5 = 6^5 a^5$
	 $\left(\dfrac{2}{3}\right)^4 = \dfrac{2^4}{3^4}$

CONCEPTS	EXAMPLES

5.2 INTEGER EXPONENTS AND THE QUOTIENT RULE

If $a \neq 0$, for integers m and n:

Zero Exponent $a^0 = 1$

Negative Exponent $a^{-n} = \dfrac{1}{a^n}$

Quotient Rule $\dfrac{a^m}{a^n} = a^{m-n}$

Negative to Positive Rules

$$\dfrac{a^{-m}}{b^{-n}} = \dfrac{b^n}{a^m} \quad (b \neq 0)$$

$$\left(\dfrac{a}{b}\right)^{-m} = \left(\dfrac{b}{a}\right)^m \quad (b \neq 0)$$

Simplify by using the rules for exponents.

$$15^0 = 1$$

$$5^{-2} = \dfrac{1}{5^2} = \dfrac{1}{25}$$

$$\dfrac{4^8}{4^3} = 4^5$$

$$\dfrac{4^{-2}}{3^{-5}} = \dfrac{3^5}{4^2}$$

$$\left(\dfrac{6}{5}\right)^{-3} = \left(\dfrac{5}{6}\right)^3$$

5.3 AN APPLICATION OF EXPONENTS: SCIENTIFIC NOTATION

To write a number in scientific notation (as $a \times 10^n$), move the decimal point to follow the first nonzero digit. If moving the decimal point makes the number smaller, n is positive. If it makes the number larger, n is negative. If the decimal point is not moved, n is 0.

Write in scientific notation.

$$247 = 2.47 \times 10^2$$
$$.0051 = 5.1 \times 10^{-3}$$
$$4.8 = 4.8 \times 10^0$$

Write without exponents.

$$3.25 \times 10^5 = 325{,}000$$
$$8.44 \times 10^{-6} = .00000844$$

5.4 ADDING AND SUBTRACTING POLYNOMIALS; GRAPHING SIMPLE POLYNOMIALS

Adding Polynomials
Add like terms.

Add.

$$\begin{array}{r} 2x^2 + 5x - 3 \\ 5x^2 - 2x + 7 \\ \hline 7x^2 + 3x + 4 \end{array}$$

Subtracting Polynomials
Change the signs of the terms in the second polynomial and add to the first polynomial.

Subtract.

$$(2x^2 + 5x - 3) - (5x^2 - 2x + 7)$$
$$= (2x^2 + 5x - 3) + (-5x^2 + 2x - 7)$$
$$= -3x^2 + 7x - 10$$

Graphing Simple Polynomials
To graph a simple polynomial equation such as $y = x^2 - 2$, plot points near the vertex. (In this chapter, all parabolas have a vertex on the x-axis or the y-axis.)

Graph $y = x^2 - 2$.

x	y
-2	2
-1	-1
0	-2
1	-1
2	2

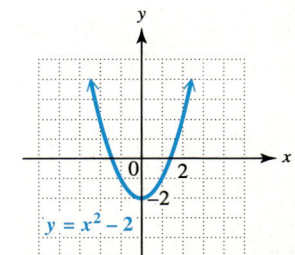

(continued)

CONCEPTS	EXAMPLES
5.5 MULTIPLYING POLYNOMIALS	
General Method for Multiplying Polynomials Multiply each term of the first polynomial by each term of the second polynomial. Then add like terms.	Multiply. $$\begin{array}{r} 3x^3 - 4x^2 + 2x - 7 \\ 4x + 3 \\ \hline 9x^3 - 12x^2 + 6x - 21 \\ 12x^4 - 16x^3 + 8x^2 - 28x \\ \hline 12x^4 - 7x^3 - 4x^2 - 22x - 21 \end{array}$$
FOIL Method for Multiplying Binomials *Step 1* Multiply the two first terms to get the first term of the answer. *Step 2* Find the outer product and the inner product and mentally add them, when possible, to get the middle term of the answer. *Step 3* Multiply the two last terms to get the last term of the answer.	Multiply. $(2x + 3)(5x - 4)$ $$2x(5x) = 10x^2$$ $$2x(-4) + 3(5x) = 7x$$ $$3(-4) = -12$$ The product of $(2x + 3)$ and $(5x - 4)$ is $10x^2 + 7x - 12$.
5.6 SPECIAL PRODUCTS	
Square of a Binomial $(x + y)^2 = x^2 + 2xy + y^2$ $(x - y)^2 = x^2 - 2xy + y^2$ **Product of the Sum and Difference of Two Terms** $(x + y)(x - y) = x^2 - y^2$	Multiply. $(3x + 1)^2 = 9x^2 + 6x + 1$ $(2m - 5n)^2 = 4m^2 - 20mn + 25n^2$ $(4a + 3)(4a - 3) = 16a^2 - 9$
5.7 DIVIDING POLYNOMIALS	
Dividing a Polynomial by a Monomial Divide each term of the polynomial by the monomial. $$\dfrac{a + b}{c} = \dfrac{a}{c} + \dfrac{b}{c}$$	Divide. $$\dfrac{4x^3 - 2x^2 + 6x - 8}{2x} = 2x^2 - x + 3 - \dfrac{4}{x}$$
Dividing a Polynomial by a Polynomial Use "long division."	Divide. $$\begin{array}{r} 2x - 5 \\ 3x + 4 \overline{\smash{)}6x^2 - 7x - 21} \\ \underline{6x^2 + 8x} \\ -15x - 21 \\ \underline{-15x - 20} \\ -1 \end{array}$$ ←Remainder The final answer is $2x - 5 + \dfrac{-1}{3x + 4}$.

Answers to Test Your Word Power
1. B; *Example:* $5x^3 + 2x^2 - 7$ 2. D; *Examples:* The term 6 has degree 0, $3x$ has degree 1, $-2x^8$ has degree 8, and $5x^2y^4$ has degree 6.
3. C; *Example:* $2a^2 - 3ab + b^2$ 4. B; *Example:* $3t^3 + 5t$ 5. A; *Examples:* -5 and $4xy^5$
6. C; *Example:* $(m + 4)(m - 3) = \underset{F}{m(m)} - \underset{O}{3m} + \underset{I}{4m} + \underset{L}{4(-3)} = m^2 + m - 12$

CHAPTER 5 REVIEW EXERCISES

[5.1] *Use the product rule, power rules, or both to simplify each expression. Write the answers in exponential form.*

1. $4^3 \cdot 4^8$
2. $(-5)^6(-5)^5$
3. $(-8x^4)(9x^3)$
4. $(2x^2)(5x^3)(x^9)$
5. $(19x)^5$
6. $(-4y)^7$
7. $5(pt)^4$
8. $\left(\dfrac{7}{5}\right)^6$
9. $(6x^2z^4)^2(x^3yz^2)^4$
10. $\left(\dfrac{2m^3n}{p^2}\right)^3$

11. Why does the product rule for exponents not apply to the expression $7^2 + 7^4$?

[5.2] *Evaluate each expression.*

12. $6^0 + (-6)^0$
13. $(-23)^0 - (-23)^0$
14. -10^0

Write each expression using only positive exponents.

15. -7^{-2}
16. $\left(\dfrac{5}{8}\right)^{-2}$
17. $(5^{-2})^{-4}$
18. $9^3 \cdot 9^{-5}$
19. $2^{-1} + 4^{-1}$
20. $\dfrac{6^{-5}}{6^{-3}}$

Simplify. Write each answer with only positive exponents. Assume that all variables represent nonzero real numbers.

21. $\dfrac{x^{-7}}{x^{-9}}$
22. $\dfrac{y^4 \cdot y^{-2}}{y^{-5}}$
23. $(3r^{-2})^{-4}$
24. $(3p)^4(3p^{-7})$
25. $\dfrac{ab^{-3}}{a^4b^2}$
26. $\dfrac{(6r^{-1})^2(2r^{-4})}{r^{-5}(r^2)^{-3}}$

[5.3] *Write each number in scientific notation.*

27. 48,000,000
28. 28,988,000,000
29. .0000000824

Write each number without exponents.

30. 2.4×10^4
31. 7.83×10^7
32. 8.97×10^{-7}

Perform each indicated operation and write the answer without exponents.

33. $(2 \times 10^{-3}) \times (4 \times 10^5)$
34. $\dfrac{8 \times 10^4}{2 \times 10^{-2}}$
35. $\dfrac{12 \times 10^{-5} \times 5 \times 10^4}{4 \times 10^3 \times 6 \times 10^{-2}}$

Each quote is taken from the source cited. Write each number that is in scientific notation in the quote without exponents.

36. The muon, a close relative of the electron produced by the bombardment of cosmic rays against the upper atmosphere, has a half-life of 2 millionths of a second (2×10^{-6} s). (Excerpt from *Conceptual Physics*, 6th edition, by Paul G. Hewitt. Copyright © by Paul G. Hewitt. Published by HarperCollins College Publishers.)

37. There are 13 red balls and 39 black balls in a box. Mix them up and draw 13 out one at a time without returning any ball . . . the probability that the 13 drawings each will produce a red ball is . . . 1.6×10^{-12}. (Weaver, Warren, *Lady Luck,* Doubleday, 1963, pp. 298–299.)

Each quote is taken from the source cited. Write each number in scientific notation.

38. An electron and a positron attract each other in two ways: the electromagnetic attraction of their opposite electric charges, and the gravitational attraction of their two masses. The electromagnetic attraction is

    4,200,000,000,000,000,000,000,000,000,000,000,000,000

    times as strong as the gravitational. (Asimov, Isaac, *Isaac Asimov's Book of Facts,* Bell Publishing Company, 1981, p. 106.)

39. The aircraft carrier USS John Stennis is a 97,000-ton nuclear powered floating city with a crew of 5000. (*Source:* Seelye, Katharine Q., "Staunch Allies Hard to Beat: Defense Dept., Hollywood," *New York Times,* in *Plain Dealer,* June 10, 2002.)

40. A googol is

    10,000,000,000,000,000,000,000,000,000,000,000,000,000, 000,000,000,000,000,000,000,000,000,000,000,000,000,000.

    The Web search engine Google is named after a googol. Sergey Brin, president and co-founder of Google, Inc., was a math major. He chose the name Google to describe the vast reach of this search engine. (*Source: The Gazette,* March 2, 2001.)

41. According to Campbell, Mitchell, and Reece in *Biology Concepts and Connections* (Benjamin Cummings, 1994, p. 230), "The amount of DNA in a human cell is about 1000 times greater than the DNA in *E. coli.* Does this mean humans have 1000 times as many genes as the 2000 in *E. coli*? The answer is probably no; the human genome is thought to carry between 50,000 and 100,000 genes, which code for various proteins (as well as for tRNA and rRNA)."

    Write each number from this quote using scientific notation.

    (a) 1000   **(b)** 2000   **(c)** 50,000   **(d)** 100,000

[5.4] *Combine like terms where possible in each polynomial. Write the answer in descending powers of the variable. Give the degree of the answer. Identify the polynomial as a* monomial, binomial, trinomial, *or* none of these.

42. $9m^2 + 11m^2 + 2m^2$

43. $-4p + p^3 - p^2 + 8p + 2$

44. $12a^5 - 9a^4 + 8a^3 + 2a^2 - a + 3$

45. $-7y^5 - 8y^4 - y^5 + y^4 + 9y$

46. $(12r^4 - 7r^3 + 2r^2) - (5r^4 - 3r^3 + 2r^2 - 1)$

47. Simplify $(5x^3y^2 - 3xy^5 + 12x^2) - (-9x^2 - 8x^3y^2 + 2xy^5)$.

CHAPTER 5 Review Exercises 379

Add or subtract as indicated.

48. Add.

$-2a^3 + 5a^2$
$\underline{3a^3 - a^2}$

49. Subtract.

$6y^2 - 8y + 2$
$\underline{5y^2 + 2y - 7}$

50. Subtract.

$-12k^4 - 8k^2 + 7k$
$\underline{k^4 + 7k^2 - 11k}$

Graph each equation by completing the table of values.

51. $y = -x^2 + 5$

x	-2	-1	0	1	2
y					

52. $y = 3x^2 - 2$

x	-2	-1	0	1	2
y					

[5.5] *Find each product.*

53. $(a + 2)(a^2 - 4a + 1)$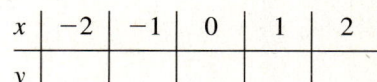

54. $(3r - 2)(2r^2 + 4r - 3)$

55. $(5p^2 + 3p)(p^3 - p^2 + 5)$

56. $(m - 9)(m + 2)$

57. $(3k - 6)(2k + 1)$

58. $(a + 3b)(2a - b)$

59. $(6k + 5q)(2k - 7q)$

60. $(s - 1)^3$

61. Find a polynomial that represents the area of the rectangle shown.

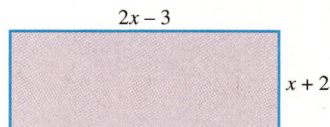

62. If the side of a square has a measure represented by $5x^4 + 2x^2$, what polynomial represents its area?

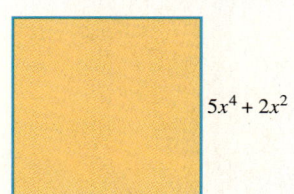

[5.6] *Find each product.*

63. $(a + 4)^2$

64. $(2r + 5t)^2$

65. $(6m - 5)(6m + 5)$

66. $(5a + 6b)(5a - 6b)$

67. $(r + 2)^3$

68. $t(5t - 3)^2$

69. Choose values for x and y to show that, in general, the following hold true.

   (a) $(x + y)^2 \neq x^2 + y^2$ **(b)** $(x + y)^3 \neq x^3 + y^3$

70. Write an explanation on how to raise a binomial to the third power. Give an example.

71. Refer to Exercise 69. Suppose that you happened to let $x = 0$ and $y = 1$. Would your results be sufficient to illustrate the truth, in general, of the inequalities shown? If not, what would you need to do as your next step in working the exercise?

72. What is the volume of a cube with one side having length $x^2 + 2$ cm?

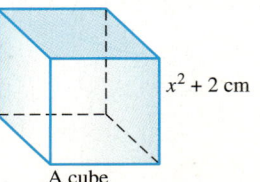

A cube

73. What is the volume of a sphere with radius $x + 1$ in.?

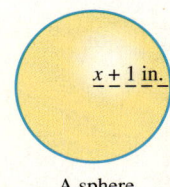

A sphere

[5.7] *Perform each division.*

74. $\dfrac{-15y^4}{9y^2}$

75. $\dfrac{6y^4 - 12y^2 + 18y}{6y}$

76. $(-10m^4n^2 + 5m^3n^2 + 6m^2n^4) \div (5m^2n)$

77. What polynomial, when multiplied by $6m^2n$, gives the product $12m^3n^2 + 18m^6n^3 - 24m^2n^2$?

78. One of your friends in class simplified $\dfrac{6x^2 - 12x}{6}$ as $x^2 - 12x$. Is this correct? If not, what error did your friend make and how would you explain the correct method of performing the division?

Perform each division.

79. $\dfrac{2r^2 + 3r - 14}{r - 2}$

80. $\dfrac{10a^3 + 9a^2 - 14a + 9}{5a - 3}$

81. $\dfrac{x^4 - 5x^2 + 3x^3 - 3x + 4}{x^2 - 1}$

82. $\dfrac{m^4 + 4m^3 - 12m - 5m^2 + 6}{m^2 - 3}$

83. $\dfrac{16x^2 - 25}{4x + 5}$

84. $\dfrac{25y^2 - 100}{5y + 10}$

85. $\dfrac{y^3 - 8}{y - 2}$

86. $\dfrac{1000x^6 + 1}{10x^2 + 1}$

87. $\dfrac{6y^4 - 15y^3 + 14y^2 - 5y - 1}{3y^2 + 1}$

88. $\dfrac{4x^5 - 8x^4 - 3x^3 + 22x^2 - 15}{4x^2 - 3}$

MIXED REVIEW EXERCISES

Perform each indicated operation. Write with positive exponents only. Assume all variables represent nonzero real numbers.

89. $5^0 + 7^0$

90. $\left(\dfrac{6r^2p}{5}\right)^3$

91. $(12a + 1)(12a - 1)$

92. 2^{-4}

93. $(8^{-3})^4$

94. $\dfrac{2p^3 - 6p^2 + 5p}{2p^2}$

95. $\dfrac{(2m^{-5})(3m^2)^{-1}}{m^{-2}(m^{-1})^2}$

96. $(3k - 6)(2k^2 + 4k + 1)$

97. $\dfrac{r^9 \cdot r^{-5}}{r^{-2} \cdot r^{-7}}$

98. $(2r + 5s)^2$

99. $(-5y^2 + 3y - 11) + (4y^2 - 7y + 15)$

100. $(2r + 5)(5r - 2)$

101. $\dfrac{2y^3 + 17y^2 + 37y + 7}{2y + 7}$

102. $(25x^2y^3 - 8xy^2 + 15x^3y) \div (10x^2y^3)$

103. $(6p^2 - p - 8) - (-4p^2 + 2p - 3)$

104. $\dfrac{5^8}{5^{19}}$

105. $(-7 + 2k)^2$

106. $\left(\dfrac{x}{y^{-3}}\right)^{-4}$

CHAPTER 5 TEST

Evaluate each expression.

1. 5^{-4}
2. $(-3)^0 + 4^0$
3. $4^{-1} + 3^{-1}$

4. Use the rules for exponents to simplify $\dfrac{(3x^2y)^2(xy^3)^2}{(xy)^3}$. Assume x and y are nonzero.

Simplify, and write the answer using only positive exponents. Assume that variables represent nonzero numbers.

5. $\dfrac{8^{-1} \cdot 8^4}{8^{-2}}$

6. $\dfrac{(x^{-3})^{-2}(x^{-1}y)^2}{(xy^{-2})^2}$

7. Determine whether each expression represents a number that is *positive, negative,* or *zero.*

   (a) 3^{-4} (b) $(-3)^4$ (c) -3^4 (d) 3^0 (e) $(-3)^0 - 3^0$ (f) $(-3)^{-3}$

8. (a) Write 45,000,000,000 using scientific notation.
   (b) Write 3.6×10^{-6} without using exponents.
   (c) Write the quotient without using exponents: $\dfrac{9.5 \times 10^{-1}}{5 \times 10^3}$.

9. A satellite galaxy of our own Milky Way, known as the Large Magellanic Cloud, is 1000 light-years across. A *light-year* is equal to 5,890,000,000,000 mi. (*Source:* "Images of Brightest Nebula Unveiled," *USA Today,* June 12, 2002.)

   (a) Write these two numbers using scientific notation.
   (b) How many miles across is the Large Magellanic Cloud?

For each polynomial, combine like terms when possible and write the polynomial in descending powers of the variable. Give the degree of the simplified polynomial. Decide whether the simplified polynomial is a monomial, binomial, trinomial, or none of these.

10. $5x^2 + 8x - 12x^2$

11. $13n^3 - n^2 + n^4 + 3n^4 - 9n^2$

12. Use the table to complete a set of ordered pairs that lie on the graph of $y = 2x^2 - 4$. Then graph the equation.

x	-2	-1	0	1	2
y					

Perform each indicated operation.

13. $(2y^2 - 8y + 8) + (-3y^2 + 2y + 3) - (y^2 + 3y - 6)$
14. $(-9a^3b^2 + 13ab^5 + 5a^2b^2) - (6ab^5 + 12a^3b^2 + 10a^2b^2)$
15. Subtract.

    $9t^3 - 4t^2 + 2t + 2$
    $9t^3 + 8t^2 - 3t - 6$

16. $3x^2(-9x^3 + 6x^2 - 2x + 1)$

17. $(t - 8)(t + 3)$

18. $(4x + 3y)(2x - y)$

19. $(5x - 2y)^2$

20. $(10v + 3w)(10v - 3w)$

21. $(2r - 3)(r^2 + 2r - 5)$

22. What polynomial expression represents the area of this square?

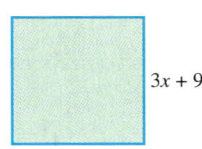

$3x + 9$

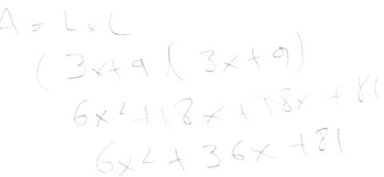

Perform each division.

23. $\dfrac{8y^3 - 6y^2 + 4y + 10}{2y}$

24. $(-9x^2y^3 + 6x^4y^3 + 12xy^3) \div (3xy)$

25. $(3x^3 - x + 4) \div (x - 2)$

CUMULATIVE REVIEW EXERCISES CHAPTERS 1–5

Write each fraction in lowest terms.

1. $\dfrac{28}{16}$

2. $\dfrac{55}{11}$

Perform each operation.

3. $\dfrac{2}{3} + \dfrac{1}{8}$

4. $\dfrac{7}{4} - \dfrac{9}{5}$

5. A contractor installs toolsheds. Each requires $1\tfrac{1}{4}$ yd³ of concrete. How much concrete would be needed for 25 sheds?

6. A retailer has $34,000 invested in her business. She finds that last year she earned 5.4% on this investment. How much did she earn?

7. List all positive integer factors of 45.

8. Find the value of

$$\dfrac{4x - 2y}{x + y}$$

if $x = -2$ and $y = 4$.

Perform each indicated operation.

9. $\dfrac{(-13 + 15) - (3 + 2)}{6 - 12}$

10. $-7 - 3[2 + (5 - 8)]$

Decide which property justifies each statement.

11. $(9 + 2) + 3 = 9 + (2 + 3)$

12. $6(4 + 2) = 6(4) + 6(2)$

13. Simplify the expression $-3(2x^2 - 8x + 9) - (4x^2 + 3x + 2)$.

Solve each equation.

14. $2 - 3(t - 5) = 4 + t$

15. $2(5h + 1) = 10h + 4$

16. $d = rt$ for r

17. $\dfrac{x}{5} = \dfrac{x - 2}{7}$

18. $\dfrac{1}{3}p - \dfrac{1}{6}p = -2$

19. $.05x + .15(50 - x) = 5.50$

20. $4 - (3x + 12) = (2x - 9) - (5x - 1)$

Solve each problem.

21. A 1-oz mouse takes about 16 times as many breaths as does a 3-ton elephant. If the two animals take a combined total of 170 breaths per minute, how many breaths does each take during that time period? (*Source:* McGowan, Christopher, *Dinosaurs, Spitfires, and Sea Dragons,* Harvard University Press, 1991.)

22. If a number is subtracted from 8 and this difference is tripled, the result is 3 times the number. Find this number, and you will learn how many times a dolphin rests during a 24-hr period.

23. One side of a triangle is twice as long as a second side. The third side of the triangle is 17 ft long. The perimeter of the triangle cannot be more than 50 ft. Find the longest possible values for the other two sides of the triangle.

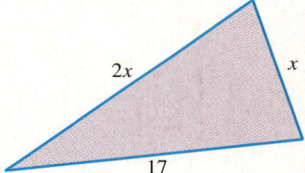

Solve each inequality.

24. $-8x \leq -80$

25. $-2(x + 4) > 3x + 6$

26. $-3 \leq 2x + 5 < 9$

27. Graph $y = -3x + 6$.

28. Consider the two points $(-1, 5)$ and $(2, 8)$.

(a) Find the slope of the line joining them.
(b) Find the equation of the line joining them.

29. Does the point $(-1, 3)$ lie within the shaded region of the graph of $y \geq x + 5$?

30. If $f(x) = x + 7$, find $f(-8)$.

Solve the system using the method indicated.

31. $y = 2x + 5$
$x + y = -4$ (Substitution)

32. $3x + 2y = 2$
$2x + 3y = -7$ (Elimination)

Evaluate each expression.

33. $4^{-1} + 3^0$

34. $2^{-4} \cdot 2^5$

35. $\dfrac{8^{-5} \cdot 8^7}{8^2}$

36. Write with positive exponents only: $\dfrac{(a^{-3}b^2)^2}{(2a^{-4}b^{-3})^{-1}}$.

37. Write in scientific notation: 34,500. 3.45×10^4

38. It takes about 3.6×10^1 sec at a speed of 3.0×10^5 km per sec for light from the sun to reach Venus. How far is Venus from the sun? (*Source: World Almanac and Book of Facts,* 2000.)

39. Graph $y = (x + 4)^2$, using the *x*-values $-6, -5, -4, -3,$ and -2 to obtain a set of points.

Perform each indicated operation.

40. $(7x^3 - 12x^2 - 3x + 8) + (6x^2 + 4) - (-4x^3 + 8x^2 - 2x - 2)$

41. $6x^5(3x^2 - 9x + 10)$

42. $(7x + 4)(9x + 3)$

43. $(5x + 8)^2$

44. $\dfrac{14x^3 - 21x^2 + 7x}{7x}$

45. $\dfrac{y^3 - 3y^2 + 8y - 6}{y - 1}$

Factoring and Applications

6

6.1 The Greatest Common Factor; Factoring by Grouping

6.2 Factoring Trinomials

6.3 More on Factoring Trinomials

6.4 Special Factoring Rules

Summary Exercises on Factoring

6.5 Solving Quadratic Equations by Factoring

6.6 Applications of Quadratic Equations

Galileo Galilei, born near Pisa, Italy, in 1564, became a professor of mathematics at the University of Pisa at age 25. He and his students conducted experiments involving the famous Leaning Tower to investigate the relationship between an object's speed of fall and its weight. (*Source: Microsoft Encarta Encyclopedia 2002.*) We use the concepts of this chapter and the formula Galileo developed from his experiments in Section 6.6.

6.1 The Greatest Common Factor; Factoring by Grouping

OBJECTIVES

1. Find the greatest common factor of a list of terms.
2. Factor out the greatest common factor.
3. Factor by grouping.

Recall from Chapter 1 that to **factor** means to write a quantity as a product. That is, factoring is the opposite of multiplying. For example,

Other **factored forms** of 12 are

$$-6(-2), \quad 3 \cdot 4, \quad -3(-4), \quad 12 \cdot 1, \quad \text{and} \quad -12(-1).$$

More than two factors may be used, so another factored form of 12 is $2 \cdot 2 \cdot 3$. The positive integer factors of 12 are

$$1, 2, 3, 4, 6, 12.$$

OBJECTIVE 1 Find the greatest common factor of a list of terms. An integer that is a factor of two or more integers is called a **common factor** of those integers. For example, 6 is a common factor of 18 and 24 since 6 is a factor of both 18 and 24. Other common factors of 18 and 24 are 1, 2, and 3. The **greatest common factor (GCF)** of a list of integers is the largest common factor of those integers. Thus, 6 is the greatest common factor of 18 and 24, since it is the largest of their common factors.

NOTE Factors of a number are also divisors of the number. The greatest common factor is actually the same as the greatest common divisor. There are many rules for deciding what numbers divide into a given number. Here are some especially useful divisibility rules for small numbers. It is surprising how many people do not know them.

A Whole Number Divisible by:	Must Have the Following Property:
2	Ends in 0, 2, 4, 6, or 8
3	Sum of its digits is divisible by 3.
4	Last two digits form a number divisible by 4
5	Ends in 0 or 5
6	Divisible by both 2 and 3
8	Last three digits form a number divisible by 8
9	Sum of its digits is divisible by 9.
10	Ends in 0

Recall from Chapter 1 that a prime number has only itself and 1 as factors. In Section 1.1 we factored numbers into prime factors. This is the first step in finding the greatest common factor of a list of numbers. We find the greatest common factor (GCF) of a list of numbers as follows.

Finding the Greatest Common Factor (GCF)

Step 1 **Factor.** Write each number in prime factored form.

Step 2 **List common factors.** List each prime number that is a factor of every number in the list. (If a prime does not appear in one of the prime factored forms, it cannot appear in the greatest common factor.)

Step 3 **Choose smallest exponents.** Use as exponents on the common prime factors the *smallest* exponent from the prime factored forms.

Step 4 **Multiply.** Multiply the primes from Step 3. If there are no primes left after Step 3, the greatest common factor is 1.

EXAMPLE 1 Finding the Greatest Common Factor for Numbers

Find the greatest common factor for each list of numbers.

(a) 30, 45

First write each number in prime factored form.

$$30 = 2 \cdot 3 \cdot 5$$
$$45 = 3 \cdot 3 \cdot 5$$

Use each prime the *least* number of times it appears in *all* the factored forms. There is no 2 in the prime factored form of 45, so there will be no 2 in the greatest common factor. The least number of times 3 appears in all the factored forms is 1, and the least number of times 5 appears is also 1. From this, the

$$\text{GCF} = 3^1 \cdot 5^1 = 15.$$

(b) 72, 120, 432

Find the prime factored form of each number.

$$72 = 2 \cdot 2 \cdot 2 \cdot 3 \cdot 3$$
$$120 = 2 \cdot 2 \cdot 2 \cdot 3 \cdot 5$$
$$432 = 2 \cdot 2 \cdot 2 \cdot 2 \cdot 3 \cdot 3 \cdot 3$$

The least number of times 2 appears in all the factored forms is 3, and the least number of times 3 appears is 1. There is no 5 in the prime factored form of either 72 or 432, so the

$$\text{GCF} = 2^3 \cdot 3^1 = 24.$$

(c) 10, 11, 14

Write the prime factored form of each number.

$$10 = 2 \cdot 5$$
$$11 = 11$$
$$14 = 2 \cdot 7$$

There are no primes common to all three numbers, so the GCF is 1.

Now Try Exercise 1.

The greatest common factor can also be found for a list of variable terms. For example, the terms x^4, x^5, x^6, and x^7 have x^4 as the greatest common factor because each of these terms can be written with x^4 as a factor.

$$x^4 = 1 \cdot x^4, \quad x^5 = x \cdot x^4, \quad x^6 = x^2 \cdot x^4, \quad x^7 = x^3 \cdot x^4$$

NOTE The exponent on a variable in the GCF is the *smallest* exponent that appears in *all* the common factors.

EXAMPLE 2 Finding the Greatest Common Factor for Variable Terms

Find the greatest common factor for each list of terms.

(a) $21m^7, -18m^6, 45m^8, -24m^5$

$$21m^7 = 3 \cdot 7 \cdot m^7$$
$$-18m^6 = -1 \cdot 2 \cdot 3^2 \cdot m^6$$
$$45m^8 = 3^2 \cdot 5 \cdot m^8$$
$$-24m^5 = -1 \cdot 2^3 \cdot 3 \cdot m^5$$

First, 3 is the greatest common factor of the coefficients 21, -18, 45, and -24. The smallest exponent on m is 5, so the GCF of the terms is $3m^5$.

(b) $x^4y^2, x^7y^5, x^3y^7, y^{15}$

$$x^4y^2 = x^4 \cdot y^2$$
$$x^7y^5 = x^7 \cdot y^5$$
$$x^3y^7 = x^3 \cdot y^7$$
$$y^{15} = y^{15}$$

There is no x in the last term, y^{15}, so x will not appear in the greatest common factor. There is a y in each term, however, and 2 is the smallest exponent on y. The GCF is y^2.

(c) $-a^2b, -ab^2$

$$-a^2b = -1a^2b = -1 \cdot 1 \cdot a^2b$$
$$-ab^2 = -1ab^2 = -1 \cdot 1 \cdot ab^2$$

The factors of -1 are -1 and 1. Since $1 > -1$, the GCF is $1ab$ or ab.

Now Try Exercises 9 and 13.

NOTE In a list of negative terms, sometimes a negative common factor is preferable (even though it is not the greatest common factor). In Example 2(c), for instance, we might prefer $-ab$ as the common factor. In factoring exercises, either answer will be acceptable.

OBJECTIVE 2 Factor out the greatest common factor. Writing a polynomial (a sum) in factored form as a product is called **factoring**. For example, the polynomial

$$3m + 12$$

has two terms, $3m$ and 12. The greatest common factor for these two terms is 3. We can write $3m + 12$ so that each term is a product with 3 as one factor.

$$3m + 12 = 3 \cdot m + 3 \cdot 4$$

Now we use the distributive property.

$$3m + 12 = 3 \cdot m + 3 \cdot 4 = 3(m + 4)$$

The factored form of $3m + 12$ is $3(m + 4)$. This process is called **factoring out the greatest common factor.**

> **CAUTION** The polynomial $3m + 12$ is *not* in factored form when written as
>
> $$3 \cdot m + 3 \cdot 4.$$
>
> The *terms* are factored, but the polynomial is not. The factored form of $3m + 12$ is the *product*
>
> $$3(m + 4).$$

EXAMPLE 3 Factoring Out the Greatest Common Factor

Factor out the greatest common factor.

(a) $5y^2 + 10y = 5y(y) + 5y(2)$ GCF = $5y$
$ = 5y(y + 2)$ Distributive property

To check, multiply out the factored form: $5y(y + 2) = 5y^2 + 10y$, which is the original polynomial.

(b) $20m^5 + 10m^4 + 15m^3$

The GCF for the terms of this polynomial is $5m^3$.

$$20m^5 + 10m^4 + 15m^3 = 5m^3(4m^2) + 5m^3(2m) + 5m^3(3)$$
$$= 5m^3(4m^2 + 2m + 3)$$

Check: $5m^3(4m^2 + 2m + 3) = 20m^5 + 10m^4 + 15m^3$, which is the original polynomial.

(c) $x^5 + x^3 = x^3(x^2) + x^3(1) = x^3(x^2 + 1)$ Don't forget the 1.

(d) $20m^7p^2 - 36m^3p^4 = 4m^3p^2(5m^4) - 4m^3p^2(9p^2)$ GCF = $4m^3p^2$
$ = 4m^3p^2(5m^4 - 9p^2)$

(e) $\dfrac{1}{6}n^2 + \dfrac{5}{6}n = \dfrac{1}{6}n(n) + \dfrac{1}{6}n(5) = \dfrac{1}{6}n(n + 5)$ GCF = $\frac{1}{6}n$

Now Try Exercises 37, 41, and 47.

> **CAUTION** Be sure to include the 1 in a problem like Example 3(c). *Always* check that the factored form can be multiplied out to give the original polynomial.

EXAMPLE 4 Factoring Out the Greatest Common Factor

Factor out the greatest common factor.

(a) $a(a + 3) + 4(a + 3)$

The binomial $a + 3$ is the greatest common factor here.

$$\underset{\text{Same}}{a(a + 3) + 4(a + 3)} = (a + 3)(a + 4)$$

(b) $x^2(x + 1) - 5(x + 1) = (x + 1)(x^2 - 5)$ Factor out $x + 1$.

Now Try Exercise 55.

OBJECTIVE 3 Factor by grouping. When a polynomial has four terms, common factors can sometimes be used to **factor by grouping.**

EXAMPLE 5 Factoring by Grouping

Factor by grouping.

(a) $2x + 6 + ax + 3a$

The first two terms have a common factor of 2, and the last two terms have a common factor of a.

$$2x + 6 + ax + 3a = 2(x + 3) + a(x + 3)$$

The expression is still not in factored form because it is the *sum* of two terms. Now, however, $x + 3$ is a common factor and can be factored out.

$$2x + 6 + ax + 3a = 2(x + 3) + a(x + 3)$$
$$= (x + 3)(2 + a)$$

The final result is in factored form because it is a *product*. Note that the goal in factoring by grouping is to get a common factor, $x + 3$ here, so that the last step is possible.

Check: $(x + 3)(2 + a) = 2x + 6 + ax + 3a$, which is the original polynomial.

(b) $2x^2 - 10x + 3xy - 15y = (2x^2 - 10x) + (3xy - 15y)$ Group terms.
$ = 2x(x - 5) + 3y(x - 5)$ Factor each group.
$ = (x - 5)(2x + 3y)$ Factor out the common factor, $x - 5$.

Check: $(x - 5)(2x + 3y) = 2x^2 + 3xy - 10x - 15y$ FOIL
$ = 2x^2 - 10x + 3xy - 15y$ Original polynomial

(c) $t^3 + 2t^2 - 3t - 6 = (t^3 + 2t^2) + (-3t - 6)$ Group terms.
$ = t^2(t + 2) - 3(t + 2)$ Factor out -3 so there is a common factor, $t + 2$; $-3(t + 2) = -3t - 6$.
$ = (t + 2)(t^2 - 3)$ Factor out $t + 2$.

Check by multiplying.

Now Try Exercises 69, 73, and 77.

CAUTION Use negative signs carefully when grouping, as in Example 5(c). Otherwise, sign errors may result. Always check by multiplying.

Use these steps when factoring four terms by grouping.

> **Factoring by Grouping**
>
> *Step 1* **Group terms.** Collect the terms into two groups so that each group has a common factor.
>
> *Step 2* **Factor within groups.** Factor out the greatest common factor from each group.
>
> *Step 3* **Factor the entire polynomial.** Factor a common binomial factor from the results of Step 2.
>
> *Step 4* **If necessary, rearrange terms.** If Step 2 does not result in a common binomial factor, try a different grouping.

EXAMPLE 6 Rearranging Terms Before Factoring by Grouping

Factor by grouping.

(a) $10x^2 - 12y + 15x - 8xy$

Factoring out the common factor of 2 from the first two terms and the common factor of x from the last two terms gives

$$10x^2 - 12y + 15x - 8xy = 2(5x^2 - 6y) + x(15 - 8y).$$

This did not lead to a common factor, so we try rearranging the terms. There is usually more than one way to do this. We try

$$10x^2 - 8xy - 12y + 15x,$$

and group the first two terms and the last two terms as follows.

$$10x^2 - 8xy - 12y + 15x = 2x(5x - 4y) + 3(-4y + 5x)$$
$$= 2x(5x - 4y) + 3(5x - 4y)$$
$$= (5x - 4y)(2x + 3)$$

Check: $(5x - 4y)(2x + 3) = 10x^2 + 15x - 8xy - 12y$ FOIL
$$= 10x^2 - 12y + 15x - 8xy$$ Original polynomial

(b) $2xy + 3 - 3y - 2x$

We need to rearrange these terms to get two groups that each have a common factor. Trial and error suggests the following grouping.

$$2xy + 3 - 3y - 2x = (2xy - 3y) + (-2x + 3)$$ Group terms.
$$= y(2x - 3) - 1(2x - 3)$$ Factor each group.
$$= (2x - 3)(y - 1)$$ Factor out the common binomial factor.

Since the quantities in parentheses in the second step must be the same, we factored out -1 rather than 1. Check by multiplying.

Now Try Exercise 81.

392 CHAPTER 6 Factoring and Applications

6.1 EXERCISES — Common Factor

For Extra Help

 Student's Solutions Manual

 MyMathLab

 InterAct Math Tutorial Software

 AW Math Tutor Center

 MathXL

Digital Video Tutor CD 11/Videotape 10

Find the greatest common factor for each list of numbers. See Example 1.

1. 40, 20, 4
2. 50, 30, 5
3. 18, 24, 36, 48
4. 15, 30, 45, 75

5. How can you check your answer when you factor a polynomial?

6. Explain how to find the greatest common factor of a list of terms. Use examples.

Find the greatest common factor for each list of terms. See Examples 1 and 2.

7. $16y, 24$
8. $18w, 27$
9. $30x^3, 40x^6, 50x^7$
10. $60z^4, 70z^8, 90z^9$
11. $12m^3n^2, 18m^5n^4, 36m^8n^3$
12. $25p^5r^7, 30p^7r^8, 50p^5r^3$
13. $-x^4y^3, -xy^2$
14. $-a^4b^5, -a^3b$
15. $42ab^3, -36a, 90b, -48ab$
16. $45c^3d, 75c, 90d, -105cd$

An expression is factored when it is written as a product, not a sum. Which of the following are not factored?

17. $2k^2(5k)$
18. $2k^2(5k + 1)$
19. $2k^2 + (5k + 1)$
20. $(2k^2 + 1)(5k + 1)$

21. Is $-xy$ a common factor of $-x^4y^3$ and $-xy^2$? If so, what is the other factor that when multiplied by $-xy$ gives $-x^4y^3$?

22. Is $-a^5b^2$ a common factor of $-a^4b^5$ and $-a^3b$?

Complete each factoring.

23. $9m^4 = 3m^2(\ \ )$
24. $12p^5 = 6p^3(\ \ )$
25. $-8z^9 = -4z^5(\ \ )$
26. $-15k^{11} = -5k^8(\ \ )$
27. $6m^4n^5 = 3m^3n(\ \ )$
28. $27a^3b^2 = 9a^2b(\ \ )$
29. $12y + 24 = 12(\ \ )$
30. $18p + 36 = 18(\ \ )$
31. $10a^2 - 20a = 10a(\ \ )$
32. $15x^2 - 30x = 15x(\ \ )$
33. $8x^2y + 12x^3y^2 = 4x^2y(\ \ )$
34. $18s^3t^2 + 10st = 2st(\ \ )$

Factor out the greatest common factor. See Examples 3 and 4.

35. $27m^3 - 9m$
36. $36p^3 + 24p$
37. $16z^4 + 24z^2$
38. $25k^4 - 15k^2$
39. $\dfrac{1}{4}d^2 - \dfrac{3}{4}d$
40. $\dfrac{1}{5}z^2 + \dfrac{3}{5}z$
41. $12x^3 + 6x^2$
42. $21b^3 - 7b^2$
43. $65y^{10} + 35y^6$
44. $100a^5 + 16a^3$
45. $11w^3 - 100$
46. $13z^5 - 80$
47. $8m^2n^3 + 24m^2n^2$
48. $19p^2y - 38p^2y^3$
49. $13y^8 + 26y^4 - 39y^2$
50. $5x^5 + 25x^4 - 20x^3$
51. $45q^4p^5 + 36qp^6 + 81q^2p^3$
52. $125a^3z^5 + 60a^4z^4 - 85a^5z^2$
53. $a^5 + 2a^3b^2 - 3a^5b^2 + 4a^4b^3$
54. $x^6 + 5x^4y^3 - 6xy^4 + 10xy$
55. $c(x + 2) - d(x + 2)$
56. $r(5 - x) + t(5 - x)$
57. $m(m + 2n) + n(m + 2n)$
58. $3p(1 - 4p) - 2q(1 - 4p)$

Students often have difficulty when factoring by grouping because they are not able to tell when the polynomial is completely factored. For example,

$$5y(2x - 3) + 8t(2x - 3)$$

is not in factored form, because it is the *sum* of two terms, $5y(2x - 3)$ and $8t(2x - 3)$. However, because $2x - 3$ is a common factor of these two terms, the expression can now be factored as

$$(2x - 3)(5y + 8t).$$

The factored form is a *product* of two factors, $2x - 3$ and $5y + 8t$.

Determine whether each expression is in factored form or is not in factored form. If it is not in factored form, factor it if possible.

59. $8(7t + 4) + x(7t + 4)$
60. $3r(5x - 1) + 7(5x - 1)$
61. $(8 + x)(7t + 4)$
62. $(3r + 7)(5x - 1)$
63. $18x^2(y + 4) + 7(y + 4)$
64. $12k^3(s - 3) + 7(s + 3)$

65. Tell why it is not possible to factor the expression in Exercise 64.

66. Summarize the method of factoring a polynomial with four terms by grouping. Give an example.

Factor by grouping. See Examples 5 and 6.

67. $p^2 + 4p + pq + 4q$
68. $m^2 + 2m + mn + 2n$
69. $a^2 - 2a + ab - 2b$
70. $y^2 - 6y + yw - 6w$
71. $7z^2 + 14z - az - 2a$
72. $5m^2 + 15mp - 2mr - 6pr$
73. $18r^2 + 12ry - 3xr - 2xy$
74. $8s^2 - 4st + 6sy - 3yt$
75. $3a^3 + 3ab^2 + 2a^2b + 2b^3$
76. $4x^3 + 3x^2y + 4xy^2 + 3y^3$
77. $1 - a + ab - b$
78. $6 - 3x - 2y + xy$
79. $16m^3 - 4m^2p^2 - 4mp + p^3$
80. $10t^3 - 2t^2s^2 - 5ts + s^3$
81. $5m - 6p - 2mp + 15$
82. $y^2 + 3x - 3y - xy$
83. $18r^2 - 2ty + 12ry - 3rt$
84. $2b^3 + 3a^3 + 3ab^2 + 2a^2b$
85. $a^5 - 3 + 2a^5b - 6b$
86. $4b^3 + a^2b - 4a - ab^4$

RELATING CONCEPTS (EXERCISES 87–90)

For Individual or Group Work

In many cases, the choice of which pairs of terms to group when factoring by grouping can be made in different ways. To see this for Example 6(b), work Exercises 87–90 in order.

87. Start with the polynomial from Example 6(b), $2xy + 3 - 3y - 2x$, and rearrange the terms as follows: $2xy - 2x - 3y + 3$. What property from Section 1.7 allows this?

88. Group the first two terms and the last two terms of the rearranged polynomial in Exercise 87. Then factor each group.

89. Is your result from Exercise 88 in factored form? Explain your answer.

90. If your answer to Exercise 89 is *no*, factor the polynomial. Is the result the same as the one shown for Example 6(b)?

91. Refer to Exercise 77. The answer given in the back of the book is $(1 - a)(1 - b)$. A student factored this same polynomial and got the result $(a - 1)(b - 1)$.

(a) Is the student's answer correct?

✎ (b) If your answer to part (a) is *yes*, explain why these two seemingly different answers are both acceptable.

✎ **92.** A student factored $18x^3y^2 + 9xy$ as $9xy(2x^2y)$. Is this correct? If not, explain the error and factor correctly.

6.2 Factoring Trinomials

OBJECTIVES

1. Factor trinomials with a coefficient of 1 for the squared term.
2. Factor such trinomials after factoring out the greatest common factor.

Using FOIL, the product of the binomials $k - 3$ and $k + 1$ is
$$(k - 3)(k + 1) = k^2 - 2k - 3. \quad \text{Multiplying}$$
Suppose instead that we are given the polynomial $k^2 - 2k - 3$ and want to rewrite it as the product $(k - 3)(k + 1)$. That is,
$$k^2 - 2k - 3 = (k - 3)(k + 1). \quad \text{Factoring}$$
Recall from the previous section that this process is called factoring the polynomial. Factoring reverses or "undoes" multiplying.

OBJECTIVE 1 **Factor trinomials with a coefficient of 1 for the squared term.** When factoring polynomials with integer coefficients, we use only integers in the factors. For example, we can factor $x^2 + 5x + 6$ by finding integers m and n such that
$$x^2 + 5x + 6 = (x + m)(x + n).$$
To find these integers m and n, we first use FOIL to multiply the two binomials on the right side of the equation:
$$(x + m)(x + n) = x^2 + nx + mx + mn.$$
By the distributive property,
$$x^2 + nx + mx + mn = x^2 + (n + m)x + mn.$$
Comparing this result with $x^2 + 5x + 6$ shows that we must find integers m and n having a sum of 5 and a product of 6.

$$x^2 + 5x + 6 = x^2 + (n + m)x + mn$$

Product of m and n is 6.
Sum of m and n is 5.

Since many pairs of integers have a sum of 5, it is best to begin by listing those pairs of integers whose product is 6. Both 5 and 6 are positive, so we consider only pairs in which both integers are positive.

Pair	Product	Sum	
1, 6	$1 \cdot 6 = 6$	$1 + 6 = 7$	
2, 3	$2 \cdot 3 = 6$	$2 + 3 = 5$	Sum is 5.

Both pairs have a product of 6, but only the pair 2 and 3 has a sum of 5. So 2 and 3 are the required integers, and

$$x^2 + 5x + 6 = (x + 2)(x + 3).$$

Check by multiplying the binomials using FOIL. Make sure that the sum of the outer and inner products produces the correct middle term.

$$(x + 2)(x + 3) = x^2 + 5x + 6$$

$$2x$$
$$3x$$
$$\overline{5x} \quad \text{Add.}$$

We can use this method of factoring only for trinomials that have 1 as the coefficient of the squared term. We give methods for factoring other trinomials in the next section.

EXAMPLE 1 Factoring a Trinomial with All Positive Terms

Factor $m^2 + 9m + 14$.

Look for two integers whose product is 14 and whose sum is 9. List the pairs of integers whose product is 14. Then examine the sums. Again, only positive integers are needed because all signs in $m^2 + 9m + 14$ are positive.

Factors of 14	Sums of Factors
14, 1	14 + 1 = 15
7, 2	7 + 2 = 9 Sum is 9.

From the list, 7 and 2 are the required integers, since $7 \cdot 2 = 14$ and $7 + 2 = 9$. Thus

$$m^2 + 9m + 14 = (m + 7)(m + 2).$$

Check by multiplying on the right side to be sure the original trinomial results.

Now Try Exercise 25.

> **NOTE** In Example 1, the answer also could have been written $(m + 2)(m + 7)$. Because of the commutative property of multiplication, the order of the factors does not matter. Always check by multiplying.

EXAMPLE 2 Factoring a Trinomial with a Negative Middle Term

Factor $x^2 - 9x + 20$.

We must find two integers whose product is 20 and whose sum is -9. Since the numbers we are looking for have a positive product and a negative sum, we consider only pairs of negative integers.

Factors of 20	Sums of Factors
−20, −1	−20 + (−1) = −21
−10, −2	−10 + (−2) = −12
−5, −4	−5 + (−4) = −9 Sum is −9.

The required integers are -5 and -4, so
$$x^2 - 9x + 20 = (x - 5)(x - 4).$$
Check: $(x - 5)(x - 4) = x^2 - 4x - 5x + 20$
$= x^2 - 9x + 20$

Now Try Exercise 29.

EXAMPLE 3 Factoring a Trinomial with Two Negative Terms

Factor $p^2 - 2p - 15$.

Find two integers whose product is -15 and whose sum is -2. If these numbers do not come to mind right away, find them (if they exist) by listing all the pairs of integers whose product is -15. Because the last term, -15, is negative, list pairs of integers with different signs.

Factors of -15	Sums of Factors
15, -1	$15 + (-1) = 14$
-15, 1	$-15 + 1 = -14$
5, -3	$5 + (-3) = 2$
-5, 3	$-5 + 3 = -2$ Sum is -2.

The required integers are -5 and 3, so
$$p^2 - 2p - 15 = (p - 5)(p + 3).$$
Check: Multiply $(p - 5)(p + 3)$ to get $p^2 - 2p - 15$.

Now Try Exercise 35.

> **NOTE** In Examples 1–3, notice that we listed factors in descending order (disregarding sign) when we were looking for the required pair of integers. This helps avoid skipping the correct combination.

As shown in the next example, some trinomials cannot be factored using only integers. We call such trinomials **prime polynomials.**

EXAMPLE 4 Deciding Whether Polynomials Are Prime

Factor each trinomial.

(a) $x^2 - 5x + 12$

As in Example 2, both factors must be negative to give a positive product and a negative sum. First, list all pairs of negative integers whose product is 12. Then examine the sums.

Factors of 12	Sums of Factors
-12, -1	$-12 + (-1) = -13$
-6, -2	$-6 + (-2) = -8$
-4, -3	$-4 + (-3) = -7$

None of the pairs of integers has a sum of -5. Therefore, the trinomial $x^2 - 5x + 12$ *cannot be factored using only integers;* it is a *prime polynomial*.

(b) $k^2 - 8k + 11$

There is no pair of integers whose product is 11 and whose sum is -8, so $k^2 - 8k + 11$ is a prime polynomial.

Now Try Exercise 31.

The procedure for factoring a trinomial of the form $x^2 + bx + c$ is summarized here.

Factoring $x^2 + bx + c$

Find two integers whose product is c and whose sum is b.

1. Both integers must be positive if b and c are positive.
2. Both integers must be negative if c is positive and b is negative.
3. One integer must be positive and one must be negative if c is negative.

EXAMPLE 5 Factoring a Trinomial with Two Variables

Factor $z^2 - 2bz - 3b^2$.

The coefficient of z in the middle term is $-2b$, so we need to find two expressions whose product is $-3b^2$ and whose sum is $-2b$. The expressions are $-3b$ and b, so

$$z^2 - 2bz - 3b^2 = (z - 3b)(z + b).$$

Check by multiplying.

Now Try Exercise 45.

OBJECTIVE 2 Factor such trinomials after factoring out the greatest common factor.
The trinomial in the next example does not have a coefficient of 1 for the squared term. (In fact, there is no squared term.) However, there may be a common factor.

EXAMPLE 6 Factoring a Trinomial with a Common Factor

Factor $4x^5 - 28x^4 + 40x^3$.

First, factor out the greatest common factor, $4x^3$.

$$4x^5 - 28x^4 + 40x^3 = 4x^3(x^2 - 7x + 10)$$

Now factor $x^2 - 7x + 10$. The integers -5 and -2 have a product of 10 and a sum of -7. The complete factored form is

$$4x^5 - 28x^4 + 40x^3 = 4x^3(x - 5)(x - 2).$$

Remember to include $4x^3$.

Check: $4x^3(x - 5)(x - 2) = 4x^3(x^2 - 7x + 10)$
$= 4x^5 - 28x^4 + 40x^3$

Now Try Exercise 53.

398 CHAPTER 6 Factoring and Applications

CAUTION When factoring, always look for a common factor first. Remember to include the common factor as part of the answer. As a check, multiplying out the factored form should always give the original polynomial.

6.2 EXERCISES

For Extra Help

- Student's Solutions Manual
- MyMathLab
- InterAct Math Tutorial Software
- AW Math Tutor Center
- MathXL
- Digital Video Tutor CD 12/Videotape 10

In Exercises 1–4, list all pairs of integers with the given product. Then find the pair whose sum is given.

1. Product: 48; Sum: -19
2. Product: 48; Sum: 14
3. Product: -24; Sum: -5
4. Product: -36; Sum: -16

5. In factoring a trinomial in x as $(x + a)(x + b)$, what must be true of a and b if the coefficient of the last term of the trinomial is negative?

6. In Exercise 5, what must be true of a and b if the coefficient of the last term is positive?

7. What is meant by a *prime polynomial*?

8. How can you check your work when factoring a trinomial? Does the check ensure that the trinomial is completely factored?

9. Which one of the following is the correct factored form of $x^2 - 12x + 32$?
   A. $(x - 8)(x + 4)$ B. $(x + 8)(x - 4)$
   C. $(x - 8)(x - 4)$ D. $(x + 8)(x + 4)$

10. What would be the first step in factoring $2x^3 + 8x^2 - 10x$?

Complete each factoring. See Examples 1–4.

11. $p^2 + 11p + 30 = (p + 5)(p+6)$
12. $x^2 + 10x + 21 = (x + 7)(\quad)$
13. $x^2 + 15x + 44 = (x + 4)(x+11)$
14. $r^2 + 15r + 56 = (r + 7)(\quad)$
15. $x^2 - 9x + 8 = (x - 1)(x-8)$
16. $t^2 - 14t + 24 = (t - 2)(\quad)$
17. $y^2 - 2y - 15 = (y + 3)(y-5)$
18. $t^2 - t - 42 = (t + 6)(\quad)$
19. $x^2 + 9x - 22 = (x - 2)(x+11)$
20. $x^2 + 6x - 27 = (x - 3)(\quad)$
21. $y^2 - 7y - 18 = (y + 2)(y-9)$
22. $y^2 - 2y - 24 = (y + 4)(\quad)$

Factor completely. If the polynomial cannot be factored, write prime. *See Examples 1–4.*

23. $y^2 + 9y + 8$ $(y+8)(y+1)$
24. $a^2 + 9a + 20$
25. $b^2 + 8b + 15$
26. $x^2 + 6x + 8$ $(x+4)(x+2)$
27. $m^2 + m - 20$
28. $p^2 + 4p - 5$
29. $y^2 - 8y + 15$ $(y-5)(y-3)$
30. $y^2 - 6y + 8$
31. $x^2 + 4x + 5$
32. $t^2 + 11t + 12$ Not Factorable
33. $z^2 - 15z + 56$
34. $x^2 - 13x + 36$
35. $r^2 - r - 30$
36. $q^2 - q - 42$
37. $a^2 - 8a - 48$
38. $m^2 - 10m - 25$ prime
39. $x^2 + 3x - 39$
40. $d^2 + 4d - 45$

41. Explain how you would factor $8 + 6x + x^2$. Commutative Property $(x+4)(x+2)$
42. Use your answer to Exercise 41 to factor $5 - 4x - x^2$. x^2-4x-5

SECTION 6.3 More on Factoring Trinomials **399**

Factor completely. See Examples 5 and 6.

43. $r^2 + 3ra + 2a^2$
44. $x^2 + 5xa + 4a^2$
45. $t^2 - tz - 6z^2$
46. $a^2 - ab - 12b^2$
47. $x^2 + 4xy + 3y^2$
48. $p^2 + 9pq + 8q^2$
49. $v^2 - 11vw + 30w^2$
50. $v^2 - 11vx + 24x^2$
51. $4x^2 + 12x - 40$
52. $5y^2 - 5y - 30$
53. $2t^3 + 8t^2 + 6t$
54. $3t^3 + 27t^2 + 24t$
55. $2x^6 + 8x^5 - 42x^4$
56. $4y^5 + 12y^4 - 40y^3$
57. $5m^5 + 25m^4 - 40m^2$
58. $12k^5 - 6k^3 + 10k^2$
59. $m^3n - 10m^2n^2 + 24mn^3$
60. $y^3z + 3y^2z^2 - 54yz^3$

61. Use the FOIL method from Section 5.5 to show that $(2x + 4)(x - 3) = 2x^2 - 2x - 12$. If you are asked to completely factor $2x^2 - 2x - 12$, why would it be incorrect to give $(2x + 4)(x - 3)$ as your answer?

62. If you are asked to completely factor the polynomial $3x^2 + 9x - 12$, why would it be incorrect to give $(x - 1)(3x + 12)$ as your answer?

Use a combination of the factoring methods discussed in this section to factor each polynomial.

63. $a^5 + 3a^4b - 4a^3b^2$
64. $m^3n - 2m^2n^2 - 3mn^3$
65. $y^3z + y^2z^2 - 6yz^3$
66. $k^7 - 2k^6m - 15k^5m^2$
67. $z^{10} - 4z^9y - 21z^8y^2$
68. $x^9 + 5x^8w - 24x^7w^2$
69. $(a + b)x^2 + (a + b)x - 12(a + b)$
70. $(x + y)n^2 + (x + y)n + 16(x + y)$
71. $(2p + q)r^2 - 12(2p + q)r + 27(2p + q)$
72. $(3m - n)k^2 - 13(3m - n)k + 40(3m - n)$
73. What polynomial can be factored as $(a + 9)(a + 4)$?
74. What polynomial can be factored as $(y - 7)(y + 3)$?

6.3 More on Factoring Trinomials

OBJECTIVES

1 Factor trinomials by grouping when the coefficient of the squared term is not 1.

2 Factor trinomials using FOIL.

Trinomials like $2x^2 + 7x + 6$, in which the coefficient of the squared term is *not* 1, are factored with extensions of the methods from the previous sections. One such method uses factoring by grouping from Section 6.1.

OBJECTIVE 1 Factor trinomials by grouping when the coefficient of the squared term is not 1. Recall that a trinomial such as $m^2 + 3m + 2$ is factored by finding two numbers whose product is 2 and whose sum is 3. To factor $2x^2 + 7x + 6$, we look for two integers whose product is $2 \cdot 6 = 12$ and whose sum is 7.

$$2x^2 + 7x + 6$$

Sum is 7.

Product is $2 \cdot 6 = 12$.

By considering pairs of positive integers whose product is 12, we find the necessary integers to be 3 and 4. We use these integers to write the middle term, $7x$, as $7x = 3x + 4x$. The trinomial $2x^2 + 7x + 6$ becomes

$$2x^2 + 7x + 6 = 2x^2 + \underbrace{3x + 4x}_{7x} + 6.$$

$$= (2x^2 + 3x) + (4x + 6) \quad \text{Group terms.}$$
$$= x(2x + 3) + 2(2x + 3) \quad \text{Factor each group.}$$

Must be the same factor

$$2x^2 + 7x + 6 = (2x + 3)(x + 2) \quad \text{Factor out } 2x + 3.$$

Check: $(2x + 3)(x + 2) = 2x^2 + 7x + 6$

In this example, we could have written $7x$ as $4x + 3x$. Factoring by grouping this way would give the same answer.

EXAMPLE 1 Factoring Trinomials by Grouping

Factor each trinomial.

(a) $6r^2 + r - 1$

We must find two integers with a product of $6(-1) = -6$ and a sum of 1.

Sum is 1.

$$6r^2 + r - 1 = 6r^2 + 1r - 1$$

Product is $6(-1) = -6$.

The integers are -2 and 3. We write the middle term, r, as $-2r + 3r$.

$$6r^2 + r - 1 = 6r^2 - 2r + 3r - 1 \quad r = -2r + 3r$$
$$= (6r^2 - 2r) + (3r - 1) \quad \text{Group terms.}$$
$$= 2r(3r - 1) + 1(3r - 1) \quad \text{The binomials must be the same.}$$
$$= (3r - 1)(2r + 1) \quad \text{Factor out } 3r - 1.$$

Check: $(3r - 1)(2r + 1) = 6r^2 + r - 1$

(b) $12z^2 - 5z - 2$

Look for two integers whose product is $12(-2) = -24$ and whose sum is -5. The required integers are 3 and -8, so

$$12z^2 - 5z - 2 = 12z^2 + 3z - 8z - 2 \quad -5z = 3z - 8z$$
$$= (12z^2 + 3z) + (-8z - 2) \quad \text{Group terms.}$$
$$= 3z(4z + 1) - 2(4z + 1) \quad \text{Factor each group; be careful with signs.}$$
$$= (4z + 1)(3z - 2). \quad \text{Factor out } 4z + 1.$$

Check: $(4z + 1)(3z - 2) = 12z^2 - 5z - 2$

(c) $10m^2 + mn - 3n^2$

Two integers whose product is $10(-3) = -30$ and whose sum is 1 are -5 and 6. Rewrite the trinomial with four terms.

$$10m^2 + mn - 3n^2 = 10m^2 - 5mn + 6mn - 3n^2 \quad\text{$mn = -5mn + 6mn$}$$
$$= 5m(2m - n) + 3n(2m - n) \quad\text{Group terms; factor each group.}$$
$$= (2m - n)(5m + 3n) \quad\text{Factor out $2m - n$.}$$

Check by multiplying.

Now Try Exercises 21, 27, and 47.

EXAMPLE 2 Factoring a Trinomial with a Common Factor by Grouping

Factor $28x^5 - 58x^4 - 30x^3$.

First factor out the greatest common factor, $2x^3$.

$$28x^5 - 58x^4 - 30x^3 = 2x^3(14x^2 - 29x - 15)$$

To factor $14x^2 - 29x - 15$, find two integers whose product is $14(-15) = -210$ and whose sum is -29. Factoring 210 into prime factors gives

$$210 = 2 \cdot 3 \cdot 5 \cdot 7.$$

Combine these prime factors in pairs in different ways, using one positive and one negative (to get -210). The factors 6 and -35 have the correct sum. Now rewrite the given trinomial and factor it.

$$28x^5 - 58x^4 - 30x^3 = 2x^3(14x^2 + 6x - 35x - 15)$$
$$= 2x^3[(14x^2 + 6x) + (-35x - 15)]$$
$$= 2x^3[2x(7x + 3) - 5(7x + 3)]$$
$$= 2x^3[(7x + 3)(2x - 5)]$$
$$= 2x^3(7x + 3)(2x - 5)$$

Check by multiplying.

Now Try Exercise 43.

CAUTION Remember to include the common factor in the final result.

OBJECTIVE 2 Factor trinomials using FOIL. In the rest of this section we show an alternative method of factoring trinomials in which the coefficient of the squared term is not 1. This method generalizes the factoring procedure explained in Section 6.2.

To factor $2x^2 + 7x + 6$ (the trinomial factored at the beginning of this section) by the method in Section 6.2, use FOIL backwards. We want to write $2x^2 + 7x + 6$ as the product of two binomials.

$$2x^2 + 7x + 6 = (\quad)(\quad)$$

The product of the two first terms of the binomials is $2x^2$. The possible factors of $2x^2$ are $2x$ and x or $-2x$ and $-x$. Since all terms of the trinomial are positive, we consider only positive factors. Thus, we have

$$2x^2 + 7x + 6 = (2x\quad)(x\quad).$$

The product of the two last terms, 6, can be factored as $1 \cdot 6, 6 \cdot 1, 2 \cdot 3$, or $3 \cdot 2$. Try each pair to find the pair that gives the correct middle term, $7x$.

$(2x + 1)(x + 6)$ Incorrect $(2x + 6)(x + 1)$ Incorrect

x, $12x$, $13x$ Add. $6x$, $2x$, $8x$ Add.

Since $2x + 6 = 2(x + 3)$, the binomial $2x + 6$ has a common factor of 2, while $2x^2 + 7x + 6$ has no common factor other than 1. The product $(2x + 6)(x + 1)$ cannot be correct.

NOTE If the original polynomial has no common factor, then none of its binomial factors will either.

Now try the numbers 2 and 3 as factors of 6. Because of the common factor of 2 in $2x + 2$, $(2x + 2)(x + 3)$ will not work, so we try $(2x + 3)(x + 2)$.

$$(2x + 3)(x + 2) = 2x^2 + 7x + 6 \quad \text{Correct}$$

$3x$, $4x$, $7x$ Add.

Thus, $2x^2 + 7x + 6$ factors as

$$2x^2 + 7x + 6 = (2x + 3)(x + 2).$$

Check by multiplying $2x + 3$ and $x + 2$.

EXAMPLE 3 Factoring a Trinomial with All Positive Terms Using FOIL

Factor $8p^2 + 14p + 5$.

The number 8 has several possible pairs of factors, but 5 has only 1 and 5 or -1 and -5. For this reason, it is easier to begin by considering the factors of 5. Ignore the negative factors since all coefficients in the trinomial are positive. If $8p^2 + 14p + 5$ can be factored, the factors will have the form

$$(\quad + 5)(\quad + 1).$$

The possible pairs of factors of $8p^2$ are $8p$ and p, or $4p$ and $2p$. Try various combinations, checking in each case to see if the middle term is $14p$.

$(8p + 5)(p + 1)$ Incorrect $(p + 5)(8p + 1)$ Incorrect

$5p$, $8p$, $13p$ Add. $40p$, p, $41p$ Add.

$(4p + 5)(2p + 1)$ Correct

$10p$, $4p$, $14p$ Add.

Since $14p$ is the correct middle term,

$$8p^2 + 14p + 5 = (4p + 5)(2p + 1).$$

Check: $(4p + 5)(2p + 1) = 8p^2 + 14p + 5$

Now Try Exercise 23.

EXAMPLE 4 Factoring a Trinomial with a Negative Middle Term Using FOIL

Factor $6x^2 - 11x + 3$.

Since 3 has only 1 and 3 or -1 and -3 as factors, it is better here to begin by factoring 3. The last term of the trinomial $6x^2 - 11x + 3$ is positive and the middle term has a negative coefficient, so we consider only negative factors. We need two negative factors because the *product* of two negative factors is positive and their *sum* is negative, as required. Use -3 and -1 as factors of 3:

$$(- 3)(- 1).$$

The factors of $6x^2$ may be either $6x$ and x, or $2x$ and $3x$. We try $2x$ and $3x$.

$(2x - 3)(3x - 1)$ Correct
$-9x$
$-2x$
$-11x$ Add.

These factors give the correct middle term, so
$$6x^2 - 11x + 3 = (2x - 3)(3x - 1).$$

Check by multiplying.

Now Try Exercise 29.

EXAMPLE 5 Factoring a Trinomial with a Negative Last Term Using FOIL

Factor $8x^2 + 6x - 9$.

The integer 8 has several possible pairs of factors, as does -9. Since the last term is negative, one positive factor and one negative factor of -9 are needed. Since the coefficient of the middle term is small, it is wise to avoid large factors such as 8 or 9. We try $4x$ and $2x$ as factors of $8x^2$, and 3 and -3 as factors of -9, and check the middle term.

$(4x + 3)(2x - 3)$ Incorrect
$6x$
$-12x$
$-6x$ Add.

Now, we try interchanging 3 and -3, since only the sign of the middle term is incorrect.

$(4x - 3)(2x + 3)$ Correct
$-6x$
$12x$
$6x$ Add.

This combination produces the correct middle term, so
$$8x^2 + 6x - 9 = (4x - 3)(2x + 3).$$

Now Try Exercise 33.

EXAMPLE 6 Factoring a Trinomial with Two Variables

Factor $12a^2 - ab - 20b^2$.

There are several pairs of factors of $12a^2$, including $12a$ and a, $6a$ and $2a$, and $3a$ and $4a$, just as there are many pairs of factors of $-20b^2$, including $20b$ and $-b$,

$-20b$ and b, $10b$ and $-2b$, $-10b$ and $2b$, $4b$ and $-5b$, and $-4b$ and $5b$. Once again, since the desired middle term is small, avoid the larger factors. Try the factors $6a$ and $2a$, and $4b$ and $-5b$.

$$(6a + 4b)(2a - 5b)$$

This cannot be correct, as mentioned before, since $6a + 4b$ has a common factor while the given trinomial has none. Try $3a$ and $4a$ with $4b$ and $-5b$.

$$(3a + 4b)(4a - 5b) = 12a^2 + ab - 20b^2 \quad \text{Incorrect}$$

Here the middle term has the wrong sign, so we interchange the signs in the factors.

$$(3a - 4b)(4a + 5b) = 12a^2 - ab - 20b^2 \quad \text{Correct}$$

Now Try Exercise 41.

EXAMPLE 7 Factoring Trinomials with Common Factors

Factor each trinomial.

(a) $15y^3 + 55y^2 + 30y$

First factor out the greatest common factor, $5y$.

$$15y^3 + 55y^2 + 30y = 5y(3y^2 + 11y + 6)$$

Now factor $3y^2 + 11y + 6$. Try $3y$ and y as factors of $3y^2$, and 2 and 3 as factors of 6.

$$(3y + 2)(y + 3) = 3y^2 + 11y + 6 \quad \text{Correct}$$

The complete factored form of $15y^3 + 55y^2 + 30y$ is

$$15y^3 + 55y^2 + 30y = 5y(3y + 2)(y + 3).$$

Check by multiplying.

(b) $-24a^3 - 42a^2 + 45a$

The common factor could be $3a$ or $-3a$. If we factor out $-3a$, the first term of the trinomial will be positive, which makes it easier to factor.

$$-24a^3 - 42a^2 + 45a = -3a(8a^2 + 14a - 15) \quad \text{Factor out } -3a.$$
$$= -3a(4a - 3)(2a + 5) \quad \text{Use FOIL.}$$

Check by multiplying.

Now Try Exercise 45.

> **CAUTION** This caution bears repeating: Remember to include the common factor in the final factored form.

The two methods for factoring trinomials are summarized here.

Factoring Trinomials of the Form $ax^2 + bx + c$

Write the middle term of the trinomial as the sum of two terms to get a polynomial with four terms. Use factoring by grouping as shown in Section 6.1.

Factor the trinomial using FOIL backwards as explained in Section 6.2 and this section.

6.3 EXERCISES

For Extra Help

 Student's Solutions Manual

 MyMathLab

 InterAct Math Tutorial Software

 AW Math Tutor Center

MathXL

 Digital Video Tutor CD 12/Videotape 10

Factor each polynomial by grouping. (The middle term of an equivalent trinomial has already been rewritten.) See Example 1.

1. $10t^2 + 5t + 4t + 2$
2. $6x^2 + 9x + 4x + 6$
3. $15z^2 - 10z - 9z + 6$
4. $12p^2 - 9p - 8p + 6$
5. $8s^2 - 4st + 6st - 3t^2$
6. $3x^2 - 7xy + 6xy - 14y^2$

7. Which pair of integers would be used to rewrite the middle term when factoring $12y^2 + 5y - 2$ by grouping?

   A. $-8, 3$ B. $8, -3$
   C. $-6, 4$ D. $6, -4$

8. Which pair of integers would be used to rewrite the middle term when factoring $20b^2 - 13b + 2$ by grouping?

   A. $10, 3$ B. $-10, -3$
   C. $8, 5$ D. $-8, -5$

Complete the steps to factor each trinomial by grouping.

9. $2m^2 + 11m + 12$
   (a) Find two integers whose product is ___ · ___ = 24 and whose sum is ___.
   (b) The required integers are ___ and ___.
   (c) Write the middle term $11m$ as ___ + ___.
   (d) Rewrite the given trinomial as ___.
   (e) Factor the polynomial in part (d) by grouping.
   (f) Check by multiplying.

10. $6y^2 - 19y + 10$
    (a) Find two integers whose product is ___ · ___ = ___ and whose sum is ___.
    (b) The required integers are ___ and ___.
    (c) Write the middle term $-19y$ as ___ + ___.
    (d) Rewrite the given trinomial as ___.
    (e) Factor the polynomial in part (d) by grouping.
    (f) Check by multiplying.

Decide which is the correct factored form of the given polynomial.

11. $2x^2 - x - 1$
    A. $(2x - 1)(x + 1)$
    B. $(2x + 1)(x - 1)$

12. $3a^2 - 5a - 2$
    A. $(3a + 1)(a - 2)$
    B. $(3a - 1)(a + 2)$

13. $4y^2 + 17y - 15$
    A. $(y + 5)(4y - 3)$
    B. $(2y - 5)(2y + 3)$

14. $12c^2 - 7c - 12$
    A. $(6c - 2)(2c + 6)$
    B. $(4c + 3)(3c - 4)$

Complete each factoring.

15. $6a^2 + 7ab - 20b^2 = (3a - 4b)(\quad)$
16. $9m^2 - 3mn - 2n^2 = (3m + n)(\quad)$
17. $2x^2 + 6x - 8 = 2(\quad)$
    $= 2(\quad)(\quad)$
18. $3x^2 - 9x - 30 = 3(\quad)$
    $= 3(\quad)(\quad)$

19. For the polynomial $12x^2 + 7x - 12$, 2 is not a common factor. Explain why the binomial $2x - 6$, then, cannot be a factor of the polynomial.

20. Factor $4k^2 + 7k - 15$ twice, using the two methods discussed in the text. Do your answers agree? Which method do you prefer?

Factor each trinomial completely. See Examples 1–7.

21. $3a^2 + 10a + 7$
22. $7r^2 + 8r + 1$
23. $2y^2 + 7y + 6$
24. $5z^2 + 12z + 4$
25. $15m^2 + m - 2$
26. $6x^2 + x - 1$
27. $12s^2 + 11s - 5$
28. $20x^2 + 11x - 3$
29. $10m^2 - 23m + 12$
30. $6x^2 - 17x + 12$
31. $8w^2 - 14w + 3$
32. $9p^2 - 18p + 8$
33. $20y^2 - 39y - 11$
34. $10x^2 - 11x - 6$
35. $3x^2 - 15x + 16$
36. $2t^2 + 13t - 18$
37. $20x^2 + 22x + 6$
38. $36y^2 + 81y + 45$
39. $24x^2 - 42x + 9$
40. $48b^2 - 74b - 10$
41. $40m^2q + mq - 6q$
42. $15a^2b + 22ab + 8b$
43. $12n^4 - 22n^3 + 10n^2$
44. $24a^4 + 10a^3 - 4a^2$
45. $15x^2y^3 + 23xy^3 - 14y^3$
46. $14a^2b^3 + 15ab^3 - 9b^3$
47. $8x^2 - 13xy - 6y^2$
48. $6x^2 - 5xy - y^2$
49. $12s^2 + 17st + 5t^2$
50. $25a^2 + 25ab + 6b^2$
51. $6m^6n + 7m^5n^2 + 2m^4n^3$
52. $12k^3q^4 - 4k^2q^5 - kq^6$
53. $5 - 6x + x^2$
54. $7 + 8x + x^2$
55. $16 + 16x + 3x^2$
56. $18 + 65x + 7x^2$
57. $-10x^3 + 5x^2 + 140x$
58. $-18k^3 - 48k^2 + 66k$
59. $12x^2 - 47x - 4$
60. $12x^2 - 19x - 10$
61. $24y^2 - 41xy - 14x^2$
62. $24x^2 + 15xy - 9y^2$
63. $36x^4 - 64x^2y + 15y^2$
64. $36x^4 + 59x^2y + 24y^2$
65. $48a^2 - 94ab - 4b^2$
66. $48t^2 - 147ts + 9s^2$
67. $10x^4y^5 + 39x^3y^5 - 4x^2y^5$
68. $14x^7y^4 - 31x^6y^4 + 6x^5y^4$
69. $36a^3b^2 - 104a^2b^2 - 12ab^2$
70. $36p^4q + 129p^3q - 60p^2q$
71. $24x^2 - 46x + 15$
72. $24x^2 - 94x + 35$
73. $24x^4 + 55x^2 - 24$
74. $24x^4 + 17x^2 - 20$
75. $24x^2 + 38xy + 15y^2$
76. $24x^2 + 62xy + 33y^2$
77. $24x^2z^4 - 113xz^2 - 35$

78. On a quiz, a student factored $3k^3 - 12k^2 - 15k$ by first factoring out the common factor $3k$ to get $3k(k^2 - 4k - 5)$. Then she wrote

$$k^2 - 4k - 5 = k^2 - 5k + k - 5$$
$$= k(k - 5) + 1(k - 5)$$
$$= (k - 5)(k + 1). \quad \text{Her answer}$$

Why was the answer marked wrong? What is the correct factored form?

If a trinomial has a negative coefficient for the squared term, such as $-2x^2 + 11x - 12$, it is usually easier to factor by first factoring out the common factor -1:

$$-2x^2 + 11x - 12 = -1(2x^2 - 11x + 12)$$
$$= -1(2x - 3)(x - 4).$$

Use this method to factor each trinomial. See Example 7(b).

79. $-x^2 - 4x + 21$
80. $-x^2 + x + 72$
81. $-3x^2 - x + 4$
82. $-5x^2 + 2x + 16$
83. $-2a^2 - 5ab - 2b^2$
84. $-3p^2 + 13pq - 4q^2$

85. The answer given in the back of the book for Exercise 79 is $-1(x + 7)(x - 3)$. Is $(x + 7)(3 - x)$ also a correct answer? Explain.

86. One answer for Exercise 80 is $-1(x + 8)(x - 9)$. Is $(-x - 8)(-x + 9)$ also a correct answer? Explain.

Factor each polynomial. Remember to factor out the greatest common factor as the first step.

87. $25q^2(m + 1)^3 - 5q(m + 1)^3 - 2(m + 1)^3$
88. $18x^2(y - 3)^2 - 21x(y - 3)^2 - 4(y - 3)^2$
89. $15x^2(r + 3)^3 - 34xy(r + 3)^3 - 16y^2(r + 3)^3$
90. $4t^2(k + 9)^7 + 20ts(k + 9)^7 + 25s^2(k + 9)^7$

6.4 Special Factoring Rules

OBJECTIVES

1. Factor a difference of squares.
2. Factor a perfect square trinomial.
3. Factor a difference of cubes.
4. Factor a sum of cubes.

By reversing the rules for multiplication of binomials from the last chapter, we get rules for factoring polynomials in certain forms.

OBJECTIVE 1 Factor a difference of squares. The formula for the product of the sum and difference of the same two terms is

$$(x + y)(x - y) = x^2 - y^2.$$

Reversing this rule leads to the following special factoring rule.

Factoring a Difference of Squares

$$x^2 - y^2 = (x + y)(x - y)$$

For example,

$$m^2 - 16 = m^2 - 4^2 = (m + 4)(m - 4).$$

As the next examples show, the following conditions must be true for a binomial to be a difference of squares.

1. Both terms of the binomial must be squares, such as

$$x^2, \quad 9y^2, \quad 25, \quad 1, \quad m^4.$$

2. The second terms of the binomials must have different signs (one positive and one negative).

EXAMPLE 1 Factoring Differences of Squares

Factor each binomial, if possible.

$$x^2 - y^2 = (x + y)(x - y)$$

(a) $a^2 - 49 = a^2 - 7^2 = (a + 7)(a - 7)$

(b) $y^2 - m^2 = (y + m)(y - m)$

(c) $z^2 - \dfrac{9}{16} = z^2 - \left(\dfrac{3}{4}\right)^2 = \left(z + \dfrac{3}{4}\right)\left(z - \dfrac{3}{4}\right)$

(d) $x^2 - 8$
Because 8 is not the square of an integer, this binomial is not a difference of squares. It is a prime polynomial.

(e) $p^2 + 16$
Since $p^2 + 16$ is a *sum* of squares, it is not equal to $(p + 4)(p - 4)$. Also, using FOIL,
$$(p - 4)(p - 4) = p^2 - 8p + 16 \neq p^2 + 16$$
and
$$(p + 4)(p + 4) = p^2 + 8p + 16 \neq p^2 + 16,$$
so $p^2 + 16$ is a prime polynomial.

Now Try Exercises 7, 9, and 11.

CAUTION As Example 1(e) suggests, after any common factor is removed, a *sum* of squares cannot be factored.

EXAMPLE 2 Factoring Differences of Squares

Factor each difference of squares.

$$x^2 - y^2 = (x + y)(x - y)$$

(a) $25m^2 - 16 = (5m)^2 - 4^2 = (5m + 4)(5m - 4)$

(b) $49z^2 - 64 = (7z)^2 - 8^2 = (7z + 8)(7z - 8)$

Now Try Exercise 13.

NOTE As in previous sections, you should always check a factored form by multiplying.

EXAMPLE 3 Factoring More Complex Differences of Squares

Factor completely.

(a) $81y^2 - 36$
First factor out the common factor, 9.
$$81y^2 - 36 = 9(9y^2 - 4) \qquad \text{Factor out 9.}$$
$$= 9[(3y)^2 - 2^2]$$
$$= 9(3y + 2)(3y - 2) \qquad \text{Difference of squares}$$

(b) $9x^2 - 4z^2 = (3x)^2 - (2z)^2 = (3x + 2z)(3x - 2z)$

(c) $p^4 - 36 = (p^2)^2 - 6^2 = (p^2 + 6)(p^2 - 6)$
Neither $p^2 + 6$ nor $p^2 - 6$ can be factored further.

(d) $m^4 - 16 = (m^2)^2 - 4^2$
$ = (m^2 + 4)(m^2 - 4)$ Difference of squares
$ = (m^2 + 4)(m + 2)(m - 2)$ Difference of squares again

Now Try Exercises 17, 21, and 25.

CAUTION Remember to factor again when any of the factors is a difference of squares, as in Example 3(d). Check by multiplying.

OBJECTIVE 2 Factor a perfect square trinomial. The expressions 144, $4x^2$, and $81m^6$ are called *perfect squares* because

$$144 = 12^2, \quad 4x^2 = (2x)^2, \quad \text{and} \quad 81m^6 = (9m^3)^2.$$

A **perfect square trinomial** is a trinomial that is the square of a binomial. For example, $x^2 + 8x + 16$ is a perfect square trinomial because it is the square of the binomial $x + 4$:

$$x^2 + 8x + 16 = (x + 4)(x + 4) = (x + 4)^2.$$

For a trinomial to be a perfect square, *two of its terms must be perfect squares.* For this reason, $16x^2 + 4x + 15$ is not a perfect square trinomial because only the term $16x^2$ is a perfect square.

On the other hand, even if two of the terms are perfect squares, the trinomial may not be a perfect square trinomial. For example, $x^2 + 6x + 36$ has two perfect square terms, x^2 and 36, but it is not a perfect square trinomial. (Try to find a binomial that can be squared to give $x^2 + 6x + 36$.)

We can multiply to see that the square of a binomial gives one of the following perfect square trinomials.

Factoring Perfect Square Trinomials

$$x^2 + 2xy + y^2 = (x + y)^2$$
$$x^2 - 2xy + y^2 = (x - y)^2$$

The middle term of a perfect square trinomial is always twice the product of the two terms in the squared binomial (as shown in Section 5.6). Use this rule to check any attempt to factor a trinomial that appears to be a perfect square.

EXAMPLE 4   Factoring a Perfect Square Trinomial

Factor $x^2 + 10x + 25$.

The term x^2 is a perfect square, and so is 25. Try to factor the trinomial as

$$x^2 + 10x + 25 = (x + 5)^2.$$

To check, take twice the product of the two terms in the squared binomial.

$$2 \cdot x \cdot 5 = 10x$$

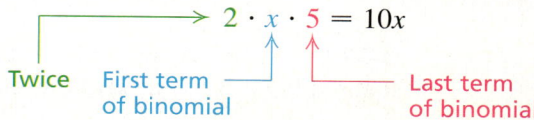

Since $10x$ is the middle term of the trinomial, the trinomial is a perfect square and can be factored as $(x + 5)^2$. Thus,

$$x^2 + 10x + 25 = (x + 5)^2.$$

Now Try Exercise 33.

EXAMPLE 5 Factoring Perfect Square Trinomials

Factor each trinomial.

(a) $x^2 - 22x + 121$

The first and last terms are perfect squares ($121 = 11^2$ or $(-11)^2$). Check to see whether the middle term of $x^2 - 22x + 121$ is twice the product of the first and last terms of the binomial $x - 11$.

$$\underbrace{2}_{\text{Twice}} \cdot \underbrace{x}_{\text{First term}} \cdot \underbrace{(-11)}_{\text{Last term}} = -22x$$

Since twice the product of the first and last terms of the binomial is the middle term, $x^2 - 22x + 121$ is a perfect square trinomial and

$$x^2 - 22x + 121 = (x - 11)^2.$$

Notice that the sign of the second term in the squared binomial is the same as the sign of the middle term in the trinomial.

(b) $9m^2 - 24m + 16 = (3m)^2 + 2(3m)(-4) + (-4)^2 = (3m - 4)^2$

$\underbrace{}_{\text{Twice}} \underbrace{(3m)}_{\text{First term}} \underbrace{(-4)}_{\text{Last term}}$

(c) $25y^2 + 20y + 16$

The first and last terms are perfect squares.

$$25y^2 = (5y)^2 \quad \text{and} \quad 16 = 4^2$$

Twice the product of the first and last terms of the binomial $5y + 4$ is

$$2 \cdot 5y \cdot 4 = 40y,$$

which is not the middle term of $25y^2 + 20y + 16$. This trinomial is not a perfect square. In fact, the trinomial cannot be factored even with the methods of the previous sections; it is a prime polynomial.

(d) $12z^3 + 60z^2 + 75z$

Factor out the common factor, $3z$, first.

$$12z^3 + 60z^2 + 75z = 3z(4z^2 + 20z + 25)$$
$$= 3z[(2z)^2 + 2(2z)(5) + 5^2]$$
$$= 3z(2z + 5)^2$$

Now Try Exercises 35, 43, and 51.

NOTE As mentioned in Example 5(a), the sign of the second term in the squared binomial is always the same as the sign of the middle term in the trinomial. Also, the first and last terms of a perfect square trinomial must be *positive*, because they are squares. For example, the polynomial $x^2 - 2x - 1$ cannot be a perfect square because the last term is negative.

Perfect square trinomials can also be factored using grouping or FOIL, although using the method of this section is often easier.

OBJECTIVE 3 Factor a difference of cubes. Just as we factored the difference of squares in Objective 1, we can also factor the **difference of cubes** using the following pattern.

Factoring a Difference of Cubes

$$x^3 - y^3 = (x - y)(x^2 + xy + y^2)$$

This pattern *should be memorized*. Multiply on the right to see that the pattern gives the correct factors.

$$\begin{array}{r} x^2 + xy + y^2 \\ x - y \\ \hline -x^2y - xy^2 - y^3 \\ x^3 + x^2y + xy^2 \\ \hline x^3 - y^3 \end{array}$$

Notice the pattern of the terms in the factored form of $x^3 - y^3$.

- $x^3 - y^3 =$ (a binomial factor)(a trinomial factor)
- The binomial factor has the difference of the cube roots of the given terms.
- The terms in the trinomial factor are all positive.
- What you write in the binomial factor determines the trinomial factor:

$$x^3 - y^3 = (x - y)(\underset{\text{First term squared}}{x^2} + \underset{\substack{\text{positive} \\ \text{product of} \\ \text{the terms}}}{xy} + \underset{\substack{\text{second term} \\ \text{squared}}}{y^2}).$$

CAUTION The polynomial $x^3 - y^3$ is not equivalent to $(x - y)^3$, because $(x - y)^3$ can also be written as

$$(x - y)^3 = (x - y)(x - y)(x - y)$$
$$= (x - y)(x^2 - 2xy + y^2)$$

but

$$x^3 - y^3 = (x - y)(x^2 + xy + y^2).$$

412 CHAPTER 6 Factoring and Applications

EXAMPLE 6 Factoring Differences of Cubes

Factor the following.

(a) $m^3 - 125$

Let $x = m$ and $y = 5$ in the pattern for the difference of cubes.

$$x^3 - y^3 = (x - y)(x^2 + xy + y^2)$$

$$m^3 - 125 = m^3 - 5^3 = (m - 5)(m^2 + 5m + 5^2) \quad \text{Let } x = m, y = 5.$$
$$= (m - 5)(m^2 + 5m + 25)$$

(b) $8p^3 - 27$

Since $8p^3 = (2p)^3$ and $27 = 3^3$,

$$8p^3 - 27 = (2p)^3 - 3^3$$
$$= (2p - 3)[(2p)^2 + (2p)3 + 3^2]$$
$$= (2p - 3)(4p^2 + 6p + 9).$$

(c) $4m^3 - 32 = 4(m^3 - 8)$
$$= 4(m^3 - 2^3)$$
$$= 4(m - 2)(m^2 + 2m + 4)$$

(d) $125t^3 - 216s^6 = (5t)^3 - (6s^2)^3$
$$= (5t - 6s^2)[(5t)^2 + 5t(6s^2) + (6s^2)^2]$$
$$= (5t - 6s^2)(25t^2 + 30ts^2 + 36s^4)$$

Now Try Exercises 59, 63, and 69.

CAUTION A common error in factoring a difference of cubes, such as $x^3 - y^3 = (x - y)(x^2 + xy + y^2)$, is to try to factor $x^2 + xy + y^2$. It is easy to confuse this factor with a perfect square trinomial, $x^2 + 2xy + y^2$. Because there is no 2 in $x^2 + xy + y^2$, it is very unusual to be able to further factor an expression of the form $x^2 + xy + y^2$.

OBJECTIVE 4 Factor a sum of cubes. A sum of squares, such as $m^2 + 25$, cannot be factored using real numbers, but a **sum of cubes** can be factored by the following pattern, *which should be memorized.*

Factoring a Sum of Cubes

$$x^3 + y^3 = (x + y)(x^2 - xy + y^2)$$

Compare the pattern for the *sum* of cubes with the pattern for the *difference* of cubes. The only difference between them is the positive and negative signs.

$$x^3 - y^3 = (x - y)(x^2 + xy + y^2) \quad \text{Difference of cubes}$$

Same sign, Opposite sign, Positive

$$x^3 + y^3 = (x + y)(x^2 - xy + y^2) \quad \text{Sum of cubes}$$

Same sign, Opposite sign, Positive

Observing these relationships should help you to remember these patterns.

EXAMPLE 7 Factoring Sums of Cubes

Factor.

(a) $k^3 + 27 = k^3 + 3^3$
$= (k + 3)(k^2 - 3k + 3^2)$
$= (k + 3)(k^2 - 3k + 9)$

(b) $8m^3 + 125 = (2m)^3 + 5^3$
$= (2m + 5)[(2m)^2 - 2m(5) + 5^2]$
$= (2m + 5)(4m^2 - 10m + 25)$

(c) $1000a^6 + 27b^3 = (10a^2)^3 + (3b)^3$
$= (10a^2 + 3b)[(10a^2)^2 - (10a^2)(3b) + (3b)^2]$
$= (10a^2 + 3b)(100a^4 - 30a^2b + 9b^2)$

Now Try Exercises 61 and 71.

The methods of factoring discussed in this section are summarized here.

Special Factorizations

Difference of squares $x^2 - y^2 = (x + y)(x - y)$

Perfect square trinomials $x^2 + 2xy + y^2 = (x + y)^2$
$x^2 - 2xy + y^2 = (x - y)^2$

Difference of cubes $x^3 - y^3 = (x - y)(x^2 + xy + y^2)$

Sum of cubes $x^3 + y^3 = (x + y)(x^2 - xy + y^2)$

Remember the *sum* of *squares* can be factored only if the terms have a common factor.

6.4 EXERCISES

For Extra Help
- Student's Solutions Manual
- MyMathLab
- InterAct Math Tutorial Software
- AW Math Tutor Center
- MathXL
- Digital Video Tutor CD 12/Videotape 11

1. To help you factor the difference of squares, complete the following list of squares.

   $1^2 = \underline{1}$ $2^2 = \underline{4}$ $3^2 = \underline{9}$ $4^2 = \underline{16}$ $5^2 = \underline{25}$
   $6^2 = \underline{36}$ $7^2 = \underline{49}$ $8^2 = \underline{64}$ $9^2 = \underline{81}$ $10^2 = \underline{100}$
   $11^2 = \underline{121}$ $12^2 = \underline{144}$ $13^2 = \underline{169}$ $14^2 = \underline{196}$ $15^2 = \underline{225}$
   $16^2 = \underline{256}$ $17^2 = \underline{}$ $18^2 = \underline{}$ $19^2 = \underline{}$ $20^2 = \underline{}$

2. The following powers of x are all perfect squares: $x^2, x^4, x^6, x^8, x^{10}$. Based on this observation, we may make a conjecture (an educated guess) that if the power of a variable is divisible by _____ (with 0 remainder), then we have a perfect square.

3. To help you factor the sum or difference of cubes, complete the following list of cubes.

   $1^3 = \underline{}$ $2^3 = \underline{}$ $3^3 = \underline{}$ $4^3 = \underline{64}$ $5^3 = \underline{125}$
   $6^3 = \underline{}$ $7^3 = \underline{}$ $8^3 = \underline{}$ $9^3 = \underline{}$ $10^3 = \underline{1000}$

4. The following powers of x are all perfect cubes: $x^3, x^6, x^9, x^{12}, x^{15}$. Based on this observation, we may make a conjecture that if the power of a variable is divisible by _____ (with 0 remainder), then we have a perfect cube.

5. Identify each monomial as a perfect square, a perfect cube, both of these, or neither of these.

   (a) $64x^6y^{12}$ **(b)** $125t^6$ **(c)** $49x^{12}$ **(d)** $81r^{10}$

6. What must be true for x^n to be both a perfect square and a perfect cube?

Factor each binomial completely. Use your answers from Exercises 1 and 2 as necessary. See Examples 1–3.

7. $y^2 - 25$
8. $t^2 - 16$
9. $p^2 - \dfrac{1}{9}$
10. $q^2 - \dfrac{1}{4}$
11. $m^2 + 64$
12. $k^2 + 49$
13. $9r^2 - 4$
14. $4x^2 - 9$
15. $36m^2 - \dfrac{16}{25}$
16. $100b^2 - \dfrac{4}{49}$
17. $36x^2 - 16$
18. $32a^2 - 8$
19. $196p^2 - 225$
20. $361q^2 - 400$
21. $16r^2 - 25a^2$
22. $49m^2 - 100p^2$
23. $100x^2 + 49$
24. $81w^2 + 16$
25. $p^4 - 49$
26. $r^4 - 25$
27. $x^4 - 1$
28. $y^4 - 16$
29. $p^4 - 256$
30. $16k^4 - 1$

31. When a student was directed to factor $x^4 - 81$ completely, his teacher did not give him full credit for the answer $(x^2 + 9)(x^2 - 9)$. The student argued that since his answer does indeed give $x^4 - 81$ when multiplied out, he should be given full credit. Was the teacher justified in her grading of this item? Why or why not?

32. The binomial $4x^2 + 16$ is a sum of squares that *can* be factored. How is this binomial factored? When can the sum of squares be factored?

Factor each trinomial completely. It may be necessary to factor out the greatest common factor first. See Examples 4 and 5.

33. $w^2 + 2w + 1$
34. $p^2 + 4p + 4$
35. $x^2 - 8x + 16$
36. $x^2 - 10x + 25$
37. $t^2 + t + \dfrac{1}{4}$
38. $m^2 + \dfrac{2}{3}m + \dfrac{1}{9}$

39. $x^2 - 1.0x + .25$ **40.** $y^2 - 1.4y + .49$ **41.** $2x^2 + 24x + 72$
42. $3y^2 - 48y + 192$ **43.** $16x^2 - 40x + 25$ **44.** $36y^2 - 60y + 25$
45. $49x^2 - 28xy + 4y^2$ **46.** $4z^2 - 12zw + 9w^2$ **47.** $64x^2 + 48xy + 9y^2$
48. $9t^2 + 24tr + 16r^2$ **49.** $-50h^2 + 40hy - 8y^2$ **50.** $-18x^2 - 48xy - 32y^2$
51. $4k^3 - 4k^2 + 9k$ **52.** $9r^3 + 6r^2 + 16r$ **53.** $25z^4 + 5z^3 + z^2$

54. In the polynomial $9y^2 + 14y + 25$, the first and last terms are perfect squares. Can the polynomial be factored? If it can, factor it. If it cannot, explain why it is not a perfect square trinomial.

Find the value of the indicated variable. (Hint: *Perform the squaring on the right side, then solve the equation using the steps given in Chapter 2.*)

55. Find a value of b so that $x^2 + bx + 25 = (x + 5)^2$.

56. For what value of c is $4m^2 - 12m + c = (2m - 3)^2$?

57. Find a so that $ay^2 - 12y + 4 = (3y - 2)^2$.

58. Find b so that $100a^2 + ba + 9 = (10a + 3)^2$.

Factor each binomial completely. Use your answers from Exercises 3 and 4 as necessary. See Examples 6 and 7.

59. $a^3 - 1$ **60.** $m^3 - 8$ **61.** $m^3 + 8$
62. $b^3 + 1$ **63.** $27x^3 - 64$ **64.** $64y^3 - 27$
65. $6p^3 + 6$ **66.** $81x^3 + 3$ **67.** $5x^3 + 40$
68. $128y^3 - 54$ **69.** $2y^3 - 16x^3$ **70.** $27w^3 - 216z^3$
71. $8p^3 + 729q^3$ **72.** $64x^3 + 125y^3$ **73.** $27a^3 + 64b^3$
74. $125m^3 - 8p^3$ **75.** $125t^3 + 8s^3$ **76.** $27r^3 + 1000s^3$

77. State and give the name of each of the five special factoring rules. For each rule, give the numbers of three exercises from this exercise set that are examples.

RELATING CONCEPTS (EXERCISES 78–81)

For Individual or Group Work

We have seen that multiplication and factoring are reverse processes. We know that multiplication and division are also related: To check a division problem, we multiply the quotient by the divisor to get the dividend. To see how factoring and division are related, **work Exercises 78–81 in order.**

78. Factor $10x^2 + 11x - 6$.

79. Use long division to divide $10x^2 + 11x - 6$ by $2x + 3$.

80. Could we have predicted the result in Exercise 79 from the result in Exercise 78? Explain.

81. Divide $x^3 - 1$ by $x - 1$. Use your answer to factor $x^3 - 1$.

Extend the methods of factoring presented so far in this chapter to factor each polynomial completely.

82. $(m + n)^2 - (m - n)^2$ **83.** $(a - b)^3 - (a + b)^3$
84. $m^2 - p^2 + 2m + 2p$ **85.** $3r - 3k + 3r^2 - 3k^2$

SUMMARY EXERCISES ON FACTORING

These mixed exercises are included to give you practice in selecting an appropriate method for factoring a particular polynomial. As you factor a polynomial, ask yourself these questions to decide on a suitable factoring technique.

Factoring a Polynomial

1. Is there a common factor? If so, factor it out.
2. How many terms are in the polynomial?

   Two terms: Check to see whether it is a difference of squares or a sum or difference of cubes. If so, factor as in Section 6.4.

   Three terms: Is it a perfect square trinomial? If the trinomial is not a perfect square, check to see whether the coefficient of the squared term is 1. If so, use the method of Section 6.2. If the coefficient of the squared term of the trinomial is not 1, use the general factoring methods of Section 6.3.

   Four terms: Try to factor the polynomial by grouping.

3. Can any factors be factored further? If so, factor them.

Factor each polynomial completely.

1. $a^2 - 4a - 12$
2. $a^2 + 17a + 72$
3. $6y^2 - 6y - 12$
4. $7y^6 + 14y^5 - 168y^4$
5. $6a + 12b + 18c$
6. $m^2 - 3mn - 4n^2$
7. $p^2 - 17p + 66$
8. $z^2 - 6z + 7z - 42$
9. $10z^2 - 7z - 6$
10. $2m^2 - 10m - 48$
11. $m^2 - n^2 + 5m - 5n$
12. $15y + 5$
13. $8a^5 - 8a^4 - 48a^3$
14. $8k^2 - 10k - 3$
15. $z^2 - 3za - 10a^2$
16. $50z^2 - 100$
17. $x^2 - 4x - 5x + 20$
18. $100n^2r^2 + 30nr^3 - 50n^2r$
19. $6n^2 - 19n + 10$
20. $9y^2 + 12y - 5$
21. $16x + 20$
22. $m^2 + 2m - 15$
23. $6y^2 - 5y - 4$
24. $m^2 - 81$
25. $6z^2 + 31z + 5$
26. $5z^2 + 24z - 5 + 3z + 15$
27. $4k^2 - 12k + 9$
28. $8p^2 + 23p - 3$
29. $54m^2 - 24z^2$
30. $8m^2 - 2m - 3$
31. $3k^2 + 4k - 4$
32. $45a^3b^5 - 60a^4b^2 + 75a^6b^4$
33. $14k^3 + 7k^2 - 70k$
34. $5 + r - 5s - rs$
35. $y^4 - 16$
36. $20y^5 - 30y^4$
37. $8m - 16m^2$
38. $k^2 - 16$
39. $z^3 - 8$
40. $y^2 - y - 56$
41. $k^2 + 9$
42. $27p^{10} - 45p^9 - 252p^8$

43. $32m^9 + 16m^5 + 24m^3$
44. $8m^3 + 125$
45. $16r^2 + 24rm + 9m^2$
46. $z^2 - 12z + 36$
47. $15h^2 + 11hg - 14g^2$
48. $5z^3 - 45z^2 + 70z$
49. $k^2 - 11k + 30$
50. $64p^2 - 100m^2$
51. $3k^3 - 12k^2 - 15k$
52. $y^2 - 4yk - 12k^2$
53. $1000p^3 + 27$
54. $64r^3 - 343$
55. $6 + 3m + 2p + mp$
56. $2m^2 + 7mn - 15n^2$
57. $16z^2 - 8z + 1$
58. $125m^4 - 400m^3n + 195m^2n^2$
59. $108m^2 - 36m + 3$
60. $100a^2 - 81y^2$
61. $64m^2 - 40mn + 25n^2$
62. $4y^2 - 25$
63. $32z^3 + 56z^2 - 16z$
64. $10m^2 + 25m - 60$
65. $20 + 5m + 12n + 3mn$
66. $4 - 2q - 6p + 3pq$
67. $6a^2 + 10a - 4$
68. $36y^6 - 42y^5 - 120y^4$
69. $a^3 - b^3 + 2a - 2b$
70. $16k^2 - 48k + 36$
71. $64m^2 - 80mn + 25n^2$
72. $72y^3z^2 + 12y^2 - 24y^4z^2$
73. $8k^2 - 2kh - 3h^2$
74. $2a^2 - 7a - 30$
75. $(m + 1)^3 + 1$
76. $8a^3 - 27$
77. $10y^2 - 7yz - 6z^2$
78. $m^2 - 4m + 4$
79. $8a^2 + 23ab - 3b^2$
80. $a^4 - 625$

RELATING CONCEPTS (EXERCISES 81–88)

For Individual or Group Work

A binomial may be *both* a difference of squares *and* a difference of cubes. One example of such a binomial is $x^6 - 1$. Using the techniques of Section 6.4, one factoring method will give the complete factored form, while the other will not. **Work Exercises 81–88 in order** to determine the method to use if you have to make such a decision.

81. Factor $x^6 - 1$ as the difference of squares.

82. The factored form obtained in Exercise 81 consists of a difference of cubes multiplied by a sum of cubes. Factor each binomial further.

83. Now start over and factor $x^6 - 1$ as the difference of cubes.

84. The factored form obtained in Exercise 83 consists of a binomial which is a difference of squares and a trinomial. Factor the binomial further.

85. Compare your results in Exercises 82 and 84. Which one of these is the completely factored form?

86. Verify that the trinomial in the factored form in Exercise 84 is the product of the two trinomials in the factored form in Exercise 82.

87. Use the results of Exercises 81–86 to complete the following statement: In general, if I must choose between factoring first using the method for difference of squares or the method for difference of cubes, I should choose the _____ method to eventually obtain the complete factored form.

88. Find the *complete* factored form of $x^6 - 729$ using the knowledge you have gained in Exercises 81–87.

418 CHAPTER 6 Factoring and Applications

6.5 Solving Quadratic Equations by Factoring

OBJECTIVES

1. Solve quadratic equations by factoring.
2. Solve other equations by factoring.

Galileo Galilei (1564–1642) developed theories to explain physical phenomena and set up experiments to test his ideas. According to legend, Galileo dropped objects of different weights from the Leaning Tower of Pisa to disprove the belief that heavier objects fall faster than lighter objects. He developed a formula for freely falling objects described by

$$d = 16t^2,$$

where d is the distance in feet that an object falls (disregarding air resistance) in t seconds, regardless of weight.

The equation $d = 16t^2$ is a *quadratic equation,* the subject of this section. A quadratic equation contains a squared term and no terms of higher degree.

Quadratic Equation

A **quadratic equation** is an equation that can be written in the form

$$ax^2 + bx + c = 0,$$

where a, b, and c are real numbers, with $a \neq 0$.

The form $ax^2 + bx + c = 0$ is the **standard form** of a quadratic equation. For example,

$$x^2 + 5x + 6 = 0, \quad 2a^2 - 5a = 3, \quad \text{and} \quad y^2 = 4$$

are all quadratic equations, but only $x^2 + 5x + 6 = 0$ is in standard form.

Up to now, we have factored *expressions,* including many quadratic expressions of the form $ax^2 + bx + c$. In this section, we see how we can use factored quadratic expressions to solve quadratic *equations.*

OBJECTIVE 1 Solve quadratic equations by factoring. We use the **zero-factor property** to solve a quadratic equation by factoring.

Zero-Factor Property

If a and b are real numbers and if $ab = 0$, then $a = 0$ or $b = 0$.

That is, if the product of two numbers is 0, then at least one of the numbers must be 0. One number *must* be 0, but both *may* be 0.

EXAMPLE 1 Using the Zero-Factor Property

Solve each equation.

(a) $(x + 3)(2x - 1) = 0$

The product $(x + 3)(2x - 1)$ is equal to 0. By the zero-factor property, the only way that the product of these two factors can be 0 is if at least one of the factors equals 0. Therefore, either $x + 3 = 0$ or $2x - 1 = 0$. Solve each of these two linear equations as in Chapter 2.

$$x + 3 = 0 \quad \text{or} \quad 2x - 1 = 0 \quad \text{Zero-factor property}$$
$$x = -3 \qquad\qquad 2x = 1 \quad \text{Add 1 to each side.}$$
$$x = \frac{1}{2} \quad \text{Divide each side by 2.}$$

The given equation, $(x + 3)(2x - 1) = 0$, has two solutions, -3 and $\frac{1}{2}$. Check these solutions by substituting -3 for x in the original equation, $(x + 3)(2x - 1) = 0$. Then start over and substitute $\frac{1}{2}$ for x.

If $x = -3$, then

$$(x + 3)(2x - 1) = 0$$
$$(-3 + 3)[2(-3) - 1] = 0 \quad ?$$
$$0(-7) = 0. \quad \text{True}$$

If $x = \frac{1}{2}$, then

$$(x + 3)(2x - 1) = 0$$
$$\left(\frac{1}{2} + 3\right)\left(2 \cdot \frac{1}{2} - 1\right) = 0 \quad ?$$
$$\frac{7}{2}(1 - 1) = 0 \quad ?$$
$$\frac{7}{2} \cdot 0 = 0. \quad \text{True}$$

Both -3 and $\frac{1}{2}$ result in true equations, so the solution set is $\{-3, \frac{1}{2}\}$.

(b) $y(3y - 4) = 0$

$$y(3y - 4) = 0$$
$$y = 0 \quad \text{or} \quad 3y - 4 = 0 \quad \text{Zero-factor property}$$
$$3y = 4$$
$$y = \frac{4}{3}$$

Check these solutions by substituting each one in the original equation. The solution set is $\{0, \frac{4}{3}\}$.

Now Try Exercises 3 and 5.

NOTE The word *or* as used in Example 1 means "one or the other or both."

In Example 1, each equation to be solved was given with the polynomial in factored form. If the polynomial in an equation is not already factored, first make sure that the equation is in standard form. Then factor.

EXAMPLE 2 Solving Quadratic Equations

Solve each equation.

(a) $x^2 - 5x = -6$

First, rewrite the equation in standard form by adding 6 to each side.

$$x^2 - 5x = -6$$
$$x^2 - 5x + 6 = 0 \quad \text{Add 6.}$$

Now factor $x^2 - 5x + 6$. Find two numbers whose product is 6 and whose sum is -5. These two numbers are -2 and -3, so the equation becomes

$$(x - 2)(x - 3) = 0. \quad \text{Factor.}$$
$$x - 2 = 0 \quad \text{or} \quad x - 3 = 0 \quad \text{Zero-factor property}$$
$$x = 2 \quad \quad \quad x = 3 \quad \text{Solve each equation.}$$

Check: If $x = 2$, then
$$x^2 - 5x = -6$$
$$2^2 - 5(2) = -6 \quad ?$$
$$4 - 10 = -6 \quad ?$$
$$-6 = -6. \quad \text{True}$$

If $x = 3$, then
$$x^2 - 5x = -6$$
$$3^2 - 5(3) = -6 \quad ?$$
$$9 - 15 = -6 \quad ?$$
$$-6 = -6. \quad \text{True}$$

Both solutions check, so the solution set is $\{2, 3\}$.

(b) $y^2 = y + 20$

Rewrite the equation in standard form.

$$y^2 = y + 20$$
$$y^2 - y - 20 = 0 \quad \text{Subtract } y \text{ and } 20.$$
$$(y - 5)(y + 4) = 0 \quad \text{Factor.}$$
$$y - 5 = 0 \quad \text{or} \quad y + 4 = 0 \quad \text{Zero-factor property}$$
$$y = 5 \quad \quad \quad y = -4 \quad \text{Solve each equation.}$$

Check these solutions by substituting each one in the original equation. The solution set is $\{-4, 5\}$.

Now Try Exercise 21.

In summary, follow these steps to solve quadratic equations by factoring.

Solving a Quadratic Equation by Factoring

Step 1 **Write the equation in standard form,** that is, with all terms on one side of the equals sign in descending powers of the variable and 0 on the other side.

Step 2 **Factor** completely.

Step 3 **Use the zero-factor property** to set each factor with a variable equal to 0, and solve the resulting equations.

Step 4 **Check** each solution in the original equation.

NOTE Not all quadratic equations can be solved by factoring. A more general method for solving such equations is given in Chapter 9.

EXAMPLE 3 Solving a Quadratic Equation with a Common Factor

Solve $4p^2 + 40 = 26p$.

Subtract $26p$ from each side and write the equation in standard form to get

$$4p^2 - 26p + 40 = 0.$$

SECTION 6.5 Solving Quadratic Equations by Factoring **421**

$$2(2p^2 - 13p + 20) = 0 \quad \text{Factor out 2.}$$
$$2p^2 - 13p + 20 = 0 \quad \text{Divide each side by 2.}$$
$$(2p - 5)(p - 4) = 0 \quad \text{Factor.}$$
$$2p - 5 = 0 \quad \text{or} \quad p - 4 = 0 \quad \text{Zero-factor property}$$
$$2p = 5 \qquad\qquad p = 4$$
$$p = \frac{5}{2}$$

Check that the solution set is $\{\frac{5}{2}, 4\}$ by substituting each solution in the original equation.

Now Try Exercise 31.

EXAMPLE 4 Solving Quadratic Equations

Solve each equation.

(a) $16m^2 - 25 = 0$

Factor the left side of the equation as the difference of squares.

$$(4m + 5)(4m - 5) = 0$$
$$4m + 5 = 0 \quad \text{or} \quad 4m - 5 = 0 \quad \text{Zero-factor property}$$
$$4m = -5 \qquad\qquad 4m = 5 \quad \text{Solve each equation.}$$
$$m = -\frac{5}{4} \qquad\qquad m = \frac{5}{4}$$

Check the two solutions, $-\frac{5}{4}$ and $\frac{5}{4}$, in the original equation. The solution set is $\{-\frac{5}{4}, \frac{5}{4}\}$.

(b) $y^2 = 2y$

First write the equation in standard form.

$$y^2 - 2y = 0 \quad \text{Standard form}$$
$$y(y - 2) = 0 \quad \text{Factor.}$$
$$y = 0 \quad \text{or} \quad y - 2 = 0 \quad \text{Zero-factor property}$$
$$y = 2 \quad \text{Solve.}$$

The solution set is $\{0, 2\}$.

(c) $k(2k + 1) = 3$

Write the equation in standard form.

$$k(2k + 1) = 3$$
$$2k^2 + k = 3 \quad \text{Distributive property}$$
$$2k^2 + k - 3 = 0 \quad \text{Subtract 3.}$$
$$(k - 1)(2k + 3) = 0 \quad \text{Factor.}$$
$$k - 1 = 0 \quad \text{or} \quad 2k + 3 = 0 \quad \text{Zero-factor property}$$
$$k = 1 \qquad\qquad 2k = -3$$
$$k = -\frac{3}{2}$$

The solution set is $\{1, -\frac{3}{2}\}$.

Now Try Exercises 37, 41, and 45.

CAUTION In Example 4(b) it is tempting to begin by dividing both sides of the equation $y^2 = 2y$ by y to get $y = 2$. Note that we do not get the other solution, 0, if we divide by a variable. (We may divide each side of an equation by a *nonzero* real number, however. For instance, in Example 3 we divided each side by 2.)

In Example 4(c) we could not use the zero-factor property to solve the equation $k(2k + 1) = 3$ in its given form because of the 3 on the right. Remember that the zero-factor property applies only to a product that equals 0.

OBJECTIVE 2 Solve other equations by factoring. We can also use the zero-factor property to solve equations that involve more than two factors with variables, as shown in Example 5. (These equations are *not* quadratic equations. Why not?)

EXAMPLE 5 Solving Equations with More Than Two Variable Factors

Solve each equation.

(a)
$$6z^3 - 6z = 0$$
$$6z(z^2 - 1) = 0 \quad \text{Factor out } 6z.$$
$$6z(z + 1)(z - 1) = 0 \quad \text{Factor } z^2 - 1.$$

By an extension of the zero-factor property, this product can equal 0 only if at least one of the factors is 0. Write and solve three equations, one for each factor with a variable.

$$6z = 0 \quad \text{or} \quad z + 1 = 0 \quad \text{or} \quad z - 1 = 0$$
$$z = 0 \qquad\qquad z = -1 \qquad\qquad z = 1$$

Check by substituting, in turn, 0, -1, and 1 in the original equation. The solution set is $\{-1, 0, 1\}$.

(b)
$$(3x - 1)(x^2 - 9x + 20) = 0$$
$$(3x - 1)(x - 5)(x - 4) = 0 \quad \text{Factor } x^2 - 9x + 20.$$
$$3x - 1 = 0 \quad \text{or} \quad x - 5 = 0 \quad \text{or} \quad x - 4 = 0 \quad \text{Zero-factor property}$$
$$x = \frac{1}{3} \qquad\qquad x = 5 \qquad\qquad x = 4$$

The solutions of the original equation are $\frac{1}{3}$, 4, and 5. Check each solution to verify that the solution set is $\{\frac{1}{3}, 4, 5\}$.

Now Try Exercises 51 and 55.

CAUTION In Example 5(b), it would be unproductive to begin by multiplying the two factors together. Keep in mind that the zero-factor property requires the *product* of two or more factors; this product must equal 0. Always consider first whether an equation is given in an appropriate form for the zero-factor property.

SECTION 6.5 Solving Quadratic Equations by Factoring   423

EXAMPLE 6 Solving an Equation Requiring Multiplication Before Factoring

Solve $(3x + 1)x = (x + 1)^2 + 5$.

The zero-factor property requires the *product* of two or more factors to equal 0. To write this equation in the required form, we must first multiply on both sides and collect terms on one side.

$$(3x + 1)x = (x + 1)^2 + 5$$
$$3x^2 + x = x^2 + 2x + 1 + 5 \quad \text{Multiply.}$$
$$3x^2 + x = x^2 + 2x + 6 \quad \text{Combine like terms.}$$
$$2x^2 - x - 6 = 0 \quad \text{Standard form}$$
$$(2x + 3)(x - 2) = 0 \quad \text{Factor.}$$
$$2x + 3 = 0 \quad \text{or} \quad x - 2 = 0 \quad \text{Zero-factor property}$$
$$x = -\frac{3}{2} \qquad\qquad x = 2$$

Check that the solution set is $\{-\frac{3}{2}, 2\}$.

Now Try Exercise 65.

6.5 EXERCISES

For Extra Help

 Student's Solutions Manual

 MyMathLab

 InterAct Math Tutorial Software

 AW Math Tutor Center

 MathXL

Digital Video Tutor CD 13/Videotape 11

Solve each equation and check your solutions. See Example 1.

1. $(x + 5)(x - 2) = 0$
2. $(x - 1)(x + 8) = 0$
3. $(2m - 7)(m - 3) = 0$
4. $(6k + 5)(k + 4) = 0$
5. $t(6t + 5) = 0$
6. $w(4w + 1) = 0$
7. $2x(3x - 4) = 0$
8. $6y(4y + 9) = 0$
9. $\left(x + \frac{1}{2}\right)\left(2x - \frac{1}{3}\right) = 0$
10. $\left(a + \frac{2}{3}\right)\left(5a - \frac{1}{2}\right) = 0$
11. $(.5z - 1)(2.5z + 2) = 0$
12. $(.25x + 1)(x - .5) = 0$
13. $(x - 9)(x - 9) = 0$
14. $(2y + 1)(2y + 1) = 0$

15. What is wrong with this "solution"?
$$2x(3x - 4) = 0$$
$$x = 2 \quad \text{or} \quad x = 0 \quad \text{or} \quad 3x - 4 = 0$$
$$x = \frac{4}{3}$$
The solution set is $\{2, 0, \frac{4}{3}\}$.

16. What is wrong with this "solution"?
$$x(7x - 1) = 0$$
$$7x - 1 = 0 \quad \text{Zero-factor property}$$
$$x = \frac{1}{7}$$
The solution set is $\{\frac{1}{7}\}$.

Solve each equation and check your solutions. See Examples 2–4.

17. $y^2 + 3y + 2 = 0$
18. $p^2 + 8p + 7 = 0$
19. $y^2 - 3y + 2 = 0$
20. $r^2 - 4r + 3 = 0$
21. $x^2 = 24 - 5x$
22. $t^2 = 2t + 15$
23. $x^2 = 3 + 2x$
24. $m^2 = 4 + 3m$
25. $z^2 + 3z = -2$
26. $p^2 - 2p = 3$
27. $m^2 + 8m + 16 = 0$
28. $b^2 - 6b + 9 = 0$
29. $8x^2 - 45x - 18 = 0$
30. $12r^2 + 19r - 10 = 0$
31. $12p^2 = 8 - 10p$

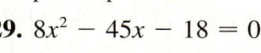

424 CHAPTER 6 Factoring and Applications

32. $18x^2 = 12 + 15x$
33. $9s^2 + 12s = -4$
34. $36x^2 + 60x = -25$
35. $y^2 - 9 = 0$
36. $m^2 - 100 = 0$
37. $16k^2 - 49 = 0$
38. $4w^2 - 9 = 0$
39. $n^2 = 121$
40. $x^2 = 400$

Solve each equation and check your solutions. See Examples 4–6.

41. $x^2 = 7x$
42. $t^2 = 9t$
43. $6r^2 = 3r$
44. $10y^2 = -5y$
45. $g(g - 7) = -10$
46. $r(r - 5) = -6$
47. $3z(2z + 7) = 12$
48. $4b(2b + 3) = 36$
49. $2(y^2 - 66) = -13y$
50. $3(t^2 + 4) = 20t$
51. $(2r + 5)(3r^2 - 16r + 5) = 0$
52. $(3m + 4)(6m^2 + m - 2) = 0$
53. $(2x + 7)(x^2 + 2x - 3) = 0$
54. $(x + 1)(6x^2 + x - 12) = 0$
55. $9y^3 - 49y = 0$
56. $16r^3 - 9r = 0$
57. $r^3 - 2r^2 - 8r = 0$
58. $x^3 - x^2 - 6x = 0$
59. $a^3 + a^2 - 20a = 0$
60. $y^3 - 6y^2 + 8y = 0$
61. $r^4 = 2r^3 + 15r^2$
62. $x^3 = 3x + 2x^2$
63. $3x(x + 1) = (2x + 3)(x + 1)$
64. $2k(k + 3) = (3k + 1)(k + 3)$
65. $x^2 + (x + 1)^2 = (x + 2)^2$
66. $(x - 7)^2 + x^2 = (x + 1)^2$
67. $(2x)^2 = (2x + 4)^2 - (x + 5)^2$
68. $5 - (x - 1)^2 = (x - 2)^2$
69. $6p^2(p + 1) = 4(p + 1) - 5p(p + 1)$
70. $6x^2(2x + 3) - 5x(2x + 3) = 4(2x + 3)$
71. $(k + 3)^2 - (2k - 1)^2 = 0$
72. $(4y - 3)^3 - 9(4y - 3) = 0$

73. Galileo's formula for freely falling objects, $d = 16t^2$, was given at the beginning of this section. The distance d in feet an object falls depends on the time elapsed t in seconds. (This is an example of an important mathematical concept, the *function*.)

  (a) Use Galileo's formula and complete the following table. (*Hint:* Substitute each given value into the formula and solve for the unknown value.)

t in seconds	0	1	2	3		
d in feet	0	16			256	576

  (b) When $t = 0$, $d = 0$. Explain this in the context of the problem.
  (c) When you substituted 256 for d and solved for t, you should have found two solutions, 4 and -4. Why doesn't -4 make sense as an answer?

TECHNOLOGY INSIGHTS (EXERCISES 74–77)

In Section 3.2 we showed how an equation in one variable can be solved with a graphing calculator by getting 0 on one side, then replacing 0 with y to get a corresponding equation in two variables. The x-values of the x-intercepts of the graph of the two-variable equation then give the solutions of the original equation.

Use the calculator screens to determine the solution set of each quadratic equation. Verify your answers by substitution.

74. $x^2 + .4x - .05 = 0$

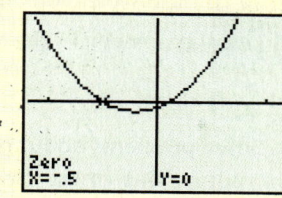 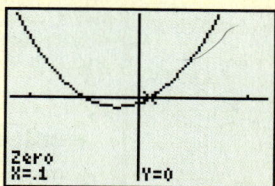

75. $2x^2 - 7.2x + 6.3 = 0$

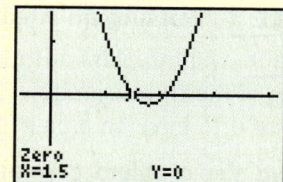 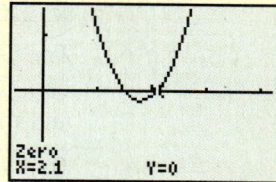

76. $2x^2 + 7.2x + 5.5 = 0$

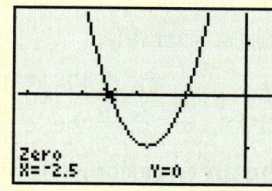 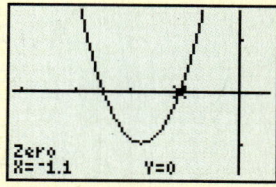

77. $4x^2 - x - 33 = 0$

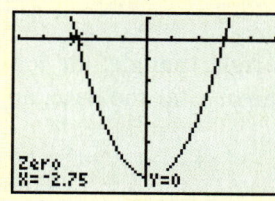 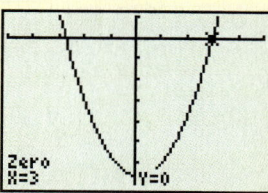

6.6 Applications of Quadratic Equations

OBJECTIVES

1. Solve problems about geometric figures.
2. Solve problems about consecutive integers.
3. Solve problems using the Pythagorean formula.
4. Solve problems using given quadratic models.

We can now use factoring to solve quadratic equations that arise in application problems. We follow the same six problem-solving steps given in Section 2.4.

Solving an Applied Problem

Step 1 **Read** the problem carefully until you understand what is given and what is to be found.

Step 2 **Assign a variable** to represent the unknown value, using diagrams or tables as needed. Write down what the variable represents. If necessary, express any other unknown values in terms of the variable.

(continued)

426 CHAPTER 6 Factoring and Applications

Step 3 **Write an equation** using the variable expression(s).
Step 4 **Solve** the equation.
Step 5 **State the answer.** Does it seem reasonable?
Step 6 **Check** the answer in the words of the original problem.

OBJECTIVE 1 Solve problems about geometric figures. Some of the applied problems in this section require one of the formulas given on the inside covers of the text.

EXAMPLE 1 Solving an Area Problem

The O'Connors want to plant a flower bed in a triangular area in a corner of their garden. One leg of the right-triangular flower bed will be 2 m shorter than the other leg, and they want it to have an area of 24 m². See Figure 1. Find the lengths of the legs.

Step 1 **Read** the problem carefully. We need to find the lengths of the legs of a right triangle with area 24 m².

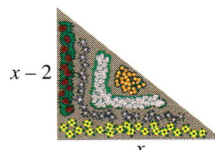

FIGURE 1

Step 2 **Assign a variable.**

Let x = the length of one leg.

Then $x - 2$ = the length of the other leg.

Step 3 **Write an equation.** The area of a right triangle is given by the formula

$$\text{area} = \frac{1}{2} \times \text{base} \times \text{height} = \frac{1}{2}bh.$$

In a right triangle, the legs are the base and height, so we substitute 24 for the area, x for the base, and $x - 2$ for the height in the formula.

$$A = \frac{1}{2}bh$$

$$24 = \frac{1}{2}x(x - 2) \quad \text{Let } A = 24, b = x, h = x - 2.$$

Step 4 **Solve.**
$$48 = x(x - 2) \quad \text{Multiply by 2.}$$
$$48 = x^2 - 2x \quad \text{Distributive property}$$
$$x^2 - 2x - 48 = 0 \quad \text{Standard form}$$
$$(x + 6)(x - 8) = 0 \quad \text{Factor.}$$
$$x + 6 = 0 \quad \text{or} \quad x - 8 = 0 \quad \text{Zero-factor property}$$
$$x = -6 \quad \text{or} \quad x = 8$$

Step 5 **State the answer.** The solutions are -6 and 8. Because a triangle cannot have a side of negative length, we discard the solution -6. Then the lengths of the legs will be 8 m and $8 - 2 = 6$ m.

Step 6 **Check.** The length of one leg is 2 m less than the length of the other leg, and the area is $\frac{1}{2}(8)(6) = 24$ m², as required.

Now Try Exercise 7.

SECTION 6.6 Applications of Quadratic Equations 427

> **CAUTION** When solving applied problems, *always* check solutions against physical facts and discard any answers that are not appropriate.

OBJECTIVE 2 Solve problems about consecutive integers. Recall from our work in Section 2.4 that consecutive integers are integers that are next to each other on a number line, such as 5 and 6, or -11 and -10. Consecutive odd integers are *odd* integers that are next to each other, such as 5 and 7, or -13 and -11. Consecutive even integers are defined similarly; for example, 4 and 6 are consecutive even integers, as are -10 and -8. The following list may be helpful.

Consecutive Integers

Let x represent the first of the integers.

Two consecutive integers	$x, x + 1$
Three consecutive integers	$x, x + 1, x + 2$
Two consecutive even or odd integers	$x, x + 2$
Three consecutive even or odd integers	$x, x + 2, x + 4$

EXAMPLE 2 Solving a Consecutive Integer Problem

The product of the second and third of three consecutive integers is 2 more than 7 times the first integer. Find the integers.

Step 1 **Read** the problem. Note that the integers are consecutive.

Step 2 **Assign a variable.**

Let $x =$ the first integer.

Then $x + 1 =$ the second integer,

and $x + 2 =$ the third integer.

Step 3 **Write an equation.**

The product of the second and third is 2 more than 7 times the first.

$$(x + 1)(x + 2) = 7x + 2$$

Step 4 **Solve.**

$$x^2 + 3x + 2 = 7x + 2 \qquad \text{Multiply.}$$
$$x^2 - 4x = 0 \qquad \text{Standard form}$$
$$x(x - 4) = 0 \qquad \text{Factor.}$$
$$x = 0 \quad \text{or} \quad x = 4 \qquad \text{Zero-factor property}$$

Step 5 **State the answer.** The solutions 0 and 4 each lead to a correct answer:

$$0, 1, 2 \quad \text{or} \quad 4, 5, 6.$$

Step 6 **Check.** Since $1 \cdot 2 = 7 \cdot 0 + 2$ and $5 \cdot 6 = 7 \cdot 4 + 2$, both sets of consecutive integers satisfy the statement of the problem.

Now Try Exercise 17.

OBJECTIVE 3 Solve problems using the Pythagorean formula. The next example requires the Pythagorean formula from geometry.

Pythagorean Formula

If a right triangle (a triangle with a 90° angle) has longest side of length c and two other sides of lengths a and b, then

$$a^2 + b^2 = c^2.$$

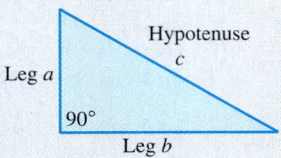

The longest side, the **hypotenuse**, is opposite the right angle. The two shorter sides are the **legs** of the triangle.

EXAMPLE 3 Using the Pythagorean Formula

Ed and Mark leave their office, with Ed traveling north and Mark traveling east. When Mark is 1 mi farther than Ed from the office, the distance between them is 2 mi more than Ed's distance from the office. Find their distances from the office and the distance between them.

Step 1 **Read** the problem again. There will be three answers to this problem.

Step 2 **Assign a variable.** Let x represent Ed's distance from the office, $x + 1$ represent Mark's distance from the office, and $x + 2$ represent the distance between them. Place these on a right triangle, as in Figure 2.

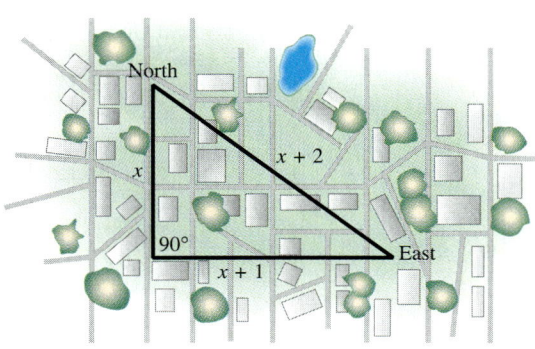

FIGURE 2

Step 3 **Write an equation.** Substitute into the Pythagorean formula.

$$a^2 + b^2 = c^2$$
$$x^2 + (x + 1)^2 = (x + 2)^2$$

Step 4 Solve.
$$x^2 + x^2 + 2x + 1 = x^2 + 4x + 4$$
$$x^2 - 2x - 3 = 0 \quad \text{Standard form}$$
$$(x - 3)(x + 1) = 0 \quad \text{Factor.}$$
$$x - 3 = 0 \quad \text{or} \quad x + 1 = 0 \quad \text{Zero-factor property}$$
$$x = 3 \quad \text{or} \quad x = -1$$

Step 5 State the answer. Since -1 cannot represent a distance, 3 is the only possible answer. Ed's distance is 3 mi, Mark's distance is $3 + 1 = 4$ mi, and the distance between them is $3 + 2 = 5$ mi.

Step 6 Check. Since $3^2 + 4^2 = 5^2$, the answers are correct.

Now Try Exercise 25.

> **CAUTION** When solving a problem involving the Pythagorean formula, be sure that the expressions for the sides are properly placed.
> $$\text{leg}^2 + \text{leg}^2 = \text{hypotenuse}^2$$

OBJECTIVE 4 Solve problems using given quadratic models. In Examples 1–3, we wrote quadratic equations to model, or mathematically describe, various situations and then solved the equations. In the final examples, we are given the quadratic models and must use them to determine data.

EXAMPLE 4 Finding the Height of a Ball

A tennis player's serve travels 180 ft per sec (125 mph). If she hits the ball directly upward, the height h of the ball in feet at time t in seconds is modeled by the quadratic equation
$$h = -16t^2 + 180t + 6.$$
How long will it take for the ball to reach a height of 206 ft?

A height of 206 ft means $h = 206$, so we substitute 206 for h in the equation.
$$206 = -16t^2 + 180t + 6 \quad \text{Let } h = 206.$$
To solve the equation, we first write it in standard form. For convenience, we reverse the sides of the equation.
$$-16t^2 + 180t + 6 = 206$$
$$-16t^2 + 180t - 200 = 0 \quad \text{Standard form}$$
$$4t^2 - 45t + 50 = 0 \quad \text{Divide by } -4.$$
$$(4t - 5)(t - 10) = 0 \quad \text{Factor.}$$
$$4t - 5 = 0 \quad \text{or} \quad t - 10 = 0 \quad \text{Zero-factor property}$$
$$t = \frac{5}{4} \quad \text{or} \quad t = 10$$

Since we found two acceptable answers, the ball will be 206 ft above the ground twice (once on its way up and once on its way down)—at $\frac{5}{4}$ sec and at 10 sec. See Figure 3.

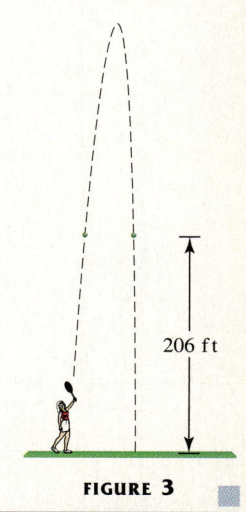

FIGURE 3

Now Try Exercise 29.

EXAMPLE 5 Modeling Increases in Drug Prices

The annual percent increase y in the amount pharmacies paid wholesalers for drugs in the years 1990–1999 can be modeled by the quadratic equation

$$y = .23x^2 - 2.6x + 9,$$

where $x = 0$ represents 1990, $x = 1$ represents 1991, and so on. (*Source: IMS Health, Retail and Provider Perspective.*)

(a) Use the model to find the annual percent increase to the nearest tenth in 1997.
In 1997, $x = 1997 - 1990 = 7$. Substitute 7 for x in the equation.

$$y = .23(7)^2 - 2.6(7) + 9 \quad \text{Let } x = 7.$$
$$y = 2.07 \quad \text{Use a calculator.}$$

To the nearest tenth, pharmacies paid about 2.1% more for drugs in 1997.

(b) Repeat part (a) for 1999.
For 1999, $x = 9$.

$$y = .23(9)^2 - 2.6(9) + 9 \quad \text{Let } x = 1999 - 1990 = 9.$$
$$y = 4.23$$

In 1999, pharmacies paid about 4.2% more for drugs.

(c) The model used in parts (a) and (b) was developed using the data shown in the table. How do the results in parts (a) and (b) compare to the actual data from the table?

Year	Percent Increase
1990	8.4
1991	7.2
1992	5.5
1993	3.0
1994	1.7
1995	1.9
1996	1.6
1997	2.5
1998	3.2
1999	4.2

From the table, the actual data for 1997 is 2.5%. Our answer, 2.1%, is a little low. For 1999, the actual data is 4.2%, which is the same as our answer in part (b).

Now Try Exercise 31.

NOTE A graph of the quadratic equation from Example 5 is shown in Figure 4. Notice the basic shape of this graph, which follows the general pattern of the data in the table—it decreases from 1990 to 1996 (with the exception of the data for 1995) and then increases from 1997 to 1999. We consider such graphs of quadratic equations, called *parabolas,* in more detail in Chapter 9.

FIGURE 4

6.6 EXERCISES

For Extra Help

 Student's Solutions Manual

 MyMathLab

 InterAct Math Tutorial Software

 AW Math Tutor Center

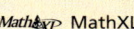 MathXL

 Digital Video Tutor CD 13/Videotape 11

1. To review the six problem-solving steps first introduced in Section 2.4, complete each statement.

   Step 1: _____ the problem carefully until you understand what is given and what must be found.

   Step 2: Assign a _____ to represent the unknown value.

   Step 3: Write a(n) _____ using the variable expression(s).

   Step 4: _____ the equation.

   Step 5: State the _____.

   Step 6: _____ the answer in the words of the _____ problem.

2. A student solves an applied problem and gets 6 or -3 for the length of the side of a square. Which of these answers is reasonable? Explain.

In Exercises 3–6, a figure and a corresponding geometric formula are given. Using x as the variable, complete Steps 3–6 for each problem. (Refer to the steps in Exercise 1 as needed.)

3.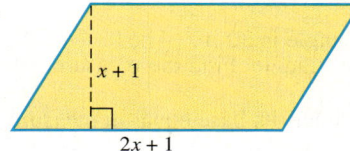

   Area of a parallelogram: $A = bh$

   The area of this parallelogram is 45 sq. units. Find its base and height.

4.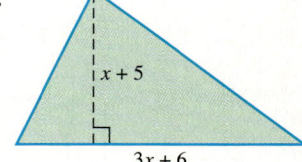

   Area of a triangle: $A = \dfrac{1}{2}bh$

   The area of this triangle is 60 sq. units. Find its base and height.

5.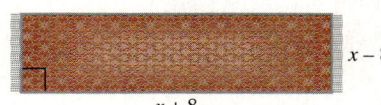

   Area of a rectangular rug: $A = LW$

   The area of this rug is 80 sq. units. Find its length and width.

6.

   Volume of a rectangular Chinese box: $V = LWH$

   The volume of this box is 192 cu. units. Find its length and width.

Solve each problem. Check your answers to be sure they are reasonable. Refer to the formulas on the inside covers. See Example 1.

7. The length of a VHS videocassette shell is 3 in. more than its width. The area of the rectangular top side of the shell is 28 in.² Find the length and width of the videocassette shell.

8. A plastic box that holds a standard audiocassette has length 4 cm longer than its width. The area of the rectangular top of the box is 77 cm². Find the length and width of the box.

9. The dimensions of a Gateway EV700 computer monitor screen are such that its length is 3 in. more than its width. If the length is increased by 1 in. while the width remains the same, the area is increased by 10 in.² What are the dimensions of the screen? (*Source:* Author's computer.)

10. The keyboard of the computer in Exercise 9 is 11 in. longer than it is wide. If both its length and width are increased by 2 in., the area of the top of the keyboard is increased by 54 in.² Find the length and width of the keyboard. (*Source:* Author's computer.)

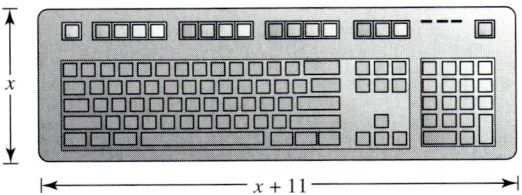

11. The area of a triangle is 30 in.² The base of the triangle measures 2 in. more than twice the height of the triangle. Find the measures of the base and the height.

12. A certain triangle has its base equal in measure to its height. The area of the triangle is 72 m². Find the equal base and height measure.

13. A 10-gal aquarium is 3 in. higher than it is wide. Its length is 21 in., and its volume is 2730 in.³ What are the height and width of the aquarium?

14. A toolbox is 2 ft high, and its width is 3 ft less than its length. If its volume is 80 ft³, find the length and width of the box.

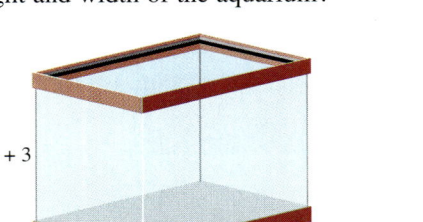
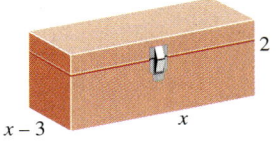

15. A square mirror has sides measuring 2 ft less than the sides of a square painting. If the difference between their areas is 32 ft², find the lengths of the sides of the mirror and the painting.

16. The sides of one square have length 3 m more than the sides of a second square. If the area of the larger square is subtracted from 4 times the area of the smaller square, the result is 36 m². What are the lengths of the sides of each square?

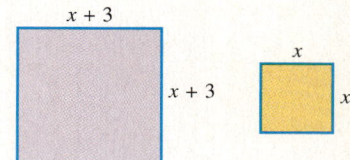

Solve each problem about consecutive integers. See Example 2.

17. The product of two consecutive integers is 11 more than their sum. Find the integers.

18. The product of two consecutive integers is 4 less than 4 times their sum. Find the integers.

19. Find three consecutive odd integers such that 3 times the sum of all three is 18 more than the product of the smaller two.

20. Find three consecutive odd integers such that the sum of all three is 42 less than the product of the larger two.

21. Find three consecutive even integers such that the sum of the squares of the smaller two is equal to the square of the largest.

22. Find three consecutive even integers such that the square of the sum of the smaller two is equal to twice the largest.

Use the Pythagorean formula to solve each problem. See Example 3.

23. The hypotenuse of a right triangle is 1 cm longer than the longer leg. The shorter leg is 7 cm shorter than the longer leg. Find the length of the longer leg of the triangle.

24. The longer leg of a right triangle is 1 m longer than the shorter leg. The hypotenuse is 1 m shorter than twice the shorter leg. Find the length of the shorter leg of the triangle.

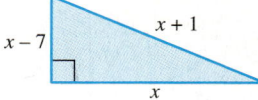
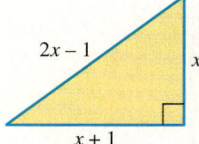

25. Tram works due north of home. Her husband Alan works due east. They leave for work at the same time. By the time Tram is 5 mi from home, the distance between them is 1 mi more than Alan's distance from home. How far from home is Alan?

26. Two cars left an intersection at the same time. One traveled north. The other traveled 14 mi farther, but to the east. How far apart were they then, if the distance between them was 4 mi more than the distance traveled east?

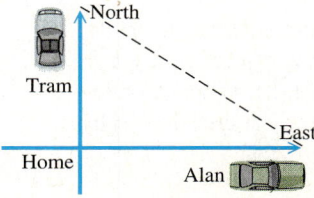
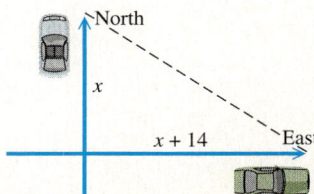

27. A ladder is leaning against a building. The distance from the bottom of the ladder to the building is 4 ft less than the length of the ladder. How high up the side of the building is the top of the ladder if that distance is 2 ft less than the length of the ladder?

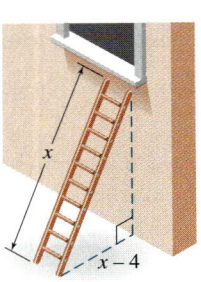

28. A lot has the shape of a right triangle with one leg 2 m longer than the other. The hypotenuse is 2 m less than twice the length of the shorter leg. Find the length of the shorter leg.

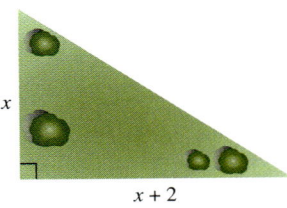

Solve each problem. See Examples 4 and 5.

29. An object propelled from a height of 48 ft with an initial velocity of 32 ft per sec after t sec has height

$$h = -16t^2 + 32t + 48.$$

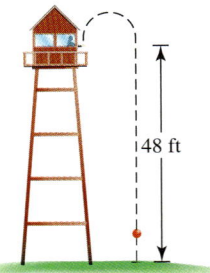

(a) After how many seconds is the height 64 ft? (*Hint:* Let $h = 64$ and solve.)
(b) After how many seconds is the height 60 ft?
(c) After how many seconds does the object hit the ground? (*Hint:* When the object hits the ground, $h = 0$.)
(d) The quadratic equation from part (c) has two solutions, yet only one of them is appropriate for answering the question. Why is this so?

30. If an object is propelled upward from ground level with an initial velocity of 64 ft per sec, its height h in feet t sec later is

$$h = -16t^2 + 64t.$$

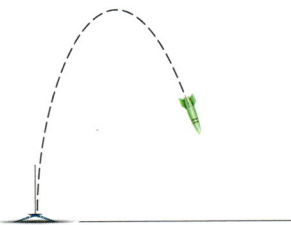

(a) After how many seconds is the height 48 ft?
(b) The object reaches its maximum height 2 sec after it is propelled. What is this maximum height?
(c) After how many seconds does the object hit the ground?
(d) The quadratic equation from part (c) has two solutions, yet only one of them is appropriate for answering the question. Why is this so?

31. The table shows the number of cellular phones (in millions) owned by Americans.

Year	Cellular Phones (in millions)
1990	5
1992	11
1994	24
1996	44
1998	62
1999	86

Source: Cellular Telecommunications Industry Association.

The quadratic equation

$$y = .813x^2 + 1.27x + 5.28$$

models the number of cellular phones y (in millions) in the year x, where $x = 0$ represents 1990, $x = 2$ represents 1992, and so on.

(a) Use the model to find the number of cellular phones to the nearest tenth of a million in 1994. How does the result compare to the actual data in the table?

(b) What value of x corresponds to 1999?

(c) Use the model to find the number of cellular phones to the nearest tenth of a million in 1999. How does the result compare to the actual data in the table?

(d) Assuming that the trend in the data continues, use the quadratic equation to approximate the number of cellular phones to the nearest tenth of a million in 2002.

RELATING CONCEPTS (EXERCISES 32–40)

For Individual or Group Work

The U.S. trade deficit represents the amount by which exports are less than imports. It provides not only a sign of economic prosperity but also a warning of potential decline. The data in the table shows the U.S. trade deficit for 1995 through 1999.

Year	Deficit (in billions of dollars)
1995	97.5
1996	104.3
1997	104.7
1998	164.3
1999	271.3

Source: U.S. Department of Commerce.

Use the data to **work Exercises 32–40 in order.**

32. How much did the trade deficit increase from 1998 to 1999? What percent increase is this (to the nearest percent)?

33. The U.S. trade deficit for the years shown in the table can be approximated by the linear equation

$$y = 40.8x + 66.9,$$

(continued)

where y is the deficit in billions of dollars. Here $x = 0$ represents 1995, $x = 1$ represents 1996, and so on. Use this equation to approximate the trade deficits in 1995, 1997, and 1999.

34. How do your answers from Exercise 33 compare to the actual data in the table?

35. The trade deficit y (in billions of dollars) can also be approximated by the quadratic equation
$$y = 18.5x^2 - 33.4x + 104,$$
where $x = 0$ again represents 1995, $x = 1$ represents 1996, and so on. Use this equation to approximate the trade deficits in 1995, 1997, and 1999.

36. Compare your answers from Exercise 35 to the actual data in the table. Which equation, the linear or quadratic one, models the data better?

37. We can also see graphically why the linear equation is not a very good model for the data. To do so, write the data from the table as a set of ordered pairs (x, y), where x represents the number of years since 1995 and y represents the trade deficit in billions of dollars.

38. Plot the ordered pairs from Exercise 37 on a graph. Recall from Chapter 3 that a linear equation has a straight line for its graph. Do the ordered pairs you plotted lie in a linear pattern?

39. Assuming that the trend in the data continues and since the quadratic equation models the data fairly well, use the quadratic equation to predict the trade deficit for the year 2000.

40. The actual trade deficit for the year 2000 was 369.7 billion dollars. (*Source:* www.census.gov)

    (a) How does the actual deficit for 2000 compare to your prediction from Exercise 39?
    (b) Should the quadratic equation be used to predict the U.S. trade deficit for years after 2000? Explain.

Chapter  Group Activity

Factoring Trinomials Made Easy

OBJECTIVE Use an organized approach to factoring by grouping.

To factor a trinomial using FOIL, we must find the outer and inner coefficients that sum to give the coefficient of the middle term. Our approach begins with a **key number,** found by multiplying the coefficients of the first and last terms. For the trinomial $6x^2 - x - 2$, for instance, the key number is -12 since $6(-2) = -12$.

Step 1 Display the factors of -12 by entering $Y_1 = -12/X$ in a graphing calculator (Screen 1) and using an automatic table (Screen 2). Factors of -12 are automatically displayed in pairs as 1, -12; 2, -6; 3, -4; 4, -3; and 6, -2 (Screen 3). You could scroll up or down to find other factors. Note that 5, -2.4 and 7, -1.714 are not factor pairs since -2.4 and -1.714 are not integers.

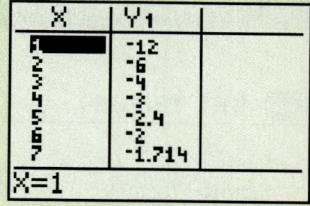

SCREEN 1   SCREEN 2   SCREEN 3

Step 2 Find the pair of factors that sum to the *middle* term coefficient, -1. We can let the calculator do this, too. Enter $Y_2 = X + -12/X$. In this case, X is one of the factors and $-12/X$ is the other, so Y_2 will give the sum (Screen 4). Look for -1 in the Y_2 column in Screen 5. (You may have to scroll up or down to find it.)

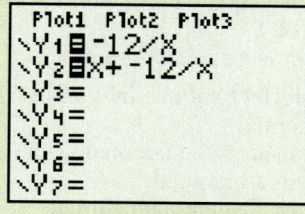
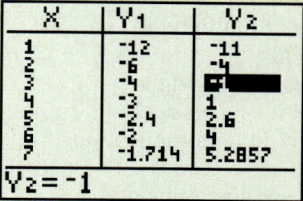

SCREEN 4   SCREEN 5

Step 3 Screen 5 shows that the coefficients of the outer and inner products are 3 and -4. Write $6x^2 - x - 2$ as $6x^2 + 3x - 4x - 2$. Using factoring by grouping,

$(6x^2 + 3x) + (-4x - 2) = 3x(2x + 1) - 2(2x + 1) = (3x - 2)(2x + 1)$.

A. To factor each trinomial given in the column heads of the following table, first find the key number, and then use a calculator to help you find the coefficients of the outer and inner products of FOIL.

Trinomial	$3x^2 - 2x - 8$	$2x^2 - 11x + 15$	$10x^2 + 11x - 6$	$4x^2 + 5x + 3$
Key Number				
Outer, Inner Coefficients				
Factor by Grouping				(*Hint:* What does it mean if the middle term coefficient is *not* listed in the Y_2 column?)

B. Factor each trinomial by grouping.

CHAPTER 6 SUMMARY

KEY TERMS

6.1 factor
factored form
common factor
greatest common factor (GCF)

6.2 prime polynomial
6.4 perfect square trinomial

6.5 quadratic equation
standard form

6.6 hypotenuse
legs

TEST YOUR WORD POWER

See how well you have learned the vocabulary in this chapter. Answers, with examples, follow the Quick Review.

1. **Factoring** is
   A. a method of multiplying polynomials
   B. the process of writing a polynomial as a product
   C. the answer in a multiplication problem
   D. a way to add the terms of a polynomial.

2. A polynomial is in **factored form** when
   A. it is prime
   B. it is written as a sum

   C. the squared term has a coefficient of 1
   D. it is written as a product.

3. A **perfect square trinomial** is a trinomial
   A. that can be factored as the square of a binomial
   B. that cannot be factored
   C. that is multiplied by a binomial
   D. where all terms are perfect squares.

4. A **quadratic equation** is a polynomial equation of
   A. degree one
   B. degree two
   C. degree three
   D. degree four.

5. A **hypotenuse** is
   A. either of the two shorter sides of a triangle
   B. the shortest side of a triangle
   C. the side opposite the right angle in a triangle
   D. the longest side in any triangle.

QUICK REVIEW

CONCEPTS	EXAMPLES
6.1 THE GREATEST COMMON FACTOR; FACTORING BY GROUPING	
Finding the Greatest Common Factor (GCF) 1. Include the largest numerical factor of every term. 2. Include each variable that is a factor of every term raised to the smallest exponent that appears in a term.	Find the greatest common factor of $$4x^2y, \quad -6x^2y^3, \quad 2xy^2.$$ $$4x^2y = 2^2 \cdot x^2 \cdot y$$ $$-6x^2y^3 = -1 \cdot 2 \cdot 3 \cdot x^2 \cdot y^3$$ $$2xy^2 = 2 \cdot x \cdot y^2$$ The greatest common factor is $2xy$.
Factoring by Grouping *Step 1* Group the terms. *Step 2* Factor out the greatest common factor in each group. *Step 3* Factor a common binomial factor from the result of Step 2. *Step 4* If necessary try a different grouping.	Factor by grouping. $$3x^2 + 5x - 24xy - 40y = (3x^2 + 5x) + (-24xy - 40y)$$ $$= x(3x + 5) - 8y(3x + 5)$$ $$= (3x + 5)(x - 8y)$$

CONCEPTS	EXAMPLES

6.2 FACTORING TRINOMIALS

To factor $x^2 + bx + c$, find m and n such that $mn = c$ and $m + n = b$.

$$x^2 + bx + c \quad \text{where } mn = c,\; m + n = b$$

Then $x^2 + bx + c = (x + m)(x + n)$.

Check by multiplying.

Factor $x^2 + 6x + 8$.

$$x^2 + 6x + 8 \quad \text{where } mn = 8,\; m + n = 6$$

$m = 2$ and $n = 4$

$$x^2 + 6x + 8 = (x + 2)(x + 4)$$

Check: $(x + 2)(x + 4) = x^2 + 4x + 2x + 8$
$= x^2 + 6x + 8$

6.3 MORE ON FACTORING TRINOMIALS

To factor $ax^2 + bx + c$:

By Grouping
Find m and n.

$$ax^2 + bx + c \quad \text{where } mn = ac,\; m + n = b$$

By Trial and Error
Use FOIL backwards.

Factor $3x^2 + 14x - 5$.

$mn = -15,\; m + n = 14$

By trial and error or by grouping,
$$3x^2 + 14x - 5 = (3x - 1)(x + 5).$$

6.4 SPECIAL FACTORING RULES

Difference of Squares
$x^2 - y^2 = (x + y)(x - y)$

Perfect Square Trinomials
$x^2 + 2xy + y^2 = (x + y)^2$
$x^2 - 2xy + y^2 = (x - y)^2$

Difference of Cubes
$x^3 - y^3 = (x - y)(x^2 + xy + y^2)$

Sum of Cubes
$x^3 + y^3 = (x + y)(x^2 - xy + y^2)$

Factor.
$$4x^2 - 9 = (2x + 3)(2x - 3)$$

$$9x^2 + 6x + 1 = (3x + 1)^2$$
$$4x^2 - 20x + 25 = (2x - 5)^2$$

$$m^3 - 8 = m^3 - 2^3 = (m - 2)(m^2 + 2m + 4)$$

$$27 + z^3 = 3^3 + z^3 = (3 + z)(9 - 3z + z^2)$$

440 CHAPTER 6 Factoring and Applications

CONCEPTS	EXAMPLES
6.5 SOLVING QUADRATIC EQUATIONS BY FACTORING	
Zero-Factor Property If a and b are real numbers and if $ab = 0$, then $a = 0$ or $b = 0$.	If $(x - 2)(x + 3) = 0$, then $x - 2 = 0$ or $x + 3 = 0$.
Solving a Quadratic Equation by Factoring *Step 1* Write in standard form. *Step 2* Factor. *Step 3* Use the zero-factor property. *Step 4* Check.	Solve $2x^2 = 7x + 15$. $$2x^2 - 7x - 15 = 0$$ $$(2x + 3)(x - 5) = 0$$ $2x + 3 = 0 \quad$ or $\quad x - 5 = 0$ $2x = -3 \qquad\qquad\quad x = 5$ $x = -\dfrac{3}{2}$ Both solutions satisfy the original equation; the solution set is $\left\{-\dfrac{3}{2}, 5\right\}$.
6.6 APPLICATIONS OF QUADRATIC EQUATIONS	
Pythagorean Formula In a right triangle, the square of the hypotenuse equals the sum of the squares of the legs. $$a^2 + b^2 = c^2$$ 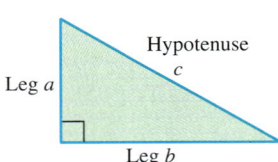	In a right triangle, one leg measures 2 ft longer than the other. The hypotenuse measures 4 ft longer than the shorter leg. Find the lengths of the three sides of the triangle. Let $x =$ the length of the shorter leg. Then $$x^2 + (x + 2)^2 = (x + 4)^2.$$ Verify that the solutions of this equation are -2 and 6. Discard -2 as a solution. Check that the sides are 6, $6 + 2 = 8$, and $6 + 4 = 10$ ft in length.

Answers to Test Your Word Power
1. B; *Example:* $x^2 - 5x - 14 = (x - 7)(x + 2)$ 2. D; *Example:* The factored form of $x^2 - 5x - 14$ is $(x - 7)(x + 2)$.
3. A; *Example:* $a^2 + 2a + 1$ is a perfect square trinomial; its factored form is $(a + 1)^2$.   4. B; *Examples:* $y^2 - 3y + 2 = 0$, $x^2 - 9 = 0$, $2m^2 = 6m + 8$ 5. C; *Example:* In Figure 2 of Section 6.6, the hypotenuse is the side labeled $x + 2$.

CHAPTER 6 REVIEW EXERCISES

[6.1] *Factor out the greatest common factor or factor by grouping.*

1. $7t + 14$
2. $60z^3 + 30z$
3. $2xy - 8y + 3x - 12$
4. $6y^2 + 9y + 4xy + 6x$

[6.2] *Factor completely.*

5. $x^2 + 5x + 6$
6. $y^2 - 13y + 40$
7. $q^2 + 6q - 27$
8. $r^2 - r - 56$
9. $r^2 - 4rs - 96s^2$
10. $p^2 + 2pq - 120q^2$
11. $8p^3 - 24p^2 - 80p$
12. $3x^4 + 30x^3 + 48x^2$
13. $p^7 - p^6q - 2p^5q^2$
14. $3r^5 - 6r^4s - 45r^3s^2$

[6.3]

15. To begin factoring $6r^2 - 5r - 6$, what are the possible first terms of the two binomial factors if we consider only positive integer coefficients?

16. What is the first step you would use to factor $2z^3 + 9z^2 - 5z$?

Factor completely.

17. $2k^2 - 5k + 2$
18. $3r^2 + 11r - 4$
19. $6r^2 - 5r - 6$
20. $10z^2 - 3z - 1$
21. $8v^2 + 17v - 21$
22. $24x^5 - 20x^4 + 4x^3$
23. $-6x^2 + 3x + 30$
24. $10r^3s + 17r^2s^2 + 6rs^3$

[6.4]

25. Which one of the following is the difference of squares?
    A. $32x^2 - 1$ B. $4x^2y^2 - 25z^2$ C. $x^2 + 36$ D. $25y^3 - 1$

26. Which one of the following is a perfect square trinomial?
    A. $x^2 + x + 1$ B. $y^2 - 4y + 9$ C. $4x^2 + 10x + 25$ D. $x^2 - 20x + 100$

Factor completely.

27. $n^2 - 49$
28. $25b^2 - 121$
29. $49y^2 - 25w^2$
30. $144p^2 - 36q^2$
31. $x^2 + 100$
32. $r^2 - 12r + 36$
33. $9t^2 - 42t + 49$
34. $m^3 + 1000$
35. $125k^3 + 64x^3$
36. $343x^3 - 64$

[6.5] *Solve each equation and check your solutions.*

37. $(4t + 3)(t - 1) = 0$
38. $(x + 7)(x - 4)(x + 3) = 0$
39. $x(2x - 5) = 0$
40. $z^2 + 4z + 3 = 0$
41. $m^2 - 5m + 4 = 0$
42. $x^2 = -15 + 8x$
43. $3z^2 - 11z - 20 = 0$
44. $81t^2 - 64 = 0$
45. $y^2 = 8y$
46. $n(n - 5) = 6$
47. $t^2 - 14t + 49 = 0$
48. $t^2 = 12(t - 3)$
49. $(5z + 2)(z^2 + 3z + 2) = 0$
50. $x^2 = 9$

[6.6] *Solve each problem.*

51. The length of a rug is 6 ft more than the width. The area is 40 ft². Find the length and width of the rug.

52. The surface area S of a box is given by
$$S = 2WH + 2WL + 2LH.$$
A treasure chest from a sunken galleon has dimensions as shown in the figure. Its surface area is 650 ft². Find its width.

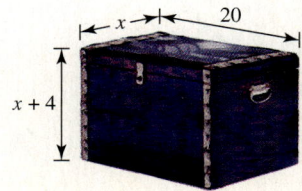

53. The length of a rectangle is three times the width. If the width were increased by 3 m while the length remained the same, the new rectangle would have an area of 30 m². Find the length and width of the original rectangle.

54. The volume of a rectangular box is 120 m³. The width of the box is 4 m, and the height is 1 m less than the length. Find the length and height of the box.

55. The product of two consecutive integers is 29 more than their sum. What are the integers?

56. Two cars left an intersection at the same time. One traveled west, and the other traveled 14 mi less, but to the south. How far apart were they then, if the distance between them was 16 mi more than the distance traveled south?

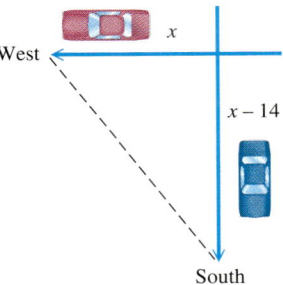

If an object is propelled upward with an initial velocity of 128 ft per sec, its height h after t sec is

$$h = 128t - 16t^2.$$

Find the height of the object after each period of time.

57. 1 sec **58.** 2 sec **59.** 4 sec

60. For the object described above, when does it return to the ground?

61. Annual revenue in millions of dollars for eBay is shown in the table.

Year	Annual Revenue (in millions of dollars)
1997	5.1
1998	47.4
1999	224.7

Source: eBay.

Using the data, we developed the quadratic equation

$$y = 67.5x^2 - 25.2x + 5.1$$

to model eBay revenues y in year x, where $x = 0$ represents 1997, $x = 1$ represents 1998, and so on. Because only three years of data were used to determine the model, we must be careful about using it to predict revenue for years beyond 1999.

(a) Use the model to predict annual revenue for eBay in 2000.

(b) The revenue for eBay through the first half of 2000 was $183.2 million. Given this information, do you think your prediction in part (a) is reliable? Explain.

MIXED REVIEW EXERCISES

62. Which of the following is *not* factored completely?

A. $3(7t)$ **B.** $3x(7t + 4)$ **C.** $(3 + x)(7t + 4)$ **D.** $3(7t + 4) + x(7t + 4)$

63. Although $(2x + 8)(3x - 4) = 6x^2 + 16x - 32$ is a true statement, the polynomial is not factored completely. Explain why and give the complete factored form.

Factor completely.

64. $z^2 - 11zx + 10x^2$

65. $3k^2 + 11k + 10$

66. $15m^2 + 20m - 12mp - 16p$

67. $y^4 - 625$

68. $6m^3 - 21m^2 - 45m$

69. $24ab^3c^2 - 56a^2bc^3 + 72a^2b^2c$

70. $25a^2 + 15ab + 9b^2$

71. $12x^2yz^3 + 12xy^2z - 30x^3y^2z^4$

72. $2a^5 - 8a^4 - 24a^3$

73. $12r^2 + 18rq - 10r - 15q$

74. $1000a^3 + 27$

75. $49t^2 + 56t + 16$

Solve.

76. $t(t - 7) = 0$

77. $x^2 + 3x = 10$

78. $25x^2 + 20x + 4 = 0$

79. The numbers of alternative-fueled vehicles, in thousands, in use for the years 1998–2001 are given in the table. Using statistical methods, we constructed the quadratic equation

$$y = .25x^2 - 25.65x + 496.6$$

to model the number of vehicles y in year x. Here we used $x = 98$ for 1998, $x = 99$ for 1999, and so on. Because only four years of data were used to determine the model, we must be particularly careful about using it to estimate for years before 1998 or after 2001.

(a) What prediction for 2002 is given by the equation?

(b) Why might the prediction for 2002 be unreliable?

ALTERNATIVE-FUELED VEHICLES

Year	Number (in thousands)
1998	384
1999	407
2000	432
2001	456

Source: Energy Information Administration, Alternatives to Traditional Fuels, 2001.

80. A lot is shaped like a right triangle. The hypotenuse is 3 m longer than the longer leg. The longer leg is 6 m longer than twice the length of the shorter leg. Find the lengths of the sides of the lot.

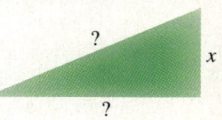

81. A pyramid has a rectangular base with a length that is 2 m more than the width. The height of the pyramid is 6 m, and its volume is 48 m³. Find the length and width of the base.

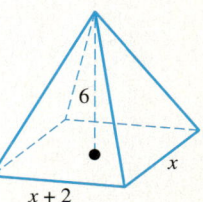

82. The product of the smaller two of three consecutive integers is equal to 23 plus the largest. Find the integers.

83. If an object is dropped, the distance d in feet it falls in t sec (disregarding air resistance) is given by the quadratic equation

$$d = 16t^2.$$

Find the distance an object would fall in the following times.

(a) 4 sec   **(b)** 8 sec

84. The floor plan for a house is a rectangle with length 7 m more than its width. The area is 170 m². Find the width and length of the house.

85. The triangular sail of a schooner has an area of 30 m². The height of the sail is 4 m more than the base. Find the base of the sail.

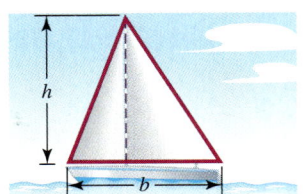

CHAPTER 6 TEST

1. Which one of the following is the correct, completely factored form of $2x^2 - 2x - 24$?
   A. $(2x + 6)(x - 4)$ **B.** $(x + 3)(2x - 8)$
   C. $2(x + 4)(x - 3)$ **D.** $2(x + 3)(x - 4)$

Factor each polynomial completely.

2. $12x^2 - 30x$
3. $2m^3n^2 + 3m^3n - 5m^2n^2$
4. $2ax - 2bx + ay - by$
5. $x^2 - 5x - 24$
6. $2x^2 + x - 3$
7. $10z^2 - 17z + 3$
8. $t^2 + 2t + 3$
9. $x^2 + 36$
10. $12 - 6a + 2b - ab$
11. $9y^2 - 64$
12. $4x^2 - 28xy + 49y^2$
13. $-2x^2 - 4x - 2$
14. $6t^4 + 3t^3 - 108t^2$
15. $r^3 - 125$
16. $8k^3 + 64$
17. $x^4 - 81$

18. Why is $(p + 3)(p + 3)$ not the correct factored form of $p^2 + 9$?

Solve each equation.

19. $2r^2 - 13r + 6 = 0$
20. $25x^2 - 4 = 0$
21. $x(x - 20) = -100$
22. $t^3 = 9t$

Solve each problem.

23. The length of a rectangular flower bed is 3 ft less than twice its width. The area of the bed is 54 ft². Find the dimensions of the flower bed.

24. Find two consecutive integers such that the square of the sum of the two integers is 11 more than the smaller integer.

25. A carpenter needs to cut a brace to support a wall stud, as shown in the figure. The brace should be 7 ft less than three times the length of the stud. If the brace will be anchored on the floor 15 ft away from the stud, how long should the brace be?

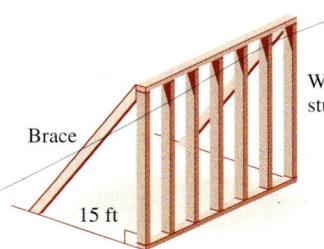

26. TV viewers have more choices than ever. The number of cable TV channels y from 1984 through 1999 can be approximated by the quadratic equation
$$y = .57x^2 + .31x + 48,$$
where $x = 0$ represents 1984, $x = 1$ represents 1985, and so on. (*Source:* National Cable Television Association.) Use the model to estimate the number of cable TV channels in 1999. Round your answer to the nearest whole number.

CUMULATIVE REVIEW EXERCISES CHAPTERS 1–6

Solve each equation.

1. $3x + 2(x - 4) = 4(x - 2)$
2. $.3x + .9x = .06$
3. $\frac{2}{3}y - \frac{1}{2}(y - 4) = 3$
4. Solve for P: $A = P + Prt$.

5. From a list of "everyday items" often taken for granted, adults were recently surveyed as to those items they wouldn't want to live without. Complete the results shown in the table if 500 adults were surveyed.

Item	Percent That Wouldn't Want to Live Without	Number That Wouldn't Want to Live Without
Toilet paper	69%	
Zipper	42%	
Frozen foods		190
Self-stick note pads		75

(Other items included tape, hairspray, pantyhose, paper clips, and Velcro.)
Source: Market Facts for Kleenex Cottonelle.

Solve each problem.

6. At the 1998 Winter Olympics in Nagano, Japan, the top medal winner was Germany with 29. Germany won 1 more silver medal than bronze and 3 more gold medals than silver. Find the number of each type of medal won. (*Source: World Almanac and Book of Facts,* 2000.)

7. In July 2000, roughly 144 million people surfed the Web from home. This was a 35% increase from the same month the previous year. How many people, to the nearest million, surfed the Web from home in July 1999? (*Source: The Gazette,* September 3, 2000.)

8. Find the measures of the marked angles.

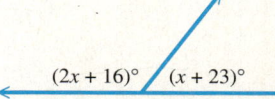

9. Fill in each blank with *positive* or *negative*. The point with coordinates (a, b) is in
    (a) quadrant II if a is _____ and b is _____.
    (b) quadrant III if a is _____ and b is _____.

Consider the equation $y = 12x + 3$. Find the following.

10. The x- and y-intercepts
11. The slope
12. The graph

446 CHAPTER 6 Factoring and Applications

13. The points on the graph show the number of U.S. radio stations in the years 1993 through 1999, along with the graph of a linear equation that models the data.

    (a) Use the ordered pairs shown on the graph to find the slope of the line to the nearest whole number. Interpret the slope.
    (b) Use the two labeled points on the graph and the slope from part (a) to write an equation of the line.

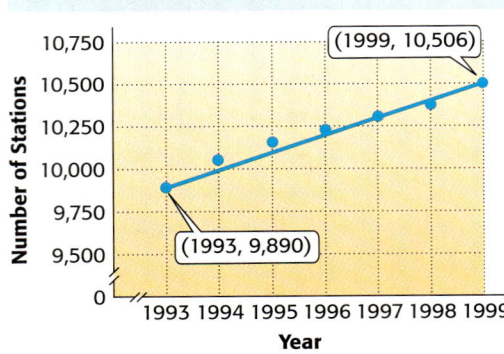

U.S. Radio Stations

Source: M Street Corporation.

14. The equation from Exercise 13(b) defines a relation R.

    (a) Find the value of R in 1999.   (b) Is the relation a function?

Solve each system of equations.

15. $4x - y = -6$
    $2x + 3y = 4$

16. $5x + 3y = 10$
    $2x + \dfrac{6}{5}y = 5$

Evaluate each expression.

17. $\left(\dfrac{3}{4}\right)^{-2}$ $\left(\dfrac{4}{3}\right)^2 = \dfrac{16}{9}$

18. $\left(\dfrac{4^{-3} \cdot 4^4}{4^5}\right)^{-1}$

Simplify each expression and write the answer using only positive exponents. Assume no denominators are 0.

19. $\dfrac{(p^2)^3 p^{-4}}{(p^{-3})^{-1} p}$

20. $\dfrac{(m^{-2})^3 m}{m^5 m^{-4}}$

Perform each indicated operation.

21. $(2k^2 + 4k) - (5k^2 - 2) - (k^2 + 8k - 6)$ 22. $(9x + 6)(5x - 3)$

23. $(3p + 2)^2$ 24. $\dfrac{8x^4 + 12x^3 - 6x^2 + 20x}{2x}$

25. To make a pound of honey, bees may travel 55,000 mi and visit more than 2,000,000 flowers. (*Source:* Home & Garden.) Write the two given numbers in scientific notation.

Factor completely.

26. $2a^2 + 7a - 4$
27. $10m^2 + 19m + 6$
28. $8t^2 + 10tv + 3v^2$
29. $4p^2 - 12p + 9$
30. $25r^2 - 81t^2$
31. $2pq + 6p^3q + 8p^2q$

Solve each equation.

32. $6m^2 + m - 2 = 0$
33. $8x^2 = 64x$

34. The length of the hypotenuse of a right triangle is twice the length of the shorter leg, plus 3 m. The longer leg is 7 m longer than the shorter leg. Find the lengths of the sides.

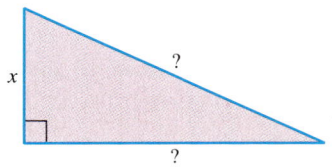

Rational Expressions and Applications

7.1 The Fundamental Property of Rational Expressions

7.2 Multiplying and Dividing Rational Expressions

7.3 Least Common Denominators

7.4 Adding and Subtracting Rational Expressions

7.5 Complex Fractions

7.6 Solving Equations with Rational Expressions

Summary Exercises on Operations and Equations with Rational Expressions

7.7 Applications of Rational Expressions

7.8 Variation

Americans have been car crazy ever since the first automobiles hit the road early in the twentieth century. Today there are about 213.5 million vehicles in the United States driving on 3.4 million mi of paved roadways. There is even a new museum devoted exclusively to our four-wheeled passion and its influence on our lives and culture. The Museum of Automobile History in Syracuse, N.Y., features some 200 years of automobile memorabilia, including rare advertising pieces, designer drawings, and Hollywood movie posters. (*Source: Home and Away,* May/June 2002.) In Exercises 75 and 76 of Section 7.4, we use a rational expression to determine the cost of restoring a vintage automobile.

448 CHAPTER 7 Rational Expressions and Applications

7.1 The Fundamental Property of Rational Expressions

OBJECTIVES

1. Find the numerical value of a rational expression.
2. Find the values of the variable for which a rational expression is undefined.
3. Write rational expressions in lowest terms.
4. Recognize equivalent forms of rational expressions.

The quotient of two integers (with denominator not 0), such as $\frac{2}{3}$ or $-\frac{3}{4}$, is called a *rational number*. In the same way, the quotient of two polynomials with denominator not equal to 0 is called a *rational expression*.

> **Rational Expression**
>
> A **rational expression** is an expression of the form
> $$\frac{P}{Q},$$
> where P and Q are polynomials, with $Q \neq 0$.

Examples of rational expressions include

$$\frac{-6x}{x^3 + 8}, \quad \frac{9x}{y + 3}, \quad \text{and} \quad \frac{2m^3}{8}.$$

Our work with rational expressions requires much of what we learned in Chapters 5 and 6 on polynomials and factoring, as well as the rules for fractions from Section 1.1.

OBJECTIVE 1 Find the numerical value of a rational expression. We use substitution to evaluate a rational expression for a given value of the variable.

EXAMPLE 1 Evaluating Rational Expressions

Find the numerical value of $\dfrac{3x + 6}{2x - 4}$ for each value of x.

(a) $x = 1$

$$\frac{3x + 6}{2x - 4} = \frac{3(1) + 6}{2(1) - 4} \quad \text{Let } x = 1.$$

$$= \frac{9}{-2} \quad \text{or} \quad -\frac{9}{2}$$

(b) $x = -2$

$$\frac{3x + 6}{2x - 4} = \frac{3(-2) + 6}{2(-2) - 4} \quad \text{Let } x = -2.$$

$$= \frac{0}{-8} \quad \text{or} \quad 0$$

Now Try Exercise 3.

OBJECTIVE 2 Find the values of the variable for which a rational expression is undefined. In the definition of a rational expression $\frac{P}{Q}$, Q cannot equal 0. The denominator of a rational expression cannot equal 0 because division by 0 is undefined.

SECTION 7.1 The Fundamental Property of Rational Expressions **449**

For instance, in the rational expression

$$\frac{3x + 6}{2x - 4} \leftarrow \text{Denominator cannot equal 0.}$$

from Example 1, the variable x can take on any real number value except 2. If x is 2, then the denominator becomes $2(2) - 4 = 0$, making the expression undefined. Thus, x cannot equal 2. We indicate this restriction by writing $x \neq 2$.

NOTE The *numerator* of a rational expression may be *any* real number. If the numerator equals 0 and the denominator does not equal 0, then the rational expression equals 0. See Example 1(b).

To determine the values for which a rational expression is undefined, use the following procedure.

Determining When a Rational Expression is Undefined

Step 1 Set the denominator of the rational expression equal to 0.
Step 2 Solve this equation.
Step 3 The solutions of the equation are the values that make the rational expression undefined.

■ **EXAMPLE 2** Finding Values That Make Rational Expressions Undefined

Find any values of the variable for which each rational expression is undefined.

(a) $\dfrac{p + 5}{3p + 2}$

Remember that the *numerator* may be any real number; you must find any value of p that makes the *denominator* equal to 0 since division by 0 is undefined.

Step 1 Set the denominator equal to 0.
$$3p + 2 = 0$$

Step 2 Solve this equation.
$$3p = -2$$
$$p = -\frac{2}{3}$$

Step 3 The given expression is undefined for $-\frac{2}{3}$, so $p \neq -\frac{2}{3}$.

(b) $\dfrac{8x^2 + 1}{x - 3}$

The denominator $x - 3 = 0$ when x is 3. The given expression is undefined for 3, so $x \neq 3$.

(c) $\dfrac{9m^2}{m^2 - 5m + 6}$

Set the denominator equal to 0, and then find the solutions of the equation $m^2 - 5m + 6 = 0$.

$$(m - 2)(m - 3) = 0 \quad \text{Factor.}$$
$$m - 2 = 0 \quad \text{or} \quad m - 3 = 0 \quad \text{Zero-factor property}$$
$$m = 2 \quad \text{or} \quad m = 3$$

The given expression is undefined for 2 and 3, so $m \neq 2$ and $m \neq 3$.

(d) $\dfrac{2r}{r^2 + 1}$

This denominator will not equal 0 for any value of r because r^2 is always greater than or equal to 0, and adding 1 makes the sum greater than 0. Thus, there are no values for which this rational expression is undefined.

Now Try Exercises 17 and 21.

OBJECTIVE 3 Write rational expressions in lowest terms. A fraction such as $\tfrac{2}{3}$ is said to be in *lowest terms*. How can "lowest terms" be defined? We use the idea of greatest common factor for this definition, which applies to all rational expressions.

> **Lowest Terms**
>
> A rational expression $\dfrac{P}{Q}$ ($Q \neq 0$) is in **lowest terms** if the greatest common factor of its numerator and denominator is 1.

The properties of rational numbers also apply to rational expressions. We use the **fundamental property of rational expressions** to write a rational expression in lowest terms.

> **Fundamental Property of Rational Expressions**
>
> If $\dfrac{P}{Q}$ ($Q \neq 0$) is a rational expression and if K represents any polynomial, where $K \neq 0$, then
>
> $$\dfrac{PK}{QK} = \dfrac{P}{Q}.$$

This property is based on the identity property of multiplication, since

$$\dfrac{PK}{QK} = \dfrac{P}{Q} \cdot \dfrac{K}{K} = \dfrac{P}{Q} \cdot 1 = \dfrac{P}{Q}.$$

The next example shows how to write both a rational number and a rational expression in lowest terms. Notice the similarity in the procedures. In both cases, we factor and then divide out the greatest common factor.

SECTION 7.1 The Fundamental Property of Rational Expressions 451

EXAMPLE 3 Writing in Lowest Terms

Write each expression in lowest terms.

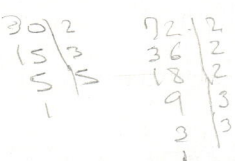

(a) $\dfrac{30}{72}$

Begin by factoring.

$$\dfrac{30}{72} = \dfrac{2 \cdot 3 \cdot 5}{2 \cdot 2 \cdot 2 \cdot 3 \cdot 3}$$

(b) $\dfrac{14k^2}{2k^3}$

Write k^2 as $k \cdot k$ and k^3 as $k \cdot k \cdot k$.

$$\dfrac{14k^2}{2k^3} = \dfrac{2 \cdot 7 \cdot k \cdot k}{2 \cdot k \cdot k \cdot k}$$

Group any factors common to the numerator and denominator.

$$\dfrac{30}{72} = \dfrac{5 \cdot (2 \cdot 3)}{2 \cdot 2 \cdot 3 \cdot (2 \cdot 3)}$$

$$\dfrac{14k^2}{2k^3} = \dfrac{7(2 \cdot k \cdot k)}{k(2 \cdot k \cdot k)}$$

Use the fundamental property.

$$\dfrac{30}{72} = \dfrac{5}{2 \cdot 2 \cdot 3} = \dfrac{5}{12}$$

$$\dfrac{14k^2}{2k^3} = \dfrac{7}{k}$$

Now Try Exercise 27.

To write a rational expression in lowest terms, follow these steps.

Writing a Rational Expression in Lowest Terms

Step 1 **Factor** the numerator and denominator completely.

Step 2 **Use the fundamental property** to divide out any common factors.

EXAMPLE 4 Writing in Lowest Terms

Write each rational expression in lowest terms.

(a) $\dfrac{3x - 12}{5x - 20} = \dfrac{3(x-4)}{5(x-4)} = \dfrac{3}{5}$

Step 1 Factor both numerator and denominator.

$$\dfrac{3x - 12}{5x - 20} = \dfrac{3(x - 4)}{5(x - 4)}$$

Step 2 Use the fundamental property.

$$= \dfrac{3}{5} \qquad \text{Divide out the common factor.}$$

The given expression is equal to $\dfrac{3}{5}$ for all values of x, where $x \neq 4$ (since the denominator of the original rational expression is 0 when x is 4).

(b) $\dfrac{2y^2 - 8}{2y + 4} = \dfrac{2(y^2 - 4)}{2(y + 2)}$ Factor. (Step 1)

$$= \dfrac{2(y + 2)(y - 2)}{2(y + 2)}$$ Be sure to factor completely.

$$= y - 2$$ Fundamental property (Step 2)

Here, $y \neq -2$ since the denominator of the original expression is 0 for this value.

452 CHAPTER 7 Rational Expressions and Applications

(c) $\dfrac{m^2 + 2m - 8}{2m^2 - m - 6} = \dfrac{(m+4)(m-2)}{(2m+3)(m-2)}$ Factor.

$\qquad\qquad\qquad= \dfrac{m+4}{2m+3}$ Fundamental property

Here, $m \neq -\dfrac{3}{2}$ and $m \neq 2$, since the denominator of the original expression is 0 for these values of m.

Now Try Exercises 33, 41, and 49.

From now on, we write statements of equality of rational expressions with the understanding that they apply only to those real numbers that make neither denominator equal to 0.

> **CAUTION** *Rational expressions cannot be written in lowest terms until after the numerator and denominator have been factored. Only common factors can be divided out,* not common terms. For example,
>
> $\dfrac{6x+9}{4x+6} = \dfrac{3(2x+3)}{2(2x+3)} = \dfrac{3}{2}$ $\qquad$ $\dfrac{6+x}{4x}$ ← Numerator cannot be factored.
>
> Divide out the common factor. $\qquad\qquad$ Already in lowest terms

EXAMPLE 5 Writing in Lowest Terms (Factors Are Opposites)

Write $\dfrac{x-y}{y-x}$ in lowest terms.

At first glance, there does not seem to be any way in which $x - y$ and $y - x$ can be factored to get a common factor. However, the denominator $y - x$ can be factored as

$$y - x = -1(-y + x) = -1(x - y).$$

Now, use the fundamental property to simplify.

$$\dfrac{x-y}{y-x} = \dfrac{1(x-y)}{-1(x-y)} = \dfrac{1}{-1} = -1$$

Now Try Exercise 83.

> **NOTE** Either the numerator or the denominator could have been factored in the first step in Example 5. Factor -1 from the numerator and confirm that the result is the same.

In Example 5, notice that $y - x$ is the opposite of $x - y$. A general rule for this situation follows.

> If the numerator and the denominator of a rational expression are opposites, as in $\dfrac{x-y}{y-x}$, then the rational expression is equal to -1.

EXAMPLE 6 Writing in Lowest Terms (Factors Are Opposites)

Write each rational expression in lowest terms.

(a) $\dfrac{2 - m}{m - 2}$

Each term in the numerator differs only in sign from the same term in the denominator, so we can factor -1 from either the numerator or the denominator.

$$\frac{2 - m}{m - 2} = \frac{-1(m - 2)}{m - 2} = -1$$

(b) $\dfrac{4x^2 - 9}{6 - 4x}$

Factor the numerator and denominator.

$$\frac{4x^2 - 9}{6 - 4x} = \frac{(2x + 3)(2x - 3)}{2(3 - 2x)}$$

$$= \frac{(2x + 3)(2x - 3)}{2(-1)(2x - 3)} \quad \text{Write } 3 - 2x \text{ as } -1(2x - 3).$$

$$= \frac{2x + 3}{2(-1)}$$

$$= \frac{2x + 3}{-2} \quad \text{or} \quad -\frac{2x + 3}{2} \qquad \frac{a}{-b} = -\frac{a}{b}$$

(c) $\dfrac{3 + r}{3 - r}$

The quantity $3 - r$ is *not* the opposite of $3 + r$ because the 3 is positive in both cases. This rational expression is already in lowest terms.

Now Try Exercises 85 and 89.

OBJECTIVE 4 Recognize equivalent forms of rational expressions. When working with rational expressions, it is important to be able to recognize equivalent forms of an expression. For example, the common fraction $-\frac{5}{6}$ can also be written $\frac{-5}{6}$ and $\frac{5}{-6}$. Consider also the rational expression

$$-\frac{2x + 3}{2}.$$

The $-$ sign representing the -1 factor is in front of the expression. The -1 factor may instead be placed in the numerator or in the denominator. Thus, some other equivalent forms of this rational expression are

Use parentheses.

$$\frac{-(2x + 3)}{2} \quad \text{and} \quad \frac{2x + 3}{-2}.$$

By the distributive property,

$$\frac{-(2x + 3)}{2} \quad \text{can also be written} \quad \frac{-2x - 3}{2}.$$

CAUTION $\frac{-2x+3}{2}$ is *not* an equivalent form of $\frac{-(2x+3)}{2}$. The sign preceding 3 in the numerator of $\frac{-2x+3}{2}$ should be $-$ rather than $+$. Be careful to apply the distributive property correctly.

EXAMPLE 7 Writing Equivalent Forms of a Rational Expression

Write four equivalent forms of the rational expression

$$-\frac{3x+2}{x-6}.$$

If we apply the negative sign to the numerator, we obtain the equivalent forms

$$\frac{-(3x+2)}{x-6} \quad \text{or, by the distributive property,} \quad \frac{-3x-2}{x-6}.$$

If we apply the negative sign to the denominator, we obtain

$$\frac{3x+2}{-(x-6)} \quad \text{or, distributing once again,} \quad \frac{3x+2}{-x+6}.$$

Now Try Exercise 93.

CAUTION Recall that $-\frac{5}{6} \neq \frac{-5}{-6}$. Thus, in Example 7, it would be incorrect to distribute the negative sign to *both* the numerator *and* the denominator. (Doing this would actually lead to the *opposite* of the original expression.)

7.1 EXERCISES

For Extra Help

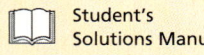 Student's Solutions Manual

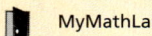 MyMathLab

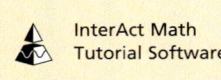 InterAct Math Tutorial Software

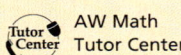 AW Math Tutor Center

 MathXL

Digital Video Tutor CD 14/Videotape 12

1. Define *rational expression* in your own words, and give an example.

Find the numerical value of each rational expression when **(a)** $x = 2$ *and* **(b)** $x = -3$. *See Example 1.*

2. $\dfrac{5x-2}{4x}$

3. $\dfrac{3x+1}{5x}$

4. $\dfrac{2x^2-4x}{3x-1}$

5. $\dfrac{x^2-4}{2x+1}$

6. $\dfrac{(-3x)^2}{4x+12}$

7. $\dfrac{(-2x)^3}{3x+9}$

8. $\dfrac{5x+2}{2x^2+11x+12}$

9. $\dfrac{7-3x}{3x^2-7x+2}$

10. Fill in each blank with the correct response: The rational expression $\frac{x+5}{x-3}$ is undefined when x is _____, so $x \neq$ _____. This rational expression is equal to 0 when $x =$ _____.

11. Why can't the denominator of a rational expression equal 0?

SECTION 7.1 The Fundamental Property of Rational Expressions

12. If 2 is substituted for x in the rational expression $\dfrac{x-2}{x^2-4}$, the result is $\dfrac{0}{0}$. An often-heard statement is "Any number divided by itself is 1." Does this mean that this expression is equal to 1 for $x = 2$? If not, explain.

Find any values for which each rational expression is undefined. See Example 2.

13. $\dfrac{12}{5y}$

14. $\dfrac{-7}{3z}$

15. $\dfrac{x+1}{x-6}$

16. $\dfrac{m-2}{m-5}$

17. $\dfrac{4x^2}{3x+5}$

18. $\dfrac{2x^3}{3x+4}$

19. $\dfrac{5m+2}{m^2+m-6}$

20. $\dfrac{2r-5}{r^2-5r+4}$

21. $\dfrac{x^2+3x}{4}$

22. $\dfrac{x^2-4x}{6}$

23. $\dfrac{3x-1}{x^2+2}$

24. $\dfrac{4q+2}{q^2+9}$

25. (a) Identify the two *terms* in the numerator and the two *terms* in the denominator of the rational expression $\dfrac{x^2+4x}{x+4}$.

(b) Describe the steps you would use to write this rational expression in lowest terms. (*Hint:* It simplifies to x.)

26. Only one of the following rational expressions can be simplified. Which one is it?

A. $\dfrac{x^2+2}{x^2}$ **B.** $\dfrac{x^2+2}{2}$ **C.** $\dfrac{x^2+y^2}{y^2}$ **D.** $\dfrac{x^2-5x}{x}$

Write each rational expression in lowest terms. See Examples 3 and 4.

27. $\dfrac{18r^3}{6r}$

28. $\dfrac{27p^2}{3p}$

29. $\dfrac{4(y-2)}{10(y-2)}$

30. $\dfrac{15(m-1)}{9(m-1)}$

31. $\dfrac{(x+1)(x-1)}{(x+1)^2}$

32. $\dfrac{(t+5)(t-3)}{(t-1)(t+5)}$

33. $\dfrac{7m+14}{5m+10}$

34. $\dfrac{8z-24}{4z-12}$

35. $\dfrac{6m-18}{7m-21}$

36. $\dfrac{5r+20}{3r+12}$

37. $\dfrac{m^2-n^2}{m+n}$

38. $\dfrac{a^2-b^2}{a-b}$

39. $\dfrac{2t+6}{t^2-9}$

40. $\dfrac{5s-25}{s^2-25}$

41. $\dfrac{12m^2-3}{8m-4}$

42. $\dfrac{20p^2-45}{6p-9}$

43. $\dfrac{3m^2-3m}{5m-5}$

44. $\dfrac{6t^2-6t}{2t-2}$

45. $\dfrac{9r^2-4s^2}{9r+6s}$

46. $\dfrac{16x^2-9y^2}{12x-9y}$

47. $\dfrac{5k^2-13k-6}{5k+2}$

48. $\dfrac{7t^2-31t-20}{7t+4}$

49. $\dfrac{x^2+2x-15}{x^2+6x+5}$

50. $\dfrac{y^2-5y-14}{y^2+y-2}$

51. $\dfrac{2x^2-3x-5}{2x^2-7x+5}$

52. $\dfrac{3x^2+8x+4}{3x^2-4x-4}$

53. $\dfrac{6x^2+13x+2}{8x^2+15x-2}$

54. $\dfrac{6x^2-29x+20}{4x^2-15x-4}$

55. $\dfrac{16x^2+24x+9}{4x^2-25x-21}$

56. $\dfrac{12x^2-5x-7}{x^2-2x+1}$

57. $\dfrac{81 - 4x^4}{9 + 29x^2 + 6x^4}$ **58.** $\dfrac{9 - 49x^4}{3 + x^2 - 14x^4}$ **59.** $\dfrac{4x^4 + 20x^2 + 25}{4x^4 - 25}$

60. $\dfrac{25x^6 + 10x^3 + 1}{25x^6 - 1}$ **61.** $\dfrac{10x^2 - 34x - 24}{10x^2 - 46x + 24}$ **62.** $\dfrac{27x^2 + 174x + 72}{27x^2 + 150x - 72}$

These exercises involve factoring by grouping (Section 6.1) and factoring sums and differences of cubes (Section 6.4). Write each rational expression in lowest terms.

63. $\dfrac{zw + 4z - 3w - 12}{zw + 4z + 5w + 20}$ **64.** $\dfrac{km + 4k + 4m + 16}{km + 4k + 5m + 20}$ **65.** $\dfrac{m^2 - n^2 - 4m - 4n}{2m - 2n - 8}$

66. $\dfrac{x^2y + y + x^2z + z}{xy + xz}$ **67.** $\dfrac{b^3 - a^3}{a^2 - b^2}$ **68.** $\dfrac{k^3 + 8}{k^2 - 4}$

69. $\dfrac{z^3 + 27}{z^3 - 3z^2 + 9z}$ **70.** $\dfrac{1 - 8r^3}{8r^2 + 4r + 2}$ **71.** $\dfrac{x^3 + 8y^3}{2x^2 - 4xy + 8y^2}$

72. $\dfrac{x^3 - 27y^3}{4x^2 + 12xy + 36y^2}$ **73.** $\dfrac{x^4 - y^4}{(x + y)^2(x^2 + y^2)}$ **74.** $\dfrac{16x^4 - 81y^4}{(2x - 3y)^2(4x^2 + 9y^2)}$

75. $\dfrac{3ac + 3ad + 6bc + 6bd}{3ac + 3ad - 15bc - 15bd}$ **76.** $\dfrac{8tr + 8tq + 4sr + 4sq}{12tr + 12tq - 4sr - 4sq}$

77. $\dfrac{-6c^2 + 12cd}{ac - 2ad + 4bc - 8bd}$ **78.** $\dfrac{-7p^2 + 42pr}{2pq - 12qr + 3rp - 18r^2}$

79. $\dfrac{x^6 + y^6}{3x^4 - 3x^2y^2 + 3y^4}$ **80.** $\dfrac{2t^4 - 8t^2r^2 + 32r^4}{2t^6 + 128r^6}$

81. Which two of the following rational expressions equal -1?

  A. $\dfrac{2x + 3}{2x - 3}$ **B.** $\dfrac{2x - 3}{3 - 2x}$ **C.** $\dfrac{2x + 3}{3 + 2x}$ **D.** $\dfrac{2x + 3}{-2x - 3}$

82. Make the correct choice for the blank: $\dfrac{4 - r^2}{4 + r^2}$ _____ equal to -1.
  (is/is not)

Write each rational expression in lowest terms. See Examples 5 and 6.

83. $\dfrac{6 - t}{t - 6}$ **84.** $\dfrac{2 - k}{k - 2}$ **85.** $\dfrac{m^2 - 1}{1 - m}$ **86.** $\dfrac{a^2 - b^2}{b - a}$

87. $\dfrac{q^2 - 4q}{4q - q^2}$ **88.** $\dfrac{z^2 - 5z}{5z - z^2}$ **89.** $\dfrac{p + 6}{p - 6}$ **90.** $\dfrac{5 - x}{5 + x}$

91. Only one of the following rational expressions is *not* equivalent to $\dfrac{x - 3}{4 - x}$. Which one is it?

  A. $\dfrac{3 - x}{x - 4}$ **B.** $\dfrac{x + 3}{4 + x}$ **C.** $-\dfrac{3 - x}{4 - x}$ **D.** $-\dfrac{x - 3}{x - 4}$

92. Make the correct choice for the blank: $\dfrac{5 + 2x}{3 - x}$ and $\dfrac{-5 - 2x}{x - 3}$ _____ equivalent rational expressions.   (are/are not)

Write four equivalent forms for each rational expression. See Example 7.

93. $-\dfrac{x+4}{x-3}$

94. $-\dfrac{x+6}{x-1}$

95. $-\dfrac{2x-3}{x+3}$

96. $-\dfrac{5x-6}{x+4}$

97. $\dfrac{-3x+1}{5x-6}$

98. $\dfrac{-2x-9}{3x+1}$

99. The area of the rectangle is represented by
$$x^4 + 10x^2 + 21.$$
What is the width? (*Hint:* Use $W = \dfrac{A}{L}$.)

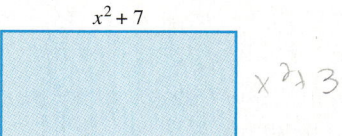

100. The volume of the box is represented by
$$(x^2 + 8x + 15)(x+4).$$
Find the polynomial that represents the area of the bottom of the box.

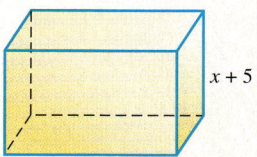

Solve each problem.

101. The average number of vehicles waiting in line to enter a sports arena parking area is approximated by the rational expression

$$\frac{x^2}{2(1-x)},$$

where x is a quantity between 0 and 1 known as the *traffic intensity*. (*Source:* Mannering, F. and W. Kilareski, *Principles of Highway Engineering and Traffic Control,* John Wiley and Sons, 1990.) To the nearest tenth, find the average number of vehicles waiting if the traffic intensity is

(a) .1 (b) .8 (c) .9.

(d) What happens to waiting time as traffic intensity increases?

102. The percent of deaths caused by smoking is modeled by the rational expression

$$\frac{x-1}{x},$$

where x is the number of times a smoker is more likely to die of lung cancer than a nonsmoker. This is called the *incidence rate*. (*Source:* Walker, A., *Observation and Inference: An Introduction to the Methods of Epidemiology,* Epidemiology Resources Inc., 1991.) For example, $x = 10$ means that a smoker is 10 times more likely than a nonsmoker to die of lung cancer. Find the percent of deaths if the incidence rate is

(a) 5 (b) 10 (c) 20.

(d) Can the incidence rate equal 0? Explain.

458 CHAPTER 7 Rational Expressions and Applications

7.2 Multiplying and Dividing Rational Expressions

OBJECTIVES

1. Multiply rational expressions.
2. Divide rational expressions.

OBJECTIVE 1 Multiply rational expressions. The product of two fractions is found by multiplying the numerators and multiplying the denominators. Rational expressions are multiplied in the same way.

Multiplying Rational Expressions

The product of the rational expressions $\frac{P}{Q}$ and $\frac{R}{S}$ is

$$\frac{P}{Q} \cdot \frac{R}{S} = \frac{PR}{QS}.$$

That is, to multiply rational expressions, multiply the numerators and multiply the denominators.

In the following example, the parallel discussion with rational numbers and rational expressions lets you compare the steps.

EXAMPLE 1 Multiplying Rational Expressions

Multiply. Write each answer in lowest terms.

(a) $\dfrac{3}{10} \cdot \dfrac{5}{9}$ (b) $\dfrac{6}{x} \cdot \dfrac{x^2}{12}$

Indicate the product of the numerators and the product of the denominators.

$$\frac{3}{10} \cdot \frac{5}{9} = \frac{3 \cdot 5}{10 \cdot 9} \qquad \frac{6}{x} \cdot \frac{x^2}{12} = \frac{6 \cdot x^2}{x \cdot 12}$$

Leave the products in factored form because common factors are needed to write the product in lowest terms. Factor the numerator and denominator to further identify any common factors. Then use the fundamental property to write each product in lowest terms.

$$\frac{3}{10} \cdot \frac{5}{9} = \frac{3 \cdot 5}{2 \cdot 5 \cdot 3 \cdot 3} = \frac{1}{6} \qquad \frac{6}{x} \cdot \frac{x^2}{12} = \frac{6 \cdot x \cdot x}{2 \cdot 6 \cdot x} = \frac{x}{2}$$

Now Try Exercise 3.

NOTE It is also possible to divide out common factors in the numerator and denominator *before* multiplying the rational expressions. For example,

$$\frac{6}{5} \cdot \frac{35}{22} = \frac{2 \cdot 3}{5} \cdot \frac{5 \cdot 7}{2 \cdot 11} \qquad \text{Identify common factors.}$$

$$= \frac{3 \cdot 7}{11} \qquad \text{Lowest terms}$$

$$= \frac{21}{11}. \qquad \text{Multiply in numerator.}$$

EXAMPLE 2 Multiplying Rational Expressions

Multiply. Write the answer in lowest terms.

$$\frac{x+y}{2x} \cdot \frac{x^2}{(x+y)^2} = \frac{(x+y)x^2}{2x(x+y)^2} \qquad \text{Multiply numerators.}$$
$$\text{Multiply denominators.}$$

$$= \frac{(x+y)x \cdot x}{2x(x+y)(x+y)} \qquad \text{Factor; identify common factors.}$$

$$= \frac{x}{2(x+y)} \qquad \text{Lowest terms}$$

Notice the quotient of factors $\frac{(x+y)x}{x(x+y)}$ in the second line of the solution. Since it is equal to 1, the final product is $\frac{x}{2(x+y)}$.

Now Try Exercise 9.

EXAMPLE 3 Multiplying Rational Expressions

Multiply. Write the answer in lowest terms.

$$\frac{x^2+3x}{x^2-3x-4} \cdot \frac{x^2-5x+4}{x^2+2x-3} = \frac{(x^2+3x)(x^2-5x+4)}{(x^2-3x-4)(x^2+2x-3)} \qquad \text{Definition of multiplication}$$

$$= \frac{x(x+3)(x-4)(x-1)}{(x-4)(x+1)(x+3)(x-1)} \qquad \text{Factor.}$$

$$= \frac{x}{x+1} \qquad \text{Lowest terms}$$

The quotients $\frac{x+3}{x+3}, \frac{x-4}{x-4}$, and $\frac{x-1}{x-1}$ all equal 1, justifying the final product $\frac{x}{x+1}$.

Now Try Exercise 43.

OBJECTIVE 2 Divide rational expressions. To develop a method for dividing rational numbers and rational expressions, consider the following problem. Suppose you have $\frac{7}{8}$ gal of milk and want to find how many quarts you have. Since 1 qt is $\frac{1}{4}$ gal, you ask yourself, "How many $\frac{1}{4}$s are there in $\frac{7}{8}$?" This would be interpreted as

$$\frac{7}{8} \div \frac{1}{4} \qquad \text{or} \qquad \frac{\frac{7}{8}}{\frac{1}{4}}$$

since the fraction bar means division.

The fundamental property of rational expressions discussed earlier can be applied to rational number values of P, Q, and K. With $P = \frac{7}{8}$, $Q = \frac{1}{4}$, and $K = 4$ (since 4 is the reciprocal of $Q = \frac{1}{4}$),

$$\frac{P}{Q} = \frac{P \cdot K}{Q \cdot K} = \frac{\frac{7}{8} \cdot 4}{\frac{1}{4} \cdot 4} = \frac{\frac{7}{8} \cdot 4}{1} = \frac{7}{8} \cdot \frac{4}{1}.$$

So, to divide $\frac{7}{8}$ by $\frac{1}{4}$, we multiply $\frac{7}{8}$ by the reciprocal of $\frac{1}{4}$, namely 4. Since $\frac{7}{8}(4) = \frac{7}{2}$, there are $\frac{7}{2}$ or $3\frac{1}{2}$ qt in $\frac{7}{8}$ gal.

The preceding discussion illustrates the rule for dividing common fractions. To divide $\frac{a}{b}$ by $\frac{c}{d}$, multiply $\frac{a}{b}$ by the reciprocal of $\frac{c}{d}$. Division of rational expressions is defined in the same way.

Dividing Rational Expressions

If $\frac{P}{Q}$ and $\frac{R}{S}$ are any two rational expressions, with $\frac{R}{S} \neq 0$, then

$$\frac{P}{Q} \div \frac{R}{S} = \frac{P}{Q} \cdot \frac{S}{R} = \frac{PS}{QR}.$$

That is, to divide one rational expression by another rational expression, multiply the first rational expression by the reciprocal of the second rational expression.

The next example shows the division of two rational numbers and the division of two rational expressions.

EXAMPLE 4 Dividing Rational Expressions

Divide. Write each answer in lowest terms.

(a) $\dfrac{5}{8} \div \dfrac{7}{16}$ **(b)** $\dfrac{y}{y+3} \div \dfrac{4y}{y+5}$

Multiply the first expression by the reciprocal of the second.

$\dfrac{5}{8} \div \dfrac{7}{16} = \dfrac{5}{8} \cdot \dfrac{16}{7}$ Reciprocal of $\frac{7}{16}$

$= \dfrac{5 \cdot 16}{8 \cdot 7}$

$= \dfrac{5 \cdot 8 \cdot 2}{8 \cdot 7}$

$= \dfrac{10}{7}$

$\dfrac{y}{y+3} \div \dfrac{4y}{y+5}$

$= \dfrac{y}{y+3} \cdot \dfrac{y+5}{4y}$ Reciprocal of $\frac{4y}{y+5}$

$= \dfrac{y(y+5)}{(y+3)(4y)}$

$= \dfrac{y+5}{4(y+3)}$

Now Try Exercise 19.

EXAMPLE 5 Dividing Rational Expressions

Divide. Write the answer in lowest terms.

$\dfrac{(3m)^2}{(2p)^3} \div \dfrac{6m^3}{16p^2} = \dfrac{(3m)^2}{(2p)^3} \cdot \dfrac{16p^2}{6m^3}$ Multiply by the reciprocal.

$= \dfrac{9m^2}{8p^3} \cdot \dfrac{16p^2}{6m^3}$ Power rule for exponents

$$= \frac{9 \cdot 16m^2p^2}{8 \cdot 6p^3m^3} \qquad \text{Multiply numerators.}$$
$$\text{Multiply denominators.}$$
$$= \frac{3}{mp} \qquad \text{Lowest terms}$$

Now Try Exercise 17.

EXAMPLE 6 Dividing Rational Expressions

Divide. Write the answer in lowest terms.

$$\frac{x^2 - 4}{(x + 3)(x - 2)} \div \frac{(x + 2)(x + 3)}{-2x}$$

$$= \frac{x^2 - 4}{(x + 3)(x - 2)} \cdot \frac{-2x}{(x + 2)(x + 3)} \qquad \text{Multiply by the reciprocal.}$$

$$= \frac{-2x(x^2 - 4)}{(x + 3)(x - 2)(x + 2)(x + 3)} \qquad \text{Multiply numerators.}$$
$$\text{Multiply denominators.}$$

$$= \frac{-2x(x + 2)(x - 2)}{(x + 3)(x - 2)(x + 2)(x + 3)} \qquad \text{Factor.}$$

$$= \frac{-2x}{(x + 3)^2} \qquad \text{Lowest terms}$$

$$= -\frac{2x}{(x + 3)^2} \qquad \frac{-a}{b} = -\frac{a}{b}$$

Now Try Exercise 35.

EXAMPLE 7 Dividing Rational Expressions (Factors Are Opposites)

Divide. Write the answer in lowest terms.

$$\frac{m^2 - 4}{m^2 - 1} \div \frac{2m^2 + 4m}{1 - m}$$

$$= \frac{m^2 - 4}{m^2 - 1} \cdot \frac{1 - m}{2m^2 + 4m} \qquad \text{Multiply by the reciprocal.}$$

$$= \frac{(m^2 - 4)(1 - m)}{(m^2 - 1)(2m^2 + 4m)} \qquad \text{Multiply numerators.}$$
$$\text{Multiply denominators.}$$

$$= \frac{(m + 2)(m - 2)(1 - m)}{(m + 1)(m - 1)(2m)(m + 2)} \qquad \text{Factor; } 1 - m \text{ and } m - 1 \text{ differ only in sign.}$$

$$= \frac{-1(m - 2)}{2m(m + 1)} \qquad \text{From Section 7.1, } \frac{1 - m}{m - 1} = -1.$$

$$= \frac{-m + 2}{2m(m + 1)} \quad \text{or} \quad \frac{2 - m}{2m(m + 1)} \qquad \text{Distribute the negative sign in the numerator.}$$

Now Try Exercise 37.

462 CHAPTER 7 Rational Expressions and Applications

In summary, follow these steps to multiply or divide rational expressions.

> **Multiplying or Dividing Rational Expressions**
>
> *Step 1* **Note the operation.** If the operation is division, use the definition of division to rewrite as multiplication.
> *Step 2* **Multiply** numerators and multiply denominators.
> *Step 3* **Factor** all numerators and denominators completely.
> *Step 4* **Write in lowest terms** using the fundamental property.
>
> *Note:* Steps 2 and 3 may be interchanged based on personal preference.

H.W = Do exercises on sybullas

7.2 EXERCISES

For Extra Help
- Student's Solutions Manual
- MyMathLab
- InterAct Math Tutorial Software
- AW Math Tutor Center
- MathXL
- Digital Video Tutor CD 14/Videotape 12

1. Match each multiplication problem in Column I with the correct product in Column II.

I

(a) $\dfrac{5x^3}{10x^4} \cdot \dfrac{10x^7}{2x}$

(b) $\dfrac{10x^4}{5x^3} \cdot \dfrac{10x^7}{2x}$

(c) $\dfrac{5x^3}{10x^4} \cdot \dfrac{2x}{10x^7}$

(d) $\dfrac{10x^4}{5x^3} \cdot \dfrac{2x}{10x^7}$

II

A. $\dfrac{2}{5x^5}$

B. $\dfrac{5x^5}{2}$

C. $\dfrac{1}{10x^7}$

D. $10x^7$

2. Match each division problem in Column I with the correct quotient in Column II.

I

(a) $\dfrac{5x^3}{10x^4} \div \dfrac{10x^7}{2x}$

(b) $\dfrac{10x^4}{5x^3} \div \dfrac{10x^7}{2x}$

(c) $\dfrac{5x^3}{10x^4} \div \dfrac{2x}{10x^7}$

(d) $\dfrac{10x^4}{5x^3} \div \dfrac{2x}{10x^7}$

II

A. $\dfrac{5x^5}{2}$

B. $10x^7$

C. $\dfrac{2}{5x^5}$

D. $\dfrac{1}{10x^7}$

Multiply. Write each answer in lowest terms. See Examples 1 and 2.

3. $\dfrac{15a^2}{14} \cdot \dfrac{7}{5a}$

4. $\dfrac{27k^3}{9k} \cdot \dfrac{24}{9k^2}$

5. $\dfrac{12x^4}{18x^3} \cdot \dfrac{-8x^5}{4x^2}$

6. $\dfrac{12m^5}{-2m^2} \cdot \dfrac{6m^6}{28m^3}$

7. $\dfrac{2(c+d)}{3} \cdot \dfrac{18}{6(c+d)^2}$

8. $\dfrac{4(y-2)}{x} \cdot \dfrac{3x}{6(y-2)^2}$

9. $\dfrac{(x-y)^2}{2} \cdot \dfrac{24}{3(x-y)}$

10. $\dfrac{(a+b)^2}{5} \cdot \dfrac{30}{2(a+b)}$

11. $\dfrac{t-4}{8} \cdot \dfrac{4t^2}{t-4}$

12. $\dfrac{z+9}{12} \cdot \dfrac{3z^2}{z+9}$

13. $\dfrac{3x}{x+3} \cdot \dfrac{(x+3)^2}{6x^2}$

14. $\dfrac{(t-2)^2}{4t^2} \cdot \dfrac{2t}{t-2}$

Divide. Write each answer in lowest terms. See Examples 4 and 5.

15. $\dfrac{9z^4}{3z^5} \div \dfrac{3z^2}{5z^3}$

16. $\dfrac{35q^8}{9q^5} \div \dfrac{25q^6}{10q^5}$

17. $\dfrac{4t^4}{2t^5} \div \dfrac{(2t)^3}{-6}$

18. $\dfrac{-12a^6}{3a^2} \div \dfrac{(2a)^3}{27a}$

19. $\dfrac{3}{2y-6} \div \dfrac{6}{y-3}$

20. $\dfrac{4m+16}{10} \div \dfrac{3m+12}{18}$

21. $\dfrac{7t+7}{-6} \div \dfrac{4t+4}{15}$ **22.** $\dfrac{8z-16}{-20} \div \dfrac{3z-6}{40}$ **23.** $\dfrac{2x}{x-1} \div \dfrac{x^2}{x+2}$

24. $\dfrac{y^2}{y+1} \div \dfrac{3y}{y-3}$ **25.** $\dfrac{(x-3)^2}{6x} \div \dfrac{x-3}{x^2}$ **26.** $\dfrac{2a}{a+4} \div \dfrac{a^2}{(a+4)^2}$

✏ **27.** Use an example to explain how to multiply rational expressions.

✏ **28.** Use an example to explain how to divide rational expressions.

Multiply or divide. Write each answer in lowest terms. See Examples 3, 6, and 7.

29. $\dfrac{5x-15}{3x+9} \cdot \dfrac{4x+12}{6x-18}$ **30.** $\dfrac{8r+16}{24r-24} \cdot \dfrac{6r-6}{3r+6}$ **31.** $\dfrac{2-t}{8} \div \dfrac{t-2}{6}$

32. $\dfrac{4}{m-2} \div \dfrac{16}{2-m}$ **33.** $\dfrac{27-3z}{4} \cdot \dfrac{12}{2z-18}$ **34.** $\dfrac{5-x}{5+x} \cdot \dfrac{x+5}{x-5}$

35. $\dfrac{p^2+4p-5}{p^2+7p+10} \div \dfrac{p-1}{p+4}$ **36.** $\dfrac{z^2-3z+2}{z^2+4z+3} \div \dfrac{z-1}{z+1}$ **37.** $\dfrac{m^2-4}{16-8m} \div \dfrac{m+2}{8}$

38. $\dfrac{2}{3-x} \div \dfrac{2x+6}{x^2-9}$ **39.** $\dfrac{2x^2-7x+3}{x-3} \cdot \dfrac{x+2}{x-1}$ **40.** $\dfrac{3x^2-5x-2}{x-2} \cdot \dfrac{x-3}{x+1}$

41. $\dfrac{2k^2-k-1}{2k^2+5k+3} \div \dfrac{4k^2-1}{2k^2+k-3}$ **42.** $\dfrac{2m^2-5m-12}{m^2+m-20} \div \dfrac{4m^2-9}{m^2+4m-5}$

43. $\dfrac{2k^2+3k-2}{6k^2-7k+2} \cdot \dfrac{4k^2-5k+1}{k^2+k-2}$ **44.** $\dfrac{2m^2-5m-12}{m^2-10m+24} \div \dfrac{4m^2-9}{m^2-9m+18}$

45. $\dfrac{m^2+2mp-3p^2}{m^2-3mp+2p^2} \div \dfrac{m^2+4mp+3p^2}{m^2+2mp-8p^2}$ **46.** $\dfrac{r^2+rs-12s^2}{r^2-rs-20s^2} \div \dfrac{r^2-2rs-3s^2}{r^2+rs-30s^2}$

47. $\dfrac{m^2+3m+2}{m^2+5m+4} \cdot \dfrac{m^2+10m+24}{m^2+5m+6}$ **48.** $\dfrac{z^2-z-6}{z^2-2z-8} \cdot \dfrac{z^2+7z+12}{z^2-9}$

49. $\dfrac{y^2+y-2}{y^2+3y-4} \div \dfrac{y+2}{y+3}$ **50.** $\dfrac{r^2+r-6}{r^2+4r-12} \div \dfrac{r+3}{r-1}$

51. $\dfrac{2m^2+7m+3}{m^2-9} \cdot \dfrac{m^2-3m}{2m^2+11m+5}$ **52.** $\dfrac{m^2+2mp-3p^2}{m^2-3mp+2p^2} \div \dfrac{m^2+4mp+3p^2}{m^2+2mp-8p^2}$

53. $\dfrac{r^2+rs-12s^2}{r^2-rs-20s^2} \div \dfrac{r^2-2rs-3s^2}{r^2+rs-30s^2}$ **54.** $\dfrac{(x+1)^3(x+4)}{x^2+5x+4} \div \dfrac{x^2+2x+1}{x^2+3x+2}$

55. $\dfrac{(q-3)^4(q+2)}{q^2+3q+2} \div \dfrac{q^2-6q+9}{q^2+4q+4}$ **56.** $\dfrac{(x+4)^3(x-3)}{x^2-9} \div \dfrac{x^2+8x+16}{x^2+6x+9}$

These exercises involve grouping symbols (Section 1.2), factoring by grouping (Section 6.1), and factoring sums and differences of cubes (Section 6.4). Multiply or divide as indicated. Write each answer in lowest terms.

57. $\dfrac{x+5}{x+10} \div \left(\dfrac{x^2+10x+25}{x^2+10x} \cdot \dfrac{10x}{x^2+15x+50} \right)$

58. $\dfrac{m-8}{m-4} \div \left(\dfrac{m^2-12m+32}{8m} \cdot \dfrac{m^2-8m}{m^2-8m+16} \right)$

59. $\dfrac{3a-3b-a^2+b^2}{4a^2-4ab+b^2} \cdot \dfrac{4a^2-b^2}{2a^2-ab-b^2}$

464 CHAPTER 7 Rational Expressions and Applications

60. $\dfrac{4r^2 - t^2 + 10r - 5t}{2r^2 + rt + 5r} \cdot \dfrac{4r^3 + 4r^2t + rt^2}{2r + t}$

61. $\dfrac{-x^3 - y^3}{x^2 - 2xy + y^2} \div \dfrac{3y^2 - 3xy}{x^2 - y^2}$

62. $\dfrac{b^3 - 8a^3}{4a^3 + 4a^2b + ab^2} \div \dfrac{4a^2 + 2ab + b^2}{-a^3 - ab^3}$

63. If the rational expression $\dfrac{5x^2y^3}{2pq}$ represents the area of a rectangle and $\dfrac{2xy}{p}$ represents the length, what rational expression represents the width?

Width

Length = $\dfrac{2xy}{p}$

The area is $\dfrac{5x^2y^3}{2pq}$.

64. If you are given the problem $\dfrac{4y + 12}{2y - 10} \div \dfrac{?}{y^2 - y - 20} = \dfrac{2(y + 4)}{y - 3}$, what must be the polynomial that is represented by the question mark?

7.3 Least Common Denominators

OBJECTIVES

1. Find the least common denominator for a group of fractions.
2. Rewrite rational expressions with given denominators.

OBJECTIVE 1 Find the least common denominator for a group of fractions. Adding or subtracting rational expressions (to be discussed in the next section) often requires a **least common denominator (LCD)**, the simplest expression that is divisible by all denominators. For example, the least common denominator for the fractions $\frac{2}{9}$ and $\frac{5}{12}$ is 36 because 36 is the smallest positive number divisible by both 9 and 12.

We can often find least common denominators by inspection. For example, the LCD for $\frac{1}{6}$ and $\frac{2}{3m}$ is $6m$. In other cases, we find the LCD by a procedure similar to that used in Chapter 6 for finding the greatest common factor.

Finding the Least Common Denominator (LCD)

Step 1 **Factor** each denominator into prime factors.

Step 2 **List each different denominator factor** the *greatest* number of times it appears in any of the denominators.

Step 3 **Multiply** the denominator factors from Step 2 to get the LCD.

When each denominator is factored into prime factors, every prime factor must be a factor of the least common denominator.

In Example 1, we find the LCD for both numerical and algebraic denominators.

EXAMPLE 1 Finding the LCD

Find the LCD for each pair of fractions.

(a) $\dfrac{1}{24}, \dfrac{7}{15}$

(b) $\dfrac{1}{8x}, \dfrac{3}{10x}$

Step 1 Write each denominator in factored form with numerical coefficients in prime factored form.

$$24 = 2 \cdot 2 \cdot 2 \cdot 3 = 2^3 \cdot 3$$
$$15 = 3 \cdot 5$$

$$8x = 2 \cdot 2 \cdot 2 \cdot x = 2^3 \cdot x$$
$$10x = 2 \cdot 5 \cdot x$$

Step 2 Find the LCD by taking each different factor the *greatest* number of times it appears as a factor in any of the denominators.

The factor 2 appears three times in one product and not at all in the other, so the greatest number of times 2 appears is three. The greatest number of times both 3 and 5 appear is one.

Here 2 appears three times in one product and once in the other, so the greatest number of times 2 appears is three. The greatest number of times 5 appears is one, and the greatest number of times x appears in either product is one.

Step 3
$$\text{LCD} = 2 \cdot 2 \cdot 2 \cdot 3 \cdot 5$$
$$= 2^3 \cdot 3 \cdot 5$$
$$= 120$$

$$\text{LCD} = 2 \cdot 2 \cdot 2 \cdot 5 \cdot x$$
$$= 2^3 \cdot 5 \cdot x$$
$$= 40x$$

Now Try Exercises 5 and 11.

EXAMPLE 2 Finding the LCD

Find the LCD for $\dfrac{5}{6r^2}$ and $\dfrac{3}{4r^3}$.

Step 1 Factor each denominator.
$$6r^2 = 2 \cdot 3 \cdot r^2$$
$$4r^3 = 2 \cdot 2 \cdot r^3 = 2^2 \cdot r^3$$

Step 2 The greatest number of times 2 appears is two, the greatest number of times 3 appears is one, and the greatest number of times r appears is three; therefore,

Step 3 $\quad\quad\quad\quad\quad\quad \text{LCD} = 2^2 \cdot 3 \cdot r^3 = 12r^3.$

Now Try Exercise 13.

EXAMPLE 3 Finding the LCD

Find each LCD.

(a) $\dfrac{6}{5m}, \dfrac{4}{m^2 - 3m}$

Factor each denominator.
$$5m = 5 \cdot m$$
$$m^2 - 3m = m(m - 3)$$

Use each different factor the greatest number of times it appears.
$$\text{LCD} = 5 \cdot m \cdot (m - 3) = 5m(m - 3)$$

Because m is not a *factor* of $m - 3$, both factors, m and $m - 3$, must appear in the LCD.

466 CHAPTER 7 Rational Expressions and Applications

(b) $\dfrac{1}{r^2 - 4r - 5}, \dfrac{3}{r^2 - r - 20}, \dfrac{1}{r^2 - 10r + 25}$

Factor each denominator.

$$r^2 - 4r - 5 = (r - 5)(r + 1)$$
$$r^2 - r - 20 = (r - 5)(r + 4)$$
$$r^2 - 10r + 25 = (r - 5)^2$$

Use each different factor the greatest number of times it appears as a factor. The LCD is

$$(r - 5)^2(r + 1)(r + 4).$$

(c) $\dfrac{1}{q - 5}, \dfrac{3}{5 - q}$

The expressions $q - 5$ and $5 - q$ are opposites of each other because

$$-(q - 5) = -q + 5 = 5 - q.$$

Therefore, either $q - 5$ or $5 - q$ can be used as the LCD.

Now Try Exercises 19, 35, and 45.

OBJECTIVE 2 Rewrite rational expressions with given denominators. Once the LCD has been found, the next step in preparing to add or subtract two rational expressions is to use the fundamental property to write equivalent rational expressions. The next example shows how to do this with both numerical and algebraic fractions.

EXAMPLE 4 Writing Rational Expressions with Given Denominators

Rewrite each rational expression with the indicated denominator.

(a) $\dfrac{3}{8} = \dfrac{15}{40}$ **(b)** $\dfrac{9k}{25} = \dfrac{18k^2}{50k}$

For each example, first factor the denominator on the right. Then compare the denominator on the left with the one on the right to decide what factors are missing. (It may sometimes be necessary to factor both denominators.)

$$\dfrac{3}{8} = \dfrac{}{5 \cdot 8} \qquad\qquad \dfrac{9k}{25} = \dfrac{}{25 \cdot 2k}$$

A factor of 5 is missing. Using the fundamental property, multiply $\dfrac{3}{8}$ by $\dfrac{5}{5}$.

Factors of 2 and k are missing. Multiply by $\dfrac{2k}{2k}$.

$$\dfrac{3}{8} = \dfrac{3}{8} \cdot \dfrac{5}{5} = \dfrac{15}{40} \qquad\qquad \dfrac{9k}{25} = \dfrac{9k}{25} \cdot \dfrac{2k}{2k} = \dfrac{18k^2}{50k}$$

$$\dfrac{5}{5} = 1 \qquad\qquad \dfrac{2k}{2k} = 1$$

Now Try Exercises 57 and 59.

EXAMPLE 5 Writing Rational Expressions with Given Denominators

Rewrite each rational expression with the indicated denominator.

(a) $\dfrac{8}{3x + 1} = \dfrac{}{12x + 4}$

Factor the denominator on the right.

$$\frac{8}{3x+1} = \frac{}{4(3x+1)} \qquad \text{Factor.}$$

The missing factor is 4, so multiply the fraction on the left by $\frac{4}{4}$.

$$\frac{8}{3x+1} \cdot \frac{4}{4} = \frac{32}{12x+4} \qquad \text{Fundamental property}$$

(b) $\dfrac{12p}{p^2+8p} = \dfrac{}{p^3+4p^2-32p}$

Factor $p^2 + 8p$ as $p(p+8)$. Compare with the denominator on the right, which factors as $p(p+8)(p-4)$. The factor $p-4$ is missing, so multiply $\dfrac{12p}{p(p+8)}$ by $\dfrac{p-4}{p-4}$.

$$\frac{12p}{p^2+8p} = \frac{12p}{p(p+8)} \cdot \frac{p-4}{p-4} \qquad \text{Fundamental property}$$

$$= \frac{12p(p-4)}{p(p+8)(p-4)} \qquad \begin{array}{l}\text{Multiply numerators.}\\ \text{Multiply denominators.}\end{array}$$

$$= \frac{12p^2-48p}{p^3+4p^2-32p} \qquad \text{Multiply the factors.}$$

Now Try Exercises 63 and 67.

NOTE In the next section we add and subtract rational expressions, which sometimes requires the steps illustrated in Examples 4 and 5. While it is beneficial to leave the denominator in factored form, we multiplied the factors in the denominator in Example 5 to give the answer in the same form as the original problem.

7.3 EXERCISES

For Extra Help

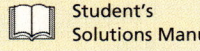
Student's Solutions Manual

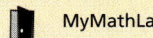
MyMathLab

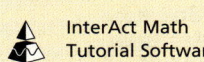
InterAct Math Tutorial Software

AW Math Tutor Center

MathXL

Digital Video Tutor CD 14/Videotape 12

Choose the correct response in Exercises 1–4.

1. Suppose that the greatest common factor of a and b is 1. Then the least common denominator for $\frac{1}{a}$ and $\frac{1}{b}$ is

    A. a **B.** b **C.** ab **D.** 1.

2. If a is a factor of b, then the least common denominator for $\frac{1}{a}$ and $\frac{1}{b}$ is

    A. a **B.** b **C.** ab **D.** 1.

3. The least common denominator for $\frac{11}{20}$ and $\frac{1}{2}$ is

    A. 40   **B.** 2   **C.** 20   **D.** none of these.

4. Suppose that we wish to write the fraction $\dfrac{1}{(x-4)^2(y-3)}$ with denominator $(x-4)^3(y-3)^2$. We must multiply both the numerator and the denominator by

    A. $(x-4)(y-3)$ **B.** $(x-4)^2$ **C.** $x-4$ **D.** $(x-4)^2(y-3)$.

Find the LCD for the fractions in each list. See Examples 1–3.

5. $\dfrac{7}{15}, \dfrac{21}{20}$

6. $\dfrac{9}{10}, \dfrac{12}{25}$

7. $\dfrac{17}{100}, \dfrac{23}{120}, \dfrac{43}{180}$

8. $\dfrac{17}{250}, \dfrac{-21}{300}, \dfrac{127}{360}$

9. $\dfrac{9}{x^2}, \dfrac{8}{x^5}$

10. $\dfrac{12}{m^7}, \dfrac{13}{m^8}$

11. $\dfrac{-2}{5p}, \dfrac{15}{6p}$

12. $\dfrac{14}{15k}, \dfrac{9}{4k}$

13. $\dfrac{17}{15y^2}, \dfrac{55}{36y^4}$

14. $\dfrac{4}{25m^3}, \dfrac{-9}{10m^4}$

15. $\dfrac{6}{21r^3}, \dfrac{8}{12r^5}$

16. $\dfrac{9}{35t^2}, \dfrac{5}{49t^6}$

17. $\dfrac{13}{5a^2b^3}, \dfrac{29}{15a^5b}$

18. $\dfrac{-7}{3r^4s^5}, \dfrac{-22}{9r^6s^8}$

19. $\dfrac{7}{6p}, \dfrac{15}{4p-8}$

20. $\dfrac{7}{8k}, \dfrac{-23}{12k-24}$

21. $\dfrac{9}{28m^2}, \dfrac{3}{12m-20}$

22. $\dfrac{15}{27a^3}, \dfrac{8}{9a-45}$

23. $\dfrac{7}{5b-10}, \dfrac{11}{6b-12}$

24. $\dfrac{3}{7x^2+21x}, \dfrac{1}{5x^2+15x}$

RELATING CONCEPTS (EXERCISES 25–28)

For Individual or Group Work

Suppose we want to find the LCD for the two common fractions

$$\dfrac{1}{24} \quad \text{and} \quad \dfrac{1}{20}.$$

In their prime factored forms, the denominators are

$$24 = 2^3 \cdot 3 \quad \text{and} \quad 20 = 2^2 \cdot 5.$$

Refer to this information as necessary, and **work Exercises 25–28 in order.**

25. What is the prime factored form of the LCD of the two fractions?

26. Suppose that two algebraic fractions have denominators $(t+4)^3(t-3)$ and $(t+4)^2(t+8)$. What is the factored form of the LCD of these?

27. What is the similarity between your answers in Exercises 25 and 26?

28. Comment on the following statement: The method for finding the LCD for two algebraic fractions is the same as the method for finding the LCD for two common fractions.

Find the LCD for the fractions in each list. See Examples 1–3.

29. $\dfrac{37}{6r-12}, \dfrac{25}{9r-18}$

30. $\dfrac{-14}{5p-30}, \dfrac{5}{6p-36}$

31. $\dfrac{5}{12p + 60}, \dfrac{17}{p^2 + 5p}, \dfrac{16}{p^2 + 10p + 25}$

32. $\dfrac{13}{r^2 + 7r}, \dfrac{-3}{5r + 35}, \dfrac{-7}{r^2 + 14r + 49}$

33. $\dfrac{3}{8y + 16}, \dfrac{22}{y^2 + 3y + 2}$

34. $\dfrac{-2}{9m - 18}, \dfrac{-9}{m^2 - 7m + 10}$

35. $\dfrac{5}{c - d}, \dfrac{8}{d - c}$

36. $\dfrac{4}{y - x}, \dfrac{7}{x - y}$

37. $\dfrac{12}{m - 3}, \dfrac{-4}{3 - m}$

38. $\dfrac{-17}{8 - a}, \dfrac{2}{a - 8}$

39. $\dfrac{29}{p - q}, \dfrac{18}{q - p}$

40. $\dfrac{16}{z - x}, \dfrac{8}{x - z}$

41. $\dfrac{3}{k^2 + 5k}, \dfrac{2}{k^2 + 3k - 10}$

42. $\dfrac{1}{z^2 - 4z}, \dfrac{4}{z^2 - 3z - 4}$

43. $\dfrac{6}{a^2 + 6a}, \dfrac{-5}{a^2 + 3a - 18}$

44. $\dfrac{8}{y^2 - 5y}, \dfrac{-2}{y^2 - 2y - 15}$

45. $\dfrac{5}{p^2 + 8p + 15}, \dfrac{3}{p^2 - 3p - 18}, \dfrac{2}{p^2 - p - 30}$

46. $\dfrac{10}{y^2 - 10y + 21}, \dfrac{2}{y^2 - 2y - 3}, \dfrac{5}{y^2 - 6y - 7}$

47. $\dfrac{-5}{k^2 + 2k - 35}, \dfrac{-8}{k^2 + 3k - 40}, \dfrac{9}{k^2 - 2k - 15}$

48. $\dfrac{19}{z^2 + 4z - 12}, \dfrac{16}{z^2 + z - 30}, \dfrac{6}{z^2 + 2z - 24}$

49. Suppose that $(2x - 5)^2$ is the LCD for two fractions. Is $(5 - 2x)^2$ also acceptable as an LCD? Why or why not?

50. Suppose that $(4t - 3)(5t - 6)$ is the LCD for two fractions. Is $(3 - 4t)(6 - 5t)$ also acceptable as an LCD? Why or why not?

RELATING CONCEPTS (EXERCISES 51–56)

For Individual or Group Work

Work Exercises 51–56 in order.

51. Suppose that you want to write $\frac{3}{4}$ as an equivalent fraction with denominator 28. By what number must you multiply both the numerator and the denominator?

52. If you write $\frac{3}{4}$ as an equivalent fraction with denominator 28, by what number are you actually multiplying the fraction?

53. What property of multiplication is being used when we write a common fraction as an equivalent one with a larger denominator? (See Section 1.7.)

54. Suppose that you want to write $\frac{2x + 5}{x - 4}$ as an equivalent fraction with denominator $7x - 28$. By what number must you multiply both the numerator and the denominator?

55. If you write $\frac{2x + 5}{x - 4}$ as an equivalent fraction with denominator $7x - 28$, by what number are you actually multiplying the fraction?

56. Repeat Exercise 53, changing "a common" to "an algebraic."

470　CHAPTER 7　Rational Expressions and Applications

Rewrite each rational expression with the indicated denominator. See Examples 4 and 5.

57. $\dfrac{4}{11} = \dfrac{20}{55}$

58. $\dfrac{6}{7} = \dfrac{}{42}$

59. $\dfrac{-5}{k} = \dfrac{}{9k}$

60. $\dfrac{-3}{q} = \dfrac{}{6q}$

61. $\dfrac{15m^2}{8k} = \dfrac{}{32k^4}$

62. $\dfrac{5t^2}{3y} = \dfrac{}{9y^2}$

63. $\dfrac{19z}{2z-6} = \dfrac{57z}{6z-18}$

64. $\dfrac{2r}{5r-5} = \dfrac{}{15r-15}$

65. $\dfrac{-2a}{9a-18} = \dfrac{-4a}{18a-36}$

66. $\dfrac{-5y}{6y+18} = \dfrac{}{24y+72}$

67. $\dfrac{6}{k^2-4k} = \dfrac{6(k+1)}{k(k-4)(k+1)}$

68. $\dfrac{15}{m^2-9m} = \dfrac{}{m(m-9)(m+8)}$

69. $\dfrac{36r}{r^2-r-6} = \dfrac{36r(r+1)}{(r-3)(r+2)(r+1)}$

70. $\dfrac{4m}{m^2-8m+15} = \dfrac{}{(m-5)(m-3)(m+2)}$

71. $\dfrac{a+2b}{2a^2+ab-b^2} = \dfrac{}{2a^3b+a^2b^2-ab^3}$

72. $\dfrac{m-4}{6m^2+7m-3} = \dfrac{}{12m^3+14m^2-6m}$

73. $\dfrac{4r-t}{r^2+rt+t^2} = \dfrac{}{t^3-r^3}$

74. $\dfrac{3x-1}{x^2+2x+4} = \dfrac{}{x^3-8}$

75. $\dfrac{2(z-y)}{y^2+yz+z^2} = \dfrac{}{y^4-z^3y}$

76. $\dfrac{2p+3q}{p^2+2pq+q^2} = \dfrac{}{(p+q)(p^3+q^3)}$

✍ **77.** Write an explanation of how to find the least common denominator for a group of denominators. Give an example.

✍ **78.** Write an explanation of how to write a rational expression as an equivalent rational expression with a given denominator. Give an example.

7.4　Adding and Subtracting Rational Expressions

OBJECTIVES

1. Add rational expressions having the same denominator.
2. Add rational expressions having different denominators.

To add and subtract rational expressions, we need the skills developed in the previous section to find least common denominators and to write equivalent fractions with the LCD.

OBJECTIVE 1　Add rational expressions having the same denominator. We find the sum of two rational expressions with the same procedure that we used for adding two fractions in Section 1.1.

SECTION 7.4 Adding and Subtracting Rational Expressions **471**

3 Subtract rational expressions.

Adding Rational Expressions

If $\frac{P}{Q}$ and $\frac{R}{Q}$ ($Q \neq 0$) are rational expressions, then

$$\frac{P}{Q} + \frac{R}{Q} = \frac{P + R}{Q}.$$

That is, to add rational expressions with the same denominator, add the numerators and keep the same denominator.

The first example shows how addition of rational expressions compares with that of rational numbers.

EXAMPLE 1 Adding Rational Expressions with the Same Denominator

Add. Write each answer in lowest terms.

(a) $\dfrac{4}{9} + \dfrac{2}{9}$ (b) $\dfrac{3x}{x+1} + \dfrac{3}{x+1}$

The denominators are the same, so the sum is found by adding the two numerators and keeping the same (common) denominator.

$$\frac{4}{9} + \frac{2}{9} = \frac{4 + 2}{9}$$
$$= \frac{6}{9}$$
$$= \frac{2}{3}$$

$$\frac{3x}{x+1} + \frac{3}{x+1} = \frac{3x + 3}{x+1}$$
$$= \frac{3(x+1)}{x+1}$$
$$= 3$$

> **Now Try Exercises 9 and 19.**

OBJECTIVE 2 Add rational expressions having different denominators. We use the following steps to add two rational expressions with different denominators. These are the same steps we used to add fractions with different denominators in Section 1.1.

Adding with Different Denominators

Step 1 **Find the least common denominator (LCD).**

Step 2 **Rewrite each rational expression** as an equivalent rational expression with the LCD as the denominator.

Step 3 **Add** the numerators to get the numerator of the sum. The LCD is the denominator of the sum.

Step 4 **Write in lowest terms** using the fundamental property.

EXAMPLE 2 Adding Rational Expressions with Different Denominators

Add. Write each answer in lowest terms.

(a) $\dfrac{1}{12} + \dfrac{7}{15}$ (b) $\dfrac{2}{3y} + \dfrac{1}{4y}$

Step 1 First find the LCD using the methods of the previous section.

$$\text{LCD} = 2^2 \cdot 3 \cdot 5 = 60 \qquad\qquad \text{LCD} = 2^2 \cdot 3 \cdot y = 12y$$

Step 2 Now rewrite each rational expression as a fraction with the LCD (either 60 or 12y) as the denominator.

$$\dfrac{1}{12} + \dfrac{7}{15} = \dfrac{1(5)}{12(5)} + \dfrac{7(4)}{15(4)} \qquad\qquad \dfrac{2}{3y} + \dfrac{1}{4y} = \dfrac{2(4)}{3y(4)} + \dfrac{1(3)}{4y(3)}$$

$$= \dfrac{5}{60} + \dfrac{28}{60} \qquad\qquad\qquad\qquad = \dfrac{8}{12y} + \dfrac{3}{12y}$$

Step 3 Since the fractions now have common denominators, add the numerators.

Step 4 Write in lowest terms if necessary.

$$\dfrac{5}{60} + \dfrac{28}{60} = \dfrac{5 + 28}{60} \qquad\qquad \dfrac{8}{12y} + \dfrac{3}{12y} = \dfrac{8 + 3}{12y}$$

$$= \dfrac{33}{60} = \dfrac{11}{20} \qquad\qquad\qquad = \dfrac{11}{12y}$$

Now Try Exercises 25 and 31.

EXAMPLE 3 Adding Rational Expressions

Add. Write the answer in lowest terms.

$$\dfrac{2x}{x^2 - 1} + \dfrac{-1}{x + 1}$$

Step 1 Since the denominators are different, find the LCD.

$$x^2 - 1 = (x + 1)(x - 1)$$

$$x + 1 \text{ is prime.}$$

The LCD is $(x + 1)(x - 1)$.

Step 2 Rewrite each rational expression as a fraction with common denominator $(x + 1)(x - 1)$.

$$\dfrac{2x}{x^2 - 1} + \dfrac{-1}{x + 1} = \dfrac{2x}{(x + 1)(x - 1)} + \dfrac{-1(x - 1)}{(x + 1)(x - 1)} \qquad \text{Multiply the second fraction by } \dfrac{x - 1}{x - 1}.$$

$$= \dfrac{2x}{(x + 1)(x - 1)} + \dfrac{-x + 1}{(x + 1)(x - 1)} \qquad \text{Distributive property}$$

Step 3 $\qquad\qquad = \dfrac{2x - x + 1}{(x + 1)(x - 1)} \qquad\qquad \text{Add numerators; keep the same denominator.}$

$$= \dfrac{x + 1}{(x + 1)(x - 1)} \qquad\qquad \text{Combine like terms.}$$

Step 4 $\quad = \dfrac{1(x+1)}{(x+1)(x-1)}$ — Identity property of multiplication

$\quad = \dfrac{1}{x-1}$ — Fundamental property

Now Try Exercise 41.

EXAMPLE 4 Adding Rational Expressions

Add. Write the answer in lowest terms.

$$\dfrac{2x}{x^2+5x+6} + \dfrac{x+1}{x^2+2x-3}$$

$= \dfrac{2x}{(x+2)(x+3)} + \dfrac{x+1}{(x+3)(x-1)}$ — Factor the denominators.

The LCD is $(x+2)(x+3)(x-1)$. Use the fundamental property.

$= \dfrac{2x(x-1)}{(x+2)(x+3)(x-1)} + \dfrac{(x+1)(x+2)}{(x+2)(x+3)(x-1)}$

$= \dfrac{2x(x-1)+(x+1)(x+2)}{(x+2)(x+3)(x-1)}$ — Add numerators; keep the same denominator.

$= \dfrac{2x^2-2x+x^2+3x+2}{(x+2)(x+3)(x-1)}$ — Multiply.

$= \dfrac{3x^2+x+2}{(x+2)(x+3)(x-1)}$ — Combine like terms.

It is usually more convenient to leave the denominator in factored form. The numerator cannot be factored here, so the expression is in lowest terms.

Now Try Exercise 43.

Rational expressions to be added or subtracted may have denominators that are opposites of each other.

EXAMPLE 5 Adding Rational Expressions with Denominators That Are Opposites

Add. Write the answer in lowest terms.

$$\dfrac{y}{y-2} + \dfrac{8}{2-y}$$

One way to get a common denominator is to multiply the second expression by -1 in both the numerator and the denominator, giving $y-2$ as a common denominator.

$\dfrac{y}{y-2} + \dfrac{8}{2-y} = \dfrac{y}{y-2} + \dfrac{8(-1)}{(2-y)(-1)}$ — Fundamental property

$= \dfrac{y}{y-2} + \dfrac{-8}{y-2}$ — Distributive property

$= \dfrac{y-8}{y-2}$ — Add numerators; keep the same denominator.

474 CHAPTER 7 Rational Expressions and Applications

If we had chosen to use $2 - y$ as the common denominator, the final answer would be $\frac{8-y}{2-y}$, which is equivalent to $\frac{y-8}{y-2}$.

Now Try Exercise 51.

OBJECTIVE 3 Subtract rational expressions. To subtract rational expressions, use the following rule.

> **Subtracting Rational Expressions**
>
> If $\frac{P}{Q}$ and $\frac{R}{Q}$ ($Q \neq 0$) are rational expressions, then
>
> $$\frac{P}{Q} - \frac{R}{Q} = \frac{P-R}{Q}.$$
>
> That is, to subtract rational expressions with the same denominator, subtract the numerators and keep the same denominator.

EXAMPLE 6 Subtracting Rational Expressions with the Same Denominator

Subtract. Write the answer in lowest terms.

$$\frac{2m}{m-1} - \frac{m+3}{m-1} = \frac{2m-(m+3)}{m-1}$$

Use parentheses around the quantity being subtracted.
Subtract numerators; keep the same denominator.

$$= \frac{2m-m-3}{m-1}$$

Distributive property

$$= \frac{m-3}{m-1}$$

Combine like terms.

Now Try Exercise 15.

CAUTION Sign errors often occur in subtraction problems like the one in Example 6. Remember that the numerator of the fraction being subtracted must be treated as a single quantity. Be sure to use parentheses after the subtraction sign.

EXAMPLE 7 Subtracting Rational Expressions with Different Denominators

Subtract. Write the answer in lowest terms.

$$\frac{9}{x-2} - \frac{3}{x} = \frac{9x}{x(x-2)} - \frac{3(x-2)}{x(x-2)}$$

The LCD is $x(x-2)$.

$$= \frac{9x - 3(x-2)}{x(x-2)}$$

Subtract numerators; keep the same denominator.

$$= \frac{9x - 3x + 6}{x(x - 2)}$$ Distributive property; Be careful with signs.

$$= \frac{6x + 6}{x(x - 2)}$$ Combine like terms.

$$= \frac{6(x + 1)}{x(x - 2)}$$ Factor the numerator.

Now Try Exercise 39.

NOTE We factored the final numerator in Example 7 to get $\frac{6(x + 1)}{x(x - 2)}$; however, the fundamental property does not apply since there are no common factors that allow us to write the answer in lower terms.

EXAMPLE 8 Subtracting Rational Expressions with Denominators That Are Opposites

Subtract. Write the answer in lowest terms.

$$\frac{3x}{x - 5} - \frac{2x - 25}{5 - x}$$

The denominators are opposites, so either may be used as the common denominator. We choose $x - 5$.

$$\frac{3x}{x - 5} - \frac{2x - 25}{5 - x} = \frac{3x}{x - 5} - \frac{(2x - 25)(-1)}{(5 - x)(-1)}$$ Fundamental property

$$= \frac{3x}{x - 5} - \frac{-2x + 25}{x - 5}$$ Multiply.

$$= \frac{3x - (-2x + 25)}{x - 5}$$ Subtract numerators; use parentheses.

$$= \frac{3x + 2x - 25}{x - 5}$$ Distributive property; Be careful with signs.

$$= \frac{5x - 25}{x - 5}$$ Combine like terms.

$$= \frac{5(x - 5)}{x - 5}$$ Factor.

$$= 5$$ Lowest terms

Now Try Exercise 53.

EXAMPLE 9 Subtracting Rational Expressions

Subtract. Write the answer in lowest terms.

$$\frac{6x}{x^2 - 2x + 1} - \frac{1}{x^2 - 1}$$

476 CHAPTER 7 Rational Expressions and Applications

We begin by factoring the denominators.
$$x^2 - 2x + 1 = (x - 1)(x - 1) \quad \text{and} \quad x^2 - 1 = (x - 1)(x + 1)$$

From the factored denominators, we identify the LCD, $(x - 1)(x - 1)(x + 1)$. We use the factor $x - 1$ twice, since it appears twice in the first denominator.

$$\frac{6x}{(x-1)(x-1)} - \frac{1}{(x-1)(x+1)}$$

$$= \frac{6x(x+1)}{(x-1)(x-1)(x+1)} - \frac{1(x-1)}{(x-1)(x-1)(x+1)} \quad \text{Fundamental property}$$

$$= \frac{6x(x+1) - 1(x-1)}{(x-1)(x-1)(x+1)} \quad \text{Subtract numerators.}$$

$$= \frac{6x^2 + 6x - x + 1}{(x-1)(x-1)(x+1)} \quad \text{Distributive property}$$

$$= \frac{6x^2 + 5x + 1}{(x-1)(x-1)(x+1)} \quad \text{Combine like terms.}$$

$$= \frac{(2x+1)(3x+1)}{(x-1)^2(x+1)} \quad \text{Factor the numerator.}$$

Verify that the final expression is in lowest terms.

Now Try Exercise 63.

7.4 EXERCISES

For Extra Help

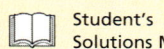 Student's Solutions Manual

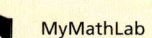 MyMathLab

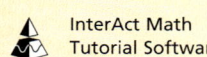 InterAct Math Tutorial Software

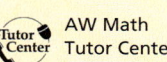 AW Math Tutor Center

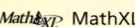 MathXL

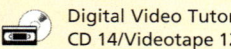 Digital Video Tutor CD 14/Videotape 12

Match the problem in Column I with the correct sum or difference in Column II.

I	II
1. $\dfrac{x}{x+6} + \dfrac{6}{x+6}$	**A.** 2
2. $\dfrac{2x}{x-6} - \dfrac{12}{x-6}$	**B.** $\dfrac{x-6}{x+6}$
3. $\dfrac{6}{x-6} - \dfrac{x}{x-6}$	**C.** -1
4. $\dfrac{6}{x+6} - \dfrac{x}{x+6}$	**D.** $\dfrac{6+x}{6x}$
5. $\dfrac{x}{x+6} - \dfrac{6}{x+6}$	**E.** 1
6. $\dfrac{1}{x} + \dfrac{1}{6}$	**F.** 0
7. $\dfrac{1}{6} - \dfrac{1}{x}$	**G.** $\dfrac{x-6}{6x}$
8. $\dfrac{1}{6x} - \dfrac{1}{6x}$	**H.** $\dfrac{6-x}{x+6}$

SECTION 7.4 Adding and Subtracting Rational Expressions **477**

Note: When adding and subtracting rational expressions, several different equivalent forms of the answer often exist. If your answer does not look exactly like the one given in the back of the book, check to see whether you have written an equivalent form. (See Example 7 of Section 7.1.)

Add or subtract. Write each answer in lowest terms. See Examples 1 and 6.

9. $\dfrac{4}{m} + \dfrac{7}{m}$ 10. $\dfrac{5}{p} + \dfrac{11}{p}$ 11. $\dfrac{5}{y+4} - \dfrac{1}{y+4}$

12. $\dfrac{4}{y+3} - \dfrac{1}{y+3}$ 13. $\dfrac{x}{x+y} + \dfrac{y}{x+y}$ 14. $\dfrac{a}{a+b} + \dfrac{b}{a+b}$

15. $\dfrac{5m}{m+1} - \dfrac{1+4m}{m+1}$ 16. $\dfrac{4x}{x+2} - \dfrac{2+3x}{x+2}$ 17. $\dfrac{a+b}{2} - \dfrac{a-b}{2}$

18. $\dfrac{x-y}{2} - \dfrac{x+y}{2}$ 19. $\dfrac{x^2}{x+5} + \dfrac{5x}{x+5}$ 20. $\dfrac{t^2}{t-3} + \dfrac{-3t}{t-3}$

21. $\dfrac{y^2 - 3y}{y+3} + \dfrac{-18}{y+3}$ 22. $\dfrac{r^2 - 8r}{r-5} + \dfrac{15}{r-5}$

23. Explain with an example how to add or subtract rational expressions with the same denominator.

24. Explain with an example how to add or subtract rational expressions with different denominators.

Add or subtract. Write each answer in lowest terms. See Examples 2, 3, 4, and 7.

25. $\dfrac{z}{5} + \dfrac{1}{3}$ 26. $\dfrac{p}{8} + \dfrac{3}{5}$ 27. $\dfrac{5}{7} - \dfrac{r}{2}$

28. $\dfrac{10}{9} - \dfrac{z}{3}$ 29. $-\dfrac{3}{4} - \dfrac{1}{2x}$ 30. $-\dfrac{5}{8} - \dfrac{3}{2a}$

31. $\dfrac{6}{5x} + \dfrac{9}{2x}$ 32. $\dfrac{3}{2x} + \dfrac{4}{7x}$ 33. $\dfrac{x+1}{6} + \dfrac{3x+3}{9}$

34. $\dfrac{2x-6}{4} + \dfrac{x+5}{6}$ 35. $\dfrac{x+3}{3x} + \dfrac{2x+2}{4x}$ 36. $\dfrac{x+2}{5x} + \dfrac{6x+3}{3x}$

37. $\dfrac{7}{3p^2} - \dfrac{2}{p}$ 38. $\dfrac{12}{5m^2} - \dfrac{5}{m}$ 39. $\dfrac{1}{k+4} - \dfrac{2}{k}$

40. $\dfrac{3}{m+1} - \dfrac{4}{m}$ 41. $\dfrac{x}{x-2} + \dfrac{-8}{x^2-4}$ 42. $\dfrac{2x}{x-1} + \dfrac{-4}{x^2-1}$

43. $\dfrac{4m}{m^2+3m+2} + \dfrac{2m-1}{m^2+6m+5}$ 44. $\dfrac{a}{a^2+3a-4} + \dfrac{4a}{a^2+7a+12}$

45. $\dfrac{4y}{y^2-1} - \dfrac{5}{y^2+2y+1}$ 46. $\dfrac{2x}{x^2-16} - \dfrac{3}{x^2+8x+16}$

47. $\dfrac{t}{t+2} + \dfrac{5-t}{t} - \dfrac{4}{t^2+2t}$ 48. $\dfrac{2p}{p-3} + \dfrac{2+p}{p} - \dfrac{-6}{p^2-3p}$

49. What are the two possible LCDs that could be used for the sum $\dfrac{10}{m-2} + \dfrac{5}{2-m}$?

50. If one form of the correct answer to a sum or difference of rational expressions is $\dfrac{4}{k-3}$, what would an alternative form of the answer be if the denominator is $3-k$?

478 CHAPTER 7 Rational Expressions and Applications

Add or subtract. Write each answer in lowest terms. See Examples 5 and 8.

51. $\dfrac{4}{x-5} + \dfrac{6}{5-x}$ **52.** $\dfrac{10}{m-2} + \dfrac{5}{2-m}$ **53.** $\dfrac{-1}{1-y} - \dfrac{4y-3}{y-1}$

54. $\dfrac{-4}{p-3} - \dfrac{p+1}{3-p}$ **55.** $\dfrac{2}{x-y^2} + \dfrac{7}{y^2-x}$ **56.** $\dfrac{-8}{p-q^2} + \dfrac{3}{q^2-p}$

57. $\dfrac{x}{5x-3y} - \dfrac{y}{3y-5x}$ **58.** $\dfrac{t}{8t-9s} - \dfrac{s}{9s-8t}$ **59.** $\dfrac{3}{4p-5} + \dfrac{9}{5-4p}$

60. $\dfrac{8}{3-7y} - \dfrac{2}{7y-3}$

In these subtraction problems, the rational expression that follows the subtraction sign has a numerator with more than one term. Be very careful with signs and find each difference. See Example 9.

61. $\dfrac{2m}{m-n} - \dfrac{5m+n}{2m-2n}$ **62.** $\dfrac{5p}{p-q} - \dfrac{3p+1}{4p-4q}$

63. $\dfrac{5}{x^2-9} - \dfrac{x+2}{x^2+4x+3}$ **64.** $\dfrac{1}{a^2-1} - \dfrac{a-1}{a^2+3a-4}$

65. $\dfrac{2q+1}{3q^2+10q-8} - \dfrac{3q+5}{2q^2+5q-12}$ **66.** $\dfrac{4y-1}{2y^2+5y-3} - \dfrac{y+3}{6y^2+y-2}$

Perform each indicated operation. See Examples 1–9.

67. $\dfrac{4}{r^2-r} + \dfrac{6}{r^2+2r} - \dfrac{1}{r^2+r-2}$ **68.** $\dfrac{6}{k^2+3k} - \dfrac{1}{k^2-k} + \dfrac{2}{k^2+2k-3}$

69. $\dfrac{x+3y}{x^2+2xy+y^2} + \dfrac{x-y}{x^2+4xy+3y^2}$ **70.** $\dfrac{m}{m^2-1} + \dfrac{m-1}{m^2+2m+1}$

71. $\dfrac{r+y}{18r^2+12ry-3ry-2y^2} + \dfrac{3r-y}{36r^2-y^2}$ **72.** $\dfrac{2x-z}{2x^2-4xz+5xz-10z^2} - \dfrac{x+z}{x^2-4z^2}$

73. Refer to the rectangle in the figure.

   (a) Find an expression that represents its perimeter. Give the simplified form.

   (b) Find an expression that represents its area. Give the simplified form.

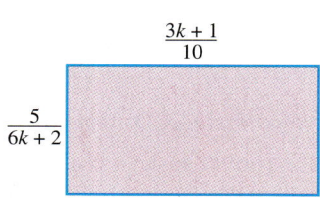

74. Refer to the triangle in the figure. Find an expression that represents its perimeter.

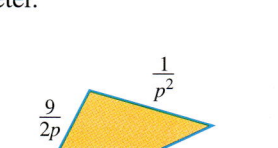

A *concours d'elegance* is *a competition in which a maximum of* 100 *points is awarded to a car based on its general attractiveness. The rational expression*

$$\frac{1010}{49(101-x)} - \frac{10}{49}$$

approximates the cost, in thousands of dollars, of restoring a car so that it will win x points.

Use this information to work Exercises 75 and 76.

75. Simplify the expression by performing the indicated subtraction.

76. Use the simplified expression from Exercise 75 to determine how much it would cost to win 95 points.

7.5 Complex Fractions

OBJECTIVES

1. Simplify a complex fraction by writing it as a division problem (Method 1).

2. Simplify a complex fraction by multiplying by the least common denominator (Method 2).

The quotient of two mixed numbers in arithmetic, such as $2\frac{1}{2} \div 3\frac{1}{4}$, can be written as a fraction:

$$2\frac{1}{2} \div 3\frac{1}{4} = \frac{2\frac{1}{2}}{3\frac{1}{4}} = \frac{2 + \frac{1}{2}}{3 + \frac{1}{4}}.$$

The last expression is the quotient of expressions that involve fractions. In algebra, some rational expressions also have fractions in the numerator, or denominator, or both.

Complex Fraction

A rational expression with one or more fractions in the numerator, denominator, or both, is called a **complex fraction.**

Examples of complex fractions include

$$\frac{2 + \frac{1}{2}}{3 + \frac{1}{4}}, \quad \frac{\frac{3x^2 - 5x}{6x^2}}{2x - \frac{1}{x}}, \quad \text{and} \quad \frac{3 + x}{5 - \frac{2}{x}}.$$

The parts of a complex fraction are named as follows.

$$\frac{\dfrac{2}{p} - \dfrac{1}{q}}{\dfrac{3}{p} + \dfrac{5}{q}}$$

← Numerator of complex fraction
← Main fraction bar
← Denominator of complex fraction

480 CHAPTER 7 Rational Expressions and Applications

OBJECTIVE 1 Simplify a complex fraction by writing it as a division problem (Method 1). Since the main fraction bar represents division in a complex fraction, one method of simplifying a complex fraction involves division.

Method 1

To simplify a complex fraction:

Step 1 Write both the numerator and denominator as single fractions.

Step 2 Change the complex fraction to a division problem.

Step 3 Perform the indicated division.

Once again, in this section the first example shows complex fractions from both arithmetic and algebra.

EXAMPLE 1 Simplifying Complex Fractions (Method 1)

Simplify each complex fraction.

(a) $\dfrac{\frac{2}{3}+\frac{5}{9}}{\frac{1}{4}+\frac{1}{12}}$ (b) $\dfrac{6+\frac{3}{x}}{\frac{x}{4}+\frac{1}{8}}$

Step 1 First, write each numerator as a single fraction.

$$\frac{2}{3}+\frac{5}{9}=\frac{2(3)}{3(3)}+\frac{5}{9} \qquad\qquad 6+\frac{3}{x}=\frac{6}{1}+\frac{3}{x}$$

$$=\frac{6}{9}+\frac{5}{9}=\frac{11}{9} \qquad\qquad =\frac{6x}{x}+\frac{3}{x}=\frac{6x+3}{x}$$

Do the same thing with each denominator.

$$\frac{1}{4}+\frac{1}{12}=\frac{1(3)}{4(3)}+\frac{1}{12} \qquad\qquad \frac{x}{4}+\frac{1}{8}=\frac{x(2)}{4(2)}+\frac{1}{8}$$

$$=\frac{3}{12}+\frac{1}{12}=\frac{4}{12} \qquad\qquad =\frac{2x}{8}+\frac{1}{8}=\frac{2x+1}{8}$$

Step 2 The original complex fraction can now be written as follows.

$$\dfrac{\frac{11}{9}}{\frac{4}{12}}=\frac{11}{9}\div\frac{4}{12} \qquad\qquad \dfrac{\frac{6x+3}{x}}{\frac{2x+1}{8}}=\frac{6x+3}{x}\div\frac{2x+1}{8}$$

Step 3 Now use the rule for division and the fundamental property.

Multiply by the reciprocal. Multiply by the reciprocal.

$$\frac{11}{9}\div\frac{4}{12}=\frac{11}{9}\cdot\frac{12}{4} \qquad\qquad \frac{6x+3}{x}\div\frac{2x+1}{8}=\frac{6x+3}{x}\cdot\frac{8}{2x+1}$$

$$= \frac{11 \cdot 3 \cdot 4}{3 \cdot 3 \cdot 4}$$

$$= \frac{11}{3}$$

$$= \frac{3(2x+1)}{x} \cdot \frac{8}{2x+1}$$

$$= \frac{24}{x}$$

Now Try Exercises 1 and 15.

EXAMPLE 2 Simplifying a Complex Fraction (Method 1)

Simplify the complex fraction.

$$\frac{\frac{xp}{q^3}}{\frac{p^2}{qx^2}} = \frac{xp}{q^3} \div \frac{p^2}{qx^2} = \frac{xp}{q^3} \cdot \frac{qx^2}{p^2} = \frac{x^3}{q^2 p}$$

Now Try Exercise 9.

EXAMPLE 3 Simplifying a Complex Fraction (Method 1)

Simplify the complex fraction.

$$\frac{\frac{3}{x+2} - 4}{\frac{2}{x+2} + 1} = \frac{\frac{3}{x+2} - \frac{4(x+2)}{x+2}}{\frac{2}{x+2} + \frac{1(x+2)}{x+2}} \quad \text{Write both second terms with a denominator of } x+2.$$

$$= \frac{\frac{3 - 4(x+2)}{x+2}}{\frac{2 + 1(x+2)}{x+2}} \quad \begin{array}{l}\text{Subtract in the numerator.} \\ \text{Add in the denominator.}\end{array}$$

$$= \frac{\frac{3 - 4x - 8}{x+2}}{\frac{2 + x + 2}{x+2}} \quad \text{Distributive property}$$

$$= \frac{\frac{-5 - 4x}{x+2}}{\frac{4 + x}{x+2}} \quad \text{Combine like terms.}$$

$$= \frac{-5 - 4x}{x+2} \cdot \frac{x+2}{4+x} \quad \text{Multiply by the reciprocal.}$$

$$= \frac{-5 - 4x}{4 + x} \quad \text{Lowest terms}$$

Now Try Exercise 33.

OBJECTIVE 2 Simplify a complex fraction by multiplying by the least common denominator (Method 2). Since any expression can be multiplied by a form of 1 to get an equivalent expression, we can multiply both the numerator and the denominator of a complex fraction by the same nonzero expression to get an equivalent complex fraction. If we choose the expression to be the LCD of all the fractions within the complex fraction, the complex fraction will be simplified. This is Method 2.

> **Method 2**
>
> To simplify a complex fraction:
>
> *Step 1* Find the LCD of all fractions within the complex fraction.
>
> *Step 2* Multiply both the numerator and the denominator of the complex fraction by this LCD using the distributive property as necessary. Write in lowest terms.

In the next example, Method 2 is used to simplify the complex fractions from Example 1.

EXAMPLE 4 Simplifying Complex Fractions (Method 2)

Simplify each complex fraction.

(a) $\dfrac{\dfrac{2}{3} + \dfrac{5}{9}}{\dfrac{1}{4} + \dfrac{1}{12}}$
(b) $\dfrac{6 + \dfrac{3}{x}}{\dfrac{x}{4} + \dfrac{1}{8}}$

Step 1 Find the LCD for all denominators in the complex fraction.

The LCD for 3, 9, 4, and 12 is 36. | The LCD for x, 4, and 8 is $8x$.

Step 2 Multiply numerator and denominator of the complex fraction by the LCD.

$$\dfrac{\dfrac{2}{3} + \dfrac{5}{9}}{\dfrac{1}{4} + \dfrac{1}{12}} = \dfrac{36\left(\dfrac{2}{3} + \dfrac{5}{9}\right)}{36\left(\dfrac{1}{4} + \dfrac{1}{12}\right)}$$

$$\dfrac{6 + \dfrac{3}{x}}{\dfrac{x}{4} + \dfrac{1}{8}} = \dfrac{8x\left(6 + \dfrac{3}{x}\right)}{8x\left(\dfrac{x}{4} + \dfrac{1}{8}\right)}$$

$$= \dfrac{36\left(\dfrac{2}{3}\right) + 36\left(\dfrac{5}{9}\right)}{36\left(\dfrac{1}{4}\right) + 36\left(\dfrac{1}{12}\right)}$$

$$= \dfrac{8x(6) + 8x\left(\dfrac{3}{x}\right)}{8x\left(\dfrac{x}{4}\right) + 8x\left(\dfrac{1}{8}\right)}$$ Distributive property

$$= \dfrac{24 + 20}{9 + 3}$$

$$= \dfrac{48x + 24}{2x^2 + x}$$

$$= \dfrac{44}{12} = \dfrac{4 \cdot 11}{4 \cdot 3}$$

$$= \dfrac{24(2x + 1)}{x(2x + 1)}$$ Factor.

$$= \dfrac{11}{3}$$

$$= \dfrac{24}{x}$$ Lowest terms

Now Try Exercises 2 and 19.

EXAMPLE 5 Simplifying a Complex Fraction (Method 2)

Simplify the complex fraction.

$$\frac{\dfrac{3}{5m} - \dfrac{2}{m^2}}{\dfrac{9}{2m} + \dfrac{3}{4m^2}} = \frac{20m^2\left(\dfrac{3}{5m} - \dfrac{2}{m^2}\right)}{20m^2\left(\dfrac{9}{2m} + \dfrac{3}{4m^2}\right)}$$

The LCD for $5m$, m^2, $2m$, and $4m^2$ is $20m^2$.

$$= \frac{20m^2\left(\dfrac{3}{5m}\right) - 20m^2\left(\dfrac{2}{m^2}\right)}{20m^2\left(\dfrac{9}{2m}\right) + 20m^2\left(\dfrac{3}{4m^2}\right)}$$

Distributive property

$$= \frac{12m - 40}{90m + 15}$$

Now Try Exercise 25.

Either of the two methods can be used to simplify a complex fraction. You may want to choose one method and stick with it to eliminate confusion. However, some students prefer to use Method 1 for problems like Example 2, which is the quotient of two fractions. They prefer Method 2 for problems like Examples 1, 3, 4, and 5, which have sums or differences in the numerators or denominators or both.

EXAMPLE 6 Deciding on a Method and Simplifying Complex Fractions

Simplify each complex fraction.

(a) $\dfrac{\dfrac{1}{y} + \dfrac{2}{y+2}}{\dfrac{4}{y} - \dfrac{3}{y+2}}$

Although either method will work, we use Method 2 here since there are sums and differences in the numerator and denominator. The LCD is $y(y + 2)$. Multiply the numerator and denominator by the LCD.

$$\frac{\dfrac{1}{y} + \dfrac{2}{y+2}}{\dfrac{4}{y} - \dfrac{3}{y+2}} \cdot \frac{y(y+2)}{y(y+2)} = \frac{1(y+2) + 2y}{4(y+2) - 3y}$$

Fundamental property

$$= \frac{y + 2 + 2y}{4y + 8 - 3y}$$

Distributive property

$$= \frac{3y + 2}{y + 8}$$

Combine like terms.

484 CHAPTER 7 Rational Expressions and Applications

(b) $\dfrac{\dfrac{x+2}{x-3}}{\dfrac{x^2-4}{x^2-9}}$

Since this is simply a quotient of two rational expressions, we use Method 1.

$$\dfrac{\dfrac{x+2}{x-3}}{\dfrac{x^2-4}{x^2-9}} = \dfrac{x+2}{x-3} \div \dfrac{x^2-4}{x^2-9}$$

$$= \dfrac{x+2}{x-3} \cdot \dfrac{x^2-9}{x^2-4} \qquad \text{Definition of division}$$

$$= \dfrac{x+2}{x-3} \cdot \dfrac{(x+3)(x-3)}{(x+2)(x-2)} \qquad \text{Factor.}$$

$$= \dfrac{x+3}{x-2} \qquad \text{Lowest terms}$$

Now Try Exercises 29 and 31.

7.5 EXERCISES

For Extra Help

- Student's Solutions Manual
- MyMathLab
- InterAct Math Tutorial Software
- AW Math Tutor Center
- MathXL
- Digital Video Tutor CD 15/Videotape 13

Note: In many problems involving complex fractions, several different equivalent forms of the answer exist. If your answer does not look exactly like the one given in the back of the book, check to see whether you have written an equivalent form.

1. Consider the complex fraction $\dfrac{\frac{1}{2} - \frac{1}{3}}{\frac{5}{6} - \frac{1}{12}}$. Answer each part, outlining Method 1 for simplifying this complex fraction.

   (a) To combine the terms in the numerator, we must find the LCD of $\frac{1}{2}$ and $\frac{1}{3}$. What is this LCD? Determine the simplified form of the numerator of the complex fraction.

   (b) To combine the terms in the denominator, we must find the LCD of $\frac{5}{6}$ and $\frac{1}{12}$. What is this LCD? Determine the simplified form of the denominator of the complex fraction.

   (c) Now use the results from parts (a) and (b) to write the complex fraction as a division problem using the symbol $\div$.

   (d) Perform the operation from part (c) to obtain the final simplification.

2. Consider the same complex fraction given in Exercise 1, $\dfrac{\frac{1}{2} - \frac{1}{3}}{\frac{5}{6} - \frac{1}{12}}$. Answer each part, outlining Method 2 for simplifying this complex fraction.

   (a) We must determine the LCD of all the fractions within the complex fraction. What is this LCD?

   (b) Multiply every term in the complex fraction by the LCD found in part (a), but do not combine the terms in the numerator and the denominator yet.

   (c) Combine the terms from part (b) to obtain the simplified form of the complex fraction.

SECTION 7.5 Complex Fractions

3. Which complex fraction is equivalent to $\dfrac{3 - \frac{1}{2}}{2 - \frac{1}{4}}$? Answer this question without showing any work, and explain your reasoning.

A. $\dfrac{3 + \frac{1}{2}}{2 + \frac{1}{4}}$ B. $\dfrac{-3 + \frac{1}{2}}{2 - \frac{1}{4}}$ C. $\dfrac{-3 - \frac{1}{2}}{-2 - \frac{1}{4}}$ D. $\dfrac{-3 + \frac{1}{2}}{-2 + \frac{1}{4}}$

4. Only one of these choices is equal to $\dfrac{\frac{1}{2} + \frac{1}{4}}{\frac{1}{3} + \frac{1}{12}}$. Which one is it? Answer this question without showing any work, and explain your reasoning.

A. $\dfrac{9}{5}$ B. $-\dfrac{9}{5}$ C. $-\dfrac{5}{9}$ D. -12

5. Describe Method 1 for simplifying complex fractions. Illustrate with the example $\dfrac{\frac{1}{2}}{\frac{2}{3}}$.

6. Describe Method 2 for simplifying complex fractions. Illustrate with the example $\dfrac{\frac{1}{2}}{\frac{2}{3}}$.

Simplify each complex fraction. Use either method. See Examples 1–6.

7. $\dfrac{-\frac{4}{3}}{\frac{2}{9}}$ **8.** $\dfrac{-\frac{5}{6}}{\frac{5}{4}}$ **9.** $\dfrac{\frac{x}{y^2}}{\frac{x^2}{y}}$ **10.** $\dfrac{\frac{p^4}{r}}{\frac{p^2}{r^2}}$

11. $\dfrac{\frac{4a^4b^3}{3a}}{\frac{2ab^4}{b^2}}$ **12.** $\dfrac{\frac{2r^4t^2}{3t}}{\frac{5r^2t^5}{3r}}$ **13.** $\dfrac{\frac{m+2}{3}}{\frac{m-4}{m}}$ **14.** $\dfrac{\frac{q-5}{q}}{\frac{q+5}{3}}$

15. $\dfrac{\frac{2}{x} - 3}{\frac{2 - 3x}{2}}$ **16.** $\dfrac{6 + \frac{2}{r}}{\frac{3r+1}{4}}$ **17.** $\dfrac{\frac{1}{x} + x}{\frac{x^2+1}{8}}$ **18.** $\dfrac{\frac{3}{m} - m}{\frac{3-m^2}{4}}$

19. $\dfrac{a - \frac{5}{a}}{a + \frac{1}{a}}$ **20.** $\dfrac{q + \frac{1}{q}}{q + \frac{4}{q}}$ **21.** $\dfrac{\frac{5}{8} + \frac{2}{3}}{\frac{7}{3} - \frac{1}{4}}$ **22.** $\dfrac{\frac{6}{5} - \frac{1}{9}}{\frac{2}{5} + \frac{5}{3}}$

23. $\dfrac{\frac{1}{x^2} + \frac{1}{y^2}}{\frac{1}{x} - \frac{1}{y}}$ **24.** $\dfrac{\frac{1}{a^2} - \frac{1}{b^2}}{\frac{1}{a} - \frac{1}{b}}$ **25.** $\dfrac{\frac{2}{p^2} - \frac{3}{5p}}{\frac{4}{p} + \frac{1}{4p}}$ **26.** $\dfrac{\frac{2}{m^2} - \frac{3}{m}}{\frac{2}{5m^2} + \frac{1}{3m}}$

27. $\dfrac{\frac{5}{x^2y} - \frac{2}{xy^2}}{\frac{3}{x^2y^2} + \frac{4}{xy}}$ **28.** $\dfrac{\frac{1}{m^3p} + \frac{2}{mp^2}}{\frac{4}{mp} + \frac{1}{m^2p}}$ **29.** $\dfrac{\frac{1}{4} - \frac{1}{a^2}}{\frac{1}{2} + \frac{1}{a}}$ **30.** $\dfrac{\frac{1}{9} - \frac{1}{m^2}}{\frac{1}{3} + \frac{1}{m}}$

486 CHAPTER 7 Rational Expressions and Applications

31. $\dfrac{\dfrac{1}{z+5}}{\dfrac{4}{z^2-25}}$
32. $\dfrac{\dfrac{1}{a+1}}{\dfrac{2}{a^2-1}}$
33. $\dfrac{\dfrac{1}{m+1} - 1}{\dfrac{1}{m+1} + 1}$

34. $\dfrac{\dfrac{2}{x-1} + 2}{\dfrac{2}{x-1} - 2}$
35. $\dfrac{\dfrac{1}{m-1} + \dfrac{2}{m+2}}{\dfrac{2}{m+2} - \dfrac{1}{m-3}}$
36. $\dfrac{\dfrac{5}{r+3} - \dfrac{1}{r-1}}{\dfrac{2}{r+2} + \dfrac{3}{r+3}}$

37. In a fraction, what operation does the fraction bar represent?

38. What property of real numbers justifies Method 2 of simplifying complex fractions?

RELATING CONCEPTS (EXERCISES 39–42)

For Individual or Group Work

To find the average of two numbers, we add them and divide by 2. Suppose that we wish to find the average of $\frac{3}{8}$ and $\frac{5}{6}$. **Work Exercises 39–42 in order,** *to see how a complex fraction occurs in a problem like this.*

39. Write in symbols: the sum of $\frac{3}{8}$ and $\frac{5}{6}$, divided by 2. Your result should be a complex fraction.

40. Simplify the complex fraction from Exercise 39 using Method 1.

41. Simplify the complex fraction from Exercise 39 using Method 2.

42. Your answers in Exercises 40 and 41 should be the same. Which method did you prefer? Why?

7.6 Solving Equations with Rational Expressions

OBJECTIVES

1. Distinguish between operations with rational expressions and equations with terms that are rational expressions.
2. Solve equations with rational expressions.
3. Solve a formula for a specified variable.

In Section 2.3 we solved equations with fractions as coefficients. By using the multiplication property of equality, we cleared the fractions by multiplying by the LCD. We continue this work here.

OBJECTIVE 1 Distinguish between operations with rational expressions and equations with terms that are rational expressions. Before solving equations with rational expressions, you must understand the difference between *sums* and *differences* of terms with rational coefficients, and *equations* with terms that are rational expressions. **Sums and differences are operations to perform, while equations are solved.**

EXAMPLE 1 Distinguishing between Operations and Equations

Identify each of the following as an operation or an equation. Then perform the operation or solve the equation.

(a) $\dfrac{3}{4}x - \dfrac{2}{3}x$

This is a difference of two terms, so it is an operation. (There is no equals sign.) Find the LCD, write each coefficient with this LCD, and combine like terms.

$$\frac{3}{4}x - \frac{2}{3}x = \frac{9}{12}x - \frac{8}{12}x \quad \text{Get a common denominator.}$$

$$= \frac{1}{12}x \quad \text{Combine like terms.}$$

(b) $\frac{3}{4}x - \frac{2}{3}x = \frac{1}{2}$

Because of the equals sign, this is an equation to be solved. Proceed as in Section 2.3, using the multiplication property of equality to clear fractions. The LCD is 12.

$$\frac{3}{4}x - \frac{2}{3}x = \frac{1}{2}$$

$$12\left(\frac{3}{4}x - \frac{2}{3}x\right) = 12\left(\frac{1}{2}\right) \quad \text{Multiply by 12.}$$

$$12\left(\frac{3}{4}x\right) - 12\left(\frac{2}{3}x\right) = 12\left(\frac{1}{2}\right) \quad \text{Distributive property; Multiply each term by 12.}$$

$$9x - 8x = 6 \quad \text{Multiply.}$$

$$x = 6 \quad \text{Combine like terms.}$$

Check: $\quad \frac{3}{4}x - \frac{2}{3}x = \frac{1}{2} \quad$ Original equation

$$\frac{3}{4}(6) - \frac{2}{3}(6) = \frac{1}{2} \quad ? \quad \text{Let } x = 6.$$

$$\frac{9}{2} - 4 = \frac{1}{2} \quad ? \quad \text{Multiply.}$$

$$\frac{1}{2} = \frac{1}{2} \quad \text{True}$$

Since a true statement results, {6} is the solution set of the equation.

Now Try Exercises 1 and 3.

The ideas of Example 1 can be summarized as follows.

> When adding or subtracting, the LCD must be kept throughout the simplification.
>
> When solving an equation, the LCD is used to multiply each side so that denominators are eliminated.

OBJECTIVE 2 Solve equations with rational expressions. When an equation involves fractions as in Example 1(b), we use the multiplication property of equality to clear the fractions. Choose as multiplier the LCD of all denominators in the fractions of the equation.

EXAMPLE 2 Solving an Equation with Rational Expressions

Solve $\dfrac{p}{2} - \dfrac{p-1}{3} = 1$.

$$6\left(\dfrac{p}{2} - \dfrac{p-1}{3}\right) = 6(1) \qquad \text{Multiply by the LCD, 6.}$$

$$6\left(\dfrac{p}{2}\right) - 6\left(\dfrac{p-1}{3}\right) = 6 \qquad \text{Distributive property}$$

$$3p - 2(p-1) = 6 \qquad \text{Multiply; use parentheses around } p-1.$$
$$3p - 2(p) - 2(-1) = 6 \qquad \text{Distributive property}$$
$$3p - 2p + 2 = 6 \qquad \text{Be careful with signs.}$$
$$p + 2 = 6 \qquad \text{Combine like terms.}$$
$$p = 4 \qquad \text{Subtract 2.}$$

Check to see that {4} is the solution set by replacing p with 4 in the original equation.

Now Try Exercise 37.

CAUTION Note that the use of the LCD here is different from its use in the previous section. Here, we use the multiplication property of equality to multiply each side of an *equation* by the LCD. Earlier, we used the fundamental property to multiply a *fraction* by another fraction that had the LCD as both its numerator and denominator. Be careful not to confuse these two methods.

Recall from Section 7.1 that the denominator of a rational expression cannot equal 0 since division by 0 is undefined. Therefore, when solving an equation with rational expressions that have variables in the denominator, *the solution cannot be a number that makes the denominator equal 0*. If it does, the proposed solution must be rejected.

EXAMPLE 3 Solving an Equation with Rational Expressions

Solve $\dfrac{x}{x-2} = \dfrac{2}{x-2} + 2$. Check the proposed solution.

Note that $x \neq 2$ here since both denominators in the given expression equal 0 if x is 2. The common denominator is $x - 2$. Solve the equation by multiplying each side by $x - 2$.

$$(x-2)\left(\dfrac{x}{x-2}\right) = (x-2)\left(\dfrac{2}{x-2} + 2\right)$$

$$(x-2)\left(\dfrac{x}{x-2}\right) = (x-2)\left(\dfrac{2}{x-2}\right) + (x-2)(2) \qquad \text{Distributive property}$$

$$x = 2 + 2x - 4$$
$$x = -2 + 2x \qquad \text{Combine like terms.}$$
$$-x = -2 \qquad \text{Subtract } 2x.$$
$$x = 2 \qquad \text{Multiply by } -1.$$

As noted, x cannot equal 2 since replacing x with 2 in the original equation causes the denominators to equal 0.

Check: $\dfrac{2}{2-2} = \dfrac{2}{2-2} + 2$?   Let $x = 2$ in the original equation.

$\dfrac{2}{0} = \dfrac{2}{0} + 2$?

Thus, 2 must be rejected as a solution, and the solution set is $\emptyset$.

Now Try Exercise 31.

While it is always a good idea to check solutions to guard against arithmetic and algebraic errors, it is *essential* to check proposed solutions when variables appear in denominators in the original equation. Some students like to determine which numbers cannot be solutions *before* solving the equation, as we did in Example 3.

The steps used to solve an equation with rational expressions follow.

Solving an Equation with Rational Expressions

Step 1 **Multiply each side of the equation by the LCD** to clear the equation of fractions.

Step 2 **Solve** the resulting equation.

Step 3 **Check** each proposed solution by substituting it in the original equation. Reject any that cause a denominator to equal 0.

EXAMPLE 4 Solving an Equation with Rational Expressions

Solve $\dfrac{2}{x^2 - x} = \dfrac{1}{x^2 - 1}$. Check the proposed solution.

Step 1 Factor the denominators to find the LCD.

$$\dfrac{2}{x(x-1)} = \dfrac{1}{(x+1)(x-1)}$$

The LCD is $x(x+1)(x-1)$. Notice that $x \neq 0, -1,$ or 1; otherwise a denominator will equal 0. Thus, 0, -1, and 1 cannot be solutions of this equation. Multiply each side of the equation by $x(x+1)(x-1)$.

$$x(x+1)(x-1)\dfrac{2}{x(x-1)} = x(x+1)(x-1)\dfrac{1}{(x+1)(x-1)}$$

Step 2 $\qquad\qquad\qquad\qquad 2(x+1) = x$

$\qquad\qquad\qquad\qquad\qquad 2x + 2 = x$ Distributive property

$\qquad\qquad\qquad\qquad\qquad\quad 2 = -x$ Subtract $2x$.

$\qquad\qquad\qquad\qquad\qquad\quad x = -2$ Multiply by -1.

490 CHAPTER 7 Rational Expressions and Applications

Step 3 The proposed solution is -2, which does not make any denominator equal 0.

Check: $\dfrac{2}{x^2 - x} = \dfrac{1}{x^2 - 1}$ Original equation

$\dfrac{2}{(-2)^2 - (-2)} = \dfrac{1}{(-2)^2 - 1}$? Let $x = -2$.

$\dfrac{2}{4 + 2} = \dfrac{1}{4 - 1}$?

$\dfrac{1}{3} = \dfrac{1}{3}$ True

Thus, the solution set is $\{-2\}$.

Now Try Exercise 43.

EXAMPLE 5 Solving an Equation with Rational Expressions

Solve $\dfrac{2m}{m^2 - 4} + \dfrac{1}{m - 2} = \dfrac{2}{m + 2}$.

Factor the first denominator on the left to find the LCD.

$$\dfrac{2m}{(m + 2)(m - 2)} + \dfrac{1}{m - 2} = \dfrac{2}{m + 2}$$

The LCD is $(m + 2)(m - 2)$. Here $m \neq -2$ or 2, so these numbers cannot be solutions of this equation. Multiply each side by the LCD.

$$(m + 2)(m - 2)\left(\dfrac{2m}{(m + 2)(m - 2)} + \dfrac{1}{m - 2}\right)$$

$$= (m + 2)(m - 2)\dfrac{2}{m + 2}$$

$$(m + 2)(m - 2)\dfrac{2m}{(m + 2)(m - 2)} + (m + 2)(m - 2)\dfrac{1}{m - 2}$$

$$= (m + 2)(m - 2)\dfrac{2}{m + 2}$$

$2m + m + 2 = 2(m - 2)$

$3m + 2 = 2m - 4$ Combine like terms; distributive property

$m + 2 = -4$ Subtract $2m$.

$m = -6$ Subtract 2.

A check verifies that $\{-6\}$ is the solution set.

Now Try Exercise 53.

EXAMPLE 6 Solving an Equation with Rational Expressions

Solve $\dfrac{1}{x - 1} + \dfrac{1}{2} = \dfrac{2}{x^2 - 1}$.

Factor the denominator on the right.

$$\dfrac{1}{x - 1} + \dfrac{1}{2} = \dfrac{2}{(x + 1)(x - 1)}$$

Notice that $x \neq 1$ or -1. Multiply each side of the equation by the LCD, $2(x + 1)(x - 1)$.

$$2(x + 1)(x - 1)\left(\frac{1}{x - 1} + \frac{1}{2}\right) = 2(x + 1)(x - 1)\frac{2}{(x + 1)(x - 1)}$$

$$2(x + 1)(x - 1)\frac{1}{x - 1} + 2(x + 1)(x - 1)\frac{1}{2} = 2(x + 1)(x - 1)\frac{2}{(x + 1)(x - 1)}$$

$$2(x + 1) + (x + 1)(x - 1) = 4$$
$$2x + 2 + x^2 - 1 = 4 \quad \text{Distributive property}$$
$$x^2 + 2x + 1 = 4 \quad \text{Combine like terms.}$$
$$x^2 + 2x - 3 = 0 \quad \text{Subtract 4.}$$
$$(x + 3)(x - 1) = 0 \quad \text{Factor.}$$
$$x + 3 = 0 \quad \text{or} \quad x - 1 = 0 \quad \text{Zero-factor property}$$
$$x = -3 \quad \text{or} \quad x = 1$$

-3 and 1 are proposed solutions. However, as noted, 1 makes an original denominator equal 0, so 1 is not a solution. Check that -3 is a solution.

Check: $\quad \dfrac{1}{x - 1} + \dfrac{1}{2} = \dfrac{2}{x^2 - 1}$ $\qquad$ Original equation

$\qquad\qquad \dfrac{1}{-3 - 1} + \dfrac{1}{2} = \dfrac{2}{(-3)^2 - 1}$? $\quad$ Let $x = -3$.

$\qquad\qquad \dfrac{1}{-4} + \dfrac{1}{2} = \dfrac{2}{9 - 1}$? $\quad$ Simplify.

$\qquad\qquad \dfrac{1}{4} = \dfrac{1}{4}$ $\qquad$ True

The solution set is $\{-3\}$.

Now Try Exercise 63.

EXAMPLE 7 Solving an Equation with Rational Expressions

Solve $\dfrac{1}{k^2 + 4k + 3} + \dfrac{1}{2k + 2} = \dfrac{3}{4k + 12}$.

Factoring each denominator gives the equation

$$\frac{1}{(k + 1)(k + 3)} + \frac{1}{2(k + 1)} = \frac{3}{4(k + 3)}.$$

The LCD is $4(k + 1)(k + 3)$, indicating that $k \neq -1$ or -3. Multiply each side by this LCD.

$$4(k + 1)(k + 3)\left(\frac{1}{(k + 1)(k + 3)} + \frac{1}{2(k + 1)}\right)$$
$$= 4(k + 1)(k + 3)\frac{3}{4(k + 3)}$$

$$4(k + 1)(k + 3)\frac{1}{(k + 1)(k + 3)} + 2 \cdot 2(k + 1)(k + 3)\frac{1}{2(k + 1)}$$
$$= 4(k + 1)(k + 3)\frac{3}{4(k + 3)}$$

$$4 + 2(k + 3) = 3(k + 1)$$
$$4 + 2k + 6 = 3k + 3 \quad \text{Distributive property}$$
$$2k + 10 = 3k + 3 \quad \text{Combine like terms.}$$
$$7 = k \quad \text{Subtract } 2k \text{ and } 3.$$

The proposed solution, 7, does not make an original denominator equal 0. A check shows that the algebra is correct, so {7} is the solution set.

Now Try Exercise 65.

OBJECTIVE 3 Solve a formula for a specified variable. Solving a formula for a specified variable was first discussed in Chapter 2. Remember to treat the variable for which you are solving as if it were the only variable, and all others as if they were constants.

EXAMPLE 8 Solving for a Specified Variable

Solve each formula for the specified variable.

(a) $F = \dfrac{k}{d - D}$ for d

We want to get d alone on one side of the equation. We begin by multiplying each side by $d - D$ to clear the fraction.

$$F = \dfrac{k}{d - D} \quad \text{Given equation}$$
$$F(d - D) = \dfrac{k}{d - D}(d - D) \quad \text{Clear the fraction.}$$
$$F(d - D) = k \quad \text{Simplify.}$$
$$Fd - FD = k \quad \text{Distributive property}$$
$$Fd = k + FD \quad \text{Add } FD.$$
$$d = \dfrac{k + FD}{F} \quad \text{or} \quad d = \dfrac{k}{F} + D \quad \text{Divide by } F.$$

(b) $\dfrac{1}{a} = \dfrac{1}{b} + \dfrac{1}{c}$ for c

The LCD of all the fractions in the equation is abc, so we multiply each side by abc.

$$abc\left(\dfrac{1}{a}\right) = abc\left(\dfrac{1}{b} + \dfrac{1}{c}\right)$$
$$abc\left(\dfrac{1}{a}\right) = abc\left(\dfrac{1}{b}\right) + abc\left(\dfrac{1}{c}\right) \quad \text{Distributive property}$$
$$bc = ac + ab$$

Since we are solving for c, we need to get all terms with c on one side of the equation. We do this by subtracting ac from each side.

$$bc - ac = ab \quad \text{Subtract } ac.$$
$$c(b - a) = ab \quad \text{Factor out } c.$$

SECTION 7.6 Solving Equations with Rational Expressions **493**

Finally, we divide each side by the coefficient of c, which is $b - a$.

$$c = \frac{ab}{b-a}$$

Now Try Exercises 77 and 83.

CAUTION Students often have trouble in the step that involves factoring out the variable for which they are solving. In Example 8(b), we had to factor out c on the left side so that we could divide both sides by $b - a$.

When solving an equation for a specified variable, *be sure that the specified variable appears alone on only one side of the equals sign in the final equation.*

7.6 EXERCISES

For Extra Help

- Student's Solutions Manual
- MyMathLab
- InterAct Math Tutorial Software
- AW Math Tutor Center
- MathXL
- Digital Video Tutor CD 15/Videotape 13

Identify as an expression or an equation. Then perform the operation or solve the equation. See Example 1.

1. $\frac{7}{8}x + \frac{1}{5}x$
2. $\frac{4}{7}x + \frac{3}{5}x$
3. $\frac{7}{8}x + \frac{1}{5}x = 1$
4. $\frac{4}{7}x + \frac{3}{5}x = 1$
5. $\frac{3}{5}x - \frac{7}{10}x$
6. $\frac{2}{3}x - \frac{7}{4}x$
7. $\frac{3}{5}x - \frac{7}{10}x = 1$
8. $\frac{2}{3}x - \frac{7}{4}x = -13$

When solving an equation with variables in denominators, we must determine the values that cause these denominators to equal 0, so we can reject these values if they appear as possible solutions. Find all values for which at least one denominator is equal to 0. Do not solve. See Examples 3–7.

9. $\frac{3}{x+2} - \frac{5}{x} = 1$
10. $\frac{7}{x} + \frac{9}{x-3} = 5$
11. $\frac{-1}{(x+3)(x-4)} = \frac{1}{2x+1}$
12. $\frac{8}{(x-8)(x+2)} = \frac{7}{3x-10}$
13. $\frac{4}{x^2 + 8x - 9} + \frac{1}{x^2 - 4} = 0$
14. $\frac{-3}{x^2 + 9x - 10} - \frac{12}{x^2 - 16} = 0$

 15. Explain how the LCD is used in a different way when adding and subtracting rational expressions compared to solving equations with rational expressions.

 16. If we multiply each side of the equation $\frac{6}{x+5} = \frac{6}{x+5}$ by $x + 5$, we get $6 = 6$. Are all real numbers solutions of this equation? Explain.

Solve each equation, and check your answers. See Examples 1(b), 2, and 3.

17. $\frac{5}{m} - \frac{3}{m} = 8$
18. $\frac{4}{y} + \frac{1}{y} = 2$
19. $\frac{5}{y} + 4 = \frac{2}{y}$
20. $\frac{11}{q} = 3 - \frac{1}{q}$
21. $\frac{3x}{5} - 6 = x$
22. $\frac{5t}{4} + t = 9$

23. $\dfrac{4m}{7} + m = 11$ **24.** $a - \dfrac{3a}{2} = 1$ **25.** $\dfrac{z-1}{4} = \dfrac{z+3}{3}$

26. $\dfrac{r-5}{2} = \dfrac{r+2}{3}$ **27.** $\dfrac{3p+6}{8} = \dfrac{3p-3}{16}$ **28.** $\dfrac{2z+1}{5} = \dfrac{7z+5}{15}$

29. $\dfrac{2x+3}{x} = \dfrac{3}{2}$ **30.** $\dfrac{5-2x}{x} = \dfrac{1}{4}$ **31.** $\dfrac{k}{k-4} - 5 = \dfrac{4}{k-4}$

32. $\dfrac{-5}{a+5} = \dfrac{a}{a+5} + 2$ **33.** $\dfrac{q+2}{3} + \dfrac{q-5}{5} = \dfrac{7}{3}$ **34.** $\dfrac{t}{6} + \dfrac{4}{3} = \dfrac{t-2}{3}$

35. $\dfrac{x}{2} = \dfrac{5}{4} + \dfrac{x-1}{4}$ **36.** $\dfrac{8p}{5} = \dfrac{3p-4}{2} + \dfrac{5}{2}$

Solve each equation, and check your answers. Be careful with signs. See Example 2.

37. $\dfrac{a+7}{8} - \dfrac{a-2}{3} = \dfrac{4}{3}$ **38.** $\dfrac{x+3}{7} - \dfrac{x+2}{6} = \dfrac{1}{6}$ **39.** $\dfrac{p}{2} - \dfrac{p-1}{4} = \dfrac{5}{4}$

40. $\dfrac{r}{6} - \dfrac{r-2}{3} = -\dfrac{4}{3}$ **41.** $\dfrac{3x}{5} - \dfrac{x-5}{7} = 3$ **42.** $\dfrac{8k}{5} - \dfrac{3k-4}{2} = \dfrac{5}{2}$

Solve each equation, and check your answers. See Examples 3–7.

43. $\dfrac{4}{x^2 - 3x} = \dfrac{1}{x^2 - 9}$ **44.** $\dfrac{2}{t^2 - 4} = \dfrac{3}{t^2 - 2t}$

45. $\dfrac{2}{m} = \dfrac{m}{5m + 12}$ **46.** $\dfrac{x}{4-x} = \dfrac{2}{x}$

47. $\dfrac{-2}{z+5} + \dfrac{3}{z-5} = \dfrac{20}{z^2 - 25}$ **48.** $\dfrac{3}{r+3} - \dfrac{2}{r-3} = \dfrac{-12}{r^2 - 9}$

49. $\dfrac{3}{x-1} + \dfrac{2}{4x-4} = \dfrac{7}{4}$ **50.** $\dfrac{2}{p+3} + \dfrac{3}{8} = \dfrac{5}{4p+12}$

51. $\dfrac{x}{3x+3} = \dfrac{2x-3}{x+1} - \dfrac{2x}{3x+3}$ **52.** $\dfrac{2k+3}{k+1} - \dfrac{3k}{2k+2} = \dfrac{-2k}{2k+2}$

53. $\dfrac{2p}{p^2 - 1} = \dfrac{2}{p+1} - \dfrac{1}{p-1}$ **54.** $\dfrac{2x}{x^2 - 16} - \dfrac{2}{x-4} = \dfrac{4}{x+4}$

55. $\dfrac{5x}{14x+3} = \dfrac{1}{x}$ **56.** $\dfrac{m}{8m+3} = \dfrac{1}{3m}$

57. $\dfrac{2}{x-1} - \dfrac{2}{3} = \dfrac{-1}{x+1}$ **58.** $\dfrac{5}{p-2} = 7 - \dfrac{10}{p+2}$

59. $\dfrac{x}{2x+2} = \dfrac{-2x}{4x+4} + \dfrac{2x-3}{x+1}$ **60.** $\dfrac{5t+1}{3t+3} = \dfrac{5t-5}{5t+5} + \dfrac{3t-1}{t+1}$

61. $\dfrac{8x+3}{x} = 3x$ **62.** $\dfrac{2}{x} = \dfrac{x}{5x-12}$

63. $\dfrac{1}{x+4} + \dfrac{x}{x-4} = \dfrac{-8}{x^2 - 16}$ **64.** $\dfrac{x}{x-3} + \dfrac{4}{x+3} = \dfrac{18}{x^2 - 9}$

65. $\dfrac{4}{3x+6} - \dfrac{3}{x+3} = \dfrac{8}{x^2+5x+6}$

66. $\dfrac{-13}{t^2+6t+8} + \dfrac{4}{t+2} = \dfrac{3}{2t+8}$

67. $\dfrac{3x}{x^2+5x+6} = \dfrac{5x}{x^2+2x-3} - \dfrac{2}{x^2+x-2}$

68. $\dfrac{m}{m^2+m-2} + \dfrac{m}{m^2-1} = \dfrac{m}{m^2+3m+2}$

69. $\dfrac{x+4}{x^2-3x+2} - \dfrac{5}{x^2-4x+3} = \dfrac{x-4}{x^2-5x+6}$

70. $\dfrac{3}{r^2+r-2} - \dfrac{1}{r^2-1} = \dfrac{7}{2(r^2+3r+2)}$

71. If you are solving a formula for the letter k, and your steps lead to the equation $kr - mr = km$, what would be your next step?

72. If you are solving a formula for the letter k, and your steps lead to the equation $kr - km = mr$, what would be your next step?

Solve each formula for the specified variable. See Example 8.

73. $m = \dfrac{kF}{a}$ for F

74. $I = \dfrac{kE}{R}$ for E

75. $m = \dfrac{kF}{a}$ for a

76. $I = \dfrac{kE}{R}$ for R

77. $I = \dfrac{E}{R+r}$ for R

78. $I = \dfrac{E}{R+r}$ for r

79. $h = \dfrac{2A}{B+b}$ for A

80. $d = \dfrac{2S}{n(a+L)}$ for S

81. $d = \dfrac{2S}{n(a+L)}$ for a

82. $h = \dfrac{2A}{B+b}$ for B

83. $\dfrac{1}{x} = \dfrac{1}{y} - \dfrac{1}{z}$ for y

84. $\dfrac{3}{k} = \dfrac{1}{p} + \dfrac{1}{q}$ for q

85. $9x + \dfrac{3}{z} = \dfrac{5}{y}$ for z

86. $\dfrac{1}{a} = \dfrac{1}{b} + \dfrac{1}{c}$ for a

SUMMARY EXERCISES ON OPERATIONS AND EQUATIONS WITH RATIONAL EXPRESSIONS

We have performed the four operations of arithmetic with rational expressions and solved equations with rational expressions. The exercises in this summary include a mixed variety of problems of these types. Since students often confuse *operations* on rational expressions with *solving equations* with rational expressions, we review the four operations using the rational expressions $\dfrac{1}{x}$ and $\dfrac{1}{x-2}$ as follows.

Add:

$\dfrac{1}{x} + \dfrac{1}{x-2} = \dfrac{1(x-2)}{x(x-2)} + \dfrac{x(1)}{x(x-2)}$ Write with a common denominator.

$= \dfrac{x-2+x}{x(x-2)}$ Add numerators; keep the same denominator.

$= \dfrac{2x-2}{x(x-2)}$ Combine like terms.

Subtract:
$$\frac{1}{x} - \frac{1}{x-2} = \frac{1(x-2)}{x(x-2)} - \frac{x(1)}{x(x-2)}$$ Write with a common denominator.
$$= \frac{x-2-x}{x(x-2)}$$ Subtract numerators; keep the same denominator.
$$= \frac{-2}{x(x-2)}$$ Combine like terms.

Multiply:
$$\frac{1}{x} \cdot \frac{1}{x-2} = \frac{1}{x(x-2)}$$ Multiply numerators and multiply denominators.

Divide:
$$\frac{1}{x} \div \frac{1}{x-2} = \frac{1}{x} \cdot \frac{x-2}{1} = \frac{x-2}{x}$$ Multiply by the reciprocal of the second fraction.

On the other hand, consider the *equation*
$$\frac{1}{x} + \frac{1}{x-2} = \frac{3}{4}.$$

Neither 0 nor 2 can be a solution of this equation, since each will cause a denominator to equal 0. We use the multiplication property of equality and multiply each side by the LCD, $4x(x-2)$, to clear fractions.

$$4x(x-2)\frac{1}{x} + 4x(x-2)\frac{1}{x-2} = 4x(x-2)\frac{3}{4}$$
$$4(x-2) + 4x = 3x(x-2)$$
$$4x - 8 + 4x = 3x^2 - 6x \quad \text{Distributive property}$$
$$0 = 3x^2 - 14x + 8 \quad \text{Get 0 on one side.}$$
$$0 = (3x-2)(x-4) \quad \text{Factor.}$$
$$3x - 2 = 0 \quad \text{or} \quad x - 4 = 0 \quad \text{Zero-factor property}$$
$$x = \frac{2}{3} \quad \text{or} \quad x = 4$$

Both $\frac{2}{3}$ and 4 are solutions since neither makes a denominator equal 0; the solution set is $\{\frac{2}{3}, 4\}$.

In conclusion, remember the following points when working exercises involving rational expressions.

Points to Remember When Working with Rational Expressions

1. The fundamental property is applied only after numerators and denominators have been *factored*.
2. When adding and subtracting rational expressions, the common denominator must be kept throughout the problem and in the final result.
3. Always look to see if the answer is in lowest terms; if it is not, use the fundamental property.

4. When solving equations, the LCD is used to clear the equation of fractions. Multiply each side by the LCD. (Notice how this differs from the use of the LCD in Point 2.)
5. When solving equations with rational expressions, reject any proposed solution that causes an original denominator to equal 0.

For each exercise, indicate "operation" if an operation is to be performed or "equation" if an equation is to be solved. Then perform the operation or solve the equation.

1. $\dfrac{4}{p} + \dfrac{6}{p}$

2. $\dfrac{x^3 y^2}{x^2 y^4} \cdot \dfrac{y^5}{x^4}$

3. $\dfrac{1}{x^2 + x - 2} \div \dfrac{4x^2}{2x - 2}$

4. $\dfrac{8}{m - 5} = 2$

5. $\dfrac{2y^2 + y - 6}{2y^2 - 9y + 9} \cdot \dfrac{y^2 - 2y - 3}{y^2 - 1}$

6. $\dfrac{2}{k^2 - 4k} + \dfrac{3}{k^2 - 16}$

7. $\dfrac{x - 4}{5} = \dfrac{x + 3}{6}$

8. $\dfrac{3t^2 - t}{6t^2 + 15t} \div \dfrac{6t^2 + t - 1}{2t^2 - 5t - 25}$

9. $\dfrac{4}{p + 2} + \dfrac{1}{3p + 6}$

10. $\dfrac{1}{x} + \dfrac{1}{x - 3} = -\dfrac{5}{4}$

11. $\dfrac{3}{t - 1} + \dfrac{1}{t} = \dfrac{7}{2}$

12. $\dfrac{6}{y} - \dfrac{2}{3y}$

13. $\dfrac{5}{4z} - \dfrac{2}{3z}$

14. $\dfrac{k + 2}{3} = \dfrac{2k - 1}{5}$

15. $\dfrac{1}{m^2 + 5m + 6} + \dfrac{2}{m^2 + 4m + 3}$

16. $\dfrac{2k^2 - 3k}{20k^2 - 5k} \div \dfrac{2k^2 - 5k + 3}{4k^2 + 11k - 3}$

17. $\dfrac{2}{x + 1} + \dfrac{5}{x - 1} = \dfrac{10}{x^2 - 1}$

18. $\dfrac{3}{x + 3} + \dfrac{4}{x + 6} = \dfrac{9}{x^2 + 9x + 18}$

19. $\dfrac{4t^2 - t}{6t^2 + 10t} \div \dfrac{8t^2 + 2t - 1}{3t^2 + 11t + 10}$

20. $\dfrac{x}{x - 2} + \dfrac{3}{x + 2} = \dfrac{8}{x^2 - 4}$

7.7 Applications of Rational Expressions

OBJECTIVES

1. Solve problems about numbers.
2. Solve problems about distance, rate, and time.
3. Solve problems about work.

When we learn how to solve a new type of equation, we are able to apply our knowledge to solving new types of applications. In Section 7.6 we solved equations with rational expressions; now we can solve applications that involve this type of equation. The six-step problem solving method of Chapter 2 still applies.

OBJECTIVE 1 Solve problems about numbers. We begin with an example about an unknown number.

498 CHAPTER 7 Rational Expressions and Applications

EXAMPLE 1 Solving a Problem about an Unknown Number

If the same number is added to both the numerator and the denominator of the fraction $\frac{2}{5}$, the result is equivalent to $\frac{2}{3}$. Find the number.

Step 1 **Read** the problem carefully. We are trying to find a number.

Step 2 **Assign a variable.** Here, we let $x =$ the number added to the numerator and the denominator.

Step 3 **Write an equation.** The fraction
$$\frac{2+x}{5+x}$$
represents the result of adding the same number to both the numerator and the denominator. Since this result is equivalent to $\frac{2}{3}$, the equation is
$$\frac{2+x}{5+x} = \frac{2}{3}.$$

Step 4 **Solve** this equation by multiplying each side by the LCD, $3(5+x)$.
$$3(5+x)\frac{2+x}{5+x} = 3(5+x)\frac{2}{3}$$
$$3(2+x) = 2(5+x)$$
$$6 + 3x = 10 + 2x \qquad \text{Distributive property}$$
$$x = 4 \qquad \text{Subtract } 2x; \text{ subtract } 6.$$

Step 5 **State the answer.** The number is 4.

Step 6 **Check** the solution in the words of the original problem. If 4 is added to both the numerator and the denominator of $\frac{2}{5}$, the result is $\frac{6}{9} = \frac{2}{3}$, as required.

Now Try Exercise 3.

OBJECTIVE 2 Solve problems about distance, rate, and time. Recall from Chapter 2 the following formulas relating distance, rate, and time.

Distance, Rate, and Time Relationship

$$d = rt \qquad r = \frac{d}{t} \qquad t = \frac{d}{r}$$

You may wish to refer to Example 5 in Section 2.7 to review the basic use of these formulas. We continue our work with motion problems here.

EXAMPLE 2 Solving a Problem about Distance, Rate, and Time

The Tickfaw River has a current of 3 mph. A motorboat takes as long to go 12 mi downstream as to go 8 mi upstream. What is the speed of the boat in still water?

Step 1 **Read** the problem again. We are looking for the speed of the boat in still water.

Step 2 **Assign a variable.** Let $x =$ the speed of the boat in still water. Because the current pushes the boat when the boat is going downstream, the speed of the boat downstream will be the sum of the speed of the boat and the speed of the current, $x + 3$ mph. Because the current slows down the boat when the boat is going upstream, the boat's speed going upstream is given by the difference between the speed of the boat and the speed of the current, $x - 3$ mph. See Figure 1.

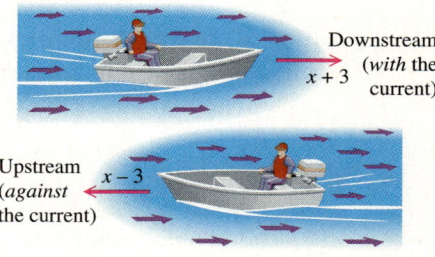

FIGURE 1

This information is summarized in the table.

	d	r	t
Downstream	12	x + 3	
Upstream	8	x − 3	

We fill in the column representing time by using the formula $t = \frac{d}{r}$. Then the time downstream is the distance divided by the rate, or

$$t = \frac{d}{r} = \frac{12}{x + 3},$$

and the time upstream is also the distance divided by the rate, or

$$t = \frac{d}{r} = \frac{8}{x - 3}.$$

The completed table follows.

	d	r	t
Downstream	12	x + 3	$\frac{12}{x+3}$
Upstream	8	x − 3	$\frac{8}{x-3}$

Times are equal.

Step 3 **Write an equation.** According to the original problem, the time downstream equals the time upstream. The two times from the table must therefore be equal, giving the equation

$$\frac{12}{x + 3} = \frac{8}{x - 3}.$$

Step 4 **Solve.** We begin by multiplying each side by the LCD, $(x + 3)(x - 3)$.

$$(x + 3)(x - 3)\frac{12}{x + 3} = (x + 3)(x - 3)\frac{8}{x - 3}$$

$$12(x - 3) = 8(x + 3)$$

$$12x - 36 = 8x + 24 \quad \text{Distributive property}$$

$$4x = 60 \quad \text{Subtract } 8x; \text{ add } 36.$$

$$x = 15 \quad \text{Divide by 4.}$$

Step 5 **State the answer.** The speed of the boat in still water is 15 mph.

Step 6 **Check.** First we find the speed of the boat going downstream, which is $15 + 3 = 18$ mph. Traveling 12 mi would take

$$t = \frac{d}{r} = \frac{12}{18} = \frac{2}{3} \text{ hr.}$$

On the other hand, the speed of the boat going upstream is $15 - 3 = 12$ mph, and traveling 8 mi would take

$$t = \frac{d}{r} = \frac{8}{12} = \frac{2}{3} \text{ hr.}$$

The time upstream equals the time downstream, as required.

Now Try Exercise 21.

OBJECTIVE 3 Solve problems about work. Suppose that you can mow your lawn in 4 hr. Then after 1 hr, you will have mowed $\frac{1}{4}$ of the lawn. After 2 hr, you will have mowed $\frac{2}{4}$ or $\frac{1}{2}$ of the lawn, and so on. This idea is generalized as follows.

Rate of Work

If a job can be completed in t units of time, then the rate of work is

$$\frac{1}{t} \text{ job per unit of time.}$$

PROBLEM SOLVING

The relationship between problems involving work and problems involving distance is a close one. Recall that the formula $d = rt$ says that distance traveled is equal to rate of travel multiplied by time traveled. Similarly, the fractional part of a job accomplished is equal to the rate of work multiplied by the time worked. In the lawn mowing example, after 3 hr, the fractional part of the job done is

$$\underbrace{\frac{1}{4}}_{\substack{\text{Rate of} \\ \text{work}}} \cdot \underbrace{3}_{\substack{\text{Time} \\ \text{worked}}} = \underbrace{\frac{3}{4}}_{\substack{\text{Fractional part} \\ \text{of job done}}}.$$

After 4 hr, $\frac{1}{4}(4) = 1$ whole job has been done.

EXAMPLE 3 Solving a Problem about Work Rates

With spraying equipment, Mateo can paint the woodwork in a small house in 8 hr. His assistant, Chet, needs 14 hr to complete the same job painting by hand. If Mateo and Chet work together, how long will it take them to paint the woodwork?

Step 1 **Read** the problem again. We are looking for time working together.

Step 2 **Assign a variable.** Let $x =$ the number of hours it will take for Mateo and Chet to paint the woodwork, working together.

Certainly, x will be less than 8, since Mateo alone can complete the job in 8 hr. We begin by making a table as shown. Remember that based on the previous discussion, Mateo's rate alone is $\frac{1}{8}$ job per hour, and Chet's rate is $\frac{1}{14}$ job per hour.

	Rate	Time Working Together	Fractional Part of the Job Done When Working Together
Mateo	$\frac{1}{8}$	x	$\frac{1}{8}x$
Chet	$\frac{1}{14}$	x	$\frac{1}{14}x$

Sum is 1 whole job.

Step 3 **Write an equation.** Since together Mateo and Chet complete 1 whole job, we must add their individual fractional parts and set the sum equal to 1.

$$\underbrace{\frac{1}{8}x}_{\text{Fractional part done by Mateo}} + \underbrace{\frac{1}{14}x}_{\text{Fractional part done by Chet}} = \underbrace{1}_{\text{1 whole job.}}$$

Step 4 **Solve.**

$$56\left(\frac{1}{8}x + \frac{1}{14}x\right) = 56(1) \quad \text{Multiply by the LCD, 56.}$$

$$56\left(\frac{1}{8}x\right) + 56\left(\frac{1}{14}x\right) = 56(1) \quad \text{Distributive property}$$

$$7x + 4x = 56$$

$$11x = 56 \quad \text{Combine like terms.}$$

$$x = \frac{56}{11} \quad \text{Divide by 11.}$$

Step 5 **State the answer.** Working together, Mateo and Chet can paint the woodwork in $\frac{56}{11}$ hr, or $5\frac{1}{11}$ hr.

Step 6 **Check** to be sure the answer is correct.

Now Try Exercise 33.

NOTE An alternative approach in work problems is to consider the part of the job that can be done in 1 hr. For instance, in Example 3 Mateo can do the entire job in 8 hr, and Chet can do it in 14 hr. Thus, their work rates, as we saw

(continued)

502 CHAPTER 7 Rational Expressions and Applications

in Example 3, are $\frac{1}{8}$ and $\frac{1}{14}$, respectively. Since it takes them x hr to complete the job when working together, in 1 hr they can paint $\frac{1}{x}$ of the woodwork. The amount painted by Mateo in 1 hr plus the amount painted by Chet in 1 hr must equal the amount they can do together. This leads to the equation

Amount by Mateo → $\frac{1}{8} + \underset{\underset{\text{Amount by Chet}}{\uparrow}}{\frac{1}{14}} = \frac{1}{x}$. ← Amount together

Compare this with the equation in Example 3. Multiplying each side by $56x$ leads to

$$7x + 4x = 56,$$

the same equation found in the third line of Step 4 in the example. The same solution results.

7.7 EXERCISES

For Extra Help

- Student's Solutions Manual
- MyMathLab
- InterAct Math Tutorial Software
- AW Math Tutor Center
- MathXL
- Digital Video Tutor CD 15/Videotape 13

Use Steps 2 and 3 of the six-step method to set up the equation you would use to solve each problem. (Remember that Step 1 is to read the problem carefully.) Do not actually solve the equation. See Example 1.

1. The numerator of the fraction $\frac{5}{6}$ is increased by an amount so that the value of the resulting fraction is equivalent to $\frac{13}{3}$. By what amount was the numerator increased?

   (a) Let $x =$ _____ . (*Step 2*)
   (b) Write an expression for "the numerator of the fraction $\frac{5}{6}$ is increased by an amount."
   (c) Set up an equation to solve the problem. (*Step 3*)

2. If the same number is added to the numerator and subtracted from the denominator of $\frac{23}{12}$, the resulting fraction is equivalent to $\frac{3}{2}$. What is the number?

   (a) Let $x =$ _____ . (*Step 2*)
   (b) Write an expression for "a number is added to the numerator of $\frac{23}{12}$." Then write an expression for "the same number is subtracted from the denominator of $\frac{23}{12}$."
   (c) Set up an equation to solve the problem. (*Step 3*)

Use the six-step method to solve each problem. See Example 1.

3. In a certain fraction, the denominator is 6 more than the numerator. If 3 is added to both the numerator and the denominator, the resulting fraction is equivalent to $\frac{5}{7}$. What was the original fraction?

4. In a certain fraction, the denominator is 4 less than the numerator. If 3 is added to both the numerator and the denominator, the resulting fraction is equivalent to $\frac{3}{2}$. What was the original fraction?

5. The numerator of a certain fraction is four times the denominator. If 6 is added to both the numerator and the denominator, the resulting fraction is equivalent to 2. What was the original fraction?

6. The denominator of a certain fraction is three times the numerator. If 2 is added to the numerator and subtracted from the denominator, the resulting fraction is equivalent to 1. What was the original fraction?

7. One-third of a number is 2 more than one-sixth of the same number. What is the number?

8. One-sixth of a number is 5 more than the same number. What is the number?

9. A quantity, its $\frac{2}{3}$, its $\frac{1}{2}$, and its $\frac{1}{7}$, added together, become 33. What is the quantity? (*Source:* Rhind Mathematical Papyrus.)

10. A quantity, its $\frac{3}{4}$, its $\frac{1}{2}$, and its $\frac{1}{3}$, added together, become 93. What is the quantity? (*Source:* Rhind Mathematical Papyrus.)

Solve each problem. See Example 5 in Section 2.7.

11. In the 2002 Winter Olympics, Catriona LeMay Doan of Canada won the 500-m speed skating event for women. Her rate was 6.6889 m per sec. What was her time (to the nearest hundredth of a second)? (*Source:* www.espn.com)

12. In the 1998 World Championships, Amy Van Dyken of the United States won the 50-m freestyle swimming event for women. Her rate was 1.9881 m per sec. What was her time (to the nearest hundredth of a second)? (*Source: Sports Illustrated 2000 Sports Almanac.*)

13. The winner of the women's 1500-m race in the 2000 Olympics was Nouria Merah-Benida of Algeria with a time of 4.085 min. What was her rate (to three decimal places)? (*Source: World Almanac and Book of Facts,* 2002.)

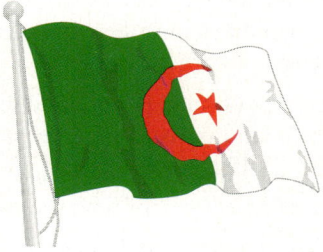

14. Gabriela Szabo of Romania won the women's 5000-m race in the 2000 Olympics with a time of 14.680 min. What was her rate (to three decimal places)? (*Source: World Almanac and Book of Facts,* 2002.)

15. The winner of the 2001 Daytona 500 (mile) race was Michael Waltrip, who drove his Chevrolet to victory with a rate of 161.794 mph. What was his time (to the nearest thousandth of an hour)? (*Source: World Almanac and Book of Facts,* 2002.)

504 CHAPTER 7 Rational Expressions and Applications

16. In 2001, Helio Castroneves drove his Reynard-Honda to victory in the Indianapolis 500 (mile) race. His rate was 131.294 mph. What was his time (to the nearest thousandth of an hour)? (*Source: World Almanac and Book of Facts, 2002.*)

Use the concepts of this section to solve each problem.

17. Suppose Stephanie walks D mi at R mph in the same time that Wally walks d mi at r mph. Give an equation relating D, R, d, and r.

18. If a migrating hawk travels m mph in still air, what is its rate when it flies into a steady headwind of 5 mph? What is its rate with a tailwind of 5 mph?

Set up the equation you would use to solve each problem. Do not actually solve the equation. See Example 2.

19. Julio flew his airplane 500 mi against the wind in the same time it took him to fly 600 mi with the wind. If the speed of the wind was 10 mph, what was the average speed of his plane? (Let x = speed of the plane in still air.)

	d	r	t
Against the Wind	500	$x - 10$	
With the Wind	600	$x + 10$	

20. Luvenia can row 4 mph in still water. She takes as long to row 8 mi upstream as 24 mi downstream. How fast is the current? (Let x = speed of the current.)

	d	r	t
Upstream	8	$4 - x$	
Downstream	24	$4 + x$	

Solve each problem. See Example 2.

21. A boat can go 20 mi against a current in the same time that it can go 60 mi with the current. The current is 4 mph. Find the speed of the boat in still water.

22. Lucinda can fly her plane 200 mi against the wind in the same time it takes her to fly 300 mi with the wind. The wind blows at 30 mph. Find the speed of her plane in still air.

23. An airplane, maintaining a constant airspeed, takes as long to go 450 mi with the wind as it does to go 375 mi against the wind. If the wind is blowing at 15 mph, what is the speed of the plane?

24. A river has a current of 4 km per hr. Find the speed of Lynn McTernan's boat in still water if it goes 40 km downstream in the same time that it takes to go 24 km upstream.

25. Kellen's boat goes 12 mph. Find the rate of the current of the river if she can go 6 mi upstream in the same amount of time she can go 10 mi downstream.

26. Kasey can travel 8 mi upstream in the same time it takes her to go 12 mi downstream. Her boat goes 15 mph in still water. What is the rate of the current?

27. The distance from Seattle, Washington, to Victoria, British Columbia, is about 148 mi by ferry. It takes about 4 hr less to travel by the same ferry from Victoria to Vancouver, British Columbia, a distance of about 74 mi. What is the average speed of the ferry?

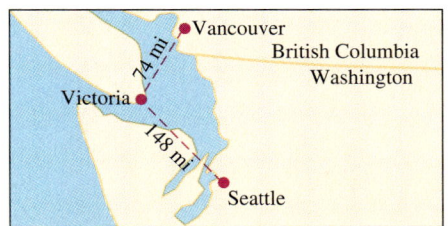

28. Driving from Tulsa to Detroit, Jeff averaged 50 mph. He figured that if he had averaged 60 mph, his driving time would have decreased 3 hr. How far is it from Tulsa to Detroit?

Use the concepts of this section to solve each problem.

29. If it takes Elayn 10 hr to do a job, what is her rate?
30. If it takes Clay 12 hr to do a job, how much of the job does he do in 8 hr?

In Exercises 31 and 32, set up the equation you would use to solve each problem. Do not actually solve the equation. See Example 3.

31. Working alone, Jorge can paint a room in 8 hr. Caterina can paint the same room working alone in 6 hr. How long will it take them if they work together? (Let x represent the time working together.)

	r	t	w
Jorge		x	
Caterina		x	

32. Edwin Bedford can tune up his Chevy in 2 hr working alone. His son, Beau, can do the job in 3 hr working alone. How long would it take them if they worked together? (Let t represent the time working together.)

	r	t	w
Edwin		t	
Beau		t	

Solve each problem. See Example 3.

33. Geraldo and Luisa Hernandez operate a small laundry. Luisa, working alone, can clean a day's laundry in 9 hr. Geraldo can clean a day's laundry in 8 hr. How long would it take them if they work together?

34. Lea can groom the horses in her boarding stable in 5 hr, while Tran needs 4 hr. How long will it take them to groom the horses if they work together?

35. A pump can pump the water out of a flooded basement in 10 hr. A smaller pump takes 12 hr. How long would it take to pump the water from the basement using both pumps?

36. Doug Todd's copier can do a printing job in 7 hr. Scott's copier can do the same job in 12 hr. How long would it take to do the job using both copiers?

37. An experienced employee can enter tax data into a computer twice as fast as a new employee. Working together, it takes the employees 2 hr. How long would it take the experienced employee working alone?

38. One roofer can put a new roof on a house three times faster than another. Working together they can roof a house in 4 days. How long would it take the faster roofer working alone?

39. One pipe can fill a swimming pool in 6 hr, and another pipe can do it in 9 hr. How long will it take the two pipes working together to fill the pool $\frac{3}{4}$ full?

40. An inlet pipe can fill a swimming pool in 9 hr, and an outlet pipe can empty the pool in 12 hr. Through an error, both pipes are left open. How long will it take to fill the pool?

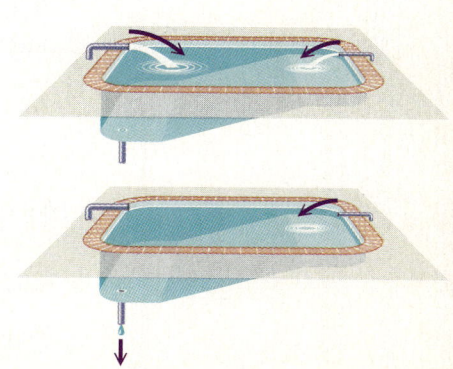

506 CHAPTER 7 Rational Expressions and Applications

41. A cold water faucet can fill a sink in 12 min, and a hot water faucet can fill it in 15 min. The drain can empty the sink in 25 min. If both faucets are on and the drain is open, how long will it take to fill the sink?

42. Refer to Exercise 40. Assume the error was discovered after both pipes had been running for 3 hr, and the outlet pipe was then closed. How much more time would then be required to fill the pool? (*Hint:* Consider how much of the job had been done when the error was discovered.)

7.8 Variation

OBJECTIVES

1. Solve direct variation problems.
2. Solve inverse variation problems.

OBJECTIVE 1 Solve direct variation problems. Suppose that gasoline costs $1.50 per gal. Then 1 gal costs $1.50, 2 gal cost 2($1.50) = $3.00, 3 gal cost 3($1.50) = $4.50, and so on. Each time, the total cost is obtained by multiplying the number of gallons by the price per gallon. In general, if k equals the price per gallon and x equals the number of gallons, then the total cost y is equal to kx. Notice that as the *number of gallons increases,* the *total cost increases.* The reverse is also true—as the *number of gallons decreases,* the *total cost decreases.*

The preceding discussion is an example of *variation.* Equations with fractions often result when discussing variation. As in the gasoline example, two variables *vary directly* if one is a constant multiple of the other.

Direct Variation

y **varies directly as** *x* if there exists a constant *k* such that
$$y = kx.$$

Also, y is said to be *proportional to x.* The k in the equation for direct variation is a numerical value, such as 1.50 in the gasoline discussion. This value is called the **constant of variation.**

EXAMPLE 1 Using Direct Variation

Suppose y varies directly as x, and $y = 20$ when $x = 4$. Find y when $x = 9$.

Since y varies directly as x, as x increases from 4 to 9, y also increases from 20 to some larger number. By the definition of direct variation, there is a constant k such that $y = kx$. We know that $y = 20$ when $x = 4$. Substituting these values into $y = kx$ and solving for k gives

$$y = kx$$
$$20 = k \cdot 4$$
$$k = 5.$$

Since $y = kx$ and $k = 5$,

$$y = 5x. \qquad \text{Let } k = 5.$$

When $x = 9$, $y = 5x = 5 \cdot 9 = 45.$ Let $x = 9$.

Thus, $y = 45$ when $x = 9$.

Now Try Exercise 19.

Solving a Variation Problem

Step 1 Write the variation equation.

Step 2 Substitute the appropriate given values and solve for *k*.

Step 3 Rewrite the variation equation with the value of *k* from Step 2.

Step 4 Substitute the remaining values, solve for the unknown, and find the required answer.

The direct variation equation $y = kx$ is a linear equation. However, other kinds of variation involve other types of equations. For example, one variable can be proportional to a power of another variable.

Direct Variation as a Power

y* varies directly as the *n*th power of *x if there exists a real number *k* such that

$$y = kx^n.$$

An example of direct variation as a power is the formula for the area of a circle, $A = \pi r^2$. Here, π is the constant of variation, and the area varies directly as the square of the radius.

EXAMPLE 2 Solving a Direct Variation Problem

The distance a body falls from rest varies directly as the square of the time it falls (disregarding air resistance). If a skydiver falls 64 ft in 2 sec, how far will she fall in 8 sec?

Step 1 If *d* represents the distance the skydiver falls and *t* the time it takes to fall, then *d* is a function of *t*, and, for some constant *k*,

$$d = kt^2.$$

Step 2 To find the value of *k*, use the fact that the skydiver falls 64 ft in 2 sec.

$d = kt^2$ Variation equation

$64 = k(2)^2$ Let $d = 64$ and $t = 2$.

$64 = 4k$

$k = 16$ Divide by 4.

Step 3 Using 16 for *k*, the variation equation becomes

$$d = 16t^2.$$

Step 4 Now let $t = 8$ to find the number of feet the skydiver will fall in 8 sec.

$d = 16(8)^2$ Let $t = 8$.

$d = 1024$

The skydiver will fall 1024 ft in 8 sec.

Now Try Exercise 41.

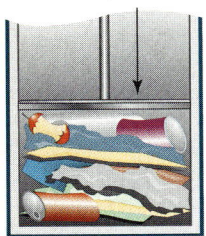

As pressure on trash increases, volume of trash decreases.

FIGURE 2

OBJECTIVE 2 Solve inverse variation problems. In direct variation, where $k > 0$, as x increases, y increases. Similarly, as x decreases, y decreases. Another type of variation is *inverse variation*. With inverse variation, where $k > 0$, as one variable increases, the other variable decreases. For example, in a closed space, volume decreases as pressure increases, as illustrated by a trash compactor. See Figure 2. As the compactor presses down, the pressure on the trash increases; in turn, the trash occupies a smaller space.

> **Inverse Variation**
>
> **y varies inversely as x** if there exists a real number k such that
>
> $$y = \frac{k}{x}.$$
>
> Also, **y varies inversely as the nth power of x** if there exists a real number k such that
>
> $$y = \frac{k}{x^n}.$$

Another example of inverse variation comes from the distance formula. In its usual form, the formula is

$$d = rt.$$

Dividing each side by r gives

$$t = \frac{d}{r}.$$

Here, t (time) varies inversely as r (rate or speed), with d (distance) serving as the constant of variation. For example, if the distance between Chicago and Des Moines is 300 mi, then

$$t = \frac{300}{r}$$

and the values of r and t might be any of the following.

$\left. \begin{array}{l} r = 50, t = 6 \\ r = 60, t = 5 \\ r = 75, t = 4 \end{array} \right\}$ As r increases, t decreases. $\left. \begin{array}{l} r = 30, t = 10 \\ r = 25, t = 12 \\ r = 20, t = 15 \end{array} \right\}$ As r decreases, t increases.

If we *increase* the rate (speed) we drive, time *decreases*. If we *decrease* the rate (speed) we drive, what happens to time?

EXAMPLE 3 Using Inverse Variation

Suppose y varies inversely as x, and $y = 3$ when $x = 8$. Find y when $x = 6$.

Since y varies inversely as x, there is a constant k such that $y = \frac{k}{x}$. We know that $y = 3$ when $x = 8$, so we can find k.

$$y = \frac{k}{x}$$

$$3 = \frac{k}{8}$$

$$k = 24$$

Since $y = \frac{24}{x}$, we let $x = 6$ and solve for y.

$$y = \frac{24}{x} = \frac{24}{6} = 4$$

Therefore, when $x = 6$, $y = 4$.

Now Try Exercise 23.

EXAMPLE 4 Using Inverse Variation

In the manufacturing of a certain medical syringe, the cost of producing the syringe varies inversely as the number produced. If 10,000 syringes are produced, the cost is $2 per syringe. Find the cost per syringe to produce 25,000 syringes.

Let x = the number of syringes produced,

and c = the cost per unit.

Here, as production increases, cost decreases and as production decreases, cost increases. Since c varies inversely as x, there is a constant k such that

$$c = \frac{k}{x}.$$

Find k by replacing c with 2 and x with 10,000.

$$2 = \frac{k}{10,000}$$

$$20,000 = k \qquad \text{Multiply by 10,000.}$$

Since $c = \frac{k}{x}$,

$$c = \frac{20,000}{25,000} = .80. \qquad \text{Let } k = 20,000 \text{ and } x = 25,000.$$

The cost per syringe to make 25,000 syringes is $.80.

Now Try Exercise 37.

510 CHAPTER 7 Rational Expressions and Applications

7.8 EXERCISES

For Extra Help

 Student's Solutions Manual

 MyMathLab

 InterAct Math Tutorial Software

 AW Math Tutor Center

 MathXL

 Digital Video Tutor CD 15/Videotape 13

Use personal experience or intuition to determine whether the situation suggests direct or inverse variation.*

1. The number of different lottery tickets you buy and your probability of winning that lottery
2. The rate and the distance traveled by a pickup truck in 3 hr
3. The amount of pressure put on the accelerator of a car and the speed of the car
4. The number of days from now until December 25 and the magnitude of the frenzy of Christmas shopping
5. The surface area of a balloon and its diameter
6. Your age and the probability that you believe in Santa Claus
7. The number of days until the end of the baseball season and the number of home runs that Sammy Sosa has
8. The amount of gasoline you pump and the amount you will pay

Determine whether each equation represents direct *or* inverse *variation.*

9. $y = \dfrac{3}{x}$ 10. $y = \dfrac{8}{x}$ 11. $y = 10x^2$ 12. $y = 2x^3$

13. $y = 50x$ 14. $y = 100x$ 15. $y = \dfrac{12}{x^2}$ 16. $y = \dfrac{10}{x^3}$

Fill in each blank with the correct response.

17. If the constant of variation is positive and y varies directly as x, then as x increases, y _____.
    (increases/decreases)

18. If the constant of variation is positive and y varies inversely as x, then as x increases, y _____.
    (increases/decreases)

Solve each problem involving direct or inverse variation. See Examples 1 and 3. (odd #)

19. If x varies directly as y, and $x = 27$ when $y = 6$, find x when $y = 2$.
20. If z varies directly as x, and $z = 30$ when $x = 8$, find z when $x = 4$.
21. If d varies directly as t, and $d = 150$ when $t = 3$, find d when $t = 5$.
22. If d varies directly as r, and $d = 200$ when $r = 40$, find d when $r = 60$.
23. If x varies inversely as y, and $x = 3$ when $y = 8$, find y when $x = 4$.
24. If z varies inversely as x, and $z = 50$ when $x = 2$, find z when $x = 25$.
25. If p varies inversely as q, and $p = 7$ when $q = 6$, find p when $q = 2$.
26. If m varies inversely as r, and $m = 12$ when $r = 8$, find m when $r = 16$.
27. If m varies inversely as p^2, and $m = 20$ when $p = 2$, find m when $p = 5$.

---

*The authors thank Linda Kodama of Kapiolani Community College for suggesting the inclusion of exercises of this type.

28. If a varies inversely as b^2, and $a = 48$ when $b = 4$, find a when $b = 7$.

29. If p varies inversely as q^2, and $p = 4$ when $q = \frac{1}{2}$, find p when $q = \frac{3}{2}$.

30. If z varies inversely as x^2, and $z = 9$ when $x = \frac{2}{3}$, find z when $x = \frac{5}{4}$.

Solve each variation problem. See Examples 1–4.

31. The interest on an investment varies directly as the rate of interest. If the interest is $48 when the interest rate is 5%, find the interest when the rate is 4.2%.

32. For a given base, the area of a triangle varies directly as its height. Find the area of a triangle with a height of 6 in., if the area is 10 in.^{2} when the height is 4 in.

33. Hooke's law for an elastic spring states that the distance a spring stretches varies directly with the force applied. If a force of 75 lb stretches a certain spring 16 in., how much will a force of 200 lb stretch the spring?

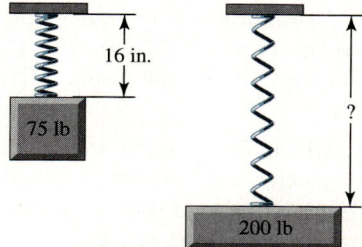

34. The pressure exerted by water at a given point varies directly with the depth of the point beneath the surface of the water. Water exerts 4.34 lb per in.^{2} for every 10 ft traveled below the water's surface. What is the pressure exerted on a scuba diver at 20 ft?

35. Over a specified distance, speed varies inversely with time. If a Dodge Viper on a test track goes a certain distance in one-half minute at 160 mph, what speed is needed to go the same distance in three-fourths minute?

36. For a constant area, the length of a rectangle varies inversely as the width. The length of a rectangle is 27 ft when the width is 10 ft. Find the width of a rectangle with the same area if the length is 18 ft.

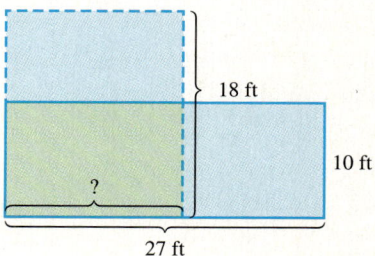

37. The current in a simple electrical circuit varies inversely as the resistance. If the current is 20 amps when the resistance is 5 ohms, find the current when the resistance is 8 ohms.

38. If the temperature is constant, the pressure of a gas in a container varies inversely as the volume of the container. If the pressure is 10 lb per ft^2 in a container with volume 3 ft^3, what is the pressure in a container with volume 1.5 ft^3?

512 CHAPTER 7 Rational Expressions and Applications

39. The force required to compress a spring varies directly as the change in the length of the spring. If a force of 12 lb is required to compress a certain spring 3 in., how much force is required to compress the spring 5 in.?

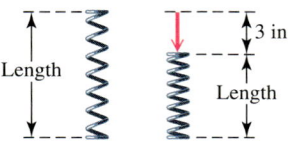

40. In the inversion of raw sugar, the rate of change of the amount of raw sugar varies directly as the amount of raw sugar remaining. The rate is 200 kg per hr when there are 800 kg left. What is the rate of change per hour when only 100 kg are left?

41. The area of a circle varies directly as the square of its radius. A circle with radius 3 in. has area 28.278 in.^{2}. What is the area of a circle with radius 4.1 in. (to the nearest thousandth)?

42. For a body falling freely from rest (disregarding air resistance), the distance the body falls varies directly as the square of the time. If an object is dropped from the top of a tower 400 ft high and hits the ground in 5 sec, how far did it fall in the first 3 sec?

43. The amount of light (measured in *foot-candles*) produced by a light source varies inversely as the square of the distance from the source. If the amount of light produced 4 ft from a light source is 75 foot-candles, find the amount of light produced 9 ft from the same source.

44. The force with which Earth attracts an object above Earth's surface varies inversely as the square of the object's distance from the center of Earth. If an object 4000 mi from the center of Earth is attracted with a force of 160 lb, find the force of attraction on an object 6000 mi from the center of Earth.

TECHNOLOGY INSIGHTS (EXERCISES 45–48)

In each table, Y_1 either varies directly or varies inversely as X. Tell which type of variation is illustrated.

45.
X	Y₁
0	0
1	2.5
2	5
3	7.5
4	10
5	12.5
6	15

X=0

46.
X	Y₁
0	0
1	3.2
2	6.4
3	9.6
4	12.8
5	16
6	19.2

X=0

47.
X	Y₁
1	24
2	12
3	8
4	6
5	4.8
6	4
7	3.4286

X=1

48.
X	Y₁
1	48
2	24
3	16
4	12
5	9.6
6	8
7	6.8571

X=1

Chapter 7 Group Activity

Buying a Car

OBJECTIVE Use a complex fraction to calculate monthly car payments.

You are shopping for a sports car and have put aside a certain amount of money each month for a car payment. Your instructor will assign this amount to you. After looking through a variety of resources, you have narrowed your choices to the cars listed in the table.

Year/Make/Model	Retail Price	Fuel Tank Size (in gallons)	Miles per Gallon (city)	Miles per Gallon (highway)
2002 Mitsubishi Eclipse	$24,197	16.4	21	28
2002 Chevy Camaro	18,415	16.8	19	31
2002 Ford Mustang	18,080	15.7	20	29
2002 Pontiac Firebird	20,050	16.8	19	31
2002 BMW-Z3	37,700	13.5	21	29
2002 Toyota Celica GTS	21,555	14.5	23	32

Source: http://aolsvc.carguides.aol.com

As a group, work through the following steps to determine which car you can afford to buy.

A. Decide which cars you think are within your budget.

B. Select one of the cars you identified in part A. Have each member of the group calculate the monthly payment for this car using a different financing option. Use the formula given below, where P is principal, r is interest rate, and m is the number of monthly payments, along with the financing options table.

Financing Options

Time (in years)	Interest Rate
4	7.0%
5	8.5%
6	10.0%

$$\text{Monthly Payment} = \frac{\frac{Pr}{12}}{1 - \left(\frac{12}{12 + r}\right)^m}$$

C. Have each group member determine the amount of money paid in interest over the duration of the loan for their financing option.

D. Consider fuel expenses.
  1. Assume you will travel an average of 75 mi in the city and 400 mi on the highway each week. How many gallons of gas will you need to buy each month?
  2. Using typical prices for gas in your area at this time, how much money will you need to have available for buying gas?

E. Repeat parts B–D as necessary until your group can reach a consensus on the car you will buy and the financing option you will use. Write a paragraph to explain your choices.

514 CHAPTER 7 Rational Expressions and Applications

CHAPTER 7 SUMMARY

KEY TERMS

7.1 rational expression
lowest terms

7.3 least common denominator (LCD)

7.5 complex fraction
7.8 direct variation

constant of variation
inverse variation

TEST YOUR WORD POWER

See how well you have learned the vocabulary in this chapter. Answers, with examples, follow the Quick Review.

1. A **rational expression** is
   A. an algebraic expression made up of a term or the sum of a finite number of terms with real coefficients and whole number exponents
   B. a polynomial equation of degree 2
   C. an expression with one or more fractions in the numerator, denominator, or both

   D. the quotient of two polynomials with denominator not 0.

2. A **complex fraction** is
   A. an algebraic expression made up of a term or the sum of a finite number of terms with real coefficients and whole number exponents

   B. a polynomial equation of degree 2
   C. a rational expression with one or more fractions in the numerator, denominator, or both
   D. the quotient of two polynomials with denominator not 0.

QUICK REVIEW

CONCEPTS	EXAMPLES

7.1 THE FUNDAMENTAL PROPERTY OF RATIONAL EXPRESSIONS

To find the value(s) for which a rational expression is undefined, set the denominator equal to 0 and solve the equation.

Find the values for which the expression $\dfrac{x-4}{x^2-16}$ is undefined.

$$x^2 - 16 = 0$$
$$(x-4)(x+4) = 0$$
$$x - 4 = 0 \quad \text{or} \quad x + 4 = 0$$
$$x = 4 \quad \text{or} \quad x = -4$$

The rational expression is undefined for 4 and -4, so $x \neq 4$ and $x \neq -4$.

Writing a Rational Expression in Lowest Terms

Write $\dfrac{x^2-1}{(x-1)^2}$ in lowest terms.

Step 1 Factor the numerator and denominator.

Step 2 Use the fundamental property to divide out common factors.

$$\frac{x^2-1}{(x-1)^2} = \frac{(x-1)(x+1)}{(x-1)(x-1)}$$
$$= \frac{x+1}{x-1}$$

There are often several different equivalent forms of a rational expression.

Give four equivalent forms of $-\dfrac{x-1}{x+2}$.

Distribute the $-$ sign in the numerator to get $\dfrac{-(x-1)}{x+2}$ or

$\dfrac{-x+1}{x+2}$; do so in the denominator to get $\dfrac{x-1}{-(x+2)}$ or $\dfrac{x-1}{-x-2}$.

CHAPTER 7 Summary

CONCEPTS	EXAMPLES

7.2 MULTIPLYING AND DIVIDING RATIONAL EXPRESSIONS

Multiplying Rational Expressions

Step 1 Multiply numerators and multiply denominators.

Step 2 Factor.

Step 3 Write in lowest terms.

Multiply. $\dfrac{3x+9}{x-5} \cdot \dfrac{x^2-3x-10}{x^2-9}$

$$= \dfrac{(3x+9)(x^2-3x-10)}{(x-5)(x^2-9)}$$

$$= \dfrac{3(x+3)(x-5)(x+2)}{(x-5)(x+3)(x-3)}$$

$$= \dfrac{3(x+2)}{x-3}$$

Dividing Rational Expressions

Step 1 Multiply the first rational expression by the reciprocal of the second.

Step 2 Multiply numerators and multiply denominators.

Step 3 Factor.

Step 4 Write in lowest terms.

Divide. $\dfrac{2x+1}{x+5} \div \dfrac{6x^2-x-2}{x^2-25}$

$$= \dfrac{2x+1}{x+5} \cdot \dfrac{x^2-25}{6x^2-x-2}$$

$$= \dfrac{(2x+1)(x^2-25)}{(x+5)(6x^2-x-2)}$$

$$= \dfrac{(2x+1)(x+5)(x-5)}{(x+5)(2x+1)(3x-2)}$$

$$= \dfrac{x-5}{3x-2}$$

7.3 LEAST COMMON DENOMINATORS

Finding the LCD

Step 1 Factor each denominator into prime factors.

Step 2 List each different factor the greatest number of times it appears.

Step 3 Multiply the factors from Step 2 to get the LCD.

Find the LCD for $\dfrac{3}{k^2-8k+16}$ and $\dfrac{1}{4k^2-16k}$.

$k^2-8k+16 = (k-4)^2$ \quad (k-4)(k-4)

$4k^2-16k = 4k(k-4)$ \quad 4k(k-4)

LCD $= (k-4)^2 \cdot 4 \cdot k$

$= 4k(k-4)^2$

Writing a Rational Expression with a Specified Denominator

Step 1 Factor both denominators.

Step 2 Decide what factors the denominator must be multiplied by to equal the specified denominator.

Step 3 Multiply the rational expression by that factor divided by itself (multiply by 1).

Find the numerator: $\dfrac{5}{2z^2-6z} = \dfrac{}{4z^3-12z^2}$.

$$\dfrac{5}{2z(z-3)} = \dfrac{}{4z^2(z-3)}$$

$2z(z-3)$ must be multiplied by $2z$.

$$\dfrac{5}{2z(z-3)} \cdot \dfrac{2z}{2z} = \dfrac{10z}{4z^2(z-3)} = \dfrac{10z}{4z^3-12z^2}$$

(continued)

516 CHAPTER 7 Rational Expressions and Applications

CONCEPTS	EXAMPLES

7.4 ADDING AND SUBTRACTING RATIONAL EXPRESSIONS

Adding Rational Expressions

Add. $\dfrac{2}{3m + 6} + \dfrac{m}{m^2 - 4}$

Step 1 Find the LCD.

$3m + 6 = 3(m + 2)$
$m^2 - 4 = (m + 2)(m - 2)$

The LCD is $3(m + 2)(m - 2)$.

Step 2 Rewrite each rational expression with the LCD as denominator.

$$= \dfrac{2(m - 2)}{3(m + 2)(m - 2)} + \dfrac{3m}{3(m + 2)(m - 2)}$$

Step 3 Add the numerators to get the numerator of the sum. The LCD is the denominator of the sum.

$$= \dfrac{2m - 4 + 3m}{3(m + 2)(m - 2)}$$

Step 4 Write in lowest terms.

$$= \dfrac{5m - 4}{3(m + 2)(m - 2)}$$

Subtracting Rational Expressions

Subtract. $\dfrac{6}{k + 4} - \dfrac{2}{k}$

Follow the same steps as for addition, but subtract in Step 3.

The LCD is $k(k + 4)$.

$$\dfrac{6k}{(k + 4)k} - \dfrac{2(k + 4)}{k(k + 4)} = \dfrac{6k - 2(k + 4)}{k(k + 4)}$$

Be careful with signs when subtracting the numerators.

$$= \dfrac{6k - 2k - 8}{k(k + 4)}$$

$$= \dfrac{4k - 8}{k(k + 4)} \quad \text{or} \quad \dfrac{4(k - 2)}{k(k + 4)}$$

7.5 COMPLEX FRACTIONS

Simplifying Complex Fractions

Simplify.

Method 1 Simplify the numerator and denominator separately. Then divide the simplified numerator by the simplified denominator.

Method 1 $\dfrac{\dfrac{1}{a} - a}{1 - a} = \dfrac{\dfrac{1}{a} - \dfrac{a^2}{a}}{1 - a} = \dfrac{\dfrac{1 - a^2}{a}}{1 - a}$

$$= \dfrac{1 - a^2}{a} \div (1 - a)$$

$$= \dfrac{1 - a^2}{a} \cdot \dfrac{1}{1 - a}$$

$$= \dfrac{(1 - a)(1 + a)}{a(1 - a)} = \dfrac{1 + a}{a}$$

Method 2 Multiply the numerator and denominator of the complex fraction by the LCD of all the denominators in the complex fraction. Write in lowest terms.

Method 2 $\dfrac{\dfrac{1}{a} - a}{1 - a} = \dfrac{\dfrac{1}{a} - a}{1 - a} \cdot \dfrac{a}{a} = \dfrac{\dfrac{a}{a} - a^2}{(1 - a)a}$

$$= \dfrac{1 - a^2}{(1 - a)a} = \dfrac{(1 + a)(1 - a)}{(1 - a)a}$$

$$= \dfrac{1 + a}{a}$$

CONCEPTS	EXAMPLES

7.6 SOLVING EQUATIONS WITH RATIONAL EXPRESSIONS

Solving Equations with Rational Expressions

Step 1 Find the LCD of all denominators in the equation.

Step 2 Multiply each side of the equation by the LCD.

Step 3 Solve the resulting equation, which should have no fractions.

Step 4 Check each proposed solution.

Solve $\dfrac{2}{x-1} + \dfrac{3}{4} = \dfrac{5}{x-1}$.

The LCD is $4(x-1)$. Note that 1 cannot be a solution.

$$4(x-1)\left(\dfrac{2}{x-1} + \dfrac{3}{4}\right) = 4(x-1)\left(\dfrac{5}{x-1}\right)$$

$$4(x-1)\left(\dfrac{2}{x-1}\right) + 4(x-1)\left(\dfrac{3}{4}\right) = 4(x-1)\left(\dfrac{5}{x-1}\right)$$

$$8 + 3(x-1) = 20$$
$$8 + 3x - 3 = 20$$
$$3x = 15$$
$$x = 5$$

The proposed solution, 5, checks. The solution set is $\{5\}$.

7.7 APPLICATIONS OF RATIONAL EXPRESSIONS

Solving Problems about Distance, Rate, Time
Use the six-step method.

Step 1 **Read** the problem carefully.

Step 2 **Assign a variable.** Use a table to identify distance, rate, and time. Solve $d = rt$ for the unknown quantity in the table.

Step 3 **Write an equation.** From the wording in the problem, decide the relationship between the quantities. Use those expressions to write an equation.

Step 4 **Solve** the equation.

Step 5 **State the answer.**

Step 6 **Check** the solution.

On a trip from Sacramento to Monterey, Marge traveled at an average speed of 60 mph. The return trip, at an average speed of 64 mph, took $\frac{1}{4}$ hr less. How far did she travel between the two cities?

Let x = the unknown distance.

	d	r	$t = \dfrac{d}{r}$
Going	x	60	$\dfrac{x}{60}$
Returning	x	64	$\dfrac{x}{64}$

Since the time for the return trip was $\frac{1}{4}$ hr less, the time going equals the time returning plus $\frac{1}{4}$.

$$\dfrac{x}{60} = \dfrac{x}{64} + \dfrac{1}{4}$$
$$16x = 15x + 240 \quad \text{Multiply by 960.}$$
$$x = 240 \quad \text{Subtract } 15x.$$

She traveled 240 mi.

The trip there took $\frac{240}{60} = 4$ hr, while the return trip took $\frac{240}{64} = 3\frac{3}{4}$ hr, which is $\frac{1}{4}$ hr less time. The solution checks.

(continued)

CONCEPTS

Solving Problems about Work

Step 1 **Read** the problem carefully.

Step 2 **Assign a variable.** State what the variable represents. Put the information from the problem in a table. If a job is done in t units of time, the rate is $\frac{1}{t}$.

Step 3 **Write an equation.** The sum of the fractional parts should equal 1 (whole job).

Step 4 **Solve** the equation.

Steps 5 and 6 **State the answer** and **check** the solution.

EXAMPLES

It takes the regular mail carrier 6 hr to cover her route. A substitute takes 8 hr to cover the same route. How long would it take them to cover the route together?

Let $x = $ the number of hours to cover the route together.

The rate of the regular carrier is $\frac{1}{6}$ job per hr; the rate of the substitute is $\frac{1}{8}$ job per hr. Multiply rate by time to get the fractional part of the job done.

	Rate	Time	Part of the Job Done
Regular	$\frac{1}{6}$	x	$\frac{1}{6}x$
Substitute	$\frac{1}{8}$	x	$\frac{1}{8}x$

The equation is $\frac{1}{6}x + \frac{1}{8}x = 1$.

The solution of the equation is $\frac{24}{7}$.

It would take them $\frac{24}{7}$ or $3\frac{3}{7}$ hr to cover the route together. The solution checks because $\frac{1}{6}\left(\frac{24}{7}\right) + \frac{1}{8}\left(\frac{24}{7}\right) = 1$ is true.

7.8 VARIATION

Solving Variation Problems

Step 1 Write the variation equation. Use
$y = kx$ or $y = kx^n$ for direct variation,
$y = \frac{k}{x}$ or $y = \frac{k}{x^n}$ for inverse variation.

Step 2 Find k by substituting the appropriate given values of x and y into the equation.

Step 3 Rewrite the variation equation with the value of k from Step 2.

Step 4 Substitute the remaining values, and solve for the unknown.

If y varies inversely as x, and $y = 4$ when $x = 9$, find y when $x = 6$.

The equation for inverse variation is

$$y = \frac{k}{x}.$$

$$4 = \frac{k}{9}$$

$$k = 36$$

$$y = \frac{36}{x} \qquad k = 36$$

$$y = \frac{36}{6} \qquad \text{Let } x = 6.$$

$$y = 6$$

Answers to Test Your Word Power

1. D; *Examples:* $-\dfrac{3}{4y}, \dfrac{5x^3}{x+2}, \dfrac{a+3}{a^2-4a-5}$ **2.** C; *Examples:* $\dfrac{\frac{2}{3}}{\frac{4}{7}}, \dfrac{x-\frac{1}{y}}{x+\frac{1}{y}}, \dfrac{\frac{2}{a+1}}{a^2-1}$

CHAPTER 7 REVIEW EXERCISES

[7.1] *Find the numerical value of each rational expression when* **(a)** $x = -2$ *and* **(b)** $x = 4$.

1. $\dfrac{4x - 3}{5x + 2}$

2. $\dfrac{3x}{x^2 - 4}$

Find the value(s) of the variable for which each rational expression is undefined.

3. $\dfrac{4}{x - 3}$

4. $\dfrac{y + 3}{2y}$

5. $\dfrac{2k + 1}{3k^2 + 17k + 10}$

6. How would you determine the values of the variable for which a rational expression is undefined?

Write each rational expression in lowest terms.

7. $\dfrac{5a^3 b^3}{15a^4 b^2}$

8. $\dfrac{m - 4}{4 - m}$

9. $\dfrac{4x^2 - 9}{6 - 4x}$

10. $\dfrac{4p^2 + 8pq - 5q^2}{10p^2 - 3pq - q^2}$

Write four equivalent forms for each rational expression.

11. $-\dfrac{4x - 9}{2x + 3}$

12. $\dfrac{8 - 3x}{3 + 6x}$

[7.2] *Multiply or divide, and write each answer in lowest terms.*

13. $\dfrac{18p^3}{6} \cdot \dfrac{24}{p^4}$

14. $\dfrac{8x^2}{12x^5} \cdot \dfrac{6x^4}{2x}$

15. $\dfrac{x - 3}{4} \cdot \dfrac{5}{2x - 6}$

16. $\dfrac{2r + 3}{r - 4} \cdot \dfrac{r^2 - 16}{6r + 9}$

17. $\dfrac{6a^2 + 7a - 3}{2a^2 - a - 6} \div \dfrac{a + 5}{a - 2}$

18. $\dfrac{y^2 - 6y + 8}{y^2 + 3y - 18} \div \dfrac{y - 4}{y + 6}$

19. $\dfrac{2p^2 + 13p + 20}{p^2 + p - 12} \cdot \dfrac{p^2 + 2p - 15}{2p^2 + 7p + 5}$

20. $\dfrac{3z^2 + 5z - 2}{9z^2 - 1} \cdot \dfrac{9z^2 + 6z + 1}{z^2 + 5z + 6}$

[7.3] *Find the least common denominator for each list of fractions.*

21. $\dfrac{4}{9y}, \dfrac{7}{12y^2}, \dfrac{5}{27y^4}$

22. $\dfrac{3}{x^2 + 4x + 3}, \dfrac{5}{x^2 + 5x + 4}$

Rewrite each rational expression with the given denominator.

23. $\dfrac{3}{2a^3} = \dfrac{}{10a^4}$

24. $\dfrac{9}{x - 3} = \dfrac{}{18 - 6x}$

25. $\dfrac{-3y}{2y - 10} = \dfrac{}{50 - 10y}$

26. $\dfrac{4b}{b^2 + 2b - 3} = \dfrac{}{(b + 3)(b - 1)(b + 2)}$

[7.4] *Add or subtract, and write each answer in lowest terms.*

27. $\dfrac{10}{x} + \dfrac{5}{x}$

28. $\dfrac{6}{3p} - \dfrac{12}{3p}$

29. $\dfrac{9}{k} - \dfrac{5}{k - 5}$

30. $\dfrac{4}{y} + \dfrac{7}{7+y}$ **31.** $\dfrac{m}{3} - \dfrac{2+5m}{6}$ **32.** $\dfrac{12}{x^2} - \dfrac{3}{4x}$

33. $\dfrac{5}{a-2b} + \dfrac{2}{a+2b}$ **34.** $\dfrac{4}{k^2-9} - \dfrac{k+3}{3k-9}$

35. $\dfrac{8}{z^2+6z} - \dfrac{3}{z^2+4z-12}$ **36.** $\dfrac{11}{2p-p^2} - \dfrac{2}{p^2-5p+6}$

[7.5]

37. Simplify the complex fraction $\dfrac{\dfrac{a^4}{b^2}}{\dfrac{a^3}{b}}$ by

    (a) Method 1 as described in Section 7.5.
    (b) Method 2 as described in Section 7.5.
    (c) Explain which method you prefer, and why.

Simplify each complex fraction.

38. $\dfrac{\dfrac{2}{3} - \dfrac{1}{6}}{\dfrac{1}{4} + \dfrac{2}{5}}$ **39.** $\dfrac{\dfrac{y-3}{y}}{\dfrac{y+3}{4y}}$ **40.** $\dfrac{\dfrac{1}{p} - \dfrac{1}{q}}{\dfrac{1}{q-p}}$

41. $\dfrac{x + \dfrac{1}{w}}{x - \dfrac{1}{w}}$ **42.** $\dfrac{\dfrac{1}{r+t} - 1}{\dfrac{1}{r+t} + 1}$

[7.6]

43. Before even beginning the solution process, how do you know that 2 cannot be a solution to the equation found in Exercise 46?

Solve each equation, and check your solutions.

44. $\dfrac{4-z}{z} + \dfrac{3}{2} = \dfrac{-4}{z}$ **45.** $\dfrac{3x-1}{x-2} = \dfrac{5}{x-2} + 1$

46. $\dfrac{3}{m-2} + \dfrac{1}{m-1} = \dfrac{7}{m^2-3m+2}$

Solve each formula for the specified variable.

47. $m = \dfrac{Ry}{t}$ for t **48.** $x = \dfrac{3y-5}{4}$ for y **49.** $p^2 = \dfrac{4}{3m-q}$ for m

[7.7] *Solve each problem.*

50. In a certain fraction, the denominator is 4 less than the numerator. If 3 is added to both the numerator and the denominator, the resulting fraction is equal to $\dfrac{3}{2}$. Find the original fraction.

51. The denominator of a certain fraction is 3 times the numerator. If 2 is added to the numerator and subtracted from the denominator, the resulting fraction is equal to 1. Find the original fraction.

52. A plane flies 350 mi with the wind in the same time that it can fly 310 mi against the wind. The plane has a still-air speed of 165 mph. Find the speed of the wind.

53. A man can plant his garden in 5 hr, working alone. His daughter can do the same job in 8 hr. How long would it take them if they worked together?

54. The head gardener can mow the lawns in the city park twice as fast as his assistant. Working together, they can complete the job in $1\frac{1}{3}$ hr. How long would it take the head gardener working alone?

[7.8] *Solve each problem.*

55. The longer the term of your subscription to *ESPN The Magazine*, the less you will have to pay per year. Is this an example of direct or inverse variation?

56. If a parallelogram has a fixed area, the height varies inversely as the base. A parallelogram has a height of 8 cm and a base of 12 cm. Find the height if the base is changed to 24 cm.

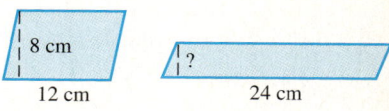

57. If y varies directly as x, and $x = 12$ when $y = 5$, find x when $y = 3$.

MIXED REVIEW EXERCISES

Perform each indicated operation.

58. $\dfrac{4}{m-1} - \dfrac{3}{m+1}$

59. $\dfrac{8p^5}{5} \div \dfrac{2p^3}{10}$

60. $\dfrac{r-3}{8} \div \dfrac{3r-9}{4}$

61. $\dfrac{\dfrac{5}{x} - 1}{\dfrac{5-x}{3x}}$

62. $\dfrac{4}{z^2 - 2z + 1} - \dfrac{3}{z^2 - 1}$

Solve.

63. $a = \dfrac{v-w}{t}$ for v

64. $\dfrac{2}{z} - \dfrac{z}{z+3} = \dfrac{1}{z+3}$

65. "If Joe can paint a house in 3 hours, and Sam can paint the same house in 5 hours, how long does it take for them to do it together?" (From the movie *Little Big League*)

66. Anne Kelly flew her plane 400 km with the wind in the same time it took her to go 200 km against the wind. The speed of the wind is 50 km per hr. Find the speed of the plane in still air.

67. In a rectangle of constant area, the length and the width vary inversely. When the length is 24, the width is 2. What is the width when the length is 12?

68. If w varies inversely as z, and $w = 16$ when $z = 3$, find w when $z = 2$.

RELATING CONCEPTS (EXERCISES 69–78)

For Individual or Group Work

In these exercises, we summarize the various concepts involving rational expressions.
Work Exercises 69–78 in order.

Let P, Q, and R be rational expressions defined as follows.

$$P = \frac{6}{x+3} \qquad Q = \frac{5}{x+1} \qquad R = \frac{4x}{x^2 + 4x + 3}$$

69. Find the value or values for which the expression is undefined.

    (a) P **(b)** Q **(c)** R

70. Find and express in lowest terms: $(P \cdot Q) \div R$.

71. Why is $(P \cdot Q) \div R$ not defined if $x = 0$?

72. Find the LCD for P, Q, and R.

73. Perform the operations and express in lowest terms: $P + Q - R$.

74. Simplify the complex fraction $\frac{P+Q}{R}$.

75. Solve the equation $P + Q = R$.

76. How does your answer to Exercise 69 help you work Exercise 75?

77. Suppose that a car travels 6 mi in $x + 3$ min. Explain why P represents the rate of the car (in miles per minute).

78. For what value or values of x is $R = \frac{40}{77}$?

CHAPTER 7 TEST

1. Find the numerical value of $\dfrac{6r+1}{2r^2 - 3r - 20}$ when

    (a) $r = -2$ **(b)** $r = 4$.

2. Find any values for which $\dfrac{3x-1}{x^2 - 2x - 8}$ is undefined.

3. Write four rational expressions equivalent to $-\dfrac{6x-5}{2x+3}$.

Write each rational expression in lowest terms.

4. $\dfrac{-15x^6 y^4}{5x^4 y}$

5. $\dfrac{6a^2 + a - 2}{2a^2 - 3a + 1}$

Multiply or divide. Write each answer in lowest terms.

6. $\dfrac{5(d-2)}{9} \div \dfrac{3(d-2)}{5}$

7. $\dfrac{6k^2 - k - 2}{8k^2 + 10k + 3} \cdot \dfrac{4k^2 + 7k + 3}{3k^2 + 5k + 2}$

8. $\dfrac{4a^2 + 9a + 2}{3a^2 + 11a + 10} \div \dfrac{4a^2 + 17a + 4}{3a^2 + 2a - 5}$

Find the least common denominator for each list of fractions.

9. $\dfrac{-3}{10p^2}, \dfrac{21}{25p^3}, \dfrac{-7}{30p^5}$

10. $\dfrac{r+1}{2r^2 + 7r + 6}, \dfrac{-2r+1}{2r^2 - 7r - 15}$

Rewrite each rational expression with the given denominator.

11. $\dfrac{15}{4p} = \dfrac{}{64p^3}$

12. $\dfrac{3}{6m - 12} = \dfrac{}{42m - 84}$

Add or subtract. Write each answer in lowest terms.

13. $\dfrac{4x+2}{x+5} + \dfrac{-2x+8}{x+5}$

14. $\dfrac{-4}{y+2} + \dfrac{6}{5y+10}$

15. $\dfrac{x+1}{3-x} + \dfrac{x^2}{x-3}$

16. $\dfrac{3}{2m^2 - 9m - 5} - \dfrac{m+1}{2m^2 - m - 1}$

Simplify each complex fraction.

17. $\dfrac{\dfrac{2p}{k^2}}{\dfrac{3p^2}{k^3}}$

18. $\dfrac{\dfrac{1}{x+3} - 1}{1 + \dfrac{1}{x+3}}$

Solve.

19. $\dfrac{2x}{x-3} + \dfrac{1}{x+3} = \dfrac{-6}{x^2-9}$

20. $F = \dfrac{k}{d-D}$ for D

Solve each problem.

21. A boat goes 7 mph in still water. It takes as long to go 20 mi upstream as 50 mi downstream. Find the speed of the current.

22. A man can paint a room in his house, working alone, in 5 hr. His wife can do the job in 4 hr. How long will it take them to paint the room if they work together?

23. If x varies directly as y, and $x = 12$ when $y = 4$, find x when $y = 9$.

24. Under certain conditions, the length of time that it takes for fruit to ripen during the growing season varies inversely as the average maximum temperature during the season. If it takes 25 days for fruit to ripen with an average maximum temperature of 80°, find the number of days it would take at 75°. Round your answer to the nearest whole number.

CUMULATIVE REVIEW EXERCISES CHAPTERS 1–7

1. Use the order of operations to evaluate $3 + 4\left(\frac{1}{2} - \frac{3}{4}\right)$.

Solve.

2. $3(2y - 5) = 2 + 5y$
3. $A = \frac{1}{2}bh$ for b
4. $\frac{2 + m}{2 - m} = \frac{3}{4}$
5. $5y \leq 6y + 8$
6. $5m - 9 > 2m + 3$
7. For the graph of $4x + 3y = -12$,
   (a) what is the x-intercept?   (b) what is the y-intercept?

Sketch each graph.

8. $y = -3x + 2$
9. $y = -x^2 + 1$

Solve each system.

10. $4x - y = -7$
    $5x + 2y = 1$
11. $5x + 2y = 7$
    $10x + 4y = 12$

Simplify each expression. Write with only positive exponents.

12. $\dfrac{(2x^3)^{-1} \cdot x}{2^3 x^5}$
13. $\dfrac{(m^{-2})^3 m}{m^5 m^{-4}}$
14. $\dfrac{2p^3 q^4}{8p^5 q^3}$

Perform each indicated operation.

15. $(2k^2 + 3k) - (k^2 + k - 1)$
16. $8x^2 y^2 (9x^4 y^5)$
17. $(2a - b)^2$
18. $(y^2 + 3y + 5)(3y - 1)$
19. $\dfrac{12p^3 + 2p^2 - 12p + 4}{2p - 2}$

20. A computer can do one operation in 1.4×10^{-7} sec. How long would it take for the computer to do one trillion (10^{12}) operations?

Factor completely.

21. $8t^2 + 10tv + 3v^2$
22. $8r^2 - 9rs + 12s^2$
23. $16x^4 - 1$

Solve each equation.

24. $r^2 = 2r + 15$
25. $(r - 5)(2r + 1)(3r - 2) = 0$

Solve each problem.

26. One number is 4 more than another. The product of the numbers is 2 less than the smaller number. Find the smaller number.

27. The length of a rectangle is 2 m less than twice the width. The area is 60 m². Find the width of the rectangle.

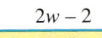

$2w - 2$

w

28. One of the following is equal to 1 for *all* real numbers. Which one is it?

   A. $\dfrac{k^2 + 2}{k^2 + 2}$ B. $\dfrac{4 - m}{4 - m}$ C. $\dfrac{2x + 9}{2x + 9}$ D. $\dfrac{x^2 - 1}{x^2 - 1}$

29. Which one of the following rational expressions is *not* equivalent to $\dfrac{4 - 3x}{7}$?

   A. $-\dfrac{-4 + 3x}{7}$ B. $-\dfrac{4 - 3x}{-7}$ C. $\dfrac{-4 + 3x}{-7}$ D. $\dfrac{-(3x + 4)}{7}$

Perform each operation and write the answer in lowest terms.

30. $\dfrac{5}{q} - \dfrac{1}{q}$

31. $\dfrac{3}{7} + \dfrac{4}{r}$

32. $\dfrac{4}{5q - 20} - \dfrac{1}{3q - 12}$

33. $\dfrac{2}{k^2 + k} - \dfrac{3}{k^2 - k}$

34. $\dfrac{7z^2 + 49z + 70}{16z^2 + 72z - 40} \div \dfrac{3z + 6}{4z^2 - 1}$

35. $\dfrac{\dfrac{4}{a} + \dfrac{5}{2a}}{\dfrac{7}{6a} - \dfrac{1}{5a}}$

36. What values of x cannot possibly be solutions of the equation $\dfrac{1}{x - 4} = \dfrac{3}{2x}$?

Solve each equation. Check your solutions.

37. $\dfrac{r + 2}{5} = \dfrac{r - 3}{3}$

38. $\dfrac{1}{x} = \dfrac{1}{x + 1} + \dfrac{1}{2}$

Solve each problem.

39. Juanita can weed the yard in 3 hr. Benito can weed the yard in 2 hr. How long would it take them if they worked together?

40. The circumference of a circle varies directly as its radius. A circle with circumference 9.42 in. has radius 1.5 in. (approximately). Find the circumference of a circle with radius 5.25 in.

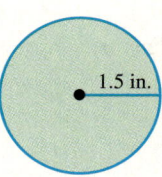

$C = 9.42$ in.

Roots and Radicals

8

8.1 Evaluating Roots

8.2 Multiplying, Dividing, and Simplifying Radicals

8.3 Adding and Subtracting Radicals

8.4 Rationalizing the Denominator

8.5 More Simplifying and Operations with Radicals

Summary Exercises on Operations with Radicals

8.6 Solving Equations with Radicals

8.7 Using Rational Numbers as Exponents

The London Eye opened on New Year's Eve in 1999. This unique Ferris wheel features 32 observation capsules and has a diameter of 135 m. Located on the bank of the Thames River, it faces the Houses of Parliament and is the fourth tallest structure in London. (*Source:* www.londoneye.com)

The formula

$$\text{sight distance} = 111.7\sqrt{\text{height of structure in kilometers}}$$

can be used to determine how far one can see (in kilometers) from the top of a structure on a clear day. (*Source: A Sourcebook of Applications of School Mathematics,* NCTM, 1980.) In Exercise 71 of Section 8.6, we use this formula to determine the truth of the claim that passengers on the London Eye can see Windsor Castle, 25 mi away.

8.1 Evaluating Roots

OBJECTIVES

1. Find square roots.
2. Decide whether a given root is rational, irrational, or not a real number.
3. Find decimal approximations for irrational square roots.
4. Use the Pythagorean formula.
5. Use the distance formula.
6. Find higher roots.

In Section 1.2, we discussed the idea of the *square* of a number. Recall that squaring a number means multiplying the number by itself.

If $a = 8$, then $a^2 = 8 \cdot 8 = 64$.
If $a = -4$, then $a^2 = (-4)(-4) = 16$.
If $a = -\frac{1}{2}$, then $a^2 = \left(-\frac{1}{2}\right)\left(-\frac{1}{2}\right) = \frac{1}{4}$.

In this chapter, we consider the opposite process.

If $a^2 = 49$, then $a = ?$.
If $a^2 = 100$, then $a = ?$.
If $a^2 = 25$, then $a = ?$.

OBJECTIVE 1 Find square roots. To find a in the three preceding statements, we must find a number that when multiplied by itself results in the given number. The number a is called a **square root** of the number a^2.

EXAMPLE 1 Finding All Square Roots of a Number

Find all square roots of 49.

To find a square root of 49, think of a number that when multiplied by itself gives 49. One square root is 7 because $7 \cdot 7 = 49$. Another square root of 49 is -7 because $(-7)(-7) = 49$. The number 49 has two square roots, 7 and -7; one is positive, and one is negative.

Now Try Exercise 7.

The **positive** or **principal square root** of a number is written with the symbol $\sqrt{}$. For example, the positive square root of 121 is 11, written

$$\sqrt{121} = 11.$$

The symbol $-\sqrt{}$ is used for the **negative square root** of a number. For example, the negative square root of 121 is -11, written

$$-\sqrt{121} = -11.$$

The symbol $\sqrt{}$ called a **radical sign,** always represents the positive square root (except that $\sqrt{0} = 0$). The number inside the radical sign is called the **radicand,** and the entire expression—radical sign and radicand—is called a **radical.**

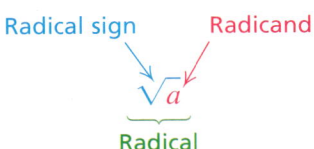

An algebraic expression containing a radical is called a **radical expression.**
Radicals have a long mathematical history. The radical sign $\sqrt{}$ has been used since sixteenth-century Germany and was probably derived from the letter R. The

Early radical symbol

radical symbol in the margin on the preceding page comes from the Latin word for root, *radix*. It was first used by Leonardo of Pisa (Fibonacci) in 1220.

We summarize our discussion of square roots as follows.

> **Square Roots of a**
>
> If a is a positive real number, then
>
> $\sqrt{a}$ is the positive or principal square root of a,
>
> and $\quad -\sqrt{a}$ is the negative square root of a.
>
> For nonnegative a,
>
> $$\sqrt{a} \cdot \sqrt{a} = (\sqrt{a})^2 = a \quad \text{and} \quad -\sqrt{a} \cdot (-\sqrt{a}) = (-\sqrt{a})^2 = a.$$
>
> Also, $\sqrt{0} = 0$.

EXAMPLE 2 Finding Square Roots

Find each square root.

(a) $\sqrt{144}$

The radical $\sqrt{144}$ represents the positive or principal square root of 144. Think of a positive number whose square is 144.

$$12^2 = 144, \quad \text{so} \quad \sqrt{144} = 12.$$

(b) $-\sqrt{1024}$

This symbol represents the negative square root of 1024. A calculator with a square root key can be used to find $\sqrt{1024} = 32$. Then, $-\sqrt{1024} = -32$.

(c) $\sqrt{\dfrac{4}{9}} = \dfrac{2}{3}$ **(d)** $-\sqrt{\dfrac{16}{49}} = -\dfrac{4}{7}$

Now Try Exercises 19, 21, and 23.

As shown in the preceding definition, when the square root of a positive real number is squared, the result is that positive real number. $\left(\text{Also, } (\sqrt{0})^2 = 0.\right)$

EXAMPLE 3 Squaring Radical Expressions

Find the *square* of each radical expression.

(a) $\sqrt{13}$

$(\sqrt{13})^2 = 13$ Definition of square root

(b) $-\sqrt{29}$

$(-\sqrt{29})^2 = 29$ The square of a *negative* number is positive.

(c) $\sqrt{p^2 + 1}$

$(\sqrt{p^2 + 1})^2 = p^2 + 1$

Now Try Exercises 27, 29, and 33.

OBJECTIVE 2 Decide whether a given root is rational, irrational, or not a real number.
All numbers with square roots that are rational are called **perfect squares.** For example, 144 and $\frac{4}{9}$ are perfect squares since their respective square roots, 12 and $\frac{2}{3}$, are rational numbers.

A number that is not a perfect square has a square root that is not a rational number. For example, $\sqrt{5}$ is not a rational number because it cannot be written as the ratio of two integers. Its decimal neither terminates nor repeats. However, $\sqrt{5}$ is a real number and corresponds to a point on the number line. As mentioned in Chapter 1, a real number that is not rational is called an *irrational number*. The number $\sqrt{5}$ is irrational. Many square roots of integers are irrational.

> If a is a positive real number that is not a perfect square, then $\sqrt{a}$ is irrational.

Not every number has a *real number* square root. For example, there is no real number that can be squared to get -36. (The square of a real number can never be negative.) Because of this, $\sqrt{-36}$ is not a real number.

> If a is a negative real number, then $\sqrt{a}$ is not a real number.

CAUTION Be careful not to confuse $\sqrt{-36}$ and $-\sqrt{36}$. $\sqrt{-36}$ is not a real number since there is no real number that can be squared to get -36. However, $-\sqrt{36}$ is the negative square root of 36, which is -6.

EXAMPLE 4 Identifying Types of Square Roots

Tell whether each square root is *rational, irrational,* or *not a real number.*

(a) $\sqrt{17}$

Because 17 is not a perfect square, $\sqrt{17}$ is irrational.

(b) $\sqrt{64}$

The number 64 is a perfect square, 8^2, so $\sqrt{64} = 8$, a rational number.

(c) $\sqrt{-25}$

There is no real number whose square is -25. Therefore, $\sqrt{-25}$ is not a real number.

Now Try Exercises 39, 41, and 47.

> **NOTE** Not all irrational numbers are square roots of integers. For example, π (approximately 3.14159) is an irrational number that is not a square root of any integer.

OBJECTIVE 3 Find decimal approximations for irrational square roots. Even if a number is irrational, a decimal that approximates the number can be found using a calculator. For example, if we use a calculator to find $\sqrt{10}$, the display will show 3.16227766, which is only an *approximation* of $\sqrt{10}$, not an exact rational value.

See Appendix A for general instructions on how to use the square root key on a calculator.

EXAMPLE 5 Approximating Irrational Square Roots

Find a decimal approximation for each square root. Round answers to the nearest thousandth.

(a) $\sqrt{11}$

Using the square root key on a calculator gives $3.31662479 \approx 3.317$, where $\approx$ means "is approximately equal to."

(b) $\sqrt{39} \approx 6.245$ Use a calculator.

(c) $-\sqrt{740} \approx -27.203$

Now Try Exercises 45 and 49.

OBJECTIVE 4 Use the Pythagorean formula. Many applications of square roots require the use of the Pythagorean formula. Recall from Section 6.6 that by this formula if c is the length of the hypotenuse of a right triangle, and a and b are the lengths of the two legs, then

$$a^2 + b^2 = c^2.$$

See Figure 1.

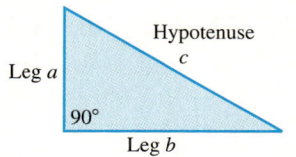

FIGURE 1

EXAMPLE 6 Using the Pythagorean Formula

Find the length of the unknown side of each right triangle with sides a, b, and c, where c is the hypotenuse.

(a) $a = 3$, $b = 4$

Use the Pythagorean formula to find c^2 first.

$$c^2 = a^2 + b^2$$
$$c^2 = 3^2 + 4^2 \quad \text{Let } a = 3 \text{ and } b = 4.$$
$$c^2 = 9 + 16 \quad \text{Square.}$$
$$c^2 = 25 \quad \text{Add.}$$

Since the length of a side of a triangle must be a positive number, find the positive square root of 25 to get c.

$$c = \sqrt{25} = 5$$

(b) $c = 9$, $b = 5$

Substitute the given values in the Pythagorean formula. Then solve for a^2.

$$c^2 = a^2 + b^2$$
$$9^2 = a^2 + 5^2 \quad \text{Let } c = 9 \text{ and } b = 5.$$
$$81 = a^2 + 25 \quad \text{Square.}$$
$$56 = a^2 \quad \text{Subtract 25.}$$

Use a calculator to find $a = \sqrt{56} \approx 7.483$.

Now Try Exercises 53 and 55.

CAUTION Be careful not to make the common mistake of thinking that $\sqrt{a^2 + b^2}$ equals $a + b$. As Example 6(a) shows, $\sqrt{9 + 16} = \sqrt{25} = 5$. However, $\sqrt{9} + \sqrt{16} = 3 + 4 = 7$. Since $5 \neq 7$, in general,

$$\sqrt{a^2 + b^2} \neq a + b.$$

The Pythagorean formula can be used to solve applied problems that involve right triangles. Use the same six problem-solving steps that we have been using throughout the text.

EXAMPLE 7 Using the Pythagorean Formula to Solve an Application

A ladder 10 ft long leans against a wall. The foot of the ladder is 6 ft from the base of the wall. How high up the wall does the top of the ladder rest?

Step 1 **Read** the problem again.

Step 2 **Assign a variable.** As shown in Figure 2, a right triangle is formed with the ladder as the hypotenuse. Let a represent the height of the top of the ladder when measured straight down to the ground.

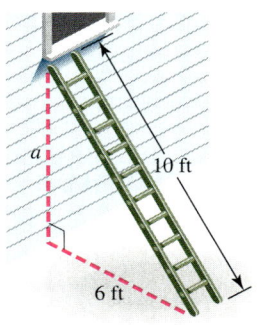

FIGURE 2

Step 3 **Write an equation** using the Pythagorean formula.

$$c^2 = a^2 + b^2$$
$$10^2 = a^2 + 6^2 \quad \text{Let } c = 10 \text{ and } b = 6.$$

Step 4 **Solve.**
$$100 = a^2 + 36 \quad \text{Square.}$$
$$64 = a^2 \quad \text{Subtract 36.}$$
$$\sqrt{64} = a$$
$$a = 8 \quad \sqrt{64} = 8$$

We choose the positive square root of 64 because a represents a length.

Step 5 **State the answer.** The top of the ladder rests 8 ft up the wall.

Step 6 **Check.** From Figure 2, we see that we must have
$$8^2 + 6^2 = 10^2 \quad ?$$
$$64 + 36 = 100. \quad \text{True}$$

The check confirms that the top of the ladder rests 8 ft up the wall.

Now Try Exercise 61.

SECTION 8.1 Evaluating Roots

OBJECTIVE 5 Use the distance formula. The Pythagorean formula is used to develop another useful formula for finding the distance between two points on the plane. Figure 3 shows two different points (x_1, y_1) and (x_2, y_2), as well as the point (x_2, y_1). The distance between (x_2, y_2) and (x_2, y_1) is $a = y_2 - y_1$, and the distance between (x_1, y_1) and (x_2, y_1) is $b = x_2 - x_1$. From the Pythagorean formula,

$$d^2 = (x_2 - x_1)^2 + (y_2 - y_1)^2.$$

Taking the square root of each side, we get the **distance formula.**

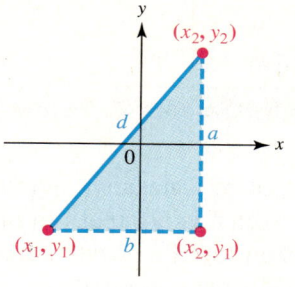

FIGURE 3

Distance Formula

The distance between the points (x_1, y_1) and (x_2, y_2) is
$$d = \sqrt{(x_2 - x_1)^2 + (y_2 - y_1)^2}.$$

EXAMPLE 8 Using the Distance Formula

Find the distance between $(-3, 4)$ and $(2, 5)$.

Use the distance formula. Choose $(x_1, y_1) = (-3, 4)$ and $(x_2, y_2) = (2, 5)$.

$$\begin{aligned} d &= \sqrt{(x_2 - x_1)^2 + (y_2 - y_1)^2} \\ &= \sqrt{(2 - (-3))^2 + (5 - 4)^2} \quad \text{Substitute.} \\ &= \sqrt{5^2 + 1^2} \\ &= \sqrt{26} \end{aligned}$$

Now Try Exercise 81.

OBJECTIVE 6 Find higher roots. Finding the square root of a number is the inverse (reverse) of squaring a number. In a similar way, there are inverses to finding the cube of a number, or finding the fourth or higher power of a number. These inverses are the **cube root**, written $\sqrt[3]{a}$, and the **fourth root**, written $\sqrt[4]{a}$. Similar symbols are used for higher roots. In general, we have the following.

$$\sqrt[n]{a}$$

The *n*th root of a is written .

In $\sqrt[n]{a}$, the number n is the **index** or **order** of the radical. It is possible to write $\sqrt[2]{a}$ instead of $\sqrt{a}$, but the simpler symbol $\sqrt{a}$ is customary since the square root is the most commonly used root.

When working with cube roots or fourth roots, it is helpful to memorize the first few *perfect cubes* ($2^3 = 8$, $3^3 = 27$, and so on) and the first few perfect fourth powers ($2^4 = 16$, $3^4 = 81$, and so on).

EXAMPLE 9 Finding Cube Roots

Find each cube root.

(a) $\sqrt[3]{8}$

Look for a number that can be cubed to give 8. Because $2^3 = 8$, $\sqrt[3]{8} = 2$.

534 CHAPTER 8 Roots and Radicals

(b) $\sqrt[3]{-8} = -2$ because $(-2)^3 = -8$.

(c) $-\sqrt[3]{216} = -6$ because $\sqrt[3]{216} = 6$, and thus $-\sqrt[3]{216} = -6$.

Now Try Exercises 87, 89, and 91.

Notice in Example 9(b) that we can find the cube root of a negative number. (Contrast this with the square root of a negative number, which is not real.) In fact, the cube root of a positive number is positive, and the cube root of a negative number is negative. *There is only one real number cube root for each real number.*

When the index of the radical is even (square root, fourth root, and so on), *the radicand must be nonnegative* to get a real number root. Also, for even indexes, the symbols $\sqrt{}, \sqrt[4]{}, \sqrt[6]{}$, and so on are used for the positive (principal) roots. The symbols $-\sqrt{}, -\sqrt[4]{}, -\sqrt[6]{}$, and so on are used for the negative roots.

EXAMPLE 10 Finding Higher Roots

Find each root.

(a) $\sqrt[4]{16} = 2$ because 2 is positive and $2^4 = 16$.

(b) $-\sqrt[4]{16}$

From part (a), $\sqrt[4]{16} = 2$, so the negative root $-\sqrt[4]{16} = -2$.

(c) $\sqrt[4]{-16}$

For a real number fourth root, the radicand must be nonnegative. There is no real number that equals $\sqrt[4]{-16}$.

(d) $-\sqrt[5]{32}$

First find $\sqrt[5]{32}$. Because 2 is the number whose fifth power is 32, $\sqrt[5]{32} = 2$. If $\sqrt[5]{32} = 2$, then
$$-\sqrt[5]{32} = -2.$$

(e) $\sqrt[5]{-32} = -2$, because $(-2)^5 = -32$.

Now Try Exercises 93, 97, and 99.

CONNECTIONS

While Pythagoras may have written the first proof of the Pythagorean relationship, there is evidence that the Babylonians knew the concept quite well. The figure on the left illustrates the formula by using a tile pattern. In the figure, the side of the square along the hypotenuse measures 5 units, while the sides along the legs measure 3 and 4 units. If we let $a = 3$, $b = 4$, and $c = 5$, the equation of the Pythagorean formula is satisfied.

$a^2 + b^2 = c^2$
$3^2 + 4^2 = 5^2$?
$25 = 25$ True

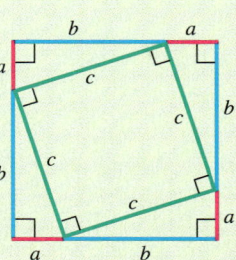

SECTION 8.1 Evaluating Roots **535**

> **For Discussion or Writing**
>
> The diagram on the right can be used to verify the Pythagorean formula. To do so, express the area of the figure in two ways: first, as the area of the large square, and then as the sum of the areas of the smaller square and the four right triangles. Finally, set the areas equal and simplify the equation.

8.1 EXERCISES

For Extra Help

 Student's Solutions Manual

 MyMathLab

 InterAct Math Tutorial Software

 AW Math Tutor Center

 MathXL

 Digital Video Tutor CD 16/Videotape 14

Decide whether each statement is true or false. If false, tell why.

1. Every positive number has two real square roots.
2. A negative number has negative square roots.
3. Every nonnegative number has two real square roots.
4. The positive square root of a positive number is its principal square root.
5. The cube root of every real number has the same sign as the number itself.
6. Every positive number has three real cube roots.

Find all square roots of each number. See Example 1.

7. 9   **8.** 16   **9.** 64   **10.** 100   **11.** 144

12. 225   **13.** $\dfrac{25}{196}$ **14.** $\dfrac{81}{400}$ **15.** 900   **16.** 1600

Find each square root. See Examples 2 and 4(c).

17. $\sqrt{1}$ **18.** $\sqrt{4}$ **19.** $\sqrt{49}$ **20.** $\sqrt{81}$ **21.** $-\sqrt{121}$

22. $-\sqrt{196}$ **23.** $-\sqrt{\dfrac{144}{121}}$ **24.** $-\sqrt{\dfrac{49}{36}}$ **25.** $\sqrt{-121}$ **26.** $\sqrt{-64}$

Find the square of each radical expression. See Example 3.

27. $\sqrt{19}$ **28.** $\sqrt{59}$ **29.** $-\sqrt{19}$ **30.** $-\sqrt{99}$

31. $\sqrt{\dfrac{2}{3}}$ **32.** $\sqrt{\dfrac{5}{7}}$ **33.** $\sqrt{3x^2 + 4}$ **34.** $\sqrt{9y^2 + 3}$

What must be true about the variable a for each statement in Exercises 35–38 to be true?

35. $\sqrt{a}$ represents a positive number.
36. $-\sqrt{a}$ represents a negative number.
37. $\sqrt{a}$ is not a real number.
38. $-\sqrt{a}$ is not a real number.

Write rational, irrational, *or* not a real number *for each number. If a number is rational, give its exact value. If a number is irrational, give a decimal approximation to the nearest thousandth. Use a calculator as necessary. See Examples 4 and 5.*

39. $\sqrt{25}$ **40.** $\sqrt{169}$ **41.** $\sqrt{29}$ **42.** $\sqrt{33}$

43. $-\sqrt{64}$ **44.** $-\sqrt{81}$ **45.** $-\sqrt{300}$ **46.** $-\sqrt{500}$

47. $\sqrt{-29}$ **48.** $\sqrt{-47}$ **49.** $\sqrt{1200}$ **50.** $\sqrt{1500}$

536 CHAPTER 8 Roots and Radicals

Work Exercises 51 and 52 without using a calculator.

51. Choose the best estimate for the length and width (in meters) of this rectangle.

A. 11 by 6   **B.** 11 by 7
C. 10 by 7   **D.** 10 by 6

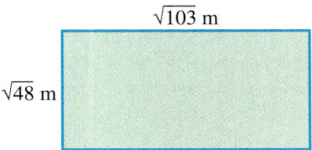

52. Choose the best estimate for the base and height (in feet) of this triangle.

A. $b = 8, h = 5$ **B.** $b = 8, h = 4$
C. $b = 9, h = 5$ **D.** $b = 9, h = 4$

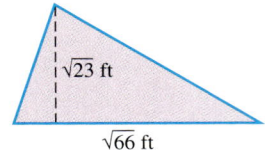

Find the length of the unknown side of each right triangle with sides a, b, and c, where c is the hypotenuse. See Figure 1 and Example 6. Give any decimal approximations to the nearest thousandth.

53. $a = 8, b = 15$ **54.** $a = 24, b = 10$ **55.** $a = 6, c = 10$

56. $b = 12, c = 13$ **57.** $a = 11, b = 4$ **58.** $a = 13, b = 9$

Use the Pythagorean formula to solve each problem. See Example 7.

59. The diagonal of a rectangle measures 25 cm. The width of the rectangle is 7 cm. Find the length of the rectangle.

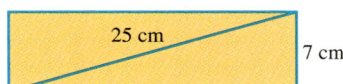

60. The length of a rectangle is 40 m, and the width is 9 m. Find the measure of the diagonal of the rectangle.

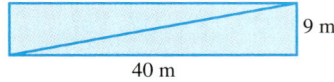

61. Tyler is flying a kite on 100 ft of string. How high is it above his hand (vertically) if the horizontal distance between Tyler and the kite is 60 ft?

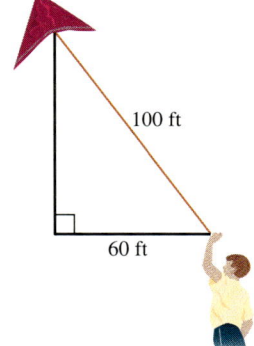

62. A guy wire is attached to the mast of a short-wave transmitting antenna. It is attached 96 ft above ground level. If the wire is staked to the ground 72 ft from the base of the mast, how long is the wire?

63. A surveyor measured the distances shown in the figure. Find the distance across the lake between points R and S.

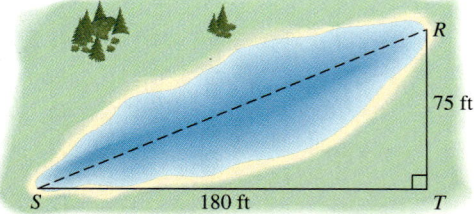

64. A boat is being pulled toward a dock with a rope attached at water level. When the boat is 24 ft from the dock, 30 ft of rope is extended. What is the height of the dock above the water?

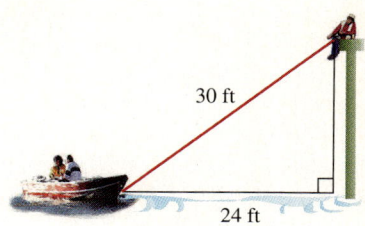

65. One of the authors of this text took this photo of a broken tree in a field near his home. The vertical distance from the base of the broken tree to the point of the break is 4.5 ft. The length of the broken part is 12 ft. How far along the ground (to the nearest tenth) is it from the base of the tree to the point where the broken part touches the ground?

66. Another of the authors recently purchased a new television set. A television set is "sized" according to the diagonal measurement of the viewing screen. The author purchased a 19-in. TV, so the TV measures 19 in. from one corner of the viewing screen diagonally to the other corner. The viewing screen is 15.5 in. wide. Find the height of the viewing screen (to the nearest tenth). (*Source:* Phillips Magnavox color television 19PR21C1.)

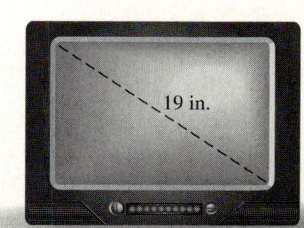

67. What is the value of x (to the nearest thousandth) in the figure?

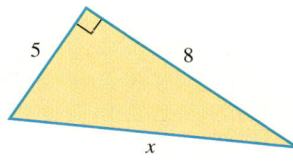

68. What is the value of y (to the nearest thousandth) in the figure?

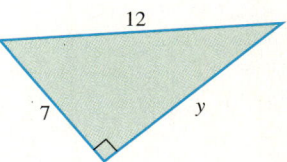

69. Use specific values for a and b different from those given in the "Caution" following Example 6 to show that $\sqrt{a^2 + b^2} \neq a + b$.

70. Why would the values $a = 0$ and $b = 1$ *not* be satisfactory in Exercise 69?

Find each square root. Express the answer as a whole number or as a square root. Do not use a calculator.

71. $\sqrt{3^2 + 4^2}$ **72.** $\sqrt{6^2 + 8^2}$ **73.** $\sqrt{8^2 + 15^2}$

74. $\sqrt{5^2 + 12^2}$ **75.** $\sqrt{(-2)^2 + 3^2}$ **76.** $\sqrt{(-1)^2 + 5^2}$

Find the distance between each pair of points. Express the answer as a whole number or as a square root. Do not use a calculator. See Example 8.

77. (5, 7) and (1, 4)   **78.** (8, 13) and (3, 1)

79. (2, 9) and (−3, −3)   **80.** (4, 6) and (−4, −9)

81. $(-1, -2)$ and $(-3, 1)$

82. $(-3, -6)$ and $(-4, 0)$

83. $\left(-\frac{1}{4}, \frac{2}{3}\right)$ and $\left(\frac{3}{4}, -\frac{1}{3}\right)$

84. $\left(\frac{2}{5}, \frac{3}{2}\right)$ and $\left(-\frac{8}{5}, \frac{1}{2}\right)$

Find each root. See Examples 9 and 10.

85. $\sqrt[3]{1}$ **86.** $\sqrt[3]{27}$ **87.** $\sqrt[3]{125}$ **88.** $\sqrt[3]{1000}$

89. $\sqrt[3]{-27}$ **90.** $\sqrt[3]{-64}$ **91.** $-\sqrt[3]{8}$ **92.** $-\sqrt[3]{343}$

93. $\sqrt[4]{625}$ **94.** $\sqrt[4]{10{,}000}$ **95.** $\sqrt[4]{-1}$ **96.** $\sqrt[4]{-625}$

97. $-\sqrt[4]{81}$ **98.** $-\sqrt[4]{256}$ **99.** $\sqrt[5]{-1024}$ **100.** $\sqrt[5]{-100{,}000}$

Use a calculator with a cube root key to find each root. Round to the nearest thousandth. $\left(\text{In Exercises 105 and 106, you may have to use the fact that if } a > 0, \sqrt[3]{-a} = -\sqrt[3]{a}.\right)$

101. $\sqrt[3]{12}$ **102.** $\sqrt[3]{74}$ **103.** $\sqrt[3]{130.6}$

104. $\sqrt[3]{251.8}$ **105.** $\sqrt[3]{-87}$ **106.** $\sqrt[3]{-95}$

RELATING CONCEPTS (EXERCISES 107–112)

For Individual or Group Work

One of the many proofs of the Pythagorean formula was given in the Connections box in this section. Here is another one, attributed to the Hindu mathematician Bhāskara. Refer to the figures and **work Exercises 107–112 in order.**

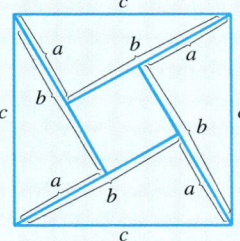 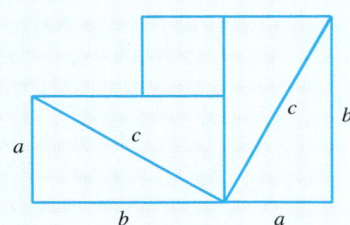

107. What is the area of the square on the left in terms of c?

108. What is the area of the small square in the middle of the figure on the left, in terms of $(b - a)$?

109. What is the sum of the areas of the two rectangles made up of triangles in the figure on the right?

110. What is the area of the small square in the figure on the right in terms of a and b?

111. The figure on the left is made up of the same square and triangles as the figure on the right. Write an equation setting the answer to Exercise 107 equal to the sum of the answers in Exercises 109 and 110.

112. Simplify the expressions you obtained in Exercise 111. What is your final result?

8.2 Multiplying, Dividing, and Simplifying Radicals

OBJECTIVES

1. Multiply radicals.
2. Simplify radicals using the product rule.
3. Simplify radicals using the quotient rule.
4. Simplify radicals involving variables.
5. Simplify higher roots.

OBJECTIVE 1 Multiply radicals. We develop several rules for finding products and quotients of radicals in this section. To illustrate the rule for products, notice that

$$\sqrt{4} \cdot \sqrt{9} = 2 \cdot 3 = 6 \quad \text{and} \quad \sqrt{4 \cdot 9} = \sqrt{36} = 6,$$

showing that

$$\sqrt{4} \cdot \sqrt{9} = \sqrt{4 \cdot 9}.$$

This result is a particular case of the more general product rule for radicals.

> **Product Rule for Radicals**
>
> For nonnegative real numbers a and b,
>
> $$\sqrt{a} \cdot \sqrt{b} = \sqrt{a \cdot b} \quad \text{and} \quad \sqrt{a \cdot b} = \sqrt{a} \cdot \sqrt{b}.$$
>
> That is, the product of two square roots is the square root of the product, and the square root of a product is the product of the square roots.

> **CAUTION** In general, $\sqrt{x + y} \neq \sqrt{x} + \sqrt{y}$. To see why this is so, let $x = 16$ and $y = 9$.
>
> $$\sqrt{16 + 9} = \sqrt{25} = 5$$
>
> but
>
> $$\sqrt{16} + \sqrt{9} = 4 + 3 = 7.$$

EXAMPLE 1 Using the Product Rule to Multiply Radicals

Use the product rule for radicals to find each product.

(a) $\sqrt{2} \cdot \sqrt{3} = \sqrt{2 \cdot 3} = \sqrt{6}$ Product rule
(b) $\sqrt{7} \cdot \sqrt{5} = \sqrt{35}$ Product rule
(c) $\sqrt{11} \cdot \sqrt{a} = \sqrt{11a}$ Assume $a \geq 0$.

> Now Try Exercises 3 and 9.

OBJECTIVE 2 Simplify radicals using the product rule. A square root radical is *simplified* when no perfect square factor remains under the radical sign. We accomplish this by using the product rule in the form $\sqrt{a \cdot b} = \sqrt{a} \cdot \sqrt{b}$.

EXAMPLE 2 Using the Product Rule to Simplify Radicals

Simplify each radical.

(a) $\sqrt{20}$

Because 20 has a perfect square factor of 4, we can write

$$\sqrt{20} = \sqrt{4 \cdot 5} \quad \text{4 is a perfect square.}$$
$$= \sqrt{4} \cdot \sqrt{5} \quad \text{Product rule}$$
$$= 2\sqrt{5}. \quad \sqrt{4} = 2$$

540 CHAPTER 8 Roots and Radicals

Thus, $\sqrt{20} = 2\sqrt{5}$. Because 5 has no perfect square factor (other than 1), $2\sqrt{5}$ is called the *simplified form* of $\sqrt{20}$. Note that $2\sqrt{5}$ represents a product, where the factors are 2 and $\sqrt{5}$.

We could also factor 20 into prime factors and look for pairs of like factors. Each pair of like factors produces one factor outside the radical in the simplified form. Therefore,

$$\sqrt{20} = \sqrt{2 \cdot 2 \cdot 5} = 2\sqrt{5}.$$

(b) $\sqrt{72}$

We begin by looking for the *largest* perfect square factor of 72. This number is 36, so

$$\begin{aligned}\sqrt{72} &= \sqrt{36 \cdot 2} & &\text{36 is a perfect square.}\\ &= \sqrt{36} \cdot \sqrt{2} & &\text{Product rule}\\ &= 6\sqrt{2}. & &\sqrt{36} = 6\end{aligned}$$

We could also factor 72 into its prime factors and look for pairs of like factors.

$$\sqrt{72} = \sqrt{2 \cdot 2 \cdot 2 \cdot 3 \cdot 3} = 2 \cdot 3 \cdot \sqrt{2} = 6\sqrt{2}$$

In either case, we obtain $6\sqrt{2}$ as the simplified form of $\sqrt{72}$. However, our work is simpler if we begin with the largest perfect square factor.

(c) $\begin{aligned}\sqrt{300} &= \sqrt{100 \cdot 3} & &\text{100 is a perfect square.}\\ &= \sqrt{100} \cdot \sqrt{3} & &\text{Product rule}\\ &= 10\sqrt{3} & &\sqrt{100} = 10\end{aligned}$

(d) $\sqrt{15}$

The number 15 has no perfect square factors (except 1), so $\sqrt{15}$ cannot be simplified further.

Now Try Exercises 13, 19, and 23.

Sometimes the product rule can be used to simplify a product, as Example 3 shows.

EXAMPLE 3 Multiplying and Simplifying Radicals

Find each product and simplify.

(a) $\begin{aligned}\sqrt{9} \cdot \sqrt{75} &= 3\sqrt{75} & &\sqrt{9} = 3\\ &= 3\sqrt{25 \cdot 3} & &\text{25 is a perfect square.}\\ &= 3\sqrt{25} \cdot \sqrt{3} & &\text{Product rule}\\ &= 3 \cdot 5\sqrt{3} & &\sqrt{25} = 5\\ &= 15\sqrt{3} & &\text{Multiply.}\end{aligned}$

Notice that we could have used the product rule to get $\sqrt{9} \cdot \sqrt{75} = \sqrt{675}$, and then simplified. However, the product rule as used here allows us to obtain the final answer without using a large number like 675.

SECTION 8.2 Multiplying, Dividing, and Simplifying Radicals 541

(b) $\sqrt{8} \cdot \sqrt{12} = \sqrt{8 \cdot 12}$ Product rule
$= \sqrt{4 \cdot 2 \cdot 4 \cdot 3}$ Factor; 4 is a perfect square.
$= \sqrt{4} \cdot \sqrt{4} \cdot \sqrt{2 \cdot 3}$ Product rule
$= 2 \cdot 2 \cdot \sqrt{6}$ $\sqrt{4} = 2$
$= 4\sqrt{6}$ Multiply.

Now Try Exercises 29 and 31.

NOTE We could also simplify Example 3(b) as follows.

$$\sqrt{8} \cdot \sqrt{12} = \sqrt{4 \cdot 2} \cdot \sqrt{4 \cdot 3}$$
$$= 2\sqrt{2} \cdot 2\sqrt{3}$$
$$= 2 \cdot 2 \cdot \sqrt{2} \cdot \sqrt{3}$$
$$= 4\sqrt{6}$$

Both approaches are correct. There is often more than one way to find such a product.

OBJECTIVE 3 Simplify radicals using the quotient rule. The quotient rule for radicals is very similar to the product rule. It, too, can be used either way.

Quotient Rule for Radicals

If a and b are nonnegative real numbers and $b \neq 0$, then

$$\sqrt{\frac{a}{b}} = \frac{\sqrt{a}}{\sqrt{b}} \quad \text{and} \quad \frac{\sqrt{a}}{\sqrt{b}} = \sqrt{\frac{a}{b}}.$$

That is, the square root of a quotient is the quotient of the square roots, and the quotient of two square roots is the square root of the quotient.

EXAMPLE 4 Using the Quotient Rule to Simplify Radicals

Simplify each radical.

(a) $\sqrt{\frac{25}{9}} = \frac{\sqrt{25}}{\sqrt{9}} = \frac{5}{3}$ Quotient rule

(b) $\frac{\sqrt{288}}{\sqrt{2}} = \sqrt{\frac{288}{2}} = \sqrt{144} = 12$ Quotient rule

(c) $\sqrt{\frac{3}{4}} = \frac{\sqrt{3}}{\sqrt{4}} = \frac{\sqrt{3}}{2}$ Quotient rule

Now Try Exercises 37, 39, and 41.

EXAMPLE 5 Using the Quotient Rule to Divide Radicals

Simplify $\dfrac{27\sqrt{15}}{9\sqrt{3}}$.

Use multiplication of fractions and the quotient rule as follows.

$$\dfrac{27\sqrt{15}}{9\sqrt{3}} = \dfrac{27}{9} \cdot \dfrac{\sqrt{15}}{\sqrt{3}} = \dfrac{27}{9} \cdot \sqrt{\dfrac{15}{3}} = 3\sqrt{5}$$

Now Try Exercise 45.

Some problems require both the product and quotient rules.

EXAMPLE 6 Using Both the Product and Quotient Rules

Simplify $\sqrt{\dfrac{3}{5}} \cdot \sqrt{\dfrac{1}{5}}$.

$$\sqrt{\dfrac{3}{5}} \cdot \sqrt{\dfrac{1}{5}} = \sqrt{\dfrac{3}{5} \cdot \dfrac{1}{5}} \qquad \text{Product rule}$$

$$= \sqrt{\dfrac{3}{25}} \qquad \text{Multiply fractions.}$$

$$= \dfrac{\sqrt{3}}{\sqrt{25}} \qquad \text{Quotient rule}$$

$$= \dfrac{\sqrt{3}}{5} \qquad \sqrt{25} = 5$$

Now Try Exercise 43.

OBJECTIVE 4 Simplify radicals involving variables. Radicals can also involve variables, such as $\sqrt{x^2}$. Simplifying such radicals can get a little tricky. If x represents a nonnegative number, then $\sqrt{x^2} = x$. If x represents a negative number, then $\sqrt{x^2} = -x$, the opposite of x (which is positive). For example,

$$\sqrt{5^2} = 5, \quad \text{but} \quad \sqrt{(-5)^2} = \sqrt{25} = 5, \quad \text{the opposite of } -5.$$

This means that the square root of a squared number is always nonnegative. We can use absolute value to express this.

$\sqrt{a^2}$

For any real number a,

$$\sqrt{a^2} = |a|.$$

The product and quotient rules apply when variables appear under the radical sign, as long as the variables represent only *nonnegative* real numbers. *To avoid negative radicands, variables under radical signs are assumed to be nonnegative in this text.* Therefore, absolute value bars are not necessary, since for $x \geq 0$, $|x| = x$.

EXAMPLE 7 Simplifying Radicals Involving Variables

Simplify each radical. Remember that we assume all variables represent nonnegative real numbers.

(a) $\sqrt{x^4} = x^2$ since $(x^2)^2 = x^4$.

(b) $\sqrt{25m^6} = \sqrt{25} \cdot \sqrt{m^6}$ Product rule
$\phantom{\sqrt{25m^6}} = 5m^3$ $(m^3)^2 = m^6$

(c) $\sqrt{8p^{10}} = \sqrt{4 \cdot 2 \cdot p^{10}}$ 4 is a perfect square.
$\phantom{\sqrt{8p^{10}}} = \sqrt{4} \cdot \sqrt{2} \cdot \sqrt{p^{10}}$ Product rule
$\phantom{\sqrt{8p^{10}}} = 2 \cdot \sqrt{2} \cdot p^5$ $(p^5)^2 = p^{10}$
$\phantom{\sqrt{8p^{10}}} = 2p^5\sqrt{2}$

(d) $\sqrt{r^9} = \sqrt{r^8 \cdot r}$
$\phantom{\sqrt{r^9}} = \sqrt{r^8} \cdot \sqrt{r}$ Product rule
$\phantom{\sqrt{r^9}} = r^4\sqrt{r}$ $(r^4)^2 = r^8$

(e) $\sqrt{\dfrac{5}{x^2}} = \dfrac{\sqrt{5}}{\sqrt{x^2}}$ Quotient rule
$\phantom{\sqrt{\dfrac{5}{x^2}}} = \dfrac{\sqrt{5}}{x}$ $x \neq 0$

Now Try Exercises 47, 53, 55, 59, and 69.

NOTE A quick way to find the square root of a variable raised to an even power is to divide the exponent by the index, 2. For example,

$$\sqrt{x^6} = x^3 \quad \text{and} \quad \sqrt{x^{10}} = x^5.$$
$\uparrow\uparrow$
$6 \div 2 = 310 \div 2 = 5$

OBJECTIVE 5 Simplify higher roots. The product and quotient rules for radicals also work for other roots. To simplify cube roots, look for factors that are *perfect cubes*. A **perfect cube** is a number with a rational cube root. For example, $\sqrt[3]{64} = 4$, and because 4 is a rational number, 64 is a perfect cube. Higher roots are handled in a similar manner.

Properties of Radicals

For all real numbers where the indicated roots exist,

$$\sqrt[n]{a} \cdot \sqrt[n]{b} = \sqrt[n]{ab} \quad \text{and} \quad \dfrac{\sqrt[n]{a}}{\sqrt[n]{b}} = \sqrt[n]{\dfrac{a}{b}} \quad (b \neq 0).$$

EXAMPLE 8 Simplifying Higher Roots

Simplify each radical.

(a) $\sqrt[3]{32} = \sqrt[3]{8 \cdot 4}$ 8 is a perfect cube.
$\phantom{\sqrt[3]{32}} = \sqrt[3]{8} \cdot \sqrt[3]{4}$ Product rule
$\phantom{\sqrt[3]{32}} = 2\sqrt[3]{4}$

544 CHAPTER 8 Roots and Radicals

(b) $\sqrt[4]{32} = \sqrt[4]{16 \cdot 2}$ 16 is a perfect fourth power.
$= \sqrt[4]{16} \cdot \sqrt[4]{2}$ Product rule
$= 2\sqrt[4]{2}$

(c) $\sqrt[3]{\dfrac{27}{125}} = \dfrac{\sqrt[3]{27}}{\sqrt[3]{125}} = \dfrac{3}{5}$ Quotient rule

Now Try Exercises 73, 79, and 83.

Higher roots of radicals involving variables can also be simplified. To simplify cube roots with variables, use the fact that for any real number a,

$$\sqrt[3]{a^3} = a.$$

This is true whether a is positive or negative. (Why?)

EXAMPLE 9   Simplifying Cube Roots Involving Variables

Simplify each radical.

(a) $\sqrt[3]{m^6} = m^2$ $(m^2)^3 = m^6$

(b) $\sqrt[3]{27x^{12}} = \sqrt[3]{27} \cdot \sqrt[3]{x^{12}}$ Product rule
$= 3x^4$ $3^3 = 27; (x^4)^3 = x^{12}$

(c) $\sqrt[3]{32a^4} = \sqrt[3]{8a^3 \cdot 4a}$ 8 is a perfect cube.
$= \sqrt[3]{8a^3} \cdot \sqrt[3]{4a}$ Product rule
$= 2a\sqrt[3]{4a}$ $(2a)^3 = 8a^3$

(d) $\sqrt[3]{\dfrac{y^3}{125}} = \dfrac{\sqrt[3]{y^3}}{\sqrt[3]{125}}$ Quotient rule
$= \dfrac{y}{5}$

Now Try Exercises 87, 89, 93, and 95.

8.2 EXERCISES

For Extra Help

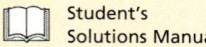
Student's Solutions Manual

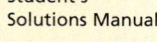
MyMathLab

InterAct Math Tutorial Software

AW Math Tutor Center

MathXL

Digital Video Tutor CD 16/Videotape 14

Decide whether each statement is true *or* false. *If false, show why.*

1. $\sqrt{(-6)^2} = -6$ F   **2.** $\sqrt[3]{(-6)^3} = -6$

Use the product rule for radicals to find each product. See Example 1.

3. $\sqrt{3} \cdot \sqrt{5}$ $\sqrt{15}$ **4.** $\sqrt{3} \cdot \sqrt{7}$ **5.** $\sqrt{2} \cdot \sqrt{11}$

6. $\sqrt{2} \cdot \sqrt{15}$ $\sqrt{30}$ **7.** $\sqrt{6} \cdot \sqrt{7}$ $\sqrt{42}$ **8.** $\sqrt{5} \cdot \sqrt{6}$

9. $\sqrt{13} \cdot \sqrt{r}, r \geq 0$ **10.** $\sqrt{19} \cdot \sqrt{k}, k \geq 0$

11. Which one of the following radicals is simplified? See Example 2.

     A. $\sqrt{47}$ **B.** $\sqrt{45}$ **C.** $\sqrt{48}$ **D.** $\sqrt{44}$

12. If p is a prime number, is $\sqrt{p}$ in simplified form? Explain your answer.

SECTION 8.2 Multiplying, Dividing, and Simplifying Radicals

Simplify each radical. See Example 2.

13. $\sqrt{45}$
14. $\sqrt{27}$
15. $\sqrt{24}$
16. $\sqrt{44}$
17. $\sqrt{90}$
18. $\sqrt{56}$
19. $\sqrt{75}$
20. $\sqrt{18}$
21. $\sqrt{125}$
22. $\sqrt{80}$
23. $\sqrt{145}$
24. $\sqrt{110}$
25. $\sqrt{160}$
26. $\sqrt{128}$
27. $-\sqrt{700}$
28. $-\sqrt{600}$

Find each product and simplify. See Example 3.

29. $\sqrt{3} \cdot \sqrt{18}$
30. $\sqrt{3} \cdot \sqrt{21}$
31. $\sqrt{12} \cdot \sqrt{48}$
32. $\sqrt{50} \cdot \sqrt{72}$
33. $\sqrt{12} \cdot \sqrt{30}$
34. $\sqrt{30} \cdot \sqrt{24}$

35. Simplify the product $\sqrt{8} \cdot \sqrt{32}$ in two ways. First, multiply 8 by 32 and simplify the square root of this product. Second, simplify $\sqrt{8}$, simplify $\sqrt{32}$, and then multiply. How do the answers compare? Make a conjecture (an educated guess) about whether the correct answer can always be obtained using either method when simplifying a product such as this.

36. Simplify the radical $\sqrt{288}$ in two ways. First, factor 288 as $144 \cdot 2$ and then simplify. Second, factor 288 as $48 \cdot 6$ and then simplify. How do the answers compare? Make a conjecture concerning the quickest way to simplify such a radical.

Use the product and quotient rules, as necessary, to simplify each radical expression. See Examples 4–6.

37. $\sqrt{\dfrac{16}{225}}$
38. $\sqrt{\dfrac{9}{100}}$
39. $\sqrt{\dfrac{7}{16}}$
40. $\sqrt{\dfrac{13}{25}}$
41. $\dfrac{\sqrt{75}}{\sqrt{3}}$
42. $\dfrac{\sqrt{200}}{\sqrt{2}}$
43. $\sqrt{\dfrac{5}{2}} \cdot \sqrt{\dfrac{125}{8}}$
44. $\sqrt{\dfrac{8}{3}} \cdot \sqrt{\dfrac{512}{27}}$
45. $\dfrac{30\sqrt{10}}{5\sqrt{2}}$
46. $\dfrac{50\sqrt{20}}{2\sqrt{10}}$

Simplify each radical. Assume that all variables represent nonnegative real numbers. See Example 7.

47. $\sqrt{m^2}$
48. $\sqrt{k^2}$
49. $\sqrt{y^4}$
50. $\sqrt{s^4}$
51. $\sqrt{36z^2}$
52. $\sqrt{49n^2}$
53. $\sqrt{400x^6}$
54. $\sqrt{900y^8}$
55. $\sqrt{18x^8}$
56. $\sqrt{20r^{10}}$
57. $\sqrt{45c^{14}}$
58. $\sqrt{50d^{20}}$
59. $\sqrt{z^5}$
60. $\sqrt{y^3}$
61. $\sqrt{a^{13}}$
62. $\sqrt{p^{17}}$
63. $\sqrt{64x^7}$
64. $\sqrt{25t^{11}}$
65. $\sqrt{x^6y^{12}}$
66. $\sqrt{a^8b^{10}}$
67. $\sqrt{81m^4n^2}$
68. $\sqrt{100c^4d^6}$
69. $\sqrt{\dfrac{7}{x^{10}}}, x \neq 0$
70. $\sqrt{\dfrac{14}{z^{12}}}, z \neq 0$
71. $\sqrt{\dfrac{y^4}{100}}$
72. $\sqrt{\dfrac{w^8}{144}}$

Simplify each radical. See Example 8.

73. $\sqrt[3]{40}$
74. $\sqrt[3]{48}$
75. $\sqrt[3]{54}$
76. $\sqrt[3]{135}$
77. $\sqrt[3]{128}$
78. $\sqrt[3]{192}$
79. $\sqrt[4]{80}$
80. $\sqrt[4]{243}$
81. $\sqrt[3]{\dfrac{8}{27}}$
82. $\sqrt[3]{\dfrac{64}{125}}$
83. $\sqrt[3]{-\dfrac{216}{125}}$
84. $\sqrt[3]{-\dfrac{1}{64}}$

Simplify each radical. See Example 9.

85. $\sqrt[3]{p^3}$ **86.** $\sqrt[3]{w^3}$ **87.** $\sqrt[3]{x^9}$ **88.** $\sqrt[3]{y^{18}}$

89. $\sqrt[3]{64z^6}$ **90.** $\sqrt[3]{125a^{15}}$ **91.** $\sqrt[3]{343a^9b^3}$ **92.** $\sqrt[3]{216m^3n^6}$

93. $\sqrt[3]{16t^5}$ **94.** $\sqrt[3]{24x^4}$ **95.** $\sqrt[3]{\dfrac{m^{12}}{8}}$ **96.** $\sqrt[3]{\dfrac{n^9}{27}}$

97. In Example 2(a) we showed *algebraically* that $\sqrt{20}$ is equal to $2\sqrt{5}$. To give *numerical support* to this result, use a calculator to do the following:

(a) Find a decimal approximation for $\sqrt{20}$ using your calculator. Record as many digits as the calculator shows.

(b) Find a decimal approximation for $\sqrt{5}$ using your calculator, and then multiply the result by 2. Record as many digits as the calculator shows.

(c) Your results in parts (a) and (b) should be the same. A mathematician would not accept this numerical exercise as *proof* that $\sqrt{20}$ is equal to $2\sqrt{5}$. Explain why.

98. On your calculator, multiply the approximations for $\sqrt{3}$ and $\sqrt{5}$. Now, predict what your calculator will show when you find an approximation for $\sqrt{15}$. What rule stated in this section justifies your answer?

99. When we multiply two radicals with variables under the radical sign, such as $\sqrt{a} \cdot \sqrt{b} = \sqrt{ab}$, why is it important to know that both a and b represent nonnegative numbers?

100. Is it necessary to restrict k to a nonnegative number to say that $\sqrt[3]{k} \cdot \sqrt[3]{k} \cdot \sqrt[3]{k} = k$? Why or why not?

The volume of a cube is found with the formula $V = s^3$, where s is the length of an edge of the cube. Use this information in Exercises 101 and 102.

101. A container in the shape of a cube has a volume of 216 cm³. What is the depth of the container?

102. A cube-shaped box must be constructed to contain 128 ft³. What should the dimensions (height, width, and length) of the box be?

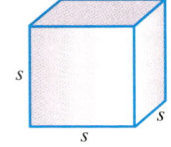

The volume of a sphere is found with the formula $V = \frac{4}{3}\pi r^3$, where r is the length of the radius of the sphere. Use this information in Exercises 103 and 104.

103. A ball in the shape of a sphere has a volume of 288π in.³. What is the radius of the ball?

104. Suppose that the volume of the ball described in Exercise 103 is multiplied by 8. How is the radius affected?

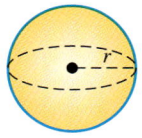

Work Exercises 105 and 106 without using a calculator.

105. Choose the best estimate for the area (in square inches) of this rectangle.

 A. 45 **B.** 72 **C.** 80 **D.** 90

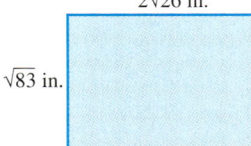

106. Choose the best estimate for the area (in square feet) of this triangle.

 A. 20 **B.** 40 **C.** 60 **D.** 80

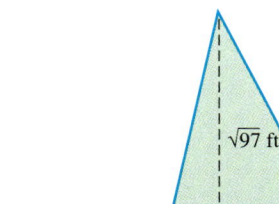

8.3 Adding and Subtracting Radicals

OBJECTIVES

1. Add and subtract radicals.
2. Simplify radical sums and differences.
3. Simplify more complicated radical expressions.

OBJECTIVE 1 Add and subtract radicals. We add or subtract radicals by using the distributive property. For example,

$$8\sqrt{3} + 6\sqrt{3} = (8 + 6)\sqrt{3} = 14\sqrt{3}.$$

Also,

$$2\sqrt{11} - 7\sqrt{11} = -5\sqrt{11}.$$

Only **like radicals,** those that are multiples of the *same root* of the *same number,* can be combined in this way. In the examples above, $8\sqrt{3}$ and $6\sqrt{3}$ are like radicals, as are $2\sqrt{11}$ and $-7\sqrt{11}$. On the other hand, examples of *unlike radicals* are

$2\sqrt{5}$ and $2\sqrt{3}$,   Radicands are different.

as well as $2\sqrt{3}$ and $2\sqrt[3]{3}$.   Indexes are different.

EXAMPLE 1 Adding and Subtracting Like Radicals

Add or subtract, as indicated.

(a) $3\sqrt{6} + 5\sqrt{6} = (3 + 5)\sqrt{6} = 8\sqrt{6}$ Distributive property

(b) $5\sqrt{10} - 7\sqrt{10} = (5 - 7)\sqrt{10} = -2\sqrt{10}$

(c) $\sqrt{7} + 2\sqrt{7} = 1\sqrt{7} + 2\sqrt{7} = (1 + 2)\sqrt{7} = 3\sqrt{7}$

(d) $\sqrt{5} + \sqrt{5} = 1\sqrt{5} + 1\sqrt{5} = 2\sqrt{5}$

(e) $\sqrt{3} + \sqrt{7}$ cannot be added using the distributive property.

Now Try Exercises 1, 3, 5, 7, and 11.

OBJECTIVE 2 Simplify radical sums and differences. Sometimes one or more radical expressions in a sum or difference must first be simplified. Any like radicals that result can then be added or subtracted.

EXAMPLE 2 Adding and Subtracting Radicals That Must Be Simplified

Add or subtract, as indicated.

(a) $3\sqrt{2} + \sqrt{8} = 3\sqrt{2} + \sqrt{4 \cdot 2}$ Factor.
$= 3\sqrt{2} + \sqrt{4} \cdot \sqrt{2}$ Product rule
$= 3\sqrt{2} + 2\sqrt{2}$ $\sqrt{4} = 2$
$= 5\sqrt{2}$ Add like radicals.

(b) $\sqrt{18} - \sqrt{27} = \sqrt{9 \cdot 2} - \sqrt{9 \cdot 3}$ Factor.
$= \sqrt{9} \cdot \sqrt{2} - \sqrt{9} \cdot \sqrt{3}$ Product rule
$= 3\sqrt{2} - 3\sqrt{3}$ $\sqrt{9} = 3$

Because $\sqrt{2}$ and $\sqrt{3}$ are unlike radicals, this difference cannot be simplified further.

(c) $2\sqrt{12} + 3\sqrt{75} = 2(\sqrt{4} \cdot \sqrt{3}) + 3(\sqrt{25} \cdot \sqrt{3})$ Product rule
$= 2(2\sqrt{3}) + 3(5\sqrt{3})$ $\sqrt{4} = 2; \sqrt{25} = 5$
$= 4\sqrt{3} + 15\sqrt{3}$ Multiply.
$= 19\sqrt{3}$ Add like radicals.

(d) $3\sqrt[3]{16} + 5\sqrt[3]{2} = 3(\sqrt[3]{8} \cdot \sqrt[3]{2}) + 5\sqrt[3]{2}$ Product rule
$= 3(2\sqrt[3]{2}) + 5\sqrt[3]{2}$ $\sqrt[3]{8} = 2$
$= 6\sqrt[3]{2} - 5\sqrt[3]{2}$ Multiply.
$= 11\sqrt[3]{2}$ Add like radicals.

Now Try Exercises 9, 13, 15, and 33.

OBJECTIVE 3   Simplify more complicated radical expressions.   When simplifying more complicated radical expressions, the order of operations (from Chapter 1) still applies.

EXAMPLE 3   Simplifying Radical Expressions

Simplify each radical expression. As before, assume all variables represent nonnegative real numbers.

(a) $\sqrt{5} \cdot \sqrt{15} + 4\sqrt{3} = \sqrt{5 \cdot 15} + 4\sqrt{3}$ Product rule
$= \sqrt{75} + 4\sqrt{3}$ Multiply.
$= \sqrt{25 \cdot 3} + 4\sqrt{3}$ 25 is a perfect square.
$= \sqrt{25} \cdot \sqrt{3} + 4\sqrt{3}$ Product rule
$= 5\sqrt{3} + 4\sqrt{3}$ $\sqrt{25} = 5$
$= 9\sqrt{3}$ Add like radicals.

(b) $\sqrt{12k} + \sqrt{27k} = \sqrt{4 \cdot 3k} + \sqrt{9 \cdot 3k}$ Factor.
$= \sqrt{4} \cdot \sqrt{3k} + \sqrt{9} \cdot \sqrt{3k}$ Product rule
$= 2\sqrt{3k} + 3\sqrt{3k}$ $\sqrt{4} = 2; \sqrt{9} = 3$
$= 5\sqrt{3k}$ Add like radicals.

(c) $3x\sqrt{50} + \sqrt{2x^2} = 3x\sqrt{25 \cdot 2} + \sqrt{x^2 \cdot 2}$ Factor.
$= 3x\sqrt{25} \cdot \sqrt{2} + \sqrt{x^2} \cdot \sqrt{2}$ Product rule
$= 3x \cdot 5\sqrt{2} + x\sqrt{2}$ $\sqrt{25} = 5; \sqrt{x^2} = x$
$= 15x\sqrt{2} + x\sqrt{2}$ Multiply.
$= 16x\sqrt{2}$ Add like radicals.

(d) $2\sqrt[3]{32m^3} - \sqrt[3]{108m^3} = 2\sqrt[3]{(8m^3)4} - \sqrt[3]{(27m^3)4}$ Factor.
$= 2(2m)\sqrt[3]{4} - 3m\sqrt[3]{4}$ $\sqrt[3]{8m^3} = 2m; \sqrt[3]{27m^3} = 3m$
$= 4m\sqrt[3]{4} - 3m\sqrt[3]{4}$ Multiply.
$= m\sqrt[3]{4}$ Subtract like radicals.

Now Try Exercises 27, 41, 51, and 59.

SECTION 8.3 Adding and Subtracting Radicals

> **CAUTION** Remember that a sum or difference of radicals can be simplified only if the radicals are *like radicals*. For example, $\sqrt{5} + 3\sqrt{5} = 4\sqrt{5}$, but $\sqrt{5} + 5\sqrt{3}$ cannot be simplified further. Also, $2\sqrt{3} + 5\sqrt[3]{3}$ cannot be simplified further.

8.3 EXERCISES

For Extra Help

 Student's Solutions Manual

 MyMathLab

 InterAct Math Tutorial Software

 AW Math Tutor Center

 MathXL

 Digital Video Tutor CD 16/Videotape 14

Add or subtract wherever possible. See Examples 1, 2, and 3(a).

1. $2\sqrt{3} + 5\sqrt{3}$
2. $6\sqrt{5} + 8\sqrt{5}$
3. $4\sqrt{7} - 9\sqrt{7}$
4. $6\sqrt{2} - 8\sqrt{2}$
5. $\sqrt{6} + \sqrt{6}$
6. $\sqrt{11} + \sqrt{11}$
7. $\sqrt{17} + 2\sqrt{17}$
8. $3\sqrt{19} + \sqrt{19}$
9. $5\sqrt{3} + \sqrt{12}$
10. $3\sqrt{23} - \sqrt{23}$
11. $\sqrt{6} + \sqrt{7}$
12. $\sqrt{14} + \sqrt{17}$
13. $-\sqrt{12} + \sqrt{75}$
14. $2\sqrt{27} - \sqrt{300}$
15. $2\sqrt{50} - 5\sqrt{72}$
16. $6\sqrt{18} - 4\sqrt{32}$
17. $5\sqrt{7} - 2\sqrt{28} + 6\sqrt{63}$
18. $3\sqrt{11} + 5\sqrt{44} - 3\sqrt{99}$
19. $9\sqrt{24} - 2\sqrt{54} + 3\sqrt{20}$
20. $2\sqrt{8} - 5\sqrt{32} + 2\sqrt{48}$
21. $5\sqrt{72} - 3\sqrt{48} - 4\sqrt{128}$
22. $4\sqrt{50} + 3\sqrt{12} + 5\sqrt{45}$
23. $\dfrac{1}{4}\sqrt{288} + \dfrac{1}{6}\sqrt{72}$
24. $\dfrac{2}{3}\sqrt{27} + \dfrac{3}{4}\sqrt{48}$
25. $\dfrac{3}{5}\sqrt{75} - \dfrac{2}{3}\sqrt{45}$
26. $\dfrac{5}{8}\sqrt{128} - \dfrac{3}{4}\sqrt{160}$
27. $\sqrt{3} \cdot \sqrt{7} + 2\sqrt{21}$
28. $\sqrt{13} \cdot \sqrt{2} + 3\sqrt{26}$
29. $\sqrt{6} \cdot \sqrt{2} + 3\sqrt{3}$
30. $4\sqrt{15} \cdot \sqrt{3} - 2\sqrt{5}$
31. $4\sqrt[3]{16} - 3\sqrt[3]{54}$
32. $5\sqrt[3]{128} + 3\sqrt[3]{250}$
33. $3\sqrt[3]{24} + 6\sqrt[3]{81}$
34. $2\sqrt[4]{48} - \sqrt[4]{243}$
35. $5\sqrt[4]{32} + 2\sqrt[4]{32} \cdot \sqrt[4]{4}$
36. $8\sqrt[3]{48} + 10\sqrt[3]{3} \cdot \sqrt[3]{18}$

37. The distributive property, which says $a(b + c) = ab + ac$ and $ba + ca = (b + c)a$, provides the justification for adding and subtracting like radicals. While we usually skip the step that indicates this property, we could not make the statement $2\sqrt{3} + 4\sqrt{3} = 6\sqrt{3}$ without it. Write an equation showing how the distributive property is actually used in this statement.

38. Refer to Example 1(e), and explain why $\sqrt{3} + \sqrt{7}$ cannot be further simplified. Confirm, by using calculator approximations, that $\sqrt{3} + \sqrt{7}$ is *not* equal to $\sqrt{10}$.

Perform each indicated operation. Assume all variables represent nonnegative real numbers. See Example 3.

39. $\sqrt{32x} - \sqrt{18x}$
40. $\sqrt{125t} - \sqrt{80t}$
41. $\sqrt{27r} + \sqrt{48r}$
42. $\sqrt{6x^2} + x\sqrt{54}$
43. $\sqrt{75x^2} + x\sqrt{300}$
44. $\sqrt{20y^2} - 3y\sqrt{5}$
45. $3\sqrt{8x^2} - 4x\sqrt{2}$
46. $\sqrt{2b^2} + 3b\sqrt{18}$
47. $5\sqrt{75p^2} - 4\sqrt{27p^2}$
48. $3\sqrt{32k} + 6\sqrt{8k}$
49. $2\sqrt{125x^2z} + 8x\sqrt{80z}$
50. $4p\sqrt{63m} + 6\sqrt{28mp^2}$
51. $3k\sqrt{24k^2h^2} + 9h\sqrt{54k^3}$
52. $6r\sqrt{27r^2s} + 3r^2\sqrt{3s}$
53. $6\sqrt[3]{8p^2} - 2\sqrt[3]{27p^2}$
54. $5\sqrt[3]{27x^2} + 8\sqrt[3]{8x^2}$
55. $5\sqrt[4]{m^3} + 8\sqrt[4]{16m^3}$
56. $5\sqrt[4]{m^4} + 3\sqrt[4]{81m^4}$
57. $2\sqrt[4]{p^5} - 5p\sqrt[4]{16p}$
58. $8k\sqrt[3]{54k} + 6\sqrt[3]{16k^4}$
59. $-5\sqrt[3]{256z^4} - 2z\sqrt[3]{32z}$
60. $10\sqrt[3]{4m^4} - 3m\sqrt[3]{32m}$

550 CHAPTER 8 Roots and Radicals

✍ **61.** In the directions for Exercises 39–60, we made the assumption that all variables represent nonnegative real numbers. However, in Exercises 53, 58, 59, and 60, variables actually *may* represent negative numbers. Explain why this is so.

62. Find the perimeter of each figure.

(a)

Rectangle with top side $7\sqrt{2}$ and right side $4\sqrt{2}$.

(b)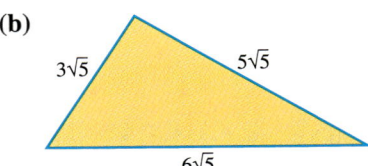

Triangle with sides $3\sqrt{5}$, $5\sqrt{5}$, and $6\sqrt{5}$.

Perform the indicated operations following the order of operations we have used throughout the book. Express all answers in simplest form.

63. $\sqrt{(-3-6)^2 + (2-4)^2}$

64. $\sqrt{(-9-3)^2 + (3-8)^2}$

65. $\sqrt{(2-(-2))^2 + (-1-2)^2}$

66. $\sqrt{(3-1)^2 + (2-(-1))^2}$

RELATING CONCEPTS (EXERCISES 67–70)

For Individual or Group Work

Adding and subtracting like radicals is no different than adding and subtracting like terms. **Work Exercises 67–70 in order.**

67. Combine like terms: $5x^2y + 3x^2y - 14x^2y$.

68. Combine like terms:
$$5(p-2q)^2(a+b) + 3(p-2q)^2(a+b) - 14(p-2q)^2(a+b).$$

69. Combine like radicals: $5a^2\sqrt{xy} + 3a^2\sqrt{xy} - 14a^2\sqrt{xy}$.

✍ **70.** Compare your answers in Exercises 67–69. How are they alike? How are they different?

8.4 Rationalizing the Denominator

OBJECTIVES

1. Rationalize denominators with square roots.
2. Write radicals in simplified form.
3. Rationalize denominators with cube roots.

OBJECTIVE 1 Rationalize denominators with square roots. Although calculators now make it fairly easy to divide by a radical in an expression such as $\frac{1}{\sqrt{2}}$, it is sometimes easier to work with radical expressions if the denominators do not contain any radicals. For example, the radical in the denominator of $\frac{1}{\sqrt{2}}$ can be eliminated by multiplying the numerator and denominator by $\sqrt{2}$, since $\sqrt{2} \cdot \sqrt{2} = \sqrt{4} = 2$.

$$\frac{1}{\sqrt{2}} = \frac{1 \cdot \sqrt{2}}{\sqrt{2} \cdot \sqrt{2}} = \frac{\sqrt{2}}{2} \qquad \text{Multiply by } \frac{\sqrt{2}}{\sqrt{2}} = 1.$$

This process of changing the denominator from a radical (irrational number) to a rational number is called **rationalizing the denominator.** The value of the radical expression is not changed; only the form is changed, because the expression has been multiplied by 1 in the form of $\frac{\sqrt{2}}{\sqrt{2}}$.

EXAMPLE 1 Rationalizing Denominators

Rationalize each denominator.

(a) $\dfrac{9}{\sqrt{6}}$

To eliminate the radical in the denominator, multiply the numerator and denominator by $\sqrt{6}$.

$$\dfrac{9}{\sqrt{6}} = \dfrac{9 \cdot \sqrt{6}}{\sqrt{6} \cdot \sqrt{6}} \qquad \text{Multiply by } \tfrac{\sqrt{6}}{\sqrt{6}} = 1.$$

$$= \dfrac{9\sqrt{6}}{6} \qquad \sqrt{6} \cdot \sqrt{6} = \sqrt{36} = 6$$

$$= \dfrac{3\sqrt{6}}{2} \qquad \text{Lowest terms}$$

(b) $\dfrac{12}{\sqrt{8}}$

The denominator could be rationalized here by multiplying by $\sqrt{8}$ in both the numerator and denominator. However, the result can be found more directly by first simplifying the denominator.

$$\sqrt{8} = \sqrt{4} \cdot \sqrt{2} = 2\sqrt{2}$$

Then multiply the numerator and denominator by $\sqrt{2}$.

$$\dfrac{12}{\sqrt{8}} = \dfrac{12}{2\sqrt{2}} \qquad \sqrt{8} = 2\sqrt{2}$$

$$= \dfrac{12 \cdot \sqrt{2}}{2\sqrt{2} \cdot \sqrt{2}} \qquad \text{Multiply by } \tfrac{\sqrt{2}}{\sqrt{2}}.$$

$$= \dfrac{12 \cdot \sqrt{2}}{2 \cdot 2} \qquad \sqrt{2} \cdot \sqrt{2} = \sqrt{4} = 2$$

$$= \dfrac{12\sqrt{2}}{4} \qquad \text{Multiply.}$$

$$= 3\sqrt{2} \qquad \text{Lowest terms}$$

Now Try Exercises 5 and 13.

NOTE In Example 1(b), we could also have rationalized the original denominator $\sqrt{8}$ by multiplying by $\sqrt{2}$, since $\sqrt{8} \cdot \sqrt{2} = \sqrt{16} = 4$.

$$\dfrac{12}{\sqrt{8}} = \dfrac{12 \cdot \sqrt{2}}{\sqrt{8} \cdot \sqrt{2}} = \dfrac{12\sqrt{2}}{\sqrt{16}} = \dfrac{12\sqrt{2}}{4} = 3\sqrt{2}$$

When simplifying radicals, be aware that there are often several approaches that yield the same correct answer.

OBJECTIVE 2 Write radicals in simplified form. A radical is considered to be in simplified form if the following three conditions are met.

552 CHAPTER 8 Roots and Radicals

> **Simplified Form of a Radical**
> 1. The radicand contains no factor (except 1) that is a perfect square (when dealing with square roots), a perfect cube (when dealing with cube roots), and so on.
> 2. The radicand has no fractions.
> 3. No denominator contains a radical.

In the following examples, we simplify radicals according to these conditions.

EXAMPLE 2 Simplifying a Radical

Simplify $\sqrt{\dfrac{27}{5}}$.

This violates condition 2. To begin, use the quotient rule for radicals.

$$\sqrt{\frac{27}{5}} = \frac{\sqrt{27}}{\sqrt{5}} \qquad \text{Quotient rule}$$

$$= \frac{\sqrt{27} \cdot \sqrt{5}}{\sqrt{5} \cdot \sqrt{5}} \qquad \text{Rationalize the denominator.}$$

$$= \frac{\sqrt{27} \cdot \sqrt{5}}{5} \qquad \sqrt{5} \cdot \sqrt{5} = 5$$

$$= \frac{\sqrt{9 \cdot 3} \cdot \sqrt{5}}{5} \qquad \text{Factor.}$$

$$= \frac{\sqrt{9} \cdot \sqrt{3} \cdot \sqrt{5}}{5} \qquad \text{Product rule}$$

$$= \frac{3 \cdot \sqrt{3} \cdot \sqrt{5}}{5} \qquad \sqrt{9} = 3$$

$$= \frac{3\sqrt{15}}{5} \qquad \text{Product rule}$$

Now Try Exercise 19.

EXAMPLE 3 Simplifying a Product of Radicals

Simplify $\sqrt{\dfrac{5}{8}} \cdot \sqrt{\dfrac{1}{6}}$.

Use both the product and quotient rules.

$$\sqrt{\frac{5}{8}} \cdot \sqrt{\frac{1}{6}} = \sqrt{\frac{5}{8} \cdot \frac{1}{6}} \qquad \text{Product rule}$$

$$= \sqrt{\frac{5}{48}} \qquad \text{Multiply fractions.}$$

$$= \frac{\sqrt{5}}{\sqrt{48}} \qquad \text{Quotient rule}$$

First simplify the denominator and then rationalize it.

$$\frac{\sqrt{5}}{\sqrt{48}} = \frac{\sqrt{5}}{\sqrt{16} \cdot \sqrt{3}} \quad \text{Product rule}$$

$$= \frac{\sqrt{5}}{4\sqrt{3}} \quad \sqrt{16} = 4$$

$$= \frac{\sqrt{5} \cdot \sqrt{3}}{4\sqrt{3} \cdot \sqrt{3}} \quad \text{Rationalize the denominator.}$$

$$= \frac{\sqrt{15}}{4 \cdot 3} \quad \text{Product rule; } \sqrt{3} \cdot \sqrt{3} = 3$$

$$= \frac{\sqrt{15}}{12} \quad \text{Multiply.}$$

Now Try Exercise 45.

EXAMPLE 4 Simplifying a Quotient of Radicals

Simplify $\dfrac{\sqrt{4x}}{\sqrt{y}}$. Assume that x and y are positive real numbers.

$$\frac{\sqrt{4x}}{\sqrt{y}} = \frac{\sqrt{4x} \cdot \sqrt{y}}{\sqrt{y} \cdot \sqrt{y}} = \frac{\sqrt{4xy}}{y} = \frac{2\sqrt{xy}}{y}$$

Now Try Exercise 59.

EXAMPLE 5 Simplifying a Radical Quotient

Simplify $\sqrt{\dfrac{2x^2y}{3}}$. Assume that x and y are nonnegative real numbers.

$$\sqrt{\frac{2x^2y}{3}} = \frac{\sqrt{2x^2y}}{\sqrt{3}} \quad \text{Quotient rule}$$

$$= \frac{\sqrt{2x^2y} \cdot \sqrt{3}}{\sqrt{3} \cdot \sqrt{3}} \quad \text{Rationalize the denominator.}$$

$$= \frac{\sqrt{6x^2y}}{3} \quad \text{Product rule}$$

$$= \frac{\sqrt{x^2}\sqrt{6y}}{3} \quad \text{Product rule}$$

$$= \frac{x\sqrt{6y}}{3} \quad \sqrt{x^2} = x, \text{ since } x \geq 0.$$

Now Try Exercise 67.

OBJECTIVE 3 Rationalize denominators with cube roots. To rationalize a denominator with a cube root, we change the radicand in the denominator to a perfect cube.

554 CHAPTER 8 Roots and Radicals

EXAMPLE 6 Rationalize Denominators with Cube Roots

Rationalize each denominator.

(a) $\sqrt[3]{\dfrac{3}{2}}$

First write the expression as a quotient of radicals. Then multiply the numerator and denominator by a sufficient number of factors of 2 to make the denominator a perfect cube. This will eliminate the radical in the denominator. Here, multiply by $\sqrt[3]{2 \cdot 2} = \sqrt[3]{2^2}$.

$$\sqrt[3]{\dfrac{3}{2}} = \dfrac{\sqrt[3]{3}}{\sqrt[3]{2}} = \dfrac{\sqrt[3]{3} \cdot \sqrt[3]{2^2}}{\sqrt[3]{2} \cdot \sqrt[3]{2^2}} = \dfrac{\sqrt[3]{3 \cdot 2^2}}{\sqrt[3]{2^3}} = \dfrac{\sqrt[3]{12}}{2} \qquad \sqrt[3]{2^3} = \sqrt[3]{8} = 2$$

Denominator is a perfect cube.

(b) $\dfrac{\sqrt[3]{3}}{\sqrt[3]{4}}$

Since $\sqrt[3]{4} \cdot \sqrt[3]{2} = \sqrt[3]{2^2} \cdot \sqrt[3]{2} = \sqrt[3]{2^3} = 2$, multiply the numerator and denominator by $\sqrt[3]{2}$.

$$\dfrac{\sqrt[3]{3}}{\sqrt[3]{4}} = \dfrac{\sqrt[3]{3} \cdot \sqrt[3]{2}}{\sqrt[3]{2^2} \cdot \sqrt[3]{2}} = \dfrac{\sqrt[3]{6}}{\sqrt[3]{2^3}} = \dfrac{\sqrt[3]{6}}{2}$$

(c) $\dfrac{\sqrt[3]{2}}{\sqrt[3]{3x^2}} \quad (x \neq 0)$

Multiply the numerator and denominator by enough factors of 3 and of x to get a perfect cube in the denominator. Here, multiply by $\sqrt[3]{3^2 x}$ (that is, $\sqrt[3]{9x}$) since $\sqrt[3]{3x^2} \cdot \sqrt[3]{3^2 x} = \sqrt[3]{(3x)^3} = 3x$.

$$\dfrac{\sqrt[3]{2}}{\sqrt[3]{3x^2}} = \dfrac{\sqrt[3]{2} \cdot \sqrt[3]{3^2 x}}{\sqrt[3]{3x^2} \cdot \sqrt[3]{3^2 x}} = \dfrac{\sqrt[3]{18x}}{\sqrt[3]{(3x)^3}} = \dfrac{\sqrt[3]{18x}}{3x}$$

Denominator is a perfect cube.

Now Try Exercises 77, 79, and 83.

> **CAUTION** A common error in a problem like the one in Example 6(a) is to multiply by $\sqrt[3]{2}$ instead of $\sqrt[3]{2^2}$. Doing this would give a denominator of $\sqrt[3]{2} \cdot \sqrt[3]{2} = \sqrt[3]{4}$. Because 4 is not a perfect cube, the denominator is still not rationalized.

8.4 EXERCISES

For Extra Help

Student's Solutions Manual

MyMathLab

InterAct Math Tutorial Software

AW Math Tutor Center

MathXL

Digital Video Tutor CD 16/Videotape 14

Rationalize each denominator. See Examples 1 and 2.

1. $\dfrac{6}{\sqrt{5}}$
2. $\dfrac{3}{\sqrt{2}}$
3. $\dfrac{5}{\sqrt{5}}$
4. $\dfrac{15}{\sqrt{15}}$

5. $\dfrac{4}{\sqrt{6}}$
6. $\dfrac{15}{\sqrt{10}}$
7. $\dfrac{8\sqrt{3}}{\sqrt{5}}$
8. $\dfrac{9\sqrt{6}}{\sqrt{5}}$

9. $\dfrac{12\sqrt{10}}{8\sqrt{3}}$
10. $\dfrac{9\sqrt{15}}{6\sqrt{2}}$
11. $\dfrac{8}{\sqrt{27}}$
12. $\dfrac{12}{\sqrt{18}}$

13. $\dfrac{6}{\sqrt{200}}$
14. $\dfrac{10}{\sqrt{300}}$
15. $\dfrac{12}{\sqrt{72}}$
16. $\dfrac{21}{\sqrt{45}}$

17. $\dfrac{\sqrt{10}}{\sqrt{5}}$
18. $\dfrac{\sqrt{6}}{\sqrt{3}}$
19. $\sqrt{\dfrac{40}{3}}$
20. $\sqrt{\dfrac{5}{8}}$

21. $\sqrt{\dfrac{1}{32}}$
22. $\sqrt{\dfrac{1}{8}}$
23. $\sqrt{\dfrac{9}{5}}$
24. $\sqrt{\dfrac{16}{7}}$

25. $\dfrac{-3}{\sqrt{50}}$
26. $\dfrac{-5}{\sqrt{75}}$
27. $\dfrac{63}{\sqrt{45}}$
28. $\dfrac{27}{\sqrt{32}}$
29. $\dfrac{\sqrt{8}}{\sqrt{24}}$

30. $\dfrac{\sqrt{5}}{\sqrt{10}}$
31. $-\sqrt{\dfrac{1}{5}}$
32. $-\sqrt{\dfrac{1}{6}}$
33. $\sqrt{\dfrac{13}{5}}$
34. $\sqrt{\dfrac{17}{11}}$

35. To rationalize the denominator of an expression such as $\dfrac{4}{\sqrt{3}}$, we multiply both the numerator and the denominator by $\sqrt{3}$. By what number are we actually multiplying the given expression, and what property of real numbers justifies the fact that our result is equal to the given expression?

36. In Example 1(a), we showed algebraically that $\dfrac{9}{\sqrt{6}} = \dfrac{3\sqrt{6}}{2}$. Support this result numerically by finding the decimal approximation of $\dfrac{9}{\sqrt{6}}$ on your calculator, and then finding the decimal approximation of $\dfrac{3\sqrt{6}}{2}$. What do you notice?

Simplify each product of radicals. See Example 3.

37. $\sqrt{\dfrac{7}{13}} \cdot \sqrt{\dfrac{13}{3}}$
38. $\sqrt{\dfrac{19}{20}} \cdot \sqrt{\dfrac{20}{3}}$
39. $\sqrt{\dfrac{21}{7}} \cdot \sqrt{\dfrac{21}{8}}$
40. $\sqrt{\dfrac{5}{8}} \cdot \sqrt{\dfrac{5}{6}}$

41. $\sqrt{\dfrac{1}{12}} \cdot \sqrt{\dfrac{1}{3}}$
42. $\sqrt{\dfrac{1}{8}} \cdot \sqrt{\dfrac{1}{2}}$
43. $\sqrt{\dfrac{2}{9}} \cdot \sqrt{\dfrac{9}{2}}$
44. $\sqrt{\dfrac{4}{3}} \cdot \sqrt{\dfrac{3}{4}}$

45. $\sqrt{\dfrac{3}{4}} \cdot \sqrt{\dfrac{1}{5}}$
46. $\sqrt{\dfrac{1}{10}} \cdot \sqrt{\dfrac{10}{3}}$
47. $\sqrt{\dfrac{17}{3}} \cdot \sqrt{\dfrac{17}{6}}$
48. $\sqrt{\dfrac{1}{11}} \cdot \sqrt{\dfrac{33}{16}}$

556 CHAPTER 8 Roots and Radicals

49. $\sqrt{\dfrac{2}{5}} \cdot \sqrt{\dfrac{3}{10}}$ 50. $\sqrt{\dfrac{9}{8}} \cdot \sqrt{\dfrac{7}{16}}$ 51. $\sqrt{\dfrac{16}{27}} \cdot \sqrt{\dfrac{1}{9}}$ 52. $\sqrt{\dfrac{5}{8}} \cdot \sqrt{\dfrac{5}{6}}$

Simplify each radical. Assume that all variables represent positive real numbers. See Examples 4 and 5.

53. $\sqrt{\dfrac{6}{p}}$ 54. $\sqrt{\dfrac{5}{x}}$ 55. $\sqrt{\dfrac{3}{y}}$ 56. $\sqrt{\dfrac{9}{k}}$

57. $\sqrt{\dfrac{16}{m}}$ 58. $\sqrt{\dfrac{2z^2}{x}}$ 59. $\dfrac{\sqrt{3p^2}}{\sqrt{q}}$ 60. $\dfrac{\sqrt{5a^3}}{\sqrt{b}}$

61. $\dfrac{\sqrt{7x^3}}{\sqrt{y}}$ 62. $\dfrac{\sqrt{4r^3}}{\sqrt{s}}$ 63. $\sqrt{\dfrac{6p^3}{3m}}$ 64. $\sqrt{\dfrac{a^3b}{6}}$

65. $\sqrt{\dfrac{x^2}{4y}}$ 66. $\sqrt{\dfrac{m^2n}{2}}$ 67. $\sqrt{\dfrac{9a^2r}{5}}$ 68. $\sqrt{\dfrac{2x^2z^4}{3y}}$

69. Which one of the following would be an appropriate choice for multiplying the numerator and the denominator of $\dfrac{\sqrt[3]{2}}{\sqrt[3]{5}}$ in order to rationalize the denominator?

   A. $\sqrt[3]{5}$ B. $\sqrt[3]{25}$ C. $\sqrt[3]{2}$ D. $\sqrt[3]{3}$

70. In Example 6(b), we multiplied the numerator and denominator of $\dfrac{\sqrt[3]{3}}{\sqrt[3]{4}}$ by $\sqrt[3]{2}$ to rationalize the denominator. Suppose we had chosen to multiply by $\sqrt[3]{16}$ instead. Would we have obtained the correct answer after all simplifications were done?

Rationalize each denominator. Assume that variables in denominators are nonzero. See Example 6.

71. $\sqrt[3]{\dfrac{1}{2}}$ 72. $\sqrt[3]{\dfrac{1}{4}}$ 73. $\sqrt[3]{\dfrac{1}{32}}$ 74. $\sqrt[3]{\dfrac{1}{5}}$

75. $\sqrt[3]{\dfrac{1}{11}}$ 76. $\sqrt[3]{\dfrac{3}{2}}$ 77. $\sqrt[3]{\dfrac{2}{5}}$ 78. $\sqrt[3]{\dfrac{4}{9}}$

79. $\dfrac{\sqrt[3]{4}}{\sqrt[3]{7}}$ 80. $\dfrac{\sqrt[3]{5}}{\sqrt[3]{10}}$ 81. $\sqrt[3]{\dfrac{3}{4y^2}}$ 82. $\sqrt[3]{\dfrac{3}{25x^2}}$

83. $\dfrac{\sqrt[3]{7m}}{\sqrt[3]{36n}}$ 84. $\dfrac{\sqrt[3]{11p}}{\sqrt[3]{49q}}$

8.5 More Simplifying and Operations with Radicals

OBJECTIVES

1 Simplify products of radical expressions.

The conditions for which a radical is in simplest form were listed in the previous section. A set of guidelines to use when you are simplifying radical expressions follows on the next page.

2 Use conjugates to rationalize denominators of radical expressions.

3 Write radical expressions with quotients in lowest terms.

Simplifying Radical Expressions

1. If a radical represents a rational number, use that rational number in place of the radical.

    Examples: $\sqrt{49} = 7$; $\sqrt{\dfrac{169}{9}} = \dfrac{13}{3}$.

2. If a radical expression contains products of radicals, use the product rule for radicals, $\sqrt[n]{a} \cdot \sqrt[n]{b} = \sqrt[n]{ab}$, to get a single radical.

    Examples: $\sqrt{5} \cdot \sqrt{x} = \sqrt{5x}$; $\sqrt[3]{3} \cdot \sqrt[3]{2} = \sqrt[3]{6}$

3. If a radicand has a factor that is a perfect square, express the radical as the product of the positive square root of the perfect square and the remaining radical factor. A similar statement applies to higher roots.

    Examples: $\sqrt{20} = \sqrt{4 \cdot 5} = \sqrt{4} \cdot \sqrt{5} = 2\sqrt{5}$;
    $\sqrt[3]{16} = \sqrt[3]{8 \cdot 2} = \sqrt[3]{8} \cdot \sqrt[3]{2} = 2\sqrt[3]{2}$.

4. If a radical expression contains sums or differences of radicals, use the distributive property to combine like radicals.

    Examples: $3\sqrt{2} + 4\sqrt{2} = 7\sqrt{2}$

    $3\sqrt{2} + 4\sqrt{3}$ cannot be simplified further.

5. Rationalize any denominator containing a radical.

    Examples: $\dfrac{5}{\sqrt{3}} = \dfrac{5 \cdot \sqrt{3}}{\sqrt{3} \cdot \sqrt{3}} = \dfrac{5\sqrt{3}}{3}$;

    $\sqrt{\dfrac{3}{2}} = \dfrac{\sqrt{3}}{\sqrt{2}} = \dfrac{\sqrt{3} \cdot \sqrt{2}}{\sqrt{2} \cdot \sqrt{2}} = \dfrac{\sqrt{6}}{2}$;

    $\sqrt[3]{\dfrac{1}{4}} = \dfrac{\sqrt[3]{1}}{\sqrt[3]{4}} = \dfrac{\sqrt[3]{1} \cdot \sqrt[3]{2}}{\sqrt[3]{4} \cdot \sqrt[3]{2}} = \dfrac{\sqrt[3]{2}}{\sqrt[3]{8}} = \dfrac{\sqrt[3]{2}}{2}$.

OBJECTIVE 1 Simplify products of radical expressions. Use the preceding guidelines.

EXAMPLE 1 Multiplying Radical Expressions

Find each product and simplify.

(a) $\sqrt{5}(\sqrt{8} - \sqrt{32})$

Start by simplifying $\sqrt{8}$ and $\sqrt{32}$.

$\sqrt{8} = 2\sqrt{2}$ and $\sqrt{32} = 4\sqrt{2}$.

Now simplify inside the parentheses.

$\sqrt{5}(\sqrt{8} - \sqrt{32}) = \sqrt{5}(2\sqrt{2} - 4\sqrt{2})$
$= \sqrt{5}(-2\sqrt{2})$ Subtract like radicals.
$= -2\sqrt{5 \cdot 2}$ Product rule
$= -2\sqrt{10}$ Multiply.

(b) $\left(\sqrt{3} + 2\sqrt{5}\right)\left(\sqrt{3} - 4\sqrt{5}\right)$

We can find the products of sums of radicals in the same way that we found the product of binomials in Chapter 5, using the FOIL method.

$$\left(\sqrt{3} + 2\sqrt{5}\right)\left(\sqrt{3} - 4\sqrt{5}\right)$$
$$= \underbrace{\sqrt{3}(\sqrt{3})}_{\text{First}} + \underbrace{\sqrt{3}(-4\sqrt{5})}_{\text{Outer}} + \underbrace{2\sqrt{5}(\sqrt{3})}_{\text{Inner}} + \underbrace{2\sqrt{5}(-4\sqrt{5})}_{\text{Last}}$$

$$= 3 - 4\sqrt{15} + 2\sqrt{15} - 8 \cdot 5 \qquad \text{Product rule}$$
$$= 3 - 2\sqrt{15} - 40 \qquad \text{Add like radicals.}$$
$$= -37 - 2\sqrt{15} \qquad \text{Combine terms.}$$

(c) $\left(\sqrt{3} + \sqrt{21}\right)\left(\sqrt{3} - \sqrt{7}\right)$
$$= \sqrt{3}(\sqrt{3}) + \sqrt{3}(-\sqrt{7}) + \sqrt{21}(\sqrt{3})$$
$$\quad + \sqrt{21}(-\sqrt{7}) \qquad \text{FOIL}$$
$$= 3 - \sqrt{21} + \sqrt{63} - \sqrt{147} \qquad \text{Product rule}$$
$$= 3 - \sqrt{21} + \sqrt{9} \cdot \sqrt{7} - \sqrt{49} \cdot \sqrt{3} \qquad \text{9 and 49 are perfect squares.}$$
$$= 3 - \sqrt{21} + 3\sqrt{7} - 7\sqrt{3} \qquad \sqrt{9} = 3; \sqrt{49} = 7$$

Since there are no like radicals, no terms can be combined.

Now Try Exercises 5, 11, and 27.

The special products of binomials discussed in Chapter 5 can be applied to radicals. Example 2 uses the rules for the square of a binomial,

$$(x + y)^2 = x^2 + 2xy + y^2 \quad \text{and} \quad (x - y)^2 = x^2 - 2xy + y^2.$$

EXAMPLE 2 Using Special Products with Radicals

Find each product.

(a) $\left(\sqrt{10} - 7\right)^2$

Follow the second pattern given above. Let $x = \sqrt{10}$ and $y = 7$.

$$\left(\sqrt{10} - 7\right)^2 = \left(\sqrt{10}\right)^2 - 2\left(\sqrt{10}\right)(7) + 7^2$$
$$= 10 - 14\sqrt{10} + 49 \qquad \left(\sqrt{10}\right)^2 = 10; 7^2 = 49$$
$$= 59 - 14\sqrt{10} \qquad \text{Combine terms.}$$

(b) $\left(2\sqrt{3} + 4\right)^2 = \left(2\sqrt{3}\right)^2 + 2\left(2\sqrt{3}\right)(4) + 4^2 \qquad x = 2\sqrt{3}; y = 4$
$$= 12 + 16\sqrt{3} + 16 \qquad \left(2\sqrt{3}\right)^2 = 4 \cdot 3 = 12$$
$$= 28 + 16\sqrt{3}$$

(c) $\left(5 - \sqrt{x}\right)^2 = 5^2 - 2(5)\left(\sqrt{x}\right) + \left(\sqrt{x}\right)^2$
$$= 25 - 10\sqrt{x} + x$$

Now Try Exercises 15, 17, and 37.

> **CAUTION** Be careful! In Examples 2(a) and (b),
> $$59 - 14\sqrt{10} \neq 45\sqrt{10} \quad \text{and} \quad 28 + 16\sqrt{3} \neq 44\sqrt{3}.$$
> *Only like radicals can be combined.*

Example 3 uses the rule for the product of the sum and difference of two terms,
$$(x + y)(x - y) = x^2 - y^2.$$

EXAMPLE 3 Using a Special Product with Radicals

Find each product.

(a) $(4 + \sqrt{3})(4 - \sqrt{3})$

Follow the pattern given above. Let $x = 4$ and $y = \sqrt{3}$.
$$(4 + \sqrt{3})(4 - \sqrt{3}) = 4^2 - (\sqrt{3})^2$$
$$= 16 - 3 \qquad 4^2 = 16; (\sqrt{3})^2 = 3$$
$$= 13$$

(b) $(\sqrt{x} - \sqrt{6})(\sqrt{x} + \sqrt{6}) = (\sqrt{x})^2 - (\sqrt{6})^2$
$$= x - 6 \qquad (\sqrt{x})^2 = x; (\sqrt{6})^2 = 6$$

Now Try Exercises 21 and 41.

Notice that the results in Example 3 do not contain radicals. The pairs of expressions being multiplied, $4 + \sqrt{3}$ and $4 - \sqrt{3}$, and $\sqrt{x} - \sqrt{6}$ and $\sqrt{x} + \sqrt{6}$, are called **conjugates** of each other.

OBJECTIVE 2 Use conjugates to rationalize denominators of radical expressions.
Conjugates similar to those in Example 3 can be used to rationalize the denominators in more complicated quotients, such as
$$\frac{2}{4 - \sqrt{3}}.$$

By Example 3(a), if this denominator, $4 - \sqrt{3}$, is multiplied by $4 + \sqrt{3}$, then the product $(4 - \sqrt{3})(4 + \sqrt{3})$ is the rational number 13. Multiplying the numerator and denominator of the quotient by $4 + \sqrt{3}$ gives
$$\frac{2}{4 - \sqrt{3}} = \frac{2(4 + \sqrt{3})}{(4 - \sqrt{3})(4 + \sqrt{3})} = \frac{2(4 + \sqrt{3})}{13}.$$

The denominator has now been rationalized; it contains no radicals.

> **Using Conjugates to Simplify a Radical Expression**
>
> To simplify a radical expression with two terms in the denominator, where at least one of those terms is a radical, multiply both the numerator and denominator by the conjugate of the denominator.

CHAPTER 8 Roots and Radicals

> **EXAMPLE 4** Using Conjugates to Rationalize Denominators

Simplify by rationalizing each denominator.

(a) $\dfrac{5}{3 + \sqrt{5}}$

We can eliminate the radical in the denominator by multiplying both the numerator and denominator by $3 - \sqrt{5}$.

$\dfrac{5}{3 + \sqrt{5}} = \dfrac{5(3 - \sqrt{5})}{(3 + \sqrt{5})(3 - \sqrt{5})}$ Multiply by the conjugate.

$= \dfrac{5(3 - \sqrt{5})}{3^2 - (\sqrt{5})^2}$ $(x + y)(x - y) = x^2 - y^2$

$= \dfrac{5(3 - \sqrt{5})}{9 - 5}$ $3^2 = 9;\ (\sqrt{5})^2 = 5$

$= \dfrac{5(3 - \sqrt{5})}{4}$ Subtract.

(b) $\dfrac{6 + \sqrt{2}}{\sqrt{2} - 5}$

$\dfrac{6 + \sqrt{2}}{\sqrt{2} - 5} = \dfrac{(6 + \sqrt{2})(\sqrt{2} + 5)}{(\sqrt{2} - 5)(\sqrt{2} + 5)}$ Multiply by the conjugate.

$= \dfrac{6\sqrt{2} + 30 + 2 + 5\sqrt{2}}{2 - 25}$ FOIL; $(x + y)(x - y) = x^2 - y^2$

$= \dfrac{11\sqrt{2} + 32}{-23}$ Combine like terms.

$= \dfrac{-11\sqrt{2} - 32}{23}$ $\dfrac{x}{-y} = \dfrac{-x}{y}$

(c) $\dfrac{4}{3 - \sqrt{x}} = \dfrac{4(3 + \sqrt{x})}{(3 - \sqrt{x})(3 + \sqrt{x})}$ Multiply by the conjugate.

$= \dfrac{4(3 + \sqrt{x})}{9 - x}$ $3^2 = 9;\ (\sqrt{x})^2 = x$

Now Try Exercises 47, 55, and 65.

OBJECTIVE 3 Write radical expressions with quotients in lowest terms.

> **EXAMPLE 5** Writing a Radical Quotient in Lowest Terms

Write $\dfrac{3\sqrt{3} + 9}{12}$ in lowest terms.

Factor the numerator and denominator, and then use the fundamental property from Section 7.1 to divide out common factors.

SECTION 8.5 More Simplifying and Operations with Radicals **561**

$$\frac{3\sqrt{3} + 9}{12} = \frac{3(\sqrt{3} + 3)}{3(4)} = 1 \cdot \frac{\sqrt{3} + 3}{4} = \frac{\sqrt{3} + 3}{4}$$

Now Try Exercise 67.

CAUTION An expression like the one in Example 5 can only be simplified by factoring a common factor from the denominator and *each* term of the numerator. For example,

$$\frac{4 + 8\sqrt{5}}{4} \neq 1 + 8\sqrt{5}.$$

First factor to get $\dfrac{4 + 8\sqrt{5}}{4} = \dfrac{4(1 + 2\sqrt{5})}{4} = 1 + 2\sqrt{5}.$

8.5 EXERCISES

For Extra Help

 Student's Solutions Manual

 MyMathLab

 InterAct Math Tutorial Software

 AW Math Tutor Center

 MathXL

Digital Video Tutor CD 17/Videotape 15

Based on the work so far in this chapter, you should now be able to mentally perform many simple operations involving radicals. In Exercises 1–4, perform the operations mentally, and write the answers without doing intermediate steps.

1. $\sqrt{49} + \sqrt{36}$
2. $\sqrt{100} - \sqrt{81}$
3. $\sqrt{2} \cdot \sqrt{8}$
4. $\sqrt{8} \cdot \sqrt{8}$

Simplify each expression. Use the five guidelines given in this section. See Examples 1–3.

5. $\sqrt{5}(\sqrt{3} - \sqrt{7})$
6. $\sqrt{7}(\sqrt{10} + \sqrt{3})$
7. $2\sqrt{5}(\sqrt{2} + 3\sqrt{5})$
8. $3\sqrt{7}(2\sqrt{7} + 4\sqrt{5})$
9. $3\sqrt{14} \cdot \sqrt{2} - \sqrt{28}$
10. $7\sqrt{6} \cdot \sqrt{3} - 2\sqrt{18}$
11. $(2\sqrt{6} + 3)(3\sqrt{6} + 7)$
12. $(4\sqrt{5} - 2)(2\sqrt{5} - 4)$
13. $(5\sqrt{7} - 2\sqrt{3})(3\sqrt{7} + 4\sqrt{3})$
14. $(2\sqrt{10} + 5\sqrt{2})(3\sqrt{10} - 3\sqrt{2})$
15. $(8 - \sqrt{7})^2$
16. $(6 - \sqrt{11})^2$
17. $(2\sqrt{7} + 3)^2$
18. $(4\sqrt{5} + 5)^2$
19. $(\sqrt{6} + 1)^2$
20. $(\sqrt{7} + 2)^2$
21. $(5 - \sqrt{2})(5 + \sqrt{2})$
22. $(3 - \sqrt{5})(3 + \sqrt{5})$
23. $(\sqrt{8} - \sqrt{7})(\sqrt{8} + \sqrt{7})$
24. $(\sqrt{12} - \sqrt{11})(\sqrt{12} + \sqrt{11})$
25. $(\sqrt{78} - \sqrt{76})(\sqrt{78} + \sqrt{76})$
26. $(\sqrt{85} - \sqrt{82})(\sqrt{85} + \sqrt{82})$
27. $(\sqrt{2} + \sqrt{3})(\sqrt{6} - \sqrt{2})$
28. $(\sqrt{3} + \sqrt{5})(\sqrt{15} - \sqrt{5})$
29. $(\sqrt{10} - \sqrt{5})(\sqrt{5} + \sqrt{20})$
30. $(\sqrt{6} - \sqrt{3})(\sqrt{3} + \sqrt{18})$
31. $(\sqrt{5} + \sqrt{30})(\sqrt{6} + \sqrt{3})$
32. $(\sqrt{10} - \sqrt{20})(\sqrt{2} - \sqrt{5})$
33. $(5\sqrt{7} - 2\sqrt{3})^2$
34. $(8\sqrt{2} - 3\sqrt{3})^2$

35. In Example 1(b), the original expression simplifies to $-37 - 2\sqrt{15}$. Students often try to further simplify expressions like this by combining the -37 and the -2 to get $-39\sqrt{15}$, which is incorrect. Explain why.

36. Find each product mentally.
    (a) $(\sqrt{x} + \sqrt{y})(\sqrt{x} - \sqrt{y})$
    (b) $(\sqrt{28} - \sqrt{14})(\sqrt{28} + \sqrt{14})$

Simplify each radical expression. Assume all variables represent nonnegative real numbers. See Examples 1–3.

37. $(7 + \sqrt{x})^2$
38. $(12 - \sqrt{r})^2$
39. $(3\sqrt{t} + \sqrt{7})(2\sqrt{t} - \sqrt{14})$
40. $(2\sqrt{z} - \sqrt{3})(\sqrt{z} - \sqrt{5})$
41. $(\sqrt{3m} + \sqrt{2n})(\sqrt{3m} - \sqrt{2n})$
42. $(\sqrt{4p} - \sqrt{3k})(\sqrt{4p} + \sqrt{3k})$

43. Determine the expression by which you should multiply the numerator and denominator to rationalize each denominator.

   (a) $\dfrac{1}{\sqrt{5} + \sqrt{3}}$ (b) $\dfrac{3}{\sqrt{6} - \sqrt{5}}$

44. If you try to rationalize the denominator of $\dfrac{2}{4 + \sqrt{3}}$ by multiplying the numerator and denominator by $4 + \sqrt{3}$, what problem arises? What should you multiply by?

Rationalize each denominator. Write quotients in lowest terms. See Example 4.

45. $\dfrac{1}{2 + \sqrt{5}}$
46. $\dfrac{1}{4 + \sqrt{15}}$
47. $\dfrac{7}{2 - \sqrt{11}}$
48. $\dfrac{38}{5 - \sqrt{6}}$
49. $\dfrac{\sqrt{12}}{\sqrt{3} + 1}$
50. $\dfrac{\sqrt{18}}{\sqrt{2} - 1}$
51. $\dfrac{2\sqrt{3}}{\sqrt{3} + 5}$
52. $\dfrac{\sqrt{12}}{2 - \sqrt{10}}$
53. $\dfrac{\sqrt{2} + 3}{\sqrt{3} - 1}$
54. $\dfrac{\sqrt{5} + 2}{2 - \sqrt{3}}$
55. $\dfrac{6 - \sqrt{5}}{\sqrt{2} + 2}$
56. $\dfrac{3 + \sqrt{2}}{\sqrt{2} + 1}$
57. $\dfrac{2\sqrt{6} + 1}{\sqrt{2} + 5}$
58. $\dfrac{3\sqrt{2} - 4}{\sqrt{3} + 2}$
59. $\dfrac{\sqrt{7} + \sqrt{2}}{\sqrt{3} - \sqrt{2}}$
60. $\dfrac{\sqrt{6} + \sqrt{5}}{\sqrt{3} + \sqrt{5}}$
61. $\dfrac{\sqrt{5}}{\sqrt{2} + \sqrt{3}}$
62. $\dfrac{\sqrt{3}}{\sqrt{2} + \sqrt{3}}$
63. $\dfrac{\sqrt{108}}{3 + 3\sqrt{3}}$
64. $\dfrac{9\sqrt{8}}{6\sqrt{2} - 6}$
65. $\dfrac{8}{4 - \sqrt{x}}$
66. $\dfrac{12}{6 + \sqrt{y}}$

Write each quotient in lowest terms. See Example 5.

67. $\dfrac{5\sqrt{7} - 10}{5}$
68. $\dfrac{6\sqrt{5} - 9}{3}$
69. $\dfrac{2\sqrt{3} + 10}{8}$
70. $\dfrac{4\sqrt{6} + 6}{10}$
71. $\dfrac{12 - 2\sqrt{10}}{4}$
72. $\dfrac{9 - 6\sqrt{2}}{12}$
73. $\dfrac{16 + 8\sqrt{2}}{24}$
74. $\dfrac{25 + 5\sqrt{3}}{10}$

Perform each operation and express the answer in simplest form.

75. $\sqrt[3]{4}(\sqrt[3]{2} - 3)$
76. $\sqrt[3]{5}(4\sqrt[3]{5} - \sqrt[3]{25})$
77. $2\sqrt[4]{2}(3\sqrt[4]{8} + 5\sqrt[4]{4})$
78. $6\sqrt[4]{9}(2\sqrt[4]{9} - \sqrt[4]{27})$
79. $(\sqrt[3]{2} - 1)(\sqrt[3]{4} + 3)$
80. $(\sqrt[3]{9} + 5)(\sqrt[3]{3} - 4)$
81. $(\sqrt[3]{5} - \sqrt[3]{4})(\sqrt[3]{25} + \sqrt[3]{20} + \sqrt[3]{16})$
82. $(\sqrt[3]{4} + \sqrt[3]{2})(\sqrt[3]{16} - \sqrt[3]{8} + \sqrt[3]{4})$

Solve each problem.

83. The radius of the circular top or bottom of a tin can with surface area S and height h is given by

$$r = \dfrac{-h + \sqrt{h^2 + .64S}}{2}.$$

What radius should be used to make a can with height 12 in. and surface area 400 in.²?

84. If an investment of P dollars grows to A dollars in 2 yr, the annual rate of return on the investment is given by

$$r = \frac{\sqrt{A} - \sqrt{P}}{\sqrt{P}}.$$

First rationalize the denominator and then find the annual rate of return (as a percent) if $50,000 increases to $58,320.

In Exercises 85 and 86, (a) give the answer as a simplified radical and (b) use a calculator to give the answer correct to the nearest thousandth.

85. The period p of a pendulum is the time it takes for it to swing from one extreme to the other and back again. The value of p in seconds is given by

$$p = k \cdot \sqrt{\frac{L}{g}},$$

where L is the length of the pendulum, g is the acceleration due to gravity, and k is a constant. Find the period when $k = 6$, $L = 9$ ft, and $g = 32$ ft per sec.

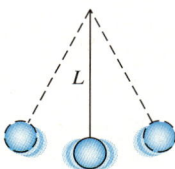

86. The velocity v of a meteorite approaching Earth is given by

$$v = \frac{k}{\sqrt{d}}$$

km per sec, where d is its distance from the center of Earth and k is a constant. What is the velocity of a meteorite that is 6000 km away from the center of Earth, if $k = 450$?

The *Golden Ratio*, the number $\dfrac{1 + \sqrt{5}}{2}$, is represented by the Greek letter phi, ϕ. It has been known since ancient times and is called the *sacred ratio* in the Rhind Papyrus from 1600 B.C. The Egyptians used ϕ to build the great pyramids, and the ancient Greeks used ϕ in art and architecture, striving for the proportion that was most pleasing to the eye.

The Golden Ratio is widespread in the star formed by connecting the vertices (corners) of a regular pentagon inscribed in a circle. In the figure shown, ABCDE forms a regular pentagon. Each of the following ratios forms the Golden Ratio, ϕ.

$$\frac{AC}{AY} \quad \text{and} \quad \frac{AY}{AX}$$

Using symmetry, you should be able to identify other similar relationships.

87. List two more ratios in the star that form the Golden Ratio. You may label additional points.

88. Use $\phi = \dfrac{1 + \sqrt{5}}{2}$ to find the following quantities. Write your answers in exact form (using radicals). Rationalize denominators when appropriate.

   (a) $\dfrac{1}{\phi}$ (b) $\phi - 1$ (c) ϕ^2 (d) $\phi + 1$

RELATING CONCEPTS (EXERCISES 89–94)

For Individual or Group Work

Work Exercises 89–94 in order, to see why a common student error is indeed an error.

89. Use the distributive property to write $6(5 + 3x)$ as a sum.
90. Your answer in Exercise 89 should be $30 + 18x$. Why can we not combine these two terms to get $48x$?
91. Repeat Exercise 14 from earlier in this exercise set.
92. Your answer in Exercise 91 should be $30 + 18\sqrt{5}$. Many students will, in error, try to combine these terms to get $48\sqrt{5}$. Why is this wrong?
93. Write the expression similar to $30 + 18x$ that simplifies to $48x$. Then write the expression similar to $30 + 18\sqrt{5}$ that simplifies to $48\sqrt{5}$.
94. Write a short explanation of the similarities between combining like terms and combining like radicals.

SUMMARY EXERCISES ON OPERATIONS WITH RADICALS

Perform all indicated operations and express each answer in simplest form. Assume that all variables represent positive numbers.

1. $5\sqrt{10} - 8\sqrt{10}$
2. $\sqrt{5}(\sqrt{5} - \sqrt{3})$
3. $(1 + \sqrt{3})(2 - \sqrt{6})$
4. $\sqrt{98} - \sqrt{72} + \sqrt{50}$
5. $(3\sqrt{5} - 2\sqrt{7})^2$
6. $\dfrac{3}{\sqrt{6}}$
7. $\sqrt[3]{16t^2} - \sqrt[3]{54t^2} + \sqrt[3]{128t^2}$
8. $\dfrac{8}{\sqrt{7} - \sqrt{5}}$
9. $\dfrac{1 + \sqrt{2}}{1 - \sqrt{2}}$
10. $(1 + \sqrt[3]{3})(1 - \sqrt[3]{3} + \sqrt[3]{9})$
11. $(\sqrt{3} + 6)(\sqrt{3} - 6)$
12. $\dfrac{1}{\sqrt{t} + \sqrt{3}}$
13. $\sqrt[3]{8x^3y^5z^6}$
14. $\dfrac{12}{\sqrt[3]{9}}$
15. $\dfrac{5}{\sqrt{6} - 1}$
16. $\sqrt{\dfrac{2}{3x}}$
17. $\dfrac{6\sqrt{3}}{5\sqrt{12}}$
18. $\dfrac{8\sqrt{50}}{2\sqrt{25}}$
19. $\dfrac{-4}{\sqrt[3]{4}}$
20. $\dfrac{\sqrt{6} - \sqrt{5}}{\sqrt{6} + \sqrt{5}}$
21. $\sqrt{75x} - \sqrt{12x}$
22. $(5 + 3\sqrt{3})^2$
23. $(\sqrt{107} - \sqrt{106})(\sqrt{107} + \sqrt{106})$
24. $\sqrt[3]{\dfrac{16}{81}}$
25. $x\sqrt[4]{x^5} - 3\sqrt[4]{x^9} + x^2\sqrt[4]{x}$

8.6 Solving Equations with Radicals

OBJECTIVES

1. Solve radical equations.
2. Identify equations with no solutions.
3. Solve equations by squaring a binomial.
4. Use a radical equation to model data.

A **radical equation** is an equation with a variable in the radicand, such as

$$\sqrt{x+1} = 3 \quad \text{or} \quad 3\sqrt{x} = \sqrt{8x+9}.$$

OBJECTIVE 1 Solve radical equations. The addition and multiplication properties of equality are not enough to solve radical equations. We need a new property, called the *squaring property*.

Squaring Property of Equality

If each side of a given equation is squared, all solutions of the original equation are *among* the solutions of the squared equation.

CAUTION Be very careful with squaring property: Using this property can give a new equation with *more* solutions than the original equation. For example, starting with the equation $x = 4$ and squaring both sides gives

$$x^2 = 4^2 \quad \text{or} \quad x^2 = 16.$$

This last equation, $x^2 = 16$, has *two* solutions, 4 or -4, while the original equation, $x = 4$, has only *one* solution, 4. Because of this possibility, checking is more than just a guard against algebraic errors when solving an equation with radicals. It is an essential part of the solution process. *All potential solutions from the squared equation must be checked in the original equation.*

EXAMPLE 1 Using the Squaring Property of Equality

Solve $\sqrt{x+1} = 3$.

Use the squaring property of equality to square both sides of the equation.

$$(\sqrt{x+1})^2 = 3^2$$
$$x + 1 = 9 \quad (\sqrt{x+1})^2 = x+1$$
$$x = 8 \quad \text{Subtract 1.}$$

Now check this potential solution in the original equation.

Check:
$$\sqrt{x+1} = 3$$
$$\sqrt{8+1} = 3 \quad ? \quad \text{Let } x = 8.$$
$$\sqrt{9} = 3 \quad ?$$
$$3 = 3 \quad \text{True}$$

Because this statement is true, {8} is the solution set of $\sqrt{x+1} = 3$. In this case the equation obtained by squaring had just one solution, which also satisfied the original equation.

Now Try Exercise 3.

EXAMPLE 2 Using the Squaring Property with a Radical on Each Side

Solve $3\sqrt{x} = \sqrt{x+8}$.

Squaring both sides gives

$$(3\sqrt{x})^2 = (\sqrt{x+8})^2$$
$$3^2(\sqrt{x})^2 = (\sqrt{x+8})^2 \quad (ab)^2 = a^2b^2$$
$$9x = x + 8 \quad (\sqrt{x})^2 = x;\ (\sqrt{x+8})^2 = x+8$$
$$8x = 8 \quad \text{Subtract } x.$$
$$x = 1. \quad \text{Divide by 8.}$$

Check:
$$3\sqrt{x} = \sqrt{x+8} \quad \text{Original equation}$$
$$3\sqrt{1} = \sqrt{1+8} \quad ?\quad \text{Let } x = 1.$$
$$3(1) = \sqrt{9} \quad ?$$
$$3 = 3 \quad \text{True}$$

The solution set of $3\sqrt{x} = \sqrt{x+8}$ is $\{1\}$.

Now Try Exercise 15.

CAUTION Do not write the final result obtained in the check in the solution set. In Example 2, the solution set is $\{1\}$, *not* $\{3\}$.

OBJECTIVE 2 Identify equations with no solutions. Not all radical equations have solutions, as shown in Examples 3 and 4.

EXAMPLE 3 Using the Squaring Property When One Side Is Negative

Solve $\sqrt{x} = -3$.

Square both sides of the equation.

$$(\sqrt{x})^2 = (-3)^2$$
$$x = 9$$

Check this potential solution in the original equation.

Check:
$$\sqrt{x} = -3$$
$$\sqrt{9} = -3 \quad ?\quad \text{Let } x = 9.$$
$$3 = -3 \quad \text{False}$$

Because the statement $3 = -3$ is false, the number 9 is not a solution of the given equation and is said to be *extraneous*. In fact, $\sqrt{x} = -3$ has no solution. The solution set is $\emptyset$.

Now Try Exercise 11.

NOTE Because $\sqrt{x}$ represents the *principal* or *nonnegative* square root of x in Example 3, we might have seen immediately that there is no solution.

Use the following steps when solving an equation with radicals.

Solving a Radical Equation

Step 1 **Isolate a radical.** Arrange the terms so that a radical is alone on one side of the equation.

Step 2 **Square both sides.**

Step 3 **Combine like terms.**

Step 4 **Repeat Steps 1–3, if necessary.** If there is still a term with a radical, repeat Steps 1–3.

Step 5 **Solve the equation.** Find all potential solutions.

Step 6 **Check.** All potential solutions *must* be checked in the original equation.

EXAMPLE 4 Using the Squaring Property with a Quadratic Expression

Solve $x = \sqrt{x^2 + 5x + 10}$.

Step 1 The radical is already alone on the right side of the equation.

Step 2 Square both sides.

$$x^2 = (\sqrt{x^2 + 5x + 10})^2$$
$$x^2 = x^2 + 5x + 10 \quad (\sqrt{x^2 + 5x + 10})^2 = x^2 + 5x + 10$$

Step 3 $\quad 0 = 5x + 10 \quad$ Subtract x^2.

Step 4 This step is not needed.

Step 5 $\quad -10 = 5x \quad$ Subtract 10.

$\qquad x = -2 \quad$ Divide by 5.

Step 6 Check this proposed solution in the original equation.

$$x = \sqrt{x^2 + 5x + 10}$$
$$-2 = \sqrt{(-2)^2 + 5(-2) + 10} \quad ? \quad \text{Let } x = -2.$$
$$-2 = \sqrt{4 - 10 + 10} \quad ? \quad \text{Multiply.}$$
$$-2 = 2 \quad \text{False}$$

Since $x = -2$ leads to a false result, the equation has no solution, and the solution set is $\emptyset$.

Now Try Exercise 21.

OBJECTIVE 3 Solve equations by squaring a binomial. The next examples use the following rules from Section 5.6:

$$(x + y)^2 = x^2 + 2xy + y^2$$

and

$$(x - y)^2 = x^2 - 2xy + y^2.$$

By these patterns, for example,

$$(x - 3)^2 = x^2 - 2x(3) + 3^2$$
$$= x^2 - 6x + 9.$$

568 CHAPTER 8 Roots and Radicals

EXAMPLE 5 Using the Squaring Property When One Side Has Two Terms

Solve $\sqrt{2x - 3} = x - 3$.

Square each side, using the preceding result to square the binomial on the right side of the equation.

$$(\sqrt{2x - 3})^2 = (x - 3)^2$$
$$2x - 3 = x^2 - 6x + 9$$

This equation is quadratic because of the x^2-term. To solve it, as shown in Section 6.5, one side of the equation must be equal to 0. Subtract $2x$ and add 3, getting

$$0 = x^2 - 8x + 12.$$
$$0 = (x - 6)(x - 2) \quad \text{Factor.}$$
$$x - 6 = 0 \quad \text{or} \quad x - 2 = 0 \quad \text{Zero-factor property}$$
$$x = 6 \quad \text{or} \quad x = 2 \quad \text{Solve.}$$

Check both of these potential solutions in the original equation.

If $x = 6$, then

$$\sqrt{2x - 3} = x - 3$$
$$\sqrt{2(6) - 3} = 6 - 3 \quad ?$$
$$\sqrt{12 - 3} = 3 \quad ?$$
$$\sqrt{9} = 3 \quad ?$$
$$3 = 3. \quad \text{True}$$

If $x = 2$, then

$$\sqrt{2x - 3} = x - 3$$
$$\sqrt{2(2) - 3} = 2 - 3 \quad ?$$
$$\sqrt{4 - 3} = -1 \quad ?$$
$$\sqrt{1} = -1 \quad ?$$
$$1 = -1. \quad \text{False}$$

Only 6 is a solution of the equation, so the solution set is {6}.

Now Try Exercise 35.

Remember to use Step 1 and isolate the radical before squaring both sides. For example, suppose we want to solve $\sqrt{9x} - 1 = 2x$. If we skip Step 1 and square both sides, we get

$$(\sqrt{9x} - 1)^2 = (2x)^2$$
$$9x - 2\sqrt{9x} + 1 = 4x^2,$$

a more complicated equation that still contains a radical. This is why we should begin by isolating the radical on one side of the equation, as shown in Example 6.

EXAMPLE 6 Rewriting an Equation before Using the Squaring Property

Solve $\sqrt{9x} - 1 = 2x$.

$$\sqrt{9x} = 2x + 1 \qquad \text{Add 1 to isolate the radical.}$$
$$(\sqrt{9x})^2 = (2x + 1)^2 \qquad \text{Square both sides.}$$
$$9x = 4x^2 + 4x + 1$$
$$0 = 4x^2 - 5x + 1 \qquad \text{Subtract } 9x.$$
$$0 = (4x - 1)(x - 1) \qquad \text{Factor.}$$
$$4x - 1 = 0 \quad \text{or} \quad x - 1 = 0 \qquad \text{Zero-factor property}$$
$$x = \frac{1}{4} \quad \text{or} \quad x = 1 \qquad \text{Solve.}$$

Check:

If $x = \frac{1}{4}$, then

$\sqrt{9x} - 1 = 2x$

$\sqrt{9\left(\frac{1}{4}\right)} - 1 = 2\left(\frac{1}{4}\right)$?

$\frac{1}{2} = \frac{1}{2}.$ True

If $x = 1$, then

$\sqrt{9x} - 1 = 2x$

$\sqrt{9(1)} - 1 = 2(1)$?

$2 = 2.$ True

Both solutions check, so the solution set is $\{\frac{1}{4}, 1\}$.

Now Try Exercise 49.

CAUTION Errors often occur when both sides of an equation are squared. For instance, in Example 6 when both sides of

$$\sqrt{9x} = 2x + 1$$

are squared, the *entire* binomial $2x + 1$ must be squared to get $4x^2 + 4x + 1$. It would be incorrect to square the $2x$ and the 1 separately to get $4x^2 + 1$.

Some radical equations require squaring twice, as in the next example.

EXAMPLE 7 Using the Squaring Property Twice

Solve $\sqrt{21 + x} = 3 + \sqrt{x}$.

$(\sqrt{21 + x})^2 = (3 + \sqrt{x})^2$ Square both sides.

$21 + x = 9 + 6\sqrt{x} + x$

$12 = 6\sqrt{x}$ Subtract 9; subtract x.

$2 = \sqrt{x}$ Divide by 6.

$2^2 = (\sqrt{x})^2$ Square both sides again.

$4 = x$

Check: If $x = 4$, then

$\sqrt{21 + x} = 3 + \sqrt{x}$ Original equation

$\sqrt{21 + 4} = 3 + \sqrt{4}$?

$5 = 5.$ True

The solution set is $\{4\}$.

Now Try Exercise 55.

OBJECTIVE 4 Use a radical equation to model data. As with linear and quadratic equations, we can also use radical equations to model real-life situations.

EXAMPLE 8 Using a Radical Equation to Model Exports

The table gives U.S. exports of electronics for selected years.

Billions of Dollars	Year
7.5	1990
19.6	1993
25.8	1994
36.4	1996

Source: U.S. Bureau of the Census; *U.S. Merchandise Trade,* series FT 900, December issue; and unpublished data.

Using the data from the table, the radical equation

$$y = 1.4\sqrt{x - 2.5} + 87.5$$

can be used to model the year y when exports were x billion dollars. Here, $y = 90$ represents 1990, $y = 93$ represents 1993, and so on.

(a) Use the equation to find the year when exports reached $50 billion.
Substitute 50 for x in the equation.

$y = 1.4\sqrt{50 - 2.5} + 87.5$ Let $x = 50$.

$y \approx 97.149$ Use a calculator.

Exports reached $50 billion in 1997.

(b) Use the equation to approximate exports of electronics in 1998.
Let $y = 98$ in the equation and solve for x.

$98 = 1.4\sqrt{x - 2.5} + 87.5$

$10.5 = 1.4\sqrt{x - 2.5}$ Subtract 87.5.

$7.5 = \sqrt{x - 2.5}$ Divide by 1.4.

$7.5^2 = (\sqrt{x - 2.5})^2$ Square both sides.

$56.25 = x - 2.5$

$58.75 = x$ Add 2.5.

This solution checks. Based on the equation, exports of electronics in 1998 were $58.75 billion.

(c) Actual exports in 1998 were $58.4 billion. How does the result of part (b) compare to the actual amount?
The approximation using the equation in part (b) is slightly high, but quite close to the actual data.

Now Try Exercise 69.

8.6 EXERCISES

For Extra Help

 Student's Solutions Manual

 MyMathLab

 InterAct Math Tutorial Software

 AW Math Tutor Center

 MathXL

Digital Video Tutor CD 17/Videotape 15

Solve each equation. See Examples 1–4.

1. $\sqrt{x} = 7$
2. $\sqrt{k} = 10$
3. $\sqrt{x + 2} = 3$
4. $\sqrt{x + 7} = 5$
5. $\sqrt{r - 4} = 9$
6. $\sqrt{k - 12} = 3$
7. $\sqrt{4 - t} = 7$
8. $\sqrt{9 - s} = 5$
9. $\sqrt{2t + 3} = 0$
10. $\sqrt{5x - 4} = 0$
11. $\sqrt{t} = -5$
12. $\sqrt{p} = -8$
13. $\sqrt{w - 4} = 7$
14. $\sqrt{t} + 3 = 10$
15. $\sqrt{10x - 8} = 3\sqrt{x}$
16. $\sqrt{17t - 4} = 4\sqrt{t}$
17. $5\sqrt{x} = \sqrt{10x + 15}$
18. $4\sqrt{x} = \sqrt{20x - 16}$
19. $\sqrt{3x - 5} = \sqrt{2x + 1}$
20. $\sqrt{5x + 2} = \sqrt{3x + 8}$
21. $k = \sqrt{k^2 - 5k - 15}$
22. $s = \sqrt{s^2 - 2s - 6}$
23. $7x = \sqrt{49x^2 + 2x - 10}$
24. $6x = \sqrt{36x^2 + 5x - 5}$
25. $\sqrt{2x + 2} = \sqrt{3x - 5}$
26. $\sqrt{x + 2} = \sqrt{2x - 5}$
27. $\sqrt{5x - 5} = \sqrt{4x + 1}$
28. $\sqrt{3m + 3} = \sqrt{5m - 1}$
29. $\sqrt{3x - 8} = -2$
30. $\sqrt{6t + 4} = -3$

31. The first step in solving the equation $\sqrt{2x + 1} = x - 7$ is to square each side of the equation. Errors often occur in solving equations such as this one when the right side of the equation is squared incorrectly. What is the square of the right side?

32. What is wrong with the following "solution"?

$-\sqrt{x - 1} = -4$
$-(x - 1) = 16$ Square both sides.
$-x + 1 = 16$ Distributive property
$-x = 15$ Subtract 1.
$x = -15$ Multiply by -1.

Solve each equation. See Examples 5 and 6.

33. $\sqrt{5x + 11} = x + 3$
34. $\sqrt{5x + 1} = x + 1$
35. $\sqrt{2x + 1} = x - 7$
36. $\sqrt{3x + 10} = 2x - 5$
37. $\sqrt{4x + 13} = 2x - 1$
38. $\sqrt{x + 1} - 1 = x$
39. $\sqrt{3x + 3} + 5 = x$
40. $\sqrt{4x + 5} - 2 = 2x - 7$
41. $\sqrt{6x + 7} - 1 = x + 1$
42. $3\sqrt{x + 13} = x + 9$
43. $2\sqrt{x + 7} = x - 1$
44. $\sqrt{4x} - x + 3 = 0$
45. $\sqrt{2x + 4} = x$
46. $\sqrt{3x + 6} = x$
47. $\sqrt{x + 9} = x + 3$
48. $\sqrt{x + 3} = x - 9$
49. $3\sqrt{x - 2} = x - 2$
50. $2\sqrt{x + 4} = x + 1$

51. Explain how you can tell that the equation $\sqrt{x} = -8$ has no real number solution without performing any algebraic steps.

52. What is wrong with the following "solution"?

$\sqrt{3x - 6} + \sqrt{x + 2} = 12$
$3x - 6 + x + 2 = 144$ Square both sides.
$4x - 4 = 144$ Combine terms.
$4x = 148$ Add 4 on both sides.
$x = 37$ Divide by 4.

Solve each equation. See Example 7.

53. $\sqrt{2x+3} + \sqrt{x+1} = 1$
54. $\sqrt{2x+1} + \sqrt{x+4} = 3$
55. $\sqrt{x+6} = \sqrt{x+72}$
56. $\sqrt{x-4} = \sqrt{x-32}$
57. $\sqrt{3x+4} - \sqrt{2x-4} = 2$
58. $\sqrt{1-x} + \sqrt{x+9} = 4$
59. $\sqrt{2x+11} + \sqrt{x+6} = 2$
60. $\sqrt{x+9} + \sqrt{x+16} = 7$

Solve each problem.

61. The square root of the sum of a number and 4 is 5. Find the number.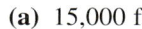
62. A certain number is the same as the square root of the product of 8 and the number. Find the number.
63. Three times the square root of 2 equals the square root of the sum of some number and 10. Find the number.
64. The negative square root of a number equals that number decreased by 2. Find the number.

Solve each problem. Give answers to the nearest tenth.

65. Police sometimes use the following procedure to estimate the speed at which a car was traveling at the time of an accident. A police officer drives the car involved in the accident under conditions similar to those during which the accident took place and then skids to a stop. If the car is driven at 30 mph, then the speed at the time of the accident is given by

$$s = 30\sqrt{\frac{a}{p}},$$

where a is the length of the skid marks left at the time of the accident and p is the length of the skid marks in the police test. Find s for the following values of a and p.

(a) $a = 862$ ft; $p = 156$ ft
(b) $a = 382$ ft; $p = 96$ ft
(c) $a = 84$ ft; $p = 26$ ft

66. A formula for calculating the distance, d, one can see from an airplane to the horizon on a clear day is

$$d = 1.22\sqrt{x},$$

where x is the altitude of the plane in feet and d is given in miles. How far can one see to the horizon in a plane flying at the following altitudes?

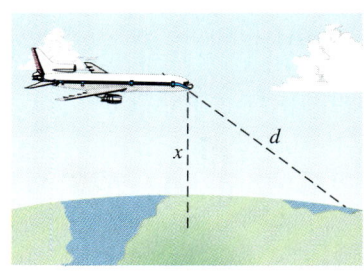

(a) 15,000 ft
(b) 18,000 ft
(c) 24,000 ft

67. A surveyor wants to find the height of a building. At a point 110.0 ft from the base of the building he sights to the top of the building and finds the distance to be 193.0 ft. How high is the building?

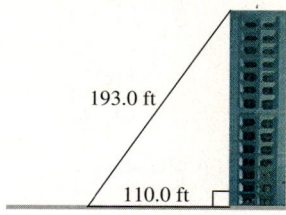

68. Two towns are separated by dense woods. To go from Town B to Town A, it is necessary to travel due west for 19.0 mi, then turn due north and travel for 14.0 mi. How far apart are the towns?

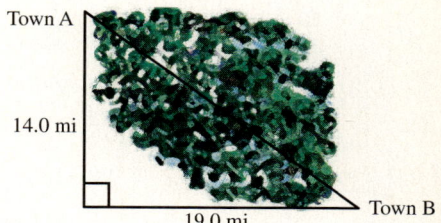

69. U.S. imports of electronics for several years are shown in the table.

Billions of Dollars	Year
11.0	1990
19.4	1993
25.9	1994
36.8	1996

Source: U.S. Bureau of the Census; *U.S. Merchandise Trade,* series FT 900, December issue; and unpublished data.

The year y when imports were x billion dollars can be modeled by the radical equation

$$y = 1.6\sqrt{x - 6} + 87,$$

where $y = 90$ represents 1990, and so on, as in Example 8.

(a) Use the equation to approximate U.S. imports of electronics in 1994 to the nearest tenth.
(b) How does the result from part (a) compare to the actual data in the table?
(c) Use the equation to predict imports for 1998 to the nearest tenth. How does your answer compare to the actual imports of $54.1 billion?
(d) Using the table for exports in Example 8 and the table for imports given above, find the differences between U.S. exports and imports of electronics in the years listed. Complete the table at the right. Then, comment on the results.

Year	Difference
1990	−3.5
1993	
1994	
1996	

70. In 1997, the number of 911 emergency calls in California had risen rapidly to 2.7 million, about a 9000% increase since 1985. The graph on the next page shows the relationship between the number of calls in millions and the years from 1985 to 1997. This relationship is closely approximated by the function defined by the radical equation

$$y = \frac{1.76 + \sqrt{.0176 + .044x}}{.022}.$$

Here, y represents the year when x million calls were made. The years are coded so that 85 represents 1985, 95 represents 1995, and so on.

574 CHAPTER 8 Roots and Radicals

911 EMERGENCY CALLS

Source: California Highway Patrol.

(a) Find the year when the number of calls reached 1 million.
(b) If $y = f(x)$, find $f(3)$. Interpret your answer in the context of this problem.
(c) Use the equation to determine how many calls were made in 1995.

On a clear day, the maximum distance in kilometers that you can see from a tall building is given by the formula

$$\text{sight distance} = 111.7 \sqrt{\text{height of building in kilometers}}.$$

(*Source: A Sourcebook of Applications of School Mathematics,* NCTM, 1980.)

Use the conversion equations 1 ft ≈ .3048 m *and* 1 km ≈ .621371 mi *as necessary to solve each problem. Round your answers to the nearest mile.*

71. As mentioned in the chapter opener, the London Eye is a unique form of a Ferris wheel that features 32 observation capsules. It is the fourth tallest structure in London, with a diameter of 135 m. (*Source:* www.londoneye.com) Does the formula justify the claim that on a clear day, passengers on the London Eye can see Windsor Castle, 25 mi away?

72. The Empire State Building opened in 1931 on 5th Avenue in New York City. The building is 1250 ft high. (The antenna reaches to 1454 ft.) The observation deck, located on the 102nd floor, is at a height of 1050 ft. (*Source:* www.esbnyc.com) How far could you see on a clear day from the observation deck?

73. The twin Petronas Towers in Kuala Lumpur, Malaysia, are listed as the tallest buildings in the world. (*Source: World Almanac and Book of Facts,* 2000.) Built in 1998, both towers are 1483 ft high (including the spires). How far would one of the builders have been able to see on a clear day from the top of a spire?

74. The Khufu Pyramid in Giza (also known as Cheops Pyramid) was built in about 2566 B.C. to a height, at that time, of 482 ft. It is now only about 450 ft high. How far would one of the original builders of the pyramid have been able to see from the top of the pyramid?

RELATING CONCEPTS (EXERCISES 75–80)

For Individual or Group Work

The most common formula for the area of a triangle is $A = \frac{1}{2}bh$, where b is the length of the base and h is the height. What if the height is not known? What if we know only the lengths of the sides? Another formula, known as *Heron's formula*, allows us to calculate the area of a triangle if we know the lengths of the sides a, b, and c. First let s equal the *semiperimeter,* which is one-half the perimeter.

$$s = \frac{1}{2}(a + b + c)$$

The area A is given by the formula

$$A = \sqrt{s(s-a)(s-b)(s-c)}.$$

For example, the familiar 3–4–5 right triangle has area

$$A = \frac{1}{2}(3)(4) = 6 \text{ square units,}$$

using the familiar formula. Using Heron's formula,

$$s = \frac{1}{2}(3 + 4 + 5) = 6.$$

Therefore,

$$A = \sqrt{6(6-3)(6-4)(6-5)}$$
$$= \sqrt{36}$$
$$= 6.$$

The area is 6 square units, as expected.

Consider the following figure, and **work Exercises 75–80 in order.**

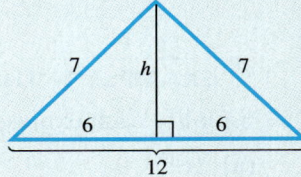

75. The lengths of the sides of the entire triangle are 7, 7, and 12. Find the semiperimeter s.

(continued)

576 CHAPTER 8 Roots and Radicals

76. Now use Heron's formula to find the area of the entire triangle. Write it as a simplified radical.

77. Find the value of h by using the Pythagorean formula.

78. Find the area of each of the congruent right triangles forming the entire triangle by using the formula $A = \frac{1}{2}bh$.

79. Double your result from Exercise 78 to determine the area of the entire triangle.

80. How do your answers in Exercises 76 and 79 compare? (*Note:* They should be equal, since the area of the entire triangle is unique.)

8.7 Using Rational Numbers as Exponents

OBJECTIVES

1. Define and use expressions of the form $a^{1/n}$.
2. Define and use expressions of the form $a^{m/n}$.
3. Apply the rules for exponents using rational exponents.
4. Use rational exponents to simplify radicals.

OBJECTIVE 1 Define and use expressions of the form $a^{1/n}$. How should an expression like $5^{1/2}$ be defined? We want to define $5^{1/2}$ so that all the rules for exponents developed earlier in this book still hold. Then we should define $5^{1/2}$ so that

$$5^{1/2} \cdot 5^{1/2} = 5^{1/2+1/2} = 5^1 = 5.$$

This agrees with the product rule for exponents from Section 5.1. By definition,

$$(\sqrt{5})(\sqrt{5}) = 5.$$

Since both $5^{1/2} \cdot 5^{1/2}$ and $\sqrt{5} \cdot \sqrt{5}$ equal 5,

$$5^{1/2} \text{ should equal } \sqrt{5}.$$

Similarly,

$$5^{1/3} \cdot 5^{1/3} \cdot 5^{1/3} = 5^{1/3+1/3+1/3} = 5^{3/3} = 5,$$

and

$$\sqrt[3]{5} \cdot \sqrt[3]{5} \cdot \sqrt[3]{5} = \sqrt[3]{5^3} = 5,$$

so

$$5^{1/3} \text{ should equal } \sqrt[3]{5}.$$

These examples suggest the following definition.

$a^{1/n}$

If a is a nonnegative number and n is a positive integer, then

$$a^{1/n} = \sqrt[n]{a}.$$

EXAMPLE 1 Using the Definition of $a^{1/n}$

Simplify each expression by first writing it in radical form.

(a) $16^{1/2}$

By the definition above, $16^{1/2} = \sqrt{16} = 4$.

(b) $27^{1/3} = \sqrt[3]{27} = 3$ **(c)** $64^{1/3} = \sqrt[3]{64} = 4$ **(d)** $64^{1/6} = \sqrt[6]{64} = 2$

Now Try Exercises 5 and 7.

OBJECTIVE 2 Define and use expressions of the form $a^{m/n}$. Now we can define a more general exponential expression like $16^{3/4}$. By the power rule, $(a^m)^n = a^{mn}$, so

$$16^{3/4} = (16^{1/4})^3 = (\sqrt[4]{16})^3 = 2^3 = 8.$$

However, $16^{3/4}$ can also be written as

$$16^{3/4} = (16^3)^{1/4} = (4096)^{1/4} = \sqrt[4]{4096} = 8.$$

— Same answer

Either way, the answer is the same. As the example suggests, taking the root first involves smaller numbers and is often easier. This example suggests the following definition for $a^{m/n}$.

$a^{m/n}$

If a is a nonnegative number and m and n are integers, with $n > 0$, then

$$a^{m/n} = (a^{1/n})^m = (\sqrt[n]{a})^m.$$

EXAMPLE 2 Using the Definition of $a^{m/n}$

Evaluate each expression.

(a) $9^{3/2} = (9^{1/2})^3 = 3^3 = 27$ **(b)** $64^{2/3} = (64^{1/3})^2 = 4^2 = 16$

(c) $-32^{4/5} = -(32^{1/5})^4 = -2^4 = -16$

Now Try Exercises 13, 15, and 21.

Earlier, a^{-n} was defined as

$$a^{-n} = \frac{1}{a^n}$$

for nonzero numbers a and integers n. This same result applies for negative rational exponents.

$a^{-m/n}$

If a is a positive number and m and n are integers, with $n > 0$, then

$$a^{-m/n} = \frac{1}{a^{m/n}}.$$

EXAMPLE 3 Using the Definition of $a^{-m/n}$

Write each expression with a positive exponent and then evaluate.

(a) $32^{-3/5} = \frac{1}{32^{3/5}} = \frac{1}{(32^{1/5})^3} = \frac{1}{2^3} = \frac{1}{8}$

(b) $27^{-4/3} = \frac{1}{27^{4/3}} = \frac{1}{(27^{1/3})^4} = \frac{1}{3^4} = \frac{1}{81}$

Now Try Exercises 25 and 27.

CAUTION In Example 3(b), a common mistake is to write $27^{-4/3}$ as $-27^{3/4}$. *This is incorrect.* The negative exponent does not indicate a negative number. Also, the negative exponent indicates the reciprocal of the *base*, not the reciprocal of the *exponent*.

OBJECTIVE 3 Apply the rules for exponents using rational exponents. All the rules for exponents given earlier still hold when the exponents are fractions.

EXAMPLE 4 Using the Rules for Exponents with Fractional Exponents

Simplify each expression. Write each answer in exponential form with only positive exponents.

(a) $3^{2/3} \cdot 3^{5/3} = 3^{2/3+5/3} = 3^{7/3}$

(b) $\dfrac{5^{1/4}}{5^{3/4}} = 5^{1/4-3/4} = 5^{-2/4} = 5^{-1/2} = \dfrac{1}{5^{1/2}}$

(c) $(9^{1/4})^2 = 9^{2(1/4)} = 9^{2/4} = 9^{1/2} = \sqrt{9} = 3$

(d) $\left(\dfrac{9}{4}\right)^{5/2} = \dfrac{9^{5/2}}{4^{5/2}} = \dfrac{(9^{1/2})^5}{(4^{1/2})^5} = \dfrac{(\sqrt{9})^5}{(\sqrt{4})^5} = \dfrac{3^5}{2^5}$

(e) $\dfrac{2^{1/2} \cdot 2^{-1}}{2^{-3/2}} = \dfrac{2^{1/2+(-1)}}{2^{-3/2}} = \dfrac{2^{-1/2}}{2^{-3/2}} = 2^{-1/2-(-3/2)} = 2^{2/2} = 2^1 = 2$

Now Try Exercises 31, 35, 39, 43, and 47.

EXAMPLE 5 Using Fractional Exponents with Variables

Simplify each expression. Write each answer in exponential form with only positive exponents. Assume that all variables represent positive numbers.

(a) $m^{1/5} \cdot m^{3/5} = m^{1/5+3/5} = m^{4/5}$

(b) $\dfrac{p^{5/3}}{p^{4/3}} = p^{5/3-4/3} = p^{1/3}$

(c) $(x^2 y^{1/2})^4 = (x^2)^4 (y^{1/2})^4 = x^8 y^2$

(d) $\left(\dfrac{z^{1/4}}{w^{1/3}}\right)^5 = \dfrac{(z^{1/4})^5}{(w^{1/3})^5} = \dfrac{z^{5/4}}{w^{5/3}}$

(e) $\dfrac{k^{2/3} \cdot k^{-1/3}}{k^{5/3}} = k^{2/3+(-1/3)-5/3} = k^{-4/3} = \dfrac{1}{k^{4/3}}$

Now Try Exercises 49, 51, 53, 55, and 57.

CAUTION Errors often occur in problems like those in Examples 4 and 5 because students try to convert the expressions to radicals. Remember that the *rules of exponents* apply here.

OBJECTIVE 4 Use rational exponents to simplify radicals. Sometimes it is easier to simplify a radical by first writing it in exponential form.

EXAMPLE 6 Simplifying Radicals by Using Rational Exponents

Simplify each radical by first writing it in exponential form.

(a) $\sqrt[6]{9^3} = (9^3)^{1/6} = 9^{3/6} = 9^{1/2} = \sqrt{9} = 3$

(b) $\left(\sqrt[4]{m}\right)^2 = (m^{1/4})^2 = m^{2/4} = m^{1/2} = \sqrt{m}$
Here it is assumed that $m \geq 0$.

> Now Try Exercises 59 and 65.

8.7 EXERCISES

For Extra Help

Student's Solutions Manual

MyMathLab

InterAct Math Tutorial Software

AW Math Tutor Center

MathXL

Digital Video Tutor
CD 17/Videotape 15

Decide which one of the four choices is not *equal to the given expression.*

1. $49^{1/2}$

    A. -7 **B.** 7 **C.** $\sqrt{49}$ **D.** $49^{.5}$

2. $81^{1/2}$

    A. 9 **B.** $\sqrt{81}$ **C.** $81^{.5}$ **D.** $\dfrac{81}{2}$

3. $-64^{1/3}$

    A. $-\sqrt{16}$ **B.** -4 **C.** 4 **D.** $-\sqrt[3]{64}$

4. $-125^{1/3}$

    A. $-\sqrt{25}$ **B.** -5 **C.** $-\sqrt[3]{125}$ **D.** 5

Simplify each expression by first writing it in radical form. See Examples 1–3.

5. $25^{1/2}$ **6.** $121^{1/2}$ **7.** $64^{1/3}$ **8.** $125^{1/3}$ **9.** $16^{1/4}$

10. $81^{1/4}$ **11.** $32^{1/5}$ **12.** $243^{1/5}$ **13.** $4^{3/2}$ **14.** $9^{5/2}$

15. $27^{2/3}$ **16.** $8^{5/3}$ **17.** $16^{3/4}$ **18.** $64^{5/3}$ **19.** $32^{2/5}$

20. $144^{3/2}$ **21.** $-8^{2/3}$ **22.** $-27^{5/3}$ **23.** $-64^{1/3}$ **24.** $-125^{5/3}$

25. $49^{-3/2}$ **26.** $9^{-5/2}$ **27.** $216^{-2/3}$ **28.** $32^{-4/5}$ **29.** $-16^{-5/4}$

30. $-81^{-3/4}$

Simplify each expression. Write answers in exponential form with only positive exponents. Assume that all variables represent positive numbers. See Examples 4 and 5.

31. $2^{1/3} \cdot 2^{7/3}$ **32.** $5^{2/3} \cdot 5^{5/3}$ **33.** $6^{1/4} \cdot 6^{-3/4}$ **34.** $12^{2/5} \cdot 12^{-1/5}$

35. $\dfrac{15^{3/4}}{15^{5/4}}$ **36.** $\dfrac{7^{3/5}}{7^{-1/5}}$ **37.** $\dfrac{11^{-2/7}}{11^{-3/7}}$ **38.** $\dfrac{4^{-2/3}}{4^{1/3}}$

39. $(8^{3/2})^2$ **40.** $(5^{2/5})^{10}$ **41.** $(6^{1/3})^{3/2}$ **42.** $(7^{2/5})^{5/3}$

43. $\left(\dfrac{25}{4}\right)^{3/2}$ **44.** $\left(\dfrac{8}{27}\right)^{2/3}$ **45.** $\dfrac{2^{2/5} \cdot 2^{-3/5}}{2^{7/5}}$ **46.** $\dfrac{3^{-3/4} \cdot 3^{5/4}}{3^{-1/4}}$

47. $\dfrac{6^{-2/9}}{6^{1/9} \cdot 6^{-5/9}}$ **48.** $\dfrac{8^{6/7}}{8^{2/7} \cdot 8^{-1/7}}$ **49.** $x^{2/5} \cdot x^{7/5}$ **50.** $y^{2/3} \cdot y^{5/3}$

51. $\dfrac{r^{4/9}}{r^{3/9}}$ **52.** $\dfrac{s^{5/6}}{s^{4/6}}$ **53.** $(m^3 n^{1/4})^{2/3}$ **54.** $(p^4 q^{1/2})^{4/3}$

55. $\left(\dfrac{a^{2/3}}{b^{1/4}}\right)^6$ **56.** $\left(\dfrac{t^{3/7}}{s^{1/3}}\right)^{21}$ **57.** $\dfrac{m^{3/4} \cdot m^{-1/4}}{m^{1/3}}$ **58.** $\dfrac{q^{5/6} \cdot q^{-1/6}}{q^{1/3}}$

Simplify each radical by first writing it in exponential form. Give the answer as an integer or a radical in simplest form. Assume that all variables represent nonnegative numbers. See Example 6.

59. $\sqrt[6]{4^3}$ **60.** $\sqrt[9]{8^3}$ **61.** $\sqrt[8]{16^2}$ **62.** $\sqrt[9]{27^3}$

63. $\sqrt[4]{a^2}$ 64. $\sqrt[9]{b^3}$ 65. $\sqrt[6]{k^4}$ 66. $\sqrt[8]{m^4}$

Use the exponential key on your calculator to find the following roots. For example, to find $\sqrt[5]{32}$, enter 32 and then raise to the 1/5 power. (The exponent 1/5 may be entered as .2 if you wish.) Refer to Appendix A if necessary. If the root is irrational, round it to the nearest thousandth.

67. $\sqrt[6]{64}$ 68. $\sqrt[5]{243}$ 69. $\sqrt[7]{84}$ 70. $\sqrt[9]{16}$

Solve each problem.

71. The formula
$$d = 1.22\sqrt{x}$$
has sometimes been used to calculate the distance in miles one can see from an airplane to the horizon on a clear day. Here, x is in kilometers.

   (a) Write the formula using a rational exponent.
   (b) Find d to the nearest hundredth if the altitude x is 30,000 ft.

72. A biologist has shown that the number of different plant species S on a Galápagos Island is related to the area of the island, A (in miles), by
$$S = 28.6A^{1/3}.$$
How many plant species would exist on such an island with the following areas?

   (a) 8 mi²   (b) 27,000 mi²

73. Explain in your own words why $7^{1/2}$ is defined as $\sqrt{7}$.
74. Explain in your own words why $7^{1/3}$ is defined as $\sqrt[3]{7}$.

RELATING CONCEPTS (EXERCISES 75–80)

For Individual or Group Work

The rules for multiplying and dividing radicals presented earlier in this chapter were stated for radicals having the same index. For example, we only multiplied or divided square roots, or multiplied or divided cube roots, and so on. Since we know how to write radicals with rational exponents and from past work know how to add and subtract fractions with different denominators, we can now multiply and divide radicals having different indexes.

Work Exercises 75–80 in order, *to see how to multiply $\sqrt{2}$ by $\sqrt[3]{2}$.*

75. Write $\sqrt{2}$ and $\sqrt[3]{2}$ using rational exponents.
76. Write the product $\sqrt{2} \cdot \sqrt[3]{2}$ using the expressions you found in Exercise 75.
77. What is the least common denominator of the rational exponents in Exercise 76?
78. Repeat Exercise 76, but write the rational exponents with the common denominator from Exercise 77.
79. Use the rule for multiplying exponential expressions with like bases to simplify the product in Exercise 78. (*Hint:* The base remains the same; do not multiply the two bases.)
80. Write the answer you obtained in Exercise 79 as a radical.

Chapter 8 Group Activity

Comparing Television Sizes

OBJECTIVE Use the Pythagorean formula to find the dimensions of different television sets.

Television sets are identified by the diagonal measurement of the viewing screen. For example, a 19-in. TV measures 19 in. from one corner of the viewing screen diagonally to the other corner.

A. The table gives some common TV sizes as well as their corresponding widths. Use the Pythagorean formula

$$a^2 + b^2 = c^2$$

to find the heights of the viewing screens. Round up to the next whole number. (*Hint:* The TV size is the hypotenuse.)

TV Set	TV Size	Width in Inches	Height in Inches
A	13-in.	10	
B	19-in.	15	
C	25-in.	21	
D	27-in.	22	
E	32-in.	25	
F	36-in.	30	

Source: Data based on information from a Sears catalog.

B. The dimensions in the table correspond only to television viewing screens. Each television set also has an outer border of $1\frac{1}{2}$ in. as well as an additional 4 in. for a control panel at the bottom of the set. Also, it is recommended that a minimum of 2 in. of space be allowed above the set for ventilation.

Consider the following entertainment centers, with dimensions for television space as given below. Find the largest television set from the table in part A that will fit in each entertainment center.

1. 27 in. by 29 in.   **2.** 28 in. by 25 in.   **3.** 20 in. by 20 in.

CHAPTER 8 SUMMARY

KEY TERMS

8.1 square root, principal square root, radicand, radical, radical expression

perfect square, cube root, fourth root, index (order)

8.2 perfect cube
8.3 like radicals
8.4 rationalizing the denominator

8.5 conjugate
8.6 radical equation, squaring property

NEW SYMBOLS

$\sqrt{\ }$ radical sign

$\approx$ is approximately equal to

$\sqrt[3]{a}$ cube root of a
$\sqrt[n]{a}$ nth root of a

$a^{1/m}$ mth root of a

TEST YOUR WORD POWER

See how well you have learned the vocabulary in this chapter. Answers, with examples, follow the Quick Review.

1. The **square root** of a number is
   A. the number raised to the second power
   B. the number under a radical sign
   C. a number that when multiplied by itself gives the original number
   D. the inverse of the number.

2. A **radical** is
   A. a symbol that indicates the nth root
   B. an algebraic expression containing a square root
   C. the positive nth root of a number
   D. a radical sign and the number or expression under it.

3. The **principal root** of a positive number with even index n is
   A. the positive nth root of the number
   B. the negative nth root of the number
   C. the square root of the number
   D. the cube root of the number.

4. **Like radicals** are
   A. radicals in simplest form
   B. algebraic expressions containing radicals
   C. multiples of the same root of the same number
   D. radicals with the same index.

5. **Rationalizing the denominator** is the process of
   A. eliminating fractions from a radical expression
   B. changing the denominator of a fraction from a radical to a rational number
   C. clearing a radical expression of radicals
   D. multiplying radical expressions.

6. The **conjugate** of $a + b$ is
   A. $a - b$
   B. $a \cdot b$
   C. $a \div b$
   D. $(a + b)^2$.

QUICK REVIEW

CONCEPTS	EXAMPLES

8.1 EVALUATING ROOTS

If a is a positive real number, then
$\sqrt{a}$ is the positive square root of a;
$-\sqrt{a}$ is the negative square root of a; $\sqrt{0} = 0$.
If a is a negative real number, then $\sqrt{a}$ is not a real number.
If a is a positive rational number, then $\sqrt{a}$ is rational if a is a perfect square. $\sqrt{a}$ is irrational if a is not a perfect square.
Each real number has exactly one real cube root.

$\sqrt{49} = 7$
$-\sqrt{81} = -9$
$\sqrt{-25}$ is not a real number.

$\sqrt{\dfrac{4}{9}}, \sqrt{16}$ are rational. $\sqrt{\dfrac{2}{3}}, \sqrt{21}$ are irrational.

$\sqrt[3]{27} = 3;\ \sqrt[3]{-8} = -2$

CONCEPTS	EXAMPLES

8.2 MULTIPLYING, DIVIDING, AND SIMPLIFYING RADICALS

Product Rule for Radicals
For nonnegative real numbers a and b,
$$\sqrt{a} \cdot \sqrt{b} = \sqrt{ab} \quad \text{and} \quad \sqrt{a \cdot b} = \sqrt{a} \cdot \sqrt{b}.$$

Quotient Rule for Radicals
If a and b are nonnegative real numbers and b is not 0,
$$\frac{\sqrt{a}}{\sqrt{b}} = \sqrt{\frac{a}{b}} \quad \text{and} \quad \sqrt{\frac{a}{b}} = \frac{\sqrt{a}}{\sqrt{b}}.$$

If all indicated roots are real, then
$$\sqrt[n]{a} \cdot \sqrt[n]{b} = \sqrt[n]{ab} \quad \text{and} \quad \frac{\sqrt[n]{a}}{\sqrt[n]{b}} = \sqrt[n]{\frac{a}{b}} \; (b \neq 0).$$

$\sqrt{5} \cdot \sqrt{7} = \sqrt{35}$
$\sqrt{8} \cdot \sqrt{2} = \sqrt{16} = 4$
$\sqrt{48} = \sqrt{16 \cdot 3} = \sqrt{16} \cdot \sqrt{3} = 4\sqrt{3}$

$\frac{\sqrt{8}}{\sqrt{2}} = \sqrt{\frac{8}{2}} = \sqrt{4} = 2; \quad \sqrt{\frac{25}{64}} = \frac{\sqrt{25}}{\sqrt{64}} = \frac{5}{8}$

$\sqrt[3]{5} \cdot \sqrt[3]{3} = \sqrt[3]{15}; \quad \frac{\sqrt[4]{12}}{\sqrt[4]{4}} = \sqrt[4]{\frac{12}{4}} = \sqrt[4]{3}$

8.3 ADDING AND SUBTRACTING RADICALS

Add and subtract like radicals by using the distributive property. *Only like radicals can be combined in this way.*

$2\sqrt{5} + 4\sqrt{5} = (2 + 4)\sqrt{5}$
$\quad\quad\quad\quad\quad = 6\sqrt{5}$
$\sqrt{8} + \sqrt{32} = 2\sqrt{2} + 4\sqrt{2}$
$\quad\quad\quad\quad\quad = 6\sqrt{2}$

8.4 RATIONALIZING THE DENOMINATOR

The denominator of a radical can be rationalized by multiplying both the numerator and denominator by a number that will eliminate the radical from the denominator.

$\frac{2}{\sqrt{3}} = \frac{2 \cdot \sqrt{3}}{\sqrt{3} \cdot \sqrt{3}} = \frac{2\sqrt{3}}{3}$

$\sqrt[3]{\frac{5}{6}} = \frac{\sqrt[3]{5} \cdot \sqrt[3]{6^2}}{\sqrt[3]{6} \cdot \sqrt[3]{6^2}} = \frac{\sqrt[3]{180}}{6}$

8.5 MORE SIMPLIFYING AND OPERATIONS WITH RADICALS

When appropriate, use the rules for adding and multiplying polynomials to simplify radical expressions.

Any denominators with radicals should be rationalized.

If a radical expression contains two terms in the denominator and at least one of those terms is a radical, multiply both the numerator and denominator by the conjugate of the denominator.

$\sqrt{6}(\sqrt{5} - \sqrt{7}) = \sqrt{30} - \sqrt{42}$
$(\sqrt{5} - \sqrt{3})(\sqrt{5} + \sqrt{3}) = 5 - 3 = 2$

$\frac{3}{\sqrt{6}} = \frac{3\sqrt{6}}{6} = \frac{\sqrt{6}}{2}$

$\frac{6}{\sqrt{7} - \sqrt{2}} = \frac{6}{\sqrt{7} - \sqrt{2}} \cdot \frac{\sqrt{7} + \sqrt{2}}{\sqrt{7} + \sqrt{2}}$
$\quad = \frac{6(\sqrt{7} + \sqrt{2})}{7 - 2}$ Multiply fractions.
$\quad = \frac{6(\sqrt{7} + \sqrt{2})}{5}$ Subtract.

(continued)

CHAPTER 8 Roots and Radicals

CONCEPTS	EXAMPLES

8.6 SOLVING EQUATIONS WITH RADICALS

Solving a Radical Equation

Step 1 Isolate a radical.

Step 2 Square both sides. (By the squaring property of equality, all solutions of the original equation are *among* the solutions of the squared equation.)

Step 3 Combine like terms.

Step 4 If there is still a term with a radical, repeat Steps 1–3.

Step 5 Solve the equation for potential solutions.

Step 6 Check all potential solutions from Step 5 in the original equation.

Solve $\sqrt{2x-3} + x = 3$.

$\sqrt{2x-3} = 3 - x$ Isolate the radical.

$(\sqrt{2x-3})^2 = (3-x)^2$ Square both sides.

$2x - 3 = 9 - 6x + x^2$

$0 = x^2 - 8x + 12$ Standard form

$0 = (x-2)(x-6)$ Factor.

$x - 2 = 0$ or $x - 6 = 0$ Zero-factor property

$x = 2$ or $x = 6$ Solve.

Verify that 2 is the only solution (6 is extraneous).
Solution set: {2}

8.7 USING RATIONAL NUMBERS AS EXPONENTS

Assume $a \geq 0$, m and n are integers, $n > 0$.

$$a^{1/n} = \sqrt[n]{a}$$

$$a^{m/n} = \sqrt[n]{a^m} = (\sqrt[n]{a})^m$$

$$a^{-m/n} = \frac{1}{a^{m/n}} \quad (a \neq 0)$$

$8^{1/3} = \sqrt[3]{8} = 2$

$(81)^{3/4} = \sqrt[4]{81^3} = (\sqrt[4]{81})^3 = 3^3 = 27$

$36^{-3/2} = \dfrac{1}{36^{3/2}} = \dfrac{1}{(36^{1/2})^3} = \dfrac{1}{6^3} = \dfrac{1}{216}$

Answers to Test Your Word Power
1. C; *Examples:* 6 is a square root of 36 since $6^2 = 6 \cdot 6 = 36$; -6 is also a square root of 36.   2. D; *Examples:* $\sqrt{144}, \sqrt{4xy^2}, \sqrt{4+t^2}$
3. A; *Examples:* $\sqrt{36} = 6$, $\sqrt[4]{81} = 3$, $\sqrt[6]{64} = 2$ 4. C; *Examples:* $\sqrt{7}$ and $3\sqrt{7}$ are like radicals; so are $2\sqrt[3]{6k}$ and $5\sqrt[3]{6k}$.
5. B; *Example:* To rationalize the denominator of $\dfrac{5}{\sqrt{3}+1}$, multiply numerator and denominator by $\sqrt{3} - 1$ to get $\dfrac{5(\sqrt{3}-1)}{2}$.
6. A; *Example:* The conjugate of $\sqrt{3} + 1$ is $\sqrt{3} - 1$.

CHAPTER 8 REVIEW EXERCISES

[8.1] *Find all square roots of each number.*

1. 49   2. 81   3. 196   4. 121   5. 225   6. 729

Find each indicated root. If the root is not a real number, say so.

7. $\sqrt{16}$ 8. $-\sqrt{36}$ 9. $\sqrt[3]{1000}$ 10. $\sqrt[4]{81}$

11. $\sqrt{-8100}$ 12. $-\sqrt{4225}$ 13. $\sqrt{\dfrac{49}{36}}$ 14. $\sqrt{\dfrac{100}{81}}$

15. If $\sqrt{a}$ is not a real number, then what kind of number must a be?

CHAPTER 8 Review Exercises **5b.**

16. Find the value of *x* in the figure.

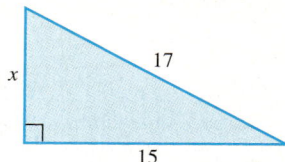

17. A Gateway EV700 computer monitor has viewing screen dimensions as shown in the figure. Find the diagonal measure of the viewing screen to the nearest tenth. (*Source:* Author's computer.)

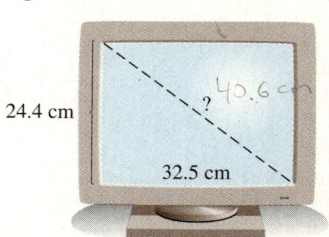

Determine whether each number is rational, irrational, *or* not a real number. *If the number is rational, give its exact value. If the number is irrational, give a decimal approximation rounded to the nearest thousandth.*

18. $\sqrt{111}$ **19.** $-\sqrt{25}$ **20.** $\sqrt{-4}$

[8.2] *Simplify each expression.*

21. $\sqrt{5} \cdot \sqrt{15}$ **22.** $-\sqrt{27}$ **23.** $\sqrt{160}$ **24.** $\sqrt[3]{-1331}$

25. $\sqrt[3]{1728}$ **26.** $\sqrt{12} \cdot \sqrt{27}$ **27.** $\sqrt{32} \cdot \sqrt{48}$ **28.** $\sqrt{50} \cdot \sqrt{125}$

Use the product rule, the quotient rule, or both to simplify each expression.

29. $-\sqrt{\dfrac{121}{400}}$ **30.**  $\sqrt{\dfrac{3}{49}}$ **31.** $\sqrt{\dfrac{7}{169}}$ **32.** $\sqrt{\dfrac{1}{6}} \cdot \sqrt{\dfrac{5}{6}}$

33. $\sqrt{\dfrac{2}{5}} \cdot \sqrt{\dfrac{2}{45}}$ **34.** $\dfrac{3\sqrt{10}}{\sqrt{5}}$ **35.** $\dfrac{24\sqrt{12}}{6\sqrt{3}}$ **36.** $\dfrac{8\sqrt{150}}{4\sqrt{75}}$

Simplify each expression. Assume that all variables represent nonnegative real numbers.

37. $\sqrt{p} \cdot \sqrt{p}$ **38.** $\sqrt{k} \cdot \sqrt{m}$ **39.** $\sqrt{r^{18}}$

40. $\sqrt{x^{10}y^{16}}$ **41.** $\sqrt{a^{15}b^{21}}$ **42.** $\sqrt{121x^6y^{10}}$

43. Use a calculator to find approximations for $\sqrt{.5}$ and $\dfrac{\sqrt{2}}{2}$. Based on your results, do you think that these two expressions represent the same number? If so, verify it algebraically.

[8.3] *Simplify, and combine terms where possible.*

44. $3\sqrt{2} + 6\sqrt{2}$ **45.** $3\sqrt{75} + 2\sqrt{27}$ **46.** $4\sqrt{12} + \sqrt{48}$

47. $4\sqrt{24} - 3\sqrt{54} + \sqrt{6}$ **48.** $2\sqrt{7} - 4\sqrt{28} + 3\sqrt{63}$ **49.** $\dfrac{2}{5}\sqrt{75} + \dfrac{3}{4}\sqrt{160}$

50. $\dfrac{1}{3}\sqrt{18} + \dfrac{1}{4}\sqrt{32}$ **51.** $\sqrt{15} \cdot \sqrt{2} + 5\sqrt{30}$

586 CHAPTER 8 Roots and Radicals

Simplify each expression. Assume that all variables represent nonnegative real numbers.

52. $\sqrt{4x} + \sqrt{36x} - \sqrt{9x}$

53. $\sqrt{16p} + 3\sqrt{p} - \sqrt{49p}$

54. $\sqrt{20m^2} - m\sqrt{45}$

55. $3k\sqrt{8k^2n} + 5k^2\sqrt{2n}$

[8.4] *Perform each indicated operation and write answers in simplest form. Assume that all variables represent positive real numbers.*

56. $\dfrac{8\sqrt{2}}{\sqrt{5}}$

57. $\dfrac{5}{\sqrt{5}}$

58. $\dfrac{12}{\sqrt{24}}$

59. $\dfrac{\sqrt{2}}{\sqrt{15}}$

60. $\sqrt{\dfrac{2}{5}}$

61. $\sqrt{\dfrac{5}{14}} \cdot \sqrt{28}$

62. $\sqrt{\dfrac{2}{7}} \cdot \sqrt{\dfrac{1}{3}}$

63. $\sqrt{\dfrac{r^2}{16x}}$

64. $\sqrt[3]{\dfrac{1}{3}}$

65. $\sqrt[3]{\dfrac{2}{7}}$

66. The radius r of a cone in terms of its volume V is given by the formula

$$r = \sqrt{\dfrac{3V}{\pi h}}.$$

Rationalize the denominator of this radical expression.

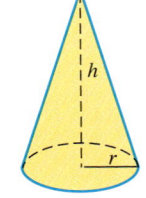

[8.5] *Simplify each expression.*

67. $-\sqrt{3}(\sqrt{5} + \sqrt{27})$

68. $3\sqrt{2}(\sqrt{3} + 2\sqrt{2})$

69. $(2\sqrt{3} - 4)(5\sqrt{3} + 2)$

70. $(5\sqrt{7} + 2)^2$

71. $(\sqrt{5} - \sqrt{7})(\sqrt{5} + \sqrt{7})$

72. $(2\sqrt{3} + 5)(2\sqrt{3} - 5)$

Rationalize each denominator.

73. $\dfrac{1}{2 + \sqrt{5}}$

74. $\dfrac{2}{\sqrt{2} - 3}$

75. $\dfrac{\sqrt{8}}{\sqrt{2} + 6}$

76. $\dfrac{\sqrt{3}}{1 + \sqrt{3}}$

77. $\dfrac{\sqrt{5} - 1}{\sqrt{2} + 3}$

78. $\dfrac{2 + \sqrt{6}}{\sqrt{3} - 1}$

Write each quotient in lowest terms.

79. $\dfrac{15 + 10\sqrt{6}}{15}$

80. $\dfrac{3 + 9\sqrt{7}}{12}$

81. $\dfrac{6 + \sqrt{192}}{2}$

[8.6] *Solve each equation.*

82. $\sqrt{m} - 5 = 0$

83. $\sqrt{p} + 4 = 0$

84. $\sqrt{k + 1} = 7$

85. $\sqrt{5m + 4} = 3\sqrt{m}$

86. $\sqrt{2p + 3} = \sqrt{5p - 3}$

87. $\sqrt{4x + 1} = x - 1$

88. $\sqrt{-2k - 4} = k + 2$

89. $\sqrt{2 - x} + 3 = x + 7$

90. $\sqrt{x} - x + 2 = 0$

91. $\sqrt{x + 4} - \sqrt{x - 4} = 2$

92. $\sqrt{5x + 6} + \sqrt{3x + 4} = 2$

[8.7] *Simplify each expression. Assume that all variables represent positive real numbers.*

93. $81^{1/2}$

94. $-125^{1/3}$

95. $7^{2/3} \cdot 7^{7/3}$

96. $\dfrac{13^{4/5}}{13^{-3/5}}$

97. $\dfrac{x^{1/4} \cdot x^{5/4}}{x^{3/4}}$

98. $\sqrt[8]{49^4}$

MIXED REVIEW EXERCISES

Simplify each expression if possible. Assume all variables represent positive real numbers.

99. $64^{2/3}$

100. $2\sqrt{27} + 3\sqrt{75} - \sqrt{300}$

101. $\dfrac{1}{5 + \sqrt{2}}$

102. $\sqrt{\dfrac{1}{3}} \cdot \sqrt{\dfrac{24}{5}}$

103. $\sqrt{50y^2}$

104. $\sqrt[3]{-125}$

105. $-\sqrt{5}(\sqrt{2} + \sqrt{75})$

106. $\sqrt{\dfrac{16r^3}{3s}}$

107. $\dfrac{12 + 6\sqrt{13}}{12}$

108. $-\sqrt{162} + \sqrt{8}$

109. $(\sqrt{5} - \sqrt{2})^2$

110. $(6\sqrt{7} + 2)(4\sqrt{7} - 1)$

111. $-\sqrt{121}$

112. $\dfrac{x^{8/3}}{x^{2/3}}$

Solve.

113. $\sqrt{x + 2} = x - 4$

114. $\sqrt{k} + 3 = 0$

115. $\sqrt{1 + 3t} - t = -3$

116. The *fall speed,* in miles per hour, of a vehicle running off the road into a ditch is given by the formula
$$S = \dfrac{2.74D}{\sqrt{h}},$$
where *D* is the horizontal distance traveled from the level surface to the bottom of the ditch and *h* is the height (or depth) of the ditch. What is the fall speed (to the nearest tenth) of a vehicle that traveled 32 ft horizontally into a ditch 5 ft deep?

RELATING CONCEPTS (EXERCISES 117–121)

For Individual or Group Work

In Chapter 3 we plotted points in the rectangular coordinate plane. In all cases our points had coordinates that were rational numbers. However, ordered pairs may have irrational coordinates as well. Consider the points $A(2\sqrt{14}, 5\sqrt{7})$ and $B(-3\sqrt{14}, 10\sqrt{7})$. Work Exercises 117–121 in order.

117. Write an expression that represents the slope of the line containing points *A* and *B*. Do not simplify yet.

118. Simplify the numerator and the denominator in the expression from Exercise 117 by combining like radicals.

119. Write the fraction from Exercise 118 as the square root of a fraction in lowest terms.

120. Rationalize the denominator of the expression found in Exercise 119.

121. Based on your answer in Exercise 120, does line *AB* rise or fall from left to right?

CHAPTER 8 TEST

On this test assume that all variables represent positive real numbers.

1. Find all square roots of 196.
2. Consider $\sqrt{142}$.
   (a) Determine whether it is rational or irrational.
   (b) Find a decimal approximation to the nearest thousandth.
3. Explain why $\sqrt{-5}$ is not a real number.

Simplify where possible.

4. $-\sqrt{27}$

5. $\sqrt{\dfrac{128}{25}}$

6. $\sqrt[3]{32}$

7. $\dfrac{20\sqrt{18}}{5\sqrt{3}}$

8. $3\sqrt{28} + \sqrt{63}$

9. $3\sqrt{27x} - 4\sqrt{48x} + 2\sqrt{3x}$

10. $\sqrt[3]{32x^2y^3}$

11. $(6 - \sqrt{5})(6 + \sqrt{5})$

12. $(2 - \sqrt{7})(3\sqrt{2} + 1)$

13. $(\sqrt{5} + \sqrt{6})^2$

14. $\sqrt[3]{16x^4} - 2\sqrt[3]{128x^4}$

Solve each problem.

15. Find the measure of the other leg of this right triangle.

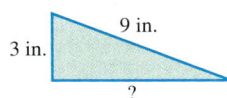

   (a) Give its length in simplified radical form.
   (b) Give the length to the nearest thousandth.

16. In electronics, the impedance Z of an alternating series circuit is given by the formula
$$Z = \sqrt{R^2 + X^2},$$
where R is the resistance and X is the reactance, both in ohms. Find the value of the impedance Z if $R = 40$ ohms and $X = 30$ ohms. (Source: Cooke, Nelson M. and Orleans, Joseph B., *Mathematics Essential to Electricity and Radio,* McGraw-Hill, 1943.)

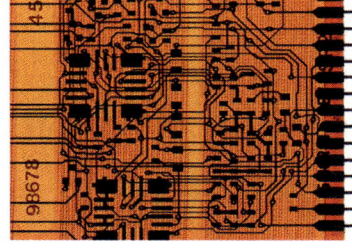

17. The radius r of a sphere in terms of its surface area S is given by the formula
$$r = \sqrt{\dfrac{S}{4\pi}}.$$

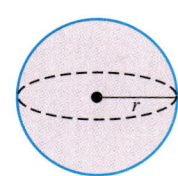

Rationalize the denominator of the radical expression.

Rationalize each denominator.

18. $\dfrac{5\sqrt{2}}{\sqrt{7}}$
19. $\sqrt{\dfrac{2}{3x}}$
20. $\dfrac{-2}{\sqrt[3]{4}}$
21. $\dfrac{-3}{4-\sqrt{3}}$

22. Write in lowest terms: $\dfrac{\sqrt{12}+3\sqrt{128}}{6}$.

Solve each equation.

23. $\sqrt{x+1} = 5 - x$
24. $3\sqrt{x-1} = 2x$
25. $\sqrt{2x+9} + \sqrt{x+5} = 2$

Simplify each expression.

26. $8^{4/3}$
27. $-125^{2/3}$
28. $5^{3/4} \cdot 5^{1/4}$
29. $\dfrac{(3^{1/4})^3}{3^{7/4}}$

30. What is wrong with the following "solution"?

$$\sqrt{2x+1} + 5 = 0$$
$$\sqrt{2x+1} = -5 \quad \text{Subtract 5.}$$
$$2x + 1 = 25 \quad \text{Square both sides.}$$
$$2x = 24 \quad \text{Subtract 1.}$$
$$x = 12 \quad \text{Divide by 2.}$$

The solution set is $\{12\}$.

CUMULATIVE REVIEW EXERCISES CHAPTERS 1–8

Simplify each expression.

1. $3(6+7) + 6 \cdot 4 - 3^2$
2. $\dfrac{3(6+7)+3}{2(4)-1}$
3. $|-6| - |-3|$

Solve each equation or inequality.

4. $5(k-4) - k = k - 11$
5. $-\dfrac{3}{4}y \leq 12$

6. $5z + 3 - 4 > 2z + 9 + z$

7. U.S. production of corn reached record levels in 2000, when .8 billion more bushels of corn were produced than in 1999. Total production for the two years was 19.6 billion bushels. How much corn was produced in each of these years? (*Source:* U.S. Department of Agriculture.)

Graph.

8. $-4x + 5y = -20$ **9.** $x = 2$ **10.** $2x - 5y > 10$

11. The graph shows a linear equation that models total federal spending on the two major-party political conventions, in millions of dollars.

(a) Use the ordered pairs shown on the graph to find the slope of the line to the nearest hundredth. Interpret the slope.

(b) Use the slope from part (a) and the ordered pair (2000, 13.5) to find the equation of the line that models the data. Write the equation in slope-intercept form.

(c) Use the equation from part (b) to project convention spending for 2004. Round your answer to the nearest tenth.

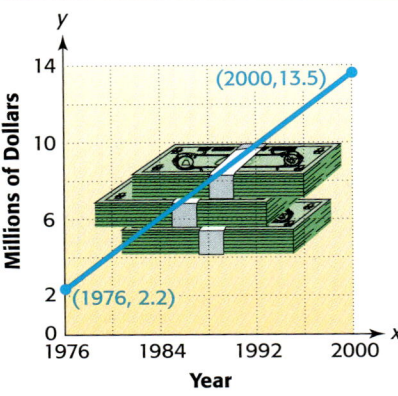

Source: Federal Election Commission.

Solve each system of equations.

12. $4x - y = 19$
$3x + 2y = -5$

13. $2x - y = 6$
$3y = 6x - 18$

14. Des Moines and Chicago are 345 mi apart. Two cars start from these cities traveling toward each other. They meet after 3 hr. The car from Chicago has an average speed 7 mph faster than the other car. Find the average speed of each car. (*Source:* State Farm Road Atlas.)

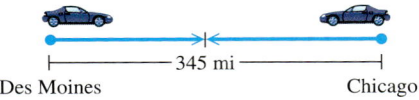

Simplify and write each expression without negative exponents. Assume that variables represent positive real numbers.

15. $(3x^6)(2x^2y)^2$ **16.** $\left(\dfrac{3^2 y^{-2}}{2^{-1} y^3}\right)^{-3}$

17. Subtract $7x^3 - 8x^2 + 4$ from $10x^3 + 3x^2 - 9$.

18. Divide: $\dfrac{8t^3 - 4t^2 - 14t + 15}{2t + 3}$.

Factor each polynomial completely.

19. $m^2 + 12m + 32$ **20.** $25t^4 - 36$

21. $12a^2 + 4ab - 5b^2$ **22.** $81z^2 + 72z + 16$

Solve each quadratic equation.

23. $x^2 - 7x = -12$ **24.** $(x + 4)(x - 1) = -6$

25. For what real number(s) is the expression $\dfrac{3}{x^2 + 5x - 14}$ undefined?

Perform each indicated operation. Express answers in lowest terms.

26. $\dfrac{x^2 - 3x - 4}{x^2 + 3x} \cdot \dfrac{x^2 + 2x - 3}{x^2 - 5x + 4}$

27. $\dfrac{t^2 + 4t - 5}{t + 5} \div \dfrac{t - 1}{t^2 + 8t + 15}$

28. $\dfrac{y}{y^2 - 1} + \dfrac{y}{y + 1}$

29. $\dfrac{2}{x + 3} - \dfrac{4}{x - 1}$

30. Simplify the complex fraction: $\dfrac{\dfrac{2}{3} + \dfrac{1}{2}}{\dfrac{1}{9} - \dfrac{1}{6}}$.

Solve each equation.

31. $\dfrac{x}{x + 8} - \dfrac{3}{x - 8} = \dfrac{128}{x^2 - 64}$

32. $A = \dfrac{B + CD}{BC + D}$ for B

33. The speed of a pulley varies inversely as its diameter. One kind of pulley, with diameter 9 in., turns at 450 revolutions per min. Find the speed of a similar pulley with diameter 10 in.

Simplify each expression. Assume all variables represent nonnegative real numbers.

34. $\sqrt{27} - 2\sqrt{12} + 6\sqrt{75}$

35. $\dfrac{2}{\sqrt{3} + \sqrt{5}}$

36. $\sqrt{200x^2y^5}$

37. $(3\sqrt{2} + 1)(4\sqrt{2} - 3)$

38. Solve $\sqrt{x + 2} = x - 10$.

Evaluate.

39. $16^{5/4}$

40. $\dfrac{8^{-7/3}}{8^{-1/3}}$

Quadratic Equations

9

9.1 Solving Quadratic Equations by the Square Root Property

9.2 Solving Quadratic Equations by Completing the Square

9.3 Solving Quadratic Equations by the Quadratic Formula

Summary Exercises on Quadratic Equations

9.4 Complex Numbers

9.5 More on Graphing Quadratic Equations; Quadratic Functions

In this chapter we develop methods of solving quadratic equations. The graphs of such equations, called *parabolas,* have many applications. For example, the Parkes radio telescope, pictured here, has a *parabolic* dish shape with diameter 210 ft and depth 32 ft. (*Source:* Mar, J. and Liebowitz, H., *Structure Technology for Large Radio and Radar Telescope Systems,* The MIT Press, 1969.) In Section 9.5 we use a graph to model this parabolic shape and find its quadratic equation.

593

9.1 Solving Quadratic Equations by the Square Root Property

OBJECTIVES

1. Solve equations of the form $x^2 = k$, where $k > 0$.
2. Solve equations of the form $(ax + b)^2 = k$, where $k > 0$.
3. Use formulas involving squared variables.

In Chapter 6 we solved quadratic equations by factoring. However, since not all quadratic equations can easily be solved by factoring, it is necessary to develop other methods. In this chapter we do just that.

Recall that a *quadratic equation* is an equation that can be written in standard form

$$ax^2 + bx + c = 0$$

for real numbers a, b, and c, with $a \neq 0$. As seen in Section 6.5, to solve the quadratic equation $x^2 + 4x + 3 = 0$ by the zero-factor property, we begin by factoring the left side and then setting each factor equal to 0.

$x^2 + 4x + 3 = 0$		
$(x + 3)(x + 1) = 0$		Factor.
$x + 3 = 0$ or $x + 1 = 0$		Zero-factor property
$x = -3$ or $x = -1$		Solve each equation.

The solution set is $\{-3, -1\}$.

OBJECTIVE 1 Solve equations of the form $x^2 = k$, where $k > 0$. We can solve equations such as $x^2 = 9$ by factoring as follows.

$x^2 = 9$		
$x^2 - 9 = 0$		Subtract 9.
$(x + 3)(x - 3) = 0$		Factor.
$x + 3 = 0$ or $x - 3 = 0$		Zero-factor property
$x = -3$ or $x = 3$		

The solution set is $\{-3, 3\}$.

We might also have solved $x^2 = 9$ by noticing that x must be a number whose square is 9. Thus, $x = \sqrt{9} = 3$ or $x = -\sqrt{9} = -3$. This is generalized as the **square root property of equations.**

Square Root Property of Equations

If k is a positive number and if $x^2 = k$, then

$$x = \sqrt{k} \quad \text{or} \quad x = -\sqrt{k},$$

and the solution set is $\{-\sqrt{k}, \sqrt{k}\}$.

NOTE When we solve an equation, we want to find *all* values of the variable that satisfy the equation. Therefore, we want both the positive and negative square roots of k.

SECTION 9.1 Solving Quadratic Equations by the Square Root Property **595**

EXAMPLE 1 Solving Quadratic Equations of the Form $x^2 = k$

Solve each equation. Write radicals in simplified form.

(a) $x^2 = 16$

By the square root property, since $x^2 = 16$,
$$x = \sqrt{16} = 4 \quad \text{or} \quad x = -\sqrt{16} = -4.$$

An abbreviation for "$x = 4$ or $x = -4$" is written $x = \pm 4$ (read "positive or negative 4"). Check each solution by substituting it for x in the original equation. The solution set is $\{-4, 4\}$.

(b) $z^2 = 5$

The solutions are $z = \sqrt{5}$ or $z = -\sqrt{5}$. The solution set may be written $\{\pm\sqrt{5}\}$.

(c)
$$5m^2 - 40 = 0$$
$$5m^2 = 40 \quad \text{Add 40.}$$
$$m^2 = 8 \quad \text{Divide by 5.}$$
$$m = \sqrt{8} \quad \text{or} \quad m = -\sqrt{8} \quad \text{Square root property}$$
$$m = 2\sqrt{2} \quad \text{or} \quad m = -2\sqrt{2} \quad \text{Simplify } \sqrt{8}.$$

The solution set is $\{\pm 2\sqrt{2}\}$.

(d) $p^2 = -4$

Since -4 is a negative number and since the square of a real number cannot be negative, there is no real number solution for this equation. (The square root property cannot be used because of the requirement that k must be positive.) The solution set is $\emptyset$.

(e) $3x^2 + 5 = 11$

First solve the equation for x^2.
$$3x^2 + 5 = 11$$
$$3x^2 = 6 \quad \text{Subtract 5.}$$
$$x^2 = 2 \quad \text{Divide by 3.}$$

Now use the square root property to get the solution set $\{\pm\sqrt{2}\}$.

Now Try Exercises 5, 7, 11, and 19.

OBJECTIVE 2 Solve equations of the form $(ax + b)^2 = k$, where $k > 0$. In each equation in Example 1, the exponent 2 appeared with a single variable as its base. We can extend the square root property to solve equations where the base is a binomial, as shown in the next example.

EXAMPLE 2 Solving Quadratic Equations of the Form $(x + b)^2 = k$

Solve each equation.

(a) $(x - 3)^2 = 16$

Apply the square root property, using $x - 3$ as the base.

596 CHAPTER 9 Quadratic Equations

$$(x - 3)^2 = 16$$
$$x - 3 = \sqrt{16} \quad \text{or} \quad x - 3 = -\sqrt{16}$$
$$x - 3 = 4 \quad \text{or} \quad x - 3 = -4 \qquad \sqrt{16} = 4$$
$$x = 7 \quad \text{or} \quad x = -1 \qquad \text{Add 3.}$$

Check both answers in the original equation.

$$(x - 3)^2 = 16 \qquad\qquad (x - 3)^2 = 16$$
$$(7 - 3)^2 = 16 \ ? \quad \text{Let } x = 7. \qquad (-1 - 3)^2 = 16 \ ? \quad \text{Let } x = -1.$$
$$4^2 = 16 \ ? \qquad\qquad (-4)^2 = 16 \ ?$$
$$16 = 16 \quad \text{True} \qquad\qquad 16 = 16 \quad \text{True}$$

The solution set is $\{7, -1\}$.

(b) $(x - 1)^2 = 6$

By the square root property,

$$x - 1 = \sqrt{6} \quad \text{or} \quad x - 1 = -\sqrt{6}$$
$$x = 1 + \sqrt{6} \quad \text{or} \quad x = 1 - \sqrt{6}.$$

Check:
$$(1 + \sqrt{6} - 1)^2 = (\sqrt{6})^2 = 6;$$
$$(1 - \sqrt{6} - 1)^2 = (-\sqrt{6})^2 = 6.$$

The solution set is $\{1 + \sqrt{6}, 1 - \sqrt{6}\}$.

Now Try Exercises 23 and 27.

NOTE The solutions in Example 2(b) may be written in abbreviated form as

$$1 \pm \sqrt{6}.$$

If they are written this way, keep in mind that *two* solutions are indicated, one with the $+$ sign and the other with the $-$ sign.

EXAMPLE 3 Solving a Quadratic Equation of the Form $(ax + b)^2 = k$

Solve $(3r - 2)^2 = 27$.

$$3r - 2 = \sqrt{27} \quad \text{or} \quad 3r - 2 = -\sqrt{27} \qquad \text{Square root property}$$
$$3r - 2 = 3\sqrt{3} \quad \text{or} \quad 3r - 2 = -3\sqrt{3} \qquad \sqrt{27} = \sqrt{9 \cdot 3} = 3\sqrt{3}$$
$$3r = 2 + 3\sqrt{3} \quad \text{or} \quad 3r = 2 - 3\sqrt{3} \qquad \text{Add 2.}$$
$$r = \frac{2 + 3\sqrt{3}}{3} \quad \text{or} \quad r = \frac{2 - 3\sqrt{3}}{3} \qquad \text{Divide by 3.}$$

The solution set is $\left\{\dfrac{2 \pm 3\sqrt{3}}{3}\right\}$.

Now Try Exercise 35.

CAUTION The solutions in Example 3 are fractions that cannot be simplified, since 3 is *not* a common factor in the numerator.

EXAMPLE 4 Recognizing a Quadratic Equation with No Real Solutions

Solve $(x + 3)^2 = -9$.

Because the square root of -9 is not a real number, the solution set is $\emptyset$.

Now Try Exercise 25.

OBJECTIVE 3 Use formulas involving squared variables.

EXAMPLE 5 Finding the Length of a Bass

The formula

$$w = \frac{L^2 g}{1200}$$

is used to approximate the weight of a bass, in pounds, given its length L and its girth g, where both are measured in inches. Approximate the length of a bass weighing 2.20 lb and having girth 10 in. (*Source: Sacramento Bee,* November 29, 2000.)

$w = \dfrac{L^2 g}{1200}$	Given formula
$2.20 = \dfrac{L^2 \cdot 10}{1200}$	$w = 2.20,\ g = 10$
$2640 = 10L^2$	Multiply by 1200.
$L^2 = 264$	Divide by 10.
$L \approx 16.25$	Use a calculator; $L > 0$.

The length of the bass is a little more than 16 in. (We discard the negative solution -16.25 since L represents length.)

Now Try Exercise 53.

9.1 EXERCISES

For Extra Help

 Student's Solutions Manual

 MyMathLab

 InterAct Math Tutorial Software

 AW Math Tutor Center

 MathXL

 Digital Video Tutor CD 18/Videotape 16

Match each equation in Column I with the correct description of its solution in Column II.

I	II
1. $x^2 = 10$	**A.** No real number solutions
2. $x^2 = -4$	**B.** Two integer solutions
3. $x^2 = \dfrac{9}{16}$	**C.** Two irrational solutions
4. $x^2 = 9$	**D.** Two rational solutions that are not integers

Solve each equation by using the square root property. Write all radicals in simplest form. See Example 1.

5. $x^2 = 81$ **6.** $z^2 = 121$ **7.** $k^2 = 14$ **8.** $m^2 = 22$

9. $t^2 = 48$ **10.** $x^2 = 54$ **11.** $z^2 = -100$ **12.** $m^2 = -64$

598 CHAPTER 9 Quadratic Equations

13. $z^2 = 2.25$
14. $w^2 = 56.25$
15. $7k^2 = 4$
16. $3p^2 = 10$
17. $5a^2 + 4 = 8$
18. $4p^2 - 3 = 7$
19. $3x^2 - 8 = 64$
20. $2t^2 + 7 = 61$

21. When a student was asked to solve $x^2 = 81$, she wrote $\{9\}$ as her answer. Her teacher did not give her full credit. The student argued that because $9^2 = 81$, her answer had to be correct. Why was her answer not completely correct?

22. Explain the square root property for solving equations, and illustrate with an example.

Solve each equation by using the square root property. Write all radicals in simplest form. See Examples 2–4.

23. $(x - 3)^2 = 25$
24. $(k - 7)^2 = 16$
25. $(z + 5)^2 = -13$
26. $(m + 2)^2 = -17$
27. $(x - 8)^2 = 27$
28. $(p - 5)^2 = 40$
29. $(3k + 2)^2 = 49$
30. $(5t + 3)^2 = 36$
31. $(4x - 3)^2 = 9$
32. $(7z - 5)^2 = 25$
33. $(5 - 2x)^2 = 30$
34. $(3 - 2a)^2 = 70$
35. $(3k + 1)^2 = 18$
36. $(5z + 6)^2 = 75$
37. $\left(\dfrac{1}{2}x + 5\right)^2 = 12$
38. $\left(\dfrac{1}{3}m + 4\right)^2 = 27$
39. $(4k - 1)^2 - 48 = 0$
40. $(2s - 5)^2 - 180 = 0$

41. Johnny solved the equation in Exercise 33 and wrote his answer as $\left\{\dfrac{5 + \sqrt{30}}{2}, \dfrac{5 - \sqrt{30}}{2}\right\}$.

Linda solved the same equation and wrote her answer as $\left\{\dfrac{-5 + \sqrt{30}}{-2}, \dfrac{-5 - \sqrt{30}}{-2}\right\}$. The teacher gave them both full credit. Explain why both students were correct, although their answers look different.

42. In the solutions found in Example 3 of this section, why is it not valid to simplify the answers by dividing out the 3s in the numerator and denominator?

Use a calculator with a square root key to solve each equation. Round your answers to the nearest hundredth.

43. $(k + 2.14)^2 = 5.46$
44. $(r - 3.91)^2 = 9.28$
45. $(2.11p + 3.42)^2 = 9.58$
46. $(1.71m - 6.20)^2 = 5.41$

RELATING CONCEPTS (EXERCISES 47–50)

For Individual or Group Work

In Section 6.4 we saw how certain trinomials can be factored as squares of binomials. Use this idea and **work Exercises 47–50 in order,** *considering the equation*

$$x^2 + 6x + 9 = 100.$$

47. Factor the left side of the equation as the square of a binomial, and write the resulting equation.

48. Use the square root property to solve the equation from Exercise 47.

49. What is the solution set of the original equation?

50. Solve the equation $4k^2 - 12k + 9 = 81$ using the method described in Exercises 47–49.

Solve each problem. See Example 5.

51. One expert at marksmanship can hold a silver dollar at forehead level, drop it, draw his gun, and shoot the coin as it passes waist level. The distance traveled by a falling object is given by

$$d = 16t^2,$$

where d is the distance (in feet) the object falls in t sec. If the coin falls about 4 ft, use the formula to estimate the time that elapses between the dropping of the coin and the shot.

52. The illumination produced by a light source depends on the distance from the source. For a particular light source, this relationship can be expressed as

$$d^2 = \frac{4050}{I},$$

where d is the distance from the source (in feet) and I is the amount of illumination in foot-candles. How far from the source is the illumination equal to 50 foot-candles?

53. The area A of a circle with radius r is given by the formula

$$A = \pi r^2.$$

If a circle has area 81π in.², what is its radius?

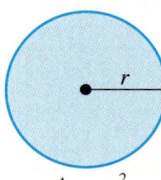

$A = \pi r^2$

54. The surface area S of a sphere with radius r is given by the formula

$$S = 4\pi r^2.$$

If a sphere has surface area 36π ft², what is its radius?

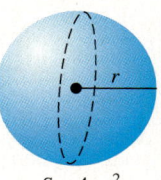

$S = 4\pi r^2$

The amount A that P dollars invested at an annual rate of interest r will grow to in 2 yr is

$$A = P(1 + r)^2.$$

55. At what interest rate will $100 grow to $110.25 in 2 yr?

56. At what interest rate will $500 grow to $572.45 in 2 yr?

600 CHAPTER 9 Quadratic Equations

9.2 Solving Quadratic Equations by Completing the Square

OBJECTIVES

1. Solve quadratic equations by completing the square when the coefficient of the squared term is 1.
2. Solve quadratic equations by completing the square when the coefficient of the squared term is not 1.
3. Simplify an equation before solving.
4. Solve applied problems that require quadratic equations.

OBJECTIVE 1 Solve quadratic equations by completing the square when the coefficient of the squared term is 1. The methods we have studied so far are not enough to solve the equation

$$x^2 + 6x + 7 = 0.$$

If we could write the equation in the form $(x + 3)^2$ equals a constant, we could solve it with the square root property discussed in the previous section. To do that, we need to have a perfect square trinomial on one side of the equation.

Recall from Section 6.4 that a perfect square trinomial has the form

$$x^2 + 2kx + k^2 \quad \text{or} \quad x^2 - 2kx + k^2,$$

where k represents a number.

EXAMPLE 1 Creating Perfect Square Trinomials

Complete each trinomial so it is a perfect square.

(a) $x^2 + \underline{\quad} + 16$

Here $k^2 = 16$, so $k = 4$ and $2kx = 2(4)(x) = 8x$. The perfect square trinomial is $x^2 + 8x + 16$.

(b) $4m^2 + \underline{\quad} + 9$

Replace x^2 by $4m^2 = (2m)^2$ and k^2 by $9 = 3^2$, to get the middle term

$$2(3)(2m) = 12m.$$

The perfect square trinomial is $4m^2 + 12m + 9$.

(c) $x^2 + 18x + \underline{\quad}$

Here the middle term $18x$ must equal $2kx$.

$$2kx = 18x$$
$$k = 9 \quad \text{Divide by } 2x.$$
$$k^2 = 81 \quad \text{Square both sides.}$$

The required trinomial is $x^2 + 18x + 81$.

(d) $4z^2 - 4z + \underline{\quad}$

Write $4z^2$ as $(2z)^2$ and set $2k(2z)$ equal to $-4z$. Then solve for k.

$$2k(2z) = -4z$$
$$4kz = -4z$$
$$k = -1 \quad \text{Divide by } 4z.$$
$$k^2 = 1 \quad \text{Square both sides.}$$

The perfect square trinomial is $4z^2 - 4z + 1$.

Now Try Exercises 1 and 3.

EXAMPLE 2 Rewriting an Equation to Use the Square Root Property

Solve $x^2 + 6x + 7 = 0.$

Start by subtracting 7 from both sides of the equation to get
$$x^2 + 6x = -7.$$

The quantity on the left-hand side of $x^2 + 6x = -7$ must be made into a perfect square trinomial. Here $2kx = 6x$, so $k = 3$ and $k^2 = 9$. The expression $x^2 + 6x + 9$ is a perfect square, since
$$x^2 + 6x + 9 = (x + 3)^2.$$

Therefore, if we add 9 to both sides of $x^2 + 6x = -7$, the equation will have a perfect square trinomial on the left side, as needed.

$$x^2 + 6x + 9 = -7 + 9 \quad \text{Add 9.}$$
$$(x + 3)^2 = 2 \quad \text{Factor; combine terms.}$$

Now use the square root property to complete the solution.

$$x + 3 = \sqrt{2} \quad \text{or} \quad x + 3 = -\sqrt{2}$$
$$x = -3 + \sqrt{2} \quad \text{or} \quad x = -3 - \sqrt{2}$$

Check by substituting $-3 + \sqrt{2}$ and $-3 - \sqrt{2}$ for x in the original equation. The solution set is $\{-3 \pm \sqrt{2}\}$.

Now Try Exercise 13.

The process of changing the form of the equation in Example 2 from
$$x^2 + 6x + 7 = 0 \quad \text{to} \quad (x + 3)^2 = 2$$
is called **completing the square.** Completing the square changes only the form of the equation. To see this, multiply out the left side of $(x + 3)^2 = 2$ and combine terms. Then subtract 2 from both sides to see that the result is $x^2 + 6x + 7 = 0$.

Look again at the original equation
$$x^2 + 6x + 7 = 0.$$

Note that 9 is the square of half the coefficient of x, 6.

$$\frac{1}{2} \cdot 6 = 3 \quad \text{and} \quad 3^2 = 9$$
↑
Coefficient of x

To complete the square in Example 2, 9 was added to each side.

EXAMPLE 3 Completing the Square to Solve a Quadratic Equation

Solve $x^2 - 8x = 5$.

To complete the square on $x^2 - 8x$, take half the coefficient of x and square it.

$$\frac{1}{2}(-8) = -4 \quad \text{and} \quad (-4)^2 = 16$$
↑
Coefficient of x

Add the result, 16, to both sides of the equation.

$$x^2 - 8x = 5 \quad \text{Given equation}$$
$$x^2 - 8x + 16 = 5 + 16 \quad \text{Add 16.}$$
$$(x - 4)^2 = 21 \quad \text{Factor the left side as the square of a binomial.}$$

Now apply the square root property.

$$(x - 4)^2 = 21$$
$$x - 4 = \pm\sqrt{21} \quad \text{Square root property}$$
$$x = 4 \pm \sqrt{21} \quad \text{Add 4.}$$

A check indicates that the solution set is $\{4 \pm \sqrt{21}\}$.

Now Try Exercise 15.

The steps to solve a quadratic equation $ax^2 + bx + c = 0$ by completing the square are summarized here.

Solving a Quadratic Equation by Completing the Square

Step 1 **Be sure the squared term has coefficient 1.** If the coefficient of the squared term is 1, go to Step 2. If it is not 1 but some other nonzero number a, divide both sides of the equation by a.

Step 2 **Put in correct form.** Make sure that all variable terms are on one side of the equation and that all constant terms are on the other.

Step 3 **Complete the square.** Take half the coefficient of the first-degree term, and square it. Add the square to both sides of the equation. Factor the variable side and combine terms on the other side.

Step 4 **Solve.** Use the square root property to solve the equation.

OBJECTIVE 2 Solve quadratic equations by completing the square when the coefficient of the squared term is not 1. To complete the square, the coefficient of the squared term must be 1 (Step 1). The next example shows what to do when this coefficient is not 1.

EXAMPLE 4 Solving a Quadratic Equation by Completing the Square

Solve $4z^2 + 24z - 13 = 0$.

Step 1 **Be sure the squared term has coefficient 1.** Divide each side by 4 so that the coefficient of z^2 is 1.

$$4z^2 + 24z - 13 = 0$$
$$z^2 + 6z - \frac{13}{4} = 0 \quad \text{Divide by 4.}$$

Step 2 **Put in correct form.** Add $\frac{13}{4}$ to each side to get the variable terms on the left and the constant on the right.

$$z^2 + 6z = \frac{13}{4}$$

Step 3 **Complete the square.** To complete the square, take half the coefficient of z and square the result: $\left[\frac{1}{2}(6)\right]^2 = 9$. Add 9 to both sides of the equation; then add the terms on the right-hand side, and factor on the left.

$$z^2 + 6z + 9 = \frac{13}{4} + 9 \quad \text{Add 9.}$$

$$z^2 + 6z + 9 = \frac{49}{4} \quad \text{Add on the right.}$$

$$(z + 3)^2 = \frac{49}{4} \quad \text{Factor on the left.}$$

Step 4 **Solve.** Use the square root property and solve for z.

$$z + 3 = \frac{7}{2} \quad \text{or} \quad z + 3 = -\frac{7}{2} \quad \text{Square root property}$$

$$z = -3 + \frac{7}{2} \quad \text{or} \quad z = -3 - \frac{7}{2} \quad \text{Subtract 3.}$$

$$z = \frac{1}{2} \quad \text{or} \quad z = -\frac{13}{2}$$

Check by substituting each solution into the original equation. The solution set is $\{\frac{1}{2}, -\frac{13}{2}\}$.

Now Try Exercise 19.

EXAMPLE 5 Solving a Quadratic Equation by Completing the Square

Solve $4p^2 + 8p + 5 = 0$.

First divide both sides by 4 to get a coefficient of 1 for the p^2-term (Step 1).

$$p^2 + 2p + \frac{5}{4} = 0$$

Add $-\frac{5}{4}$ to both sides (Step 2).

$$p^2 + 2p = -\frac{5}{4}$$

The coefficient of p is 2. Take half of 2, square the result, and add it to both sides: $[\frac{1}{2}(2)]^2 = 1$. The left-hand side can then be factored as a perfect square (Step 3).

$$p^2 + 2p + 1 = -\frac{5}{4} + 1 \quad \text{Add 1.}$$

$$(p + 1)^2 = -\frac{1}{4} \quad \text{Factor; add.}$$

We cannot use the square root property to solve this equation (Step 4) because the square root of $-\frac{1}{4}$ is not a real number. The solution set is $\emptyset$.

Now Try Exercise 21.

OBJECTIVE 3 Simplify an equation before solving. Sometimes an equation must be simplified before completing the square. The next example illustrates this.

EXAMPLE 6 Simplifying an Equation Before Completing the Square

Solve $(x + 3)(x - 1) = 2$.

$$(x + 3)(x - 1) = 2$$
$$x^2 + 2x - 3 = 2 \qquad \text{Use FOIL.}$$
$$x^2 + 2x = 5 \qquad \text{Add 3.}$$
$$x^2 + 2x + 1 = 5 + 1 \qquad \text{Add } \left[\tfrac{1}{2}(2)\right]^2 = 1.$$
$$(x + 1)^2 = 6 \qquad \text{Factor on the left; add on the right.}$$
$$x + 1 = \sqrt{6} \quad \text{or} \quad x + 1 = -\sqrt{6} \qquad \text{Square root property}$$
$$x = -1 + \sqrt{6} \quad \text{or} \quad x = -1 - \sqrt{6} \qquad \text{Subtract 1.}$$

The solution set is $\{-1 \pm \sqrt{6}\}$.

Now Try Exercise 27.

NOTE The solutions given in Example 6 are *exact*. In applications, decimal solutions are more appropriate. Using the square root key of a calculator, $\sqrt{6} \approx 2.449$. Evaluating the two solutions gives

$$x \approx 1.449 \quad \text{and} \quad x \approx -3.449.$$

OBJECTIVE 4 Solve applied problems that require quadratic equations. There are many practical applications of quadratic equations. The next example illustrates an application from physics.

EXAMPLE 7 Solving a Velocity Problem

If a ball is propelled into the air from ground level with an initial velocity of 64 ft per sec, its height s (in feet) in t sec is given by the formula

$$s = -16t^2 + 64t.$$

How long will it take the ball to reach a height of 48 ft?

Since s represents the height, let $s = 48$ in the formula to get

$$48 = -16t^2 + 64t.$$

We solve this equation for time, t, by completing the square. First we divide each side by -16. We also reverse the sides of the equation.

$$-3 = t^2 - 4t \qquad \text{Divide by } -16.$$
$$t^2 - 4t = -3 \qquad \text{Reverse the sides.}$$
$$t^2 - 4t + 4 = -3 + 4 \qquad \text{Add } \left[\tfrac{1}{2}(-4)\right]^2 = 4.$$
$$(t - 2)^2 = 1 \qquad \text{Factor.}$$
$$t - 2 = 1 \quad \text{or} \quad t - 2 = -1 \qquad \text{Square root property}$$
$$t = 3 \quad \text{or} \quad t = 1 \qquad \text{Add 2.}$$

You may wonder how we can get two correct answers for the time required for the ball to reach a height of 48 ft. The ball reaches that height twice, once on the way up and again on the way down. So it takes 1 sec to reach 48 ft on the way up, and then after 3 sec, the ball reaches 48 ft again on the way down.

Now Try Exercise 37.

9.2 EXERCISES

For Extra Help

 Student's Solutions Manual

 MyMathLab

 InterAct Math Tutorial Software

 AW Math Tutor Center

MathXL

 Digital Video Tutor CD 18/Videotape 16

Complete each trinomial so it is a perfect square. See Example 1.

1. $k^2 + \underline{} + 25$
2. $9x^2 + \underline{} + 4$
3. $z^2 + 14z + \underline{}$
4. $4a^2 - 32a + \underline{}$

5. Which one of the following steps is an appropriate way to begin solving the quadratic equation
$$2x^2 - 4x = 9$$
by completing the square?

A. Add 4 to both sides of the equation. B. Factor the left side as $2x(x - 2)$.
C. Factor the left side as $x(2x - 4)$. D. Divide both sides by 2.

6. In Example 3 of Section 6.5, we solved the quadratic equation
$$4p^2 - 26p + 40 = 0$$
by factoring. If we were to solve by completing the square, would we get the same solution set, $\{\frac{5}{2}, 4\}$?

Find the constant that should be added to each expression to make it a perfect square. See Examples 1–3.

7. $x^2 + 4x$
8. $z^2 + 16z$
9. $k^2 - 5k$
10. $m^2 - 9m$
11. $r^2 + \frac{1}{2}r$
12. $s^2 - \frac{1}{3}s$

Solve each equation by completing the square. See Examples 2 and 3.

13. $x^2 - 4x = -3$
14. $p^2 - 2p = 8$
15. $x^2 + 2x - 5 = 0$
16. $r^2 + 4r + 1 = 0$
17. $z^2 + 6z + 9 = 0$
18. $k^2 - 8k + 16 = 0$

Solve each equation by completing the square. See Examples 4–6.

19. $4x^2 + 4x = 3$
20. $9x^2 + 3x = 2$
21. $2p^2 - 2p + 3 = 0$
22. $3q^2 - 3q + 4 = 0$
23. $3a^2 - 9a + 5 = 0$
24. $6b^2 - 8b - 3 = 0$
25. $3k^2 + 7k = 4$
26. $2k^2 + 5k = 1$
27. $(x + 3)(x - 1) = 5$
28. $(z - 8)(z + 2) = 24$
29. $-x^2 + 2x = -5$
30. $-x^2 + 4x = 1$

*Solve each equation by completing the square. Give **(a)** exact solutions and **(b)** solutions rounded to the nearest thousandth.*

31. $3r^2 - 2 = 6r + 3$
32. $4p + 3 = 2p^2 + 2p$
33. $(x + 1)(x + 3) = 2$
34. $(x - 3)(x + 1) = 1$

35. In using the method of completing the square to solve $2x^2 - 10x = -8$, a student began by adding the square of half the coefficient of x (that is, $\left[\frac{1}{2}(-10)\right]^2 = 25$) to both sides of the equation. He then encountered difficulty in his later steps. What was his error? Explain the steps needed to solve the problem, and give the solution set.

36. The equation $x^4 - 2x^2 = 8$ can be solved by completing the square, even though it is not a quadratic equation. What number should be added to both sides of the equation so that the left side can be factored as the square of a binomial? Solve the equation for its real solutions.

Solve each problem. See Example 7.

37. If an object is propelled upward on the surface of Mars from ground level with an initial velocity of 104 ft per sec, its height s (in feet) in t sec is given by the formula $s = -13t^2 + 104t$. How long will it take for the object to be at a height of 195 ft?

38. How long will it take the object in Exercise 37 to return to the surface? (*Hint:* When it returns to the surface, $s = 0$.)

39. If an object is propelled upward from ground level on Earth with an initial velocity of 96 ft per sec, its height, s (in feet), in t sec is given by the formula $s = -16t^2 + 96t$. How long will it take for the object to be at a height of 80 ft? (*Hint:* Let $s = 80$.)

40. How long will it take the object described in Exercise 39 to be at a height of 100 ft? Round your answers to the nearest tenth.

41. A farmer has a rectangular cattle pen with perimeter 350 ft and area 7500 ft². What are the dimensions of the pen? (*Hint:* Use the figure to set up the equation.)

42. The base of a triangle measures 1 m more than three times the height of the triangle. Its area is 15 m². Find the lengths of the base and the height.

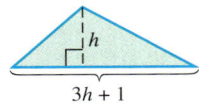

43. Two cars travel at right angles to each other from an intersection until they are 17 mi apart. At that point one car has gone 7 mi farther than the other. How far did the slower car travel? (*Hint:* Use the Pythagorean formula.)

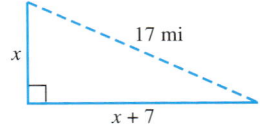

44. Two painters are painting a house in a development of new homes. One of the painters takes 2 hr longer to paint a house working alone than the other painter. When they do the job together, they can complete it in 4.8 hr. How long would it take the faster painter alone to paint the house? (Give your answer to the nearest tenth.)

RELATING CONCEPTS (EXERCISES 45–48)

For Individual or Group Work

We have discussed "completing the square" so far only in an algebraic sense. However, this procedure can literally be applied to a geometric figure so that it becomes a square. For example, to complete the square for $x^2 + 8x$, begin with a square having a side of length x. Add four rectangles of width 1 to the right side and to the bottom. To "complete the square," fill in the bottom right corner with 16 squares of area 1.

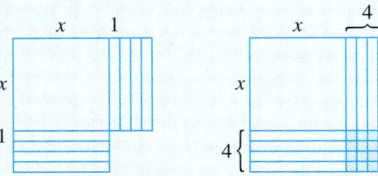

Work Exercises 45–48 in order.

45. What is the area of the original square?

46. What is the area of the figure after the 8 rectangles are added?

47. What is the area of the figure after the 16 small squares are added?

48. At what point did we "complete the square"?

9.3 Solving Quadratic Equations by the Quadratic Formula

OBJECTIVES

1. Identify the values of *a*, *b*, and *c* in a quadratic equation.
2. Use the quadratic formula to solve quadratic equations.
3. Solve quadratic equations with only one solution.
4. Solve quadratic equations with fractions.

We can solve any quadratic equation by completing the square, but the method is tedious. In this section we complete the square on the general quadratic equation $ax^2 + bx + c = 0$ to get the *quadratic formula*, a formula that gives the solutions for any quadratic equation. (Note that $a \neq 0$, or we would have a linear, not a quadratic, equation.)

OBJECTIVE 1 Identify the values of *a*, *b*, and *c* in a quadratic equation. The first step in solving a quadratic equation by this new method is to identify the values of *a*, *b*, and *c* in the standard form of the quadratic equation.

EXAMPLE 1 Determining Values of *a*, *b*, and *c* in Quadratic Equations

Match the coefficients of each quadratic equation with the letters *a*, *b*, and *c* of the standard quadratic equation

$$ax^2 + bx + c = 0.$$

(a) $2x^2 + 3x - 5 = 0$

In this example, $a = 2$, $b = 3$, and $c = -5$.

(b) $-x^2 + 2 = 6x$

First rewrite the equation with 0 on the right side to match the standard form $ax^2 + bx + c = 0$.

$$-x^2 + 2 = 6x$$
$$-x^2 - 6x + 2 = 0 \qquad \text{Subtract } 6x.$$

Here, $a = -1$, $b = -6$, and $c = 2$. (Notice that the coefficient of x^2 is understood to be -1.)

(c) $5x^2 - 12 = 0$

The *x*-term is missing, so write the equation as

$$5x^2 + 0x - 12 = 0.$$

Then $a = 5$, $b = 0$, and $c = -12$.

(d) $(2x - 7)(x + 4) = -23$

Write the equation in standard form.

$$(2x - 7)(x + 4) = -23$$
$$2x^2 + x - 28 = -23 \quad \text{Use FOIL.}$$
$$2x^2 + x - 5 = 0 \quad \text{Add 23.}$$

Now, identify the values: $a = 2$, $b = 1$, and $c = -5$.

Now Try Exercises 1, 7, and 9.

OBJECTIVE 2 Use the quadratic formula to solve quadratic equations. To develop the quadratic formula, we follow the steps given in the previous section for completing the square on $ax^2 + bx + c = 0$. For comparison, we also show the corresponding steps for solving $2x^2 + x - 5 = 0$ (from Example 1(d)).

Step 1 Make the coefficient of the squared term equal to 1.

$$2x^2 + x - 5 = 0 \qquad\qquad ax^2 + bx + c = 0$$
$$x^2 + \frac{1}{2}x - \frac{5}{2} = 0 \quad \text{Divide by 2.} \qquad x^2 + \frac{b}{a}x + \frac{c}{a} = 0 \quad \text{Divide by } a.$$

Step 2 Get the variable terms alone on the left side.

$$x^2 + \frac{1}{2}x = \frac{5}{2} \quad \text{Add } \tfrac{5}{2}. \qquad x^2 + \frac{b}{a}x = -\frac{c}{a} \quad \text{Subtract } \tfrac{c}{a}.$$

Step 3 Add the square of half the coefficient of x to both sides, factor the left side, and combine terms on the right.

$$x^2 + \frac{1}{2}x + \frac{1}{16} = \frac{5}{2} + \frac{1}{16} \quad \text{Add } \tfrac{1}{16}. \qquad x^2 + \frac{b}{a}x + \frac{b^2}{4a^2} = -\frac{c}{a} + \frac{b^2}{4a^2}$$

$$\text{Add } \tfrac{b^2}{4a^2}.$$

$$\left(x + \frac{1}{4}\right)^2 = \frac{41}{16} \quad \begin{array}{l}\text{Factor;}\\ \text{add on}\\ \text{right.}\end{array} \qquad \left(x + \frac{b}{2a}\right)^2 = \frac{b^2 - 4ac}{4a^2}$$

Factor; add on right.

Step 4 Use the square root property to complete the solution.

$$x + \frac{1}{4} = \pm\sqrt{\frac{41}{16}} \qquad\qquad x + \frac{b}{2a} = \pm\sqrt{\frac{b^2 - 4ac}{4a^2}}$$

$$x + \frac{1}{4} = \pm\frac{\sqrt{41}}{4} \qquad\qquad x + \frac{b}{2a} = \pm\frac{\sqrt{b^2 - 4ac}}{2a}$$

$$x = -\frac{1}{4} \pm \frac{\sqrt{41}}{4} \qquad\qquad x = -\frac{b}{2a} \pm \frac{\sqrt{b^2 - 4ac}}{2a}$$

$$x = \frac{-1 \pm \sqrt{41}}{4} \qquad\qquad x = \frac{-b \pm \sqrt{b^2 - 4ac}}{2a}$$

The final result in the column on the right is called the **quadratic formula.** *It is a key result that should be memorized.* Notice that there are two values, one for the $+$ sign and one for the $-$ sign.

Quadratic Formula

The solutions of the quadratic equation $ax^2 + bx + c = 0$, $a \neq 0$, are

$$x = \frac{-b + \sqrt{b^2 - 4ac}}{2a} \quad \text{and} \quad x = \frac{-b - \sqrt{b^2 - 4ac}}{2a}$$

or, in compact form, $\quad x = \dfrac{-b \pm \sqrt{b^2 - 4ac}}{2a}.$

CAUTION Notice that the fraction bar is under $-b$ as well as the radical. When using this formula, be sure to find the values of $-b \pm \sqrt{b^2 - 4ac}$ first, then divide those results by the value of $2a$.

EXAMPLE 2 Solving a Quadratic Equation by the Quadratic Formula

Use the quadratic formula to solve $2x^2 - 7x - 9 = 0$.

Match the coefficients of the variables with those of the standard quadratic equation

$$ax^2 + bx + c = 0.$$

Here, $a = 2$, $b = -7$, and $c = -9$. Substitute these numbers into the quadratic formula, and simplify the result.

$$x = \frac{-b \pm \sqrt{b^2 - 4ac}}{2a}$$

$$x = \frac{-(-7) \pm \sqrt{(-7)^2 - 4(2)(-9)}}{2(2)} \quad \text{Let } a = 2, b = -7, c = -9.$$

$$x = \frac{7 \pm \sqrt{49 + 72}}{4}$$

$$x = \frac{7 \pm \sqrt{121}}{4}$$

$$x = \frac{7 \pm 11}{4} \quad \sqrt{121} = 11$$

Find the two separate solutions by first using the plus sign, and then using the minus sign:

$$x = \frac{7 + 11}{4} = \frac{18}{4} = \frac{9}{2} \quad \text{or} \quad x = \frac{7 - 11}{4} = \frac{-4}{4} = -1.$$

Check by substituting each solution into the original equation. The solution set is $\{-1, \frac{9}{2}\}$.

Now Try Exercise 19.

EXAMPLE 3 Rewriting a Quadratic Equation Before Using the Quadratic Formula

Solve $x^2 = 2x + 1$.

Find a, b, and c by rewriting the equation $x^2 = 2x + 1$ in standard form (with 0 on one side). Add $-2x - 1$ to each side of the equation to get

$$x^2 - 2x - 1 = 0.$$

Then $a = 1$, $b = -2$, and $c = -1$. Substitute these values into the quadratic formula.

$$x = \frac{-b \pm \sqrt{b^2 - 4ac}}{2a}$$

$$x = \frac{-(-2) \pm \sqrt{(-2)^2 - 4(1)(-1)}}{2(1)} \quad \text{Let } a = 1, b = -2, c = -1.$$

$$x = \frac{2 \pm \sqrt{4 + 4}}{2}$$

$$x = \frac{2 \pm \sqrt{8}}{2}$$

$$x = \frac{2 \pm 2\sqrt{2}}{2} \quad \sqrt{8} = \sqrt{4} \cdot \sqrt{2} = 2\sqrt{2}$$

Write the solutions in lowest terms by factoring $2 \pm 2\sqrt{2}$ as $2(1 \pm \sqrt{2})$ to get

$$x = \frac{2(1 \pm \sqrt{2})}{2} = 1 \pm \sqrt{2}. \quad \text{Factor, then divide out.}$$

The solution set is $\{1 \pm \sqrt{2}\}$.

Now Try Exercise 21.

OBJECTIVE 3 Solve quadratic equations with only one solution. In the quadratic formula, the quantity under the radical, $b^2 - 4ac$, is called the **discriminant.** When the discriminant equals 0, the equation has just one rational number solution. In this case, the trinomial $ax^2 + bx + c$ is a perfect square.

EXAMPLE 4   Solving a Quadratic Equation with Only One Solution

Solve $4x^2 + 25 = 20x$.

Write the equation as

$$4x^2 - 20x + 25 = 0. \quad \text{Subtract } 20x.$$

Here, $a = 4$, $b = -20$, and $c = 25$. By the quadratic formula,

$$x = \frac{-(-20) \pm \sqrt{(-20)^2 - 400}}{8} = \frac{20 \pm 0}{8} = \frac{5}{2}.$$

In this case, $b^2 - 4ac = 0$, and the trinomial $4x^2 - 20x + 25$ is a perfect square. There is just one solution in the solution set $\{\frac{5}{2}\}$.

Now Try Exercise 17.

NOTE The single solution of the equation in Example 4 is a rational number. If all solutions of a quadratic equation are rational, the equation can be solved by factoring as well.

58. An astronaut on the moon throws a baseball upward. The height h of the ball, in feet, x seconds after he throws it, is given by the equation
$$h = -2.7x^2 + 30x + 6.5.$$
After how many seconds is the ball 12 ft above the moon's surface?

59. A rule for estimating the number of board feet of lumber that can be cut from a log depends on the diameter of the log. To find the diameter d required to get 9 board ft of lumber, we use the equation
$$\left(\frac{d-4}{4}\right)^2 = 9.$$
Solve this equation for d. Are both answers reasonable?

60. An old Babylonian problem asks for the length of the side of a square, given that the area of the square minus the length of a side is 870. Find the length of the side. (*Source:* Eves, Howard, *An Introduction to the History of Mathematics,* Sixth Edition, Saunders College Publishing, 1990.)

RELATING CONCEPTS (EXERCISES 61–66)

For Individual or Group Work

In Chapter 6 we presented methods for factoring trinomials. Some trinomials cannot be factored using integer coefficients, however. There is a way to determine beforehand whether a trinomial of the form $ax^2 + bx + c$ can be factored using the discriminant, $b^2 - 4ac$. **Work Exercises 61–66 in order.**

61. Each of the following trinomials is factorable. Find the discriminant for each one.
  (a) $18x^2 - 9x - 2$ (b) $5x^2 + 7x - 6$
  (c) $48x^2 + 14x + 1$ (d) $x^2 - 5x - 24$

62. What do you notice about the discriminants you found in Exercise 61?

63. Factor each of the trinomials in Exercise 61.

64. Each of the following trinomials is not factorable using the methods of Chapter 6. Find the discriminant for each one.
  (a) $2x^2 + x - 5$ (b) $2x^2 + x + 5$ (c) $x^2 + 6x + 6$ (d) $3x^2 + 2x - 9$

65. Are any of the discriminants you found in Exercise 64 perfect squares?

66. Make a conjecture (an educated guess) concerning when a trinomial of the form $ax^2 + bx + c$ is factorable. Then use your conjecture to determine whether each trinomial is factorable. (Do not actually factor.)
  (a) $42x^2 + 117x + 66$ (b) $99x^2 + 186x - 24$ (c) $58x^2 + 184x + 27$

SUMMARY EXERCISES ON QUADRATIC EQUATIONS

Four algebraic methods have now been introduced for solving quadratic equations written in the form $ax^2 + bx + c = 0$. The table shows some advantages and some disadvantages of each method.

Method	Advantages	Disadvantages
1. Factoring	It is usually the fastest method.	Not all equations can be solved by factoring. Some factorable polynomials are difficult to factor.
2. Square root property	It is the simplest method for solving equations of the form $(ax + b)^2 = $ a number.	Few equations are given in this form.
3. Completing the square	It can always be used. (Also, the procedure is useful in other areas of mathematics.)	It requires more steps than other methods.
4. Quadratic formula	It can always be used.	It is more difficult than factoring because of the $\sqrt{b^2 - 4ac}$ expression.

Solve each quadratic equation by the method of your choice.

1. $s^2 = 36$
2. $x^2 + 3x = -1$
3. $x^2 - \dfrac{100}{81} = 0$
4. $81t^2 = 49$
5. $z^2 - 4z + 3 = 0$
6. $w^2 + 3w + 2 = 0$
7. $z(z - 9) = -20$
8. $x^2 + 3x - 2 = 0$
9. $(3k - 2)^2 = 9$
10. $(2s - 1)^2 = 10$
11. $(x + 6)^2 = 121$
12. $(5k + 1)^2 = 36$
13. $(3r - 7)^2 = 24$
14. $(7p - 1)^2 = 32$
15. $(5x - 8)^2 = -6$
16. $2t^2 + 1 = t$
17. $-2x^2 = -3x - 2$
18. $-2x^2 + x = -1$
19. $8z^2 = 15 + 2z$
20. $3k^2 = 3 - 8k$
21. $0 = -x^2 + 2x + 1$
22. $3x^2 + 5x = -1$
23. $5z^2 - 22z = -8$
24. $z(z + 6) + 4 = 0$
25. $(x + 2)(x + 1) = 10$
26. $16x^2 + 40x + 25 = 0$
27. $4x^2 = -1 + 5x$
28. $2p^2 = 2p + 1$
29. $3m(3m + 4) = 7$
30. $5x - 1 + 4x^2 = 0$
31. $\dfrac{r^2}{2} + \dfrac{7r}{4} + \dfrac{11}{8} = 0$
32. $t(15t + 58) = -48$
33. $9k^2 = 16(3k + 4)$
34. $\dfrac{1}{5}x^2 + x + 1 = 0$
35. $x^2 - x + 3 = 0$
36. $4m^2 - 11m + 8 = -2$

37. $-3x^2 + 4x = -4$ **38.** $z^2 - \frac{5}{12}z = \frac{1}{6}$ **39.** $5k^2 + 19k = 2k + 12$

40. $\frac{1}{2}n^2 - n = \frac{15}{2}$ **41.** $k^2 - \frac{4}{15} = -\frac{4}{15}k$ **42.** $(x + 2)(x - 4) = 16$

43. If $D > 0$ and $\frac{5 + \sqrt{D}}{3}$ is a solution of $ax^2 + bx + c = 0$, what must be another solution of the equation?

44. How many real solutions are there for a quadratic equation that has a negative number as its radicand in the quadratic formula?

9.4 Complex Numbers

OBJECTIVES

1. Write complex numbers as multiples of *i*.
2. Add and subtract complex numbers.
3. Multiply complex numbers.
4. Write complex number quotients in standard form.
5. Solve quadratic equations with complex number solutions.

As shown earlier in this chapter, some quadratic equations have no real number solutions. For example, the numbers

$$\frac{4 \pm \sqrt{-4}}{2},$$

which occurred in the solution of Example 5 in the previous section, are not real numbers, because -4 appears as the radicand. For every quadratic equation to have a solution, we need a new set of numbers that includes the real numbers. This new set of numbers is defined using a new number *i* having the properties given below.

$$i = \sqrt{-1} \quad \text{and} \quad i^2 = -1$$

OBJECTIVE 1 Write complex numbers as multiples of *i*. We can write numbers like $\sqrt{-4}, \sqrt{-5},$ and $\sqrt{-8}$ as multiples of *i*, using the properties of *i* to define any square root of a negative number as follows. For any positive real number *b*,

$$\sqrt{-b} = i\sqrt{b}.$$

EXAMPLE 1 Simplifying Square Roots of Negative Numbers

Write each number as a multiple of *i*.

(a) $\sqrt{-4} = i\sqrt{4} = i \cdot 2 = 2i$
(b) $\sqrt{-5} = i\sqrt{5}$
(c) $\sqrt{-8} = i\sqrt{8} = i \cdot 2 \cdot \sqrt{2} = 2i\sqrt{2}$

Now Try Exercises 1 and 3.

CAUTION It is easy to mistake $\sqrt{2}i$ for $\sqrt{2i}$, with the *i* under the radical. For this reason, it is customary to write the *i* factor first when it is multiplied by a radical. For example, we usually write $i\sqrt{2}$ rather than $\sqrt{2}i$.

616 CHAPTER 9 Quadratic Equations

Numbers that are nonzero multiples of *i* are *imaginary numbers*. The *complex numbers* include all real numbers and all imaginary numbers.

> **Complex Number**
>
> A **complex number** is a number of the form $a + bi$, where a and b are real numbers. If $b \neq 0$, $a + bi$ is also an **imaginary number.**

For example, the real number 2 is a complex number since it can be written as $2 + 0i$. Also, the imaginary number $3i = 0 + 3i$ is a complex number. Other imaginary complex numbers are

$$3 - 2i, \quad 1 + i\sqrt{2}, \quad \text{and} \quad -5 + 4i.$$

In the complex number $a + bi$, a is called the **real part** and b (*not bi*) is called the **imaginary part.***

A complex number written in the form $a + bi$ (or $a + ib$) is in **standard form.** Figure 1 shows the relationships among the various types of numbers discussed in this book. (Compare this figure to Figure 4 in Chapter 1.)

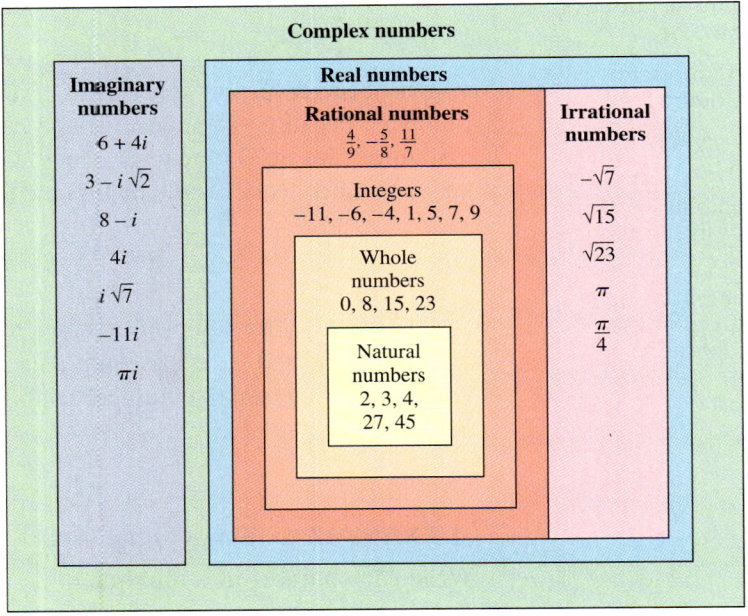

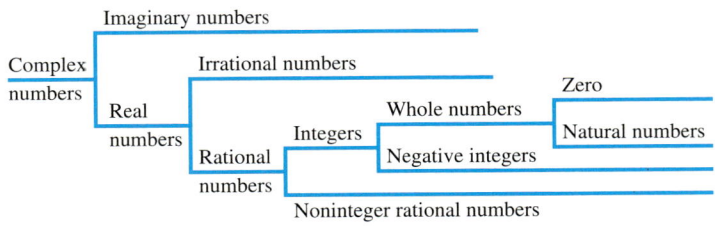

FIGURE 1

---

*Some texts *do* refer to *bi* as the imaginary part.

OBJECTIVE 2 Add and subtract complex numbers. Adding and subtracting complex numbers is similar to adding and subtracting binomials.

> **Adding and Subtracting Complex Numbers**
>
> To add complex numbers, add their real parts and add their imaginary parts.
>
> To subtract complex numbers, change the number following the subtraction sign to its negative, and then add.

The properties of real numbers from Section 1.7 (commutative, associative, etc.) also hold for operations with complex numbers.

EXAMPLE 2   Adding and Subtracting Complex Numbers

Add or subtract.

(a) $(2 - 6i) + (7 + 4i) = (2 + 7) + (-6 + 4)i = 9 - 2i$

(b) $3i + (-2 - i) = -2 + (3 - 1)i = -2 + 2i$

(c) $(2 + 6i) - (-4 + i)$

Change $-4 + i$ to its negative, and then add.

$$(2 + 6i) - (-4 + i) = (2 + 6i) + (4 - i) \quad -(-4 + i) = 4 - i$$
$$= (2 + 4) + (6 - 1)i \quad \text{Commutative, associative, and distributive properties}$$
$$= 6 + 5i$$

(d) $(-1 + 2i) - 4 = (-1 - 4) + 2i = -5 + 2i$

Now Try Exercises 9 and 11.

OBJECTIVE 3 Multiply complex numbers. We multiply complex numbers as we do polynomials. Since $i^2 = -1$ by definition, whenever i^2 appears, we replace it with -1.

EXAMPLE 3   Multiplying Complex Numbers

Find each product.

(a) $3i(2 - 5i) = 6i - 15i^2$ Distributive property
$$= 6i - 15(-1) \quad i^2 = -1$$
$$= 6i + 15$$
$$= 15 + 6i \quad \text{Standard form}$$

(b) $(4 - 3i)(2 + 5i) = 4(2) + 4(5i) + (-3i)2 + (-3i)5i$ Use FOIL.
$$= 8 + 20i - 6i - 15i^2$$
$$= 8 + 14i - 15(-1)$$
$$= 8 + 14i + 15$$
$$= 23 + 14i$$

(c) $(1 + 2i)(1 - 2i) = 1 - 2i + 2i - 4i^2$
$= 1 - 4(-1)$
$= 1 + 4$
$= 5$

Now Try Exercises 15 and 19.

OBJECTIVE 4 Write complex number quotients in standard form. The quotient of two complex numbers is expressed in standard form by changing the denominator into a real number. For example, to write

$$\frac{8 + i}{1 + 2i}$$

in standard form, the denominator must be a real number. As seen in Example 3(c), the product $(1 + 2i)(1 - 2i)$ is 5, a real number. This suggests multiplying the numerator and denominator of the given quotient by $1 - 2i$ as follows.

$$\frac{8 + i}{1 + 2i} = \frac{8 + i}{1 + 2i} \cdot \frac{1 - 2i}{1 - 2i}$$

$$= \frac{8 - 16i + i - 2i^2}{1 - 4i^2} \quad \text{Multiply.}$$

$$= \frac{8 - 16i + i - 2(-1)}{1 - 4(-1)} \quad i^2 = -1$$

$$= \frac{10 - 15i}{5} \quad \text{Combine terms.}$$

$$= \frac{5(2 - 3i)}{5} \quad \text{Factor.}$$

$$= 2 - 3i \quad \text{Standard form}$$

We used a similar method to rationalize some radical expressions in Chapter 8. The complex numbers $1 + 2i$ and $1 - 2i$ are *conjugates*. That is, the **conjugate** of the complex number $a + bi$ is $a - bi$. Multiplying the complex number $a + bi$ by its conjugate $a - bi$ gives the real number $a^2 + b^2$.

Product of Conjugates

$$(a + bi)(a - bi) = a^2 + b^2$$

That is, the product of a complex number and its conjugate is the sum of the squares of the real and imaginary parts.

EXAMPLE 4 Dividing Complex Numbers

Write each quotient in standard form.

(a) $\dfrac{-4 + i}{2 - i}$

Multiply numerator and denominator by $2 + i$, the conjugate of the denominator.

$$\frac{-4 + i}{2 - i} \cdot \frac{2 + i}{2 + i} = \frac{-8 - 4i + 2i + i^2}{4 - i^2}$$

$$= \frac{-8 - 2i - 1}{4 - (-1)} \qquad i^2 = -1$$

$$= \frac{-9 - 2i}{5}$$

$$= -\frac{9}{5} - \frac{2}{5}i \qquad \text{Standard form}$$

(b) $\dfrac{3 + i}{-i}$

Here, the conjugate of $0 - i$ is $0 + i$, or i.

$$\frac{3 + i}{-i} \cdot \frac{i}{i} = \frac{3i + i^2}{-i^2}$$

$$= \frac{-1 + 3i}{-(-1)} \qquad i^2 = -1; \text{ commutative property}$$

$$= -1 + 3i$$

Now Try Exercise 21.

OBJECTIVE 5 Solve quadratic equations with complex number solutions. Quadratic equations that have no real solutions do have complex solutions, as shown in the next examples.

EXAMPLE 5 Solving a Quadratic Equation with Complex Solutions (Square Root Method)

Solve $(x + 3)^2 = -25$.

Use the square root property.

$$(x + 3)^2 = -25$$

$x + 3 = \sqrt{-25}$ or $x + 3 = -\sqrt{-25}$

$x + 3 = 5i$ or $x + 3 = -5i$ $\sqrt{-25} = 5i$

$x = -3 + 5i$ or $x = -3 - 5i$.

The solution set is $\{-3 \pm 5i\}$.

Now Try Exercise 27.

EXAMPLE 6 Solving a Quadratic Equation with Complex Solutions (Quadratic Formula)

Solve $2p^2 = 4p - 5$.

620 CHAPTER 9 Quadratic Equations

Write the equation $2p^2 = 4p - 5$ as $2p^2 - 4p + 5 = 0$. Then $a = 2$, $b = -4$, and $c = 5$. The solutions are

$$p = \frac{-(-4) \pm \sqrt{(-4)^2 - 4(2)(5)}}{2(2)}$$

$$p = \frac{4 \pm \sqrt{16 - 40}}{4}$$

$$p = \frac{4 \pm \sqrt{-24}}{4}.$$

Since $\sqrt{-24} = i\sqrt{24} = i \cdot \sqrt{4} \cdot \sqrt{6} = i \cdot 2 \cdot \sqrt{6} = 2i\sqrt{6}$,

$$p = \frac{4 \pm 2i\sqrt{6}}{4}$$

$$p = \frac{2(2 \pm i\sqrt{6})}{2(2)} \qquad \text{Factor out 2.}$$

$$p = \frac{2 \pm i\sqrt{6}}{2} \qquad \text{Lowest terms}$$

$$p = \frac{2}{2} \pm \frac{i\sqrt{6}}{2} \qquad \text{Separate into real and imaginary parts.}$$

$$p = 1 \pm \frac{\sqrt{6}}{2}i. \qquad \text{Standard form}$$

The solution set is $\left\{1 \pm \frac{\sqrt{6}}{2}i\right\}$.

Now Try Exercise 35.

9.4 EXERCISES

For Extra Help

 Student's Solutions Manual

 MyMathLab

 InterAct Math Tutorial Software

 AW Math Tutor Center

 MathXL

Digital Video Tutor CD 19/Videotape 17

Write each number as a multiple of i. See Example 1.

1. $\sqrt{-9}$ **2.** $\sqrt{-36}$ **3.** $\sqrt{-20}$ **4.** $\sqrt{-27}$
5. $\sqrt{-18}$ **6.** $\sqrt{-50}$ **7.** $\sqrt{-125}$ **8.** $\sqrt{-98}$

Add or subtract as indicated. See Example 2.

9. $(2 + 8i) + (3 - 5i)$ **10.** $(4 + 5i) + (7 - 2i)$
11. $(8 - 3i) - (2 + 6i)$ **12.** $(1 + i) - (3 - 2i)$
13. $(3 - 4i) + (6 - i) - (3 + 2i)$ **14.** $(5 + 8i) - (4 + 2i) + (3 - i)$

Find each product. See Example 3.

15. $(3 + 2i)(4 - i)$ **16.** $(9 - 2i)(3 + i)$ **17.** $(5 - 4i)(3 - 2i)$
18. $(10 + 6i)(8 - 4i)$ **19.** $(3 + 6i)(3 - 6i)$ **20.** $(11 - 2i)(11 + 2i)$

Write each quotient in standard form. See Example 4.

21. $\dfrac{17 + i}{5 + 2i}$ **22.** $\dfrac{21 + i}{4 + i}$ **23.** $\dfrac{40}{2 + 6i}$

24. $\dfrac{13}{3 + 2i}$ **25.** $\dfrac{i}{4 - 3i}$ **26.** $\dfrac{-i}{1 + 2i}$

Solve each quadratic equation for complex solutions by the square root property. Write solutions in standard form. See Example 5.

27. $(a + 1)^2 = -4$ **28.** $(p - 5)^2 = -36$ **29.** $(k - 3)^2 = -5$
30. $(x + 6)^2 = -12$ **31.** $(3x + 2)^2 = -18$ **32.** $(4z - 1)^2 = -20$

Solve each quadratic equation for complex solutions by the quadratic formula. Write solutions in standard form. See Example 6.

33. $m^2 - 2m + 2 = 0$ **34.** $b^2 + b + 3 = 0$ **35.** $2r^2 + 3r + 5 = 0$
36. $3q^2 = 2q - 3$ **37.** $p^2 - 3p + 4 = 0$ **38.** $2a^2 = -a - 3$
39. $5x^2 + 3 = 2x$ **40.** $6x^2 + 2x + 1 = 0$ **41.** $2m^2 + 7 = -2m$
42. $4z^2 + 2z + 3 = 0$ **43.** $r^2 + 3 = r$ **44.** $4q^2 - 2q + 3 = 0$

Exercises 45 and 46 refer to quadratic equations having real number coefficients.

45. Suppose you are solving a quadratic equation by the quadratic formula. How can you tell, before completing the solution, whether the equation will have solutions that are not real numbers?

46. Refer to the solutions in Examples 5 and 6, and complete the following statement: If a quadratic equation has imaginary solutions, they are _____ of each other.

Answer true *or* false *to each of the following. If false, say why.*

47. Every real number is a complex number.

48. Every imaginary number is a complex number.

49. Every complex number is a real number.

50. Some complex numbers are imaginary.

9.5 More on Graphing Quadratic Equations; Quadratic Functions

OBJECTIVES

1. Graph quadratic equations of the form $y = ax^2 + bx + c$ ($a \neq 0$).
2. Use a graph to determine the number of real solutions of a quadratic equation.
3. Use a quadratic function to solve an application.

In Section 5.4 we graphed the quadratic equation $y = x^2$. By plotting points, we obtained the graph of a parabola, shown here in Figure 2.

x	y
3	9
2	4
1	1
0	0
-1	1
-2	4
-3	9

FIGURE 2

622 CHAPTER 9 Quadratic Equations

Recall that the lowest point on this graph is called the **vertex** of the parabola. (If the parabola opens downward, the vertex is the highest point.) The vertical line through the vertex is called the **axis,** or **axis of symmetry.** The two halves of the parabola are mirror images of each other across this axis.

We have also seen that other quadratic equations such as $y = -x^2 + 3$ and $y = (x + 2)^2$ have parabolas as their graphs. See Figures 3 and 4.

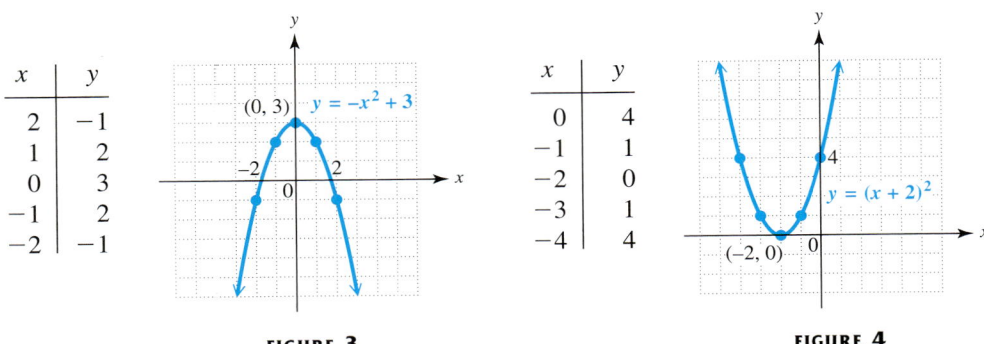

FIGURE 3   FIGURE 4

OBJECTIVE 1 Graph quadratic equations of the form $y = ax^2 + bx + c$ $(a \neq 0)$.
Every equation of the form
$$y = ax^2 + bx + c,$$
with $a \neq 0$, has a graph that is a parabola. As the preceding graphs suggest, the vertex is the most important point to locate when you are graphing a quadratic equation. The next example shows how to find the vertex in a general case.

EXAMPLE 1 Graphing a Parabola by Finding the Vertex and Intercepts

Graph $y = x^2 - 2x - 3$.

First find the vertex of the graph. *Because of its symmetry, if a parabola has two x-intercepts, the x-value of the vertex is exactly halfway between them.* See Figure 3, for example. Therefore, begin by finding the x-intercepts. Let $y = 0$ in the equation and solve for x.

$$0 = x^2 - 2x - 3$$
$$x^2 - 2x - 3 = 0 \quad \text{Reverse sides.}$$
$$(x + 1)(x - 3) = 0 \quad \text{Factor.}$$
$$x + 1 = 0 \quad \text{or} \quad x - 3 = 0 \quad \text{Zero-factor property}$$
$$x = -1 \quad \text{or} \quad x = 3$$

There are two x-intercepts, $(-1, 0)$ and $(3, 0)$.

As mentioned above, the x-value of the vertex is halfway between the x-values of the two x-intercepts. Thus, it is half their sum.

$$x = \frac{1}{2}(-1 + 3) = 1$$

Find the corresponding y-value by substituting 1 for x in the equation.

$$y = 1^2 - 2(1) - 3 = -4$$

The vertex is $(1, -4)$. The axis is the line $x = 1$. To find the y-intercept, substitute $x = 0$ in the equation.
$$y = 0^2 - 2(0) - 3 = -3$$
The y-intercept is $(0, -3)$.

Plot the three intercepts and the vertex. Find additional ordered pairs as needed. For example, if $x = 2$,
$$y = 2^2 - 2(2) - 3 = -3,$$
leading to the ordered pair $(2, -3)$. A table with all these ordered pairs is shown with the graph in Figure 5.

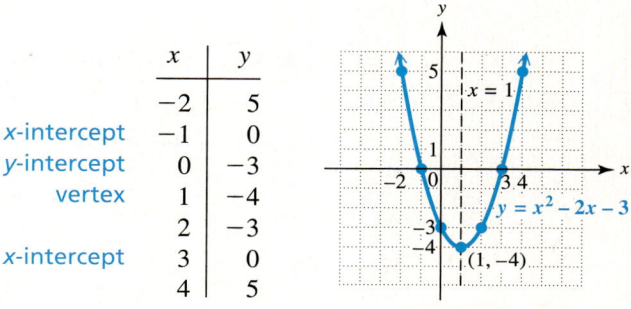

	x	y
	-2	5
x-intercept	-1	0
y-intercept	0	-3
vertex	1	-4
	2	-3
x-intercept	3	0
	4	5

FIGURE 5

Now Try Exercise 7.

We can generalize from Example 1. The x-coordinates of the x-intercepts for the equation $y = ax^2 + bx + c$, by the quadratic formula, are
$$x = \frac{-b + \sqrt{b^2 - 4ac}}{2a} \quad \text{and} \quad x = \frac{-b - \sqrt{b^2 - 4ac}}{2a}.$$
Thus, the x-value of the vertex is
$$x = \frac{1}{2}\left(\frac{-b + \sqrt{b^2 - 4ac}}{2a} + \frac{-b - \sqrt{b^2 - 4ac}}{2a}\right)$$
$$x = \frac{1}{2}\left(\frac{-b + \sqrt{b^2 - 4ac} - b - \sqrt{b^2 - 4ac}}{2a}\right)$$
$$x = \frac{1}{2}\left(\frac{-2b}{2a}\right)$$
$$x = -\frac{b}{2a}.$$
For the equation in Example 1, $y = x^2 - 2x - 3$, $a = 1$ and $b = -2$. Thus, the x-value of the vertex is
$$x = -\frac{b}{2a} = -\frac{-2}{2(1)} = 1,$$
which is the same x-value for the vertex we found in Example 1. (*The x-value of the vertex is $x = -\frac{b}{2a}$ even if the graph has no x-intercepts.*) A procedure for graphing quadratic equations follows.

> **Graphing the Parabola** $y = ax^2 + bx + c$
>
> *Step 1* **Find the vertex.** Let $x = -\frac{b}{2a}$, and find the corresponding y-value by substituting for x in the equation.
>
> *Step 2* **Find the y-intercept.**
>
> *Step 3* **Find the x-intercepts** (if they exist).
>
> *Step 4* **Plot** the intercepts and the vertex.
>
> *Step 5* **Find and plot additional ordered pairs** near the vertex and intercepts as needed, using symmetry about the axis of the parabola.

EXAMPLE 2 Graphing a Parabola

Graph $y = x^2 - 4x + 1$.

The x-value of the vertex is

$$x = -\frac{b}{2a} = -\frac{-4}{2(1)} = 2.$$

The y-value of the vertex is

$$y = 2^2 - 4(2) + 1 = -3,$$

so the vertex is $(2, -3)$. The axis is the line $x = 2$. Now find the intercepts. Let $x = 0$ in $y = x^2 - 4x + 1$ to get the y-intercept $(0, 1)$. Let $y = 0$ to get the x-intercepts. If $y = 0$, the equation is $0 = x^2 - 4x + 1$, which cannot be solved by factoring. Use the quadratic formula to solve for x.

$$x = \frac{4 \pm \sqrt{16 - 4}}{2} \quad \text{Let } a = 1, b = -4, c = 1.$$

$$x = \frac{4 \pm \sqrt{12}}{2}$$

$$x = \frac{4 \pm 2\sqrt{3}}{2} \quad \sqrt{12} = 2\sqrt{3}$$

$$x = \frac{2(2 \pm \sqrt{3})}{2} = 2 \pm \sqrt{3}$$

Use a calculator to find that the x-intercepts are $(3.7, 0)$ and $(.3, 0)$ to the nearest tenth. Plot the intercepts, vertex, and the additional points shown in the table. Connect these points with a smooth curve. The graph is shown in Figure 6.

x	y
-1	6
0	1
$2 - \sqrt{3} \approx .3$	0
1	-2
2	-3
3	-2
$2 + \sqrt{3} \approx 3.7$	0
4	1
5	6

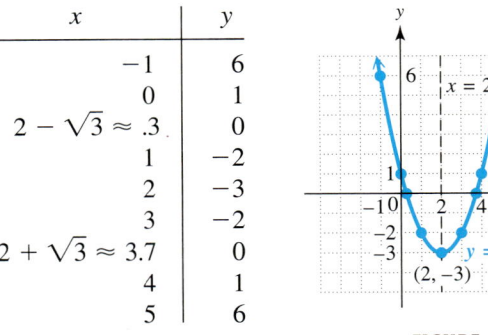

FIGURE 6

Now Try Exercise 9.

SECTION 9.5 More on Graphing Quadratic Equations; Quadratic Functions **625**

OBJECTIVE 2 Use a graph to determine the number of real solutions of a quadratic equation. It can be verified by the vertical line test (Section 3.6) that the graph of an equation of the form $y = ax^2 + bx + c$ is the graph of a function. A function defined by an equation of the form $f(x) = ax^2 + bx + c$ ($a \neq 0$) is called a **quadratic function**. The domain (possible x-values) of a quadratic function is all real numbers, or $(-\infty, \infty)$; the range (the resulting y-values) can be determined after the function is graphed. In Example 2, the domain is $(-\infty, \infty)$ and the range is $[-3, \infty)$.

In Example 2, we found that the x-intercepts of the graph of $y = x^2 - 4x + 1$ (where $y = 0$) are

$$2 - \sqrt{3} \approx .3 \quad \text{and} \quad 2 + \sqrt{3} \approx 3.7.$$

This means that $2 - \sqrt{3} \approx .3$ and $2 + \sqrt{3} \approx 3.7$ are also the solutions of the equation $0 = x^2 - 4x + 1$.

---

x-Intercepts of the Graph of a Quadratic Function

The real number solutions of a quadratic equation $ax^2 + bx + c = 0$ are the x-values of the x-intercepts of the graph of the corresponding quadratic function defined by $f(x) = ax^2 + bx + c$.

Since the graph of a quadratic function can intersect the x-axis in two, one, or no points, this result shows why some quadratic equations have two, some have one, and some have no real solutions.

EXAMPLE 3 Determining the Number of Real Solutions from Graphs

(a) Figure 7 shows the graph of $f(x) = x^2 - 3$. The equation $0 = x^2 - 3$ has two real solutions, $\sqrt{3}$ and $-\sqrt{3}$, which correspond to the x-intercepts. The solution set is $\{\pm\sqrt{3}\}$.

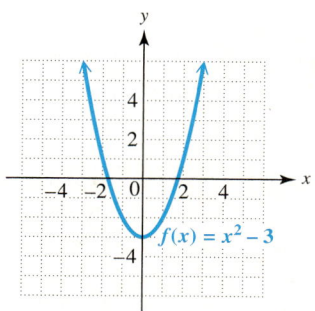

FIGURE 7

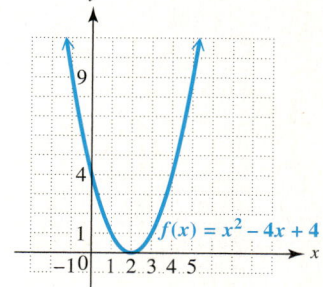

FIGURE 8

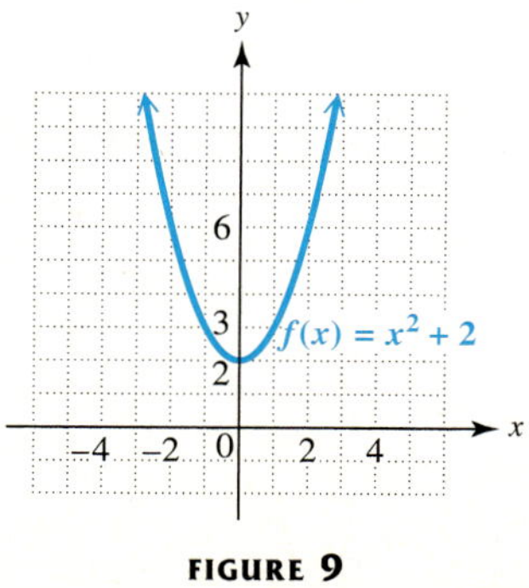

(b) Figure 8 shows the graph of $f(x) = x^2 - 4x + 4$. The corresponding equation $0 = x^2 - 4x + 4$ has one real solution, 2, which is the x-value of the x-intercept of the graph. The solution set is $\{2\}$.

(c) Figure 9 shows the graph of $f(x) = x^2 + 2$. The equation $0 = x^2 + 2$ has no real solutions, since there are no x-intercepts. The solution set over the domain of real numbers is $\emptyset$. (The equation *does* have two imaginary solutions, $i\sqrt{2}$ and $-i\sqrt{2}$.)

FIGURE 9

Now Try Exercises 13, 15, and 17.

OBJECTIVE 3 Use a quadratic function to solve an application. Parabolic shapes are found all around us. Satellite dishes that receive television signals are becoming more popular each year. Radio telescopes use parabolic reflectors to track incoming signals. The final example discusses how to describe a cross section of a parabolic dish using a quadratic function.

EXAMPLE 4 Finding the Equation of a Parabolic Satellite Dish

The Parkes radio telescope has a parabolic dish shape with diameter 210 ft and depth 32 ft. (*Source:* Mar, J., and H. Liebowitz, *Structure Technology for Large Radio and Radar Telescope Systems,* The MIT Press, 1969.) Figure 10(a) shows a diagram of such a dish, and Figure 10(b) shows how a cross section of the dish can be modeled by a graph, with the vertex of the parabola at the origin of a coordinate system. Find the equation of this graph.

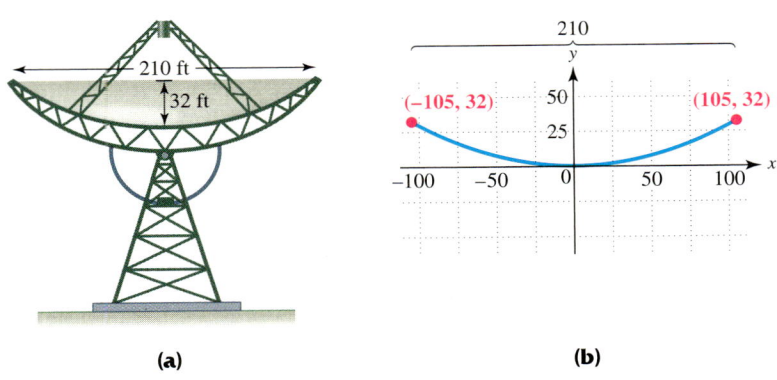

(a) (b)

FIGURE 10

Because the vertex is at the origin, the equation will be of the form
$$y = ax^2.$$
As shown in Figure 10(b), one point on the graph has coordinates (105, 32). Letting $x = 105$ and $y = 32$, we can solve for a.

$y = ax^2$	General equation
$32 = a(105)^2$	Substitute for *x* and *y*.
$32 = 11,025a$	$105^2 = 11,025$
$a = \dfrac{32}{11,025}$	Divide by 11,025.

Thus the equation is $y = \dfrac{32}{11,025}x^2$.

Now Try Exercise 35.

9.5 EXERCISES

For Extra Help

 Student's Solutions Manual

 MyMathLab

 InterAct Math Tutorial Software

 AW Math Tutor Center

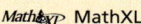 MathXL

 Digital Video Tutor CD 19/Videotape 17

1. In your own words, explain what is meant by the vertex of a parabola.

2. In your own words, explain what is meant by the line of symmetry of a parabola that opens upward or downward.

Give the coordinates of the vertex and sketch the graph of each equation. See Examples 1 and 2.

3. $y = x^2 - 6$
4. $y = -x^2 + 2$
5. $y = (x + 3)^2$
6. $y = (x - 4)^2$
7. $y = x^2 + 2x + 3$
8. $y = x^2 - 4x + 3$
9. $y = x^2 - 8x + 16$
10. $y = x^2 + 6x + 9$
11. $y = -x^2 + 6x - 5$
12. $y = -x^2 - 4x - 3$

Decide from each graph how many real solutions $f(x) = 0$ has. Then give the solution set (of real solutions). See Example 3.

13.
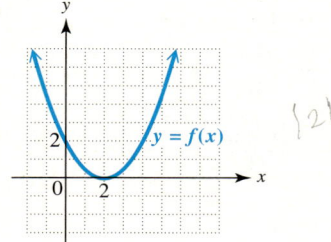

14.
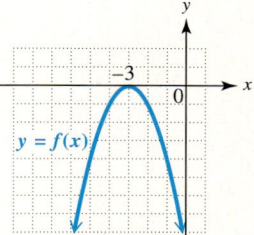

15.
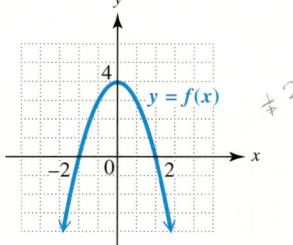

16.
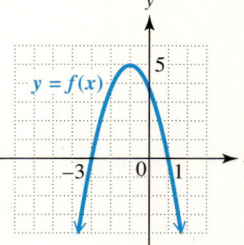

17.
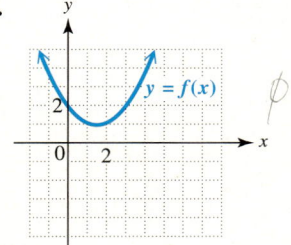

18.
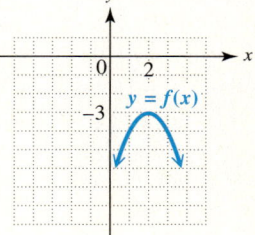

19. Based on your work in Exercises 3–12, what seems to be the direction in which the parabola $y = ax^2 + bx + c$ opens if $a > 0$? if $a < 0$?

20. How many real solutions does a quadratic equation have if its corresponding graph has (a) no *x*-intercepts, (b) one *x*-intercept, (c) two *x*-intercepts? See Examples 1–3.

628 CHAPTER 9 Quadratic Equations

> **TECHNOLOGY INSIGHTS** (EXERCISES 21–24)
>
> *The connection between the solutions of an equation and the x-intercepts of its graph enables us to solve quadratic equations using a graphing calculator. With the equation in the form $ax^2 + bx + c = 0$, enter $ax^2 + bx + c$ as Y_1, and then direct the calculator to find the x-intercepts of the graph. (These are also referred to as* zeros *of the function.) For example, to solve $x^2 - 5x - 6 = 0$ graphically, refer to the three screens below. The displays at the bottoms of the lower two screens show the two solutions, -1 and 6.*

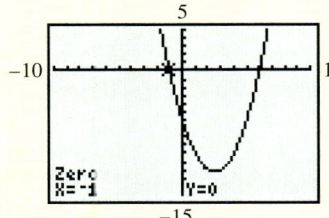 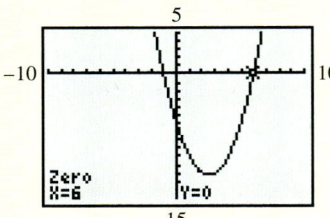

Determine the solution set of each quadratic equation by observing the corresponding screens. Then verify your answers by solving the quadratic equation using the method of your choice.

21. $x^2 - x - 6 = 0$

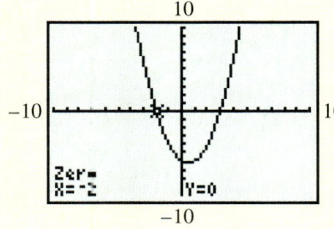 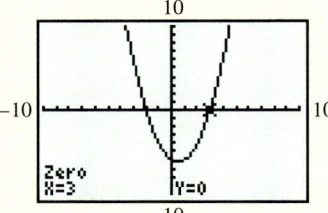

22. $x^2 + 6x + 5 = 0$

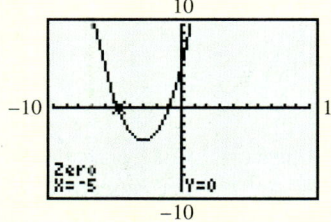 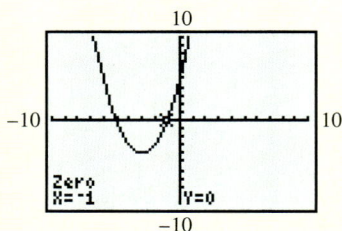

SECTION 9.5 More on Graphing Quadratic Equations; Quadratic Functions **629**

23. $2x^2 - x - 3 = 0$

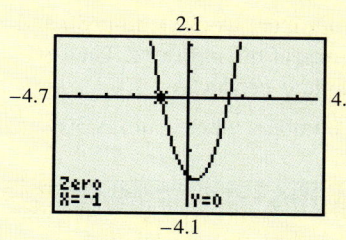 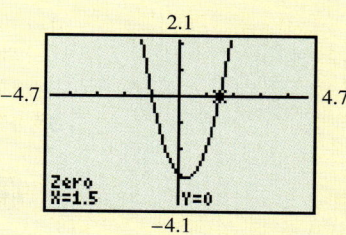

24. $4x^2 - 11x - 3 = 0$

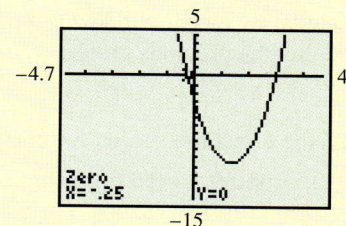 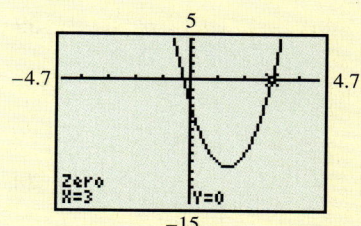

Find the domain and range of each function graphed in the indicated exercise.

25. Exercise 13   **26.** Exercise 14   **27.** Exercise 15
28. Exercise 16   **29.** Exercise 17   **30.** Exercise 18

Given $f(x) = 2x^2 - 5x + 3$, find each of the following.

31. $f(0)$ **32.** $f(1)$ **33.** $f(-2)$ **34.** $f(-1)$

Use a quadratic function to solve each problem. See Example 4.

35. The U.S. Naval Research Laboratory designed a giant radio telescope that had diameter 300 ft and maximum depth 44 ft. The graph depicts a cross section of this telescope. Find the equation of this parabola. (*Source:* Mar, J., and H. Liebowitz, *Structure Technology for Large Radio and Radar Telescope Systems,* The MIT Press, 1969.)

 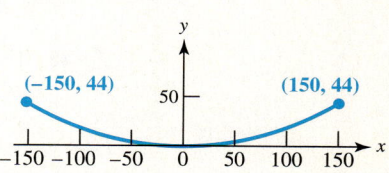

36. Suppose the telescope in Exercise 35 had diameter 400 ft and maximum depth 50 ft. Find the equation of this parabola.

37. Find two numbers whose sum is 80 and whose product is a maximum. (*Hint:* Let x represent one of the numbers. Then $80 - x$ represents the other. A quadratic function represents their product.)

38. Find two numbers whose sum is 300 and whose product is a maximum.

RELATING CONCEPTS (EXERCISES 39–44)

For Individual or Group Work

We can use a graphing calculator to illustrate how the graph of $y = x^2$ can be transformed using arithmetic operations. **Work Exercises 39–44 in order.**

39. In the standard viewing window of your calculator, graph the following one at a time, leaving the previous graphs on the screen as you move along.

$$Y_1 = x^2 \quad Y_2 = 2x^2 \quad Y_3 = 3x^2 \quad Y_4 = 4x^2$$

Describe the effect the successive coefficients have on the parabola.

40. Repeat Exercise 39 for the following.

$$Y_1 = x^2 \quad Y_2 = \frac{1}{2}x^2 \quad Y_3 = \frac{1}{4}x^2 \quad Y_4 = \frac{1}{8}x^2$$

41. In the standard viewing window of your calculator, graph the following pair of parabolas on the same screen.

$$Y_1 = x^2 \quad Y_2 = -x^2$$

In your own words, describe how the graph of Y_2 can be obtained from the graph of Y_1.

42. In the standard viewing window of your calculator, graph the following parabolas on the same screen.

$$Y_1 = -x^2 \quad Y_2 = -2x^2 \quad Y_3 = -3x^2 \quad Y_4 = -4x^2$$

Make a conjecture about what happens when the coefficient of x^2 is negative.

43. In the standard viewing window of your calculator, graph the following one at a time, leaving the previous graphs on the screen as you move along.

$$Y_1 = x^2 \quad Y_2 = x^2 + 3 \quad Y_3 = x^2 - 6$$

Describe the effect that adding or subtracting a constant has on the parabola.

44. Repeat Exercise 43 for the following.

$$Y_1 = x^2 \quad Y_2 = (x + 3)^2 \quad Y_3 = (x - 6)^2$$

45. The equation $x = y^2 - 2y + 3$ has a parabola with a horizontal axis of symmetry as its graph. Sketch the graph of this equation by letting y take on the values $-1, 0, 1, 2,$ and 3. Then write an explanation of why this is not the graph of a function.

Chapter 9 Group Activity

How Is a Radio Telescope Like a Wok?

OBJECTIVE Determine appropriate domains and ranges when graphing parabolas on a graphing calculator.

In Example 4 of Section 9.5 we explored the parabolic shape of a radio telescope. This activity uses that example and another to show the importance of setting appropriate domains and ranges on a graphing calculator. One student should write the answers to the questions while the other one does the graphing on a calculator. When you start part B, switch tasks.

A. In Example 4 of Section 9.5, an equation was found to represent the shape of the Parkes radio telescope. Using this equation, graph the parabola on a graphing calculator.

1. Use a standard viewing window, that is, Xmin = −10, Xmax = 10, Ymin = −10, Ymax = 10. Describe your graph. Does its shape look similar to the picture of the telescope in Section 9.5?

2. Change the domain and range settings to Xmin = −50, Xmax = 50, Ymin = −5, Ymax = 50. How does the graph look now? You may want to change Xscl and Yscl to 5.

3. Continue to adjust the domain and range settings until the calculator graph is close in shape to the picture of the telescope in Section 9.5. What are your settings?

B. A wok used in oriental cooking has a parabolic shape. Just as the shape of a parabola focuses radio waves from space, a wok focuses heat and oil for cooking.

1. Find an equation that would model the cross section of a wok with diameter 14 in. and depth 4 in.

2. Graph this equation using a standard viewing window on a calculator. Describe the graph of the equation. Describe the differences between this graph and the graph in part A, Exercise 1.

3. Adjust the settings for domain and range until the graph looks similar to the cross section of a wok.

4. How are a wok and a radio telescope alike?

5. How is the range (or domain) of a function different from the range (or domain) setting on a graphing calculator?

CHAPTER 9 SUMMARY

KEY TERMS

9.1 quadratic equation
9.2 completing the square
9.3 discriminant
9.4 complex number

imaginary number
real part
imaginary part

standard form (of a complex number)
conjugate (of a complex number)

9.5 parabola
vertex
axis (of symmetry)
quadratic function

NEW SYMBOLS

$\pm$ positive or negative (plus or minus)

i $i = \sqrt{-1}$ and $i^2 = -1$

TEST YOUR WORD POWER

See how well you have learned the vocabulary in this chapter. Answers, with examples, follow the Quick Review.

1. A **quadratic equation** is an equation that can be written in the form
   A. $Ax + By = C$
   B. $ax^2 + bx + c = 0$
   C. $Ax + B = 0$
   D. $y = mx + b$.

2. A **complex number** is
   A. a real number that includes a complex fraction
   B. a nonzero multiple of i
   C. a number of the form $a + bi$, where a and b are real numbers
   D. the square root of -1.

3. An **imaginary number** is
   A. a complex number $a + bi$ where $b \neq 0$
   B. a number that does not exist
   C. a complex number $a + bi$ where $b = 0$
   D. any real number.

4. A **parabola** is the graph of
   A. any equation in two variables
   B. a linear equation
   C. an equation of degree three
   D. a quadratic equation.

5. The **vertex** of a parabola is
   A. the point where the graph intersects the y-axis
   B. the point where the graph intersects the x-axis
   C. the lowest point on a parabola that opens up or the highest point on a parabola that opens down
   D. the origin.

6. The **axis** of a vertical parabola is
   A. either the x-axis or the y-axis
   B. the vertical line through the vertex
   C. the horizontal line through the vertex
   D. the x-axis.

QUICK REVIEW

CONCEPTS	EXAMPLES

9.1 SOLVING QUADRATIC EQUATIONS BY THE SQUARE ROOT PROPERTY

Square Root Property of Equations
If k is positive, and if $x^2 = k$, then
$$x = \sqrt{k} \quad \text{or} \quad x = -\sqrt{k}.$$

Solve $(2x + 1)^2 = 5$.

$2x + 1 = \sqrt{5}$ or $2x + 1 = -\sqrt{5}$
$2x = -1 + \sqrt{5}$ or $2x = -1 - \sqrt{5}$
$x = \dfrac{-1 + \sqrt{5}}{2}$ or $x = \dfrac{-1 - \sqrt{5}}{2}$

Solution set: $\left\{ \dfrac{-1 \pm \sqrt{5}}{2} \right\}$

CHAPTER 9   Summary   **633**

CONCEPTS	EXAMPLES

9.2 SOLVING QUADRATIC EQUATIONS BY COMPLETING THE SQUARE

Completing the Square

Step 1   If the coefficient of the squared term is 1, go to Step 2. If it is not 1, divide each side of the equation by this coefficient.

Step 2   Make sure that all variable terms are on one side of the equation and all constant terms are on the other.

Step 3   Take half the coefficient of x, square it, and add the square to each side of the equation. Factor the variable side and combine terms on the other side.

Step 4   Use the square root property to solve the equation.

Solve $2x^2 + 4x - 1 = 0$.

$x^2 + 2x - \dfrac{1}{2} = 0$ Divide by 2.

$x^2 + 2x = \dfrac{1}{2}$ Add $\tfrac{1}{2}$.

$x^2 + 2x + 1 = \dfrac{1}{2} + 1$ Add $\left[\tfrac{1}{2}(2)\right]^2 = 1$.

$(x + 1)^2 = \dfrac{3}{2}$ Factor; add.

$x + 1 = \sqrt{\dfrac{3}{2}} = \dfrac{\sqrt{6}}{2}$ or $x + 1 = -\sqrt{\dfrac{3}{2}} = -\dfrac{\sqrt{6}}{2}$

$x = -1 + \dfrac{\sqrt{6}}{2}$ or $x = -1 - \dfrac{\sqrt{6}}{2}$

$x = \dfrac{-2 + \sqrt{6}}{2}$ or $x = \dfrac{-2 - \sqrt{6}}{2}$

Solution set: $\left\{\dfrac{-2 \pm \sqrt{6}}{2}\right\}$

9.3 SOLVING QUADRATIC EQUATIONS BY THE QUADRATIC FORMULA

Quadratic Formula
The solutions of $ax^2 + bx + c = 0$, $a \neq 0$, are

$$x = \dfrac{-b \pm \sqrt{b^2 - 4ac}}{2a}.$$

The discriminant for the quadratic equation is $b^2 - 4ac$.

Solve $3x^2 - 4x - 2 = 0$.

$x = \dfrac{-(-4) \pm \sqrt{(-4)^2 - 4(3)(-2)}}{2(3)}$ $a = 3, b = -4, c = -2$

$x = \dfrac{4 \pm \sqrt{16 + 24}}{6}$

$x = \dfrac{4 \pm \sqrt{40}}{6} = \dfrac{4 \pm 2\sqrt{10}}{6}$ Discriminant: 40

$x = \dfrac{2(2 \pm \sqrt{10})}{2(3)} = \dfrac{2 \pm \sqrt{10}}{3}$ Factor; then divide out 2.

Solution set: $\left\{\dfrac{2 \pm \sqrt{10}}{3}\right\}$

9.4 COMPLEX NUMBERS

The number $i = \sqrt{-1}$ and $i^2 = -1$.
For the positive number b,
$$\sqrt{-b} = i\sqrt{b}.$$

$\sqrt{-19} = i\sqrt{19}$

(continued)

CONCEPTS	EXAMPLES
Addition Add complex numbers by adding the real parts and adding the imaginary parts.	Add: $(3 + 6i) + (-9 + 2i)$. $(3 + 6i) + (-9 + 2i) = (3 - 9) + (6 + 2)i$ $\qquad\qquad\qquad\qquad\quad = -6 + 8i$
Subtraction To subtract complex numbers, change the number following the subtraction sign to its negative and add.	Subtract: $(5 + 4i) - (2 - 4i)$. $(5 + 4i) - (2 - 4i) = (5 + 4i) + (-2 + 4i)$ $\qquad\qquad\qquad\qquad\quad = (5 - 2) + (4 + 4)i$ $\qquad\qquad\qquad\qquad\quad = 3 + 8i$
Multiplication Multiply complex numbers in the same way polynomials are multiplied. Replace i^2 with -1.	Multiply: $(7 + i)(3 - 4i)$. $(7 + i)(3 - 4i)$ $= 7(3) + 7(-4i) + i(3) + i(-4i)$ FOIL $= 21 - 28i + 3i - 4i^2$ $= 21 - 25i - 4(-1)$ $i^2 = -1$ $= 21 - 25i + 4$ $= 25 - 25i$
Division Divide complex numbers by multiplying the numerator and the denominator by the conjugate of the denominator.	Divide: $\dfrac{2}{6 + i}$. $\dfrac{2}{6 + i} = \dfrac{2}{6 + i} \cdot \dfrac{6 - i}{6 - i}$ $\quad\;\; = \dfrac{2(6 - i)}{36 - i^2}$ $\quad\;\; = \dfrac{12 - 2i}{36 + 1}$ $\quad\;\; = \dfrac{12 - 2i}{37}$ $\quad\;\; = \dfrac{12}{37} - \dfrac{2}{37}i$ Standard form
Complex Solutions A quadratic equation may have nonreal, complex solutions. This occurs when the discriminant is negative. The quadratic formula will give complex solutions in such cases.	Solve for all complex solutions of $$x^2 + x + 1 = 0.$$ Here, $a = 1$, $b = 1$, and $c = 1$. $x = \dfrac{-1 \pm \sqrt{1^2 - 4(1)(1)}}{2(1)}$ $x = \dfrac{-1 \pm \sqrt{1 - 4}}{2}$ $x = \dfrac{-1 \pm \sqrt{-3}}{2}$ $x = \dfrac{-1 \pm i\sqrt{3}}{2}$ Solution set: $\left\{-\dfrac{1}{2} \pm \dfrac{\sqrt{3}}{2}i\right\}$

CONCEPTS	EXAMPLES

9.5 MORE ON GRAPHING QUADRATIC EQUATIONS; QUADRATIC FUNCTIONS

To graph $y = ax^2 + bx + c$,

Step 1 Find the vertex: $x = -\frac{b}{2a}$; find y by substituting this value for x in the equation.

Graph $y = 2x^2 - 5x - 3$.

$$x = -\frac{b}{2a} = -\frac{-5}{2(2)} = \frac{5}{4}$$

$$y = 2\left(\frac{5}{4}\right)^2 - 5\left(\frac{5}{4}\right) - 3$$

$$= 2\left(\frac{25}{16}\right) - \frac{25}{4} - 3$$

$$= \frac{25}{8} - \frac{50}{8} - \frac{24}{8} = -\frac{49}{8}$$

The vertex is $\left(\frac{5}{4}, -\frac{49}{8}\right)$.

Step 2 Find the y-intercept.

$y = 2(0)^2 - 5(0) - 3 = -3$

The y-intercept is $(0, -3)$.

Step 3 Find the x-intercepts (if they exist).

$$0 = 2x^2 - 5x - 3$$
$$0 = (2x + 1)(x - 3)$$
$$2x + 1 = 0 \quad \text{or} \quad x - 3 = 0$$
$$2x = -1 \quad \text{or} \quad x = 3$$
$$x = -\frac{1}{2} \quad \text{or} \quad x = 3$$

The x-intercepts are $\left(-\frac{1}{2}, 0\right)$ and $(3, 0)$.

Step 4 Plot the intercepts and the vertex.

Step 5 Find and plot additional ordered pairs near the vertex and intercepts as needed.

x	y
$-\frac{1}{2}$	0
0	-3
1	-6
$\frac{5}{4}$	$-\frac{49}{8}$
2	-5
3	0

$y = 2x^2 - 5x - 3$, vertex $\left(\frac{5}{4}, -\frac{49}{8}\right)$

Answers to Test Your Word Power

1. B; *Examples:* $z^2 + 6z + 9 = 0$, $y^2 - 2y = 8$, $(x + 3)(x - 1) = 5$ 2. C; *Examples:* -5 (or $-5 + 0i$), $7i$ (or $0 + 7i$), $\sqrt{2} - 4i$
3. A; *Examples:* $2i$, $-13i$, $3 + i\sqrt{6}$ 4. D; *Examples:* See Figures 2–9 in Section 9.5.   5. C; *Example:* The graph of $y = (x + 3)^2$ has vertex $(-3, 0)$, which is the lowest point on the graph.   6. B; *Example:* The axis of the graph of $y = (x + 3)^2$ is the line $x = -3$.

CHAPTER 9 REVIEW EXERCISES

[9.1] *Solve each equation by using the square root property. Give only real number solutions. Express all radicals in simplest form.*

1. $z^2 = 144$
2. $x^2 = 37$
3. $m^2 = 128$
4. $(k + 2)^2 = 25$
5. $(r - 3)^2 = 10$
6. $(2p + 1)^2 = 14$
7. $(3k + 2)^2 = -3$
8. $(5x + 3)^2 = 0$

[9.2] *Solve each equation by completing the square. Give only real number solutions.*

9. $m^2 + 6m + 5 = 0$
10. $p^2 + 4p = 7$
11. $-x^2 + 5 = 2x$
12. $2z^2 - 3 = -8z$
13. $5k^2 - 3k - 2 = 0$
14. $(4a + 1)(a - 1) = -7$

Solve each problem.

15. If an object is thrown upward on Earth from a height of 50 ft, with an initial velocity of 32 ft per sec, then its height after t sec is given by $h = -16t^2 + 32t + 50$, where h is in feet. After how many seconds will it reach a height of 30 ft?

16. Find the lengths of the three sides of the right triangle shown.

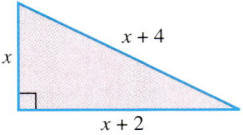

17. What must be added to $x^2 + 3x$ to make it a perfect square?

[9.3]
18. Consider the equation $x^2 - 9 = 0$.
    (a) Solve the equation by factoring.
    (b) Solve the equation by the square root property.
    (c) Solve the equation by the quadratic formula.
    (d) Compare your answers. If a quadratic equation can be solved by both the factoring and the quadratic formula methods, should you always get the same results? Explain.

Solve each equation by using the quadratic formula. Give only real number solutions.

19. $x^2 - 2x - 4 = 0$
20. $3k^2 + 2k = -3$
21. $2p^2 + 8 = 4p + 11$
22. $-4x^2 + 7 = 2x$
23. $\frac{1}{4}p^2 = 2 - \frac{3}{4}p$
24. $3x^2 - x - 2 = 0$

25. Why is this not the statement of the quadratic formula for $ax^2 + bx + c = 0$?

$$x = -b \pm \frac{\sqrt{b^2 - 4ac}}{2a}$$

[9.4] *Perform each indicated operation.*

26. $(3 + 5i) + (2 - 6i)$
27. $(-2 - 8i) - (4 - 3i)$
28. $(6 - 2i)(3 + i)$
29. $(2 + 3i)(2 - 3i)$
30. $\dfrac{1 + i}{1 - i}$
31. $\dfrac{5 + 6i}{2 + 3i}$

32. What is the conjugate of the real number a?

33. Is it possible to multiply a complex number by its conjugate and get an imaginary product? Explain.

Find the complex solutions of each quadratic equation.

34. $(m + 2)^2 = -3$
35. $(3p - 2)^2 = -8$
36. $3k^2 = 2k - 1$
37. $h^2 + 3h = -8$
38. $4q^2 + 2 = 3q$
39. $9z^2 + 2z + 1 = 0$

[9.5] *Identify the vertex and sketch the graph of each equation.*

40. $y = -3x^2$
41. $y = -x^2 + 5$
42. $y = (x + 4)^2$
43. $y = x^2 - 2x + 1$
44. $y = -x^2 + 2x + 3$
45. $y = x^2 + 4x + 2$

Decide from the graph how many real number solutions the equation $f(x) = 0$ has. Determine the solution set (of real solutions) for $f(x) = 0$ from the graph.

46.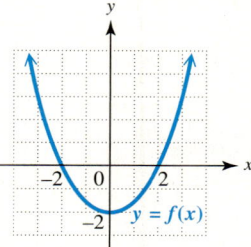
47.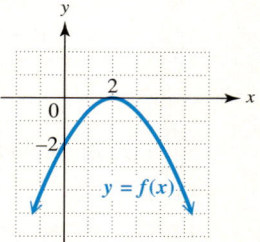
48.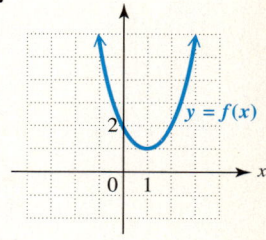

49. Give the domain and range of the graphs in Exercises 46–48.

50. Refer to Example 4 and Exercise 35 in Section 9.5. Suppose that a telescope has diameter 200 ft and maximum depth 30 ft. Find the equation for a cross section of the parabolic dish.

MIXED REVIEW EXERCISES

Solve by any method. Give only real number solutions.

51. $(2t - 1)(t + 1) = 54$
52. $(2p + 1)^2 = 100$
53. $(k + 2)(k - 1) = 3$
54. $6t^2 + 7t - 3 = 0$
55. $2x^2 + 3x + 2 = x^2 - 2x$
56. $x^2 + 2x + 5 = 7$
57. $m^2 - 4m + 10 = 0$
58. $k^2 - 9k + 10 = 0$
59. $(3x + 5)^2 = 0$
60. $\frac{1}{2}r^2 = \frac{7}{2} - r$
61. $x^2 + 4x = 1$
62. $7x^2 - 8 = 5x^2 + 8$

63. Becky and Brad are the owners of Cole's Baseball Cards. They have found that the price p, in dollars, of a particular Brad Radke baseball card depends on the demand d, in hundreds, for the card, according to the function defined by $p = -(d - 6)^2 + 10$. What demand produces a price of $6 for the card?

64. Find the vertex of the parabola from Exercise 63. Give the corresponding demand and price.

CHAPTER 9 TEST

Items marked * require knowledge of complex numbers.

Solve by using the square root property.

1. $x^2 = 39$
2. $(z + 3)^2 = 64$
3. $(4x + 3)^2 = 24$

Solve by completing the square.

4. $x^2 - 4x = 6$
5. $2x^2 + 12x - 3 = 0$

Solve by the quadratic formula.

6. $5x^2 + 2x = 0$
7. $2x^2 + 5x - 3 = 0$
8. $3w^2 + 2 = 6w$
*9. $4x^2 + 8x + 11 = 0$
10. $t^2 - \frac{5}{3}t + \frac{1}{3} = 0$

Solve by the method of your choice.

11. $p^2 - 2p - 1 = 0$
12. $(2x + 1)^2 = 18$
13. $(x - 5)(2x - 1) = 1$
14. $t^2 + 25 = 10t$

Solve each problem.

15. If an object is propelled vertically into the air from ground level on Earth with an initial velocity of 64 ft per sec, its height s (in feet) after t sec is given by the formula
$$s = -16t^2 + 64t.$$
After how many seconds will the object reach a height of 64 ft?

16. Find the lengths of the three sides of the right triangle.

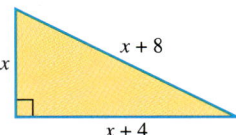

*Perform each indicated operation.

17. $(3 + i) + (-2 + 3i) - (6 - i)$
18. $(6 + 5i)(-2 + i)$
19. $(3 - 8i)(3 + 8i)$
20. $\dfrac{15 - 5i}{7 + i}$

Identify the vertex and sketch the graph of each equation.

21. $y = x^2 - 6x + 9$
22. $y = -x^2 - 2x - 4$
23. $f(x) = x^2 + 6x + 7$

24. Refer to the equation in Exercise 23, and do the following:
    (a) Determine the number of real solutions of $x^2 + 6x + 7 = 0$ by looking at the graph.
    (b) Use the quadratic formula to find the exact values of the real solutions. Give the solution set.
    (c) Use a calculator to find approximations for the solutions. Round your answers to the nearest thousandth.
25. Find two numbers whose sum is 400 and whose product is a maximum.

CUMULATIVE REVIEW EXERCISES CHAPTERS 1–9

Note: **This cumulative review exercise set can be considered a final examination for the course.**

Perform each indicated operation.

1. $\dfrac{-4 \cdot 3^2 + 2 \cdot 3}{2 - 4 \cdot 1}$
2. $|-3| - |1 - 6|$
3. $-9 - (-8)(2) + 6 - (6 + 2)$
4. $-4r + 14 + 3r - 7$
5. $13k - 4k + k - 14k + 2k$
6. $5(4m - 2) - (m + 7)$

Solve each equation.

7. $x - 5 = 13$
8. $3k - 9k - 8k + 6 = -64$
9. $\dfrac{3}{5}t - \dfrac{1}{10} = \dfrac{3}{2}$
10. $2(m - 1) - 6(3 - m) = -4$

Solve each problem.

11. Find the measures of the marked angles.

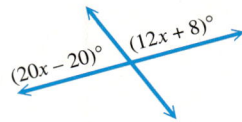

12. The perimeter of a basketball court is 288 ft. The width of the court is 44 ft less than the length. What are the dimensions of the court?

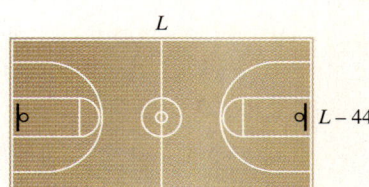

13. Solve the formula $P = 2L + 2W$ for L.

Solve each inequality and graph the solution set.

14. $-8m < 16$
15. $-9p + 2(8 - p) - 6 \geq 4p - 50$

Graph the following.

16. $2x + 3y = 6$ **17.** $y = 3$ **18.** $2x - 5y < 10$

19. Find the slope of the line through $(-1, 4)$ and $(5, 2)$.

20. Write an equation of a line with slope 2 and y-intercept $(0, 3)$. Give it in the form $Ax + By = C$.

Solve each system of equations.

21. $\begin{aligned} 2x + y &= -4 \\ -3x + 2y &= 13 \end{aligned}$

22. $\begin{aligned} 3x - 5y &= 8 \\ -6x + 10y &= 16 \end{aligned}$

23. Based on the prices in a 2002 Best Buy ad, you can purchase 3 GE phones and 2 Motorola phones for $69.95. You can purchase 2 GE phones and 3 Motorola phones for $79.95. Find the price for a single phone of each model.

24. Graph the solution set of the system of inequalities.

$$2x + y \leq 4$$
$$x - y > 2$$

Simplify each expression. Write answers with positive exponents.

25. $(3^2 \cdot x^{-4})^{-1}$

26. $\left(\dfrac{b^{-3}c^4}{b^5c^3}\right)^{-2}$

27. $\left(\dfrac{5}{3}\right)^{-3}$

Perform each indicated operation.

28. $(5x^5 - 9x^4 + 8x^2) - (9x^2 + 8x^4 - 3x^5)$

29. $(2x - 5)(x^3 + 3x^2 - 2x - 4)$

30. $\dfrac{3x^3 + 10x^2 - 7x + 4}{x + 4}$

31. (a) The number of possible hands in contract bridge is about 6,350,000,000. Write this number in scientific notation.
(b) The body of a 150-lb person contains about 2.3×10^{-4} lb of copper. Write this number without using exponents.

Factor.

32. $16x^3 - 48x^2y$

33. $2a^2 - 5a - 3$

34. $16x^4 - 1$

35. $25m^2 - 20m + 4$

Solve.

36. $x^2 + 3x - 54 = 0$

37. The length of a rectangle is 2.5 times its width. The area is 1000 m². Find the length.

Simplify each expression as much as possible.

38. $\dfrac{2}{a - 3} \div \dfrac{5}{2a - 6}$

39. $\dfrac{1}{k} - \dfrac{2}{k - 1}$

40. $\dfrac{2}{a^2 - 4} + \dfrac{3}{a^2 - 4a + 4}$

41. $\dfrac{\dfrac{1}{a} + \dfrac{1}{b}}{\dfrac{1}{a} - \dfrac{1}{b}}$

Solve.

42. $\dfrac{1}{x+3} + \dfrac{1}{x} = \dfrac{7}{10}$

43. Jake's boat goes 12 mph in still water. Find the speed of the current of the river if he can go 6 mi upstream in the same time that he can go 10 mi downstream.

Simplify each expression as much as possible.

44. $\sqrt{100}$

45. $\dfrac{6\sqrt{6}}{\sqrt{5}}$

46. $\sqrt[3]{\dfrac{7}{16}}$

47. $3\sqrt{5} - 2\sqrt{20} + \sqrt{125}$

48. $\sqrt[3]{16a^3b^4} - \sqrt[3]{54a^3b^4}$

49. Solve $\sqrt{x+2} = x - 4$.

50. Simplify.
   (a) $8^{2/3}$ (b) $-16^{1/4}$

Solve each quadratic equation using the method indicated. Give only real solutions.

51. $(3x+2)^2 = 12$ (square root property)
52. $-x^2 + 5 = 2x$ (completing the square)
53. $2x(x-2) - 3 = 0$ (quadratic formula)
54. $(4x+1)(x-1) = -3$ (any method)

*Solve each problem. (Items marked * require knowledge of complex numbers.)*

55. In a right triangle, the lengths of the sides are consecutive integers. Use the Pythagorean formula to find these lengths.

*56. Write in standard form.
   (a) $(-9 + 3i) + (4 + 2i) - (-5 - 3i)$ (b) $\dfrac{-17 - i}{-3 + i}$

*57. Find the complex solutions of $2x^2 + 2x = -9$.

58. Graph the quadratic function defined by $f(x) = -x^2 - 2x + 1$ and identify the vertex. Give the domain and range.

Appendix A
An Introduction to Calculators

There is little doubt that the appearance of handheld calculators three decades ago and the later development of scientific and graphing calculators have changed the methods of learning and studying mathematics forever. For example, computations with tables of logarithms and slide rules made up an important part of mathematics courses prior to 1970. Today, with the widespread availability of calculators, these topics are studied only for their historical significance.

Calculators come in a large array of different types, sizes, and prices. *For the course for which this textbook is intended, the most appropriate type is the scientific calculator,* which costs $10–$20.

In this introduction, we explain some of the features of scientific and graphing calculators. However, remember that calculators vary among manufacturers and models, and that while the methods explained here apply to many of them, they may not apply to your specific calculator. *This introduction is only a guide and is not intended to take the place of your owner's manual.* Always refer to the manual in the event you need an explanation of how to perform a particular operation.

Scientific Calculators

Scientific calculators are capable of much more than the typical four-function calculator that you might use for balancing your checkbook. Most scientific calculators use *algebraic logic.* (Models sold by Texas Instruments, Sharp, Casio, and Radio Shack, for example, use algebraic logic.) A notable exception is Hewlett-Packard, a company whose calculators use *Reverse Polish Notation* (RPN). In this introduction, we explain the use of calculators with algebraic logic.

Arithmetic Operations To perform an operation of arithmetic, simply enter the first number, press the operation key, $(+)$, $(-)$, $(\times)$, or $(\div)$, enter the second number, and then press the $(=)$ key. For example, to add 4 and 3, use the following keystrokes.

$$(4)\ (+)\ (3)\ (=)\ \boxed{7}$$

Change Sign Key The key marked $(+/-)$ allows you to change the sign of a display. This is particularly useful when you wish to enter a negative number. For example,

643

to enter −3, use the following keystrokes.

$$\boxed{3} \quad \boxed{+/-} \quad \boxed{-3}$$

Memory Key Scientific calculators can hold a number in memory for later use. The label of the memory key varies among models; two of these are $\boxed{M}$ and $\boxed{STO}$. The $\boxed{M+}$ and $\boxed{M-}$ keys allow you to add to or subtract from the value currently in memory. The memory recall key, labeled $\boxed{MR}$, $\boxed{RM}$, or $\boxed{RCL}$, allows you to retrieve the value stored in memory.

Suppose that you wish to store the number 5 in memory. Enter 5, then press the key for memory. You can then perform other calculations. When you need to retrieve the 5, press the key for memory recall.

If a calculator has a constant memory feature, the value in memory will be retained even after the power is turned off. Some advanced calculators have more than one memory. Read the owner's manual for your model to see exactly how memory is activated.

Clearing/Clear Entry Keys The keys $\boxed{C}$ or $\boxed{CE}$ allow you to clear the display or clear the last entry entered into the display. In some models, pressing the $\boxed{C}$ key once will clear the last entry, while pressing it twice will clear the entire operation in progress.

Second Function Key This key, usually marked $\boxed{2nd}$, is used in conjunction with another key to activate a function that is printed *above* an operation key (and not on the key itself). For example, suppose you wish to find the square of a number, and the squaring function (explained in more detail later) is printed above another key. You would need to press $\boxed{2nd}$ before the desired squaring function can be activated.

Square Root Key Pressing $\boxed{\sqrt{}}$ or $\boxed{\sqrt{x}}$ will give the square root (or an approximation of the square root) of the number in the display. On many newer scientific calculators, the square root key is pressed *before* entering the number, while other calculators use the opposite order. Experiment with your calculator to see which method it uses. For example, to find the square root of 36, use the following keystrokes.

$$\boxed{\sqrt{}} \; \boxed{3} \; \boxed{6} \quad \boxed{6} \quad \text{or} \quad \boxed{3} \; \boxed{6} \; \boxed{\sqrt{}} \quad \boxed{6}$$

The square root of 2 is an example of an irrational number (Chapter 8). The calculator will give an approximation of its value, since the decimal for $\sqrt{2}$ never terminates and never repeats. The number of digits shown will vary among models. To find an approximation for $\sqrt{2}$, use the following keystrokes.

$$\boxed{\sqrt{}} \; \boxed{2} \quad \boxed{1.4142136} \quad \text{or} \quad \boxed{2} \; \boxed{\sqrt{}} \quad \boxed{1.4142136}$$

An approximation for $\sqrt{2}$

Squaring Key The $\boxed{x^2}$ key allows you to square the entry in the display. For example, to square 35.7, use the following keystrokes.

$$\boxed{3} \; \boxed{5} \; \boxed{.} \; \boxed{7} \; \boxed{x^2} \quad \boxed{1274.49}$$

The squaring key and the square root key are often found on the same key, with one of them being a second function (that is, activated by the second function key previously described).

Reciprocal Key The key marked (1/x) is the reciprocal key. (When two numbers have a product of 1, they are called *reciprocals*. See Chapter 1.) Suppose that you wish to find the reciprocal of 5. Use the following keystrokes.

$$\boxed{5} \quad \boxed{1/x} \quad \boxed{0.2}$$

Inverse Key Some calculators have an inverse key, marked (INV). Inverse operations are operations that "undo" each other. For example, the operations of squaring and taking the square root are inverse operations. The use of the (INV) key varies among different models of calculators, so read your owner's manual carefully.

Exponential Key The key marked (x^y) or (y^x) allows you to raise a number to a power. For example, if you wish to raise 4 to the fifth power (that is, find 4^5, as explained in Chapter 1), use the following keystrokes.

$$\boxed{4} \quad \boxed{x^y} \quad \boxed{5} \quad \boxed{=} \quad \boxed{1024}$$

Root Key Some calculators have a key specifically marked ($\sqrt[x]{}$) or ($\sqrt[y]{}$); with others, the operation of taking roots is accomplished by using the inverse key in conjunction with the exponential key. Suppose, for example, your calculator is of the latter type and you wish to find the fifth root of 1024. Use the following keystrokes.

$$\boxed{1} \quad \boxed{0} \quad \boxed{2} \quad \boxed{4} \quad \boxed{INV} \quad \boxed{x^y} \quad \boxed{5} \quad \boxed{=} \quad \boxed{4}$$

Notice how this "undoes" the operation explained in the exponential key discussion.

Pi Key The number π is an important number in mathematics. It occurs, for example, in the area and circumference formulas for a circle. By pressing the (π) key, you can display the first few digits of π. (Because π is irrational, the display shows only an approximation.) One popular model gives the following display when the (π) key is pressed.

$$\boxed{3.1415927} \quad \text{An approximation for } \pi$$

Methods of Display When decimal approximations are shown on scientific calculators, they are either *truncated* or *rounded*. To see how a particular model is programmed, evaluate 1/18 as an example. If the display shows .0555555 (last digit 5), it truncates the display. If it shows .0555556 (last digit 6), it rounds the display.

When very large or very small numbers are obtained as answers, scientific calculators often express these numbers in scientific notation (Chapter 5). For example, if you multiply 6,265,804 by 8,980,591, the display might look like this:

$$\boxed{5.6270623 \ 13}$$

The 13 at the far right means that the number on the left is multiplied by 10^{13}. This means that the decimal point must be moved 13 places to the right if the answer is to be expressed in its usual form. Even then, the value obtained will only be an approximation: 56,270,623,000,000.

Graphing Calculators

While you are not expected to have a graphing calculator to study from this book, we include the following as background information and reference should your course or future courses require the use of graphing calculators.

Basic Features In addition to the typical keys found on scientific calculators, graphing calculators have keys that can be used to create graphs, make tables, analyze data, and change settings. One of the major differences between graphing and scientific calculators is that a graphing calculator has a larger viewing screen with graphing capabilities. The screens below illustrate the graphs of $Y = X$ and $Y = X^2$.

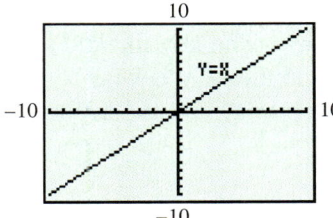 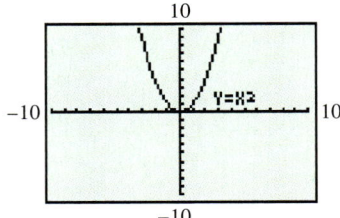

If you look closely at the screens, you will see that the graphs appear to be jagged rather than smooth, as they should be. The reason for this is that graphing calculators have much lower resolution than computer screens. Because of this, graphs generated by graphing calculators must be interpreted carefully.

Editing Input The screen of a graphing calculator can display several lines of text at a time. This feature allows you to view both previous and current expressions. If an incorrect expression is entered, an error message is displayed. The erroneous expression can be viewed and corrected by using various editing keys, much like a word-processing program. You do not need to enter the entire expression again. Many graphing calculators can also recall past expressions for editing or updating. The screen on the left shows how two expressions are evaluated. The final line is entered incorrectly, and the resulting error message is shown in the screen on the right.

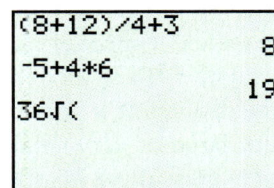 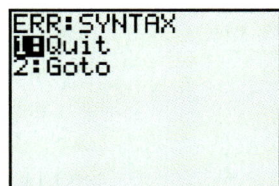

Order of Operations Arithmetic operations on graphing calculators are usually entered as they are written in mathematical expressions. For example, to evaluate $\sqrt{36}$ you would first press the square root key, and then enter 36. See the left screen at the top of the next page. The order of operations on a graphing calculator is also important, and current models insert parentheses when typical errors might occur. The open parenthesis that follows the square root symbol is automatically entered by the calculator so that an expression such as $\sqrt{2 \times 8}$ will not be calculated incorrectly as $\sqrt{2} \times 8$. Compare the two entries and their results in the screen on the right.

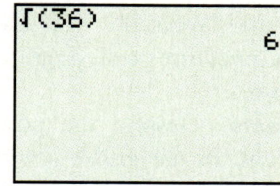 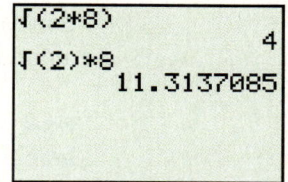

Viewing Windows The viewing window for a graphing calculator is similar to the viewfinder in a camera. A camera usually cannot take a photograph of an entire view of a scene. The camera must be centered on some object and can capture only a portion of the available scenery. A camera with a zoom lens can photograph different views of the same scene by zooming in and out. Graphing calculators have similar capabilities. The xy-coordinate plane is infinite. The calculator screen can only show a finite, rectangular region in the plane, and it must be specified before the graph can be drawn. This is done by setting both minimum and maximum values for the x- and y-axes. The scale (distance between tick marks) is usually specified as well. Determining an appropriate viewing window for a graph is often a challenge, and many times it will take a few attempts before a satisfactory window is found.

The screen on the left shows a standard viewing window, and the graph of $Y = 2X + 1$ is shown on the right. Using a different window would give a different view of the line.

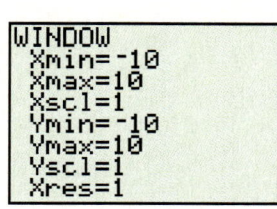 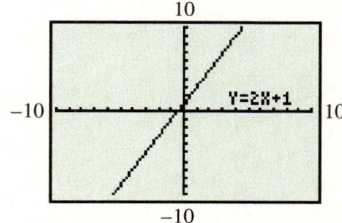

Locating Points on a Graph: Tracing and Tables Graphing calculators allow you to trace along the graph of an equation and display the coordinates of points on the graph. See the screen on the left below, which indicates that the point $(2, 5)$ lies on the graph of $Y = 2X + 1$. Tables for equations can also be displayed. The screen on the right shows a partial table for this same equation. Note the middle of the screen, which indicates that when $X = 2$, $Y = 5$.

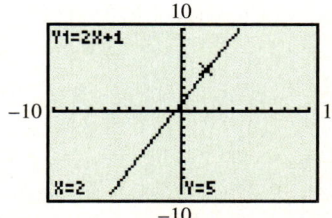 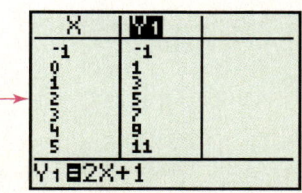

Additional Features There are many features of graphing calculators that go far beyond the scope of this book. These calculators can be programmed, much like computers. Many of them can solve equations at the stroke of a key, analyze statistical data, and perform symbolic algebraic manipulations. Calculators also provide the opportunity to ask "What if . . . ?" more easily. Values in algebraic expressions can be altered and conjectures tested quickly.

Final Comments Despite the power of today's calculators, they cannot replace human thought. *In the entire problem-solving process, your brain is the most important component.* Calculators are only tools and, like any tool, they must be used appropriately in order to enhance our ability to understand mathematics. Mathematical insight may often be the quickest and easiest way to solve a problem; a calculator may neither be needed nor appropriate. By applying mathematical concepts, you can make the decision whether or not to use a calculator.

Appendix B
Review of Decimals and Percents

OBJECTIVES

1. Add and subtract decimals.
2. Multiply and divide decimals.
3. Convert percents to decimals and decimals to percents.
4. Find percentages by multiplying.

OBJECTIVE 1 Add and subtract decimals. A **decimal** is a number written with a decimal point, such as 4.2. The operations on decimals—addition, subtraction, multiplication, and division—are reviewed in the next examples.

EXAMPLE 1 Adding and Subtracting Decimals

Add or subtract as indicated.

(a) $6.92 + 14.8 + 3.217$

Place the numbers in a column, with decimal points lined up, then add. If you like, attach 0s to make all the numbers the same length; this is a good way to avoid errors.

Decimal points lined up

```
   6.92           6.920
  14.8      or   14.800
+  3.217        +  3.217
 _____          _____
  24.937          24.937
```

(b) $47.6 - 32.509$

Write the numbers in a column, attaching 0s to 47.6.

```
  47.6                  47.600
 -32.509    becomes    -32.509
                        _____
                        15.091
```

(c) $3 - .253$

```
  3.000
 - .253
 _____
  2.747
```

Now Try Exercises 1 and 5.

OBJECTIVE 2 Multiply and divide decimals. Multiplication and division of decimals are similar to the same operations with whole numbers.

EXAMPLE 2 Multiplying Decimals

Multiply.

(a) 29.3×4.52

Multiply as if the numbers were whole numbers. To find the number of decimal places in the answer, add the numbers of decimal places in the factors.

$$\begin{array}{r} 29.3 \\ \times\ 4.52 \\ \hline 5\ 86 \\ 14\ 6\ 5 \\ 117\ 2 \\ \hline 132.4\ 36 \end{array}$$

29.3 — 1 decimal place in first factor
× 4.52 — 2 decimal places in second factor
1 + 2 = 3
132.4 36 — 3 decimal places in answer

(b) 7.003×55.8

$$\begin{array}{r} 7.003 \\ \times\ 55.8 \\ \hline 5\ 602\ 4 \\ 35\ 015 \\ 350\ 15 \\ \hline 390.767\ 4 \end{array}$$

7.003 — 3 decimal places
× 55.8 — 1 decimal place
3 + 1 = 4
390.767 4 — 4 decimal places in answer

Now Try Exercises 11 and 15.

EXAMPLE 3 Dividing Decimals

Divide: $279.45 \div 24.3$.

Move the decimal point in 24.3 one place to the right, to get 243. Move the decimal point the same number of places in 279.45. By doing this, 24.3 is converted into the whole number 243.

$$243.\overline{)2794.5}$$

Bring the decimal point straight up and divide as with whole numbers.

$$\begin{array}{r} 11.5 \\ 243.\overline{)2794.5} \\ \underline{243} \\ 364 \\ \underline{243} \\ 121\ 5 \\ \underline{121\ 5} \\ 0 \end{array}$$

Now Try Exercise 19.

OBJECTIVE 3 Convert percents to decimals and decimals to percents. One of the main uses of decimals is in percent problems. The word **percent** means "per one hundred." Percent is written with the sign %. One percent means "one per one hundred."

Percent

$$1\% = .01 \quad \text{or} \quad 1\% = \frac{1}{100}$$

EXAMPLE 4 Converting between Decimals and Percents

Convert.

(a) 75% to a decimal
Since 1% = .01,
$$75\% = 75 \cdot 1\% = 75 \cdot .01 = .75.$$

The fraction form 1% = $\frac{1}{100}$ can also be used to convert 75% to a decimal.

$$75\% = 75 \cdot 1\% = 75 \cdot \frac{1}{100} = .75$$

(b) 2.63 to a percent
$$2.63 = 263 \cdot .01 = 263 \cdot 1\% = 263\%$$

Now Try Exercises 23 and 33.

OBJECTIVE 4 Find percentages by multiplying. A part of a whole is called a **percentage.** For example, since 50% represents $\frac{50}{100} = \frac{1}{2}$ of a whole, 50% of 800 is half of 800, or 400. Multiply to find percentages, as in the next example.

EXAMPLE 5 Finding Percentages

Find each percentage.

(a) 15% of 600
The word *of* indicates multiplication here. For this reason, 15% of 600 is found by multiplying.
$$15\% \cdot 600 = .15 \cdot 600 = 90$$

(b) 125% of 80
$$125\% \cdot 80 = 1.25 \cdot 80 = 100$$

(c) What percent of 52 is 7.8?
We can translate this sentence to symbols word by word.

What percent   of   52   is   7.8?
$$p \cdot 52 = 7.8$$
$$52p = 7.8$$
$$p = .15 \quad \text{Divide both sides by 52.}$$
$$p = 15\% \quad \text{Change to percent.}$$

Now Try Exercises 39 and 43.

EXAMPLE 6 Using Percent in a Consumer Problem

A DVD movie with a regular price of $18 is on sale at 22% off. Find the amount of the discount.

The discount is 22% of $18. Change 22% to .22 and use the fact that *of* indicates multiplication.
$$22\% \cdot 18 = .22 \cdot 18 = 3.96$$

The discount is $3.96.

Now Try Exercise 55.

APPENDIX B EXERCISES

Perform each indicated operation. See Examples 1–3.

1. $14.23 + 9.81 + 74.63 + 18.715$
2. $89.416 + 21.32 + 478.91 + 298.213$
3. $19.74 - 6.53$
4. $27.96 - 8.39$
5. $219 - 68.51$
6. $283 - 12.42$
7. $48.96 + 37.421 + 9.72$
8. $9.71 + 4.8 + 3.6 + 5.2 + 8.17$
9. $8.6 - 3.751$
10. $27.8 - 13.582$
11. 39.6×4.2
12. 18.7×2.3
13. 42.1×3.9
14. 19.63×4.08
15. $.042 \times 32$
16. 571×2.9
17. $24.84 \div 6$
18. $32.84 \div 4$
19. $7.6266 \div 3.42$
20. $14.9202 \div 2.43$
21. $2496 \div .52$
22. $.56984 \div .034$

Convert each percent to a decimal. See Example 4(a).

23. 53%
24. 38%
25. 129%
26. 174%
27. 96%
28. 11%
29. .9%
30. .1%

Convert each decimal to a percent. See Example 4(b).

31. .80
32. .75
33. .007
34. 1.4
35. .67
36. .003
37. .125
38. .983

Respond to each statement or question. Round your answer to the nearest hundredth if appropriate. See Example 5.

39. What is 14% of 780?
40. Find 12% of 350.
41. Find 22% of 1086.
42. What is 20% of 1500?
43. 4 is what percent of 80?
44. 1300 is what percent of 2000?
45. What percent of 5820 is 6402?
46. What percent of 75 is 90?
47. 121 is what percent of 484?
48. What percent of 3200 is 64?
49. Find 118% of 125.8.
50. Find 3% of 128.
51. What is 91.72% of 8546.95?
52. Find 12.741% of 58.902.
53. What percent of 198.72 is 14.68?
54. 586.3 is what percent of 765.4?

Solve each problem. See Example 6.

55. A retailer has $23,000 invested in her business. She finds that she is earning 12% per year on this investment. How much money is she earning per year?

56. Harley Dabler recently bought a duplex for $144,000. He expects to earn 16% per year on the purchase price. How many dollars per year will he earn?

57. For a recent tour of the eastern United States, a travel agent figured that the trip totaled 2300 mi, with 35% of the trip by air. How many miles of the trip were by air?

58. Capitol Savings Bank pays 3.2% interest per year. What is the annual interest on an account of $3000?

59. An ad for steel-belted radial tires promises 15% better mileage when the tires are used. Alexandria's Escort now goes 420 mi on a tank of gas. If she switched to the new tires, how many extra miles could she drive on a tank of gas?

60. A home worth $77,000 is located in an area where home prices are increasing at a rate of 12% per year. By how much would the value of this home increase in 1 year?

61. A family of four with a monthly income of $2000 spends 90% of its earnings and saves the rest. Find the *annual* savings of this family.

Appendix C
Sets

OBJECTIVES

1. List the elements of a set.
2. Learn the vocabulary and symbols used to discuss sets.
3. Decide whether a set is finite or infinite.
4. Decide whether a given set is a subset of another set.
5. Find the complement of a set.
6. Find the union and the intersection of two sets.

OBJECTIVE 1 List the elements of a set. A **set** is a collection of things. The objects in a set are called the **elements** of the set. A set is represented by listing its elements between **set braces, { }**. The order in which the elements of a set are listed is unimportant.

EXAMPLE 1 Listing the Elements of Sets

Represent each set by listing the elements.

(a) The set of states in the United States that border on the Pacific Ocean is
$$\{California, Oregon, Washington, Hawaii, Alaska\}.$$

(b) The set of all counting numbers less than $6 = \{1, 2, 3, 4, 5\}$.

Now Try Exercises 1 and 3.

OBJECTIVE 2 Learn the vocabulary and symbols used to discuss sets. Capital letters are used to name sets. To state that 5 is an element of
$$S = \{1, 2, 3, 4, 5\},$$
write $5 \in S$. The statement $6 \notin S$ means that 6 is not an element of S.

A set with no elements is called the **empty set,** or the **null set.** The symbols $\emptyset$ or { } are used for the empty set. If we let A be the set of all cats that fly, then A is the empty set.
$$A = \emptyset \quad \text{or} \quad A = \{ \}$$

CAUTION Do not make the common error of writing the empty set as $\{\emptyset\}$.

In any discussion of sets, there is some set that includes all the elements under consideration. This set is called the **universal set** for that situation. For example, if the discussion is about presidents of the United States, then the set of all presidents of the United States is the universal set. The universal set is denoted U.

OBJECTIVE 3 Decide whether a set is finite or infinite. In Example 1, there are five elements in the set in part (a), and five in part (b). If the number of elements in a set is either 0 or a counting number, then the set is **finite.** On the other hand, the set of natural numbers, for example, is an **infinite** set, because there is no final natural number. We can list the elements of the set of natural numbers as
$$N = \{1, 2, 3, 4, \ldots\},$$

655

where the three dots indicate that the set continues indefinitely. Not all infinite sets can be listed in this way. For example, there is no way to list the elements in the set of all real numbers between 1 and 2.

EXAMPLE 2 Distinguishing between Finite and Infinite Sets

List the elements of each set, if possible. Decide whether each set is finite or infinite.

(a) The set of all integers
One way to list the elements is $\{\ldots, -2, -1, 0, 1, 2, \ldots\}$. The set is infinite.

(b) The set of all natural numbers between 0 and 5
$\{1, 2, 3, 4\}$ The set is finite.

(c) The set of all irrational numbers
This is an infinite set whose elements cannot be listed.

Now Try Exercise 11.

Two sets are equal if they have exactly the same elements. Thus, the set of natural numbers and the set of positive integers are equal sets. Also, the sets

$$\{1, 2, 4, 7\} \quad \text{and} \quad \{4, 2, 7, 1\}$$

are equal. The order of the elements does not make a difference.

OBJECTIVE 4 Decide whether a given set is a subset of another set. If all elements of a set A are also elements of another set B, then we say A is a **subset** of B, written $A \subseteq B$. We use the symbol $A \not\subseteq B$ to mean that A is not a subset of B.

EXAMPLE 3 Using Subset Notation

Let $A = \{1, 2, 3, 4\}$, $B = \{1, 4\}$, and $C = \{1\}$. Then

$$B \subseteq A, \quad C \subseteq A, \quad \text{and} \quad C \subseteq B,$$

but

$$A \not\subseteq B, \quad A \not\subseteq C, \quad \text{and} \quad B \not\subseteq C.$$

Now Try Exercises 21 and 25.

The set $M = \{a, b\}$ has four subsets: $\{a, b\}$, $\{a\}$, $\{b\}$, and $\emptyset$. The empty set is defined to be a subset of any set. How many subsets does $N = \{a, b, c\}$ have? There is one subset with 3 elements: $\{a, b, c\}$. There are three subsets with 2 elements:

$$\{a, b\}, \quad \{a, c\}, \quad \text{and} \quad \{b, c\}.$$

There are three subsets with 1 element:

$$\{a\}, \quad \{b\}, \quad \text{and} \quad \{c\}.$$

There is one subset with 0 elements: $\emptyset$. Thus, set N has eight subsets. The following generalization can be made.

Number of Subsets of a Set

A set with n elements has 2^n subsets.

APPENDIX C Sets **657**

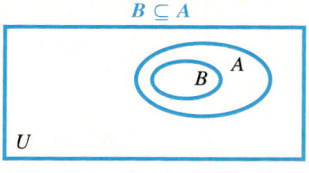

FIGURE 1

To illustrate the relationships between sets, **Venn diagrams** are often used. A rectangle represents the universal set, U. The sets under discussion are represented by regions within the rectangle. The Venn diagram in Figure 1 shows that $B \subseteq A$.

OBJECTIVE 5 Find the complement of a set. For every set A, there is a set A', the **complement** of A, that contains all the elements of U that are not in A. The shaded region in the Venn diagram in Figure 2 represents A'.

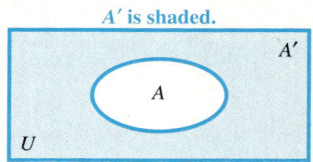

FIGURE 2

EXAMPLE 4 Determining Complements of a Set

Given $U = \{a, b, c, d, e, f, g\}$, $A = \{a, b, c\}$, $B = \{a, d, f, g\}$, and $C = \{d, e\}$, then
$$A' = \{d, e, f, g\}, \quad B' = \{b, c, e\}, \quad \text{and} \quad C' = \{a, b, c, f, g\}.$$

Now Try Exercises 45 and 47.

OBJECTIVE 6 Find the union and the intersection of two sets. The **union** of two sets A and B, written $A \cup B$, is the set of all elements of A together with all elements of B. Thus, for the sets in Example 4,
$$A \cup B = \{a, b, c, d, f, g\}$$
and
$$A \cup C = \{a, b, c, d, e\}.$$

In Figure 3 the shaded region is the union of sets A and B.

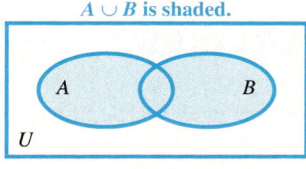 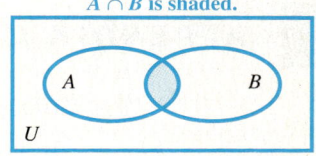

FIGURE 3   FIGURE 4

EXAMPLE 5 Finding the Union of Two Sets

If $M = \{2, 5, 7\}$ and $N = \{1, 2, 3, 4, 5\}$, then
$$M \cup N = \{1, 2, 3, 4, 5, 7\}.$$

Now Try Exercise 55.

The **intersection** of two sets A and B, written $A \cap B$, is the set of all elements that belong to both A and B. For example if,
$$A = \{\text{Jose, Ellen, Marge, Kevin}\}$$
and
$$B = \{\text{Jose, Patrick, Ellen, Sue}\},$$
then
$$A \cap B = \{\text{Jose, Ellen}\}.$$

The shaded region in Figure 4 represents the intersection of the two sets A and B.

EXAMPLE 6 Finding the Intersection of Two Sets

Suppose that $P = \{3, 9, 27\}$, $Q = \{2, 3, 10, 18, 27, 28\}$, and $R = \{2, 10, 28\}$. List the elements in each set.

(a) $P \cap Q = \{3, 27\}$ (b) $Q \cap R = \{2, 10, 28\} = R$ (c) $P \cap R = \emptyset$

Now Try Exercises 49 and 51.

Sets like P and R in Example 6 that have no elements in common are called **disjoint sets**. The Venn diagram in Figure 5 shows a pair of disjoint sets.

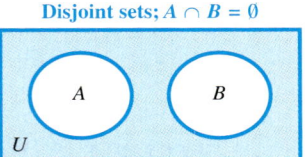

Disjoint sets; $A \cap B = \emptyset$

FIGURE 5

EXAMPLE 7 Using Set Operations

Let $U = \{2, 5, 7, 10, 14, 20\}$, $A = \{2, 10, 14, 20\}$, $B = \{5, 7\}$, and $C = \{2, 5, 7\}$. Find each set.

(a) $A \cup B = \{2, 5, 7, 10, 14, 20\} = U$ (b) $A \cap B = \emptyset$
(c) $B \cup C = \{2, 5, 7\} = C$ (d) $B \cap C = \{5, 7\} = B$
(e) $A' = \{5, 7\} = B$

Now Try Exercises 53 and 57.

APPENDIX C EXERCISES

List the elements of each set. See Examples 1 and 2.

1. The set of all natural numbers less than 8
2. The set of all integers between 4 and 10
3. The set of seasons
4. The set of months of the year
5. The set of women presidents of the United States
6. The set of all living humans who are more than 200 years old
7. The set of letters of the alphabet between K and M
8. The set of letters of the alphabet between D and H
9. The set of positive even integers
10. The set of all multiples of 5
11. Which of the sets described in Exercises 1–10 are infinite sets?
12. Which of the sets described in Exercises 1–10 are finite sets?

Tell whether each statement is true or false.

13. $5 \in \{1, 2, 5, 8\}$
14. $6 \in \{1, 2, 3, 4, 5\}$
15. $2 \in \{1, 3, 5, 7, 9\}$
16. $1 \in \{6, 2, 5, 1\}$
17. $7 \notin \{2, 4, 6, 8\}$
18. $7 \notin \{1, 3, 5, 7\}$
19. $\{2, 4, 9, 12, 13\} = \{13, 12, 9, 4, 2\}$
20. $\{7, 11, 4\} = \{7, 11, 4, 0\}$

Let

$$A = \{1, 3, 4, 5, 7, 8\}, \quad B = \{2, 4, 6, 8\},$$
$$C = \{1, 3, 5, 7\}, \quad D = \{1, 2, 3\},$$
$$E = \{3, 7\}, \quad \text{and} \quad U = \{1, 2, 3, 4, 5, 6, 7, 8, 9, 10\}.$$

Tell whether each statement is true *or* false. *See Examples 3, 5, 6, and 7.*

21. $A \subseteq U$
22. $D \subseteq A$
23. $\emptyset \subseteq A$
24. $\{1, 2\} \subseteq D$
25. $C \subseteq A$
26. $A \subseteq C$
27. $D \subseteq B$
28. $E \subseteq C$
29. $D \not\subseteq E$
30. $E \not\subseteq A$
31. There are exactly 4 subsets of E.
32. There are exactly 8 subsets of D.
33. There are exactly 12 subsets of C.
34. There are exactly 16 subsets of B.
35. $\{4, 6, 8, 12\} \cap \{6, 8, 14, 17\} = \{6, 8\}$
36. $\{2, 5, 9\} \cap \{1, 2, 3, 4, 5\} = \{2, 5\}$
37. $\{3, 1, 0\} \cap \{0, 2, 4\} = \{0\}$
38. $\{4, 2, 1\} \cap \{1, 2, 3, 4\} = \{1, 2, 3\}$
39. $\{3, 9, 12\} \cap \emptyset = \{3, 9, 12\}$
40. $\{3, 9, 12\} \cup \emptyset = \emptyset$
41. $\{4, 9, 11, 7, 3\} \cup \{1, 2, 3, 4, 5\} = \{1, 2, 3, 4, 5, 7, 9, 11\}$
42. $\{1, 2, 3\} \cup \{1, 2, 3\} = \{1, 2, 3\}$
43. $\{3, 5, 7, 9\} \cup \{4, 6, 8\} = \emptyset$
44. $\{5, 10, 15, 20\} \cup \{5, 15, 30\} = \{5, 15\}$

Let

$$U = \{a, b, c, d, e, f, g, h\}, \quad A = \{a, b, c, d, e, f\},$$
$$B = \{a, c, e\}, \quad C = \{a, f\},$$
and $\quad D = \{d\}.$

List the elements in each set. See Examples 4–7.

45. A'
46. B'
47. C'
48. D'
49. $A \cap B$
50. $B \cap A$
51. $A \cap D$
52. $B \cap D$
53. $B \cap C$
54. $A \cup B$
55. $B \cup D$
56. $B \cup C$
57. $C \cup B$
58. $C \cup D$
59. $A \cap \emptyset$
60. $B \cup \emptyset$
61. Name every pair of disjoint sets among sets A–D above.

Appendix D
Mean, Median, and Mode

OBJECTIVES

1. Find the mean of a list of numbers.
2. Find a weighted mean.
3. Find the median.
4. Find the mode.

OBJECTIVE 1 Find the mean of a list of numbers. Making sense of a long list of numbers can be difficult. So when we analyze data, one of the first things to look for is a *measure of central tendency*—a single number that we can use to represent the entire list of numbers. One such measure is the *average* or **mean**. The mean can be found with the following formula.

Finding the Mean (Average)

$$\text{mean} = \frac{\text{sum of all values}}{\text{number of values}}$$

EXAMPLE 1 Finding the Mean (Average)

David had test scores of 84, 90, 95, 98, and 88. Find his mean (average) score.

Use the formula for finding the mean. Add up all the test scores and then divide the sum by the number of tests.

$$\text{mean} = \frac{84 + 90 + 95 + 98 + 88}{5} \quad \leftarrow \text{Sum of test scores} \\ \leftarrow \text{Number of tests}$$

$$\text{mean} = \frac{455}{5}$$

$$\text{mean} = 91 \quad \text{Divide.}$$

David has a mean (average) score of 91.

Now Try Exercise 1.

EXAMPLE 2 Applying the Mean (Average)

The sales of photo albums at Sarah's Card Shop for each day last week were $86, $103, $118, $117, $126, $158, and $149. Find the mean daily sales of photo albums.

To find the mean, add all the daily sales amounts and then divide the sum by the number of days (7).

$$\text{mean} = \frac{\$86 + \$103 + \$118 + \$117 + \$126 + \$158 + \$149}{7} \quad \leftarrow \text{Sum of sales} \\ \leftarrow \text{Number of days}$$

$$\text{mean} = \frac{\$857}{7}$$

$$\text{mean} \approx \$122.43 \quad \text{Nearest cent}$$

Now Try Exercise 7.

OBJECTIVE 2 Find a weighted mean. Some items in a list of data might appear more than once. In this case, we find a **weighted mean,** in which each value is "weighted" by multiplying it by the number of times it occurs.

EXAMPLE 3 Finding a Weighted Mean

The table shows the amount of contribution and the number of times the amount was given (frequency) to a food pantry. Find the weighted mean.

Contribution Value	Frequency	
$ 3	4	← 4 people each contributed $3.
$ 5	2	
$ 7	1	
$ 8	5	
$ 9	3	
$10	2	
$12	1	
$13	2	

In most cases, the same amount was given by more than one person: for example, $3 was given by four people, and $8 was given by five people. Other amounts, such as $12, were given by only one person.

To find the mean, multiply each contribution value by its frequency. Then add the products. Next, add the numbers in the *frequency* column to find the total number of values, that is, the total number of people who contributed money.

Value	Frequency	Product
$ 3	4	($3 · 4) = $12
$ 5	2	($5 · 2) = $10
$ 7	1	($7 · 1) = $ 7
$ 8	5	($8 · 5) = $40
$ 9	3	($9 · 3) = $27
$10	2	($10 · 2) = $20
$12	1	($12 · 1) = $12
$13	2	($13 · 2) = $26
Totals	**20**	**$154**

Finally, divide the totals.

$$\text{mean} = \frac{\$154}{20} = \$7.70$$

The mean contribution to the food pantry was $7.70.

Now Try Exercise 11.

A common use of the weighted mean is to find a student's *grade point average (GPA),* as shown in the next example.

EXAMPLE 4 Applying the Weighted Mean

Find the GPA (grade point average) for a student who earned the following grades last semester. Assume A = 4, B = 3, C = 2, D = 1, and F = 0. The number of credits determines how many times the grade is counted (the frequency).

Course	Credits	Grade	Credits · Grade
Mathematics	4	A (= 4)	4 · 4 = 16
Speech	3	C (= 2)	3 · 2 = 6
English	3	B (= 3)	3 · 3 = 9
Computer Science	2	A (= 4)	2 · 4 = 8
Theater	2	D (= 1)	2 · 1 = 2
Totals	14		41

It is common to round grade point averages to the nearest hundredth, so

$$\text{GPA} = \frac{41}{14} \approx 2.93.$$

Now Try Exercise 15.

OBJECTIVE 3 Find the median. Because it can be affected by extremely high or low numbers, the mean is often a poor indicator of central tendency for a list of numbers. In cases like this, another measure of central tendency, called the *median*, can be used. The **median** divides a group of numbers in half; half the numbers lie above the median, and half lie below the median.

Find the median by listing the numbers *in order* from *smallest* to *largest*. If the list contains an *odd* number of items, the median is the *middle number*.

EXAMPLE 5 Finding the Median

Find the median for this list of prices.

$$\$7, \$23, \$15, \$6, \$18, \$12, \$24$$

First arrange the numbers in numerical order from smallest to largest.

Smallest → 6, 7, 12, 15, 18, 23, 24 ← Largest

Next, find the middle number in the list.

6, 7, 12, *15*, 18, 23, 24

Three are below.   Three are above.
Middle number

The median price is $15.

Now Try Exercise 19.

If a list contains an *even* number of items, there is no single middle number. In this case, the median is defined as the mean (average) of the *middle two* numbers.

EXAMPLE 6 Finding the Median

Find the median for this list of ages, in years.

$$74, 7, 15, 13, 25, 28, 47, 59, 32, 68$$

First arrange the numbers in numerical order from smallest to largest. Then, because the list has an even number of ages, find the middle *two* numbers.

Smallest → 7, 13, 15, 25, <u>28, 32</u>, 47, 59, 68, 74 ← Largest

Middle two numbers

The median age is the mean of the two middle numbers.

$$\text{median} = \frac{28 + 32}{2} = \frac{60}{2} = 30 \text{ yr}$$

Now Try Exercise 21.

OBJECTIVE 4 Find the mode. Another statistical measure is the **mode,** which is the number that occurs *most often* in a list of numbers. For example, if the test scores for 10 students were

$$74, 81, 39, 74, 82, 80, 100, 92, 74, \text{ and } 85,$$

then the mode is 74. Three students earned a score of 74, so 74 appears more times on the list than any other score. It is *not* necessary to place the numbers in numerical order when looking for the mode, although that may help you find it more easily.

A list can have two modes; such a list is sometimes called **bimodal.** If no number occurs more frequently than any other number in a list, the list has *no mode.*

EXAMPLE 7 Finding the Mode

Find the mode for each list of numbers.

(a) 51, 32, 49, 51, 49, 90, 49, 60, 17, 60

The number 49 occurs three times, which is more often than any other number. Therefore, 49 is the mode.

(b) 482, 485, 483, 485, 487, 487, 489, 486

Because both 485 and 487 occur twice, each is a mode. This list is binomial.

(c) 10,708; 11,519; 10,972; 12,546; 13,905; 12,182

No number occurs more than once. This list has no mode.

Now Try Exercises 29, 31, and 33.

Measures of Central Tendency

The **mean** is the sum of all the values divided by the number of values. It is the mathematical *average.*

The **median** is the middle number in a group of values that are listed from smallest to largest. It divides a group of numbers in half.

The **mode** is the value that occurs most often in a group of values.

APPENDIX D EXERCISES

Find the mean for each list of numbers. Round answers to the nearest tenth when necessary. See Example 1.

1. Ages of infants at the child care center (in months): 4, 9, 6, 4, 7, 10, 9

2. Monthly electric bills: $53, $77, $38, $29, $49, $48

3. Final exam scores: 92, 51, 59, 86, 68, 73, 49, 80

4. Quiz scores: 18, 25, 21, 8, 16, 13, 23, 19

5. Annual salaries: $31,900; $32,850; $34,930; $39,712; $38,340; $60,000

6. Numbers of people attending baseball games: 27,500; 18,250; 17,357; 14,298; 33,110

Solve each problem. See Examples 2 and 3.

7. The Athletic Shoe Store sold shoes at the following prices: $75.52, $36.15, $58.24, $21.86, $47.68, $106.57, $82.72, $52.14, $28.60, $72.92. Find the mean shoe sales amount.

8. In one evening, a waitress collected the following checks from her dinner customers: $30.10, $42.80, $91.60, $51.20, $88.30, $21.90, $43.70, $51.20. Find the mean dinner check amount.

9. The table shows the face value (policy amount) of life insurance policies sold and the number of policies sold for each amount by the New World Life Company during one week. Find the weighted mean amount for the policies sold.

Policy Amount	Number of Policies Sold
$ 10,000	6
$ 20,000	24
$ 25,000	12
$ 30,000	8
$ 50,000	5
$100,000	3
$250,000	2

10. Detroit Metro-Sales Company prepared the following table showing the gasoline mileage obtained by each of the cars in their automobile fleet. Find the weighted mean to determine the miles per gallon for the fleet of cars.

Miles per Gallon	Number of Autos
15	5
20	6
24	10
30	14
32	5
35	6
40	4

Find each weighted mean. Round answers to the nearest tenth when necessary. See Example 3.

11.
Quiz Scores	Frequency
3	4
5	2
6	5
8	5
9	2

12.
Credits per Student	Frequency
9	3
12	5
13	2
15	6
18	1

13.

Hours Worked	Frequency
12	4
13	2
15	5
19	3
22	1
23	5

14.

Students per Class	Frequency
25	1
26	2
29	5
30	4
32	3
33	5

Find the GPA (grade point average) for students earning the following grades. Assume A = 4, B = 3, C = 2, D = 1, and F = 0. Round answers to the nearest hundredth. See Example 4.

15.

Course	Credits	Grade
Biology	4	B
Biology Lab	2	A
Mathematics	5	C
Health	1	F
Psychology	3	B

16.

Course	Credits	Grade
Chemistry	3	A
English	3	B
Mathematics	4	B
Theater	2	C
Astronomy	3	C

17. Look again at the grades in Exercise 15. Find the student's GPA in each of these situations.

  (a) The student earned a B instead of an F in the 1-credit class.
  (b) The student earned a B instead of a C in the 5-credit class.
  (c) Both (a) and (b) happened.

18. List the credits for the courses you're taking at this time. List the lowest grades you think you will earn in each class and find your GPA. Then list the highest grades you think you will earn and find your GPA.

Find the median for each list of numbers. See Examples 5 and 6.

19. Number of e-mail messages received: 9, 12, 14, 15, 23, 24, 28

20. Deliveries by a newspaper distributor: 99, 108, 109, 123, 126, 129, 146, 168, 170

21. Students enrolled in algebra each semester: 328, 549, 420, 592, 715, 483

22. Number of cars in the parking lot each day: 520, 523, 513, 1283, 338, 509, 290, 420

23. Number of computer service calls taken each day: 51, 48, 96, 40, 47, 23, 95, 56, 34, 48

24. Number of gallons of paint sold per week: 1072, 1068, 1093, 1042, 1056, 205, 1009, 1081

The table lists the cruising speed and distance flown without refueling for several types of larger airplanes used to carry passengers. Use the table to answer Exercises 25–28.

Type of Airplane	Cruising Speed (miles per hour)	Distance without Refueling (miles)
747-400	565	7650
747-200	558	6450
DC-9	505	1100
DC-10	550	5225
727	530	1550
757	530	2875

Source: Northwest Airlines *WorldTraveler.*

25. What is the mean distance flown without refueling, to the nearest mile?

26. Find the mean cruising speed.

27. (a) Find the median distance.

(b) Is the median distance similar to the mean distance from Exercise 25? Explain why or why not.

28. (a) Find the median cruising speed.

(b) Is the median speed similar to the mean speed from Exercise 26? Explain why or why not.

Find the mode(s) for each list of numbers. Indicate whether a list is bimodal or has no mode. See Example 7.

29. Number of samples taken each hour: 3, 8, 5, 1, 7, 6, 8, 4, 5, 8

30. Monthly water bills: $21, $32, $46, $32, $49, $32, $49, $25, $32

31. Ages of retirees (in years): 74, 68, 68, 68, 75, 75, 74, 74, 70, 77

32. Patients admitted to the hospital each week: 30, 19, 25, 78, 36, 20, 45, 85, 38

33. The number of boxes of candy sold by each child: 5, 9, 17, 3, 2, 8, 19, 1, 4, 20, 10, 6

34. The weights of soccer players (in pounds): 158, 161, 165, 162, 165, 157, 163, 162

Appendix E
Solving Quadratic Inequalities

OBJECTIVE

1 Solve quadratic inequalities and graph their solutions.

OBJECTIVE 1 Solve quadratic inequalities and graph their solutions. A **quadratic inequality** is an inequality that involves a second-degree polynomial. Examples of quadratic inequalities include

$$2x^2 + 3x - 5 < 0, \qquad x^2 \leq 4, \qquad \text{and} \qquad x^2 + 5x + 6 > 0.$$

Examples 1 and 2 show how to solve such inequalities.

EXAMPLE 1 Solving a Quadratic Inequality Including Endpoints

Solve $x^2 - 3x - 10 \leq 0$.

To begin, we find the solutions of the corresponding quadratic *equation*.

$$x^2 - 3x - 10 = 0$$
$$(x - 5)(x + 2) = 0 \qquad \text{Factor.}$$
$$x - 5 = 0 \quad \text{or} \quad x + 2 = 0 \qquad \text{Zero-factor property}$$
$$x = 5 \quad \text{or} \quad x = -2 \qquad \text{Solve each equation.}$$

Since 5 and -2 are the only values that satisfy $x^2 - 3x - 10 = 0$, all other values of x will make $x^2 - 3x - 10$ either less than 0 (< 0) or greater than 0 (> 0). The values $x = 5$ and $x = -2$ determine three regions (intervals) on the number line, as shown in Figure 1. Region A includes all numbers less than -2, Region B includes the numbers between -2 and 5, and Region C includes all numbers greater than 5.

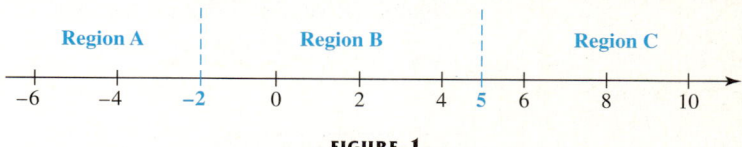

FIGURE 1

All values of x in a given region will cause $x^2 - 3x - 10$ to have the same sign (either positive or negative). We test one value of x from each region to see which regions satisfy $x^2 - 3x - 10 \leq 0$. First, are the points in Region A part of the solution set? As a trial value, we choose any number less than -2, say -6.

$$x^2 - 3x - 10 \leq 0 \qquad \text{Original inequality}$$
$$(-6)^2 - 3(-6) - 10 \leq 0 \quad ? \qquad \text{Let } x = -6.$$
$$36 + 18 - 10 \leq 0 \quad ? \qquad \text{Simplify.}$$
$$44 \leq 0 \qquad \text{False}$$

Since $44 \leq 0$ is false, the points in Region A do not belong to the solution set of inequality.

670 APPENDIX E Solving Quadratic Inequalities

What about Region B? We try the value $x = 0$ in the original inequality.

$$0^2 - 3(0) - 10 \leq 0 \quad ? \quad \text{Let } x = 0.$$
$$-10 \leq 0 \quad \text{True}$$

Since $-10 \leq 0$ is true, the points in Region B do belong to the solution set. Try $x = 6$ to check Region C.

$$6^2 - 3(6) - 10 \leq 0 \quad ? \quad \text{Let } x = 6.$$
$$36 - 18 - 10 \leq 0 \quad ? \quad \text{Simplify.}$$
$$8 \leq 0 \quad \text{False}$$

Since $8 \leq 0$ is false, the points in Region C do not belong to the solution set.

The points in Region B are the only ones that satisfy $x^2 - 3x - 10 < 0$. As shown in Figure 2, the solution set includes the points in Region B together with the endpoints -2 and 5 since the original inequality uses $\leq$. The solution set is written as the interval $[-2, 5]$.

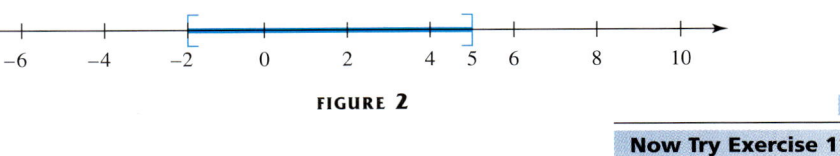

FIGURE 2

Now Try Exercise 11.

To summarize, we use the following steps to solve a quadratic inequality.

Solving a Quadratic Inequality

Step 1 **Write the inequality as an equation and solve it.**

Step 2 **Determine intervals.** Use the solutions of the equation to divide a number line into intervals.

Step 3 **Test each interval.** Substitute a number from each interval into the original inequality to determine the intervals that satisfy the inequality.

Step 4 **Consider the endpoints separately.** The solutions from Step 1 are included in the solution set only if the inequality is $\leq$ or $\geq$.

■ **EXAMPLE 2** Solving a Quadratic Inequality Excluding Endpoints

Solve $-x^2 - 5x - 6 < 0$.

It will be easier to factor if we multiply both sides by -1. We must also remember to reverse the direction of the inequality symbol:

$$x^2 + 5x + 6 > 0.$$

Step 1 Factoring the expression in the corresponding equation $x^2 + 5x + 6 = 0$, we get $(x + 2)(x + 3) = 0$. The solutions of the equation are -2 and -3.

Step 2 These numbers determine three regions on a number line. See Figure 3. This time, these points will not belong to the solution set because only values of x that make $x^2 + 5x + 6$ *greater than* 0 are solutions.

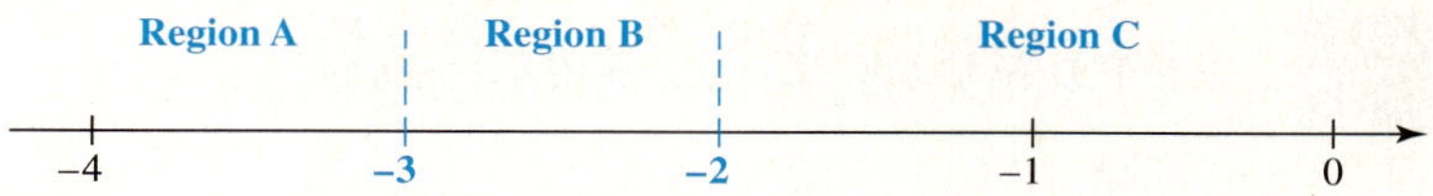

APPENDIX E Solving Quadratic Inequalities **671**

```
         Region A    |    Region B    |           Region C
    ─────┼────────────┼────────────────┼────────────────┼─────▶
        -4           -3               -2               -1              0
```
FIGURE 3

Step 3 To decide whether the points in Region A belong to the solution set, we select any number in Region A, such as -4, and substitute.

$$-x^2 - 5x - 6 < 0 \qquad \text{Original inequality}$$
$$-(-4)^2 - 5(-4) - 6 < 0 \quad ? \quad \text{Let } x = -4.$$
$$-16 + 20 - 6 < 0 \quad ? \quad \text{Simplify.}$$
$$-2 < 0 \qquad \text{True}$$

Since $-2 < 0$ is true, all the points in Region A belong to the solution set of the inequality.

For Region B, we choose a number between -3 and -2, say $-2\frac{1}{2}$ or $-\frac{5}{2}$.

$$-\left(-\frac{5}{2}\right)^2 - 5\left(-\frac{5}{2}\right) - 6 < 0 \quad ? \quad \text{Let } x = -\frac{5}{2}.$$
$$-\frac{25}{4} + \frac{25}{2} - 6 < 0 \quad ? \quad \text{Simplify.}$$
$$\frac{1}{4} < 0 \qquad \text{False}$$

Since $\frac{1}{4} < 0$ is false, no point in Region B belongs to the solution set.

For Region C, we try the number 0.

$$-0^2 - 5(0) - 6 < 0 \quad ? \quad \text{Let } x = 0.$$
$$-6 < 0 \qquad \text{True}$$

Since $-6 < 0$ is true, the points in Region C belong to the solution set.

Step 4 The solutions include all values of x less than -3, together with all values of x greater than -2, as shown in the graph in Figure 4. Using inequalities, the solutions may be written as $x < -3$ or $x > -2$. Any number that satisfies *either* of these inequalities will be a solution, because *or* means "either one or the other or both." In interval notation, we write the solution set using the *union symbol* $\cup$ for *or*:

$$(-\infty, -3) \cup (-2, \infty).$$

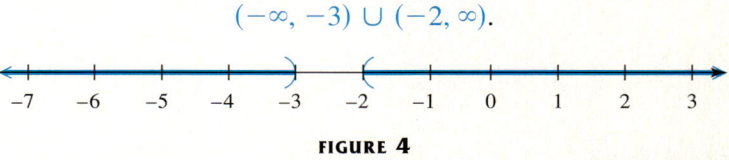

FIGURE 4

Now Try Exercise 9.

CAUTION There is no shortcut way to write the solutions $x < -3$ or $x > -2$.

APPENDIX E EXERCISES

1. Which of these inequalities is not a quadratic inequality?
   A. $5x^2 - x + 7 < 0$ B. $x^2 + 4 > 0$ C. $3x + 5 \geq 0$ D. $(x - 1)(x + 2) \leq 0$

2. To solve a quadratic inequality, we use the _____ of the corresponding equation to determine _____ on the number line.

To prepare for using test values in the various regions, answer true *or* false *depending on whether the given value of x satisfies or does not satisfy the inequality.*

3. $(2x + 1)(x - 5) \geq 0$
   (a) $x = -\frac{1}{2}$ (b) $x = 5$ (c) $x = 0$ (d) $x = -6$

4. $(x - 6)(3x + 4) < 0$
   (a) $x = -3$ (b) $x = 6$ (c) $x = 0$ (d) $x = -\frac{4}{3}$

Solve each inequality and graph the solution set. See Examples 1 and 2.

5. $(a + 3)(a - 3) < 0$
6. $(b - 2)(b + 2) > 0$
7. $(a + 6)(a - 7) \geq 0$
8. $(z - 5)(z - 4) \leq 0$
9. $m^2 + 5m + 6 > 0$
10. $x^2 - 3x + 2 < 0$
11. $z^2 - 4z - 5 \leq 0$
12. $3p^2 - 5p - 2 \leq 0$
13. $5m^2 + 3m - 2 < 0$
14. $2k^2 + 7k - 4 > 0$
15. $6r^2 - 5r < 4$
16. $6r^2 + 7r > 3$
17. $q^2 - 7q < -6$
18. $2k^2 - 7k \leq 15$
19. $6m^2 + m - 1 > 0$
20. $30r^2 + 3r - 6 \leq 0$
21. $12p^2 + 11p + 2 < 0$
22. $a^2 - 16 < 0$
23. $9m^2 - 36 > 0$
24. $r^2 - 100 \geq 0$
25. $r^2 > 16$
26. $m^2 \geq 25$

Each given inequality is not quadratic, but it may be solved in a similar manner. Solve and graph each inequality. (Hint: Because these inequalities correspond to equations with three solutions, they determine four regions on a number line.)

27. $(a + 2)(3a - 1)(a - 4) \geq 0$
28. $(2p - 7)(p - 1)(p + 3) \leq 0$
29. $(r - 2)(r^2 - 3r - 4) < 0$
30. $(m + 5)(m^2 - m - 6) > 0$

An object is propelled upward from ground level with an initial velocity of 256 ft per second. After t sec, its height h is given by the equation $h = -16t^2 + 256t$, where h is in feet.

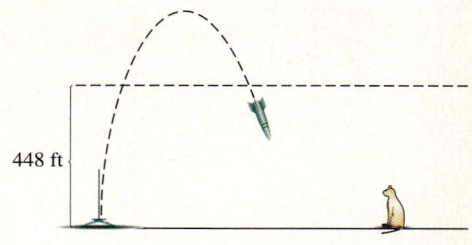

31. The object reaches a height of 448 ft when $h = 448$ in the equation. Use the method of Section 6.5 to solve the resulting equation and determine the times at which the object reaches this height.

32. The object is more than 448 ft above the ground when $h > 448$. This is given by the inequality $-16t^2 + 256t > 448$. Solve this inequality to find the time interval over which the object is more than 448 ft above the ground.

33. By solving $-16t^2 + 256t < 448$ and realizing that t must be greater than or equal to 0, determine the time intervals over which the object is less than 448 ft above the ground. (*Note:* We must also have $t \leq 16$.)

34. Explain the restrictions on t described in Exercise 33.

Answers to Selected Exercises   A-3

37. $\dfrac{2 \cdot 2 \cdot 2 \cdot \cancel{3}^{1} \cdot \cancel{3}^{1}}{\cancel{3}_{1} \cdot \cancel{3}_{1} \cdot 5 \cdot 5} = \dfrac{8}{25}$ 39. equivalent   41. not equivalent   43. not equivalent   45. equivalent   47. not equivalent   49. equivalent

51. A fraction is in lowest terms when the numerator and the denominator have no common factors other than 1. Some examples are $\dfrac{1}{2}, \dfrac{3}{8}$, and $\dfrac{2}{3}$.   53. $\dfrac{7}{8}$ 55. $\dfrac{2}{1} = 2$

Section 0.11   (page 0-61)

Exercises   1. $\dfrac{3}{8}$ 3. $\dfrac{5}{12}$ 5. $\dfrac{3}{4}$ 7. $\dfrac{1}{4}$ 9. $\dfrac{5}{12}$ 11. $\dfrac{9}{32}$ 13. $\dfrac{2}{5}$ 15. $\dfrac{13}{32}$ 17. $\dfrac{21}{128}$ 19. 5   21. 40   23. 21   25. $13\dfrac{1}{2}$ 27. $31\dfrac{1}{2}$
29. 240   31. $189\dfrac{1}{3}$ 33. 400   35. 810

Section 0.12   (page 0-66)

Exercises   1. $\dfrac{3}{2}$ 3. $\dfrac{5}{8}$ 5. $\dfrac{6}{5}$ 7. $\dfrac{1}{4}$ 9. $\dfrac{1}{3}$ 11. $2\dfrac{5}{8}$ 13. $\dfrac{9}{20}$ 15. 4   17. 6   19. $\dfrac{13}{16}$ 21. $\dfrac{4}{5}$ 23. 18   25. $22\dfrac{1}{2}$ 27. $\dfrac{1}{14}$

Section 0.13   (page 0-71)

Exercises   1. Exact: $8\dfrac{1}{8}$; Estimate: $3 \cdot 3 = 9$ 3. Exact: $4\dfrac{1}{2}$; Estimate: $2 \cdot 3 = 6$ 5. Exact: 4; Estimate: $3 \cdot 1 = 3$ 7. Exact: 50; Estimate: $8 \cdot 6 = 48$ 9. Exact: $49\dfrac{1}{2}$; Estimate: $5 \cdot 2 \cdot 5 = 50$ 11. Exact: 12; Estimate: $3 \cdot 2 \cdot 3 = 18$ 13. Exact: $\dfrac{1}{3}$; Estimate: $3 \div 8 = \dfrac{3}{8}$ 15. Exact: $\dfrac{5}{6}$; Estimate: $3 \div 3 = 1$ 17. Exact: $3\dfrac{3}{5}$; Estimate: $9 \div 3 = 3$ 19. Exact: $\dfrac{3}{7}$; Estimate: $1 \div 2 = \dfrac{1}{2}$
21. Exact: $\dfrac{3}{10}$; Estimate: $2 \div 6 = \dfrac{1}{3}$ 23. Exact: $\dfrac{17}{18}$; Estimate: $6 \div 6 = 1$

Sections 0.7–0.13   Test (page 0-73)

1. $\dfrac{5}{6}$ 2. $\dfrac{3}{8}$ 3. $\dfrac{2}{3}, \dfrac{6}{7}, \dfrac{1}{4}, \dfrac{5}{8}$ 4. $\dfrac{20}{3}$ 5. $23\dfrac{3}{5}$ 6. 1, 2, 3, 6, 9, 18   7. $3^2 \cdot 7$ 8. $2^5 \cdot 3$ 9. $2^2 \cdot 5^3$ 10. $\dfrac{3}{4}$ 11. $\dfrac{5}{6}$
12. $\dfrac{56}{84} = \dfrac{\cancel{2}^{1} \cdot \cancel{2}^{1} \cdot 2 \cdot \cancel{7}^{1}}{\cancel{2}_{1} \cdot \cancel{2}_{1} \cdot 3 \cdot \cancel{7}_{1}} = \dfrac{2}{3}$ 13. Multiply fractions by multiplying the numerators and multiplying the denominators. Divide two fractions by using the reciprocal of the second fraction (divisor) and multiplying.   14. $\dfrac{3}{20}$ 15. 16   16. $\dfrac{9}{10}$ 17. $15\dfrac{3}{4}$ 18. $\dfrac{1}{9}$
19. Estimate: $4 \cdot 4 = 16$; Exact: $14\dfrac{7}{16}$ 20. Estimate: $2 \cdot 4 = 8$; Exact: $7\dfrac{17}{18}$ 21. Estimate: $10 \div 2 = 5$; Exact: $4\dfrac{4}{15}$
22. Estimate: $9 \div 2 = 4\dfrac{1}{2}$; Exact: $5\dfrac{1}{10}$

Section 0.14   (page 0-77)

Exercises   1. $\dfrac{4}{5}$ 3. $\dfrac{5}{6}$ 5. $\dfrac{1}{3}$ 7. $\dfrac{6}{5}$ 9. $\dfrac{1}{3}$ 11. $\dfrac{13}{20}$ 13. $\dfrac{11}{15}$ 15. $\dfrac{3}{2}$ 17. $\dfrac{11}{27}$ 19. $\dfrac{3}{8}$ 21. $\dfrac{6}{11}$ 23. $\dfrac{3}{5}$ 25. $\dfrac{8}{7}$ 27. $\dfrac{1}{5}$ 29. $\dfrac{7}{6}$
31. $\dfrac{11}{10}$

A-4 Answers to Selected Exercises

Section 0.15 (page 0-84)

Exercises 1. 4 3. 15 5. 36 7. 14 9. 30 11. 100 13. 20 15. 72 17. 36 19. 120 21. 180 23. 144 25. $\frac{8}{24}$ 27. $\frac{18}{24}$

29. $\frac{20}{24}$ 31. 2 33. 15 35. 28 37. 55 39. 32 41. 72

Section 0.16 (page 0-88)

Exercises 1. $\frac{5}{6}$ 3. $\frac{8}{9}$ 5. $\frac{3}{4}$ 7. $\frac{39}{40}$ 9. $\frac{23}{36}$ 11. $\frac{14}{15}$ 13. $\frac{29}{36}$ 15. $\frac{17}{20}$ 17. $\frac{23}{30}$ 19. $\frac{7}{12}$ 21. $\frac{23}{48}$ 23. $\frac{3}{8}$ 25. $\frac{1}{2}$ 27. $\frac{7}{15}$ 29. $\frac{1}{6}$

31. $\frac{19}{45}$ 33. $\frac{3}{40}$ 35. $\frac{17}{48}$ 37. $\frac{23}{24}$ yd³

Section 0.17 (page 0-94)

Exercises 1. *Estimate:* 6 + 3 = 9; *Exact:* $8\frac{5}{6}$ 3. *Estimate:* 7 + 4 = 11; *Exact:* $11\frac{1}{2}$ 5. *Estimate:* 1 + 4 = 5; *Exact:* $4\frac{5}{24}$

7. *Estimate:* 25 + 19 = 44; *Exact:* $43\frac{2}{3}$ 9. *Estimate:* 34 + 19 = 53; *Exact:* $52\frac{1}{10}$ 11. *Estimate:* 23 + 15 = 38; *Exact:* $38\frac{5}{28}$

13. *Estimate:* 11 + 18 + 16 = 45; *Exact:* $43\frac{5}{6}$ 15. *Estimate:* 12 − 10 = 2; *Exact:* $2\frac{1}{8}$ 17. *Estimate:* 13 − 1 = 12; *Exact:* $11\frac{7}{15}$

19. *Estimate:* 28 − 6 = 22; *Exact:* $22\frac{7}{30}$ 21. *Estimate:* 17 − 7 = 10; *Exact:* $10\frac{3}{8}$ 23. *Estimate:* 19 − 6 = 13; *Exact:* $12\frac{19}{20}$

25. *Estimate:* 16 − 11 = 5; *Exact:* $5\frac{7}{8}$ 27. $9\frac{3}{8}$ 29. $11\frac{1}{2}$ 31. $3\frac{5}{6}$ 33. $6\frac{11}{12}$ 35. $8\frac{1}{8}$ 37. $\frac{5}{6}$

Sections 0.14–0.17 Test (page 0-96)

1. $\frac{3}{4}$ 2. $\frac{1}{2}$ 3. $\frac{2}{5}$ 4. $\frac{1}{6}$ 5. 12 6. 30 7. 108 8. $\frac{5}{8}$ 9. $\frac{23}{36}$ 10. $\frac{5}{24}$ 11. $\frac{1}{40}$ 12. *Estimate:* 8 + 5 = 13; *Exact:* $12\frac{1}{2}$

13. *Estimate:* 16 − 12 = 4; *Exact:* $4\frac{11}{15}$ 14. *Estimate:* 19 + 9 + 12 = 40; *Exact:* $40\frac{29}{60}$ 15. *Estimate:* 24 − 18 = 6; *Exact:* $5\frac{5}{8}$

Section 0.18 (page 0-102)

Exercises 1. 7; 0; 4 3. 5; 1; 8 5. 4; 7; 0 7. 1; 6; 3 9. 1; 8; 9 11. 6; 2; 1 13. 410.25 15. 6.5432 17. 5406.045 19. $\frac{7}{10}$

21. $13\frac{2}{5}$ 23. $\frac{1}{4}$ 25. $\frac{33}{50}$ 27. $10\frac{17}{100}$ 29. $\frac{3}{50}$ 31. $\frac{41}{200}$ 33. $5\frac{1}{500}$ 35. $\frac{343}{500}$ 37. five tenths 39. seventy-eight hundredths

41. one hundred five thousandths 43. twelve and four hundredths 45. one and seventy-five thousandths 47. 6.7 49. 0.32

51. 420.008 53. 0.0703 55. 75.030

Section 0.19 (page 0-107)

Exercises 1. 16.9 3. 0.956 5. 0.80 7. 3.661 9. 794.0 11. 0.0980 13. 49 15. 9.09 17. 82.0002

Section 0.20 (page 0-111)

Exercises 1. 17.48 3. 23.013 5. 7.763 7. 77.006 9. 20.104 11. 0.109 13. 330.86895 15. (a) 24.75 in. (b) 3.95 in.

17. (a) 62.27 in. (b) 0.39 in.

Section 0.21 (page 0-113)

Exercises 1. 0.1344 3. 159.10 5. 15.5844 7. $34,500.20 9. 43.2 11. 0.432 13. 0.0432 15. 0.00432 17. 0.0000312
19. 0.000006 21. Multiplying by 10, decimal point moves one place to the right; by 100, two places to the right; by 1000, three places to the right.

Section 0.22 (page 0-119)

Exercises 1. 3.9 3. 0.47 5. 400.2 7. 36 9. 0.06 11. 6000 13. 25.3 15. 516.67 (rounded) 17. 24.291 (rounded)
19. 10,082.647 (rounded) 21. Dividing by 10, decimal point moves one place to the left; by 100, two places to the left; by 1000, three places to the left.

Section 0.23 (page 0-123)

Exercises 1. 0.5 3. 0.75 5. 0.3 7. 0.9 9. 0.6 11. 0.875 13. 2.25 15. 14.7 17. 3.625 19. 0.333 (rounded) 21. 0.833 (rounded) 23. 1.889 (rounded) 25. (a) A proper fraction is less than 1, so it cannot be equivalent to a mixed number. (b) $\frac{5}{9}$ means $5 \div 9$ or $9\overline{)5}$ so correct answer is 0.556 (rounded). This makes sense because both the fraction and decimal are less than 1.
26. (a) $2.035 = 2\frac{35}{1000} = 2\frac{7}{200}$, not $2\frac{7}{20}$. (b) Adding the whole number part gives $2 + 0.35$, which is 2.35, not 2.035. To check, $2.35 = 2\frac{35}{100} = 2\frac{7}{20}$. 27. Just add the whole number part to 0.375. So $1\frac{3}{8} = 1.375$; $3\frac{3}{8} = 3.375$; $295\frac{3}{8} = 295.375$.
28. It works only when the fraction part has a one-digit numerator and a denominator of 10, a two-digit numerator and a denominator of 100, and so on. 29. $\frac{2}{5}$ 31. $\frac{5}{8}$ 33. $\frac{7}{20}$ 35. 0.35 37. $\frac{1}{25}$ 39. $\frac{3}{20}$ 41. 0.2 43. $\frac{9}{100}$

Sections 0.18–0.23 Test (page 0-125)

1. $18\frac{2}{5}$ 2. $\frac{3}{40}$ 3. sixty and seven thousandths 4. two hundred eight ten-thousandths 5. 725.6 6. 0.630 7. $1.49 8. $7860
9. *Estimate*: $8 + 80 + 40 = 128$; *Exact*: 129.2028 10. *Estimate*: $80 - 4 = 76$; *Exact*: 75.498 11. *Estimate*: $6 \cdot 1 = 6$; *Exact*: 6.948
12. *Estimate*: $20 \div 5 = 4$; *Exact*: 4.175 13. 839.762 14. 669.004 15. 0.0000483 16. 480 17. 2.625

Section 0.24 (page 0-134)

Exercises 1. $\frac{8}{9}$ 3. $\frac{2}{1}$ 5. $\frac{1}{3}$ 7. $\frac{8}{5}$ 9. $\frac{3}{8}$ 11. $\frac{9}{7}$ 13. $\frac{6}{1}$ 15. $\frac{5}{6}$ 17. $\frac{8}{5}$ 19. $\frac{1}{12}$ 21. $\frac{5}{16}$ 23. $\frac{4}{1}$ 25. $\frac{1}{2}$ 27. $\frac{36}{1}$ 29. Answers will vary. One possibility is stocking cards of various types in the same ratios as those in the table. 31. Comparing the violin to piano, guitar, organ, clarinet, and drums gives ratios of $\frac{1}{11}, \frac{1}{10}, \frac{1}{3}, \frac{1}{2}$, and $\frac{2}{3}$, respectively. 33. Answers will vary. Possibilities include: guitars are less expensive than drums and easier to carry around; more guitar players than drummers are needed in a band. 35. $\frac{5 \text{ cups}}{3 \text{ people}}$
37. $\frac{3 \text{ feet}}{7 \text{ seconds}}$ 39. $\frac{1 \text{ person}}{2 \text{ dresses}}$ 41. $\frac{5 \text{ letters}}{1 \text{ minute}}$ 43. $\frac{\$21}{2 \text{ visits}}$ 45. $\frac{18 \text{ miles}}{1 \text{ gallon}}$ 47. $12 per hour or $12/hour 49. 5 eggs per chicken or 5 eggs/chicken 51. 1.25 pounds/person 53. $103.30/day 55. 4 ounces for $0.89 57. 15 ounces for $3.15 59. 18 ounces for $1.79

Section 0.25 (page 0-140)

Exercises 1. $\dfrac{\$9}{12 \text{ cans}} = \dfrac{\$18}{24 \text{ cans}}$ 3. $\dfrac{200 \text{ adults}}{450 \text{ children}} = \dfrac{4 \text{ adults}}{9 \text{ children}}$ 5. $\dfrac{120}{150} = \dfrac{8}{10}$ 7. true 9. true 11. false 13. true 15. true
17. false 19. true 21. false 23. false 25. true 27. false 29. true 31. true 33. false 35. false
37. $\dfrac{16 \text{ hits}}{50 \text{ at bats}} = \dfrac{128 \text{ hits}}{400 \text{ at bats}}$ $\begin{array}{l} 50 \cdot 128 = 6400 \\ 16 \cdot 400 = 6400 \end{array}$ Diagonal products are *equal* so the proportion is *true*; they hit equally well.

Section 0.26 (page 0-147)

Exercises 1. 4 3. 2 5. 88 7. 91 9. 5 11. 10 13. 24.44 (rounded) 15. 50.4 17. 17.64 (rounded) 19. 1 21. $3\dfrac{1}{2}$
23. 0.2 or $\dfrac{1}{5}$ 25. 0.005 or $\dfrac{1}{200}$ 27. Find diagonal products: $20 \neq 30$, so the proportion is false.

$$\dfrac{6\tfrac{2}{3}}{4} = \dfrac{5}{3} \text{ or } \dfrac{10}{6} = \dfrac{5}{3} \text{ or } \dfrac{10}{4} = \dfrac{7.5}{3} \text{ or } \dfrac{10}{4} = \dfrac{5}{2}$$

28. Find diagonal products: $192 \neq 180$, so the proportion is false.

$$\dfrac{6.4}{8} = \dfrac{24}{30} \text{ or } \dfrac{6}{7.5} = \dfrac{24}{30} \text{ or } \dfrac{6}{8} = \dfrac{22.5}{30} \text{ or } \dfrac{6}{8} = \dfrac{24}{32}$$ 29. 22.5 hours 31. $7.20 33. 42 pounds 35. $273.45
37. 10 ounces (rounded) 39. 5 quarts

Sections 0.24–0.26 Test (page 0-149)

1. $\dfrac{4}{5}$ 2. $\dfrac{20 \text{ miles}}{1 \text{ gallon}}$ 3. $\dfrac{\$1}{5 \text{ minutes}}$ 4. $\dfrac{15}{4}$ 5. $\dfrac{1}{80}$ 6. $\dfrac{9}{2}$ 7. 13 ounces of Brand Z for $1.29 8. false 9. true 10. 25 11. 2.67
(rounded) 12. 325 13. $10\dfrac{1}{2}$ 14. 576 words 15. 3.6 ounces 16. 23.8 grams (rounded)

Section 0.27 (page 0-156)

Exercises 1. 3 3. 8 5. 5280 7. 2000 9. 60 11. 2 13. 2 15. 76,000 to 80,000 lb 17. 27 19. 112 21. 10 23. $1\dfrac{1}{2}$ or 1.5
25. $\dfrac{1}{4}$ or 0.25 27. $1\dfrac{1}{2}$ or 1.5 29. $2\dfrac{1}{2}$ or 2.5 31. $\dfrac{1}{2}$ day or 0.5 day 33. 5000 35. 17 37. 44 ounces 39. 216 41. 28
43. 518,400 45. 48,000 47. (a) pound/ounces (b) quarts/pints or pints/cups (c) minutes/hours or seconds/minutes (d) feet/inches
(e) pounds/tons (f) days/weeks 49. 174,240 51. 800 53. 0.75 or $\dfrac{3}{4}$ 55. 51.3 ft/sec 57. 17.0 mi/hr 59. 20.5 mi/hr

Section 0.28 (page 0-163)

Exercises 1. 1000; 1000 3. $\dfrac{1}{1000}$ or 0.001; $\dfrac{1}{1000}$ or 0.001 5. $\dfrac{1}{100}$ or 0.01; $\dfrac{1}{100}$ or 0.01 7. 700 cm 9. 0.040 m or 0.04 m
11. 9400 m 13. 5.09 m 15. 40 cm 17. 910 mm 19. 0.0000056 km 21. 16.7 m/sec 23. 24.4 m/sec 25. 90 km/hr 27. 162 km/hr

Section 0.29 (page 0-168)

Exercises 1. Unit for your answer (g) is in numerator; unit being changed (kg) is in denominator so it will divide out. The unit fraction is
$\dfrac{1000 \text{ g}}{1 \text{ kg}}$. 3. 15,000 mL 5. 3 L 7. 0.925 L 9. 0.008 L 11. 4150 mL 13. 8 kg 15. 5200 g 17. 850 mg 19. 30 g 21. 0.598 g
23. 0.06 L 25. 0.003 kg 27. 990 mL

Answers to Selected Exercises A-7

Section 0.30 (page 0-171)

Exercises 1. 21.8 yd 3. 262.4 ft 5. 4.8 m 7. 5.3 oz 9. 111.6 kg 11. 30.3 qt 13. (a) about 0.2 oz (b) probably not 15. about 31.8 L 17. about 1.3 cm

Sections 0.27–0.30 Test (page 0-172)

1. 36 qt 2. 15 yd 3. 2.25 or $2\frac{1}{4}$ hr 4. 0.75 or $\frac{3}{4}$ ft 5. 56 oz 6. 7200 min 7. 22 ft/sec 8. 13.3 m/sec 9. 2.5 m 10. 4600 m 11. 0.5 cm 12. 0.325 g 13. 16,000 mL 14. 400 g 15. 1055 cm 16. 0.095 L 17. 1.8 m (rounded) 18. 56.3 kg (rounded) 19. 13 gal 20. 5.0 mi (rounded)

Section 0.31 (page 0-176)

Exercises 1. 0.15 3. 0.60 or 0.6 5. 0.25 7. 1.40 or 1.4 9. 0.055 11. 1.00 or 1 13. 0.005 15. 0.0035 17. 80% 19. 58% 21. 1% 23. 12.5% 25. 37.5% 27. 200% 29. 370% 31. 3.12% 33. 416.2% 35. 0.28% 37. Answers will vary. Some possibilities are: No common denominators are needed with percents. The denominator is always 100 with percent, which makes comparisons easier to understand. 39. 0.45 41. 0.18 43. 3.5% 45. 200% 47. 0.5% 49. 1.536

Section 0.32 (page 0-181)

Exercises 1. $\frac{1}{4}$ 3. $\frac{3}{4}$ 5. $\frac{17}{20}$ 7. $\frac{5}{8}$ 9. $\frac{1}{16}$ 11. $\frac{1}{6}$ 13. $\frac{1}{15}$ 15. $\frac{1}{200}$ 17. $1\frac{1}{5}$ 19. $3\frac{3}{4}$ 21. 50% 23. 80% 25. 25% 27. 37% 29. 62.5% 31. 87.5% 33. 36% 35. 46% 37. 15% 39. 83.3% (rounded) 41. 55.6% (rounded) 43. 14.3% (rounded) 45. $\frac{1}{2}$; 50% 47. $\frac{7}{8}$; 0.875 49. $\frac{4}{5}$; 80% 51. 0.167 (rounded) 16.7% (rounded) 53. $\frac{1}{4}$; 25% 55. $\frac{1}{8}$; 0.125 57. 0.667 (rounded); 66.7% (rounded) 59. 0.4; 40% 61. 0.08; 8% 63. 0.005; 0.5%

Section 0.33 (page 0-188)

Exercises 1. 50 3. 150 5. 70 7. 25% 9. 150% 11. 33.3% (rounded) 13. 26 15. 21.6 17. 26.5% (rounded) 19. 115 21. 0.4% 23. 2.5% 25. 0.3% 27. 10; unknown; 60 29. 75; $800; $600 31. 25; $970; unknown 33. 20; unknown; 12 35. 50; 68; 34 37. unknown; $296; $177.60 39. 3.25; unknown; 54.34 41. 0.68; $487; unknown

Section 0.34 (page 0-196)

Exercises 1. 16 guests 3. 1836 military personnel 5. 4.8 ft 7. 315 files 9. 819 trucks 11. $3.28 13. 1530 tables 15. 182 homes 17. $21.60 19. 320 e-mails 21. 160 hay bales 23. 550 students 25. 1360 graduates 27. 2800 29. 50% 31. 52% 33. 8% 35. 1.5% 37. 18.6% (rounded) 39. 9.2%

Section 0.35 (page 0-202)

Exercises 1. 135 hamburgers 3. 675 garments 5. 83.2 quarts 7. 700 tablets 9. 1029.2 meters 11. $4.16 13. 160 patients 15. 325 salads 17. 680 circuits 19. 1080 people 21. 300 gallons 23. 50% 25. 76% 27. 125% 29. 1.5% 31. 250% 33. You must first change the fraction in the percent to a decimal, then divide the percent by 100 to change it to a decimal.

$2\frac{1}{2}\% = 2.5\% = 0.025 \leftarrow$ 2.5% as a decimal

↑ ↑

Change $2\frac{1}{2}$ to 2.5.

A-8 Answers to Selected Exercises

Sections 0.31–0.35 Test (page 0-203)

1. 0.65   2. 80%   3. 175%   4. 87.5% or $87\frac{1}{2}$%   5. 3.00 or 3   6. 0.0005   7. $\frac{1}{8}$ 8. $\frac{1}{400}$ 9. 60%   10. 62.5% or $62\frac{1}{2}$%
11. 250%   12. 800 sacks   13. 20%   14. $18,500   15. $2854.20   16. $628

CHAPTER 1 THE REAL NUMBER SYSTEM

Section 1.1 (page 10)

Exercises 1. true 3. false; The fraction $\frac{17}{51}$ is written in lowest terms as $\frac{1}{3}$. 5. false; *Product* refers to multiplication, so the product of 8 and 2 is 16. 7. prime 9. composite; $2 \cdot 2 \cdot 2 \cdot 2 \cdot 2 \cdot 2$ 11. composite; $2 \cdot 7 \cdot 13 \cdot 19$ 13. neither 15. composite; $2 \cdot 3 \cdot 5$
17. composite; $2 \cdot 2 \cdot 5 \cdot 5 \cdot 5$ 19. composite; $2 \cdot 2 \cdot 31$ 21. prime 23. $\frac{1}{2}$ 25. $\frac{5}{6}$ 27. $\frac{3}{10}$ 29. $\frac{6}{5}$ 31. C 33. $\frac{24}{35}$
35. $\frac{6}{25}$ 37. $\frac{6}{5}$ or $1\frac{1}{5}$ 39. $\frac{232}{15}$ or $15\frac{7}{15}$ 41. $\frac{10}{3}$ or $3\frac{1}{3}$ 43. 12 45. $\frac{1}{16}$ 47. $\frac{84}{47}$ or $1\frac{37}{47}$ 49. To multiply two fractions, multiply their numerators to get the numerator of the product and multiply their denominators to get the denominator of the product. For example, $\frac{2}{3} \cdot \frac{8}{5} = \frac{2 \cdot 8}{3 \cdot 5} = \frac{16}{15}$. To divide two fractions, replace the divisor with its reciprocal and then multiply. For example, $\frac{2}{5} \div \frac{7}{9} = \frac{2}{5} \cdot \frac{9}{7} = \frac{2 \cdot 9}{5 \cdot 7} = \frac{18}{35}$. 51. $\frac{2}{3}$ 53. $\frac{8}{9}$ 55. $\frac{43}{8}$ or $5\frac{3}{8}$ 57. $\frac{2}{3}$ 59. $\frac{17}{36}$ 61. $\frac{11}{12}$
63. 6 cups 65. 34 dollars 67. $\frac{9}{16}$ in. 69. $618\frac{3}{4}$ ft 71. $5\frac{5}{24}$ in. 73. $\frac{1}{3}$ cup 75. $\frac{1}{20}$ 77. more than $1\frac{1}{25}$ million
79. (a) $\frac{1}{2}$ (b) $\frac{1}{4}$ (c) $\frac{1}{3}$ (d) $\frac{1}{6}$

Section 1.2 (page 20)

Exercises 1. false; $4 + 3(8 − 2) = 4 + 3 \cdot 6 = 4 + 18 = 22$. The common error leading to 42 is adding 4 to 3 and then multiplying by 6. One must follow the order of operations. 3. false; The correct interpretation is $4 = 16 − 12$. 5. 49 7. 144 9. 64 11. 1000
13. 81 15. 1024 17. $\frac{16}{81}$ 19. .000064 21. Write the base as a factor the number of times indicated by the exponent. For example, $6^3 = 6 \cdot 6 \cdot 6 = 216$. 23. 32 25. $\frac{49}{30}$ or $1\frac{19}{30}$ 27. 12 29. 42 31. 95 33. 90 35. 14 37. 9

39. -19 **41.** 112 **43.** -24 **45.** -9 **47.** 8.55 **49.** $16 \leq 16$; true **51.** $61 \leq 60$; false **53.** $0 \geq 0$; true **55.** $45 \geq 46$; false **57.** $66 > 72$; false **59.** $2 \geq 3$; false **61.** $3 \geq 3$; true **63.** $15 = 5 + 10$ **65.** $9 > 5 - 4$ **67.** $16 \neq 19$ **69.** $2 \leq 3$ **71.** Seven is less than nineteen; true **73.** Three is not equal to six; true **75.** Eight is greater than or equal to eleven; false **77.** $30 > 5$ **79.** $3 \leq 12$ **81.** is younger than **83.** The inequality symbol $\geq$ implies a true statement if 12 equals 12 *or* if 12 is greater than 12. **85.** 1998, 1999 **87.** (a) $16.96 (b) approximately 31.5% **89.** $3 \cdot (6 + 4) \cdot 2 = 60$ **91.** $10 - (7 - 3) = 6$ **93.** $(8 + 2)^2 = 100$

Section 1.3 (page 27)

Exercises 1. 10 **3.** $12 + x$; 21 **5.** no **7.** $2x^3 = 2 \cdot x \cdot x \cdot x$, while $2x \cdot 2x \cdot 2x = (2x)^3$. **9.** The exponent 2 applies only to its base, which is x. (The expression $(4x)^2$ would require multiplying 4 by $x = 3$ first.) **11.** Answers will vary. Two such pairs are $x = 0, y = 6$ and $x = 1, y = 4$. To determine them, choose a value for x, substitute it into the expression $2x + y$, and then subtract the value of $2x$ from 6. **13.** (a) 13 (b) 15 **15.** (a) 20 (b) 30 **17.** (a) 64 (b) 144 **19.** (a) $\frac{5}{3}$ (b) $\frac{7}{3}$ **21.** (a) $\frac{7}{8}$ (b) $\frac{13}{12}$ **23.** (a) 52 (b) 114 **25.** (a) 25.836 (b) 38.754 **27.** (a) 24 (b) 28 **29.** (a) 12 (b) 33 **31.** (a) 6 (b) $\frac{9}{5}$ **33.** (a) $\frac{4}{3}$ (b) $\frac{13}{6}$ **35.** (a) $\frac{2}{7}$ (b) $\frac{16}{27}$ **37.** (a) 12 (b) 55 **39.** (a) 1 (b) $\frac{28}{17}$ **41.** (a) 3.684 (b) 8.841 **43.** $12x$ **45.** $x + 7$ **47.** $x - 2$ **49.** $7 - x$ **51.** $x - 6$ **53.** $\frac{12}{x}$ **55.** $6(x - 4)$ **57.** No, it is a connective word that joins the two factors: the number and 6. **59.** yes **61.** no **63.** yes **65.** yes **67.** yes **69.** $x + 8 = 18$; 10 **71.** $16 - \frac{3}{4}x = 13$; 4 **73.** $2x + 1 = 5$; 2 **75.** $3x = 2x + 8$; 8 **77.** expression **79.** equation **81.** equation **83.** 128.02 ft **85.** 187.08 ft

Section 1.4 (page 36)

Exercises 1. 1,198,000 **3.** 925 **5.** -5074 **7.** -11.35 **9.** 4 **11.** 0 **13.** One example is $\sqrt{12}$. There are others. **15.** true **17.** true **19.** (a) 3, 7 (b) 0, 3, 7 (c) $-9, 0, 3, 7$ (d) $-9, -1\frac{1}{4}, -\frac{3}{5}, 0, 3, 5.9, 7$ (e) $-\sqrt{7}, \sqrt{5}$ (f) All are real numbers. **21.** The *natural numbers* are the numbers with which we count. An example is 1. The *whole numbers* are the natural numbers with 0 also included. An example is 0. The *integers* are the whole numbers and their negatives. An example is -1. The *rational numbers* are the numbers that can be represented by a quotient of integers, such as $\frac{1}{2}$. The *irrational numbers,* such as $\sqrt{2}$, cannot be represented as a quotient of integers. The *real numbers* include all positive numbers, negative numbers, and zero. All the numbers discussed are real. **23.** [number line: $-6, -4, -2, 0, 2$] **25.** [number line: $-6, -4, -2, 0, 2, 4$] **27.** [number line: $-3\frac{4}{5}, -1\frac{5}{8}, \frac{1}{4}, 2\frac{1}{2}$; $-4, -2, 0, 2, 4$] **29.** (a) A (b) A (c) B (d) B **31.** (a) 2 (b) 2 **33.** (a) -6 (b) 6 **35.** 6 **37.** -12 **39.** 3 **41.** -12 **43.** -8 **45.** 3 **47.** $|-3|$ or 3 **49.** $-|-6|$ or -6 **51.** $|5 - 3|$ or 2 **53.** true **55.** true **57.** true **59.** false **61.** true **63.** false **65.** softwood plywood, 1998 to 1999 **67.** paving mixtures and blocks, 1998 to 1999 In Exercises 69–73, answers will vary. **69.** $\frac{1}{2}, \frac{5}{8}, 1\frac{3}{4}$ **71.** $-3\frac{1}{2}, -\frac{2}{3}, \frac{3}{7}$ **73.** $\sqrt{5}, \pi, -\sqrt{3}$ **75.** This is not true. The absolute value of 0 is 0, and 0 is not positive. A more accurate way of describing absolute value is to say that *absolute value is never negative*, or *absolute value is always nonnegative*.

Section 1.5 (page 46)

Exercises 1. negative 3. negative 5. To add two numbers with the same sign, add their absolute values and keep the same sign for the sum. For example, $3 + 4 = 7$ and $-3 + (-4) = -7$. To add two numbers with different signs, subtract the smaller absolute value from the larger absolute value, and use the sign of the number with the larger absolute value. For example, $6 + (-4) = 2$ and $(-6) + 4 = -2$. 7. -8 9. -12 11. 2 13. -2 15. 4 17. 12 19. 5 21. 2 23. -9 25. 0 27. $\frac{1}{2}$ 29. $-\frac{19}{24}$ 31. $-\frac{3}{4}$ 33. -7.7 35. -8 37. 0 39. -20 41. -3 43. -4 45. -8 47. -14 49. 9 51. -4 53. 4 55. $\frac{3}{4}$ 57. $-\frac{11}{8}$ or $-1\frac{3}{8}$ 59. $\frac{15}{8}$ or $1\frac{7}{8}$ 61. 11.6 63. -9.9 65. 10 67. -5 69. 11 71. -10 73. 22 75. -2 77. -6 79. -12 81. $\frac{11}{4}$ 83. -2 85. $5\frac{1}{2}$ 87. $-\frac{43}{42}$ 89. $\frac{17}{12}$ 91. -20.01 93. -1.7 95. $-.8$ 97. 9.91 99. -14.1 101. -5.90617 103. $-5 + 12 + 6; 13$ 105. $[-19 + (-4)] + 14; -9$ 107. $[-4 + (-10)] + 12; -2$ 109. $[8 + (-18)] + 4; -6$ 111. $4 - (-8); 12$ 113. $-2 - 8; -10$ 115. $[9 + (-4)] - 7; -2$ 117. $[8 - (-5)] - 12; 1$ 119. -3.4 (billion dollars) 121. -2.7 (billion dollars) 123. 50,395 ft 125. 1345 ft 127. 136 ft 129. 45°F 131. -58°F 133. 27 ft 135. $+31,900$ ft 137. $-\$107$

Section 1.6 (page 60)

Exercises 1. greater than 0 3. greater than 0 5. less than 0 7. greater than 0 9. equal to 0 11. 12 13. -12 15. 120 17. -33 19. -165 21. $\frac{5}{12}$ 23. $-\frac{1}{6}$ 25. 6 27. $-32, -16, -8, -4, -2, -1, 1, 2, 4, 8, 16, 32$ 29. $-40, -20, -10, -8, -5, -4, -2, -1, 1, 2, 4, 5, 8, 10, 20, 40$ 31. $-31, -1, 1, 31$ 33. 3 35. -5 37. 7 39. -6 41. $\frac{32}{3}$ or $10\frac{2}{3}$ 43. -4 45. 0 47. undefined 49. -11 51. -2 53. 35 55. 6 57. -18 59. 67 61. -8 63. $-\frac{22}{225}$ 65. $\frac{59}{120}$ 67. -16.86 69. -2.13 71. 22.75 73. 3 75. 7 77. 4 79. -3 81. 10 83. 2 85. -36 87. 5.25 89. $\frac{3}{8}$ 91. 47 93. 72 95. $-\frac{78}{25}$ 97. 0 99. -23 101. 2 103. $9 + (-9)(2); -9$ 105. $-4 - 2(-1)(6); 8$ 107. $(1.5)(-3.2) - 9; -13.8$ 109. $12[9 - (-8)]; 204$ 111. $\frac{-12}{-5 + (-1)}; 2$ 113. $\frac{15 + (-3)}{4(-3)}; -1$ 115. $2(8 + 9); 34$ 117. $.20(-5 \cdot 6); -6$ 119. $\frac{x}{3} = -3; -9$ 121. $x - 6 = 4; 10$ 123. $x + 5 = -5; -10$ 125. $8\frac{2}{5}$ 127. 2 129. (a) 6 is divisible by 2. (b) 9 is not divisible by 2. 131. (a) 64 is divisible by 4. (b) 35 is not divisible by 4. 133. (a) 2 is divisible by 2 and $1 + 5 + 2 + 4 + 8 + 2 + 2 = 24$ is divisible by 3. (b) While 0 is divisible by 2, $2 + 8 + 7 + 3 + 5 + 9 + 0 = 34$ is not divisible by 3. 135. (a) $4 + 1 + 1 + 4 + 1 + 0 + 7 = 18$ is divisible by 9. (b) $2 + 2 + 8 + 7 + 3 + 2 + 1 = 25$ is not divisible by 9.

Summary Exercises on Operations with Real Numbers (page 63)

1. -16 2. 4 3. 0 4. -24 5. -17 6. 76 7. -18 8. 90 9. 38 10. 4 11. -5 12. 5 13. $-\frac{7}{2}$ or $-3\frac{1}{2}$ 14. 4 15. 13 16. $\frac{5}{4}$ or $1\frac{1}{4}$ 17. 9 18. $\frac{37}{10}$ or $3\frac{7}{10}$ 19. 0 20. 25 21. 14 22. 0 23. -4 24. $\frac{6}{5}$ or $1\frac{1}{5}$ 25. -1 26. $\frac{52}{37}$ or $1\frac{15}{37}$ 27. $\frac{17}{16}$ or $1\frac{1}{16}$ 28. $-\frac{2}{3}$ 29. 3.33 30. 1.02 31. -13 32. 0 33. 24 34. -7 35. 37 36. -3 37. -1 38. $\frac{1}{2}$ 39. $-\frac{5}{13}$ 40. 5

Answers to Selected Exercises

Section 1.7 (page 70)

Exercises 1. -12; commutative property   3. 3; commutative property   5. 7; associative property   7. 8; associative property   9. (a) B (b) F (c) C (d) I (e) B (f) D, F (g) B (h) A (i) G (j) H   11. commutative property   13. associative property   15. associative property   17. inverse property   19. inverse property   21. identity property   23. commutative property   25. distributive property   27. identity property   29. distributive property   31. identity property   33. 150   35. 2010   37. 400   39. 1400   41. 0   43. 7   45. 3   47. 11   49. 0   51. $-.38$ 53. 1   55. -2 57. 0   59. Subtraction is not associative. 61. The expression following the first equals sign should be $-3(4) - 3(-6)$. The student forgot that 6 should be preceded by a $-$ sign. The correct work is $-3(4-6) = -3(4) - 3(-6) = -12 + 18 = 6$.   63. 85   65. $4t + 12$ 67. $-8r - 24$ 69. $-5y + 20$ 71. $-16y - 20z$ 73. $8(z + w)$ 75. $7(2v + 5r)$ 77. $24r + 32s - 40y$ 79. $-24x - 9y - 12z$ 81. $5(x + 3)$ 83. $\frac{4}{5}(x - y + t)$ 85. $m - n$ 87. $1.5(xy - ab + mn)$ 89. $-2p - 10q + 5r + 8s$ 91. $-67r - 670s + 6700t$ 93. $-4t - 3m$ 95. $5c + 4d$ 97. $3q - 5r + 8s$ 99. Answers will vary; for example, "putting on your socks" and "putting on your shoes."   101. 0   102. $-3(5) + (-3)(-5)$ 103. -15 104. We must interpret $(-3)(-5)$ as 15, since it is the additive inverse of -15.

Section 1.8 (page 76)

Exercises 1. $4r + 11$ 3. $5 + 2x - 6y$ 5. $-7 + 3p$ 7. $2 - 3x$ 9. -12 11. 5   13. 1   15. -1 17. 74   19. like   21. unlike   23. like   25. unlike   27. $17y$ 29. $-6a$ 31. $13b$ 33. $7k + 15$ 35. $-4y$ 37. $2x + 6$ 39. $14 - 7m$ 41. $-17 + x$ 43. $23x$ 45. $9y^2$ 47. $-14p^3 + 5p^2$ 49. $8x + 15$ 51. $5x + 15$ 53. $-4y + 22$ 55. $-16y + 63$ 57. $4r + 15$ 59. $12k - 5$ 61. $-2k - 3$ 63. $4k - 7$ 65. $-23.7y - 12.6$ 67. $-53.8x - 21y - 72.4$ 69. $112y - 22x$ 71. $-6.99s - 45.89t$ 73. $107a - 66b - 66c$ 75. $-165x - 77y + 672$ 77. $(x + 3) + 5x$; $6x + 3$ 79. $(13 + 6x) - (-7x)$; $13 + 13x$ 81. $2(3x + 4) - (-4 + 6x)$; 12   83. 2, 3, 4, 5   84. 1   85. (a) 1, 2, 3, 4 (b) 3, 4, 5, 6 (c) 4, 5, 6, 7   86. The value of $x + b$ also increases by 1 unit.   87. (a) 2, 4, 6, 8 (b) 2, 5, 8, 11 (c) 2, 6, 10, 14   88. m 89. (a) 7, 9, 11, 13 (b) 5, 8, 11, 14 (c) 1, 5, 9, 13; In comparison, we see that while the values themselves are different, the number of units of increase is the same as the corresponding parts of Exercise 87.   90. m 91. Apples and oranges are examples of unlike fruits, just like x and y are unlike terms. We cannot add x and y to get an expression any simpler than $x + y$; we cannot add, for example, 2 apples and 3 oranges to obtain 5 fruits that are all alike.   93. Wording will vary. One example is "the difference between 9 times a number and the sum of the number and 2."

Chapter 1 Review Exercises (page 84)

1. $\frac{3}{4}$ 3. $\frac{9}{40}$ 5. 625   7. $.0000000032$ 9. 27   11. 39   13. true   15. false   17. $5 + 2 \neq 10$ 19. 30   21. 14   23. $x + 6$ 25. $6x - 9$ 27. yes   29. $2x - 6 = 10$; 8   31. [number line from -4 to 4 with points at $-\frac{1}{2}$ and 2.5]   33. rational numbers, real numbers   35. -10 37. $-\frac{3}{4}$ 39. true   41. true   43. (a) 9 (b) 9   45. (a) -6 (b) 6   47. 12   49. -19 51. -6 53. -17 55. -21.8 57. -10 59. -11 61. 7   63. 10.31   65. 2   67. $(-31 + 12) + 19$; 0   69. $-4 - (-6)$; 2   71. -2 73. $\$26.25$ 75. $-\$29$ 77. It gained 4 yd.   79. 36   81. $\frac{1}{2}$ 83. -20 85. -24 87. 4   89. $-\frac{3}{4}$ 91. -1 93. 1   95. -18 97. 125   99. $-4(5) - 9$; -29 101. $\frac{12}{8 + (-4)}$; 3   103. $8x = -24$; -3 105. 32   107. identity property   109. inverse property   111. associative property   113. distributive property   115. $7(y + 2)$ 117. $3(2s + 5y)$ 119. $25 - (5 - 2) = 22$ and $(25 - 5) - 2 = 18$. Because different groupings lead to different results, we conclude that in general subtraction is not associative.   121. $11m$ 123. $16p^2 + 2p$ 125. $-2m + 29$ 127. C   129. A   131. 16   133. $\frac{8}{3}$ or $2\frac{2}{3}$ 135. 2   137. $-\frac{3}{2}$ or $-1\frac{1}{2}$ 139. $-\frac{28}{15}$ or $-1\frac{13}{15}$ 141. $8x^2 - 21y^2$ 143. When dividing 0 by a nonzero number, the quotient will be 0. However, dividing a number by 0 is undefined.   145. $5(x + 7)$; $5x + 35$

Answers to Selected Exercises

Chapter 1 Test (page 89)

1. $\dfrac{7}{11}$ 2. $\dfrac{241}{120}$ or $2\dfrac{1}{120}$ 3. $\dfrac{19}{18}$ or $1\dfrac{1}{18}$ 4. (a) 492 million (b) 861 million   [1.2] 5. true   [1.4] 6. ←•—•—•—•—•—•—•→ $-4\ -2\ \ 0\ \ 2\ \ 4$

rational numbers, real numbers   8. If -8 and -1 are both graphed on a number line, we see that the point for -8 is to the *left* of the point for -1. This indicates $-8 < -1$.   [1.6] 9. $\dfrac{-6}{2+(-8)}$; 1   [1.1, 1.4–1.6] 10. 4   11. $-\dfrac{17}{6}$ or $-2\dfrac{5}{6}$ 12. 2   13. 6   14. 108   15. 3

16. $\dfrac{30}{7}$ or $4\dfrac{2}{7}$ [1.3, 1.5, 1.6] 17. 6   18. 4   [1.4–1.6] 19. -70 20. 3   21. 7000 m   22. 15   23. (a) $25.1 billion

(b) $-$$11.3 billion (c) $5.0 billion (d) $-$$2.2 billion   [1.7] 24. B   25. D   26. E   27. A   28. C   29. distributive property

30. (a) -18 (b) -18 (c) The distributive property assures us that the answers must be the same, because $a(b+c) = ab + ac$ for all a, b, c.
[1.8] 31. $21x$ 32. $15x - 3$

CHAPTER 2 LINEAR EQUATIONS AND INEQUALITIES IN ONE VARIABLE

Section 2.1 (page 98)

Exercises   1. A and C   3. The addition property of equality says that the same number (or expression) added to each side of an equation results in an equivalent equation. Example: $-x$ can be added to each side of $2x + 3 = x - 5$ to get the equivalent equation $x + 3 = -5$.
5. $\{12\}$ 7. $\{31\}$ 9. $\{-3\}$ 11. $\{4\}$ 13. $\{-9\}$ 15. $\{-10\}$ 17. $\{-13\}$ 19. $\{10\}$ 21. $\{6.3\}$ 23. $\{-16.9\}$ 25. $\{-6\}$
27. $\{-2\}$ 29. $\{4\}$ 31. $\{0\}$ 33. $\{-2\}$ 35. $\{-7\}$ 37. A and B; A sample answer might be, "A linear equation in one variable is an equation that can be written using only one variable term with the variable to the first power."   39. $\{13\}$ 41. $\{-4\}$ 43. $\{0\}$ 45. $\left\{\dfrac{7}{15}\right\}$
47. $\{7\}$ 49. $\{-4\}$ 51. $\{13\}$ 53. $\{29\}$ 55. $\{18\}$ 57. $\{12\}$ 59. Answers will vary. One example is $x - 6 = -8$.
61. $3x = 2x + 17; \{17\}$ 63. $7x - 6x = -9; \{-9\}$

Section 2.2 (page 104)

Exercises   1. The multiplication property of equality says that the same nonzero number (or expression) multiplied on each side of the equation results in an equivalent equation. Example: Multiplying each side of $7x = 4$ by $\dfrac{1}{7}$ gives the equivalent equation $x = \dfrac{4}{7}$.   3. C
5. To get x alone on the left side, divide each side by 4, the coefficient of x.   7. $\dfrac{3}{2}$ 9. 10   11. $-\dfrac{2}{9}$ 13. -1 15. 6   17. -4
19. .12   21. -1 23. $\{6\}$ 25. $\left\{\dfrac{15}{2}\right\}$ 27. $\{-5\}$ 29. $\{-4\}$ 31. $\left\{-\dfrac{18}{5}\right\}$ 33. $\{12\}$ 35. $\{0\}$ 37. $\{40\}$ 39. $\{-12.2\}$
41. $\{-48\}$ 43. $\{72\}$ 45. $\{-35\}$ 47. $\{14\}$ 49. $\{18\}$ 51. $\left\{-\dfrac{27}{35}\right\}$ 53. $\{-12\}$ 55. $\left\{\dfrac{3}{4}\right\}$ 57. $\{3\}$ 59. $\{-5\}$ 61. $\{7\}$
63. $\{0\}$ 65. $\left\{-\dfrac{3}{5}\right\}$ 67. Answers will vary. One example is $\dfrac{3}{2}x = -6$.   69. $4x = 6; \left\{\dfrac{3}{2}\right\}$ 71. $\dfrac{x}{-5} = 2; \{-10\}$

Section 2.3 (page 112)

Exercises   1. *Step 1:* Clear parentheses and combine like terms, as needed. *Step 2:* Use the addition property to get all variable terms on one side of the equation and all numbers on the other. Then combine like terms. *Step 3:* Use the multiplication property to get the equation in the form $x = $ a number. *Step 4:* Check the solution. Examples will vary.   3. D   5. $\{-1\}$ 7. $\{5\}$ 9. $\left\{-\dfrac{5}{3}\right\}$ 11. $\left\{\dfrac{4}{3}\right\}$ 13. $\{5\}$

Answers to Selected Exercises   A-13

15. ∅   **17.** {all real numbers}   **19.** {1}   **21.** ∅   **23.** {5}   **25.** {0}   **27.** $\left\{-\dfrac{7}{5}\right\}$ **29.** {120}   **31.** {6}   **33.** {15,000}   **35.** {8}
37. {0}   **39.** {4}   **41.** {20}   **43.** {all real numbers}   **45.** ∅   **47.** $11 - q$ **49.** $x + 7$ **51.** $a + 12$; $a - 5$ **53.** $\dfrac{t}{5}$

Summary Exercises on Solving Linear Equations   (page 114)

1. {−5}   **2.** {4}   **3.** {−5.1}   **4.** {12}   **5.** {−25}   **6.** {−6}   **7.** {−3}   **8.** {−16}   **9.** {7}   **10.** $\left\{-\dfrac{96}{5}\right\}$ **11.** {5}   **12.** {23.7}
13. {all real numbers}   **14.** {1}   **15.** {−6}   **16.** ∅   **17.** {6}   **18.** {3}   **19.** ∅   **20.** $\left\{\dfrac{7}{3}\right\}$ **21.** {25}   **22.** {−10.8}   **23.** {3}
24. {7}   **25.** {2}   **26.** {all real numbers}   **27.** {−2}   **28.** {70}   **29.** $\left\{\dfrac{14}{17}\right\}$ **30.** $\left\{-\dfrac{5}{2}\right\}$

Section 2.4   (page 122)

Connections   (page 122)   Polya's Step 1 corresponds to our Steps 1 and 2. Polya's Step 2 corresponds to our Step 3. Polya's Step 3 corresponds to our Steps 4 and 5. Polya's Step 4 corresponds to our Step 6. Trial and error or guessing and checking fit into Polya's Step 2, devising a plan.

Exercises   **1.** The procedure should include the following steps: read the problem carefully; assign a variable to represent the unknown to be found, and write down variable expressions for any other unknown quantities; translate into an equation; solve the equation; state the answer; check your solution.   **3.** D; there cannot be a fractional number of cars.   **5.** 3   **7.** 6   **9.** −3   **11.** California: 59 screens; New York: 48 screens   **13.** Democrats: 45; Republicans: 55   **15.** U2: $109.7 million; 'N Sync: $86.8 million   **17.** wins: 61; losses: 21
19. 1950 Denver nickel: $14.00; 1945 Philadelphia nickel: $12.00   **21.** ice cream: 44,687.9 lb; topping: 537.1 lb   **23.** 18 prescriptions
25. peanuts: $22\dfrac{1}{2}$ oz; cashews: $4\dfrac{1}{2}$ oz   **27.** Airborne Express: 3; Federal Express: 9; United Parcel Service: 1   **29.** gold: 10; silver: 13; bronze: 11   **31.** 36 million mi   **33.** A and B: 40°; C: 100°   **35.** $k - m$ **37.** no   **39.** $x - 1$ **41.** 18°   **43.** 39°   **45.** 50°
47. 68, 69   **49.** 10, 12   **51.** 101, 102   **53.** 10, 11   **55.** 18   **57.** 15, 17, 19   **59.** $2.78 billion; $3.33 billion; $3.53 billion

Section 2.5   (page 133)

Exercises   **1. (a)** The perimeter of a plane geometric figure is the distance around the figure.   **(b)** The area of a plane geometric figure is the measure of the surface covered or enclosed by the figure.   **3.** four   **5.** area   **7.** perimeter   **9.** area   **11.** area   **13.** $P = 26$
15. $A = 64$ **17.** $b = 4$ **19.** $t = 5.6$ **21.** $I = 1575$ **23.** $B = 14$ **25.** $r = 2.6$ **27.** $A = 50.24$ **29.** $V = 150$ **31.** $V = 52$
33. $V = 7234.56$ **35.** about 154,000 ft²   **37.** perimeter: 13 in., area: 10.5 in.²   **39.** 132.665 ft²   **41.** 23,800.10 ft²
43. length: 36 in.; volume: 11,664 in.³   **45.** 48°, 132°   **47.** 51°, 51°   **49.** 105°, 105°   **51.** $t = \dfrac{d}{r}$ **53.** $b = \dfrac{A}{h}$ **55.** $d = \dfrac{C}{\pi}$
57. $H = \dfrac{V}{LW}$ **59.** $r = \dfrac{I}{pt}$ **61.** $h = \dfrac{2A}{b}$ **63.** $h = \dfrac{3V}{\pi r^2}$ **65.** $b = P - a - c$ **67.** $W = \dfrac{P - 2L}{2}$ or $W = \dfrac{P}{2} - L$ **69.** $m = \dfrac{y - b}{x}$
71. $y = \dfrac{C - Ax}{B}$ **73.** $r = \dfrac{M - C}{C}$

Section 2.6   (page 141)

Exercises   **1. (a)** C   **(b)** D   **(c)** B   **(d)** A   **3.** $\dfrac{6}{7}$ **5.** $\dfrac{18}{55}$ **7.** $\dfrac{5}{16}$ **9.** $\dfrac{4}{15}$ **11.** $\dfrac{3}{1}$ **13.** 17-oz size   **15.** 64-oz can   **17.** 500-count
19. 28-oz size   **21.** A ratio is a comparison, while a proportion is a statement that two ratios are equal. For example, $\dfrac{2}{3}$ is a ratio and
$\dfrac{2}{3} = \dfrac{8}{12}$ is a proportion.   **23.** true   **25.** false   **27.** true   **29.** {35}   **31.** {7}   **33.** $\left\{\dfrac{45}{2}\right\}$ **35.** {2}   **37.** {−1}   **39.** {5}

A-14 Answers to Selected Exercises

41. $\left\{-\dfrac{31}{5}\right\}$ **43.** $67.50 **45.** $28.35 **47.** 4 ft **49.** 6.875 fluid oz **51.** $670.48 **53.** 50,000 fish **55.** (a) $\dfrac{26}{100} = \dfrac{x}{350}$; (b) 54 ft **63.** $270
$91 million (b) $112 million; $11.2 million (c) $119 million **57.** 4 **59.** 1 **61.** (a)

65. $287 **67.** 30 **68.** (a) $5x = 12$ (b) $\left\{\dfrac{12}{5}\right\}$ **69.** $\left\{\dfrac{12}{5}\right\}$ **70.** Both methods give the same solution set.

Section 2.7 (page 152)

Exercises **1.** 35 mL **3.** $350 **5.** $14.15 **7.** C **9.** (a) 375,000 (b) 575,000 (c) 275,000 **11.** 4.5% **13.** D **15.** 160 gal
17. $53\dfrac{1}{3}$ kg **19.** 4 L **21.** $13\dfrac{1}{3}$ L **23.** 25 mL **25.** $5000 at 3%; $1000 at 5% **27.** $40,000 at 3%; $110,000 at 4%
29. 25 fives **31.** fives: 84; tens: 42 **33.** 20 lb **35.** A **37.** 530 mi **39.** 4.059 hr **41.** 7.91 m per sec **43.** 8.42 m per sec
45. 10 hr **47.** $2\dfrac{1}{2}$ hr **49.** 5 hr **51.** northbound: 60 mph; southbound: 80 mph **53.** 50 km per hr; 65 km per hr

Section 2.8 (page 166)

Connections **(page 166)** The revenue is represented by $5x - 100$. The production cost is $125 + 4x$. The profit is represented by $R - C = (5x - 100) - (125 + 4x) = x - 225$. The solution of $x - 225 > 0$ is $x > 225$. To make a profit, more than 225 cassettes must be produced and sold.

Exercises **1.** Use a parenthesis if the symbol is $<$ or $>$. Use a square bracket if the symbol is $\leq$ or $\geq$. **3.** $x > -4$ **5.** $x \leq 4$
7. $(-\infty, 4]$ **9.** $(-\infty, -3)$ **11.** $(4, \infty)$ **13.** $[8, 10]$
15. $(0, 10]$ **17.** It would imply that $3 < -2$, a false statement. **19.** $[1, \infty)$
21. $[5, \infty)$ **23.** $(-\infty, -11)$ **25.** It must be reversed when multiplying or dividing by a negative number. **27.** His method is incorrect. He divided *by* the positive number 6. The sign of the number divided *into* does not matter.
29. $(-\infty, 6)$ **31.** $[-10, \infty)$ **33.** $(-\infty, -3)$
35. $(-\infty, 0]$ **37.** $(20, \infty)$ **39.** $[-3, \infty)$
41. $[-5, \infty)$ **43.** $(-\infty, 1)$ **45.** $(-\infty, 0]$
47. $[4, \infty)$ **49.** $(-\infty, 32)$ **51.** $\left[\dfrac{5}{12}, \infty\right)$
53. $(-21, \infty)$ **55.** $-1 < x < 2$ **57.** $-1 < x \leq 2$ **59.** $[-1, 6]$
61. $\left(-\dfrac{11}{6}, -\dfrac{2}{3}\right)$ **63.** $(1, 3)$ **65.** $[-26, 6]$

67. $[-3, 6]$ 69. $\left[-\dfrac{24}{5}, 0\right]$ 71. $\{4\}$ 72. $(4, \infty)$

73. $(-\infty, 4)$ 74. The graph would be all real numbers.   75. The graph would be all real numbers.

76. If a point on the number line satisfies an equation, the points on one side of that point will satisfy the corresponding less-than inequality, and the points on the other side will satisfy the corresponding greater-than inequality.   77. 83 or greater   79. all numbers greater than 16   81. It is never more than 86° Fahrenheit.   83. 32 or greater   85. 15 min   87. 9.5 gal   89. $x \geq 500$

Chapter 2 Review Exercises   (page 175)

1. $\{6\}$ 3. $\{7\}$ 5. $\{11\}$ 7. $\{5\}$ 9. $\{5\}$ 11. $\left\{\dfrac{64}{5}\right\}$ 13. {all real numbers}   15. {all real numbers}   17. $\emptyset$
19. Democrats: 75; Republicans: 45   21. Seven Falls: 300 ft; Twin Falls: 120 ft   23. 11, 13   25. $A = 28$ 27. $V = 904.32$
29. $h = \dfrac{2A}{b+B}$ 31. 100°; 100°   33. diameter: approximately 19.9 ft; radius: approximately 9.95 ft; area: approximately 311 ft²
35. Not enough information is given. We also need the value of B.   37. $\dfrac{5}{14}$ 39. $\dfrac{1}{12}$ 41. $\left\{-\dfrac{8}{3}\right\}$ 43. $6\dfrac{2}{3}$ lb   45. 375 km
47. 25.5-oz size   49. 3.75 L   51. 8.2 mph   53. $2\dfrac{1}{2}$ hr   55. $(-\infty, 7)$ 57. $[-3, \infty)$
59. $[3, \infty)$ 61. $(-\infty, -5)$ 63. $\left[-2, \dfrac{3}{2}\right]$ 65. 88 or more
67. $\{7\}$ 69. $(-\infty, 2)$ 71. $\{70\}$ 73. $\emptyset$ 75. Since $-(8 + 4x) = -1(8 + 4x)$, the first step is to distribute -1 over *both* terms in the parentheses, to get $3 - 8 - 4x$ on the left side of the equation. The student got $3 - 8 + 4x$ instead. The correct solution set is $\{-2\}$.
77. Golden Gate Bridge: 4200 ft; Brooklyn Bridge: 1595 ft   79. 8 qt   81. 44 m

Chapter 2 Test   (page 179)

[2.1–2.3] 1. $\{-6\}$ 2. $\{21\}$ 3. $\emptyset$ 4. $\{30\}$ 5. {all real numbers}   [2.4] 6. wins: 116; losses: 46   7. Hawaii: 4021 mi²; Maui: 728 mi²; Kauai: 551 mi²   8. 50°   [2.5] 9. (a) $W = \dfrac{P - 2L}{2}$ or $W = \dfrac{P}{2} - L$ (b) 18   10. 75°, 75°   [2.6] 11. $\{6\}$ 12. $\{-29\}$
13. 8 slices for $2.19   14. 2300 mi   [2.7] 15. $8000 at 3%; $14,000 at 4.5%   16. 4 hr   [2.8] 17. $(-\infty, 4]$
18. $(-2, 6]$ 19. 83 or more   20. When an inequality is multiplied or divided by a negative number, the direction of the inequality symbol must be reversed.

Cumulative Review Exercises Chapters 1–2   (page 181)

[1.1] 1. $\dfrac{3}{4}$ 2. $\dfrac{37}{60}$ 3. $\dfrac{48}{5}$ [1.2] 4. $\dfrac{1}{2}x - 18$ 5. $\dfrac{6}{x + 12} = 2$ [1.4] 6. true   [1.5–1.6] 7. 11   8. -8 9. 28
[1.3] 10. $-\dfrac{19}{3}$ [1.7] 11. distributive property   12. commutative property   [1.8] 13. $2k - 11$ [2.1–2.3] 14. $\{-1\}$ 15. $\{-1\}$
16. $\{-12\}$ [2.6] 17. $\{26\}$ [2.5] 18. $y = \dfrac{24 - 3x}{4}$ 19. $n = \dfrac{A - P}{iP}$ [2.8] 20. $(-\infty, 1]$
21. $(-1, 2]$ [2.4] 22. 4 cm; 9 cm; 27 cm   [2.5] 23. 12.42 cm   [2.6] 24. $\dfrac{25}{6}$ or $4\dfrac{1}{6}$ cups
[2.7] 25. 40 mph; 60 mph

A-16 Answers to Selected Exercises

CHAPTER 3 LINEAR EQUATIONS AND INEQUALITIES IN TWO VARIABLES; FUNCTIONS

Section 3.1 (page 192)

Exercises **1.** Ohio (OH): about 680 million eggs; Iowa (IA): about 550 million eggs   **3.** Indiana (IN) and Pennsylvania (PA); about 490 million eggs each   **5.** from 1975 to 1980; about $.75   **7.** The price of a gallon of gas was decreasing.   **9.** does; do not   **11.** II   **13.** 3   **15.** A linear equation in one variable can be written in the form $Ax + B = C$, where $A \neq 0$. Examples are $2x + 5 = 0$, $3x + 6 = 2$, and $x = -5$. A linear equation in two variables can be written in the form $Ax + By = C$, where A and B cannot both equal 0. Examples are $2x + 3y = 8$, $3x = 5y$, and $x - y = 0$.   **17.** yes   **19.** yes   **21.** no   **23.** yes   **25.** yes   **27.** no   **29.** No, the ordered pair (3, 4) represents the point 3 units to the right of the origin and 4 units up from the x-axis. The ordered pair (4, 3) represents the point 4 units to the right of the origin and 3 units up from the x-axis.   **31.** 17   **33.** -5 **35.** -1 **37.** -7 **39.** b **41.** 8; 6; 3   **43.** -9; 4; 9   **45.** 12; 12; 12   **47.** A: (2, 5); B: (-2, 2); C: (-5, 5); D: (-6, -2); E: (0, -1)   **49.–55.** 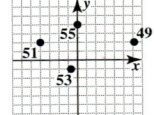   **57.** negative; negative   **59.** positive; negative

61. -3; 6; -2; 4   **63.** -3; 4; -6; $-\dfrac{4}{3}$ **65.** -4; -4; -4; -4 **67.** The points in each graph appear to lie on a straight line.   **69.** (a) (5, 45) (b) (6, 50)   **71.** (a) (1996, 53.3), (1997, 52.8), (1998, 52.1), (1999, 51.6) (b) (1995, 54.0) means that in 1995, the graduation rate for 4-yr college students within 5 yr was 54.0%. (c) 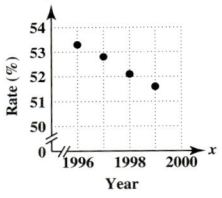 (d) The points appear to lie on a straight line. Graduation rates for 4-yr college students within 5 yr are decreasing.   **73.** (a) 170; 154; 138; 122 (b) (20, 170), (40, 154), (60, 138), (80, 122) (c) 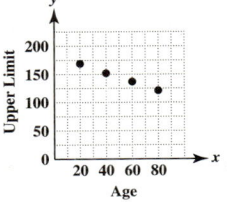 The points lie in a linear pattern.

Section 3.2 (page 204)

Connections (page 202)   **1.** $3x + 4 - 2x - 7 - 4x - 3 = 0$ **2.** $5x - 15 - 3(x - 2) = 0$

Exercises   **1.** 5; 5; 3 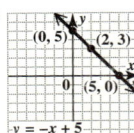   **3.** 1; 3; -1 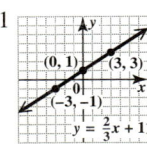   **5.** -6; -2; -5 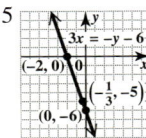   **7.** C   **9.** D   **11.** B

Answers to Selected Exercises   A-17

13. (12, 0); (0, −8)   **15.** (0, 0); (0, 0)   **17.** (4, 0); (0, −10)   **19.** (4, 0); none   **21.** $y = 0$; $x = 0$ **23.**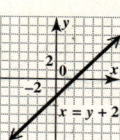

25. 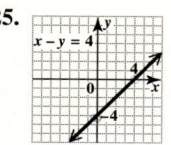   **27.** 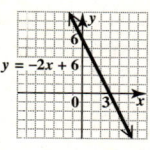   **29.** 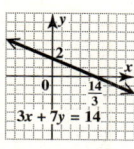   **31.** 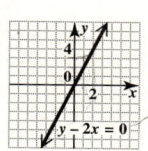   **33.** 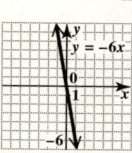   **35.**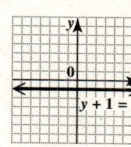

37.   **39.** 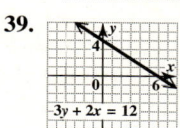   **41.**   **43.** 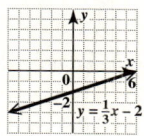   **45.** 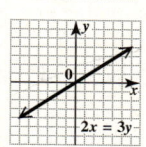   **47.**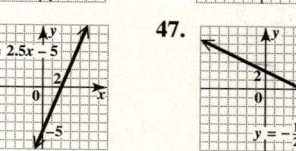

49.   **51.** 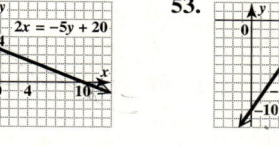   **53.**   **55.**   **57.** 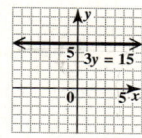   **59.**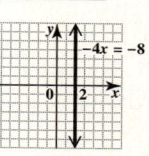

61. 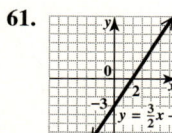   **63.** 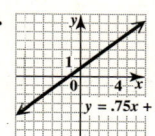   **65.**   **67.** 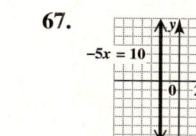   **69.**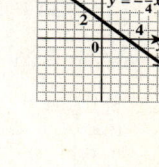

71. (a) 151.5 cm, 174.9 cm, 159.3 cm (c) 24 cm; 24 cm   **73.** (a) 130 (b) 133 (c) They are quite close.   **75.** between 133 and 162
77. (a) 1993: 49.8 gal; 1995: 51.4 gal; 1997: 53 gal (b) 1993: 50.1 gal; 1995: 51.6 gal; 1997: 53 gal (c) Corresponding values are quite close.
79. (a) $30,000 (b) $15,000 (c) $5000 (d) After 5 yr, the SUV has a value of $5000.

Section 3.3   (page 215)

Exercises   **1.** Rise is the vertical change between two different points on a line. Run is the horizontal change between two different points on a line.   **3.** 4   **5.** $-\frac{1}{2}$ **7.** 0   **9.** Yes; it doesn't matter which point you start with. Both differences will be the negatives of the differences in Exercise 8, and the quotient will be the same.   **In Exercises 11 and 13, sketches will vary.**   **11.** The line must rise from left to right.   **13.** The line must be horizontal.   **15.** Because he found the difference $3 - 5 = -2$ in the numerator, he should have subtracted in the same order in the denominator to get $-1 - 2 = -3$. The correct slope is $\frac{-2}{-3} = \frac{2}{3}$.   **17.** $\frac{5}{4}$ **19.** $\frac{3}{2}$ **21.** −3   **23.** 0
25. undefined   **27.** $-\frac{1}{2}$ **29.** 5   **31.** $\frac{1}{4}$ **33.** $\frac{3}{2}$ **35.** $\frac{3}{2}$ **37.** 0   **39.** undefined   **41.** −3; $\frac{1}{3}$ **43.** (a) negative (b) zero
45. (a) positive (b) negative   **47.** (a) zero (b) negative   **49.** A   **51.** $-\frac{2}{5}$; $-\frac{2}{5}$; parallel   **53.** $\frac{8}{9}$; $-\frac{4}{3}$; neither   **55.** $\frac{3}{2}$; $-\frac{2}{3}$; perpendicular   **57.** 5; $\frac{1}{5}$; neither   **59.** $\frac{3}{10}$ **61.** 232 thousand or 232,000   **62.** positive; increased   **63.** 232,000 students   **64.** −1.66
65. negative; decreased   **66.** 1.66 students per computer   **67.** The change for each year is .1 billion (or 100,000,000) ft², so the graph is a straight line.   **69.** $\frac{2}{5}$ **71.** (0, 4)

Section 3.4   (page 226)

Connections   (page 222)   **1.** $3500, $5000; $35,000, $50,000   **2.** the loss in value each year; the depreciation when the item is brand new ($D = 0$)

A-18 Answers to Selected Exercises

Exercises 1. D 3. B 5. The slope m of a vertical line is undefined, so it is not possible to write the equation of the line in the form $y = mx + b$. 7. $y = 3x - 3$ 9. $y = -x + 3$ 11. $y = 4x - 3$ 13. $y = 3$ 15. $x = 0$ 17.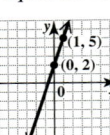

19. 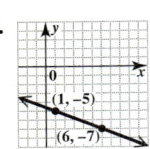 21. 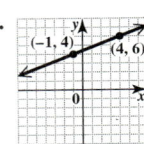 23. 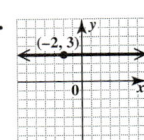 25. 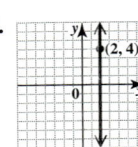 27. the x-axis 29. $y = -4x - 1$

31. $y = \frac{2}{3}x + \frac{19}{3}$ 33. $y = \frac{3}{4}x + 4$ 35. $y = \frac{5}{2}x - 4$ 37. $y = x$ (There are other forms as well.) 39. $y = x + 6$
41. $y = -\frac{3}{5}x - \frac{11}{5}$ 43. $y = \frac{1}{2}x + 2$ 45. $y = -\frac{1}{3}x + \frac{22}{9}$ 47. (0, 32), (100, 212) 48. $\frac{9}{5}$ 49. $F - 32 = \frac{9}{5}(C - 0)$
50. $F = \frac{9}{5}C + 32$ 51. $C = \frac{5}{9}(F - 32)$ 52. 86° 53. 10° 54. −40° 55. $y = \frac{3}{4}x - \frac{9}{2}$ 57. $y = -2x - 3$
59. (a) $400 (b) $.25 (c) $y = .25x + 400$ (d) $425 (e) 1500 61. (a) (1, 10,017), (3, 11,025), (5, 12,432), (7, 13,785), (9, 15,380)
(b) Yes (c) $y = 725.8x + 8847.6$ or $y = 725.8x + 8847.8$ (depending on the point used) (d) $18,283

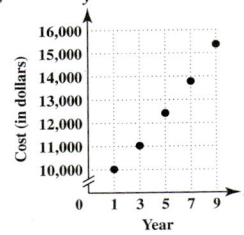

63. $y = -3x + 6$ 65. $Y_1 = \frac{3}{4}X + 1$

Section 3.5 (page 235)

Connections (page 235) 1. Answers will vary. 2. (a) $(5, \infty)$ (b) $(-\infty, 5)$ (c) $\left[3, \frac{11}{3}\right)$

Exercises 1. false 3. true 5. > 7. ≤ 9. < 11. 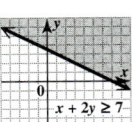 13. 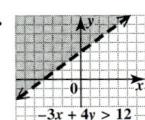 15.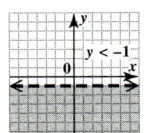

17. Use a dashed line if the symbol is < or >. Use a solid line if the symbol is ≤ or ≥. 19. 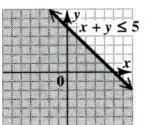 21.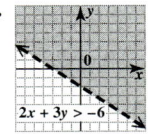

23. 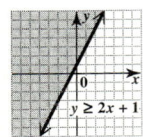 25. 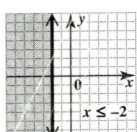 27. 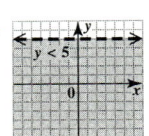 29. 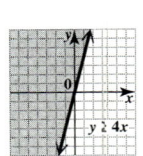 31. Every point in quadrant IV has a positive x-value and a negative y-value. Substituting into $y > x$ would imply that a negative number is greater than a positive number, which is always

Answers to Selected Exercises **A-19**

false. Thus, the graph of $y > x$ cannot lie in quadrant IV.   **33.** A   **35.** C   **37.** (a) 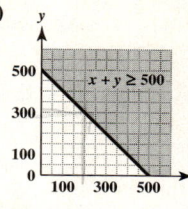   (b) (500, 0), (200, 400) Answers

may vary.   **39.** (a) 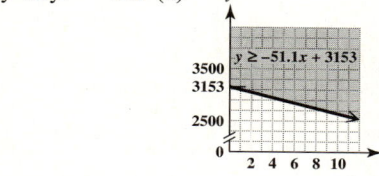   (b) (0, 3153), (2, 3050.8), (10, 2642) Answers may vary.

Section 3.6 (page 244)

Exercises   **1.** 3; 3; (1, 3)   **3.** 5; 5; (3, 5)   **5.** The graph consists of the four points (0, 2), (1, 3), (2, 4), and (3, 5).
7. (a) domain: $\{-4, -2, 0\}$; range: $\{3, 1, 5, -8\}$ (b) not a function   **9.** (a) domain: $\{A, B, C, D, E\}$; range: $\{2, 3, 6, 4\}$ (b) function
11. (a) domain: $\{-4, -2, 0, 2, 3\}$; range: $\{-2, 0, 1, 2, 3\}$ (b) not a function   **13.** function   **15.** not a function   **17.** function
19. function   **21.** not a function   **23.** domain: $(-\infty, \infty)$; range: $(-\infty, \infty)$ **25.** domain: $(-\infty, \infty)$; range: $[2, \infty)$ **27.** domain: $[0, \infty)$;
range: $[0, \infty)$ **29.** (2, 4)   **30.** $(-1, -4)$ **31.** $\frac{8}{3}$ **32.** $f(x) = \frac{8}{3}x - \frac{4}{3}$ **33.** (a) 11 (b) 3 (c) -9 **35.** (a) 4 (b) 2 (c) 14
37. (a) 2 (b) 0 (c) 3   **39.** (a) 4 (b) 2   **41.** {(1970, 9.6), (1980, 14.1), (1990, 19.8), (2000, 28.4)}; yes   **43.** $g(1980) = 14.1$;
$g(1990) = 19.8$ **45.** For the year 2002, the function indicates 30.3 million foreign-born residents in the United States.
46. (a) 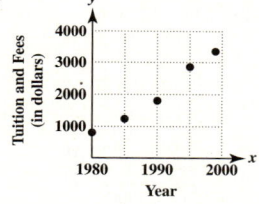   domain: {1980, 1985, 1990, 1995, 1999}; range: {804, 1242, 1809, 2860, 3356}; yes (b) 2860; 1999

47. $y = 134.32x - 265{,}150$ **48.** 1990: $2147; 1995: $2818   **49.** $y = 161.8x - 319{,}931$ **50.** 1980: $433; 1999: $3507; The results
from Exercise 48 are a little better.   **51.** 4   **53.** 1   **55.** 1   **57.** $y = x + 1$ **59.** The domain of a function is the set of all possible
replacements for x; the range is the set of all possible values of y.

Chapter 3 Review Exercises (page 253)

1. (a) $1.05 per gal; $1.75 per gal (b) $.70 per gal; about 67% (c) between April and June 2000; about $.40 per gal (d) August–October
1999 and February–April 2000   **3.** 2; $\frac{3}{2}$; $\frac{14}{3}$ **5.** 7; 7; 7   **7.** no   **9.** I   Exercises 9 and 11   **11.** none   **13.** I or III

15. $\left(\frac{8}{3}, 0\right)$; (0, 4) 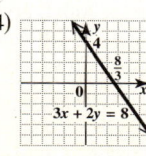   **17.** $-\frac{1}{2}$ **19.** 3   **21.** $\frac{3}{2}$ **23.** $\frac{3}{2}$ **25.** parallel   **27.** neither   **29.** $y = -\frac{1}{2}x + 4$

A-20 Answers to Selected Exercises

31. $y = \frac{2}{3}x + \frac{14}{3}$ **33.** $y = -\frac{1}{4}x + \frac{3}{2}$ **35.** $x = \frac{1}{3}$; It is not possible to express this equation in the form $y = mx + b$.

37. 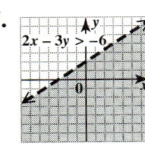 **39.** not a function; domain: $\{-2, 0, 2\}$; range: $\{4, 8, 5, 3\}$ **41.** not a function **43.** function **45.** (a) 8 (b) -1

47. (a) 5 (b) 2 **49.** C, D **51.** D **53.** B **55.** $(0, 0); (0, 0); -\frac{1}{3}$ 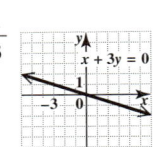 **57.** $y = -\frac{1}{4}x - \frac{5}{4}$ **59.** $y = -\frac{4}{7}x - \frac{23}{7}$

61. 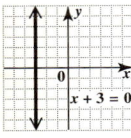 **62.** about $1.5 billion **63.** It will have negative slope since the total spent on video rentals is decreasing over these years. **64.** (1996, 11.1), (2000, 9.6) **65.** $y = -.375x + 759.6$ **66.** $-.375$; Yes, the slope is negative. **67.** 10.7; 10.4; 10.0 **68.** The actual amounts are fairly close to those given by the equation. **69.** $8.9 billion **70.** Answers will vary.

Chapter 3 Test (page 257)

[3.1] **1.** $-6, -10, -5$ **2.** no [3.2] **3.** To find the x-intercept, let $y = 0$, and to find the y-intercept, let $x = 0$. **4.** x-intercept: $(2, 0)$; y-intercept: $(0, 6)$ **5.** x-intercept: $(0, 0)$; y-intercept: $(0, 0)$ **6.** x-intercept: $(-3, 0)$; y-intercept: none **7.** x-intercept: none; y-intercept: $(0, 1)$ **8.** x-intercept: $(4, 0)$; y-intercept: $(0, -4)$

[3.3] **9.** $-\frac{8}{3}$ **10.** -2 **11.** undefined **12.** $\frac{5}{2}$ **13.** 0 [3.4] **14.** $y = 2x + 6$ **15.** $y = \frac{5}{2}x - 4$ **16.** $y = -9x + 12$ [3.5] **17.** **18.** [3.1–3.3] **19.** The slope is positive since food and drink sales are increasing. **20.** $(0, 43), (30, 376); 11.1$ **21.** 1990: $265 billion; 1995: $320.5 billion **22.** In 2000, food and drink sales were $376 billion. [3.6] **23.** (a) not a function (b) function; domain: $\{0, 1, 2\}$; range: $\{2\}$ **24.** not a function **25.** 1

Cumulative Review Exercises Chapters 1–3 (page 258)

[1.1] **1.** $\frac{301}{40}$ or $7\frac{21}{40}$ **2.** 6 [1.5] **3.** 7 [1.6] **4.** $\frac{73}{18}$ or $4\frac{1}{18}$ [1.2–1.6] **5.** true **6.** -43 [1.7] **7.** distributive property [1.8] **8.** $-p + 2$ [2.5] **9.** $h = \frac{3V}{\pi r^2}$ [2.3] **10.** $\{-1\}$ **11.** $\{2\}$ [2.6] **12.** $\{-13\}$ [2.8] **13.** $(-2.6, \infty)$

14. $(0, \infty)$ **15.** $(-\infty, -4]$ [2.4, 2.5] **16.** high school diploma: $22,895; bachelor's

degree: $40,478   **17.** 13 mi   **[3.1] 18. (a)** 89.45; 81.95; 78.20   **(b)** In 1980, the winning time was 85.7 sec.
19. (a) $7000   **(b)** $10,000   **(c)** about $30,000   **[3.2] 20.** $(-4, 0); (0, 3)$ **[3.3] 21.** $\frac{3}{4}$ **[3.2] 22.**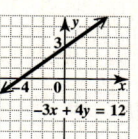
[3.3] 23. perpendicular   **[3.4] 24.** $y = 3x - 11$ **25.** $y = 4$

CHAPTER 4 SYSTEMS OF LINEAR EQUATIONS AND INEQUALITIES

Section 4.1 (page 267)

Exercises   **1. (a)** B   **(b)** C   **(c)** D   **(d)** A   **3.** no   **5.** yes   **7.** yes   **9.** no   **11.** A; The ordered pair solution must be in quadrant II, and $(-4, -4)$ is in quadrant III.   **13.** $\{(4, 2)\}$ **15.** $\{(0, 4)\}$ **17.** $\{(4, -1)\}$

In Exercises 19–27, we do not show the graphs.   **19.** $\{(1, 3)\}$ **21.** $\{(0, 2)\}$ **23.** ∅ (inconsistent system)   **25.** infinite number of solutions (dependent equations)   **27.** $\{(4, -3)\}$ **29.** If the coordinates of the point of intersection are not integers, the solution will be difficult to determine from a graph.   **31. (a)** neither   **(b)** intersecting lines   **(c)** one solution   **33. (a)** dependent   **(b)** one line   **(c)** infinite number of solutions   **35. (a)** inconsistent   **(b)** parallel lines   **(c)** no solution   **37. (a)** neither   **(b)** intersecting lines   **(c)** one solution   **39.** 40   **41.** (40, 30)   **43.** 1995–1997   **45. (a)** 1989–1997   **(b)** 1997; NBC; 17%   **(c)** 1989: share 20%; 1998: share 16%   **(d)** NBC and ABC; (2000, 16)   **(e)** Viewership has generally declined during these years.   **47.** B   **49.** A   **51.** $\{(-1, 5)\}$ **53.** $\{(.25, -.5)\}$

Section 4.2 (page 277)

Exercises   **1.** No, it is not correct, because the solution set is $\{(3, 0)\}$. The y-value in the ordered pair must also be determined.   **3.** $\{(3, 9)\}$
5. $\{(7, 3)\}$ **7.** $\{(0, 5)\}$ **9.** $\{(-4, 8)\}$ **11.** $\{(3, -2)\}$ **13.** infinite number of solutions   **15.** $\left\{\left(\frac{1}{4}, -\frac{1}{2}\right)\right\}$ **17.** ∅   **19.** infinite number of solutions   **21. (a)** A false statement, such as $0 = 3$, occurs.   **(b)** A true statement, such as $0 = 0$, occurs.   **23.** $\{(2, 6)\}$
25. $\{(2, -4)\}$ **27.** $\{(-2, 1)\}$ **29.** $\left\{\left(13, -\frac{7}{5}\right)\right\}$ **31.** infinite number of solutions   **33.** To find the total cost, multiply the number of bicycles (x) by the cost per bicycle ($400), and add the fixed cost ($5000). Thus, $y_1 = 400x + 5000$ gives this total cost (in dollars).
34. $y_2 = 600x$ **35.** $y_1 = 400x + 5000$, $y_2 = 600x$; solution set: $\{(25, 15,000)\}$ **36.** 25; 15,000; 15,000
37. $\{(2, 4)\}$ 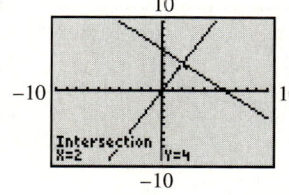   **39.** $\{(1, 5)\}$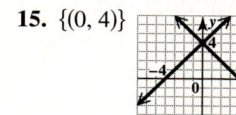

41. $\{(5, -3)\}$; The equations to input are $Y_1 = \frac{5 - 4X}{5}$ and $Y_2 = \frac{1 - 2X}{3}$.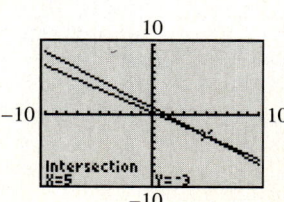

43. Adjust the viewing window so that it does appear.

Section 4.3 (page 283)

Exercises **1.** false; Multiply by -3. **3.** true **5.** $\{(4, 6)\}$ **7.** $\{(-1, -3)\}$ **9.** $\{(-2, 3)\}$ **11.** $\{(\frac{1}{2}, 4)\}$ **13.** $\{(3, -6)\}$ **15.** $\{(0, 4)\}$ **17.** $\{(0, 0)\}$ **19.** $\{(7, 4)\}$ **21.** $\{(0, 3)\}$ **23.** $\{(3, 0)\}$ **25.** $\{(-\frac{2}{3}, \frac{17}{2})\}$ **27.** $\{(-\frac{32}{23}, -\frac{17}{23})\}$ **29.** $\{(-\frac{6}{5}, \frac{4}{5})\}$ **31.** $\{(\frac{1}{8}, -\frac{5}{6})\}$ **33.** $\{(11, 15)\}$ **35.** infinite number of solutions **37.** $\emptyset$ **39.** $1141 = 1991a + b$ **40.** $1465 = 1999a + b$ **41.** $1991a + b = 1141$; $1999a + b = 1465$; solution set: $\{(40.5, -79{,}494.5)\}$ **42.** $y = 40.5x - 79{,}494.5$ **43.** 1424.5 (million); This is less than the actual figure. **44.** Since the data do not lie in a perfectly straight line, the quantity obtained from an equation determined in this way will probably be "off" a bit. We cannot put too much faith in models such as this one, because not all sets of data points are linear in nature.

Summary Exercises on Solving Systems of Linear Equations (page 286)

1. (a) Use substitution since the second equation is solved for y. (b) Use elimination since the coefficients of the y-terms are opposites. (c) Use elimination since the equations are in standard form with no coefficients of 1 or -1. Solving by substitution would involve fractions. **2.** The system on the right is easier to solve by substitution because the second equation is already solved for y. **3.** (a) $\{(1, 4)\}$ (b) $\{(1, 4)\}$ (c) Answers will vary. **4.** (a) $\{(-5, 2)\}$ (b) $\{(-5, 2)\}$ (c) Answers will vary. **5.** $\{(3, 12)\}$ **6.** $\{(-3, 2)\}$ **7.** $\{(\frac{1}{3}, \frac{1}{2})\}$ **8.** $\emptyset$ **9.** $\{(3, -2)\}$ **10.** $\{(-1, -11)\}$ **11.** infinite number of solutions **12.** $\{(9, 4)\}$ **13.** $\{(\frac{45}{31}, \frac{4}{31})\}$ **14.** $\{(4, -5)\}$ **15.** $\emptyset$ **16.** $\{(-4, 6)\}$ **17.** $\{(-3, 2)\}$ **18.** $\{(\frac{22}{13}, -\frac{23}{13})\}$ **19.** $\{(5, 3)\}$ **20.** $\{(2, -3)\}$ **21.** $\{(24, -12)\}$ **22.** $\{(10, -12)\}$ **23.** $\{(3, 2)\}$ **24.** $\{(-4, 2)\}$

Section 4.4 (page 292)

Exercises **1.** D **3.** B **5.** C **7.** the second number; $x - y = 48$; The two numbers are 73 and 25. **9.** KISS: 120; Dave Matthews Band: 43 **11.** (a) *Harry Potter and the Sorcerer's Stone*: $294 million; *Shrek*: $268 million (b) $241 million (c) $803 million **13.** (a) 45 units (b) Do not produce; the product will lead to a loss. **15.** ones: 46; tens: 28 **17.** 2 copies of *Harry Potter and the Sorcerer's Stone*; 5 Backstreet Boys CDs **19.** $2500 at 4%; $5000 at 5% **21.** Japan: $17.19; Switzerland: $13.15 **23.** 40% solution: 80 L; 70% solution: 40 L **25.** 30 lb at $6 per lb; 60 lb at $3 per lb **27.** 30 barrels at $40 per barrel; 20 barrels at $60 per barrel **29.** 45 mph, 75 mph **31.** car leaving Cincinnati: 55 mph; car leaving Toledo: 70 mph **33.** boat: 10 mph; current: 2 mph **35.** plane: 470 mph; wind: 30 mph **37.** Roberto: 17.5 mph; Juana: 12.5 mph

Section 4.5 (page 300)

Exercises **1.** C **3.** B **5.** 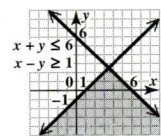 **7.** 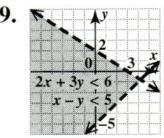 **9.** 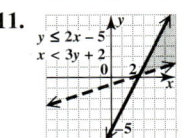 **11.** 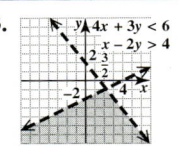 **13.**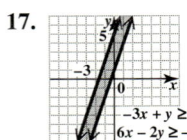

15. **17.** 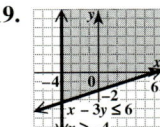 **19.** 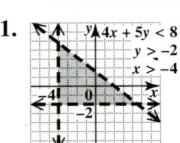 **21.** 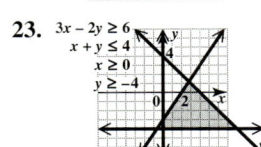 **23.** **25.** D **27.** A

Chapter 4 Review Exercises   (page 306)

1. yes   **3.** {(3, 1)}   **5.** ∅   **7.** No, two lines cannot intersect in exactly three points.   **9.** {(3, 5)}   **11.** His answer was incorrect since the system has infinitely many solutions (as indicated by the true statement 0 = 0).   **13.** C   **15.** {(7, 1)}   **17.** infinite number of solutions   **19.** {(−4, 1)}   **21.** {(9, 2)}   **23.** Answers will vary.   **25.** Subway: 13,247 restaurants; McDonald's: 13,099 restaurants   **27.** Texas Commerce Tower: 75 stories; First Interstate Plaza: 71 stories   **29.** length: 27 m; width: 18 m   **31.** 25 lb of $1.30 candy; 75 lb of $.90 candy   **33.** $7000 at 3%; $11,000 at 4%   **35.** 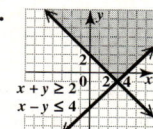   **37.** 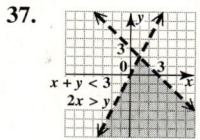   **39. (a)** years 0 to 6   **(b)** year 6; about $650

41. infinite number of solutions   **43.** {(−4, 15)}   **45.**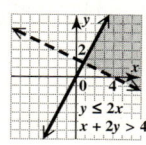

47. Statue of Liberty: 5.5 million; Vietnam Veterans Memorial: 3.8 million   **49.** small bottles: 102; large bottles: 44

Chapter 4 Test   (page 309)

[4.1] **1. (a)** x = 8 or 800 parts **(b)** $3000   **2. (a)** no **(b)** no **(c)** yes   **3.** {(4, 1)}   [4.2] **4.** {(1, −6)}   **5.** {(−35, 35)}   [4.3] **6.** {(5, 6)}   **7.** {(−1, 3)}   **8.** {(−1, 3)}   **9.** ∅   **10.** {(0, 0)}   **11.** {(−15, 6)}   [4.1–4.3] **12.** infinite number of solutions   **13.** It has no solution.   [4.4] **14.** Memphis and Atlanta: 371 mi; Minneapolis and Houston: 671 mi   **15.** Magic Kingdom: 15.4 million; Disneyland: 13.9 million   **16.** 25% solution: $33\frac{1}{3}$ L; 40% solution: $16\frac{2}{3}$ L   **17.** slower car: 45 mph; faster car: 60 mph   [4.5] **18.** 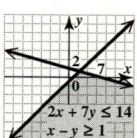   **19.** 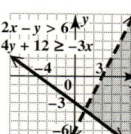   **20.** B

Cumulative Review Exercises Chapters 1–4   (page 311)

[1.6] **1.** −1, 1, −2, 2, −4, 4, −5, 5, −8, 8, −10, 10, −20, 20, −40, 40   [1.3] **2.** 1   [1.7] **3.** commutative property   **4.** distributive property   **5.** inverse property   [1.6] **6.** 46   [2.3] **7.** $\left\{-\frac{13}{11}\right\}$ **8.** $\left\{\frac{9}{11}\right\}$ [2.5] **9.** $T = \frac{PV}{k}$ [2.8] **10.** (−18, ∞)   **11.** $\left(-\frac{11}{2}, \infty\right)$ [2.7] **12.** 2010; 1813; 62.8%; 57.2%   [2.4] **13.** not guilty: 105; guilty: 95   [2.5] **14.** 46°, 46°, 88°   **15.** width: 8.25 in.; length: 10.75 in.   [3.2] **16.** 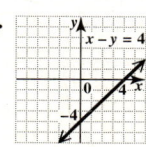   **17.** 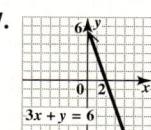   [3.3] **18.** $-\frac{4}{3}$ **19.** $-\frac{1}{4}$ [3.4] **20.** $y = \frac{1}{2}x + 3$ **21.** $y = 2x + 1$ **22. (a)** x = 9 **(b)** y = −1   [4.1–4.3] **23.** {(−1, 6)}   **24.** {(3, −4)}   **25.** {(2, −1)}   **26.** ∅   [4.4] **27.** 405 adults and 49 children   **28.** 19 in., 19 in., 15 in.   **29.** 20% solution: 4 L; 50% solution: 8 L   [4.5] **30.**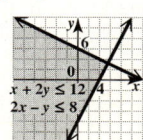

CHAPTER 5 EXPONENTS AND POLYNOMIALS

Section 5.1 (page 318)

Exercises 1. false 3. false 5. w^6 7. $\frac{1}{4^4}$ 9. $(-7x)^4$ 11. $\left(\frac{1}{2}\right)^6$ 13. In $(-3)^4$, -3 is the base, while in -3^4, 3 is the base. $(-3)^4 = 81$, while $-3^4 = -81$. 15. base: 3; exponent: 5; 243 17. base: -3; exponent: 5; -243 19. base: $-6x$; exponent: 4 21. base: x; exponent: 4 23. $5^2 + 5^3$ is a sum, not a product. $5^2 + 5^3 = 25 + 125 = 150$. 25. 5^8 27. 4^{12} 29. $(-7)^9$ 31. t^{24} 33. $-56r^7$ 35. $42p^{10}$ 37. $-30x^9$ 39. The product rule does not apply. 41. The product rule does not apply. 43. 4^6 45. t^{20} 47. 7^3r^3 49. $5^5x^5y^5$ 51. 5^{12} 53. -8^{15} 55. $8q^3r^3$ 57. $\frac{1}{2^3}$ 59. $\frac{a^3}{b^3}$ 61. $\frac{9^8}{5^8}$ 63. $\frac{5^5}{2^5}$ 65. $\frac{9^5}{8^3}$ 67. $2^{12}x^{12}$ 69. -6^5p^5 71. $6^5x^{10}y^{15}$ 73. x^{21} 75. $2^2w^4x^{26}y^7$ 77. $-r^{18}s^{17}$ 79. $\frac{5^3a^6b^{15}}{c^{18}}$ 81. This is incorrect. Using the product rule, it is simplified as follows: $(10^2)^3 = 10^{2 \cdot 3} = 10^6 = 1{,}000{,}000$. 83. $12x^5$ 85. $6p^7$ 87. $125x^6$ 89. $-a^4, -a^3, -(-a)^3, (-a)^4$; One way is to choose a positive number greater than 1 and substitute it for a in each expression. Then arrange the terms from smallest to largest. 91. $304.16 93. $1843.88

Section 5.2 (page 327)

Exercises 1. 1 3. 1 5. -1 7. 0 9. 0 11. C 13. F 15. E 17. 2 19. $\frac{1}{64}$ 21. 16 23. $\frac{49}{36}$ 25. $\frac{1}{81}$ 27. $\frac{8}{15}$ 29. 5^3 31. $\frac{5^3}{3^2}$ or $\frac{125}{9}$ 33. 5^2 35. x^{15} 37. 6^3 39. $2r^4$ 41. $\frac{5^2}{4^3}$ 43. $\frac{p^5}{q^8}$ 45. r^9 47. $\frac{yz^2}{4x^3}$ 49. $a+b$ 51. $(x+2y)^2$ 53. 1 54. $\frac{5^2}{5^2}$ 55. 5^0 56. $1 = 5^0$; This supports the definition of 0 as an exponent. 57. 7^3 or 343 59. $\frac{1}{x^2}$ 61. $\frac{64x}{9}$ 63. $\frac{x^2z^4}{y^2}$ 65. $6x$ 67. $\frac{1}{m^{10}n^5}$ 69. $\frac{1}{xyz}$ 71. x^3y^9 73. $\frac{a^{11}}{2b^5}$ 75. $\frac{108}{y^5z^3}$ 77. $\frac{9z^2}{400x^3}$ 79. The student attempted to use the quotient rule with unequal bases. The correct way to simplify this expression is $\frac{16^3}{2^2} = \frac{(2^4)^3}{2^2} = \frac{2^{12}}{2^2} = 2^{10} = 1024$.

Summary Exercises on the Rules for Exponents (page 329)

1. $\frac{6^{12}x^{24}}{5^{12}}$ 2. $\frac{r^6s^{12}}{729t^6}$ 3. $10^5x^7y^{14}$ 4. $-128a^{10}b^{15}c^4$ 5. $\frac{729w^3x^9}{y^{12}}$ 6. $\frac{x^4y^6}{16}$ 7. c^{22} 8. $\frac{1}{k^4t^{12}}$ 9. $\frac{11}{30}$ 10. $y^{12}z^3$ 11. $\frac{x^6}{y^5}$ 12. 0 13. $\frac{1}{z^2}$ 14. $\frac{9}{r^2s^2t^{10}}$ 15. $\frac{300x^3}{y^3}$ 16. $\frac{3}{5x^6}$ 17. x^8 18. $\frac{y^{11}}{x^{11}}$ 19. $\frac{a^6}{b^4}$ 20. $6ab$ 21. $\frac{61}{900}$ 22. 1 23. $\frac{343a^6b^9}{8}$ 24. 1 25. -1 26. 0 27. $\frac{27y^{18}}{4x^8}$ 28. $\frac{1}{a^8b^{12}c^{16}}$ 29. $\frac{x^{15}}{216z^9}$ 30. $\frac{q}{8p^6r^3}$ 31. x^6y^6 32. 0 33. $\frac{343}{x^{15}}$ 34. $\frac{9}{x^6}$ 35. $5p^{10}q^9$ 36. $\frac{7}{24}$ 37. $\frac{r^{14}t}{2s^2}$ 38. 1 39. $8p^{10}q$ 40. $\frac{1}{mn^3p^3}$ 41. -1 42. $\frac{3}{40}$

Section 5.3 (page 334)

Connections (page 333) 1. The Erzincan earthquake was ten times as powerful as the Sumatra earthquake. 2. The Adana earthquake had one hundredth the power of the San Francisco earthquake. 3. The Erzincan earthquake was about 50.12 times as powerful as the Adana earthquake. 4. "+3" corresponds to a factor of 1000 times stronger; "−1" corresponds to a factor of one-tenth.

Answers to Selected Exercises   A-25

Exercises   **1.** A   **3.** C   **5.** in scientific notation   **7.** not in scientific notation; 5.6×10^6 **9.** not in scientific notation; 8×10^1 **11.** not in scientific notation; 4×10^{-3} **13.** It is written as the product of a power of 10 and a number whose absolute value is between 1 and 10 (inclusive of 1). Some examples are 2.3×10^{-4} and 6.02×10^{23}.   **15.** 5.876×10^9 **17.** 8.235×10^4 **19.** 7×10^{-6} **21.** 2.03×10^{-3} **23.** 750,000   **25.** 5,677,000,000,000   **27.** 6.21   **29.** .00078   **31.** .000000005134   **33.** 600,000,000,000   **35.** 15,000,000   **37.** 60,000   **39.** .0003   **41.** 40   **43.** .000013   **45.** 4.7E−7   **47.** 2E7   **49.** 1E1   **51.** 1×10^{10} **53.** 2,000,000,000   **55.** $\$6.03 \times 10^8$ **57.** $\$5.2466 \times 10^{10}$ **59.** about 15,300 sec   **61.** about 9.2×10^8 acres   **63.** about 3.3   **65.** about .276 lb

Section 5.4   (page 344)

Exercises   **1.** 7; 5   **3.** 8   **5.** 26   **7.** 0   **9.** 1; 6   **11.** 1; 1   **13.** 2; −19, −1   **15.** 3; 1, 8, 5   **17.** $2m^5$ **19.** $-r^5$ **21.** cannot be simplified   **23.** $-5x^5$ **25.** $5p^9 + 4p^7$ **27.** $-2xy^2$ **29.** already simplified; 4; binomial   **31.** $11m^4 - 7m^3 - 3m^2$; 4; trinomial   **33.** x^4; 4; monomial   **35.** 7; 0; monomial   **37.** (a) 36 (b) −12   **39.** (a) 14 (b) −19   **41.** (a) −3 (b) 0   **43.** $5x^2 - 2x$ **45.** $5m^2 + 3m + 2$ **47.** $\frac{7}{6}x^2 - \frac{2}{15}x + \frac{5}{6}$ **49.** $6m^3 + m^2 + 12m - 14$ **51.** $3y^3 - 11y^2$ **53.** $4x^4 - 4x^2 + 4x$ **55.** $15m^3 - 13m^2 + 8m + 11$ **57.** Answers will vary.   **59.** $5m^2 - 14m + 6$ **61.** $4x^3 + 2x^2 + 5x$ **63.** $-11y^4 + 8y^2 + y$ **65.** $a^4 - a^2 + 1$ **67.** $5m^2 + 8m - 10$ **69.** $-6x^2 - 12x + 12$ **71.** $13.01x^2 + 4.93x - .56$ **73.** $1.5a^3 - 11.5a^2 + 3.8a - 22.5$ **75.** $-2.5x^2 + 4.3x - 6.8$ **77.** $2y^3 - y^2 + y - 1$ **79.** $\frac{1}{2}x^2 - \frac{13}{9}x + \frac{1}{3}$ **81.** −10   **83.** $4b - 5c$ **85.** $6x - xy - 7$ **87.** $-3x^2y - 15xy - 3xy^2$ **89.** $14.1x - 5.3y - 10.1z$ **91.** $17.10s + 17.24st - .35$ **93.** $-\frac{1}{6}x^2y^3 + \frac{11}{8}x^3y^2 + \frac{5}{9}x^4y$ **95.** $-\frac{1}{4}a^2 + \frac{13}{12}b^2$ **97.** $-a^3 + 2.8a^2 + .8a + 8.5$ (or $-a^3 + \frac{14}{5}a^2 + \frac{4}{5}a + \frac{17}{2}$)   **99.** $8x^2 + 8x + 6$ **101.** (a) $23y + 5t$ (b) approximately 37.26°, 79.52°, and 63.22°   **103.** $-6x^2 + 6x - 7$ **105.** $-7x - 1$ **107.** 0, −3, −4, −3, 0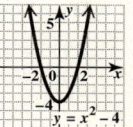

109. 7, 1, −1, 1, 7 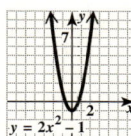   **111.** 0, 3, 4, 3, 0 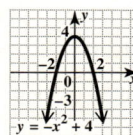   **113.** 4, 1, 0, 1, 4 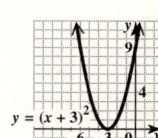   **115.** 175 ft   **117.** 4; $5.00

118. 6; $27   **119.** 2.5; 130   **120.** 459.87 billion

Section 5.5   (page 353)

Exercises   **1.** (a) B (b) D (c) A (d) C   **3.** $15y^{11}$ **5.** $30a^9$ **7.** $15pq^2$ **9.** $-18m^3n^2$ **11.** $6m^2 + 4m$ **13.** $-6p^4 + 12p^3$ **15.** $-16z^2 - 24z^3 - 24z^4$ **17.** $6y^3 + 4y^4 + 10y^7$ **19.** $28r^5 - 32r^4 + 36r^3$ **21.** $6a^4 - 12a^3b + 15a^2b^2$ **23.** $21m^5n^2 + 14m^4n^3 - 7m^3n^5$ **25.** $12x^3 + 26x^2 + 10x + 1$ **27.** $81a^3 + 27a^2 + 11a + 2$ **29.** $20m^4 - m^3 - 8m^2 - 17m - 15$ **31.** $6x^6 - 3x^5 - 4x^4 + 4x^3 - 5x^2 + 8x - 3$ **33.** $5x^4 - 13x^3 + 20x^2 + 7x + 5$ **35.** $3x^5 + 18x^4 - 2x^3 - 8x^2 + 24x$ **37.** $m^2 + 12m + 35$ **39.** $x^2 - 25$ **41.** $12x^2 + 10x - 12$ **43.** $9x^2 - 12x + 4$ **45.** $10a^2 + 37a + 7$ **47.** $12 + 8m - 15m^2$ **49.** $20 - 7x - 3x^2$ **51.** $8xy - 4x + 6y - 3$ **53.** $15x^2 + xy - 6y^2$ **55.** $6y^5 - 21y^4 - 45y^3$ **57.** $-200r^7 + 32r^3$ **59.** $3y^2 + 10y + 7$ **61.** $6p^2 - \frac{5}{2}pq - \frac{25}{12}q^2$ **63.** $x^2 + 14x + 49$ **65.** $a^2 - 16$ **67.** $4p^2 - 20p + 25$ **69.** $25k^2 + 30kq + 9q^2$ **71.** $m^3 - 15m^2 + 75m - 125$ **73.** $8a^3 + 12a^2 + 6a + 1$ **75.** $56m^2 - 14m - 21$ **77.** $-9a^3 + 33a^2 + 12a$ **79.** $81r^4 - 216r^3s + 216r^2s^2 - 96rs^3 + 16s^4$ **81.** $6p^8 + 15p^7 + 12p^6 + 36p^5 + 15p^4$ **83.** $-24x^8 - 28x^7 + 32x^6 + 20x^5$ **85.** $14x + 49$ **87.** $\pi x^2 - 9$ **89.** $30x + 60$ **90.** $30x + 60 = 600$ **91.** 18   **92.** 10 yd by 60 yd   **93.** $2100   **94.** 140 yd   **95.** $1260   **96.** (a) $30kx + 60k$ (dollars) (b) $6rx + 32r$ **97.** Suppose that we want to multiply $(x + 3)(2x + 5)$. The letter F stands for *first*. We multiply the two first terms to get $2x^2$. The letter O represents *outer*. We next multiply the two outer terms to get $5x$. The letter I represents *inner*. The product of the two inner terms is $6x$. L stands for *last*. The product of the two last terms is 15. Very often the outer and inner products are like terms, as they are in this case. So we simplify $2x^2 + 5x + 6x + 15$ to get the final product, $2x^2 + 11x + 15$.

Section 5.6 (page 359)

Exercises 1. (a) $4x^2$ (b) $12x$ (c) 9 (d) $4x^2 + 12x + 9$ 3. $m^2 + 4m + 4$ 5. $r^2 - 6r + 9$ 7. $x^2 + 4xy + 4y^2$ 9. $25p^2 + 20pq + 4q^2$ 11. $16a^2 + 40ab + 25b^2$ 13. $49t^2 + 14ts + s^2$ 15. $36m^2 - \frac{48}{5}mn + \frac{16}{25}n^2$ 17. $9t^3 - 6t^2 + t$ 19. $-9y^2 + 48y - 64$ 21. To find the product of the sum and the difference of two terms, we find the difference of the squares of the two terms. For example, $(5x + 7y)(5x - 7y) = (5x)^2 - (7y)^2 = 25x^2 - 49y^2$. 23. $a^2 - 64$ 25. $4 - p^2$ 27. $4m^2 - 25$ 29. $9x^2 - 16y^2$ 31. $25y^2 - 9x^2$ 33. $169r^2 - 4z^2$ 35. $81y^4 - 4$ 37. $81y^2 - \frac{4}{9}$ 39. $25q^3 - q$ 41. $x^3 + 3x^2 + 3x + 1$ 43. $t^3 - 9t^2 + 27t - 27$ 45. $r^3 + 15r^2 + 75r + 125$ 47. $8a^3 + 12a^2 + 6a + 1$ 49. $81r^4 - 216r^3t + 216r^2t^2 - 96rt^3 + 16t^4$ 51. $(a + b)^2$ 52. a^2 53. $2ab$ 54. b^2 55. $a^2 + 2ab + b^2$ 56. They both represent the area of the entire large square. 57. 1225 58. $30^2 + 2(30)(5) + 5^2$ 59. 1225 60. They are equal. 61. 9999 63. $39,999$ 65. $399\frac{3}{4}$ 67. $\frac{1}{2}m^2 - 2n^2$ 69. $9a^2 - 4$ 71. $\pi x^2 + 4\pi x + 4\pi$ 73. $x^3 + 6x^2 + 12x + 8$

Section 5.7 (page 368)

Connections (page 368) 1. -104 2. -104 3. They are both -104. 4. The answers should agree.

Exercises 1. $6x^2 + 8$; 2; $3x^2 + 4$ 3. $3x^2 + 4$; 2 (These may be reversed.); $6x^2 + 8$ 5. The first is a polynomial divided by a monomial, covered in Objective 1. This section does not cover dividing a monomial by a polynomial of several terms. 7. $30x^3 - 10x + 5$ 9. $4m^3 - 2m^2 + 1$ 11. $4t^4 - 2t^2 + 2t$ 13. $a^4 - a + \frac{2}{a}$ 15. $4x^3 - 3x^2 + 2x$ 17. $1 + 5x - 9x^2$ 19. $\frac{12}{x} + 8 + 2x$ 21. $\frac{4x^2}{3} + x + \frac{2}{3x}$ 23. $9r^3 - 12r^2 + 2r + \frac{26}{3} - \frac{2}{3r}$ 25. $-m^2 + 3m - \frac{4}{m}$ 27. $4 - 3a + \frac{5}{a}$ 29. $\frac{12}{x} - \frac{6}{x^2} + \frac{14}{x^3} - \frac{10}{x^4}$ 31. $6x^4y^2 - 4xy + 2xy^2 - x^4y$ 33. $5x^3 + 4x^2 - 3x + 1$ 35. $-63m^4 - 21m^3 - 35m^2 + 14m$ 37. 1423 38. $(1 \times 10^3) + (4 \times 10^2) + (2 \times 10^1) + (3 \times 10^0)$ 39. $x^3 + 4x^2 + 2x + 3$ 40. They are similar in that the coefficients of powers of ten are equal to the coefficients of the powers of x. They are different in that one is a constant while the other is a polynomial. They are equal if $x = 10$ (the base of our decimal system). 41. $x + 2$ 43. $2y - 5$ 45. $p - 4 + \frac{44}{p + 6}$ 47. $r - 5$ 49. $6m - 1$ 51. $2a - 14 + \frac{74}{2a + 3}$ 53. $4x^2 - 7x + 3$ 55. $4k^3 - k + 2$ 57. $5y^3 + 2y - 3$ 59. $3k^2 + 2k - 2 + \frac{6}{k - 2}$ 61. $2p^3 - 6p^2 + 7p - 4 + \frac{14}{3p + 1}$ 63. $r^2 - 1 + \frac{4}{r^2 - 1}$ 65. $y^2 - y + 1$ 67. $a^2 + 1$ 69. $x^2 - 4x + 2 + \frac{9x - 4}{x^2 + 3}$ 71. $x^3 + 3x^2 - x + 5$ 73. $\frac{3}{2}a - 10 + \frac{77}{2a + 6}$ 75. The process stops when the degree of the remainder is less than the degree of the divisor. 77. $x^2 + x - 3$ units 79. $5x^2 - 11x + 14$ hr 81. A is correct; B is incorrect. 82. A is incorrect; B is correct. 83. A is correct; B is incorrect. 84. Because any power of 1 is 1, to evaluate a polynomial for 1 we simply add the coefficients. If the divisor is $x - 1$, this method does not apply, since $1 - 1 = 0$ and division by 0 is undefined.

Chapter 5 Review Exercises (page 377)

1. 4^{11} 3. $-72x^7$ 5. 19^5x^5 7. $5p^4t^4$ 9. $6^2x^{16}y^4z^{16}$ 11. The expression is a *sum* of powers of 7, not a *product*. 13. 0 15. $-\frac{1}{49}$ 17. 5^8 19. $\frac{3}{4}$ 21. x^2 23. $\frac{r^8}{81}$ 25. $\frac{1}{a^3b^5}$ 27. 4.8×10^7 29. 8.24×10^{-8} 31. $78,300,000$ 33. 800 35. $.025$ 37. $.0000000000016$ 39. 9.7×10^4; 5×10^3 41. (a) 1×10^3 (b) 2×10^3 (c) 5×10^4 (d) 1×10^5 43. $p^3 - p^2 + 4p + 2$; degree 3; none of these 45. $-8y^5 - 7y^4 + 9y$; degree 5; trinomial 47. $13x^3y^2 - 5xy^5 + 21x^2$ 49. $y^2 - 10y + 9$

Answers to Selected Exercises   **A-27**

51. 1, 4, 5, 4, 1 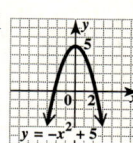   53. $a^3 - 2a^2 - 7a + 2$ 55. $5p^5 - 2p^4 - 3p^3 + 25p^2 + 15p$ 57. $6k^2 - 9k - 6$

59. $12k^2 - 32kq - 35q^2$ 61. $2x^2 + x - 6$ 63. $a^2 + 8a + 16$ 65. $36m^2 - 25$ 67. $r^3 + 6r^2 + 12r + 8$ 69. (a) Answers will vary. For example, let $x = 1$ and $y = 2$. $(1 + 2)^2 \neq 1^2 + 2^2$, because $9 \neq 5$. (b) Answers will vary. For example, let $x = 1$ and $y = 2$. $(1 + 2)^3 \neq 1^3 + 2^3$, because $27 \neq 9$.   71. In both cases, $x = 0$ and $y = 1$ lead to 1 on each side of the inequality. This would not be sufficient to show that *in general* the inequality is true. It would be necessary to choose other values of x and y.   73. $\frac{4}{3}\pi(x + 1)^3$ or $\frac{4}{3}\pi x^3 + 4\pi x^2 + 4\pi x + \frac{4}{3}\pi$ in.^{3} 75. $y^3 - 2y + 3$ 77. $2mn + 3m^4n^2 - 4n$ 79. $2r + 7$ 81. $x^2 + 3x - 4$ 83. $4x - 5$
85. $y^2 + 2y + 4$ 87. $2y^2 - 5y + 4 + \frac{-5}{3y^2 + 1}$ 89. 2   91. $144a^2 - 1$ 93. $\frac{1}{8^{12}}$ 95. $\frac{2}{3m^3}$ 97. r^{13} 99. $-y^2 - 4y + 4$
101. $y^2 + 5y + 1$ 103. $10p^2 - 3p - 5$ 105. $49 - 28k + 4k^2$

Chapter 5 Test   (page 381)

[5.1, 5.2] 1. $\frac{1}{625}$ 2. 2   3. $\frac{7}{12}$ 4. $9x^3y^5$ 5. 8^5 6. x^2y^6 7. (a) positive (b) positive (c) negative (d) positive (e) zero (f) negative   [5.3] 8. (a) 4.5×10^{10} (b) .0000036 (c) .00019   9. (a) 1×10^3; 5.89×10^{12} (b) 5.89×10^{15} mi   [5.4] 10. $-7x^2 + 8x$; 2; binomial   11. $4n^4 + 13n^3 - 10n^2$; 4; trinomial   12. $4, -2, -4, -2, 4$ 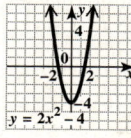   13. $-2y^2 - 9y + 17$
14. $-21a^3b^2 + 7ab^5 - 5a^2b^2$ 15. $-12t^2 + 5t + 8$ [5.5] 16. $-27x^5 + 18x^4 - 6x^3 + 3x^2$ 17. $t^2 - 5t - 24$ 18. $8x^2 + 2xy - 3y^2$ [5.6] 19. $25x^2 - 20xy + 4y^2$ 20. $100v^2 - 9w^2$ [5.5] 21. $2r^3 + r^2 - 16r + 15$ [5.6] 22. $9x^2 + 54x + 81$ [5.7] 23. $4y^2 - 3y + 2 + \frac{5}{y}$ 24. $-3xy^2 + 2x^3y^2 + 4y^2$ 25. $3x^2 + 6x + 11 + \frac{26}{x - 2}$

Cumulative Review Exercises Chapters 1–5   (page 382)

[1.1] 1. $\frac{7}{4}$ 2. 5   3. $\frac{19}{24}$ 4. $-\frac{1}{20}$ 5. $31\frac{1}{4}$ yd^3 [1.6] 6. $1836   7. 1, 3, 5, 9, 15, 45   8. -8 9. $\frac{1}{2}$ 10. -4
[1.7] 11. associative property   12. distributive property   [1.8] 13. $-10x^2 + 21x - 29$ [2.1–2.3] 14. $\left\{\frac{13}{4}\right\}$ 15. ∅
[2.5] 16. $r = \frac{d}{t}$ [2.6] 17. $\{-5\}$ [2.1–2.3] 18. $\{-12\}$ 19. $\{20\}$ 20. {all real numbers}   [2.4] 21. mouse: 160; elephant: 10
22. 4   [2.8] 23. 11 ft and 22 ft   [2.8] 24. $[10, \infty)$ 25. $\left(-\infty, -\frac{14}{5}\right)$ 26. $[-4, 2)$ [3.2] 27.

[3.3, 3.4] 28. (a) 1 (b) $y = x + 6$ [3.5] 29. no   [3.6] 30. -1 [4.2] 31. $\{(-3, -1)\}$ [4.3] 32. $\{(4, -5)\}$

[5.1, 5.2] **33.** $\frac{5}{4}$ **34.** 2 **35.** 1 **36.** $\frac{2b}{a^{10}}$ [5.3] **37.** 3.45×10^4 **38.** about 10,800,000 km [5.4] **39.**

40. $11x^3 - 14x^2 - x + 14$ [5.5] **41.** $18x^7 - 54x^6 + 60x^5$ **42.** $63x^2 + 57x + 12$ [5.6] **43.** $25x^2 + 80x + 64$
[5.7] **44.** $2x^2 - 3x + 1$ **45.** $y^2 - 2y + 6$

CHAPTER 6 FACTORING AND APPLICATIONS

Section 6.1 (page 392)

Exercises 1. 4 **3.** 6 **5.** First, verify that you have factored completely. Then multiply the factors. The product should be the original polynomial. **7.** 8 **9.** $10x^3$ **11.** $6m^3n^2$ **13.** xy^2 **15.** 6 **17.** factored **19.** not factored **21.** yes; x^3y^2 **23.** $3m^2$ **25.** $2z^4$
27. $2mn^4$ **29.** $y + 2$ **31.** $a - 2$ **33.** $2 + 3xy$ **35.** $9m(3m^2 - 1)$ **37.** $8z^2(2z^2 + 3)$ **39.** $\frac{1}{4}d(d - 3)$ **41.** $6x^2(2x + 1)$
43. $5y^6(13y^4 + 7)$ **45.** no common factor (except 1) **47.** $8m^2n^2(n + 3)$ **49.** $13y^2(y^6 + 2y^2 - 3)$ **51.** $9qp^3(5q^3p^2 + 4p^3 + 9q)$
53. $a^3(a^2 + 2b^2 - 3a^2b^2 + 4ab^3)$ **55.** $(x + 2)(c - d)$ **57.** $(m + 2n)(m + n)$ **59.** not in factored form; $(7t + 4)(8 + x)$
61. in factored form **63.** not in factored form; $(y + 4)(18x^2 + 7)$ **65.** The quantities in parentheses are not the same, so there is no common factor in the two terms $12k^3(s - 3)$ and $7(s + 3)$. **67.** $(p + 4)(p + q)$ **69.** $(a - 2)(a + b)$ **71.** $(z + 2)(7z - a)$
73. $(3r + 2y)(6r - x)$ **75.** $(a^2 + b^2)(3a + 2b)$ **77.** $(1 - a)(1 - b)$ **79.** $(4m - p^2)(4m^2 - p)$ **81.** $(5 - 2p)(m + 3)$
83. $(3r + 2y)(6r - t)$ **85.** $(a^5 - 3)(1 + 2b)$ **87.** commutative property **88.** $2x(y - 1) - 3(y - 1)$ **89.** No, because it is not a product. It is the difference between $2x(y - 1)$ and $3(y - 1)$. **90.** $(2x - 3)(y - 1)$; yes **91.** (a) yes (b) When either one is multiplied out, the product is $1 - a + ab - b$.

Section 6.2 (page 398)

Exercises 1. 1 and 48, -1 and -48, 2 and 24, -2 and -24, 3 and 16, -3 and -16, 4 and 12, -4 and -12, 6 and 8, -6 and -8; the pair with a sum of -19 is -3 and -16. **3.** 1 and -24, -1 and 24, 2 and -12, -2 and 12, 3 and -8, -3 and 8, 4 and -6, -4 and 6; the pair with a sum of -5 is 3 and -8. **5.** a and b must have different signs, one positive and one negative. **7.** A prime polynomial is one that cannot be factored using only integers in the factors. **9.** C **11.** $p + 6$ **13.** $x + 11$ **15.** $x - 8$ **17.** $y - 5$ **19.** $x + 11$
21. $y - 9$ **23.** $(y + 8)(y + 1)$ **25.** $(b + 3)(b + 5)$ **27.** $(m + 5)(m - 4)$ **29.** $(y - 5)(y - 3)$ **31.** prime **33.** $(z - 7)(z - 8)$
35. $(r - 6)(r + 5)$ **37.** $(a + 4)(a - 12)$ **39.** prime **41.** Factor $8 + 6x + x^2$ directly to get $(2 + x)(4 + x)$. Alternatively, use the commutative property to write the trinomial as $x^2 + 6x + 8$ and factor to get $(x + 2)(x + 4)$, an equivalent answer. **43.** $(r + 2a)(r + a)$
45. $(t + 2z)(t - 3z)$ **47.** $(x + y)(x + 3y)$ **49.** $(v - 5w)(v - 6w)$ **51.** $4(x + 5)(x - 2)$ **53.** $2t(t + 1)(t + 3)$
55. $2x^4(x - 3)(x + 7)$ **57.** $5m^2(m^3 + 5m^2 - 8)$ **59.** $mn(m - 6n)(m - 4n)$ **61.** $(2x + 4)(x - 3)$ is incorrect because $2x + 4$ has a common factor of 2, which must be factored out. **63.** $a^3(a + 4b)(a - b)$ **65.** $yz(y + 3z)(y - 2z)$ **67.** $z^8(z - 7y)(z + 3y)$
69. $(a + b)(x + 4)(x - 3)$ **71.** $(2p + q)(r - 9)(r - 3)$ **73.** $a^2 + 13a + 36$

Section 6.3 (page 405)

Exercises 1. $(2t + 1)(5t + 2)$ **3.** $(3z - 2)(5z - 3)$ **5.** $(2s - t)(4s + 3t)$ **7.** B **9.** (a) 2, 12, 24, 11 (b) 3, 8 (Order is irrelevant.)
(c) $3m, 8m$ (d) $2m^2 + 3m + 8m + 12$ (e) $(2m + 3)(m + 4)$ (f) $(2m + 3)(m + 4) = 2m^2 + 11m + 12$ **11.** B **13.** A **15.** $2a + 5b$
17. $x^2 + 3x - 4; x + 4, x - 1$ or $x - 1, x + 4$ **19.** The binomial $2x - 6$ cannot be a factor because it has a common factor of 2, which the polynomial does not have. **21.** $(3a + 7)(a + 1)$ **23.** $(2y + 3)(y + 2)$ **25.** $(3m - 1)(5m + 2)$ **27.** $(3s - 1)(4s + 5)$

29. $(5m - 4)(2m - 3)$ **31.** $(4w - 1)(2w - 3)$ **33.** $(4y + 1)(5y - 11)$ **35.** prime   **37.** $2(5x + 3)(2x + 1)$
39. $3(4x - 1)(2x - 3)$ **41.** $q(5m + 2)(8m - 3)$ **43.** $2n^2(6n - 5)(n - 1)$ **45.** $y^3(15x - 7)(x + 2)$ **47.** $(8x + 3y)(x - 2y)$
49. $(12s + 5t)(s + t)$ **51.** $m^4n(3m + 2n)(2m + n)$ **53.** $(5 - x)(1 - x)$ **55.** $(4 + 3x)(4 + x)$ **57.** $-5x(2x + 7)(x - 4)$
59. $(12x + 1)(x - 4)$ **61.** $(24y + 7x)(y - 2x)$ **63.** $(18x^2 - 5y)(2x^2 - 3y)$ **65.** $2(24a + b)(a - 2b)$ **67.** $x^2y^5(10x - 1)(x + 4)$
69. $4ab^2(9a + 1)(a - 3)$ **71.** $(12x - 5)(2x - 3)$ **73.** $(8x^2 - 3)(3x^2 + 8)$ **75.** $(4x + 3y)(6x + 5y)$ **77.** $(xz^2 - 5)(24xz^2 + 7)$
79. $-1(x + 7)(x - 3)$ **81.** $-1(3x + 4)(x - 1)$ **83.** $-1(a + 2b)(2a + b)$ **85.** Yes, $(x + 7)(3 - x)$ is equivalent to $-1(x + 7)(x - 3)$ because $-1(x - 3) = -x + 3 = 3 - x$.   **87.** $(m + 1)^3(5q - 2)(5q + 1)$ **89.** $(r + 3)^3(5x + 2y)(3x - 8y)$

Section 6.4   (page 414)

Exercises   **1.** 1; 4; 9; 16; 25; 36; 49; 64; 81; 100; 121; 144; 169; 196; 225; 256; 289; 324; 361; 400   **3.** 1; 8; 27; 64; 125; 216; 343; 512; 729; 1000   **5. (a)** both of these **(b)** a perfect cube **(c)** a perfect square **(d)** a perfect square   **7.** $(y + 5)(y - 5)$
9. $\left(p + \frac{1}{3}\right)\left(p - \frac{1}{3}\right)$ **11.** prime   **13.** $(3r + 2)(3r - 2)$ **15.** $\left(6m + \frac{4}{5}\right)\left(6m - \frac{4}{5}\right)$ **17.** $4(3x + 2)(3x - 2)$
19. $(14p + 15)(14p - 15)$ **21.** $(4r + 5a)(4r - 5a)$ **23.** prime   **25.** $(p^2 + 7)(p^2 - 7)$ **27.** $(x^2 + 1)(x + 1)(x - 1)$
29. $(p^2 + 16)(p + 4)(p - 4)$ **31.** The teacher was justified. $x^2 - 9$ can be factored as $(x + 3)(x - 3)$. The complete factored form is $(x^2 + 9)(x + 3)(x - 3)$.   **33.** $(w + 1)^2$ **35.** $(x - 4)^2$ **37.** $\left(t + \frac{1}{2}\right)^2$ **39.** $(x - .5)^2$ **41.** $2(x + 6)^2$ **43.** $(4x - 5)^2$
45. $(7x - 2y)^2$ **47.** $(8x + 3y)^2$ **49.** $-2(5h - 2y)^2$ **51.** $k(4k^2 - 4k + 9)$ **53.** $z^2(25z^2 + 5z + 1)$ **55.** 10   **57.** 9
59. $(a - 1)(a^2 + a + 1)$ **61.** $(m + 2)(m^2 - 2m + 4)$ **63.** $(3x - 4)(9x^2 + 12x + 16)$ **65.** $6(p + 1)(p^2 - p + 1)$
67. $5(x + 2)(x^2 - 2x + 4)$ **69.** $2(y - 2x)(y^2 + 2xy + 4x^2)$ **71.** $(2p + 9q)(4p^2 - 18pq + 81q^2)$ **73.** $(3a + 4b)(9a^2 - 12ab + 16b^2)$
75. $(5t + 2s)(25t^2 - 10ts + 4s^2)$ **77.** $x^2 - y^2 = (x + y)(x - y)$ Difference of Squares; See Exercises 7–10, 13–22, and 25–30.
$x^2 + 2xy + y^2 = (x + y)^2$ Perfect Square Trinomial; See Exercises 33, 34, 37, 38, 41, 47, 48, 50. $x^2 - 2xy + y^2 = (x - y)^2$ Perfect Square Trinomial; See Exercises 35, 36, 39, 40, 42–46, 49. $x^3 - y^3 = (x - y)(x^2 + xy + y^2)$ Difference of Cubes; See Exercises 59, 60, 63, 64, 68–70, 74. $x^3 + y^3 = (x + y)(x^2 - xy + y^2)$ Sum of Cubes; See Exercises 61, 62, 65–67, 71–73, 75, 76.   **78.** $(5x - 2)(2x + 3)$
79. $5x - 2$ **80.** Yes. If $10x^2 + 11x - 6$ factors as $(5x - 2)(2x + 3)$, then when $10x^2 + 11x - 6$ is divided by $2x + 3$, the quotient should be $5x - 2$.   **81.** $x^2 + x + 1$; $(x - 1)(x^2 + x + 1)$ **83.** $-2b(3a^2 + b^2)$ **85.** $3(r - k)(1 + r + k)$

Summary Exercises on Factoring   (page 416)

1. $(a - 6)(a + 2)$ **2.** $(a + 8)(a + 9)$ **3.** $6(y - 2)(y + 1)$ **4.** $7y^4(y + 6)(y - 4)$ **5.** $6(a + 2b + 3c)$ **6.** $(m - 4n)(m + n)$
7. $(p - 11)(p - 6)$ **8.** $(z + 7)(z - 6)$ **9.** $(5z - 6)(2z + 1)$ **10.** $2(m - 8)(m + 3)$ **11.** $(m + n + 5)(m - n)$ **12.** $5(3y + 1)$
13. $8a^3(a - 3)(a + 2)$ **14.** $(4k + 1)(2k - 3)$ **15.** $(z - 5a)(z + 2a)$ **16.** $50(z^2 - 2)$ **17.** $(x - 5)(x - 4)$
18. $10nr(10nr + 3r^2 - 5n)$ **19.** $(3n - 2)(2n - 5)$ **20.** $(3y - 1)(3y + 5)$ **21.** $4(4x + 5)$ **22.** $(m + 5)(m - 3)$
23. $(3y - 4)(2y + 1)$ **24.** $(m + 9)(m - 9)$ **25.** $(6z + 1)(z + 5)$ **26.** $(5z + 2)(z + 5)$ **27.** $(2k - 3)^2$ **28.** $(8p - 1)(p + 3)$
29. $6(3m + 2z)(3m - 2z)$ **30.** $(4m - 3)(2m + 1)$ **31.** $(3k - 2)(k + 2)$ **32.** $15a^3b^2(3b^3 - 4a + 5a^3b^2)$ **33.** $7k(2k + 5)(k - 2)$
34. $(5 + r)(1 - s)$ **35.** $(y^2 + 4)(y + 2)(y - 2)$ **36.** $10y^4(2y - 3)$ **37.** $8m(1 - 2m)$ **38.** $(k + 4)(k - 4)$
39. $(z - 2)(z^2 + 2z + 4)$ **40.** $(y - 8)(y + 7)$ **41.** prime   **42.** $9p^8(3p + 7)(p - 4)$ **43.** $8m^3(4m^6 + 2m^2 + 3)$
44. $(2m + 5)(4m^2 - 10m + 25)$ **45.** $(4r + 3m)^2$ **46.** $(z - 6)^2$ **47.** $(5h + 7g)(3h - 2g)$ **48.** $5z(z - 7)(z - 2)$
49. $(k - 5)(k - 6)$ **50.** $4(4p - 5m)(4p + 5m)$ **51.** $3k(k - 5)(k + 1)$ **52.** $(y - 6k)(y + 2k)$ **53.** $(10p + 3)(100p^2 - 30p + 9)$
54. $(4r - 7)(16r^2 + 28r + 49)$ **55.** $(2 + m)(3 + p)$ **56.** $(2m - 3n)(m + 5n)$ **57.** $(4z - 1)^2$ **58.** $5m^2(5m - 3n)(5m - 13n)$
59. $3(6m - 1)^2$ **60.** $(10a + 9y)(10a - 9y)$ **61.** prime   **62.** $(2y + 5)(2y - 5)$ **63.** $8z(4z - 1)(z + 2)$ **64.** $5(2m - 3)(m + 4)$
65. $(4 + m)(5 + 3n)$ **66.** $(2 - q)(2 - 3p)$ **67.** $2(3a - 1)(a + 2)$ **68.** $6y^4(3y + 4)(2y - 5)$ **69.** $(a - b)(a^2 + ab + b^2 + 2)$
70. $4(2k - 3)^2$ **71.** $(8m - 5n)^2$ **72.** $12y^2(6yz^2 + 1 - 2y^2z^2)$ **73.** $(4k - 3h)(2k + h)$ **74.** $(2a + 5)(a - 6)$
75. $(m + 2)(m^2 + m + 1)$ **76.** $(2a - 3)(4a^2 + 6a + 9)$ **77.** $(5y - 6z)(2y + z)$ **78.** $(m - 2)^2$ **79.** $(8a - b)(a + 3b)$
80. $(a^2 + 25)(a + 5)(a - 5)$ **81.** $(x^3 - 1)(x^3 + 1)$ **82.** $(x - 1)(x^2 + x + 1)(x + 1)(x^2 - x + 1)$ **83.** $(x^2 - 1)(x^4 + x^2 + 1)$
84. $(x - 1)(x + 1)(x^4 + x^2 + 1)$ **85.** The result in Exercise 82 is completely factored.   **86.** Show that $x^4 + x^2 + 1 = (x^2 + x + 1)(x^2 - x + 1)$.   **87.** difference of squares   **88.** $(x - 3)(x^2 + 3x + 9)(x + 3)(x^2 - 3x + 9)$

Section 6.5 (page 423)

Exercises 1. $\{-5, 2\}$ **3.** $\left\{\dfrac{7}{2}, 3\right\}$ **5.** $\left\{-\dfrac{5}{6}, 0\right\}$ **7.** $\left\{0, \dfrac{4}{3}\right\}$ **9.** $\left\{-\dfrac{1}{2}, \dfrac{1}{6}\right\}$ **11.** $\{-.8, 2\}$ **13.** $\{9\}$ **15.** Set each *variable* factor equal to 0, to get $2x = 0$ or $3x - 4 = 0$. The solution set is $\left\{0, \dfrac{4}{3}\right\}$. **17.** $\{-2, -1\}$ **19.** $\{1, 2\}$ **21.** $\{-8, 3\}$ **23.** $\{-1, 3\}$ **25.** $\{-2, -1\}$ **27.** $\{-4\}$ **29.** $\left\{-\dfrac{3}{8}, 6\right\}$ **31.** $\left\{-\dfrac{4}{3}, \dfrac{1}{2}\right\}$ **33.** $\left\{-\dfrac{2}{3}\right\}$ **35.** $\{-3, 3\}$ **37.** $\left\{-\dfrac{7}{4}, \dfrac{7}{4}\right\}$ **39.** $\{-11, 11\}$ **41.** $\{0, 7\}$ **43.** $\left\{0, \dfrac{1}{2}\right\}$ **45.** $\{2, 5\}$ **47.** $\left\{-4, \dfrac{1}{2}\right\}$ **49.** $\left\{-12, \dfrac{11}{2}\right\}$ **51.** $\left\{-\dfrac{5}{2}, \dfrac{1}{3}, 5\right\}$ **53.** $\left\{-\dfrac{7}{2}, -3, 1\right\}$ **55.** $\left\{-\dfrac{7}{3}, 0, \dfrac{7}{3}\right\}$ **57.** $\{-2, 0, 4\}$ **59.** $\{-5, 0, 4\}$ **61.** $\{-3, 0, 5\}$ **63.** $\{-1, 3\}$ **65.** $\{-1, 3\}$ **67.** $\{3\}$ **69.** $\left\{-\dfrac{4}{3}, -1, \dfrac{1}{2}\right\}$ **71.** $\left\{-\dfrac{2}{3}, 4\right\}$ **73. (a)** 64; 144; 4; 6 **(b)** No time has elapsed, so the object hasn't fallen (been released) yet. **(c)** Time cannot be negative. **75.** $\{1.5, 2.1\}$ **77.** $\{-2.75, 3\}$

Section 6.6 (page 431)

Exercises 1. Read; variable; equation; Solve; answer; Check, original **3.** Step 3: $45 = (2x + 1)(x + 1)$; Step 4: $x = 4$ or $x = -\dfrac{11}{2}$; Step 5: base: 9 units; height: 5 units; Step 6: $9 \cdot 5 = 45$ **5.** Step 3: $80 = (x + 8)(x - 8)$; Step 4: $x = 12$ or $x = -12$; Step 5: length: 20 units; width: 4 units; Step 6: $20 \cdot 4 = 80$ **7.** length: 7 in.; width: 4 in. **9.** length: 13 in.; width: 10 in. **11.** base: 12 in.; height: 5 in. **13.** height: 13 in.; width: 10 in. **15.** mirror: 7 ft; painting: 9 ft **17.** $-3, -2$ or $4, 5$ **19.** 7, 9, 11 **21.** $-2, 0, 2$ or 6, 8, 10 **23.** 12 cm **25.** 12 mi **27.** 8 ft **29. (a)** 1 sec **(b)** $\dfrac{1}{2}$ sec and $1\dfrac{1}{2}$ sec **(c)** 3 sec **(d)** The negative solution, -1, does not make sense since t represents time, which cannot be negative. **31. (a)** 23.4 million; The result using the model is less than 24 million, the actual number for 1994. **(b)** 9 **(c)** 82.6 million; The result is less than 86 million, the actual number for 1999. **(d)** 137.6 million **32.** 107 billion dollars; 65% **33.** 1995: 66.9 billion dollars; 1997: 148.5 billion dollars; 1999: 230.1 billion dollars **34.** The answers using the linear equation are not at all close to the actual data. **35.** 1995: 104 billion dollars; 1997: 111.2 billion dollars; 1999: 266.4 billion dollars **36.** The answers in Exercise 35 are fairly close to the actual data. The quadratic equation models the data better. **37.** (0, 97.5), (1, 104.3), (2, 104.7), (3, 164.3), (4, 271.3) **38.** no **39.** 399.5 billion dollars **40. (a)** The actual deficit is about 30 billion dollars less than the prediction. **(b)** No, the equation is based on data for the years 1995–1999. Data for later years might not follow the same pattern.

Chapter 6 Review Exercises (page 440)

1. $7(t + 2)$ **3.** $(x - 4)(2y + 3)$ **5.** $(x + 3)(x + 2)$ **7.** $(q + 9)(q - 3)$ **9.** $(r + 8s)(r - 12s)$ **11.** $8p(p + 2)(p - 5)$ **13.** $p^5(p - 2q)(p + q)$ **15.** r and $6r$, $2r$ and $3r$ **17.** $(2k - 1)(k - 2)$ **19.** $(3r + 2)(2r - 3)$ **21.** $(v + 3)(8v - 7)$ **23.** $-3(x + 2)(2x - 5)$ **25.** B **27.** $(n + 7)(n - 7)$ **29.** $(7y + 5w)(7y - 5w)$ **31.** prime **33.** $(3t - 7)^2$ **35.** $(5k + 4x)(25k^2 - 20kx + 16x^2)$ **37.** $\left\{-\dfrac{3}{4}, 1\right\}$ **39.** $\left\{0, \dfrac{5}{2}\right\}$ **41.** $\{1, 4\}$ **43.** $\left\{-\dfrac{4}{3}, 5\right\}$ **45.** $\{0, 8\}$ **47.** $\{7\}$

49. $\left\{-\dfrac{2}{5}, -2, -1\right\}$ **51.** length: 10 ft; width: 4 ft **53.** length: 6 m; width: 2 m **55.** 6, 7 or −5, −4 **57.** 112 ft **59.** 256 ft
61. (a) $537 million **(b)** No, the prediction seems high. If eBay revenues in the last half of 2000 are comparable to those for the first half of the year, annual revenue in 2000 would be about $366 million. **63.** The factor $2x + 8$ has a common factor of 2. The complete factored form is $2(x + 4)(3x - 4)$. **65.** $(3k + 5)(k + 2)$ **67.** $(y^2 + 25)(y + 5)(y - 5)$ **69.** $8abc(3b^2c - 7ac^2 + 9ab)$
71. $6xyz(2xz^2 + 2y - 5x^2yz^3)$ **73.** $(2r + 3q)(6r - 5)$ **75.** $(7t + 4)^2$ **77.** $\{-5, 2\}$ **79. (a)** 481,000 vehicles **(b)** The estimate may be unreliable because the conditions that prevailed in the years 1998–2001 may have changed, causing either a greater increase or a greater decrease in the numbers of alternative-fueled vehicles. **81.** length: 6 m; width: 4 m **83. (a)** 256 ft **(b)** 1024 ft **85.** 6 m

Chapter 6 Test (page 444)

[6.1–6.4] **1.** D **2.** $6x(2x - 5)$ **3.** $m^2n(2mn + 3m - 5n)$ **4.** $(2x + y)(a - b)$ **5.** $(x + 3)(x - 8)$ **6.** $(2x + 3)(x - 1)$
7. $(5z - 1)(2z - 3)$ **8.** prime **9.** prime **10.** $(2 - a)(6 + b)$ **11.** $(3y + 8)(3y - 8)$ **12.** $(2x - 7y)^2$ **13.** $-2(x + 1)^2$
14. $3t^2(2t + 9)(t - 4)$ **15.** $(r - 5)(r^2 + 5r + 25)$ **16.** $8(k + 2)(k^2 - 2k + 4)$ **17.** $(x^2 + 9)(x + 3)(x - 3)$ **18.** The product $(p + 3)(p + 3) = p^2 + 6p + 9$, which does not equal $p^2 + 9$. The binomial $p^2 + 9$ is a prime polynomial. [6.5] **19.** $\left\{6, \dfrac{1}{2}\right\}$
20. $\left\{-\dfrac{2}{5}, \dfrac{2}{5}\right\}$ **21.** $\{10\}$ **22.** $\{-3, 0, 3\}$ [6.6] **23.** 6 ft by 9 ft **24.** $-2, -1$ **25.** 17 ft **26.** 181

Cumulative Review Exercises Chapters 1–6 (page 445)

[2.1–2.3] **1.** $\{0\}$ **2.** $\{.05\}$ **3.** $\{6\}$ [2.5] **4.** $P = \dfrac{A}{1 + rt}$ [2.7] **5.** 345; 210; 38%; 15% [2.4] **6.** gold: 12; silver: 9; bronze: 8
[2.7] **7.** 107 million [2.5] **8.** 110° and 70° [3.1] **9. (a)** negative, positive **(b)** negative, negative [3.2] **10.** $\left(-\dfrac{1}{4}, 0\right), (0, 3)$
[3.3] **11.** 12 [3.2] **12.** [3.3, 3.4] **13. (a)** 103; A slope of 103 means that the number of radio stations increased by about 103 stations per year. **(b)** $y = 103x - 195{,}389$ [3.6] **14. (a)** 10,508 **(b)** yes [4.1] **15.** $\{(-1, 2)\}$ **16.** ∅ [5.1, 5.2] **17.** $\dfrac{16}{9}$
18. 256 **19.** $\dfrac{1}{p^2}$ **20.** $\dfrac{1}{m^6}$ [5.4] **21.** $-4k^2 - 4k + 8$ [5.5] **22.** $45x^2 + 3x - 18$ [5.6] **23.** $9p^2 + 12p + 4$
[5.7] **24.** $4x^3 + 6x^2 - 3x + 10$ [5.3] **25.** 5.5×10^4; 2.0×10^6 [6.2, 6.3] **26.** $(2a - 1)(a + 4)$ **27.** $(2m + 3)(5m + 2)$
28. $(4t + 3v)(2t + v)$ [6.4] **29.** $(2p - 3)^2$ **30.** $(5r + 9t)(5r - 9t)$ [6.3] **31.** $2pq(3p + 1)(p + 1)$ [6.5] **32.** $\left\{-\dfrac{2}{3}, \dfrac{1}{2}\right\}$
33. $\{0, 8\}$ [6.6] **34.** 5 m, 12 m, 13 m

CHAPTER 7 RATIONAL EXPRESSIONS AND APPLICATIONS

Section 7.1 (page 454)

Exercises **1.** A rational expression is a quotient of two polynomials, such as $\dfrac{x^2 + 3x - 6}{x + 4}$. One can think of this as an algebraic fraction.
3. (a) $\dfrac{7}{10}$ **(b)** $\dfrac{8}{15}$ **5. (a)** 0 **(b)** -1 **7. (a)** $-\dfrac{64}{15}$ **(b)** undefined **9. (a)** undefined **(b)** $\dfrac{8}{25}$ **11.** Division by 0 is undefined, so if the

denominator of a rational expression equals 0, the expression is undefined. **13.** 0 **15.** 6 **17.** $-\dfrac{5}{3}$ **19.** $-3, 2$ **21.** never undefined **23.** never undefined **25. (a)** numerator: x^2, $4x$; denominator: x, 4 **(b)** First factor the numerator, getting $x(x+4)$, then divide the numerator and denominator by the common factor of $x+4$ to get $\dfrac{x}{1}$ or x. **27.** $3r^2$ **29.** $\dfrac{2}{5}$ **31.** $\dfrac{x-1}{x+1}$ **33.** $\dfrac{7}{5}$ **35.** $\dfrac{6}{7}$ **37.** $m - n$ **39.** $\dfrac{2}{t-3}$ **41.** $\dfrac{3(2m+1)}{4}$ **43.** $\dfrac{3m}{5}$ **45.** $\dfrac{3r-2s}{3}$ **47.** $k-3$ **49.** $\dfrac{x-3}{x+1}$ **51.** $\dfrac{x+1}{x-1}$ **53.** $\dfrac{6x+1}{8x-1}$ **55.** $\dfrac{4x+3}{x-7}$ **57.** $\dfrac{9-2x^2}{1+3x^2}$ **59.** $\dfrac{2x^2+5}{2x^2-5}$ **61.** $\dfrac{5x+3}{5x-3}$ **63.** $\dfrac{z-3}{z+5}$ **65.** $\dfrac{m+n}{2}$ **67.** $-\dfrac{b^2+ba+a^2}{a+b}$ **69.** $\dfrac{z+3}{z}$ **71.** $\dfrac{x+2y}{2}$ **73.** $\dfrac{x-y}{x+y}$ **75.** $\dfrac{a+2b}{a-5b}$ **77.** $-\dfrac{6c}{a+4b}$ **79.** $\dfrac{x^2+y^2}{3}$ **81.** B, D **83.** -1 **85.** $-(m+1)$ **87.** -1 **89.** already in lowest terms **91.** B Answers may vary in Exercises 93, 95, and 99. **93.** $\dfrac{-(x+4)}{x-3}, \dfrac{-x-4}{x-3}, \dfrac{x+4}{-(x-3)}, \dfrac{x+4}{-x+3}$ **95.** $\dfrac{-(2x-3)}{x+3}, \dfrac{-2x+3}{x+3}, \dfrac{2x-3}{-(x+3)}$, $\dfrac{2x-3}{-x-3}$ **97.** $-\dfrac{3x-1}{5x-6}, \dfrac{-(3x-1)}{5x-6}, \dfrac{-3x+1}{-5x+6}, \dfrac{3x-1}{-5x+6}$ **99.** x^2+3 **101. (a)** 0 **(b)** 1.6 **(c)** 4.1 **(d)** The waiting time also increases.

Section 7.2 (page 462)

Exercises **1. (a)** B **(b)** D **(c)** C **(d)** A **3.** $\dfrac{3a}{2}$ **5.** $-\dfrac{4x^4}{3}$ **7.** $\dfrac{2}{c+d}$ **9.** $4(x-y)$ **11.** $\dfrac{t^2}{2}$ **13.** $\dfrac{x+3}{2x}$ **15.** 5 **17.** $-\dfrac{3}{2t^4}$ **19.** $\dfrac{1}{4}$ **21.** $-\dfrac{35}{8}$ **23.** $\dfrac{2(x+2)}{x(x-1)}$ **25.** $\dfrac{x(x-3)}{6}$ **27.** Suppose I want to multiply $\dfrac{a^2-1}{6} \cdot \dfrac{9}{2a+2}$. I start by factoring where possible: $\dfrac{(a+1)(a-1)}{6} \cdot \dfrac{9}{2(a+1)}$. Next, I divide out common factors in the numerator and denominator to get $\dfrac{a-1}{2} \cdot \dfrac{3}{2}$. Finally, I multiply numerator times numerator and denominator times denominator to get the final product, $\dfrac{3(a-1)}{4}$. **29.** $\dfrac{10}{9}$ **31.** $\dfrac{3}{4}$ **33.** $\dfrac{9}{2}$ **35.** $\dfrac{p+4}{p+2}$ **37.** -1 **39.** $\dfrac{(2x-1)(x+2)}{x-1}$ **41.** $\dfrac{(k-1)^2}{(k+1)(2k-1)}$ **43.** $\dfrac{4k-1}{3k-2}$ **45.** $\dfrac{m+4p}{m+p}$ **47.** $\dfrac{m+6}{m+3}$ **49.** $\dfrac{y+3}{y+4}$ **51.** $\dfrac{m}{m+5}$ **53.** $\dfrac{r+6s}{r+s}$ **55.** $\dfrac{(q-3)^2(q+2)^2}{q+1}$ **57.** $\dfrac{x+10}{10}$ **59.** $\dfrac{3-a-b}{2a-b}$ **61.** $-\dfrac{(x+y)^2(x^2-xy+y^2)}{3y(y-x)(x-y)}$ or $\dfrac{(x+y)^2(x^2-xy+y^2)}{3y(x-y)^2}$ **63.** $\dfrac{5xy^2}{4q}$

Section 7.3 (page 467)

Exercises **1.** C **3.** C **5.** 60 **7.** 1800 **9.** x^5 **11.** $30p$ **13.** $180y^4$ **15.** $84r^5$ **17.** $15a^5b^3$ **19.** $12p(p-2)$ **21.** $28m^2(3m-5)$ **23.** $30(b-2)$ **25.** $2^3 \cdot 3 \cdot 5$ **26.** $(t+4)^3(t-3)(t+8)$ **27.** The similarity is that 2 is replaced by $t+4$, 3 is replaced by $t-3$, and 5 is replaced by $t+8$. **28.** The procedure used is the same. The only difference is that for algebraic fractions, the factors may contain variables, while in common fractions, the factors are numbers. **29.** $18(r-2)$ **31.** $12p(p+5)^2$ **33.** $8(y+2)(y+1)$ **35.** $c-d$ or $d-c$ **37.** $m-3$ or $3-m$ **39.** $p-q$ or $q-p$ **41.** $k(k+5)(k-2)$ **43.** $a(a+6)(a-3)$ **45.** $(p+3)(p+5)(p-6)$ **47.** $(k+3)(k-5)(k+7)(k+8)$ **49.** Yes, because $(2x-5)^2 = (5-2x)^2$. **51.** 7 **52.** 1 **53.** identity property of multiplication **54.** 7 **55.** 1 **56.** identity property of multiplication **57.** $\dfrac{20}{55}$ **59.** $\dfrac{-45}{9k}$ **61.** $\dfrac{60m^2k^3}{32k^4}$ **63.** $\dfrac{57z}{6z-18}$ **65.** $\dfrac{-4a}{18a-36}$ **67.** $\dfrac{6(k+1)}{k(k-4)(k+1)}$ **69.** $\dfrac{36r(r+1)}{(r-3)(r+2)(r+1)}$ **71.** $\dfrac{ab(a+2b)}{2a^3b+a^2b^2-ab^3}$ **73.** $\dfrac{(t-r)(4r-t)}{t^3-r^3}$

75. $\dfrac{2y(z-y)(y-z)}{y^4-z^3y}$ or $\dfrac{-2y(y-z)^2}{y^4-z^3y}$ **77.** *Step 1:* Factor each denominator into prime factors. *Step 2:* List each different denominator factor the greatest number of times it appears in any of the denominators. *Step 3:* Multiply the factors in the list to get the LCD. For example, the least common denominator for $\dfrac{1}{(x+y)^3}$ and $\dfrac{-2}{(x+y)^2(p+q)}$ is $(x+y)^3(p+q)$.

Section 7.4 (page 476)

Exercises 1. E   **3.** C   **5.** B   **7.** G   **9.** $\dfrac{11}{m}$ **11.** $\dfrac{4}{y+4}$ **13.** 1   **15.** $\dfrac{m-1}{m+1}$ **17.** b   **19.** x   **21.** $y-6$ **23.** To add or subtract rational expressions with the same denominators, combine the numerators and keep the same denominator. For example, $\dfrac{3x+2}{x-6}+\dfrac{-2x-8}{x-6}=\dfrac{x-6}{x-6}$. Then write in lowest terms. In this example, the sum simplifies to 1.   **25.** $\dfrac{3z+5}{15}$ **27.** $\dfrac{10-7r}{14}$ **29.** $\dfrac{-3x-2}{4x}$
31. $\dfrac{57}{10x}$ **33.** $\dfrac{x+1}{2}$ **35.** $\dfrac{5x+9}{6x}$ **37.** $\dfrac{7-6p}{3p^2}$ **39.** $\dfrac{-k-8}{k(k+4)}$ **41.** $\dfrac{x+4}{x+2}$ **43.** $\dfrac{6m^2+23m-2}{(m+2)(m+1)(m+5)}$ **45.** $\dfrac{4y^2-y+5}{(y+1)^2(y-1)}$
47. $\dfrac{3}{t}$ **49.** $m-2$ or $2-m$ **51.** $\dfrac{-2}{x-5}$ or $\dfrac{2}{5-x}$ **53.** -4 **55.** $\dfrac{-5}{x-y^2}$ or $\dfrac{5}{y^2-x}$ **57.** $\dfrac{x+y}{5x-3y}$ or $\dfrac{-x-y}{3y-5x}$
59. $\dfrac{-6}{4p-5}$ or $\dfrac{6}{5-4p}$ **61.** $\dfrac{-(m+n)}{2(m-n)}$ **63.** $\dfrac{-x^2+6x+11}{(x+3)(x-3)(x+1)}$ **65.** $\dfrac{-5q^2-13q+7}{(3q-2)(q+4)(2q-3)}$ **67.** $\dfrac{9r+2}{r(r+2)(r-1)}$
69. $\dfrac{2(x^2+3xy+4y^2)}{(x+y)(x+y)(x+3y)}$ or $\dfrac{2(x^2+3xy+4y^2)}{(x+y)^2(x+3y)}$ **71.** $\dfrac{15r^2+10ry-y^2}{(3r+2y)(6r-y)(6r+y)}$ **73.** (a) $\dfrac{9k^2+6k+26}{5(3k+1)}$ (b) $\dfrac{1}{4}$ **75.** $\dfrac{10x}{49(101-x)}$

Section 7.5 (page 484)

Exercises 1. (a) 6; $\dfrac{1}{6}$ **(b)** 12; $\dfrac{3}{4}$ **(c)** $\dfrac{1}{6}\div\dfrac{3}{4}$ **(d)** $\dfrac{2}{9}$ **3.** Choice D is correct, because every sign has been changed in the fraction.
5. Method 1 indicates to write the complex fraction as a division problem, and then perform the division. For example, to simplify $\dfrac{\frac{1}{2}}{\frac{2}{3}}$, write $\dfrac{1}{2}\div\dfrac{2}{3}$. Then simplify as $\dfrac{1}{2}\cdot\dfrac{3}{2}=\dfrac{3}{4}$.   **7.** -6 **9.** $\dfrac{1}{xy}$ **11.** $\dfrac{2a^2b}{3}$ **13.** $\dfrac{m(m+2)}{3(m-4)}$ **15.** $\dfrac{2}{x}$ **17.** $\dfrac{8}{x}$ **19.** $\dfrac{a^2-5}{a^2+1}$ **21.** $\dfrac{31}{50}$
23. $\dfrac{y^2+x^2}{xy(y-x)}$ **25.** $\dfrac{40-12p}{85p}$ **27.** $\dfrac{5y-2x}{3+4xy}$ **29.** $\dfrac{a-2}{2a}$ **31.** $\dfrac{z-5}{4}$ **33.** $\dfrac{-m}{m+2}$ **35.** $\dfrac{3m(m-3)}{(m-1)(m-8)}$ **37.** division
39. $\dfrac{\frac{3}{8}+\frac{5}{6}}{2}$ **40.** $\dfrac{29}{48}$ **41.** $\dfrac{29}{48}$ **42.** Answers will vary.

Section 7.6 (page 493)

Exercises 1. expression; $\dfrac{43}{40}x$ **3.** equation; $\left\{\dfrac{40}{43}\right\}$ **5.** expression; $-\dfrac{1}{10}x$ **7.** equation; $\{-10\}$ **9.** $-2, 0$ **11.** $-3, 4, -\dfrac{1}{2}$
13. $-9, 1, -2, 2$ **15.** When adding and subtracting, the LCD is retained as part of the answer. When solving an equation, the LCD is used as the multiplier in applying the multiplication property of equality.   **17.** $\left\{\dfrac{1}{4}\right\}$ **19.** $\left\{-\dfrac{3}{4}\right\}$ **21.** $\{-15\}$ **23.** $\{7\}$ **25.** $\{-15\}$
27. $\{-5\}$ **29.** $\{-6\}$ **31.** $\emptyset$ **33.** $\{5\}$ **35.** $\{4\}$ **37.** $\{1\}$ **39.** $\{4\}$ **41.** $\{5\}$ **43.** $\{-4\}$ **45.** $\{-2, 12\}$ **47.** $\emptyset$ **49.** $\{3\}$
51. $\{3\}$ **53.** $\{-3\}$ **55.** $\left\{-\dfrac{1}{5}, 3\right\}$ **57.** $\left\{-\dfrac{1}{2}, 5\right\}$ **59.** $\{3\}$ **61.** $\left\{-\dfrac{1}{3}, 3\right\}$ **63.** $\{-1\}$ **65.** $\{-6\}$ **67.** $\left\{-6, \dfrac{1}{2}\right\}$ **69.** $\{6\}$

A-34 Answers to Selected Exercises

71. Transform the equation so that the terms with k are on one side and the remaining term is on the other.   **73.** $F = \dfrac{ma}{k}$ **75.** $a = \dfrac{kF}{m}$

77. $R = \dfrac{E - Ir}{I}$ or $R = \dfrac{E}{I} - r$ **79.** $A = \dfrac{h(B + b)}{2}$ **81.** $a = \dfrac{2S - ndL}{nd}$ or $a = \dfrac{2S}{nd} - L$ **83.** $y = \dfrac{xz}{x + z}$ **85.** $z = \dfrac{3y}{5 - 9xy}$ or $z = \dfrac{-3y}{9xy - 5}$

Summary Exercises on Operations and Equations with Rational Expressions   (page 497)

1. operation; $\dfrac{10}{p}$ **2.** operation; $\dfrac{y^3}{x^3}$ **3.** operation; $\dfrac{1}{2x^2(x + 2)}$ **4.** equation; $\{9\}$ **5.** operation; $\dfrac{y + 2}{y - 1}$ **6.** operation; $\dfrac{5k + 8}{k(k - 4)(k + 4)}$

7. equation; $\{39\}$ **8.** operation; $\dfrac{t - 5}{3(2t + 1)}$ **9.** operation; $\dfrac{13}{3(p + 2)}$ **10.** equation; $\left\{-1, \dfrac{12}{5}\right\}$ **11.** equation; $\left\{\dfrac{1}{7}, 2\right\}$

12. operation; $\dfrac{16}{3y}$ **13.** operation; $\dfrac{7}{12z}$ **14.** equation; $\{13\}$ **15.** operation; $\dfrac{3m + 5}{(m + 3)(m + 2)(m + 1)}$ **16.** operation; $\dfrac{k + 3}{5(k - 1)}$

17. equation; $\emptyset$ **18.** equation; $\emptyset$ **19.** operation; $\dfrac{t + 2}{2(2t + 1)}$ **20.** equation; $\{-7\}$

Section 7.7   (page 502)

Exercises   **1. (a)** the amount   **(b)** $5 + x$ **(c)** $\dfrac{5 + x}{6} = \dfrac{13}{3}$ **3.** $\dfrac{12}{18}$ **5.** $\dfrac{12}{3}$ **7.** 12 **9.** $\dfrac{1386}{97}$ **11.** 74.75 sec   **13.** 367.197 m per min

15. 3.090 hr   **17.** $\dfrac{D}{R} = \dfrac{d}{r}$ **19.** $\dfrac{500}{x - 10} = \dfrac{600}{x + 10}$ **21.** 8 mph   **23.** 165 mph   **25.** 3 mph   **27.** 18.5 mph   **29.** $\dfrac{1}{10}$ job per hr

31. $\dfrac{1}{8}x + \dfrac{1}{6}x = 1$ or $\dfrac{1}{8} + \dfrac{1}{6} = \dfrac{1}{x}$ **33.** $4\dfrac{4}{17}$ hr   **35.** $5\dfrac{5}{11}$ hr   **37.** 3 hr   **39.** $2\dfrac{7}{10}$ hr   **41.** $9\dfrac{1}{11}$ min

Section 7.8   (page 510)

Exercises   **1.** direct   **3.** direct   **5.** direct   **7.** inverse   **9.** inverse   **11.** direct   **13.** direct   **15.** inverse   **17.** increases   **19.** 9

21. 250   **23.** 6   **25.** 21   **27.** $\dfrac{16}{5}$ **29.** $\dfrac{4}{9}$ **31.** $40.32   **33.** $42\dfrac{2}{3}$ in.   **35.** $106\dfrac{2}{3}$ mph   **37.** $12\dfrac{1}{2}$ amps   **39.** 20 lb

41. 52.817 in.^{2} **43.** $14\dfrac{22}{27}$ foot-candles   **45.** direct   **47.** inverse

Chapter 7 Review Exercises   (page 519)

1. (a) $\dfrac{11}{8}$ **(b)** $\dfrac{13}{22}$ **3.** 3   **5.** $-5, -\dfrac{2}{3}$ **7.** $\dfrac{b}{3a}$ **9.** $\dfrac{-(2x + 3)}{2}$ **11.** (Answers may vary.) $\dfrac{-(4x - 9)}{2x + 3}, \dfrac{-4x + 9}{2x + 3}, \dfrac{4x - 9}{-(2x + 3)}$,

$\dfrac{4x - 9}{-2x - 3}$ **13.** $\dfrac{72}{p}$ **15.** $\dfrac{5}{8}$ **17.** $\dfrac{3a - 1}{a + 5}$ **19.** $\dfrac{p + 5}{p + 1}$ **21.** $108y^4$ **23.** $\dfrac{15a}{10a^4}$ **25.** $\dfrac{15y}{50 - 10y}$ **27.** $\dfrac{15}{x}$ **29.** $\dfrac{4k - 45}{k(k - 5)}$

31. $\dfrac{-2 - 3m}{6}$ **33.** $\dfrac{7a + 6b}{(a - 2b)(a + 2b)}$ **35.** $\dfrac{5z - 16}{z(z + 6)(z - 2)}$ **37. (a)** $\dfrac{a}{b}$ **(b)** $\dfrac{a}{b}$ **(c)** Answers will vary.   **39.** $\dfrac{4(y - 3)}{y + 3}$ **41.** $\dfrac{xw + 1}{xw - 1}$

43. It would cause the first and third denominators to equal 0.   **45.** $\emptyset$ **47.** $t = \dfrac{Ry}{m}$ **49.** $m = \dfrac{4 + p^2q}{3p^2}$ **51.** $\dfrac{2}{6}$ **53.** $3\dfrac{1}{13}$ hr

55. inverse   **57.** $\dfrac{36}{5}$ **59.** $8p^2$ **61.** 3   **63.** $v = at + w$ **65.** $1\dfrac{7}{8}$ hr   **67.** 4   **69. (a)** -3 **(b)** -1 **(c)** $-3, -1$ **70.** $\dfrac{15}{2x}$

71. If $x = 0$, the divisor R is equal to 0, and division by 0 is undefined.   **72.** $(x + 3)(x + 1)$ **73.** $\dfrac{7}{x + 1}$ **74.** $\dfrac{11x + 21}{4x}$ **75.** $\emptyset$

Answers to Selected Exercises   A-35

76. We know that -3 is not allowed because P and R are undefined for $x = -3$.   **77.** Rate is equal to distance divided by time. Here, distance is 6 mi, and time is $x + 3$ min, so rate $= \dfrac{6}{x+3}$, which is the expression for P.   **78.** $\dfrac{6}{5}, \dfrac{5}{2}$

Chapter 7 Test   (page 522)

[7.1] **1.** (a) $\dfrac{11}{6}$ (b) undefined   **2.** $-2, 4$ **3.** (Answers may vary.) $\dfrac{-(6x-5)}{2x+3}, \dfrac{-6x+5}{2x+3}, \dfrac{6x-5}{-(2x+3)}, \dfrac{6x-5}{-2x-3}$ **4.** $-3x^2y^3$

5. $\dfrac{3a+2}{a-1}$ [7.2] **6.** $\dfrac{25}{27}$ **7.** $\dfrac{3k-2}{3k+2}$ **8.** $\dfrac{a-1}{a+4}$ [7.3] **9.** $150p^5$ **10.** $(2r+3)(r+2)(r-5)$ **11.** $\dfrac{240p^2}{64p^3}$ **12.** $\dfrac{21}{42m-84}$

[7.4] **13.** 2   **14.** $\dfrac{-14}{5(y+2)}$ **15.** $\dfrac{-x^2+x+1}{3-x}$ or $\dfrac{x^2-x-1}{x-3}$ **16.** $\dfrac{-m^2+7m+2}{(2m+1)(m-5)(m-1)}$ [7.5] **17.** $\dfrac{2k}{3p}$ **18.** $\dfrac{-2-x}{4+x}$

[7.6] **19.** $\left\{-\dfrac{1}{2}\right\}$ **20.** $D = \dfrac{dF-k}{F}$ or $D = \dfrac{k-dF}{-F}$ [7.7] **21.** 3 mph   **22.** $2\dfrac{2}{9}$ hr   [7.8] **23.** 27   **24.** 27 days

Cumulative Review Exercises Chapters 1–7   (page 524)

[1.2, 1.5, 1.6] **1.** 2   [2.3] **2.** {17}   [2.5] **3.** $b = \dfrac{2A}{h}$ [2.6] **4.** $\left\{-\dfrac{2}{7}\right\}$ [2.8] **5.** $[-8, \infty)$ **6.** $(4, \infty)$ [3.1] **7.** (a) $(-3, 0)$

(b) $(0, -4)$ [3.2] **8.** [graph: $y = -3x + 2$]   [5.4] **9.** [graph: $y = -x^2 + 1$]   [4.1–4.3] **10.** $\{(-1, 3)\}$ **11.** ∅   [5.1, 5.2] **12.** $\dfrac{1}{2^4 x^7}$ **13.** $\dfrac{1}{m^6}$

14. $\dfrac{q}{4p^2}$ [5.4] **15.** $k^2 + 2k + 1$ [5.1] **16.** $72x^6 y^7$ [5.6] **17.** $4a^2 - 4ab + b^2$ [5.5] **18.** $3y^3 + 8y^2 + 12y - 5$

[5.7] **19.** $6p^2 + 7p + 1 + \dfrac{3}{p-1}$ [5.3] **20.** 1.4×10^5 sec   [6.3] **21.** $(4t+3v)(2t+v)$ **22.** prime

[6.4] **23.** $(4x^2+1)(2x+1)(2x-1)$ [6.5] **24.** $\{-3, 5\}$ **25.** $\left\{5, -\dfrac{1}{2}, \dfrac{2}{3}\right\}$ [6.6] **26.** -2 or -1 **27.** 6 m   [7.1] **28.** A   **29.** D

[7.4] **30.** $\dfrac{4}{q}$ **31.** $\dfrac{3r+28}{7r}$ **32.** $\dfrac{7}{15(q-4)}$ **33.** $\dfrac{-k-5}{k(k+1)(k-1)}$ [7.2] **34.** $\dfrac{7(2z+1)}{24}$ [7.5] **35.** $\dfrac{195}{29}$ [7.6] **36.** 4, 0

37. $\left\{\dfrac{21}{2}\right\}$ **38.** $\{-2, 1\}$ [7.7] **39.** $1\dfrac{1}{5}$ hr   [7.8] **40.** 32.97 in.

CHAPTER 8 ROOTS AND RADICALS

Section 8.1   (page 535)

Connections   (page 535)   The area of the large square is $(a+b)^2$ or $a^2 + 2ab + b^2$. The sum of the areas of the smaller square and the four right triangles is $c^2 + 2ab$. Set these equal to each other and subtract $2ab$ from each side to get $a^2 + b^2 = c^2$.

Exercises   **1.** true   **3.** false; Zero has only one square root.   **5.** true   **7.** $-3, 3$ **9.** $-8, 8$ **11.** $-12, 12$ **13.** $-\dfrac{5}{14}, \dfrac{5}{14}$

15. $-30, 30$ **17.** 1   **19.** 7   **21.** -11 **23.** $-\dfrac{12}{11}$ **25.** not a real number   **27.** 19   **29.** 19   **31.** $\dfrac{2}{3}$ **33.** $3x^2 + 4$

35. a must be positive.   **37.** a must be negative.   **39.** rational; 5   **41.** irrational; 5.385   **43.** rational; -8 **45.** irrational; -17.321

47. not a real number **49.** irrational; 34.641 **51.** C **53.** $c = 17$ **55.** $b = 8$ **57.** $c \approx 11.705$ **59.** 24 cm **61.** 80 ft **63.** 195 ft **65.** 11.1 ft **67.** 9.434 **69.** Answers will vary. For example, if $a = 2$ and $b = 7$, $\sqrt{a^2 + b^2} = \sqrt{2^2 + 7^2} = \sqrt{53}$, while $a + b = 2 + 7 = 9$. Therefore, $\sqrt{a^2 + b^2} \neq a + b$ because $\sqrt{53} \neq 9$. **71.** 5 **73.** 17 **75.** $\sqrt{13}$ **77.** 5 **79.** 13 **81.** $\sqrt{13}$ **83.** $\sqrt{2}$ **85.** 1 **87.** 5 **89.** -3 **91.** -2 **93.** 5 **95.** not a real number **97.** -3 **99.** -4 **101.** 2.289 **103.** 5.074 **105.** -4.431 **107.** c^2 **108.** $(b - a)^2$ **109.** $2ab$ **110.** $(b - a)^2 = a^2 - 2ab + b^2$ **111.** $c^2 = 2ab + (a^2 - 2ab + b^2)$ **112.** $c^2 = a^2 + b^2$

Section 8.2 (page 544)

Exercises **1.** false; $\sqrt{(-6)^2} = \sqrt{36} = 6$ **3.** $\sqrt{15}$ **5.** $\sqrt{22}$ **7.** $\sqrt{42}$ **9.** $\sqrt{13r}$ **11.** A **13.** $3\sqrt{5}$ **15.** $2\sqrt{6}$ **17.** $3\sqrt{10}$ **19.** $5\sqrt{3}$ **21.** $5\sqrt{5}$ **23.** cannot be simplified **25.** $4\sqrt{10}$ **27.** $-10\sqrt{7}$ **29.** $3\sqrt{6}$ **31.** 24 **33.** $6\sqrt{10}$ **35.** $\sqrt{8} \cdot \sqrt{32} = \sqrt{8 \cdot 32} = \sqrt{256} = 16$. Also, $\sqrt{8} = 2\sqrt{2}$ and $\sqrt{32} = 4\sqrt{2}$, so $\sqrt{8} \cdot \sqrt{32} = 2\sqrt{2} \cdot 4\sqrt{2} = 8 \cdot 2 = 16$. Both methods give the same answer, and the correct answer can always be obtained using either method. **37.** $\dfrac{4}{15}$ **39.** $\dfrac{\sqrt{7}}{4}$ **41.** 5 **43.** $\dfrac{25}{4}$ **45.** $6\sqrt{5}$ **47.** m **49.** y^2 **51.** $6z$ **53.** $20x^3$ **55.** $3x^4\sqrt{2}$ **57.** $3c^7\sqrt{5}$ **59.** $z^2\sqrt{z}$ **61.** $a^6\sqrt{a}$ **63.** $8x^3\sqrt{x}$ **65.** x^3y^6 **67.** $9m^2n$ **69.** $\dfrac{\sqrt{7}}{x^5}$ **71.** $\dfrac{y^2}{10}$ **73.** $2\sqrt[3]{5}$ **75.** $3\sqrt[3]{2}$ **77.** $4\sqrt[3]{2}$ **79.** $2\sqrt[3]{5}$ **81.** $\dfrac{2}{3}$ **83.** $-\dfrac{6}{5}$ **85.** p **87.** x^3 **89.** $4z^2$ **91.** $7a^3b$ **93.** $2t\sqrt[3]{2t^2}$ **95.** $\dfrac{m^4}{2}$ **97.** (a) 4.472135955 (b) 4.472135955 (c) The numerical results are not a proof because both answers are approximations and they might differ if calculated to more decimal places. **99.** The product rule for radicals requires that both a and b must be nonnegative. Otherwise $\sqrt{a}$ and $\sqrt{b}$ would not be real numbers (except when $a = b = 0$). **101.** 6 cm **103.** 6 in. **105.** D

Section 8.3 (page 549)

Exercises **1.** $7\sqrt{3}$ **3.** $-5\sqrt{7}$ **5.** $2\sqrt{6}$ **7.** $3\sqrt{17}$ **9.** $7\sqrt{3}$ **11.** cannot be added using the distributive property **13.** $3\sqrt{3}$ **15.** $-20\sqrt{2}$ **17.** $19\sqrt{7}$ **19.** $12\sqrt{6} + 6\sqrt{5}$ **21.** $-2\sqrt{2} - 12\sqrt{3}$ **23.** $4\sqrt{2}$ **25.** $3\sqrt{3} - 2\sqrt{5}$ **27.** $3\sqrt{21}$ **29.** $5\sqrt{3}$ **31.** $-\sqrt[3]{2}$ **33.** $24\sqrt[3]{3}$ **35.** $10\sqrt[3]{2} + 4\sqrt[3]{8}$ **37.** $2\sqrt{3} + 4\sqrt{3} = (2 + 4)\sqrt{3} = 6\sqrt{3}$ **39.** $\sqrt{2x}$ **41.** $7\sqrt{3r}$ **43.** $15x\sqrt{3}$ **45.** $2x\sqrt{2}$ **47.** $13p\sqrt{3}$ **49.** $42x\sqrt{5z}$ **51.** $6k^2h\sqrt{6} + 27hk\sqrt{6k}$ **53.** $6\sqrt[3]{p^2}$ **55.** $21\sqrt[3]{m^3}$ **57.** $-8p\sqrt[3]{p}$ **59.** $-24z\sqrt[3]{4z}$ **61.** The variables may represent negative numbers, because negative numbers have cube roots. **63.** $\sqrt{85}$ **65.** 5 **67.** $-6x^2y$ **68.** $-6(p - 2q)^2(a + b)$ **69.** $-6a^2\sqrt{xy}$ **70.** The answers are alike because the numerical coefficient of the three answers is the same: -6. Also, the first variable factor is raised to the second power, and the second variable factor is raised to the first power. The answers are different because the variables are different: x and y, then $p - 2q$ and $a + b$, and then a and $\sqrt{xy}$.

Section 8.4 (page 555)

Exercises **1.** $\dfrac{6\sqrt{5}}{5}$ **3.** $\sqrt{5}$ **5.** $\dfrac{2\sqrt{6}}{3}$ **7.** $\dfrac{8\sqrt{15}}{5}$ **9.** $\dfrac{\sqrt{30}}{2}$ **11.** $\dfrac{8\sqrt{3}}{9}$ **13.** $\dfrac{3\sqrt{2}}{10}$ **15.** $\sqrt{2}$ **17.** $\sqrt{2}$ **19.** $\dfrac{2\sqrt{30}}{3}$ **21.** $\dfrac{\sqrt{2}}{8}$ **23.** $\dfrac{3\sqrt{5}}{5}$ **25.** $\dfrac{-3\sqrt{2}}{10}$ **27.** $\dfrac{21\sqrt{5}}{5}$ **29.** $\dfrac{\sqrt{3}}{3}$ **31.** $-\dfrac{\sqrt{5}}{5}$ **33.** $\dfrac{\sqrt{65}}{5}$ **35.** 1; identity property for multiplication **37.** $\dfrac{\sqrt{21}}{3}$ **39.** $\dfrac{3\sqrt{14}}{4}$ **41.** $\dfrac{1}{6}$ **43.** 1 **45.** $\dfrac{\sqrt{15}}{10}$ **47.** $\dfrac{17\sqrt{2}}{6}$ **49.** $\dfrac{\sqrt{3}}{5}$ **51.** $\dfrac{4\sqrt{3}}{27}$ **53.** $\dfrac{\sqrt{6p}}{p}$ **55.** $\dfrac{\sqrt{3y}}{y}$ **57.** $\dfrac{4\sqrt{m}}{m}$ **59.** $\dfrac{p\sqrt{3q}}{q}$ **61.** $\dfrac{x\sqrt{7xy}}{y}$ **63.** $\dfrac{p\sqrt{2pm}}{m}$ **65.** $\dfrac{x\sqrt{y}}{2y}$ **67.** $\dfrac{3a\sqrt{5r}}{5}$ **69.** B **71.** $\dfrac{\sqrt[3]{4}}{2}$ **73.** $\dfrac{\sqrt[3]{2}}{4}$ **75.** $\dfrac{\sqrt[3]{121}}{11}$ **77.** $\dfrac{\sqrt[3]{50}}{5}$ **79.** $\dfrac{\sqrt[3]{196}}{7}$ **81.** $\dfrac{\sqrt[3]{6y}}{2y}$ **83.** $\dfrac{\sqrt[3]{42mn^2}}{6n}$

Section 8.5 (page 561)

Exercises 1. 13   3. 4   5. $\sqrt{15} - \sqrt{35}$ 7. $2\sqrt{10} + 30$ 9. $4\sqrt{7}$ 11. $57 + 23\sqrt{6}$ 13. $81 + 14\sqrt{21}$ 15. $71 - 16\sqrt{7}$ 17. $37 + 12\sqrt{7}$ 19. $7 + 2\sqrt{6}$ 21. 23   23. 1   25. 2   27. $2\sqrt{3} - 2 + 3\sqrt{2} - \sqrt{6}$ 29. $15\sqrt{2} - 15$ 31. $\sqrt{30} + \sqrt{15} + 6\sqrt{5} + 3\sqrt{10}$ 33. $187 - 20\sqrt{21}$ 35. Because multiplication must be performed before addition, it is incorrect to add -37 and -2. Since $-2\sqrt{15}$ cannot be simplified, the expression cannot be written in a simpler form, and the final answer is $-37 - 2\sqrt{15}$.   37. $49 + 14\sqrt{x} + x$ 39. $6t - 3\sqrt{14t} + 2\sqrt{7t} - 7\sqrt{2}$ 41. $3m - 2n$ 43. (a) $\sqrt{5} - \sqrt{3}$ (b) $\sqrt{6} + \sqrt{5}$ 45. $-2 + \sqrt{5}$ 47. $-2 - \sqrt{11}$ 49. $3 - \sqrt{3}$ 51. $\dfrac{-3 + 5\sqrt{3}}{11}$ 53. $\dfrac{\sqrt{6} + \sqrt{2} + 3\sqrt{3} + 3}{2}$ 55. $\dfrac{-6\sqrt{2} + 12 + \sqrt{10} - 2\sqrt{5}}{2}$ 57. $\dfrac{-4\sqrt{3} - \sqrt{2} + 10\sqrt{6} + 5}{23}$ 59. $\sqrt{21} + \sqrt{14} + \sqrt{6} + 2$ 61. $-\sqrt{10} + \sqrt{15}$ 63. $3 - \sqrt{3}$ 65. $\dfrac{8(4 + \sqrt{x})}{16 - x}$ 67. $\sqrt{7} - 2$ 69. $\dfrac{\sqrt{3} + 5}{4}$ 71. $\dfrac{6 - \sqrt{10}}{2}$ 73. $\dfrac{2 + \sqrt{2}}{3}$ 75. $2 - 3\sqrt[3]{4}$ 77. $12 + 10\sqrt[3]{8}$ 79. $-1 + 3\sqrt[3]{2} - \sqrt[3]{4}$ 81. 1   83. 4 in.   85. (a) $\dfrac{9\sqrt{2}}{4}$ sec (b) 3.182 sec   87. Answers will vary. Some possibilities are $\dfrac{DB}{DY}$ and $\dfrac{EB}{EX}$.   89. $30 + 18x$ 90. They are not like terms.   91. $30 + 18\sqrt{5}$ 92. They are not like radicals.   93. Make the first term $30x$, so that $30x + 18x = 48x$; make the first term $30\sqrt{5}$, so that $30\sqrt{5} + 18\sqrt{5} = 48\sqrt{5}$.   94. When combining like terms, we add (or subtract) the coefficients of the common factors of the terms: $2xy + 5xy = 7xy$. When combining like radicals, we add (or subtract) the coefficients of the common radical terms: $2\sqrt{ab} + 5\sqrt{ab} = 7\sqrt{ab}$.

Summary Exercises on Operations with Radicals (page 564)

1. $-3\sqrt{10}$ 2. $5 - \sqrt{15}$ 3. $2 - \sqrt{6} + 2\sqrt{3} - 3\sqrt{2}$ 4. $6\sqrt{2}$ 5. $73 - 12\sqrt{35}$ 6. $\dfrac{\sqrt{6}}{2}$ 7. $3\sqrt[3]{2t^2}$ 8. $4\sqrt{7} + 4\sqrt{5}$ 9. $-3 - 2\sqrt{2}$ 10. 4   11. -33 12. $\dfrac{\sqrt{t} - \sqrt{3}}{t - 3}$ 13. $2xyz^2\sqrt[3]{y^2}$ 14. $4\sqrt[3]{3}$ 15. $\sqrt{6} + 1$ 16. $\dfrac{\sqrt{6x}}{3x}$ 17. $\dfrac{3}{5}$ 18. $4\sqrt{2}$ 19. $-2\sqrt[3]{2}$ 20. $11 - 2\sqrt{30}$ 21. $3\sqrt{3x}$ 22. $52 + 30\sqrt{3}$ 23. 1   24. $\dfrac{2\sqrt[3]{18}}{9}$ 25. $-x^2\sqrt[3]{x}$

Section 8.6 (page 571)

Exercises 1. {49}   3. {7}   5. {85}   7. {−45}   9. $\left\{-\dfrac{3}{2}\right\}$ 11. ∅   13. {121}   15. {8}   17. {1}   19. {6}   21. ∅   23. {5}   25. {7}   27. {6}   29. ∅   31. $x^2 - 14x + 49$ 33. {−2, 1}   35. {12}   37. {3}   39. {11}   41. {−1, 3}   43. {9}   45. {8}   47. {9}   49. {2, 11}   51. Since $\sqrt{x}$ must be greater than or equal to 0 for any replacement for x, it cannot equal -8, a negative number.   53. {−1}   55. {9}   57. {4, 20}   59. {−5}   61. 21   63. 8   65. (a) 70.5 mph (b) 59.8 mph (c) 53.9 mph   67. 158.6 ft   69. (a) $25.1 billion (b) Actual imports for 1994 were $25.9 billion, so the result using the equation is a little low. (c) $53.3 billion; The result using the equation is a little low. (d) .2, −.1, −.4; Exports exceeded imports only in 1993.   71. yes; 26 mi   73. 47 mi   75. $s = 13$ units   76. $6\sqrt{13}$ sq. units   77. $h = \sqrt{13}$ units   78. $3\sqrt{13}$ sq. units   79. $6\sqrt{13}$ sq. units   80. They are both $6\sqrt{13}$.

Section 8.7 (page 579)

Exercises 1. A   3. C   5. 5   7. 4   9. 2   11. 2   13. 8   15. 9   17. 8   19. 4   21. -4 23. -4 25. $\dfrac{1}{343}$ 27. $\dfrac{1}{36}$ 29. $-\dfrac{1}{32}$ 31. $2^{8/3}$ 33. $\dfrac{1}{6^{1/2}}$ 35. $\dfrac{1}{15^{1/2}}$ 37. $11^{1/7}$ 39. 8^3 41. $6^{1/2}$ 43. $\dfrac{5^3}{2^3}$ 45. $\dfrac{1}{2^{8/5}}$ 47. $6^{2/9}$ 49. $x^{9/5}$ 51. $r^{1/9}$ 53. $m^2 n^{1/6}$ 55. $\dfrac{a^4}{b^{3/2}}$ 57. $m^{1/6}$ 59. 2   61. 2   63. $\sqrt{a}$ 65. $\sqrt[3]{k^2}$ 67. 2   69. 1.883   71. (a) $d = 1.22x^{1/2}$ (b) 211.31 mi   73. Because $(7^{1/2})^2 = 7$ and $(\sqrt{7})^2 = 7$, they should be equal, so we define $7^{1/2}$ to be $\sqrt{7}$.   75. $\sqrt{2} = 2^{1/2}$ and $\sqrt[3]{2} = 2^{1/3}$ 76. $2^{1/2} \cdot 2^{1/3}$ 77. 6   78. $2^{3/6} \cdot 2^{2/6}$ 79. $2^{5/6}$ 80. $\sqrt[6]{2^5}$ or $\sqrt[6]{32}$

Chapter 8 Review Exercises (page 584)

1. $-7, 7$ 3. $-14, 14$ 5. $-15, 15$ 7. 4 9. 10 11. not a real number 13. $\dfrac{7}{6}$ 15. a must be negative. 17. 40.6 cm
19. rational; -5 21. $5\sqrt{3}$ 23. $4\sqrt{10}$ 25. 12 27. $16\sqrt{6}$ 29. $-\dfrac{11}{20}$ 31. $\dfrac{\sqrt{7}}{13}$ 33. $\dfrac{2}{15}$ 35. 8 37. p 39. r^9
41. $a^7 b^{10}\sqrt{ab}$ 43. Yes, because both approximations are .7071067812. 45. $21\sqrt{3}$ 47. 0 49. $2\sqrt{3} + 3\sqrt{10}$ 51. $6\sqrt{30}$
53. 0 55. $11k^2\sqrt{2n}$ 57. $\sqrt{5}$ 59. $\dfrac{\sqrt{30}}{15}$ 61. $\sqrt{10}$ 63. $\dfrac{r\sqrt{x}}{4x}$ 65. $\dfrac{\sqrt{98}}{7}$ 67. $-\sqrt{15} - 9$ 69. $22 - 16\sqrt{3}$ 71. -2
73. $-2 + \sqrt{5}$ 75. $\dfrac{-2 + 6\sqrt{2}}{17}$ 77. $\dfrac{-\sqrt{10} + 3\sqrt{5} + \sqrt{2} - 3}{7}$ 79. $\dfrac{3 + 2\sqrt{6}}{3}$ 81. $3 + 4\sqrt{3}$ 83. $\emptyset$ 85. $\{1\}$ 87. $\{6\}$
89. $\{-2\}$ 91. $\{5\}$ 93. 9 95. 7^3 or 343 97. $x^{3/4}$ 99. 16 101. $\dfrac{5 - \sqrt{2}}{23}$ 103. $5y\sqrt{2}$ 105. $-\sqrt{10} - 5\sqrt{15}$
107. $\dfrac{2 + \sqrt{13}}{2}$ 109. $7 - 2\sqrt{10}$ 111. -11 113. $\{7\}$ 115. $\{8\}$ 117. $\dfrac{10\sqrt{7} - 5\sqrt{7}}{-3\sqrt{14} - 2\sqrt{14}}$ or $\dfrac{5\sqrt{7} - 10\sqrt{7}}{2\sqrt{14} + 3\sqrt{14}}$ 118. $-\dfrac{5\sqrt{7}}{5\sqrt{14}}$
119. $-\sqrt{\dfrac{1}{2}}$ 120. $-\dfrac{\sqrt{2}}{2}$ 121. It falls from left to right.

Chapter 8 Test (page 588)

[8.1] 1. $-14, 14$ 2. (a) irrational (b) 11.916 3. There is no real number whose square is -5. [8.2] 4. $-3\sqrt{3}$ 5. $\dfrac{8\sqrt{2}}{5}$
6. $2\sqrt[3]{4}$ 7. $4\sqrt{6}$ [8.3] 8. $9\sqrt{7}$ 9. $-5\sqrt{3x}$ [8.2] 10. $2y\sqrt[3]{4x^2}$ [8.5] 11. 31 12. $6\sqrt{2} + 2 - 3\sqrt{14} - \sqrt{7}$
13. $11 + 2\sqrt{30}$ [8.3] 14. $-6x\sqrt[3]{2x}$ [8.1] 15. (a) $6\sqrt{2}$ in. (b) 8.485 in. 16. 50 ohms [8.4] 17. $r = \dfrac{\sqrt{\pi S}}{2\pi}$ 18. $\dfrac{5\sqrt{14}}{7}$
19. $\dfrac{\sqrt{6x}}{3x}$ 20. $-\sqrt[3]{2}$ [8.5] 21. $\dfrac{-12 - 3\sqrt{3}}{13}$ 22. $\dfrac{\sqrt{3} + 12\sqrt{2}}{3}$ [8.6] 23. $\{3\}$ 24. $\left\{\dfrac{1}{4}, 1\right\}$ 25. $\{-4\}$ [8.7] 26. 16
27. -25 28. 5 29. $\dfrac{1}{3}$ [8.6] 30. 12 is not a solution. A check shows that it does not satisfy the original equation.

Cumulative Review Exercises Chapters 1–8 (page 589)

[1.2] 1. 54 2. 6 [1.4] 3. 3 [2.3] 4. $\{3\}$ [2.8] 5. $[-16, \infty)$ 6. $(5, \infty)$ [2.4] 7. 1999: 9.4 billion bushels; 2000: 10.2 billion bushels [3.2] 8.

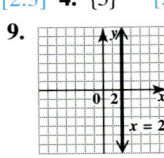

9.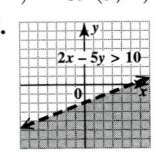

[3.5] 10.

[3.3, 3.4] 11. (a) .47; Convention spending increased $.47 million per year. (b) $y = .47x - 926.5$ (c) $15.4 million [4.1–4.3] 12. $\{(3, -7)\}$ 13. infinite number of solutions
[4.4] 14. from Chicago: 61 mph; from Des Moines: 54 mph [5.1] 15. $12x^{10}y^2$ [5.2] 16. $\dfrac{y^{15}}{2^3 \cdot 3^6}$ or $\dfrac{y^{15}}{5832}$
[5.4] 17. $3x^3 + 11x^2 - 13$ [5.7] 18. $4t^2 - 8t + 5$ [6.2–6.4] 19. $(m + 8)(m + 4)$ 20. $(5t^2 + 6)(5t^2 - 6)$
21. $(6a + 5b)(2a - b)$ 22. $(9z + 4)^2$ [6.5] 23. $\{3, 4\}$ 24. $\{-2, -1\}$ [7.1] 25. $-7, 2$ [7.2] 26. $\dfrac{x + 1}{x}$
27. $(t + 5)(t + 3)$ or $t^2 + 8t + 15$ [7.4] 28. $\dfrac{y^2}{(y + 1)(y - 1)}$ 29. $\dfrac{-2x - 14}{(x + 3)(x - 1)}$ [7.5] 30. -21 [7.6] 31. $\{19\}$
32. $B = \dfrac{CD - AD}{AC - 1}$ [7.8] 33. 405 revolutions per min [8.3] 34. $29\sqrt{3}$ [8.5] 35. $-\sqrt{3} + \sqrt{5}$ [8.2] 36. $10xy^2\sqrt{2y}$
[8.5] 37. $21 - 5\sqrt{2}$ [8.6] 38. $\{16\}$ [8.7] 39. 32 40. $\dfrac{1}{64}$

CHAPTER 9 QUADRATIC EQUATIONS

Section 9.1 (page 597)

Exercises 1. C 3. D 5. $\{\pm 9\}$ 7. $\{\pm\sqrt{14}\}$ 9. $\{\pm 4\sqrt{3}\}$ 11. $\emptyset$ 13. $\{\pm 1.5\}$ 15. $\left\{\pm\dfrac{2\sqrt{7}}{7}\right\}$ 17. $\left\{\pm\dfrac{2\sqrt{5}}{5}\right\}$ 19. $\{\pm 2\sqrt{6}\}$ 21. According to the square root property, -9 is also a solution, so her answer was not completely correct. The solution set is $\{\pm 9\}$. 23. $\{-2, 8\}$ 25. $\emptyset$ 27. $\{8 \pm 3\sqrt{3}\}$ 29. $\left\{-3, \dfrac{5}{3}\right\}$ 31. $\left\{0, \dfrac{3}{2}\right\}$ 33. $\left\{\dfrac{5 \pm \sqrt{30}}{2}\right\}$ 35. $\left\{\dfrac{-1 \pm 3\sqrt{2}}{3}\right\}$ 37. $\{-10 \pm 4\sqrt{3}\}$ 39. $\left\{\dfrac{1 \pm 4\sqrt{3}}{4}\right\}$ 41. Johnny's first solution, $\dfrac{5 + \sqrt{30}}{2}$, is equivalent to Linda's second solution, $\dfrac{-5 - \sqrt{30}}{-2}$. This can be verified by multiplying $\dfrac{5 + \sqrt{30}}{2}$ by 1 in the form $\dfrac{-1}{-1}$. Similarly, Johnny's second solution is equivalent to Linda's first one. 43. $\{-4.48, .20\}$ 45. $\{-3.09, -.15\}$ 47. $(x + 3)^2 = 100$ 48. $\{-13, 7\}$ 49. $\{-13, 7\}$ 50. $\{-3, 6\}$ 51. about $\dfrac{1}{2}$ sec 53. 9 in. 55. 5%

Section 9.2 (page 605)

Exercises 1. $10k$ 3. 49 5. D 7. 4 9. $\dfrac{25}{4}$ 11. $\dfrac{1}{16}$ 13. $\{1, 3\}$ 15. $\{-1 \pm \sqrt{6}\}$ 17. $\{-3\}$ 19. $\left\{-\dfrac{3}{2}, \dfrac{1}{2}\right\}$ 21. $\emptyset$ 23. $\left\{\dfrac{9 \pm \sqrt{21}}{6}\right\}$ 25. $\left\{\dfrac{-7 \pm \sqrt{97}}{6}\right\}$ 27. $\{-4, 2\}$ 29. $\{1 \pm \sqrt{6}\}$ 31. (a) $\left\{\dfrac{3 \pm 2\sqrt{6}}{3}\right\}$ (b) $\{-.633, 2.633\}$ 33. (a) $\{-2 \pm \sqrt{3}\}$ (b) $\{-3.732, -.268\}$ 35. The student should have divided both sides of the equation by 2 as his first step. The correct solution set is $\{1, 4\}$. 37. 3 sec and 5 sec 39. 1 sec and 5 sec 41. 75 ft by 100 ft 43. 8 mi 45. x^2 46. $x^2 + 8x$ 47. $x^2 + 8x + 16$ 48. It occurred when we added the 16 squares.

Section 9.3 (page 611)

Exercises 1. $a = 3, b = 4, c = -8$ 3. $a = -8, b = -2, c = -3$ 5. $a = 3, b = -4, c = -2$ 7. $a = 3, b = 7, c = 0$ 9. $a = 1, b = 1, c = -12$ 11. $a = 9, b = 9, c = -26$ 13. If a were 0, the equation would be linear, not quadratic. 15. No, because $2a$ should be the denominator for $-b$ as well. The correct formula is $x = \dfrac{-b \pm \sqrt{b^2 - 4ac}}{2a}$. 17. $\{2\}$ 19. $\{-13, 1\}$ 21. $\left\{\dfrac{-6 \pm \sqrt{26}}{2}\right\}$ 23. $\left\{-1, \dfrac{5}{2}\right\}$ 25. $\{-1, 0\}$ 27. $\left\{0, \dfrac{12}{7}\right\}$ 29. $\{\pm 2\sqrt{6}\}$ 31. $\left\{\pm\dfrac{2}{5}\right\}$ 33. $\left\{\dfrac{6 \pm 2\sqrt{6}}{3}\right\}$ 35. $\emptyset$ 37. $\emptyset$ 39. $\left\{\dfrac{-5 \pm \sqrt{61}}{2}\right\}$ 41. (a) $\left\{\dfrac{-1 \pm \sqrt{11}}{2}\right\}$ (b) $\{-2.158, 1.158\}$ 43. (a) $\left\{\dfrac{1 \pm \sqrt{5}}{2}\right\}$ (b) $\{-.618, 1.618\}$ 45. $\left\{-\dfrac{2}{3}, \dfrac{4}{3}\right\}$ 47. $\left\{\dfrac{-1 \pm \sqrt{73}}{6}\right\}$ 49. $\emptyset$ 51. $\{1 \pm \sqrt{2}\}$ 53. $\left\{-1, \dfrac{5}{2}\right\}$ 55. $r = \dfrac{-\pi h \pm \sqrt{\pi^2 h^2 + \pi S}}{\pi}$ 57. 3.5 ft 59. $\{16, -8\}$; Only 16 ft is a reasonable answer. 61. (a) 225 (b) 169 (c) 4 (d) 121 62. Each is a perfect square. 63. (a) $(3x - 2)(6x + 1)$ (b) $(x + 2)(5x - 3)$ (c) $(8x + 1)(6x + 1)$ (d) $(x + 3)(x - 8)$ 64. (a) 41 (b) -39 (c) 12 (d) 112 65. no 66. If the discriminant is a perfect square the trinomial is factorable. (a) yes (b) yes (c) no

A-40 Answers to Selected Exercises

Summary Exercises on Quadratic Equations (page 614)

1. $\{\pm 6\}$ 2. $\left\{\dfrac{-3 \pm \sqrt{5}}{2}\right\}$ 3. $\left\{\pm \dfrac{10}{9}\right\}$ 4. $\left\{\pm \dfrac{7}{9}\right\}$ 5. $\{1, 3\}$ 6. $\{-2, -1\}$ 7. $\{4, 5\}$ 8. $\left\{\dfrac{-3 \pm \sqrt{17}}{2}\right\}$ 9. $\left\{-\dfrac{1}{3}, \dfrac{5}{3}\right\}$
10. $\left\{\dfrac{1 \pm \sqrt{10}}{2}\right\}$ 11. $\{-17, 5\}$ 12. $\left\{-\dfrac{7}{5}, 1\right\}$ 13. $\left\{\dfrac{7 \pm 2\sqrt{6}}{3}\right\}$ 14. $\left\{\dfrac{1 \pm 4\sqrt{2}}{7}\right\}$ 15. $\emptyset$ 16. $\emptyset$ 17. $\left\{-\dfrac{1}{2}, 2\right\}$
18. $\left\{-\dfrac{1}{2}, 1\right\}$ 19. $\left\{-\dfrac{5}{4}, \dfrac{3}{2}\right\}$ 20. $\left\{-3, \dfrac{1}{3}\right\}$ 21. $\{1 \pm \sqrt{2}\}$ 22. $\left\{\dfrac{-5 \pm \sqrt{13}}{6}\right\}$ 23. $\left\{\dfrac{2}{5}, 4\right\}$ 24. $\{-3 \pm \sqrt{5}\}$
25. $\left\{\dfrac{-3 \pm \sqrt{41}}{2}\right\}$ 26. $\left\{-\dfrac{5}{4}\right\}$ 27. $\left\{\dfrac{1}{4}, 1\right\}$ 28. $\left\{\dfrac{1 \pm \sqrt{3}}{2}\right\}$ 29. $\left\{\dfrac{-2 \pm \sqrt{11}}{3}\right\}$ 30. $\left\{\dfrac{-5 \pm \sqrt{41}}{8}\right\}$ 31. $\left\{\dfrac{-7 \pm \sqrt{5}}{4}\right\}$
32. $\left\{-\dfrac{8}{3}, -\dfrac{6}{5}\right\}$ 33. $\left\{\dfrac{8 \pm 8\sqrt{2}}{3}\right\}$ 34. $\left\{\dfrac{-5 \pm \sqrt{5}}{2}\right\}$ 35. $\emptyset$ 36. $\emptyset$ 37. $\left\{-\dfrac{2}{3}, 2\right\}$ 38. $\left\{-\dfrac{1}{4}, \dfrac{2}{3}\right\}$ 39. $\left\{-4, \dfrac{3}{5}\right\}$
40. $\{-3, 5\}$ 41. $\left\{-\dfrac{2}{3}, \dfrac{2}{5}\right\}$ 42. $\{-4, 6\}$ 43. $\dfrac{5 - \sqrt{D}}{3}$ 44. There are no real solutions.

Section 9.4 (page 620)

Exercises 1. $3i$ 3. $2i\sqrt{5}$ 5. $3i\sqrt{2}$ 7. $5i\sqrt{5}$ 9. $5 + 3i$ 11. $6 - 9i$ 13. $6 - 7i$ 15. $14 + 5i$ 17. $7 - 22i$ 19. 45
21. $3 - i$ 23. $2 - 6i$ 25. $-\dfrac{3}{25} + \dfrac{4}{25}i$ 27. $\{-1 \pm 2i\}$ 29. $\{3 \pm i\sqrt{5}\}$ 31. $\left\{-\dfrac{2}{3} \pm i\sqrt{2}\right\}$ 33. $\{1 \pm i\}$
35. $\left\{-\dfrac{3}{4} \pm \dfrac{\sqrt{31}}{4}i\right\}$ 37. $\left\{\dfrac{3}{2} \pm \dfrac{\sqrt{7}}{2}i\right\}$ 39. $\left\{\dfrac{1}{5} \pm \dfrac{\sqrt{14}}{5}i\right\}$ 41. $\left\{-\dfrac{1}{2} \pm \dfrac{\sqrt{13}}{2}i\right\}$ 43. $\left\{\dfrac{1}{2} \pm \dfrac{\sqrt{11}}{2}i\right\}$ 45. If the discriminant $b^2 - 4ac$ is negative, the equation will have solutions that are not real. 47. true 49. false; For example, $3 + 2i$ is a complex number but it is not real.

Section 9.5 (page 627)

Exercises 1. The vertex of a parabola is the lowest or highest point on the graph. 3. $(0, -6)$ 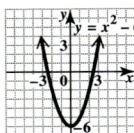 5. $(-3, 0)$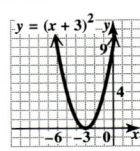

7. $(-1, 2)$ 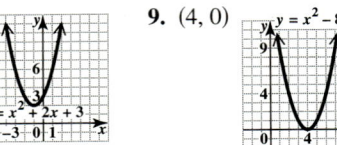 9. $(4, 0)$ 11. $(3, 4)$ 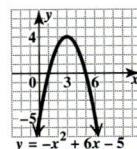 13. one real solution; $\{2\}$

15. two real solutions; $\{\pm 2\}$ 17. no real solutions; $\emptyset$ 19. If $a > 0$, it opens upward, and if $a < 0$, it opens downward. 21. $\{-2, 3\}$
23. $\{-1, 1.5\}$ 25. $(-\infty, \infty)$; $[0, \infty)$ 27. $(-\infty, \infty)$; $(-\infty, 4]$ 29. $(-\infty, \infty)$; $[1, \infty)$ 31. 3 33. 21 35. $y = \dfrac{11}{5625}x^2$
37. 40 and 40 39. In each case, there is a vertical "stretch" of the parabola. It becomes narrower as the coefficient gets larger.
40. In each case, there is a vertical "shrink" of the parabola. It becomes wider as the coefficient gets smaller. 41. The graph of Y_2 is obtained by reflecting the graph of Y_1 across the x-axis. 42. When the coefficient of x^2 is negative, the parabola opens downward.

43. By adding a positive constant k, the graph is shifted k units upward. By subtracting a positive constant k, the graph is shifted k units downward. **44.** Adding a positive constant k moves the graph k units to the left. Subtracting a positive constant k moves the graph k units to the right. **45.** 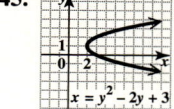 This is not the graph of a function because for many x-values, there are two y-values.

The graph does not pass the vertical line test.

Chapter 9 Review Exercises (page 636)

1. $\{\pm 12\}$ **3.** $\{\pm 8\sqrt{2}\}$ **5.** $\{3 \pm \sqrt{10}\}$ **7.** $\emptyset$ **9.** $\{-5, -1\}$ **11.** $\{-1 \pm \sqrt{6}\}$ **13.** $\left\{-\dfrac{2}{5}, 1\right\}$ **15.** 2.5 sec **17.** $\left(\dfrac{3}{2}\right)^2$ or $\dfrac{9}{4}$

19. $\{1 \pm \sqrt{5}\}$ **21.** $\left\{\dfrac{2 \pm \sqrt{10}}{2}\right\}$ **23.** $\left\{\dfrac{-3 \pm \sqrt{41}}{2}\right\}$ **25.** The $-b$ term should be above the fraction bar. **27.** $-6 - 5i$ **29.** 13

31. $\dfrac{28}{13} - \dfrac{3}{13}i$ **33.** No, the product $(a + bi)(a - bi) = a^2 + b^2$ will always be the sum of the squares of two real numbers, which is a real number. **35.** $\left\{\dfrac{2}{3} \pm \dfrac{2\sqrt{2}}{3}i\right\}$ **37.** $\left\{-\dfrac{3}{2} \pm \dfrac{\sqrt{23}}{2}i\right\}$ **39.** $\left\{-\dfrac{1}{9} \pm \dfrac{2\sqrt{2}}{9}i\right\}$ **41.** vertex: $(0, 5)$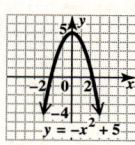

43. vertex: $(1, 0)$ **45.** vertex: $(-2, -2)$ **47.** one; $\{2\}$ **49.** in Exercise 46, domain: $(-\infty, \infty)$;

range: $[-2, \infty)$; in Exercise 47, domain: $(-\infty, \infty)$; range: $(-\infty, 0]$; in Exercise 48, domain: $(-\infty, \infty)$; range: $[1, \infty)$ **51.** $\left\{-\dfrac{11}{2}, 5\right\}$

53. $\left\{\dfrac{-1 \pm \sqrt{21}}{2}\right\}$ **55.** $\left\{\dfrac{-5 \pm \sqrt{17}}{2}\right\}$ **57.** $\emptyset$ **59.** $\left\{-\dfrac{5}{3}\right\}$ **61.** $\{-2 \pm \sqrt{5}\}$ **63.** 400 or 800

Chapter 9 Test (page 638)

[9.1] **1.** $\{\pm\sqrt{39}\}$ **2.** $\{-11, 5\}$ **3.** $\left\{\dfrac{-3 \pm 2\sqrt{6}}{4}\right\}$ [9.2] **4.** $\{2 \pm \sqrt{10}\}$ **5.** $\left\{\dfrac{-6 \pm \sqrt{42}}{2}\right\}$ [9.3] **6.** $\left\{0, -\dfrac{2}{5}\right\}$ **7.** $\left\{-3, \dfrac{1}{2}\right\}$

8. $\left\{\dfrac{3 \pm \sqrt{3}}{3}\right\}$ [9.4] **9.** $\left\{-1 \pm \dfrac{\sqrt{7}}{2}i\right\}$ [9.3] **10.** $\left\{\dfrac{5 \pm \sqrt{13}}{6}\right\}$ [9.1–9.3] **11.** $\{1 \pm \sqrt{2}\}$ **12.** $\left\{\dfrac{-1 \pm 3\sqrt{2}}{2}\right\}$

13. $\left\{\dfrac{11 \pm \sqrt{89}}{4}\right\}$ **14.** $\{5\}$ **15.** 2 sec **16.** 12, 16, 20 [9.4] **17.** $-5 + 5i$ **18.** $-17 - 4i$ **19.** 73 **20.** $2 - i$

[9.5] **21.** vertex: $(3, 0)$ **22.** vertex: $(-1, -3)$ **23.** vertex: $(-3, -2)$ **24.** (a) two

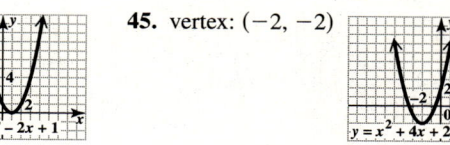

(b) $\{-3 \pm \sqrt{2}\}$ (c) $-3 - \sqrt{2} \approx -4.414$ and $-3 + \sqrt{2} \approx -1.586$ **25.** 200 and 200

Answers to Selected Exercises

Cumulative Review Exercises Chapters 1–9 (page 639)

[1.2] **1.** 15　[1.5] **2.** −2　[1.6] **3.** 5　[1.8] **4.** $-r + 7$　**5.** $-2k$　**6.** $19m - 17$　[2.1] **7.** {18}　[2.2] **8.** {5}　[2.3] **9.** $\left\{\dfrac{8}{3}\right\}$

10. {2}　[2.5] **11.** 100°, 80°　**12.** width: 50 ft; length: 94 ft　**13.** $L = \dfrac{P - 2W}{2}$ or $L = \dfrac{P}{2} - W$

[2.8] **14.** $(-2, \infty)$　**15.** $(-\infty, 4]$　[3.2] **16.**　**17.**

[3.5] **18.**　[3.3] **19.** $-\dfrac{1}{3}$　[3.4] **20.** $2x - y = -3$　[4.1–4.3] **21.** {(−3, 2)}　**22.** ∅　[4.4] **23.** GE: $9.99;

Motorola: $19.99　[4.5] **24.**　[5.1, 5.2] **25.** $\dfrac{x^4}{3^2}$ or $\dfrac{x^4}{9}$　**26.** $\dfrac{b^{16}}{c^2}$　**27.** $\dfrac{3^3}{5^3}$ or $\dfrac{27}{125}$　[5.4] **28.** $8x^5 - 17x^4 - x^2$

[5.5] **29.** $2x^4 + x^3 - 19x^2 + 2x + 20$　[5.7] **30.** $3x^2 - 2x + 1$　[5.3] **31.** (a) 6.35×10^9 (b) .00023　[6.1] **32.** $16x^2(x - 3y)$

[6.3] **33.** $(2a + 1)(a - 3)$　[6.4] **34.** $(4x^2 + 1)(2x + 1)(2x - 1)$　**35.** $(5m - 2)^2$　[6.5] **36.** {−9, 6}　[6.6] **37.** 50 m　[7.2] **38.** $\dfrac{4}{5}$

[7.4] **39.** $\dfrac{-k - 1}{k(k - 1)}$　**40.** $\dfrac{5a + 2}{(a - 2)^2(a + 2)}$　[7.5] **41.** $\dfrac{b + a}{b - a}$　[7.6] **42.** $\left\{-\dfrac{15}{7}, 2\right\}$　[7.7] **43.** 3 mph　[8.1] **44.** 10

[8.4] **45.** $\dfrac{6\sqrt{30}}{5}$　**46.** $\dfrac{\sqrt[3]{28}}{4}$　[8.3] **47.** $4\sqrt{5}$　**48.** $-ab\sqrt[3]{2b}$　[8.6] **49.** {7}　[8.7] **50.** (a) 4 (b) −2　[9.1] **51.** $\left\{\dfrac{-2 \pm 2\sqrt{3}}{3}\right\}$

[9.2] **52.** $\{-1 \pm \sqrt{6}\}$　[9.3] **53.** $\left\{\dfrac{2 \pm \sqrt{10}}{2}\right\}$　[9.2, 9.3] **54.** ∅　**55.** 3, 4, 5　[9.4] **56.** (a) $8i$ (b) $5 + 2i$　**57.** $\left\{-\dfrac{1}{2} \pm \dfrac{\sqrt{17}}{2}i\right\}$

[9.5] **58.** vertex: $(-1, 2)$; domain: $(-\infty, \infty)$; range: $(-\infty, 2]$

APPENDIXES

Appendix B (page 652)

1. 117.385　**3.** 13.21　**5.** 150.49　**7.** 96.101　**9.** 4.849　**11.** 166.32　**13.** 164.19　**15.** 1.344　**17.** 4.14　**19.** 2.23
21. 4800　**23.** .53　**25.** 1.29　**27.** .96　**29.** .009　**31.** 80%　**33.** .7%　**35.** 67%　**37.** 12.5%　**39.** 109.2　**41.** 238.92
43. 5%　**45.** 110%　**47.** 25%　**49.** 148.44　**51.** 7839.26　**53.** 7.39%　**55.** $2760　**57.** 805 mi　**59.** 63 mi　**61.** $2400

Answers to Selected Exercises **A-43**

Appendix C (page 658)

1. {1, 2, 3, 4, 5, 6, 7}   **3.** {winter, spring, summer, fall}   **5.** ∅   **7.** {L}   **9.** {2, 4, 6, 8, 10, ...}   **11.** The sets in Exercises 9 and 10 are infinite sets.   **13.** true   **15.** false   **17.** true   **19.** true   **21.** true   **23.** true   **25.** true   **27.** false   **29.** true   **31.** true   **33.** false   **35.** true   **37.** true   **39.** false   **41.** true   **43.** false   **45.** {g, h}   **47.** {b, c, d, e, g, h}   **49.** {a, c, e} = B   **51.** {d} = D   **53.** {a}   **55.** {a, c, d, e}   **57.** {a, c, e, f}   **59.** ∅   **61.** B and D; C and D

Appendix D (page 665)

1. 7   **3.** 69.8 (rounded)   **5.** $39,622   **7.** $58.24   **9.** $35,500   **11.** 6.1   **13.** 17.2   **15.** 2.60   **17. (a)** 2.80 **(b)** 2.93 (rounded) **(c)** 3.13 (rounded)   **19.** 15   **21.** 516   **23.** 48   **25.** 4142 mi   **27. (a)** 4050 mi **(b)** The median is somewhat different from the mean; the mean is more affected by the high and low numbers.   **29.** 8   **31.** 68 and 74; bimodal   **33.** no mode

Appendix E (page 672)

1. C   **3. (a)** true **(b)** true **(c)** false **(d)** true   **5.** $(-3, 3)$ **7.** $(-\infty, -6] \cup [7, \infty)$ **9.** $(-\infty, -3) \cup (-2, \infty)$ **11.** $[-1, 5]$ **13.** $\left(-1, \dfrac{2}{5}\right)$ **15.** $\left(-\dfrac{1}{2}, \dfrac{4}{3}\right)$ **17.** $(1, 6)$ **19.** $\left(-\infty, -\dfrac{1}{2}\right) \cup \left(\dfrac{1}{3}, \infty\right)$ **21.** $\left(-\dfrac{2}{3}, -\dfrac{1}{4}\right)$ **23.** $(-\infty, -2) \cup (2, \infty)$ **25.** $(-\infty, -4) \cup (4, \infty)$ **27.** $\left[-2, \dfrac{1}{3}\right] \cup [4, \infty)$ **29.** $(-\infty, -1) \cup (2, 4)$ **31.** 2 sec and 14 sec   **33.** between 0 and 2 sec or between 14 and 16 sec

Glossary

A

absolute value The absolute value of a number is the distance between 0 and the number on a number line. (Section 1.4)

addition property of equality The addition property of equality states that the same number can be added to (or subtracted from) both sides of an equation to obtain an equivalent equation. (Section 2.1)

addition property of inequality The addition property of inequality states that the same number can be added to (or subtracted from) both sides of an inequality without changing the solution set. (Section 2.8)

additive inverse (opposite) Two numbers that are the same distance from 0 on a number line but on opposite sides of 0 are called additive inverses. (Section 1.4)

algebraic expression Any collection of numbers or variables joined by the basic operations of addition, subtraction, multiplication, or division (except by 0), or the operation of raising to powers or taking roots is called an algebraic expression. (Section 1.3)

area Area is a measure of the surface covered by a two-dimensional (flat) figure. (Section 2.5)

associative property of addition The associative property of addition states that the way in which numbers being added are grouped does not change the sum. (Section 1.7)

associative property of multiplication The associative property of multiplication states that the way in which numbers being multiplied are grouped does not change the product. (Section 1.7)

axis (axis of symmetry) The axis of a parabola is the vertical or horizontal line through the vertex of the parabola. (Sections 5.4, 9.5)

B

base The base is the number that is a repeated factor when written with an exponent. (Sections 1.2, 5.1)

bimodal A group of numbers with two modes is bimodal. (Appendix D)

binomial A binomial is a polynomial with exactly two terms. (Section 5.4)

boundary line In the graph of a linear inequality, the boundary line separates the region that satisfies the inequality from the region that does not satisfy the inequality. (Section 3.5)

C

coefficient (numerical coefficient) A coefficient is the numerical factor of a term. (Sections 1.8, 5.4)

combining like terms Combining like terms is a method of adding or subtracting like terms by using the properties of real numbers. (Section 1.8)

common factor An integer that is a factor of two or more integers is called a common factor of those integers. (Section 6.1)

commutative property of addition The commutative property of addition states that the order of numbers in an addition problem can be changed without changing the sum. (Section 1.7)

commutative property of multiplication The commutative property of multiplication states that the product in a multiplication problem remains the same regardless of the order of the factors. (Section 1.7)

complement of a set The set of elements in the universal set that are not in a set A is the complement of A, written A'. (Appendix C)

complementary angles (complements) Complementary angles are angles whose measures have a sum of 90°. (Section 2.4)

completing the square The process of adding to a binomial the number that makes it a perfect square trinomial is called completing the square. (Section 9.2)

complex fraction A complex fraction is an expression with one or more fractions in the numerator, denominator, or both. (Section 7.5)

complex number A complex number is any number that can be written in the form $a + bi$, where a and b are real numbers. (Section 9.4)

components In an ordered pair (x, y), x and y are called the components of the ordered pair. (Section 3.6)

composite number A composite number has at least one factor other than itself and 1. (Section 1.1)

conditional equation A conditional equation is true for some replacements of the variable and false for others. (Section 2.3)

conjugate The conjugate of $a + b$ is $a - b$. (Section 8.5)

conjugate of a complex number The conjugate of a complex number $a + bi$ is $a - bi$. (Section 9.4)

consecutive integers Two integers that differ by 1 are called consecutive integers. (Section 2.4)

consistent system A system of equations with a solution is called a consistent system. (Section 4.1)

constant of variation In the equation $y = kx$, the number k is called the constant of variation. (Section 7.8)

contradiction A contradiction is an equation that is never true. It has no solutions. (Section 2.3)

coordinate on a number line Each number on a number line is called the coordinate of the point that it labels. (Section 1.4)

coordinates of a point The numbers in an ordered pair are called the coordinates of the corresponding point. (Section 3.1)

cross products The cross products in the proportion $\frac{a}{b} = \frac{c}{d}$ are ad and bc. (Section 2.6)

cube root A number b is the cube root of a if $b^3 = a$. (Section 8.1)

D

decimal A decimal is a number written with a decimal point. (Appendix B)

degree A degree is a basic unit of measure for angles in which one degree (1°) is $\frac{1}{360}$ of a complete revolution. (Section 2.4)

G-1

degree of a polynomial The degree of a polynomial is the greatest degree of any of the terms in the polynomial. (Section 5.4)

degree of a term The degree of a term is the sum of the exponents on the variables in the term. (Section 5.4)

denominator The number below the fraction bar in a fraction is called the denominator. It shows the number of equal parts in a whole. (Section 1.1)

dependent equations Equations of a system that have the same graph (because they are different forms of the same equation) are called dependent equations. (Section 4.1)

descending powers A polynomial in one variable is written in descending powers of the variable if the exponents on the terms of the polynomial decrease from left to right. (Section 5.4)

difference The answer to a subtraction problem is called the difference. (Section 1.1)

difference of cubes The difference of cubes, $x^3 - y^3$, can be factored as $x^3 - y^3 = (x - y)(x^2 + xy + y^2)$. (Section 6.4)

difference of squares The difference of squares, $x^2 - y^2$, can be factored as the product of the sum and difference of two terms, or $x^2 - y^2 = (x + y)(x - y)$. (Section 6.4)

direct variation y varies directly as x if there exists a real number (constant) k such that $y = kx$. (Section 7.8)

discriminant The discriminant is the quantity under the radical, $b^2 - 4ac$, in the quadratic formula. (Section 9.3)

disjoint sets Sets that have no elements in common are disjoint sets. (Appendix C)

distributive property For any real numbers a, b, and c, the distributive property states that $a(b + c) = ab + ac$ and $(b + c)a = ba + ca$. (Section 1.7)

domain The set of all first components (x-values) in the ordered pairs of a relation is the domain. (Section 3.6)

E

elements (members) Elements are the objects that belong to a set. (Section 1.3, Appendix C)

elimination method The elimination method is an algebraic method used to solve a system of equations in which the equations of the system are combined so that one or more variables is eliminated. (Section 4.3)

empty set (null set) The empty set, denoted by { } or ∅, is the set containing no elements. (Section 2.3, Appendix C)

equation An equation is a statement that two algebraic expressions are equal. (Section 1.3)

equivalent equations Equivalent equations are equations that have the same solution set. (Section 2.1)

exponent (power) An exponent is a number that indicates how many times a factor is repeated. (Sections 1.2, 5.1)

exponential expression A number or letter (variable) written with an exponent is an exponential expression. (Sections 1.2, 5.1)

extremes of a proportion In the proportion $\frac{a}{b} = \frac{c}{d}$, the a- and d-terms are called the extremes. (Section 2.6)

F

factor A factor of a given number is any number that divides evenly (without remainder) into the given number. (Sections 1.1, 6.1)

factored A number is factored by writing it as the product of two or more numbers. (Section 1.1)

factored form An expression is in factored form when it is written as a product. (Section 6.1)

factoring Writing a polynomial as the product of two or more simpler polynomials is called factoring. (Section 6.1)

factor by grouping Factoring by grouping is a method for grouping the terms of a polynomial in such a way that the polynomial can be factored even though its greatest common factor is 1. (Section 6.1)

factor out the greatest common factor Factoring out the greatest common factor is the process of using the distributive property to write a polynomial as a product of the greatest common factor and a simpler polynomial. (Section 6.1)

FOIL FOIL is a method for multiplying two binomials $(A + B)(C + D)$. Multiply **F**irst terms AC, **O**uter terms AD, **I**nner terms BC, and **L**ast terms BD. Then combine like terms. (Section 5.5)

formula A formula is a mathematical expression in which letters are used to describe a relationship. (Section 2.5)

fourth root A number b is a fourth root of a if $b^4 = a$. (Section 8.1)

function A function is a set of ordered pairs (relation) in which each value of the first component x corresponds to exactly one value of the second component y. (Section 3.6)

function notation Function notation $f(x)$ represents the value of the function at x, that is, the y-value that corresponds to x. (Section 3.6)

G

graph of a number The point on a number line that corresponds to a number is its graph. (Section 1.4)

graph of an equation The graph of an equation is the set of all points that correspond to all of the ordered pairs that satisfy the equation. (Section 3.2)

graphing method The graphing method for solving a system of equations requires graphing all equations of the system on the same axes and locating the ordered pairs at their intersection. (Section 4.1)

greatest common factor (GCF) The greatest common factor of a list of integers is the largest common factor of those integers. The greatest common factor of a polynomial is the largest term that is a factor of all terms in the polynomial. (Sections 1.1, 6.1)

grouping symbols Grouping symbols are parentheses ( ), square brackets [ ], or fraction bars. (Section 1.2)

H

hypotenuse The hypotenuse is the longest side in a right triangle. It is the side opposite the right angle. (Section 6.6)

I

identity An identity is an equation that is true for all replacements of the variable. It has an infinite number of solutions. (Section 2.3)

identity element for addition Since adding 0 to a number does not change the number, 0 is called the identity element for addition. (Section 1.7)

identity element for multiplication Since multiplying a number by 1 does not change the number, 1 is called the identity element for multiplication. (Section 1.7)

Glossary G-3

identity property The identity properties state that the sum of 0 and any number equals the number, and the product of 1 and any number equals the number. (Section 1.7)

imaginary number A complex number $a + bi$ with $b \neq 0$ is called an imaginary number. (Section 9.4)

imaginary part The imaginary part of the complex number $a + bi$ is b. (Section 9.4)

inconsistent system An inconsistent system of equations is a system with no solution. (Section 4.1)

independent equations Equations of a system that have different graphs are called independent equations. (Section 4.1)

index (order) In a radical of the form $\sqrt[n]{a}$, n is called the index or order. (Section 8.1)

inequality An inequality is a statement that two expressions are not equal. (Section 1.2)

inner product When using the FOIL method to multiply two binomials $(A + B)(C + D)$, the inner product is BC. (Section 5.5)

integers The set of integers is $\{\ldots, -3, -2, -1, 0, 1, 2, 3, \ldots\}$. (Section 1.4)

intersection The intersection of two sets A and B, written $A \cap B$, is the set of elements that belong to both A and B. (Appendix C)

interval An interval is a portion of a number line. (Section 2.8)

interval notation Interval notation is a simplified notation that uses parentheses ( ) and/or brackets [ ] to describe an interval on a number line. (Section 2.8)

inverse property The inverse properties state that a number added to its opposite is 0 and a number multiplied by its reciprocal is 1. (Section 1.7)

inverse variation y varies inversely as x if there exists a real number (constant) k such that $y = \frac{k}{x}$. (Section 7.8)

irrational number An irrational number cannot be written as the quotient of two integers but can be represented by a point on the number line. (Section 1.4)

L

least common denominator (LCD) Given several denominators, the smallest expression that is divisible by all the denominators is called the least common denominator. (Sections 1.1, 7.3)

legs of a right triangle The two shorter sides of a right triangle are called the legs. (Section 6.6)

like radicals Like radicals are terms that have multiples of the same root of the same number or expression. (Section 8.3)

like terms Terms with exactly the same variables raised to exactly the same powers are called like terms. (Sections 1.8, 5.4)

linear equation in one variable A linear equation in one variable can be written in the form $Ax + B = C$, where A, B, and C are real numbers, with $A \neq 0$. (Section 2.1)

linear equation in two variables A linear equation in two variables is an equation that can be written in the form $Ax + By = C$, where A, B, and C are real numbers and A and B are not both 0. (Section 3.1)

linear inequality in one variable A linear inequality in one variable can be written in the form $Ax + B < C$, $Ax + B \leq C$, $Ax + B > C$, or $Ax + B \geq C$, where A, B, and C are real numbers, with $A \neq 0$. (Section 2.8)

linear inequality in two variables A linear inequality in two variables can be written in the form $Ax + By < C$ or $Ax + By > C$ (or with $\leq$ or $\geq$), where A, B, and C are real numbers and A and B are not both 0. (Section 3.5)

line of symmetry The axis of a parabola is a line of symmetry for the graph. It is a line that can be drawn through the graph in such a way that the part of the graph on one side of the line is an exact reflection of the part on the opposite side. (Sections 5.4, 9.5)

lowest terms A fraction is in lowest terms when there are no common factors in the numerator and denominator (except 1). (Sections 1.1, 7.1)

M

mean The mean (average) of a group of numbers is the sum of the numbers divided by the number of numbers. (Appendix D)

means of a proportion In the proportion $\frac{a}{b} = \frac{c}{d}$, the b- and c-terms are called the means. (Section 2.6)

median The median divides a group of numbers listed in order from smallest to largest in half; that is, half the numbers lie above the median and half lie below the median. (Appendix D)

mixed number A mixed number includes a whole number and a fraction written together and is understood to be the sum of the whole number and the fraction. (Section 1.1)

mode The mode is the number (or numbers) that occurs most often in a group of numbers. (Appendix D)

monomial A monomial is a polynomial with only one term. (Section 5.4)

multiplication property of equality The multiplication property of equality states that the same nonzero number can be multiplied by (or divided into) both sides of an equation to obtain an equivalent equation. (Section 2.2)

multiplication property of inequality The multiplication property of inequality states that both sides of an inequality may be multiplied (or divided) by a positive number without changing the direction of the inequality symbol. Multiplying (or dividing) by a negative number reverses the inequality symbol. (Section 2.8)

multiplicative inverse (reciprocal) The multiplicative inverse of a nonzero real number a is $\frac{1}{a}$. (Section 1.6)

N

natural numbers The set of natural numbers includes the numbers used for counting: $\{1, 2, 3, 4, \ldots\}$. (Sections 1.1, 1.4)

negative number A negative number is located to the left of 0 on a number line. (Section 1.4)

number line A number line is a line with a scale that is used to show how numbers relate to each other. (Section 1.4)

numerator The number above the fraction bar in a fraction is called the numerator. It shows how many of the equivalent parts are being considered. (Section 1.1)

numerical coefficient The numerical factor in a term is its numerical coefficient. (Sections 1.8, 5.4)

O

ordered pair An ordered pair is a pair of numbers written within parentheses in which the order of the numbers is important. (Section 3.1)

origin The point at which the x-axis and y-axis of a rectangular coordinate system intersect is called the origin. (Section 3.1)

G-4 Glossary

outer product When using the FOIL method to multiply two binomials $(A + B)(C + D)$, the outer product is AD. (Section 5.5)

P

parabola The graph of a second-degree (quadratic) equation in two variables is called a parabola. (Sections 5.4, 9.5)

parallel lines Parallel lines are two lines in the same plane that never intersect. (Section 3.3)

percent Percent, written with the sign %, means per one hundred. (Appendix B)

percentage A percentage is a part of a whole. (Appendix B)

perfect cube A perfect cube is a number with a rational cube root. (Section 8.1)

perfect square A perfect square is a number with a rational square root. (Section 8.1)

perfect square trinomial A perfect square trinomial is a trinomial that can be factored as the square of a binomial. (Section 6.4)

perimeter The perimeter of a two-dimensional figure is a measure of the distance around the outside edges of the figure, or the sum of the lengths of its sides. (Section 2.5)

perpendicular lines Perpendicular lines are two lines that intersect to form a right (90°) angle. (Section 3.3)

plot To plot an ordered pair is to locate it on a rectangular coordinate system. (Section 3.1)

point-slope form A linear equation is written in point-slope form if it is in the form $y - y_1 = m(x - x_1)$, where m is slope and (x_1, y_1) is a point on the line. (Section 3.4)

polynomial A polynomial is a term or a finite sum of terms in which all coefficients are real, all variables have whole number exponents, and no variables appear in denominators. (Section 5.4)

polynomial in x A polynomial containing only the variable x is called a polynomial in x. (Section 5.4)

positive number A positive number is located to the right of 0 on a number line. (Section 1.4)

prime number A natural number (except 1) is prime if it has only 1 and itself as factors. (Section 1.1)

prime polynomial A prime polynomial is a polynomial that cannot be factored using only integer coefficients. (Section 6.2)

principal root (principal nth root) For even indexes, the symbols $\sqrt{}$, $\sqrt[4]{}$, $\sqrt[6]{}$, ..., $\sqrt[n]{}$, are used for nonnegative roots, which are called principal roots. (Section 8.1)

product The answer to a multiplication problem is called the product. (Section 1.1)

product of the sum and difference of two terms The product of the sum and difference of two terms is the difference of the squares of the terms, or $(x + y)(x - y) = x^2 - y^2$. (Section 5.6)

proportion A proportion is a statement that two ratios are equal. (Section 2.6)

Pythagorean formula The Pythagorean formula states that the square of the length of the hypotenuse of a right triangle equals the sum of the squares of the lengths of the two legs. (Section 6.6)

Q

quadrant A quadrant is one of the four regions in the plane determined by a rectangular coordinate system. (Section 3.1)

quadratic equation A quadratic equation is an equation that can be written in the form $ax^2 + bx + c = 0$, where a, b, and c are real numbers, with $a \neq 0$. (Sections 6.5, 9.1)

quadratic formula The quadratic formula is a general formula used to solve any quadratic equation. (Section 9.3)

quadratic function A function defined by an equation of the form $f(x) = ax^2 + bx + c$, for real numbers a, b, and c, with $a \neq 0$, is a quadratic function. (Section 9.5)

quadratic inequality A quadratic inequality can be written in the form $ax^2 + bx + c < 0$ or $ax^2 + bx + c > 0$ (or with $\leq$ or $\geq$), where a, b, and c are real numbers, with $a \neq 0$. (Appendix E)

quotient The answer to a division problem is called the quotient. (Section 1.1)

R

radical A radical sign with a radicand is called a radical. (Section 8.1)

radical equation A radical equation is an equation with a variable in the radicand. (Section 8.6)

radical expression A radical expression is an algebraic expression that contains radicals. (Section 8.1)

radical sign The symbol $\sqrt{}$ is called a radical sign. (Section 8.1)

radicand The number or expression under a radical sign is called the radicand. (Section 8.1)

range The set of all second components (y-values) in the ordered pairs of a relation is the range. (Section 3.6)

ratio A ratio is a comparison of two quantities with the same units. (Section 2.6)

rational expression The quotient of two polynomials with denominator not 0 is called a rational expression. (Section 7.1)

rationalizing the denominator The process of removing radicals from a denominator so that the denominator contains only rational numbers is called rationalizing the denominator. (Section 8.4)

rational numbers Rational numbers can be written as the quotient of two integers, with denominator not 0. (Section 1.4)

real numbers Real numbers include all numbers that can be represented by points on the number line, that is, all rational and irrational numbers. (Section 1.4)

real part The real part of a complex number $a + bi$ is a. (Section 9.4)

reciprocal Pairs of numbers whose product is 1 are called reciprocals of each other. (Section 1.1)

rectangular (Cartesian) coordinate system The x-axis and y-axis placed at a right angle at their zero points form a rectangular coordinate system, also called the Cartesian coordinate system. (Section 3.1)

relation A relation is a set of ordered pairs. (Section 3.6)

right angle A right angle measures 90°. (Section 2.4)

rise Rise is the vertical change between two points on a line, that is, the change in y-values. (Section 3.3)

run Run is the horizontal change between two points on a line, that is, the change in x-values. (Section 3.3)

S

scatter diagram A scatter diagram is a graph of ordered pairs of data. (Section 3.1)

scientific notation A number is written in scientific notation when it is expressed in the form $a \times 10^n$, where $1 \leq |a| < 10$ and n is an integer. (Section 5.3)

set A set is a collection of objects. (Section 1.3, Appendix C)

set-builder notation Set-builder notation is used to describe a set of numbers without actually having to list all of the elements. (Section 1.4)

signed numbers Signed numbers are numbers that can be written with a positive or negative sign. (Section 1.4)

simplified radical A simplified radical meets three conditions:
1. The radicand has no factor raised to a power greater than or equal to the index.
2. The radicand has no fractions.
3. No denominator contains a radical.

(Section 8.4)

slope The ratio of the change in y to the change in x along a line is called the slope of the line. (Section 3.3)

slope-intercept form A linear equation is written in slope-intercept form if it is in the form $y = mx + b$, where m is the slope and $(0, b)$ is the y-intercept. (Section 3.4)

solution of an equation A solution of an equation is any replacement for the variable that makes the equation true. (Section 1.3)

solution of a system A solution of a system of two equations is an ordered pair (x, y) that makes both equations true at the same time. (Section 4.1)

solution set The solution set is the set of all solutions of a particular equation. (Section 2.1)

solution set of a linear system The solution set of a linear system of equations includes all ordered pairs that satisfy all the equations of the system at the same time. (Section 4.1)

solution set of a system of linear inequalities The solution set of a system of linear inequalities includes all ordered pairs that make all inequalities of the system true at the same time. (Section 4.5)

square of a binomial The square of a binomial is the sum of the square of the first term, twice the product of the two terms, and the square of the last term, that is, $(x + y)^2 = x^2 + 2xy + y^2$ or $(x - y)^2 = x^2 - 2xy + y^2$. (Section 5.6)

square root The opposite of squaring a number is called taking its square root; that is, a number b is a square root of a if $b^2 = a$. (Section 8.1)

square root property The square root property states that if $x^2 = k$, $k > 0$, then $x = \sqrt{k}$ or $x = -\sqrt{k}$. (Section 9.1)

squaring property The squaring property is used to solve equations that include square roots. It states that if both sides of a given equation are squared, all solutions of the original equation are among the solutions of the squared equation. (Section 8.6)

standard form of a complex number The standard form of a complex number is $a + bi$. (Section 9.4)

standard form of a linear equation A linear equation in two variables written in the form $Ax + By = C$, with A and B not both 0, is in standard form. (Section 3.4)

standard form of a quadratic equation A quadratic equation written in the form $ax^2 + bx + c = 0$, where $a \neq 0$, is in standard form. (Section 6.5)

straight angle A straight angle measures 180°. (Section 2.4)

subscript notation Subscript notation is a way of indicating nonspecific values, such as x_1 and x_2. (Section 3.3)

subset If all elements of set A are in set B, then A is a subset of B, written $A \subseteq B$. (Appendix C)

substitution method The substitution method is an algebraic method for solving a system of equations in which one equation is solved for one of the variables and the result is substituted in the other equation. (Section 4.2)

sum The answer to an addition problem is called the sum. (Section 1.1)

sum of cubes The sum of cubes, $x^3 + y^3$, can be factored as $x^3 + y^3 = (x + y) \cdot (x^2 - xy + y^2)$. (Section 6.4)

supplementary angles (supplements) Supplementary angles are angles whose measures have a sum of 180°. (Section 2.4)

system of linear equations (linear system) A system of linear equations consists of two or more linear equations to be solved at the same time. (Section 4.1)

system of linear inequalities A system of linear inequalities consists of two or more linear inequalities to be solved at the same time. (Section 4.5)

T

table of values A table of values is an organized way of displaying ordered pairs. (Section 3.1)

term A term is a number, a variable, or the product or quotient of a number and one or more variables raised to powers. (Section 1.8)

terms of a proportion The terms of the proportion $\frac{a}{b} = \frac{c}{d}$ are a, b, c, and d. (Section 2.6)

three-part inequality An inequality that says that one number is between two other numbers is called a three-part inequality. (Section 2.8)

trinomial A trinomial is a polynomial with exactly three terms. (Section 5.4)

U

union The union of two sets A and B, written $A \cup B$, is the set that includes all elements of A together with all elements of B. (Appendix C)

universal set The set that includes all elements under consideration is the universal set, written U. (Appendix C)

unlike terms Unlike terms are terms that do not have the same variable or the variables are not raised to the same powers. (Section 1.8)

V

variable A variable is a symbol, usually a letter, used to represent an unknown number. (Section 1.3)

vary directly (is proportional to) y varies directly as x if there exists a nonzero number (constant) k such that $y = kx$. (Section 7.8)

vary inversely y varies inversely as x if there exists a real number (constant) k such that $y = \frac{k}{x}$. (Section 7.8)

Venn diagram A Venn diagram represents the relationships between sets. (Appendix C)

vertex The point on a parabola that has the smallest y-value (if the parabola opens upward) or the largest y-value (if the parabola opens downward) is called the vertex of the parabola. (Sections 5.4, 9.5)

vertical angles When two intersecting lines are drawn, the angles that lie opposite each other have the same measure and are called vertical angles. (Section 2.5)

vertical line test The vertical line test states that any vertical line drawn through the graph of a function must intersect the graph in at most one point. (Section 3.6)

volume The volume of a three-dimensional figure is a measure of the space occupied by the figure. (Section 2.5)

W

weighted mean A weighted mean is one in which each number is weighted by multiplying it by the number of times it occurs. (Appendix D)

whole numbers The set of whole numbers is {0, 1, 2, 3, 4, . . . }. (Sections 1.1, 1.4)

X

x-axis The horizontal number line in a rectangular coordinate system is called the x-axis. (Section 3.1)

x-intercept A point where a graph intersects the x-axis is called an x-intercept. (Section 3.2)

Y

y-axis The vertical number line in a rectangular coordinate system is called the y-axis. (Section 3.1)

y-intercept A point where a graph intersects the y-axis is called a y-intercept. (Section 3.2)

Z

zero-factor property The zero-factor property states that if two numbers have a product of 0, then at least one of the numbers must be 0. (Section 6.5)

Index

A

Absolute value, 34
  definition of, 34
  evaluating, 35
Addition
  associative property of, 64
  commutative property of, 64
  of complex numbers, 617
  of decimals, 649
  of fractions, 6
  with grouping symbols, 42
  identity element for, 66
  identity property of, 66
  inverse for, 67
  of like terms, 337
  of multivariable polynomials, 342
  of negative numbers, 39
  on a number line, 39
  of polynomials, 340
  properties of, 64
  of radicals, 547
  of rational expressions, 471
  of real numbers, 38
  of signed numbers, 39
  summary of properties of, 69
  word phrases for, 43
Addition property
  of equality, 95
  of inequality, 158
Additive identity element, 66
Additive inverse, 33, 41, 67
  definition of, 33
Algebraic expressions, 23
  distinguishing from equations, 26
  evaluating, 23, 56
  from word phrases, 24, 76, 112
  simplifying, 73
Angles
  complementary, 119
  measure of, 119, 130
  right, 119
  straight, 120, 130
  supplementary, 119
  vertical, 130
Approximately equal to, 531
  symbol for, 531
Area of geometric figures, 128, 320, 361, 599
Associative properties, 64
  distinguishing from commutative, 65
Average, 62
Axes of a coordinate system, 188
Axis
  of a parabola, 343, 622
  of symmetry, 622

B

Bar graph, 19, 184
  interpreting, 184
Base of an exponential expression, 14, 314
Basic principle of fractions, 3
Bhāskara, 538
Bimodal, 664
Binomials, 339
  higher powers of, 358
  multiplication by FOIL method, 351
  squares of, 356
  steps to multiply by FOIL method, 352
Boundary line, 231
Brackets, 15, 16
Business profit, determining, 247

C

Calculators, 643
  arithmetic operations on, 643
  graphing, 646
  introduction to, 643
  keys on, 643, 644, 645
  viewing window of, 647
Cartesian coordinate system, 188
  plotting points on, 189
Celsius-Fahrenheit relationship, 228
Central tendency, measures of, 661
Change of sign key on calculators, 643
Circle
  area of, 599
  graph of, 9, 184
Classifying polynomials, 339
Clear entry key on calculators, 644
Coefficient, 74, 337
Combining like terms, 74, 337
Combining variables, 337
Common denominator, 6
Common factor, 386
Commutative properties, 64
  distinguishing from associative, 65
Comparing floor areas of houses, 79
Comparing television sizes, 581
Complement of a set, 657
  symbol for, 657
Complementary angles, 119
Completing the square method for solving quadratic equations, 601
  steps to solve by completing the square, 602
Complex fractions, 479
  steps to simplify, 480, 482
Complex numbers, 616
  chart of, 616
  conjugate of, 618
  imaginary part of, 616
  operations on, 617, 618
  real part of, 616
  standard form of, 616, 618
Components of an ordered pair, 238
Composite number, 2
Compound interest, 320
Conditional equation, 111
Conjugates
  of complex numbers, 618
  multiplication of, 559, 618
  of radicals, 559
Consecutive integers, 121, 427
  even, 121, 427
  odd, 121, 427
Consistent system, 265
Constant of variation, 506
Contradiction, 111
Coordinate system, 188
  Cartesian, 188
  origin of, 189
  quadrants of, 188
  rectangular, 188
Coordinates of a point, 29, 189
Cost
  net, 222
  unit, 137
Cross multiplication, 138
Cross products, 138
Cube(s)
  of a number, 14
  perfect, 543
  sum of, 412
  volume of, 546
Cube root, 533
  symbol for, 534

I-1

D

Data, interpreting, 35
Data set, 224
Decimal approximation of an irrational number, 530
Decimal numbers, 649
  arithmetic operations on, 649
  converting to percents, 651
  linear equations with, 110
  operations on, 649, 650
Decimal numeration system, 369
Degree, 119
  of a polynomial, 338
  of a term, 338
Denominator, 2, 449
  common, 6
  least common, 6, 464
  rationalizing, 550
Dependent equations, 265
  elimination method for solving, 283
  substitution method for solving, 275
Depreciation, 207, 221
  straight-line, 222
Descartes, René, 188, 189
Descending powers, 338
Determining business profit, 247
Difference, 7, 44
  of cubes, 411
  of squares, 407
Direct variation, 506
  as a power, 507
Discriminant of quadratic formula, 610
Distance formula, 533
  for falling objects, 418
Distance, rate, and time problems, 150, 291
Distance, rate, and time relationship, 149, 239, 291, 498
Distributive property, 68
Divisibility tests for numbers, 62, 386
Division
  of complex numbers, 618
  of decimals, 650
  definition of, 53
  of fractions, 5
  long, 363
  of polynomials, 362, 364
  of radicals, 541
  of rational expressions, 459
  of real numbers, 53
  of signed numbers, 54
  word phrases for, 58
  by zero, 53
Domain, 238, 241
  of a quadratic equation, 625
Double negative rule, 34

E

Earthquake, intensity of, 333
Elements of a set, 25, 655
Elimination method
  for solving dependent equations, 283
  for solving inconsistent systems, 283
  for solving linear systems, 279
  steps to solve by, 280
Empty set, 111, 655
  symbols for, 111, 655
Endpoints of a quadratic inequality, 669
Equal sets, 656
Equality
  addition property of, 95
  multiplication property of, 100
  squaring property of, 565
Equation(s), 25, 94. *See also* System of equations
  conditional, 111
  dependent, 265
  for depreciation, 207, 221
  distinguishing from expressions, 26, 486
  equivalent, 94
  extraneous solution of, 566
  from word statements, 59
  graph of, 197
  of a horizontal line, 200
  independent, 265
  linear in one variable, 94
  linear in two variables, 185, 186, 199
  with no solutions, 566
  quadratic, 342, 418, 594
  with radicals, 565
  with rational expressions, 486
  simplifying, 97, 104
  slope of, 212
  solution set of, 94
  solutions of, 25, 94
  square root property of, 594
  of a vertical line, 200
Equation of a line, 220
  point-slope form of, 222
  slope-intercept form of, 220
  standard form of, 224
  steps to find, 226
Equilibrium demand, 269
Equilibrium supply, 269
Equivalent equations, 94
Equivalent forms for a rational expression, 453
Even consecutive integers, 121, 427
Exponential expressions, 14, 314
  base of, 14, 314
  evaluating, 14
Exponential key on calculators, 645
Exponents, 14, 314
  application of, 330
  fractional, 576
  integer, 321
  negative, 322
  power rules for, 315, 316, 317
  product rule for, 314
  quotient rule for, 324
  rational, 576
  and scientific notation, 330
  summary of rules for, 317, 325
  zero, 321
Expressions, 418
  algebraic, 23
  distinguishing from equations, 26, 486
  exponential, 14, 314
  from word phrases, 76
  quadratic, 418
  radical, 528
  rational, 448
  simplifying, 73
  terms of, 74, 337
Extraneous solution, 566
Extremes of a proportion, 138

F

Factoring, 290
  difference of cubes, 411
  difference of squares, 407
  greatest common factor, 386
  by grouping, 390
  perfect square trinomials, 409
  sum of cubes, 412
Factoring method for solving quadratic equations, 418, 594
Factoring trinomials, 394
  by grouping, 399
  made easy, 436
  in two variables, 403
  using FOIL, 394, 401
Factors, 2, 386
  common, 386
  greatest common, 3, 386
  of integers, 52
  of a number, 2, 386
  prime, 2
Fahrenheit-Celsius relationship, 228
Fibonacci, 529
Finite set, 655
  distinguishing from infinite, 656
First-degree equation. *See* Linear equations
FOIL, 351, 394, 401
  inner product of, 352
  outer product of, 352

Formulas
    to evaluate variables, 128
    geometry, 128
Fourth root, 533
Fraction bar, 2
Fractional exponents, 576
    definition of, 576
    rules for, 578
Fractions, 2
    basic principle of, 3
    complex, 479
    improper, 2
    least common denominator of, 464
    linear equations with, 109
    lowest terms of, 3
    mixed numbers, 6
    operations on, 4, 5, 6, 8
    proper, 2
    quadratic equations with, 611
    reciprocals of, 5, 53, 67
    solving linear systems with, 275
Frequency, 662
Function, 192, 238
    domain of, 241
    input-output, 239
    linear, 220
    notation, 242
    quadratic, 625
    range of, 241
    vertical line test for, 240
Fundamental property of rational
    expressions, 450
$f(x)$ notation, 242

G

Galileo Galilei, 385, 418, 424
Geometry applications, 128
Geometry formulas, 128, 320, 361
Golden ratio, 563
Graph(ing)
    bar, 19, 184
    calculators, 646
    circle, 9, 184
    of an equation, 197
    of horizontal lines, 200
    of inequalities, 158
    line, 184
    of linear equations, 196, 197
    of linear inequalities, 158, 232
    of a natural number, 31
    of numbers, 29
    of ordered pairs, 188
    of parabolas, 342, 343
    pie, 9, 184
    polynomials, 337

of quadratic equations, 342, 343, 621
of rational numbers, 31
region of, 231
of straight lines, 201
of vertical lines, 200
Graphical method
    for solving linear equations, 202, 263
    for solving linear inequalities, 298
Graphing calculator method
    for solving linear equations, 202
    for solving linear inequalities, 299
    for solving linear systems, 266
    for solving quadratic equations, 425, 628
Greater than, 17, 33, 157
    definition of, 33
Greater than or equal to, 17, 157
Greatest common factor, 3, 386
    factoring out, 388
    of numbers, 386
    steps to find, 387
    for variable terms, 388
Grouping
    factoring by, 390
    factoring trinomials by, 399
Grouping symbols, 15
    addition with, 42
    subtraction with, 42

H

Heron's formula, 575
Higher powers of binomials, 358
Horizontal line, 200
    equation of, 200
    graph of, 200
    slope of, 211
Hypotenuse of a right triangle, 428

I

i, 615
Identity element, 66
Identity properties, 66
Imaginary numbers, 616
Imaginary part of a complex number, 616
Improper fraction, 2
Inconsistent system, 265
    elimination method for solving, 283
    substitution method for solving, 274
Independent equations, 265
Index of a radical, 533
Inequalities, 17, 157
    addition property of, 158
    applied problems using, 163
    graphs of, 158

linear in one variable, 158
linear in two variables, 230
multiplication property of, 160
quadratic, 669
solving linear, 157
symbols of, 17, 19, 157
three-part, 164
Infinite set, 655
    distinguishing from finite, 656
Infinity, 157
Inner product, 352
Input of an ordered pair, 239
Input-output machine, 192, 239
Integers, 29
    consecutive, 121, 427
    consecutive even, 121, 427
    consecutive odd, 121, 427
    as exponents, 321
    factors of, 52
Intensity of an earthquake, 333
Intercepts, 198, 220
    of a linear equation, 198
    of a parabola, 622
    of a quadratic equation, 622
    of a quadratic function, 625
Interest, compound, 320
Interest problems, 145, 147
Interpreting graphs, 184
Intersection of sets, 657
    symbol for, 657
Interval notation, 157
Inverse
    additive, 33, 41, 67
    multiplicative, 53, 67
Inverse key on calculators, 645
Inverse properties, 67
Inverse variation, 508
    as a power, 508
Irrational numbers, 31, 530
    conjugate of, 559
    decimal approximation of, 530

K

Khowarizmi, 104
Kilometer-mile relationship, 170

L

Least common denominator, 6, 464
    steps to find, 464
Legs of a right triangle, 428
Leonardo da Pisa, 529
Less than, 17, 32, 157
    definition of, 32
Less than or equal to, 17, 157
Like radicals, 547

Like terms, 74, 337
   addition of, 337
   combining, 74, 337
Line(s)
   equations of, 220
   horizontal, 200
   intercepts of, 220
   number, 29, 157
   parallel, 213
   perpendicular, 213
   slope of, 208
   of symmetry, 343
   vertical, 200
Line graph, 184
   interpreting, 185
Linear equations in one variable, 94
   applications of, 114, 115
   with decimal coefficients, 110
   with fractions, 109
   geometric applications of, 128
   with infinitely many solutions, 111
   with no solutions, 111
   solving, 106
   steps to solve, 106
Linear equations in two variables, 184, 186, 199. *See also* System of linear equations
   calculator graphing of, 202
   graphing calculator method for solving, 202
   graphing of, 196, 197
   intercepts of, 198
   point-slope form of, 222
   slope of, 212
   slope-intercept form of, 220
   solution of, 186
   standard form of, 224
   summary of types of, 224
   systems of, 262
   use to model, 202
Linear function, 220
Linear inequalities in one variable, 158
   graph of, 158
   solution of, 159
   steps to solve, 161
Linear inequalities in two variables, 230
   boundary line of, 231
   calculator graphing of, 234
   graph of, 232
   solution of, 231
   steps to graph, 233
   system of, 297
Long division, 363
Lowest terms
   of a fraction, 3
   of a rational expression, 450

M

Mathematical model, 28
Mean, 661
   weighted, 662
Means of a proportion, 138
Measure of an angle, 119, 130
Measures of central tendency, 661
   mean, 661
   median, 663
   mode, 664
   summary of, 664
Median, 663
Memory key on calculators, 644
Mile-kilometer relationship, 170
Mixed number, 2
Mixture problems, 118, 145, 289
   steps to solve, 146
Mode, 664
Model(s)
   mathematical, 28
   quadratic, 429
   using a linear equation to, 202
Money problems, 148
Monomial, 74, 339
Motion problems, 150, 291
Multiplication
   associative property of, 64
   of binomials, 351
   commutative property of, 64
   of complex numbers, 617
   of conjugates, 559, 618
   of decimals, 647
   FOIL method of, 351
   of fractions, 4
   identity element for, 66
   identity property of, 66
   inverse for, 67
   of a monomial and a polynomial, 348
   of polynomials, 349
   properties of, 64
   of radicals, 539
   of rational expressions, 458
   of real numbers, 51
   of signed numbers, 51
   of sum and difference of two terms, 357
   summary of properties of, 69
   word phrases for, 57
   by zero, 50
Multiplication property
   of equality, 100
   of inequality, 160
Multiplicative identity element, 66
Multiplicative inverse, 53, 67

Multivariable polynomial, 342
   addition of, 342
   subtraction of, 342

N

Natural numbers, 2, 29
   negative of, 29
   opposite of, 29
Negative exponents, 322
   changing to positive, 323
Negative infinity, 157
   symbol for, 157
Negative numbers, 24, 30
   addition of, 39
   as exponents, 322
   simplifying square roots of, 615
Negative of a number, 29
Negative slope, 209
Negative square roots, 528
Net cost, 222
Not equal, 17
nth root, 533
Null set, 111, 655
   symbols for, 111, 655
Number(s). *See also* Real numbers
   absolute value of, 34
   complex, 616
   composite, 2
   cube of, 14
   decimal, 649
   divisibility tests for, 62, 386
   factors of, 2, 386
   fractions, 2
   graph of, 29
   greatest common factor of, 386
   imaginary, 616
   integers, 29
   irrational, 31, 530
   mixed, 2
   natural, 2, 29
   negative, 24, 30
   opposite of, 29, 33, 41, 67
   ordering of, 32
   perfect square, 530
   positive, 30
   prime, 2, 386
   prime factors of, 2
   rational, 30
   real, 31
   reciprocal of, 5, 53, 57
   signed, 30
   square of, 14, 528
   square roots of, 528
   whole, 2, 29

Number line, 29, 157
    addition on, 39
    graphing intervals on, 157
    graphing a number on, 29
    subtraction on, 41
Numerator, 2, 449
Numerical coefficient, 74, 337
Numerical expressions
    evaluating, 55
    from word phrases, 43, 57, 58

O

Odd consecutive integers, 121, 427
Opposite of a number, 29, 33, 41, 67
Order
    of operations, 15, 42
    of a radical, 533
Ordered pairs, 186, 238
    completing, 187
    components of, 238
    graph of, 188
    input of, 239
    output of, 239
    plotting, 188
    table of, 187
Ordering of real numbers, 32
Origin, 189
Outer product, 352
Output of an ordered pair, 239

P

Pairs, ordered, 186, 238
Parabola, 343, 593
    axis of, 343, 622
    axis of symmetry of, 622
    graph of, 342, 343
    graphing by calculator, 631
    intercepts of, 622
    line of symmetry of, 343
    steps to graph, 624
    vertex of, 343, 622
Parabolic dish, 593
Parallel lines, 213
    slopes of, 213
Parentheses, 15
Percents, 650
    converting to decimals, 650
Perfect cube, 543
Perfect square, 530
Perfect square trinomial, 409, 600
    factoring of, 409
Perimeter of a geometric figure, 129
Perpendicular lines, 213
    slopes of, 213

Pi (π), 134, 530
Pi key on calculators, 645
Pie chart, 9, 184
Pisa, Leonardo da, 529
Plane, 189
Plotting points, 189
Plus or minus symbol, 595
Point-slope form, 222
Polya, George, 122
    four-step method for problem solving, 122
Polynomial(s), 313
    binomial, 339
    classifying, 339
    degree of, 338
    of degree two, 343
    in descending powers, 338
    division by a monomial, 362
    division by a polynomial, 364
    evaluating, 339
    factoring summary, 416
    graphing, 337
    graphing equations defined by, 342
    long division of, 363
    monomial, 339
    multiplication by a monomial, 348
    multivariable, 342
    numerical coefficients of, 337
    operations on, 340, 341, 349, 362, 364
    prime, 396
    terms of, 339
    trinomial, 338
    in x, 338
Positive exponents, 321
Positive numbers, 30
Positive slope, 209
Positive square roots, 528
Power rules for exponents, 315, 316, 317
Powers, 14, 314
    descending, 338
Price per unit, 137
Prime factors of a number, 2
Prime number, 2, 386
Prime polynomials, 396
Principal square root, 566
Product, 2, 51, 57
    of the sum and difference of two terms, 357
Product rule
    for exponents, 314
    for radicals, 539
Profit, determining, 247
Proper fraction, 2

Properties
    of radicals, 543
    of real numbers, 64
Proportions, 138
    applications of, 140
    cross multiplication of, 138
    cross products of, 138
    extremes of, 138
    means of, 138
    solving, 140
    terms of, 138
Pyramid, volume of, 134
Pythagorean formula, 428, 531
    application of, 532
    converse of, 329
    proof of, 534, 538

Q

Quadrants, 188
Quadratic equations, 342, 418, 594
    applications of, 425, 604
    completing the square method for solving, 601
    with complex number solutions, 619
    factoring method for solving, 418, 594
    with fractions, 611
    graphing of, 342, 343, 621
    intercept, 622
    with no real solution, 597
    with one solution, 610
    quadratic formula method for solving, 608
    square root method for solving, 595
    standard form of, 418, 594
    steps to solve, 420
    steps to solve an applied problem, 425
    summary of solution methods, 614
Quadratic expression, 418
Quadratic formula, 607, 609
    discriminant of, 610
Quadratic formula method for solving quadratic equations, 608
Quadratic functions, 625
    application of, 626
    domain of, 625
    intercepts of, 625
    range of, 625
Quadratic inequalities, 669
    endpoints of, 669
    steps for solving, 670
Quadratic models, 429
Quotient, 5, 53, 58
Quotient rule
    for exponents, 324
    for radicals, 541

R

Radical equation, 565
  steps to solve, 567
  using to model data, 569
Radical expressions, 528
  squaring of, 529
  summary of simplifying, 557
  using conjugates to simplify, 559
Radical sign, 528
Radicals, 528
  conjugates of, 559
  equations with, 565
  index of, 533
  like, 547
  operations on, 539, 541, 547
  order of, 533
  product rule for, 539
  properties of, 543
  quotient rule for, 541
  in simplified form, 552
  simplified form of, 541, 542
  simplifying, 539
  simplifying higher roots, 543
  simplifying using rational exponents, 578
  unlike, 547
  with variables, 542
Radicand, 528
Range, 238, 241
  of a quadratic equation, 625
Rate of change, 218
Rate of work, 500
Ratio, 137
  from word phrases, 137
  golden, 563
  sacred, 563
Rational expressions, 448
  applications of, 497
  with denominator zero, 449
  equations with, 486
  equivalent forms for, 453
  evaluating, 448
  fundamental property of, 450
  in lowest terms, 450
  with numerator zero, 449
  operations on, 458, 459, 471, 474
  solving an equation with, 487
  steps for division of, 462
  steps for multiplication of, 462
  summary of operations on, 495
  undefined values for, 449
Rational numbers, 30
  graph of, 31
  using as exponents, 576

Rationalizing denominators, 550
Reading graphs, 184
Real numbers, 31. *See also* Numbers
  absolute value of, 34
  additive inverse of, 33
  operations on, 38, 41, 51, 53
  opposites of, 33
  order of operations of, 42
  ordering of, 32
  properties of, 64
  reciprocals of, 53
  sets of, 31
  summary of operations on, 63
Real part of a complex number, 616
Reciprocal key on calculators, 645
Reciprocals of fractions, 5, 53, 67
Rectangular box, volume of, 134
Rectangular coordinate system, 188
Region of a graph, 231
Relation, 238
  domain of, 238
  range of, 238
Rhind Papyrus, 104, 563
Richter, Charles F., 333
Richter scale, 333
Right angle, 119
Right triangle, 428
  hypotenuse of, 428
  legs of, 428
Rise, 208
Root key on calculators, 645
Roots
  cube, 533
  fourth, 533
  nth, 533
  principal, 566
  square, 528, 530
Run, 208

S

Sacred ratio, 563
Salvage value, 222
Scatter diagram, 191
Scientific notation, 330
  application of, 330
  on calculators, 332
  and exponents, 330
  steps to write a number in, 330
Second-degree equation. *See* Quadratic equations
Second function key on calculators, 644
Semiperimeter, 575
  of a triangle, 575

Set(s), 25, 655
  complement of, 657
  elements of, 25, 655
  empty, 111, 655
  equal, 656
  finite, 655
  infinite, 655
  intersection of, 657
  null, 111, 655
  of real numbers, 31
  subset of, 656
  union of, 657
  universal, 655
Set braces, 25, 655
Set-builder notation, 30
Sight distance, 527, 574
Signed numbers, 30
  interpreting data with, 44
  operations on, 39, 40, 41, 51, 52, 54
  summary of operations with, 56
Similar triangles, 143
Simple interest problems, 147
Simplified form of a radical, 541, 542
Simplifying
  algebraic expressions, 73
  radicals, 539
Six-step method for solving applied problems, 115
Slope, 208
  formula for, 210
  from an equation, 212
  of horizontal lines, 211
  negative, 209
  notation for, 208
  of parallel lines, 213
  of perpendicular lines, 213
  positive, 209
  of vertical lines, 211
Slope-intercept form, 220
Solution set
  of an equation, 94
  of system of linear equations, 262
  of system of linear inequalities, 297
Solutions of an equation, 25, 94
Solving for a specified variable, 131, 492
Special factorizations, summary of, 413
Sphere
  surface area of, 599
  volume of, 134, 546
Square root key on calculators, 644
Square root method for solving quadratic equations, 595
Square root property of equations, 594

Square roots, 528, 530
  of *a*, 529
  approximation of, 531
  negative, 528
  of a number, 528
  positive, 528
  principal, 566
  symbol for, 528
Squares
  of binomials, 356
  completing, 601
  difference of, 407
  of a number, 14, 528
Squaring key on calculators, 644
Squaring of radical expressions, 529
Squaring property of equality, 565
Standard form
  of complex numbers, 616, 618
  of a linear equation, 224
  of a quadratic equation, 418, 594
Stock market, 1
Stopping distance, 347
Straight angle, 120, 130
Straight-line depreciation, 222
Straight lines, summary of graphs of, 201
Subscript notation, 208
Subset of a set, 656
  symbol for, 656
Substitution method
  for solving dependent equations, 275
  for solving inconsistent systems, 274
  for solving linear systems, 272
  steps to solve by, 273
Subtraction
  of complex numbers, 617
  of decimals, 649
  definition of, 41
  of fractions, 8
  with grouping symbols, 42
  of a multivariable polynomial, 342
  on the number line, 41
  of polynomials, 341
  of radicals, 547
  of rational expressions, 474
  of real numbers, 41
  of signed numbers, 41
  word phrases for, 44
Sum, 6
  of cubes, 412
Supplementary angles, 119
Supply and demand, 269
Surface area of a sphere, 599
Symbols of inequality, 17, 19, 157
  statements with, 18

Symmetry line of a parabola, 343
System of linear equations, 262
  alternative method for solving, 282
  to analyze data, 302
  applications of, 287
  choosing a method to solve, 285
  consistent, 265
  elimination method for solving, 279
  with fractions, 275
  graphical method for solving, 263
  graphing calculator method for solving, 266
  inconsistent, 265
  with no solution, 264
  solution of, 262
  solution set of, 262
  steps for solving, 287
  steps to solve by elimination, 280
  steps to solve by graphing, 264
  steps to solve by substitution, 273
  substitution method for solving, 272
  summary of outcomes, 265
System of linear inequalities, 297
  graphical method for solving, 298
  graphing calculator method for solving, 299
  solution of, 297
  solution set of, 297
  steps to solve, 297

T

Table of data, 35
  interpreting, 35
Table of ordered pairs, 187
Terms, 74
  combining, 74, 337
  degree of, 338
  of an expression, 74, 337
  like, 74, 337
  numerical coefficient of, 74, 337
  of a polynomial, 339
  of a proportion, 138
  unlike, 74, 338
Tests for divisibility, 62, 386
Three-part inequalities, 164
Translating words into equations, 59
Triangles
  right, 428
  semiperimeter of, 575
  similar, 143
Trinomials, 338
  factoring of, 394
  perfect square, 409, 600

U

Undefined rational expressions, 449
Union of sets, 657
  symbol for, 657
Unit cost, 137
Unit pricing, 137
Universal set, 655
Unlike radicals, 547
Unlike terms, 74, 338

V

Variables, 23
  combining, 337
  formulas to evaluate, 128
  radicals with, 542
  solving for specified, 131, 492
Variation, 506
  constant of, 506
  direct, 506
  inverse, 508
Velocity problem, 604
Venn diagrams, 657
Vertex of a parabola, 343, 622
  solving for, 623
Vertical angles, 130
Vertical line, 200
  equation of, 200
  graph of, 200
  slope of, 211
Vertical line test for functions, 240
Volume, 134
  of a cube, 546
  of a pyramid, 134
  of a rectangular box, 134
  of a sphere, 134, 546

W

Weighted mean, 662
Whole numbers, 2, 29
Word phrases
  for addition, 43
  to algebraic expressions, 24, 76, 112
  for division, 58
  to expressions, 76
  for multiplication, 57
  to numerical expressions, 43, 57, 58
  to ratios, 137
  for subtraction, 44
Word statements to equations, 59
Words to symbols conversions, 18, 43
Work rate problems, 500

X

x-axis, 188
x-intercepts, 198
   of a parabola, 622

Y

y-axis, 188
y-intercept, 198, 220
   of a parabola, 622

Z

Zero
   division by, 53
   multiplication by, 50
Zero denominator in a rational expression, 449
Zero exponent, 321
Zero-factor property, 418

Photo Credits: **p. 1** © Alan Schein Photography/CORBIS; **p. 10** © Anthony Cooper; Ecoscene/CORBIS; **p. 12 top** © Alan Schein Photography/CORBIS; **p. 12 bottom** © 2002 PhotoDisc; **p. 22** © Keith Wood/Getty Images; **p. 28** © Pete Saloutos/CORBIS; **p. 36** © 2002 PhotoDisc; **p. 45** © Charles Gupton/CORBIS; **p. 48 top** © James Leynse/CORBIS SABA; **p. 48 bottom** © James L. Amos/CORBIS; **p. 87** Rob Tringali Jr./SportsChrome USA; **p. 90** AP/Wide World Photos; **p. 93** © Karl Weatherly/CORBIS; **p. 116** AP/Wide World Photos; **p. 123 left** AP/Wide World Photos; **p. 123 right** © White/Packert/Getty Images; **p. 124** AP/Wide World Photos; **p. 125** © Karl Weatherly/CORBIS; **p. 137** Courtesy of John Hornsby; **p. 140** Courtesy of John Hornsby; **p. 142** Jeff Greenberg/PhotoEdit; **p. 147** © 2002 Eyewire; **p. 150** © Reuters New Media Inc./CORBIS; **p. 154** © Paul Conklin/PhotoEdit; **p. 155 left** Greg Crisp/SportsChrome USA; **p. 155 right** Brian Spurlock/SportsChrome USA; **p. 156** © Karl Weatherly/CORBIS; **p. 164** © Gary Conner/PhotoEdit; **p. 170** © Karl Weatherly/CORBIS; **p. 177** © Douglas Peebles/CORBIS; **p. 179** © 2002 PhotoDisc; **p. 183** © 2002 PhotoDisc; **p. 190** © 2002 PhotoDisc; **p. 196** © 2002 PhotoDisc; **p. 203** © 2002 PhotoDisc; **p. 206** © 2002 PhotoDisc; **p. 219** © Alan Schein Photography/CORBIS; **p. 224** © 2002 PhotoDisc; **p. 229** © 2002 PhotoDisc; **p. 235** © Wolfgang Kaehler/CORBIS; **p. 239** © Michael Newman/PhotoEdit; **p. 246** © Gary Buss/Getty Images; **p. 257** © Kate Powers/Getty Images; **p. 259** © 2002 PhotoDisc; **p. 261** THE KOBAL COLLECTION/WARNER BROS/MOUNTAIN, PETER; **p. 278** © 2002 PhotoDisc; **p. 285** © Chuck Savage/CORBIS; **p. 287** © David Young-Wolff/PhotoEdit; **p. 288** © Michael Newman/PhotoEdit; **p. 293** THE KOBAL COLLECTION/WARNER BROS/MOUNTAIN, PETER; **p. 294** © 2002 PhotoDisc; **p. 295** AP/Wide World Photos; **p. 302** © FORESTIER YVES/CORBIS SYGMA; **p. 307 top** © Tony Freeman/PhotoEdit; **p. 307 bottom** © 2002 PhotoDisc; **p. 309** © 2002 PhotoDisc; **p. 310** © Kelly-Mooney Photography/CORBIS; **p. 313** Credit Line Henryk T. Kaiser/IndexStock; **p. 320** © Jim Cummins/Getty Images; **p. 332** Courtesy of NASA; **p. 333** © Roger Ressmeyer/CORBIS; **p. 335** Courtesy of NASA; **p. 336** © Digital Vision; **p. 337** THE KOBAL COLLECTION/20TH CENTURY FOX/PARAMOUNT; **p. 344** Courtesy of NASA; **p. 347** © 2002 PhotoDisc; **p. 348** © Ron Levine/Getty Images; **p. 378** © Reuters New Media Inc./CORBIS; **p. 381** Courtesy of NASA; **p. 385** © 2002 PhotoDisc; **p. 418** © 2002 PhotoDisc; **p. 430** © 2002 PhotoDisc; **p. 435 top** © 2002 PhotoDisc; **p. 435 bottom** © Bettmann/CORBIS; **p. 442** © 2002 PhotoDisc; **p. 443** AP/Wide World Photos; **p. 447** © Neil Rabinowitz/CORBIS; **p. 479** © Neil Rabinowitz/CORBIS; **p. 503 top left** AP/Wide World Photos; **p. 503 top right** © Reuters New Media Inc./CORBIS; **p. 503 bottom** © Duomo/CORBIS; **p. 507** © Steve Fitchett/Getty Images; **p. 509** © Robert Brenner/PhotoEdit; **p. 511** © 2002 PhotoDisc; **p. 513** © Douglas Slone/CORBIS; **p. 521** © Joesph Sohm; ChromoSohm Inc./CORBIS; **p. 527** © Ben Wood/CORBIS; **p. 537** Courtesy of John Hornsby; **p. 563 top** © Bruce Ayres/Getty Images; **p. 563 bottom** Courtesy of NASA; **p. 570** James R. Holland/Stock Boston; **p. 572** Henryk T. Kaiser/IndexStock; **p. 574 top** © Ben Wood/CORBIS; **p. 574 middle** © 2002 PhotoDisc; **p. 574 bottom** © 2002 PhotoDisc; **p. 575** © 2002 PhotoDisc; **p. 580** © Craig Lovell/CORBIS; **p. 587** AP/Wide World Photos; **p. 588** David Shopper/Stock Boston; **p. 589** Chuck Pefley/Stock Boston; **p. 593** John Sarkissian/Parkes Observatory; **p. 597** Arles Gupton/Stock Boston; **p. 599** © John Van Hasselt/CORBIS SYGMA; **p. 604** © Eyewire; **p. 613** Courtesy of NASA; **p. 626** John Sarkissian/Parkes Observatory; **p. 637** AP/Wide World Photos; **p. 638** AP/Wide World Photos; **p. 641** © Neil Rabinowitz/CORBIS

Videotape and CD Index

Text/Video/CD Section	Exercise Numbers	Text/Video/CD Section	Exercise Numbers
Section 1.1	65	Section 6.1	none
Section 1.2	5, 9	Section 6.2	33
Section 1.3	3	Section 6.3	none
Section 1.4	17	Section 6.4	13, 17, 25, 45, 59
Section 1.5	49	Section 6.5	none
Section 1.6	25, 95	Section 6.6	7, 29
Section 1.7	none		
Section 1.8	none	Section 7.1	33, 49
		Section 7.2	49
Section 2.1	7, 13, 19, 25, 37, 55, 49, 63	Section 7.3	none
		Section 7.4	none
Section 2.2	27, 31, 45, 49, 47, 61, 69	Section 7.5	none
		Section 7.6	none
Section 2.3	23, 47, 53	Section 7.7	23, 35
Section 2.4	7, 19, 37	Section 7.8	21, 23, 25, 27
Section 2.5	17, 23, 37, 45, 49, 71, 73		
		Section 8.1	77
Section 2.6	25, 29, 61	Section 8.2	none
Section 2.7	11, 21, 27, 31, 33, 39	Section 8.3	none
Section 2.8	67, 77	Section 8.4	none
		Section 8.5	none
Section 3.1	35	Section 8.6	none
Section 3.2	5, 31	Section 8.7	53
Section 3.3	31		
Section 3.4	45	Section 9.1	29, 55
Section 3.5	none	Section 9.2	3, 13, 23, 27, 39
Section 3.6	41, 43	Section 9.3	none
		Section 9.4	43
Section 4.1	7, 11	Section 9.5	41
Section 4.2	19		
Section 4.3	13		
Section 4.4	9, 21		
Section 4.5	none		
Section 5.1	79		
Section 5.2	none		
Section 5.3	none		
Section 5.4	31, 67		
Section 5.5	21, 49		
Section 5.6	47, 51		
Section 5.7	23, 29, 45, 65		

Formulas

Figure	Formulas	Illustration
Square	Perimeter: $P = 4s$ Area: $A = s^2$	
Rectangle	Perimeter: $P = 2L + 2W$ Area: $A = LW$	
Triangle	Perimeter: $P = a + b + c$ Area: $A = \frac{1}{2}bh$	
Circle	Diameter: $d = 2r$ Circumference: $C = 2\pi r$ $C = \pi d$ Area: $A = \pi r^2$	
Parallelogram	Area: $A = bh$ Perimeter: $P = 2a + 2b$	
Trapezoid	Area: $A = \frac{1}{2}(B + b)h$ Perimeter: $P = a + b + c + B$	